COGS, SMALL COGS AND BOATS

The thirteenth- until sixteenth- century Dutch and Flemish archaeological finds from the Hanseatic shipbuilding tradition seen in a broader perspective

Karel Vlierman

General view from the stern with the repositioned frames, keelson and riders of the Rutten Cog.

COGS, SMALL COGS AND BOATS

Cogghen, Kleene Cogghen ende Schuten

Quoted from a letter by a German merchant in Bruges addressing the entire Hanseatic League in 1401

SPA UITGEVERS

Made possible by:

Ms. H.C. van der Velde-Doyer

Ms. B. Espeet

Mr. P.B. Zwiers

Stichting professor van Winter fonds

Stichting Hanzefonds

P.J.R. Modderman Stichting

provincie Overijssel

Provincie Flevoland

Stichting voor de Middeleeuwse Archeologie

Rijksdienst voor het Cultureel Erfgoed, Ministerie van Onderwijs, Cultuur en Wetenschap

Ruud van Beek Stichting

Stichting Promotie Archeologisch Onderzoek SPAO

Stadsarchief Kampen

Kampen Archeologie

NISS

Buro de Brug

Prins Bernhard Cultuurfonds

Stichting Promotie Archeologie SPA

Vaderlandsch Fonds ter aanmoediging van 's-lands zeedienst

Admiraal van Kinsbergen Fonds

Vereeniging de Prins Hendrik Stichting

Directie der Oostersche Handel en Reederijen

Colophon:

ISBN:	978-90-8932-048-3
Author:	Karel Vlierman
Editors:	Hemmy Clevis, Hans Janssen and Jos Koldeweij
Translators:	Xandra Bardet, Hemmy Clevis, Stefanie Hoss, Christine Jefferis, Joy Phillips, Sasja van der Vaart-Verschoof, Jeroen Vermeersch
Graphic Design:	Hidde Heikamp
Lay-out:	Hidde Heikamp
Plates:	Coen Bouwstra, Hidde Heikamp
Drawings:	All drawings and impressions are made by Karel Vlierman
Photographs:	Unless otherwise by Karel Vlierman and the Rijksdienst Cultureel Erfgoed
Print:	Huvé Graphics

SPA UITGEVERS

SPA uitgevers
Assendorperstraat 174-4
8012 CE Zwolle
The Netherlands
info@spa-uitgevers.nl
www.spa-uitgevers.nl

Contents

Preface

In 2006 the *Canon van Nederland* ('Canon of Dutch History') was compiled, presenting 50 'windows' on the history of the Netherlands. The seventh topic from the distant past is entitled 'The Hanseatic League, 1356 - ca 1450'; it discusses the trading network that brought wealth and international contacts to Northwest-European cities and regions. The updated *Canon* of 2020 provided the chapter on the Hanseatic League with a subtitle: 'Cooperation pays'. After all, it was collaboration among a large group of merchants and cities that proved the formula of success. All these cities had a waterfront: on the North Sea, the Baltic, an inland sea or a (major) river. Waterborne transport lay at the basis of this successful alliance, and this was carried out with characteristic ships: the cogs. The cog, between 15 and 25 m long, was the late-medieval bulk carrier, in which products such as salt, grain, timber, fish, wine and beer, pelts, beeswax, wool and cloth were shipped over long distances.
Since the late 19th century, historians and others have attempted to visualise the by then legendary cog. There are some images on civic seals of Hanseatic cities, and in paintings and illuminated manuscripts, but these provide only limited information. More evidence did not emerge until 1962, when near Bremen the submerged wreck of a cog was discovered, excavated and rebuilt. Numerous questions could now be answered, about dimensions, construction, tonnage, any crew's quarters and number of crew members. But such questions cannot be fully answered on the basis of a single, partially preserved shipwreck. Herein lies the great significance of *Cogghen, kleene cogghen ende schuten*. Karel Vlierman, who has devoted his life and career to the many recovered wrecks of ships that had foundered in the former Zuiderzee, identified some twenty of them as cogs, and in several others recognised cog-like traits. Vlierman has studied the mostly disjointed wreckage in meticulous detail. His deep understanding of this combined material, his shipwright's eye and draughtsman's talent have allowed him to largely reconstruct all of these vessels on paper in a reliable manner. The constructions in one wreck would show him how the detached timbers of another sunken vessel must once have fitted together. Comparison of the wrecks and broken parts also clarified a great deal about the shipbuilding techniques of the period and how these evolved. Besides, remnants of the ships' inventories were found in and around the wrecks. This material too Vlierman draws into his magisterial study: careful analysis of all cogs from the unparalleled wealth of the Zuiderzee soil archive and supremely detailed 2-D reconstruction drawings of these ships. All in all, in this study Karel Vlierman has realised what promises to be an enduring and defining standard work on the cog and its variants and evolution.

The Hanseatic League owed its flourishing to collaboration and to the cog. Likewise, the present publication - comprising two volumes and a folder of large reconstruction drawings to a scale of 1:10 - is the result of many years of intensive collaboration. In 1999, Thijs Maarleveld persuaded Karel to write a doctoral thesis on a subject in the field of maritime archaeology, upon which Karel decided to bring together and publish in a comprehensive overview the immense expertise he had gained with regard to the cogs of the Hanseatic era. The PhD ceremony, originally to be held at Leiden University, is now to take place at the Radboud University in Nijmegen. As an unexpected bonus, 2015-'16 saw the IJssel cog lifted from the river at Kampen, and Karel Vlierman was invited to do reconstruction drawings for this ship. His study by now was very voluminous; for putting it into print, Karel found SPA Publishers, where Hemmy Clevis went to great lengths to realise its publication. For a long time it seemed impossible to fund the project, among other things as an aftereffect of the banking crisis. Some years later, especially on the strength of commitments made by private sponsors, the mega-project was finally given the green light after, which set the ball rolling. In due course, funds too were disposed to step in. At this point a dilemma arose: printed or digital publication? Some funds were prepared to contribute on condition that the publication be digital. Ms Els van Eijk van Heslinga pointed out the possibility of thus integrating the study into the *Maritiem Portal* (website on Dutch maritime history). The 'Huygens Instituut voor Nederlandse Geschiedenis' (Huygens Institute for the History of the Netherlands of the Royal Netherlands Academy of Arts and Sciences) would be happy to cooperate in such a scheme. Anyone who closely examines the drawings will be fascinated by Karel Vlierman's approach. Everything is drawn and recorded in such detail that one can calculate how many nails and sintel clamps must have been used in the construction of each hull. The drawings effectively convey the impressive degree of craftsmanship with which these cogs had been built. The author is convinced that these drawings come into their own best of all in a printed version. The collaborative efforts of many parties, listed elsewhere in this publication, have as their happy outcome that this work will now be available in multiple formats: in Dutch as a PhD thesis, as a printed commercial edition in English, and in a digital version via the *Maritiem Portaal*, with all of the drawings magnifiable to show every detail.

With this mega-project, Karel Vlierman has set a new standard for examining, measuring, describing and presenting cogs and other shipwrecks. But having done so, he has gone the extra mile. He contrasted the cog-building tradition with other shipbuilding traditions, describing the similarities and differences. Step by step he explains how a cog was built; he classifies them; and by comparing them he is able to draw remarkable conclusions which would not have been possible by investigating just a single ship. The examination of ships' inventories and cargoes provided him with insight- in the case of some small cogs - into the number of crew members and what the vessels were carrying. With the aid of dendrochronology, but equally by his analysis of the typology of sintel clamps, he is able to date the ships. Dendrochronology and ballast may also offer clues about the provenance of the ship and the likely geographical origin of the ballast stones taken aboard. Bone analysis offers a glimpse of the provisions carried on voyages. In one case even an 'individual' on board can be described on the basis of finds: a traveller who had probably been a leper.

Karel Vlierman's study opens new horizons in what has been published on cogs. It offers arresting insight into the cargo vessels of the illustrious Hanseatic league, but also shows that this vessel type could not be upgraded into a larger ship to carry even more freight. The cog was part of the Hanseatic era, and was to be succeeded by other mighty vessel types such as those we know from the days of the Dutch East India Company.

Lex Heerma van Voss
Director, Huygens Institute for the History of the Netherlands

Acknowledgements

In the summer of 1999, what sparked this research project was the entirely unexpected question by fellow maritime archaeologist Professor Thijs Maarleveld whether, if given the opportunity, I would be interested in writing a PhD thesis on a maritime-archaeological subject. This would be an honor and a welcome opportunity to conclude my by then thirty-year career at the Netherlands Institute of Maritime and Underwater Archaeology (NISA) and its predecessors. Having responded positively to the suggestion, I gave it little further thought. So I was most surprised when on 16 November of that year, I received a letter from Leiden University's Faculty of Archaeology, via Professor L.P. Louwe Kooijmans, inviting me to submit a PhD thesis. Soon after, I opted for the intriguing late-medieval (Hanseatic) cog as my subject, and embarked on my research.
By collating Heinsius' publication of 1956, the discovery of a large late-medieval shipwreck in the river Weser near Bremen in 1962, and images on late-medieval civic seals, it had proved possible to identify characteristic features, and to link the historical designation of *Cogghe* (cog) to previously known and new archaeological finds.
The thirteen shipwrecks with cog features discovered - and in some cases excavated - in the Netherlands between 1944 and 1999 (by the time of writing some eighteen to twenty) had given rise to just four (concise and) in part dated excavation reports. Guided by personal interest, the principal reasons for opting for the full working-out of the excavation documentation and, if possible, the reconstruction on paper of these Dutch cog finds, which are major European maritime heritage, are fourfold:

A fascination with the Middle Ages, and medieval ships and shipbuilding in particular;
The NISA's ca 450 files of shipwrecks excavated or archaeologically assessed in the former Zuiderzee area. They date from ca 1250-1900, and ca 150 of them would more than merit elaboration and publication. Roughly half of these vessels predate 1600. The full range of large to small craft of Hanseatic times are represented;

- The new national policy regarding maritime archaeology, introduced in 1999/2000, made it clear that as of 2000 NISA would no longer be performing any excavations, and that there would be precious little opportunity for working-out, analyzing and publishing previous excavations. Excavating shipwrecks would in the future be left to 'the Market'. Yet the Market would largely lack, and indeed still lacks the required experience and specialist knowledge;
- Publication of the research results of all cog finds from the Low Countries by someone who in many cases had been immediately involved with their excavation and analysis would be a substantial contribution, of supra-national significance, to our knowledge about this historically best-known West-European and imaginal medieval vessel type and shipbuilding tradition.

A reorganization at the ROB and NISA entailed my unexpected early retirement as of July 1, 2002. As a result, the cog project was put on hold for three years. But over the 2005-2015 decade, I made standardized (new) descriptions of all Dutch and Flemish cog finds, ship by ship, and prepared drawings and drawn reconstructions of the vessels and all items found within them.

First of all I wish to thank Professor T.J. Maarleveld for his advocacy in allowing me to undertake this PhD research. I wish to thank Emeritus Professor L.P. Louwe Kooijmans for the opportunity of being admitted to the PhD trajectory, and, together with co-examiner Dr J.-M. A.W. Morel, for their early advice with respect to setting up the PhD study. To Emeritus Professor H.L. Janssen I am much indebted for his willingness to take over the role of supervisor from Professor Louwe Kooijmans and, together with his wife Ms drs. M. van Gent, offering me guidance in structuring the manuscript. Also for his substantial and textual comments, guidance and advice regarding substance and text. Unfortunately, it was not possible for Professor Janssen to continue as my supervisor. I thank my two current supervisors, Emeritus Professor A.M. Koldeweij and Professor Maarleveld, for jointly taking over from Professor Janssen; and Professor Koldeweij and the University of Nijmegen for allowing the ceremony to take place at Nijmegen University. Dr H.H.C.C. Clevis and Emeritus Professor Janssen I thank for their willingness to function as co-examiners.

I am grateful to Professor J.G.A. Bazelmans (*Rijksdienst voor het Cultureel Erfgoed*, RCE) for the faith he placed in my completing the manuscript, and for the funds he made available in 2009 and 2010 to have the standardized construction drawings developed and to have the drawings and figures of the Nijkerk-II and the Spakenburg-I cogs, as well as those of the Ens, Marknesse and Kuinre cogs, printed, by Evers Litho & Druk at Lelystad. In 2013, however, it became clear that the RCE's written commitment to funding and carrying out the thesis' publication was to be reneged on.

I am immensely grateful to Dr Clevis (SPA Publishers, Zwolle) for his tireless commitment to realizing this voluminous publication. Without his assessment and opinion that the 'cog book' simply *had* to be published, I would not be submitting this PhD thesis.

Drs. H. Dessens, Dr E.S. van Eijck van Heslinga, Professor A.F. Heerma van Voss, Mr T. Fruithof, I thank them all for their support in realizing the publication.

Further I am indebted to the following bodies and private individuals for their financial support: Prins Bernhard Cultuurfonds, the Samenwerkende Maritieme Fondsen, the Rijksdienst voor het Cultureel Erfgoed, the Stichting Professor van Winterfonds, the Stichting Hanzefonds, the Stichting Promotie Archeologie, the Stichting Promotie Archeologisch Onderzoek, the Ruud van Beek Stichting, the Modderman Stichting, the Province of Overijssel, the Province of Flevoland, the Municipality of Kampen, Ms H.C. van der Velde-Doyer, Mr P.B. Zwiers and Ms B. Espeet. I am grateful to my friends Hubert de Witte and Paul Zwiers for agreeing to be my *paranimfen* at the graduation ceremony.

I thank Mr H. Heikamp (SPA Publishers, Zwolle) for laying out the book and drawings, and Mr C. Bouwstra for 'vectorising' the drawings. Mr Heikamp's work has surpassed the result that I envisaged from the very start. SPA Publishers and the employees of the municipal archaeology department of Zwolle too I want to thank for their special and friendly cooperation. Also I have learnt a great deal about how a voluminous text with a lot of pictures and many outsized drawings is transformed into a book. I am grateful to the *Huygens Instituut voor Nederlandse Geschiedenis* for making the entire work digitally available through the 'Maritiem Portaal'.
I am grateful to Mr drs. F. Laarman (RCE) for performing the analyses of bone remains, and to Dr S. Comis for her part in studying the sailcloth from the 'Myosotis' excavation site in Kampen. I thank Professor E. Jansma and her team at RING for making available the tree-ring datings. Drs. W. Waldus and the *Archeologisch Diensten Centrum* (ADC) I thank for supplying photos and drawings of the Kampen (IJssel) cog. Ms E. Mook-Kamps performed the pollen analyses, for which I am grateful. I thank my ex-colleagues and other employees at the NISA for their help and their always cordial welcome at my former work location.

I thank Mr A. van den Heuvel, master carpenter with the firm of Sars Houtbouw in the cog shipyard at Kampen, for his contribution in building the replica of the Nijkerk-II cog (the Kampen Cog) and for his permission to reproduce the construction drawings that he made for Sars Houtbouw. And I thank Mr K. Sars for his remarks and additions to the text, and for making available his color slides documenting the various construction phases of the Kampen Cog. My gratitude of course includes all those who made available photographic material for this book. For translating the thesis into English I thank Drs A.C. Bardet, Ms J. Phillips, Ms C. Jefferis, Dr Dr S. Hoss, Dr S. van der Vaart-Verschoof and Drs. J. Vermeersch.

Above all, I owe the greatest possible gratitude to my dear wife Joke, who throughout my career as a maritime archaeologist and the many years devoted to this study, remained unreservedly supportive. Without her support, the 'cog project' would never have been completed, even though she would at regular intervals remind me that there is more to life than research into medieval ships. I dedicate this study to her, and to Angélique and Peter, to Dilâ, and to my two mentors in the 1950s, now both long departed: M. Gerrits (headmaster at Eesveen) and Dr F.D. Carels (school doctor in Overijssel).
All responsibility for the data presented here, the text and the drawings, and for the conclusions derived from them, is entirely my own.

Karel Vlierman

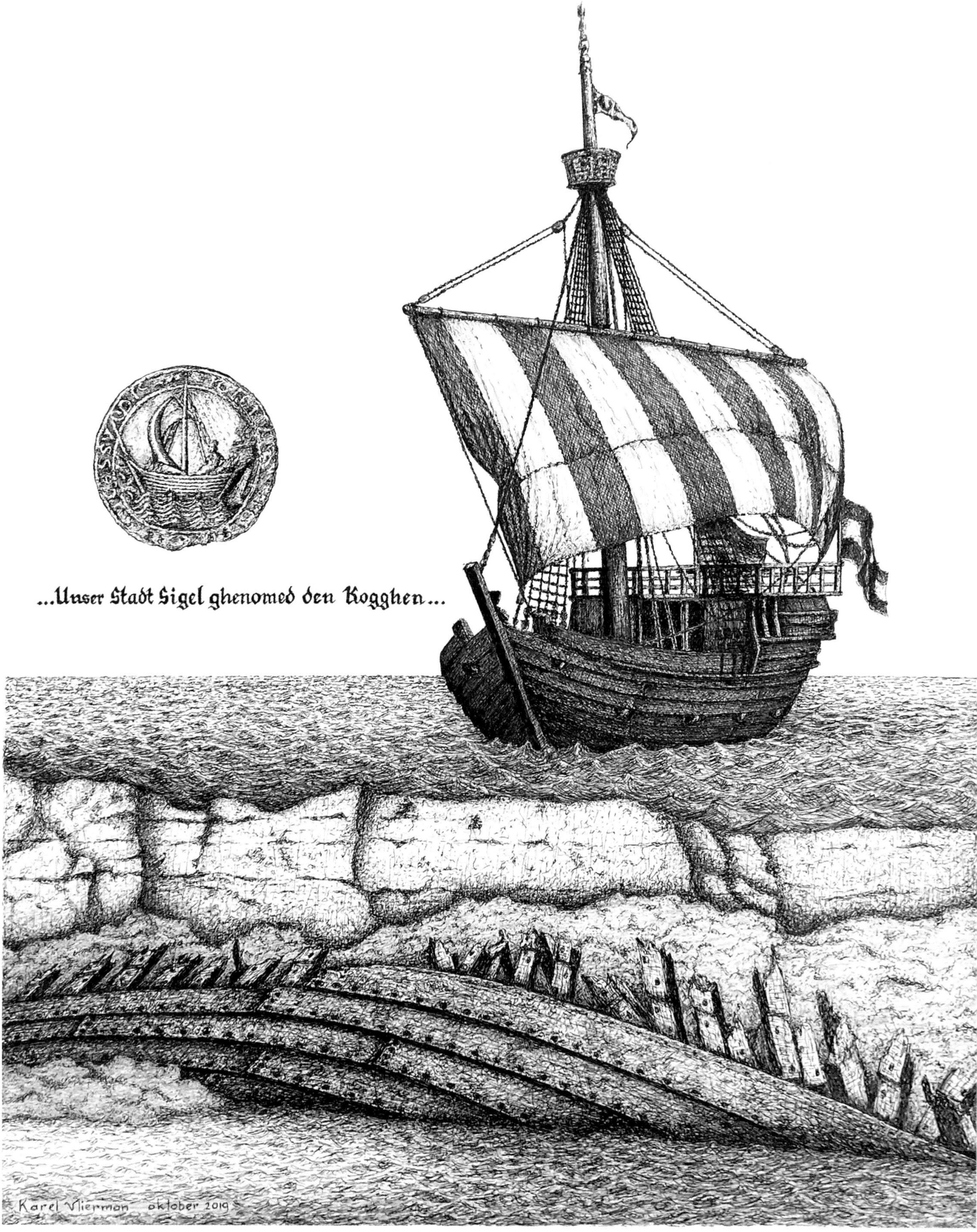
...Unser Stadt Sigel ghenomed den Kogghen...
Karel Vlierman oktober 2019

The cog: introduction[1]

Introduction

In late-medieval documentary sources, the designation 'cog' in the context of cargo shipping in Hanseatic times, as well as the word 'heerkogge' and contemporary images of sailing vessels on seals, have through the ages prompted researchers to wonder about this apparent vessel type. The seals are mainly civic seals of northwest-European ports situated on the lower reaches of the major rivers flowing into the Baltic and Noordzee.

These were the sources available to Heinsius when he wrote Das Schiff der Hanzischen Frühzeit (1956). Apart from the hulls on the seals, which show various details, in fact little could be said with certainty, given the rather scant further evidence regarding the structure and interior layout, or the building of these vessels. Yet when in 1962 the wreck of a large, late-14th-century cargo ship was found in the river Weser near Bremen, Heinsius' study did present sufficient information to immediately bring to mind a cog. An unusual circumstance was that this find comprised the complete starboard side, about half of the port side, the crossbeam constructions and large parts of the aftercastle - a virtually entirely reconstructable vessel. The image of a ship on a seal from Stralsund (Germany) shows a hull that almost fully matches the ship from Bremen. The letter to which this seal was attached mentions '*Unser Stadt Sigel ghenomed den Kogghen*...' ('our city's seal known as The Cog...') The seal dates from 1329; the letter, from 1483. This is the only historical source combining an image of a cog and a description referring to it.

The Bremen wreck was dismantled at the site and transferred in parts to the Deutsches Schiffahrtsmuseum at Bremerhaven. In the 1970s, the vessel was rebuilt there, from the keel plank and the stern and stem constructions upwards, plank by plank, and subsequently impregnated. At fairly regular intervals throughout the period of rebuilding and conservation, articles appeared about the ship, as progressively more evidence emerged about how it had been constructed. Lahn (1992) published about the rebuilding of the Bremen Cog, giving detailed attention to its technical aspects.

In the period preceding the recovery of the Bremen Cog and even during World War II, some shipwrecks had been encountered in the Netherlands, Denmark and Sweden, which had set researchers surmising that they might be cogs. However, these had been far less completely preserved, were small, or were not yet examined closely enough to allow anything definite to be said about them. Another limitation was that nautical archaeology was still in its infancy and hardly anything had been published about investigated vessels.

In most cases the shipwrecks were chance discoveries. The multiple wrecks in the harbour of Kalmar (Sweden) and in the Roskilde Fiord (Denmark) were exceptions to the rule. In the Netherlands, the situation was an entirely different one. Hundreds of shipwrecks were recorded and investigated during the first cultivation of the 'soft' sea-floor sediments of the former Zuiderzee (transformed into Lake IJsselmeer in 1932) from which large polders - now the province of Flevoland - were reclaimed between 1942 and 1968. This area is a 'ship cemetery' unique in the world, with wrecks dating from ca 1250 to 1900. From the start, a permanent team performed regular nautical-archaeological research in these polders, and until ca 2000, annually several shipwrecks were excavated and documented. Two wrecks of fairly large late-medieval cargo ships, excavated in 1981 and 1983, were found to have structural features noted also in the Bremen cog.

Prompted by these finds, visits to and fro sparked a growing interest in the cog also on an international level. Moreover, it became increasingly clear that nautical archaeology might contribute significantly to further understanding of the cogs known from documentary sources. Through comparison, wrecks excavated earlier in the Flevoland polders could now provisionally be labelled as 'cogs' or 'cog-like vessels'. Various prominent as well as novice researchers enthusiastically (re-)immersed themselves in the cog phenomenon. They came up with further, albeit still tentative new insights and conclusions about, for example, the ships' construction, range of dimensions, function, nomenclature, origin, dating and provenance. With researchers far

1. Translated by Xandra Bardet.

from agreeing on these topics. In hindsight, it can be said that at the time, terminology was used quite loosely with regard to the 'cog' concept. Even in the more recent literature it is unclear what exactly is meant by a 'cog', with different authors employing differing definitions.

Reinders (1985a) compared the nine then known wrecks from the Flevoland polders that might be cogs, formulated some characteristics, inferred that there had probably been a cog-building tradition producing various large and small types, and wondered whether they should all be termed 'cogs'. Also the dating of the various wrecks still was uncertain, so it was not yet possible to establish whether the cog had seen any evolution. Crumlin-Pedersen (2000) states that the name 'cog' for large, seagoing vessels like the Bremen cog by the turn of the century had become generally accepted and is appropriate as an archaeological term. Besides, he remarks that just a few wreck finds have been exhaustively investigated and published. And also that on the basis of the details presented of otherwise still unpublished vessels it is hardly possible to offer more than a first impression of the dimensions, dating, provenance and types of the vessel.

Meanwhile (by mid 2017) about 20 wreck finds have come to light in the Low Countries, with features (in the underwater body) that match those of the Bremen cog. These are over two-thirds of all known cog finds in Europe. A number of other, mostly smaller vessels display (some of) these traits as well. In the present study, a total of almost thirty shipwrecks from the Low Countries are described in detail and presented together with the excavation and reconstruction drawings. Almost certainly, each of these was built according to the cog-building tradition, in the ca 1250-1500 period, and in the (heartland of the) Hanseatic territories.

The mention of the designation 'cog' in 12th- and 13th-century sources and the dating of some foreign shipwrecks suggest that the method of building the large cogs evolved in the mid-12th century and was further developed in the latter half of that century. In the preceding period, but also during the heyday of the Hanseatic League, seagoing vessels outside the Hanseatic territories were built by a different method: fully clinker-built and riveted. This approach was rooted in the Nordic (Scandinavian or Viking) shipbuilding tradition.

The thirty wrecks from the Low Countries with their variation in dimensions probably represent the full range of vessels from the Hanseatic period and the Hanseatic territories. Therefore they can collectively be regarded as a reliable source for investigating what might be definitely called a cog, how these ships were built, what (functional) types can be distinguished, and what other kinds of vessel, smaller craft especially, may also be reckoned to this shipbuilding tradition.

The Kampen (IJssel) cog, built around 1420, is the youngest among the large cogs known so far. On the grounds of its deviating construction and dimensions, it may well be considered a 'key find'. Most probably it is an example from the final phase of the cog-building tradition.

No definite explanation can be given of the meaning and etymology of the word 'cog'. Maybe it is simply a Middle-German and/or Middle-Dutch designation or descriptive term equivalent to the English 'coaster'. A coaster, according to the Dutch definition, is a seagoing commercial vessel up to 500 register tons.

The historical sources also offer us various other names of vessel types. Archaeological ship finds that with certainty can be associated with any of these names have so far not been found. The same goes for the designation 'heerkogge' ('military cog'). In this case too, it is not clear whether we are dealing with a relatively small, rowed (river) craft to carry soldiers, with a man-of-war, or with an armed commercial cog. The relatively frequent mention in the sources of a 'cog' at any rate is congruent with the substantial number of excavated and structurally similar vessels to which the word almost certainly refers. Moreover, given the available historical and archaeological sources, there is no credible alternative interpretation.

The wrecks recovered from the river IJssel at Kampen in 2015 and 2016

The manuscript of the present publication, treating all cogs from the Low Countries, was completed in late 2014. The texts in chapter 23 about the Kampen (IJssel) cog, and about the small, cog-tradition crafts encountered with the cog (Sections 26.2 and 26.3) were at the time based on the results of the Inventariserend Veldonderzoek Onderwater-waarderend by ADC-Maritiem (Waldus *et al.* 2012) supplemented with my own observations and information obtained during a visit to the diving pod (on 23 October 2012) at the invitation of the researchers. Also it incorporated relevant findings from my own research on all the other cogs.

On 29 September 2015 the wreck of Object 14, initially typified as a punt, was salvaged. Immediately an offer was made to do reconstruction drawings of the vessel. Before the year was out, the set of drawings was ready, and this was followed by collaboration on the description and report on the boat. On 22 October 2015, Object 3, a barge-like wreck, was lifted from the river IJssel. No reconstruction drawings were made of this vessel.

The wreck of the Kampen (IJssel) cog was lifted from the water on 10 February 2016. I was privileged to be among the first who in a small cage were lowered by a crane into the wreck as it lay on a pontoon at the IJssel quayside in Kampen. On that occasion I could observe for myself the complexity of the wreck resulting from distortions and the detached timbers that still lay inside it, and consider what this implied for documenting and reconstructing the vessel. The researchers at that point assumed that it would be possible to achieve 3-D reconstructions by means of the planned 3-D photography and special computer programs.

After initially being invited to offer advice for a few days during the lifting and documenting of the cog, I was asked to join the research team once the wreck had been placed in the specially constructed conservation building at Lelystad. On the basis of many years' experience in documenting shipwrecks in Flevoland and reconstructing on paper all other cogs in the Low Countries, it soon became clear to me that 3D-reconstruction would probably be possible only if - with manually gathered additional data - 2D-reconstruction drawings could first be made. In the contract with the client (Rijkswaterstaat, the Ministry of Waterways and Public Works), no account had been taken of the possible need to prepare both 2D- and 3D-reconstructions. So additional to this contract, new requests had to be made from various subsidising agencies. In the summer and early autumn of 2016, an opportunity to gather the minimally necessary data was created on the understanding that the 2D-reconstruction drawings could be made. In January 2017 a start was made on this reconstruction; the full set of drawings was completed by 30 November 2017. A separate section describes how the 2D-reconstructions of the cog and the boat were arrived at. Waldus *et al.* (2019) discuss the 3D-reconstruction eventually made on the basis of the 2D-drawings.

The various chapters of the manuscript completed in late 2014 in fact remained virtually unchanged. The analysis and the conclusions are based on the then known evidence, and were retained as the state of research at that moment. Additional and/or new data and insights that the Kampen cog produced were incorporated in the summary of the rewritten Chapter 23, discussed in the sections on the boats and the barge in sections 26.1 to 26.4, added as final paragraphs to the Summary and Conclusions (Chapter 31), and outlined below.

Types

Among the investigated 13th- to mid-16th-century wrecks, a number of ships can now be identified with certainty as (large, seagoing) cogs. The vessels from a second group can be classified under the heading of 'small (seaworthy) cogs'; they differ from the first group in their deviating dimensions and height of the sides, but have a comparable length-beam ratio of 3.2:1 or a little less (measured fore-to-aft at gunwale level and across the greatest beam). A potential third group, also of small cogs, distinguishes itself from the others by slightly different, more elongated lines. They might be typified as small cogs for inland navigation. The somewhat different hull shapes may relate to specific functions, to the area served by these ships, and presumably also to their later date, after ca 1400 (evolution of the shipbuilding method). A fourth group, of generally smaller craft with a low freeboard, can on the strength of their typical constructions best be put under the collective heading of 'boats in the cog-building tradition'.

The terms cogs, small cogs and boats (in Middle Dutch: *Cogghen, Kleene Cogghen ende Schuten*), as in the title of the present book, are quoted from a letter by a German merchant in Bruges addressing the entire Hanseatic League in 1401.

Definitions

The range of vessel types identified within the cog tradition makes it impossible for a single, concise definition to cover all cogs. The differing forms of the most prominent structural features among the large and the small cogs are too great; they require explanation. Besides, there are typical (general) features that need to be highlighted to show how cog building was distinct from other shipbuilding traditions. Some of the investigated wrecks in terms of dimensions do not quite match the two definitions formulated below. This may be due to traditions in the region of provenance, particular functions and/or date of construction.

Cogs (*Cogghen*)

The (large) cog, right from the late 13th century, was a high-sided vessel, generally over 20 m long, with straight, inclined stern and stem and everted sides. From aside, it displays a marked sheer, which forward rises higher than aft. The ship has a single mast, carries a square sail and is kept on course by a stern rudder. The underwater body is sharp both aft and forward, and with flowing lines merges from there into the almost horizontal midships bottom. The transition from the bottom to the sides (the turn of the bilge) too has flowing lines, sometimes a little angular. From above, the outline of the gunwales resembles a laurel leaf in shape. From the underwater body with the sharp stern and bow, the lines of the strakes upwards and towards the stern and stem (through time) steadily became 'rounder'. At similar distances from the stern and stem, a cog is always 10-20 cm narrower in the after- than in the forebody. The length-width ratio at gunwale level is always 3.2:1 or a little less. The sides, over 4 m high amidships, are marked by a number of crossbeams protruding through the shell. They sit a little above halfway up the sides, and are aligned parallel to the sheer. Inside the shell, heavy standing knees were placed onto these beams, to which the strakes of the uppermost 1.5-2 m of the everted sides were fastened. To make up for the differences in elevation resulting from the sheer, filling rider beams of differing heights were placed on the crossbeams and under the knees, resulting in a near-horizontal deck level above them. Thick treenails joined the various components of the crossbeam construction together and to the shell planking. For further raising the sides in the 15th century, additional vertical reinforcements of the upper strakes were put in between the crossbeam knees in the form of heavy vertical riders. Crossbeams above the 'main deck' indicate that the vessel probably had more than one deck level. Inasmuch as can now be established, the 'main deck' was in the first instance supported by two long stringers to port and starboard of the centre line, which were fixed to just one crossbeam construction. Between these stringers and the sides and between the various crossbeam constructions, short stringers were placed, also longitudinally. Rabbets in the upper surfaces of the stringers hold short, transverse deck planks. In the afterbody, at different levels, sat short platforms with longitudinal planking, from which the windlass and the rudder were operated. With the exception of the long stringers, it seems that all further components of the decks were removable. The 3-3.5 m long spaces framed by the various crossbeam constructions and the sides could thus be opened up simply and completely for loading and unloading the ship. There are indications that intermediate decks too might be inserted, also consisting of loose parts. The cogs of the 13th and first half of the 14th century may have had freestanding castles in the stern and the bow. The second half of the 14th century probably saw the emergence of a more poop-like castle, a superstructure integrated with the hull. In the 15th century some (poop-like) castles probably for the first time were provided with a roof structure.

Small Cogs (*Cleene Cogghen*)

The 'small cog' is a vessel 11-17 m long with straight, inclined stern and stem and everted sides. From aside, it displays a marked sheer which rises higher towards the bow than to the stern. In some (river) cogs the sheer is less pronounced. The ship has a single mast, carries a square sail, is kept on course by a stern rudder and has no leeboards. There are indications that from about the mid 14th century (some) smaller cogs had a mast that could be lowered. The somewhat forward position of the mast may mean that these vessels carried a spritsail. So far, none of the evidence has suggested that these cogs could be rowed. The underwater body has a sharp stern and bow, but shows a flowing transition to the almost flat bottom. From the late 14th century, the more-or-less flat part of the bottom of (river) cogs becomes longer and the sides are built somewhat steeper. The turn of the bilge may be flowing, rounded or angular. From above, the outline of the gunwales resembles that of a laurel leaf. At similar distances from the stern and stem, the small cog always is 10-20 cm narrower aft than forward. The length-width ratio at the gunwales always is 3.2:1 or a little less. In the case of 'river' cogs, it is a little more. Depending on their length, small cogs amidships are ca 1.5 to 3 m high. The hull basically comprises three sections: the afterbody, the midship and the forebody. These are separated by comparatively heavy (topside) beams, resting on thick, waterway-like gangways and on frames that locally do not extend up to the gunwales. The gangways often run on into the after- and foredeck. The foremost beam usually is the most massive and functions also as the sail beam, supporting the mast. The after beam generally carries a heavy windlass. In some 'river' cogs, a third heavy beam may rest upon the gangways amidships. The connection between the beams and the sides may be reinforced with iron rosebolts. Onto these two or three transverse beams were fixed short standing knees to which the upper strakes of the sides were treenailed. In the afterbody, a deck was built. The longitudinal deck planks rest upon the heavy beam and on some lighter deck beams. At the sides, these planks are fitted around the frames and to the shell. As early as the first half of the 14th century there is evidence of an oblique coaming on the inboard edge of the gangways. To date, there is no archaeological evidence to suggest that the holds of late-medieval small cargo vessels were closed with hatches. Most probably they were open. However, a not widely known drawing in the Kampen City Digest (*Digestum Vetus*) dating from 1456 shows six ships (probably including two small cogs), three of which are provided with a convex structure above the hold. The foredeck is formed by one or more stringers supporting transverse deck planking. On the after- and foredecks, the low bulwark is fixed to a few short standing knees. The components of the decks are held together with treenails and iron nails; the gangways, held with treenails onto the frames and shell strakes.

- Cogs and boats have straight, inclined sterns and stems. The rake of the bow is usually greater than that of the stern. Sometimes the rakes (in smaller ships) differ very little.

- The bottom is carvel-built , with the aid of temporary framing. These frames were to be later replaced by permanent ones. Towards the stern and bow, the bottom strakes transition from carvel to clinker constructions. In 13th-century cogs the bottom strakes or even all strakes of the underwater body are fully carvel-built. There is a brother-and-sister pattern; that is to say that, in principle, in the equivalent positions in the hull to port and to starboard, planks are used that were sawn adjacently from the same tree. The side strakes are joined from stern to stem in clinker fashion. The overlapping planks/strakes are held together with iron nails twice beaten sideways and driven back into the timber (double-clenching).

General features of cogs

As an outcome of this study, the definitions of the various vessel types in the cog-building tradition can be elaborated with the following general features.

- Cogs are not built on a keel or keel beam, but have a keel composed of a thick keel plank which at both ends is extended with a knee-shaped timber, a (stern or stem) hook. The hooks also constitute the lowermost parts of the stern- and stemposts. The various timbers are held together by flat scarfs. In some cases there was no keel plank, but just two long hooks scarfed together. The boats after ca 1400 have a keel plank with a hook-like stern and stem fixed onto the ends of the keel plank, extending a little beyond the after and forward ends of the plank. The keel plank of these boats is of the same thickness as the other bottom planks.

- The strakes of the underwater body of cogs are nailed into rabbets in the hooks (and in the lower parts of the stern- and stemposts). The strakes of the sides end on the mitred rear surface of the sternpost and are nailed onto it. On the bow this is also the case on the similarly mitred front of the stempost. Here the fragile hood ends are protected by a (massive) false (or outer) stem, which covers them and is fixed to the stempost. Through time, the side strakes in the small cogs end in rabbets progressively higher up the sides of the stempost. In the boats, the rabbets extend upwards to hold even the top strake.

- The shell strakes are composed of a few or multiple sawn planks. The planks of a strake are joined by oblique ‘flat scarfs’ secured with double-clenched nails - in principle with two rows of nails: one driven from the inside, the other from the outside. The plank ends at the scarfs are one to several centimetres thick, and stand proud of the shell both inside and out.

- The seams of the hull were rendered watertight with so-called sintel clamped moss caulking. Into the seams, first reamed into a V shape, strands of moss (*Sphagnum*) were crammed, which were covered with split oak laths ca 1.5 cm wide and 0.5 cm thick ('moss laths') which in their turn were secured with forged iron clamps (sintels). The sintel clamps were driven into the strakes on either side of the seam. Between ca 1150 and 1500 the shape of these sintel clamps evolved from flat, narrow clamps to clamps with an oval central part (the shape of 'cigar bands' or 'butterflies'). The various consecutive shapes can be dated to periods of 50-75 years, some to even shorter phases. After ca 1500 the method is applied less and less in large and small cogs, and over the first half of the 16th century finally disappears, together with the cog-building tradition.

- The hull was constructed in phases, according to the bottom first/shell first principle. The connections between the shell planks and the frames are made with treenails: wooden pegs 2.5-3 cm in diameter. Until the end of the 14th century, the fastenings between the massive crossbeams, bulkheads, standing knees and the upper parts of the sides were executed with treenails (3-5 cm thick) and the occasional heavy iron bolt. After this time, heavy iron bolts (30 mm thick) were increasingly used in crucial fastenings.

- Above the keel plank and on the frames lies a long keelson with a mast step. Together with the open ceiling, this ensures the longitudinal stiffness on the inside of the hull. In smaller cogs, the mast is set into a cavity in a frame, or in a short separate mast step on the frames.

- The rigidity on the inside of the hull in smaller cogs is provided especially by (heavy) bilge deals, a few ceiling strakes and the heavy gangway planks. Sometimes the hold has a fully closed underfoot ceiling.

- At any rate until the end of the 14th century, the cogs sailed with square, tarred sails of coarsely woven wool. Some were woven in bands of different colours.

- The ships and boats in the cog tradition were built from oak timber and are the earliest vessels in northwestern Europe -based on archaeological finds only- to be steered with a stern rudder.

Summarising conclusions

The present study of the thirty-odd shipwrecks in the cog-building tradition recovered in the Low Countries allows the following conclusions.

- Two types of cog can be distinguished: cogs and small cogs. Besides, there are boats of various sizes which clearly display features and methods typical of the cog-building tradition. The small cogs may be subdivided into seagoing and river cogs. The latter type is not quite clear-cut because the differences from the seagoing type are small and may possibly reflect slowly evolving methods of construction.

- It is likely that the speed of evolution of the construction style of large vessels in the 15th century differs by shipbuilding region and may relate to local economic conditions. For instance, in the case of the Kampen (IJssel) cog, built after ca 1420 (the youngest known cog), the shell seams still are closed with clamped moss caulking on the inside and outside, while for cogs around 1350 there already is a growing trend to apply caulking to the inside seams only. On the other hand, this hull is much higher than in any other cog (5.85 m instead of a good 4 m amidships), with vertical riders up the inside and outside of the bulwarks and two wales on the outside of the shell to strengthen these unusually high parts. Also, greater numbers of heavy iron fastenings were used.

- The (large) cogs were seagoing vessels. So, it appears, were some of the smaller ones, suited especially for the Zuiderzee inlet and the Frisian Waddenzee. Other small cogs, given their sheer, seem less appropriate for maritime navigation and probably were intended for inland transport.

- Given the long period during which cogs remained largely unchanged (ca 1200-1500), they must have been vessels constructively well-suited to their purpose. In the period concerned, the (large) cogs would have been a prominent sight on the northwest-European waterways, and certainly would have met the contemporary requirements of capacity and possibilities for transport of various goods and commodities.

- The loadbed of branches, twigs and straw in the forward part of the hold, identified with certainty in the recent (2016-2017) investigation of the Kampen (IJssel) cog, indicates that people went to great lengths to keep the cargo in the hold well-stowed and dry. Rainwater, spray and leakage would seep in through the chinks in the deck and along the frames, and through the limber holes in the floor timbers on the bottom would be led to a pump. Through the pump and a drain, the water would be discharged overboard. The earliest scuppers on a watertight deck so far are known from a *waterschip* of the mid 16th century.

- Watertight packaging of cargo at this time could only be achieved with wooden barrels. On the Kampen (IJssel) cog, some individual items could be loaded and unloaded through an opening in the port side, at the level of the crossbeam constructions.

- Also in the Kampen (IJssel) cog, the first clear evidence was encountered of removable between-decks. Whether crew or passenger quarters were provided there - or elsewhere in the ship - is impossible to tell.

- A built-in hearth and a domed, brick-built oven on the 'main deck' to starboard in the afterbody, and the locally somewhat unusual construction of the deck, suggest an 'open' galley. It is hardly possible to establish whether any of the ships served a military purpose. Aboard the Kampen (IJssel) cog some stone (slingshot) balls were found. Some spearheads turned up in the Dronten-I and Swifterbant cogs; and in the the Nijkerk-I cog, an axe that probably should be classified as a battleaxe.

- A shipyard dated to ca 1175-1375, discovered in 2014 on the Burgel channel in Kampen, is the first Hanseatic-period shipyard to be discovered in the Netherlands. A daring hypothesis states that the Burgel most probably was the main channel of the IJssel during the earliest (at any rate 12th-century) phase of the settlement of Kampen. Implying that the earliest dwellings and the predecessor of the Bovenkerk church were built on the right-hand bank of the river, in the Oversticht seigneury.

The present publication describes all of the archaeological evidence and ensuing reconstructions of ships and their contents in words and images. Most chapters also include one or more pen-drawn impressions of the ship and its inventory. Drawing together all of the Low Countries' cog finds known by mid 2017 makes it possible to extract more evidence from the wrecks recovered in a fragmentary state. Thanks to the opportunities for comparison, it proved possible in many cases to reliably reconstruct greater parts of these vessels than initially assumed. Any available iconographic and other (documentary) sources were employed to achieve maximum result. By means of this publication the combined data will be made available for further research by nautical archaeologists, shipwrights with an interest in medieval shipbuilding, historians, modelmakers and other interested people. Hopefully this study will prompt further research into this fabled medieval ship. The separate chapters treating individual cogs and the shipyard at Kampen can be read as stand-alone reports.

After its spectacular lifting, the research on the Kampen (IJssel) cog over the 2015-2018 period has made it clear that as a result of the defective national government policy with regard to Dutch nautical archaeology, adopted at the close of the 20th century, it in fact is no longer possible to make reliable 2D-reconstruction drawings of archaeological ship finds. Hence the creation of reliable 3D-reconstructions, more suited to the digital era, has been rendered virtually impossible.

The present study of this highly interesting and internationally unique group of medieval cog ships should therefore be a wake-up call: an urgent plea for the Government's reinstatement of the Nederlands Instituut voor Scheeps- en onderwaterArcheologie (NISA, Netherlands Institute for Ship- and underwater Archaeology) under the Rijksdienst voor het Cultureel Erfgoed (RCE, Cultural Heritage Agency of the Netherlands), which was closed down in 2016. It needs to be revived as an independent research institute, and indeed expanded. Such an establishment is truly indispensable for the Netherlands as a great, if not the greatest, seafaring nation in world history. Almost a hundred shipwrecks of supreme interest still lie embedded in the former Zuiderzee, that urgently require excavation. The preservation conditions of most of these wrecks are so dire that in about 50 years' time nothing will be left of them; the evidence of the life and work of our seafaring ancestors and the ships in which they sailed, preserved for centuries in the Zuiderzee sea floor, will be inexorably lost. Besides, at least 150 files relating to unique late- and post-medieval shipwrecks, already excavated, need analysis and publication. And this is without even mentioning the dozens of submerged (VOC) shipwrecks of our national maritime heritage, that within and beyond the territorial waters of the Netherlands also await our attention.

Two methods for building: the Atlantic completely clinker built *nef* an the cog.

Introduction[1]

1.1 Two methods for building seagoing ships

In 1962 the wreck of a large, high-sided, late-14th-century cargo ship was recovered in the river Weser at Bremen (Germany) (figs 1.1a and b). Since then, certain archaeological finds of late-medieval shipwrecks in northwestern Europe have been labelled as cogs. In documentary sources of the 1200-1450 period, this designation mainly refers to large, seagoing cargo ships of the Hanseatic era. Before 1962 little was known about this vessel type. From the mid-19th century, which saw the rise of an interest in these topics, it was only civic seals and other documentary sources that offered historians with a maritime interest any evidence of what such ships might have looked like. Hardly anything was known about their construction and dimensions. Then Reinders compared a number of wrecks from the Zuiderzee area with the Bremen cog and concluded that they belonged to the same shipbuilding tradition.[2]

Since the 19th century, wrecks of other seagoing ships, dating from the preceding period (ca 1000-1200), were known from northwestern Europe and particularly from Scandinavia. They are slender vessels, built in the Nordic style, with relatively low sides, rooted in the Viking shipbuilding style and influenced by traditions along the coast of the southern Baltic.[3] By ca 1980 some thirty such finds were known, dating from 800-1200. The ships of Gokstad and Oseberg (Norway) and Roskilde (Denmark) are the most famous examples. About half of them may be cargo vessels.[4] In recent decades further finds have emerged in Scandinavia, elaborate research has been conducted into construction and dimensions, and thanks to the replica construction of various ships much knowledge could be gained about their handling properties and how to sail these replicas.

In the Netherlands, so far just a few ships' timbers have been found of vessels built according to the Nordic or Viking shipbuilding method. Four finds date from ca 750-1050. Just one dates from the 13th century; some pieces of shell, from the 14th and/or early 15th century. A further two vessels were found that must have been built in the 15th century; four others, in the 16th century. They are briefly discussed in section 3.2 and Chapter 27. Although we cannot be totally certain, it seems very likely that in these Dutch finds we are dealing with vessels built in the former Frisia, England, Scandinavia or the Baltic area.

Fig. 1.1a. The reconstructed Bremen Cog in the Deutsches Schiffahrtsmuseum in Bremerhaven. Exterior of the starboard side with the stem (and false stem). (Photo: Deutsches Schiffahrtsmuseum).

Fig. 1.1b. Interior of the Bremen Cog viewed from the port side of the forebody. (Photo: Deutsches Schiffahrtsmuseum).

1. Translated by Xandra Bardet.
2. Reinders 1985a.
3. Crumlin-Pedersen 1997, 20-31.
4. McGrail 1981, 31.

Crumlin-Pedersen[5] presents an overview of the nine ships found in Scandinavia and classified as cargo vessels of the Nordic tradition from the 1000-1200 period, also referring to future publications. He compares their data with those of the Bremen cog, and with the available evidence of the eighteen wrecks then known (of which nine in the Netherlands) that structurally match the Bremen wreck. He remarks that just a small number of these ships excavated between 1943 and 1993 has been fully published, making it possible only to present an initial impression of the dimensions, dating and provenance of the type. Over the past two decades further finds have come to light around the Baltic; in the Netherlands, by now some twenty ship finds with these features. They date from ca 1250-1500.

Insofar as currently can be established on the basis of archaeological finds, northwestern Europe between ca 1000 and 1450 saw two different methods by which seagoing ships were built.[6]

1. The Nordic method, which is rooted in Viking shipbuilding and which in Scandinavian wooden vessels has continued even up to today.

2. The building method believed to be that of the cog, which seems to have evolved in the second half of the 12th century; the last large ships built fully according to this tradition probably sailed around 1500.

Section 3.5 discusses the most prominent differences between ships of the Nordic and the cog traditions.

Besides the regularly encountered 'cog', the documentary sources also mention names of other vessel types. Yet none of these sources provides sufficient information to reveal to any extent the nature, shape and dimensions of these ships, and whether they were built in the Nordic, cog, or any other style. Judging by 13th-century English civic seals which probably depict the frequently mentioned Atlantic nef, this ship seems to be an example of the Nordic building method. Unfortunately, no archaeological finds have yet been linked to the *nef* or to the name of any other late-medieval vessel type.[7]

Northwestern Europe by now has a substantial number of medieval extended logboats, boats and barges. They were used on the rivers and other inland waterways. In a number of Dutch and Flemish craft (some) structural (detail) features have been identified that also occur in the Bremen cog and related vessel finds. These are briefly discussed in chapter 26. This has (by 2016) raised the number of vessels in the Low Countries that can be related to cog-style construction to ca 30.

1.2 Aims and significance of the study

Aims

By the turn of the century, a small number of northwest-European wreck finds were generally considered to be, like the Bremen find, (large) cogs. However, most of the other Dutch finds are smaller vessels, and some researchers doubt whether they should be labelled cogs at all. The designations 'cog' and 'cog-like' in reference to these wrecks, as remarked above, are based on a number of typical structural features that were identified in the underwater body of the Bremen cog. What came as a surprise at the start of the study was the observation that - with the exception of a few finds built according to the Nordic method - all other 13th-to 16th-century wreck finds in the former Zuiderzee area displayed the features typical of the large cogs. So there was no need for selection. So far, publications have appeared about just a few of these ships, and some of those are outdated and/or very succinct. On the basis of these publications and the present details of existing cog finds, what tends to happen in the case of a new find is that with respect to the dimensions, the type, dating, function and provenance just an initial impression is presented, together with a reference to the published vessels. The principal aim of the present study is to change this situation, creating greater clarity by working out and publishing the available data relating to each individual known cog wreck from the Low Countries, including the previously published wrecks. The emphasis thereby lies on the technical aspects, the hull shape and the constructions of the vessels. These comprise a good two-thirds of all European cog finds. Elaborating on the series of excavation reports by the Nederlands Instituut voor Scheeps- en onderwaterArcheologie (NISA)[8] and adhering especially to the format of the publication on the Almere W13 cog,[9] it seemed desirable to create as detailed as possible a reference study, while developing a standard format for research reports and (re-)construction drawings.

5. Crumlin-Pedersen 2000.
6. Vessels from the Slavonic and Wendish territories are here reckoned to the Nordic tradition. Owing to the lack of archaeological finds, there is no concrete evidence that the scapha and the barca of Carolingian times (Asaert 1974, 14) still sailed in the Ottonian period.
7. The large 16th-century, armed cargo or warship wreck found in lot U34 in Oostelijk Flevoland is tentatively identified as a hulc (Reinders & Oosting 1989, 106-122).
8. The NISA, founded in 1995, was slimmed down in the new millennium and passed into the department of Ship Archaeology of the State Service of Cultural Heritage end was closed down in mid 2016.
9. Hocker & Vlierman 1996; the Almere-I cog, Chapter 16.

Frame of reference

As the subtitle indicates and as has been explained above, the starting point for the present study were the archaeological wrecks of 13th- to 16th-century vessels built in the tradition of the (Hanseatic) cog that have been uncovered in the Netherlands and Flanders. These shipwrecks are compared with each other and placed in a wider perspective. This means that the excavated cogs and vessels with cog-like features are the principal focus, and that the highly detailed (scale 1:10) reconstruction drawings, based on the recovered remains, are an essential part. This systematic analysis of the eighteen investigated shipwrecks will not only provide an anchor for further historical and more theory-oriented research, but will also allow us to better assess previously published views on 'the cog' in terms of ideology or historiography. The emphasis in this investigation has been on concrete, material sources, which generated the research questions enumerated below. To conclude this two-volume study, I shall in the Epilogue (see p. 835) review the project as a whole, and point by point formulate concise answers to these fifteen questions, while also providing a more detailed account of my chosen starting point.

Research questions

Comparison of the data in the excavation files of all 13th- to 16th-century shipwrecks from the area of the former Zuiderzee, as noted above, revealed that virtually all were built with features like those encountered in the Bremen cog. I documented the twenty-plus wrecks of large and small(er) ships with these traits. The features are described in section 2.4. The questions are:

1. *Do they belong to the same construction type and were they built to serve a particular geographical area?*

2. *Are there specific (functional) types, to be identified by, for instance, dimensions, shape or internal layout?*

3. *Were they all (in a particular area) built following a generally applied tradition, and what do different types look like?*

4. *How were they built?*

5. *Can we identify a particular sequence in their construction?*

6. *In the past, these ships have been labelled as 'cogs' or described as 'cog-like' on the basis of incomplete documentation. Was this justified?*

7. *From the literature we know that in the Baltic area too, several wrecks have been identified as cogs. Do their structural features match those of the finds in the Low Countries?*

8. *Were these vessels demonstrably built in the Baltic area, and if so, can we spot regional differences?*

9. *Why did the (seagoing, large) cog become such an apparently dominant type both in the written sources and in archaeology? Was this due to the nature or volume of the cargoes, the seas it sailed and/or the ports of call?*

10. *What are the dates of the ships (years/periods in which they were built and foundered); and is there any evolution in their construction? If so, did this occur simultaneously across north-western Europe/the Hanseatic territories?*

11. *Where did the recovered ships come from (where were they built)?*

Other topics too figured in the research, such as:

12. *Obtaining more certainty about the already mooted suggestion that the datable evolution in the shape of the sintel clamps and the application of the method of clamped moss caulking*[10] *are directly linked to the area and the period in which the cogs sailed.*

13. *Earlier researchers already wondered whether the cog was a new vessel type that arose in a certain brief period, or instead a long-established type of ship gradually evolving in the emerging Hanseatic territories - questions to which this research might provide answers, or at any rate additional data and/ or new insights.*

14. *The region where the cog and its building tradition may have arisen, when this took place, and the possible underlying reasons are further intriguing aspects, and secondary aims of this study. Any relation to the Cru- sades and the Hanseatic League should not be ignored in this context.*

15. *The standardised description also of the objects found inside the cogs may provide a better insight into the ship's equipment and inventory, and greater clarity about the number of crew aboard the various ships.*

Chapter 4, 'The medieval cog, the state of research' describes to what extent the historical and archaeological sources have through time already provided answers to some of the questions listed above. 'Analysis and Conclusions' will clarify whether those answers still hold and agree with the results of the present study.

10. Vlierman 1996a, 54.

Significance

From the start, it was obvious that first the data of all relevant shipwrecks should be worked out into individual excavation and research reports, before there could be any comparative research within the group as a whole. Without any doubt, the descriptions and (re-)construction drawings of the eighteen wrecks treated in detail in this study provide enhanced insights into the construction (sequence) of the vessels, and supplement the previously known technical data and existing publications. This increased insight is in part due to the experience gained in the remaking of the Nijkerk-II cog, the most complete Dutch cog wreck (discussed in Chapter 6): the replica Kamper Cog. The Nijkerk-II wreck was selected as the basic example for comparing the data of the various vessels, and together with the Kampen Cog constitutes the reference/research model. As progressively more ships were worked out and the distinction between large and small cogs became increasingly clear, it was obvious that in the treatment of any subsequent wreck greater attention was given to the specific constructions and components of the vessels from either of these two groups.

The questions at the start of the investigation were in part answered, while new questions arose as well. Both are important for future research. We should also gain a growing insight into the kind of ships in which the important and increasingly voluminous transportation of goods within that whole took place.

Analysing and publishing in a standardised format the descriptions and images of more than two-thirds of all European finds with cog features will considerably increase our knowledge about the late-medieval cog. Besides, more clarity can be achieved about the significance, the function and the form of the various types. The research outcome also contributes to a clearer and more realistic understanding of the ships, shipbuilding and waterborne transport in western and northwestern Europe in the ca 1150-1450 period, as well as offering a good impression of the wealth of the Dutch maritime cultural heritage, unparallelled worldwide, and enhancing our insight into the practice of Dutch nautical archaeology as it evolved between 1942 and ca 2000. Figures 1.2 and 1.3 indicate the findspots of the large and small cogs and the boats with cog features discussed in this study.

Shipwrecks in the Zuiderzee area

Between 1945 and the early 1970s the field drawings of the uncovered wrecks, together with the usually modest number of photos and slides plus the generally succinct descriptions or scanty notes, were preserved in the files of the ships in question. Only in 1975, after the appointment of H.R. Reinders as archaeologist and head of the department of Nautical Archaeology of the Rijksdienst voor de IJsselmeerpolders (RIJP, IJsselmeerpolders Development Authority) and P.B. Zwiers as shipbuilding engineer, did nautical archaeology properly take off - from excavation and reconstruction to publication. I personally became increasingly involved in the working-out of the assembled field data and the earliest publications of excavated wrecks. These activities taught me how complex ships are, and how difficult it is during excavations to correctly decide what data should be recorded - or not - to allow reconstruction of the vessel in question. Through the collective working-out of the data[11] it became clear what should be given particular attention in future shipwreck excavations, and what needed recording to optimise the possibilities of reconstruction. During my official employment at NISA (1969-2002) the amount of recorded data per wreck gradually increased about tenfold. This meant that the files on the whole contain far more evidence about the ships excavated in the 1990s than about the previously investigated vessels. Since I had been involved with a large number of ship excavations over that period, and was able to contribute to improving the necessary documentation, I am in a position, retrospectively, to speak authoritatively about details of the older excavations. For anyone who did not take part in the excavation and documentation of a wreck, this would be very difficult or indeed nigh-on impossible.

Evaluating investigation

For the shipwrecks discovered in the reclaimed polders of Flevoland, this meant in practical terms that up to the mid 1970s they were excavated in random order. An important point was that any wreck would have to be removed before the reclaimed land could be given out to a tenant farmer. The numerous ship finds in the Noordoostpolder, but especially the fast-growing number of finds in Oostelijk and Zuidelijk Flevoland eventually made it impossible to maintain this procedure. Therefore a second excavation team was set up in the second half of the 1960s. So far, some 450 shipwrecks have been recorded in the area of the former Zuiderzee. They date roughly from the 1250-1900 period. Owing to the expanded excavation capacity and the excavation of more ships every year, the number of new finds continued to grow relentlessly, and ways were sought to protect them in situ against rapid further decay. Some fifteen special and fairly complete

11. Modderman initiated this in 1945; after which the first wreck to be actually reconstructed on paper was the *waterschip* from lot W10 in Oostelijk Flevoland, but only in its excavation report of 1978 (Reinders *et al.*).

Fig. 1.2. One of the earliest maps of the Zuiderzee (mid-16th century) with the findspots of the cogs and boats with cog features and some vessels of the Nordic shipbuilding tradition. The numbers correspond with the chapters in which the respective wrecks are discussed (Brussels Atlas 1573 of Christaan Sgroten).

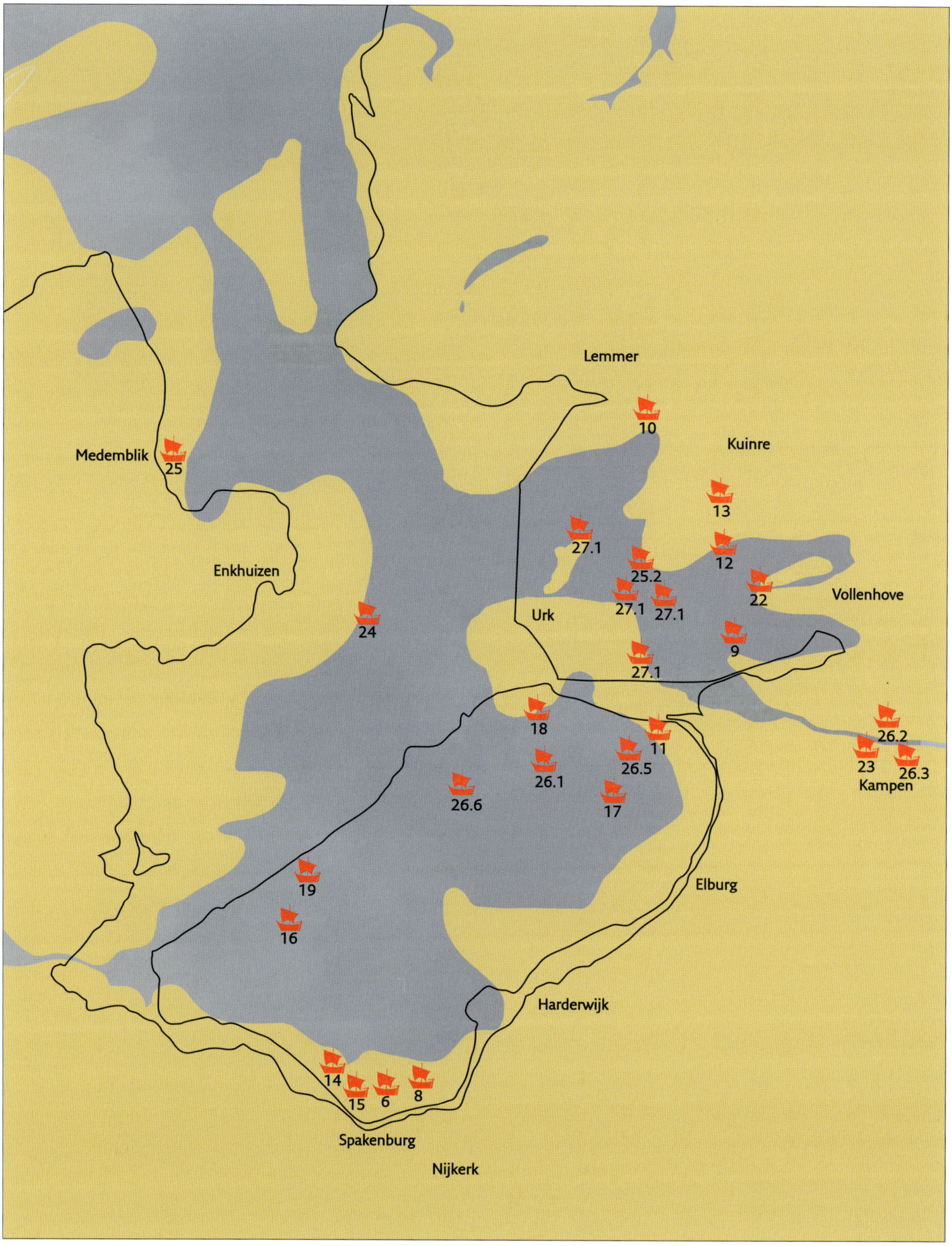

Fig. 1.3. Map of the Netherlands showing the areas of land and water in the 11th and 12th centuries with the findspots of the cogs, small cogs and boats with cog features and some vessels of the Nordic shipbuilding tradition. The numbers correspond with the chapters in which the respective wrecks are discussed.

wrecks are now protected by a newly developed method.[12] For most shipwrecks this is not possible, because of local soil conditions. Since decades there have been about 70 wrecks that, to prevent the evidence of these 14th- to 20th-century ships from being irretrievably lost, should be excavated as soon as possible. They still lie in situ. Driving around the IJsselmeerpolders, one may here and there spot a post with a blue-and-white chequered pattern topped by a little red sailing ship. The post indicates that one of these shipwrecks still lies buried in this field. Among them are the cog finds marked in Table 1 with w.o. (evaluating investigation) performed.

From the mid 1970s, a reported ship find prompted an immediate, more elaborate (evaluating) investigation and ensuing documentation. Until then it was usually deemed sufficient to record the location, a brief description of what had come to light during discovery and to secure any movable finds. But from that time onwards, test trenches were dug across the wreck and the stern and stem timbers were uncovered; and files were created in which the observations were recorded in detail, complemented with survey and construction drawings, section drawings, photos and slides. In this way we recorded what type of ship it probably was, how complete it seemed to be (e.g. presence of decks, stern and stem), its apparent antiquity (from the mid 1980s also by means of tree-ring dating), the local soil stratigraphy around the wreck, and possibilities and recommendations for protection in situ or for early excavation. Then also a programme was intiated to perform similar evaluating investigations also of the wrecks known since World War II. This gradually produced a solid, substantive body of data covering the then known 100 to 150 shipwrecks and their scientific potential. Up till the late 1970s, the timbers of excavated ships would after completion of an excavation be burnt. Since then, most timbers of suitable ships have in principle been stored in a field depot. Besides the large numbers of wrecks discovered during the first cultivation of the reclaimed sea floor, also many historical wrecks have been located elsewhere in Dutch waters. In a manner adapted to underwater conditions, these too are subjected to evaluating investigations.

The new Dutch national policy with reference to nautical archaeology

Around 2000, the changed government policy as regards Dutch nautical archaeology meant that NISA's painstakingly amassed expertise and experience over the 1945-2000 period with respect to this globally unique maritime heritage would actually be no further expanded, but as it were 'laid up'. The 'market' was to take over, and of necessity started in many cases to reinvent the wheel. Besides, new regulations were introduced with which (evaluating) research reports were to comply. In general, these reports largely consist of - doubtless invaluable - administrative data. For the actual subject of research - the archaeological investigation, analysis of the collected data, reconstruction of ships and publication of the results - there has seemed to be ever less scope, or none at all - a most troubling development. To me this was a major reason for focusing on these late-medieval cog finds, that are of international importance to nautical archaeology; besides, I felt a responsibility for securing, making accessible and interpreting the data recorded during the excavations of these ships which so excite the imagination. It is a (full) group of wrecks excavated by NISA, the first, and most probably also the last, to be extensively investigated, described and drawn, and published in this way. This study is also intended as a manual for (future) nautical archaeologists, lay enthusiasts involved in nautical archaeology and any further interested parties, to give them an idea of the complexity of nautical archaeology and of how good results can be achieved by an 'old-fashioned' approach. Given the unique nature of this National Maritime Heritage, we should, for any further wreck that comes to light, content ourselves with nothing less.

1.3 Sources

The excavation and evaluating investigation of wrecks, and the Kampen Cog

With the exception of the above-mentioned wrecks of the Nordic shipbuilding tradition, the source material for this study is formed by all late-medieval shipwrecks with structural features typical of the large cogs, which have been excavated and/or undergone evaluating investigation by the Netherlands Institute for Ship- and underwater Archaeology (NISA) and its predecessors in the polders of the former Zuiderzee. Also included are two wrecks discovered and salvaged by the Archeologische Dienst Waasland (ADW, Waasland Archaeological Agency) in Flanders (Belgium), near Doel on the river Scheldt. Briefly, attention will be given to the evaluating investigation and salvaging of the shipwrecks in the river IJssel at Kampen by ADC Maritiem on behalf of Rijkswaterstaat Programmadirectie Ruimte voor de Rivier (RWS PDR); to the wreck discovered on behalf of Rijkswaterstaat-Directie IJsselmeergebied during the dredging of the shipping channel in the Noordelijke Aanloop Houtribsluizen (Lake IJssel); to the evaluating investigation of the wreck in the Lake IJssel 'Zeebad' at Medemblik; and

12. Vlierman 1997a, 15-20.

Tabel 1. Cog-finds in the Low Countries

Lot/findspot	excavation in	Name	Construction date
Zo 43	1981	Nijkerk-I cog	1225-1250
Nq 75	1949	Ens cog	1225-1275
Na 57	1985	Rutten cog	1275-1285
On 5	1976	Dronten-I cog	1300-1325
Zo 36	1983	Nijkerk-II cog	1325-1350
Nm 107	1944	Marknesse cog	1340-1350
Nm 133	1943	Kuinre cog	1350-1375
Zn 42II	1979	Spakenburg-I cog	1385-1400
Zn 43	1979	Spakenburg-II cog	1415-1430
Zw 13	1986	Almere-I cog	ca 1400
Om 61	1995	Dronten-II cog	1425-1475

Supplemented with the not yet fully excavated and/or not yet fully documented or elaborated finds from the Netherlands and Flanders.

Og 77	w.o.*	Swifterbant cog	1300-1325
Nt 25	w.o.*	Kraggenburg cog	1310-1320
Doel, Belgium	2000	Doel-I cog	1325-1330
Zc 46	w.o.*	Almere-II cog	1325-1350
Doel, Belgium	2002	Doel-II cog	1330-1350
Kampen (IJssel)	w.o.*/2016	Kampen (IJssel) cog	1400-1450
Noordelijke Aanloop	w.o.*	Enkhuizerzand cog	1470-1485
Houtribsluizen (IJsselmeer)			
Medemblik (Strandbad)	w.o.*		1450-1500
Ng 37	1955		15th century?

w.o. = evaluating investigation. These wrecks are still present in their find spot.*

to a wreck excavated in the Noordoostpolder in 1955 (Table 1). Also ship finds with some cog-like features are discussed. The study of sintel clamped moss caulking published in 1996 will here be summarised and updated. The construction of the Kampen Cog, by 'Sars Houtbouw' carpentry firm, was an important source for gaining a better understanding of how the medieval shipwrights went about their work and the probable rationale for their decisions. The building experience, early sea trials and the Ommelandvaart (voyage commemorating the Baltic trade route) in 2004 will be described in a separate chapter (7).

Possibilities

It would be too great a digression to describe all pertinent actions in the reconstruction of a ship. It will suffice to establish that despite the limitations mentioned below, reconstruction on the basis of the field drawings, photos and written documentation is possible. The experiences and solutions adopted in the construction of the Kampen Cog are helpful too. In this study, drawing together and working out all cog finds in the Low Countries has made it possible to say more about the less fully preserved vessels. In other words, what was observed in one wreck may be 'transferred' to another, less complete wreck. Besides, chances are very small that these fragmentary wrecks might be eligible for stand-alone publication. In short, not only are they now being published in a full overview, but through mutual comparison in a combined study they gain added value.

Apart from dating the foundering of a vessel on the basis of datable items from its inventory and/or the disturbance of the soil profile, dating of its building and identification of the provenance of the timber (or the ship) is of interest. In many cases dendrochrological analysis may provide answers. To achieve a reliable result, ideally some twenty samples should be analysed. It is important to record from which part of the ship a sample comes, and whether it might have been part of a repair. For a reliable outcome, samples should have at least eighty tree rings and preferably include some sapwood. Timber dated before ca 1300 can in principle also reveal a ship's region of origin. It was around this date that in certain regions the locally available timber for constructing ships and buildings ran out, and wood began to be imported from other parts.

Limitations

The possibilities for attaining the aims outlined above are limited. In most cases a wreck comprises little more than the underwater body with fragments of the lower parts of the sides. In just a few cases also higher parts of the hull have survived. The wreck of the Bremen cog is the most fully salvaged and reconstructed large cargo ship (figs 1.1a and b). In the Netherlands, the Nijkerk-II cog is the most completely recovered ship. Among the smaller craft, the most complete are the Marknesse- and the Almere-I cogs. Other limitations for reconstructing hull lines and dimensions are the facts that the hulls often have collapsed, and that the timbers through the centuries may have become distorted, for instance through pressure in the soil. Usually nothing has survived of the mast, the spar or sprit, the rigging or the sails of (late-medieval) shipwrecks. A few rare instances excepted, hardly anything concrete can be said about these. The study therefore is largely limited to the ships' hulls and the artefacts found within them.

A limitation of a very different kind is the way in which the various shipwrecks have been documented. In general it may be said that the data on the wrecks from the Lake IJssel polders through the years have been reasonably to well documented in field drawings, photos and slides. A description of the excavated ship is sometimes present; in other cases, no more than a few notes. It was not until 1975 that a wreck was excavated with the stated intention of subsequently publishing an excavation report by those involved in the investigation.[13] On the working-out of the excavation data, the following remarks may be made:

- In the ships excavated between 1944 and 2002 we are of course dealing with heterogeneous documentation. The wrecks were excavated and drawn by different teams in years far apart. Besides, a part is played by what knowledge, insight, interest and (research) intentions prevailed when the data were recorded, and under what conditions;

- In working out the excavation data (the drawings), account should be taken of shrinkage and/or expansion of the wet and often degraded ships' timbers. Even during the excavation! This shrinking or expanding depends on the depth position and condition of the wood, of the water table, and of the soil structure. Within individual wrecks great differences in wood quality may be encountered. In other words, the dimensions of a particular timber as recorded at the beginning of the excavation may a few weeks later be different, while also deviating from its original dimensions.

These limitations having been taken into account, and all available possibilities and new insights having been applied, it can be decided whether any previously published evidence still stands, and whether the inferences are upheld or need revision. Also, the analyses may prompt new views, for instance on the origin, the provenance and the dating of the vessel type or shipbuilding tradition.

1.4 Structure of this study

After the Introduction, Part I (Chapters 2 to 4) in a nutshell describes the origins of waterborne transport and the context in which the emergence and evolution of the cog/the cog-building tradition probably took place. To put it differently: the state of research, i.e. what was archaeologically and historically known at the time when the present study was undertaken. But first of all, Dutch nautical archaeology in the IJsselmeerpolders/the former Zuiderzee, up to ca 2000 will be discussed in the sections of this chapter.
Then follows (Chapter 2) a description of the soil stratigraphy of the lake-/sea bottom of the former Flevomeer/Almere/Zuiderzee/IJsselmeer, and its effect on wreck formation, state of conservation, distortion and completeness. This is followed by the history of archaeological research on the shipwrecks interpreted as cogs, and the characteristics of the cog(-building tradition) among the wrecks excavated in the Low Countries. Also mentioned are the late-medieval shipwrecks from the Flevoland polders (and beyond) that do not belong in the cog-building tradition. Chapter 3 deals with the archaeologically known river craft up to 1300, and the archaeologically and historically documented west- and north-European seagoing vessels of the period up to 1200. The ships of the Mediterranean and the Normans in southern Italy are the next subject. This is followed by the Crusades, the socio-economic developments and the emerging seaborne trade between the Mediterranean and the North Sea area and around the Baltic. The final section deals with the northwest-European seagoing vessels of the 13th to 16th centuries. Chapter 4, 'The medieval cog, the state of research', deals with the historiographically recorded findings.

In Part II (Chapters 5 to 7), chapter 5 focuses on the investigation and the research/reference model (the reference wreck Nijkerk-II, and its replica the Kampen cog). This is followed by chapters 6 and 7, which are the reports on

13. Reinders *et al.* 1978

the Nijkerk-II wreck and the building of the Kampen cog. From the specified data of this ship, a standard descriptive model was drawn up. This archaeological and technical description is consistently adhered to in the descriptions of the other wrecks. At the same time this descriptive model is a checklist for the validity of the reference model.

In the chapters of Part III (Chapters 8 tot 28), the other wrecks are treated ship by ship and in chronological order. In the wrecks which so far have only received evaluating investigation and the wrecks from Doel, naturally there is only succinct, summary information about what at this point is known about these ships. Two wrecks provided supplementary evidence, the vessel excavated in lot G37 in the Noordoostpolder in 1955, and the wreck in the 'Zeebad' at Medemblik, which in 1982 underwent evaluating investigation. For a long time during the proceedings there was doubt as to whether these wrecks too could be identified as cogs. The answer now seems to be that they can. For practical reasons, however, they could not be treated in detail here. The subsequent chapters focus on a number of excavated boats in which certain cog features were recognised. Next on the agenda are the late-medieval wrecks encountered in the Flevoland polders and elsewhere in the Netherlands, that display some cog-like features but cannot be reckoned to the cog-building tradition. The final chapter of Part III describes the discovery of a presumably medieval shipyard at Kampen and some finds from it, and gives a brief review of potential cog-building centres in the Hanseatic territories.

Part IV (Chapers 29 up to 31) presents the Results and Analysis, the Origin and Decline of the cog and the Summary and Conclusions. Here first the prominent differences between the vessels of the Nordic or Viking tradition and the cogs are discussed and the supplemented characteristics of the latter are summed up. Also there are references to what is known about all other vessel types mentioned in documentary sources. After this are unfolded the characteristic structural features, the construction method and the building sequence of the large and small cogs. Following this, it is explained where the origin of the cog should probably be sought, why the vessel type barely if at all evolved over a period of 300 years, what made it such a success, and what might have been reasons for its demise. Also the progressive research into the method of sintel clamped moss caulking is discussed, as well as the confirmation of the typological and chronological sequencing of the sintels (iron clamps). This caulking method is a characteristic trait of cog building. A medieval miniature which, very unusually, depicts the application of sintel clamped moss caulking to a Crusader ship (an early cog?) is considered from an archaeological perspective. The miniature is familiar to nautical historians and archaeologists, but has never before been interpreted in this light. In some other, early 15th-century miniatures very realistic structural details are shown on the depicted boats. Also attention is given to the layout, fitting-out and inventories of the cogs and to their external connections, trade and the Hanseatic network. The historically mentioned *nef* a vessel type of the Atlantic and Channel coasts of England and France is discussed, as there is a potential relationship with (the emergence of) the cog-building tradition. The *waterschip* of the Zuiderzee is also included in the review.

The section research questions (Section 1.2) examines the wrecks from the Baltic identified as cogs: whether their structural features match those of the cogs of the Low Countries, whether they were demonstrably built in the Baltic region and whether, for example, there are regional differences - these are obvious questions. Crumlin-Pedersen[14] presented an overview of the then known cog finds from Denmark, Sweden, Germany and the Netherlands. He discusses what other researchers and he himself have so far discovered and established about the vessel type, about a possible prototype of the seagoing cogs and about the exchange of technical features between the Scandinavian ships and cogs. In the same article he warned that most wrecks still awaited publication, and that the evidence presented about their construction etc. was no more than an initial impression (see section 1.1). There seems little point in repeating Crumlin-Pederson's work. Indeed I have too little data on the various wrecks from the North. Therefore I shall limit myself to all wrecks of the cog-building tradition in the Low Countries. If a particular wreck, a construction or a structural feature calls for it, there will be an occasional reference to a foreign wreck.

In general, we may state that among the cog finds from Scandinavia and the Baltic there will certainly be one or more that match the finds in the Netherlands. This also goes for wreck finds in the Low Countries in which Nordic influences are evident. Besides, there are wrecks that resemble cogs, yet these are clinker-built throughout and often built from conifer wood. As has already been done, they might be labelled 'Baltic cogs'. Crumlin-Pedersen also writes about the Gedesby ship, a vessel built in the Nordic tradition with typical structural traits and constructions which must have been borrowed from the cog-building tradition. The vessel dates from ca 1300. There is a sailing replica.

My aim in this study is to present realistic and responsible information, as detailed as possible, about the shape, construction method and building sequence of a specific category of shipwrecks, which in the Low Countries is the

14. Crumlin-Pedersen 2000.

most numerous among the discovered and investigated late-medieval wrecks. The conclusion is that, on the basis of comparison with the Bremen cog, these wrecks can all be reckoned to the same shipbuilding tradition. The difference in dimensions indicates that we are dealing with cogs, small cogs and boats. The results of this study will contribute to the debate about this research topic which on an international level still is far from settled. Working out the cog finds from the Baltic area in a comparable manner should make it possible to answer (several of) the questions enumerated above.

Many nautical terms are explained in the Glossary of ship's terms, and are indicated in the drawings of a cog's hull (Plates 1.1 and 1.2).

State of research

I

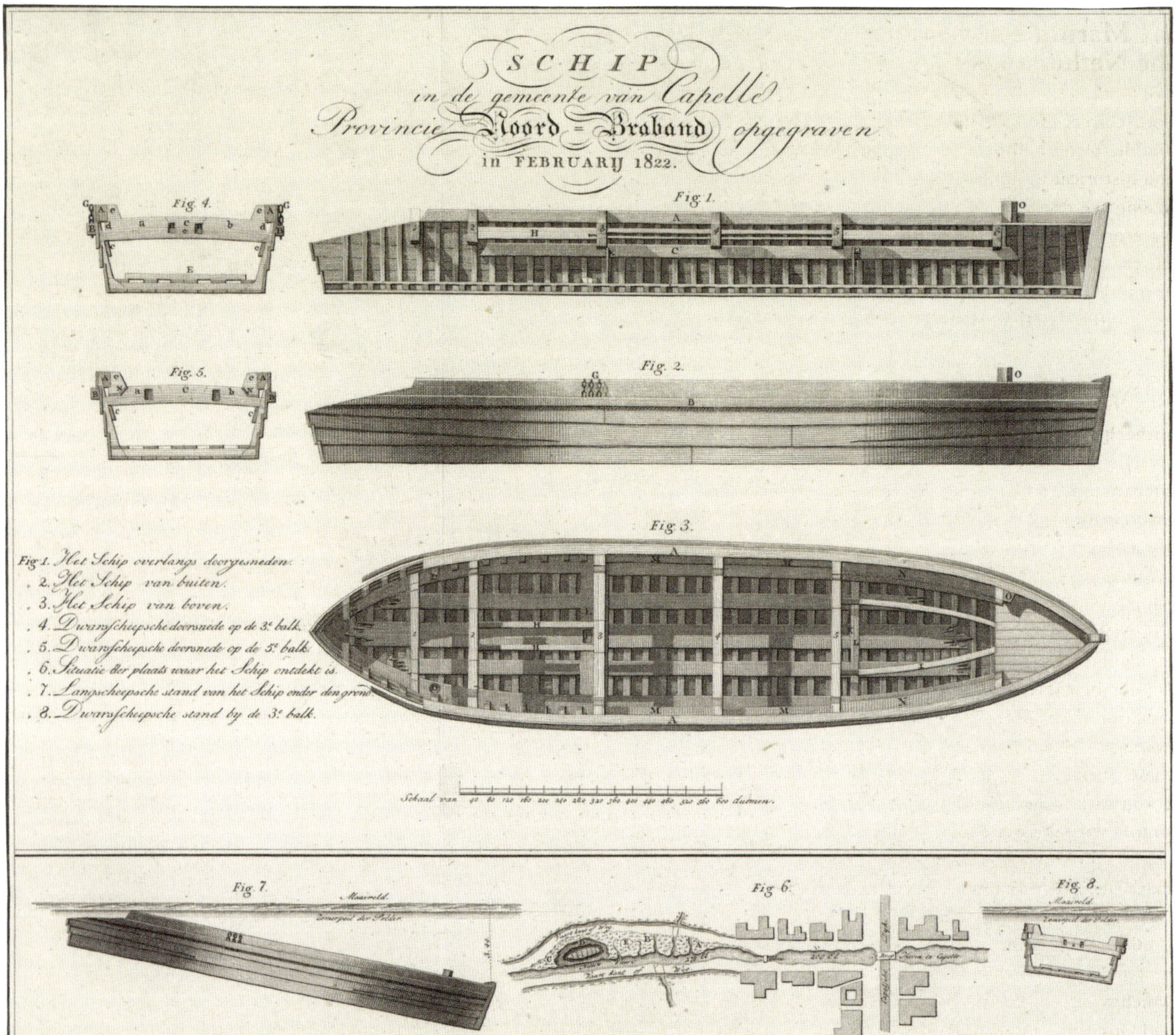
SCHIP
in de gemeente van Capelle
Provincie Noord = Braband opgegraven
in FEBRUARIJ 1822.
Fig. 1.
Fig. 2.
Fig. 3.
Fig. 4.
Fig. 5.
Fig 1. Het Schip overlangs doorgesneden.
2. Het Schip van buiten.
3. Het Schip van boven.
4. Dwarsscheepsche doorsnede op de 3e balk.
5. Dwarsscheepsche doorsnede op de 5e balk.
6. Situatie der plaats waar het Schip ontdekt is.
7. Langscheepsche stand van het Schip onder den grond.
8. Dwarsscheepsche stand by de 3e balk.
Schaal van 40 80 120 160 200 240 280 320 360 400 440 480 520 560 600 duimen.
Fig. 7.
Fig 6.
Fig. 8.

Previous history and context[1]

2.1 Maritime archaeological research in the Netherlands

For a correct interpretation of the data collected it is advisable to consider first the archaeological, pedological and historical information available at the start of and during the research. Of particular importance are the socio-economic situations/developments, trade, transport, the crusades and the ships involved as well as other means of transport in Western Europe and the Mediterranean region in the period up to ca 1200.

In the Konst- en Letterbode of 1822 C.J. Glavimans published the excavation that he carried out under the authority of the minister of the Navy on a shipwreck at Capelle in the Langstraat in the province of Noord Brabant. Judged by present standards it was an excellent investigation to which the archaeologist C.J.C. Reuvens also contributed.[2] The discovery and salvage of what was originally assumed to be the Roman or Early Medieval 'Ship of Utrecht' in 1930 is a more familiar example of a shipwreck which received attention in the past. These incidental finds may be regarded as the start of maritime archaeological research in the Netherlands. After the reclamation of the Wieringermeer polder in 1930, several dozen post-medieval shipwrecks were uncovered, though these received only limited attention. They literally did not fit into archaeological thinking at that time. They simply were not old enough.[3]

With the construction in 1932 of the Afsluitdijk between the provinces of Noord Holland and Friesland the Zuiderzee was dammed. In this inland sea of over 500,000 hectares in the northwest and centre of the Netherlands which was renamed IJsselmeer, the Noordoostpolder (1942) and Oostelijk- and Zuidelijk Flevoland (1957 and 1968 respectively) were drained in addition to the Wieringermeer polder which had already been reclaimed in 1930. This amounted to a total of 166,000 hectares of new land.[4] Plans for a fifth polder (Markerwaard) were shelved.

During the Middle Ages and also in later centuries the Zuiderzee was an important part of a complex Northwest-European traffic network. In the same period, ships were by far the most important means of transport – certainly until the invention of the train. The Netherlands Institute for Ship- and underwaterArchaeology/ROB (NISA) and its predecessors, and also the present Department of Maritime Archaeology of the (RCE) the Cultural Heritage Agency of the Netherlands[5] has registered almost 450 shipwrecks in these reclaimed areas of former seabed since 1942. They date mainly from ca 1250-1900. Over 350 of these have now been excavated. It is a ship graveyard that is unique in the world and comparatively simple to excavate. A considerable number of wrecks contained (virtually) complete inventories. The objects belonging to these inventories provide insight into the life and work on board ships of our seafaring ancestors, while the vessels themselves often supply hitherto unknown information about types of ships and building methods throughout the centuries. For the majority of ships the period in which they foundered can be fairly precisely dated (within 25 years or less). As a result, they can be regarded as genuine time capsules.[6] Approximately seventy wrecks date from ca 1250-1600. The finds from the former Zuiderzee area are mostly inland freight vessels, fishing boats and work barges. In addition, several larger, seagoing cargo ships and armed vessels were discovered. The shipwreck finds prior to ca 1500 are, almost without exception, all representatives of the same tradition, which will henceforth be referred to as the cog-building tradition.

Fairly soon after the reclamation of the first sections of the Northeast Polder archaeological artefacts were collected near Kuinre. In July 1941 the Biological Archaeological Institute of the Rijksuniversiteit Groningen (BAI) was charged with the archaeological investigation of the Zuiderzee region. As well as the discovery of the location of the first castle of Kuinre, the first shipwrecks and ship remains were also found in this area.[7] The discovery and publication in 1944/1945 of a small 14th-century freight

1. Translated by Christine Jefferis.
2. Reinders 1987, 9.
3. Vlierman 2002, 120.
4. Vlierman 1997a, 15 and fig.1.
5. The department was closed down in 2016.
6. Vlierman 1983, 1 and 1997a, 15.
7. Modderman 1945; chapter 13, the Kuinre cog.

vessel with cargo and inventory[8] made it clear that regular marine archaeological investigation would be required in the Northeast Polder and in future polders in the former Zuiderzee region. The drawing method devised and carried out by the BAI for recording the ship excavated by P.J.R. Modderman and its position in the soil became almost standard procedure after the Second World War under the direction of G.D. van der Heide. The drawing (scale 1:20) consisted of one or more top views, a longitudinal section across the keel and stern/stem combined with the interior view of one of the sides, various cross sections of the frames/ribs, drawings of parts found separately and important for reconstruction and cross/transverse sections together with the disturbed soil profile. Photographs of the situations encountered were also taken. This is basically the documentation procedure still followed. Since the beginning of the 1970s additional new techniques have been applied for this purpose (stereo photogrammetry, drawing with a 2D and 3D pantograph). In 1981 it was decided that all parts of a shipwreck would be drawn on a scale of 1:10.

2.2 Structure of the Zuiderzee bed

In order to fully understand the difference in completeness and the varying degrees of preservation of the wrecks a brief description of the soil structure will now follow. The Pleistocene surface whose depth determines the preservation of the wrecks was taken as the starting-point.

Roughly speaking, the Pleistocene surface which mainly consists of sand, apart from the odd outcrop of boulder clay, extends along the eastern edge of the Northeast Polder at 3 to 4 metres below NAP (Normal Amsterdam Level). On the west side of this polder its depth is ca 8-9 metres below NAP. In the Noordoostpolder and the Oostelijk Flevoland area around the IJssel/Vecht delta its depth on the east side is 4-9 metres and near the island of Urk it is 8-10 metres below NAP. Along the southeast coast of the former Zuiderzee between Muiderberg and Elburg the Pleistocene sand surface is ca 1 metre below NAP, while between the Ketel- and the Hollandse Brug it is ca 10-11 metres below NAP. In the IJssel/Vecht delta which fans out farther towards the west the Pleistocene surface along the coast of Noord Holland at Enkhuizen and Marken lies at approximately 12 metres below NAP, while near Scharwoude it is about 18 metres. At Wieringen in the Wieringermeerpolder the Pleistocene surface is ca 4 metres and west of Medemblik ca 14 metres below NAP. In addition to these differences in height/depth the Pleistocene landscape is transected by the channels of the Steenwijker Aa and the Vecht in the southern half of the Noordoostpolder, under the Ketelmeer and in the northern part of Oostelijk Flevoland by that of the IJssel (and possibly also the Vecht) and in Zuidelijk Flevoland by a channel system starting at Zeewolde, then running west-north-west and disappearing under Noord Holland just north of Edam. Roughly in the middle of Zuidelijk Flevoland the channel of the Eem flows into this channel system. Various channels were also observed in the Wieringermeerpolder.

Fig. 2.1. The disturbed soil profile around a shipwreck (Os19, 17th century). The white layer is the shell layer Zu IV, dating from the early 17th century.

This Pleistocene surface/landscape remained virtually unaltered until ca 2000 BC or slightly later. In Prehistoric times people lived on the highest parts and on river dunes, as evidenced by the finds of the Swifterbant culture in particular, in the northwest of Oostelijk Flevoland and in the south of the Noordoostpolder. The prehistoric and also the medieval finds in the vicinity of the cog ships discussed in this study are briefly described in the relevant chapters.

The rise and fall of the sea level in various periods, the silting up of the channels with marine clays, the closing of connections with the Noordzee and the subsequent wetting by water from the rivers in the later Zuiderzee area leads to the development of peat in the period between ca 2000-1750 BC.

We are not sufficiently aware of the fact that it is thanks to this growth of peat that the Netherlands possesses a huge and unique piece of Pleistocene/prehistoric landscape dating from ca 4000 years ago. With the exception of the prehistoric surface which has been damaged by canals, ditches and agriculture (deep ploughing) this landscape is still ex-

8. Modderman 1945; chapter 12, the Marknesse cog.

tant undisturbed and in fact deserves UNESCO protection in its entirety as exceptional natural and cultural heritage.

The growth of the peat bogs continued for centuries. Around the beginning of this era this region consisted of a dynamic system of lakes and peat islands and peninsulas, the *Flevo Lacus* or Lake Flevo (Flevomeer). From the north, east and south these lakes were fed by peat channels, streams and rivers. After a layer of clay was deposited over the peat, the region once again became suitable for habitation in the Carolingian period. Under the influence of the North Sea large sections of the peat area were subsequently cleared. There were many storm surges particularly in the 12th century resulting in the formation of the Zuiderzee as we now know it and the water gradually began to turn brackish. The layers of peat and clay which had been eroded but also parts of the top of the Pleistocene sand formed the material for new sedimentation layers which covered the original seabed. These 'freshwater' deposits developed between the beginning of this era and AD 1600 the Almere layer, as it was named by soil scientists. Around ca 1600, but at any rate in the first quarter of the 17th century, the influence (ebb and flow) of the Noordzee gradually increased and the entire Zuiderzee became salty. The clay deposited from then on was marine. After the construction of the Afsluitdijk, the Zuiderzee was renamed IJsselmeer and gradually became freshwater again. Shortly after the reclamation of the polders thin layers of IJsselmeer deposits were observed.

Depending on the location (close to the coast or in the middle of the sea), the height/depth of the Pleistocene and whether or not there were any remains of the original Holocene peat layer on top of the Pleistocene, the Almere and Zuiderzee deposits were either complete or are entirely or partially lacking. Wiggers[9] describes the sedimentation history, erosion and relief development of the soil of the Noordoostpolder and names and dates the various layers of the Sloef and Zuiderzee deposits. Ente *et al.* describe the soil of Oostelijk Flevoland, and substitute the name Sloef layers for Almere layers.[10] The (rough) dates of the various sedimentation layers are derived from archaeological finds, mainly shipwrecks.

The degree of completeness and the state of preservation of the various wrecks is strongly influenced by the place where the ship sank. Also the way in which this (process) took place will have affected preservation, while in some cases the construction of the ship's hull will be to blame for the fact that certain parts of the ship are missing. This is occasionally due to breaking-up.

In principle it may be stated that the sedimentation layer on top of the Pleistocene surface formed a 'soft' seabed. Depending on the height/depth of the Pleistocene sand this seabed is a half to several metres thick. It may be assumed that a sunken ship sank fairly quickly into this 'soft' bed entirely or partially and that this could, in theory, be down as far as or into the Pleistocene sand. In the case of thick layers of clay or peat, the wreck could as it were stay 'floating' in the sedimentation layers and/or sometimes completely disappear in them. If there is a thin sedimentation layer or a hard sandy layer containing shells, the ship sinks down to or into the Pleistocene sand or on to the layer of shells. The latter does not affect the cogs because all of these sank into the 'soft' Almere layers or into peat.

If there is a thin layer of sedimentation, the ship sticks out above the seabed. These projecting parts were subsequently eroded by wave action, currents, fishing activities and breaking and/or drifted away. If a wreck permanently sticks out above the seabed, floating ice and later post-worm damage may also be responsible for the disappearance of ship parts.

After a wreck has come to rest in the seabed, the hole caused by the ship sinking into it fills up again with sedimentation layers. The first undisturbed layer to accumulate over this filled in 'pit' can be given an *ante-quem* date of the moment the ship sank. This date can in many cases be derived from the ship's inventory. And reciprocally, the sedimentation layers thus dated can serve as a rough date of sinking for wrecks in which no dateable finds were encountered. Instances are known in which the wrecking process of a ship must have occurred in a continuous short period, but also in which there must have been different processes in periods sometimes far apart. The shell layer Zu IV which is the oldest Zuiderzee deposit indicates the border between the 'fresh' Almere deposits and the salty clay or sand sedimentation layers of the Zuiderzee. Wrecks above which this layer of shells is undisturbed therefore date from before ca 1600.

2.3 Archaeological cog research/investigation

Early on, wrecks were interpreted as cogs without all the characteristics being analysed in detail. Consequently, concrete answers to many of the still open questions could not be given. This might to some extent be possible if

9. Wiggers 1955.
10. Ente, Koning & Koopstra 1986.

archaeological discoveries identifiable as cogs had been made, but this was not the case up to the Second World War. P.J.R. Modderman was the first to regard a ship find from 1944 as a cog, indicating that this was a small freight vessel and that the type of ship, on the basis of the historical data, was not uniform but had undergone development.[11] The publication of the ship went for years unnoticed by later international researchers. Nothing new could yet be reported about the 'big' Hanse ship, while the criteria to which the construction of a vessel had to comply in order to be designated a cog had not yet been established. The 'Modderman cog' made it clear to the historical and archaeological world of the Netherlands that shipwrecks were a highly interesting field of research with regard to the life and work of our medieval (seafaring) ancestors. The wreck thus gave rise to regular marine archaeological research in the Netherlands. It was in fact the first wreck in the only recently reclaimed Northeast Polder to be seriously investigated. Several years later Åkerlund described and drew the late- and post-medieval finds from the harbour of Kalmar, Sweden.[12]

In 1962 a large shipwreck was discovered in the Weser near Bremen. On the basis of the data in Heinsius'[13] book which had been published only a few years before, it was established with a high degree of certainty that this must have been a (Hanse) cog. The ship was salvaged in parts and reconstructed and preserved in the Deutsches Schiffahrtsmuseum in Bremerhaven (figs 1.1a and b). Dendrochronological analysis showed that the vessel must have been built in/after 1380.[14]

Other less completely preserved and not, or not yet fully excavated and/or published medieval wrecks in Denmark (Kolding, Vejby and Kollerup) and in the Netherlands (Northeast Polder Q 75, the Ens cog, chapter 9) were involved in the further identification of the ship type in the 1960s and 70s on the basis of the characteristics of (the underwater ship of) the Bremen cog. The discovery of the wreck on lot Q 75 featured conspicuously in the discussion on the development of the cog, on the basis of the then assumed date.[15] Crumlin-Pedersen assumed that the cog developed from a Frisian, cog-like *waddenschip* and is not therefore a revolutionary type.[16] Ellmers was of the same opinion, but was under the impression that evidence for this had yet to be found.[17]

A number of characteristic constructional features of these archaeological finds were also found in the wreck on lot N 5 in Oostelijk Flevoland in 1973.[18] However, in addition to similarities there were also differences with regard to the previous finds. The analysis of the excavation data by the former department of Maritime Archaeology of the IJsselmeerpolders (RIJP) was still at an early stage. The information about the considerable number of ships that were excavated per year was stored in files. After Modderman's publication in 1945 it was not until 1978 that an excavation report on a wreck discovered in the IJsselmeer polders appeared for the first time.[19] The wreck on lot N 5 together with two other vessels was the second publication of (modest) excavation reports and the first in which several, in this case late-medieval, finds were presented together and in which attention was paid to similarities and differences.[20]

The excavation of the wreck on lot Oz 43 in Flevoland and the discovery of a second vessel on lot Oz 36 in the same year sparked focused (international) attention. The records of the late-medieval vessels excavated in the IJsselmeer polders were subsequently examined and as a result, several wrecks were attributed to this shipbuilding tradition. The excavations on lots Oz 43 and Oz 36 in Flevoland in 1981 and 1983 respectively[21] and on lot A 57 in the Noordoostpolder in 1985,[22] made it clear that these, like the Bremen cog, were large cogs.

The Kollerup cog is the first find from Denmark to be (apparently) fully published (drawings of the preserved parts, a preliminary reconstruction of the ship and a description of the finds discovered in the wreck).[23] Reinders studied the characteristic data available for the nine medieval vessels from the IJsselmeer polders which were then eligible and compared them to the data available at the time for the Bremen cog. He presented the provisional dates of these ships in a brief overview.[24] Due to insufficient certainty regarding the dates of the wrecks at that stage of

11. Modderman 1945, 80; the Marknesse cog, chapter 12.
12. Åkerlund 1951.
13. Heinsius 1956.
14. Lahn 1992.
15. Crumlin-Pedersen 1965 and 1979; Ellmers 1972, 1975 and 1979.
16. Crumlin-Pedersen 1965.
17. Ellmers 1972.
18. Reinders *et al.* 1980, 14 and 15; the Dronten-I cog, chapter 11.
19. Reinders *et al.* 1978.
20. Reinders *et al.* 1980, 44 and 45.
21. Nijkerk-I and Nijkerk-II cog, chapters 8 and 6.
22. Rutten cog, chapter 10.
23. Andersen 1983.
24. Reinders 1985a.

the investigation he came to the conclusion that it was not possible to explain the differences as being due to the development of the cog in the 13th and 14th centuries and queried whether the term cog was relevant in all cases. From the beginning of the 1980s dendrochronological research developed in the Netherlands giving rise to possibilities of dating the construction of the ships and the provenance of the timber used.

Van de Moortel published the Dutch wreck of lot Nz 43[25] and presented a reconstruction model scale 1:10 made by herself from cardboard and the hull analyses derived from it. Lahn[26] published the book on the Bremen cog with drawings of the reconstructed ship's hull. In the same year (1992) the components of the equipment and inventory from a small 15th-century cog were presented.[27] Hocker & Vlierman[28] published the wreck of this small cog from Zuidelijk Flevoland. In addition to the description of the ship, the report presents the field drawings, reconstruction drawings and an oak model scale 1:10 and the inventory discovered in the vessel is described and analysed in detail.

Until the start of this study only reports of finds and brief articles had been devoted to the remaining Dutch and international cog finds. In 1996 a research report was published on the characteristic caulking method (sintel clamped moss caulking) for ships from the Hanseatic region and from the period in which the Hanse was active,[29] while in the following year two articles on the cog and other medieval vessels were published in which there is speculation about the origin of the cog, among other things.[30] In 1998 an article appeared on the reconstruction of a Hanse ship in Kampen.[31]

Weski[32] interpreted what had been published up to then about cog finds and came to the conclusion that there must have been an IJsselmeer type. In his survey *To be or not to be a cog: the Bremen Cog in perspective*, Crumlin-Pedersen rightly remarked that only few cog finds had so far been published and that, on the basis of the details of these ships which have already been presented, it is at most only possible to give a first impression of the dimensions, date, origin and types of ships.[33] This study provides more information in this respect about the Dutch finds.

The unusual find of an early-14th-century Hanse cog at Doel in Vlaanderen, Belgium in September 2000[34] provided an unexpected and first opportunity for researching a completely accessible exterior of the hull of a medieval vessel and ultimately gaining more knowledge about the 'workhorse' of the Middle Ages. It is very likely that the discovery of the wreck in the IJsselmeer[35] and that in the IJssel off Kampen[36] will provide further and possibly so far new insight into the large cogs of the 15th century.

2.4 Characteristics of the cog

Already in the case of the first wreck published Modderman writes that it was a small cargo vessel, probably of the cog type.[37] The dimensions of the various vessels discovered later showed that there were in fact large and small ships. Section IV explains to which of these groups the vessels can be assigned according to a Middle Dutch description.

Fig. 2.2. The overturned cog found in the Deurganck dock under construction at Doel.

25. Van de Moortel 1991; Spakenburg-II cog, chapter 15.
26. Lahn 1992.
27. Vlierman 1992b, 10-22.
28. Hocker & Vlierman 1996; Almere-I cog, chapter 16.
29. Vlierman 1996a.
30. Vlierman 1997b, 71-95 and 1997c, 41-52, respectively.
31. Vlierman & Tahey 1998, 30-40.
32. Weski 1999.
33. Crumlin-Pedersen 2000, 236.
34. Doel-I cog, chapter 20.
35. Enkhuizerzand cog, chapter 24.
36. Waldus *et al.* 2012; the Kampen(IJssel) cog, chapter 23.
37. Modderman 1945, 80.

Characteristics

Thanks to the archaeological find at Bremen in the river Weser in 1962, it was possible to assemble realistic information about the construction and shape of the hull of a large late-medieval cog. This information was tested against several Danish and Dutch finds. By about the year 2000 a number of striking similarities in shape, construction and building method had already been accepted as characteristics of the cog, at any rate with regard to large seagoing ships. These characteristics are:

- A keel plank with stern/stem hooks and straight, falling stern/stemposts.
- An underwater body V-shaped fore and aft and relatively flat amidships.
- A carvel built section/surface becoming a clinker built construction fore and aft.
- The planks of the carvel built section are joined together by means of ribs.
- Clinker-built and sloping sides.
- The overlapping planks are secured together with nails double clenched and driven back into the wood.
- The hull seams are filled with sintel clamped moss caulking.
- Beams that protrude through the hull with heavy knees on them forming the intersection at deck level and above.

Almost all the wrecks from the Zuiderzee area and several Flemish finds had these characteristics. They date from ca 1250-1450. The rough preselection of the ships to be discussed in this study consisted of the wrecks of seven or eight large and ten small(er) vessels which were all classed as cogs or coglike. All of them were described and drawn in detail, as far as possible. During this process it became gradually clearer that both the large and the smaller ships were built in the same tradition. This corroborated Luns' and Reinders' prior conclusion.[38] Not only, therefore, was there 'the cog', but more probably a 'cog-building tradition'. A cargo vessel excavated in the Noordoostpolder in 1955 and a appreciatively investigated excavated wreck in the IJsselmeer near Medemblik also appeared to fit the criteria for cogs when this study was nearing completion. They could not be discussed in detail, but are briefly dealt with in chapter 25. In table 1 (Section 1.2) all the cog finds discussed are listed. Figure 29.1 (Chapter 29) shows the contours of all ship hulls on the same scale.

Other vessels (with some of the characteristics) may also be included in the cog-building tradition. These are at least ten inland vessels, some of which were found in Antwerp at the end of the 19th century, some in Oostelijk Flevoland and recently in the IJssel at Kampen. A number of them have been published. Most may be regarded as barges built in the cog-building tradition. They are briefly discussed in sections 26.1 through 26.8.

2.5 Other late-medieval shipwrecks

In the former Zuiderzee region a small number of late-medieval shipwrecks were also discovered which can with a high degree of certainty be classed as belonging to other ('foreign') building traditions, on the basis of their construction. In section 1.1 finds of the Norse/Northern or Viking shipbuilding method are mentioned. In sections 3.2 and 3.5 finds of seagoing ships up to ca 1200 are briefly described and the differences outlined between the building traditions of these vessels and those of cogs. From this, one can see how specific the cog/the cog-building tradition is. The *Hollandse Waterschip* of the Zuiderzee is also discussed. A comparatively large group of fishing boats consists of two types, a medieval completely clinker-built type and a completely carvel-built post-medieval variant. For a correct insight in the significance and scope of the cog-building tradition these 15th- and 16th-century vessels are briefly discussed in part III, chapter 27. The most striking differences with cogs and probable provenance are indicated. Some of these wrecks do still share one or more features similar to those of the cog, probably as a result of the changes that took place in shipbuilding in the 15th/16th century. The transition from (partially) clinker to carvel building and the large cogs with one mast and one square sail which had remained unchanged for a number of centuries and which were succeeded by larger ships with several masts and more sails plus the disappearance of the clamped moss caulking method are the most conspicuous changes. We shall return to this in part IV.

2.6 River vessels with cog features

From elsewhere in the Netherlands 13th-century boats are known that may be regarded as river vessels. Apart from several features from the cog-building tradition, the design of these ships also reveals elements that may clearly be ascribed to a different building tradition. They are briefly discussed in section 26.9. In section 26.10 attention is briefly given to a small cargo vessel/barge from the Oosterschelde. The boat dates from ca 1500, differs

38. Luns 1985, 57-58; Reinders 1985a, 3.

clearly from the construction of vessels belonging to the cog-building tradition and is discussed for this reason. In part IV these finds are again dealt with in comparison with the cog-building tradition.

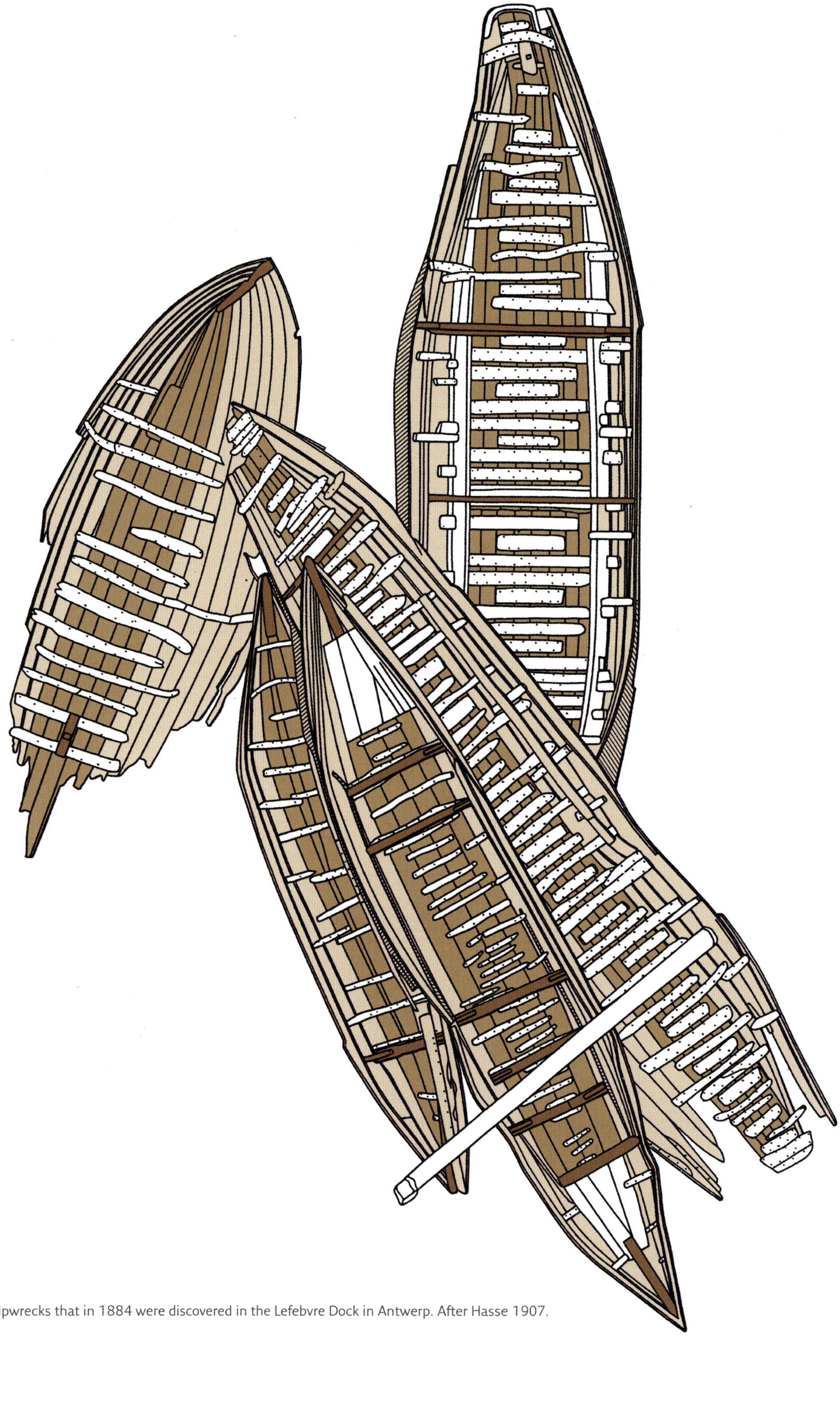

The five shipwrecks that in 1884 were discovered in the Lefebvre Dock in Antwerp. After Hasse 1907.

Trade and ships in the Middle Ages[1]

Core areas

Within the Roman Empire there was intensive (long distance) trade and transport. This was also the case in the then relatively densely populated northwestern part, the Low Countries. However, during the period between ca 300-600 the latter area was very sparsely populated. The population increased in some places during the Carolingian period and a core area developed in the Meuse basin between Aachen, Dinant, Namur, Tongres and Maastricht (fig. 3.1).[2] In this period the inhabitants relied mainly on agriculture and stockbreeding and were self-sufficient. In the post-Roman period until (well) into the Carolingian period there was probably only a limited degree of long-distance and overseas trade and transport. Archaeological finds, particularly pottery, show that long-distance trade and transport did take place within the Merovingian and Carolingian empires. In the 7th-9th centuries Utrecht and Nijmegen were the centres of religion and power in the Low Countries. Important trade settlements and ports were Dorestad, Witla, Domburg, Deventer and the much smaller town of Medemblik. After the Viking raids in the late 8th and early 9th century, Dorestad's importance declined. After the Carolingian period the present-day territory of the Netherlands was part of the German Empire.

In the Ottonian period (10th-11th century) the number of trading places increased. Nijmegen and Utrecht had developed into trading centres with an even higher status; Tiel, Vlaardingen, Staveren, Zutphen and Dokkum were the new centres on rivers and on the coast. In this period Deventer was the largest town in the present-day territory of the Netherlands and an important centre for overland transport and the shipping of goods from and to the Westphalian and Lower Saxon hinterland.[3] These large centres had a modest number of inhabitants by current standards. Seldom or never would a situation have existed in which dozens of river and seagoing vessels would have moored or cast off in one of the harbours at the same time or within a short period. In addition, trade over land would have been at least as extensive and important until well into the 12th century. Dorestad and later Tiel and Vlaardingen were at their peak major ports of transhipment of goods between the German Rhineland and England.

In the 11th-13th century another core area developed in the southwest part of the Low Countries (Flanders and Artesia, which included the cities of Bruges, Ghent, Courtrai, Doornik, Atrecht, St Omars and Nieuwpoort). The 12th-13th century saw almost everything change in this region and heralded explosive growth. The earliest developments occurred in Artesia and in the south of Flanders. Improved agricultural methods resulted in a better yield and an increasing percentage of the population started to engage in non-agrarian activities. In the 11th and 12th centuries there was comparative order and stability under the rule of the territorial leaders. The expansion of the towns went hand in hand with the development of trade and transport. However, it was becoming more and more difficult to satisfy the first necessities of the sharply increasing population locally or regionally

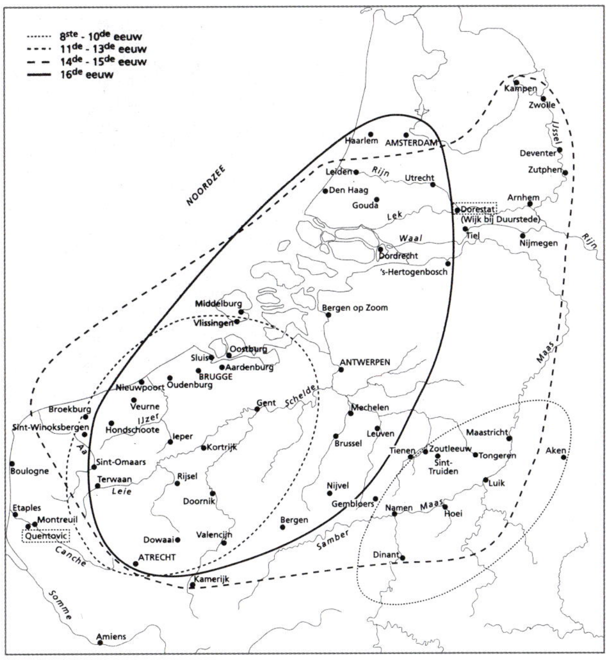

Fig. 3.1. The successive dominant regions in the Low Countries, 8th to16th century. After Blockmans 2010.

1. Translated by Christine Jefferis.
2. Blockmans 2010, 9-22.
3. Bartels 2001, 8; Vlierman 2002, 137.

and ever larger quantities had to be brought in from afar. The headstart in the core area had far-reaching effects on the neighbouring districts and the British isles.[4] The Flemings had acquired a good reputation for their capacity for work and technical knowhow. A large number emigrated to England (Scotland/Wales, Pembroke and Kent) and at the request of local princes also to Germany (the middle course of the Weser, Harz and Thüringen).[5] Trade between England and the continent flourished. At the beginning of the 11th century London and York were important trade centres in England and attracted an increasing number of merchants from Germany, Flanders, the valley of the Meuse and France. Towards the end of the 10th century there was probably even an English merchant fleet.

In the 10th, but mainly in the 11th century, Bruges was already known from the written sources as a busy centre with overseas contacts and a thriving cloth industry.[6] Flemish cloth (of English wool) for example was exported to every corner of Europe. The most obvious trade route was by ship over the Zwin and from other Flemish towns over the Schelde, over the waterways of Zeeland and further upstream via the Meuse and Rhine.[7] There was also a land route to Cologne and via the *Königstrassen* through Swabia and over the Alps. Via the annual fairs in the Champagne region and the rivers in eastern France traders travelled to the Mediterranean area as far as Italy and vice versa. The 11th century also saw the emergence of the ports of Dover, Hastings, Hythe, Romney and Sandwich (the so-called Cinque Ports). Their economic importance to England resulting from trade with the continent was great and was evident from the fact that merchants from these towns enjoyed exemption from toll throughout England. The importance of English trade increased further as a result of contacts with Scandinavia and the Baltic, when England became part of the great kingdom of Canute the Great of Denmark. Irish trade also flourished, not only through contacts with Wales and the west coast of England, but also with Norway and the northern isles.[8] At the beginning of the 11th century trade also started to develop in France and to integrate with that of the northern seas. The Loire and Garonne areas saw a revival. Transport over the North Sea was mainly carried out by the North Sea coastal population of England, the Low Countries and Germany.

Although Blockmans describes it as such and shows it on the map (fig. 3.1), there is, without discussing it in detail but for example for economic and political reasons, certainly no question of a third core area in the Low Countries in the 14th-15th century. Interests in the various regions differed considerably. In addition to Flanders and Artesia, Blockmans also includes in this area the first core area along the Meuse, present-day Zeeland, Brabant, Limburg, Zuid-Holland, Utrecht, Gelderland, the eastern Zuiderzee coast and the IJssel region in Overijssel. In these 'new areas' existing trade settlements grew as did the towns founded mainly at the end of the 12th and the beginning of the 13th century. Some of these genuinely flourished. Due to the creativity of townspeople and farmers a society was created in the Netherlands between 1100-1560 that in almost all respects was at the peak of European economic development.[9]

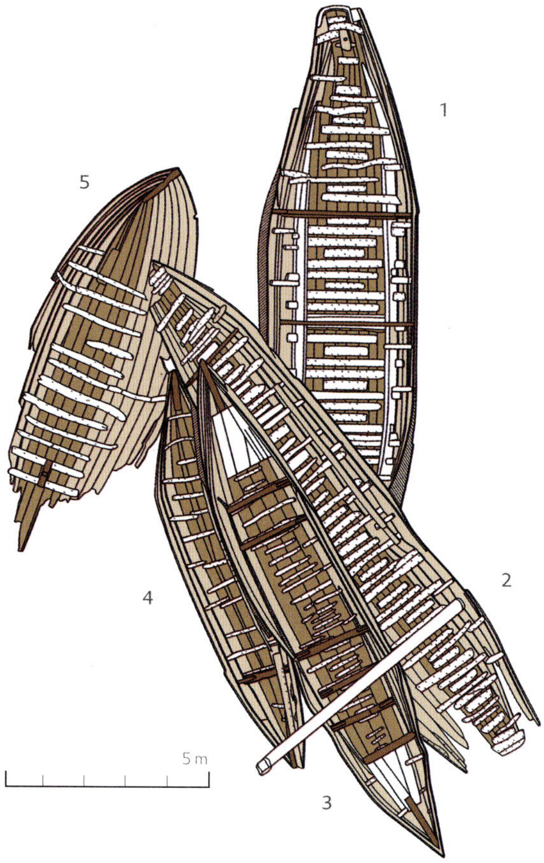

Fig. 3.2. The five shipwrecks that in 1884 were discovered in the Lefebvre Dock in Antwerp. After Hasse 1907.

3.1. Inland vessels up to ca 1300

Transportation by water was carried out by relatively primitively built river barges and barge-like plank boats on the continent between the Roman period and ca 1300. The by

4. Blockmans 2010, 23-24.
5. De Maesschalck 2012, 139.
6. Hillewaert 2011, 118.
7. Blockmans 2010, 107.
8. Asaert 1974, 19-21.
9. Blockmans 2010, 22.

now sizeable number of barges from the second to fourth centuries discovered and excavated has shown that they were not all built in the South German or Central German Rhine area and used for transporting building materials and other goods to encampments in the Low Countries. The shipyards of some of these boats may have been located in the present-day territory of the Netherlands. The vessels, which in certain cases were over 30 metres long and built of 6-10 cm thick planks, have sides joined at right angles to the carvel-built bottom with L-shaped bilge parts. The heavily constructed vessels were hardly suitable for sailing upstream. Several of them were discarded for this reason at their destination point and rediscovered centuries later during archaeological investigations in the vicinity of *castella*.

Most of the medieval and much more lightly built river barges and barge-like plank boats discovered up to now date from the 10th to the beginning of the 14th century, and a few from the Carolingian period. The so far earliest medieval example (4th quarter of the 7th century) was found in 2013 in the Catharijnesingel in Utrecht.[10] In contrast to the Roman boats which were designed with vertically built sides, the medieval ones have sloping sides.[11] From the Roman period on, they were completely open, had a flat bottom ending at both ends in a rising aft and forward and carvel or clinker built sides. Other examples of these are also the 13th-, but more likely 12th-century vessels 1 and 2 from Antwerp (fig. 3.2). In one case only was a mast step found, evidence that this was probably a 'genuine' sailing ship. It is not clear of the other barges how they were propelled, as no evidence was found. Floating downstream controlled with the help of steering oars, towing, punting and rowing, possibly with the addition of a small sail, are the most likely possibilities. The barges or parts of them discovered were used as bank reinforcement or revetment. The cog and the two small vessels at Kampen in the IJssel were probably used to dam a river branch/river channel. Many others will have been broken up for use as building or firewood.

3.2 West and North European seagoing vessels up to ca 1200

Not much is known about the type of vessels used to sail across the Channel and the North Sea between the Roman period and ca 1200. Historical sources report that in the 5th century Angles, Jutes and Saxons crossed the North Sea to the British Isles and landed there. These ships were most probably clinker built. From the written sources it appears that in the 7th century English ships crossed the Channel to the Seine, Brittany, the mouth of the Loire and more to the south, the Charente. The 7th-century oared vessel discovered at Sutton Hoo (in Suffolk) was a completely clinker-built ship whose construction was similar to that of a Baltic vessel, the ship found at Nydam in Denmark which dated from 400.[12] Ships also went to the Netherlands delta. In 716 there were trade relations with the Frisians, particularly the trade centre of Dorestad.

At the end of the 8th century the aggressive raids of the Norsemen/Vikings began, in their renowned streamlined and completely clinker-built and clenched warships. The long, narrow ships had a low deck level and curved, high sterns/stems. They were kept on course by a steering oar and had a square sail. They could also be rowed with oars (sometimes several dozen).[13] These ships, which are well known thanks to various Scandinavian finds, will not be discussed further here. In retaliation to the Viking raids Charlemagne and Louis the Pious organized within about ten years a maritime defence system between the estuaries of the Rhine and the Garonne. The main body of the fleet was stationed in Boulogne to defend the Strait of Calais. The estuaries of the Rhine, Seine, Loire and Garonne were safeguarded by smaller squadrons. Along the coast of Gaul two carvel-built types of ship were used for this purpose, the *scapha* and the *barca* which already existed in the 6th century.[14] For the latter defence fleet, ships were built in Boulogne and Ghent, among other places. In 811 the emperor himself came to Ghent to personally inspect construction progress.[15] The 7th-century carvel-built vessel with cross beams protruding through the shell Port Berteau II (Charente-Maritime) possibly reflects the type of ship we have in mind.[16]

In the last quarter of the 9th century the Scheldt estuary and adjoining Flemish territory were under Viking rule by an army that had crossed over from England. What his ships looked like is difficult to say, but from the records they did not resemble the Danish Viking ships nor the Frisian vessels. The new ships of King Alfred the Great of England are according to the Anglo-Saxon Chronicle of 897 almost twice as long as the others, some had sixty oars, some more.[17] As already stated in the introduction, only four ship (timber) finds of seagoing vessels are known from the period ca 750-1200 in the Netherlands. These were built in the Nordic or Viking shipbuilding tradition.

10. Brouwers *et al.* 2015, 19 and fig.12.
11. Vlierman 1996b, 101-106, fig. 73.
12. Johnstone 1974, 102-114.
13. See, for example, Crumlin-Pedersen 1997, 310, the Hedeby 1 wreck; ca 30 m long, 2.5 m wide midships ,1.5 m high and stern/stem of 4 m.
14. Asaert 1974, 14.
15. Henderikx 1995, 82.
16. Westphal 2014,12; Das Lochbuch 40, 12.
17 Crumlin-Pedersen 1997, 18.

Despite the fact that there is regular mention in the sources of Frisian vessels, until a few years ago it was not at all clear in what types of ships the Frisians sailed. Thanks to archaeological discoveries there now appears to be more clarity on this point. In the former medieval harbour of Dorestad a number of fragments of a clinker-built and clenched vessel were discovered during the large-scale excavations from 1967 to 1977. The fragments were subjected to dendrochronological analysis which yielded a felling date of after AD 739 ± 8. The provenance of the timber was Oost-Friesland.[18] In 1996 several fragments of a vessel built in the Viking or Nordic tradition were found in the former harbour of Tiel (Dorestad's medieval successor). The distinctive features are:

- The planks are not sawn but split and shaped and smoothened with a (T-shaped) carpenter's axe. The planks are usually no thicker than 1.5-2.5 cm.
- The planks are joined together in the ship's hull with iron rivets.
- The wooden pins affixing the frames/ribs to the shell have a thick rounded head projecting slightly from the outside of the ship.
- The ribs are comparatively far apart: c. 90 cm.
- On the scarves and the lands (the parts where two planks are fixed together) animal hair (often wool soaked in tar) was used to make the seams watertight (caulking).
- Some planks (from foreign/international finds) are decorated with grooves on the outside and sometimes on the inside. In certain ships this can also be observed on the ribs.[19]

Dendrochronological analysis of these items from Tiel produced a felling date for the timber of between AD 971-1008/1018. The timber was dated with the GB London calendar. Secondarily used ship's timber from this shipbuilding tradition in Vlaardingen dates from AD 961-973 and originates from the Netherlands, Northwest Germany or Schleswig-Holstein. The provenance of a second vessel, whose timber must have been felled after AD 1043 is England.[20] With regard to seagoing vessels, it now appears that, on the basis of current knowledge and insight, the ships mainly or almost exclusively used in the English and Flemish coastal areas and in what used to be known as Frisia were built in the Nordic/(English) or Viking tradition, in other words completely clinker-built and clenched.

At the beginning of the 10th century the age of the Vikings came to an end and authority over the northern seas passed into the hands of Anglo-Saxon England and Ottonian Germany. Trade with Scandinavia lost almost all its importance as a result of problems in Southern Russia which cut off the old connection between Northern Europe and the East. At the beginning of the 11th century merchants from Hamburg and Bremen in particular had an increasing share in Noordzee shipping. The land route between Hamburg and Lübeck, but also between Schleswig and the Noordzee coast through Holstein were still important in the 12th century.

On the Baltic side, sailing ships at that time were certainly built in the Nordic tradition. The 11th and 12th centuries initially saw little change in seagoing vessels. The over 15 m long and completely clinker built and clenched cargo vessels found in Denmark, Sweden and Norway date from 1000-1200.[21] According to Asaert, the so-called *knorr* or *knar*, which belonged to this group of ships, was closely related to the 9th-century Frisian ship.[22] The ships depicted on the famous Bayeux tapestry also clearly show that William the Conqueror crossed over to England in 1066 in ships built in the Nordic tradition. Several younger ship finds from Rotterdam and Dordrecht, which would appear to suggest a southern or northern North Sea or Baltic Sea provenance, provide evidence that seagoing vessels were still (?) built in the Nordic tradition until into the 13th century. They will again be discussed in part IV.

3.3 Mediterranean ships and the Normans in Southern Italy

For centuries prior to the Early Middle Ages the Mediterranean region had known a tradition of intensive seaborne freight traffic, particularly in the Roman period. Not only were the Mediterranean and the Black Sea navigated, but ships already sailed via the Strait of Gibraltar and the Atlantic Ocean to the south of England, for instance, for tin. In certain cases, very large ships were used. On the extremely detailed 3rd-century 'Torlonia relief' depicting two large cargo vessels in the harbour of Rome, one can see that even then the hull was constructed with crossbeams protruding through the shell.[23] They were observed in the wreck of the cargo ship of L'Anse des Laurons (Var, France) dating from the Late Roman Imperial period. In this case,

18. Vlierman 1996b, 83-87 and appendix 8; Vlierman 2002, 135 and 138; RING January 1997 and August 2001.
19. Vlierman 1998, 56-58.
20. Vlierman 1997d; Vlierman 2004, 22-29.
21. Crumlin-Pedersen 2000, 243 table 2.
22. Asaert 1974,18.
23. Gardiner 1996, 48.

the beams were not under, but on/over the deck.[24] Roman Mediterranean vessels were carvel built in a specific way (*mortice and tenon* construction). Finds from Britain (Blackfriars 1 in London, St. Peter Port on Guernsey and Barland's Farm in Wales) and the boat from Bruges are examples of the Roman-Celtic tradition of seagoing shipbuilding. These are carvel-built vessels whose shell planks were fastened to the ribs with long nails (through wooden pegs), which were bent over twice on the inside of the ship and driven back into the wood.[25] Due to the incompleteness of these wrecks it is unclear whether they had through-beams.

Several researchers regard the Mediterranean region as the origin of certain building elements (through-beams, stern rudder). It has been stated above that the former was at any rate in use in the Roman period. The stern rudder was probably familiar to Arab peoples or Saracens, of which the 'Moors' from North Africa conquered the Visigoth Kingdom in Spain at the beginning of the 8th century. Quite soon after, they also occupied the coastal area of the Languedoc. In the first half of the 9th century the Arabs seized Palermo, set fire to Rome and shortly after 900 conquered Sicily and Southern Italy. In 1009 the first Norman raids took place in Italy. In 1043 they took control of Southern Italy (Apulia) from the Arabs. In 1059 the Pope signed the Treaty of Melfi with the Normans. From then on the Normans concentrated their military strength to capture the Byzantine strongholds in Italy. They conquered Sicily between 1061-1091.

The Normans undoubtedly sailed to the Mediterranean in their well-known clinker-built ships. On arrival, the ships would have been viewed by local shipbuilders and almost certainly also by Arab shipbuilders not only with surprise but also with admiration with regard to the degree of craftsmanship with which they were constructed. Consequently, it would appear almost a matter of course for certain constructions and means and methods of fastening (adapted) to be adopted and applied to existing ships as well as those still to be built. The planks of Arab vessels, for instance, were still sewn together, and their method of shipbuilding might be regarded as primitive, certainly in comparison with the Norman ships. It goes almost without saying that the Normans, from the occupied areas in the south of Italy, must have had a (substantial) influence on the further development of shipbuilding in Italy and particularly so on that of the already large ports in the north. There is evidence that at the time of the First Crusade (1096-1099), the Italians, at any rate temporarily, led the way with the dimensions of their ships and use of technical skills.[26] It seems likely that they maintained and/or adopted and improved the Arab stern rudder, and used it increasingly. The Norman shipbuilders would certainly have tried out the stern rudder on their ships, probably adopted it to some extent and subsequently introduced it into their original territory, the Atlantic coastal region of France, Flanders and England.[27] However, how the 'new' Italian ships were built remains unknown. Hardly any archaeological shipwrecks are known from the period in question, and if any exist, there still appears to be little interest in them in Italy. The great maritime cities of Genoa, Pisa and Venice underwent a huge development in the 12th century and around the middle of the century already had large (trading) fleets.[28]

3.4 Crusades and sea trade in Europe

Since the 11th century, the northern half of Italy was part of the Holy Roman empire. Towards the end of that century the armies of the French from the royal domain under the command of the king's brother formed the army for the First Crusade (1096-1099). The counts of Normandy and Flanders, the North French, the Lotharingians and the Germans under Godfrey of Bouillon joined them, as did eventually the Normans from Southern Italy and Sicily. Some of these Crusaders embarked in Bari and Brindisi on ships that were chartered in Genoa, and set sail for Constantinople. During the regular presence in Italy and fierce confrontations between the German rulers and the popes in the first half of the 12th century, and particularly also during the preparations for and implementation of the Second Crusade (1147-1149) led by king Louis VII of France and the German king Konrad III, much experience was gained concerning large-scale transport of troops by sea and the transportation of the necessary equipment and provisions. Possibly the crossing of this army was not carried out in a single operation, so that the ships returned to collect the remaining army units. During the return voyages use would certainly have been made of the cargo capacity of the ships and goods already seized or obtained through trade shipped to Italy and beyond.

This resulted in increasingly busy trade through Europe and a growing trading area in Europe north of the Alps of goods which were often completely unknown up to

24. Hockman 1994, 62.
25. The boat of Bruges reviewed, Vlierman in preparation b.
26. Pryor 1992, 30.
27. The earliest depictions of these are the 12th-century vessels with stern rudders on a relief in Winchester Cathedral and on a baptismal font in Zedelgem; Heinsius 1956, 48-49. Chapter 4, fig. 4.4a and b.
28. Erbstösser 1976, 173-174.

then. A side issue was that rulers of the countries passed through started to levy toll in more places for their own benefit.[29] In the northwest and south of Europe two economic centres developed in the 11th century, in which the most important cities included Bruges in Flanders and Genoa and Venice in Italy. Most of the trade between these trading centres was originally via the land routes on the continent, undoubtedly due to the aversion to and fear of the ocean. This fear gradually disappeared as Flemish and English trade with Gascony increased and the Italian cities established ever-closer ties with west Morocco. During the Crusades more and more sea voyages were undertaken between the Mediterranean and the Noordzee.[30]

As a result of the growing number of tollhouses which hindered goods transport in the inland parts of Europe, the sea started to gain importance as a traffic route. The Mediterranean Sea resumed the role of connection that it had during the Roman Empires. The Crusades, with the conquest of the Syrian coast, brought the Italian cities of Genoa, Pisa and Venice great activity in the East. Northern Italy's increasing wealth did not escape the authority in the north and led to the increasing economic importance of the German Empire (north of the Alps) after the Roncalian decrees of 1159.[31] Cologne began to make more use of its own ships for trade instead of hiring them from the Frisians as was previously the case. This change coincided with Cologne's participation in the Second Crusade. The ships from this city on the Rhine were the largest proportion of German ships that met the English and Flemish squadrons at Dartmouth. The Cologne ships also took part in the siege of Lisbon in 1147 before sailing on to the Holy Land.[32]

During the regular and assertive presence of Frederick I, who was crowned emperor in 1155, and his armies in the northern part of Italy, Italian Mediterranean shipbuilding will have undergone stronger innovating growth in the third and fourth quarters of the 12th century, not least as preparation for the Third Crusade (1189-1192). These developments undoubtedly also became known in the part of the Empire northwest of the Alps at roughly the same time. Several large Italian vessels (as examples?) probably even sailed to the North Sea area. At the instigation of emperor Barbarossa and the bishop of Cologne and also as a result of various privileges bestowed by the emperor, it would, in my view, appear likely that Crusader ships were built or that ships were equipped as such in the trade centres in the core area near the Flemish coast and in the Lower Rhine area.

Henry the Lion, (duke of Bavaria and Saxony, 1142-1180), the Saxon cousin of emperor Frederick I (Barbarossa) and initially his right hand, had a great deal of power in the North German coastal area where the emperor's influence was small. In 1159 he had the city of Lübeck rebuilt after a devastating fire. At the same time this meant opening up the areas east of the Elbe as part of the German expansion eastwards (also referred to as a Crusade) and the foundation of ports according to Lübeck (or derived from it), Magdeburg or Kulmer law along the southern Baltic coast. The foundation of the new port of Lübeck gave German merchants from Westphalia and Saxony, but also those from Holstein direct access to the Baltic. It was in this period that the Hanseatic League or Hanse came into being. Prior to this there were contacts between merchants from the Rhineland and from the Netherlands-Frisian coastal area with the Scandinavian and Russian Baltic merchants via the isthmus at Haithabu/Schleswig.[33]

The merchants from the west did not themselves sail on the Baltic from the 10th to the middle of the 12th century. Protected by Henry the Lion's privileges they first sailed to Gotland after the middle of the 12th century. The Germans succeeded in taking control of the Baltic trade by the end of the 12th century and virtually eliminating Scandinavian and Russian competition. This success was mainly due to their better relations and trading partners in the West, as well as to more capital, though not to their ships (cogs?) because they were not technically as good.[34] As can be seen from the 13th-century seals of Lübeck and the late-12th and early-13th century ship finds from the Schlei, Denmark and Sweden, the ships that sailed from this city and the surrounding area were almost certainly still clinker-built and clenched, in the Nordic or Viking tradition. Bolin[35] was under the impression that up to the middle of the 13th century the merchants of Lübeck transported their goods in Wendic or Nordic ships, or ships from other countries. In his view, cogs did not appear in the Baltic until the Ommelandvaart (sailing around Jutland to the Baltic) in the middle of the 13th century. According to Heinsius[36], little attention was paid to Bolin's assertion, though neither was it refuted

29. In 1200 there were for example 19 tollhouses along the Rhine, and a century later, 35 (Asaert 1974, 35).
30. Asaert 1974, 32-34.
31. Wahl 1959, 90-92.
32. Asaert 1974, 25-26.
33. The first impetus for the economic rise of the new German territories east of the Elbe is partly due to Dutch colonists and Flemish merchants (Asaert 1974, 36).
34. Henn 2010, 13-15.
35. Bolin 1933.
36. Heinsius 1956, 247.

by German historians. This is still the case at present. Knowledge of the innovative shipbuilding method for large vessels in the Mediterranean region and the experiences with it during the Second Crusade was undoubtedly put to use in (North)west Europe. It will have contributed to the improvement and further development of the fleets on the Artesian/Flemish/English coasts, while in the Lower Rhine area it will also have led to improving and enlarging existing ships, but also to building a new type of seaworthy cargo vessel in order to trade over sea with the Mediterranean area. The pragmatic Henry the Lion, who had returned from Italy with an empty purse,[37] saw unprecedented economic opportunities in the Baltic region, and he introduced this innovative Mediterranean method of shipbuilding in the third quarter of the 12th century, probably starting in and around Lübeck. In the newly established German ports along the Baltic coast an early start was probably made in building shipyards and constructing (large) cargo ships for the Baltic Sea. Whether this led during Henry's rule that lasted until 1180 to the building of cog ships as discussed in this study cannot be said. The oldest (Danish) cog finds (Kollerup, Kolding and Skagen) consist of ships that were probably built between ca 1150-1200 in South(west) Jutland.[38] The oldest depictions of cogs on city seals from the North German and Polish coastal areas date from about the middle of the 13th century (Stralsund, Wismar, Elbing and Danzig). On the seal of Barth from ca 1300[39] a Nordic/Slavic ship is also shown, but with a stern rudder.

3.5 Northwest European sea ships ca 1200-1500

It was already stated in section 1.1 that with regard to (seagoing) ships from the period up to ca 1200 we have to resort mainly to archaeological finds. The information available from historical sources is limited and the extent to which vessels depicted on miniatures are realistic is questionable.
Generally speaking, it may be said that these are more symbolic representations. Flatman[40] (only) shows four 10th- and 12th-century miniatures depicting probably reliable hull forms.

After ca 1200 more historical sources are available and the number of names of ship types also increases.[41] Asaert[42] discusses several of these names (the *balengier*, the *bark*, *bargie of baerdze*, the *buis*, the *galei*, the *heude*, the *hulk*, the *karveel*, the *kogge*, the *kraaier*, the *kraak*, the *nef* and the *nao*). In the introduction he describes how a ship type can be defined: according to function, place of origin, building method or method of propulsion, and he gives examples of these. From his descriptions one can infer that they were mainly based on information from the written sources. For want of published archaeological finds in the 1970s, this was hardly surprising. He shows two familiar, late-15th-century ship portraits by master WA, both of which bear the name of the ship type (a *baerdze* and a *kraeck*). However, the problem still remains that so far no archaeological finds are known which can be linked with certainty to one of these two types, nor to those depicted by master WA. In the case of the large vessel from lot U 34 in Eastern Flevoland mentioned above, it is cautiously thought to be a *hulk*. This ship was built in the first half of the 16th century, the period in which most probably no cogs were still sailing. The fact that the name 'cog' can possibly still be linked to a substantial number of late-medieval ship finds appears to be a striking exception. In the next chapter that state of research into the cog will be discussed. The main differences established between ships of the Nordic and the cogbuilding traditions are listed below.

The Nordic shipbuilding tradition up to ca 1200

- A (T-shaped) keel with curved (and high rising) stern/stem, sometimes with a transitional/connecting part between the shell planks and the stern/stem.
- A hull completely clinker built from keel to gunwale of generally narrow strakes fastened together with rivets. The planks are split from a treetrunk and usually no thicker than 1.5-2.5 cm.
- A shallow groove was gouged out on the land between the overlapping planks in which, before the planks were riveted/clenched together, a thinly braided animal hair thread soaked in tar was laid.
- The hull is comparatively low and broad midships, whereas the ships rise sharply towards stern/stem.
- The hull, built entirely according to the *shell first* principle, is kept in shape with generally rather thin and narrow frames attached every 80-90 cm to the shell planks by means of wooden treenails. The treenails on the outside usually have a somewhat rounded head which projects through the shell.
- About halfway up the gunwale height deck beams were fitted and provided/covered with loose lying deck planks lengthwise in the ship.

37. Wahl 1959, 193.
38. Dokkedal 1996; Westphal 1999; Hocker 2000.
39. Ewe 1972.
40. Flatman 2009, figures 19, 22, 30 and 138.
41. Heinsius 1956, 39-69.
42. Asaert 1974, 101-113.

- Between this deck and the edge of the gunwale area are crossbeams attached to the sides with rising and/or lodging (horizontal) knees to provide the necessary transverse rigidity in the hull.
- Longitudinal rigidity is reinforced by a long keelson overlying the keel and bottom frames inside the ship.
- The ships were kept on course with a steering oar.

The building tradition of the (large) cogs between ca 1200 and 1500

- A (short) keelplank with a knee-shaped hook at both ends forming the back, respectively front part of the keelplank as well as the bottom parts of the rising stern/stempost. The hooks are attached to the keel plank with slanting scarves. The straight and relatively sharply falling stern/stempost are (also) attached to the hooks with slanting scarves.
- A carvel built surface almost horizontal or slightly rising towards the turn of the bilge whose lands fore and aft twist increasingly in the sharply built parts of the hull and turn into a clinker-built construction. The carvel-built part is held together during construction with temporary frames and subsequently floor timbers are added.
- The sides are completely clinker built from stern to stem, according to the *shell first* principle. The overlapping strakes were joined with nails double-clenched and driven back into the wood.
- The strakes of the sides run as far as the middle of the back of the sternpost and the middle of the front of the stempost. On the fore side the plank ends are protected and covered with a false stem.
- Before the frames/ribs were put in, the seams of the overlapping strakes on the inside were made watertight with sintel clamped moss caulking. This method of caulking was carried out on these seams and on the seams of the carvel-built part of the surface on the outside of the hull shortly before the ship was launched.
- Depending on the dimensions of the ship, the floor timbers in the sides were lengthened with one or more futtocks. The ca 15-20-cm thick frames/ribs are set every ca 40-50 cm, and attached to them in principle with two wooden treenails per plank width. The first futtocks placed on the floor timbers are usually slightly less wide and thick. The second futtocks often even less.
- The transverse and longitudinal rigidity at deck level was achieved by a number of crossbeams protruding through the shell with rising knee constructions on them. The strakes of the sides were attached to these up to the top edge of the gunwale. The longitudinal rigidity at deck level consisted of stringers, and this also applied to the planks of the open ceiling and the long keelson.
- The ships were held on course with a stern rudder.

The late-15th-century document that mentions the name 'cog': "*...unser Stad Sigel ghenomed den Kogghen wylyken laten henghe yn dussen Unsen breff...*" (... the seal of our city, known as the Cog, which we affix to this missive of ours ...), and the civic seal of Stralsund (fourth from the left). Stadtarchiv der Hansestadt Stralsund.

The medieval cog: state of research[1]

The object/topic of this study is the technical details, the various aspects of the construction and appearance of the archaeological finds from the Low Countries designated as cogs plus the analysis and ensuing conclusions. One cannot, of course, disregard research previously carried out by others with regard to the iconographical and written late-medieval sources on the ship, the attempts to link these sources to the archaeological evidence, the development of the building tradition and the meaning of the name 'cog'.

4.1 Historiographical research

The late-medieval cog has long been an intriguing topic for maritime historians. Seals, miniatures, several early paintings and the regular appearance of the name cog in written reports mainly referring to the Hanse (trade), were the only sources until the middle of the 20th century. City seals and some miniatures depicting ships provided the most important information about the shape of the hull, rigging and sail plan of the ships that appear to be eligible for classification as cogs. The written sources also contain names and descriptions of other late-medieval types of vessels. One good example is an informative anchor(ing) instruction for the Zwin from 1401 from a German merchant in Bruges to the entire Hanse. Not only the names of ship types are mentioned, but some other ships are referred to with the place of origin, and from the descriptions it can also be inferred that there were vessels of different dimensions.

The reason for writing the instruction was the silting up of the Zwin which made the passage to Damme increasingly difficult. Many ships transferred their cargo at the roads of Sluis. Not all ships were allowed to anchor there so many did so in the Zwin, thus hampering passage. This required organization. The mayor, the Schepenen (sheriffs) and the Council of Bruges concluded, among other things:

'*...Item dat alle manieren van schepen, hoe gedanich zy zin, de ind verscreven Zwen komen zullen, dats te wetene tuusschen der Slus en de Mude, men legge langhes den strome, te wetene de cogghen, schepen, crayers, evers ende barghen dre schepe dicke deen neffens den anderen, pleyten, cleene cogghen, schepen van Brabant, de Hollandesche, Zeelandesche unde van anderen vre(m)den steden viere schepe dicke deen neffens den anderen, ende de schuten ende soyen 6 schepe dicke deen neffens den anderen, wel gheancert voren ende bachten; ten dien heynden, dat zij niet ne drayen metter hogher vloet noch mit der ebbe dwers den vorscreven Zwene, up de boete...*' [2]

There were, then, three categories: the first being the large seaworthy ships which were able to sail via the North Sea and the Channel to the Zwin. These types were used by the merchants of the Hanse, among others. The second category includes smaller types of ships for the transshipment of goods from the sea ships and for inland shipping. The last category, the small barges, were used for transshipment and local transport of goods. The name cog appears twice. In the first category as a large sea or Hanse ship, in the second as a smaller variant and probably intended more for inland navigation.

Other types of ships

In addition to the *cogghe*, other written sources such as toll registers and bills mention the *heercogghe,* as well as other names of ships. The types described by Asaert have already been mentioned in the preceding section, but even more names are also known. River shipping refers to names such as *drubbort (dryeboerden, druppert, drobbert) hoegeboerden*, *schep van vyeff boerden*, *lordannen* (pinewood ships), *gasthuis*, *swarten schippen* and *swelis, sameroes, beyerschiff, Reesischeraachen, beitelaachen*, *Oberländer.*[3] Alongside the names of ship types in bills and toll registers many technical shipbuilding terms can be found. For example, the toll regulations for Dordrecht of 1287 mention among the many types of cargo ... *elken doerbalcte schepen* ... Unfortunately no mention is made of the type of ship this refers to.[4] Neither was a distinction made on the basis of

1. Translated by Christine Jefferis.
2. ...that all types of vessels entering the Zwin must be moored in the direction of the current; depending on their size, three (such as cogs), four (such as small cogs) or six (such as boats) and anchored fore and aft to prevent them from turning with the tide and blocking passage through the Zwin. Luns 1985, 35-36; Höhlbaum 1879, HUB II.
3. Reinders 1983, 34.
4. Luns 1985, 35-36.

the technical construction. This means that the typology that we reconstruct on the basis of the material sources was not applied in the Middle Ages or that (most) medieval people were not even able to define it.[5] Moreover, we do not know whether the construction of, for instance, cogs in the Middle Ages was also considered a characteristic of this type of ship, or whether the name was used to designate a (large) ship. With regard to the river ships that sailed on the Rhine, Waal and IJssel between ca 1360-1560, Weststrate observes that hardly anything is known about them, except for the names mentioned in the historical sources.[6] And indeed, not a single archaeological river ship find has so far been linked with (any) certainty to a late-medieval name. To return to the cog, this would appear to be a different situation, though this is still not absolutely certain.

The heercogghe

In addition to the name *cogghe* or cog, there is regular reference to a *heercogghe.* Luns discusses what other researchers (for instance Ellmers 1972) have written about this. In most cases, the name *heercogghe* is linked with the word Kogschuld. A hypothesis is put forward that the kogschuld was a sum of money to pay off the *heervaart* (kind of military service) obligation in times of peace. Luns then gives various examples of the payment of *koggeld* (cog money) for the heervaart or for the equipment of heerkoggen or other arms. In his view, the paucity of written sources does not permit this to be anything more than a suggestion. In his appendix V, Luns mentions sources in which, in addition to the name *heercogghe,* there is reference to oars/rowlocks and rowers: in the year 1283 twenty oars and fifteen men; in 1361 rowlocks; in 1362 sixteen oars; in 1371 thirty-five oars; and in the same year, thirty-one oars.[7] These cases apparently refer to vessels that were rowed. Further indications as to the shape and size etc. are lacking. Flatman[8] shows a rare illustration from the English Queen Mary's Psalter from ca 1325-1353. The ships rowed by two men and on which two knights in armour are attacking each other with lances are not cogs, but almost certainly vessels built in the Nordic(/English) tradition. No ship/cog finds are known which might cautiously point to a rowed heerkogge. In the case of the Rutten cog, this may have been a large warship/heerkogge, whereas the Nijkerk-II cog may have been converted in order to accommodate soldiers, for example.[9]

Where, in what period and from what shipbuilding tradition(s) did the cog develop?

Hagedorn observes that the word *kogge* is of Frisian origin and that the earliest mentions of the word come from this region. In 867 the name *Cokingi* was used for Frisians.[10] However, the underwater body of the cog which he assumes to be V-shaped makes the origin of this ship type along the shallow Frisian coast unlikely. Hagedorn solved

Fig. 4.1. Armed knights in combat with lances on small, rowed (river)craft. (Queen Mary's Psalter, England, ca 1325-1353) after Flatman 2009, 79

5. H.L. Janssen, written communication August 2016.
6. Weststrate 2008, 253-254. More archaeological river ship finds are known than those mentioned by Weststrate. Especially vessels older than ca 1300. See for this § 2.6, Van Holk 2001 and Vlierman 1996b, 1997b, 2000 and 2002. In a recent survey (Brouwer *et al.* 2015, 6-24) recently discovered finds have been included.
7. Luns 1985, 23-25 and 70.
8. Flatman 2009, 79 fig. 70.
9. Chapters 10 and 6 respectively.
10. Hagedorn 1914, 23.

this problem by arguing that the provenance of the cog should be sought on the French coasts. The type of ship in use there, the *nef*, he claims was at some stage fitted with a Frisian stern rudder. Luns observes that Hagedorn's hypothesis invites the necessary criticism. Heinsius comes to the conclusion that the cog in the Northwest German Baltic region must have originated in the second half of the 12th century, that according to several historical sources at that time the Germans appeared on the sea with remarkably large ships and that the type must have been built according to a housebuilding method.[11] Asaert[12] writes about the *nef* among other things "in addition to the general and vague meaning of vessel, the names also refer to certain types. The *nef* of the 11th and later centuries is the large Atlantic cargo ship, mainly between France and England, an area to which the cog had not at first penetrated ..... The difference between these two ships is not clear. Both are round, low in shape and clinker built... It develops simultaneously and analogously with the cog. In the 13th century a distinction was still made between the two vessels. Cogs and nefs are mentioned in the same texts, they sail in the same fleets. Cog and nef begin to influence each other so much that their names are only a matter of usage and the cog becomes the large medieval cargo ship, the northern version of the nef, the Atlantic translation/version." Ellmers and Crumlin-Pedersen assume that the cog developed from a Frisian, coglike *waddenschip*, and that it must have originated in the northwest part of Germany. The former adds the remark that no evidence for this has been found. Reinders[13] states that no shipwrecks have yet been found to corroborate a development of the cog from a Frisian *waddenschip* in the 13th century. This is still applicable even now (2016).

The name *kogge/cog*

That the German word *Kocke* (*kogge/cog*) referred to a distinctive, large, seagoing type of ship could be inferred from *Parzival,* which mentions this and several other names of ship types around 1200. In *Willehalm* by the same author (Wolfram van Eschenbach) the name occurs in the meaning of a large transport ship during the Crusades, while in three cases it also appears to designate a kind of river ship. Almost contemporaneously the name 'Kokke' crops up in South German literature in the sense of a large sea ship, while this name of a ship type is mentioned in documents/records from 1206 and 1210 in the North Sea region. Heinsius does not wish to devote too much attention to the possibility of the name cog designating a river ship.[14] Other researchers have also pored over the name cog and have given their views on its meaning and origin. Luns discussed in detail what the well-known authors Hagedorn, Vogel, Heinsius, Crumlin-Pedersen, Ellmers and others have written about the name cog (in relation to archaeological finds) and complemented this with data from his own (sources) research. He came to the conclusion that there was a simultaneous development of a clearly recognizable ship type on the 13th-and 14th-century city seals and the emergence of the word cog for a large merchant vessel in the written sources of the same period. He was also of the opinion that the smaller vessels with similar characteristics should be called cogs as well and that there was probably a building tradition. Furthermore he concluded that archaeological investigation and historical research into bills in particular would considerably improve insight in medieval shipping and shipbuilding.[15]

As already mentioned at the beginning of this section, it is uncertain whether the name *kogge/cog* refers to a certain type of ship. It could quite well be a collective term. By way of comparison: one cannot infer from the modern name or description *kustvaarder*/coaster to what type of ship it refers and/or its dimensions. The possibility cannot even be excluded that the word *kogge* at that time may also have had one or more completely different meanings. We shall probably never find out for certain.[16] Nor is it yet completely certain which iconographic images of ships from the period ca 1200-1500 may be linked to the name cog. Ewe portrayed all the seals from Northwest Europe at his disposal, described them with their corresponding dates and published the data in his indispensable *Schiffe auf Siegeln*. On a substantial number of the depicted seals of towns along the south coast of the Baltic and from the Low Countries are ships with straight, rising sterns/stems, one mast and one or two castles.[17]

However, after the discovery of the Bremen cog, attention was drawn to the fact that the shape of the hulls of a number of the ships depicted on seals of Hanse cities was very similar to the hull shape of this vessel. This also applies to the wrecks of large cargo vessels discovered later and discussed in parts II and III of this study. The smaller ships with similar constructions discussed in this study were built in the same tradition. This does not mean, however, that they were given the same name. The striking (high) shape of the hull with steep upright and raked sterns/

11. Heinsius 1956, 117.
12. Asaert 1974, 111 and 112.
13. Reinders 1985a, 12.
14. Heinsius 1956, 72; Wolfram van Eschenbach (c. 1170-c.1220).
15. Luns 1985.
16. See also below under Seals and miniatures in a footnote the remark concerning the ship type of the late 19th-century cargo ship Lutina.
17. Ewe 1972.

stems and a stern rudder and the generally clearly visible clamps/and sintel clamped moss caulking in the lapstrake seams could in my opinion have been enough for the average medieval person to classify all ships with these characteristics under the same name.

What did the cog look like?

The most complete and the first survey of what could be inferred from historical and iconographic data about the cog up to the middle of the 1950s is by P. Heinsius.[18] It may be concluded that Heinsius, on the basis of historical sources and despite the lack of archaeological information, provides the most precise description of what a large Hanse cog must have looked like. It is evident that on the basis of these sources and with respect to the building, construction, design and the main measurements, in many cases he was not able to come to the correct descriptions and conclusions. This also applies to the above-mentioned description by Asaert[19] (a ship completely clinker built, round and low in shape). This immediately shows the importance of archaeological sources.

Heinsius came to the following description: *...die Koggen waren ein geradkieliger, hochbordiger Fahrzeugtyp mit ziemlich geraden und recht steilen Steven. Sie führten ein grosses viereckiges, mit zwei Schoten dirigierter, an einer beweglichen Rah befestigter Segel. Ausserdem zeigen einige Reliefs auf Siegeln, dass die Fahrzeuge oft ein recht gedrungener Form hatten. Trotzdem wahr ihr Unterwasserschiff verhältnismassig scharf gebaut und zum Segeln gut geeignet. Landschafftliche Sonderformen zeigen eine leichte Varierung dieses Schiffs.*[20] Heinsius goes on to write that there is no more doubt as to the building method of the cog: *...Nachdem uns aber die auf den Lübecker Siegel erkannte Bauweise durch verschiedene Quellen bestätigt ist, kann kein Zweifel mehr bestehen, dass beim Bau der ersten Koggen mit dem Annageln der Planken, entgegen dem sonstigen Brauch, von oben begonnen wurde. ..... Est ist aber nun weiter möglich, mit dem Aneinandernageln der Planken von oben anzufangen wenn zuerst das Spantengerüst aufgestellt und festgefugt ist. Wir können nunmehr in den Koggen Fahrzeuge sehen, bei denen der Bau des Spantengerüstes vor dem bau der Aussenhaut errichtet wurde...*[21]

Between the excavations of two cog wrecks in the IJsselmeer Polders (and with the information on several excavated wrecks from 1944, 1949 and 1976), several staff members of the Department of Maritime Archaeology of the RIJP paid a visit to the Deutsches Schiffahrtsmuseum in Bremerhaven in 1982. The wreck of the Bremen cog which had been completely reconstructed there and placed in a conservation tank was at that time dry and we had the opportunity of examining the ship inside and outside. With respect to the building method of the ship and the wrecks in the Netherlands, we had long held a different view from that of Heinsius. We are of the opinion that with cogs there must have been a *shell first* method or the striking combination of carvel and clinker building. With this method, the entire shell was built in phases from the keel to the gunwale. On completion of each phase, the frame parts were attached as reinforcement, to get the required shape and strengthen it. When building the bottom several (temporary) support frames were used. Heinsius visited the excavation of the Nijkerk-I cog in 1981, and in 1986 published a second, revised edition of *Das Schiff der Hansischen Frühzeit.* In the afterword he concludes *zum stand der Forschung bis 1983* that the building method on frames which he had previously assumed was not applicable.[22]

Baltic cogs

The name cog is also used for some completely clinker built and often of softwood late-medieval shipwrecks from the Baltic region. Taking into account the characteristics listed in § 2.4, this is probably incorrect. Some authors have found a way of getting around this by calling these wrecks Baltic cogs. Apart from the vessels from the Low Countries discussed in this study, in certain cases there do appear to be typical cogs built in the Baltic area that show more resemblance to the Nordic or Scandinavian shipbuilding tradition. These completely clinker built Baltic ships are not further discussed here. The same goes for the cog wrecks from the Baltic whose construction probably (largely) corresponds to the ships from the Low Countries discussed in this study. The main reason for not including these finds here is the limited information on these ships available and published[23] compared with the detailed descriptions etc. of the cogs in this study and the impracticability of processing these shipwrecks myself in a similar fashion and adding them to this already voluminous study.

18. Heinsius, 1956.
19. Asaert 1974, 112.
20. ...the cogs were a straight keeled, high gunwale type of vessel with rather straight and steep sterns/stems. They carried a large square sail fixed to a moveable spar and controlled by two sheets.
In addition, several reliefs on seals show that the vessels were often quite compact in shape. Nevertheless, the underwater body was comparatively sharp and well adapted for sailing. Slight regional variations of this ship can be seen. Heinsius 1956, 247-248.
21. ...after confirmation by various sources of the recognised building method shown on the Lübeck seal, there can be no more doubt that when constructing the first cogs, the planks were nailed from the top downwards contrary to the usual custom.
However, it is also possible to start nailing the planks together starting from the top when the frame has first been built. Cogs can now be seen to be vessels in which the frame was built before the shell. Idem, 111.
22. Heinsius 1986, 263.
23. See the remark by Crumlin-Pedersen in § 2.3.

Cogs on medieval seals

According to Ellmers, Heinsius knew of two early seals depicting ships designated in the medieval sources as cogs (La Rochelle and Lübeck, figs 4.1a en b).[24] Ellmers writes of the La Rochelle seal that it may on no account be used as a certain depiction of a cog, but he thinks that this may however be the case with the Lübeck seals.[25] Hulst is of the opinion that the latter cannot be assumed to portray cogs either.[26] He bases this on the interpretation of the Latin texts in which the word cog is mentioned. These do not state, as Heinsius says, that the ship on the seal of La Rochelle is a cog, but that it resembles one (*ad similitudinem cochae*). In the case of the Lübeck seal of 1226, *coqko sive liburna* is written, a cog or a liburna. According to Hulst, the latter is a lightly built and fast ship. The image does in fact resemble a lightly built ship in the Slavic/Scandinavian tradition.[27] The ship depicted could never be said to have a flat, carvel-built cog bottom. Besides which, it has a curved stem and a steering oar. With regard to steering, it has meanwhile been established that the Skagen cog (DK) built around c. 1200 and the oldest Dutch cog finds dating from the middle of the 13th century were equipped with a stern rudder. The sterns with gudgeons belonging to the Dutch finds show that this was certainly not a proto-design, but a construction that was long in use and worked well. Taking into consideration the characteristics already put forward by others, I am also of the opinion that the ships on the La Rochelle and Lübeck seals are certainly not cogs.

.Fig. 4.2a. The civic seal of La Rochelle (12th century). After Heinsius 1956.

There are seals and miniatures on which cogs are almost certainly depicted, but none of these bear the name cog. The identification of the wreck found near Bremen in 1962 as a cog, was based initially on Heinsius' work.[28] A late 15th-century document was subsequently 'discovered' which mentions a cog, while the attached seal of the town of Stralsund shows a ship very similar to the external appearance of the Bremen wreck.[29] The earliest date for the

Fig. 4.2b. The civic seal of Lübeck (1224). After Ewe 1972.

Fig. 4.2c. The civic seal of Stralsund (1329). After Ewe 1972.

24. Ellmers 2002, 160.
25. Ellmers 1972, 58-59; Ellmers 2002, 163.
26. Hulst 1987, 49-55.
27. Wiechell 1978.
28. Heinsius 1956.
29. Fliedner 1964, 5.

seal is 1329. It is, in fact, the only specimen in which the name cog and a contemporaneous image are found together (although there may be a difference of ca 150 years). The text of the document of 1483 refers to the ship type by name ....*unser Stad Sigel ghenomed den Kogghen wylyken laten henghe yn dussen Unsen breff*....(fig. 4.2c).[30] Even this source does not provide 100% certainty that the Bremen find is a cog. It cannot therefore be proven that the designation of a wreck as a cog is correct in all cases. Historical sources provide insufficient evidence for this.[31] The same problem applies to almost all ship finds in the former Zuiderzee area (and elsewhere). Even with familiar names of more recent ship types such as *praam, aak* and *tjalk* etc. it is often the archaeologically observed differences with the published and generally accepted characteristics and descriptions that make it difficult to establish with certainty the type of vessel in question.[32]

In addition to the Stralsund seal, Fliedner and Ewe should in fact have mentioned the Elbing seal from ca 1350 (figs 4.3a and b).[33] The shape of the ship's hull is almost the same on both seals. The Bremen cog was built with cross-beams protruding through the shell. This is not the case on the Stralsund seal. The Elbing seal from 1350 shows beam fenders (plus cross-beams protruding through the shell). Furthermore in the highest but one strake fore and aft there are rectangular holes (or also beam guides) visible and against the topmost strake there is a channel for the attachment of the standing rigging (the ropes supporting the mast). It is remarkable that on only three of the seals depicted by Ewe on which, judging by the Stralsund seal, a cog is portrayed, are protruding cross-beams shown; on the Harderwijk seal from 1263, on that of Damme from c. 1300 and on the second seal of Elbing from ca 1350.[34] Are perhaps large, seagoing cogs represented here? If there are rose bolts instead of beams protruding through the shell of the ship on the Harderwijk seal, this would probably mean a small cog.

The cog on the Elbing seal from ca 1350 shows two rows, one above the other, of four cross-beams protruding through the shell plus beam guides.[35] These two rows of beams suggest that this ship must have had two deck levels with a relatively small space in between. This was probably intended initially as extra stowage space for fragile or valuable goods, for example, but possibly also for quartering soldiers. One should, however, note that on insignia depicting a warship[36] soldiers and weapons are always portrayed, whereas this is not the case with city seals. On the other hand, the trading cogs did carry soldiers in turbulent times. Space therefore had to be created for this extra complement with kit/arms. For this reason

Fig. 4.3a. The second seal of Elbing (ca 1350). After Ewe 1972.

Fig. 4.3b. Idem.

30. ... our townseal named the cog hanging on our letter... Fliedner 1964, 5.
31. Reinders 1985a, 31.
32. Zwiers & Vlierman 1988, 70. The inland cargo vessel in question, 'Lutina', wrecked on 20 November 1888, displayed both characteristic elements of the Overijsselse praam and of the Hasselter aak, while the descendants of the skipper spoke of their grandfather's *tjalk* during a reunion in the Museum voor Scheepsarcheologie in Ketelhaven.
33. Fliedner 1964, 5.
34. Luns 1985, 37; on determining the amount of toll in Dordrecht, in 1287 the presence of cross-beams was probably one of the factors involved '....elcken doerbalcten scepe....'.
35. Ewe 1972, 50 en 122.
36. Van Beuningen & Koldeweij 1993, 247, fig. 774-576.

it does not appear inconceivable that an extra deck level was built in and that this necessary provision was already taken into account when building a warship or a trading cog if it was likely that it would have to be able to carry soldiers.

Many medieval sources mention the cog as being the largest cargo ship of the German Hanse towns. It is, therefore, not illogical to assume that this type of ship was depicted on the seals of these towns. Moreover, there is hardly another alternative. But there is no irrefutable evidence.[37] Furthermore, a large number of characteristics of the Bremen cog made it possible to establish that they differed from the wrecks dating from the period directly preceding which were built in the Scandinavian and West Slavic traditions and from vessels such as river barges and the Utrecht boat in Northwest Europe. The differences mainly relate to the bottom and stern/stem constructions and the way the shell planks were attached together.[38] The relatively large number of known shipwrecks with characteristics of the Bremen cog is as it were a confirmation of the equally frequent appearance in the written historical sources of the name cog in the Hanseatic period.

Fig. 4.4a. The baptismal font (of Tournai stone) in the church of St Laurence at Zedelgem near Bruges (12th century). The carvings include a type of ship known as a nef, with a stern rudder. After Asaert 1974, 105.

Fig. 4.4b. The baptismal font in Winchester Cathedral (12th century). After Asaert 1974, 106.

37. Hulst 1987, 49-56.
38. Crumlin-Pedersen 1997, 20-31 and Crumlin-Pedersen 2000, 232.

English seals

Depictions of ships on the seals of English towns from the 12th/13th centuries often show ship hulls with through-beams. These are certainly not cogs. In most cases these are almost certainly vessels built in the Nordic tradition (Hythe, Yarmouth, Winchelsea, Sandwich, Dunwich and Poole). The Dunwich seal dates from ca 1200, that of Yarmouth from ca 1300. The others date from the 13th century. On five of the English seals the ship has a straight stern with a stern rudder and a curved stem. The same goes for the vessels portrayed on a miniature from the *Roman de Troie* (see part IV), on the relief sculpture in Winchester Cathedral (GB) and on the baptismal font in the church of St Laurentius in Zedelgem (B; figs 4.4a en b). The type of ship is probably an (adapted) *nef.* The ships on the seals of Rye (13th century), on the second seal of Southampton (13th century) and the third Council seal from ca 1400, of Tenby (second half 14th century), of the Admiral Earl of Rutland and Cork (a personal seal?, 1391-1398) and on the third seal of Amsterdam (around 1400) could perhaps represent cogs,[39] though they are more likely to be a *nef* or *hulk,* for example.[40]

The Atlantic *nef*

Little is known about the above-mentioned *nef*, the ship very probably depicted on various 13th-century English city seals. As Asaert has already observed, *nef* is only a vague designation, meaning nothing more than ship. Apart from these images, we do not know precisely what this name in the historical sources implies with regard to shape, size and construction. No archaeological evidence from the 12th-14th century has so far been found.

Zimmermann has reconstructed a 13th-century *nef* (on paper) on the basis of the images on English city seals (Romney, Hastings, Southampton, Pevensey, Rey, Winchelsea, Sandwich, Dunwich, Hythe, Yarmouth, Faversham and Dover) and of the city seals of Nieuwpoort and San Sebastian on the continent.[41] He also made use of the information on the archaeological finds of the vessels from Gokstad, Oseberg, Kalmar and Roskilde. The reconstructed ship is ca 22 m long, has a maximum width of ca 5.1 m and a deck level midships of ca 3 m and is entirely clinker built (in the Nordic or Viking ship tradition). In his view, the ship was sailed, could be rowed and was kept on course with a steering oar (fig. 4.5). The stern/stem projected at deck level ca 5 m above the keel line, while the construction athwartships was carried out

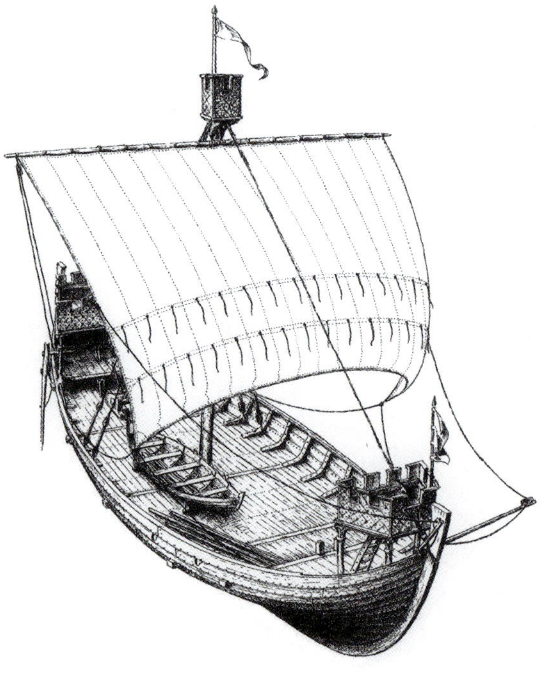

Fig. 4.5. The 13th-century Scandinavian-type ship reconstructed by Zimmermann: the *nef* de cinque ports. After Zimmermann 1982.

Fig. 4.6. The 13th-century civic seal of Ipswich (UK). After Zimmermann 1982/Ewe 1972.

39. Ewe 1972.
40. See section 27.3.
41. Zimmermann 1982, 105-111.

with through-beams piercing the shell at deck level.[42] The reconstructed 'Viking' ship/this *nef* had a direct connection with the so-called 'Cinque Ports', a confederation of coastal towns on the Channel coast of Southeast England already mentioned in historical sources in 1158. In 1189, this confederation placed 33 ships at the disposal of Richard the Lionheart for the Third Crusade. In the 13th century the fleet of the confederation was deployed during various piracy and defence campaigns against, in particular, the French. The ship type, the *nef*, was probably first intended as a trading vessel.[43] However, it seems more likely that the possibility of rowing was reserved for the typical, relatively low built Viking ship. This was not possible with the considerably (larger? and) higher ships with through-beams. Zimmermann is also of the opinion that the *nef* was equipped with a stern rudder under the influence of the 'superior and ubiquitous Hanse cogs' in the 13th and 14th centuries.[44] However, it is far more probable that the opposite occurred. The oldest, 13th-century seal of Ipswich shows a *nef* with a curved stem and a straight stern with a stern rudder (fig. 4.6). This type of ship will again be discussed in the analysis.

In the written source from 1401 referred to previously, mention is made, among other things, of *COGGHEN* (large seagoing cogs), *CLEENE COGGHEN* (small cogs) and *SCHUTEN* (boats).[45] These late-medieval names 'cover' the contents of this manuscript and have, for this reason, been used as a title. Only after all the eligible wrecks had been (completely) drawn and described could they be classified into one of these three groups of ships. In part IV, section 29.2 the groups are explained in detail. And in the same part IV, in addition to bringing up to date the research dating from 1996 into the clamped moss caulking method, several less known late-medieval miniatures and a sculpture are also discussed. They are interpreted and described with the current maritime archaeological knowledge regarding cogs. They show the full spectrum of ships from the building tradition, the characteristic clamped moss caulking method and exceptionally realistic constructions of barges from the cogbuilding tradition in the early 15th century.

42. Zimmermann 1982, Plan 5.
43. Zimmermann 1982.
44. Zimmermann 1982, 68.
45. Luns 1985, 36; Höhlbaum 1879, HUB II.

Research and reference model

II

5

The research, the 'reference' wreck Nijkerk-II and the Kampen cog[1]

5.1 The research

For a long time the findings and conclusions of the maritime historians about the cog could not be tested against a real ship. Only after the Second World War have archaeological finds become known that could possibly be related to the cog. The discovery of a large late-medieval cargo vessel in the Weser near Bremen 1962 changed all this once and for all. The world of maritime archaeology started to stir, and several previously excavated wrecks and new finds could now be designated as cogs or coglike, on the basis of the specific characteristics of the Bremen wreck. Nevertheless, a lot of questions remained unanswered. The few published wrecks with these characteristics provided insufficient information. In order to throw more light in this matter, the files of all the wrecks excavated in the former Zuiderzee area by the NISA and its predecessors were examined and the ships considered eligible were selected for the research planned. This amounted to over two-thirds of all the cogs found in Europe so far, including the wrecks discovered in 2000 and 2002 at Doel in Flanders and the wreck of the Kampen (IJssel) cog evaluated in 2011 and recovered in 2016. It was immediately decided that the data from all these ships should, as far as possible, be completely processed in the form of excavation reports in which the vessels, their construction, sequence of building and the artefacts encountered are described and depicted in detail. Only then could questions be answered and could existing ideas about the cog be established as correct. A realistic picture could also be created of the large and small vessels from the cogbuilding tradition.

The drawings made during the excavations (top views, longitudinal and cross sections and side views) have been prepared for publication and serve as justification for my own reconstruction drawings of the ships which are also presented. The reconstructions were not restricted to what was actually found, but an attempt was made to sketch reliable images of the entire ship hulls and what the deck lay out may have looked like. The (shape) reconstructions may be regarded as realistic and the main measurements for at least 95%. Figure 29.1 (Chapter 29) shows the contours of all reconstructed ships scale 1:100.

The reports and drawings are presented below in parts II and III and in the appendices in a standardized form and ship by ship. The vessels described represent the full spectrum of (large) seagoing cogs, inland and Zuiderzee cogs and barges built according to the specific cog tradition. The ships discussed sailed at any rate on the Zuiderzee and/or the Scheldt between ca 1250 and 1450. Finally, in part IV the excavation results are given and tested against the data applicable up to now and conclusions from the literature and an indication is also given as to whether the research has led to other insights and/or conclusions. The reports on the building of the Kampen cog and the discovery in 2014 of the probable site of a late-medieval shipbuilding yard in the Hanse town of Kampen are valuable additions.

5.2 The reference wreck Nijkerk-II and the Kampen Cog

It has already been mentioned in § 1.2 that a reference wreck was taken as a standard model for the description and drawing of all cog wrecks still to be investigated. The Nijkerk-II was selected because it was a (large) cog, the wreck of this ship was the most completely preserved and the wreck and its parts were extensively documented during the excavation.[2] Moreover, a reconstruction model (scale 1:10) had already been made and the ship itself reconstructed full size at Kampen almost simultaneously with the model.

During the renewed start of the research in 2005 it was originally assumed that reconstruction drawings of the ship had also been made, so that I could concentrate

The excavation of the Nijkerk-II cog and the experimental replica built on the basis of the archaeological data.

1. Translated by Christine Jefferis.
2. That the cog in the IJssel at Kampen was even more complete was not known until the end of 2015.

straight away on the description of the vessel. However, while describing the ship questions quickly arose that could only be answered with the help of reconstruction drawings. On inquiry at NISA however, it appeared that these had never been made apart from the bottom with stern/stem hooks, keel plank and frame parts belonging on it, keelson and ceiling planks (Plates 6.III A, B and C). Meanwhile it had already become clear that describing a ship find/a ship would only be advisable on the basis of definitive reconstruction drawings. In the case of earlier reconstructions it appeared that the original shape of a ship often differed from what was thought during excavations and later presuppositions and assumptions. Armed with only little experience of reconstructing on paper the hulls of a few boats and ship fragments, the Nijkerk-II wreck proved to be a huge challenge. More than two years of intensive work was required to obtain the result (reconstructing, devising a standard design for the drawings, the actual drawing itself and compiling a standard description. All the cog finds in this study were then first drawn and subsequently described in a similar fashion.

The report on the experimental reconstruction of the ship, the Kampen cog, was also described according to this standard description. This (also) makes clear what was found of the original ship and how the shipbuilders completed the ship as authentically as possible on the basis of the information available. The reference wreck, the model and the reconstructed cog provided an extremely detailed source of information on how the ship must have been built and together formed the reference/research model for the other cogs. The research subsequently carried out and presented here and the reconstruction on paper of the Nijkerk-II showed that, from approximately deck level, the ship must originally have looked different from the now sailing Kampen cog.

The Nijkerk-II cog, impression based on the reconstruction drawings.

The Nijkerk-II cog[1]

6.1 Find conditions

6.1.1 Discovery and investigation

During the excavation in 1981 of the Nijkerk-I cog, in lot Oz43 in the province of Flevoland (Chapter 8), it was found that this vessel must have sunk before 1300, on or just off the then coastline, a very unusual situation. Indeed almost all of the ca 450 wrecks recorded in the IJsselmeer area seem to have foundered in 'open' sea. This situation brought to mind a large amount of *Kugeltopf* sherds that had been recovered in a nearby field in the early 1970s.[2] On 23 September 1981, the lots Oz35 and Oz36 were searched for the spot where these potsherds had reportedly been gathered. Apart from some further *Kugeltopf* sherds, ship's timbers were discovered in two adjacent reclamation ditches in lot Oz36.

Subsequently an evaluation was carried out between 28 September and 10 October 1981 (figs 6.2 and 6.3), during which it was decided that these most probably were the fairly complete, but dispersed remains of a cargo vessel with cog-like features. The date of the ship, on the basis of some potsherds found with the wreck, might be 13th century. The wreck lay in Zuidelijk Flevoland, in lot Oz36, some 140 m from the ditch separating lots Oz35 and Oz36, and 220 m from the ditch along the foot of the dike (figs 6.1, 6.2a and b and 6.4).[3] In view of the shallow position of some already badly degraded timbers, it was decided to cover the wreck site with a protective layer of soil in November of that same year, and to plan an excavation for 1983.

Simultaneously with the application of this protective layer, about 130 m south of the wreck a trench was cut for a

Fig. 6.1. The (splayed) hull during excavation, after removal of the detached timbers, viewed from the stern.

1. Translated by Xandra Bardet.
2. Vlierman, 1985a.
3. The vessel lay in a north-south direction, the bow pointing south, 1 km northwest of the Nijkerker Sluis (locks) near the old Zeedijk (sea wall); Ordnance map 26D, coordinates 159.382/474.990, with the highest-lying parts at 2.5 m below NAP. According to the Hydrographic Map of 1921, the former depth of the sea was ca 1.7 m at mean low tide.

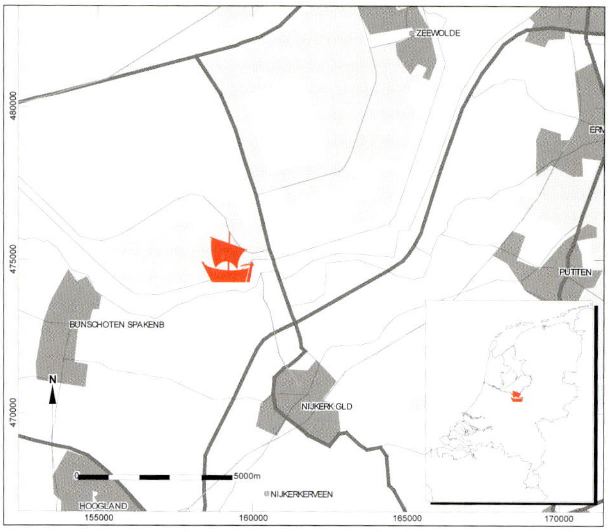

Fig. 6.2a. Ordnance map indicating the site of the cog.

section documenting the soil profile around the vessel and any traces of habitation. A highly informative section of a channel filled in in phases was encountered, into which waste (mostly pottery and bones) from a medieval settlement had been dumped. In the spring of 1982, employees of the pedological department of the RIJP performed a series of borings, establishing the course of this channel and a second one. In fact, little more than the bottom of these former drainage channels had survived. On the whole, their depth beneath the modern surface is slight. Only where their course meanders do deeper areas occur, in one place up to as much as 3 m beneath the modern surface. The widths of the channel remnants, yet to be measurable, range between 7 and 40 m. In September of that same year, an intensive survey of the extent of the habitation traces was carried out with the assistance of volunteers from the

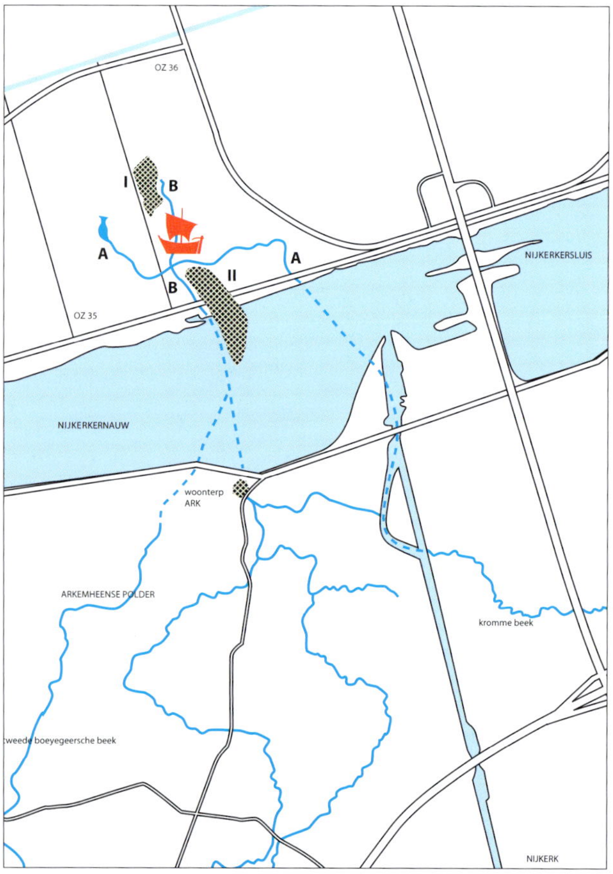

Fig. 6.2b. Fields Oz 35 and Oz36 with surroundings, the site, the course of the two channels (A and B) and the approximate locations of the Carolingian (I) and 12th to 14th century (II) locations of the settlement known as Ark.

Fig. 6.3 Test trench with crossbeam 2 during exploratory investigations in October 1981. The head of the beam (with the two oblique rabbets) lies exactly over the spot where the beam protruded through the shell. The missing part of the shell was damaged and destroyed in the digging of a ditch.

antiquarian society 'Arent thoe Boecop' of Elburg.[4] Some of the potsherds recovered during these surveys will be discussed in section 6.1.3.

The excavation and documentation of what would turn out to be a cog, and the cutting of two section trenches was carried out by the department of Nautical Archaeology of the RIJP, between 27 April and 28 October 1983. At the start, two dams have been hold in reserve (to study the profile), crosswise, around which the four excavation areas were opened up. This liberated the loose timbers overlying the actual wreck (fig. 6.5). Only after the removal of the dams could a full overview be achieved. (fig. 6.6 and Plate 6.I). Given the wreck's shallow position, the fragility of the wood and the partly chaotic distribution of the wreck's components, all the digging was performed manually. Two trenches - a north-south section down the length of the cog and including channel A to the south of it, and an east-west section transversely across the wreck and channel B to the west of it - served to provide insight into any relation between the ship and habitation (fig. 6.7). After the loose parts had been drawn and numbered, they were lifted and put into a water basin. Then the wreck was uncovered more fully and cleaned. Only now did the actual condition of the hull become clear (fig. 6.8). An overview drawing was completed (Plate I), and the then National Survey Service (Delft) took stereoscopic photogrammetric images from an airforce helicopter. The usual sections across the hull were not drawn, because the scattered and almost flat-lying timbers, possibly warped in the soil, were unlikely to provide any useful evidence to assist reconstruction. Moreover, it had already been decided to draw the parts to a scale of 1:10, which at any rate would record detailed information about every plank and frame timber. The findspots of all other artefacts were marked in the overview drawing (Plate 6.II B) with numbers corresponding to the numbers in the finds list (Plate 6.A). The finds themselves are discussed in section 6.5.

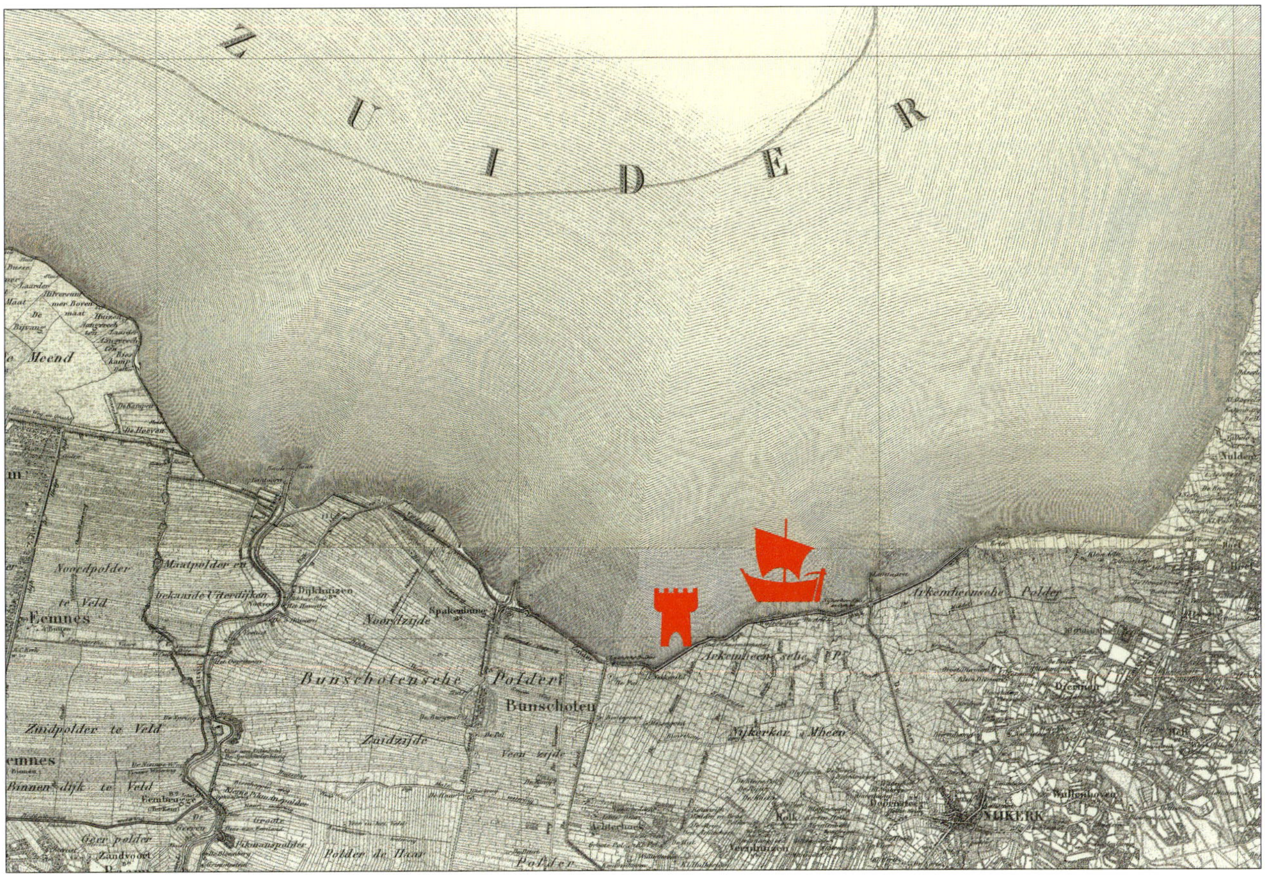

Fig. 6.4. Ordnance map of the Kingdom of the Netherlands (scale 1:50,000) of 1861, indicating the sites of the cog and Hulkestein castle.

4. Vlierman 1985a, 13-16.

Fig. 6.5. View of the two eastern quadrants of the excavated area under investigation.

Fig. 6.6. View with the detached timbers *in situ*, including (at the centre) the larger of the two crossbeam constructions (beam 2; see also fig. 6.31a)

6.1.2 Geology and stratigraphy

The Pleistocene surface and overlying deposits

Since the soil profile had been barely affected (at the time of discovery, the reclamation ditches were still present, the land lay as yet uncultivated and there had been almost no agricultural activity), it proved possible during the surveys and the excavations to obtain a good impression of the irregular local soil structure. This irregularity is caused by the very uneven upper surface of the Pleistocene sand, whose elevation in this southernmost part of Flevoland ranges between 0.25 and 1.5 m below the modern surface. Remnants of the Holocene peat deposit have survived mainly on the deeper-lying surfaces of this sand, which largely still has its A horizon. Within the channels, which also cut through this peat, a variety of sediment layers can be seen, reflecting a filling-in and building-up process occurring in phases, and comprising humic Almere clay, alternating with thin layers of sand (containing potsherds), thin layers of redeposited peat, chunks of peat, driftwood and plant remains. This complex is sealed by a layer of Zuiderzee/IJsselmeer clay deposits, 15 to 35 cm thick. Judging by the soil structure of the adjacent 'old land' (former shore), we may assume that the peat originally had a substantial thickness. Over this peat, a layer of clay was deposited in the Early Middle Ages, on which people were able to settle but presumably no earlier than the 8th or 9th century.[5] The elevation of this clay-on-peat land, dissected by drainage channels, will have been roughly the same as that of the adjacent 'old land', the Arkemheense Polder. Which, globally speaking, now roughly corresponds to NAP or slightly above it. It is likely that the higher Pleistocene outcrops (ca 2.1 m below NAP) on the east and west side of channel B, and north and south of channel A in lot Oz36 (fig. 6.2) were noticeable also in the contemporary clay-on-peat landscape, and in the Middle Ages were favoured for habitation.

Channel A

Channel remnant A remained open at least up to the end of the Neolithic. The humic Almere clay that settled in the channel in phases, suggests that subsequently, into the Middle Ages, parts of it still carried water. This applies especially to the course of the channel in lot Oz35 and the stretch in lot Oz36 west of the cog. Remarkable are the absence of medieval sherds in the channel fill to the south and east of the cog and the lack of humic clay (fig. 6.7a). This prompts the tentative conclusion that this last part of the channel had already (largely) silted up by the Middle Ages.[6]

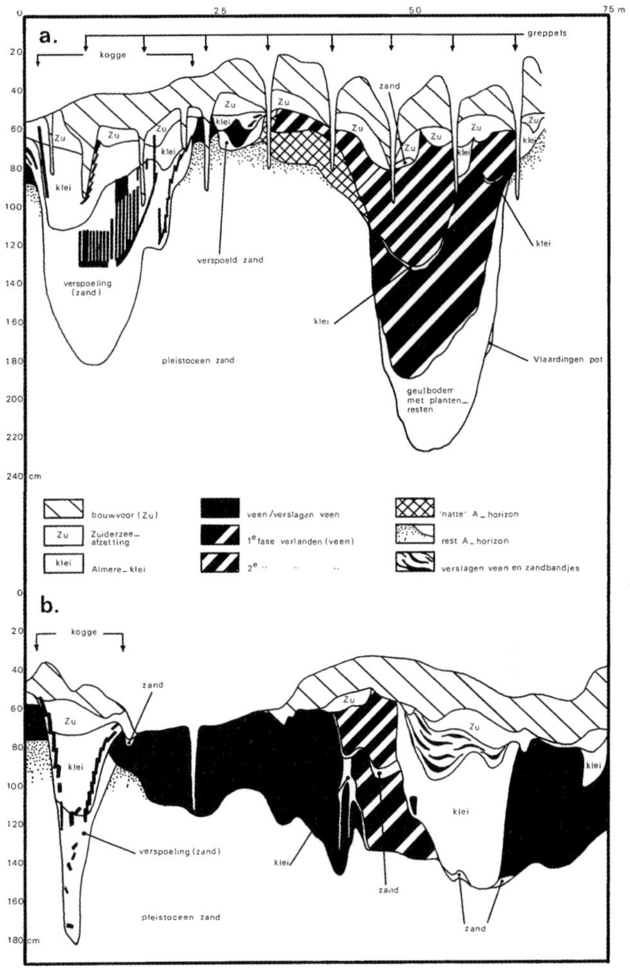

Fig. 6.7. North-south and east-west soil profiles.

Fig. 6.8. The hull *in situ*, after removal of the detached timbers, viewed from the stern. In the background, the digging of a section to examen the soil profile of a trench across channel A, parallel in orientation to the cog.

5. Vlierman 2002, 137.
6. Vlierman 1985a, 33.

Channel B

The location and fill of channel B to the north of the cog are marked by some peculiarities. The channel cuts into the Pleistocene in two places and there are indications that this channel may have (partly) been cut artificially.[7] Its fill of humic Almere clay indicates that the channel may have been open for several centuries. Maybe the channel came to a dead end in the peatland, having been used for peat-cutting and peat transportation (fig. 6.7b).

Position of the cog with respect to the channels

At about 160 m from the ditch along the foot of the modern dike, channel A, which locally runs east-west, crosses channel B, which runs south-north (fig. 6.2b). The deepest observed stretch of channel remnant lies just south of this 'intersection' in channel B, and is largely filled with clay from the Almere deposition phase.[8] The clay overlies the layer with the 12th-/13th-century sherds, and should probably be reckoned to the Alc deposits. The section across channel A and lengthwise across the ship, shows a watercourse filled in by peat development in two phases, above which there are no Almere clay deposits. The transverse section across the cog contains a similar pattern on the east side of channel B along about a third of the channel's width. The remainder, about two thirds of the channel, is filled in with Almere clay; this clay-filled part of channel B retains its depth (ca 1.5 m below the modern surface) up to about 20 m north of the cog, then becomes significantly shallower, and towards the end deepens again. The cog lies about 35 m north of channel remnant A and a similar distance east of channel B. The wreck had worked itself through a remnant of the Holocene peat and into the Pleistocene sand. It is no longer possible to ascertain how thick the peat was when the ship was lost - in other words, how deep the water was locally, because the peat was eroded away in the Late Middle Ages. Yet is seems highly likely that at the fateful moment the southern stretch of channel B and the western stretch of channel A constituted a navigable channel out to sea and that their 'intersection' formed a more or less natural harbour basin, ca 150 m long and 50 m wide. The cog lay in the northeastern corner of this basin.

6.1.3 Medieval habitation

During the various surveys hundreds of potsherds were picked up from the surface of lot Oz36; most of them lay between the cog and the dike. It was mostly *Kugeltopf* ware, with the odd piece of proto-stoneware, as well as some Paffrath and Pingsdorf-style ware. This material is dated to between the 12th and the late 13th or early 14th century. The *Kugeltopf* ware from the section trench in channel B to the south of the channels' 'intersection' also is of 12th-/13th-century date. East of the cog some sherds were found on the surface that may be dated to the mid-14th century at the latest while in 1970, when the shipping channel in the Nijkerkernauw was widened, sherds were dredged up that dated from the 12th to the 15th century. The findspot of these sherds lies to the south of the site in lot Oz36, and may be reckoned to the same habitation.

The suggestion uttered at the time, that the sherds might relate to the castle of Hulkestein, built in 1427 and demolished (by the sea) in the second half of the 16th century, but whose location then was unknown, could be ruled out in 1989.[9] The castle had stood about 1.5 km southwest of the medieval settlement. Mostly within and around the northern end of channel B, various fragments of Badorf-style and *Kugeltopf* ware as well as wall daub were found, which indicate that the site was inhabited even in Carolingian times. The absence of 10th- and 11th-century pottery suggests that in those days the site may have been unoccupied. The prehistoric finds discovered during the surveys and excavations are left outside the scope of this discussion.

Spatial and stratigraphical distribution of the finds

The 9th- and 12th-/13th-century sherds (fig. 6.9), which were locally found at the bottom of the filled-in channel remnants, demonstrate that at the time they were still open. Above this in the channels, a (thick) deposit of humic Almere clay was deposited containing chunks and plaques of peat and (drift-)wood. On this clay deposit lay thin layers of sand with Carolingian or 12th-/13th-century potsherds with abraded edges, alternating with thin bands of peat debris. Without doubt, these sandy layers derive from the occupation surface on the original clay-on-peat land adjacent to the channels. Over this sediment deposit, and beside the channels, there were occasional small pits in the remaining peat, or if such no longer survived, immediately in the Pleistocene sand, filled with the earliest Zuiderzee deposit (the seashell layer Zu IV). The ensemble is sealed below a deposit of Zuiderzee-/IJsselmeer clay, 15 to 35 cm thick. The chunks of peat and the driftwood, the sandy layers with sherds and the thin deposits of peat debris indicate that the (remnant) channels were probably filled in catastrophically in a period of strong erosion of the land surface. Subsequently both erosion and sedimentation appear to have taken a calmer course.[10] The sand with (abraded) potsherds, deposited in the channel remnants in alternation with thin peaty layers, differs in composition from the Pleistocene sand in the immediate

7. Vlierman 1985a, 19 and 33.
8. AD 0-1600, Ente *et al.* 1986, 130.
9. RAAP-Notitie 43, 1992.
10. Vlierman 1985a, 33; fig. 27a en b.

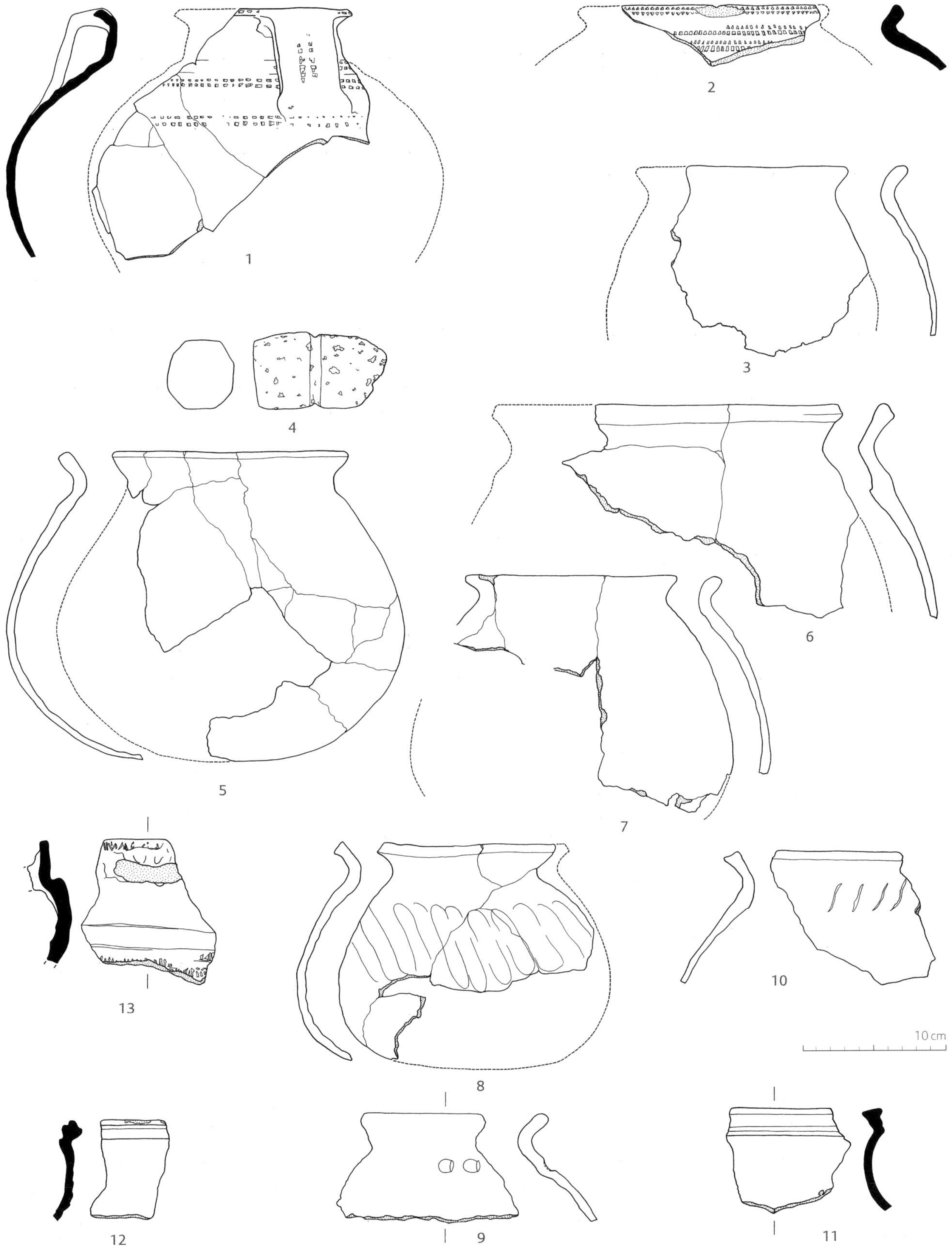

Fig. 6.9. Ceramic finds from the 9th century and the 12th-to 14th-century sites of the village of *Ark*.

vicinity. It seems likely that this sand had been imported from elsewhere (the Veluwe region) to consolidate the habitation surface on the clay-on-peat soil underlying the settlement.[11] Such a levelling-up of the site with imported sand was identified also in excavations at Elburg, where at any rate in the western part of the new town, relocated (in part) between 1392 and 1396, 40 to 60 cm of sand was laid down.[12] Yet the sand may also have been deposited during severe storms as the Almere lake was transformed into the marine Zuiderzee. The Carolingian pottery was encountered predominantly at the end of channel remnant B and on a high Pleistocene outcrop on the west side of this channel between channel A and the end of channel B (figs 6.2, 6.10 and 6.11).[13] This outcrop measures ca 400 x 150 m and rises to 0.25 m below the modern surface (ca 2.1 m below NAP). Presumably there were a few 9th-century farms on the northern half of this outcrop (fig. 6.10). The southernmost part of lot Oz36, between channels A and B and the part situated north of channel A, contains a Pleistocene outcrop similar in height. The presence of the waste in the deepest part of channel B at a good 100 m southwest of the wreck and the sherds at the surface around and especially to the south of it suggest that the 12th-/13th-century settlement probably lay on the southeast side of channel B. The ceramics from the Nijkerkernauw shipping channel support this hypothesis and moreover indicate that the settlement extended over at least 400 m in a north-south direction (figs 6.2 and 6.11).

During the excavations much attention was given to dating the foundering of the cog, and the possibly related dating of the youngest settlement traces. From the profile dams above the wreck and the sections in its vicinity it became clear that the potsherds found within them did not match up with a normal chronostratigraphy. The medieval sherds found above, within and around the wreck date from the 9th to 14th centuries. Curiously, the oldest sherds do not lie at the base of the section but in the top, and indeed prehistoric, Carolingian and 12th- to 14th-century sherds may appear in the same deposit layer (fig. 6.7).
It took some effort to work out how this was possible, but the eventual explanation was quite simple. In the immediate vicinity of the cog site, there had been two distinct settlement sites, in the 9th and 12th-14th century, respectively. The farms (of the younger settlement) had most probably been built on a sandy platform in a clay-on-peat landscape. As usual, much of the domestic waste ended up scattered around the farmsteads, the oldest sherds in the base of the sandy occupation layers and the youngest on top. The vessel lay close to the younger settlement site.

In the course of the wrecking process also the surrounding land was, in phases, devoured by the sea, layer by layer and peaty chunk by peaty chunk. First to be washed into the bottom of the channels and to end up among the wreckage would have been the topmost occupation layers with their potsherds and bones. The successively exposed older layers were progressively washed away and, alternating with thin layers of peat debris, produced further layers of sediment. The process repeated itself several times until in some parts the peat had completely been eroded and finally Neolithic potsherds from the Pleistocene sand were deposited on top of the earlier layers covering the wreck.

Fig. 6.10. Hypothetical reconstruction of the landscape around the Carolingian settlement of *Archi* (=*Ark*) on the western shore and at the end of the (man-made) channel B and the partially silted-up eastern part of channel A.

Fig. 6.11. Hypothetical reconstruction of the landscape around the 12th- to 14th-century settlement of *(H)erca* or *Erke* (=*Ark*) on the eastern shore of channel B with the cog in the northeast corner of the 'harbour basin', the (largely) silted-up channel A and the eastern course of channel B and the open connection to the Zuiderzee via the southernmost part of channel B in field Oz36 and the westerly course of channel A in field Oz35. And the partially silted-up eastern part of channel A.

11. Vlierman 1985a, 36.
12. Vlierman 1981, 6.
13. Vlierman 1985a, 24.

There where the channel remnant bottom passed from Pleistocene sand to peat, and where on the adjacent old shore the peat had been washed away to expose the Pleistocene sand, the earliest new sedimentation probably did not occur until after the 16th century: the seashell layer Zu IV. It was found to contain potsherds of prehistoric as well as Carolingian, 12th- to 14th-century and 17th-/18th-century age. If the ship had lain (moored?) close to the still inhabited settlement, theoretically the finds belonging to the ship's inventory and the youngest sherds from the settlement should both lie in the bottommost position at the wreck site dating from the same period. However, none of the finds offers 'hard' dating evidence to demonstrate the parallel demise of a functioning ship and the settlement within a few decades.

Pottery

The pottery from among the settlement waste has been re-examined. It consists mainly of *Kugeltopf* sherds together with some fragments of Badorf, Pingsdorf, Andenne and Siegburg wares. Their dates range from the 9th century to the early 14th century, but mainly the 13th century, and can be specified as follows:

Badorf ware. Represented is Dorestad type W IIB (figs 6.9-1 and -2), in use between ca AD 800 and 900.[14]

Kugeltopf ware. The ovoid *Kugeltopf* (fig. 6.9-3) too may be reckoned to the Carolingian period. The vessel corresponds with Dorestad type H IA.[15] The slight number of sherds of this period and their limited distribution point to a modest-sized settlement (a few farms), which probably was quite short-lived. 10th- and 11th-century pottery seems to be lacking.

Pingsdorf ware. Just a few small sherds of Pingsdorf ware came to light. Some may belong to an early production phase, the others are more similar to proto-stoneware and most were found on the surface among 12th-/13th-century material.

Kugeltopf ware. The bulk of the pottery consists of *Kugeltopf* sherds. They display a variety of familiar rim profiles. Most rims have a rounded section and are of a soft fabric; others are blue-grey in colour and have a hard fabric. Janssen[16] calls the latter '*pseudo-Paffrath*' and dates them to the first half of the 13th century. The other *Kugeltopf* ware is harder to date. Just a few of them display traits that can be attributed to a particular period. A large fragment decorated with finger swipes (fig. 6.9-8) and some comparable sherds of other specimens according to Verhoeven[17] stem from as early as the late 12th century, but mainly from the 13th. Amersfoort probably is one of the easternmost findspots; east of the river Eem they seem to occur only sporadically.[18] This findspot may therefore be added to Verhoeven's distribution map, which also goes for the *Kugeltopf* pots with brushed decoration, fingertip impressions and incisions (figs 6.9-9 and -10). With the exception of the finger swipes, all mentioned decoration styles are amply represented also in a broad collection of sherds that in the 1950s were recovered from the base of the eastern sea wall of the former island of Schokland. Among them are also hundreds of (rim) sherds of undecorated *Kugeltopf* vessels, Pingsdorf and Paffrath ware, Andenne pots, and proto-stoneware. Janssen and Nijhof[19] believe that *Kugeltopf* vessels with finger swipes are a frequent occurrence specially in the second half of the 13th century, while this kind of decoration continues in use at least into the early 14th century. *Kugeltopf* vessels with brushed decoration (see section 6.5.2) are typically 13th-century. They come into general use around 1290, but still regularly turn up in contexts of the first half of the 14th century.[20] In the Dronten-I cog (Chapter 11, figs. 11.29-5 and 11.31) a small specimen was found together with three pitchers of 'near-stoneware' and a small redware cooking pot. On the basis of the stoneware, the foundering of that ship is dated to the first half of the 14th century, most probably the 1320s.[21]

Kugeltopf sherds with incisions and brushed decoration have been found in archaeological excavations at Elburg too.[22] A moulded rim-/shoulder sherd of a fairly large *Kugeltopf* vessel (neck diameter ca 20 cm, fig. 6.9-6) is of an uncommon type and has its closest match in vessels found at IJlst (Friesland).[23] Several examples of such large *Kugeltopf* vessels are available from the sea-wall base at Schokland as well.

Andenne ware. The few, small Andenne-ware sherds are dated to the 12th century by Janssen and Nijhof,[24] while the earliest might even be from the late 11th century and the youngest from the early 13th century.

14. Van Es & Verwers 1980, 73, 74 and 78.
15. Van Es & Verwers 1980, 112-113.
16. Janssen 1981, 171.
17. Verhoeven 1998, 91.
18. Verhoeven 1998, 110.
19. Janssen & Nijhof, pers. comm. 2000.
20. Janssen & Nijhof, pers. comm. 2000..
21. Vlierman 1983, 14.
22. Vlierman 1981, figs 2.10 and 2.11.
23. Verhoeven 1998, fig. 59-4.
24. Janssen & Nijhof, pers. comm. 2000.

Proto-stoneware. The also slight number of proto-stoneware sherds (figs 6.9-11 and -12) are easy to place in terms of chronology: ca 1225-1275.[25]

Blue-grey ware. Blue-grey ware is represented by a prominent neck sherd with handle attachment of a large pitcher, pronounced rilling and rouletted ornamentation (fig. 6.9-13). The pitcher is of an unknown type, but is reminiscent of imitation-stoneware pitchers from e.g., Maastricht[26] and Schinveld.[27] A late 13th-/early 14th-century date seems the most likely.

Siegburg stoneware. A mere four fragments of Siegburg stoneware (find numbers 81-159, 82-13, 83-46 and 83-91) may be regarded as typically 14th-century. Such products are regarded as domestic ceramics, in use between ca 1310 and 1370, and not afterwards.[28] The pottery found within and above the cog wreck during excavation, which in part should be attributed to the settlement, is described in section 6.5.

Historical and cultural landscape evidence regarding Ark and the Arkemheense Polder

Might it be possible, by using historical records, to relate these recovered traces to a medieval settlement which fell prey to the waves as the Almere lake expanded and transformed into the Zuiderzee? This in fact seems to be so. The historical name of *Almere* is mentioned between 755 and 1340 at the very latest.[29] We are probably dealing with the Carolingian settlement of *Archi* and the 12th- to 14th-century *(H)erca* or *Erce* (=*Ark*). Heidinga[30] writes that according to Slicher van Bath the occupation history of the Veluwe region divides into two periods of intensive reclamation: the Carolingian period and the 11th-12th century, with an intervening period of stasis between ca 950 and 1050. Blommesteijn *et al.*[31] present an archaeological-historical survey of the occupation history of the Veluwe up to 1200. The following is a quotation from the chapter '*De Werdense Hof te Putten en Nederzettingen op Noord-west Veluwe*':

"From the earliest written sources relating to the Werdense Hof at Putten, it is evident that in the modern municipality of Nijkerk already a number of farmsteads are known. The donation by Folkerus in 855 is mentioned as very important. With this donation the focus of the possessions of [the monastic institution of] Werden in the Veluwe region came to lie in the area of Putten." *Archi* was one of these settlements. In the donation charter of 855 and another of 806 there are in fact three clusters: at Nijkerk (including Ark), Putten and Ermelo. These charters explicitly mention reclamations: *comprehensiones* The earliest and comprehensive list of revenues of the demesne (*curtis*) at Putten, dating from the mid-12th century, still mentions some familiar place names besides some new ones. The old names are:

Putte (2 x)	*Ermelo*
(H)erca	*Hezzera*
Niutlo	*Thri*

A second revenue list of the demesne of Putten, dating from the mid-13th century, features the familiar names:

Ermele	*Putte*	*Dri*
Nordere	*Husszere*	*Becsteren*
Halvinchusen (2 x)	*Buchorst* (2 x)	*Erke*

Among these old names, Nulde has disappeared. Also the 12th-century reference to Holthusen ('Woodland Village') no longer occurs. One of the new toponyms, Ekelenvelde ('Ekelen Field'), might be linked to pastureland near Ark. From the 14th century we know of two further properties in the Ark district, though the village of Ark itself must have been engulfed by the sea in the same century. After 1360-1370 it is no more heard of. The sea in relation to the Nijkerk area is also mentioned by De Zeeuw.[32] It was not until 1356 that Duke Reinald III of Guelders granted the people of this area permission to build a dike across the Arkemheen. This was much to the advantage of the duke and to the people of Nijkerk, as every winter the settlement and his landed property would be flooded.

Immediately to the south of the current old sea wall beside the Arkerweg road, in the continuation of the findspot of medieval sherds in the Nijkerkernauw and in lot Oz36 in Flevoland, lies the dwelling mound called *Ark*. The Arkerweg towards Nijkerk becomes the Bremerseweg, off which there are three more dwelling mounds, then bends, beyond which it is called Korte Holk. On this stretch lie two further dwelling mounds. It seems likely that these mounds were built before the construction of the sea wall, but no datings are known. Besides the documentary evidence about the presence of the dwelling mounds, and evidence of a clearly observable sandy elevation in a ditch cutting, the spoil from which yielded 14th- and 15th-century sherds, Hulst draws attention to a remarkable relic in the cultural landscape in part of the Arkemheense Polder: its irregular block parcelling.[33]

25. Janssen & Nijhof, pers. comm. 2000.
26. Hupperetz & Nijhof 1995, 1-2.
27. Bruijn 1966, figs 53 and 54.
28. Janssen, pers. comm. 2000.
29. Wiggers 1955, 81.
30. Heidinga 1984, 184.
31. Blommesteijn *et al.* 1977, 140-148.
32. De Zeeuw 1974, 7 and 8.
33. Hulst in: Vlierman 1985a, 36-37.

This seems to be restricted to the Pleistocene Wullenhove-Diermen area, which extends into the Flevoland polder. A feature of this area is the large number of small lowland brooks that drain it. These watercourses join up in the Ark area to form a main stream running northwest within Flevoland. It was along these streams that the parcelling pushed into the peatland. Presumably the settlement of *Ark* in Carolingian times can be regarded as a result of this reclamation drive, with the field boundaries too dating from this period. No archaeological evidence, e.g. remains of farmhouses, survives to support this hypothesis, because the land outside the sea wall has since been eroded away by the sea. At any rate the area of irregular block parcelling must be the core area of the reclamations in the Arkemheense Polder. Adjoining this nucleus lie areas with more regular parcelling, first in the form of regular blocks, then in strips. This succession, possibly concluded even in the 12th/13th century, is very clear towards Nekkeveld (13th-century *Enkelenvelde*). Probably this Nekkeveld originally extended further northward, beyond the later sea wall. From these younger reclamations derives a rim sherd of Elmpt ware, from the 13th or 14th century.

Van Dooren[34] argues that in the Carolingian toponym *Archi* the suffix -i indicates a (rural?) settlement; even though in this period the peatlands were not regarded as desirable areas for settling. In his opinion, this can only mean that we are dealing with a peatland reclamation which by Dutch standards is quite early. He believes the name *Ark* to derive from *archa*, which indicates a dam. However, *arca* may also mean a boundary bank, while *ark(e)* or *erk(e)* may refer to a small sluice or a man-made drainage ditch. Dam, sluice or drainage ditch would all point to hydraulic engineering activities by the inhabitants of the central part of the northern peatlands of Nijkerk-Arkemheen. The prominent block parcelling surviving in the centre of the Arkemheen area must, according to Van Dooren,[35] be younger than Hulst assumes, and he expects there to be old reclamation features in the peat underlying the clay cover. According to my view this last part is very unlikely for the medieval period (see section 6.1.2).

On the basis of the revenue lists, Van Dooren[36] reports that a small hamlet of four farms (excluding any reclamations belonging to Elten or Paderborn) had arisen. The heads of households bore the names Nameco, Oza, Oenga and Oenga the Elder. They were obliged annually to pay the Werden monastery 6, 8, 4 and 4 pennies, respectively. Also one Gozwin is mentioned, who possesses a piece of land and manages a fishery. The mid-13th-century list of Werden's estates contains no further stipulations on any fishery, but the four farmsteads do recur. With respect to the permit for dike building granted by Reinald III on 28 March 1356, Van Dooren[37] believes that this dike-building charter should be considered a supplement, aiming to improve an existing situation: he believes that the various personal names that include *'dijk'* predating 1356 point to previously undertaken hydraulic engineering activities.

Coastal erosion of the area outside the sea wall at the Arkemheen and the weather in the 12th to 14th centuries

The 9th- and 12th- to 14th-century settlement of *Ark* apparently lay on or close to the shores of the then Lake Almere, later the Zuiderzee. Little is known about the details of the coastal erosion of the unendiked land near the Arkemheen, but a few general remarks, both geological and historical, may be made. The dwelling mounds, the imported sand at the settlement sites in lot Oz36 and along the Bremenseweg, and the construction of the dike represent protective engineering measures taken by the inhabitants of the area. Recurring storm floods from the mid-12th century each time will have washed away new areas of land - witness the fill (incorporating chunks of peat) of the channel remnants - which meant that people were forced to repeat these activities time and again, presumably having to move their homes progressively further south. Van Dooren's observation about personal names including *'dijk'* predating the 1356 dike-building charter indeed points in this direction.

For a rough idea of what the situation was around the disappearance of Ark, we should look back taking account of some definite facts. The latest sea wall does not follow the same course as the oldest dike, but almost perfectly follows that of 1750. It is known that the dike was moved landward along various stretches between 1650 and 1750, but there are no records relating to the preceding centuries. The maps from Blaeu's Atlas of 1631 and 1662 within the dike show Hulkestein castle, which actually was demolished in the second half of the 16th century. The location of the castle is now in the Nijkerkernauw, close to the shipping channel, about 350 m off the dike enclosing Flevoland. Assuming that Blaeu presents the correct, original, endiked location of the castle site, this means that the then sea wall would have run roughly along the course of the current Flevoland dike. With the possible meanings of *archi*, *ark(e)* and *erc(e)* as suggested by Van Dooren, I believe that potentially the earliest hydraulic engineering took place even further north. The dark-coloured dike and dwelling-mound bases that for many

34. Van Dooren 1986, 25-26.
35. Van Dooren 1986, 87.
36. Van Dooren 1986, 15.
37. Van Dooren 1986, 62-63.

years in winter were visible in the ploughsoil around the former island of Schokland, and from which the fields P14 and P33 yielded the medieval potsherds, have in 2003 been made visible throughout the year in the form of low banks and mounds in grassland to the east of this World Heritage monument. A similar pattern of dikes and mounds shifting and expanding through time probably marked the landscape around the village of *Ark* as well, until the land was washed away by the sea.

As for the mentioned storms and coastal erosion it is worth checking 14th-century sources for evidence on the subject. Little is known about the weather in the Middle Ages. Moreover, it is uncertain how reliable any reports may be. As regards the drowning of the village of *Ark* and the foundering of the cog, a few documented storm floods will be enumerated here. From the mid-12th century on, several serious storm floods have been recorded. Buisman[38] writes that they occurred north of the great rivers, in 1163/1164, 1170, and probably also 1173. In 1196 they caused utter devastation in the northern and central Netherlands, playing a part in the transformation of Lake Almere into the Zuiderzee. The flood of 1196 probably was the final major event in the process. In 1262 and 1268 serious storm floods occurred in the coastal zone, as was the case in 1287 and 1288. In 1290/91 and 1300/01 mention is made of severe river floods.[39] In 1334 the south of the country experienced a major storm flood which also affected the north. As a result of the frequent disasters caused by both high river water and storm floods, the 1341-1344 period may well be called catastrophic.[40] In 1356 and 1357 too there were severe storm floods, and 1362 saw the worst storm of the century.[41] The engulfing and rendering uninhabitable of (the part in field Oz36 of) 12th- to 14th-century *Ark* was probably due to the storm floods in the 1300-1356 period. The younger dating of the pottery dredged from the Nijkerkernauw (up to the 15th century) suggests that there it may have taken place at a later date. The site of the Carolingian hamlet may have been drowned even before 1300.

6.1.4 Condition and wreck formation

Condition

It was especially thanks to the groundwater that seeped into the polder under the dike, that the timbers of the cog in comparison to many other wrecks in the IJsselmeer polders[42] was so well preserved, despite it shallow position. Locally, the water table during much of the year is just a few decimetres below the surface. The collapsed and now almost horizontal wreck comprised the near-complete bottom with the keel plank and the stern and stem hooks, the flat and V-shaped floor timbers, the keelson with a step for the heel of the mast and the ceiling boards, the stempost toppled forward, with the port side almost up to deck level and the forward half of the starboard side almost up to the gunwale, and the after part of this side ripped away amidships. The sternpost separated from the stern hook lay partly beneath the starboard shell. As is usual in wrecks in the former Zuiderzee area, the mast and spars are missing. Generally some blocks and lengths of rope of the rigging, and sometimes the rudder may be recovered, usually still attached to the ship's hull, but this was not the case with the Nijkerk-II cog.

Wreck formation

Ships that foundered in the Zuiderzee after a while would sink into the soft sea floor, but never farther than (just into) the firm Pleistocene sand. Since the Pleistocene sand along the Veluwe coast rises to just below the sea floor, the layer of soft sediment there is largely absent, and as a result the wrecks of most vessels foundering there are beaten to pieces. Sometimes at most a bottom and a few bits and pieces survived. The force of the waves would disjoint the hull and the separated parts would float away. Also people living along the coast will have played their part in demolishing any parts that emerged above the water. It is clear that the foundering of the Nijkerk-II cog was linked to coastal erosion and the drowning of a small settlement. A series of observations in the excavations can be explained from this perspective and suggest that it was not a stranding, but a 'harbour' situation. Probably it was a written-off and/or moored vessel. Some vertically placed posts to the east of it suggest the latter possibility (see Plate 6.I). Already it has been noted that a comparatively large part of this vessel has survived. The remains of the starboard shell show that this part must have been complete at the time of foundering. This is less certain for the port side: here the top four to six strakes are missing, and all of the upper futtocks of the midships and forebody. One would expect at least some of these parts to be found loose on top of the wreckage, but they may in fact have been washed away at an earlier stage or salvaged from the hulk. Furthermore just two of the presumably eight, or more probably nine crossbeams have been found, while no parts were encountered of a deck and any fore- and aftercastle.

38. Buisman 1995, 596-597.
39. Buisman 1995, 601.
40. Buisman 1995, 632.
41. Buisman 1995, 633.
42. Vlierman 1997a, 19.

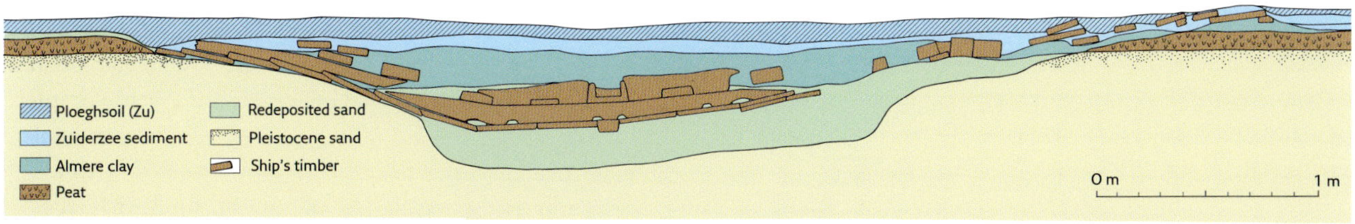

Fig. 6.12. The roughly 2 m wide washed-out disturbance of the soil profile below the ship, from which it is evident that the hull had moved on several occasions, sometimes laterally, in the course of the wreck-formation process. A partially reconstructed profile/transverse section at frame 18.

Therefore it seems credible that the vessel was already being dismantled and partially broken, moored in what may have been a 'harbour basin' near the (site of) the late-medieval settlement of Ark (fig. 6-11), and therefore not in open water. The vessel lay there relatively sheltered, until water and wind attacked the adjacent land. The disturbance of the soil profile beneath the wreck (figs 6.7 and 6.12) shows that the vessel had been 'shaking' repeatedly and with intervals, and that the keel (and hence the entire ship) had shifted sideways a few times over a distance of about 2 m. The dismantling of the ship and the loose parts of the afterbody (Plate 6.I) suggest that the vessel at the time of its definitive demise at any rate still consisted of the complete hull up to the bottommost protruding crossbeams, the adjoining part of the starboard side and two crossbeams in the afterbody with above them the adjoining shell of the port side. A 'shelf clamb' on the inside of the shell across the hole of the foremost crossbeam to starboard indicates that the crossbeam was no longer present at the time of foundering (Section 6.2.6 and Plate 6.V.B). The starboard side subsequently must have (partly) broken a few metres forward of the mast, causing the bow and the forward part of the starboard side to topple forward and to port, in the first instance dragging along the top part of the starboard side of the midships and afterbody, consisting of six planks, over a good four metres. The lower part of the side must have come away from the bottom more or less simultaneously, initially dragging along the stern and eventually coming to rest about a metre beside the bottom.

The break in the starboard shell may in part have been the result of the pre-existing absence of crossbeams in the midships and forebody. During this collapse of the hull, also the two crossbeams and the entirely broken-off and displaced piece of port shell with the upper part of the stern would have become detached. The port shell broke at the seam between strakes B and C and fell sideways, maybe simultaneously with the breaking of the starboard side and the bow falling forward and to port. The position of almost all detached timbers lying across the wreck suggests that they were ripped from their original context and displaced by northerly storms. A detached, near-complete piece of a load-bed of branches twigs and straw or reed immediately aft of the mast (Section 6.2.2), covered with wattle, straw and a layer of sand, indicates that the after half of the ship up to the moment of definitive wrecking must have remained (almost) intact, as otherwise the sand and the lighter wattle fragments certainly would have been washed from their context.

6.2 The ship

6.2.1 Documentation and reconstruction

Documentation

By around 1980 the then department of Maritime Archaeology of the RIJP had already been working for years on improving excavation techniques and the documentation of the near-continuous excavations of shipwrecks. A more or less standard procedure had been developed.[43] The documentation of wrecks at the time comprised the drawing (scale 1:20) of one or more top views (of decks, ceiling, frame and shell), a longitudinal section along the keel and bows, in some cases complemented with an interior view of one of the sides, several cross sections and detailed drawings of constructions or detached parts. The survey drawings were coloured in to clarify which parts were detached and which parts of the hull belonged to the bottom, the sides and the ceiling, etc. (Plate 6.1). The overviews and sections were hand-drawn on site. As a base line, a tape measure was strung over the stem and stern; and at right angles to it some transverse tapes, varying in number according to the vessel. To replace this time-consuming and fragile measuring system involving tapes, the Technical Department of the RIJP developed an aluminium measuring rod 6 m long with two easy-to-move stainless steel feet, which for the most prevalent sizes of wrecks (up to 4.5 m wide) could

43. Vlierman 1997a, 19-20.

be deployed to measure in the cross sections. This device would have been useful at any excavation in the foregoing decade. Yet, as things will go, the first wreck excavation at which this labour-saving device might have come in handy was the Nijkerk-II, for which it proved useless.

It was the Nijkerk-I cog in lot Oz43 (Chapter 8) of which in 1981 all components were for the first time drawn to scale 1:10. This was prompted by the condition of the wreck, the desire to record all details better and on the basis of various experiences in Britain, Denmark and Sweden in documenting ships' timbers there. These experiences and the hypothetical reconstructions based on them had been presented at the symposium 'Sources and Techniques in Boat Archaeology' at Greenwich in September 1976.[44] In short, the methods discussed there boil down to the drawing of plank templates from which, together with the available and appropriate evidence from cross-sections (frames), as reliable as possible a hull reconstruction may be obtained. The condition of the wood and the construction of the Nijkerk-I cog was such that especially the flat-lying sides no longer reliably reflected the shape of the hull. Even the frames attached to areas of shell had warped through lying flat and had lost their original shape. The drawing of plank templates therefore seemed the best solution. To record as closely as possible the shape of the surviving frames, the floor timbers and futtocks belonging together were lifted off the ship's bottom and sides, and joined together horizontally in a levelled area of the excavation. After their detachment the warping of the futtocks would largely correct itself. In this way a number of reliable cross-sections could be measured in and drawn (fig. 8.13 and Plate 8.IIC).

Of the Nijkerk-II cog, the plank templates and the frame timbers were drawn to the scale of 1:10, which has been the standard ever since 1981. To this end, drawing tables 5 m long and 60 cm wide were installed, marked with a 10x10-cm grid. Onto this a timber would be laid, to be measured with T-squares and rulers, and drawn. All details, including the exact place and dimensions of any treenails and nails, treenail wedges, nailhole plugs, sintel clamps, markings and scarfs and the presence of sapwood were indicated, while the thickness of the plank or other timber was added in writing in various places (figs 6.13, 6.14 and 6.15). Of the planks only the inside was drawn; of the frame timbers, both sides: a rear and a bottom view. The position and the angle of the wooden treenails and iron nails is indicated with broken lines in the rear views of the frame members. The stern- and stem hook, keel plank, stern- and stempost, keelson and crossbeam constructions were also drawn in detail.

The overview drawing with written notes in it (scale 1:20; Plate 6.1) was important during reconstruction for determining the original position of detached parts and in reconstructing the process of wreck formation. The drawing also offers a good and instructive overview of the situation as encountered, in a way that was not, nor could be recorded photographically: the situation as drawn never actually occurred at any one point in time. After being measured in and drawn, all loose parts were lifted. Only afterwards could the timbers of the wreck still lying 'in articulation' be uncovered, cleaned and drawn. During the excavation a list was compiled of all loose parts, which were coded, in several cases mentioning their (presumed) position in the ship's construction.

Since 1981 also the description and numbering of all parts of a wreck are performed according to a fixed system and sequence commonly employed in shipbuilding. The numbering and description of the shell and ceiling planks is done from the keel(plank) upwards to both sides and from stern to bow. The codes G for *gang* (strake) and W for *wegering* (ceiling) are followed by A, B etc. (in this case up to O) and sequential numbers (depending on the number of planks per strake) plus the addition of sb (for starboard) or bb (for port). For example, the planks of the lowermost starboard shell strake are given the codes GA1sb, GA2sb, etc.. If the plank consists of several broken parts, or if it has been sawn through for lifting, the code will be extended to for example GA1-1sb, GA1-2sb, or GA2-1sb, GA2-2sb. The same goes for the ceiling planks. In the description of the wreck in this chapter, the G and W are omitted. The frame timbers are numbered as follows: the lowermost floor timber on the rising part of the stern hook is labelled 0, and frame numbers ascend towards the bow up to (in this wreck) 42. Aft of 0, the frames (above the stern) are numbered -1 to -9 (Plate 6.IV B). Between frames 25 and 26 to starboard, merely a short futtock was found above the hole of the protruding crossbeam, so it was numbered 25½. Between 18 and 19 too an extra futtock was fitted (18½), which starts above ceiling plank B. The futtock(s) extending from the floor timbers are given the same numbers as the frames, for instance 5. The first (i.e. lower) futtock on the starboard side then is coded 5Asb; the second (i.e. upper), 5Bsb. The other parts of the ship are similarly coded, or identified as keel(plank), stem hook, deck beam 1, etc., with the addition sb or bb, or fore or aft, depending on where the marking is placed (the aft- or forwards side of the plank). Also, the dimensions of the timbers are noted in metres and centimetres, and those of iron parts in millimetres.

44. Crumlin-Pedersen 1977, 163-179; Evans 1977, 179-191; Cederlund 1977, 191-201; McKee 1977, 205-215.

Reconstruction

With the Nijkerk-II cog the bottom was made the base line for the reconstruction, assuming an entirely straight and horizontal keel from fore to aft, including the stern- and stem hook. The bottom was found to be almost complete *in situ* and had largely retained its original shape, thanks to the still articulated frame members. As a result, the detached floor timbers that were recovered could be fitted snugly into the sharper parts of the fore and aft parts of the bottom. Also the complete stern- and stem hook and the lowermost parts of the stern- and stempost were still present, so that the first three phases of the ship's construction could be fully reconstructed (Section 6.3.2 and beyond). The reconstructed lines may be regarded as very reliable. After the field research, the frame timbers, the stern- and stem hook and the stern- and stempost could also be simply fitted together, thanks to the drawings and the corresponding treenail and nail holes marked on them. This allowed the complete bottom of planks and frame timbers in articulation to be reconstructed 'on paper'. The planks of the bottom together form a 'mosaic'. They had been sawn or hewn into shape, which is particularly evident in the garboards (Plate 6.III A). Reconstruction of the sides 'on paper' was abandoned after a few attempts. The fragments had been pulled apart and warped to such an extent that it was decided to use the available data (the drawn plank templates, lengths and widths of the planks and futtocks, etc.) only for building the 1:10 scale

Fig. 6.13. Cleaning a frame.

Fig. 6.14. Measuring and drawing a frame and a plank.

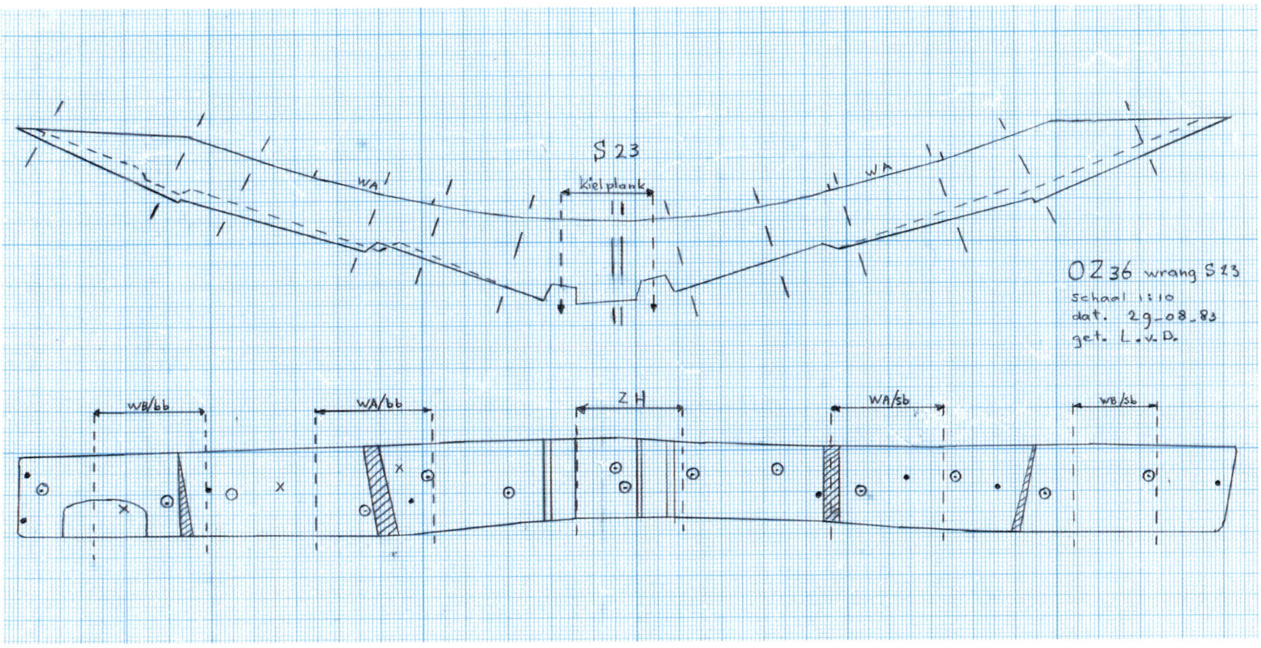

Fig. 6.15. A (field) drawing of a V-shaped floor timber to scale 1:10.

model and in building the full-sized reconstruction of the Kampen Cog, which was taking place at roughly the same time. The two recovered crossbeams are of essential importance for reconstructing the correct hull lines at deck level. No other cross beams were preserved, but at the time six transverse beams protruding from the sides were assumed on the basis of structural features to accommodate them, the traces they left and the assumption of roughly equidistant beams across the midships (Section 6.2.2).

During the attempts at reconstructing the sides 'on paper', it soon became clear that two cross-sections with crossbeams were insufficient to achieve a reliable shape of the full hull at deck level. This conclusion may in part have resulted from an erroneous premise: at the time, it was assumed that crossbeam 2 in the afterbody should be crossbeam 6 in the forebody, and that crossbeam 1 should be crossbeam 2. As a result, the reconstructed widths of the hull, both in the model and in the Kampen Cog, are incorrect at the level of the crossbeams (and upwards). In the context of the present study, which is also intended as a plea for continuous attention to and improvement of the methods, this was sufficient reason to undertake a second reconstruction attempt on paper. In due course it was established that the afterbody especially must have been considerably broader at the level of crossbeams and gunwales; that the vessel numbered fifteen strakes from keel plank to top of the gunwale rather than the previously assumed fourteen; that at any rate there must have been a seventh, an eighth and probably also a ninth crossbeam projecting through the shell; and that the cog in the final years of its working life had featured both an aftercastle and a forecastle. For the detailed information on the reconstruction of the sides, on the crossbeams, the hull lines and the castles, the reader is referred to sections 6.2.2, 6.3 and 6.4.2 and to the building of the Kampen Cog (Chapter 7).

6.2.2 Construction and layout of the hull

The keel plank
Including the stern- and stem hook, the keel is about 12.3 m long. The keel plank has a length of 6.97 m, with a thickness throughout of 11.5 cm, and is attached to the hooks with scarfs. The scarfs run from fore to aft, from top to bottom. The sloping flat scarf (measured along the keel line) is 57 cm long in the forebody and 50 cm in the afterbody. The ends are fixed to the hooks with iron nails: the forward ends with five nails from inside outwards and six from the outside inwards; and the after ends, the other way around. The heads of the nails always are on the thinner part of the scarf, the nail itself then would emerge through the thicker plank section, where it was flattened, or else, depending on its length, turned back into this thicker timber. Turning back a nail point into a thin timber nib would certainly have split the wood.

The keel plank on its upper surface is 22 cm wide at its forward end, gradually narrows, and towards the stern measures 20 cm. It is slightly trapezoid in section, the bottom surface being 20 and 17 cm wide, respectively. Below frame 12, at 7.09 m and 7.02 m from the front of the stem hook and measured along the keel line, two transverse grooves are scratched into the top. The foremost one is about midway along the keelson. Also under frames 17 and 18 are scratched grooves, at 5.01 and 4.49 m from the front of the stem hook, respectively. The grooves almost certainly relate to the measurements made at the start of the ship's construction. The sides of the keel plank were not rabbeted to accommodate the garboards. These A strakes were placed against it by a butt seam.

Stern hook
The horizontal part of the stern hook had broken at about 1.65 m from the rear of the false (also known as outer) stem (fig. 6.16a and b). Its forward part was still connected to the keel plank. The break has been abraded by the continuous movement of water and its scraping on the sandy sea floor. Its reconstructed length, including the scarf, is 3.04 m; including the false stem, 3.2 m. The horizontal part gradually narrows towards the stern, and at 1.42 m from the back of the stern hook is 13 cm wide and thick. From there the upper surface runs upward at an angle along a length of 11 cm. On the inside of the rising part its width is 9 cm and its thickness 22 cm. The back of the stern hook is 7 cm wide; the width of the back of the false stern is 6 cm. The section of the stern construction is trapezoid and in the sides of the hook, from the part slanting upward in the horizontal part there is a rabbet for the garboards, which starts flush and ultimately is 3 cm deep at the bottom. The rising part, measured at right angles from the keel line, is about 80 cm high. From its highest (aftmost) point, a flat scarf runs downward, which connects the stern hook with the sternpost.

Sternpost
The (true) sternpost, of which 2.8 m were preserved, at its base was attached to the stern hook, and only on the inner side, by five nails. Combined with the rising part of the hook, its surviving length is 3.1 m. The sternpost is trapezoid in section, 25 cm thick at the bottom and 21 cm at the top end. The width on the inner side increases

Fig. 6.16a. The stern hook with the sternpost and the gudgeons. The photo shows the recovered part of the starboard side; on it, the thin (filler) plank supporting the gudgeons

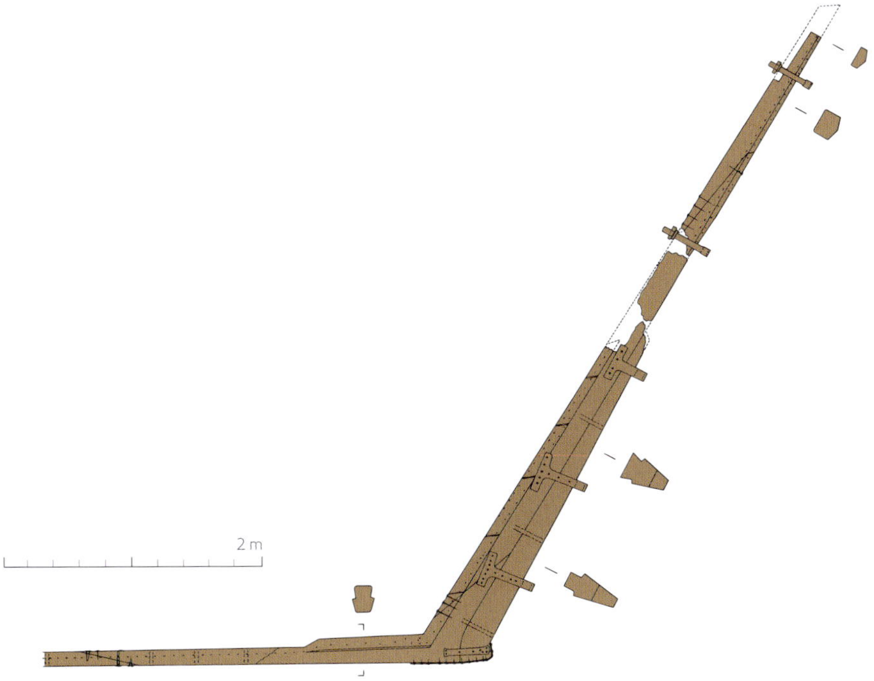

Fig 6. 16b. Reconstruction drawing to scale 1:10 of the port side of the stern construction with the detached upper part of the sternpost.

upwards from strake to strake: at the garboards it is 12 cm, then upwards 15, 17 and 18 cm. On the outside, the sternpost is 14 cm wide. On both sides a rabbet about 10 cm wide and 3 cm deep had been cut out for fixing the hood ends of the strakes. The stern stands at a rake of 120° to the keel line. Beside this part of the sternpost lay a piece broken off from it, about 60 cm long. Originally its length will have been 65-70 cm. The adjoining part was found loose in the afterbody with attached to it fragments of several shell strakes of the port side and a small floor timber (8; see Plate 6.I). This part of the stern is 2 m long, about 20 cm wide, 18 cm thick at the base, and at 47 cm from the top it is 16 cm thick. From the last-mentioned point, its forward surface recedes 5 cm. It consists of the broken-off top of the sternpost and a second part connected upwards with it by a flat scarf. The very top of the stern, which had been joined to it by a hooked scarf, was never recovered. On the rear, the top part of the sternpost was finished in a mitred shape, from which it is evident that the upper strakes (from strake G) ran on outside the stern and had been nailed to its rear surface. The upper end of the scarf had been fixed to the inner side of the sternpost with two nails, the lower end with four nails. In the upper (recessed) part of the stern is a treenail hole with a diameter of 3.5 cm and at 1.88 m from the top is a round hole with a diameter of 4 cm: the top two suspension points of the rudder. The overall length of the stern, measured from the keel line, must have been about 6 m.

The false stern

To create sufficient space on the sternpost for attaching the braces of the three gudgeons, an extra beam was fastened to the rear of it: the false stern. Its surviving length is 2.9 m. At the base of this beam it is 7 cm wide, at a height of 2 m it is 9 cm, and in these places it has a thickness of 22 and 7.5 cm, respectively. The missing top section tapered away or ended with a thickness of a few centimetres at the top of strake G; the latter seems the more likely. The original length of the false stern would have been about 3.35 m. The surviving piece was fixed to the sternpost with three blind treenails. Two of these had a diameter of 3.5 cm; the middle one, 3 cm. The false stern is broader than the back of the sternpost. In order to attach the braces of the gudgeons with a close fit, a thin plank was nailed to the starboard side of the sternpost (as appears from the nail holes), which originally must have covered almost all of the visible part of the sternpost. It is not clear whether this is a construction made at the time when the ship was built. Neither in the field, nor in the drawing of the stern did anything suggest that the false stern was ever replaced. Yet this does seem the most likely explanation. It is unlikely that a stern should be composed like this in a new-built ship, as such a construction is not a robust one.

The sternpost and false stern still bear two wrought-iron gudgeons and a fragment of a third: the lowermost three of a total of five original attachments for the stern rudder. The flat braces, attached on either side of the sternpost, are T-shaped, 60-70 mm wide, 20 mm thick at the rear and 10 mm at their forward end. Into the rear part, protruding from the stern in a loop, a ring was wrought with an opening 70 mm in diameter, the hole that held the pintle. The bottommost gudgeon was attached to the stern with four nails through the horizontal parts of the braces, and with seven nails through the vertical wings. The second gudgeon (143, figs 6.17, 6.53-143 and 6.54) was attached with three and five nails, respectively, and the third one with two and five nails, respectively. The fourth suspension point of the rudder (144, figs 6.17, 6.53-144 and 6.54) was found among the detached parts; the top one was never found. It is a thick, round iron bolt (diameter ca 30 mm) with a ring at one end that had held the pintle of the rudder. The ring was bent and broken. The hole for the pintle has a diameter of about 45 mm, and hence is narrower than that of the three gudgeons. The bolt passed through the stern in the above-mentioned hole at about 2.1 m below the top, and at the back had a slot to hold a securing pin. Neither the end of the bolt nor the pin were recovered. Various late-medieval examples of such a (uppermost) rudder suspension are known from the former Zuiderzee area. To reinforce the connection between the false stern and the stern hook, a flat metal strip 55 mm wide and 8 mm thick was nailed across the heel of the stern. This strip is 662 mm long (650 mm on the keel side), bends around the heel of the

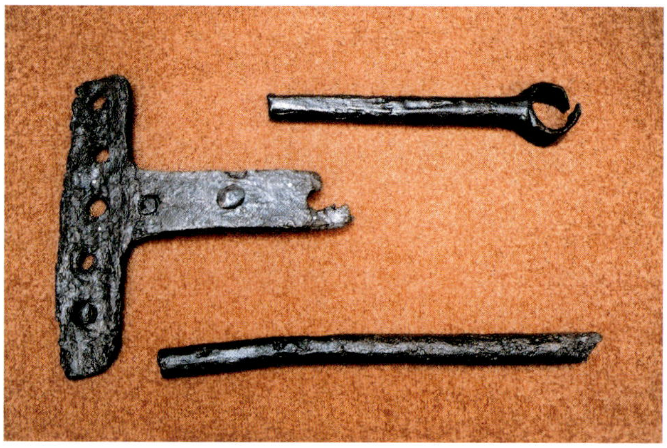

Fig. 6.17. The starboard brace of the second gudgeon from below, the broken bolt with the (broken) suspension ring for the pintle, and the bolt through the stempost and false stem.

stern and continues to the beginning of the back. The iron piece is fixed with 14 nails. To the same end, a flat iron strip was nailed to both sides of the stern construction at 10 cm from the base of the keel, 350 mm long, 50-70 mm wide and 8 mm thick.

Rudder

In section 6.1.4 it was already noted that the rudder is missing. From the iron gudgeons it is evident that the ship had a stern rudder. The bent and broken ring that held the rudder's pintle, the broken end of its bolt, and the difference in diameter between this bolt (ca 30 mm) and that of the worn-out hole (35-40 mm) that held it, suggest that the rudder may have hung on until the upper part of the port shell and the stern became detached, and may have drifted away together with them. So far, no cog has been found with its rudder. For the shape of the stern rudder, our reconstruction has made use of images on seals and archaeological finds of rudders that, especially with regard to dating, approximate cogs as closely as possible. The rudders of early-16th-century fishing vessels (*waterschepen*) are particularly appropriate.

The stem hook

The base of the stem hook (fig. 6.18) has been abraded to an angle along 70 cm at the front; removing 13 or 14 cm from the front of its rising part. This must be due to frequent scraping on the sea floor or beaching on a sandy shore, and the ship's probably quite advanced age by the time it was wrecked. This degree of abrasion had never before been encountered in the hundreds of younger, late- and post-medieval wrecks excavated in the IJsselmeer polders. The bow rakes at an angle of about 130°. The length of the horizontal section of the hook is 3.2 m at most, its width at the top is 22 cm, its thickness at the scarf is 12.5 cm, and in a forward direction gradually increases, up to 16.5 cm at 1.73 m from the scarf. There the upper surface slopes upwards and its height becomes 22 cm. At the angle with the upper face of the rising hook part the original height (before the abrasion) had been 28 cm. The horizontal part of the stem hook, like the keel plank, is trapezoid in section. From the rising part to 1.73 m forward of the scarf, there is a rabbet to hold the garboards, which is 3 cm deep. About 20 cm off the front of the stem, the rabbet runs upward, parallel to the front of the stem. The highest point of the stem hook stands about 80 cm above the line of the keel (measured vertically). This is also the outermost point of the scarf joining it to the stempost, which runs inward and downward.

Stempost

The scarf to the stem hook supports the stempost, which is secured by two sets of three thick, blind nails from within and from the outside. Of this stempost 5.5 m survives, the overall preserved stem length including the stem hook is 6.4 m. On the inside, it is rectangular in section down to the rabbet. The first six strakes of the bottom and sides are held in a 3 cm deep rabbet. The part of the stempost that projects ahead of the strakes is trapezoid in section, with the narrow side forward. From strake D upwards, the front is mitred progressively less sharply. The front of the stempost holds three blind treenails, two with a diameter of 2 cm and one with a diameter of 2.5-3 cm. At the top of strake F the stem still was visible over a width of 7 cm; from strake G upwards this was no longer the case. From there on, the port and starboard strakes extended across the mitred front of the stempost, where they formed a butt join on the ridge. At the level of the sixth to eighth strakes (F-H) on the port side, a timber triangular in section is fixed to the side of the stempost with two nails. Presumably this was a repair to provide extra support for the (vulnerable) ends of the strakes.

False stem

Against the front of the stempost a false stem was fitted, of which 1.9 m has survived. It begins 5 cm below the top of the scarf connecting the stem hook and the stempost, where it was 9 cm wide and 10 cm thick. The underwater part is badly worn down, leaving the lowermost 20 cm of the front quite rounded. The false stem is trapezoid in section throughout, widening and thickening upwards, and at its (topmost surviving) end is 12 cm wide and 13 cm thick. The missing part had, from strake G on, covered and protected the cross-cut ends of the strakes. The base of the false stem was attached with two small nails to the top of the stem hook. A little above it and in a vertical line, with three thick nails close together and then nails at successive intervals of 50, 60, 80 and 80 cm, to the front of the stempost. At 1 m above the base of the false stem, the stempost holds a blind treenail with a diameter of 2.5-3 cm, which does not penetrate into the false stem. Because of the absence of the broken top of the false stem it is unclear whether this also goes for any blind treenails in the upper part of the stempost. Under the bow, an iron bolt (figs 6.17, 6.53-145 and 6.54) was found. It still has a length of 415 mm, at the intact end it was 27 mm thick and at the other, broken end, 30 mm. Judging by the length of the bolt, it presumably was a fastening device that had passed through the false stem and the top of the stempost (fig. 6.18) at a level above the preserved timbers. The blind treenail(s) in the stempost may be evidence of an earlier and since replaced false stem.

Bottom and sides

The strakes of the shell are each composed of three or four planks, which are interconnected by flat scarfs. Plate 6.V B shows the recovered shell parts. The bottom is complete. Of the sides, the starboard side has survived more fully than the port side. The few (apparently) square frame heads in the forebody suggest that the topmost strake had been attached here. This means that the vessel on either side of the keel plank had numbered thirteen strakes (A to M). On the port side, just the strakes A to H and parts of I survived. Remnants of a fourteenth strake (N) were found aft on the starboard shell and on the broken-off piece of port shell with the upper part of the sternpost; it was presumed that this was a 'drop strake' which never continued all the way to the bow. During the final phase of the definitive reconstruction, the global location of a shell plank found loose in the stern (AY, Plate 6.I) could be established: most probably a part of strake O to port in the afterbody. A number of other findings, of which at first it was unclear how they should be interpreted, now were found to be explainable. They demonstrated that the N strakes must have continued fully aft to fore. Strake O probably stopped short of the stern and stem, about a metre.

One of the clues pointing to this is the beam found *in situ* that had been fixed vertically on the outside of the ship forward of the gap in strake M in the forebody. Assuming that the upper strakes had a roughly similar width, this means that there must have been a short and presumably oblique plank (P) above frames 35 to 38. The others are the square head of frame 25 and immediately above it the evidence of a square hole for a crossbeam protruding through the shell. The top of this beam, in terms of height above the keel line, would match the bottom of the rider beam on crossbeam 2 in the afterbody, with in the upper side four notches for longitudinal stringers (Section 6.2.2). From close scrutiny of the excavation photos, it also became clear that the aforementioned 'square' frame heads in the bow were badly damaged. It is imaginable that in the course of documentation in the field, with the insights of the time, these 'square' frame heads had been misinterpreted. The upper parts of these frames may have broken off in the process of wreck formation. Protruding from the sea floor, they were eroded away. The bottom very gradually merges into the sides; a 'proper' turn of the bilge is hard to identify, and is evident only amidships, from the angles at which the C- and D strakes were laid. The upper surface of the garboards is flush with that of the keel plank; the latter does not protrude above the garboards.

Bottom

The bottom comprises three strakes (A, B and C) on either side of the keel plank (Plate 6.III A). The bottom is carvel-laid, but towards the stern and stem (aft of floor timber 8 and forward of floor timber 22) becomes clinker-built. Broad planks (51-60 cm) were used with an original thickness of 4-5 cm. The lengths of the scarfs range from 34 to 50 cm, the joins usually being fixed with two rows of six nails: one row from the outside, and one row from the inside. The points of the nails were bent over twice and driven back into the timber. Sometimes there are five or seven nails, or a few extra between the two rows. The ends of the overlapping planks are 1-2 cm thick and project from the shell on the inside and the outside. For the sake of strength, the scarfs in adjacent planks are staggered. The so-called 'brother-and-sister' pattern on either side of the keel plank was applied. During the excavation and on-site documentation, it was not observed whether this system had been fully followed through, i.e. that two adjoining planks sawn from the same tree were used in similar positions to port and starboard.

In the garboards (A), the fore and aft planks are a good 5 m long; the middle plank, about 3 m. The aft plank to port (A3), however, is shorter: about 4 m. This strake was complemented with a fourth plank of just 1.65 m (A4), the only place in the bottom that deviates from the pattern of planks and scarfs. It may be assumed that initially a single long plank was prepared, which broke during bending or for some other reason was deemed to be partially inadequate, maybe because of the presence of a knot. The garboards are spiked into the rabbets of the stern and stem hooks with iron nails; and, in two places, ahead of the rabbets, onto the sides of the horizontal part of the hook. The B- and C strakes each are composed of four planks. In the B strakes, like the garboards, 5 m planks were used fore and aft; and between these, two planks ca 2.5 m long. The C strakes comprise planks measuring about 4.5 m fore and 5.7 m aft; and in between, planks of 2.65 m and 3.2 m. The scarfs run downwards fore to aft, which is an obvious construction given the vessel's direction of travel.

On strake C between frames 7 and 8 to starboard, a notable area of wear was observed. The remaining thickness of the plank of strake B on that location is only 0.5 cm. The adjacent plank of strake B was not recovered, but probably had a similarly worn area. Maybe this is where the foot of the bilge pump was located. Section 6.2.2 will return to this matter. In the carvel-laid part, the planks are interconnected only by the floor timbers, without any joins between the adjacent planks and the keel plank.

Where the carvel-laid bottom transitions into the clinker-built side, the lower edges of the planks of the B- and C strakes are about 4 cm wider, so as to achieve the necessary overlap (the 'land'). From there towards the stern

Fig. 6.18a. The stem hook and the adjoining stempost fragment. Viewed from below, from the keel.

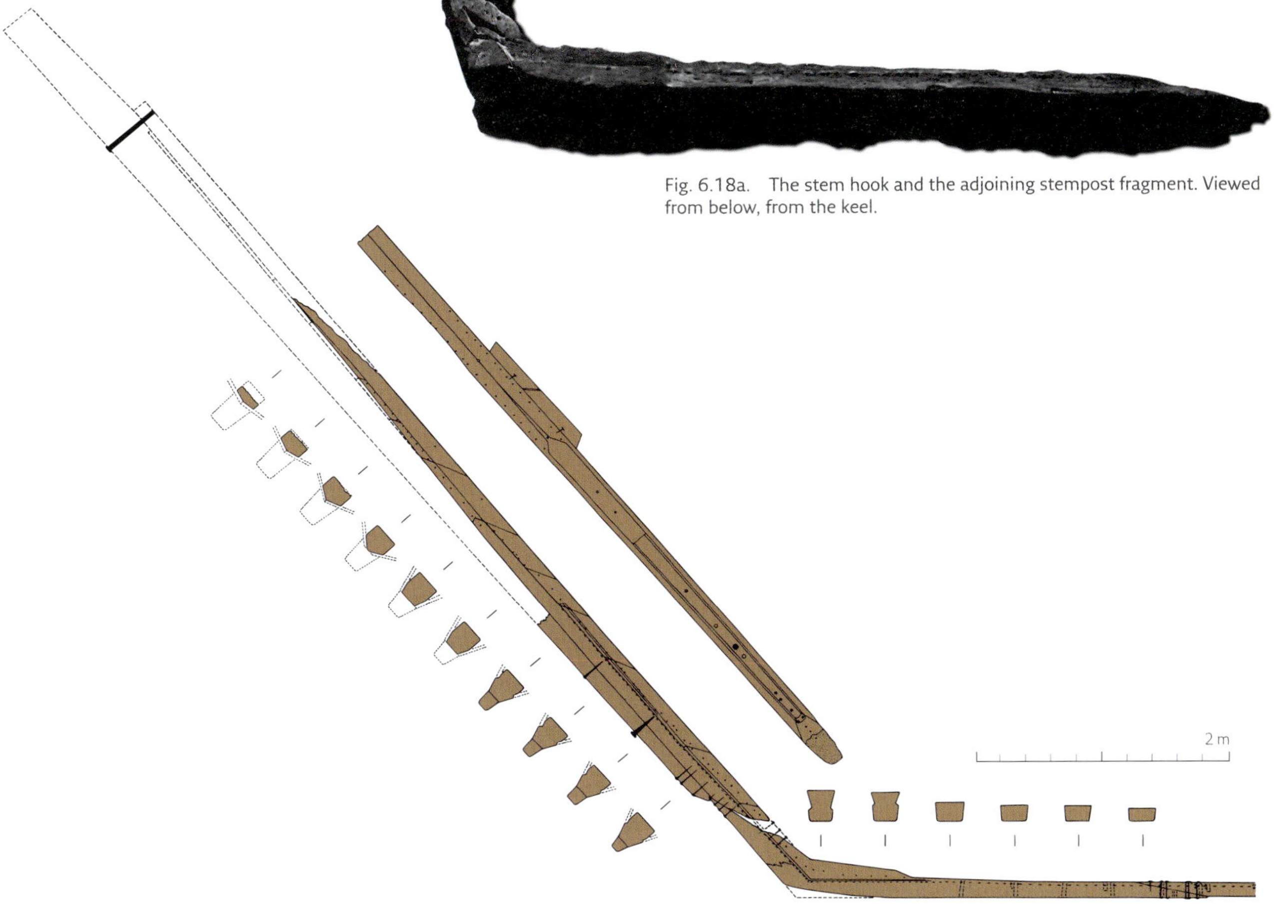

Fig. 6.18b. Reconstruction drawing of the port face of the stem hook, the stempost, and the false stem, plus the frontal view of the recovered parts of the stempost and false stem.

and stem, the planks were positioned so as to overlap each other and secured with double-clenched nails (turned back into the planks), at regular intervals of 14-18 cm. The ends of the planks towards the stern and stem become narrower and thinner and the hood ends are fixed with nails into the rabbets in the stern and stem hooks, and higher up in the stern- and stempost.

The seams between the carvel-laid planks are closed with sintel clamped moss caulking only from below, as are the seams between the garboards and the keel plank. The clinker-built sides were caulked in this way both on the inside and the outside. The seams between the B- and C strakes of the carvel-built bottom are on the inside covered by 15 cm wide, and 1.5-2 cm thick laths, with a flattened, more-or-less oval to diamond-shaped section. Both to port and starboard they consist of three parts: two lengths of 2.6 m and one short piece of 0.45 and 0.6 m, respectively. The pieces are interconnected with short (6-10 cm) flat scarfs, and every 15-20 cm they are nailed to both strakes on either side of the seam.They are so-called 'bilge laths', in this case serving to prevent any opening-up of the carvel seams in the bottom, which might occur as a result of shrinkage of the clinker-laid planks of the sides higher up. (see also Chapter 7). These 'bilge laths' were put in place before the floor timbers were inserted. The flush-laid bottom (between floor timbers 8 and 22) 'nowhere has a dead flat'. The first two strakes on either side of the keel plank (the garboards A and the bottom strakes B) throughout are positioned in an even, smooth and gently ascending line towards the turn of the bilge. The third strake (C) sits at a slight angle to the previous two.

Starboard side

The foremost 4.5 m of the starboard side is the most fully preserved. From bottom strake C all strakes (up to and including strake M) still were interconnected in clinker fashion and thus provided incontestable evidence on the fashioning and the number of strakes for reconstructing the sides. The midships and after parts consist of two large fragments: the lower part (strakes D-H) which lay almost in its original place longitudinally but was displaced sideways by about a metre, and an upper part (strakes I-N), which partially underlay it and had shifted forward a good 4 m (Plate 6.I). The two parts became separated from each other and from the bottom by the lengthways breaking off of strakes D and I. The breaks occurred here because the most scarfs between floor timbers and the lowermost futtocks were situated above strake D; and the scarfs between the lower and upper futtocks, above strakes H and I. Above strake D especially this was a weakness in a structurally important part of the hull. The forward part of the side still numbered ten strakes (D-M). The fact that the top parts of futtocks 25½, 26, 29 and 30 project above strake M confirms that the fourteenth strake (N), of which fragments survived in the after part to both port and starboard, initially continued to frame 30 or beyond, while from the foregoing it is evident that even a strake O existed. The strakes of the

Fig. 6.19. The starboard shell with the forebody, with the two square holes in strake J (fourth strake from the right) for the (missing) crossbeams 5 and 6 at frames 25½ and 29.

Fig. 6.20. The starboard afterbody shell with the square hole in strake K for crossbeam 1 (in the foreground to the right of the repair plank), the small square hole for the drain, and to the right of this the damaged area where crossbeam 2 passed through the shell.

gunwales were composed of three or four planks of varying lengths, and were 42-52 cm wide. Insofar as could be established, there are no true 'brother and sister' planks to port and starboard, but planks of differing lengths arranged more or less arbitrarily. From strake G upwards, planks of more than 7 m long might be used. The shortest is 0.8 m long (strake E at the bow). The lengths of the flat scarfs connecting the planks range from 34 to 48 cm, while here too the plank ends project from the shell on the in- and outside. The overlapping plank ends are joined in the same way as the flat scarfs in the bottom. All strakes are fixed in clinker fashion from fore to aft, also with twice-bent nails turned back into the timber, placed at regular intervals of 14-18 cm. In some places they are more closely spaced (Section 6.4.1). The planks are straight, to which we shall return in sections 6.4.1 and 6.3.4. Yet they do not all have clean-cut, straight sides, as a result of which the width (or height) of the land between scarfed planks varies between 6 and 9 cm. This is the case especially along the metres closest to the stern and stem.

In strake J between frames 25 and 26 and at 29, there are two large, more-or-less square holes (fig. 6.19), through which the foremost two crossbeams had protruded. In strake K a similar square hole appears, at frame -1 (fig. 6.20). In the same strake, immediately above frame 2 there also seems to have been such a hole in the shell. The dimensions of the hole above frame 29 originally were about 20x20 cm; those of the hole at frame 25½, about 23x23 cm. The hole at frame -1 to starboard measures about 16x16 cm, while in the port shell it is about 18x20 cm. The relatively slight damage to the edges of the two holes for the protruding crossbeams to starboard in the forebody (at 25½ and 29) and the fact that this part of the starboard side has remained almost undamaged and still articulated from top to bottom, support the idea that the beams were deliberately hewn or sawn through before the final disintegration of the ship. Besides, the shelf timber G found *in situ* and fixed to the futtocks closes off one of the openings (Plate 6.V B). This was probably done to strengthen the weak part of the shell that resulted from removal of the crossbeam, but it may also indicate an adjustment or change to the layout of the ship. The two observed square-finished edges of an opening in strake L, directly above the square-finished top of frame 25, indicate that this gap too had once held a shell-penetrating crossbeam. It seems highly likely that this had been the case too above frame 19 or between 18 and 19 (the likely place for a mast beam), although no shell planking survived here.

Between 2 and 3 there is a small square hole (10x10 cm) in strake J (fig. 3.20), which probably is the opening through which the wooden drain passed (see section 6.2.2). The sieving of sand from the hold yielded a square oaken 'lid' or 'cover' 13x13x3 cm). At its back, part of the edge has been cut away, leaving a square of almost 10x10 cm, standing proud 1.5 cm (figs 6.53-114 and 6.54). A little off-centre in this square is a small, 6-mm hole containing a remnant of iron, probably belonging to a pin with a ring or a cotter pin hole. The small square part fits exactly into the scuttle in strake J and may very well have served to close it when for some reason the drain was not in use. Between frames 34 and 35 in strake M (and in strakes N and O) there is a trapezoid opening in the shell, about 45 cm wide at the bottom, and at the top - where it coincides with the top of strake M - about 62 cm. A similar gap was observed in the Doel-I cog, in a very similar position in the bulwark.

In 1985 a wall painting depicting a cog was discovered in a cellar vault of the town hall of Wismar, and it shows just such a gap. The image dates from the 14th or 15th century. Although the opening in the painting is shown more amidships, we may assume that a similar feature is depicted. The image (fig. 6.21) offers no clues as to the function of the hole. A likely explanation is that the crew could push a gangplank through it. An opening for dropping and weighing an anchor is another possibility but seems less plausible. Close to the stern in strake K, both

Fig. 6.21. The opening in the gunwale of a cog, 14th-/15th-century fresco in the town hall of Wismar, Germany (after Förster 2000).

to port and starboard, there are holes 5 cm across, through which a rope could be passed for securing the rudder. The detached shell plank AY, given its findspot in the afterbody, must have been one of the first timbers to have broken off as the ship disintegrated. Most probably it is part of strake O between frames -1 and 8 on the port side.

Port side

The surviving part of the port side (strakes D-I) provides little more evidence than the starboard side. Three or four planks per strake were used. The side remnant, preserved almost in one piece, lay in its correct place lengthwise. Bottom strake B especially had broken longitudinally, which caused the entire piece to collapse sideways. On this side, the scarfs between the floor timbers and the lowermost futtocks were located mainly above strake C. The after part of the shell had presumably broken off previously (strakes J-N), taking along the upper part of the sternpost (fig. 6.22). In it, strake K above frame -1 displays the hole corresponding to its starboard equivalent, through which the aftmost crossbeam 1 once protruded. The absence of the rest of the upper part of the port shell with its upper futtocks and ceiling planks, may, as was suggested above, have drifted away, but may also be the result of demolition and deliberate removal of the midship and forward crossbeams.

Caulking: sintel clamped moss caulking

Before the V-shaped floor timbers and futtocks were fixed into the shell of the ship, the seams were caulked on the interior with moss. This was done also to the seams in the clinker-built parts of the bottom aft and fore, and between the stern- and stempost and the hood ends of the shell planks. The seams in the carvel-laid part of the bottom were treated only externally with sintel clamped moss caulking; this was not done until the hull was completed. The same goes for the seams on the outside of the clinker-built sides and those between the plank ends and the stern and stem on the outside. The cross-cut ends of the planks at the scarfs were not provided with moss caulking neither on the inside nor on the outside. By contrast, this probably *was* the case around the protruding crossbeams on the outside of the shell. Also cracks in the shell planks were externally closed in this way. Between overlapping plank ends (in the scarfs), a thin layer of moss would be put, and also under the repair planks on the inside and outside of the shell. Whether under the 'bilge laths' also moss had been used was not recorded during the excavation; but it is likely.

The moss was held in the seams with laths of oak wood (*Quercus* sp.), which in their turn were kept in place with iron sintel clamps. This method is known as sintel clamped moss caulking (fig. 6.23).[45] The origin of this caulking method should probably be sought in the Celto-Roman inland shipping tradition in and (immediately) north and south of the Alps. This method was not part of the Nordic or Viking shipbuilding tradition. The sintel clamped moss caulking associated with the cog-building tradition, i.e. the development from sintel nail to sintel clamp, probably occurred in the Lower Rhine region and (later) in northern Germany, i.e. the Hanseatic region.[46] The sintel clamps used in the Nijkerk-II cog are of the D type, variant 2. This type was in use between about 1260 and 1350; variant 2 probably from around 1300.[47] The clamps are not large. The roughly oval centre part is 35-50 mm in length. Overall, about 1100 m of moss caulking was applied to the shell seams. the average distance between the clamps of about 7 cm (centre to centre) means that in this ship some 15,000 to 16,000 sintel clamps were used.[48]

Identification and pollen analysis of the moss caulking

The caulking itself consists of a single species of moss: sphagnum moss (*Sphagnum cuspidatum*), which usually is gathered from a wet, boggy or peaty habitat such as heathland, peatbog, ditches and reedland.[49] *Sphagnum cuspidatum* will grow abundantly both in oligotrophic peatbogs and in the less eutrophic parts of fenland in the low-lying parts of the Netherlands. For reasons of convenience and economy, the provenance will not have been too far from the shipyards. Pollen analysis of the moss caulking made it clear that the sample is a pure, be it degraded moss preparation, and contains little pollen (trees 63%, herbs 62%). The tree pollen divides into: alder 18%, hazel 8%, birch 3%, oak 4%, beech, willow, lime and elm 0.5%. The herb pollen divides as follows: heather 36%, cereals 11%, grasses and cyperus grasses 5%, arable weeds 4% (including mugwort, *cruciferae* and narrowleaf plantain). Local: peat moss 120%, bracken and dryopteris ferns 1%. Undetermined: 0%. This means that the moss grew in oligotrophic conditions in a large peatland adjacent to reclaimed land, heathland and trees of many types, growing in an open landscape.[50]

45. Vlierman 1996a; fig. 6.23.
46. Idem., 54.
47. Idem., 58 and 77.
48. Idem., 31.
49. Cappers *et al.* 2000, 584.
50. Mook-Kamps 1997, sample 17; see also Appendix.

Fig. 6.22. The piece of after port shell found detached on top of the wreck with the upper end of the sternpost with transversely across it frame-8.

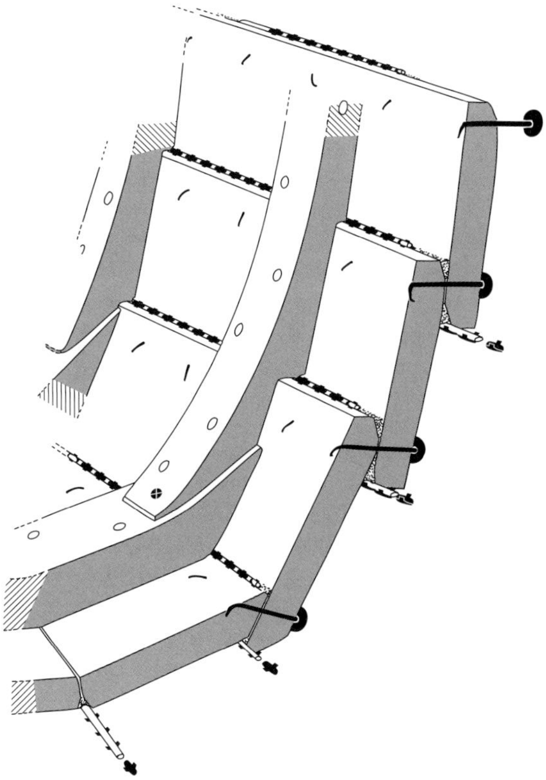

Fig. 6.23. Clamped moss caulking (moss, moss lath and sintel clamps) in the seams of the carvel-built bottom (only on the outside), on the inside and outside of the shell in the seams of the clinker-built sides and the manner of attachment of the shell planks with double-clenched nails.

Frames, floor timbers, V-shaped timbers and futtocks
The bottom and sides are held in shape with frames (fig. 6.24), composed of straight floor timbers on the carvel-laid part of the bottom, and V-shaped floor timbers in the narrowing fore- and afterbody and on the stern- and stempost. The floor timbers are extended up the sides with lower and upper futtocks (Plate 3.IV A). Possibly some (short) upper futtocks had been extended with a third futtock. Between the futtocks of 15 and 18 to port (beside the mast) an extra futtock had been fitted (17½). Between 25 and 26 this is the case above crossbeam 5 to starboard (25½). This was done to fill the space between the frame and futtocks placed somewhat aft in the afterbody and the somewhat forward position of the frame and futtocks in the forebody. The floor timbers are quite massive: 18-24 cm wide and 15-20 cm thick or high; the V-shaped floor timbers are up to 30 cm high. The broadest floor timbers are found in the mast-step area to give the construction extra strength to support the mast. The lower futtocks generally have a width of 10-15 cm and a thickness or height of 8-15 cm, while the upper futtocks are rarely wider or thicker than 10-11 cm. The V-shaped floor timbers of frames 1, 2 and 29 and 30 above the stern- and stem hook each have a single limber hole, and the floor timbers of frames 4 to 26 each have two, above the transition from the keel plank to the garboards, while the straight floor timbers in 8-22 have additional limber holes also above the middle of the B strakes in the bottom. These limber holes are about 5-6 cm high, and about 11 cm wide; they have an inverted U shape with the sides somewhat everted. In some V-shaped floor timbers they are triangular.

The floor timbers 10-24 to starboard either extended to the join of bilge strake C and strake D, or ended above the middle of strake D; to port they ended above strakes B and C, floor timber 15 being the sole exception. The scarf joining it with the lower futtock there is on the port side above strake E. The adjacent floor timbers both to port and starboard are barely if at all staggered in length with respect to each other, but generally end at the same level, where they have a flat scarf with the lower futtock. This creates a structural weakness at the transition from the bottom to the sides, which is all the more remarkable given the great attention given to the composition of the bottom. Maybe the shipbuilders ran out of suitable longer floor timbers. The V-shaped floor timbers in the afterbody and forebody in general are asymmetrical and rarely cover more than the garboards (A) and the immediately adjacent strakes (B). Just here and there, and generally alternating between port and starboard, does the longer arm extend to strake C. The V-shaped floor timbers of frames 0 to -9 and 26 to 36 (and beyond) are not fixed to the rising ends of the stern- and stem hook or to the stern- and stempost. The somewhat less massive lower futtocks are fixed to the ends of the heavy floor timbers. The scarfs between the lower and upper futtocks between -2 and 31 sit above strakes G, H or I and are almost always staggered. This shows greater consideration for the strength of the construction. In as far as observed, the bow and stern floor timbers (frames -3 to -8 and 32 to 39) seem to have had just single futtocks on both sides.

The missing V-shaped floor timbers 40 to 42 probably continued right up to the gunwale. In the short upper futtocks with square lower ends (above frames -1, 2, 25½ and 29) are treenail holes with diameters of 5 cm for fixing the upright parts of the knees on top of the protruding crossbeams. At frame 10 on the starboard side the strakes J, K and L also were found to have treenail holes with this diameter (Section 6.2.2). In some (thin) ends of frame timbers also nails instead of treenails might be used to secure the scarfs, to join up frame parts. In frame 0, at starboard strake L, there is a horizontal groove 4 cm wide and 1.5-2 cm deep, most probably a rabbet to accommodate the outer plank of a small deck between the aftermost two crossbeams. In frame 26, just above strake F and below the fifth ceiling plank, a spacious triangular rabbet was cut which continued right to the shell and in which sat a remnant of a ca 4 cm thick plank. Initially it was thought that this held a transverse timber of the hull, doubling as a support for a platform, of which the plank remnant was a part. In reconstructing the cross-section of the hull it became clear that the rabbet must have held a vertical support of a forecastle (Section 6.2.2 and Plates 6.IV A and B and 6.V A and B).

Keelson with mast step

Above the keel plank and on the floor timbers (4 to 26) lies a massive keelson with a broadened mast-step area (fig. 6.25). The timber is 9.12 m long, has a maximum width (at the mast step) of 40 cm and a maximum thick-

Fig. 6.24. The V-shaped and straight floor timbers and lowermost futtocks on the bottom and the port side, viewed from the bow. The V-shaped floor timbers and futtocks of the forebody and the starboard shell have already been removed.

ness of 22 cm. It was made from a gnarled tree trunk. The lowermost part of the trunk, i.e. the foremost 5 m of the keelson including the widened mast-step part, is roughly rectangular in section. The after part too is rectangular, but still displays more of the (squared) shape of the tree trunk, with its natural curve remaining recognizable. The broadest part, the mast step with the hole to take the heel of the mast, is 2 cm higher than the forward part of the keelson. The upper surface of the keelson from there runs in a straight line to about 2.85 m aft of the hole. There we see a clear distinction from the after part of the keelson, which starts here 1.5 cm above the foregoing part. Taking into account on the shape of the trunk, its lower surface where it crosses the floor timbers has been somewhat cut away in sixteen places so that it partially engages with these frame timbers. The keelson was fixed to most of the frames with a single, oblique, blind treenail (diameter 2.5 cm). Some treenails penetrate (in)to the keel plank.

The ca 2.25 m long and 40 cm wide mast-step part contains the hole for the heel of the mast. The dimensions of the hole are: length 48 cm, width 21 cm, depth 13 cm. A small, 6 cm thick plank was (at some later time) placed into the hole aft of the mast, and this was probably also the case forward of the mast. The latter may have served to jam the mast fast into the hole. Probably the ship initially had a mast with a heel of about 42x21 cm; during its final years maybe one of about 30x21 cm. In the bottom of the hole a cross was carved out (fig. 6.26), to starboard connecting to a drilled hole, which opened above the bottom just forward of frame 18. The carved cross and the drilled hole served to drain away (rain)water, so as to stop the base of the mast and the mast step from rotting. Near the little plank, the top surface beside the hole on both sides shows cutmarks hewn obliquely towards the centre, which indicate the probable earlier presence of yet another differently-shaped infill or structure. The stepped upper surface of the keelson at 2.85 m aft of the hole for the mast may point to some vertical construction (a bulkhead?) at this spot. On the rear end of the keelson rested a plank of about 4 cm thick, which in a sense formed an extension of the keelson and continued to the rear of V-shaped floor timber 1.

Riders

To take the sidewards pressure of the mast, riders were fixed to the massive floor timbers 17-19 on either side of the mast-step part of the keelson (fig. 6.27 and Plates 6.III C and 6.IV B). These 'extra' floor timbers lie with their upper surfaces flush with the keelson and run (almost) horizontally to the outer edge of the bilge ceiling strake (B). On the underside they are notched so as to engage with the first ceiling plank and the bilge ceiling strake. They are fixed with three treenails each (diameters

Fig. 6.25. The irregularly shaped keelson with the broader mast-step part.

Fig. 6.26. The hole for the heel of the mast with the carved cross, the drainhole for water, the small plank set into the hole to the rear, and the hewing traces belonging to an earlier construction at the spot and the hewn rabbet for a small plank in the front of the hole.

2.5 cm), most of them blind, to the floor timbers, with a few extending down to or into the bottom strakes. The ends of some of the knee riders are fixed with a nail onto the bilge ceiling strake and/or the keelson, probably to hold them in place during the boring of the holes and insertion of the treenails.

Mast support?

Beside the mast step, to port, a knee-shaped piece of conifer wood (130, fig. 6.28 and Plate 6.I) was found on the roundwood floor (Section 6.5.3). The knee, roughly squared on all sides, forms an angle of almost 90°. The short arm is 45 cm long, 8.5 cm wide and 5.5 cm thick. The longer arm is 72 cm long and 8 cm wide and thick. In the short arm are two nails with heads 2.5 cm across. The timber may have been attached to the keelson, immediately aft of the hole for the heel of the mast, where it may have served as a support to which the bottom end of the mast could have been tied for holding down the mast in its hole. However, no attachment points have been observed on the keelson. A somewhat similar knee was found in its original position in the Nijkerk-I cog (see section 8.2.2). Alternatively, the knee may have served to support some kind of partition (Section 6.2.2). As already mentioned in section 6.1.4, nothing survived of the mast or other spars.

Ceiling and roundwood floor, ceiling up to the crossbeams

The vessel had a ceiling consisting of six strakes (A-F) on either side of the keelson and continuing to the six lowermost protruding crossbeams (figs 6.8 and 6.27, Plate 6.III C). The ceiling planks were not placed side-by-side, but with a space of about a plank's width. Such open ceilings generally occur (only) in late-medieval vessels and principally should be regarded as parts of the longitudinal frame of the ship and not as a covering of the frame to - for instance - hold a cargo. The bilge ceiling strakes (B and C) are more massive, and ran between floor timbers 9 and 24, though some may have been longer (the ends of the ceiling planks to port had been destroyed by the digging of a reclamation ditch; starboard ceiling plank C was missing altogether). Together with the keelson they form the principal parts of the longitudinal frame and ensure the internal rigidity of the bottom. The thick B and C ceiling planks lie above the turn of the bilge and the base of the sides, where they were placed to counter the weakness caused by the more-or-less aligned scarfs between the floor timbers and the lower futtocks. Such massive ceiling planks are also known as 'lateral keelsons.' They are about 6.5 m long, 25-30 m wide and 11-12 cm thick, and on their underside have deep or shallow notches to accommodate the transverse timbers in the same way as the keelson does. In putting two of these heavy bilge ceiling strakes into both sides, the shipwright clearly did acknowledge the vulnerability of the turn of the bilge, while in positioning of the scarfs joining the floor timbers to the lower futtocks he seems to have disregarded this problem. In the middle of the bottom, between the keelson and the bilge ceiling strakes, there are thinner ceiling planks (A strakes), starting between frames 6 and 7, and continuing to frame 26. They are a good 8 m long; about 30 cm wide; and 5-7 cm thick.

Fig. 6.27. The broader mast-step area of the keelson with the hole for the heel of the mast, the buttress timbers overlying the floor timbers, which are set across the A ceiling strakes and continue onto the B ceiling boards.

Both the first ceiling planks beside the keelson and the bilge ceiling strakes had been extended fore and aft with shorter pieces, up to 30-70 cm short of the stern and stem. This probably went for all ceiling planks. The observed flat scarfs indicate that the ceiling strakes were composed of three to six planks. With the exception of the fifth ceiling strake (E), which was 35-40 cm wide, the others are no more than 30 cm wide. The foremost and after planks in some cases are just 20 cm wide. The narrow planks usually were fixed to each of the underlying frame timbers with single treenails, sometimes two (diameter 2.5 cm). The ceiling planks were pegged in a single go, together with the shell strakes and frame timbers. The A ceiling planks immediately beside the keelson and the fourth and fifth planks (D and E) usually are secured with two treenails per frame, but sometimes with just one. The only recovered fragments of the sixth ceiling plank (F), with single ones. There where the plank insufficiently engaged with the floor timber, also nails were used. These nails may initially have served to hold the planks in place while the treenail holes were drilled and the treenails

Fig. 6.28. The detached mast support[?] encountered beside the rear of the hole for the heel of the mast, on the roundwood floor aft of the mast.

Fig. 6.29. The mast support[?] on the cleft stems of the roundwood floor to port.

inserted. The massive B bilge ceiling strakes despite their important function were fixed with just one treenail per frame member. The equally massive C ceiling planks usually with two, sometimes one. Remarkable about these heavy bilge ceiling strakes also are the thick nails, which were countersunk in carved-out cavities on the edges of the planks in some places: four on the port side and six on the starboard side (also Section 6.4.1). These nails too may initially have served to fix the planks before the treenail holes were bored.

Roundwood floor

In the course of the excavation a curious 'floor' appeared immediately aft of the mast. It was about 2.8 m long and covered the full width (a good 3 m) of the bottom, up to and including the B bilge ceiling strakes (fig. 6.28 and Plate 6.I). The 'floor' was constructed as follows: between and on the frame timbers had been laid a large number of thin stems and branches of probably Alder (*Alnus* sp.) and Oak, and thin conifer logs cleft in three or four (probably Scots pine, *Pinus sylvestris*). The alder branches, with a diameter of 5-6 cm, on which in many cases the bark was still present, differed in length (0.5-2.5 m); the cleft logs, with an original diameter of about 20 cm, generally were no longer than a metre. They all lay transversely in the ship, and displayed clear traces of hewing and splitting (133-135; figs 6.28, 6.29 and 6.30). On the starboard side on the front of these branches and stakes lay a piece of woven willow withies (*Salix*), of about 80x80 cm, maybe part of a discarded basket or the like. On top of this, and on the rest of the roundwood floor, was a layer of straw and on top of this a layer of (coarse) sand. The step in the after surface of the keelson, rising by 1.5 cm at 2.85 m

Fig. 6.30. Parts of some of the preserved, cleft and cut saplings and branches of the roundwood floor.

from the mast (as mentioned in section 6.2.2), does not coincide with a crossbeam at deck level, but is about 50 cm forward of crossbeam 3. The roundwood floor starts quite abruptly aft of the hole for the mast heel, so that possibly, and also in view of the recovered knee of conifer wood (probably silver fir, *Abies alba*), there may have been a partition here too. So there may have been a separate (accommodation) area (about 2.85x3m) with a roundwood floor just aft of the mast.

Ceiling above the crossbeams

On the starboard shell of the forebody was a seventh ceiling plank (strake G). it is in fact a beam (a shelf clamb), 3.3 m long, 12-14 cm wide and 4-8 cm thick. It had been fixed to the futtocks from frame 28 to beyond 34, with just three treenails and two nails. The beam had been joggled where it crossed some of the futtocks, so that it engaged with them. It is quite puzzling that this timber almost entirely obstructed the hole in the shell for the foremost protruding crossbeam. The crossbeam and the shelf clamb cannot possibly have been part of the ship simultaneously. This shelf clamb must have been fitted after the foremost crossbeam (no 6) had been removed (see section 6.1.4). A somewhat similar shelf clamb was found detached, overlying the starboard shell in the afterbody (GG in Plate 6.I). It is curved and 4.85 m long, 18 cm wide and 8-10 cm thick. It too has rectangular notches in its fore and aft ends laterally where it crossed the futtocks. Yet this beam had been fixed in place by eleven treenails (ø 2.5 cm) and four nails, of which one obliquely. Its shape suggests that it was an extension piece of a B or C ceiling timber. One of the treenail holes, roughly midway along the beam, has a deviating, oval shape measuring 4.5x5.5 cm. Maybe the treenail had been the last by which the beam remained attached to the hull, and the peghole was worn wider by the to-and-fro motion of the wreckage. Overlying the shell strakes K and L of the starboard shell in the afterbody, fragments were found of two ceiling strakes (H and I) on the futtocks (Plate 6.I). The notches in the outer face of the upstanding knees which sat on top of the protruding crossbeams (Section 6.2.2, figs 6.31 and 6.32, and Plate 6.IV B) indicate that above the protruding crossbeams as many as three ceiling planks (H-J) had been fixed. The notches in the top ends of these knees suggest that there had been a fourth (thick) plank or inwale along the gunwale. In all, therefore, the ship probably had ten ceiling strakes on either side of the keelson (A-J), including the inwale, but not counting shelf clamb G.

Crossbeams with knees, beam-head guides, stringers, deck(s), windlasses, pump and drain/latrine, holes for crossbeams with knees and (other transverse) connecting beams

The transverse rigidity of the ship at deck level was ensured by a series of transverse beams that protruded through relatively large, square openings in the shell. Some of these holes and the function of the crossbeams have been mentioned earlier in section 6.2.2. These openings, as well as the square-cut top of frame 2 to starboard and the end of a complete crossbeam with knees found *in situ*, indicate that protruding crossbeams were located at any rate at -1, 2, 25½ and 29 (figs 6.19 and 6.20, Plate 6.I). Treenails with diameters of 4-5 cm were observed only in the knees set atop the crossbeams to connect them with the futtocks and the shell strakes of the upper part of the side. Such treenails were also found in strakes K and L at the spot of 10 to starboard, which suggests a fifth cross-

Fig. 6.31a. The rear of crossbeam 1 *in situ* with the sawn-off head of the knee.

Fig. 6.31b. The front of the sawn-off starboard knee and the end of the crossbeam with the oblique rabbets for the shell planks.

beam. If we assume that the crossbeams in the midships section were more-or-less regularly spaced, this implies that there must have been a sixth crossbeam as well, at 17. The relevant shell parts were no longer present. The holes in the shell for the crossbeams in the afterbody (at -1 and 2) are in strake K; the others are a strake lower, in strake J.

It has already been remarked that the square top of frame 25 and the edges of a square opening in strake L immediately above it also indicate a crossbeam. The hole would have been about 20x20 cm. Also it seems obvious that there would have been a crossbeam immediately before the mast. The shell planks here are missing, but it is almost certain that such a crossbeam passed through strake L above 19 or between 18 and 19. Given the numbering of timbers from stern to bow and from below upwards, the sail beam must be crossbeam 7, and the beam above 25 must be crossbeam 8. There are treenail holes with diameters between 3.5 and 4 cm in strakes J and K at frames -5 and -6 in the stern; those at -6 may in fact relate to the attachment of standing knees of an (in this position necessary) crossbeam-and-knee construction, be it one not protruding through the shell. Here probably was a bit as well. It cannot be ascertained whether this beam sat beneath or above the platform - probably the former. The treenail holes in -5 almost certainly relate to the attachment of the after stanchions of the aftercastle which were probably interconnected across the top of the after deck (Section 6.2.2 and Plate 6.IV B). The gap in a detached ceiling plank (ii, Plate 6.I), which very probably comes from the port shell in the afterbody, once enclosed a crossbeam, probably crossbeam 1. Although no 5 cm thick treenails have been recorded in or forward of frame 29 in the bow, it may be assumed that some support was needed there for a deck. Below in this section we shall return to this matter.

Given the distance to the bow, there must have been another transverse connection forward of 29 (the location of the foremost documented protruding crossbeam) - maybe one similar to that in the bow of the Nijkerk-I cog (Section 8.2.2). The only potential evidence of this is a notch in the top of the C ceiling plank above 31 and some extra treenails in the lower futtock of 31 on the starboard side. But more probably we are in fact dealing with the attachment of the foremost stanchion of the forecastle (Section 6.2.2). A ninth protruding crossbeam does seem a possibility above the squared top of 35. The beam would then have rested on the top of the L strakes and the frame heads with just a rabbet, and would have protruded through the (later cut?) opening in the shell (Sections 6.2.2 and 6.3.3 and Plates 6.IV A and B, and 6.V B and C). In the A and B ceiling timbers above 28 to starboard too, some extra treenails were observed (Plate 6.I). There is nothing to suggest that this vessel ever had so-called 'side stems' on either side of the stempost, fixed on the inside across the ceiling and frame timbers.

Crossbeam 1

The complete aftmost transverse connection, comprising crossbeam 1 and two knees, was found to starboard of the midship section, roughly longitudinally to the ship (fig. 6.31 and Plates 6.I and 6.VIII). It was a crossbeam protruding through the shell on either side, carrying standing knees both on the starboard and the port side. These knees were joined together by a vertical flat scarf. The heads of the knees did not protrude above the gunwale. Crossbeam 1 has a maximum length of 4.7 m, is 20x20 cm across, and at 10 cm from both ends has a rab-

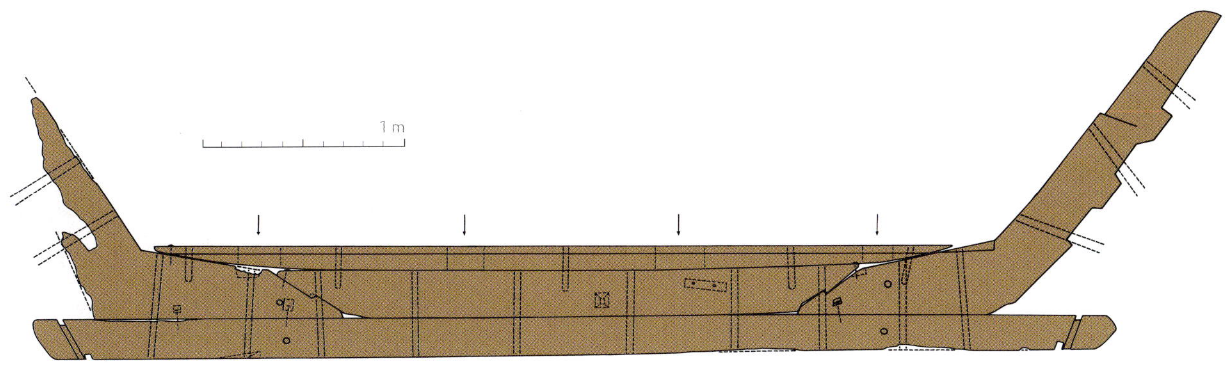

Fig. 6.32 The rear view of crossbeam construction 2 with the location of the four stringers (A to D).

bet around its upper surface and sides, 3-4 cm wide and 3 cm deep. The rabbets were made in the sides where the beam pierced the shell. On its lower surface is a similar rabbet, yet at a plank's width further inward, owing to the clinker-style construction of the shell. This bottom rabbet engages with the top of strake J, while the beam itself emerges through strake K. The remaining core of the beam, i.e. the size of the holes in the shell into which it was fitted, measures about 16x16 cm on one end of the beam and 18x18 cm on the other end. These dimensions almost perfectly match those of the gaps in the K strakes at frame -1. The smaller hole is to starboard. Crossbeam 1 was recovered with its aft side forward, probably having drifted forward at the time when the port shell became detached in the afterbody. Laths had been nailed to the knees both on the front and rear. The forward lath still is almost complete. It consists of two pieces, with a combined length of 4.2 m, their height ca 4.5 cm, their thickness 3 cm. The lath was applied about 5 cm below the top of the horizontal part of the knees, and fixed with fifteen nails.

In the final reconstruction, a detached lath (NN) found in the afterbody could yet be placed. This lath, 7 cm high and 4 cm thick, had been fixed about 5 cm below the one mentioned above, where it served to support deck planks between crossbeams 1 and 2. The laths had held the deck-plank ends down and in their place. Of the lath on the rear of the knees, just two fragments were found, on the port knee, at about 14 cm below the top. A well-finished flat scarf at the end of one of the pieces shows that this plank too had consisted of at least two parts. Originally these planks, which together extended at least to the starboard knee, had been attached to the rear of the knee by six or seven nails. No nail holes were observed in the starboard knee, maybe because the wood was too degraded. Yet it may be assumed that this lath, like the one on the front of the knee, had extended across the full width of the ship. The lath on the rear of the knee had served to hold in place the planking of the small after platform, that rested on the crossbeam. No nail holes were observed in the upper surface of these laths, that might point to a connection with (deck) planks. Curious and unexplained is the absence of the uppermost 20-25 cm of the starboard knee. Its head had been sawn off at the place of a treenail connecting it with the upper futtock of frame -1 and strake N. Another point of interest is the wide space between the outward side of the knees and the inside of the shell. We shall return to this in section 6.3.2.

Crossbeam 2

The second full transverse connection, consisting of crossbeam 2 with a standing knee on both ends, the two joined by a connecting beam, and a rider beam covering the horizontal surfaces, lay to starboard at the very spot where it had protruded through the shell. The crossbeam was without doubt still connected with the shell when in the process of wreck formation the top part of the starboard shell of the midship section and stern was pushed 4 m forward (Section 6.1.4, figs 6.3, 6.6 and 6.32 and Plate 6.I). The starboard lower futtock of frame 2 is finished square on top and thus indicates where (reckoned from the stern) the second protruding crossbeam had been positioned. Yet the necessary square opening to accommodate it in strake K is lacking, having been dug through by a reclamation ditch. The entire crossbeam construction had toppled forward and was found *in situ*, with the after side facing up. Crossbeam 2 has a length of 5.6 m and in section is about 20x20 cm. At 17 cm from either end, as in crossbeam 1 a rabbet was cut in the upper and lateral surfaces, 3.5 cm wide and 1-4 cm deep (Plate 3.IV B). At the underside of the beam is a similar rabbet, 3 cm deep, a plank's width further inward. This rabbet engaged with the top of shell strake J, while the sides of the square hole in strake K fitted into the top and lateral rabbets (also Section 6.4.1). The gaps in the shell for this crossbeam must have been at least 14x14 cm.

In the underside of the crossbeam, at about 80 cm from the shell, an oblique rabbet was present: evidence of a stanchion? On the upper surface of the beam, starting at about 35 cm from either end, are two heavy knees. The horizontal parts of these are 25-35 cm high and 13-15 cm wide. They each were fixed to the underlying beam by three 4-cm-diameter treenails and some massive and countersunk nails. Three of these thicker treenails also fastened the vertical parts of these knees to the upper shell, right through the ceiling planks and futtocks. Between the knees, joined by a hooked scarf, a connecting beam was fixed: 3.07 m long, 25 cm high and 15 cm thick. This was fastened to the knees and the crossbeam by six 3-cm-thick treenails. On the upper surface of the horizontal part of the knees and the intervening beam, a rider beam was mounted with five or six blind, 3-cm-thick treenails. This beam at the top of its rear side has a rabbet 6 cm wide and 4 cm high, which to starboard runs on into the knee. In the rabbet lay the foremost ends of the planks of a small, slightly elevated *'tweendeck'*, that extended to crossbeam 1. In the rabbet no nail holes were observed that might have secured the planking. On the rear face of the crossbeam and the connecting beam, to port, several nail holes were observed, which might relate to a vertical partition in the hold. On the front of the connecting beam between the two knees, to starboard, is a clamp, fixed at a slight angle with two nails. Slightly off-centre on the front of this beam is a countersunk nail.

On the front of the rider beam four rabbets can be observed. These rabbets are 3 cm deep at the top, and towards the underside of the beam become 7 cm deep, so their rear surface slopes downwards and inwards. The middle two are 24 and 19 cm wide with an interspace of 85 cm, the outer two have widths of 15 and 20 cm. The distance between the latter two and the outer edges of the two middle rabbets is about 80 cm. These four bevelled rabbets were without doubt created to accommodate stringers, which rested on the knees and the connecting beam between them, and thence extended forward.
The rabbets hold no treenails or holes of massive iron nails, so we must assume that the ends of the stringers rested loose upon the connecting beam and were held in place by the rider beam, which overlay them. The outward sides of the two knees were joggled to fit the ceiling planks and the inwale. The space from the shell (about 35 cm) is very large to accommodate a futtock. We shall return to this in section 6.3.2. The knees extended upward to the top of strake O. On the back of the vertical part of the port knee there was a bevelled notch, about 1 m above the crossbeam, which is a fairly obvious place to put a belaying rack: between the knees of crossbeams 1 and 2. Both to port and starboard, in the crossbeam as well as in the knees, treenails 3 cm in diameter were inserted one above the other and in a longitudinal direction. Surely these must have secured the foremost pair of stanchions supporting the aftercastle (Section 6.2.2).

Crossbeams 3 to 8 (or 9)
Of course little can be said with certainty about the lost crossbeams 3 to 8 (or 9). All we know is that the square gaps in the starboard shell for crossbeams 5, 6 and 8 measured about 23x23 cm, 20x20 cm and 20 x20 cm, respectively. With rabbets 3 cm deep for engaging with the shell planks, these beams must have been about 29x29 and 26x26 cm in diameter. For crossbeams 3, 4 and 7 we may assume similar dimensions. A clue to the location of crossbeam 7 is the extra futtock (17½) between frames 17 and 18 on the port side. Certainly there will have been standing knees (with interconnecting beams) on crossbeams 3 to 6, witness the 4-5 cm thick treenails in the surviving frame timbers and shell planks. Whether this was also the case with crossbeams 7 and 8, inserted at a higher level, is impossible to tell, though it is quite likely. The knees must have been fixed onto the deck. Starboard strake L has survived between frames 3 and 15. Along this stretch, there is no evidence of a third crossbeam at this level, which implies that crossbeam 3 (passing through strake J at 10) with knees, a connecting beam and possibly a rider beam, must have been elevated by about 1 m. This was necessary to support the stringers. Whether the described constructions on crossbeams 4 and 5 were elevated to meet the underside of the stringers, despite the presence of crossbeams 7 and 8 is hard to tell, but seems quite likely. The distance between crossbeam 6 and the bow does suggest the original presence of a ninth crossbeam, at frame 35 (Section 6.3.2).

Guide
Under the starboard forebody of the ship, in the redeposited sediment, appeared a piece of wood, triangular in section and 95 cm long. Two of its three sides are flat and straight at one end and measure 21 and about 18 cm in width, respectively; the third side is convex. The other end tapers to a point about 2 cm thick (fig. 6.33). The broadest, convex side holds two nails; roughly midway along the length is a third. A small part of the broadest end had been finished flat on the convex side. The wood species was not identified, but it certainly is not oak wood; its red colour suggests that it might be beech wood. During the construction of the Kampen Cog, Mr A.J. van den Heuvel recognised it as a 'guide', a piece of wood tapering forward into a point, to be fixed to the exterior of the hull with its thicker end up against the front of a protruding crossbeam (probably number 5) to protect the latter. Its function was to stop the protruding end from getting caught, for instance during berthing manoeuvres.

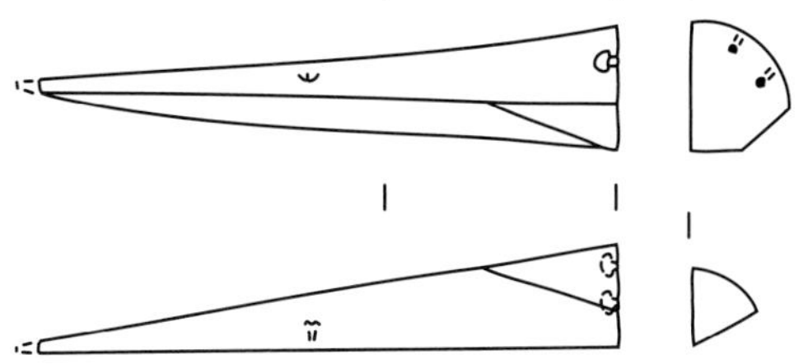

Fig. 6.33. The beam-head guide found detached under the forebody.

Such beam-head guides have also been found in the 15th-century, fully clinker-built wrecks in the Aber-Wrac'h river in Brittany (France)[51] and at Vedby Hage in the Storstrømmen between Sjæland and Falster in Denmark,[52] while they are known also from the Norwegian cog find 'Ny Hellesund'.[53] In the Doel-I cog, various specimens were found *in situ* in 2000. The shape of these guides can be recognised on the hulls of vessels depicted on many medieval seals. Yet until the recovery of the Bremen cog it was unclear what they represented, since it had not been understood that the beams projected through the shell. In this instance, archaeology proved able to provide the answer. Some good examples of guides shown on seals are the second seal of Elblag (Poland) of ca 1350 (a cog), and the English seals of Rye (ca 1400), Sandwich (13th century) and Southampton (second seal, 13th century), which show what probably are fully clinker-built vessels, '*nefs*'. (fig. 6.34).[54]

Fig. 6.34. Civic seals of Eblag (= Elbing, Poland), ca 1350; Rye, ca 1400; Sandwich, 13th century; and Southampton, 13th century (3 x England), which show such beam-head guides. After Ewe 1972.

51. L'Hour & Veyrat 1994, 174-176.
52. Gøthche & Myrhøj 1996, 14 en 15.
53. Line Dokkedal, pers. comm. 2000.
54. Ewe 1972, 122 no. 42, 185 no. 168, 186 no. 170 and 191 no. 180.

Stringers and deck(s)

No beams or planks were recovered that could be attributed with certainty to a deck or decks. For a long time it was thought that the ship might just have had platforms fore and aft, maybe connected by thick planks along the gunwales, with stringers providing extra longitudinal stiffness. The Kampen Cog was provided with a continuous fore-to-aft deck (see Chapter 7). Evidence for an after deck are the laths fixed to the front and rear faces of crossbeam 1, the rabbet in frame 0 to starboard, and the rabbet on the rear of crossbeam 2. The ca 1.5 m long platform for the helmsman, between the stern and crossbeam 1, sat ca 10 cm lower than the small deck between crossbeams 1 and 2. The latter small deck, just 70 cm long, most probably accommodated the great windlass for hoisting and lowering the square sail. The lower afterdeck rested on a V-shaped floor timber at its rear end, and forward was supported by crossbeam 1. The planks of the higher after platform rested on a lath fixed to the front of crossbeam 1 and were held in a rabbet on the rear side of the rider beam on crossbeam 2.

Remarkably, the decking planks, probably placed longitudinally, seem not to have been fixed in place but were simply held down with laths on the front and rear of crossbeam 1. The openings for four stringers in the front of the rider beam on crossbeam 2 are the only recovered evidence of a midships construction at that level. Given the similar elevation with respect to the keel line of these stringers and the top of crossbeam 8, we may assume that this vessel during the final phase of its working life had four stringers, about 9.6 m long, fitted between crossbeams 2 and 8. As mentioned above, these stringers must have been supported by the knees, connecting beams, rider beams and crossbeams. The deck was almost horizontal or followed the sheer of the ship. Judging by the non-fixing of the ends of the stringers to crossbeam 2, it seems quite plausible and maybe more realistic that shorter stringers were installed between crossbeams 2 and 3 and between 3 upto 7. Some of these, especially those between crossbeams 2 and 3, may have been removable. This might well have been the case also between crossbeams 7 and 8.

Further, it seems almost certain that the ship had hatches between the midship stringers between crossbeams 2 and 3 and between 7 and 8. In the reconstruction drawings the stringers are shown as if they were all removable (Plates 6.IV A and 6.V B). A transverse beam (sail/mast beam) through strake L at 19 would seem a logical feature, though no concrete evidence was encountered. Having two crossbeams at different levels and close together (17 and 19) is not unlike the documented situation further forward (25 and 25½). Between the stringers, and between the outer stringers and the shell, very probably (removable) transverse decking planks were placed. Whether this midship deck continued forward from crossbeam 8 with the same sheer cannot be ascertained, but it is to be expected. Quite remarkable is the total absence of scuppers in the shell at deck level. This means that all rainwater would have to be pumped overboard. The shelf beam G can only have been inserted after crossbeam 6 was removed. Presumably its sole purpose was to somewhat remedy the thereby weakened construction.

Windlasses

On the after deck, probably between and against the inside of the foremost stanchions of the aftercastle, there must have been a massive windlass for hoisting and lowering the spar with the square sail; its drum must have been about 2.5 m long. For operating the anchor(s) there must have been a windlass or anchor bitt on the foredeck as well. This is fairly certain to have stood on and againstg crossbeam construction 6. Still it is imaginable that it was constructed against crossbeam 9.

Bilge pump

The vessel without doubt will have been provided with one or two (fixed) pumps, but no wooden pump tube was recovered. Yet about 2 m aft of the mast, under the hull, a round, oakwood 'lid', 3 cm thick, was found, with an original diameter of 16.5 cm (figs 6.35, 6.53-116 and Plates 6.IV A and B and 6.V B). It has two opposing lugs, 2 cm wide. The length of the lugs is unknown because they have both broken off, but this would have been 1-4 cm. At the centre is a hole 3.4 cm across, with around it, set in a square, four small holes with a diameter of 1.2 cm. One of these holes is closed up with an octagonal wooden peg, jammed in with a wedge. Presumably this is the 'lid' of the top of a wooden pump tube. The central hole would have been for the pump rod. The two lugs probably served for its attachment with nails into the top

Fig. 6.35. "Lid" of a pump barrel? Reconstruction of a pump barrel and lid.

of the tube. The purpose of the four small holes is unclear. Maybe they served as a vent when the pump rod was pulled up. Section 6.2.2 discusses a thin-worn area that was observed in bottom strake C (and B) between frames 7 and 8. The worn area may have been caused by the foot of the pump having rested here. The most likely place for a pump on the bottom however is the space between floor timbers 2 and 4, just forward of the higher horizontal part of the stern hook. There never was an 3 floor timber. At the equivalent place forward, on the stem hook, there was not, nor had there been, any v-shaped floor timber 27; so this may have been the location of a second pump tube.

Drain

A curious elongated and trapezoid 'box' with a rectangular section (Plates 6.IV A and B and 6.V B) without doubt also relates to the pump. It was found detached in the afterbody, aft of the keelson (Plate 6.I). The box is 1.75 m long, at one end 45 cm wide and 25 cm high; at the other, 8 cm wide and 7 cm high (fig. 6.36). It was made of planks 3 cm thick. The top, under and rear planks are of oak; the sides are of conifer wood. The broader end plank has a gap in the top, 35 cm long, while the sides at this spot remain at the same height (25 cm). The side planks each are attached to the bottom plank with twelve nails, the top plank with ten nails on either side in the top edges of the side planks. The rear plank is attached from below with three nails, and with two nails into the side planks. The top edge of this plank displays a concave, worn area midway along its length. In the bottom plank there are on one side three blind treenail holes with a diameter of 1.5 cm, one of them still holding a square piece of peg. Above it in the side of the 'box' sit two nails with flat heads (ca 25 mm across) and an original length of ca 100 mm. The two seams between the bottom plank and the sides are somewhat V-shaped and caulked with moss. The narrow end of the box is open: a gap of 4x3 cm. Probably this was the receptacle into which the water from the pump would be spilled and from which it was released overboard. The opening in the top of the 'box' and the worn area on the upper edge of the rear plank suggest that the drain was also used by the crew for relieving themselves.

In section 6.2.2 it was already surmised that the 10x10 cm hole in strake J to starboard between frames 2 and 3 might have been the hole through which the drain protruded; at closer consideration, this seems as good as certain. This would mean that the drain had been fixed against or just forward of crossbeam 2. On the front of this crossbeam, to starboard, there is a clamp set at a slight angle. The distance from the side of this clamp to the exterior of the shell at the hole in strake J is 1.74 m. The thin end of the 1.75-m-long drain would then have protruded from the hull by 1 cm.

This angled clamp may thus have held a small plank that served as a foot support in front of the latrine. The above-mentioned worn area in the bottom is about 2 m forward of this, which makes a direct link between the drain and the pump not very likely in this construction. A pump tube between frames 2 and 4 would in the longitudinal direction end almost up against the drain, but would have stood leaning towards the gunwale to have the bilge water emptying directly into the drain. If the pump tube stood vertically, there must have been a simple gut-

Fig. 6.36a. The complete drain/latrine *in situ* above the afterbody.

4Fig. 6.36b. The drain beside the excavation area.

ter towards the drain. Yet in fact a very different situation and function may have prevailed. The available evidence does not allow a better explanation. Still we may assume that the square opening in the top was (also) used as a latrine. A crew member could in the envisaged situation be able to sit on the wear-hollowed edge of the rear wall of the drain, with his feet resting on a small plank fixed below between two stringers. A somewhat similar drain was encountered in the late-14th-century cog of Bremen, which however had a separate, almost square receptacle, while the vessel also had a proper latrine.[55] among the sandy sediment in the bilge, a wooden lid was found that might have served to close the 10x10cm scuttle in the shell when the drain was not in use (Section 6.5.3 and fig. 53-114).

Castles, the aftercastle
In the first instance, no evidence was encountered of a fore- or aftercastle. Yet in crossbeam 2 and in the knees attached to it, on either side of the ship, two treenails can be seen one above the other. These treenails, with a diameter of 3 cm, are in the longitudinal direction of the ship, and directly above the top of strake F. They must have served to fasten the forward pair of stanchions that supported an aftercastle. The bases of these stanchions must have been attached to frame -1. The part of this frame found to starboard was badly damaged, while the corresponding part to port was never found. Therefore it is unclear whether the posts stood in a rabbet and/or were treenailed to the frame. The posts had been set against the rear of crossbeam 2 (on the front were stringers). The inward faces of the two stanchions may have supported the bitts of the windlass for operating the square sail. The thick treenails in -5, as noted before, point to the attachment of the after stanchions of the castle. They were probably interconnected by a deck beam that ran across the afterdeck. They would have stood about 2.4 m above the small afterdeck. The deck of the aftercastle would have been about 3.2 m long, with a width of 3.65 m (aft) and 4.5 m (forward) respectively, at about 5.7 m above the keel line. Plates 6.IV A and B and 6.V B show what the castle may have looked like.

Forecastle
At the transition of starboard strakes F and G a hewn-out gap is to be seen in frame 26. During the final reconstruction, it was concluded that this gap was likely to have held a vertical post. The gap was immediately forward of the (missing) crossbeam construction 5. To this beam may have been attached the after stanchions of the forecastle, comparable to the forward supports of the aftercastle. The semicircular repair on the outside of strake G, directly below the position of the stanchion, indicates that the shell had broken here under the pressure of (excessive) weight and poorly judged construction (a V-shaped notch had been cut into the frame right down to the shell). Such a construction would certainly not have been applied in a new-built ship, so we may assume that this was a subsequent alteration. The underside of the forecastle deck was about 6.3 m above the keel line and some 2.65 m above the afterdeck. Presumably the castle had been about 3.2 m long, with a width of 5.6 m (aft) and 4.3 m (forward) (Plates 6.IV A and B, and 6.V B). A good example of what the vessel may have looked like from aside can be seen on the second civic seal of the town of Damme of ca 1300 (fig. 6.37).

Repairs
The repairs observed on the hull can be divided into two groups: one group that fairly certainly were made even during the building of the ship; the other, as later additions. Repairs from the building phase will be closed-up cracks, which usually appear (roughly down the middle,) longitudinally in a plank or in a narrow curve around a knot. They would be cut out open in a V-shape on the exterior of the shell, and closed up with sintel clamped moss caulking. Plate 6.III A shows the caulked splits in the bottom strakes. In the planks of the sides similar repairs were observed in various places. The repairs on the outside of the port shell are marked in plate 6.V.C. Such cracks in the wood might occur in the course of the ship's construction, as the timber dried out and shrank. They are a familiar feature also in other cog finds.
A different or additional method of repair is that of fixing

Fig. 6.37. The seal of Damme (Flanders, Belgium), ca 1300. After Ewe 1972.

55. Lahn 1992, 162-163 and fig. 152.

a thin plank, usually 1-1.5 cm thick, to the interior of the shell. This would be nailed over a crack that had been caulked on the exterior. The most likely explanation of this procedure is that people were not altogether confident about a crack closed with clamped moss caulking being able to withstand the pressure of water from outside, and therefore felt it required further measures. For instance, on the inside of starboard strake H three thin planks were nailed over a caulked-and-sintel clamped crack, which have a joint length of 4.3 m, are 18 cm wide and 1.5 cm thick. Another example is a 2.7 m long and 18 cm wide plank on the inside of strake I in the forebody, below which, however, the crack had not been closed up with sintel clamped moss caulking. Between the planks, nailed to the shell, and the strake is a thin layer of *Sphagnum* moss. Strake F in the port shell has three repair planks (fig. 6.38). Starboard strake J amidships also holds such a plank (fig. 6.39). And aft of and below protruding cross-beam 1, two further small planks sit on the inside of the starboard shell.

Fig. 6.38. The inside of the port shell (without futtocks) displaying various repair planks, viewed from the stern.

With the exception of these two examples, which were affixed in areas between frames, all such planks sit between the shell and the frame. Therefore most must have been put in place during construction, before the frame was built in. It was not established in every case whether there were cracks in the shell underneath these thin planks. The small planks nailed here and there onto the outside of the shell may well have been added while the ship was in use. A second type of repair are the patches inserted into the shell strakes. These are semi-circular in shape, with long bevelled edges. Their seams were closed with sintel clamped moss caulking. Among other places, such repairs were found in starboard strakes H and L and port strakes D, J and K, in the fore- and afterbody, respectively. They seem to be repairs carried out while the ship was in use. The only construction that may have seen any replacement by way of repair, is the stern. In section 6.1.1 it was already noted that this might be the case with the thicker false stern, given the filling-out planks on one side of where it adjoins the true sternpost. Indeed it may not be entirely ruled out that the false stem on the bow was at some time replaced. Besides, it is as good as certain that the starboard shelf clamb (G) in the forebody replaced a previous construction (Section 6.2.2).

Timber use and processing

With the exception of the beam-head guide, all recovered fixed parts of the ship are of oak (*Quercus* sp.). The planks of the shell and the ceiling on the whole are rift-sawn. Cutmarks show that they were squared with an axe. No

Fig. 6.39. Repair plank on the inside of starboard strake J in the midships area.

sapwood was found on the edges of the planks forming the underwater body. As remarked before, much attention had been given to the laying of the bottom. In the clinker planking of the sides and especially the higher parts (above the water line), the shipwright was somewhat less particular about his timber. Indeed there we see some sapwood on the edges. How the planks were used, with the heart turned inward or outward, or alternatingly, was not recorded in the excavation. On the inside of the bottom planks, twisted towards the stern and stem, a thin charred layer was observed beneath the frames only at plank F1, Which might be a (first?) indication that the planks of cogs were furnaced in order to make them amenable to bending. The stern- and stem hooks, the knees on the crossbeams, the keelson and the floor timbers were made from naturally-grown trunks and crooks, hewn into the desired shape with a squaring axe and/or an adze. On these parts, little if any sapwood was encountered. The lower futtocks were hewn from thick limbs or thin trunks, with here and there the circumference of the tree or branch still recognisable, including some sapwood. The generally thinner upper futtocks show that here the builders were even less particular about the ultimate shape and finish.

Fastenings and nail-hole plugs of temporary frames

The frames are fixed to the shell, the keel plank and the stem- and stern hooks with oak treenails (diameter 2.5-3 cm). Generally with two treenails per plank width, and single ones in the case of narrow planks. The outside ends of the treenails are somewhat thicker. In individual planks they are aligned in a somewhat staggered way, creating a kind of zigzag pattern of treenails on the frames and strakes. A rough estimate of the overall number of treenails used in this vessel is about 3000. The treenails used for fixing the crossbeams were 4-5 cm thick. To what extent, or whether at all, any wedges had been driven into the treenails to tighten them was not recorded. Frames 4 to 26 (Plate 6.III A) on the whole were fixed to the keel or hook by single treenails. Frames 8, 14, 15, 17 and 21 had two, and in 12 there were three. The second treenails in frames 14 and 15 are blind ones, which also goes for the third treenail in 12. Besides, single blind treenails of conifer wood with a diameter of 2 cm were used at frame 8, between 10 and 11, between 12 and 13, beneath 13, at 20 and beneath 21. Between 21 and 22 there are a further two, with a diameter of 1 cm, in the scarf joining the keel plank and the stem hook. These thin treenails are nail-hole plugs and served to close up nail holes, particularly those left by small planks nailed on in the process of construction. These would be temporarily fixed transversely across the keel plank and two shell strakes to hold them tightly together while a permanent frame was being inserted nearby. Holes with a diameter of 2 cm may sooner derive from thin treenails. Such holes were found around the future locations of floor timbers 8, 13, 17 and 22. Without doubt these are treenail holes that show where during construction the bottom planks were fixed to temporary floor planks. These temporary floor timbers probably ran across the full width of the bottom. At the scarfs in the garboards it seems that during construction two planks about 0.8 m long had been nailed on, which held the A strakes tightly in place against the keel. For the temporary fixing of the next two bottom strakes (B) probably two planks ca 1.6 m long were used, fastened at the foremost and aftmost scarfs. After most floor timbers had been put in and treenailed, the temporary frame timbers and planks large and small could be removed, their nail holes plugged with pegs to avoid wood rot, and the last of the permanent floor timbers inserted.

The clinker joins between the planks in the sides were fastened with iron nails 110-130 mm long (fig. 6.40). The nails have a more or less round head with a convex upper surface forged into four faces, and a diameter of about 25-30 mm. The average space between the nails is 16 cm. For joining the strake planks of the ship's hull, in all some 5000 nails would have been used. Although no concrete evidence of this was found (or had survived), we may assume that the nails in the clinker-built shell had been driven from the outside inwards into pre-drilled holes. This would stop the plank from splitting along the row of nails. At any rate this precaution seems necessary for the shell planks, and for the frame planks at least advisable. For the underlying timbers, a drill with a smaller diameter may have been used, as the nails tapered towards the point. In the 16th-century painting of a ship under construction (fig. 6.41), two braces are prominently in

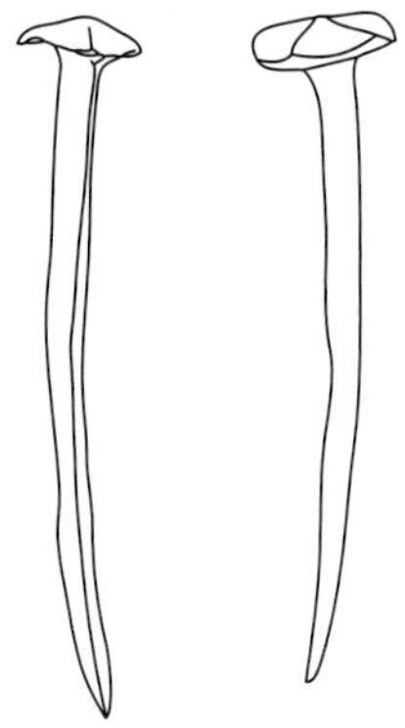

Fig. 6.40. Two nails from the clinker construction of the sides, a so-called 'cog nail' or double-clenched nail.

Fig. 6.41. The fixing with *boeitangen* of a bilge strake in the course of the 'shell-first' construction of a small vessel. Anonymous painting by the 'Master of Gouda', 1565. Stedelijk Museum het Catharina-Gasthuis, Gouda.

view, holding drill bits of possibly different diameters. During the application of the nails, first the utmost tip as it emerged on the other side would be hammered into a right angle, before the nail was fully driven through the planks. After this, the now emerging end would also be hammered sideways, thereby driving the tip back into the wood (fig. 6.42). The result somewhat resembles 'iron stitching'.[56] Such 'cog nails' are also known as double-clenched nails.

For joining the components of the crossbeam constructions, not just treenails but also some thick iron nails had been used. In the fastening of stern-/stem hooks to stern-/stempost, false stern/stem to stern-/stempost and in the bilge deals too, some of these massive nails were used. The iron strips on the underside and sides of the bottom end of the stern; the iron gudgeons; the never recovered rudder fittings; the two top bolts with a ring for suspending the rudder (the upper of which was not recovered); and the bolt through the stempost; these were the only larger and massive iron components and fastenings in the vessel. For fixing the oak laths onto the seams, use was made of sintel clamps of the D type, variant 2 (figs 6.53 and 6.54). They are not large: the roughly oval centre part is just 35-50 mm long. In the vessel 15,000 to 16,000 such clamps were used (Section 6.2.2).

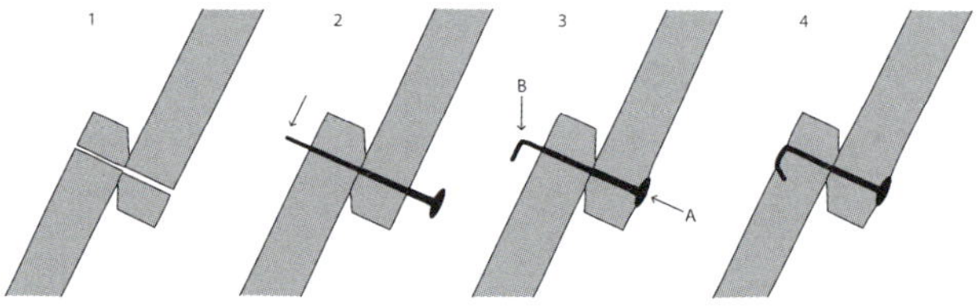

Fig. 6.42. The insertion of double-clenched nails: holes drilled (1); nail partially driven in and its tip beaten sideways(2); nail fully driven in (3A); the tip again beaten sideways (3B); the double-clenched nail (4).

56. Vlierman 1996a, 54.

6.3 Construction data

In the foregoing sections prominent features in the ship's construction and observations were discussed and in part interpreted. On the basis of which various conclusions can be drawn about the ways in which the ship was built and used: the ship's biography. About some phases in its life quite a lot may be deduced with the means available to us, and less about others. The phases in this vessel's life may be listed as follows:

1. Design
2. Construction
3. Use
4. Abandonment and demolition
5. Wreck formation and embedding

6.3.1 Design

We have no design drawings or blueprints or detailed descriptions of the building of cogs, nor is it likely that any ever existed. The surviving medieval images, especially on seals and in illuminated manuscripts, do provide some general data about the shape of the hull above water, about any fore- and aftercastles or both, about the rudder, about the mast and the rigging, and about the square sail. Also they may show various details. Until a few decades ago, it was unclear how reliable the evidence of these medieval images was - especially regarding the proportions (e.g. the ship in relation to the depicted crew), and the degree to which the seal maker had had to make allowances for the small round seal and the limited space into which to compress the ship's hull. Certain details shown on the hull were a mystery as to what they represented (the projecting crossbeams and/or guides). Section 4.1 presents the definition of a cog as formulated by P. Heinsius on the basis of iconographic evidence.[57] The cog wrecks found and investigated since then generally support this definition but also present variations. The wrecks in fact yielded a great deal of new, unknown evidence about the lines, the underwater body, the (sequence of) construction, the interior finish and partitioning, the protruding crossbeams, various details of construction, and the caulking method typical of cogs: sintel clamped moss caulking. The Nijkerk-II cog allowed the following observations and conclusions derived from them.

Lines

The lines and plank drawings (Plates 6.V B and C) show a remarkably high-sided hull with a straight keel and little sheer, and with straight, inclined stern and stem. The gudgeons show that the ship had a stern rudder. The stem is raked at an angle of 130° with respect to the keel line; the stern at an angle of 120°. The bottom, or more precisely, the underwater body, has a near-symmetrical, laurel-leaf-shaped outline. It is very sharp at the base of the stern and stem, but widens upwards because of the everted sides, so that the gunwales display a somewhat elongated half-walnut outline. As a result, the frames in the bow and stern are somewhat S-shaped. Only amidships is there a discernible rounded transition from the bottom to the sides: the turn of the bilge. Further fore and aft, this becomes ever less distinct, because of the torsion of the aft- and foremost planks and the transition from carvel- to clinker-style joining of the strakes. The A- and B strakes in the carvel-built part of the bottom, which 'in no part is dead flat', have a similar, rising and sleek line from the keel plank to the turn of the bilge. The C strakes are joined to these at a slight upward angle.

The next strake, D, which is the first to run from stern to stem in clinker-built fashion, in its turn also adjoins strake C at a slight upward angle. The subsequent strakes (E-O) amidships are fixed together clinker-style, in an almost straight, everted line (ca 115°). This design produces a slender underwater body, while at deck and gunwale level the ship has - or appears to have - a voluminous fore- and afterbody, and an exceedingly spacious midships section. The forebody is somewhat wider than the afterbody. After the first trial sailings with the Kampen Cog, it became clear what this implied for the handling of the ship, about which more will follow in chapter 7. There are indications that this cog had castles fore and aft. The design, the lines and the construction of the ship reveal well-established shipbuilding knowledge and expertise. In other words, this is definitely not a prototype, but an example of a fully evolved vessel type.[58]

Measurement system, partitioning and tonnage

Before people started building the Nijkerk-II cog, the shipwright must have agreed with the client how large the ship was to be. Specifications for a cog to be built have never been encountered in historical records. Presumably global agreements would have been made, in part guided by the proportions known to the shipwright and the available timber - especially for the keel plank, the stern and stem hook, the stern- and stempost and the crossbeams. For the actual shape and dimensions of cogs we therefore have to rely on archaeological finds.

57. Heinsius 1956, 247 and 248.
58. Vlierman 1997b, 91.

Measurement system

The only observed marks that most probably relate to measurements at the start of construction, are four grooves scratched into the wood across the full width of the keel plank (Section 6.2.2). Two of these, 7 cm apart, are side-by-side under floor timber 12; the two others were found under frames 17 and 18. The locations of the scratches do not correspond with e.g. the midpoint along the keel plank. During the final reconstruction on paper, the distance (measured along the keel line) between the aftermost groove underlying frame 12 and the rear of the sternpost was found to match the (reconstructed) length of the rear of the sternpost, measured from its scarf with the stern hook to the top of strake O. The same goes for the foremost groove (on the keelplank) underneath and the front of the stempost: his distance exactly corresponds with the length of the stempost. The sternpost originally was 4.95 m long; the stempost, 6.95 m. Be it that this length is just up to the top of the gunwale at the bow. The false stem on the bow is certain to have projected well above this and must have been about 8 m long. If the above hypothesis to explain these grooves is correct, they must have been scratched into the keel plank after the keel plank and the stern- and stem hook were in place. The two other grooves, 53 cm apart, may well have marked the place where the shipwright (initially) planned to put the mast. In the situation as excavated, the hole for the mast was 10 cm further forward in the keelson, while also it is not 53 but 47 cm long.

The reconstructed bottom (Plates 6.III A to C) may be considered very reliable, so the measurements of the bottom are well-established. Thanks to the recovery of the complete stern and stem hook and large parts of the stern- and stemposts, it also proved possible to determine the exact rake of the latter. The full lengths of the stern- and stempost could at first not be established, because the top end of the stempost was not recovered and about halfway along the sternpost a section is missing whose precise length was difficult to determine. As described above, these measurements could eventually be inferred after all. The exact measurements of the two recovered crossbeams are known; those of the six or seven others were derived from the definitive reconstructions of the thirteen drawn

Table 1. Principal dimensions of the Nijkerk-II cog

Length of the keel, including stern- and stem hook	12.3 m
Length of the bottom (between the rising part of the stern- and stem hook)	11.0 m
Greatest width of the bottom (seam between C and D strakes at frame 17)	3.4 m
Overall length at gunwale level	20.0 m
Overall width at gunwale level at frames 17-18	8.7 m
Height of the stern (keel line to top gunwale level)	5.1 m
Height of the stem (keel line to top gunwale level)	6.2 m
Height of the side amidships (keel line to top of gunwale level at frame 17)	4.3 m
Length of stern	6.0 m
Length of stem	8.0 m
Length of crossbeam 1 (to outside of shell at frame -1)	4.5 m (3.7 m)*
Length of crossbeam 2 (to outside of shell at frame 2	5.2 m (4.8 m)*
Length of crossbeam 3 (to outside of shell at frame 10)	6.6 m (6.3 m)*
Length of crossbeam 4 (to outside of shell at frame 17)	7.0 m (6.7 m)*
Length of crossbeam 5 (to outside of shell at frame 25½)	6.2 m (6.1 m)*
Length of crossbeam 6 (to outside of shell at frame 29)	5.5 m (5.3 m)*
Length of crossbeam 7 (to outside of shell at frame 18)	7.9 m
Length of crossbeam 8 (to outside of shell at frame 25)	7.2 m
Length of crossbeam 9 (to outside of shell at frame 35)	4.4 m
Height of the hold	3.0 m

Length-width ratio of bottom	3.2:1
Length-width ratio at gunwale level	2.3:1

** The measuremenths shown in brackets are the global comparable size of the beam in the Kanpen cog. All measurements rounded to the nearest 10 cm.*

cross-sections (frames -1, 2, 4, 6, 8, 10, 12, 17, 19, 21, 23, 26 and 29) and the top view composed from them (Plate 6.IV A). The cross-sections of frames -1, 2, 10, 17, 19, 26 and 29 have been further elaborated and expanded (Plate 6.IV B).

For the height of the sides, which towards the top are increasingly incomplete, assumptions were made for the reconstruction. On the one hand this was because of the missing parts, on the other because of the sometimes poor preservation of the surviving fragments. Because of shrinkage through drying-out or swelling through water absorption, the original widths of the planks were no longer certain. Nor could the lengths of the planks of the strakes be established, because parts were missing. The discovery of loose plank AY (Plate 6.I), the discovery that crossbeams must also have passed through strake L and the insight that roughly level with its top there must have been a deck, and the vertical beam on the outside beside the opening in the gunwale of the forebody - these all allowed the conclusion that the sides from stern to stem did not number fourteen but fifteen strakes and that the ship not only featured an aftercastle, but (from a certain point in time) also a forecastle. The top strake (O) presumably did not continue all the way to the stern and stem.

In general it may be said that the reconstruction presented in this study (Pl tes 6.IV and V) very closely approaches the original hull lines, layout and dimensions of the Nijkerk-II cog. This simultaneously means that as a result of the above-mentioned misinterpretation (in the first reconstruction on paper) concerning the original place of the two recovered crossbeam constructions, the shape and dimensions of the scale model and the actual Kampen Cog contain errors. These concern the level at which the lower crossbeams were inserted and the parts above it - especially in the afterbody. Also no O strakes, and no seventh, eighth and ninth protruding crossbeams were envisaged either in the scale model or in the Kampen Cog. However, the location of the aftercastle built onto the Kampen Cog almost corresponds with the revised situation presented here. In this reconstruction too it is assumed that its width diminishes towards the stern. The forecastle similarly narrows towards the bow. The deviating hull shape of the Kampen Cog barely affects its handling properties, since the underwater body (draught about 2 m) remains almost unchanged. The presence or absence of a forecastle and the extra strake would have been of little consequence. Sailing trials (on the North Sea) anyway did make it clear that the Kampen Cog did require extra ballast on such trips.

Layout and tonnage

The very sharp bow and stern sections were not well suited to stowing cargo. Therefore it seems quite likely that on the underside of the foremost beams supporting the aftercastle and on the after beams of the forecastle, at about 2 and 1.6 m above the keel line respectively, transverse beams were fixed onto which a deck was laid. In the afterward, such a deck could have had a surface of up to 1.4 x 3.4 m; in the forward of up to about 2.7 x 4.4 m. The largest part of the ship's interior may in any case be reckoned to the hold. In determining the length of the hold we may assume that this was situated roughly between crossbeam 2 (frame 2) - nail holes on the rear of the connecting beam and crossbeam 2 suggest a vertical partition there - and crossbeam 5 (frame 26). The hold thus was about 10 m long, had a greatest width below the deck (at crossbeam 4 and between the frames) of 7.5 m; at crossbeam 2 this was 5.1 m and at crossbeam 5 it was 6.6 m. In these places, the width of the floor between the middle of the C ceiling planks was 3.6 m, 2.5 m and 3.2 m, respectively. This amounts to an average breadth of the hold of about 4.75 m. The space (headroom) between the bottom of the deck and the top of the ceiling covering the bottom was about 3 m. This means that the volume of the hold would have been about 142 m^3 (10x4.75x3 m). If there was a separate crew compartment aft of the mast on the roundwood floor, about 2.8 m long, 3 m wide (widening upwards to 7.5 m; 5 m average) and 3 m high (2.8x5x3 = 42 cu m), then the available cargo hold would be about 100 m^3, divided between a forward and an after hold. The vessel therefore had a gross cargo capacity of about 100 tonnes. This still excludes the space above the presumed small decks fore and aft. Of course, space was also needed for storing the tools of the trade, (anchor) cables, etc., and night accommodation for the crew.

Selection of materials

As is usual with late-medieval shipwrecks in the Zuiderzee area, the Nijkerk II-cog was built entirely in oak (*Quercus* sp.). The roughly 3000 treenails and the caulking laths too are of this wood species. A few thin pegs, the detached knee at the mast step and the sides of the drain are of conifer wood. For the stern- and stempost, the keel plank, the keelson, the bilge ceiling strakes, the crossbeams, the stringers, the straight floor timbers and some of the V-shaped floor timbers on the carvel-laid part of the bottom, the shipbuilder used (large parts of) tree trunks. For the stern= and stem hook, the massive knees on the crossbeams and the V-shaped floor timbers in the stern and bow, he carefully selected suitable, naturally crooked parts of trunks and limbs. For the lower and upper futtocks, parts of thinner trunks and limbs were chosen, while the planks were sawn from trunks. Judging by other late-medieval ship finds, the decking too would most probably have been of oak, while little can be said about the wood of the mast and spar, except that Scots pine (*Pinus sylvestris*) also known as deal, is the most likely.[59] It never became clear whether a full 'brother-and-sister' patterning was striven for in laying the shell strakes.

The gudgeons, the metal plates, the bolts, the nails and the sintel clamps were made of wrought iron. The numbers of nails and clamps (about 5,000 and 15,000-16,000, respectively) are relatively large. Apart from the gudgeons, the bolts of the upper two suspension points of the rudder, the (not recovered) rudder fittings and the bolt through the bow, the ship's hull probably contained few if any further heavy iron structural or fastening elements.

6.3.2 Construction: techniques and sequence

In his historical approach of the construction (sequence) of a cog and the (construction) techniques employed, Heinsius states that the shell planks were joined in clinker (or lapstrake) fashion.[60] After due deliberation, given the two apparently different ways in which this was represented on seals, he arrived at the conclusion that this was not done in an overlapping fashion, as was known from the Viking shipbuilding tradition then already known from archaeological finds, but the other way around. Such a method would imply that the planks were applied from the top downwards, which is possible only in a frame-first rather than a shell-first approach. As to the exact execution of the keel, the stern, stem and crossbeams, barely anything could be said at the time. However, it had been established that in some images the shell planks seemed to extend beside the stern- and stempost. The archaeological finds, including the Nijkerk-II cog, have confirmed that the above-water part of the shell was indeed entirely clinker-built. Yet not, as Heinsius believed, in an inverted clinker fashion, but with simple overlaps like roofing tiles. The midships part of the bottom, by contrast, was found to be carvel-built. The curious construction of the keel with the stern- and stem hook and the posts rising from them, and the combined carvel/clinker-built shell, which above a certain height indeed extends alongside the stern and stem, is so far known only from the late-medieval vessels of the cog-building tradition. The stern- and stem hook/keel plank/stern- and stempost construction is without doubt due to the shipbuilders' reluctance to having a seam between the thin keel plank and the stem- and sternposts resting upon it.[61] Such a construction is indeed quite vulnerable if the ship is to be beached on a sandy shore. The crossbeams protruding through the shell are known also from other shipbuilding traditions.

Once the global dimensions of the desired vessel were agreed and the necessary timber was available, construction of the cog could start. This certainly took place in several phases. In each, particular techniques were employed and a predetermined result was envisaged. Only after completion would the next phase be initiated. Examination of the wreck remnants and testing the presumably followed procedures by building a scale model, and above all by building a full-sized replica, resulted in reconstruction of the following highly likely building phases and sequence, and of how they probably were executed.

Construction phase 1: Keel plank, stern- and stem hook, stern- and stempost and temporary frames

After the keel plank and stern- and stem hooks were shaped, they were laid on fairly high launching blocks and joined together. Such blocks were necessary so as to allow enough space later to finish the exterior of the bottom and to drive in the treenails from outside for fastening the floor and ceiling timbers to the shell strakes; to caulk the exterior seams in the bottom; and to ream open any drought cracks arising during construction and close them with sintel clamped moss caulking. Then the stern- and stemposts were shaped, placed onto their respective hooks and joined to them. Between the surfaces to be scarfed together a layer of moss felt would be inserted. It has not been established whether this moss caulking was first impregnated with tar. Installing the stern- and stempost could not have taken place at a later stage, because no strakes other than the garboards could be spiked to the rising part of the hooks. Most probably the stern- and stempost were supported with struts to keep them secure at the correct rake. The aim and importance of this was to achieve a stable fore-and-aft centreline along the keel. On either side of this the two sides of the vessel should be built as evenly as possible. Also the definitive length of the ship could be fixed in this way, from which widths would be derived according to established ratios. This method is employed in the building of authentic Giethoorn punts even today.

After this, it was a matter of laying the bottom. In order to hold the planks of the carvel-built part of the bottom - which were not to be fastened to the keel plank or to each other - together and in the right position, it was necessary to fix a number of transverse connections to the keel plank. These provisional floor timbers served as temporary

59. Vlierman 2007, 21.
60. Heinsius 1956, 109-111.
61. Vlierman 1997b, 84.

frames. Small pegs with a diameter of 2 cm were observed in the keel plank and some bottom planks, roughly at the place of the definitive floor timbers 8, 13, 17 and 22; these almost certainly reveal where such temporary frames once were attached. The aftmost and foremost of these temporary frames were found at the transitions between the carvel- and clinker-built parts of the bottom; the two others were evenly spaced between them. These temporary frames probably extended a little beyond the local full width of the bottom to be built, their lower surfaces having been shaped to rise towards the sides and, in the case of the outer two, to the stern and stem; they must have been supported at their ends to hold them in the desired position. The objective of this building phase was to set up a sound base for constructing a symmetrical and stable bottom, which was essential and decisive for the further construction of the ship and the eventual outcome.

Construction phase 2: Bottom strakes, bilge laths and small repair planks

The second phase entailed the sawing of the bottom planks, and hewing and bending them into the required shape. Among the timbers of the Nijkerk-II cog just one (non-bottom) plank shows traces to indicate that planks were burned in order to achieve the desired curvature. During the excavations, much attention was given to identifying any traces of burning. In most shipwrecks found in the IJsselmeerpolders (prov. Flevoland), especially in vessels dating from about the mid-16th-century onwards, the charred inner surface of the strongly curved planks in the bows often still is in clear evidence, at any rate beneath the frames. Probably at least the aftmost and foremost planks in the strakes up to F or G had been burned. Even if they were used as 'greenwood', in other words, recently felled and sawn, without having been watered, it would have been a tremendous effort to bend them without burning, especially the aftmost and foremost planks of the A-, B- and C strakes. The ends of each of these planks would have to be twisted some 60 to 80°. The near-absence of traces of charring may be explained in part by the fact that both in the Nijkerk-I and the Nijkerk-II cogs the final metres of the planks toward the stem and stern become progressively thinner. In the bow of the Nijkerk-I cog even down to a thickness of just 1 cm! This could mean that after burning, and the fitting and cutting into the desired shape, the plank was thinned on the inside with an adze, which removed any traces of charring. However, at the part where the greatest torsion was exerted, the planks still were 4-5 cm thick, so we may assume that here nothing was removed with an adze. In constructing the Nijkerk-II replica, the planks were duly burned, which was found to be quite feasible without significant charring (Chapter 7).

After the planks of both garboards had been made fully fitting, the scarfs had been made and the sides of the planks had been partially bevelled from below in order later to take the sintel clamped moss caulking, the midship garboard planks were pressed against the keel plank and the underside of the temporary frames. They were presumably held in place with small struts on the floor, while here and there also a treenail or nail might have been used. These planks were followed by their aft and fore extensions. Within the scarfs, moss felt was used; and the scarfs were secured with two rows of nails: one from within and one from the outside. Wherever possible, these planks too had been tightly held to the temporary frames with small struts. In the strongly twisted aft and fore parts of the garboards this was not possible (there were no temporary frames here), but this would probably not be altogether necessary, because their bottom surface and cross-cut hood ends were to be nailed into the previously cut rabbets in the stern- and stem hook. The same procedure was followed with the B- and C strakes, be it that these hood ends were fixed into the rabbets in the stern- and stempost, and the aft- and foremost planks of the clinker strakes were broader (for the overlap) and on the inside were chamfered over the full width of the land (overlap). This chamfering was also done to the exterior land on the upper edge of the previously placed A (and then B) strakes. Before these clinker parts were spiked together, also the upper/outer edge of the preceding strake would be bevelled, creating a V-shaped seam between the two planks, which was later to be filled from the inside with sintel clamped moss caulking. The same was done with the lower/inner edge of the upper strake, forming the seam for caulking the outside of the shell. At the stern and stem, the plank ends were thin and almost carvel-built, leaving the outside of the shell almost smooth. Towards the midships the clinker-built construction became progressively more apparent. Beyond a few metres from the stern and stem, the overlapping lands of the planks matched average plank thickness on the exterior.

Two clinker plank sections to be joined together would be pressed together with so-called boeitangen (large wooden pincers, figs 6.41 and 7.5a and b); and, presumably after the plank to be attached was provided with nail holes, it was spiked to the previously placed strake. The small nail hole plugs found here and there suggest that the carvel-built strakes sometimes closed up insufficiently and that this problem was resolved by employing small additional temporary frames: for example, short planks fixed across several strakes and/or the keel plank. During this - presumably relatively protracted and (most) difficult - phase of construction, the bottom planks would dry out and might develop cracks. Also builders had to take into account the shrinkage to be expected from the clinker-built sides that were yet to be put onto the carvel-built bottom, potentially causing the seam between the B- and C strakes to open up. The following action then consisted of placing the bilge laths at this part of the seam and repair planks were fixed to the inside of the shell over the new drought cracks, with a layer of moss luting in between. After the entire bottom had been successfully laid, a start could be made on consolidating the shape and lines of the bottom by means of straight and V-shaped floor timbers, i.e. the third phase of construction.

Construction phase 3. Straight and V-shaped floor timbers, keelson, bottom ceiling boards and the caulking of seams in the clinker-built parts of the bottom
In the first instance, the massive floor timbers for the hold (4-26) were now made. The tree-trunk parts selected for these purposes were hewn to a rectangular cross-section, and on the underside cut to match the local profile of the underlying bottom. Then limber holes were cut and the floor timber could be put in its proper position. Treenail holes were drilled in from above, after which the treenails could be driven in from below. Maybe this was initially done only for the treenail holes and treenails between the ceiling planks to be put in subsequently. First the floor timbers to be placed between the temporary frames were inserted. Probably the builders started in the mast-step area, working towards the stern and bow. After which the temporary frames were taken out, the treenail and nail holes closed up, and the definitive frames put in place. So the entire carvel-built bottom was provided with frames. Now first the seams of the clinker-built after and fore parts of the bottom were to be filled with sintel clamped moss caulking. Only then could the floor timbers right up to the stern and stem be inserted. These V-shaped floor timbers, from 2 to the stern and from 28 to the bow (there never were any 3 and 27 floor timbers), were not fastened to the hooks or the stern- and stempost, and most probably were put in at a later date (in construction phase 5). Subsequently the keelson and the A ceiling planks were shaped and treenailed in place. The massive bilge ceiling strakes (B) could not yet be fitted in this phase, because some of the floor timbers first needed to be extended by the lower futtocks. The keelson was fixed with blind treenails; the ceiling planks were treenailed to the floor timbers and the shell strakes in a single go. It seems reasonable to place the keelson and the ceiling planks between the frames 2 and 26, as otherwise the heavy keelson would have to be lifted unnecessarily high across the shell and lower futtocks. But even more significantly, it was a component essential to the rigidity of the underwater body, of supreme importance to the further construction of the ship.

Construction phase 4: Shell strakes up to crossbeam level, caulking of seams, lower futtocks, ceiling boards up to crossbeam level and riders
In this phase the almost straight shell planks were prepared strake by strake, and with the aid of boeitangen (large wooden pincers) clamped to the preceding strake which had been fully spiked to the hull. This was done more-or-less simultaneously on both sides, so the ship took shape upwards in an even fashion. Before the first plank of a strake (probably the aftmost) was fixed to the preceding strake and the stern- or stempost, the scarf for its connection to the next plank had to be adzed and the upper outer edge of the previous strake would be bevelled: the future seam to hold the sintel clamped moss caulking on the inside of the shell. This bevelling was done also to the lower/inner edge of the strake to be attached, the future V-shaped caulking seam on the outside of the shell. After this, the plank was fixed with the typical, twice-bent nails that were driven back into the timber (fig. 6.42), while the plank ends were slotted and spiked fast into the rabbets in the stern- and stempost (strakes D to G). All subsequent strakes (F to N) ended at or on the (rear or front, respectively, of the) stern- and stempost and were nailed to them. The uppermost strake (O) probably ended about a metre clear of the stern and two metres clear of the bow. Once the sides were built up to and including strake J, and also the after planks of the K strakes had been put in place, the interior seams of the hull were filled with clamped moss caulking. This was done also with the seam between the stem- and sternpost and the plank ends on the inside.

Any cracks in the planks were covered on the inside with small planks and a layer of moss luting in between. After this, the lower futtocks could be inserted. These were the futtocks on either side of the hold (frames 4 to 26). These frame timbers, made from thin trunks or thick branches were hewn to a roughly square cross-section. The outward surface would then be marked out and joggled to match the stepped profile of the clinker-built shell. Their lower and upper ends were cut at an angle, for scarfing to the floor timbers and the upper futtocks, respectively. In contrast to all other (recovered) frames, the oblique ends of the starboard upper futtocks of 19 and 21 (and more) interfaced with the outward rather than the inward faces of the lower futtocks. Maybe this relates to the (belated) insertion of crossbeams 7 and 8 (see also Sections 6.3.2 and 6.3.4). The upper ends of the lower futtocks of frames -1, 2, 10, 17, 19, 25, 29 and 35 were finished square, because the undersides of the transverse beams were to rest upon them there. Frame 25½ had a futtock only above the transverse beam. The futtocks probably were first fixed with a nail through the scarf and the land to the floor timber, and from the outside through the land and the scarf on the upper side. These last nails were observed in various places as extra nails between those of the clinker construction of the two strakes. Subsequently the futtocks probably were connected to the shell by a few treenails (between the ceiling boards). After the bilge ceiling strakes (B and C) and after them the three other ceiling boards had been made to fit and been nailed fast, they could in one go be treenailed to the futtocks and the shell. To complete this phase, the riders previously hewn into shape were placed across the A and B ceiling boards and fixed with a few nails, on port and starboard, beside the mast-step part of the keelson. Then the treenail holes were drilled and the knee riders were fastened onto the floor timbers with blind treenails.

Construction phase 5: crossbeams, shell strakes up to the gunwale, caulking the seams, upper futtocks, V-shaped floor timbers, futtocks and transverse connections in the aft and forward, remaining ceiling boards and inwale

The four lowermost crossbeams were laid upon the square heads of the futtocks concerned, which reached up to or almost up to the top of the I strakes. At the two aftmost crossbeams this is one strake higher. Into the crossbeams, rabbets had been cut right around. The rabbet on the underside engaged with the upper edge of strake I or J, respectively. Subsequently square gaps were cut into the planks of the next strake, which fitted around the upper and lateral rabbets of the crossbeam, thus entirely enclosing the crossbeam (fig. 6.43). The crossbeams were, two at a time. Enclosed by single planks on their under and upper sides. Now the planks up and including strake N could be prepared, attached and caulked on the inside. Upon this followed the upper futtocks between 4 and 26, and the ceiling boards to be mounted upon them.

Probably this was followed by the moment when the floor timbers in the stern and bow were made to fit and treenailed to the shell, when the futtocks and transverse connections in the stern and bow were inserted, and finally the last of the ceiling boards were put in. Indeed it was necessary during construction to keep the weight of the stern and bow to a minimum, to stop these parts from tilting forward or backwards, in other words, threatening to affect the shape of the ship at deck level and putting the entire construction under undue stress. With the shape being sufficiently rigid and secured by means of the enclosed crossbeams and the ceiling boards, this risk was averted. Strake O was probably (later, simultaneously with the insertion of crossbeams 7, 8 and 9) mounted onto strake N as a removable wash strake with short 'futtocks' fitting between the frame heads. This can be inferred from the few treenail holes found in pairs, one obliquely above the other, in the plank fragment on the port aft gunwale. In the afterbody a small number of thick treenails were observed, which presumably relate to the attachment of knee-shaped transverse connections that did not protrude through the shell, probably to hold up the rear stanchions of the aftercastle. For constructional reasons it seems very likely that the ship, both in the aft and forward, had transverse timbers between crossbeam 1 and the stern and between crossbeam 6 and the stem. Yet no actual evidence was found for this. The same goes for the question whether the vessel was provided with side-stems.

Construction phase 6: crossbeam knees, stringers, windlasses and castles

In order to keep in place the everted sides (strakes up to N) that rose about 1.6 m above crossbeams 1 to 6, and to add the necessary lateral rigidity to the ship, two heavy knees were placed on each of the protruding crossbeams, whose upright parts had been joggled to accommodate the ceiling boards and the inwales to engage with them, and were fastened to the upper futtocks and the shell. The distance between the shell rabbets in crossbeams 1 and 2 and the outward faces of their knees, however, is such that there must have been additional futtocks in between (see section 6.3.2). The attachment of these knees was executed with 4-5 cm thick treenails. The two recovered crossbeam constructions show that the aftmost (hence shortest)

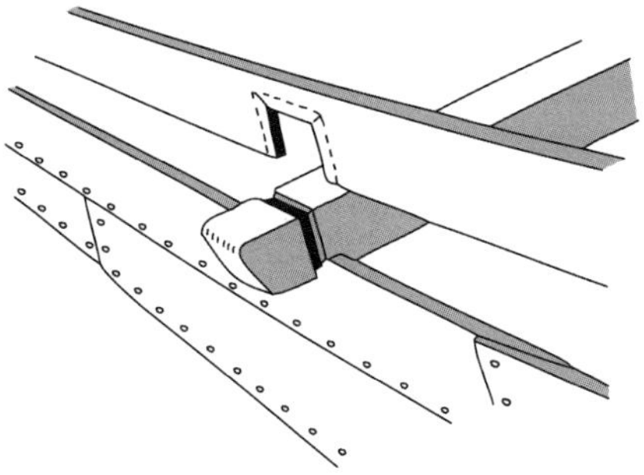

Fig. 6.43. The insertion of a crossbeam.

crossbeam carried just the knees, scarfed together roughly midway along the crossbeam; while the other had a connecting beam between the knees. Besides, these knees and the connecting beam were topped by a rider beam, into which on its front surface several rabbets had been cut. These had held the ends of the (unrecovered) stringers which ran from here to crossbeam 3. It seems most likely that such sets of four stringers had run between the other crossbeams too. Besides it is quite possible that several of these beams were removable to allow voluminous cargo to be hoisted in and out. Whether there ever were stringers between the foremost crossbeam and the bow cannot be established, though it is likely. These probably were fixed. It cannot be ruled out that the stringers (from crossbeam 3) initially sat at a lower level, more closely following the sheer. This construction phase must also have seen the installation of the stanchions supporting the castles and the windlass(es). Be it that the forecastle in particular is most likely to have been added at a later date.

Construction phase 7: decks (with gangways), false stems and exterior caulking

Just a few small details were found that point to the presence of an afterdeck, and a small, 10 cm higher deck forward of it. This concerns the areas between the stern and crossbeam construction 1 and between crossbeam constructions 1 and 2. Besides, the rabbets for four stringers on the front of crossbeam construction 2 point to massive longitudinal connections which at any rate extended to the next crossbeam construction. On the basis of the presence of the opening in the shell for crossbeam 8 we may assume that during its final phase of use the ship had a near-horizontal deck extending from crossbeam 2 (actually from crossbeam 1) to crossbeam 8. It was continuous up to the bow and in this part of the ship ran fairly parallel to the sheer. For a long time it was thought to be a fore- and an afterdeck joined by a gangway-like construction along the sides. Whether this was the original state of affairs can no longer be ascertained. Yet we may assume that in this phase the false stem was installed to cover and protect the cross-cut ends of the shell strakes, while at the other end the false stern was added and provided with gudgeons. At the end of this construction phase, the seams and any cracks occurring in the shell during construction were externally closed with sintel clamped moss caulking and maybe tarred (locally) - though few if any traces of tar were encountered on the wreck.

Construction phase 8: rudder, belaying racks, pump and drain, mast and rigging, spar and sail, guides

The rudder with its fittings, prepared separately, would have been hung onto the stern probably after launching. The mast too would be erected at this stage, after which the rigging could be put in place. The belaying racks, which probably sat between the upright parts of the knees on the crossbeams, the pump tube, the drain/latrine, the guides and the ship's and crew's further equipment will also have been fitted and finished at this final stage. Once the sail was in place, the ship would be ready to sail. The final two construction phases proposed here are barely if at all deduceable from the evidence of the Nijkerk-II wreck, and should be considered hypothetical.

Alterations and adaptations to the cog?

In the foregoing the expression 'initially' and 'in the first instance' were used in reference to certain structural features. This is to say that the described situation may deviate from the situation immediately after the cog's construction. The most notable features are the hole for crossbeam 6 which was covered up with a shelf clamb; the hole cut into frame 26 for placing the rear stanchion for the forecastle; and the different direction of the scarf connecting the upper futtocks of 19 and 21 to the lower futtocks (and maybe all other futtocks between 18 and 25?). Also the curiously large distance between the outer face of the crossbeam knees and the shell suggests that the construction may originally have been different. The repair on the exterior of the shell, immediately below the obliquely cut hole in 26, suggests that a significant weight had pressed on this spot over a long time (years). Also it is clear that the full severance of the frame to provide a solid base for the stanchion contributed to a far from robust construction, which certainly would not have been the original shipwright's choice. Also, one may wonder whether crossbeams 7 and 8 passing through strake L, and crossbeam 9 through strake M, and the washboard strake (O) indeed were original features. The lack of further evidence however means that no all-encompassing final answer can be given on these points. Nonetheless it seems useful to formulate our thoughts on the subject.

For instance, it seems quite possible that the hull initially was no higher than strake M and had an extra drop strake (N) only in the afterbody (from the stern to e.g. crossbeam 3). The vessel would then certainly have featured an after- and a foredeck. Whether the midships was entirely covered over remains an open question. Gangway-like, broad and heavy boards along both sides, together with some stringers to aid longitudinal rigidity, are possible as well.

These would have sat between and atop the knees and the connecting and rider beams on the crossbeams, and may have followed the sheer of the ship. The cog may have had an aftercastle right from the start. It seems that the hull was adapted a good time after the ship's construction, maybe to serve a new purpose (Section 6.3.4). It is probably then that the forecastle was added, and crossbeams 7, 8 and 9 were inserted. In this new situation, the stringers between the various crossbeams presumably were put in on a virtually horizontal level or slightly following the sheer, between crossbeams 2 and 8. The crossbeams with knees (3, 4, 5 and 6) were probably levelled-up with extra timbers to support the deck.

The spaces between the stringers and between the stringers and the sides were most probably covered over with short planks, fixed transversely to the ship (the gangways or outer waterways). It seems plausible that this deck construction continued from crossbeam 8 to the bow, and followed the sheer of the vessel. The decking between the stern and crossbeam 1, and between crossbeams 1 and 2, was placed longitudinally. Because of the elevated deck, it became necessary to raise the bulwarks as well. To this end, strake N was extended up to the bow and raised with a wash strake (strake O), and the sternpost was given non-recovered extension. Strake O presumably did not continue right up to the stems. The reconstruction drawings (Plates 6.IV and V) show the final, hypothetical situation. Plates 6.V A and C indicate which parts were recovered. Several starboard futtocks in the midships are left blank. These detached futtocks were in fact recovered, but it never became clear which belonged exactly where. Detailed tree-ring analysis on appropriate timbers could bring more clarity as regards these hypotheses.

6.3.3 Characteristics

Reinders has described a number of typical cog(-like) structural features.[62] The recorded and analysed data of the Nijkerk-II cog produced the following list of characteristic features, which may be regarded as typical traits of a cog.

Plank keel

The vessel does not have a high keel, but a relatively narrow plank keel that slightly narrows from fore to aft, has over twice the thickness of the bottom planks, is wider at the top than at the bottom (trapezoid in cross-section) and which does not continue to below the stern and stem, but - in the manner of a connecting beam - is scarfed to the horizontal arms of the stern- and stem hook.

Stern- and stem hook

The aft and fore hooks were hewn from naturally grown crooks and like the keel plank are trapezoid in section. They form the aftmost and foremost parts of the keel and the lowermost part of the stern and stem. The top of the rising arm is finished obliquely from outside to inside and from top to bottom, forming the scarf connecting it to the stern- or stempost.

Stern- and stempost and false stern and stem

The stern- and stempost are joined to the hooks by scarfs secured with (two rows of) nails. In the stern, the nails were driven in from the inside; on the stem, from the outside. The stempost, viewed from aside, progressively narrows towards the top, while the false stem progressively broadens; thus the combined timbers achieve a more-or-less even thickness throughout. The stem probably stood well proud of the gunwales. The sternpost ended roughly at gunwale level and viewed from aside had a fairly even thickness. Because of the shape of the false stem attached to it, which was broadest at its base and ended somewhat short of the top, the stem as a whole however becomes elongated-triangular in side elevation. The plank ends of the lower strakes (A to F) are spiked into rabbets in the sides of the stern- and stem hook and stern- and stempost. The subsequent strakes (from G upwards) continue onto the sides of the stern- and stemposts, obscuring the actual posts. The cross-cut ends of the strakes on the bow were covered over and protected by the false stem, also trapezoid in section, which most probably rose above the stempost by a good metre. This by contrast is not the case with the false stern, where the plank ends of the port and starboard strakes meet on its mitred rear face.

Carvel-laid and clinker-built bottom

The most notable points about the bottom are its laurel-leaf-shaped outline which converges very sharply towards the stern and stem, the carvel-laid midships and the clinker-built after- and forebody. Besides, the bottom 'is flat at no point' and on either side of the keel comprises three broad strakes, each composed of several planks. These planks were probably arranged in the bottom according to a 'brother-and-sister' pattern, and joined together with scarfs fixed with double rows of nails. One row from the inside outward; the other, from the outside inward. The plank ends at the scarfs protrude a few centimetres from the shell, both on the inside and the outside. It was only after the insertion of the heavy floor timbers and ceiling boards on the carvel-built part of the bottom and the subsequent treenailing of the keelson and the first ceiling boards that the underwater body consisted of a tight and rigid construction, onto which the rest of the ship could

62. Reinders 1985a, 31.

be built. The building of the ship therefore was a 'bottom-first' approach in the case of the bottom, and a 'shell-first' matter when it came to the sides, whereas the Viking shipbuilding tradition was fully shell-first.

A long keelson and an 'open ceiling' with heavy bilge ceiling strakes

The keelson is a long, heavy inner keel with a broadened mast-step part, into which the hole for the mast heel is cut. The B and C bilge ceiling strakes too are executed quite massively and are more or less rectangular in section. The keelson and bilge ceiling strakes where necessary are notched to accommodate and engage with the underlying floor timbers. The ceiling boards are spaced about a plank's width apart, forming a so-called 'open ceiling'. The uppermost (J) may have been a thicker plank or an inwale. Keelson and ceiling timbers together ensure the longitudinal rigidity in the frame of the ship.

Clinker-built sides and 'double-clenching'

The strakes upward of strake D are fully clinker-built. The form and the securing of the scarfs between the constituent planks are the same as those in the bottom. The strake planks are attached laterally with regularly spaced nails. The nails were twice hammered into an angle and finally driven back into the timber: 'double clenching' (fig. 6.42). This method of construction makes it possible to add the planks strake by strake, without frames. This was done in two phases. The first phase comprised the shell from the bottom to the lowermost crossbeams. Most probably this was followed by the sides between frames 2 and 26 being provided with futtocks, crossbeams and ceiling planks; and after this the second phase: shell planks, futtocks, and ceiling boards of the above-decks part. This 'sandwich' method produced a rigid structure. Only afterwards were the futtocks and ceiling boards in the fore- and afterbody inserted. The lower futtocks are of a lighter execution than the straight and V-shaped floor timbers; the upper futtocks are lighter than the lower ones, while the finish becomes progressively less meticulous from the bottom upwards.

Caulking method

All clinker seams were caulked both on the inside and the outside of the shell. This was not the case with the cross-cut parts of the planks in the scarfs. Some evidence was found that sintel clamped moss caulking was applied also around the crossbeams where they protruded through the shell. The seams of the flush-laid part of the bottom were caulked only externally. The method employed for watertight closure of the shell seams, known since the mid-1990s as 'sintel clamped moss caulking', is typical of the cog-building tradition.[63] The materials used in this method (moss, laths and sintel clamps) were applied on the inside of the clinker-built part of the shell before the futtocks were put in. The caulking of the exterior seams was done shortly before the ship was launched.

Crossbeam constructions, deck(s), castles

Remarkable features are the crossbeams protruding through the shell. They determine and consolidate the shape and rigidity of the vessel at about deck level. The knees on the crossbeams serve to support the bulwarks and provide rigidity to the above-deck part of the ship. So do the stringers, be it for longitudinal strength. The heads of the crossbeams, protruding from the shell, were protected on the front by pointed and rounded, tapering pieces of timber, fenders also known as beam-head guides. These were not made from oak wood. In its final phase of use, the cog probably had a deck at one level from crossbeam 1 forward, which in the forebody followed the sheer of the ship. The vessel had a slightly lower afterdeck or platform for the helmsman. Initially the sides may have been one strake lower, without crossbeams protruding through strakes L and M; and the ship may have had a fore- and afterdeck connected along the gunwales by means of gangway planks. The ship also may have had an aftercastle from the start. The forecastle almost certainly was a later addition.

Fastenings

Not only the 2.5-3 cm thick treenails usually employed in wooden shipbuilding were used for fixing the shell planks and ceiling boards to the frame; but also thicker treenails with a diameter of 4-5 cm were used, only in the crossbeam constructions and the stern- and stempost and false stern and stem. The shell strakes are 'stitched' laterally to the adjacent strakes with iron nails 110-130 mm long. The moss caulking and the laths covering it were held in place with iron sintel clamps. Apart from the gudgeons and the rudder fittings, barely any other large or heavy iron connections or fastenings were used in the vessel.

A floor of branches and twigs

Just aft of the mast the vessel had a (separate) 'dwelling floor' consisting of branches and split short tree trunks, which lay among and across the floor timbers, the bottom ceiling boards and the keelson. On this lay a layer of straw and then a layer of coarse sand. It is not certain that this floor was part of the ship's original appointments.

63. Vlierman 1996a.

6.3.4 Use

Function

Apart from the dimensions and the lines and layout of the ship, there are in fact no concrete clues as to its function. Despite this limitation, it seems plausible that the cog was built as a cargo vessel, and that for those days it was a medium-sized to sizeable ship. It was definitely suitable as a sea-going vessel and it can be cautiously inferred from the recovered ballast rocks (section 6.7) that it visited the Baltic area at some point: almost certainly these rocks could have been gathered only on the southwest coast of Sweden. We may assume that the cog served for cargo transportation in that area. The little redware jar (Section 6.5.3 and figs 6.63-5 and 6.64), an as a type entirely unknown utensil in the Low Countries, displays a fair degree of similarity to 13th-/14th-century pottery from Saintonge, Charente-Maritime, in southwestern France.[64] The jug almost certainly was part of the ship's inventory and must have broken during its use on board. The sherds eventually slipped down among the frames and judging by the wear on some of them must have lain there for a considerable length of time. The apparent subsequent alteration of the cog (extra crossbeams through strakes L and M, a continuous deck at a single level, a forecastle and raised gunwales) may indicate that the ship was given a different or additional function. The forecastle especially, but also the near-horizontal deck (except the foremost part) are adjustments that may have been made to make the ship more defendable or to deploy it for warfare. In the latter context we might remember the elevation to dukedom of Count Reinald II of Guelders in 1339 by Emperor Louis of Bavaria. This was a reward for the efforts made by Reinald, whose wife was an English princess, during the One-Hundred-Years' War between France and England. These efforts may have included the provision of ships and soldiers.

Wear, tear and repairs

The notable wear to the stem hook and the false stem points to many landings on (sandy) shores and a lengthy sojourn in the water. This can also be inferred from the round, abraded corners of the keel plank, while the almost total absence of tar remnants on the shell too may be an indication of long exposure to water. Yet this may merely reflect lack of maintenance during the final years of the ship's remaining afloat. From these observations it may be concluded that the vessel was definitely not in its prime when it was abandoned. Some repairs too, the adaptation/alteration, and the possible replacement of the false stern and stem suggest the same. The results of the tree-ring analysis (Section 6.8 and Appendix) support this conclusion: there were one or two building phases, or one building phase and one or more repair phases. An age of twenty to thirty years by the time of the vessel's foundering is virtually certain, but may have been yet greater.

6.3.5 Abandonment and breaking

The presumably considerable age of the ship and its potential loss of seaworthiness may have caused it to be laid up. The findspot and the completeness of the wreck show that the vessel must have arrived here afloat, and may even have lain moored in the 'harbour basin' beside the late-medieval settlement of *Ark*. It seems unlikely that this was its home port. The historical sources relate to a farming settlement. Nor have any archaeological finds come to light that might point to a trading community. A few kilometres further inland, the town of Nijkerk, which had expanded around the new church built in 1222, might have been a likelier candidate for commerce, but nothing of this kind was recorded over the 13th and 14th centuries. We know that by the first quarter of the 15th century it was a fair-sized town, which in 1463 was to gain its city charter.

The area northwest of Nijkerk was of strategic importance to the counts (from 1339 the dukes) of Guelders, witness the building of Hulkestein castle in 1427. Nijkerk, and in any case Ark, from the start were situated on navigable waterways close to and with good access to Lake Almere (later the Zuiderzee marine inlet), and close to the mouth of the river Eem and the border with Utrecht territory. This could well have been a Guelders cog, which in its final years served as a watch ship or outpost at the mouth of the Eem, during Guelders' turbulent times in the mid 14th century. The ship was eventually written off and laid up at the site where it was excavated (its winter port?). The bulwarks on the port side, the castles, the midships deck and foredeck, the stringers and the foremost four and upper three crossbeam constructions were removed. Possibly these massive timbers were reused for instance for building or repairing a sluice or a farmhouse in the vicinity, which had been damaged as a result of flooding. The abandoned ship may in its final years even have served as a fisherman's shelter or a storage place for fishing gear (Section 6.5.2).

6.3.6 Wreck formation and embedding

It has already been noted that the ship lay close to the settlement of Ark and that there, probably already partially dismantled, it must have been subjected to severe storms accompanied by flooding of the surrounding coun-

64. Rieth 1979, 117-129.

tryside. In several episodes the ship was ripped apart, the sides collapsed, after which detached parts were washed away. Also any farmhouses still standing in the vicinity would have been destroyed; while the surrounding clay-on-peat land was eroded layer by layer by the action of the waves, was washed about and then sedimented on top of the wreck and in the channels. This process was probably repeated several times in the course of a few decades. Moreover, it seems quite likely that here and there in the vicinity of the wreck some peat, up to just below the water surface, managed to persist for a considerable period.

The recovered panel of woven withies, presumably a kind of keepnet for fish (Section 6.5.2), which given its position in lacustrine Almere sediments is certainly of late-medieval antiquity, shows that after the sea floor settled, there was water over and immediately around the wreck for a considerable length of time. Yet the panel may also have been associated with the cog when it was still afloat. The stratigraphic position of this 'keepnet' indicates that after its foundering initially large parts of the cog must have protruded from the sea floor. These timbers eventually dispersed or collapsed and became covered with sediment. Only after ca 1600, once the peat around the wreck had been washed away almost completely, and the submerged channel remnants had become filled in to approximately the same level as the surrounding sea floor, did a deposit of marine clay cover the wreck, rendering it invisible both under water and in the field after the draining of the Zuidelijk Flevoland polder, until its chance discovery.

6.4 Model

6.4.1 The NISA standard method for building scale models

For building a scale model of the cog, we initially employed the well-tested NISA standard building method developed by J. van der Zee in collaboration with the researchers (author and other excavators). The principle is that all components are fashioned on the basis of the scale 1:10 drawings, using the same kind of wood as employed for the original timbers, and put together as follows:

- The keel(plank), the hooks and the stern and stem are joined up (with glue).
- A set of frame templates is constructed on the basis of evidence from the transverse sections, combining as many as possible cross sections to fit the inside of the shell, which are cut out in the form of bulkheads. These are fixed to the keel in the appropriate places and rigidly interconnected along the top.
- The planks of the bottom and side strakes are shaped one by one, bent if necessary and glued together where the scarfs would be. This is done both with the flush-laid strakes and with those to be mounted overlapping with each other. Once the entire hull is completed in this way, the set of frame templates is removed.
- Now the frames are made, starting with the floor timbers and then the futtocks up the sides. One by one they are shaped to fit the hull and glued into place.
- Then the keelson and the ceiling boards are inserted and the model is provided with its treenails and metal nails. The treenail and nail holes are made with a dentist's drill. Toothpicks were used by way of treenails, and small copper nails instead of iron ones. Iron nails have a tendency to rust or turn 'blue' in response to the lye in the oak wood, which will eventually stain or even split the wood.
- Finally the deck beams, the decking planks, and whatever else was found of the upper parts of the ship are put in place. The gudgeons and metal fittings of the rudder and other metalwork are cut from sheet copper and fixed to the stern.

The model is never elaborated beyond what can be inferred from the archaeological evidence, be it that parts found on one side may be mirrored and copied on the other side as well. Besides, for constructional reasons, a few deck timbers or crossbeams were put in, but only if such have been recovered or if there is clear evidence of their former presence. To complete the model, all recovered parts are coloured, while merely reconstructed parts are left blank.

6.4.2 The 1:10 scale model of the Nijkerk-II cog

After the first three phases described above were completed in the usual way in the construction of the model of the Nijkerk-II cog (fig. 6.44), a start was made on shaping the floor timbers and futtocks of the most fully preserved frames, on the basis of the scale drawings. The floor timbers were glued to the bottom. However, the futtocks, next to be inserted, were found not to fit, despite being apparently faithful copies. During the attempts to place these futtocks, the still flexible hull was found to be losing its shape, taking on a shape that could not possibly reflect the original lines. On the one hand, this could be due to the not quite reliable hull shape of the long-buried wreck, caused by warping, and the sometimes poor condition of the frames; on the other, possibly mistaken assumptions regarding the lengths of some strakes, because of these, upwards progressively greater proportions were missing, along the entire length. A third reason is that at the time it was erroneously assumed that the recovered crossbeams were numbers 2 and 6. At any rate the hull must have had, and should now have, a flowing outline. Therefore the shell was largely dismantled and remade. Once more the shape turned out unsatisfactory and would not allow further construction of the model. The serious distortion of the hull by the attempts to place several frames anew made it clear that these still were not correct, and that a different building method should be adopted. This after all was not a matter of merely putting together a threedimensional version of the reconstruction drawing of the ship as a whole, as had been the case with earlier scale models; this attempt had two aims, which were essential to the technical investigation. These two aims were:

1. Achieving an accurate reconstruction of the shape.
2. Developing insights into the construction process.

It was an attempt as closely as possible to reconstruct the medieval building method and sequence, based on the technical excavation data; in other words, the technical impossibilities needed to be ruled out, after which a few possibilities remained. Some of these were more plausible than others. While it should be remembered that there are fundamental differences between building a 1:10 scale model and a full-sized replica. The model-building was in part undertaken because of the initiative to build a replica of this cog at Kampen. After the initial frustrating experiences in building the scale model, it therefore was decided to start afresh and to build up both the model and the replica in phases, the model always keeping at least a full phase ahead. The model was to function as the shipyard model. Every time a solution for a particular construction problem was found or opted for, this would be applied to the scale model. The outcome would be discussed with the builders of the Kampen Cog, maybe tested at full scale and if necessary adjusted.

On the basis of the 1:10 scale drawings, first the keel plank, hooks and stern- and stempost were glued together, and fixed to a rigid support. The stern and stem were left longer that they should eventually become. Then the four floor timbers 8, 13, 17 and 22 were shaped, which globally occupied the places of the temporary frames, before being fastened to the keel plank. These were followed by the garboard planks (strake A) on either side of the keel plank. The foremost and aftmost planks of

Fig. 6.44. The first attempt at building a scale model (1:10) of the

these strakes were heated with hot air so they could be bent into shape, and then glued to the keel plank and the hooks.
This procedure was repeated for the planks of the other bottom strakes (B and C). The seams in the clinker-built parts of these strakes were fixed with little nails; and upon its completion the bottom was sanded down where necessary, in preparation for the next phase. Even in reconstructing the bottom on paper, it was evident how beautifully the various components fitted together, and what the underwater body must have looked like. The same was observed in the building of the model and the replica. How the laying of the bottom strakes and the subsequent construction phases apparently were actually executed was described in the above analyses (Section 6.3.2). The experience gained in building the Kampen Cog replica will be discussed in chapter 7. At the next stage all the floor timbers to be fitted onto the carvel-laid part of the bottom were shaped and glued into place. These were followed by the keelson and the lowermost (A) ceiling boards (fig. 6.45). The underwater body, composed sandwich-style, now had a very rigid, stiff con-

Nijkerk-II cog.
Fig. 6.45. The underwater body of the model: the bottom, the stern and stem, the floor timbers, the keelson, and the A strakes of the ceiling.

struction. In fact, it was an almost complete vessel, be it with rather low sides. This method is now known as the 'bottom first' method.

The almost straight plank templates of the side strakes and the clinker joining suggested a relatively simple construction method. Therefore it was investigated whether, starting with the D strake, the sides might not be built onto the C strakes of the underwater body plank by plank, subsequently adding strake upon strake, without the need of temporary filling frames. A fairly obvious medieval construction method, which was used also in the Viking shipbuilding tradition, be it right from the keel upwards (the 'shell-first' method), of which some experience had been gained by replica builders in Denmark. In building the model, this was of course not difficult. Taking in to account the dimensions of the two recovered crossbeams, and the locations where at the time they were thought to belong (crossbeams 2 and 6), it was possible to globally assess the flare of the sides at these points. Starting from the ratio of this dimension to the width of the bottom below it, now also the dimensions of the other, missing beams could be reconstructed, so that there too the incline of the sides could be established. After this, the lengths of the strakes could be determined up to strake K. The forebody was broader than the afterbody. Then the lower futtocks were shaped and inserted, which neither in the model nor the replica encountered any substantial problems. The required adjustments of the futtocks to the hull shape were deemed acceptable. In other words, the hull retained its shape and gained in rigidity.

Generally speaking, the crossbeams would be fitted at this stage. Yet in this case this was postponed, as it still was uncertain whether both beams indeed belonged in the places assigned to them: there was a difference between the length across the gunwales and the reconstructed length at the site of crossbeam 6. Also from the small number of surviving upper futtocks, whose originals on the whole were in poor condition, it was hard to estab-

Fig. 6.46. The completed (shipyard) model (scale 1:10) of the Nijkerk-II cog. Showing part of the ceiling, frames and crossbeams; the recovered timbers are coloured.

lish how the original lines of the upper sides had been - continuing in a direct, flowing line up to the gunwale, or maybe somewhat inturned. After a few trials, it was decided to complete the shell of the model by the first method, and only then to shape and place the crossbeams. Once the shell was ready, the recovered crossbeams were found to fit in none of the six locations, and the lengths of the crossbeams for the model were adjusted. The recovered knee constructions (with connecting and rider beams) on both crossbeams were not copied in wood, because they would not fit into the model. Naturally, in the replica Kampen Cog the crossbeams were placed first. To complete the model, the upper futtocks and the starboard ceiling boards were put in. The model has fifty frames.

Apart from the small nails fixing the clinker-built ends of the bottom strakes, initially no treenails, nails or clamps were used in the model, as it served as a shipyard model. After the recovered timbers (apart from the knees and connecting/rider beams on the crossbeams) had been copied and glued into the hull, it was exhibited at the shipyard while the Kampen Cog was being finished, while also serving to discuss and occasionally try out the next steps to be taken in building the replica. Besides, there initially was a plan to make an entirely new model after completion of the replica, incorporating all of the findings and necessary adjustments, before completing the model in accordance with the NISA model-building protocol. Eventually this plan was abandoned and the existing model was finished in a manner that seemed acceptable and presentable (fig. 6.46).

6.5 Inventory

6.5.1 Introduction and distribution of the finds

Section 6.1.3 describes a number of potsherds that are relatable to the late-medieval settlement of Ark. It mentions that two trenches were dug to obtain insight into any relationship between the ship and the settlement, on the basis of the potsherds found in the stratigraphy of the sections. The sherds found at different depths in the baulks and in the levels of the excavation will be discussed here. Thereby a distinction will be made between, on the one hand, those found on the surface, in the plough soil, in the marine Zuiderzee sediments and in the top of the lacustrine Almere clay deposit, and on the other, those found in the deepest part of the ship's hull, or indeed below it where parts of the ship had broken away. These latter finds all lay within a layer of redeposited sand. The former group simply was waste from the nearby settlement. These sherds, as well as the many bones that lay among them, ended up in the deposits covering the wreck as the sea swallowed up the surrounding land. Most lay in the oldest Zuiderzee sediment, the shell layer Zu IV. This means that they may have ended up there even after ca 1600. During the wreck formation process a layer of redeposited sand formed in the bottom of the wreck. This was covered by a deposit of Almere clay (fig. 6.12), of which just ca 20 cm remains.

It is possible and indeed quite likely that this layer of Almere clay initially was thicker. The top of it was eroded away in the 16th century. The sherds and bones that lay within it gradually sank downward as the finer particles were washed away, and ended up in the shell layer. This might explain the unabraded edges of the half small *Kugeltopf* (figs 6.47-10 and 6.48a and b) found partially embedded in the Almere clay (so maybe at the level where it first came to rest), with the rim of the pot reaching the sea floor for a short period because of the unabraded edges. On most of the other *Kugeltopf* sherds there is considerable abrasion of the edges, due to prolonged washing to-and-fro among the seashells. Given their stratigraphic position, the finds from the sand deposit are as good as certain to belong to the ship itself, or in any case predate the definitive wreck formation. Plate 6.II B shows the accession numbers of the finds; they correspond with those in the finds list and with the serial numbers in the figures.

6.5.2 Finds attributable to the late-medieval settlement of Ark

Pottery

In the sections overlying the cog and in the excavation, over seventy sherds were found above the Almere sediments. In fourteen cases they were of prehistoric origin, three date from Carolingian times. These finds will not be further discussed here. Pingsdorf ware is represented with sixteen sherds, including base fragments with pinched footrings (figs 6.47-37 to -40). There are three sherds of Paffrath ware; one of these is a foot of a small pot or jug (fig. 6.47-47). Proto-stoneware is present in ten fragments from different items, including three rim sherds (figs 6.47-41 to -43). Twenty-six sherds derive from various *Kugeltopf* pots. Two rim sherds have a hole drilled through the narrowest part of the neck (figs 6.47-54 and -55). Another, a rim sherd of a large *Kugeltopf* pot with a rim diameter of 25 cm (fig. 6.47-60), has diagonal finger strokes on the shoulder; while a wall sherd is decorated with broad,

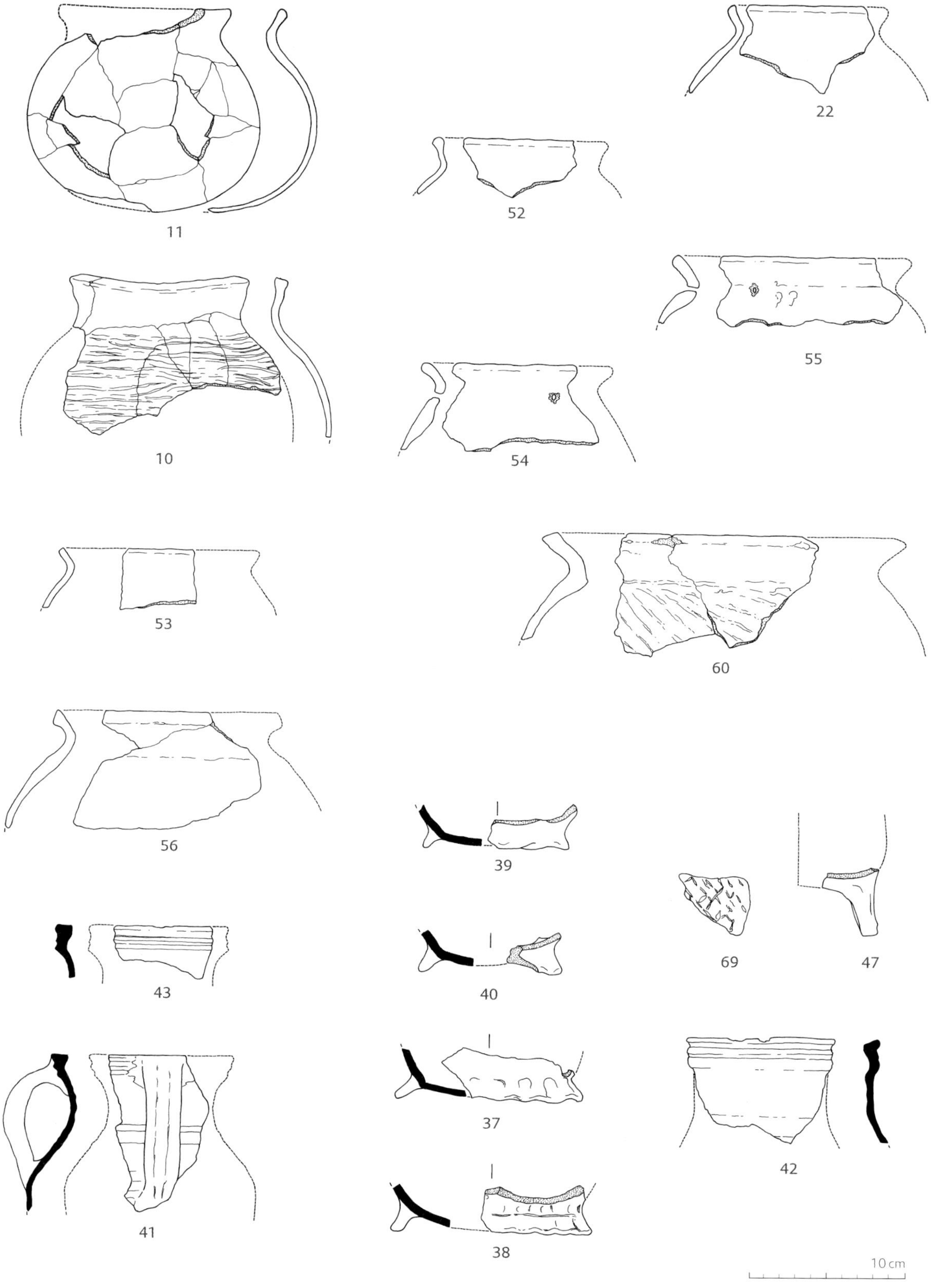

Fig. 6.47. Potsherds from the late-medieval settlement of Ark found in the cog excavation pit.

Fig. 6.48a. Rim of a small *Kugeltopf* vessel with brushed surface treatment (10) *in situ* in the section.

Fig. 6.48b. After the restauration.

crosswise scratches (fig. 6.47-69). Of two small *Kugeltopf* pots, about half has survived. One, with brushstroke decoration (figs 6.47-10 and 6.48), was found above the afterbody in the profile dam and with its rim in the marine Zu IV deposit. The other (fig. 6.47-11) partially sat in the same layer, while the remainder of both pots lay in the top of the underlying Almere-clay deposit.

These sherds can all be dated to the 13th century, the Pingsdorf material possibly to the (late) 12th century, while *Kugeltopf* ware with brushstroke decoration occurs into the early 14th century (see section 6.1.3). A fairly flat redware sherd with ochre-yellow lead glaze on the inside and soot on the outside (72) probably belongs to a flat, 15th-century frying pan. The findspot, among the frames of a higher-lying part of the wreck, but still in the Zu IV shell layer, confirms this younger date. The sherd definitely does not belong to a frying pan from the cog. The sand from the bilge was sieved. This brought to light some very small sherds of proto-stoneware. Still, it cannot be ruled out that these are contaminations originating from the excavation of overlying sand. Also they may have sunk into the ground as a result of movement of timbers protruding from the sea floor. A third possibility is that they were in the coarse sand that covered the roundwood floor. At any rate it is most unlikely that they came from jugs that belonged to the ship's inventory.

Tools

Beside the stem hook, outside the hull to starboard, an iron axe with a handle fragment was found (figs 6.49-137 and 6.50). It is a so-called broadaxe, a right-hand hewing axe.[65] Weighing about 900 gr, the axe has a triangular blade, a fairly straight cutting edge and a triangular haft eye with somewhat convex sides and a hammer-like butt. The blade has a chisel-like bevel only on the right, and the head is offset somewhat to the left, so we may assume that it is a right-hand broadaxe. The axe had a handle of beech wood (*Fagus sylvatica*). Despite the nearby breaks in the first two starboard bottom strakes, it seems very unlikely that the axe was washed out of the wreck. The findspot is very close to that of crossbeams 5, 6 and 8. Therefore it is imaginable that it was lost for instance during the chopping-off of the starboard heads of the protruding crossbeams, having fallen into the water alongside the ship. Beside the starboard futtocks of frame 24, a block of wood was recovered at the very start of the excavation, which was described a possibly hewn-off head of a protruding crossbeam (K in Plate 6.1).

65. Vlierman 1985b, 46-48.

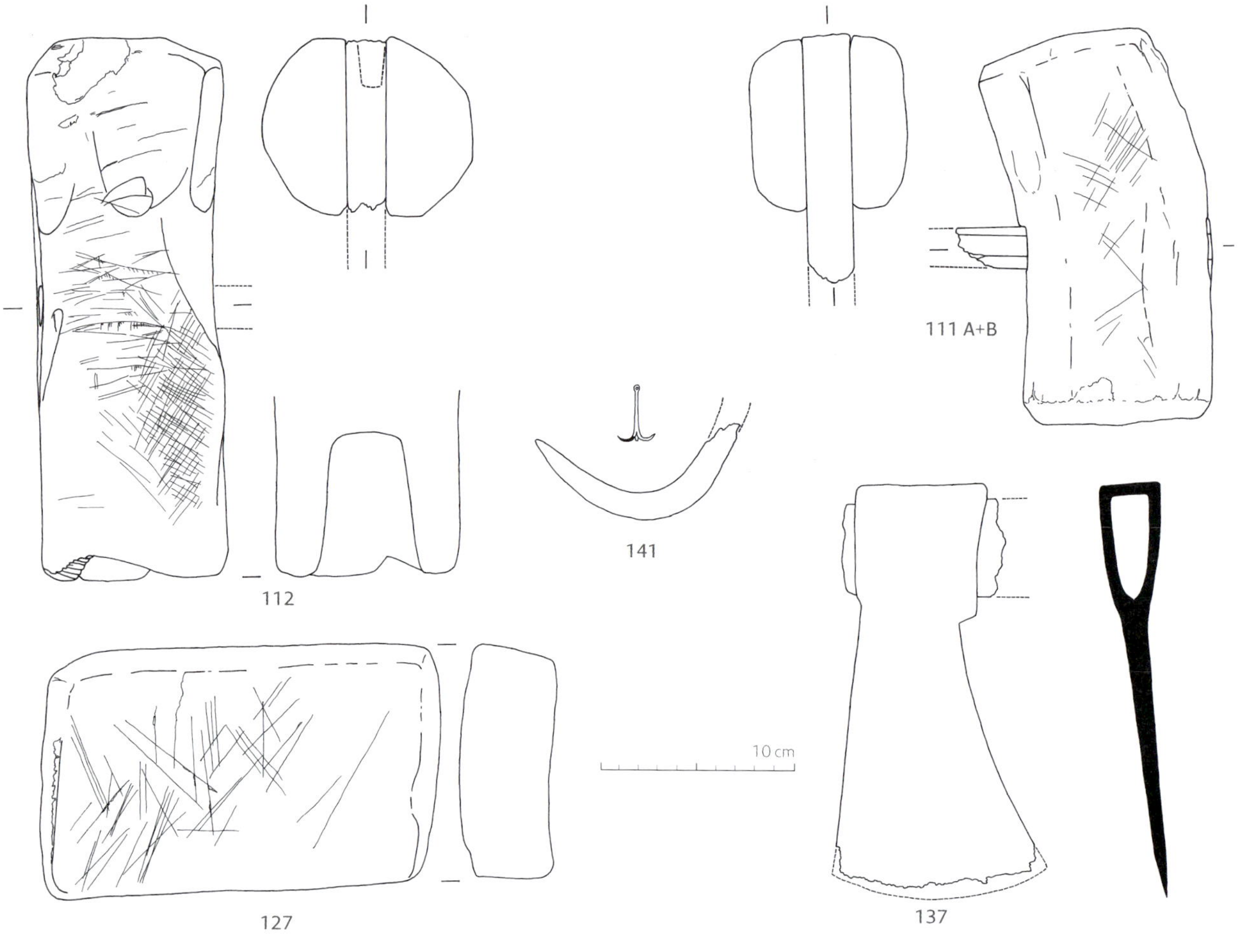

Fig. 6.49. Fisherman's gear(?): Axe (137), sledgehammers (111 and 112), small plank with cutmarks (127), and the tip of a grapnel (141).

Fig. 6.50a & b. The axe *in situ* to starboard beside the stem hook and after the restauration.

Below the midships and at the bow end of the keelson, two wooden (Alder? *Alnus glutinosa*) sledgehammers were found in the redeposited sand. One (figs 6.49-111 and 6.51) has a slightly curved shape. The hammer head still contains part of the oak handle. The edges of the hammer surface are rounded through use. On the sides some cutmarks can be seen. The other sledgehammer (figs 6.49-112 and 6.51) is longer and straight, while in section it is hewn into a roughly octagonal shape. This hammer head too contains part of a wooden handle, fixed into it with a wooden wedge. The edges of one hammer surface are rounded through use. The other end across its entire width and roughly across the middle has an obliquely cut groove, with a hole drilled into the centre, 7.5 cm deep and 4.5 cm in diameter. The edges around the hole are much worn; and various cutmarks are to be seen centrally on both of the sides. Sledgehammers would primarily have been used for driving in posts.

It may be no coincidence that the hole in the hammer has the same diameter as the stakes found with the 'keep net' to be discussed below. It would be practical for the hammer to be placed with its hole on top of the stake which was thus pressed down into position. Then it could be hit with the other hammer to drive the stake into the hard sandy subsoil that was ubiquitous here (on a map of the Zuiderzee of 1852 the area in question is indicated as '*Areker Hard*'). If the stake ends were hammered directly, they might easily splinter and break. Below the stern a small board of oak wood was found, measuring 20x12x5 cm (figs 6.49-127 and 6.51). One side is slightly concave and has quite a few cutmarks. It seems quite likely that the two sledgehammers, the board and the 'keepnet' were part of a fisherman's gear. The cutmarks may relate to the gutting of fish. In this context it makes sense to remember the name of Gozwin, mentioned in section 6.1.3, who in Carolingian times possessed a piece of land near Ark and managed a local fishing water. The management of fishing waters no longer features in the 13th-century records, but in those days, and later too, there surely was good fishing here. The axe, the hammers and the board must slightly predate the definitive wrecking of the ship. They are unlikely to have been part of the ship's equipment.

A killick (anchor) and a 'keepnet'

Above and beside the afterbody, three objects were recovered that most probably do not relate to the ship itself, but to fishing activities. On the wreck, in the port side of the afterbody, close to the drain (Plate 6.I) lay a heavy, subrectangular (40x20x17 cm), granite erratic boulder

Fig. 6.51. The two sledgehammers and the small plank with cutmarks.

with remains of an iron eye (298). The stone lay on the Zu IV shell layer, and therefore must date from well after 1600. Such stones were used as anchor stones for small (fishing) craft, and are known as killicks. Usually such anchoring sites lay close to the shore. The stone may also have been lost on a fishing ground. On the upper edge of the port side and embedded in the Zu IV shell layer, lay a broken-off iron point (fig. 6.49-141). This may have been part of a small grapnel. The stratigraphic dating of both finds indicates that there can be no connection either with late-medieval Ark or the cog.

At first, a link seemed fairly obvious between the killick and the piece of woven withies (*Salix* sp.), together with three wooden stakes 4.5 cm thick and originally a good 2 metres long (Plate 6.I and fig. 6.52). However, the basketry lay flat on top of a redeposited sandy layer and was covered over with a deposit of Almere clay and a chunk of redeposited peat. Also a large piece of driftwood lay on top of it, which had warped the basketry. This implies that this presumed 'keep net' for holding live caught fish must be considerably older than the killick and may have been installed at the stern of the berthed cog in the 'harbour basin'. The surviving piece of basketry was 1.2 m high and 1.4 m wide, part of it lying folded upon itself. Its overall height was at least 1.3 m; its length or width more than 1.9 m. The basketry was made of withies 0.5-1 cm thick with interspaces of 2-2.5 cm, therefore unsuitable for holding eels. The withies had been twisted together at roughly 30-cm intervals in strands of three withies 0.3 to 0.5 cm thick. A remnant of the fifth woven connection was found beneath the basketwork. The way in which the stakes lay close to the basketwork showed very clearly that they belonged together. Probably this was a 'keepnet', which had been held in place with three or more stakes. Two stakes lay upon the basketwork, the

Fig. 6.52. The fish cage made of willow withies, close to the afterbody.

third beneath it. The shape of the keep net could not be established; the position of the three stakes suggests a triangular shape, but this should be evident from the basketwork, which it is not. A round shape therefore seems more logical and is practicable with three stakes set 60 cm apart. A keepnet or a cage for holding live caught fish can even today be observed in some places where there is good and regularly used fishing water. There is no dating for the basketwork. Given the above considerations, a mid-14th-century age and a connection with a fisherman (from the settlement of *Ark*) seem most likely.

Bones

Large numbers of bones were found in the excavation of the wreck. Judging by their stratigraphic distribution, they can in part be interpreted as redeposited (butchering) waste from the settlement. A smaller proportion must derive from meat consumed aboard the vessel. All bones were examined (see Appendix). In section 6.5.3 the research results are linked to the context data in relation to the other finds and the ship, and the inferences these allow.

6.5.3 Ship's equipment and inventory

For the purpose of systematically presenting the large assemblage of utensils from the late-19th-century cargo ship *Lutina*,[66] Reinders drew up a subdivision scheme.[67] Research into the equipment and inventory of a small, early-15th-century cog made it clear that late-medieval ships require a different approach from 19th-century (cargo) ships.[68] In Part IV, Results and Analysis, this point will be elaborated. The items from this and the other cogs in the present study will be classified according to the following categories and in this order.

66. Zwiers & Vlierman, 1988.
67. Reinders, 1985b.
68. Vlierman 1992b, 10-22; Idem. 1993, 69-76 and Idem. in: Hocker & Vlierman 1996.

Table 2. Distribution of finds

Category	Forebody	Hold	Afterbody
1. Ship's equipment	anchor?	sail, cordage, sheave, spoke, 'lids', drain	-
2. Operational equipment	-	spades, load hook	-
3. Administration and navigation	-		compass needle?
4. Tools	-	marlinespike, spoon auger, tar pot, needle	marlinespike
5. Household items	-	candle holder	-
6. Cooking, eating and drinking	-	bronze Grapen, cooking pots, treen dish, pottery and stoneware jugs	-
7. Victuals	barrel	stave bowls, sticks, hazelnut shells, bones	-
8. Personal belongings, weaponry	coin	belt, footwear, knives, sheath	jacket?

1. Ship's equipment
2. Operational equipment
3. Administration and navigation
4. Tools
5. Household items
6. Food preparation, eating and drinking
7. Victuals
8. Personal belongings and weaponry

Ship's equipment

The stern carried three iron gudgeons for the suspension of the rudder (fig. 6.16). The starboard brace of the middle gudgeon was detached and features in the finds list (figs 6.53-143 and 6.54). The loose fourth rudder attachment, found beneath the ship, consisted of a thick iron bolt with a ring (fig. 6.53-144) which had been attached through the hole in the stern at the level of strake I. The ring, which had held the pintle, was worn through at the back, while the starboard half of the ring had broken off. On the inside of the sternpost the bolt had broken at the hole for the securing pin; the end with the pin and the fifth, comparable and uppermost attachment device were never recovered. The stempost and the false stem had probably been joined together at the top by a thick iron bolt (figs 6.53-145 and 6.54). Iron nails of differing sizes were used on the hull (figs 6.53-146A, -147, -149-3, -149-5, -149-10, -150A, -151B, -152A and -153B and 6.54). To secure the caulking, sintel clamps of various sizes were employed (figs 6.53-156A-C and -157F and 6.54). For fastening the shell strakes to the frames and the ceiling boards, oak treenails were used (figs 6.53-121 and -122, and 6.54). The first one illustrated probably was a spare treenail; the other seems to have been used. Clear to see are the somewhat thicker ends by which the treenail would be jammed into the outside of the shell plank. A small oakwood lid (figs 6.53-114A and B, and 6.54) probably served to temporarily close the scuttle through which the drain emptied overboard. On the inside there must have been an iron eyelet or a bolt with a hole through which a bolt or pin could be passed to secure it in place. Beneath the stern lay a worked and broken-off piece of oak wood (fig. 6.53-128). It may be a frame head that once stood proud of the gunwale. On the top it has an oblique notch on one side; a cross was carved into the oblique surface. It is reminiscent of (part of) a bitt, but it lacks abrasion marks left by rope in the appropriate area. A small piece appears to have been sawn off at the broken and shipworm-eaten end. Indeed the piece may have belonged to a very different part of the ship, e.g. an ornamental end on a beam of the castle. The round, almost 3 cm thick oaken 'lid' with broken lugs (figs 6.35, 6.53-116 and 6.54) probably was the cover of a wooden pump tube. The function of the rather rough-hewn, elongated piece of beech wood (fig. 6.53-132) is unclear. Maybe it was a spoke for operating a windlass. The small wooden (probably Ash, *Fraxinus excelsior*) sheave of a block (figs 6.53-113 and 6.54) was found between the floor timbers beside the keelson. There is heavy use-wear both on the single flange beside the groove that held the rope, and on the sides of the axis hole.

The rope fragments (228 and fig. 6.53-229) were found in the lengthways of the ship inside some longitudinal limber holes in the floor timbers and between the floor timbers of the hold. They are of a four-strand, Z-twisted line with a 1.8 cm diameter, so-called shroud-laid rope. The fragments probably belong to a single, or a few lengths of line that had been strung through the aligned limber holes that ran in the underside of the floor timbers. Each of the (presumably four) lines could at times be tugged to and fro a bit, to prevent the limber holes from clogging up. This was important especially for the

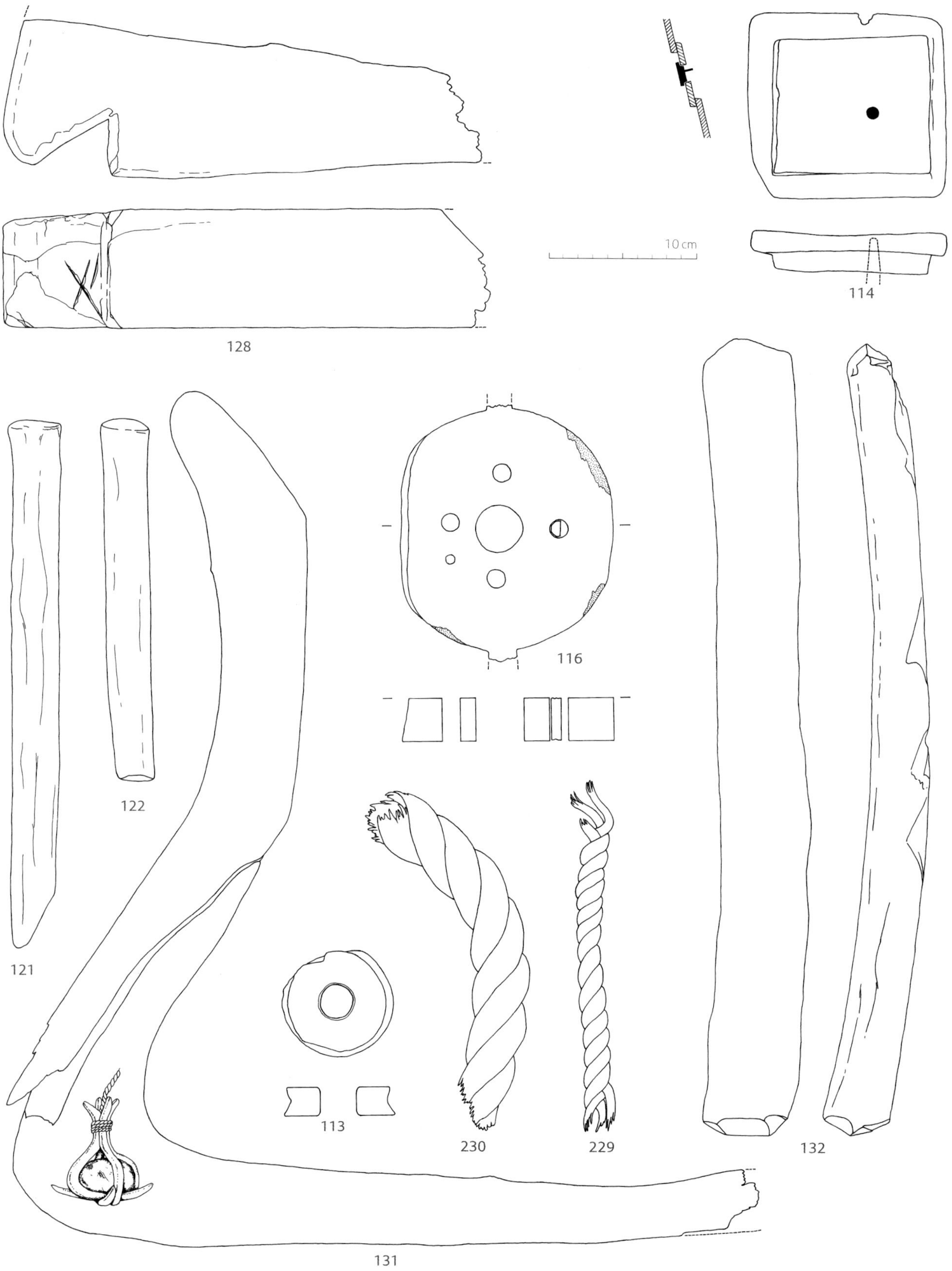

Fig. 6.53a. Ship's equipment. Wooden part of an anchor (131), spoke? (132), sheave of a block (113), lid of a pump barrel? (116), cover for a drainage opening in the shell (114), (spare) treenail (121, 122), part of a frame head? (128), two lengths of rope (229, 230),

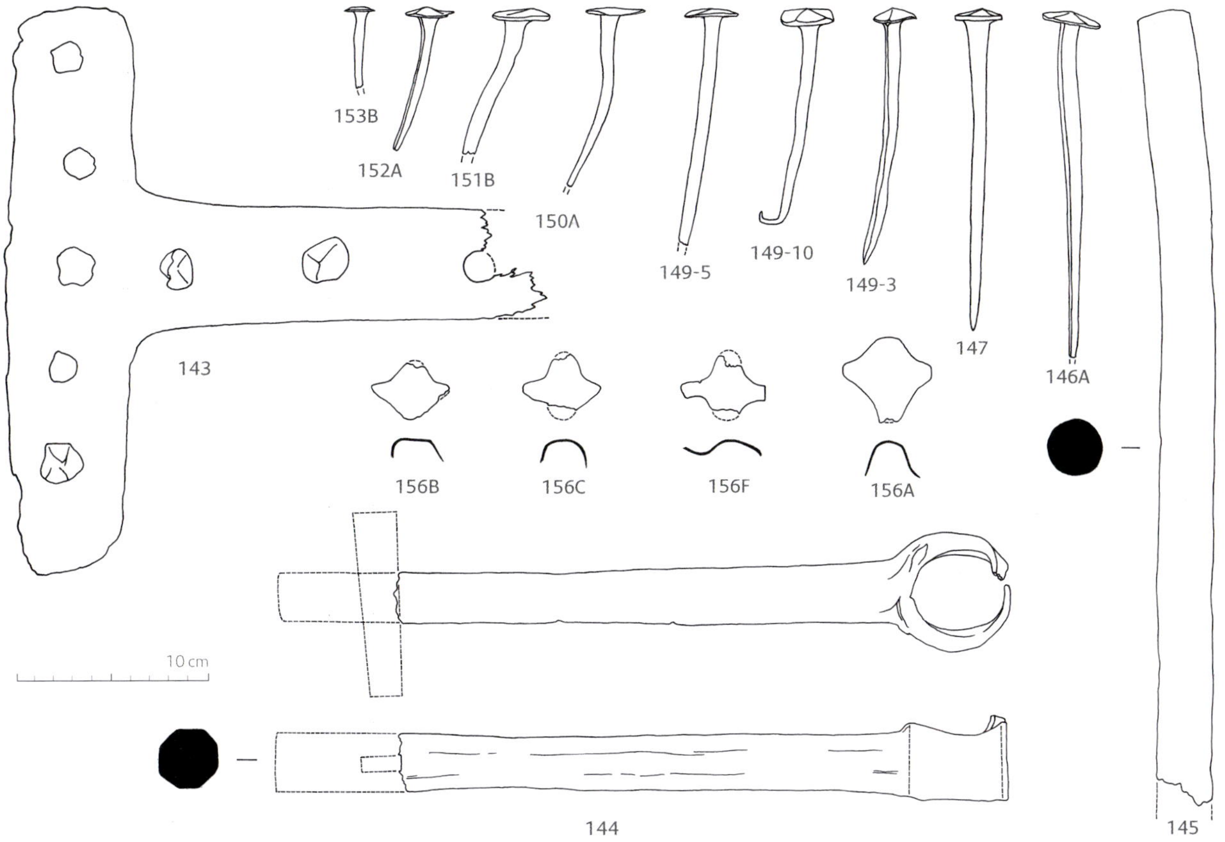

Fig. 6.53b. Ship's equipment. Bolt with eye for a helm (144), bolt of the rudder pin (145), wing of the middle gudgeon (143), several nails (146A, 147, 149-3, -5 and -10, 150A, 151B and 153B and sintels (156A to -C end 157F).

Fig. 6.54a. Ship's equipment. Gudgeon, bolts, nails and sintel clamps.

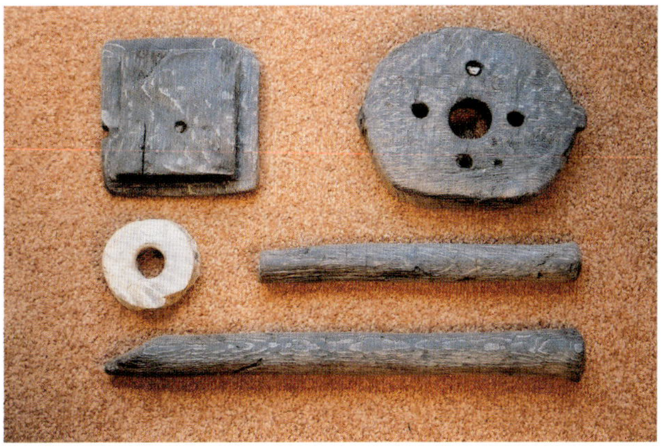

Fig. 6.54b. 'Lids' of pump barrel and drain, (spare) treenails and sheave of a block..

limber holes in the floor timbers under the roundwood floor. The piece of Z-twisted rope, so-called cable-laid rope (fig. 6.53-230), probably was part of the rigging. The cable has a diameter of 3.5 cm, is composed of three S-twisted shroud-laid ropes, which in their turn consist of four Z-twisted strands. Besides, three pieces of line (228, 229, 230) were recovered. The function of an naturally inward-curving L-shaped, worn-down and hewn piece of oak wood (fig. 6.53-131) has remained unclear. It may have been one of three (or four) more-or-less comparable pieces of wood that together enclosed a heavy anchor stone while being held together at the top and below, and whose lower ends pointed outwards. One of the late-medieval shipwrecks in the harbour of Kalmar (Sweden) contained an anchor whose wooden parts enclosed a boulder.[69] In the Almere-I cog a chopped-off, pointed piece of oak wood was found, which too may have belonged to a somewhat similar anchor.[70]

Highly unusual is the recovery of three textile fragments. Late-medieval textiles are rarely found in archaeological contexts. Two pieces measure about 70x55 cm (226A and fig. 6.55-226B), the third piece (227) is about 40x30 cm. The fairly uneven woollen fabrics were woven in so-called basket or plain weave on a broad handloom. The weave consists of S-twisted, coarsely spun strands, twined into Z-twisted yarn 2-5 mm thick. One piece (226B) has a selvage on one side. This side has a border about 18 cm wide of dark brown wool, followed by a 3 cm wide strip of a paler shade. Although this is impossible to ascertain, is seems that there is a repeating pattern. There are ample traces of tar on the fragments. These most probably are the remains of a sail. Until the present study, nothing was known about the material used for cog sails.

Viking ships are known probably to have had mainly woollen sails. Crumlin-Pedersen,[71] judging by more recent woollen sails, writes that a small sail of 20 sq m requires a fabric of about 500 gr/m^2, while for a sail of 60 to 130 m^2 this should be about 1000 gr/m^2, if it is to be tarred for protection and wind proofing. Given the dimensions of the hull, it may be assumed that the Nijkerk-II cog carried a sail of about 150 m^2. The pieces of woollen fabric with tar remnants from the Nijkerk-II cog currently have a weight of about 1200 gr/m^2, which amounts to about 180 kg for a 150 m^2 sail. Large and small fragments of fabric have been found also in the Nijkerk-I cog, the Marknesse cog, the Spakenburg-I and -II cogs and the Doel-I cog (Chapters 8, 12, 14, 15 and 20, respectively).

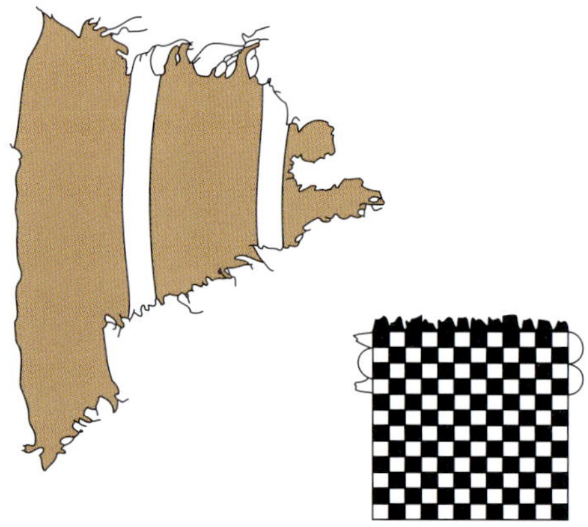

Fig. 6.55a & b. Ship's equipment. Two pieces of a dark brown, woven woollen sail (226A and B) with selvedge and one or two paler bands of warp in plain weave. Sail fragment (226B).

69. Åkerlund 1951, 35, 36 and Fig.18 and Pl.5,b.
70. Hocker & Vlierman, 1996.
71. Crumlin-Pedersen 1997, 190.

Table 3. Ship's equipment

Category	Item	Number	Findspot
1.1 Ground tackle	anchor/killick?	131	under the bow
1.2 Sails	sail	226A and B, 227	hold
1.3 Cordage	line	228-230	hold
1.4 Blocks	sheave	113	hold
1.5 Windlass	spoke?	132	under the stern
1.6 Pump, drain	lids, drain	114, 116, 117	under the hold and in the stern
1.7 Navigation aids	-	-	-
1.8 Boats	-	-	-
1.9 Spare parts	treenail, nails, clamps	121-126, 146-154, 156-158	hold

Operational equipment

Just a few items were recovered that may be reckoned to the ship's operational equipment. A special find is the complete spade made from a single piece of ash wood (*Fraxinus excelsior*) (figs 6.56-107 and 6.57). The fairly straight spade has a blade 15 cm wide and 20 cm high. On both sides the fairly flat blade is sheathed in a single iron sheet, folded around the cutting edge and tapering upwards on both the front and back. The two sides were riveted together with a single rivet passing through the blade. The cutting edge is worn to an angle through use. The spade has a fairly thick, round handle (4.5 cm in diameter) and a closed grip 12.5 cm wide. Its exact function is difficult to determine, because its design makes it barely suitable for either digging or shovelling. Maybe it was intended for peat-cutting. Yet a miniature (fig. 6.56-107A) shows that in the High Middle Ages such spades were indeed used for tilling the land. This spade sat in the hold among the frames and behind a bilge ceiling strake to port. Belonging to shovels are the fragments of presumably ash wood (figs 6.56-109 and -110) that were found to starboard beside the stern in the redeposited sand. These are fragments of two blades, which originally were 15-25 cm wide and 25-30 cm high. They had slightly upturned sides, making the blades somewhat concave at the front. Presumably these are so-called grain shovels. The model remained virtually unchanged even into the 20th century (figs 6.56-109A). The iron load hook (fig. 6.56-140) was found in the redeposited sand to starboard of the midships. Two such hooks would be fixed to the ends of a rope, sometimes with a ring midway along the rope, and were used especially for hoisting barrels.

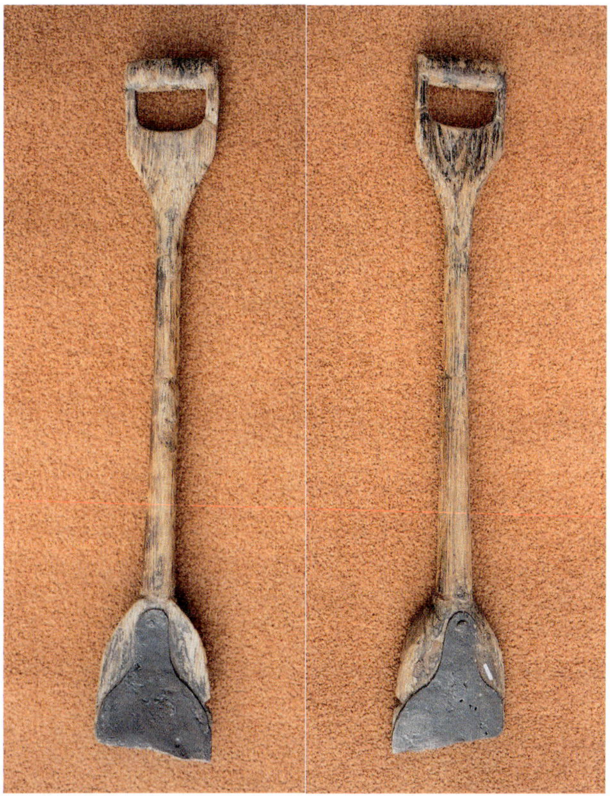

Fig. 6.57a & b. Rear and frontal views of the wooden spade (107).

107

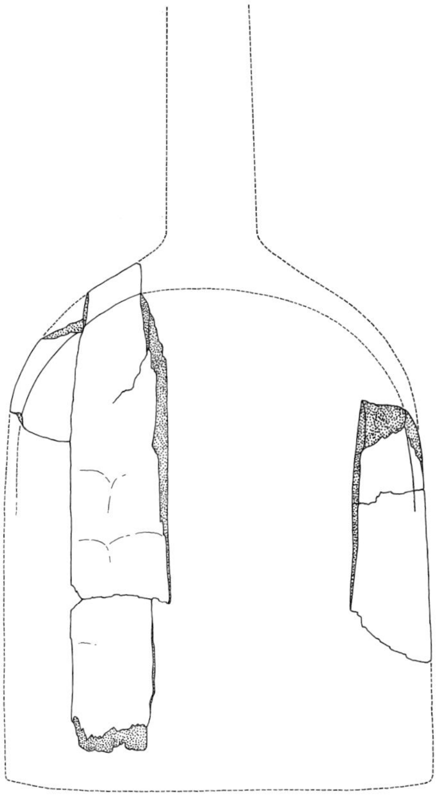

109

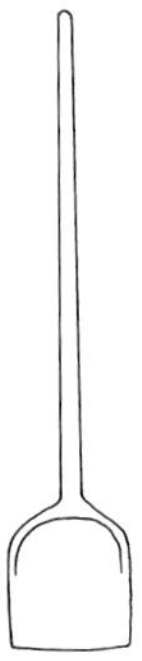

109A

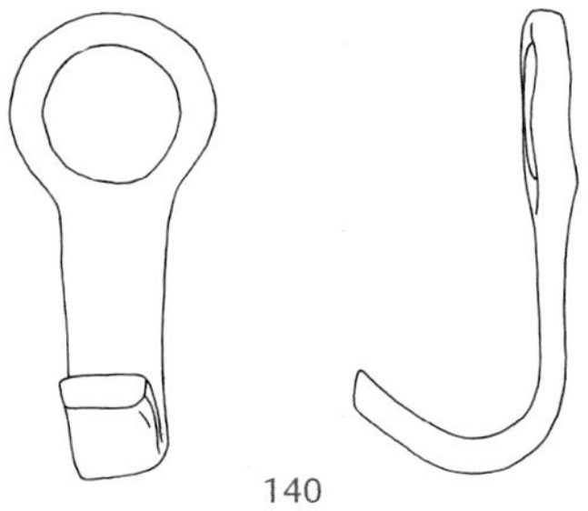

140

107A

Fig. 6.56. Operational equipment. Wooden spade (107); detail of a late-12th-century miniature showing a similar type of wooden spade made from a single piece of wood with the lower part of the blade sheathed in iron (107A; after Groten 1990, 73); grain shovel (109); and load hook (140).

Table 4. Operational equipment

Category	Item	Number	Findspot
2.1 (Un)loading gear	spade, grain shovels, load hook	107, 109, 110, 140	hold
2.2 Additional equipment	-	-	-

Administration and navigation

Aboard late-medieval vessels no great amount of administration will have been carried out. The skipper probably kept notes on what he was carrying, what the goods fetched at their destination, and the cost of the return cargo. Of course he would also carry cash on board. In the case of barrels and other solid packaging or items, these are likely to have been marked with crayon for identification. Often fragments of crayon are the sole evidence of administration or the marking of cargo to be found aboard shipwrecks. Besides, owner's marks carved into barrels are regularly encountered, which sometimes can be linked to provenance. Aboard the Nijkerk-II nothing of this kind was found. Navigation aids too will have been very limited in the Middle Ages; indeed it is questionable whether any were present in every ship. Actually most shipping passed along the mainland coast and north of the Frisian Isles, which allowed navigation by following onshore landmarks. Often sailors would make notes of where they were on particular days. Ships sailing across the North Sea and beyond are likely to have had a compass on board. A piece of thin copper plate with a sharp point and a flaring end may have been part of a compass needle (figs 6.58-160 and 6.59). In the middle of the broader end is a small hole, while the item has an impressed groove there too and seems to have broken. The needle was found under a floor timber, aft of the roundwood floor.

Tools

Just four items can be classified as tools. The broken iron spoon auger (figs 6.58-138 and 6.59) is a carpentry tool. The diameter of the bit is about 10 mm, so it was certainly not intended for cutting treenail holes into the shell. It may in fact have been lost at an early stage of construction, when it served to make holes for the pegs holding the temporary frames. The hand drill for using this bit was not recovered. Two marlinespikes (figs 6.58-167, -168 and 6.59) and a bone needle (figs 6.58-170 and 6.59) are tools for splicing rope and presumably for mending the sail. Marlinespike 167 was cut from a red-deer antler (*Cervus elaphus*). At the place where it was cut from the tine, about two-thirds of the edge is bevelled. The remaining part was cut off flat over about half the length of the spike. Given the traces, the point was sharpened quite roughly, probably with a knife; the very tip has broken off. The other marlinespike (168) was made from the upper tine of a forked antler of a roe deer (*Capreolus capreolus*). The other tine of the fork had probably been sawn off parallel to the main tine which itself is fairly rough and from two sides cut or broken off obliquely. The point had been slightly sharpened with a knife, the very tip had broken off. The needle (170) was made from a bone splinter. The broad, thin, somewhat concave end had broken off at the drilled eye. The pointed other end is quite robust. The needle seems very suitable for, for instance, mending a woollen sail.

Table 5. Administration and navigation

Category	Item	Number	Findspot
3.1 Ship's papers	-	-	-
3.2 Stationery	-	-	-
3.3 Financial/commercial matters	-	-	-
3.4 Orientation and dead reckoning	compass needle(?)	160	afterbody
3.5 Timekeeping	-	-	-
3.6 Signalling and distress signals	-	-	-

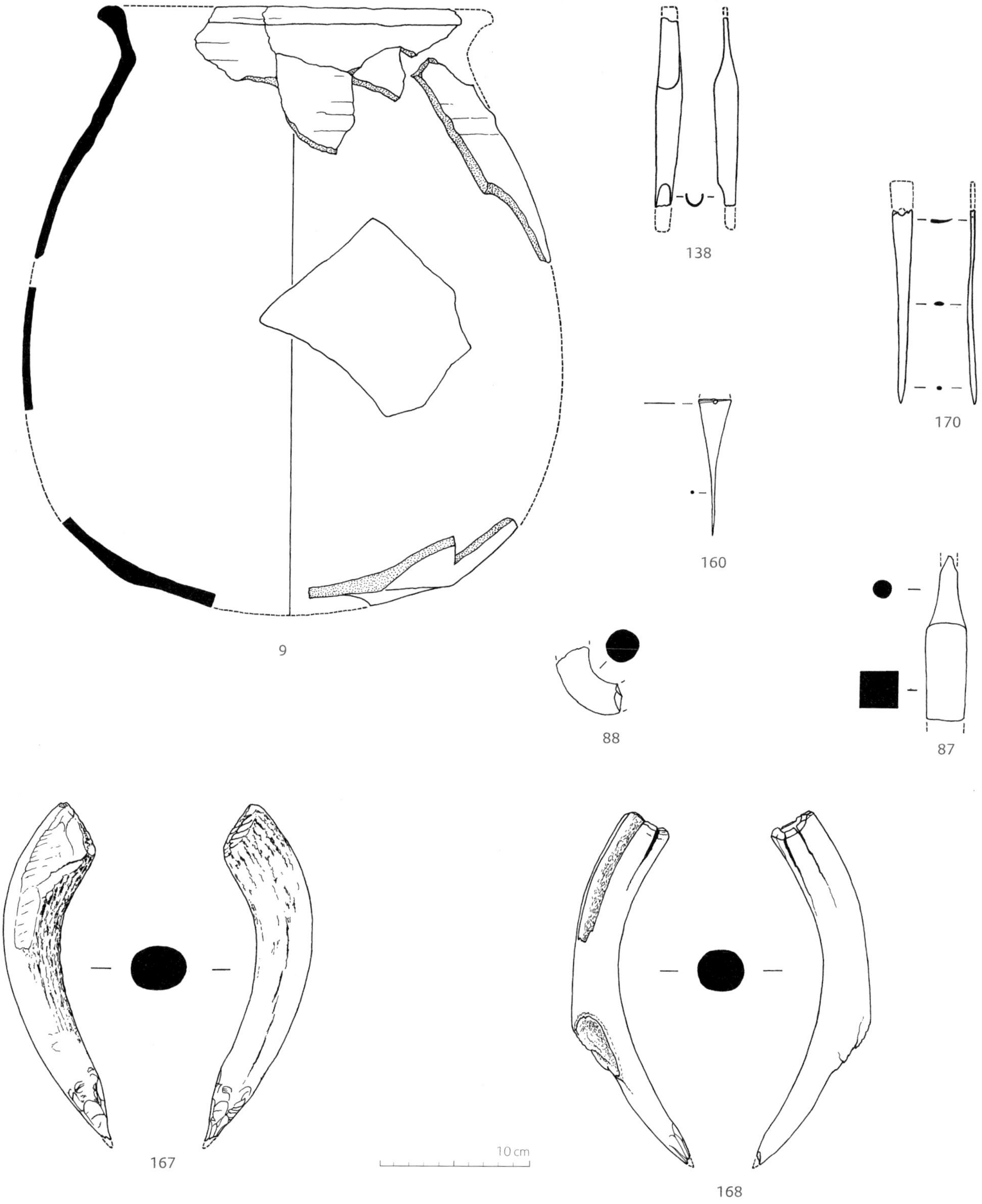

Fig. 6.58. Navigation and tools. Tar pot (9); tool handle? (87, 88); spoon auger (138), compass needle? (160), marlinspikes (167, 168), and (sailcloth) needle (170).

Fig. 6.59. Navigation and tools. Sherds of the tar pot, tool handles?, the spoon auger, the compass needle, the two marlinspikes and the (sailcloth) needle.

Fig. 6.60. The application of a novel caulking method: sintel clamped moss caulking, and the final preparations of a voyage by *nef/*(Mediterranean) cog. Detail of a miniature from the *Histoire Ancienne jusqu'à César* (Folio 31r of MS Douce 353, Bodleian Library, Oxford).

Table 6. Tools

Category	Item	Number	Findspot
4.1 Carpentry	spoon auger	138	the hull
4.2 Caulking and maintenance	tar pot	9	hold/under the hull
4.3 Splicing, sailmaking	marlinespikes, (sail), needle	167, 168, 170	under the hull
4.4 Miscellaneous	?, ?	87, 88	under the hull

No tools were found that can be associated with caulking. Yet scattered between the floor timbers under the roundwood floor, in the midships and under the wreck, sherds were found of a fairly large, blue-grey cooking pot (figs 6.58-9 and 6.59), with traces of soot and tar on the inside and the outside of some base sherds. Most probably this was a pot for melting the tar that was needed for maintaining the ship and the sail. An illustration in the *Histoire Ancienne jusqu'a César* (fig. 6.60) most probably depicts tar being melted and stirred with a stick, in two *Kugeltopf*-like pots over wood fires. The illustration will be more elaborately discussed in section 29.5.2. Two worked pieces of wood from the cog may also have belonged to tools. Of one piece, cut to a square cross-section (2x2 cm), one end is thinner and round in section (figs 6.58-87 and 6.59), and a fragment of a curved piece of wood, round in section and sanded smooth (figs 6.58-88 and 6.59) has likewise been classified under the heading of 'Miscellaneous'.

Household items

Aboard (small) late-medieval vessels there probably was no clear separation between working or cargo space and the crew's quarters.[72] Specifically domestic items, such as furniture, therefore are hardly to be expected. Nor have there been any indications of bedding space. This in fact applies to virtually all the hundreds of pre-mid-19th-century vessels excavated so far in the former Zuiderzee area. It is possible that a separate accommodation space was partitioned off in the Nijkerk-II cog (Sections 6.2.2 and 6.2.3).

The only recovered domestic item is a beechwood candle holder in the shape of an hourglass (figs 6.61 and 6.62-82). The original piece of wood, 18.5 cm high, had a diameter of 8.5 cm at the bottom and 6-6.5 cm at the top. Both at the top and the bottom 3 cm, its round shape was maintained. A little above the middle, the diameter is 3 cm, and above and below this waist the circumference was carved into eight faces. Into one of the faces an owner's mark or decoration had been cut. Into the centre of the

72. Hocker & Vlierman 1996, 61.

top, an iron candle holder had been inserted (diameter 15 mm), driven into the wood and similar to a sintel clamp lacking one spike and a bit of the central part. On one side the lower half is badly damaged, while on the other side part of the bottom edge was charred. From a wreck that was found in 2000 in the close vicinity of the Poeler cog in the Wismar Bight, a somewhat similar candle holder was recovered (fig. 6.61-82A). That wreck was tree-ring dated to 1486. Sanitation must have included the drain, which most probably doubled as a latrine.

Food preparation, eating and drinking; fireplace

Apart from a few ballast stones on which traces of soot were identified (Section 6.7), and a few pieces of coal (299-300), no finds indicated that the vessel had a fireplace. But originally this must certainly have been the case. The only known fireplace aboard a cog so far was encountered in the Nijkerk-I cog (Section 8.5.2 and fig. 8.9). It is a square wooden tub with a closed base and filled with loam, ca 80x80 cm and 20 cm high.[73]

The Nijkerk-II cog almost certainly had a similar fireplace, but it may have been filled with sand rather than loam, on which a floor of tiles or bricks was laid. Such fireplaces are known in several versions from other late-medieval ships in the former Zuiderzee area. The cog may in fact have had a yet different fireplace. Within the sandy base on the roundwood floor may have been a surface paved with erratic cobbles on which an open fire was burnt. In the wreck of a large seagoing and armed ship in field U34 in Eastern Flevoland, dating from the first half of the 16th century, a somewhat similar roundwood floor was found (NISA file). In the layer of sand lay a 'cobbled pavement' of erratics. A substantial number of floor tiles were found dispersed at the front of this pavement, which suggests that the ship had a built-in, tiled fireplace of considerable dimensions. No fire-tending tools were found in the cog. Coal was never before identified as a fuel in the late-medieval wrecks of the Zuiderzee/IJsselmeer polders.

Cooking ware

In the redeposited sand below the ship, to port, a broken-off foot of a small bronze *Grapen* vessel was recovered (fig. 6.63-164), together with some other small non-ferro metal items. Scattered about the redeposited sand in the bow and midship section, lay a dozen small sherds of a redware cooking pot. This pot had had a somewhat flattened base with three feet, a globular belly and a decoration of double rills and grooves on the shoulder. One sherd has a handle attachment (figs 6.63-6 and 6.64). On the outside, the shoulder was partially covered with greenish lead glaze; on the inside, there is a splash of blackish-green lead glaze on the bottom. The base sherds are sooty on the exterior. H.L. Janssen dates this globular pot with rilling to the second half, or possibly the middle of the 14th century.[74] Six small redware sherds (7 and 8), partially splashed with greenish lead glaze, belong to two different small pots. One of them shows traces of soot.

Table 7. Household items

Category	Item	Number	Findspot
5.1 Lighting	candle holder	82	midships
5.2 Storage	-	-	-
5.3 Sanitation	drain	-	afterbody

73. Vlierman 1992a, 51.
74. Janssen, pers. comm. 2000.

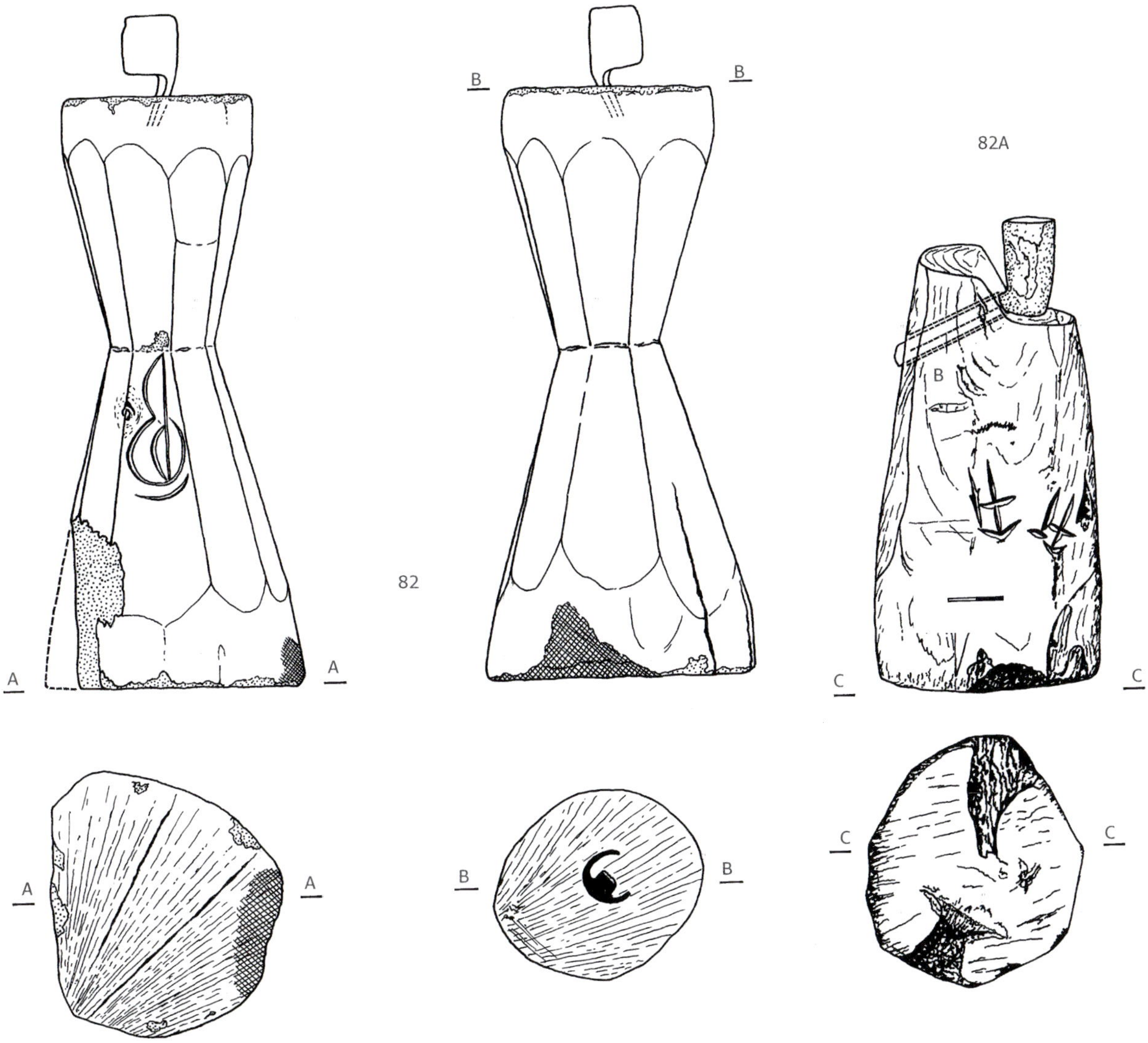

Fig. 6.61. Candlestick (82) and detail of the owner's mark.

Fig. 6.62. Household items. Candlestick (82) and the candlestick from a wreck in the Wismar Bight (82A; after Förster 2000, 49, *Abb.* 3).

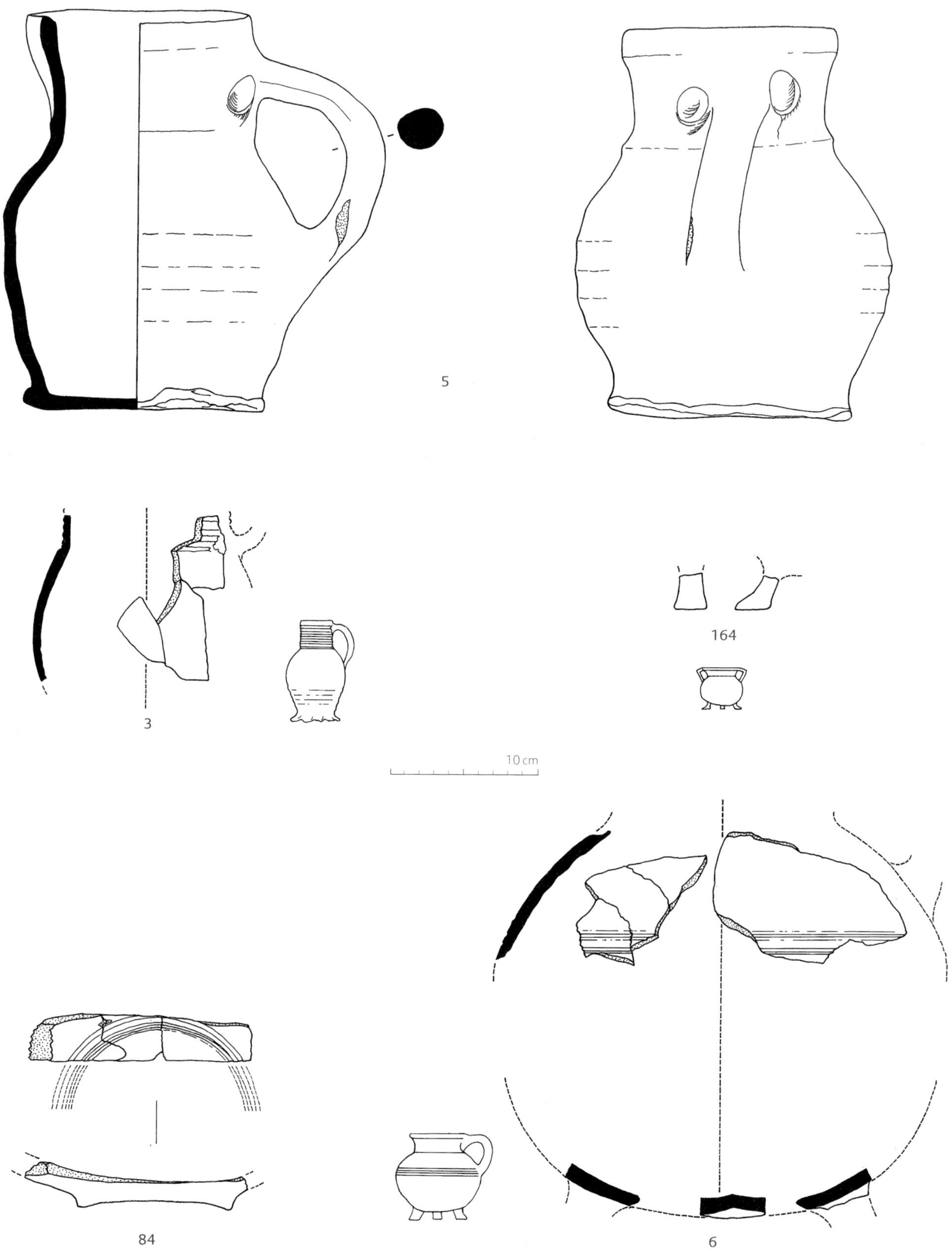

Fig. 6.63. Cooking, eating and drinking. Small stoneware jug (3), small redware jug (5), redware cooking pot (6), lath-turned wooden dish (84) and foot of a bronze *Grapen* pot.

Eating and drinking ware

Scattered about the redeposited sand below the wreck were three joining pieces of a small lath-turned wooden dish (fig. 6.63-84). It has a footring and rising flanges. On the inside, at the transition from the flat to the flange, it has two narrow ornamental grooves and a slightly wider one.

Drinking ware is represented by a redware jug (figs 6.63-5 and 6.64).[75] and fragments of three different stoneware jugs (1 and 2, fig. 6.63-3). The redware jug is of a type quite unknown in northwestern Europe. It has a flat base, a globular belly and a wide neck with a fairly large pouring lip. Opposite the pouring lip it has an obliquely placed roll handle with a pair of finger imprints at the top. On the shoulder are a patch and some spatters of yellow-green lead glaze. The jug has a pale, yellowish to reddish ceramic body of a rather soft fabric. It is definitely not a Dutch or Flemish product, but does display some traits that are (faintly) reminiscent of pottery from various regions, including the squat form, the peculiar handle attachment, and the shape of the base. What is certain is that these traits are typical of the late 13th and early 14th centuries.[76] Between 1971 and 1973, the riverbed of the river Charente at Port-Berteau, a little southeast of La Rochelle (Charente-Maritime, France)[77] yielded a batch of ceramics that must have been produced between ca 1250 and 1350 at the pottery centre of Saintonge, some way upstream. It included a number of jugs that display a marked similarity to the specimen from the Nijkerk-II cog. Of stoneware jug 1, three neck/rim sherds were found; of jug 2, one neck and two belly sherds; and of jug 3, three belly sherds and a belly-neck sherd with a handle attachment. All three are products of Siegburg (Germany), which according to H.L. Janssen were still unknown around 1350 and should be dated to the third or fourth quarter of the 14th century.[78]

Fig. 6.64a. Cooking, eating and drinking. Sherds of small stoneware jugs 1, 2 and 3; redware jug (5), sherds of redware cooking pot (6) and foot of a bronze *Grapen* (164).

Fig. 6.64b. The redware jug (5).

Table 8. Cooking, eating and drinking

Category	Item	Number	Findspot
6.1 Fireplace	-	-	-
6.2 Fire-tending tools	-	-	-
6.3 Fuel	coal	299, 300	midships and forebody
6.4 Cooking equipment	cooking pots, bronze Grapen	6, 7, 8, 164	midships
6.5 Eating and drinking gear	drinking jugs, treen dish	1, 2, 3, 5, 84	midships, under hull

75. Vlierman 2010, 193 fig. 7.
76. H.L. Janssen and F. Verhaeghe, pers. comm. 2000.
77. Rieth 1979, 117-129.
78. Janssen, pers. comm. 2000.

Victuals

During long voyages, a ship naturally needed to carry supplies of food and fresh water. However, there are several indications that this vessel had been laid up for a considerable time in the 'harbour basin', so hopes of finding provisions or their containers on board are rather unrealistic. Yet various objects have been found that may be food-related. Therefore it was decided to put them under this heading. Some may in fact derive from a former cargo. A half base with a diameter of 30 cm and some stave fragments may have belonged to a single, small oak barrel (figs 6.65-98 and -99, and 6.66). A fragment of a partially lathe-turned wooden stopper (fig. 6.65-101), four small stoppers (figs 6.65-102 to -105 and 6.66), a wooden stopper almost square in section (fig. 6.65-106) and a large and possibly conifer-wood stopper (figs 6.65-119) without doubt served to close large and small barrels. A base fragment and five small staves (fig. 6.65-90) belong to a (conifer-?) wooden dish. The base is about 18 cm across, the stave dish was 6 cm high, while the rim was 14 to 15 cm across. A sixth small stave, 5.1 cm long, belongs to a smaller bowl (figs 6.65-90G and 6.66). Excavations in the Hanseatic city of Lübeck (and elsewhere) have yielded several such stave bowls.[79] Further, dozens of small, thin (broken) oakwood sticks were found throughout the ship (91-97, figs 6.65-93 and 6.66). They are cut to a roughly round diameter, and originally 12-15 cm long. The thicker ends are up to 0.8 cm thick, while the other end tapers to a point. Such sticks were also found in abundance in the Nijkerk-I cog (Section 8.5.2), in the Rutten cog (section 10.5.2) and the Doel-I cog (Section 20.5). For many years they were simply known as 'cog sticks' and no-one had a clue as to their purpose. However, a chance discovery in 2002 provided a quite plausible explanation.

Fig. 6.66. Victuals. Fragment of a barrel base and fragments of staves, staves of dishes, various stoppers, and 'cog sticks'.

In the southern valleys of the Alps, even today there are extensive forests of sweet chestnut (*Castanea sativa*). For example, in the Val Bregaglia, southwest of St. Moritz (Switzerland), close to the Italian frontier; here lies the hillside village of Soglio, one of the most authentic villages in the region. Not only is it a tourist hotspot, it is also famous even today for the various products that the villagers have since centuries been making from the sweet chestnut. Formerly, one of the ways chestnuts were prepared is slowly drying them over an open fire, after which they were ground and mixed with milk. This made a kind of gruel: polenta (nowadays polenta is usually made from maize). This gruel was typically the food of poor peasants. The chestnuts were also traded and transported, packaged in small batches. To this end, people used simple, home-made baskets. These were made from small oval planks ca 1 cm thick (nowadays hardboard is used for

Fig. 6.67a. Souvenir basket for packaging sweet chestnuts from Soglio (Switzerland).

Fig. 6.67b. Rounded, flat shingles on the wall of a farmhouse in Haute-Savoie, France.

79. Neugebauer 1975, 117-121.

Fig. 6.65. Victuals. Fragment of a barrel base and fragments of staves (98, 99), various stoppers (101-106, 119), staves of small bowls (90) with reconstruction, and 'cog sticks' (91-97).

such souvenirs), measuring ca 25x10 cm. Holes are drilled along the edge at intervals of about 3 cm. Into these are driven 15-cm sticks, through which are woven thin twigs to form the sides (fig.6.67a). Small planks of roughly this size used to be made in huge quantities as a winter cottage industry in the Alpine region, the Mediterranean and in the Pyrenees, in and west of the Haute Savoie and the Jura. They served as shingles for covering roofs and house walls. They were made with a wedge, a broad chisel and a hammer from greenwood, therefore still moist trunks, that were previously sawn to lengths of about 25 cm. The exposed lower end of the shingle might be rounded (fig. 6.67b). Depending on what was available, people used wood of spruce, chestnut, beech, oak or acacia.

Presumably these shingles, which were easy to make and would have been kept in store, would also be used for other purposes, e.g. for making these baskets. The sticks in the modern souvenir baskets are identical to those from the cogs. Hence it is quite likely that the sticks had been made for such baskets. In olden days, nuts and chestnuts were an important (supplementary) part of people's diet, and because they preserve well are very suitable for taking along on sea voyages. The provenance of sweet chestnuts is southern Europe, but also in western and parts of central Europe they became native, having escaped since Roman times. In these more northerly parts, sweet chestnuts ripen less well and those on sale here nowadays mostly come from southern France or Italy. It seems far from unlikely that sweet chestnuts and nuts were carried as victuals on Crusaders' ships sailing between Italy and the Holy Land. So that as a result people in northwestern Europe became (re-)acquainted with this food, as was the case with many other products and technical inventions. Indeed sweet chestnuts together with hazelnuts and walnuts may have been transported through the Hanseatic network to these northern parts and beyond, in just such baskets. In the redeposited sand in the bow section, one whole and eight half hazelnut shells (304) came to light.

During the excavation of the cog 255 bones were recovered. On the basis of their stratigraphic distribution, they were classified into seven groups:

1. left over from meals eaten aboard
2. as 1 or from the settlement
3-5. very probably as 1
6. as 1
7. scattered around the site, presumably settlement waste

The bones in group 1 were gathered in the redeposited sand inside and below the wreck and are certain to have been inside the ship when it finally sank. Their examination (see Appendix) tells us that they came from cattle and from sheep or goat, and that 71 % of them display cutmarks. This is a result to be expected from meals consumed aboard a ship. 43% of the bones are abraded, which indicates that they were washed about for a long time during the wreck-formation process, and were worn down through contact with the sand and timbers. The bones of groups 2 to 6 all come from the Almere clay deposit and must (in part) have been on board (on the deck?) or were washed into the ship at an early stage of wreck formation. One of Laarman's conclusions (see Appendix) is that group 4 is the largest, and that besides considerable numbers of cattle, sheep/goat and pig bones, it also includes a few bones of horse, red deer, chicken, duck, goose, black-throated diver and crane - species hardly to be expected aboard a ship. Also the presence of odd bones of pets (dog and cat) in groups 2, 3, 4 and 6 seems rather curious. In a wreck, one would sooner expect to find a complete skeleton of a cat or dog.

However, considering the explanation of the complex stratigraphic distribution of the finds (Section 6.1.3), and the fact that almost certainly the cog in its final years was no longer a working vessel but probably served as a fisherman's shelter or storage space, the appearance of these

Table 9. Victuals

Category	Item	Number	Findspot
7.1 Fresh-water supply	barrel, stoppers	98, 99, 101-106, 119	under the hull
7.2 Supply of beverages	-	-	-
7.3 Livestock	-	-	-
7.4 Food provisions	stave bowls, sticks, hazelnut shells, bones	90, 91-97, 304, 231-288	under the hull

species is quite acceptable. In this period we are dealing not with the victuals aboard a ship, but with a more varied pattern of (fresh) food in a land-based situation. The fisherman who is presumed to have dwelt in the cog, or at any rate to have regularly occupied it, will surely have bagged the occasional duck, goose or black-throated diver, and maybe poached a red deer. The bones of dog and cat probably came from drowned pets. The large numbers of bones suggests that redeposited settlement waste and the remains of animals drowned in floods account for a large proportion of them.

Personal belongings and weaponry

The category of personal belongings and weaponry is unlikely to be represented in a vessel that probably had not seen any service for a long time. An almost straight and in cross-section almost oval root-wood knife handle and an iron blade that probably belonged to it (figs 6.68-86 and -139 and 6.69), part of another root-wood knife handle (figs 6.68-85 and 6.69) and fragments of a decorated leather knife sheath (figs 6.70 and 6.72-213A and B) in fact are the only (incomplete) items that can be reckoned to this category. A fragment of thin copper plate with a punched decoration (figs 6.68-162 and 6.69) and a small, presumably pewter button (figs 6.68-163 and 6.69) may come from a knife- or dagger-sheath and a belt, respectively. In this category have been placed also a round copper plaque with a hole (figs 6.68-161 and 6.69) and a small, broken, turned bone rod with some ornamental grooves (figs 6.68-169 and 6.69) and a thin plaque of lead or pewter pierced through with holes (figs 6.68-165 and 6.69), items the function or meaning of which could not be established. Some of the holes in the last-named item are somewhat lozenge-shaped and may have been made with an awl.

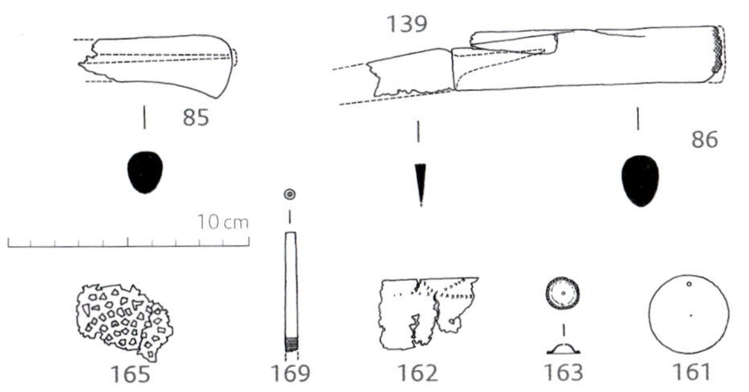

Fig. 6.68. Personal equipment/weaponry. Knives (85, 86, 139), sheath fitting? (162), button (163), round copper plaque (161), lead or pewter plaque with (awl?) holes (165), and bone rod (169).

The relatively large numbers of leather fragments, large and small, mainly of footwear, naturally were initially classified as personal belongings. In all cases, these were parts of (heavily) worn and discarded objects. These fragments were kept on board to be cut up and reused for a variety of purposes. A couple of right and left soles probably belonged to a pair (figs 6.70 and 6.72-191 and -192); two others (figs 6.70 and 6.72-193B and -194) probably did not. A strip had been cut out of the last-mentioned piece. The heel part of a sole composed of two pieces (figs 6.70, 6.71 and 6.72-211) had been worked. A rough round hole was cut into it, while the round heel end had been made angular. Maybe it was intended as a suspension patch for e.g. a handle on some object. In all, 23 soles or parts of soles, and fifteen fragments of stitched welts were found. Four fairly completely preserved parts of uppers (figs 6.71 and 6.72-171A, B and C, -172A and B, -174A and B, and -175A), judging by the stitching on the top edges, come from an ankle boot or a higher model (see

Fig. 6.69. Personal equipment/weaponry. Knives, sheath fitting? (162), button, round copper plaque, lead or pewter plaque with holes, and bone rod.

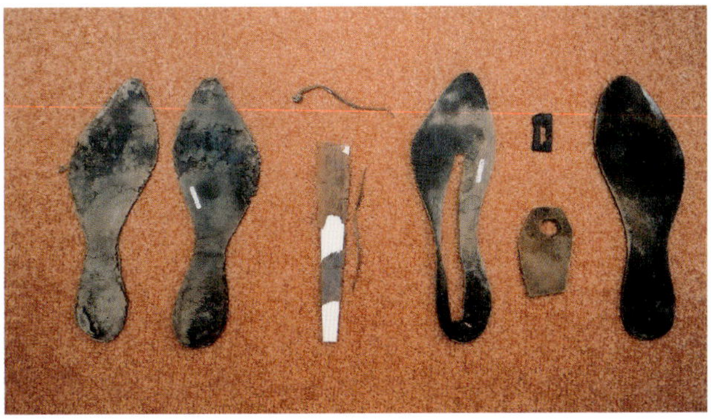

Fig. 6.70. Personal equipment/weaponry. Shoe soles (191, 192, 193B, 194), sole fragment with a hole (211), knife sheath (213), shoelace (223) and belt fragment (224).

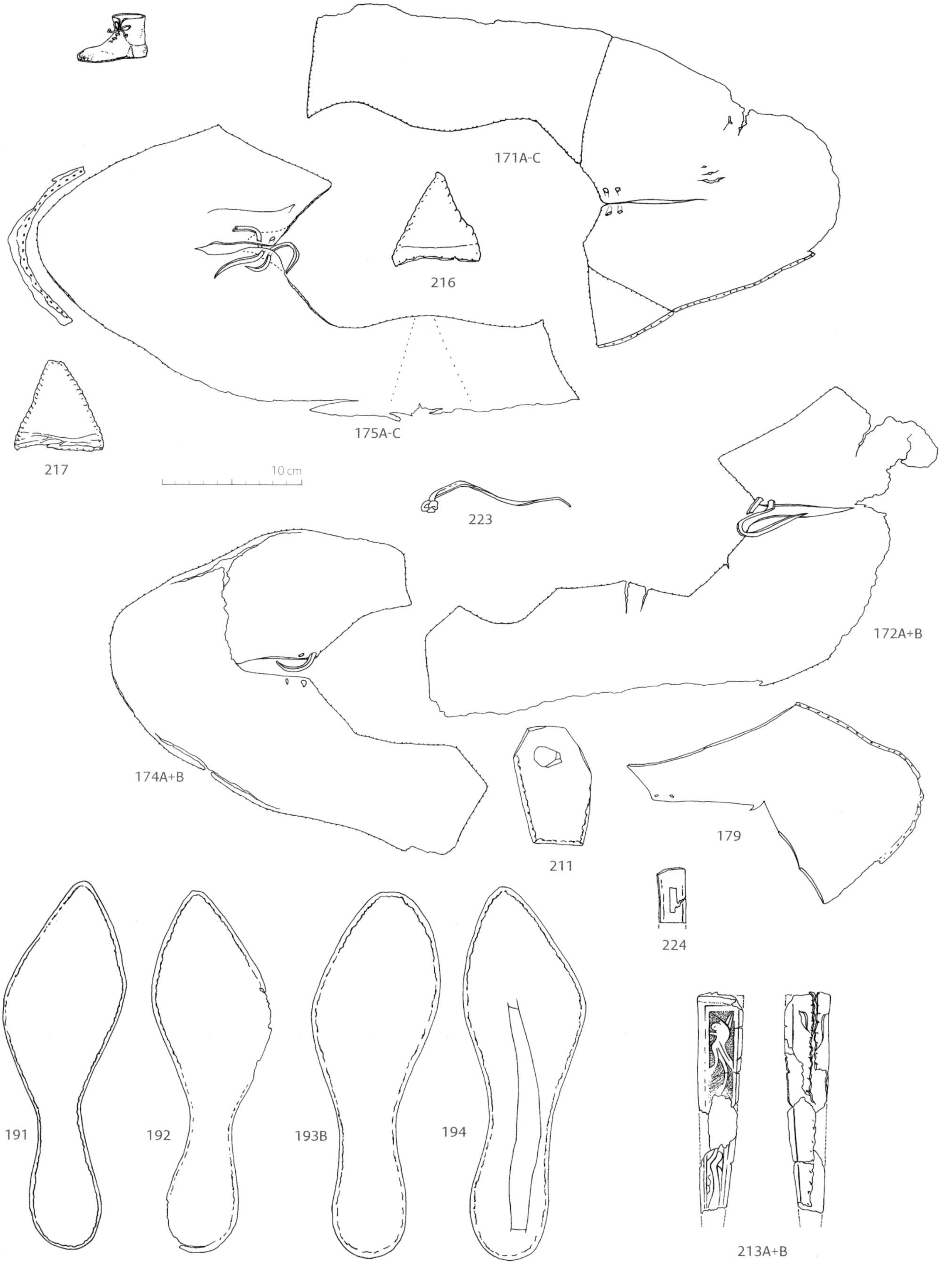

Fig. 6.72. Personal equipment/weaponry. Vamps of shoes (171A-C, 172A and B, 174A and B, 175A-C), shoe soles (191, 192, 193B, 194), sole fragment with a hole (211), knife sheath (213A and B), shoelace (223) and belt fragment (224).

reconstruction sketch, fig. 6.72).They are comparable to the mid-14th-century type DDT708 from Dordrecht,[80] The first- and the last-mentioned, still contain remnants of the laces. At the end of a separately found lace, a knot had been made (fig. 6.70 and 6.72-223). Further, two heel stiffeners were found (figs 6.17 and 6.72-216 and -217) and one tongue of a shoe (215). The bulk of the leather finds were fragments of various sizes, of uppers of different shoes. It seems that in all cases they were shoes of an ankle model or high type with a sole stitched on with welts. Parts of some had been cut off for re-use. A comparatively large piece of leather (187) cannot have been part of a shoe or boot. Maybe it belonged to a jacket of some kind. A small, rectangular piece of leather with an also rectangular hole in it (figs 6.70 and 6.72-224) is the cut-off end of a belt. The quality of the leather and its tanning is remarkably diverse. No personal weapons were found in or near the wreck.

A small coin (fig. 673-308) was found in the redeposited sand at the fore end of the keel plank. Judging by its findspot, it is likely to have been lost aboard the ship. After cleaning, the coin was identified at Rijksmuseum Het Koninklijk Penningkabinet; it is described in section 6.8.

Fig. 6.71. Personal equipment/weaponry. Shoe vamps (172, 174, 175, 179) and two heel stiffeners (216, 217).

Fig. 6.73. Counterfeit coin of Reinald II or III (308).

Table 10. Personal belongings and weaponry

Category	Item	Number	Findspot
8.1 Garments	button, jacket?, belt	163, 187, 224	afterbody, hold
8.2 Footwear	shoes	171-186, 188-223, 225	in and under the wreck
8.3 Sewing kit	-	-	-
8.4 Tools, weapons	knives, sheath fitting?, button, knife sheath	85, 86, 139, 162, 163, 213	in and under the wreck
8.5 Cash	coin	308	forebody
8.6 Miscellaneous	copper plaque, pierced plaque, bone rod	161, 165, 169	to port beside the wreck

80. Goubitz *et al.* 2001, 189 fig. 7.

6.6 Cargo and merchandise

Scattered within and below the afterbody, various fragments of a round, flat stone were found (288-1 to -19, fig. 6.74). The rock is a very fine, greenish-grey and thinly laminated schist and originally had a diameter of ca 46 cm. Along its edges it is just 0.6 cm thick, while towards the centre it is thicker. Its greatest measured thickness is 1.3 cm. Both sides of the disc-like stone bear a dense pattern of grooves 1.5 mm wide and 1 mm deep. It seems that these grooves were scratched in in 20x10 cm areas of parallel grooves, with the grooves in adjoining areas running at a different angle. Where the areas meet, the grooves overlap, sometimes crosswise. The dense pattern of grooves suggests that the piece may have been intended as a millstone. It must have broken before it could be used, as it lacks any use-wear. Judging by its thickness and its crumbly texture, it seems unsuitable for milling. Besides, such an item is rather anomalous aboard a ship, which prompted the idea that this may have been part of an erstwhile cargo. Two small, unworked pieces of rock (figs 6.74-292-1 and -2) are very similar to the rock of the 'millstone'. Section 6.5.3 describes some parts of objects that may in fact derive from cargo.

6.7 Ballast?

Among the V-shaped and straight floor timbers of the forebody, on and immediately beside the stem hook and keel plank and among the redeposited sand under the forebody, about 300 roughly fist-sized erratic boulders were found. Large(r) erratics are regularly found as ballast in shipwrecks in the former Zuiderzee area. The place in the ship occupied by the stones suggests that they may be ballast. Yet their small number and the size of the stones cast doubt on this interpretation. Assuming a weight of 500-750 gr per stone, the maximum weight of the stones is 225 kg in all. In terms of ballast, this would have barely any effect on this vessel. Of course it is possible that these rocks were part of a much larger batch. On some quartzite and quartzitic sandstones faint traces of soot were observed during the excavation. Such remnants may have been present on more stones before being rubbed away in the course of wreck formation. The presence of the soot traces may mean that the rocks once may have paved a fireplace. The potential form of such a fireplace was discussed in section 6.5.3. The total number of stones may cover an area of ca 2.5 m^2. This is too much for paving a fireplace that consists of a ca 80x80x20 cm, sand-filled wooden box. Without sand, the whole batch might fill a box measuring ca 100x100x25 cm. A fireplace embedded in the sand covering the roundwood floor seems another possibility.

The boulders were all strongly rounded (rolled smooth) and remarkably intact. They were identified by the late W.Tj. Hellinga.[81] His written report goes as follows:

Granite	104	
Syenite	1	
Diorite	1	
Gabbro	1	
Aplite	1	
Sandstone	3	
Fissure-vein quartz	4	
Quartzite	21	
Schist	27	
Gneiss	100	
Garnet gneiss	1	
Granite Bohuslän	1	environs of Gothenburg
Lysekill granite	1	idem
Augen gneiss	5	environs of Trollhättan
Quartzitic sandstone	13	southern Sweden
Flint	8	idem
Småland granite	7	Småland (southeastern Sweden)
Fissure-vein quartz	1	idem
Basalt	3	idem
East-Baltic granite	1	
Total	304	

The material is definitely not of Dutch origin. It is too homogeneous, the stones too undamaged, while the rock types are very different from Dutch associations of erratics. The total lack of limestone means that the batch cannot be from Denmark. Remarkably, the identifiable material comes exclusively from southwestern, central-southern and southeastern Sweden. Also the large proportion of gneisses points to this provenance. The strong rounding indicates that the stones were rolled over a considerable distance. The principal directions of transport by ice and meltwater are north-south and northeast-southwest. The inclusion of various Småland rocks is explained by the latter course. From the above it can be concluded with near-total certainty that the material derives from the very southwest of Sweden (Malmö, Trelleborg).' In addition to Hellinga's report it may be remarked that the strongly rounded and in some cases almost polished stones will not be (solely) the result of transportation by ice and/or a meltwater river over a distance of 500-700 km, but probably also of subsequent

81. Hellinga 1984, report on the identification of erratics, file Zo 36 RCE-Lelystad.

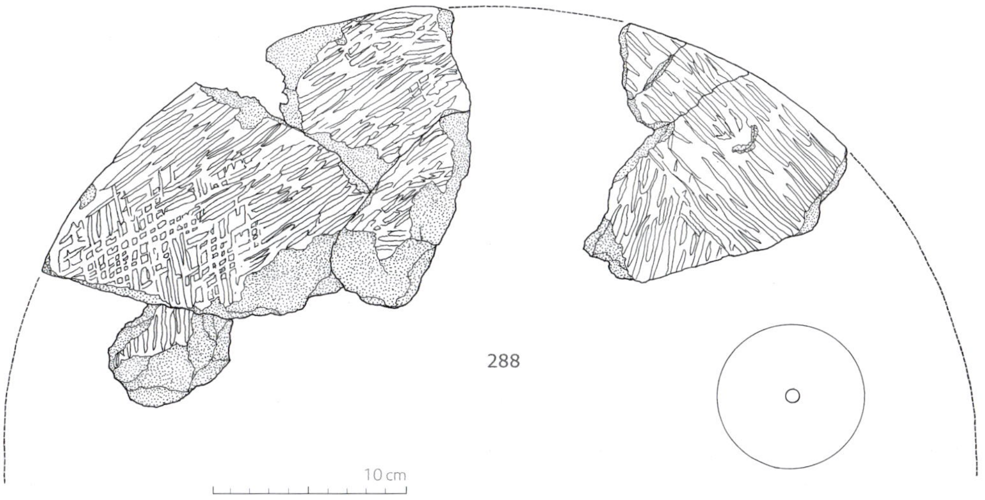

292-1

292-2

Fig. 6.74. Pieces of a millstone (288) and two fragments of rock (292-1 and -2).

rolling to and fro on a rocky coast for thousands of years. It therefore seems most likely that the stones were gathered on the coast of the above-mentioned region.

6.8 Dating

Ark

Despite the very carefully examined soil profiles, it proved impossible to arrive at a global dating of the foundering of the cog on the basis of the local sedimentary layers. This has been possible in the case of most other wrecks in the IJsselmeer polders. Here the oldest datable deposit is the Zu IV shell layer, which was deposited in the first or second quarter of the 17th century.[82] In this southernmost part of the Zuiderzee it may in fact have been a little later. The Almere clay deposited between this layer and the cog is barely stratified, while in this area also a considerable part of the Almere deposit is absent. In the top of this Almere clay layer and partially poking into the Zu IV layer, two large fragments of small *Kugeltöpfe* were found, one of them with brushstroke decoration. They are settlement waste, both dating from the 13th century. The decorated specimen may actually be early-14th-century. The bulk of the recovered potsherds from the settlement suggests that the settlement's heyday was in the 13th century. On the surface, and at a good distance from the wreck and the channels, just four sherds of Siegburg stoneware were found, domestic pottery of ca 1310-1370. They are the only sherds that can be definitely dated to the 14th century. On this basis it can be concluded that in the second quarter, and certainly by the middle of the 14th century, all permanent occupation at the settlement had ceased. A confirmation of this can be seen in the dike-building charter granted by Reinald

82. Vlierman 2006, 151.

III in the spring of 1356. The charter may have been a response to the historically documented, severe storm of that year. The construction of the dike implies that the unendiked land was now being sacrificed to the encroaching sea. This matches the historical evidence, which indicates that Ark was engulfed in the 14th century, and is no longer mentioned in the records of the 1360s. The small, Carolingian-period settlement somewhat further north had probably disappeared through shoreline erosion (well) before 1300.

Finds from the cog
A number of globally datable finds were recovered from the sand deposit in the bilge. For example, a small coin was found at the fore end of the keelson (fig. 6.73). It was probably lost aboard the ship. It was examined by the Rijksmuseum Het Koninklijk Penningkabinet,[83] which produced the following data:

In the first instance, it seems to be a quarter Great of Guelders, issued by Duke Reinald II (1327-1343) or Reinald III (1343-1361). Such a coin is formally described as follows:
Guelders
Reinald II or Reinald III
Quarter Great, struck at Harderwijk
obverse: Guelders lion rampant with circumscription:
+ REIN : DVX : GHELRIE
reverse: long-armed cross with circumscription:
MON-ETA-DEH-ERD
catalogue: Van der Chijs, Gelderland, Pl. II, nos 6 and 7.

A problem is that the circumscriptions on this specimen do not quite match up with this description. In as far as they are legible, they read:
obverse: + REI(N) : DVX : O(N)ELRIE
reverse: MO(N)-AC-DE(N)-ERD
The letters in brackets are poorly legible.

The small deviation, underscored in this rendering, suggest that this is a contemporary, i.e. 14th-century, counterfeit coin. The advantage to the forger would have been the smaller amount of silver that went into the counterfeit coin compared to a genuine Quarter Great (lower content and lower weight). The dating value of this coin however remains the same as that of a proper Quarter Great: (second half of the) 14th century.
The redware jug of unknown type must, judging by various traits, be dated to the late 13th or early 14th century. Besides, the ship held sherds of a redware cooking pot. Its globular shape and ornamental grooves indicate that it is a type belonging in the second half or maybe even middle of the 14th century, whereas the (sherds of) Siegburg stoneware jugs of three different types were as yet unknown in 1350 and must date from the second or third quarter of the 14th century. This means that it is as good as certain that the vessel cannot have (definitively) sunk before ca 1360. Also it is unlikely to have happened much later. The cog's foundering in the third quarter of the 14th century is therefore the most plausible scenario. A number of clues point to the possibility of the ship already having been partly dismantled by the time it sank, having been written off and probably laid up at this spot for some time already. The (afterbody of the) ship may well in this final phase have been used by a fisherman as a (temporary) shelter or for storing his tackle.

Dendrochronology
In 1983, at the end of the excavation, eleven wood samples were taken for the purpose of any future dendrochronological analysis. At that time facilities for such research still were unavailable in the Netherlands, while in the case of most shipwrecks in the IJsselmeer polders a fairly close dating of the ship's foundering could be derived from the associated finds.[84] Also the condition of the ship's timbers and the degree of repair often allow an acceptably close assessment of the ship's age at the time of its demise. Together with the financial aspect, these were the reasons for not yet attaching a high priority to having tree-ring analysis performed in Germany. During excavation it was decided that to preserve the ship all timbers were to be reburied below the water table, to allow proper conservation in the future; therefore samples were taken only from detached parts, whose original position in the hull could not yet be established at the time, and from some ceiling boards. Hence without a clear, predefined question regarding construction phases or repairs. The full results of the tree-ring investigations by RING[85] are presented in the appendix. Part of the report reads as follows: 'None of the samples could be dated individually with the reference chronologies available to us. All samples except m. 12 and 13 could, however, be cross-dated and were averaged to compile an object curve, kkostd. This curve was subsequently dated in absolute terms as follows:

83. H.W. Jacobi, written comm., 25 October 1983.
84. Vlierman 1983, 3 and 4.
85. Jansma & Hanraets, written report, March 1994.

Table 11. Object curve

Dendro name	Reference chronology	Dating	gl.	t.	ÓC
kkostd	Apeldoorn	AD 1173-1334	68.7	5.53	0.001kkostd
kkostd	E. Netherlands/ Westphalia	AD 1173-1334	61.8	4.91	0.01

Table 12. Dating of the individual samples relative to the reference chronologies

Dendro name	Reference chronology	Dating
kko070	-	AD 1238-1305
kko100	-	AD 1258-1334
kko110	-	AD 1220-1312
kko120	-	undatable
kko130	-	undatable
kko140	-	AD 1199-1333
kko151	-	AD 1247-1325
kko160	-	AD 1202-1311
kko171	-	AD 1233-1318
kko180	-	AD 1173-1322
kko190	-	1236-1325 AD

Table 13. Interpretation of the datings

Object	End date	Felling date
Oz36 (cog)	1334+1(the last ring is always omitted)	AD 1135/1336

Given the position of the sapwood boundary, it is unlikely that the wood samples come from simultaneously felled trees. In chronological order, while assuming an average number of sapwood rings of 27+7, we list the estimated felling date of the wood samples that contain sapwood as follows:

Table 14. Estimated felling dates of the samples with sapwood

Dendro name	End dating	Felling date
kko171	1319	AD 1319-132
kko180	1323	AD 1323-1331
kko190	1326	AD 1326-1339
kko151	1326	AD 1326-1339
kko140	1334	AD 1334-1342
kko100	1335	AD 1335-1342

Although these estimates do in part overlap, it is clear that the felling date of kko171 cannot coincide with those of kko100 and kko140, and probably also is incompatible with those of kko151 and kko190; this means that we are dealing with at least two 'building phases', or at least one repair or alteration episode, maybe more.

In the eventual reconstruction, the original position in the ship of several of the sampled detached timbers could be determined. Of others, it is possible to identify their global position, on the basis of their shape and findspot in the wreck. Dendro sample kko171 comes from floor timber M, which could be identified as part of frame 28.

Kko 180 comes from a frame fragment from the afterbody to port, level with strakes J and K. The felling date of kko171 suggests that construction may have started (quite shortly) after 1326. The V-shaped floor timber can in fact have been put in only during (the final part of) construction phase 3 (Section 6.3.2). No samples were taken from timbers used in the first two construction phases, so a slightly earlier start cannot be entirely ruled out. Of the afterbody it was concluded above that this almost certainly was built throughout the construction period, up to strake N, and that wash strake O was not added until (much) later. The felling date of both samples suggest an overall construction period of at least 5 years, which seems rather long. Sample kko190 is of an upper futtock from the after half of the ship and kko151 probably comes from the starboard lower futtock of frame 11. The felling dates of both timbers indicate that they can only have been used after AD 1339. In the first case this seems reasonable, but that the builders should have waited for several years before installing upper futtocks is rather implausible.

A more likely procedure might be that around ca 1340 the crossbeams 7, 8 and 9 were installed, requiring the system of futtocks and frames to be adjusted. Still, the identical felling date of kko151 is hard to explain in this vein, since it seems unlikely that a lower futtock should be replaced. The substantial number of cracks closed up with sintel clamped moss caulking in the underside of the bottom during construction suggests a relatively long construction period. Kko 100 is from an upper futtock with a sloping, straight top end. It is very probably the starboard upper futtock of frame 36 in the forebody. The felling date of this piece suggests that it was installed a considerable time after construction (after AD 1342). The same dating of starboard bilge ceiling strake B in the forebody strikes one as peculiar. But it is well possible that during construction of the forecastle also the foremost part of the bilge celing strake was replaced by a more massive timber (compare the removal of crossbeam 6, after which its hole in the shell was covered with a 'shelf clamb'). More extensive tree-ring research on designated parts of the wreck might provide more clarity.

Radiocarbon dating

The Centre for Isotope Research of the University of Groningen analysed one sample. It was a piece with sapwood from the V-shaped floor timber of frame 28 in the bow (GrN-205770); the outcome was a date of 710 ± 20 BP, global AD 1300.[86]

To summarise, we can come to the following conclusion.

On the basis of the results obtained from tree-ring analysis so far, construction of the cog was started in the beginning of the second quarter of the 14th century and may at first have seen some delay, for instance owing to unavailability of the required timber. Around 1340 the vessel was adjusted or altered (instalment of crossbeams 7, 8 and 9, and a fully closed deck on the stringers). Some years later the forecastle was added - though this may have been done as early as ca 1340 too. The youngest dated samples will have been repairs (e.g. the repair in the shell below the spot where the forecastle stanchion had rested). After this repair the cog may have seen further service for a good ten to twenty years before being written off. The crossbeams from the midships and forebody, large parts of the deck, the stringers and the castle(s), as well as the port inwale of the midships and forebody may have been subsequently dismantled, while for some years a fisherman continued to make use of the afterbody. The eventual sinking and disintegration of the hull and the washing away of the final traces of occupation at the village of Ark may be linked to the storm of 1356 mentioned in section 6.3.1, and the severest storm of the 14th century in 1362. The sherds of the redware cooking pot and of the stoneware jugs indicate that this is most unlikely to have happened before ca 1350.

6.9 Summary

Ark

Earthenware and stoneware sherds and a substantial amount of bones belonging to butchering waste, found in the sediment layers overlying the cog and in the remnants of the catastrophically silted-up drainage channels, have demonstrated that between the 12th and the 14th centuries a farming settlement flourished in the vicinity of the wreck of the Nijkerk-II cog. Apart from these finds, all further traces of the settlement have disappeared, together with the sea engulfing a deposit of clay on peat at least 2-2.5 m thick, in several phases. Very probably the heavy storms in 1356 and 1362, known from historical records, were the coups de grâce for the hamlet. The charter, granted by duke Reinald III of Guelders in 1356, to build a dike, probably just a little south of the wreck site, supports this assumption. 300-400 m further north, an earlier, Carolingian settlement had arisen beside a channel that may have resulted from peat-cutting. Presumably this hamlet comprised a few farmsteads, but during the 10th and 11th centuries there almost certainly was no more occupation at (either of) the two settlements.

86. J.N. Lanting, written report, 8 July 1994.

The location of both sites; their names still remembered in the adjacent 'old' land (the Arkemheense Polder); and the historical records of the Werdense Hof at Putten, all suggest that we must be dealing with the Carolingian settlement *Archi* and the 12th- to 14th-century *(H)erca* or *Erke* = *Ark*, whose name no longer occurs in records of 1360-1370. The cog lay (moored, laid up) in a 'harbour basin' a few hundred metres north of the site of the younger settlement.

Ship

The entirely oak-built ship had a striking hull shape, a typical construction and some characteristic features. It carried a single mast, which stood in a cavity in a massively constructed part of a long keelson, probably had an aftercastle from the start, and was equipped with a stern rudder. On the basis of Heinsius' definition, of comparison with vessels depicted on various seals and described in historical sources as cogs, and on the assumption that the Bremen Cog indeed *is* a cog, we may conclude that this ship is a cog dating from the heyday of the Hanseatic League. The ship, which was probably built as a cargo vessel, many years later was altered: provided with a forecastle and possibly at the same time with a continuous deck, which may relate to a (partially?) new (military?) function.

The lines drawing or plank plan shows a striking, high-sided hull with a straight keel and little sheer and with straight, raked stern and stem and everted sides. In the stern and bow, the frames have a faint S-shape. The forebody is slightly higher than the afterbody. The stern and stem rise at rakes of 120° and 130°, respectively. The strakes of the visible, above-water part of the ship are clinker-built and from strake G upwards extend to the rear and front of the stern- and stempost, respectively. The cross-cut hood ends of the strakes on the bow are covered and protected by a false stem; on the mitred sternpost, the port and starboard hood ends meet. The planks of the bottom (strakes A to C) below the hold are laid flush, carvel-style, and are interconnected only by the treenails that hold them to the floor timbers. The bottom strakes as they run into the bow and stern gradually transition into a clinker-built construction. The hood ends, like those of the subsequent strakes D to G, are nailed into rabbets in the hooks and the lower half of the stern- and stempost.

Strake D is the first fully clinker-built strake, attached with double-clenched nails that were twice hammered sideways and driven back into the plank. The frames consist of straight or V-shaped floor timbers most of which were extended up the sides with lower and upper futtocks. Of these, the floor timbers are the most massive; the upper futtocks are the lightest. Across the frames a so-called 'open ceiling' was laid, mainly intended to internally strengthen the fore-and-aft construction; two bilge ceiling strakes on either side were more heavily executed than the other ceiling boards. Initially the ship probably had six crossbeams protruding through the shell: four through strake J and two (in the afterbody) through strake K.

The crossbeams protruding through strakes L and M were probably inserted simultaneously with the installation of the forecastle and the full decking of the ship. Viewed from above, the reconstructed hull displays a striking underwater body, shaped like a laurel leaf and sharply converging towards both ends. At deck and gunwale level, however, the outline is more like a somewhat stretched half-walnut, with the forebody a little wider than the afterbody. Together with the crossbeams protruding through the shell, whose heads were protected by guides mounted on their forward sides, the stern- and stem hook are the most characteristic structural elements. They constitute the foremost and the aftmost parts of the plank keel, as well as the bases for the stern- and stempost.

The bulwarks, rising about 1.5 m above the (lower six) crossbeams and everted, were held in place by heavy upright knees. These knees together with the longitudinal stringers provided a rigid structure to support the above-deck part of the ship. The seams between the shell planks were rendered watertight with clamped moss caulking. This typical sintel clamped caulking method evolved as part of the cog-building tradition and disappeared almost simultaneously with the demise of the Hanseatic League. The design, the lines and the construction of the vessel are evidence of well-established tradition and expertise. This is definitely not a prototype, but an example of a fully developed vessel type. The ship's principal (reconstructed) dimensions are:

- Length over the stern and stem at gunwale level — 20.0 m
- Largest width at gunwale level — 8.7 m
- Height of the stern (keel line to the top gunwale) — 5.1 m
- Height of the stem (keel line to the top gunwale) — 6.2 m
- Height of the side amidships (keel line to the top gunwale) — 4.3 m

From the detailed analysis of the wreck, the building of a scale model and the near-simultaneous and in part experimental construction of a full-sized replica (the Kampen Cog, see Chapter 7), it transpired that the field drawings needed to be reexamined and that a paper-based reconstruction was required. Misinterpretation of the original location of the two recovered crossbeam constructions had meant that the available evidence was considered insufficient to allow such a reconstruction. The ultimate reconstruction on paper shows what the ship must have looked like and what its original dimensions must have been. The building sequence and (eight) construction phases could be identified. Among other things this revealed that we were dealing with the initial almost complete construction of the underwater body ('bottom-first' approach). This was followed by the shell strakes ('shell-first' approach), then the futtocks and ceiling boards up to crossbeam level. The final phase of construction saw the fitting of crossbeams, further shell strakes and futtocks, ceiling boards up to gunwale level, the aftercastle and the fore- and afterdecks. The V-shaped floor timbers, futtocks and ceiling boards in the after- and forebody very probably were not fitted until the lower six crossbeams, the shell strakes above them and the futtocks and ceiling boards between frames 2 and 26 were in place. The building of the forecastle, the addition of the three upper crossbeams, the fitting of a probably continuous deck and

Fig. 6.75. Deposition of the keelson (and the cog's other timbers) below the water table at the site of the wreck excavation.

the placing of a wash strake atop the original uppermost strake probably occurred some ten to fifteen years into the ship's career. From the position of the detached parts of the wreck as excavated (see survey drawing, Plate 6.I) it could be inferred how the ship must have disintegrated in stages, and that this most probably occurred largely as a result of northerly storms. This study also describes the documentation methods followed during and after excavation. The wreck, taken apart into sections, was reburied below the water table, for preservation at the site (fig. 6.75).

Equipment and inventory

A first step towards the classification of items found aboard late-medieval vessels that belonged to the ship's equipment and inventory was made in the publication of the finds from the Almere-I cog.[87] In the present study, all objects from the cogs, which like any other shipwreck with its contents must be considered "as an inseparable whole, and as the only available source for studying the history of different vessel types and their use",[88] have been classified under the following headings and in this order:

1. Ship's equipment
2. Operational equipment
3. Administration and navigation
4. Tools
5. Household items
6. Food preparation, eating and drinking
7. Victuals
8. Personal belongings and weaponry

If indeed the cog had been lying in the 'harbour basin' for a considerable time, derelict and partially dismantled, before finally sinking and disintegrating, then it is small wonder that few if any items of value remained on board that with any certainty might be attributed to its (original) equipment and inventory. The few (parts of) objects that potentially once belonged to it are described under the various headings and in the respective tables. For the category of ship's equipment there are the three pieces of a woven woollen and tarred sail, which in fact is a most exceptional find. Thanks to these pieces it can be established for the first time that cogs, like the Viking ships, carried woollen sails. Confirmation for this finding is provided by some of the other vessels discussed in this study. Another remarkable find is the wooden drain or slop chute, which probably also served as a latrine. The category of operational equipment is represented only by some fragments of wooden grain shovels and a load hook.

87. Vlierman, 1992b.
88. Reinders 1986, 2.

The straight wooden spade, found intact, whose blade was partially sheathed with iron plating, seems suitable only for peat-cutting. Therefore the question remains whether this item actually belonged with the ship. Almost inevitably, no items were recovered that belonged to the cog's administration and navigation. A broken compass needle, if that is what it is, might yet represent this category. The small spoon auger (lost during the ship's construction?) belongs in the category of carpentry tools. The sherds of a tar pot can with some certainty be reckoned to the ship's tool kit, as can the two marlinespikes, although the latter may also have been part of a fisherman's gear. Any household items are hardly to be expected aboard a medieval cargo ship. Therefore the curious hourglass-shaped beechwood candleholder may equally have been a possession of the final occupant. The same may go for the above-mentioned (peat-cutter/spade.

Some soot traces on the erratic boulders in the forebody are the only traces that might relate to a fireplace; a few pieces of coal represent fuel used on board. It is the first time that this type of fuel has been found in a medieval shipwreck in the Netherlands. Originally the cog almost certainly would have had a fireplace. A broken foot of a small bronze cooking pot is of a *Grapen*, which once belonged to the galley equipment on board. We cannot be equally certain about the sherds of a redware cooking pot. This pot, datable to the second half or possibly the middle of the 14th century, may in fact have belonged to the final occupant. The same goes for the sherds of three different small jugs of Siegburg stoneware. They are of a type that was still unknown around 1350, and was not in common use until the third and fourth quarters of that century. The unusual and in these parts typologically quite unknown small redware jug must, on the grounds of particular features, be attributed to the late 13th or early 14th century, and almost certainly was an item from the ship's inventory. The little jug is definitely not from the Netherlands or any of the immediately adjacent countries, and must have been brought along aboard the ship. It shows several features associated with the pottery centre of Saintonge, southeast of La Rochelle (France). The fragment of a small, lathe-turned treen dish too almost certainly belongs to the ship's inventory. Of special interest is the possible function of the sticks that for many years were known as 'cog sticks': roughly cut, thin and about 15 cm long, with pointed ends. They are close parallels to sticks that traditionally have been part of small woven baskets in which (in northern Italy) sweet chestnuts were packaged and traded. Further, in the category of victuals, parts of a small barrel, stave bowls, bones of consumed meat, and hazelnut shells were recovered. The final category, that of personal belongings and weaponry, is represented only by a few parts of discarded footwear and parts of some knives and knife sheaths. The erratic boulders in the ship, serving as ballast or contained in a box as a substrate for a fireplace, can in fact only have been collected on the coast of southern Sweden, from which we may with some caution infer that the vessel at some time visited the Baltic Sea.

Dating

The (counterfeit) coin of Reinald II or III was presumably lost within the ship and, together with the stoneware and the sherds of the redware cooking pot, provides the most important dating clue for the definitive foundering of the cog. This almost certainly occurred in the third quarter of the 14th century - possibly as a result of the 1362 storm. Dendrochronology revealed that the necessary wood for the ship was felled in the adjacent Veluwe region, in the eastern Netherlands and/or in Westphalia, and that construction of the cog must have started in the later 1320s/ early 1330s. Construction probably did not take place over a 'short' span of time, but may have seen some delay. It seems that around 1340 the vessel was altered, provided with a forecastle and a continuous deck. Afterwards the cog may well have sailed for at least another ten to twenty years. It will always remain an open question whether this was a Guelders ship, and whether the alteration of ca 1340 and the presence of the jug that may come from southwestern France might in any way be linked to Reinald II's efforts in the 'One Hundred Years War', and/or military action on the Zuiderzee in the mid-14th century. Eventually the cog was laid up and probably partially dismantled. A number of fishery attributes and a 'keepnet' plus some earlier-mentioned items, suggest that in its final years the ship may have served as a storage space and/or shelter for a fisherman. It never became clear whether the curious roundwood floor aft of the mast was part of the ship or maybe related to the hulk's possible function as a dwelling.

Fig. 7.1. The Kampen Cog under sail on the river IJssel.

The experimental reconstruction of the Nijkerk-II: the Kampen Cog[1,2]

In 2000, at the start of this study on all the archaeological cog finds in the Low Countries, there was a very strong desire to have the most extensive reference possible, as a point of comparison for the various shipwrecks and relevant data. The Nijkerk-II cog was chosen as the base wreck for this purpose. The reason was that it is the most complete and extensively documented large cog excavated in the Netherlands. In addition, NISA had already built a reconstruction model (1:10 scale) of the cog. In the same period, a replica of the ship had been built in Kampen as an experimental reconstruction. The starting point for the construction of the Kampen Cog was the bottom of the Nijkerk-II cog, which had been reconstructed on paper by J.-M.A.W. Morel.[3]

At the Netherlands Institute for Ship and Underwater Archaeology (NISA), there had long been a desire to recreate an excavated late-medieval ship in order to experimentally determine how vessels from this specific period of maritime history were built. It goes without saying that all sorts of aspects of these wrecks already gave rise to quite a few ideas as to how the building process must have gone and what the vessels must have looked like. Recreating a life-size ship could provide concrete answers and insights. The frame of reference was provided not only by the original archaeological data on the shipwreck, but also the model that was built, the findings during the construction of that model, and the 1:1 recreated/reconstructed vessel, the Kampen Cog (fig. 7.1).

No report has ever been published about the construction of the Kampen Cog. It was an important reason, besides the above, to include the report written predominantly by the ship-builder and apprentice trainer (A.J. van den Heuvel) in this study. The fairly detailed report has been structured according to the previous excavation reports (Chapters 8 through 20) to facilitate comparison with the Nijkerk-II report. The report, as well as the summary and conclusion, set out the answers to some long-standing questions and the likely sequence of their construction. It also indicates the extent to which the construction readjusted or confirmed existing ideas, or refuted them, and discusses e.g. how problems were solved. In 2005/2006, as part of the present study, a complete reconstruction of the ship's hull was drawn on the basis of the archaeological data, and all the excavation data on the Nijkerk-II were presented in a report (Chapter 6 and Plates 6.I through 6.V).

Until the reconstruction of the Kampen Cog, experimental nautical archaeology research in the Netherlands had been limited to recreating the Mesolithic tree-trunk canoe of Pesse (Drents Museum, Assen), the reconstruction for Archeon of the planking river barge and dating back to Roman times known as Zwammerdam 6, and a salmon scow (Nz 66-II, Zuidelijk Flevoland) by De Boer & Sars. Besides that, the Roman-era vessels De Meern I and II and a few historical ships have been reconstructed. The Batavia, built by Willem Vos in Lelystad, is the most famous and impressive example of this. The construction of the Batavia was based on historical information and the reconstruction drawings made from that evidence. The experience gained while rebuilding/reconstructing the Nijkerk-II cog made an essential contribution to the interpretation of how this ship was built and in which sequence.

7.1 The preceding sequence of events

In 1989, an initiative was launched in Kampen to build a Hanseatic cog.[4] Partly because of this well-known trading vessel, Kampen became a prominent city in the Late Middle Ages. Led by Alderman H. Bastiaan, a working group was assembled, comprising representatives of the municipality and the business community, to explore whether such a project was feasible and what would be involved. Besides garnering political support, choosing a form of organisation, finding ways to obtain financial resources, finding a suitable building location, and determining who should be hired as the master ship-builder, the chosen approach was highly desirable from the perspective of Dutch nautical archaeology: a scientifically and historically sound building plan. Discussions on the subject took place with NISA and other experts in the field.

1. Translated by Joy Phillips.
2. By A.J. van den Heuvel & K. Vlierman, supplemented by K. Sars.
3. Director of NISA at that time.
4. Van den Berg *et al.* 2007, 12-13.

Fig. 7.1. The Kampen Cog under sail on the river IJssel. Photo: Richard Tennekes.

Morel studied the various files on the cogs excavated in the IJsselmeer polders and concluded that the shipwreck from lot Oz 36 in Zuidelijk Flevoland (the Nijkerk-II cog, Chapter 6) could provide the best and most complete information for the realisation of such a project. The articles of association were signed on 18 December 1991, and the Kampen Cog Foundation (*Stichting Kamper Kogge*) was established. In addition to the board of the foundation, a project office was set up to act as a sounding board for the construction, and a technical committee conducted extensive research into sound building practices. Sars Houtbouw (master ship-builder Kees Sars and ship-builder and apprentice trainer Anton van den Heuvel) was hired as the construction company. Sars drew up a budget, a work plan and a training plan. The European Social Fund (ESF) invested money in the project to train job-seekers and provide them with work experience. On average, there were about twelve participants working on the construction process at any given time, a total of sixty people during 3.5 years of construction. Once it was certain that construction was going to take place in Kampen, Morel started extrapolating from the excavation drawings of the Nijkerk-II cog, and J. van der Zee (model builder at NISA) started working on a reconstruction model at 1:10 scale. The construction of this model is executed in phases (Section 6.2.4) and served as a shipyard model for the construction and for determining the sequence of work for building the Kampen Cog.

Neither the model nor the Kampen Cog presented any problems in terms of building the bottom of the ship. Large parts of the upper parts of the sides were missing from the shipwreck. Despite the detailed 1:10 drawings of the parts that were found, it proved impossible to reconstruct the sides on paper at that time. In the early stages, there were some difficulties building the sides of the model; it was extremely difficult to achieve what would presumably have been the original shape of the hull based on the frame structures that were drawn and recreated. The frames may have warped somewhat over the centuries in the soil. Morel and Van der Zee consulted at great length on the best way to continue work on and finalise the construction of the model and the Kampen Cog. It was then decided to deviate from the initial plan, which had been to start with the frames and build from there; instead, they first constructed the shell of the sides and then put the frames in place. This 'shell-first' method had previously been used to build replicas of the Viking ships – which were, it should be noted, fully clinker-built. During the reconstruction of the cog, it soon became clear that this method must also have been used for the clinker-built planking parts of the bottom near the stern and stem and sides of cogs. One confirmation that this construction method was used is the fact that the clinker-built shell strakes were sealed with sintel clamped moss caulking on the inside of the ship. On the wrecked cogs excavated thus far, these seams were sealed with sintel clamped moss caulking behind the frames, extending from stern to stem with no interruptions. This means that the sintel clamped moss caulking was applied before the frames were put in place. The part of the ship that was not built shell-first was the carvel-built part of the bottom. Those planks were fitted into place on either side of the keel plank, under four 'temporary floor timbers' that had been placed on the keel plank beforehand, and effectively served as a kind of matrix for the ship's bottom.

No significant difficulties were encountered during the construction of the sides; when the strakes were added, both the model and the replica took on the intended hull shape almost automatically. However, it became clear at a certain point while building the model that the lengths of the two crossbeams found in the wreck did not match the dimensions in the afterbody of the model in the places where they belonged. The lengths of the beams have been adjusted in the model and in the Kampen Cog. The reason for the difference in length could be that a choice had to be made early on regarding the angle of the stern and stem. That choice determines the lines of the rest of the shell. The knees found on the two original beams have been omitted in the model (fig. 6.46); the structure has been adjusted (automatically) in the reconstructed vessel. The missing parts of the shell were supplemented on both the model and the Kampen Cog. The decks and other items at deck height and above, plus the rigging and sail, were derived from other archaeological finds and historical information. The Bremen cog was the main inspiration. As problems emerged, they were resolved and reshaped by the builders themselves, opting for the most obvious method and in consultation with the technical (advisory) committee.

Construction officially commenced on 17 June 1994. The Kampen Cog was launched on 1 November 1997, and the final structural elements were fitted onto the cog on 14 March 1998. In the following sections, there will be a brief summary of the basic principles used in the construction process, followed by a description of the elements added in each phase of construction, with occasional references to the original ship, and an account of the initial sailing tests, the experiences gained in those tests, and the hull features and operating and sailing characteristics, plus the installation of an engine.

7.2 Basic principles of construction

A number of important basic principles were selected before starting to build the ship. It had to be built as authentically as possible, it had to be easy to maintain, and it had to be durable, sustainable and safe. With these principles in mind, there was an interesting and fascinating task for the builders to combine the sometimes conflicting interests in the different areas in an elegant, acceptable way. In principle, as problems arose, the builders always sought to choose the most authentic solutions available to them.

Materials

The following materials were used for the ship:

- Construction timber: oak *(Quercus spec.)* and Douglas fir *(Pseudostuga Menzisi)*.
- Fastenings: wooden treenails, wrought-iron gudgeons and rudder fittings, bolts, nails and sintel clamps.
- Caulking material: moss (*Sphagnum spec.*), tarred hemp fibre
- Preserving agents: boiled linseed oil, raw linseed oil. brown tar, carbolineum, Tencofix (bituminous sealant) and Tencolak (glossy top coat especially for boats).
- Mast and spar: Douglas fir (*Pseudostuga Menzisi*).

Tools used

Electric, motorised and manually operated tools were all used in the construction process. The hand tools used on the project include several items that were developed and/or made specifically for this purpose: the oak *boeitangen* or planking tongs, the oak 'caulking iron', and wrought-iron tongs for adding the sintel clamps. A fire rack was built to hot-mould the planks into shape, and a self-propelled crane was present at the shipyard to handle the heavy lifting and manoeuvre the various pieces into position.

Construction method and sequence

The construction method and sequence of the various activities were as follows:

- Setting up and sealing the keel plank with stern and stem hooks, sternpost and stempost.
- Installing four temporary timbers where the hold would be, and making and placing the bottom strakes.
- Making and installing the floor timbers and planks in the hold and removing/replacing or adjusting the temporary timbers.
- Making and placing the keelson and mast step.
- Manufacturing and installing the planks/strakes of

the sides up to the level of the crossbeams protruding through the shell.
- Applying sintel clamped moss caulking on the inside of the vessel in the seams of the clinker-built planking that had been completed up to that point.
- Making and placing the lowest futtocks on the already fitted floor timbers and the timbers in the afterbody and forebody up to the start of the stern and stem.
- Making and placing the six crossbeams protruding through the shell.
- Manufacturing and installing the planks/strakes above the crossbeams.
- Waterproofing the inside of the clinker-built shell strakes above the crossbeams with sintel clamped moss caulking, and making and placing the remaining futtocks in the afterbody and forebody, as well as the top futtocks at the level of the hold and the heavy knees and intermediate pieces on the crossbeams.
- Making and installing the splashboard.
- Making the deck and the aftercastle, the belaying rack and winch, etc., the rudder, the mast, the spar and all the other items such as ropes, blocks etc. to prepare the ship for sailing.
- Sealing the outside seams of the carvel-built bottom and the clinker-built composite sides with sintel clamped moss caulking, and caulking the deck seams and tarring the ship (to the extent that it had not already been done in a previous phase).

7.3 Construction

Construction phase 1: the piles, the keel plank, the stern hook and stem hook, the sternpost and stempost, and the temporary floor timbers

Before the actual construction could commence, the piles were built: a temporary foundation pier on two concrete stop beams, consisting of stacked blocks of wood, on which the ship would be built. Railway sleepers were used for this purpose. Six piles were constructed, distributed evenly along the length of the keel plank including the stern and stem hooks (figs 7.2a and b). The piles were 1 m high, leaving sufficient working space under the bottom of the ship. Then the keel plank and the stern- and stem hooks were put in place. Stern- and stem hooks are made from wood that naturally grew in a curved shape, and are intended as a transition between the keel plank and the stern and stem. The pieces of wood selected for the stern- and stem hooks were shaped according to the measurements on the scale drawings of the originals, except for the rabbets in which the ends of the shell planks had to be fastened. These rabbets were 'coarsely' carved out and then cut to fit later in the work to accommodate the shell planks. The keel plank and the stern- and stem hooks were fastened to each other with sloping flat scarfs, 60 cm and nailed into place.

It was decided to waterproof all seams in the vessel using the original method with sintel clamped moss caulking. However, it proved far from simple to find the required quantity of moss, with the consequence that this material was not yet available during the first phase of construction. For that reason, a layer of tarred hemp fibre was applied between the scarf surfaces of the keel plank and stern- and stem hooks and between the hooks and the sternpost and stempost. During the official keel laying by Morel, three nails from the original, 14th-century cog were ceremoniously hammered into the aftmost scarf of the keel plank. The assembly of the keel plank and stern- and stem hooks was temporarily attached to the underlying concrete stop beams with threaded studs. This was done to prevent the total structure of the bottom from being pushed up at a later stage by the upward force of the support posts, which press against the floor timbers when the planks are placed.

When Morel reconstructed the ship's bottom in the drawings, he discovered that small wooden pegs made of pinewood (known as nailhole plugs) had been hammered into the planks at frames 8, 13, 17 and 22. They formed the basis for the theory regarding the sequence in which the ship's bottom was constructed. Although they were closed later, these nailholes marked where connecting nails were placed during construction, to temporarily fasten the first three carvel-built shell strakes (A, B and C) on either side of the keel to the temporary floor timbers. This method solves an important practical problem: maintaining the shape of the carvel-built section of the ship's bottom, since those planks were not fastened to each other. Following the same method, which was mostly likely medieval, four temporary floor timbers were added. In building the Kampen Cog, it was decided to place them where frames 11, 17, 21 and 24 would be added later, attaching an extra temporary timber in the spaces between temporary floor timbers 11 and 24 and the stern and stem. The strakes fit around these extra timbers without any tension. In order to fasten the plank ends of the B- and C strakes, the sternpost and stempost were placed on the stern hook and stem hook respectively. The flat scarfs between the stern- and stem hooks and stempost and sternpost were 65 cm long, and the sections were nailed together. At this stage, the sternpost and stempost had a square cross-section; later on, rabbets would be added, finishing by shaping the posts at an oblique angle (making them trapezoidal in cross-section). The stern and stem were set vertically with a plumb line and braced with three struts each.

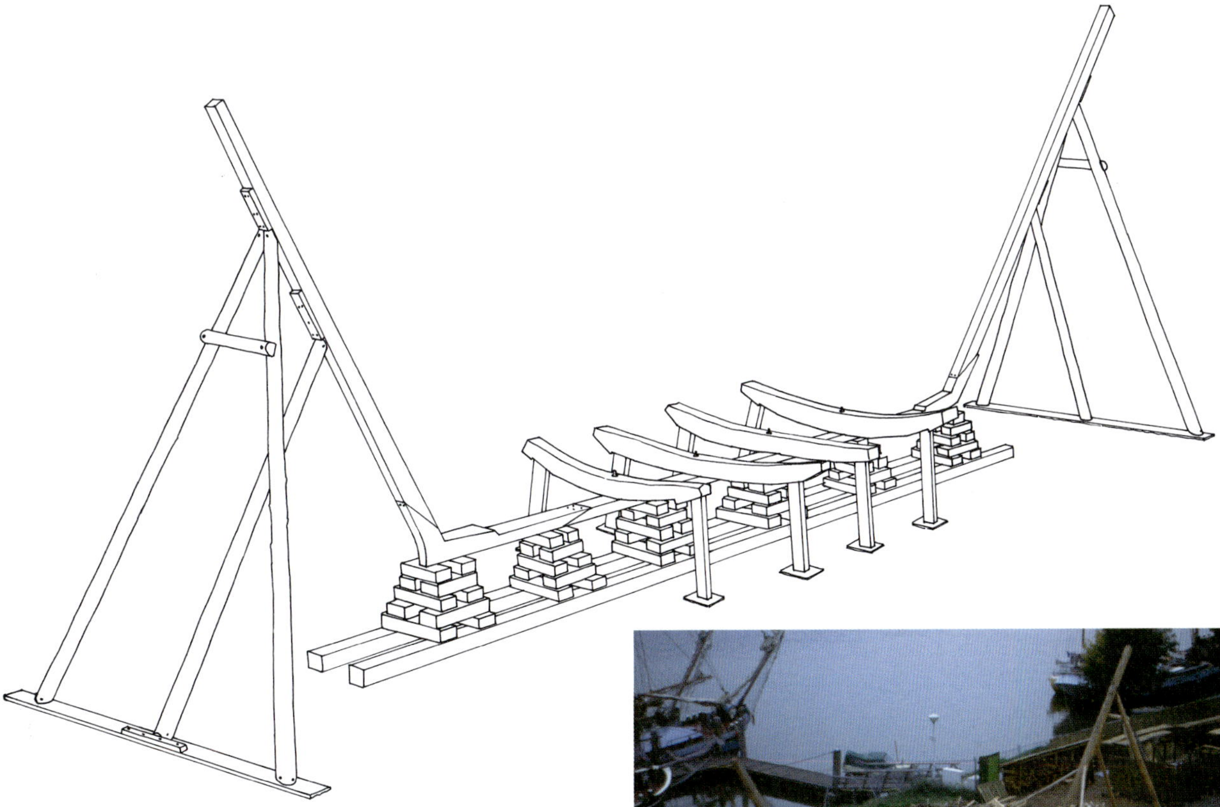

Fig. 7.2a. Keel plank, stern hook and sternpost, stem hook and stempost, temporary floor timbers, stocks and the struts supporting the stern and stem. Drawing: A. van den Heuvel/Sars Houtbouw '97.

Fig. 7.2b. The same as in A, with the garboards fitted. Photo: Sars Houtbouw.

Construction phase 2: the bottom of the ship, the lowest strake of the sides, the (V-shaped) floor timbers on the bottom, the sintel clamping of the seams on the inside of the clinker-built planks, and the installation of the keelson

Bottom of the ship

The underside of the first shell strakes (A) fell neatly abaft and forward into the rabbets in the stern and stem hooks. The middle section was carvel-built against the keel plank. The parts that slot into the rabbets were fastened with blind nails. The seams between the A strakes and the subsequently placed B strakes, and between the B strakes and C strakes, were clinker-built abaft and forward, and carvel-built against each other in between. From the locations where the strakes visibly twist towards the stern and stem, the upper strakes were left wider, and the overlapping sections on the overlap of the A- and B strakes and the B- and C strakes were chamfered (fig. 7.3) and fastened to the stern and stem hooks, sternpost and stempost with blind nails through the cross-cut ends of the planks. The upper edge of the strake was cut to be chamfered on the outside, while the lower edge of the following strake was similarly

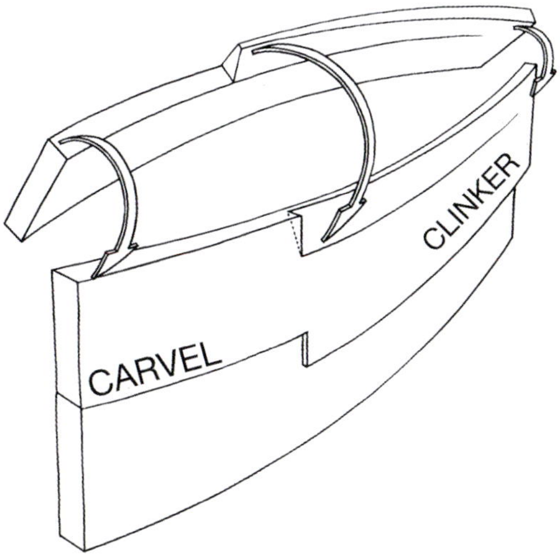

Fig. 7.3. The manner in which the transition between the carvel- and clinker-built parts of a seam between two bottom strakes is constructed. Drawing: A. van den Heuvel/Sars Houtbouw '97.

Fig. 7.4a. Burning a shell plank. Photo: Sars Houtbouw.

Fig. 7.4b. A newly burned plank being lifted into position. Photo: Sars Houtbouw.

chamfered on the inside before placement; the V-shaped seams created after assembling both planks formed a caulking seam which would be sealed later on the inside and outside with sintel clamped moss caulking.

Hot-moulding and installing the planks

The twisted planks abaft and forward in the underwater body were hot-moulded over an open wood fire. Before the sawn plank could be held over the fire for hot-moulding, all the dimensions had to be compiled. The measurements for the shape of the curve were taken by holding a taut string at the place where the plank would be installed on top of the previous strake, tensioned from the start to the finish of the new plank. Then a measurement was taken every metre (measured along the string), at right angles from the rope to the top of the previous strake. The angle of the plank was also measured every metre using a spiling batten or carpenter's bevel. Using the dimensions of the curve and the spiling, the fire rack was set to the shape of the desired plank.

On the day that the planks would be bent, the fire was lit early in the morning to obtain a long strip of glowing coals. The idea was to have an even bed of coals with minimal flames, so the plank would not char. A garden hose equipped with a gun nozzle was kept ready to spray down the underside of the plank regularly (fig. 7.4a). This was necessary, since the wood would otherwise dry out too quickly and/or crack due to the intense heat. The firewood was chopped, dried 'scrap wood' from ship elements that had been roughly shaped beforehand. The importance of using dried wood became apparent when it was necessary to use semi-dried wood on one occasion. The fire produced billowing clouds of smoke, causing everyone's eyes to sting and tear up. The entire process of hot-moulding a plank took an average of two and a half hours. The time required over the fire depended primarily on the thickness of the plank and how curved it needed to be, but also – albeit to a lesser extent – on the quality of the wood (e.g. the number of knots, the crossgrain of the fibres). The hardness and toughness of the various planks also played a role.

Once the plank had the desired shape, it had to be placed at the intended location on the ship as soon as possible. The plank was removed from its restraints of tongs, weights, levers and chains and trotted over to the ship, carried by five or six men, where it was clamped into place with wooden *boeitangen* or planking tongs (fig. 7.4b) and glue clamps, held in shape while it cooled down.

The planking tongs (an old form of glue clamps), which were in any case used as early as the 16th century (see the Nijkerk-II cog, fig. 6.41; fig. 7.5a and b), were made of oakwood rods of 90x8x8 cm. The clamping force was achieved by striking a wedge between the two arms. Once the plank had cooled down, the scarfs and the overlap could be drawn and cut. The flat scarfs were 40 cm long. The section on the outside of the shell ended at a thickness of 1.5 cm, while the section on the inside (of the next plank of the strake) was 1 cm thick at the end. In many cases, the overlap was chamfered slightly so the next plank up could be added at a slightly different angle. After this, every 15 cm and in the middle of the overlap, the nailholes in the outer plank were pre-drilled in order to insert the double-clenched nails or '*naainagels*'.

The double-clenching method used for these nails (folding the tip down twice and hammering it back into the wood) is illustrated in chapter 6, fig. 6.42. The scarfs were 'clenched' together with two rows of nails: one row from the outside in, the other from the inside out. During this process, which can almost be described as riveting, a hammer weighing five kilograms was held against the nailhead when the tip was folded down the second time, to prevent it from shooting back out. On the Kampen Cog, the nails were inserted at an angle of 45° in relation to the wood grain. This has never been established in shipwrecked cogs. If memory serves, the more than one hundred wrecks worked on to date have always had nails hammered in at right angles to the plank width. Moreover, it has been assumed that holes were drilled beforehand (Section 6.2.2 and fig. 6.42). Bad spots and knots in the planks were immediately removed and replaced by an insert, in a waffle-like pattern known as honeycomb. In other places, in accordance with the original ship, a board has been nailed on the inside to cover spots where the wood was of inferior quality.

Bilge strakes
Above the carvel-built section of the seams between the B- and C strakes, planks 18 cm wide and 2 cm thick were fitted under the temporary floor timbers. They were nailed onto the B- and C strakes on either side of the seam. These planks were intended to prevent the seam from pulling open at those spots due to the difference in potential shrinkage between the rigid clinker-built construction of the sides above that and the carvel-built planks of the bottom that were only fastened to the floor timbers with wooden treenails. This shrinkage problem was observed during the reconstruction of the Bremen cog in Kiel and referred to as the 'Kieler Ritze' or 'Kiel crack'.[5]

The floor timbers in the carvel-built section of the bottom
In addition to the A-, B- and C strakes of the bottom, the lowest strakes of the sides (the D strakes) were also added. After that, the roughly pre-shaped floor timbers above this section of the carvel-built ship's bottom could be measured to the shell using a water level (or plumb line) and then cut to fit and put in place. This measuring instrument is a modern version of the 'plumb bob' described by Nicolaas Witsen in his book entitled *Shipbuilding and Management* (1671/1690). The difference is that the principle of the ancient instrument is based on a plumb line, while the modern version is based on a spirit level. There was a practical

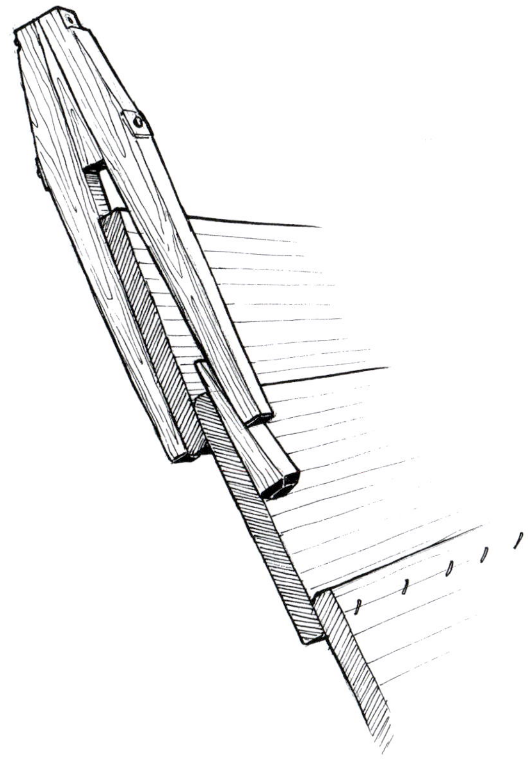

Fig. 7.5a. *Boeitang* on a section of shell. Drawing: A. van den Heuvel/Sars Houtbouw '97.

Fig. 7.5b. *Boeitangen* on the shell. Photo: Sars Houtbouw

5. Baykowski, 1991.

reason for also installing the D strakes at this stage. The scarfs on the ends of the floor timbers were in a number of cases above the C- and D strakes (frames 8 through 24). Moreover, the D strakes were also part of the carvel-built underwater body. The subsequent E strakes were the first to be completely clinker-built.

Sintel clamped moss caulking in the seams of the clinker-built planking of the underwater body

Before it was possible to start applying the timbers in the aft and forward parts of the clinker-built section of the ship's bottom up to the start of the stern and stem, the sintel clamped moss caulking first had to be applied on the inside. At the special request of NISA in relation to ongoing research on the sintel clamped moss caulking in late-medieval shipwrecks[6], this authentic method was used with particular care and attention in caulking the Kampen Cog. Until then, a number of questions had effectively remained unanswered, such as: How were the moss, the moss laths and the sintel clamps applied? How much time would it have taken? What would it have required in terms of craftsmanship?

Such questions were also put to a smith regarding the production of the approx. 16,000 sintel clamps and the wrought-iron nails. In the case of the sintel clamps, it was decided to have them punched. The dry moss *(Sphagnum)* was hammered into the V-shaped seams, which were approx. 2 cm wide and deep, using an oak 'caulking iron' made on site and a wooden hammer. The moss laths were about 1.5 cm wide and 0.5 cm thick and made from strips of oak, which were laid on the moss. They were held in place by initially securing them with a few sintel clamps, and then fastened more securely by adding the other sintel clamps every 7 cm or so. Various methods were tried in order to find out how the sintel clamps were applied in the Late Middle Ages, probably above the seam between two clinker-built shell planks fastened to each other. While the clamps were still straight, one end (ear) was hammered into the wood and bent. Then the other end was hammered into the other plank. In another method, the folded 'ear' was bent beforehand, after which the other, straight end was held against the upper plank. The carpenter then hit the bent side with a hammer, causing it to sink into the wood. Then the curved side of the sintel clamp was hit from above to drive it into the lower plank.

Since these practices proved fairly difficult, efforts were made to develop a tool to assist them. Shipyard workers developed a 'sintel iron', a caulking iron with an opening in the middle (similar to the slot in a claw hammer) that could be used to pull out nails. Using a hammer and this 'sintel iron', it was possible to perform the aforementioned actions properly. The 'sintel iron' left impressions on the sintel clamps that had never been observed on any of the thousands of authentic sintel clamps examined previously. Moreover, a 'sintel iron' was never found. It is therefore questionable whether such a tool would have been used in the Low Countries in the Late Middle Ages.

On the vessels on the Danube, a long, thin hammer was used which had a triangular groove in the flat end (similar to a claw hammer). This tool was used both for hammering in and removing the *Klampeln* (sintel nails).[7] The hammer displayed on the miniature from the *Roman de Troie* (figs 6.60 and 29.11a). bears some similarity to the item. The wooden 'caulking iron' is a similar case. Caulking irons made from wrought iron are almost a standard feature in shipwrecks, and a wooden specimen has never before been found in the Netherlands, to the best of my knowledge. However, it is known that wide wooden chisels were used for caulking the characteristic river barge and barge-like planking boats (*Plätten, Zillen* etc.) in the river basin of the Rhine and Danube, but those seams were generally much wider.

Figure 7.6a through e shows a number of the actions described above and the tools that were recorded while applying the sintel-clamped moss caulking in the Kampen Cog. The carvel-built seams of the bottom and clinker-built seams of the sides on the outside of the shell were sealed with sintel clamped moss caulking shortly before the launch. After drying, the moss used *(Sphagnum spec.)* was stripped of irregularities (heather twigs, pine needles, blueberries and blueberry twigs, etc.). Approximately 25-30% of the original volume was removed in this way, leaving approx. 2 m^3 of moss compressed, enough to caulk a vessel of approx. 20x7.5 m.[8]

Fig. 7.6a-e. a. Securing a moss lath with sintel clamps.In the foreground the inserted moss caulking is seen before being covered with a lath. Right in front is the wooden 'caulking iron' with which the strands of moss were driven into the seam. The top edges of the strakes show the double-clenched nails by which the strakes are fastened clinker-style. b. Driving a sintel clamp into the lower edge of a clinker seam. The other 'wing', to be driven into the lower edge of the upper plank, has been previously bent over. c. The sintel clamp fixed into the lower strake.
d. The sintel also fixed into the upper strake. e. The interior of the starboard forebody with sintel clamped moss caulking.

6. Vlierman 1996a.
7. Wöginger 2007, 15.
8. Cappers *et al.* 1997/1998, 585.

a.

b.

c.

d.

e.

Keelson with mast step
After all the various floor timbers had been placed on the bottom of the ship, the keelson could be made and then fastened to frames 4 through 26. The keelson found in the wreckage was made from a jagged tree, of which the lowest, thickest part of the trunk was used as the front. The first approx. 2.5 m has been chopped into a more or less square cross-section, followed by a section that was left wider and thicker; this was the mast step, in which the hole for the mast heel would later be cut. The aft section retained the uneven, tapering shape of the trunk and was square-hewn here and there. Notches were cut into the underside of the keelson at the level of some floor timbers, allowing the remaining pieces to 'hang' between the floor timbers (fig. 7.7). These notches also provided more longitudinal rigidity. In principle, the keelson was fastened to the frames with blind wooden treenails. For the Kampen Cog, a trunk was sought which was as similar in shape as possible to the trunk used in the Nijkerk-II cog. It was similarly worked and put into place.

Construction phase 3: adding the shell strakes up to the crossbeams, sintel clamping the seams on the inside, and making and installing the lowest futtocks and crossbeams

The strakes from the sides to the crossbeams
As indicated above, the first strakes of the sides (D) were already fastened into place during the previous phase. After this, the planks of the next five strakes (E through I) and the aft planks of the J strakes were cut to fit and put into place (figs 7.8a and b), in principle always starting at the stern. After the first plank was nailed into place, the next plank was marked off, cut to fit, and nailed into place until the entire strake had been assembled from stern to stem, and then the next one was started. A thin layer of Tencofix was smeared onto the overlap of the planks before nailing them down. After all the planks in these parts of the sides were assembled, the entire structure was strutted to hold the relatively thin, somewhat flexible shell in shape.

Sintel clamping the clinker-built seams
The D strakes were chamfered along the top edge on the outside. On the inside of the ship, the same was done to the lower edge of the planks of the E- strakes. As a result, each time a plank was fitted onto the overlap, it created V-shaped seams about 2 cm wide and deep on the inside and the outside of the shell. After the sides were completed up to the crossbeams, the seams were sealed on the inside with sintel clamped moss caulking (fig. 7.6).

Fig. 7.7. The carvel-built part of the underwater body with floor timbers, and the still unconnected keelson on its side, with notches for the floor timbers. Photo: Sars Houtbouw.

Making and placing the lowest futtocks and protruding crossbeams
For the shape of the futtocks, branches or trunks that had grown crooked with a straight stretch were preferred amidships, while S-shaped and straight pieces were preferred in the afterbody, and crooked pieces of wood sufficed in the forebody. At certain frames in the wreck, the lowest futtocks are made of several pieces. The reason for this is probably that there was no suitable piece of wood available to fashion them from one piece. A centreline was first used to determine the location of the futtocks. A chain mould was used to determine the shape of the shell at that location. Then a suitable piece of wood was selected. The piece was roughly cut to the shape of the mould, and then placed in the right position in the hull and marked off. In ancient times, the wood would then have been cut to fit using a saw, axe and adze, but in modern times, an electric planer was used to make the outside of the frame section fit the inside of the shell, leaving the surfaces of the boards a thumb-width longer. This created

Fig. 7.8a. Overview towards the forebody, showing the floor timbers on the bottom, and the strakes of the sides built 'shell-first' (with fetter cramps).
Photo: Sars Houtbouw.

Fig. 7.8b. The clinker-built strakes of the starboard afterbody. A futtock is being shaped to fit. Photo: Sars Houtbouw.

a 'limber hole' above the clinker-built overlap between two shell strakes fastened one above the other. The scarf with the top of the floor timber on which it will rest was marked off and cut to fit. The top of the futtock was chamfered (the scarf with the upper futtock) or squared off if a crossbeam will be resting on it.

In principle, the hearts of the tree trunks and branches were placed on the shell side. However, this has the (possible) consequence that the wood will change shape, causing the contact surfaces of the shell planks to deform or swell up as they dry. Another consequence could be that the futtocks will push themselves loose from the shell over time. To prevent this from happening, all surfaces of the frames were in principle hollowed out a few millimetres. A recess was created at the places where the double-clenched nails would touch the contact surface. Whether this method of providing a slightly concave surface and adding recesses was also used on the Nijkerk-II cog was not observed or noticed. This method was certainly observed in a number of other shipwrecks from later periods.

After the futtock was finished, it was fastened into place with several 130-150 mm long 8x8 mm nails, which were hammered through the shell into the frame section from the outside, between the double-clenched nails of the clinker-built fastenings of the strakes. They are immediately noticeable because they halve the regular intervals of approx. 15 cm between the double-clenched nails. A number of studs temporarily ensured that the nearly fresh-cut pieces of wood did not warp and that the shell sections stayed neatly against the carefully fitted surfaces. The studs were placed where wooden treenails would later be placed. The wooden treenails connect all three structural layers (shell plank, frame and ceiling board) and were therefore only applied after the ceiling boards had been cut to fit and fastened to the frames with a few nails.

The protruding crossbeams

Above the scattered wreck, two intact crossbeams were found, with the knee sections still attached. They had become detached from their original construction and were 20x20 cm in cross-section. The holes found in the shell from the missing beams indicate that these beams would have been about 25x25 cm or slightly more. The measurements used for the Kampen Cog followed 25x25 cm. The beams of the Bremen and Doel cog are about the same dimensions. Three of the forward four crossbeams rested on squared-off frames and on the top of strake I, while the two aft protruding crossbeams rested on frame heads and the top of strake J. The second forward crossbeam was placed between frames 25 and 26 and only rested on the top of strake I. The builders were unable to determine why this anomalous fastening point was used.[9]

Not far from the ends of the crossbeams, a rabbet about 3 cm deep and 4 cm wide was made all the way around. The top of the I strake and J strake respectively were enclosed in the lowest rabbet. The rabbets in the sides were slightly staggered and placed in the sloping part of the shell. After placing the protruding crossbeams, the next shell planks (J above the forward four beams and K above the aft two) were marked off and cut to fit (fig. 7.9a and b). Square recesses about 20x20 cm were made in the underside of these planks, so they could slide over the beams from above into the rabbets. At that point, the beams were fully enclosed. The planks were then fastened to the underlying strake with double-clenched nails. This construction of crossbeams protruding from the shell and sealed in place absorbs the (transverse) pressure and tensile forces. Each shell plank encloses two crossbeams: one on the aft two, one on the middle two, and one on the forward two.

9. For a possible explanation, see the Nijkerk-II cog in section 6.2.2 and Plates 6.IVA and B.

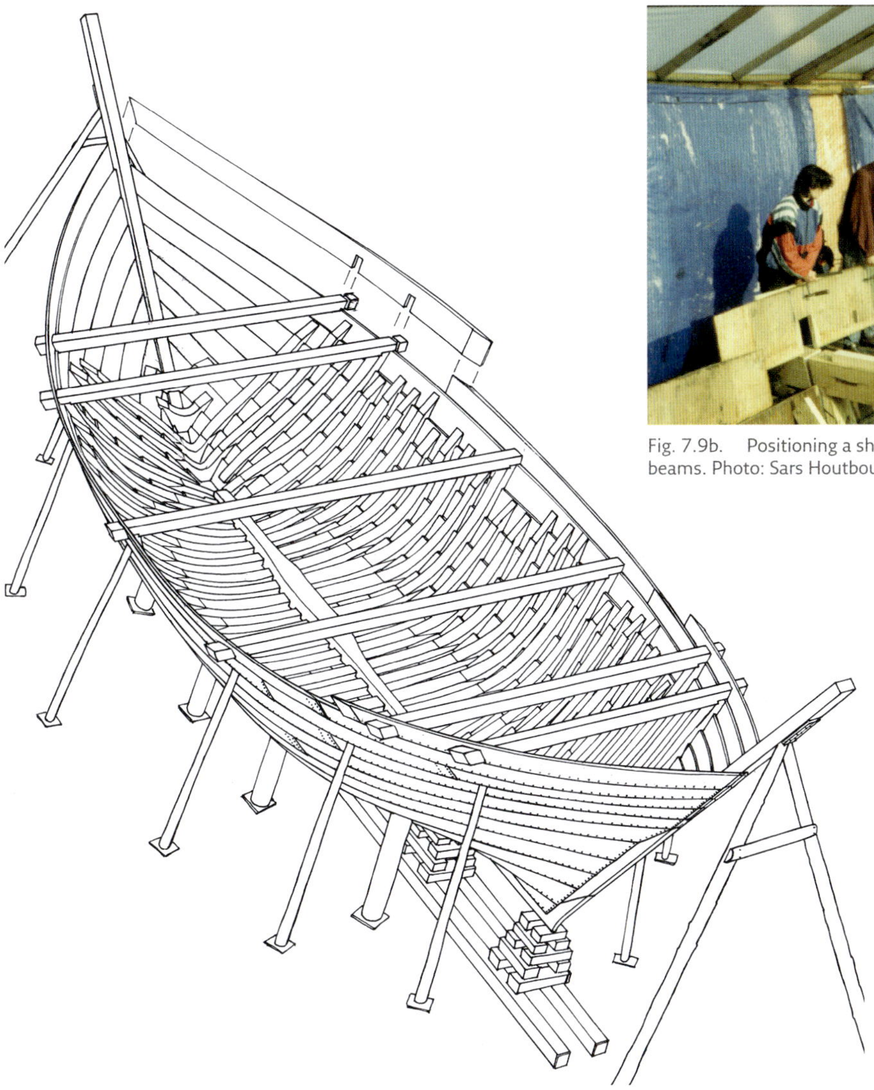

Fig. 7.9a. Top view of the Kampen Cog with floor timbers, the lower futtocks and the protruding crossbeams with the lower side strakes. Drawing: A. van den Heuvel/Sars Houtbouw '97.

Fig. 7.9b. Positioning a shell plank shaped to enclose protruding crossbeams. Photo: Sars Houtbouw.

Construction phase 4: the gunwale to the washboard, sintel clamping the seams, the top series of futtocks, the floor timbers on the stern and stem and the washboard

The gunwale, sintel clamps and futtocks

Then the J strakes were completed, and then the K-, L- and M strakes were put into place, as well as the N strakes from the stern up to and including frame 10. After sintel clamping the seams on the inside, the top futtocks were marked off, cut to fit and fastened into place.

The floor timbers on the stern and stem

After the series of lower futtocks of frames 1 through 30 had been installed, the crossbeams had been assembled and the gunwale with futtocks had been installed in this part of the ship, the remaining floor timbers and futtocks in the afterbody and forebody were marked off, cut to fit and assembled. The builders had waited to install these frames until the ship's hull had reached sufficient structural rigidity at the location of the hold. Due to the total weight, earlier installation of the generally quite thick floor timbers on the raked stern and stem (but not fastened onto them) would have posed a risk to maintaining the shape of the ship's hull.

The washboard

The part of the N strakes from frame 10 forward was constructed as a washboard (fig. 7.10), in accordance with a complete construction found in the Bremen cog. The planks were fitted with relatively thin, short timber, which protruded at the bottom and fell between the

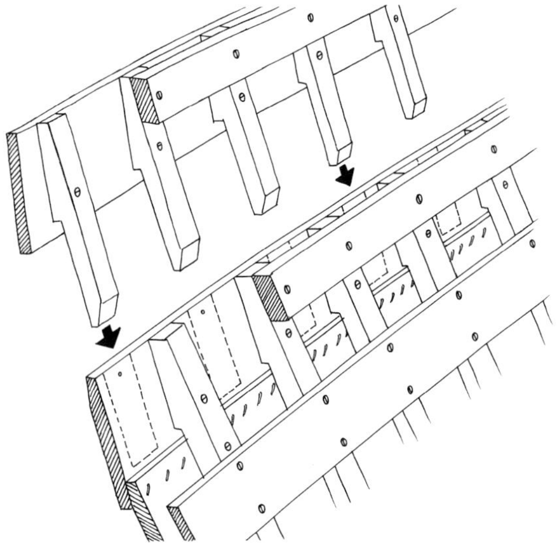

Fig. 7.10. The construction of the washboard, and the manner in which the top timbers of the washboard slot in between the frames. Drawing: A. van den Heuvel/Sars Houtbouw '97.

top futtocks of the frames, and were fitted with a ceiling board on the inside. The washboard ended on the futtock of frame 36, which protruded above the gunwale.[10] The washboard construction was reinforced with a ceiling beam. The structural elements were fastened together with wooden treenails. The whole structure was assembled in such a way that it could be used for experimentation involving taking it apart and putting it back in place, as well as sailing under various conditions. The entire construction was secured with nails driven through the overlap of the N- and M strakes into the heads of the futtocks. A lath 8 cm wide and 1.5 cm thick was placed between the nail heads and the splashboard. If the splashboard had to be removed, this lath could be split open and removed, creating room below the nail heads to insert a crowbar and pull out the nails. No indications of a washboard with such a lath was found in the shipwreck; it is a solution that the builders came up with for the practical problem of how to remove the washboard (fig. 7.11).

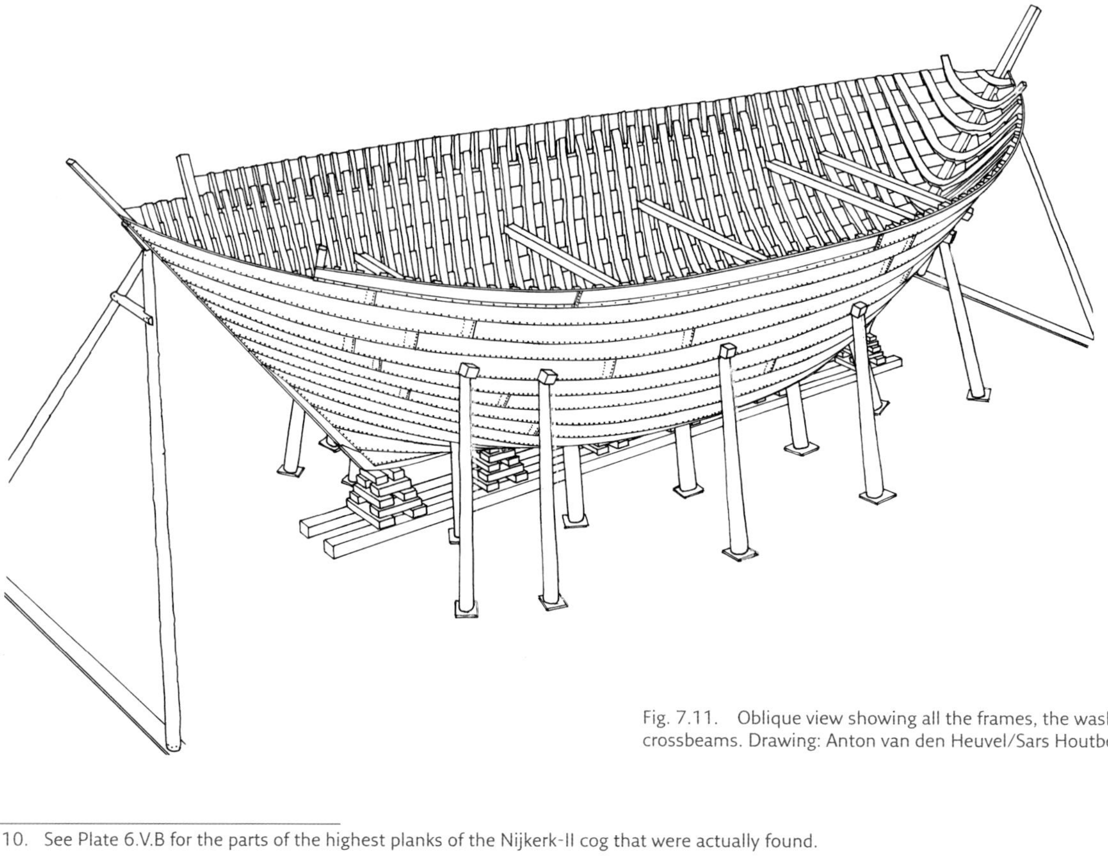

Fig. 7.11. Oblique view showing all the frames, the washboards and crossbeams. Drawing: Anton van den Heuvel/Sars Houtbouw '97l.

10. See Plate 6.V.B for the parts of the highest planks of the Nijkerk-II cog that were actually found.

Construction phase 5: the ceiling

The main function of the internal planking or ceiling is to provide rigidity to the ship's hull and to support cargo. Without this 'sandwiching', the ship would be flexible and too weak. The 'open ceiling' (approximately a strake width was left open between every two ceiling strakes) consists of eleven strakes fastened into place on the frames lengthwise in the ship. Each ceiling strake consists of a number of sections that are fastened together with sloping flat scarfs. The strakes continue to the floor timbers at the stern and stem. There are two types of ceiling strakes: ceiling beams and ceiling boards (fig. 7.12). The beams were notched at the bottom so they would fall over the frames, while the boards were placed on the lined insides of the frames. The ceiling beams are placed at the locations where the scarfs are in the frames; the boards are in between.

The ceiling beams

The B- and C ceiling beams were installed side by side at the bilges. They are located above the places where the staggered scarfs are located in the frames and are about 13 cm thick. A major advantage, and probably the most important reason, is that the differences in thickness of the floor timbers and the futtocks placed on top of them can be compensated by the beam. In other words, the notches vary in depth and are sometimes not necessary at all. The reason that some scarfs in the floor timbers/futtocks are thin or thick is mainly due how the grain runs in the pieces of wood used. In one case, the foot of the futtock is thicker and stands on a nearly straight floor timber end; in the other case, the timber has a crooked end on which the futtock could be placed with a short base. This problem only occurs with the frames around the bilges, because the variation in the hull shape is greatest there. Selecting the required pieces of wood was therefore also the most time-consuming task.

The G ceiling beams were installed below the two aft crossbeams and below the four forward crossbeams. The shelf clamps were not only notched at the location of the frames, but also on the top at the location of the crossbeams, which they partly enclose. The I- and J ceiling beams were installed against the inside of frames of the gunwale. They were fastened into place at the M- and N shell strakes. Two beams were stacked here, since the lowered gunwale would

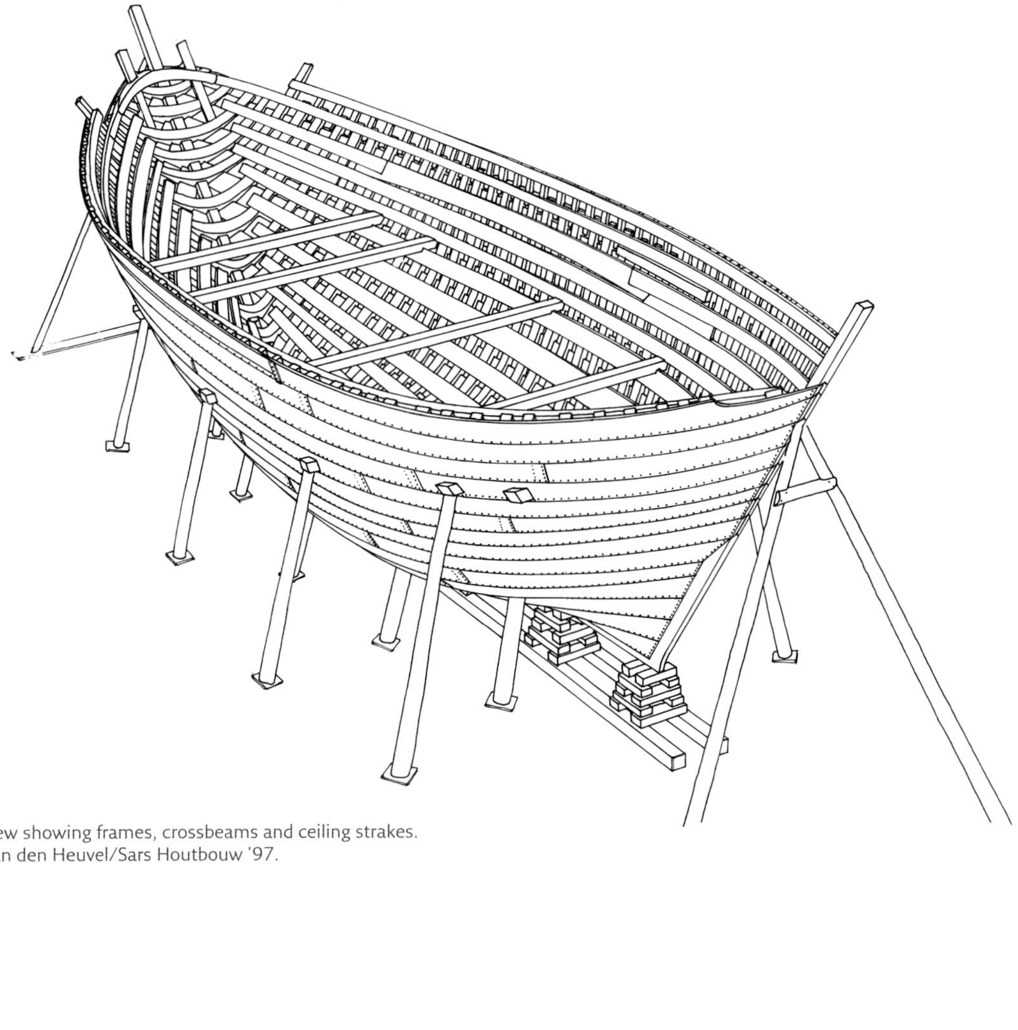

7.12. Oblique view showing frames, crossbeams and ceiling strakes. Drawing: Anton van den Heuvel/Sars Houtbouw '97.

not be strong enough on its own when the splashboard and the top ceiling beam were removed. These three beams in the ceiling strakes consist of thick boards that have not been notched at the locations of the frames.

The ceiling boards
Ceiling strakes A, D, E, F and H were made of boards 4-5 cm thick.

After cutting a ceiling beam or board to fit, it was fastened into place with several nails. Then the 'sandwich' of shell planks, frames and ceiling boards were permanently fastened together with wooden treenails. In principle, two treenails were used at the intersecting surface of a shell plank and a frame section. They were placed in a staggered pattern. The treenails were pounded in from the outside. They have a slightly curved head that rises slightly from the shell. On the inside, the protruding treenail was split and a wedge was driven into it.

Construction phase 6: The false stern and stem, the side supports, the deck knees, the deck, the deck construction, the deck layout and the castle

The false stern and stem
An outer or 'false' stern and stem were placed over the sternpost and stempost respectively, which had already been put into position, and the ends of the shell planks that had been attached to them. The lower 1.9 m of the original false stem was recovered. The rounded underside starts several centimetres below the seam between the stem hook and stempost and was fastened against the top of the hook with two blind nails. Above that, it was fastened with several slightly thicker nails in a vertical line placed close together, and then with even thicker iron nails at intervals of 50, 60, 80 and 80 cm respectively on the front of the inner stem or stempost. The false stem serves to cover the starboard and port shell strakes, which end against each other in front of the stempost and reinforce the relatively thin inner stem. The apex serves as a fastening point for the forestay. The false stem is secured with cotter bolts, which pass through all the structural parts of the stem. The lower 2.9 m of the original false stern was recovered. It can be seen here that the interfacing surface of both the sternpost and stempost is curved at right angles to the keel. The purpose of this is to prevent water from entering the scarf and causing damage while sailing. A somewhat similar function is served by the 'curved seam' between the false stern and the rudder. This form of opening prevents floating debris in the water from becoming trapped between the false stern and the rudder, thus preventing the rudder from moving.

The Bremen cog was used as a reference for reconstructing the following elements, since they were either not found with the Nijkerk II cog or were not (or no longer) available.

The side stems
No evidence of side stems was found in the shipwreck. Such evidence was similarly lacking in the Doel-I cog that was discovered and salvaged later. It was decided to install them on the Kampen Cog, following the example of the Bremen Cog. These long, wide boards/beams were 12 cm thick, with notches on the underside, and were fastened into place over the V-shaped floor timbers of the bow. They were fastened onto the V-shaped floor timbers at a position 70 cm from the centre of the middle of the stempost, with wooden treenails 4 cm thick. The function of breasthooks is to provide rigidity to the bow and to absorb the pressure of the two stringers – the two heavy beams in the deck running forward from the mast.

The knee sections on the crossbeams
The knee sections together with the crossbeams protruding through the shell, on which they are fastened, provide the transverse connections of the hull construction, and above all carry the weight and provide the connection between the deck construction and the upper gunwale. The horizontal parts of the knees lie on the crossbeam, while the vertical parts hold the (strongly) everted sides in place, as it were. The vertical part of each knee is notched, enclosing the ceiling boards and aligning against the front of the futtock. The two sections that come together amidships are connected by a vertical nibbed scarf. They are fastened together with wooden treenails and iron nails. The knee sections are 13 cm thick on average; in principle, two sections were taken from one piece of trunk/branch, one placed on the starboard side and the other to port. The knee sections were fastened to the crossbeams and the sides with wooden treenails 4 cm thick. On the Nijkerk-II cog, only four knee sections were found, which were fastened onto crossbeams 1 and 2. The construction on crossbeam 2 differs from the other knee sections in the cog replica. This is where the horizontal parts of the knee sections are connected by an intermediate beam. The parts were put together with horizontal hooked scarfs. A beam was placed over this construction which covers the scarfs and continues until the vertical parts. A rabbet was made in the beam on the top back in which the deck planks would be laid. Four recesses were made on the front in which longitudinal strength members or stringers were placed. The construction of the other knee sections and connecting beams was derived from the sections on crossbeam 1 and from the crossbeam constructions in the Bremen cog.

The deck and deck construction

The deck (fig. 7.13a and b). consists of a number of elements, nearly all of which are supported by the horizontal parts of the knee sections. Indications on the deck beam constructions 1 and 2 that were found made it clear that the part abaft of beam 2 was different from the part forward of it. The planks in the stern were placed lengthwise in the ship, while the planks in front of the beam were athwartships. A wooden rail was nailed against the back of the knee sections on crossbeam 1 of the Nijkerk-II cog, on which the deck planks were laid between this beam and the stern. This deck section was 10 cm lower than the rest of the deck. It was not possible to determine how the deck planks were supported beyond that. In the Bremen cog, two small crossbeams were used for this purpose, which were fastened onto the ceiling boards with an oblique surface. This solution was applied. A similar wooden rail was placed against the forward side of beam 1, on which rested the ends of the short deck planks between crossbeams 1 and 2; the forward ends were fastened into place in the aforementioned rabbet in the back of the lower deck beam on the knee sections of beam 2. Longitudinal strength members or stringers were placed in the four rabbets cut into the front of this beam. The size of these beams corresponds to those of the Bremen cog. The beams had an inverted T-shape in cross-section. The short

Fig. 7.13b. The hull, viewed from the stern, showing the side stems and stringers. Photo: Sars Houtbouw.

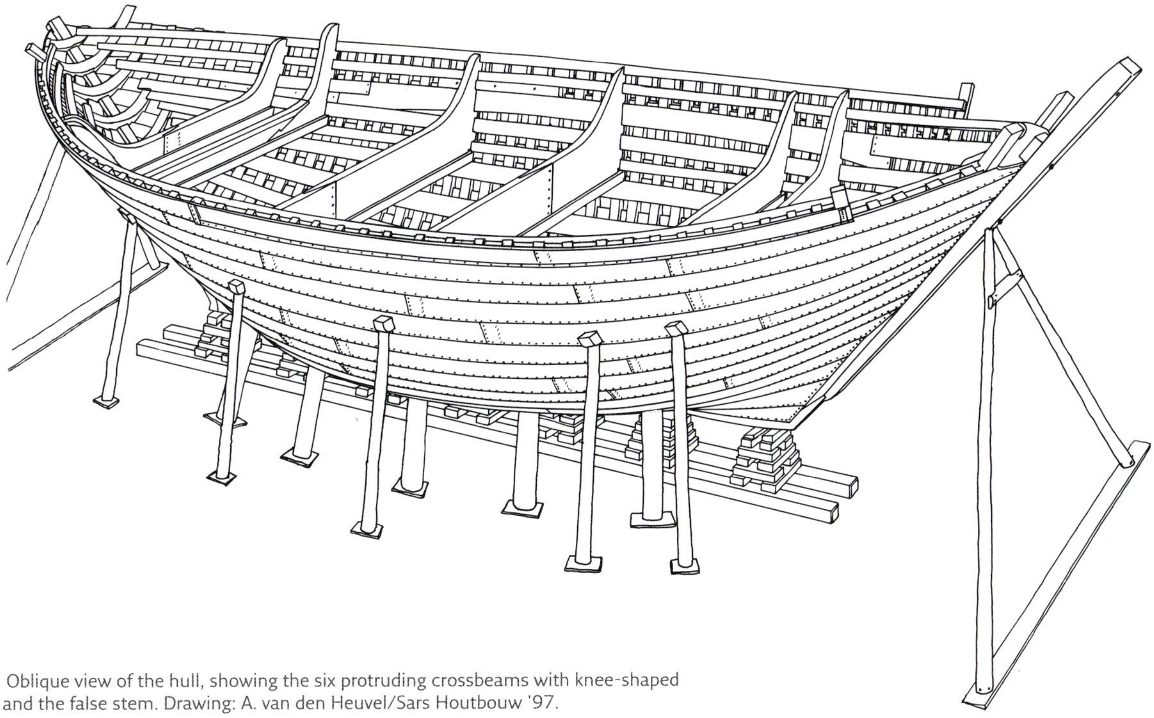

Fig. 7.13a. Oblique view of the hull, showing the six protruding crossbeams with knee-shaped beam riders and the false stem. Drawing: A. van den Heuvel/Sars Houtbouw '97.

Fig. 7.13c. The deck and topside of the clinker-built *waterschip* wreck (first half of the 16th century) in lot Mz 22 in Zuidelijk Flevoland, reconstructed beside the excavation. In the foreground is the after part of the fish well deck, (*bundeken,*) with a topside beam and the gangways to the left and right. The afterdeck with on it a massive topside beam composed of two knees. In its forward face are two oblique rabbets to hold the planks of a roof. A windlass (lacking its drum), two massive bitts with belaying racks on them and inwales.

deck planks laid athwartships rested on the rabbets created on both sides in this manner, between two stringers or between a stringer and the sides of the ship. The rest of the deck was laid according to the deck construction found in the Bremen Cog.

The two stringers in the middle, from beam 4 forward, were thicker, since these are elements in the construction that had to direct the sailing forces from the mast and the forces exerted by the waves to the bow and the deck respectively. These are therefore not 'short' beams between every two crossbeam constructions, but long stringers that extend from 50 cm abaft of beam 4 until the side supports. They are notched at the knee bulkheads of the protruding crossbeams 4, 5 and 6. These two stringers are 20 wide, ranging in height from 38 cm at bulkhead 4 to 15 cm at the side supports.

The following elements can be specified in sequence of distribution of the sailing forces: the mast, the fissing plates, the fissing beam, two chocks with a breasthook on the mast pressure beams ending with a breasthook on the side stem, the sidestems that are notched over the V-shaped floor timbers, and finally the shell planks that are against these timbers. All these structural elements were fastened together with wooden treenails 3-4 cm thick. The deck planks are 4 cm thick and made of oak (remnant of heart sections from shell planks). The widths and tapered shape depended on the available wood and vary in width between 10-35 cm. The ends were placed in the rabbets cut out of the top corners of the stringers and have a overlay of 5 cm. The planks have an overlay against the sides on a wooden rail that is nailed against the ceiling.

Very little is known about cog decks from archaeological sources. A number of planks from the main deck were found in the Bremen cog; in the Nijkerk-II cog, only the aforementioned indications on crossbeams 1 and 2 were found. The question of whether or not the decks/deck planks were nailed in place also remains unresolved. The wreck of the Nijkerk-II cog indicates that they were not, since no nails or nailholes were found at the locations where they would have been placed. The deck planks on the Bremen Cog do seem to have been partly nailed in place, but it should be noted here that it is assumed that the vessel was still in construction when it sank. Important indications for the construction of such late medieval decks, have also been derived from the decks of an early-16th century *'waterschip'* on lot Mz 22 in Zuidelijk Flevoland (fig. 7.13c).[11] On the replica of the Nijkerk-II cog, the deck planks of the middle three sections between crossbeams 2 and 3 as well as 3 and 4 were not nailed into place. These planks were made into hatches that were nailed together with three clamps on the underside. These hatches and the underlying stringers are removable, creating a large loading aperture. The deck seams were caulked with hemp fibres and pitch. The hatches were caulked separately.

The anchor bitts

The anchor bitts is a heavy construction placed at crossbeams 5 and 6. The construction consists of two vertical beams installed against the back of protruding crossbeam 5 and against the outer edges of the middle stringers (the lower ends resting on the shell between two frames), while on the front these beams were installed between the knee bulkheads of crossbeams 5 and 6. Attached to these two beams, there is a strongback at about 1 m above the deck. The strongback and the two vertical beams are 25x25 cm in diameter, and the knees are 15 cm thick. The parts were fastened together with wooden treenails. The construction was designed in such a way that the cog could be towed by it. It was decided not to use the bitts as found on the Bremen Cog, since no indication of such a construction was found on the Nijkerk-II cog.

The small or mast bitts (belaying rack)

The mast bitts are a simple construction of two vertical posts, connected by a horizontal beam, which is fastened into place against the outer edges of the central stringers and against the back of crossbeam 5. Seven holes were drilled in the composite horizontal beam made from two smaller beams (the belaying rack), which is fastened together with wooden treenails around the beams; the corvine nails were inserted into these seven holes. The lines from the mast can be belayed here.

The windlass cylinder

No evidence of a windlass cylinder was found during the excavation of the Nijkerk-II cog. However, several nailholes 3 cm in diameter were observed in crossbeam 2 and the knee sections fastened onto it, about 90 cm from the sides. It was decided to place the cheeks of the windlass cylinder here. The cheeks were also the forward posts that supported the castle. The dimensions and design of the windlass cylinder on the Bremen Cog were chosen as a reference and adapted to the Kampen Cog.

The castle

During the construction of the Kampen Cog, no evidence had been found in the documentation on the Nijkerk-II cog that would indicate the presence of a castle. It was only during the reconstruction of the vessel in drawings in 2005/2006 and the writing of the report for this study that it became clear that the vessel most likely even had an aftercastle and a forecastle.[12] The technical (advisory) committee and the builders held the opinion that the ship should be given an aftercastle. The builders determined the lines and the fastenings. The result was a simple platform on four posts (fig. 7.14a and b). The two aft posts were knees; the lower ends were slanted, cut to fit and fastened into place on frame -2. The rising sections were perfectly perpendicular. The other two posts were straight and placed against the aft side of frame and crossbeam 2. A square cross frame was made to fit over the four posts, with the boards of the castle deck nailed onto it. A simple construction of posts and struts was fastened against the sides of the frame, with the sidings nailed on to it. On the forward side,

Fig. 7.14a. Overview of the afterbody at deck level with the rear stanchions (pale wood) to support the castle. Photo: Sars Houtbouw.

Fig. 7.14b. The aftercastle.

11. File Mz 22, RCE-Lelystad.
12. See section 6.2.2 and plates 6.IV A, 6.IV B, 6.V A and 6.V B.

openings were left to starboard and port through which it was possible to enter or exit the castle deck. The castle deck could be accessed via ladders from the deck. The upper edge of the sidings was crenellated to form battlements. The overall concept for the castle was derived from the cog on the seal of Damme (West Flanders, Belgium) dating from ca 1300 (see the Nijkerk-II cog, fig. 6.37). In retrospect, the battlements could have been slightly wider and the crenels could have been narrower.

Construction phase 7: the rudder, the mast and spar, and adding sintel clamps to the outside of the ship's hull

The rudder

No rudder was found with the Nijkerk-II cog. Rudders have seldom been found in shipwrecks in the IJsselmeer polders. A rudder found on its own near the Barbizon Hotel on Prins Hendrikkade in Amsterdam served as a model for the rudder of the Kampen Cog. This reused specimen was found underneath a house as part of a foundation, the walls of which have been dated to the 14th/15th century. It was made of an oak board 60 cm wide and 12-16 cm thick. A piece was added on the back to widen it. The rudder was fitted with wrought-iron rudder fittings and pintles and then suspended from the rudder pintles inserted into the wrought-iron gudgeons, which were attached to the stern construction. The rudder is operated with a tiller approx. 3 m long made of ash, which was secured with a dowelled joint.

The mast and the spar

Masts and spars are essentially never found with shipwrecks in the former Zuiderzee region. To the best of my knowledge, of the nearly 450 registered shipwrecks, only one instance is known, found near Lelystad, in which the lower part of the (broken-off) mast is probably still in place. No parts of a mast or spar were found with the wreck of the Nijkerk-II cog. The dimensions of both elements were designed and calculated by yacht-builder G. Kroes and Professor W. Draijer PhD, both members of the Technical Committee. The mast was made of a 94-year-old Douglas fir *(Pseudostuga menzisi)* from the Ardennes region of Belgium. The mast has a total height of 22.4 m, and a diameter of 46 cm at the base and where it passes through the deck, and 32 cm at a height of 18.8 m (rigging). A Scots pine *(Pinus sylvestris)* would have been a preferable choice for the mast, but no good-quality trunk was available at that time. The spar was also made of Douglas fir and was 13.1 m long, with a diameter of 26 cm in the middle and 14 cm at both ends.

7.4 Design and execution of the sail, rigging and blocks

The sail

Jan van Gammeren, rigging master of Sars Houtbouw and former rigging master of the Batavia shipyard, designed the sailing and rigging plan in collaboration with the master builder. The sail for the Kampen Cog has a total surface area of 144 m^2. The upper half of this is known as the oversail; the lower half consists of three bonnets. The bonnets can be removed piece by piece, reducing the sail area. The sail and spar are hoisted or lowered using handspikes, which are inserted into the windlass cylinder. Initially, a great deal of effort was put into attempts to weave sailcloth from hemp thread, but it did not yield the desired results. The sail was then made from factory-produced flax cloth and assembled from twenty lengths of fabric 62 cm wide. The heavy sewing, hemming of the leech, and finishing was done by Petra Mullecom (sailmaker at Sars Houtbouw). Several pieces of woven textiles were found in the wreckage of the Nijkerk-II cog which indicate that the cog (and several other cogs in this study) was equipped with a sail made from wool.[13]

The ropes

The guiding principle used for the rigging was that it had to be as simple as possible. Anything that was considered even slightly superfluous was therefore omitted. The ropes for the ship were twisted from tarred hemp, and consisted of:

- one forestay (Ø 5.5 cm)
- six inwale stays (Ø 5.5 cm)
  The stays were tensioned using deadeyes and lanyards.
- two shrouds (Ø 1.6 cm)
- two braces (Ø 1.6 cm)
- two sheets (Ø 1.6 cm)
- the halyard and the emergency halyard (Ø 2.4 cm)

Whether or not bowlines are advisable will be decided during sailing. The leech line and sleeve consisted of rope and sailcloth to which two jam cleats were attached to ensure that the spar does not get stuck while it is being lowered.

The blocks

Not much is known about the types of blocks that cogs were equipped with. The oldest and well-preserved block of a shipwreck from the former Zuiderzee area dates from ca 1400 and was found in the Almere-I cog (figs 7.15a, b and c).[14] It is most likely what is known as a brace block. This 32.5 cm high block was cut out of a single piece

13. See section 6.5.3 and figures 6.55 a through e.
14. Vlierman 1992b, 12; Hocker & Vlierman 1996, 53-55 and figures 26 and 27; chapter 16, section 16.5.2.

of ash wood *(Fraxinus spec.)*, with a disc of quarter-sawn pear wood (*Pyrus spec.*) and an axle made from a boxwood branch (*Buxus spec.*). Tensile tests with a replica made it clear that these types of wood would be very suitable for the various components and very capable of withstanding the expected forces. The deadeyes installed on the Kampen Cog were derived from specimens found on various 16th-century shipwrecks in the IJsselmeer polders. They are flat and pear-shaped with the narrow side sawn off straight. Three holes were drilled in the wide part: two in a row and the middle one slightly lower/higher. A groove was cut in the side to enclose the thick rope, for example the end of a stay. The deadeyes were made of oak. Such blocks are also known as 'deadhead blocks'. These blocks are reeved together with rope in pairs (fig. 7.16).

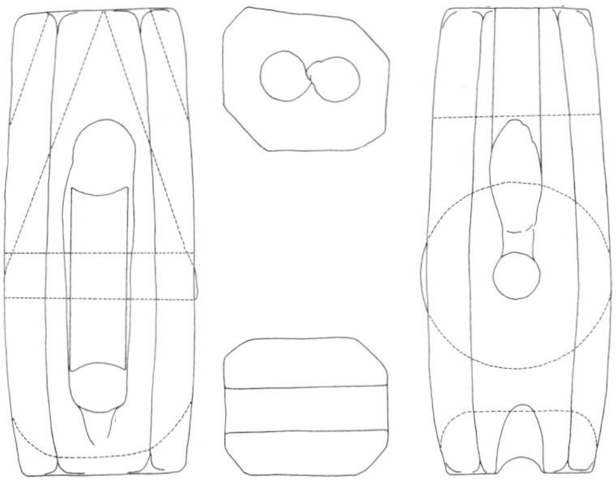

Fig. 7.16. Shrouds with deadeyes and the forestay. Photo: Kamper Kogge, p. 81.

Fig. 7.15a. The brace block from the Almere-I wreck, first quarter of the 15th century. Photo: Hocker & Vlierman 1996.
Fig. 7.15b. The brace block, cross-sections. Drawing: Hocker & Vlierman 1996.
Fig. 7.15c. The brace block, fastening.

7.5 Ballast

In order to sail the cog, the ship has to be ballasted. Lead and boulders were used for this purpose. Lead was melted and poured into moulds in a casting assembly at the cog yard. The lead bars weighed 200-600 kg. They were laid between the lowest ceiling strakes on the frames to keep them from shifting. The rest of the ballast consisted of boulders, as frequently found in shipwrecks in the IJsselmeer polders.

7.6 Construction and sailing data

Dimensions:	
Length from stern to stem	ca 21.6 m
Broadest width of the gunwale	ca 7.6 m
Inside of the ship from deck to bottom	ca 3.4 m
Draught	ca 1.9 m
Weight	45 tonnes
Loading capacity	60 tonnes
Mast:	
Length	22.4 m
Diameter at deck height	46 cm
Weight	2,500 kg
Spar:	
Length	13.1 m
Largest diameter (centre)	26 cm
Sail:	
Flax cloth	144 m²
Weight	122 kg
Ballast:	
Lead bars	20 tonnes
Boulders	20 tonnes
total	40 tonnes
Wood required for the cog:	
Oak from Denmark, France and the Netherlands	150 m³

Before installing the planks, the quality of the wood was examined. Normal quality was selected for the part of the ship that would be permanently under water, better quality along the water line, and normal quality again above that. The inferior quality was used for the ceiling boards, while the best quarter-sawn wood from the heart sections of the shell trunks was used in the deck.

Blocks:	
Blocks of ashwood and oak	25
Nails (double-clenched nails):	
Forged by hand by blacksmith	
Hengeveld from Kampen	10,000 nails
Sintel clamps:	
Die-cut sheet iron	16,000 clamps

7.7 Launch, first sailing tests and experiences

The launch

The launch took place on 1 November 1997 in the presence of a large group of invited guests, interested parties and the press (figs 7.17a, b and c). The largest mobile crane in the Netherlands (supplied by the Mammoet firm) lifted the Kampen Cog into the IJssel River, only 100 m from the place where the wreckage of a large cog from the first half of the 15th century had been discovered in the riverbed of the IJssel in April 2011.[15] After that, the deck was put into place. Then the windlass cylinder was installed. On Saturday 14 March 1998, the final structural elements were fitted onto the ship. Then the cog was rigged and the sail was fastened into place on the spar. In winter 1999/2000, a floor was installed in the hold. The ship was ready to sail.

During a visit to the National Museum's Institute of Maritime Archaeology (NMU) in Roskilde, Denmark to give several lectures, Ole Crumlin-Pedersen – coordinator of the Centre for Maritime Archaeology at the National Museum (NMF) – informed me in May 1998 that of all the cogs recreated until then, the Kampen Cog was the most authentically designed and constructed.

The first sailing tests and experiences

After the cog had been towed out of Kampen to past the Ketelbrug bridge, I had the pleasure of being aboard for the second sailing test on the IJsselmeer. It was an extraordinary experience for someone who is essentially only familiar with ships when they are in the ground. Being able to sail along on the replica of a late-medieval vessel that I myself had discovered and excavated made that voyage even more extraordinary. With this study in mind, which was already underway, it was possible to record extremely valuable impressions of how such a vessel handles on the water and approximately what it would probably have been like to live and work on board the cog. Despite the idea I had at the time that a ship of approx. 20 m was not very large by today's standards, I had to conclude that the crew was able

15. Waldus *et al.* 2012, chapter 23.

Fig. 7.17a. Exterior of the forebody, shortly before the launch. Photo: Sars Houtbouw.

Fig. 7.17b. Exterior of the forebody, shortly before the launch. Photo: Sars Houtbouw.

Fig. 7.17c. The Kampen Cog immediately after launching on 1 November 1997. Photo: Sars Houtbouw.

Fig. 7.18. The Kampen Cog on the river IJssel off Kampen, setting sail for the Ommelandvaart (voyage into the Baltic) on 7 June 2004. Photo: Van den Berg 2007.

to move around easily aboard a cog and that the hold below deck was quite spacious. Sleeping aboard the ship was the only part that would not have been pleasant; in essence, there is no data on that for the late-medieval period.

We only sailed on the wind with the oversail (i.e. without bonnets), because it was still almost entirely unknown how the vessel would respond. The sailing speed was measured. At some point, the ship sailed for some time in this manner at wind force 4-5 Bft. at a speed of about 4 knots without any problems. First-hand experience showed that the tiller was easy to operate in this situation and that the vessel barely made any leeway movement. It was a good sailing ship, as had already been predicted during the launch[16] by Werner Lahn, who had reassembled and built the Bremen cog in the German Maritime Museum (*Deutsches Schiffarhtsmuseum*, DSM) in Bremerhaven.

7.8 Characteristics of the hull, vessel and sails and the Baltic voyage in 2004

Characteristics of the hull, vessel and sails

From that point on, the vessel was taken out onto the IJsselmeer on practice trips on a regular basis. To bring the cog to the attention of the general public, the ship went to a number of events across the country. Invitations were also received for special occasions. Amongst others, the cog attended Sail 2000 in Amsterdam, was present at receptions for the queen in Doesburg and Deventer, and showed up at Botterdag events and Open Monument events. Some experience was gained during these activities, but even more during the practice trips on the IJsselmeer. As Lahn predicted, the hull shape had indeed guaranteed that the vessel would provide a stable sailing experience. All sorts of minor adjustments to e.g. the rigging or the way in which operations were carried out made the ship increasingly manageable and controllable.

The Baltic voyage: Ommelandvaart 2004

During the various sailing activities, a number of crew members talked about taking the cog on a trip to the Baltic Sea. The reason: everything about the cog was scientifically sound, from excavation to construction, but it had never been sailed as it would have been in the Middle Ages: taking the vessel on an unmotorised trip around Denmark to the Baltic Sea. That, too, need to be scientifically proven.[17] And so the idea for a voyage to the Baltic was born: the Ommelandvaart or 'voyage around the land'. On 11 December 2002, a meeting was organised in the Taveerne at the cog wharf at which the plans were presented. Many volunteers signed up straight away, leading to the Ommelandvaart 2004 project. After

16. Werner Lahn, personal statement, 1 November 1997.
17. Van de Berg *et al.* 2007, 24.

that, a large organisation was set up that would carry out the preparations. The first practice trip with twelve crew members took place on 18 April 2004. Just 10 minutes after the sail was hoisted and the cog sailed out of the harbour, after the command forward had been given the sail started flapping and two crew members were injured by the sheets. This sailing manoeuvre revealed many problems that required a solution. Before embarking on the actual voyage, the ship had to be fitted out (sleeping berths, etc.).

On 7 June 2004, the Kampen Cog and its escort ship the Carlotte departed from Kampen (fig. 7.18) to Stavoren, continuing the next day to the island of Terschelling. On 11 June, the reconstructed Nijkerk-II truly headed out to sea for the first time. Under poor weather conditions and wind force 8 to 9 Bft., the vessels reached Helgoland on 13 June. The prevailing wind and the time schedule of the voyage forced the skippers to decide not to sail along the west coast of Jutland to Skagen, but via the Kiel Canal and the Baltic Sea. The ship was in Skagen from 18 to 26 June and then headed off via Helsingborg (fig. 7.19) to Malmö in Sweden. Via Hiddensee, they sailed on to Stralsund, where the cog was moored from 6 to 9 July. Then the voyage continued via Rostock and Warnemünde to Wismar, and from there to Lübeck. From there, they sailed along Svendborg, Æroskøbing to Kiel. Via the Kiel Canal, the voyage continued to Bremerhaven, heading back toward the Dutch island of Terschelling on 31 July. The final leg of the journey home to Kampen took place on 2 August. For the preparations and the experiences, contacts, highs and lows gained during the voyage, it suffices to refer to the ship's log from the Ommelandvaart.[18]

The evaluation and reflection on the Baltic voyage describes how the crew experienced the experience of sailing a replica of the late-medieval ship, especially on open water.[19] But how did the ship sail at sea? Once it passed the sandbanks above Terschelling, the ship was released from the towing lines and was able to show its true strengths at wind force SW 4-5 Bft., diagonally from astern. Sailing on a choppy sea, the cog travelled 100 miles within 24 hours, an average speed of 4 knots, which corresponds to the wind force. The skipper said, "as hard as the wind blows, that's how fast it goes".

Steerability

Steering the ship proved possible with one person at the helm up to 5 Bft. At wind force 6 Bft. and more, the lateral pressure of the water on the rudder was so great that

Fig. 7.19. The Kampen Cog passing Helsingborg castle on the Sont, en route from Skagen to Malmö on 26 June 2004. Photo: Van den Berg 2007.

it was not possible to sail without a tackle. With a lot of wind and high waves from behind, steering the cog was a difficult undertaking.

Sailing on the wind

The cog could not sail on the wind at an angle sharper than 68°. As the wind increased and the sail area decreased, the ship's became increasingly powerless to take action. In strong winds (7 Bft. and more) with sail reeved as much as possible (all bonnets furled), the cog could only sail before the wind. In weaker winds and without significant waves, the ship could only make minor gains against the wind. And such gains were soon lost on longer stretches when manoeuvres had to be made, such as wearing. The primary causes of this sailing behaviour are:
the lines of the underwater body: short, wide and carvel-built with a short waterline (the length-width ratio of the cog is 3:1);
the lack of a keel;
the high sideboards of the ship catch a lot of wind, and that increases in percentage as the ship carries less sail.
From this, it can be concluded that the Kampen Cog does not sail well when running upwind. Running downwind does not go as well if reeving becomes necessary. It was clear why the ship and crew would sometimes have had to wait a long time for a suitable wind.

Sailing before the wind

It has already been noted that the cog could handle any courses at an angle greater than 68° from the direction of the wind. And the more the cog sailed before the wind, the more effective it sailed. A German study showed that a cog on this course exhibits superior performance compared to other cross-rigged vessels and can be compared

18. Van de Berg *et al.* 2007, 36-83.
19. Van de Berg *et al.* 2007, 84-86.

to sailing yachts in terms of efficiency. The Kampen Cog was placed along the hydromechanical bar and analysed.[20] With regard to sailing performance, a simplified velocity prediction programme or VPP (a speed forecast for various wind speeds/directions) was used to determine the sailing speeds of the cog for a number of wind speeds and wind directions. The results are shown in table 1.

The hull shape and sailing performance were also compared to the ships of the Dutch East India Company (VOC) that sailed later. The investigation concluded: The cog is simple in shape. The length-width ratio in particular gives it more resistance than other ships. The underwater design of VOC ships had also clearly been developed further, but the cog performed better before the wind compared to VOC ships.

Tacking
The most practical way to tack (ready about) is wearing. After some practice, this manoeuvre was executed smoothly by the crew. The disadvantage of wearing is the loss of height, since the ship needs a great deal of room to perform this manoeuvre.

Manoeuvrability
When space is limited, the cog is very restricted in its mobility. When sailing before the wind, the ship responds quickly and well to the rudder.

Stability
The compact construction and lack of a keel makes the cog very uncomfortable for the crew at sea. The movements of rolling and pitching follow each other very closely (5-7 sec.). This places a considerable physical strain on the crew. Despite this rocking behaviour, the ship exhibited rough, but safe sailing performance at sea in stormy conditions on the North Sea. The high bulwark kept the deck dry as a bone while rocking as much as 10°.

7.9 Conservation and management of the Kampen Cog and the installation of an engine

The conservation and management of the cog is handled by the "Kamper Kogge" Foundation in Kampen. Many people were and are involved in the construction of the replica, in the organisation and execution of the Baltic voyage (Ommelandvaart), and in all other sailing trips and maintenance. Volunteers played a major role in all these activities. The Kampen Cog has been sailing for about twenty years by now and has become a familiar sight at all sorts of events in the Netherlands and abroad. Its limited manoeuvrability consistently caused problems in harbours and on rivers etc. and made the presence of an engine on board more and more desirable. Despite the fact that this would affect the authenticity of the vessel, it was decided to install an engine, both for safety reasons (at the insistence of the Shipping Inspectorate and in compliance with prevailing regulations) and for cost reasons (an accompanying [towing] vessel always had to be present during sailing trips).

Tabel 1. Performance

Wind force	Course	Drift angle	Speed (in knots)
Bft. 6 (25 kn)	(to sail) before the wind	0	7
	(to sail) large	3,3	7
	A beam wind	5	6
Bft. 4 (14 kn)	(to sail) before the wind	0	4,5
	(to sail) large	2	4,5
	A beam wind	2	3
Bft. 2-3 (7 kn)	(to sail) before the wind	0	2,5
	(to sail) large	2	2
	A beam wind	2	1,5

7.10 Summary and conclusions

Summary
In 1989, an initiative was launched in Kampen to build a Hanseatic cog. In consultation with the Netherlands Institute for Ship and Underwater Archaeology (NISA) and other experts, a construction plan was chosen that was scientifically and historically sound. The data on the cog excavated in 1983 in southern Flevoland on lot Oz 36 (the Nijkerk-II cog, Chapter 6) was selected as the reference model, since that vessel had been a large cog, and because this shipwreck was more intact than the others.

Between June 1994 and March 1998, a replica or experimental reconstruction of the ship was built. The underwater body could be reconstructed relatively easily, thanks to all the preserved parts. The sides, on the other hand, presented problems, since large parts were missing from the wreckage, and the frames that were found had probably been warped during the centuries they had spent in the ground. Presumably, the reason why that the underwater body was reconstructed on paper beforehand, but the sides were not, was for example because the necessary

20. Allema & Hubregtse 2008, 34-38.

Fig. 7.20. The application of sintel clamped moss caulking (moss, laths and iron sintel clamps) to the seams of the Kampen Cog. June 1995.

time was lacking. The building team therefore started building up the sides during the process, with the result that the guiding principle they had started from – building on frames – soon encountered a number of problems. This had already been established when building the scale model as well.

After consultation, it was then decided to continue the rest of the hull as an experimental archaeological project. With the simultaneous construction of a model at a 1:10 scale, with the construction of the model always staying at least one phase farther along than the replica, it became clear that cogs were not built on frames; in fact, the sides used a 'shell first' construction process, which was carried out in phases. At the end of the first phase, it became clear that the protruding crossbeam structures 1 and 2 that had been found did not fit in their originally designated location in the replica; they were too short. Construction continued and the beam dimensions were adjusted. In the report on the Nijkerk-II (Chapter 6, Section 6.3.1, Table 1), the general dimensions of the beams in the Kampen Cog are shown in brackets behind the dimensions of those beams in question in the reconstruction drawings from 2005/2006.

It had been established on multiple occasions in the original cog finds that the clinker-built seams in the sides had been sealed on the inside with sintel clamped moss caulking in a single swoop, from stern to stem, and that the frames were not installed until after that. In relation to research on sintel clamped moss caulking[21], the seams were caulked according to this method, just as they had been on the original ship, to gain insight into how the various elements had been combined and how much time and manpower was required to do the work (fig. 7.20). After some experimentation and practice, it turned out that a few people could sintel clamp about two seams from stern to stem in a single working day. The part of the ship above the crossbeams protruding through the shell plus the deck and deck layout was derived largely from historical sources, such as seals and other archaeological finds, discussed with the technical (advisory) committee, and designed and built by master builder Kees Sars and ship-builder and apprentice trainer Anton van den Heuvel (Sars Houtbouw) at the Koggewerf or "cog wharf" in Kampen. The European Social Fund (ESF) invested money in the project to train job-seekers and provide them with work experience. The cog hull was launched on 1 November 1997, after which construction on the rest of the ship was finished.

Sailing on practice trips and taking part in all sorts of events revealed that various (minor) adjustments were needed in order to sail the ship well and safely. Adaptation and thus deviation from the original ship was necessary to prepare the vessel for an organised sailing voyage to the Baltic Sea, along a traditional route known as the Ommelandvaart or "voyage around the land", to gain insight into how such a late-medieval (Hanseatic) cargo vessel would behave at sea under all kinds of circumstances. The Baltic trip (Ommelandvaart) took place between 7 June and 2 August 2004. After the ship sailed for 1720 nautical miles on this journey, overcoming all sorts of obstacles and encounter highs, but also sad lows, it can be stated that the Kampen Cog proved itself to be a paragon of late-medieval seaworthiness. The booklet about the Kampen Cog and the Ommelandvaart 2004 project offers a decent impression of the experiences with the ship. Unfortunately, the Baltic sailing trip in 2004, which was prompted largely by a desire to conduct scientific research on sailing at sea, following the excavation and the construction of the replica, did not result in a scientific report.

Conclusions

In addition to the above observations, it can be concluded that the reconstruction of the Nijkerk-II cog provided valuable and often confirmatory information with regard to the interpretation of archaeological cog finds, revealing more about how these ships were built and about the various stages of the construction process. This was confirmed over and over during reconstruction on paper of the other cog finds in this study. Along the way, it also became increasingly clear that there must have been a cog-building tradition in the core area of the Hanseatic League during 1200-1450, and that such vessels had already been built for that long when the Nijkerk-II cog was constructed in the first half of the 14th century. During the excavation of the wreckage, it had already been concluded that it was definitely not a prototype, but rather a prime example of a hull shape and construction method that was already fully developed at that time. The construction of the Kampen Cog also clarified how and in some cases why certain constructions were used and how they were executed. The construction of the ship also led to a greater understanding of how the late-medieval shipbuilders worked and how they managed to achieve these results with the resources and tools at their disposal – certainly in comparison with, for example, the tools and resources that are available to us today. In any case, the medieval shipbuilders who built the cogs by sight exhibited incredible skill and craftsmanship. The builders of the Kampen Cog definitely deserve to be showered with compliments!

21. Vlierman, 1996a.

Fig. 7.21. The Kampen cog on the high seas.

More cogs, other vessel types and a medieval shipyard

III

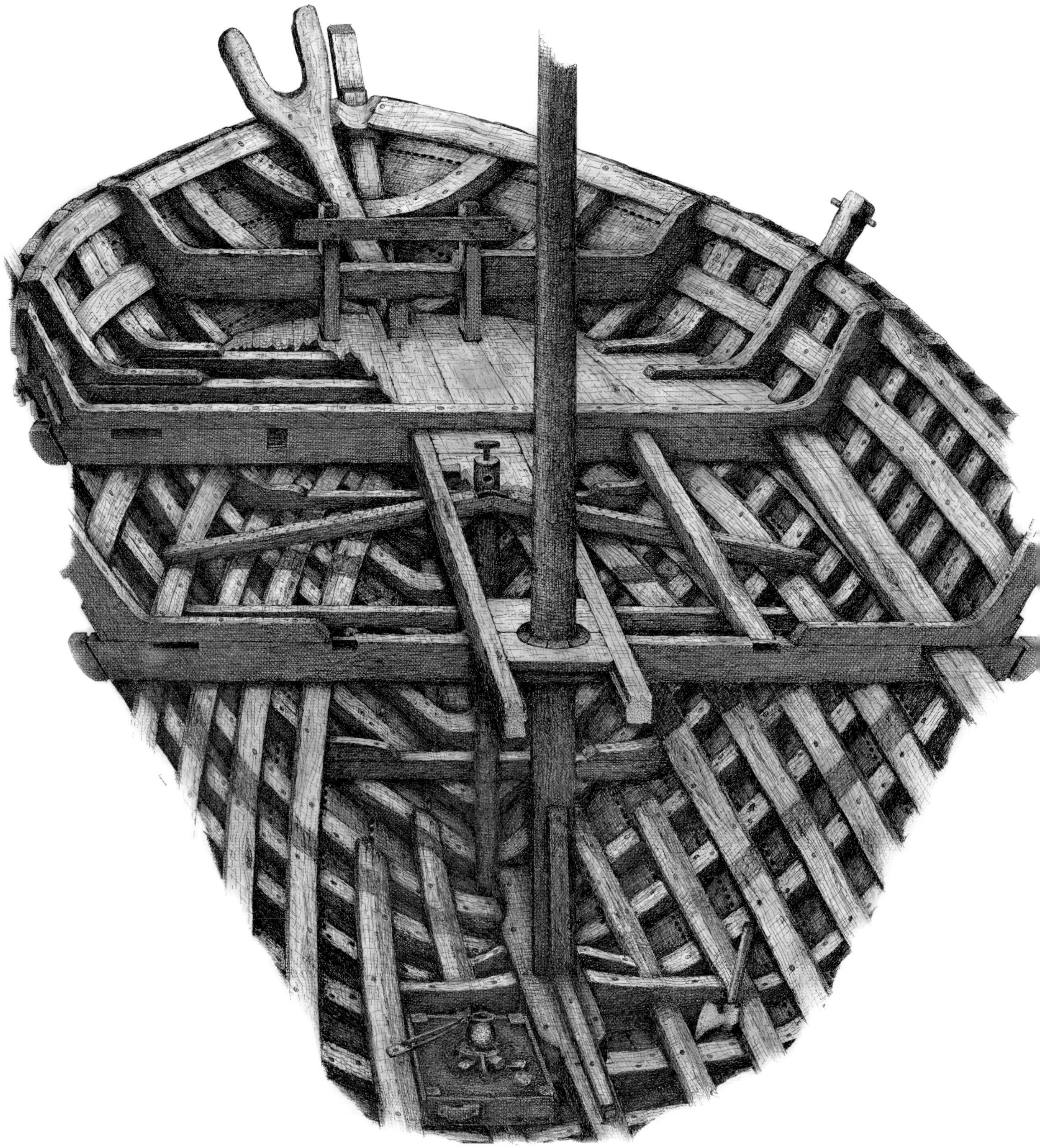

Nijkerk-I cog: impression of the forebody.

The Nijkerk-1 cog[1]

8.1 Find conditions

8.1.1 Discovery and investigation

Not far from the Nijkerkerbrug (bridge) in Flevoland, an exploratory investigation was carried out on 3 November 1980. The digging of a boundary ditch necessary to the future use of the field (part pasture, part forestry) had some time previously brought to light a shipwreck. The exploratory trenches revealed that a presumably near-complete, carvel-laid vessel bottom was present, which, with a slight list, had sunk some way into the Pleistocene sand. Lying virtually horizontally to the southwest of the bottom, was one of the clinker-built sides, which had broken at the turn of the bilge, but still remained connected to the bottom. During the investigation no movable finds were recovered. The soil profile above the wreck, the sintel clamps on the caulked seams, and structural parallels to a wreck excavated in 1976[2] suggested that this was a late-medieval, possibly 14th-century cargo vessel. Early frost put an abrupt end to the exploratory investigation. The suspected vessel type (a cog), the side surviving probably up to a high level, combined with the poor condition (especially of the side) and the wreck's shallow position prompted the decision that since *in-situ* protection was not feasible, the ship should be excavated as soon as possible. The cog Nijkerk-I to be discussed here lay in lot Oz43 in the Zuidelijk Flevoland polder (figs 8.2, 8.3 and 8.4).[3]

The excavation and field documentation took place between 1 May and 10 November 1981. In view of the fra-

Fig. 8.1. The excavated wreck viewed from the stern.

1. Translation by Xandra Bardet
2. Reinders *et al.* 1980, 6-15; the Dronten-I cog, chapter 11.
3. 200 metres from the centre of the roadside ditch along the Erkemederweg and 60 m from the centreline of the roadside ditch along the Erkemederpad. The ship lay oriented NW-SE with the bow to the NW, about 2 km NE of the Nijkerkersluis (locks) in the Zeedijk (sea wall). Ordnance map sheet 26G, coordinates 161.500/475.700, with the highest-lying parts at 2.6 m below NAP (ordnance datum). The local water depth formerly was ca 1.6 m: Hydrographic map 1921, low water mean.

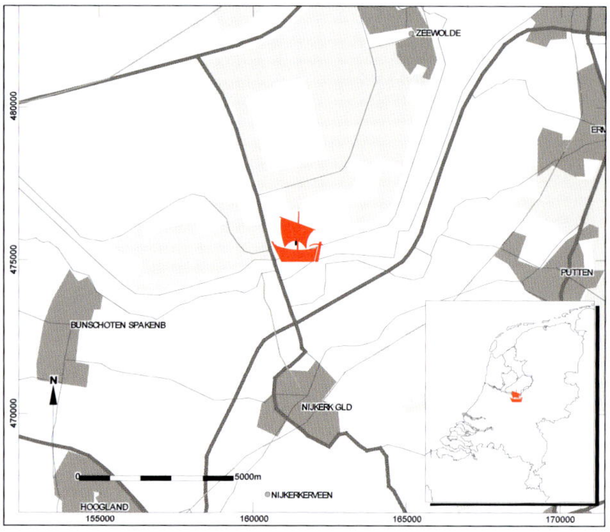

Fig. 8.2. The findspot of the cog.

Fig. 8.3. Ordnance map of the Kingdom of the Netherlands (scale 1:50,000), 1861, showing the findspot of the Nijkerk-I cog.

Fig. 8.4. The excavated wreck viewed from the bow.

Fig. 8.5. The site during excavation viewed from the bow.

Fig. 8.6. P. Heinsius, H.L. Janssen and his wife, and K. Vlierman at the excavation on the 12th of June 1981. Photo: Roelfsema.

gility of the timbers and their shallow position, no mechanical digger was deployed and all excavation was done manually (fig. 8.5). The NISA employees were assisted by volunteers from two antiquarian societies: 'Archeologische Werkgemeenschap Nederland' (AWN, IJsselmeerpolders division), and 'Oudheidkundige Vereniging Arent toe Boecop' from Elburg. For the first time, also two students/interns joined a shipwreck dig in Flevoland. To begin with, the trenches of 1980 were opened up again, as well as some small pits to determine the overall spread of the wreckage and to assess the shape and size of the area to be uncovered. A profile dam was reserved over the forebody. Initially an area of 22.4 x 7.5 m was opened up. For uncovering the starboard side, which had been dislodged and displaced in the course of the wreck-formation process, the area had to be ladder-like expanded to the southeast up to a width of 14.5 m.

After the loose timbers found above and beside the wreck had been marked with numbers *in situ*, recorded in a survey drawing, lifted and transferred to a previously prepared water basin, the wreck itself was cleaned and documented. During the fieldwork a great deal of puzzling took place over the original position of these detached timbers. Also as fully as possible the parts of frames were matched up, reconstructed and documented. For the first time in Flevoland, all the shell planks were drawn to a scale of 1:10. The findspots of all recovered finds are recorded in plate 8.I A with their find/accession numbers, as they are in the text and find list.

The excavation was visited by many interested visitors. A special guest was Dr P. Heinsius (author of *Das Schiff der Hansischen Frühzeit*) on 12 June (fig. 8.6). He referred to the wreck as "a spectacular find, probably the oldest and

Fig. 8.7a & b. The whirlwind-wrecked office and tool huts, 6 October 1981.

most complete cog so far [1981]". In the autumn, prof. dr. D. Ellmers and W. Lahn (Deutsches Schiffahrtsmuseum, Bremerhaven) and O. Crumlin-Pedersen (Vikingmuseum Roskilde, Denmark) paid a visit. Thanks to their expertise and experience in documenting and reconstructing medieval ship remains, a great deal of valuable advice was obtained. In the course of the excavation the impression arose that the vessel probably dated from the 13th century.

The author during the excavation also looked for the spot on lots Oz35 and Oz36 where in the mid 1970s medieval potsherds had been picked up, as a possible and presumably contemporaneous habitation not far from the cog site. Not just this spot was reidentified, but in the reclamation ditches still present at that time, indeed a second, even more complete cog was discovered.[4] During the excavation an unusual incident occurred in the late afternoon of 6 October when a tornado passed the site, by Dutch standards a very severe one. As a result, the office and tool caravans were overturned and wrecked (fig. 8.7), the tent in which the timbers were drawn was blown away, and a range of minor damage. One drawing table was recovered ca 100 m from the site. Amazingly, the protective plastic sheeting still covered the wreck, untouched by the storm which must have passed the excavation within yards. At the Nijkerkerbrug bridge, pleasure craft had been lifted from the water and thrown onto caravans at a nearby campsite, and near the Moerdijkbrug bridge a passenger aircraft crashed as a result of this same storm.

8.1.2 Geology and stratigraphy

The Pleistocene surface and its covering deposits

Apart from the ploughsoil (about 20 cm thick), the boundary ditch cutting obliquely across the wreck, and the drainage ditches across the fore- and afterbody, the soil profile was barely disturbed. The Pleistocene surface in this area is quite irregular, its elevation varying between 0.55 and 1 m below the surface within the excavation. The forebody had settled up to 1 m deep into the Pleistocene sand. Overlying the sand, the lowermost part of the original and at least 2-m-thick Holocene peat layer was still present, in some places right up to - and originally extending into - the ploughsoil. Parts of the sides lay on top of the peat. Nothing was known about the age of the peat in this area. Initially, peat growth at the time of - or until shortly before - the foundering of the ship had seemed possible; yet a (much) greater age was more likely. During the excavation samples were taken from the section for pollen analysis and radiocarbon dating. The pollen spectrum from the mineral soil shows the evolution of Late-Atlantic forest under relatively dry conditions. There are signs of a podzol forming, which however grew progressively wetter because of a rising water table. Peat growth occurred during the '*Cardium* transgression' phase. The earliest dating of the peat is 4302 BP; the youngest, 3883 BP.[5] This leaves a gap of at least 3500 years before the (marine) clay sediments were deposited on the peat.[6]

On the remainder of the peat there was just a thin layer of strongly humic Almere sediment around the sides of the excavation. Immediately outside the excavation, the peat layer in most places extended up to the tilth, so presumably there is sedimentary material of the Almere phase only in the cavity created by the ship in the peat and the Pleistocene sand. Seashells were encountered in the clayey ploughsoil. But there no longer was any suggestion of a distinct shell layer, so the Zuiderzee deposit cannot have been thicker than 15 cm. The partially reconstructed section across the forebody (fig. 8.8) shows the hollowed-out cavity in the Pleistocene sand with inside it first of all a stratified

4. Vlierman 1985a; the Nijkerk-II cog, chapter 6.
5. GrN -13035 and GrN -13033, respectively.
6. Van Smeerdijk 1989, 477 and 486; for the full report, see appendix.

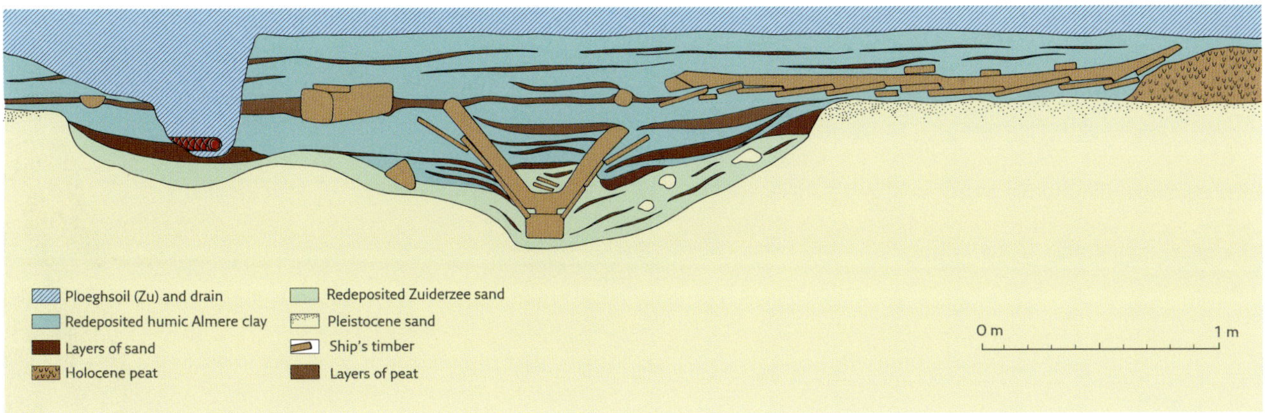

Fig. 8.8. Soil profile across the forebody.

deposit of redeposited Pleistocene sand with bands of redeposited peat and a few clods of sand containing iron-pan, while in other places chunks of displaced peat were encountered as well. Above it is strongly humic Almere clay shot through with sandy bands, thin bands of peaty matter and much driftwood, the bulk of which must derive from the eroded peat. The Almere deposit defies further subdivision. Like the Nijkerk-II cog (Section 6.1.2), this vessel most probably did not sink under (fully) open-water conditions. Section 8.1.4 will focus in more detail on the wreck-formation process. The ship's foundering cannot be closely dated by geological means. The Almere deposit, ca 25 cm thick, at any rate indicates that it must have occurred well before the salinisation of the area, which took place from about 1600:[7] definitely no later than 1500.

8.1.3 Medieval habitation

In the vicinity of the Nijkerk-II cog[8] evidence has been found of the 12th- to 14th-century farming settlement of *(H)erca* or *Erce* (i.e., *Ark*), while there was evidence of occupation in Carolingian times too.[9] Evidence of habitation in the immediate vicinity of the Nijkerk-I wreck is lacking. Nor could it be determined whether the ship had sunk in, for instance, a former peatland channel, or whether any channel remnant was present nearby. Yet as indicated in section 8.1.2, some lumps of sand with iron pan were found in the redeposited material around the ship, which might tentatively be interpreted as evidence of a channel bottom, while the uneven depth of the Pleistocene surface within the excavation might potentially be a further indication of the same. Indeed, judging by old maps it seems quite credible that the brook which ran northwards from the Oud Aller estate once made a westward curve as it crossed the former peatland north of the modern sea wall. In the map showing the depths of the Pleistocene sand below the modern surface[10] there is a deep (1-1.5 m) and ca 300 m wide depression in the Pleistocene sand at, and extending east of, the site of the cog. Some 400 m towards the Nijkerkerbrug bridge, the sand appears at a depth of just 0.5 m; 800 m eastwards this is also the case. At about 1 km north of the cog, the subsoil displays a depression in the Pleistocene surface more than 3 m deep, extending in a roughly east-west direction.

Which is an obvious spot for the original and natural discharge of the above-mentioned brook/peatland channel, and for the Schuiten Beek which runs a little further east.[11] Also the Middelbeek stream in the higher interior may, before the building of the sea wall (Zeedijk) in 1356, might have followed a slightly different course. For example, through the cog depression towards the above-mentioned depression, before emptying into the river Eem. In 1977, during the excavation of an 18th-century cargo vessel on lot Pz33 close to the Nulderdijk,[12] redeposited medieval potsherds were encountered in the top of the soil profile which had been disturbed by the wreck. This site lies about 2.5 km east of that of the Nijkerk-1 cog. In the early 1980s, an almost intact *Kugeltopf* with traces of soot was found in the cutting of a ditch about 3.5-4 km north of the cog.[13] This kind of *Kugeltopf* is known as 'pseu-

7. See the Marknesse cog section 12.1.2.
8. More than 2 km away west- southwest from the above mentioned find of the Nijkerk-II in section 6.1.3.
9. Vlierman, 1985a; section 6.1.3.
10. RIJP-scientific department of soil science, January 1981, stb.nr. 113.107.
11. Some years ago the presumable channel remnant of the Schuiten Beek was visible in the ditch cutting along the Nulderpad, from the Spiekweg road almost to the point where the path bends southwards (author's own observation).
12. File RCE, Lelystad.
13 Vlierman 1985d, 29-32.

do-Paffrath' ware, a fairly hard, grey, hand-formed ware that is dated to the first half of the 13th century.[14] These finds may actually have a link to the hamlet of Nulde (*Niuthlo*) which is still mentioned in the mid-12th-century revenue lists of the demesne of Wierden, but a century later no more. The likely location of this hamlet, which can be roughly identified, in the meantime had become part of the expanding Lake Almere.[15]

8.1.4. Condition and wreck formation

Condition

In the course of the early days of the excavation, it became clear that the side observed during the exploratory investigation was the port side. The oblique hood ends of the lowermost four strakes, originally spiked to the stem hook and stempost, were found to be still present. Those further 'up' had been damaged or destroyed by the cutting of a drainage ditch. The port side, of which the after part was largely missing, in the forebody still comprised nine strakes. The starboard side had broken off at the turn of the bilge, become disconnected from the bottom, and had floated away a few metres. The fragment lay beside and aft of the afterbody and in its forward part still numbered eight strakes. Above and beside the bottom and the port shell to port and starboard lay various timbers that had become removed from their original position. These included some transverse constructions with their associated knees (Plate 8.I). A remarkable find was the fire place, a square wooden box still partly filled with loam (fig. 8.9, Plate 8.I and Section 8.5.2). The underwater body was found to be almost entirely present, including the (broken) stern hook. The stem hook had broken close to the scarf with the keel plank, and initially was thought to be lost. However, at the end of the excavation the stem hook was found, in two pieces, lying beneath the forebody. In the bottom, all floor timbers were still present in their original position or lying close by.

Fig. 8.9. The fireplace: a low wooden box filled with loam.

On the floor timbers and above the keel plank a low keelson had been fixed with a broadened and somewhat thicker part, the mast step, containing a cavity to receive the heel of the mast. On both sides the mast step was supported by a pair of riders. Both to starboard and port, a heavy ceiling board lay on the frames, while on either side (parts of) three ceiling strakes were found. As a result of the high position in the soil, and the cutting of drainage ditches and the boundary ditch, the sides had sustained serious damage in three areas. The timbers damaged in the process had been lost (Plate 8.I A). Above, it was already mentioned that the sides of the vessel especially were in poor condition as a result of their shallow burial. The bottom with the frames was generally in better shape, because between ca 0.6 and 1.8 m below the surface it lay under the water table, which as a result of seepage water entering under the *randmeer* dike is relatively high here.[16] Moreover, a large part of the wreck was surrounded (and covered) by a 20-60 cm thick peat layer, which worked as a sponge, keeping the underlying layers and the humic clay/sand fill above the wreck quite moist.

Wreck formation and the weather in the second half of the 13th century

The recovered wreckage, the peat overlying the Pleistocene sand, the flush sediments-filled wreck site and the track that the bottom of the vessel scraped in the ground during the early stages of wreck formation, provide sufficient evidence for a reconstruction of the process. Fig. 8.10a shows a general view of the excavation area after removal of the ploughsoil (the top 20-30 cm). The peat is still evident on the west, south and east sides of the wreck. The wreck lies embedded in layers of Almere-clay and sand intercalated with layers of redeposited peat, chunks of peat, pieces of driftwood and detached ship's timbers. The parts of the sides still connected mostly rested on a final remnant of the peat. Figure 8.10b shows the same situation, but this time at about 60 cm below the surface. Interestingly, the peat now is absent beside the wreck to the southwest, because of a local rise in the sand surface. The depth of the Pleistocene surface on the south side is about 60 cm, at the southeast to northeast sides this ranges between 70 and

14. Janssen 1981, 171.
15. See Nijkerk-II cog, section 6.1.3.
16. A more or less comparable situation as nearby the Nijkerk-II cog (section 6.1.4).

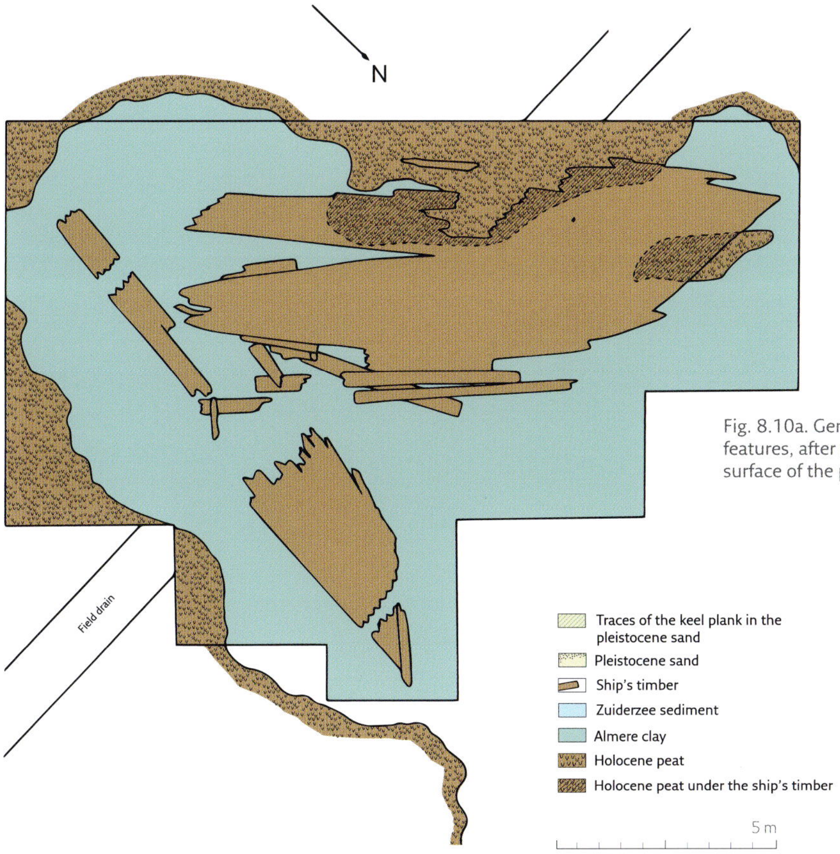

Fig. 8.10a. General view of the excavated area with soil features, after removal of the topsoil the remaining original surface of the peat.

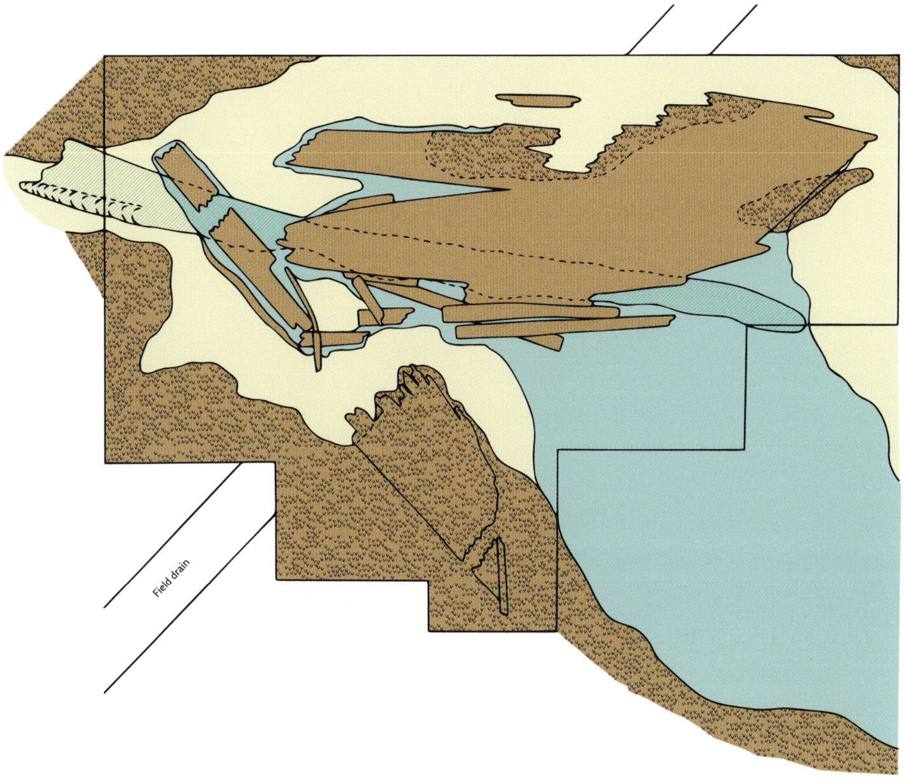

Fig. 8.10b. General view of the excavated area with the soil features at ca 60 cm below the surface.

95 cm below the surface. At about 20 m north of the site, this is over 100 cm. On the north side of the forebody is a roughly 9 m wide flush sediment-filled cavity which extends at least 10 m in a northerly direction. This 'channel' was scraped into the peat and the Pleistocene sand by the vessel's keel plank; the feature is 20-30 cm deep.

Most probably, one or more severe northerly storms pushed the ship, stern pointing south, onto the peaty (underwater) coast or shore, and progressively worked itself into it. Within this channel there was a trace obliquely running under the bottom, ca 22 m long, 2-3 m wide and ca 15 cm deep, scraped into the Pleistocene sand even more deeply. This trace without a doubt marks the place where (the keel plank of) the cog initially ran aground and/or came to a standstill. The wavelike patterns in the southeasternmost part of this feature must be due to the wreck eventually shifting in fits and starts back to the northwest by about 2 m. The keel plank is 13 m long and 0.3 m wide. Taking this into account, it is evident that the ship must have moved about quite significantly both lengthways and sideways before it finally came to rest. The broken stern and stem hooks and the broken-off stem hook lying beneath the bottom show that the ship must have disintegrated under very rough conditions. In the process, the ship changed its orientation somewhat and also much of the surrounding peat became detached, was washed away or remained in large chunks, some of which ultimately ended up in the sedimentary deposit overlying the wreck. In the fill of the scar, at the original location of the stem hook, a textile fragment, fragments of a grind-/whetstone and a shoe were found. They must have been washed out of the ship when the stem hook was broken and displaced. They are described in section 8.5.

The condition of the ship before disaster struck and the subsequent process of disintegration can also be reliably reconstructed. Yet it should be noted that there are two potential scenarios. In one, we may be dealing with a still complete vessel, that became stranded on a clay-on-peat shore of a channel or a coastal bay. The alternative is that this was a written-off, incomplete ship which like the Nijkerk-II cog had already been abandoned in a peatland channel before being ripped from its berth and drifting further inland. After the hulk could no longer be pushed further into the peat and the sand, northerly storms and the associated waves and currents subsequently proceeded to break up large areas of the peat around the wreck, and detached various structural elements and the sides. In the course of this, the hulk also rotated a bit with respect to the position in which it had initially become stranded.

The detached timbers above the bottom, lying on the first sedimentary layers of the flush sediments and the height of the preserved sides show that the the deeper-lying forebody must have remained intact the longest. The absence of large parts of the afterbody and of the entire 'above-deck' part of the sides may also be explained in two ways, or as a combination of the two. One is that they became detached from the hull and then floated away; the alternative involves human agency. In other words, the above-water parts of the ship that lay stranded on a shore and possibly had already sunk, were salvaged by local people for use as construction timber and firewood. The head hewn from starboard futtock 18B (fig. 8.11 and Plates 8.I A and B, futtock K), which was found beside the ship, separate from the shell and the detached futtock, distinctly argues in favour of the latter. Also the position beside the underwater body of the shell planks from high-up strakes J2, K1-1, K1-2 and L1 (Plates 8.I A and B) confirms that the highest preserved parts of the starboard side had already become detached from the hull and had sunk at the site before the side broke at the turn of the bilge and drifted off a few metres.

It may have been residents of the nearby village of Ark who dismantled the hulk. The people of Nulde, a hamlet about equally distant, but east of the cog site, are unlikely candidates because, as mentioned above, this settlement probably no longer existed by the time of the ship's foundering. As a result of the pounding waves wrecking the ship, and/or human agency, the free-hanging starboard

Fig. 8.11. The detached, hewn-off head of starboard futtock 18b.

side would have been the first to gradually become further detached, folding over and ultimately breaking away and landing obliquely behind and beside the wreck. The port side must initially have been jammed up against the peaty shore, and probably did not fall away until (much) later, after the gradual erosion of the peat by waves and currents. Eventually, both the starboard and port sides came to rest horizontally on top of a peat remnant.

That the sides broke at the turn of the bilge directly follows from the weak construction of scarfs just there joining the floor timbers to the futtocks, fragile once the crossbeams at deck level had gone. As for the fireplace, the 'tower' of the original loam fill, still present in the middle of the fire box and eroded into a round shape, shows that this item must have been exposed to open water for a long time. Only under such conditions would the swirling water be able to slowly dissolve the solid loam (fig. 8.12). Also the ample amount of driftwood around and inside the wreck indicates that for a long time the wreck must have trapped bits of floating wood and peat, until the wreckage was finally covered over with sediment. The total wreck-formation process will have taken a long time, presumably decades, and undoubtedly occurred in episodes. Section 8.8 explains that the foundering of the cog should be dated to the second half of the 13th century. In 1262 and 1268 severe storm floods affected the coastal zone of the Low Countries, as was the case in 1287 and 1288. In 1290/1291 and 1300/1301, these parts saw serious river floodings.[17]

Fig. 8.12. The fireplace: a low, square wooden box, in which a cone of loam with a small part of the burnt surface has survived the swirling water.

8.2 The ship

During the investigation and drawing of reconstructions, naturally much attention was given to the data of the reference wreck (Nijkerk II). Besides parallels, also differences were observed. Answers to several questions were found in the Swedish Kalmar-I wreck.[18] Also parallels and differences with regard to other 11th- to 14th-century vessel finds were looked into (see section 8.8).

8.2.1. Documentation and reconstruction

Documentation

Thanks to the horizontal position of the preserved parts of the sides and the frame timbers found with them, the forebody of the ship can be relatively fully reconstructed. However, the situation as found *in situ* also made it clear that the flat-lying and poorly preserved sides and the futtocks attached to them had been distorted, and that in terms of reconstruction it would provide inadequate or at any rate unreliable evidence, if the findings were to be recorded, as up till then was usual at the nautical archaeology department of the RIJP, in field sections only. Indeed at about 1-m intervals the most informative transverse sections were drawn as well as a longitudinal section along the keel plank and hooks, so as to record the wreck's position in the soil profile (Plates 8.II A and B). Also the Geodesic Service at Delft took stereo photogrammetric images from an airforce helicopter (fig. 8.13). Besides, a survey drawing was made of all the loose timbers and the underlying wreck (Plate 8.1A). This drawing proved crucial in our attempt in the field to put back the detached timbers, and in reconstructing the ship from the scale 1:10 drawings of individual parts.

The chapter on the Nijkerk-II cog (Section 6.2.1) already explained the rationale for drawing all shell planks in detail to scale 1:10 besides the usual plans and sections. The recovered loose crossbeams with knees and a few V-shaped floor timbers on the bottom and a further one on the stempost (32) were temporarily placed back onto the collapsed starboard side, in order to gain an impression of the shape of the forebody (figs 8.14a and b). The above-mentioned distortion of the frame timbers on the sides however necessitated an additional approach. During the puzzling, some frame timbers had been lifted off the shell, and laid on their sides to avoid breakage. The wood was found still to be so resilient that the frames more or less resumed their original shapes (and no longer fitted onto the sides when put back where they came from). In

17. Buisman 1995, 601.
18. Åkerlund 1951, 27-50, 151-152 and plates 1 t/m 9.

Fig. 8.13. General view of the excavation from the airforce helicopter. Photo: Meetkundige Dienst Delft.

Fig. 8.14a. The transverse constructions with knees DV4 and DV5 returned to their original positions on the starboard shell (aft of frames 25 and 27).

Fig. 8.14b. The detached third-highest floor timber (frame 32) replaced onto the stempost

Fig. 8.15a. Frames reconstituted beside the wreck. Frame 27 with transverse connection DV5 and the port knee of DV10 above it.

Fig. 8.15b. Frame 25 with transverse connection DV4.

Fig. 8.15c. Frame 17.

the light of this unexpected phenomenon, it was decided to lift all parts of the more complete frames and to reconstruct, draw and photograph the individual frames on a flat area of the excavation floor (figs 8.15a, b and c). Plate 8.II C shows drawings of all the frames as reconstructed in the field, complemented with sections of the recovered shell planks and crossbeams.

Reconstruction
The frames, reconstructed as fully as possible in the field, proved to be crucial evidence in the ultimate reconstruction of the shape and dimensions of the hull. Though it did emerge in the final drawing process that the upper widths of the frames/the vessel could not all be accurate (which came as no surprise). The absence of plank ends (upwards of strake E in the afterbody) meant that even the data from the shell planks drawn to scale 1:10 could not supply full certainty about the lengths of the strakes from stern to stem along the frames. Reconstructing the plan, the lines of the sides, and the cross sections (Plates 8.IV A, B, C and D) was a laborious process, requiring many revisions. However, the lines and dimensions presented here may be regarded as over 95% reliable. Ultimately it seemed the most plausible that the vessel numbered fifteen strakes on either side of the keel plank. Given the absence of the top three, the widths of those strakes is unknown and hence the height of the gunwale is an estimate. A series of short planks were found, belonging together, which might represent decking.
There is no actual evidence of a deck or of cuddies fore and aft. Yet it is almost certain that the latter must have been the case, and that connecting the two there would have been gangways of some kind. The height of the gunwale (ca 3.35 m amidships) in fact necessitates those. Section 8.2.2 discusses the crossbeams, deck beams and decks.

8.2.2 Construction and layout of the hull

Plank keel
The overall length of the plank keel, including the stern and stem hooks, is 13.3 m. The plank itself is 9.5 m long, 7 cm thick almost throughout, and joined to the hooks with flat scarfs. At the scarf with the stern hook, its width is 20 cm, it gradually widens forward, and between frames 16 and 20 is at its widest (33 cm). Further forward, the keel plank gradually narrows again and at the scarf with the stem hook is 28 cm wide. In the drawing of the keel plank it was not recorded whether its section was rectangular. Judging by the sections of the horizontal parts of the hooks, it probably was slightly trapezoid, with the broader side uppermost. In the areas mentioned above, the bottom of the keel plank would have been about 2 cm narrower than the top. The flat scarfs run downward fore to aft, with lengths of 28 cm (aft) and 37 cm (fore). The keel plank and the stern hook were joined at the scarf with two rows of iron nails, four nails in each. One row driven in from above; the other, from below. The nails were double-clenched. The join to the stem hook shows the same pattern. But here there is an extra nail between the two rows, driven in from below; besides, a treenail belonging to the fastening of frame 23 extends into the scarf.

Stern hook
The stern hook (figs 8.16a, b and c) still was joined to the keel plank. Its rising part was broken just above the lowermost gudgeon, the top end was damaged by ploughing. The remaining part still is 73 cm high (reckoned perpendicularly from the keel line), having originally been about 140 cm. At the top it would have had a flat scarf running top-down and from the outside inwards, connecting it with the sternpost. The horizontal part is 1.88 m long. The flat scarf does not taper to the end but has a nib 5 cm thick. The total thickness of the horizontal part is 9 cm here, so the end of the hook at the scarf stood proud of the 7-cm keel plank by 2 cm. At the scarf, the bottom surface of the stern hook is 18 cm wide and at its after end about 5 cm. Towards the stern, its thickness gradually increases and at the angle towards the rising part becomes ca 20 cm. The width of the stern hook at the same spot on the inboard side is 13 cm and of its uppermost preserved part is ca 12 cm. The rear of the rising part is about 5 cm wide; its thickness, about 17 cm. The rear surface of the stern hook is not a straight line. The lower part has a rake of 108°, narrowing by about 2 cm at the lowermost rudder-pintle, and then continues upwards in a straight line with a rake of 117°. The section of the stern hook is trapezoid throughout.

On the sides of the stern hook, between frames 1 and 2, are obliquely cut rabbets into which the hood ends of the shell strakes were fitted. These rabbets are 2 cm deep at most. At 17 cm above the keel line, the rising part of the stern hook supports the lowermost wrought-iron gudgeon for suspending the rudder. The straight braces of the gudgeon are about 550 mm long, 50 mm wide and ca 10 mm thick. A ring was forged into the 90 mm wide loop at the rear. Together with the gudgeon loop, the thickness of the ring is about 20 mm. The hole to take the pintle has a diameter of ca 50 mm. The braces extend across the outside of the garboards and on either side are fixed to the stern hook and stern hook/garboard with at least two massive nails on either side. A loose gudgeon was recovered near the stern. It is one of the gudgeons that had been affixed higher up to the strakes and sternpost. Its braces are not straight but curved, which reflects the limited scope for attachment to the sternpost. The part of the gudgeon with the eye for the pintle is 160 mm long, measured in a straight line, and projected 8 cm from the stern. Close to the stern the gudgeon forms a

90° angle, so the braces from there are perpendicular to the pintle. The braces were at least 400-450 mm long, 45 mm wide and 7 mm thick, and at their forward end spread about 12 cm (i.e. the inboard width of the sternpost plus the thicknesses of the shell planks' hood ends at this spot). Each of the braces has at least four nail holes, 14 mm in diameter. The pintle eye was made by fixing a forged ring into the loop formed by the two braces. The iron ring was 6-7 mm thick. The thickness of the part of the gudgeon that protrudes from the stern is 18 mm, so the combined thickness is 24-25 mm. The hole for the pintle has a diameter of 45 mm. The forged ring inside the gudgeon is

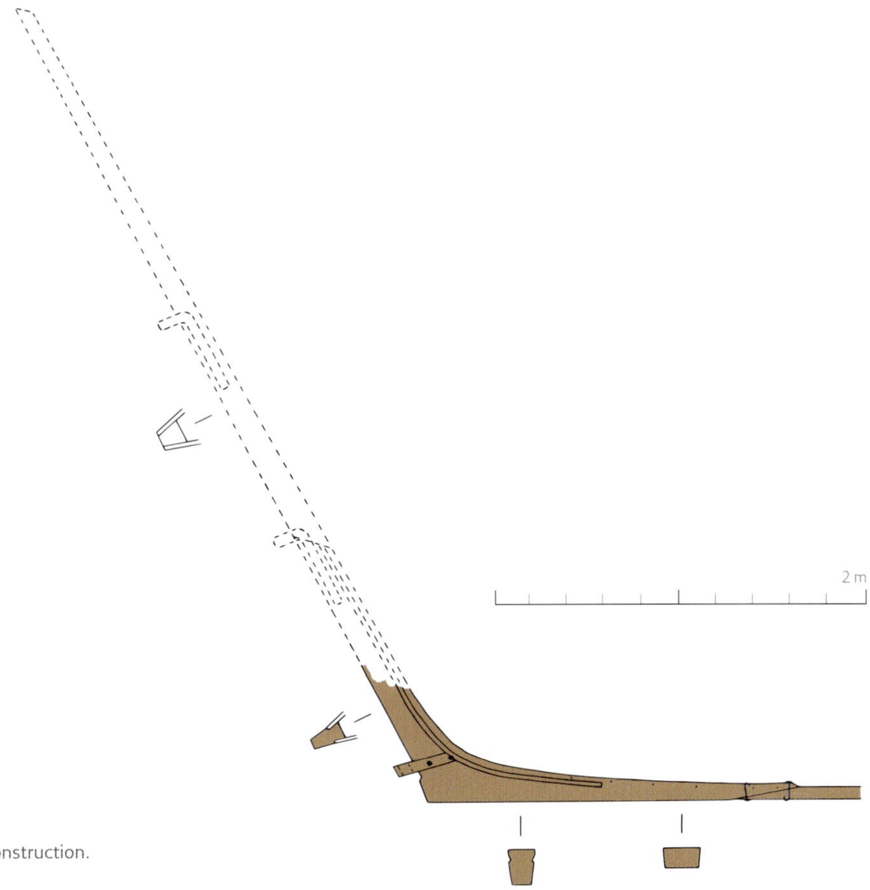

Fig. 8.16a. The reconstructed stern construction.

Fig. 8.16c. The sloping scarf joining the keel plank with the higher horizontal part of the stern hook and the starboard garboard.

Fig. 8.16b. The broken stern hook (*in situ*) with the lowermost gudgeon and the port garboard.

Fig. 8.17a. The dislodged stem hook.

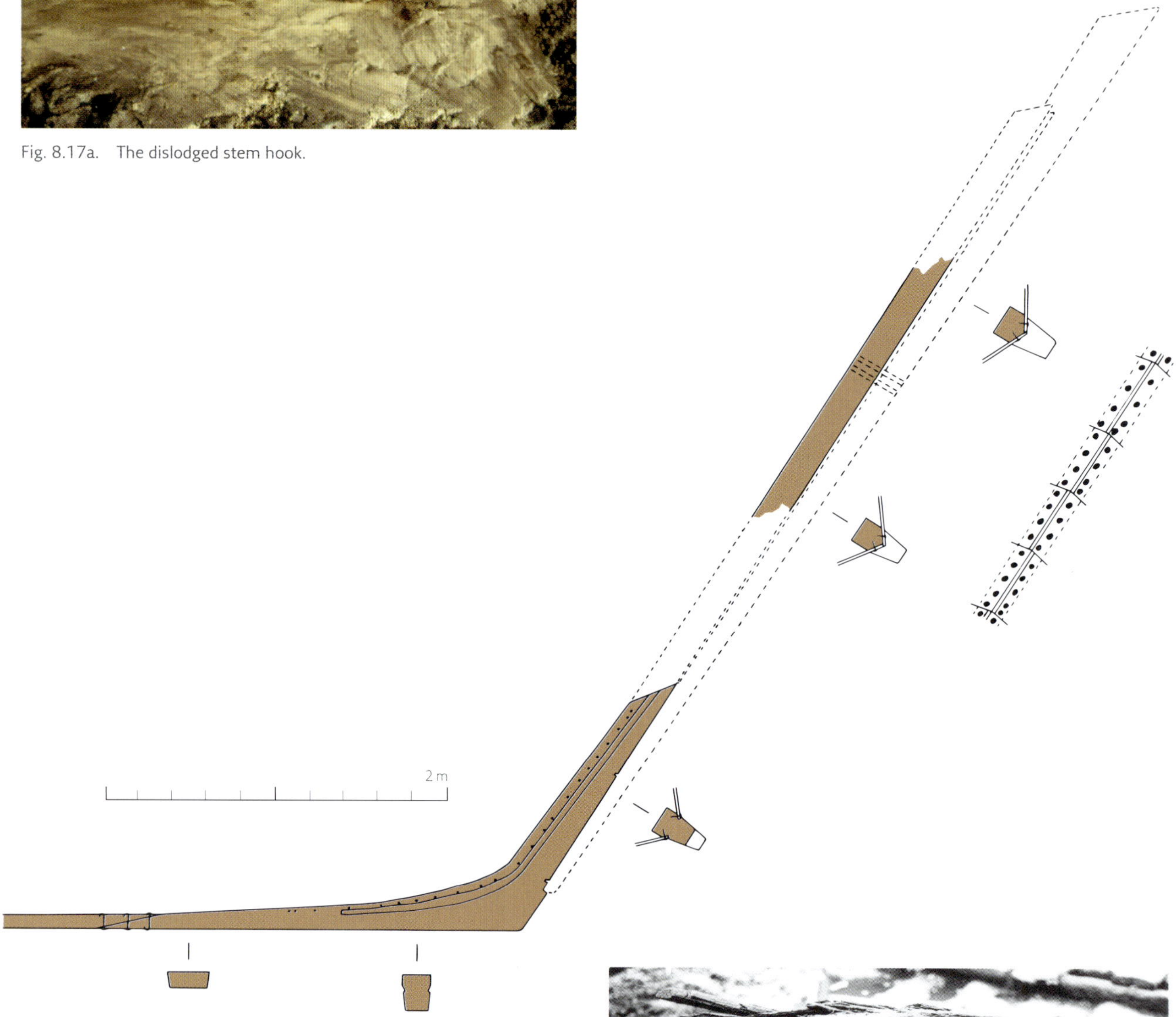

Fig. 8.17b. The reconstructed stem construction.

Fig. 8.17c. The hood ends of the starboard strakes on the front of the stempost.

open (worn away?) towards the sternpost. The iron of the gudgeon shows no distinct traces of abrasion by the pintle, so we cannot be sure whether the braces were angled upwards or downwards when they were mounted on the sternpost. The latter seems the more likely (Plate 8.III A).

Stern post

The sternpost, which originally extended the stern hook, was not recovered.

False stern

No false stern had been attached to the stern hook and sternpost.

Rudder

The two gudgeons found on and near the stern hook show that the vessel had a stern rudder. According to Heinsius,[19] this at the time (1981) was the earliest archaeological evidence of a medieval vessel with a stern rudder. Nothing was recovered of the rudder itself.

Stem hook

The aftmost horizontal part of the stem hook was still connected to the keel plank. The broken-off remainder of the horizontal part of the stem hook lay beneath the bottom, where also the rising part of the stem hook was found. The stem hook thus is complete, be it in three pieces. The horizontal part is 2.45 m long in total, and somewhat abraded over the foremost 70 cm of its underside (figs 8.17a and b): by 2-3 cm at the front. This most probably is use-wear from landing on sandy shores over many years. The rising part has a rake of 122°, rises to 1.43 m (perpendicularly) above the keel line, and at the top has a flat scarf which runs from above to below, from fore to aft. The scarf is 29 cm long. The width of the horizontal arm at the top of its scarf is 28 cm, and gradually narrows towards the bow. At the transition to the rising part of the stem hook, its width is 14 cm, while the inner surface of the rising part elsewhere is 15 cm. The section of the stem hook is trapezoid throughout, with the widest surface on the inboard side.

The bottom of the horizontal part at the scarf with the keel plank is 26 cm wide, and 5 cm at the bow. The front of the rising arm of the stem hook is 5 cm wide over the lowermost 23 cm. Then the front surface is offset by 3 cm, leaving the flat front of the higher part to widen to a good 6 cm. At 5 and 73 cm upwards from the offset, 2-cm-deep and 1.5-cm-wide sawcuts can be observed. Their significance has remained unclear. The offset part of the stem hook was almost certainly designed for attaching a false stem. However, no nail- or treenail holes have been observed on the front of the stem hook. For fastening the stempost, two nails had been driven into the front of the stem hook just below its end. From the rear of frame 26 and along the top edge of the stem hook an oblique rabbet had been cut with a maximum depth of 2 cm. Into this were spiked the hood ends of the garboards and of the other bottom strakes.

Stempost

Just part of the stempost has survived (fig. 8.17c). The piece was joined to the ends of strakes G to K of the starboard side (Plate 8.I) and was about 90 cm long. It is a post rectangular in section, of which part had been mitred on the front, and which at strake G is about 15 cm wide. The straight sides are 10 cm thick. Including the mitred front, the overall thickness is 16.5 cm. At its top end, at strake K, its width is 18 cm; the thickness of the sides is 10 cm and its overall thickness including the mitre is 15 cm. The thickness of the top of the stem hook is about 16 cm, so the stempost slightly narrows upwards. On the mitred sides of the front, various nail holes are visible for fastening the hood ends. The plank ends of equivalent strakes to starboard and port alternatingly abutted on their counterparts.[20] At the level of strake J, three blind treenails were observed on the inside surface, almost certainly treenails for fastening the false stem.

False stem

The false stem was not recovered, but originally it fully covered the stempost and the ends of the shell strakes nailed onto it (Plate 8.IV A and figs 8.17a and b).

Bottom and sides

The shell of the vessel is built up of strakes each composed of three or four planks. Plates 8.III A and B and fig. 8.18 show the shell planks and a general view of the parts recovered during the excavation. The planks are joined to each other by flat scarfs. The bottom was preserved almost entirely, just a few hood ends at the stern and stem are missing. Of the bottom, the turn of the bilge and the sides overall (the remains of) thirteen strakes (A-M) were found, on either side of the keel plank. It is highly remarkable that the A to G strakes in principle are carvel-built. In the afterbody, strake F is the first strake that over a short distance (1.4 m) was joined to strake E in clinker fashion - in the sense that the two adjacent planks were chamfered and that the F strake on the outside slightly overlapped strake E and was nailed onto it. On the inside the strakes either appear to be carvel-built, or plank edges protrude just 0.5-1 cm in order to locally attain a mini-

19. Heinsius, 12 June 1981, pers. comm.
20. This was also observed in the Doel-I cog, section 20.3.

Fig. 8.18. General view of the shell after removal of all frame timbers. In the foreground, the ripped-off and dislodged starboard shell.

mally required shell thickness, as has also been observed in some other areas of the shell. In the forebody, it is strake C which is the first to have been fastened to strake B in clinker fashion over a short distance (0.5 m). As was strake D along 0.9 m onto strake C; strake E onto D over 2.2 m; strake F onto E over 2.4 m, and strake G onto strake F over 2.6 m. The overlapping parts of these clinker-built strake sections are also chamfered. Up to the seam between strakes D and E they are flush on the inside. We can only speak of true clinker construction, i.e. with 5-7 cm wide, flat lands and from stern to stem, in the joining of the H strakes to the G strakes. A distinct turn of the bilge is difficult to indicate. The bottom curves upwards from the keel plank to the sides and from the midships to the fore- and afterbody, and smoothly transitions into the sides. At the hold, the more-or-less flat part of the bottom, the transition to the sides is fairly rounded and is in fact formed by strakes E and F.

Bottom

Beside the keel plank the bottom is made up on either side of four strakes (A-D). To starboard, these strakes comprised four planks. On the port side, strake A has three planks; B has four; C, three; and D, four. The bottom is carvel-built from stern to bow. The planks of the bottom strakes are fairly straight and relatively narrow (Plates 8.I B, 8.III B and 8.IV B). The A strakes are 27-28 cm wide; A3bb is wider at the bow (39 cm). The B strakes are a little wider than the A strakes (30-33 cm). Towards the stern, they gradually become narrower than 25 cm or more. At the bow this is 30 cm to starboard and 20 cm to port. Still the combined width of the A and B strakes is roughly identical on both sides. The C strakes vary in width between 20 and 29 cm with the narrower ends towards the stern, while the D strakes tend to be a little wider (22-23 cm). The thickness of the planks is very uneven throughout the bottom and ranges from 1.5 to 4 cm. They are thinnest in the afterbody: in many places just 1.5-2.5 cm. Amidships we see both planks that are 2.5 and (locally) 4 cm thick, while in some places this also is the difference in thickness between opposite edges of one and the same plank. For the bottom, planks with lengths of 3 to 4.5 m were used. Just two other planks, A3bb and C2bb, are longer: 5.4 and 6.5 m, respectively. The parts of the A strakes where they were nailed into the rabbets of the stern hook, had been chamfered. In the stem hook, this is not the case; here the full plank thicknesses were maintained.

The sloping flat scarfs that join the planks of the strakes are 30-40 cm long. The ends of the scarfed planks stand proud both inside and outside. They have straight nibs or are chamfered and are 1.2-2 cm thick (fig. 8.19). The fastening of the planks at the scarfs always involves two rows of three or four nails, sometimes with a single extra nail. One row driven in from the outside, the other from the inside. The nail points are twice beaten sideways and driven back into the timber: 'double clenching'.[21] No indications were found of a so-called 'brother-and-sister pattern' being applied in the building of this ship.[22] For the sake of strength, the scarfs in adjacent strakes were staggered. The planks of the bottom are fixed side-to-side only by the floor timbers. The short, clinker-built plank ends of the C and D strakes at the bow, which incidentally on the inside were flush with one another, have just a 2-3 cm overlap (the 'land'). For fastening them to each other, just a few double-clenched nails were used.

Apart from the (usually) two drilled treenail holes per plank width for the strakes' attachment to the floor timbers, also thinner treenails (2 cm across) were observed in some places in the bottom. A little forward of frame 8 in port strake A, and in starboard strake C; under or just aft of frame 13 in the keel plank, in port strake C and in starboard strake B; under or just aft of frame 16 in port strakes B, C and D, and in starboard strakes B and C; and below

Fig. 8.19. The keel plank with the stern hook and bottom planks with scarfs in the afterbody.

and aft of frame 20 in port strakes A and C, and forward of frame 19 in starboard strakes A and B. These small treenails without doubt relate to the early stages of the ship's construction, which we shall return to in section 8.4.1.

Starboard side

Most parts of the starboard side were recovered, despite their no longer forming a connected whole. With the exception of the missing parts of certain strakes between the stern and frames 1 and 4 (destroyed by the cutting of the boundary ditch and drainage ditches), the side is present from strake E to K. Along the foremost 7.5 m also strake L is preserved, as well as a small fragment of strake M. Because of the absence of certain parts it is impossible to ascertain how many planks each strake comprised. Strake E numbers three planks: 3, 6.5 and 6.5 m long, respectively. Strake F at any rate consists of four planks. Strake G probably of three, of which those at either end were about 6.5 m long. For strakes H to K probably also three planks were used. The middle plank of strake K is about 8 m long, while the foremost measures almost 9 m.

This can also be inferred from the square-cut futtock head below the upper-edge of the K strake. The head of futtock 30 to port is also flat, probably to accommodate another protruding crossbeam. In the forebody at 25, there are openings in the top of the L strakes which relate to a protruding crossbeam. Besides, the flat scarfs on the heads of the starboard futtocks of frames 24 and 27-29 indicate that upper futtocks were to follow there. The frames of the vessel were constructed from straight or V-shaped floor timbers plus two or three futtocks on either side. Hence the sides must have been higher. Originally it is likely to have numbered fifteen strakes (A-O) between keel plank and gunwale (see Plate 8.IV A). The length of the scarfs and their fastening are similar to those in the bottom strakes.

Yet in one place there is a deviation. Plank L3-1 on the outside at the scarf to plank L2 and at 23 and 24 does not have a straight end but a tongue-shaped one (fig. 8.20a). Its edges are bevelled. This probably was a decorative feature on the outside of the hull. At 22, shell plank L1 to port shows the same (fig. 8.20b). The pattern of nails in the scarf of plank K1-2sb suggests that also the formerly nailed-on but now missing end of plank K2-1 had such a finish, while port strake I at 25 displayed the same pattern. Apart from these decorative features there are some other striking details in the uppermost preserved strakes. In planks K1-1, K1-2 and K2-1 between frames 1 and 2, between 16 and 17 and between 23 and 24 there are oval

21. For the double-clenching method, see the Nijkerk-II cog, fig. 6.42.
22. For a description and the method, see e.g. the Spakenburg-I cog, section14.2.2 and fig. 14.17.

Fig. 8.20a. The tongue-shaped end of starboard shell plank L3-1.

Fig. 8.20b. The tongue-shaped end of port shell plank L1-1.

holes of 6.5x3.5 cm, 6x3 cm, and 7.5 x 4 cm, respectively. In strake L2, immediately below the tip of the tongue-shaped plank end joined to L3-1, a similar hole of 6.5x3 cm had been made, but this one had been closed up with a wooden plug. Figure 8.21 shows the outer face of the plank on which besides the closed-up hole also the adze marks of the flat scarf are clearly visible. In the port shell too some such oval holes are found. Initially the function of these holes was a mystery. However, there does seem to be an explanation; this will be discussed in section 8.2.2.

Plank K2-2 at 30 in its upper edge displays two blind treenails, about 5 cm long and 2.5 cm diameter. In the middle of the plank between these treenails is another, with a different, 3-cm diameter. The three treenails very probably relate to the fastening of a transverse connection. However, in the corresponding place to port, no evidence of such treenails was found (probably this end of the transverse connection was fastened to a futtock). In a few places, as in the bottom planks, treenails with a diameter of 2 cm may well be linked to the construction of the ship (temporary frames). They were encountered in shell planks H2, J3 and K2 (Plate 8.III B). In the upper edge of plank L3-1sb was a notch above 25, which was 7 cm deep and 20-22 cm wide. The intact edges are chamfered. The surviving remnant of plank M1 starts at the forward end of this gap. Together with the square-finished top of 2 at 8 cm below the bottom of the hole and the equivalent situation to port, it is clear that we are here without any doubt dealing with the place where a heavy crossbeam had protruded through the shell (fig. 8.22a and b). Furthermore, in some places treenail holes are seen that are linked to other transverse constructions within the hull. The transverse constructions are treated in section 8.2.2.

Fig. 8.21. The exterior of starboard shell plank L2 with the closed-up oval opening between frames 23 and 24, and part of a of a clamped-moss-caulked crack.

Port side

The strakes E to I all have survived from 0, 1 or 2 up to the stern. From strakes G, H and I (small) pieces are missing, between 15 and 18 and at the foremost ends, as a result of the cutting of the boundary ditch and the drainage ditches. Of strakes J, K and L only parts between 11 and 30 have survived, while of strake M just a 3-m length was left. On this side too, strakes were made up of three or four planks. Just as in the bottom, there is no evidence of a 'brother-and-sister' arrangement of planks in the starboard and port sides. For a further description of the port shell, apart from some details, see the description of the starboard shell. The opening for the protruding crossbeam above 25 was mentioned above. So was the tongue-shaped scarf in the same plank (L1-1), be it that to port it appears between frames 21 and 22 rather than between 23 and 24. However, a deviating point in the starboard shell is that

Fig. 8.22a. The square head of starboard futtock S25 and the lower edge of the hole for protruding crossbeam DB5.

Fig. 8.22b The same to port.

Fig. 8.23a. The two oval openings in port shell strake K1-1 between frames S23 and S24. At the upper opening, small nail holes are observed on the outside, probably belonging to a 'spout' through which bilge water could be pumped out.

Fig. 8.23b. The inboard side of the oval hole; below it, a small plank fitted into the frame.

also the beginning of plank I3 where it is scarfed to I2-2 has a tongue-shaped finish on the outside. The scarf between the last two planks of the K strake is lost, so nothing can be said about this. Yet it is quite possible that, as on the starboard side, the tongue-shaped decoration had been applied here too. The oval holes described on the starboard side were also observed on the port side. In plank K1-1 there are two such holes, one above the other, between 23 and 24. The lower measures 8x3 cm and was closed with a wooden plug and sintel-clamped from within. The one above it is open, measures 8.5 x 4 cm and on the outside displays various small nail holes around its perimeter (figs 8.23a and b). In section 8.2.2 its likely function and significance will be discussed. Such an oval hole (6.5x3.5 cm) is to be seen also just aft of 30 in plank I3. This hole cannot be linked to any pump. Sooner one might imagine an anchor cable passing through it. The hole is just 2.3 m above the keel line, and cannot have been much above the waterline, so it is a strange place for a scupper. In the port side, a small treenail (2 cm thick) was observed only in strake G1 where probably it had once fixed a temporary frame (Plate 8.III B).

Caulking method

All seams of the hull were closed with sintel-clamped moss caulking, both on the inside and the outside; in this manner also occasional short cracks were closed on the outside (Plate 8.II B). All the sintel clamped moss caulking (or just sintel nail clamps, on the carvel seams) on the inside was applied before the frames were fitted. We shall return to this in section 8.4.1. A thin layer of moss had been laid within the scarfs joining the shell planks. The

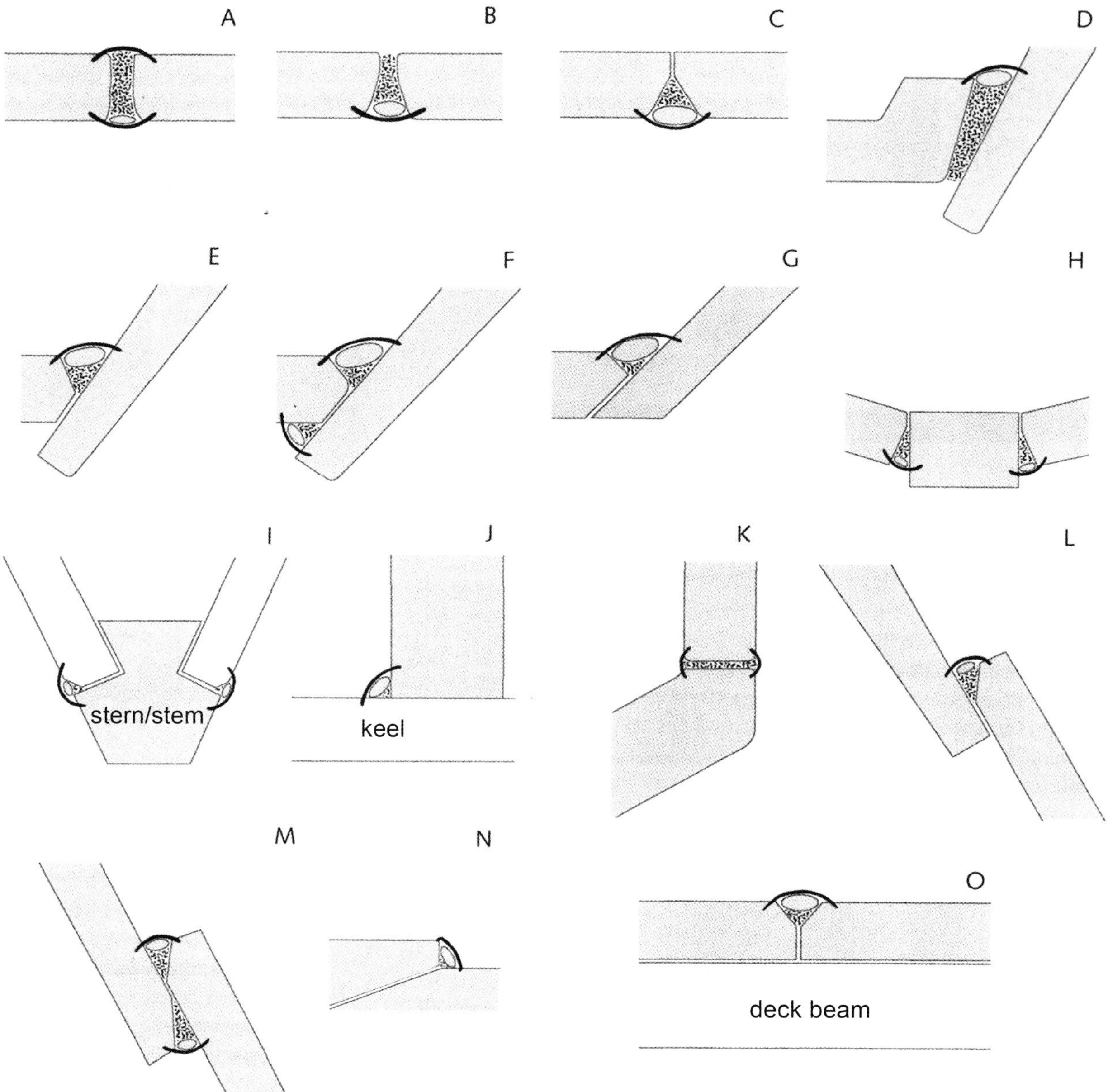

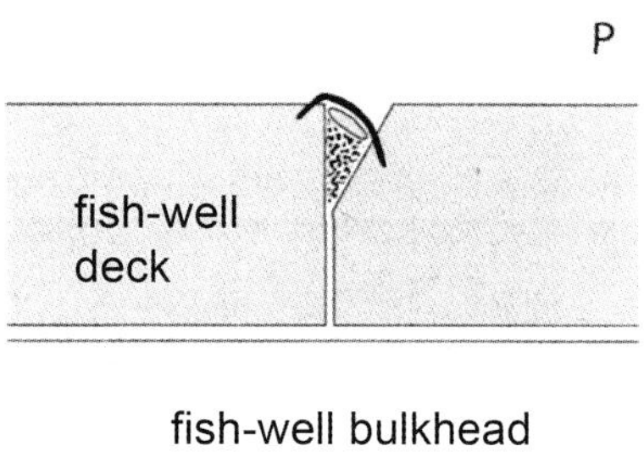

Fig. 8.24. Variants of seams closed with sintel (nail) clamped moss caulking in late medieval ships. After Vlierman 1996a.

Fig. 8.25. Bottom shell planking with (impressions of) sintel nail clamps on the inboard side of carvel seams, also beneath the frames.

moss used comprised at least three species: *Sphagnum, Drepanocladus aduncus* and presumably *Cratoneuron commutatum* ; a sample (85) analysed by E. Mook-Kamps (University of Groningen) was found to be devoid of pollen.[23] The carvel seams between the keel plank and the A strakes (garboards) and between the subsequent strakes right up to the seams between the F and G strakes gape a little on the outside, where the planks sit at a slight angle with respect to their neighbours. This creates V-shaped seams. In places where two planks almost fully abut, one or both are a little bevelled, thus creating a triangular gap of unequal sides. These gaps run from the outside to the inside of the shell, so the planks will at most slightly touch on the inboard surface. This probably is the transition between the chevron-shaped bottom seam which on the inside gapes up to 0.5 cm, and the seam which was reamed into a V-shape for just half or three-quarters of the plank thickness on the out-/underside, with the rest of the plank thickness abutting: variants B and C (fig. 8.24).[24]

In the Nijkerk-1 cog the gaping carvel seams on the underside are about 0.5 cm wide. The opening gradually widens (up to 1.5 cm) over the outer 0.5 cm of plank thickness. Into this groove the 1-1.5 cm wide and a few mm thick moss laths were laid, by which the moss caulking was held in place. The oaken moss laths were secured with sintel nail clamps. Most remarkably, along the carvel seams on the inside of the shell too sintel nails are present, at 15 cm intervals. The two wings of these sintel nail clamps were driven into the planks on either side at a good centimetre beside the seam. Here no moss laths were used. The sintel nail clamps are found also beneath the frames, so they were put in before the latter were fitted (fig. 8.25). This point will be elaborated in section 8.4.1. This clamping of carvel seams at considerable intervals on the inside of the ship, with only sintel (nail) clamps holding down the moss caulking and on the outside of the hull the moss caulking and the moss laths so far is known only from one or a few vessels from Antwerp, found in 1884 and from the dugout boat found at Meinerswijk in 1976.[25] Yet these differ from the vessel described here in that these examples have fully open seams.[26]

The 'clinker-built' but on the outside carvel-like seams at the stern and bow of strakes C to G are closed with sintel nail clamped moss caulking. Figures 8.26a, b and c show some details of cracks caulked on the outside, of an area of shell with seams thus treated on the inside, and the sintel nail clamped moss caulking on the outside of clinker-joined planking.

Fig. 8.26a. Cracks closed with sintel nail clamped moss caulking on the outside of a shell plank.

Fig. 8.26b. An area of starboard shell in the forebody with some futtocks and a ceiling strake. Clear to see are the sintel nail clamps and moss laths on the clinker seams.

Fig. 8.26c. An exterior area of clinker-built shell with sintel nail clamps, moss lath and a convex, protruding treenail for fastening a frame.

23. Cappers *et al.*, 1997/98, 582-283; written report by E. Mook-Kamps.
24. Vlierman 1996a, 26-28 and fig. 11.
25. Hasse 1907; Reinders 1983 and Vlierman 1996a, 26-28 and fig. 11.
26. Vlierman 1996a, 26-28, variant A.

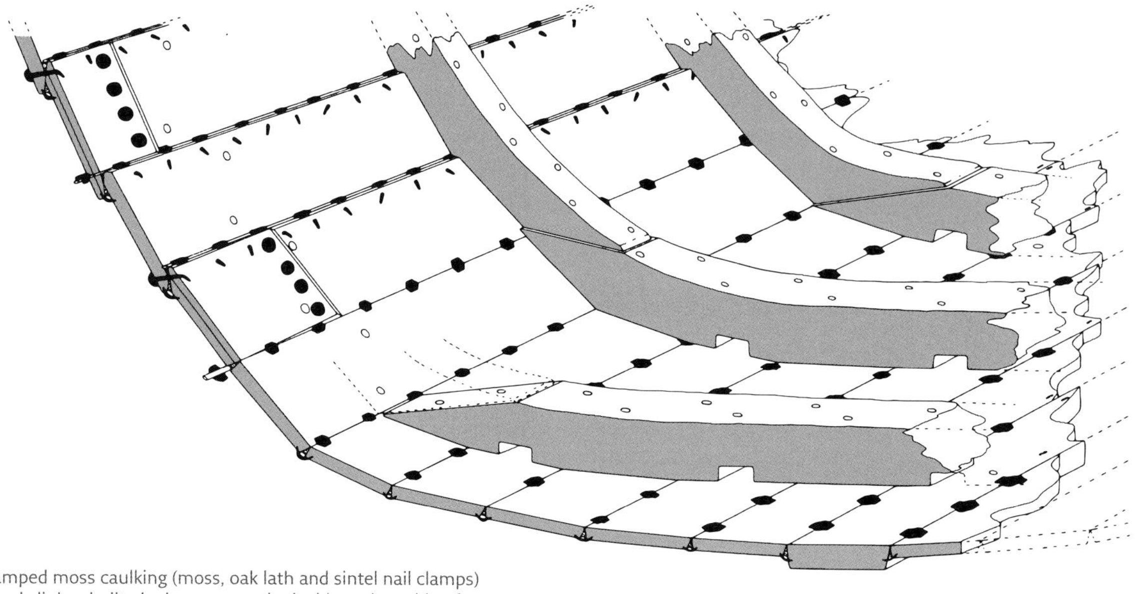

Fig. 8.27. Clamped moss caulking (moss, oak lath and sintel nail clamps) of the carvel- and clinker-built plank seams on the inside and outside of the shell, and just sintel nail clamps on the inside of the carvel seams; and the joinning of clinker shell planks with double-clenched nails.

As mentioned before, the clinker-built strakes of the sides were sintel nail clamped both on the inside and on the outside (fig. 8.27). In the shell seams, singly twisted strips of moss caulking were inserted, covered with moss laths and secured with sintel nail clamps. The seams between the stern- and stem hooks and stern- and stemposts and the hood ends of the shell, as well as the repaired cracks in the shell were made watertight in the same way.

The scarfed ends of the joined shell planks were not filled with sintel nail clamped moss caulking. On the seams on the inside and outside of the vessel, and also along the stern and stem, and the seams between the keel plank and the garboards, and probably also around the protruding crossbeams, clamped moss caulking (moss, laths and sintelnail clamps) was applied with varying spaces (4-12 cm) between the clamps. The sintel nail clamps on the inside of the carvel seams sit at intervals of 15 cm. In all, about 800 m of seams were caulked in this way. With an average interval of 8 cm, the number of sintel nail clamps amounts to some 10,000. The carvel seams on the inside add up to 195 m. With 15 cm intervals, this means a further 1,400 sintel nail clamps. This does not include those on the exterior repair seams. Altogether we may assume that the vessel contained at least some 12,000 sintel nail clamps. They are of the type B/C(2), vary considerably in size (45-80 mm long and 18-36 mm wide) and can typologically be dated to the second quarter of the 13th century (figs 8.44 and 8.45-12).[27]

Frames: (V-shaped) floor timbers and futtocks

The bottom and the sides are in the first instance held together by frames (ribs) (figs 8.28a, b and c). The frames are composed of straight floor timbers on the more-or-less flat part of the bottom, and V-shaped ones in the increasingly sharp parts towards the stern and stem. The sloping flat scarfs joining these floor timbers to the lower futtocks alternate in height on the port and starboard sides, and also are staggered in height in the successive frames towards the stern and stem. The scarfs are at the level of the third to the seventh strake (C-G). The scarfs between the lower and second futtocks are almost always at the level of the ninth or tenth strake (I or J); while the scarfs joining the second and third futtocks (or the top of the second futtock that ends below a protruding crossbeam) are at the level of the twelfth or thirteenth strake (L or M). The heads of the third futtocks or the knees on the crossbeams emerge a little above the gunwale.

The (V-shaped) floor timbers are the most massive parts and vary a good deal in width (18-32 cm). An irregular sequential pattern of broader and narrower timbers was applied. The height of the floor timbers is 20-23 cm (fig. 8.29). The frames are placed at intervals of about 50 cm (space and room). The futtocks are on the whole lighter than the floor timbers. It is impossible to tell whether the upper futtocks in their turn were thinner and narrower than the lowermost. The top futtocks of 2 and 27 may have functioned also as bitts (Plates 8.IV A, B and C).

27. Vlierman 1996a, 73.

Fig. 8.28a. Floor timbers (straight and V-shaped) and futtocks, the keelson and the lowermost ceiling strakes on the bottom in the afterbody and on part of the port shell. b. Floor timbers (straight and V-shaped) and futtocks, the keelson with mast step, and the lowermost ceiling boards on the bottom of the forebody and part of the port shell. c. Timber futtocks and ceiling strakes on the starboard shell of the forebody, with a fragment of the stempost.

Fig. 8.29. General view of the frames on the bottom, viewed from the stern.

In the floor timbers were five rectangular limber holes: above the keel plank and and above the second and fourth bottom strakes on either side of it. The V-shaped floor timbers had triangular holes above the seam between the keel plank or hooks and the garboards. Also there are some with a single limber hole above this seam to port of starboard. The V-shaped floor timbers are not fastened to either of the hooks or the stern- or stemposts. The ship probably had a total of forty frames (-1 to -5 and 0 to 34).

Keelson with mast step area

Above the keel plank a keelson 8.03 m long lay across the floor timbers 5 up to and including 22 (figs 8.30a and b). Towards the stern, the keelson is 22 cm wide. It gradually widens to 24 cm at the rear of the mast-step area. At frames 17 to 20, the keelson is 50 cm wide and 25 cm thick, *i.e.* a few cm thicker than the rest of it. Above the forward half of 18 and up to 19, is a hewn-out cavity for the heel of the mast, 35 cm long and 26 cm wide. The foremost part of the keelson at the mast-step section is 30 cm wide; and towards the front, 22 cm. In the hole for the heel of the mast sits a small, 4-cm-thick plank, the top of which shows abrasion on the side of the mast. The rear 4 cm of the hole is less deep than the rest (8 instead of 12 cm). Roughly in the middle of the hole for the mast, a drainage tunnel had been drilled, which empties into the space between the frames 18 and 19. Transversely across its mouth, a 15 cm long groove had been carved to drain off to the bottom of the ship any rainwater and spray that ran down the mast. A second such drainage hole is seen in the starboard rear corner of the mast cavity (fig. 8.31 and Plate 8.III C).

In the underside of the keelson notches were made to engage with the floor timbers. Yet these display an irregular pattern and do not fit perfectly with all of the frames. The lengths of keelson that sit between the frames are irregularly shaped. The keelson is fastened to the frames with blind treenails (Plate 4.III C, the yellow treenails). Besides, there are different treenails too (the green ones in the same plate). The second drainage hole in the cavity for the heel of the mast, which ends above frame 18 and hence has no function, together with the poorly engaging underside of the keelson and the other (green) treenails, indicate that the keelson originally belonged to a different vessel. The parts that fall between the frames in that vessel had been made somewhat concave, as may still be seen between frames 11 and 12 (fig. 8.32 and Plate 8.III C). At frames 7 and 8, the keelson has a cavity on the starboard side (fig. 8.33). Possibly this relates to a pump barrel in the earlier ship, which stood between the frames. This reused keelson almost certainly is the first and only keelson that this vessel ever knew.

Fig. 8.30a & b. Frontal view of the keelson, showing the broadened mast-step area with the hole for the heel of the mast, and the riders against and across the lowermost ceiling strakes. b. The mast step, viewed from the starboard forebody. Aft of the mast is the mast support, broken in antiquity.

Fig. 8.31. The hewn-out hole for the mast heel. The water drainage hole with carved-out drainage rill in the centre; the primary drainage hole immediately beside the 4-cm-wide elevation in the rear of the mast hole; and the small, rounded, abraded plank in the forward part of the mast hole.

Fig. 8.32. The keelson removed from the wreck.

Fig. 8.33. The notch in the keelson which may mark the place of a pump barrel in the vessel to which the keelson originally belonged.

Fig. 8.34. Reconstruction of the broken mast support.

Knee-shaped mast partner

Immediately aft of the hole for the heel of the mast, on the centre line of the keelson, stood a knee (Plate 8.III C and 8.IV A; fig. 8.34). Its vertical part had broken off and lay beside its horizontal part on the keelson. The upper tip had broken off at a treenail hole and was no longer present. On the rear the knee had been bevelled stepwise (ornamentally?). The knee had been fastened to the keelson by two blind treenails. The forward face of the vertical arm was exactly flush with the rear of the hole for the heel of the mast; be it that between the mast and the supporting knee there must have stood a 4-cm-thick and probably 26-cm-wide plank, which occupied the shallower part of the mast hole. At 90 cm above the keelson, this plank had been fastened to the partner knee with a treenail, and at the top it most probably fitted into a rabbet in the sail beam (Plates 8.IV A and 8.IV C, Section 18). The knee with the plank, together with the abraded plank at the front of the mast hole, suggest that the mast was occasionally lowered, with the supporting knee and the plank serving to guide the mast. All of this means that the mast was set up from the forebody, and almost certainly was raised or lowered by means of a winch on the afterdeck, between two stringers (for this see also Section 8.8).

Riders

The mast-step part of the keelson was supported by just two pairs of short riders to take the lateral pressure exerted by the mast with the sail (figs 8.30a and b; Plate 8.IV C, Section 18). The after pair (on 18) are notched on the underside so as to engage with the first ceiling strakes beside the keelson. The other two riders to port ends up against the second ceiling strake; to starboard, just onto this strake. The riders were fastened to the floor timber with blind treenails, and have a virtually flat top surface, which rises a little towards the mast step. The top edges are chamfered. The riders on frame 19 are shorter. They extend to the first ceiling strakes, stand a little proud of them and have a faintly concave top line which continues to the top of the mast step area of the keelson.

Ceiling

The vessel had an open ceiling and up to the protruding crossbeams numbered seven ceiling strakes (A to G). Above the crossbeams there most probably were two further strakes (H and I; Plates 8.IV A and C and figs 8.28a to c). The strakes were meant especially to provide additional longitudinal stiffness. The first ceiling strakes beside the keelson were adze-finished on the upper surface and somewhat roughly roof shaped. The B and C ceiling strakes are missing altogether, probably because the hull broke at the turn of the bilge in the course of the wreck-formation process. The B strakes (and possibly also the C strakes) probably also had been adze-trimmed to a roughly roof shaped section and may have been thicker than the other ceiling strakes. This would have been done with an eye to their important function of strengthening the hull above the turn of the bilge, sitting more or less above the scarfs joining the floor timbers and the lowermost futtocks. Possibly they were joggled on the underside to engage with the frames.

The first ceiling strakes ran from a little forward of frame 4 to 23. The B ceiling strakes probably started forward of 5 and ended aft of frame 23 because of transverse constructions; the C and D strakes probably ran from forward of frame 3 on to 29. The E and F strakes ran from frame 0 and -1 to 29 and 30, respectively; while the seventh ceiling strake (G) probably started forward of frame -1 and ended at 26. The eighth strake (H) began and ended presumably behind the deck knees on frames -1 and 30. The highest would have run along the gunwale from 5 right onto 34. Apart from the A (and B?) ceiling strakes they probably each comprised multiple planks. In as far as could be established, the planks had been fastened to the frames with blind treenails, hence not until the shell with its frames had been completed. No indications were found to suggest that the spaces between the keelson and the first ceiling strake, between the first and second, and between the second and third ceiling strake had been closed with planking. So the ship had no continuous ceiling.

Transverse connections

In the forebody some transverse constructions were found, which had been inserted relatively low in the hull immediately against the sides or onto and against the frames Plate 8.I A). Of the transverse construction immediately aft of frame 5 (DV4), the beam was at the level of strake F and to port ended in a standing knee. On the starboard side, a separate knee had been fixed to the beam. The angled end of the beam and the knee had been fastened to the sides with treenails, while to starboard the beam had been fastened with a treenail to 25 (Plates 8.I A and 8.IV A). The transverse connection on frame 27 (DV5) also sat at the level of the F strakes (figs 8.14a and 8.35a, b and c; Plates 8.II C and 8.IV A and C). It consists of a beam with knees on it at either end. The entire construction is fastened to and against the futtocks with blind treenails as well as treenails passing through a futtock and the shell, and a few iron nails. The vertical parts of the knees engage with the third (C) ceiling strakes. At the level of these strakes there are holes in the knees in a fore-and-aft direction, whose function is unclear. A notch in the fifth (E) ceiling strake, at the level of the top of shell strake I, indicates that a transverse connection (DV10) had been there too.

The standing knees probably extended to the top of shell strake L and presumably 'supported' a deck beam below the foredeck. Knees were probably also mounted on the deck, that extended to the top of the gunwale (Plates 8.IV A, B and C). The lowermost starboard futtock of frame 23 (fig. 8.36) is knee-shaped and on its underside is flat over a length of 60 cm. The top of the V-shaped floor timbers, at the level of the top of the D shell strakes, had supported a beam (DV2), on which rested knee-shaped futtocks (Plates 8.IV A and C). At the level of the J shell strakes there is a notch in the top edge of ceiling strake F to starboard; beside it is a more-or-less vertical treenail (Plate 8.I A). Hence another transverse construction (DV9) must

Fig. 8.35a. Viewed 'down' from the gunwale, the starboard shell in the forebody with the replaced V-shaped floor timbers and the transverse connections against frame 25 and on frame 27.

Fig. 8.35b. The reconstructed frame 25 with on its after surface the transverse construction fastened to the shell.

Fig. 8.35c. The starboard knee of transverse construction DV5, *in situ*, with the hole against the ceiling strake.

have sat immediately aft of 23 at this level, structurally comparable to that aft of 25. At the level of shell strake L there very probably also was a transverse connection (DV11), attached to the inside of 30. A broken part of it was recovered to starboard beside the forebody. The beam has a rectangular opening (15 x 6 cm) through which presumably a small post had passed, which supported the foremost crossbeam (DB6) midway, and with its heel on curved floor timber 30 rested on the stempost (Plates 8.IV A and C). It seems as good as certain that the heads of V-shaped floor timber 5 in the afterbody, at the level of shell strakes D and E, were flat and that there too, analogous to 23 in the forebody, a transverse construction (DV1) had rested which had been fastened to the shell with knee-shaped futtocks. Through the absence of higher parts of the afterbody, we cannot be sure how many and where exactly the transverse constructions had been inserted. Yet it seems quite plausible that their numbers and locations in the afterbody were roughly analogous to those in the forebody (Plates 8.IV A and C; DV3, DV6, DV7 and DV8).

Fig. 8.36. The starboard shell with remnants of the open ceiling in the forebody, and (centre right) the knee-shaped lowermost futtock of frame 23 with a 60 cm long, straight underside.

Crossbeams with knees

In the shell of the forebody some openings can be seen in the upper edges of the L strakes, above the square futtock heads of 25. To port even a fragment of an M strake was present, so it was evident that here had been a roughly square, ca 18x18 cm gap in the shell (figs 8.22a and b). The port futtock of 30 ends in a square head a few centimetres below the upper edge of shell strake M. From the presence of these square futtock heads and the hole in the shell above frame 25 it may be inferred that the vessel had protruding crossbeams. Most probably four on the upper edge of strake L and passing through strake M, at frames 4, 11, 18 and 25 (DB2, DB3, DB4 and DB5); one at frame -1 through the join between the K and L strakes (DB1); and one at frame 30 through the join between the M and N strakes (DB6). It may be assumed that on the inside of the vessel these beams had carried knees to which the strakes of the 'above deck'-part of the sides were treenailed. Further, it seems likely that the horizontal parts of these knees were 60-100 cm long, with those at the forward edge of the afterdeck probably ending at the cheeks of the windlass. Those at the after edge of the foredeck may have been interconnected by a deck timber (Plates 8.IV A, B and C).

Beam-head guides

It cannot be ascertained whether on the outside of the shell these protruding crossbeams were protected by so-called guides or fenders on their forward face. The only piece of M strake (from the port side of the forebody) did not show any evidence of such a device.

Stringers, deck(s) and gangways

The only source of evidence that might tell whether the vessel had been longitudinally strengthened at 'deck level' with stringers, are the protruding crossbeams. None of these were recovered. Yet it seems entirely plausible that there were some. Plates 8.IV A, B and C show what kind of construction might be envisaged. The sail beam (DB4) was almost certainly reinforced by two stringers that extended to DB5 with an interspace of 50-60 cm. It is between these stringers that the mast would be raised and lowered. Between crossbeams DB2 and DB3 and between DB3 and DB4, there possibly were more such stringers, aligned with this forward pair. Judging by the distance between these stringers and the sides of the ship (2.5-3 m), it seems well possible that the shipwright put in further (removable?) stringers at roughly mid distance. At deck level, the mast was kept in position by an inserted plank, part of which was recovered. The fragment is 63 cm wide and was probably mortised into the stringers. Its original full length would have been about 70 cm; its thickness, ca 10 cm. The mast hole in the king plank had a maximum diameter of 40 cm. The mast probably was thinner at this spot, so it could be set fast into the king plank with wedges (Plates 8.IV A and C). It seems rather

Fig. 8.37a, b and c. Nine small planks, found detached in the forebody. The notches in three of them indicate that they probably were fitted around the futtocks and against the shell, so they may have been parts of a deck or floor.

unlikely that the openings between the after- and foredeck and the gangways were closed while sailing.

Given the height of the ship, it is certain to have had decks both fore and aft. What exact form they took is hard to tell. For the reconstructions in the plates it has been assumed that the ship was built with two decks, each about 3.5 m long, with longitudinal planking. Of the latter, the outermost planks were notched to allow for the futtocks; they ran right up to the shell. The afterdeck most probably rested on crossbeams DB1 and DB2, the foredeck on DB5, transverse construction DV4 and on frame 32. In the stern, the deck would have required at least one further load-bearing deck beam at frame 2; in the bow, a least two (at frames 27 and 29). A number of loose small planks (fig. 8.37a, b and c), apparently belonging together, were found but it is unclear where they belonged. They may have been outermost deck planks, which with notches fitted around the frames and against the shell of the hull. Alternatively, they could have formed a floor at the fireplace. For movement between the fore- and afterdeck, it is almost certain (given the height of the sides) that gangways were constructed along the bulwarks. For the reconstruction drawings it has been assumed that these were thick planks, ca 35 cm wide, running from DB2 to DB5.[28] Yet the planks may also have been joggled to fit around the futtocks and to the shell.

Windlass, and support for the lowered spar (and mast)
Above, it has already been suggested that the spar (with the sail), and maybe also the mast, could be lowered and that this must have been done with the aid of a windlass in the afterbody. Of the windlass no part has been recovered. In sections 8.IV A, B and C its possible shape and location are indicated. The cheeks of the windlass were probably treenailed to the rear of crossbeam DB2 and transverse construction DV7. A rabbet had been cut into both sides of frame 4 (fig. 8.38) in which very probably vertical posts were set to support DV7. A piece was cut out of the first port ceiling board to make space for them. These posts did not stand aligned with the windlass cheeks. Beside the bow there presumably was a forked support to take the lowered mast and spar.

Fig. 8.38. The rectangular rabbet to port in frame 4 and the narrowed end of the first ceiling board.

28. Evidence for such 'gangway planks' was found on the crossbeams of the Doel-I cog near Antwerp, Flanders (Belgium) (section 20.3) and in a crossbeam of the cog from Kolding, Denmark.

Pump and drain

Apart from the notch in the keelson that may relate to a pump barrel in the ship to which the keelson had initially belonged (Section 8.2.2), in fact no indications for a pump were observed. The oval holes in the starboard and port shell and the obliquely fixed little plank in the port shell between 23 and 24 under one of these holes however do suggest that we are dealing with scuppers for bilge water. If such is indeed the case, then very probably a pump barrel stood between 23 and 24 above the scarf between the keel plank and the stem hook, and reached to deck level. In the top of this barrel there may have been a hole on either side. The pumped-up water was guided to port or starboard through a simple wooden gutter towards a sump above the oblique little plank and between te futtocks, and spilled overboard through the scupper.

The interpretation of the nail holes around one of the oval scupper holes on the outside of the shell (fig. 8.23a) for a long time remained a mystery. However, the image on the title page of the *Maritime Laws of Wisby* of 1535, and the picture reproduced on the cover of the book *Westeuropese scheepvaart in de Middeleeuwen* [29] (figs 8.39a and b), appear to provide an explanation. In the middle of the side of the ship shown in fig. 8.39a, there is a round, hoselike feature at the centre of an area enclosed by three small planks. Most probably we are here dealing with a piece of leather 'hose', that was nailed to the shell and was protected by the three planks. The other picture shows a late-medieval 'hulc' ship with two such scupper hoses. With such drains, the bilge water could be easily disposed of, while it was almost impossible for water to enter from outside. The other - closed-up - oval holes suggest that the construction and location of this drainage device was adjusted or replaced in several instances.

Fig. 8.39a. The title page of the *Maritime Law of Wisby*, 1535. Stiftung Hanseatisches Wirtschaftsarchiv (S/132). Handelskammer Hamburg (photo M. Zapf).

29. Brand & Knol 2010, 160, and Asaert 1974, respectively.

Fig. 8.39b. A late-medieval 'hulc' with two spouts in the side. Österreichisches Nationalbibliothek, Vienna, Cad. 2538, *fol.* 109 *recto*. After Asaert 1974.

Castles
It is as good as certain that the vessel had no castles.

Repairs
Plank GA1bb contains a small rectangular patch measuring 36x16 cm, a repair whose seams are closed from the outside with sintel nail clamped moss caulking. At frame 21, the outside of plank C3bb bore a nailed-on repair plank (Plate 8.II A). Forward of frame 5, on the inside of the shell, at right angles across the seam joining port strakes C and D, a small plank was nailed on measuring 22x4 cm. This plank may in fact have been put there during construction, as a temporary connection to hold fast the seam between bottom strakes C and D, while later on there seemed no need to remove it (Plate 8.IV A). Such small planks were also found on the bottom of the Dronten-I cog (Chapter 11, Section 11.2.2) and in the Swifterbant cog (Section 18.1). Between frame 18 and 20, on the outside of plank H3-1bb sat a small, nailed-on repair plank, 1.5 cm thick. The method for applying caulking to such planks was discussed above, in section 8.2.2. The cracks on the outside of the shell planks are closed with sintel nail clamped moss caulking (Plate 8.III B). These are drought cracks which opened up during the ship's construction. So these may be regarded as (very early) repairs.

Wood selection, fastenings, timber use and processing
All recovered parts of the vessel are of oak (*Quercus* sp.). The shell planks are tangential-sawn/cut. Many of the planks employ the full width of the trunk; others less than half a width (fig. 8.40). In a few planks, it was investigated how the tree-ring pattern manifested itself in section and how the planks had been positioned. Most were used with the heartwood on the inboard side. Another curious fact is the ample presence of sapwood, especially since this was the case even in the carvel-built part of the shell (Plate 8.III B). Clearly the shipwright had not been too choosey. In fact, by leaving the sapwood on, he could employ its property of more readily absorbing water. In the seams of the bottom of a late-medieval punt from the river Regge near Hellendoorn, no caulking was used to render the craft watertight, but the swollen sapwood took care of this.[30] Despite these being sawn planks, it is noticeable that especially the carvel-built planking of the underwater body is of very uneven thickness (1.5-4 cm), both transversely and longitudinally. Along the first few metres from the stern the planks are thinnest (1.5-2.5 cm). In some places distinct adze marks were observed. The only plausible explanation for the irregular plank thicknesses is that the outside of the carvel-built underwater body (the keel plank up to the G strakes) had been faired after the frames were inserted.

At the plank ends it was clear to see that the lands of the flat, chamfered scarfs had been fashioned with an adze. The hooks, (V-shaped) frame timbers, crossbeams, curved ribs and knees were made from naturally grown trunks and limbs. The surfaces of the timbers visible within the ship had been squared; wherever the trunk or limb was naturally thinner, the original round shape of the timber - including the sapwood - had been retained (fig. 8.41). The shell planks bore no traces to show that they were furnaced to give them the desired curves. Nor is there any obvious sign of the shipwright having applied a so-called 'brother-and-sister' pattern.

Fastenings
In a number of planks of the bottom and the sides, some oak treenails (2 cm thick) were observed between the frames. Without doubt, these were fastenings for temporary frames used during the construction of the ship (Plate 8.III B). The shell planks in principle are always fastened to the frame with two oaken treenails each, with a 2.5 cm diameter. They are somewhat staggered with respect to adjacent strakes (fig.8.25). Although this was not recorded exactly, it seems very likely that treenails with slightly broader heads were used (figs 8.44 and 8.45-17 to 45-20). They had been driven in from the outside. Their heads at first protruded by a few millimetres, which in most cases have worn away. On the inboard side, no wedges were found in the treenails. In the carvel-laid underwater body,

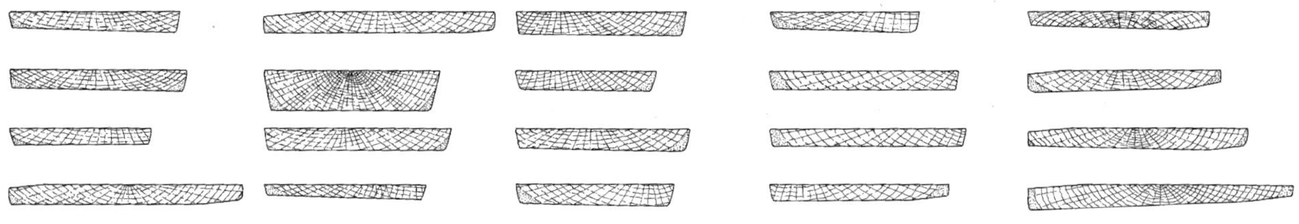

Fig. 8.40. Some sections across the keel and shell planks with their schematic tree-ring patterns and the way they were placed in the ship's hull (the heartwood side facing upwards, inboard).

30. Vlierman 1996b, 59-60.

Fig. 8.41. Futtocks and ceiling strakes on the starboard shell of the fore-body. In some of the squared futtocks, clearly parts of the original round trunks or limbs can be recognised.

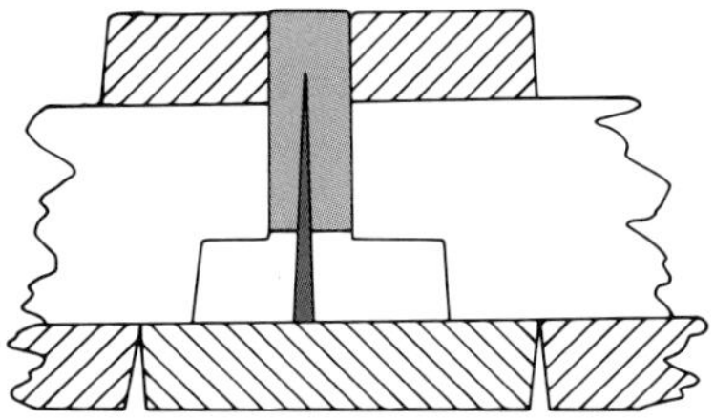

Fig. 8.42. In a ceiling board above frame 8, the wedged treenail that penetrated into a limber hole.

any protruding treenails were probably smoothed down together with the adze-fairing of the planking. Curiously, no wedges seem to have been driven in afterwards.

The boards of the ceiling strakes, the keelson and the riders were fastened to the frames with blind treenails. For securing the knees to the transverse constructions/crossbeams and to the shell, on the whole somewhat thicker treenails were used (3 cm thick; the heads ca 3.5 cm). Overall, about 3000 treenails will have been used in the vessel. An arresting detail is the solution found by a carpenter to a hole for a blind treenail erroneously drilled too far. The hole passed through a ceiling strake above frame 8 and entered into a limber hole. To avoid the treenail being driven down to the bottom strake and blocking the limber hole (thus creating a blockage to accumulating water and debris), a long narrow wedge was inserted from below into a shorter treenail. When the treenail with the wedge were driven in, the wedge struck the bottom and the short treenail was jammed fast into the floor timber above the limber hole. The ingenious carpenter however forgot to cut away the protruding wedge, thereby frustrating the object of the exercise (fig. 8.42).

For the clinker (parts of) shell strakes, iron nails 80-100 mm long were used, with flat, irregularly round heads or with slightly faceted forged heads (diameter 22-30 mm; figs 8.44 and 8.45-10). They were driven in from the outside, and on the inside beaten sideways and then back into the wood (double-clenched). On the whole the nails are spaced at irregular intervals. Besides, there are nails which are closer together (in the underwater body up to strake I, spacing 6-10 cm) and nails set further apart (14-22 cm; clinker strakes J to O). The part of the vessel that (usually) was below water thus is marked by a denser pattern of nails. Including the nails in the plank scarfs and the fastening of the hood ends of the strakes to the stern and stem, the vessel is likely to have contained some 3,100-3,250 nails. Presumably the gudgeons and pintles and their braces were the only major wrought-iron fastenings in the ship. For securing the moss laths and the moss caulking, about 12,000 sintel nail clamps of the B/C(2) type were used (figs 8.44 and 8.45-12).

8.3 Construction data

Above, the recovered parts, structural elements and observations have been discussed and in part interpreted. With some effort, a reliable reconstruction of the hull and layout of the vessel have been made on the basis of these data (Plates 8.IV A, B and C). From this reconstruction and other evidence a few things can be inferred about the lines and the building, the use, the foundering or abandonment and dismantling of the ship, and the formation of the wreck and its embedding in the soil.

8.3.1 Design

It is highly questionable whether the Nijkerk-1 cog was ever made to a design. It is quite unlikely. We may assume that in those days shipwrights built vessels according to familiar rules of thumb, relating to ratios of length to width and height, etc.

Lines

From the lines or planks drawings (Plates 8.III A and 8.IV D) it is evident that the vessel from lot Oz43 had a relatively high hull, with a straight keel, no pronounced sheer and with a straight, raked stern and stem. The ship was steered with a stern rudder, witness the gudgeons. The stern had a rake of 117° with respect to the keel line; the stem had a rake of 122°. The bottom (the keel plank plus the A- to D strakes) is symmetrical and in the after- and forebody sharply built. Towards the midships the progressively wider, everted sides rise from an almost flat bottom, with a slight rise towards the turn of the bilge. Nor is the bottom dead flat longitudinally. The cog nowhere has a sharp transition from the bottom to the sides. The bilge strakes and lower strakes of the sides (up to strake G) together form a round to lightly curved line.

The clinker-built higher parts of the sides amidships rise in a straight, everted line (118°) towards the top of the gunwales. At gunwale level, the ship has a relatively slender shape, which is not very different from the outline of the bottom. Yet the after- and forebody at the gunwales are in fact wider, with the ship forward of the mast is a little broader than aft. Only in the sharp parts of the fore- and afterbody do the frames on the sides display a slight S-shape. Most remarkably in the hull of the Nijkerk-1 cog, it is not only the more-or-less flat part of the bottom that is carvel-built, but actually the A to G strakes were built fully in this style. Just short stretches of these strakes are bevelled towards the stern and stem, where they appear to be joined in clinker fashion. Only from strake H up are the sides constructed stern-to-stem in clinker style (with a land of ca 6 cm).

System of measurement

In the ship no marks were observed which during construction might have marked, for instance, the centre line or the place of the mast. The thin treenails between the frames, mentioned in section 8.2.2, are the only traces that indicate the position of temporary framing. Nor do the reconstructed lengths of the bottom or between the stern and stem at gunwale height, for instance, form multiples of the lengths of the stern or stem.

Table 1. Principal dimensions of the Nijkerk-1 cog*

Length of the keel (including the stern- and stem hook)	13.2 m
Length of the bottom (between the rising part of the stern- and stem hooks	12.9 m
Greatest width of the bottom (seams between the D and E strakes at frame 15)	2.7 m
Overall length at gunwale level	18.3 m
Overall width at gunwale level (at frames 18-20)	6.9 m
Height of the stern (keel line to top gunwale level)	4.2 m
Height of the stem (keel line to top gunwale level)	3.4 m
Height of side amidships (keel line to top gunwale level at frame 13)	4.8 m
Length of the stern	4.7 m
Length of the stem	5.7 m
Height of the hold	2.1 m

Length-width ratio of the bottom	4.8:1
Length-width ratio at gunwale level	2.7:1

**All measurements rounded to the nearest 10 cm.*

Layout and tonnage
The midship section between transverse constructions DV1 and DV2, and between the V-shaped floor timbers of frames 5 and 23, may be regarded as the hold in any case; but maybe the hold extended to crossbeams DB2 and DB5 (Plates 8.IV A and B). The sharply built stern and forebody were less suited to stowage. In one area of this hold, close to the mast, was the fireplace, and around it without any doubt would have been the crew quarters. If indeed the hold was enclosed by DB2 and DB5 (between the after- and foredecks) this means that it was 9.5 m long. Its greatest width at the gangways at DB2 was 4.8 m; at DB4, 5.8 m; and at DB5, 5.20 m. The widths of the bottom ceiling (the outer edges of the D-shell-strakes) in these places are 1.8 m, 2.6 m, and 1.8 m, respectively. This amounts to an average width of ca 3.65 m. The depth of the hold between the top of the first ceiling board and the gangways was about 2.1 m. The volume of the hold therefore will be 9.5 x 3.65 x 2.1 = 72.8 cu m, i.e. a good 70 gross tons. It is doubtful that any vertical partitions were present in the hold, fore or aft. The stern and bow almost certainly were provided with decks, interconnected by removable or fixed gangways.

Choice of materials
Exclusively oak (*Quercus* sp.) was used for the vessel. This includes the treenails and the moss laths. For the large, straight timbers such as stern- and stemposts, (sawn) shell planks, floor timbers, keelson and crossbeams, tree trunks were used. For the hooks, V-shaped and curved floor timbers and knees, the wood came from trunks with limbs. Nothing can be said with certainty about the mast and yard. Most probably pine was used for these. The fastenings other than treenails were wrought-iron nails and sintel nail clamps.

8.3.2 Construction: techniques and sequence
The Nijkerk-1 cog has a keel plank, extended fore and aft with hooks. The rising part of the stem hook extends 1.45 m above the keel line. The top of the stern hook has broken off and disappeared; but probably it too originally rose ca 1.40 m above the keel line. The shell strakes A to G, almost entirely carvel-built, during the first two phases of construction probably would have required a slightly different approach than if large parts of these strakes were to be attached clinker-style.

Construction phase 1: keel plank, hooks and temporary frames
The keel plank and the hooks were probably joined together on fairly high blocks, by flat scarfs. In the keel plank, 2 cm thick treenails were observed in just three places, probably indicating the positions of temporary frames (between frames 10 and 11, under 20 and between 22 and 23). In the port A strake, and in the starboard A and B strakes, there also are such small treenails, at 20 or a little forward of it. On the strength of these treenails it may be assumed that during the first phase of construction a small timber had been fastened here to temporarily hold in place the A and B strakes. Aft, just one small treenail under frame 6 in port strake A was observed. At frame 16, the port strakes B, C and D contain small treenails that point to temporary fastenings of these planks. To starboard, they were just in strakes B and C. Around frame 13, there must have been something similar for strakes C and D.

Construction phase 2: bottom strakes, sintelnail clamps on the interior on the carvel seams, straight and V-shaped floor timbers of frames 2 to 27 and the transverse constructions DV1 and DV2.
Durig the fitting of the planks, the chamfering of the scarfs, the joining of the shell planks into strakes and the fastening of the hood ends to the hooks, every strake would have been temporarily fastened to the temporary frames or held tight against its neighbours by struts on the outside. The strakes were attached from aft to fore. At an early stage, the shipwright would have started on fashioning the permanent floor timbers. Before these were inserted, sintelnail clamps were put across the seams on the inside, roughly at 15 cm intervals, to hold the planks together. This must have been done before insertion of the frames, as the clamps were found also beneath the latter. Presumably during this construction phase only the (V-shaped) floor timbers 2 to 29 were inserted; and very probably also the crossbeams of transverse constructions DV1 and DV2 were put onto the V-shaped floor timbers of 5 and 23, in order by means of frames to provide some initial rigidity to the bottom.

Construction phase 3: stern- and stempost, sides up to strake I, sintel nail clamps on carvel seams and sintel nail clamped moss caulking in the clinker seams on the inside, V-shaped floor timbers of frames 1 to -1 and of 28 and 29, lowermost futtocks from 2 to 27, knees/futtocks on DV1 and DV2, transverse constructions with knees DV3, DV4 and DV5; ceiling strakes A to E; and the keelson.
After insertion of the stern- and stemposts, which were supported with struts, and the strakes, which almost entirely were attached carvel-style (up to strake G), the outside of the shell built so far will have been faired with an adze. From strake H up it became easier to add strakes, as from there on they could be simply nailed clinker-style to the preceding strake. Once the transverse connections with

knees, DV1 to DV5, were inserted, the ceiling strakes A to E could be attached onto the frames. The keelson too would have been inserted at this stage, or a little earlier.

Construction phase 4: strakes of the sides up to strake L; sintel nail clamped moss caulking on the inside; second futtocks on the frames and at the crossbeams; and the first futtocks on 0, -1 and crossbeam DB1
The frames 4, 11, 18 and 25 were provided with second futtocks, by which the crossbeams could be supported. In the top edge of the aftmost planks of the K strakes, there are notches at frame -1, to take the heads of crossbeam DB1. The same was done to the lower edge of the L strakes just above. The crossbeam was provided with rabbets to engage with the notches in the planks.[31] The enclosed beam head was surrounded with sintel clamped moss caulking to seal it. After the L strakes were put in place, notches were cut into their top edges to accept crossbeams DB2 to DB5. Then the seams until the L-strakes on the inside were closed with sintel nail clamped moss caulking. After copy the shell and shaping of the second futtocks, these too were fitted in. Finally the transverse constructions with knees (DV6 to DV10) and protruding crossbeam DB1 were put in place.

Construction phase 5: crossbeams DB2 to DB6; strakes of the sides up to the gunwales; sintel nail clamped moss caulking in the seams on the inside; top futtocks and uppermost V-shaped floor timbers in the stern and bow; knees on crossbeams DB3 and DB4, and the uppermost ceiling boards
Crossbeams DB2 to DB6 are placed into the notches in the upper edges of the L strakes. subsequently the strakes up to te gunwale strakes were placed and caulked. Now the uppermost futtocks, the knees on DB3 and DB4, and the still lacking floor timbers in the stern and bow could be shaped and inserted. To complete this phase, the top two ceiling strakes were put in.

Construction phase 6: stringers, gangway planks or fixed gangways, deck beams, windlass, anchor bitt, decks, knees on the decks above DB1, DB2, DB5, DB6 and above the deck beams on frames 2, 27 and against 29; pump, and a support for the lowered yard and mast
Between deck beams DB2 and DB5, the stringers and the gangways or gangway planks were inserted. Also the windlass, the anchor bitt, and the deck beams will have been fashioned and put in at this stage. These were followed by the deck planking and then the knees supported by the transverse and deck beams. In this or the next phase, the pump barrel, the drains and the support for the lowered yard and mast were put in place.

Construction phase 7: sintel nail clamped moss caulking in all seams and cracks on the outside of the shell; the false stem; the rudder and the mast
After phase 6, in which the hull and the interior were completed, it was time for the finishing touches. This involved installing the false stem, and closing all seams and cracks in the outside of the shell with sintel nail clamped moss caulking. The rudder and the mast are likely to have been placed only after the launching of the ship. Once the rigging and the sail with its yard were up, the ship was ready to sail.

8.3.3 Characteristics

The analysed structural elements and data resulted in the following reconstructed features of the Nijkerk-1 cog.

Keel plank
The ship has a keel plank which is narrowest at the stern hook and widest around the mast step. Forward the keel plank also narrows. Its section is rectangular to slightly trapezoid (the widest side uppermost). The keel plank is joined to the hooks by flat, chamfered scarfs. Its thickness is roughly equivalent to two plank thicknesses.

Stern- and stem hook
The hooks are trapezoid in section with the broadest side inward. The quite steeply rising parts are remarkably high (ca 1.4 m) and are joined to the stern- and stempost by a flat and chamfered scarf.

Stern- and stempost and false stem
The sternpost gradually tapers towards the top, where it is just 7 cm wide. The stempost by contrast remains roughly even in width. On the bow, the ship had a false stem, that covered the hood ends of the shell planks which were nailed to the front of the stempost. There was no false stern.

Carvel-built underwater body
The A to G shell strakes are basically attached entirely carvel-style. In some of the strakes, the plank edges were bevelled over relatively short distances towards the stern and stem and nailed together. This makes them appear attached in clinker fashion. Strake E is the first strake that was bevelled on its underside and runs on towards the rear and front of the sternpost and stempost. The preceding strakes had all ended in rabbets in the hooks. In order to give the underwater body rigidity as soon as possible, not just the temporary frames, but also as many permanent floor timbers as possible as well as beams of the lowermost two transverse constructions were fitted and inserted.

31. See the Nijkerk-II cog, fig. 6.43.

Clinker-built sides and double-clenching
Upwards of strake H the shell strakes are fastened clinker-style. The hood ends, one above the other, slightly differ in length where they are attached to the roof shaped front of the stempost: the starboard plank extends across the port plank by 1-2 cm (the latter plank's thickness) and vice versa, alternatingly. In the underwater body, the double-clenched nails of clinker-style fastenings are set closer together than in the strakes above.

Caulking method
Sintelnail clamps were put on all carvel seams on the inside, at intervals of ca 15 cm. They were (in the first instance) intended to hold together the planks of the bottom during construction. The seams are V-shaped in section, with the plank edges only just meeting on the inboard side. From the outside the seams were closed with sintel nail clamped moss caulking. In the apparently clinker-built oblique joining of the strake ends in the otherwise carvel-built underwater body, the caulking was done in a similar way. The clinker-built seams of the sides are provided with sintelnail clamped moss caulking both on the inside and outside, in gaps of V-shaped section.

Transverse connections, crossbeam constructions, stringers and decks
During the construction of the sides, the hull was reinforced immediately with futtocks and in the fore- and afterbody with transverse connections fixed onto the futtocks by naturally grown knees. At deck level, there are six crossbeams protruding through the shell, between which there probably were longitudinal stringers connecting the fore- and afterdeck. The vessel's fore- and afterdecks almost certainly carried a windlass and an anchor bitt.

Fastenings
For the structural attachment of the shell planks to the frames, 2.5 cm thick treenails were used with slightly thicker heads on the outside. For attaching the knees to the transverse constructions and crossbeams and to the shell, 3 cm thick treenails were used. The wrought-iron nails and double-clenched fastenings used in the ship are 80-100 mm long and have flat or slightly faceted heads. Very few larger iron fastenings (rudder fittings and gudgeons) were used in the ship.

8.3.4 Use

Function
No concrete evidence has been found regarding the ship's function. It dimensions, shape and relatively large hold (ca 70 gross tons) suggest that we are almost certainly dealing with a cargo vessel. The ship was definitely suitable for maritime voyages. The nature and provenance of the ballast stones found on board (see section 8.7) reveal that they were not collected as erratics in the Netherlands, but most probably in southern Denmark. The axe found in the Nijkerk-1 cog has no parallel in the Netherlands, but does in Denmark[32] (Section 8.5.2 and figs 8.47a, 8.48 and 8.49).

Wear and repairs
The underside of the stem hook is abraded over a length of ca 70 cm. This must be the result of frequent landings on sandy shores. From the rounded (use) wear on the ceiling boards, the keelson and the frames, it further is evident that the ship was far from new when it finally sank. The reused keelson, and the scuppers for the pump (drains) which changed position or were closed up, and a few repair planks set into and on the shell strakes also suggest that the vessel was at least several decades old when it finally met its end. The multiple drought cracks in the shell planks (closed with sintel nail clamped moss caulking) on the one hand show that the shipwright would have done better to use just half or less of tangential-sawn planks. Presumably the available timber was not thick enough to provide the minimum desired width for shell planks. On the other hand, it may also indicate a prolonged period of construction, while probably also poorly seasoned wood was used.

8.3.5 Abandonment/dismantling
The substantial age of the vessel, which maybe was no longer seaworthy, could have been a reason for abandoning it. It seems less likely that the ship got stranded or foundered at or near the wreck's findspot, although there is no hard evidence for this. Part of the ship probably stood out above the waves for a long time after it sank. The futtock (frame 18B) found beside the hull and its chopped-off head may be considered concrete evidence that part of the ship was dismantled. Besides, it is notable that none of the large structural elements suitable for reuse were recovered, while some timbers belonging higher up in the ship did survive.

8.3.6 Wreck formation and embedding
Section 8.1.4 elaborately (and reliably) reconstructs the wreck-formation process and the wreck's embedding. This phase, including the ship's partial demolition, presumably took decades. A stoneware sherd, found in the disturbed topsoil (Almere deposit) overlying the ship, must have

32. Vlierman 1985, 22-25 en 44.

ended up there long after the ship sank and became largely covered by sediment. The type of stoneware jug to which the sherd belonged was in use around the mid 14th century.[33] The dating of the construction and the demise of the cog is discussed in section 8.8.

8.4 Model

So far, no scale model of the Nijkerk-1 cog has been built.

8.4.1 Trial models

Already quite soon after the excavation, an attempt was made to build a model on the basis of the drawn plank templates and the frames reconstructed in the field. This was executed in model-aircraft plywood (fig. 8.43b). This material was selected for its flexibility; but the result was unsatisfactory. This was caused not just by excessive flexibility but also by the lack of clarity at the time as to whether a cog had a straight keel (including the hooks) or one with a rise towards the stern and stem. For these reasons, also a half-model of the forebody and two scale models of the bottom with floor timbers and some V-shaped floor timbers were built in oak. Of the latter two, one was executed with a straight keel throughout, and the other with a keel rising towards the stern and stem (figs 8.43a). A visit to the Deutsches Schiffahrtsmuseum at Bremerhaven in 1982, to see the Bremen Cog which at that moment was dry and exposed in its conservation tank, showed that there too people had struggled with the same dilemma during its rebuilding. Indeed the largely completed bottom/underwater body had had to be taken apart again, because the originally adopted starting point (a keel rising towards the stern and stem) was found to be erroneous. The building of the trial models of the bottom of the Nijkerk-1 cog too had shown us that the medieval cog featured a dead straight keel: the keelson could be fitted only into the straight-keeled model.

Fig. 8.43 a & b. Oak model of the bottom with floor timbers, keelson, buttress timbers and a keel plank that rises towards the stern and stem. The keelson failed to engage with the floor timbers, from which it was evident that the keel plank had not been curved but straight. b. Trial model of the Nijkerk-1 cog executed in model-aircraft plywood.

33. H.L. Janssen, pers. comm., July 2011.

8.5 Inventory

8.5.1 Distribution of the finds

Within the wreck and in the soil profile disturbed by the wreck, a modest number of items were encountered. The objects that might be reckoned to the equipment or inventory of the ship were all found in the bottommost, sandy sedimentary layer within the wreck and in the redeposited sand beneath the forebody. The latter can be explained by the stempost and hook being ripped from the hull in the process of wreck formation, and the objects that initially lay within the ship being washed out. The stoneware sherd (4) and the skull crown with horn cores (41) were encountered within the Almere-Zuiderzee layers in the redeposited sediment overlying the wreck and hence have no direct link with it. The inventory numbers of the items are shown at their findspots in plate 8.I A and correspond with the numbers in the text, in the captions and in the find list. The bones found in and around the wreck were investigated by F. Laarman (Appendix). The finds marked with an asterisk were not preserved.

8.5.2 Ship's equipment and inventory

For the sake of systematics, the finds from the wreck are classified under the eight categories listed in section 6.5.3 (regarding the Nijkerk-II cog). From the table (below) showing the distribution of the finds, their small number

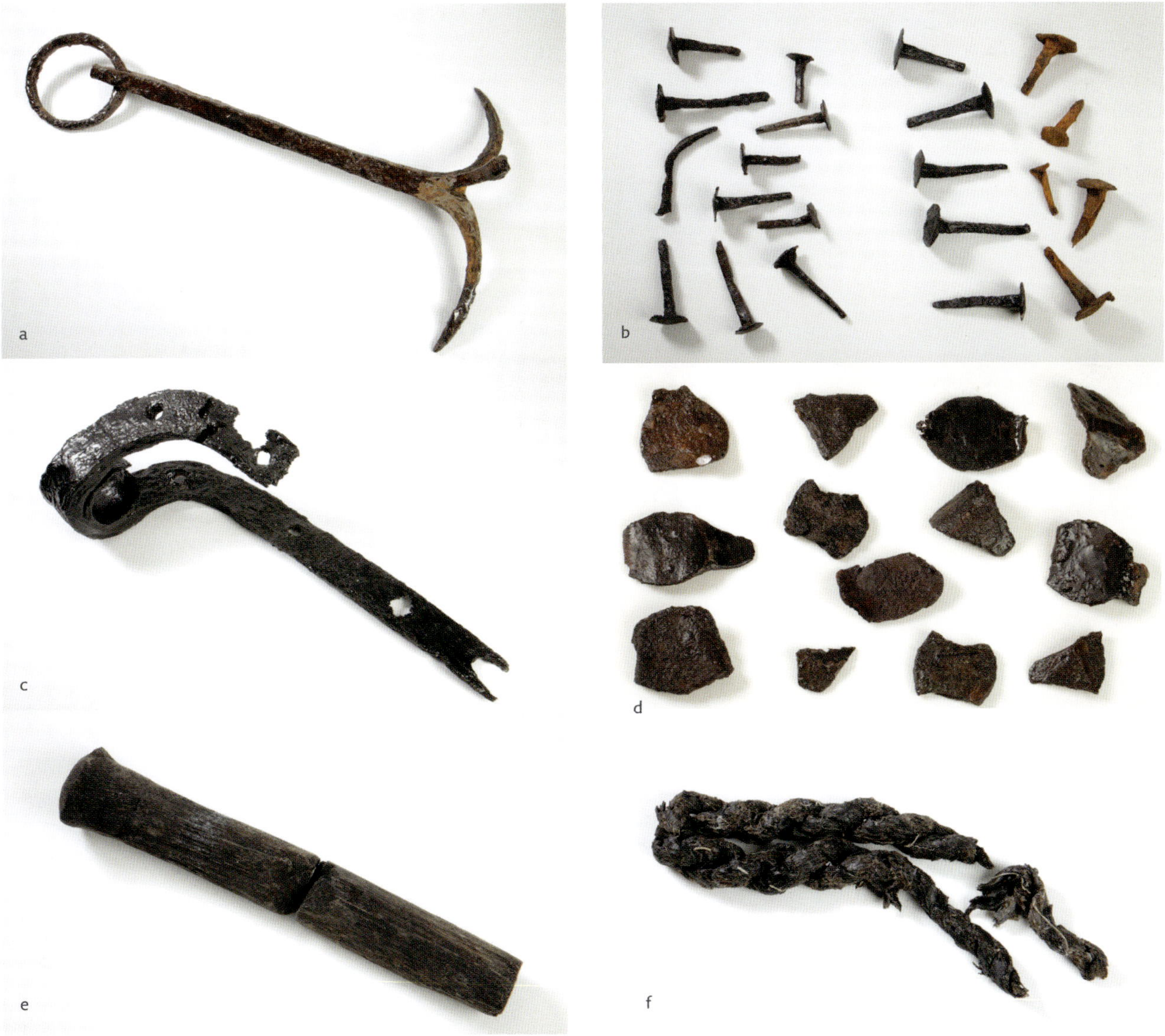

Fig. 8.44. a. Grapnel anchor. b. Nails. c. Gudgeon. d. Sintel nail clamps. e. Treenail. f. Piece of rope.

Table 2. Distribution of items

Category	Forebody	Hold	Afterbody
1. Ship's equipment	sail, cordage	cordage	grapnel?, cordage
2. Operational equipment	-	-	-
3. Administration and navigation	-	compass box?	-
4. Tools	grind-/whetstone	axe	grindstone
5. Household items	-	-	-
6. Cooking, eating and drinking	stoneware jug	stoneware jug, firebox	*Kugeltopf*, peat
7. Victuals	'cog sticks', bone	'cog sticks', bone	stopper, 'cog sticks', bone, hazelnut
8. Personal belongings, weaponry	scrap of leather	sole, shoe*	

** no longer present*

is also evident. It is clear that we cannot truly speak of an inventory or equipment. Given the findspot of the cog, which probably lay half-sunk, presumably easily accessible over a long period and part-dismantled, it is of course unlikely that people should have left anything of value inside the wreck.

Ship's equipment

The remains of the lowermost gudgeon were found attached to the stern hook. Beside the stern hook lay a detached gudgeon and two of the nails by which it had been fastened (figs 8.44 and 8.45-11a, b and c). This probably is the gudgeon that had sat a good metre above the lowermost. The nails of the clinker fastening of the shell planks, the sintelnail clamps and the treenails (figs 8.45-10, -12, -16 and -24), like the gudgeons, are part of the hull and are therefore not listed in the table under 'ship's equipment'.

Although the three-armed grapnel (figs 8.44 and 8.45-7) does feature in the table, it is debatable whether it did belong to this ship. The grapnel is distorted at the bottom and was caught on the after end of the port A ceiling board, at the rabbet in frame 8 which once had held a small upright post. The grapnel may well have belonged to a fisherman who set his nets over the submerged wreck, and lost the grapnel anchor when it became caught in the wreckage. What definitely does belong in the table, is the piece of an ancient sail (64, figs 8.46a and b). When in the wreck-formation process the stem hook and stempost became detached from the hull, it was swept from the hold and came to rest in the disturbed sand beside the forebody. As with the sail fragments in the Nijkerk-II cog,[34] we may here too speak of a remarkable find. Up until this study, it had been a mystery what kind of sail the cogs had carried. The fragment is about 60 cm high and 80 cm wide, and is woven from wool. On one side it has a selvedge; the top, bottom and other side are frayed, so the original width and length of the woven material canot be determined. The fabric is rather irregular and fairly loose, woven in plain weave on a broad loom. The fabric consists of S-twisted, irregularly spun strands, twined with a Z twist into yarn 2-3 mm thick. The fabric was woven from brown wool. At the selvedge, the first eight yarns of warp are of a paler, yellowish colour. This is followed by a 23-cm band of entirely brown wool. Then six paler warp threads, followed by a 30 cm band of brown ones. Another six pale warp threads, and then a brown band whose width can no longer be determined (>18 cm). If this too was a 30-cm band, plus another selvedge of paler warp threads, the fabric coming off the loom would have had the following width: 30+30+23+8 to 10 cm (of the paler stripes) = ca 90-95 cm.

The fragment was probably scrubbed with sand, so to speak, and stripped of any visible tar remnants. It looks clean; yet after more than seven centuries in the ground, its tarry aroma is still quite potent. Crumlin-Pedersen[35] says about the Viking ships that they carried woollen sails, and that, judging by recent woollen sails, a small, 20 sq m sail would require about 500 gr of woollen fabric per sq m. For a sail of 60-130 sq m the weight should be about double this if it was to be protected and made wind-tight with tar. The 'clean' piece from the Nijkerk-1 cog weighs about 750 gr per sq m. Tarred, it must have weighed between 1000 and 1500 gr per sq m. The square sail of the Nijkerk-I is likely to have been at least 80 sq m.

The fragments of cordage (figs 8.45-75, -77, -78 and -80), which turn up scattered among the (V-shaped) floor timbers on the bottom, certainly belong to the ship. Although this has not been definitively established, it seems

34. See section 6.5.3.
35. Crumlin-Pedersen 1997, 190.

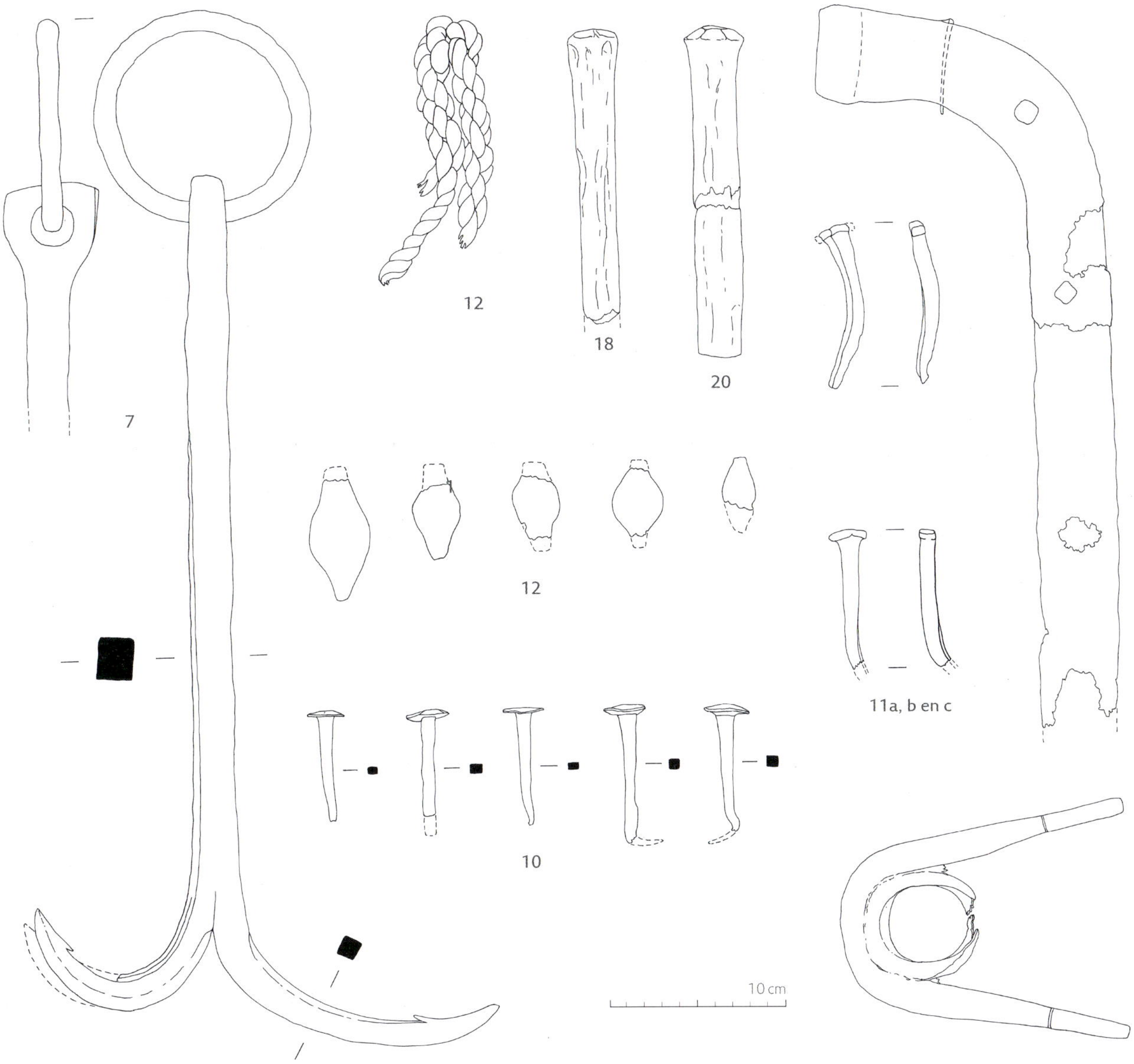

Fig. 8.45. Ship's fittings and equipment: gudgeon with fastening nails, nails of the shell, sintel nail clamps, rope and grapnel anchor.

Fig. 8.46a & b. The fragment of a woven brown woollen sail with bands of warp in a paler shade. b. The same, detail.

Table 3. Ship's equipment

Category	Item	Number	Findspot
1.1 Ground tackle	grapnel	11 A, B and C	afterbody
1.2 Sails	sail	64	forebody
1.3 Cordage	line	70 - 82	after- and forebody, midships
1.4 Blocks	-	-	-
1.5 Windlass	-	-	-
1.6 Pump, drain	-	-	-
1.7 Navigation aids	-	-	-
1.8 Boats	-	-	-

that they are fragments of lines that ran through the limber holes in the frames.[36] By tugging these ropes to and fro from time to time, the limber holes could be cleared of any obstructions, to allow water to be pumped from the bottom. The rope is about 1.8 cm thick, made of two strands with a Z twist. The material from which it was made has not yet been determined.

Working equipment

No finds were recovered that may be reckoned to the category of working equipment.

Documents and navigational equipment

Inside the hull a small plank with a dovetail joint was found, 14.3 cm long, >6.5 cm wide (broken), and 0.8 cm thick (fig. 8.45-29). It may have been part of a compass box.

Tools

Inside the wreck just three, or possibly four tools were found. One is an axe with handle fragments (figs 8.47a and 8.48-8); the others are two grindstones and possibly part of a tar brush or deck mop. The axe lay on port strake F of the flat-lying side between frames 13 and 14, and (originally) most probably had been tucked behind bilge deal C. The axe is of an unusual model and sparked the research report *'Bijlen uit laat-middeleeuwse schepen'*.[37] The axe has a roughly triangular blade and a slightly sagging flowing and a sloping rising obliquely to the top of the cutting edge which extends well above the eye. The cutting edge has a symmetrical section, is slightly curved and 143 mm long. The top line is 200 mm long. The blade has a curved bottom line: the steep rear line of the blade in a smooth arc merges into the underline of the neck and continues into the underside of the cheeks of the eye. The eye has a flat, hammer-like butt. The cheeks extend beyond the actual eye at the top and bottom, thus providing extra support/strength to the shaft or handle attachment. They also contribute to the ornamental aspect of the axe. The axe weighs 1100 gr. The handle is preserved only fragmentarily, had a length >30 cm, and is made of ash wood (*Fraxinus excelsior*). The top of the handle had been secured into the eye by means of a nail ca 100 mm long. The axe type, probably a battleaxe, has no known parallels in the Netherlands. If indeed it is a battleaxe, it should in fact belong in the category of 'Personal equipment/ weaponry'.

Yet a number of similar axes are known from Denmark (fig. 8.49). From excavations at the medieval castle of Næsholm on the Danish island of Sjælland, some 5 km west of Nykoping, three axes are recorded that display great similarity to the axe from the Nijkerk-1 cog. The only difference from Number N. 570 is the length from the cutting edge to the rear of the eye. The Danish spec-

Table 4. Administration and navigation

Category	Item	Number	Findspot
3.1 Ship's papers		-	-
3.2 Stationary	-	-	-
3.3 Financial/commercial matters	-	-	-
3.4 Orientation and dead reckoning	compass box?	29	midship section
3.5 Timekeeping	-	-	-
3.6 Signalling and distress signals	-	-	-

36. In the Rutten cog (Chapter 10) this was established with certainty during the excavation.
37. Vlierman 1985b, 22-25 and 43-45.

Fig. 8.47. a. Axe with fragments of the handle and a nail by which it was secured into the handle. b. Whetstone fragments. c. Granite grindstone.

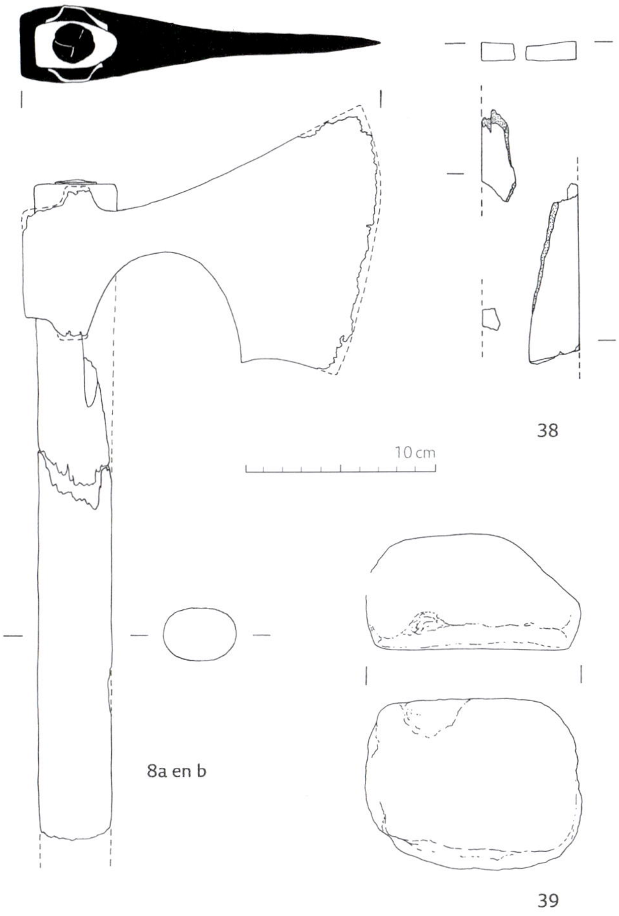

Fig. 8.48. Axe with handle, whetstone fragments and grindstone.

imen is 30 mm shorter. In other respects its design is virtually identical. The angle of the cutting edge with respect to the eye, the direction and the curve of the under and top line, and even the protruding cheeks on the sides of the eye show great similarity. The second specimen of this type is number N. 567. This axe however is less shapely and lacks the cheek extensions on the eye. Its maximum length is almost identical to that of the Flevoland find. The third axe from Næsholm bears the number N. 568. This axe shows a striking similarity to the axe (8) from the cog, described above. One might even speak of a sibling axe. Just a more distinct transition from blade to eye (a slight kink) and a cutting edge that runs more parallel to the eye, are the most notable differences in the lateral aspect of the axes. The Danish axes too have a symmetrical cutting edge. La Cour believes axes N. 567 and N. 570 to have been ordinary tools. But N. 568 reminds him of a battleaxe. *'I modsaetning til de overige funde Økser, der alle ma opfattes som arbejdokser, er det taenkelig, at N. 568 er en stridokse'*. Næsholm Castle was destroyed short-

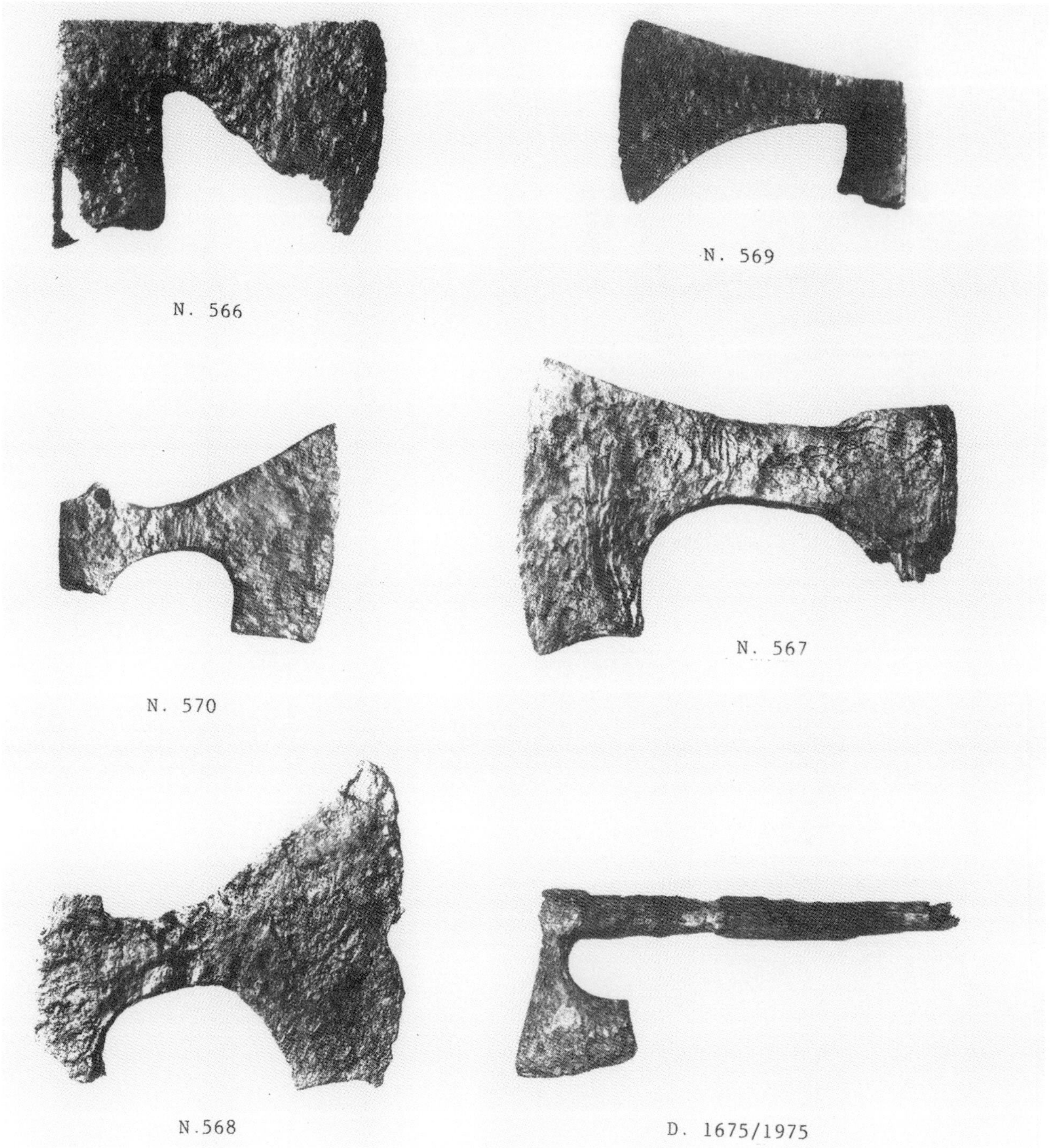

Fig. 8.49. Five axe-heads from Næsholm Castle in Denmark (accession numbers N. 566-N. 570) and an axe with handle (accession number D. 1675/1975) from the presumed wreck site of a cog at Vigsø, northwestern Jutland, Denmark (photo Nationalmuseet, Copenhagen; Wennar Warsen 1979). After Vlierman 1985 b.

ly after 1340 and was never rebuilt.[38] Besides the axes from Næsholm, the Nationalmuseet in Copenhagen has a fourth axe (fig. 8.49-D. 1675/1975) that shows similarities to the axe from the Nijkerk-1 cog, in this case with a large part of its handle. The axe was found at Vigsø, at the most northwesterly point of Jutland. Just behind the dunes, there is a great sand- and gravel-extraction site, where in 1974 two fine bronze *aquamaniles*, in the shape of horsemen, were found. Investigations in subsequent years brought to light a large number of iron-oxide concretions containing further items, including the axe, iron nails, a gudgeon and the upper pintle with an eye for holding a rudder. Of the actual ship no timbers were found. By the shape of the nails and the dating of the finds, it is for now assumed that a cog was lost here around 1300.[39] The Vigsø axe in its shape most closely matches axe N. 567 from Næsholm.

Two other finds that may be reckoned to the tools are four small pieces of a broken grindstone or whetstone, and a grindstone (figs 8.47, 8.48-38 and -39). The grind-/whetstone is of quartz phyllite, a type of rock often used for this purpose in the Middle Ages. These tend to be relatively long, narrow and thin tools.[40] The other grindstone is a granite erratic. The potential fourth find in this category is a small, broken plank fragment with a ragged scrap of fabric wound around it (31A and B). The fragment is of a quite loose, woollen weave. It consists of S-twisted, thin (ca 1 mm thick) strands, two of them Z-twisted into yarn 1.5-2 mm thick. The spaces between the threads are ca 4x4 mm. The fabric has a brown colour and smells of tar. It may have been part of a deck mop or a tar brush.

Domestic implements

In the wreck no objects were found that belong under the heading of '*domestic implements*'.

Cooking, eating, drinking and fireplace

A remarkable find is the oaken box mentioned above in section 8.1.4; it measures 84x84 cm and is 22 cm high. Inside it was an abraded 'cone' of loam, with which it originally had been entirely filled (figs 8.12, 8.50 and 8.51). The uppermost remaining bits of loam are hard and coloured red as a result of heat. It is a movable (weight ca 80-90 kg) fireplace, of a type found more than once in vessels excavated in Flevoland,[41] but usually filled with sand beneath a 'floor' of bricks or small tiles set in lime mortar. The box from the Nijkerk-1 cog is the earliest example known so far, and the only one filled with loam. Later ones mostly come from larger fishing vessels known as *waterschepen*.[42] On these fireplaces, an open, peat-fuelled fire would be tended, with the cooking vessels usually set on an iron tripod. The youngest archaeological example dates from the second half of the 17th century,[43] while such boxes are known to have still been used on *kwak* vessels from Volendam in the early 20th century. The box stood beside the (location of the) mast on the frames/ceiling strakes, embedded in the sandy sediment. This appears to have been roughly its original location. In boat III of the Herlgeandsholmen excavations in the centre of Stockholm, a somewhat similar box with a clay or loam fill was encountered. It was found in the afterbody of a vessel built in the Nordic tradition, presumably built in the 14th century.[44]

The box from the cog was made by joining the 22 cm wide and 2.5 cm thick planks of the four sides with pairs of treenails (2 cm across) to four 23 cm uprights ca 8x10 cm in section (figs 8.51 and 8.52). The uprights stood

Table 5. Tools

Category	Item	Number	Findspot
4.1 Carpentry	axe, grind-/whetstone, grindstone	8A and B, 38, 39	midship, fore- and afterbody
4.2 Caulking and maintenance	deck mop/tar brush?	31 A and B	forebody
4.3 Splicing, sailmaking	-	-	-
4.4 Miscellaneous	-	-	-

38. La Cour 1961, 233 and 241.
39. Crumlin-Pedersen 1979, 23 figs 2.5 and 2.6; Landberg 1980, 6.
40. Cf. the grind-/whetstone from the Marknesse cog, section 12.5.4, figs 12.26 and 12.28b.
41. Vlierman 1992, 51-52.
42. See e.g. Reinders *et al.* 1978, 25- 26 and fig. 5.
43. Shipwreck (*waterschip*) Zo 69, depot RCE, Lelystad.
44. Varenius 1985, 437 and fig. 34.3.

proud of the sides by 1 cm. In the planks are a few treenail holes, which suggest some earlier use. Maybe the four pieces were originally cut from a single plank. In two opposing sides, at 2 cm from the underside, there are three rectangular openings of 6, 8.5 and 7 cm length and a good 1 cm wide. Through these passed narrow planks, which each had been secured to the sides with single nails driven in from below. At right angles to these three 'supports' were laid three loose planks, 19, 22.5 and 24 cm wide and 2 cm thick. They formed the base, from side to side between the uprights. The remaining gaps between the uprights and the other sides had been filled in with shorter, narrower planks. Nailed to the outside of one of the sides was a wooden clamp (27x8x6 cm), into which a rectangular vertical tunnel (6x3 cm) had been cut. On the opposite side of the box, diagonally across from the clamp, two nails and two small treenail holes (1 cm across) were observed. This presumably was the attachment spot of a second, matching clamp. In the stern, a similar clamp was found, but the nail holes in it do not match up with those in the side of the box. Probably the original pair of clamps held a couple of planks or rods supporting a simple roasting spit. No tripod or firetongs were found; but in the afterbody some small pieces of fen peat (67) were recovered among the curved floor timbers. The box was in very poor condition and has not survived; the detached clamp and a sample of the loam fill (68) were preserved.

Fig. 8.50. The fireplace consisting of a loam-filled wooden box, *in situ*, to port beside the mast step.

Cooking utensils

To port, beside the stern hook and aft of frame S4, a fairly large rim-and-body fragment of a *Kugeltopf* vessel was found (figs 8.53d and 8.54-5).[45] The traces of soot indicate that the pot was used for cooking. The originally ca 24 cm high pot with a rim diameter of ca 18 cm is of a type indigenous to the northern Netherlands.[46] The only other ceramic item identifiable as cooking utensil is a small sherd of another *Kugeltopf* vessel (6). Typologically the large pot cannot be (closely) dated.

Tableware

In the sieving of the sandy sediment from the bottom of the wreck and from the flush sediment below the forebody, a number of small proto-stoneware sherds were recovered. They come from three different, small (drinking) jugs. Numbers 1A and 1B are a base sherd with a pinched foot and a moulded neck sherd from the same jug. 2A, B and C are rim sherds of the second jug, and 3A, B and C are shoulder fragments and a body sherd from a spot just above the pinched foot of the third (figs 8.53 and 8.54). Jug 1 is definitely datable typologically to the second

Fig. 8.51. The wooden fireplace box after removal of the loam fill.

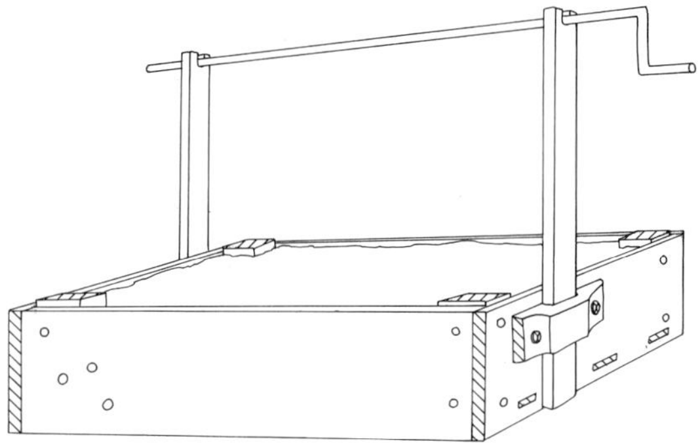

Fig. 8.52. (Re-)construction drawing of the fireplace box.

45. Vlierman 2010, 191 fig.3.
46. H.L. Jansen, pers. comm., July 2011.

Fig. 8.53. a. Base sherd of a proto-stoneware jug. b. Small proto-stoneware sherd. c. Sherd of Siegburg stoneware. d. *Kugeltopf* potsherds. e. 'Cog sticks'.

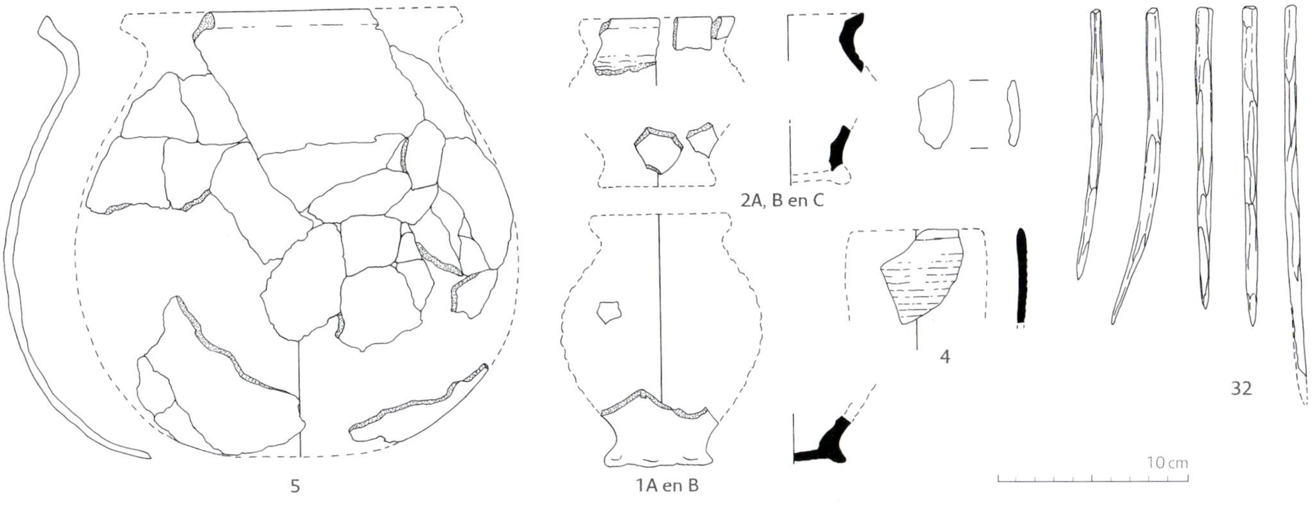

Fig. 8.54. Sherds of proto-stoneware, sherd of Siegburg ware, *Kugeltopf* sherds, and 'cog sticks'.

quarter of the 13th century; the two others, around the middle or third quarter of that century.[47] Given their findspots, the proto-stoneware sherds and the *Kugeltopf* without doubt were part of the inventory of the ship in its sailing days. They are the remains of broken crockery, the other parts of which were probably disposed of overboard. In ships excavated in Flevoland, this phenomenon has been recognised in several instances. Usually it is one or a few sherds of crockery that in most cases can be dated to an earlier period than the other ware still in use when the ship foundered.[48] The now dry sea floor which is Flevoland, has through the years yielded thousands of sherds or incomplete pieces of crockery which had been thrown overboard.

Victuals

A number of 'cog sticks' were found inside the wreck (figs 8.53e and 8.54-12). Section 6.5.3 on the Nijkerk-II cog describes the presumable function of the 15-20 cm long and 0.5-0.8 cm thick sticks whittled to a point, that were stuck into regularly spaced holes around the edge of a wooden base, and into which simple basketry was woven. Such little baskets were used for transporting sweet chestnuts or nuts. In the ship numerous half hazelnut shells (65 and 66) were found. A number of bones were recovered inside the wreck and in the flush sediment beside it. The skull crown of a cow (41) lay in the disturbed sediment above the ship and certainly does not derive from meat eaten on board. Probably the bone was at a later date washed into a depression in the sea floor above the wreck. The other bones may represent meat prepared and eaten by the sailors. They display no rounding from being washed to and fro. They were investigated and described by F. Laarman (Appendix). They yielded a predictable outcome.

Personal belongings/armament

The finds attributable to this category are a scrap of leather and some shoe fragments (figs 8.55-34 to -37). The scrap of leather (34) may have been part of a jacket or the like. Bits have been cut off for reuse. Along the straight edge are possibly two buttonholes with impressions of a button around them. Number 35 is an outer sole of a flat-heeled shoe or boot. The stitchmarks along the edge show that the sole had been stitched to the vamp. Fragments of a vamp and loose bits of welt (36) once were part of footwear. Nothing can be said about the shoe that was found beside the forebody (37), as it was not preserved. All of the recovered leather belonged to discarded items.

Table 6. Cooking, eating and drinking

Category	Item	Number	Findspot
6.1 Fireplace	(box)*, clamp, loam	13, 68	midship, afterbody
6.2 Fire-tending tools	-	-	-
6.3 Fuel	fen peat	67	afterbody
6.4 Cooking equipment	*Kugeltopf*	5, 6	afterbody
6.5 Eating and drinking gear	stoneware jugs	1, 2 and 3	midship, forebody

Table 7. Victuals

Category	Item	Number	Findspot
7.1 Fresh-water supply	-	-	-
7.2 Supply of beverages	-	-	-
7.3 Livestock	-	-	-
7.4 Food provisions	sticks, hazelnuts, bones	12, 65, 66, 41 to 63	forebody, midship, afterbody

47 H.L. Janssen, pers. comm., July 2011.
48 See e.g. Schröder 1994, VIII.

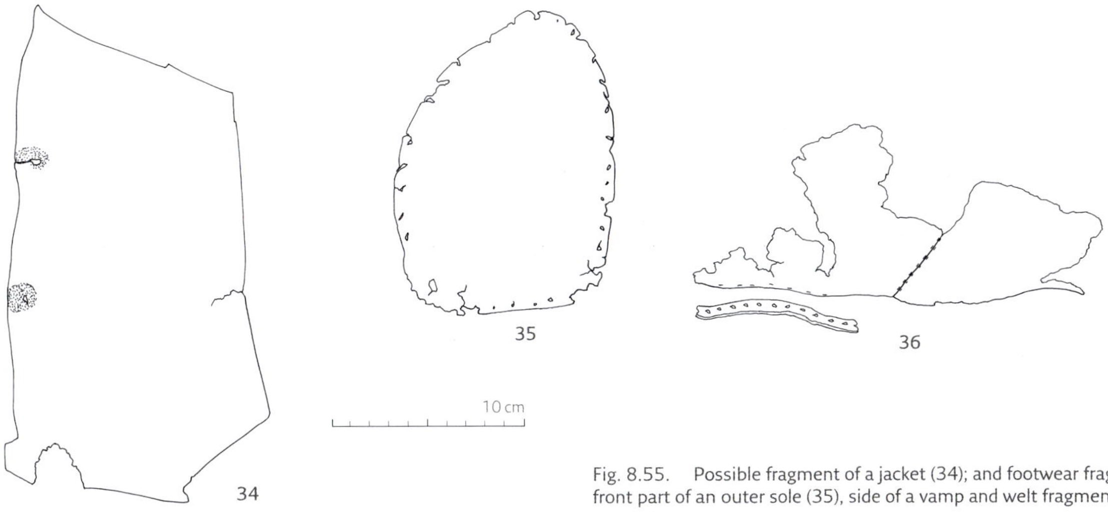

Fig. 8.55. Possible fragment of a jacket (34); and footwear fragments: front part of an outer sole (35), side of a vamp and welt fragment (36).

Table 8. Personal belongings and weaponry

Category	Item	Number	Findspot
8.1 Garments	jacket fragment(?)	34	midship section
8.2 Footwear	shoe (fragment)	35, 36, 37	fore- and afterbody
8.3 Sewing kit	-	-	-
8.4 Tools/weaponry	-	-	-
8.5 Cash	-	-	-
8.6 Miscellaneus	-	-	-

8.6 Cargo and merchandise

No evidence of cargo or merchandise was found in the wreck.

8.7 Ballast?

It is especially in the stern, that between the frames above the hook and the first three bottom strakes on either side of it, a total of 74 erratic boulders were found. They are fairly small, just 10-15 cm in diameter. Their size and small number beg the question whether they indeed were ballast, but formerly there may of course have been many more. W. Tj. Hellinga[49] identified the rocks, drawing the following conclusion:

Granite	21
Småland granite	3
Diorite	2
Hornblendite	1
Basalt	2
Småland volcanite	1
Secondary quartz	3
Sandstone	2
Chalk	3
Flint	22 (!)
Mylonite	2
Gneiss	4
Eye gneiss	1
Hälleflinta	1
Quartzite	6
Total	74

49. Written report, April 1983, file Zo 43 RCE, Lelystad

Under normal conditions, such a small number of stones would not allow any inferences. In this case, however, there is the unusual circumstance that 30(!) percent of the stones are flint, and that there also are three angular pieces of Danish chalk. The implication of the latter is that we cannot be dealing with northern erratics gathered in The Netherlands. The chalk, being a soft rock, would at any rate have been rounded (if not entirely disintegrated). Since the flint has its provenance in the same Danish rock formation as the chalk, and even this small number of specimens includes some rocks typical of Småland in southern Sweden, the conclusion must be that the batch was imported from southern Denmark. To be more precise, its provenance was definitely not at the latitude of Bornholm or more northerly, as no rock types from those areas were encountered.

8.8 Dating

Soil profile

The disturbance of the soil profile offers no clues as to the date of the ship's demise. The wreck has worked its way into a (submerged) peaty shore and into the underlying Pleistocene sand. The depression created by the ship subsequently was partly filled with the eroded peat and sand, followed by Almere clay. There is no question of a complete Almere sedimentary sequence. The Zuiderzee deposit (on the peat and above the wreck) never was any thicker than 15 cm. Of the homogeneous Almere deposit all we can say is that it sedimented before ca 1600; while the thickness of the layer suggests that this may well have been before 1500.

Finds

The neck sherd of a bottle of Siegburg stoneware (figs 8.53c and 8.54-4), which cannot be directly linked to the wreck, was found in the top of the Almere deposit of the disturbed sediment. The type of jug was in use around the mid 14th century. This means that the ship had been on the sea floor for a considerable period by that time and that the depression in the peat caused by the wreck had already been largely filled up with sediment. The *Kugeltopf* sherds and the proto-stoneware sherds can unproblematically be reckoned to the inventory of the ship as a working vessel. The large *Kugeltopf* fragment is not closely datable, but may be classified as a probably 13th-century model.[50] The foot of a small proto-stoneware jug is typologically dated to the second quarter of the 13th century. The sherds of the two other proto-stoneware jugs date from the middle or the third quarter of that same century.

Sintelnail clamps and nails

The sintelnail clamps used in the ship to secure the moss caulking are of the B/C2 type, which is dated to the second quarter of the 13th century.[51] The wrought-iron nails in the vessel have either flat, round, or somewhat faceted heads. There is some evidence that nails with flat, round heads are 13th-century or older, and that nails with somewhat convex heads forged into facets begin to appear in the early 13th century.[52]

Dendrochronology

So far, no dendrochronological analysis has been performed. At the time of the excavation (1981), collecting samples for this purpose was not yet standard procedure at the RIJP; indeed, facilities for such research were barely available in the Netherlands.[53] For many years there have been plans to assess the state of conservation or decay of the timbers reburied below the water table some decades ago. When this happens, also some suitable samples for tree-ring analysis will be gathered. The outcome of that study will show whether the presumable time of construction (second quarter of the 13th century) is correct, and where the timber, or the ship, originated.

Radiocarbon analysis

A sample for radiocarbon analysis was taken from the chopped-off futtock head of frame 18B. The result was a date of 830±30 BP. Using published calibration curves, and through a trajectory of plus or minus two standard deviations, this comes down to ca 1160-1260, with a confidence of 95 percent.[54] The sample had already been treated with PEG, of which we do not know whether it affects the dating outcome.

Similarities to and differences from some other 11th- to 14th-century ship finds

The possibilities of meaningful comparison with other ship finds in the Netherlands are very limited. For the purpose of dating, the most appropriate wrecks are two cogs presented in this study: the 13th-century cogs from Ens and Rutten (Chapters 9 and 10). The wreck at Ens is a far smaller and lighter craft than the Nijkerk-1 cog. The Rutten ship is (much) larger and more massively built. Besides, little more than the underwater body of both was recovered, so comparison is possible only as re-

50. Cf. for instance the finds with the Nijkerk-II cog, section 6.1.3, fig. 6.9.
51. Vlierman 1996a, 73.
52. See also the Ens cog, section 9.8. Similarities with other 12th-/13th-century ship (wood) finds.
53. Vlierman 1983, 4.
54. Lanting, written comm., 2 November 1993; GrN-20057.

gards the bottom and details of their construction. Of all three vessels it is certain that the carvel seams were open, V-shaped, over the full thickness of the planks. These seams were closed with sintelnail clamped moss caulking on the outside. In the Nijkerk-1 cog the carvel seams also have sintelnail clamps on the inside; this is not the case in the two others.

In the Ens cog, the A to E strakes are carvel-built all the way from stern to stem. In the Nijkerk-1 cog, this goes for the A to G strakes in principle whereas the Rutten cog has only its A and B strakes fully carvel-built while its C strakes have properly clinker-joined ends of at least 3 m. Besides the already mentioned differences in size among the three vessels, the provenance of the ships - and hence traditions varying in details - probably also played a part in the differences in carpentry. Although nothing of the transverse constructions was recovered, it may be assumed that the Ens cog only had them inserted in the interior of the hull. The Rutten cog without doubt was built with protruding crossbeams. This ship is likely also to have had transverse constructions fitted wholly inside the shell, as was the case in the Nijkerk-1 cog. Transverse connections low down in the hull as part of a frame, as in frames 5 and 23 in the Nijkerk-1 cog, were observed also in the lower parts of the after- and forebody of the Doel-I cog found near Antwerp, which was built after 1326 (Section 20.3)[55] and in the Vejby cog in Denmark, which dates from the second half of the 14th century.[56] Indications of the same were encountered also in the Dronten-I cog (Section 11.2.2).

An almost identical combination of transverse constructions and protruding crossbeams was thus far known only from wreck I at Kalmar, Sweden.[57] Although this vessel was just ca 11 m long, 4.55 m wide and ca 2 m high amidships, it constituted an important reference for the lines and construction of the Nijkerk-1 cog. Yet we should at once note many marked differences too. The Kalmar-I find has a typically Scandinavian/Nordic, T-shaped keel, was built clinker-style from keel to gunwale, fastened with rivets, and just the top strakes by means of thin treenails.[58] The ship has a curved stem. Ellmers[59] remarks that the plank fastenings in the Kalmar-I ship are executed with double-clenched nails, and that Åkerlund in his day was still unaware that this is a typical feature of cogs. It is unclear who of the two authors is right. At any rate it is the earliest of 25 finds of ships (timbers) that were recovered in 1933/'34 during the clearing-out of the silted-up old harbour, now known as Slottsfjärden (castle inlet). The presumably roughly similar construction dates (mid 13th century) can be considered fascinating evidence of a mixture of different building traditions.

In his reconstruction, P.K. Andersen assumes that the Kollerup cog, presumably built in the late 12th century and lost in the early 13th century,[60] contained only transverse constructions mounted inside the shell, onto which short decks were built fore and aft. The vessel, a good 20 m long and 4.5 m wide, and a good 2 m high amidships, did not have a long keelson with a broader part for the mast step, but instead a hole for the heel of the mast in a (moulded) V-shaped floor timber. Which is somewhat similar to the primary mast step in the Marknesse cog and the mast step in the Dronten-I cog.[61] Remarkable also is, with the exception of the 1.7-4 m long clinker-built ends towards the stern and stem of strakes C to H, while the central parts of these strakes are carvel-built. However, with the exception of the clinker-built fastening of shell strake F to the E strake in the Kalmar-I vessel a construction (thus far) unknown in Dutch finds. Westphal[62] reports the (externally) sintelnail clamped moss caulking of carvel seams and cracks, but fails to mention the presence or absence of any clamped moss caulking on clinker seams, or of sintelnail clamps on the inside of seams.

The four prominent tongue-shaped scarf ends on the outside of the forebody of the Nijkerk-I are unparalleled on any Dutch vessel. They are however known from Long-Ship Wreck I from the harbour of Haithabu (Hedeby) in Schleswig-Holstein. The ship probably sank between AD 990 and 1010, while dendrochronological analysis has shown thet the timber must have been felled in Schleswig-Holstein, with Haithabu (Hedeby) as the focus of the likely region of origin.[63] Notable also is the similarity between the mast support/knee on the keelson and the original shape of the (underside of the) keelson of the Nijkerk-1 cog on the one hand, and on the other, the keelsons/mast supports of various vessels from Haithabu (Hedeby) and Schleswig, and the various treenail forms with thicker heads.[64] The presumable possibility of lowering the mast and resting it on a forked support at the bow, together with the yard and the sail, is confirmed

55. At frames 1, 3, 26 and 28.
56. Crumlin-Pedersen 1979, 25-29 fig. 2.9.
57. Åkerlund 1951, 27-50, 151-152 and plates 1 to 9.
58. Westphal 1999, 152.
59. Ellmers 1979, 11.
60. Westphal 1999, 101-115.
61. Section 12.2.2; Reinders *et al.*1980, 11 fig.6, appendix 1; section 11.2.2.
62. Westphal 1999, Abb. 3.
63. Crumlin-Pedersen 1997, 81-95 figs 4.8 and 4.9.
64. Crumlin-Pedersen 1997 , 119, fig. 5.19 and 124 and fig. 5.25.

by the find of such a support with the Kalmar wreck,[65] but is also documented on civic seals and especially the 13th-century ones of ports along the southern shores of the Baltic, and of Stavoren and Harderwijk in the Netherlands (fig. 8.56).

Despite the (as yet) unavailability of tree-ring analyses, the foregoing does allow some conclusions with regard to dating the building and foundering of the ship, and to its likely region of origin.

The type of sintelnail clamp was common in the second quarter of the 13th century. The V-section carvel seams, which cover the entire thickness of the shell planks with the adjacent shell planks only just touching on the inside, may be regarded as a transitional form between caulked-seam variants B and C.[66] Which, judging by other - younger and slightly older - finds of ship's timbers, may well have evolved in this period. The use of nails both with flat and slightly faceted heads also suggests a date in the first half/second quarter of the 13th century. The working life of the vessel having been around the middle or third quarter of the 13th century is confirmed by the proto-stoneware sherds.

The similar tongue-shaped ornamentation of some plank scarfs on both the Nijkerk-1 cog and the Long Ship from the harbour of Haithabu/Hedeby which presumably was built in that very region, the similarities to and differences from the Kalmar ship as described above, and the (in part erratic) ballast stones that must have been gathered in southern Denmark, as well as the typical axe, which has counterparts only in Denmark, indicates that we are dealing with a vessel that almost certainly visited Denmark, and may well have been built in northwestern Germany (Schleswig-Holstein) or adjacent southern Jutland.

The ship's demise is harder to date closely. Given the observed abrasion, the vessel is certain to have sailed for several decades. Its abandonment or foundering is most likely to have occurred in the third or fourth quarter of the 13th century.

8.9 Summary

Ship

The characteristic construction and typical structural features (keel plank, hooks, raked, straight stern and stem, protruding crossbeams, (a large) part carvel-built, and partly clinker-built, the closing of the shell seams with sintelnail clamped moss caulking, and stern rudder) and the exclusive use of oak timber make it clear that we are dealing with a vessel built in the cog tradition. The remarkable transverse connections in the after- and fore-body and the length-width ratios of the bottom and the gunwales, suggest a region of origin in an area where the Nordic shipbuilding tradition (the Kalmar-I ship) and that of the cogs influenced each other. Thanks to the presence of large parts of the sides, and these preserved up to a fairly high level, and to the evidence that the ship was provided with protruding crossbeams, the Nijkerk-1 cog presents the first and earliest archaeological evidence and example of a probably medium-sized (Hanseatic) cargo ship built up high above the crossbeams, with a capacity of ca 70 gross tons. The Kollerup cog has a somewhat greater length of 20 m and a gunwale height of just ca 2 m. The Nijkerk-1 cog is 3.4 m high amidships, while the Nordic ship from Kalmar would have had a roughly similar height at the gunwale, with a similar length to our Flevoland ship. The late-12th-century Kollerup cog does not yet have crossbeams protruding through the shell; the mid-13th-century Kalmar-I ship does.

The lines/planks drawing shows a hull with a straight keel, little sheer and straight, raked stern and stem (117° and 122°, respectively) and everted sides (118°).
The first seven strakes of the bottom (A-G) are in principle carvel-built; short lengths towards the stern and stem appear clinker-built from the outside, and are indeed fastened with double-clenched nails, but the strake ends apparently overlapping here are merely chamfered. In the nailed-together after and fore parts of the underwater body (up to the seam joining strakes H and I) the nails are closer together (6-10 cm) than in the strakes higher up (14-22 cm). The H strakes are the first to be fully nailed on from stern to stem, while the F strakes are the first to run all the way from the rear of the stern post to the front of the stem post. The hood ends on the bow were covered by a false stem. The stern had no false stern. The shell is closed with sintelnail clamped moss caulking both on the inside and the outside. Moss was pressed from the outside into the V-shaped seams of the carvel-built bottom, that cover the full thickness of the planks; the moss was covered with oak laths held down by sintelnail clamps. This applied also to the (many) drought cracks in the shell planks, which on the one hand suggest a prolonged building process, and on the other, the use of poorly seasoned wood.
The planks were sawn, and no traces were found of their being furnaced in order to bend them. The clinker con-

65. Åkerlund 1951, plates 7b and 7c.
66. Vlierman 1996a, 27 fig. 11; fig 4.24

a

b

c

d

e

Fig. 8.56. Five examples of 13th-century civic seals that show a forked mast/yard support beside or on the bow: a. Stavoren, first half of the 13th century; b. Wismar, ca. 1250; c. Stralsund, 1265; d. second seal of Stralsund, 1278; e. Harderwijk, ca 1280 (after Ewe 1972).

nections, the hood ends of the strakes at the hooks and stern= and stemposts, and the seams around the heads of the protruding crossbeams were also sealed with sintelnail clamped moss caulking. The spacing of the sintelnail clamps is very irregular (4-12 cm). As an extra, sintelnail clamps were placed across the carvel seams on the inside of the shell, at 15 cm intervals.This cannot have served for keeping in place the moss caulking, but presumably principally to hold the planking tightly together during construction: they were inserted before the frames were inserted. The shell on either side of the keel plank numbers fifteen strakes (A to O) and the hull is principally held in shape by frames consisting of straight and crooked floor timbers extended on both sides with two or three futtocks. The futtocks are somewhat less massive than the floor timbers. In all, the ship numbered forty frames. The ten transverse connections (DV1 to DV10) up to strake J ensured good transverse rigidity in the below-decks part of the fore- and afterbody. The six protruding crossbeams (DB1 to DB6) and transverse connection DV11, together with the stringers, provided transverse rigidity at deck level. These beams carried knees to which the strakes of the bulwarks were fastened. Whether the protruding heads of the crossbeams were (already) provided with wooden fenders or guides on their forward side, is unclear.

In the underwater body the vessel has a notable, almost symmetrical shape, and at the gunwale the forebody of the ship is just 10 cm wider than the afterbody. The cog has a nearly round turn of the bilge, and nowhere a sharp, angular transition from bottom to side. From the midship section, the strakes gently rise towards stern and stem, as do the strakes from the keel plank to the turn of the bilge/the sides. In short, the bottom at no point is dead flat. Just a few frames in the sharpest parts of the bow and stern have a faint S-profile. The long keelson with its broader mast step and knee-shaped mast support originally belonged in another ship. The original joggling on its underside to engage with the frames had been adjusted in order to fit into this ship. The reused keelson however seems to be the only one to have ever been part of this vessel.

The mast support on the rear of the hole for the heel of the mast and the probable presence of closely spaced stringers suggest that the mast could be lowered. The mast was guided from the bow into the hole for the heel by means of a plank that was fastened to the mast support in the mast hole and to the sail beam. This lowering and raising of the mast would be done by means of a heavy windlass on the front of the afterdeck. The mast with the sail and the yard may have been lowered onto a forked support beside the stem. This is inferred from the recovery of such a support in the Kalmar-I wreck and especially from 13th-century civic seals showing cogs from some southern Baltic ports and, in the Netherlands, the towns of Stavoren and Harderwijk. For moving between the fore and after decks, the high sides necessitated some kind of gangway construction. These probably were heavy planks that were shaped to engage with the crossbeams. Presumably the hold was open to the sky between the fore- and afterdeck and the gangways.

The principal (reconstructed) dimensions of the ship were as follows:

- Length over the stern and stem at gunwale level — 18.3 m
- Largest width at gunwale level — 6.9 m
- Height of the stern (keel line to the top gunwale) — 4.2 m
- Height of the stem (keel line to the top gunwale) — 4.8 m
- Height of the side amidships (keel line to the top gunwale) — 3.4 m

The painstaking reconstruction of the hull may be considered reliable and will reflect the original shape and dimensions by more than 95%. The unusually extensive carvel construction of the underwater body, as well as the V-shaped seams between planks touching only at their inner edges and closed with sintel nail clamped moss caulking, may be regarded as typically 13th-century.

The ship had become stranded on a (submerged) peaty shore - or had been abandoned there - before drifting further inland or being pressed into the peat, and during a probably lengthy wreck-formation process gradually fell apart. Human intervention (partial dismantling) seems almost certainly to have occurred and will have speeded the wreck-formation process. Unusual structural features, ballast stones that can only have been gathered in southern Denmark and a fine (battle-?)axe that has parallels only in Denmark tell us that the ship almost certainly visited Denmark and that it is likely to have been built in northwestern Germany (Schleswig-Holstein) or southern Jutland. Dendrochronological analysis, as yet unperformed, will be able to shed more light on this point and on the presumed date of construction (second quarter of the 13th century).

Inventory
Hardly any (parts of) objects were found that might be attributed to the ship's inventory. This is an understandable situation, since the half-submerged wreck probably was accessible over a long period, during which it was partly dismantled and any item of value will have been carried off. The fine axe must have been overlooked, probably because it sat jammed behind a bilge strake between two frames. The square, loam-filled box formed the movable fireplace on which food was prepared over a peat fire.
It seems that a couple of wooden supports could be attached to the box, onto which a simple roasting spit might be placed. Various bones of cattle and sheep or goat were found. This type of fireplace is known from several (fishing) vessels from the former Zuiderzee area, and was used even into the 20th century. This fire box from the Nijkerk-1 cog is the earliest example known so far, and the only one filled with loam. From Stockholm a loam-filled box is known which probably dates from the mid 14th century. In the younger examples the box is filled with sand, covered by a floor of bricks or tiles set in lime mortar.

What should also be noted is the presumable presence of thin lines (rope) running longitudinally through the limber holes and which could be occasionally pulled to and fro to remove any obstruction. Other finds of interest are the thin, so-called 'cog sticks', ca 15-20 cm long and cut to a point, which probably belonged to small baskets in which nuts or chestnuts were shipped. The proto-stoneware sherds of the three small drinking jugs indicate that the ship must have sailed around the middle and in the third quarter of the 13th century. The vessel's ultimate demise must have occurred probably in the third or fourth quarter of that century. The fragment of brown sail, woven from wool with paler bands due to lighter-coloured warp threads, is a truly exceptional find.

The Ens cog[1]

9.1 Find conditions

9.1.1 Discovery and investigation

In 1946, during the digging of the Neushoorntocht along lot Q75 in the Noordoostpolder (NOP), the remains of a ship were found in the talus of this watercourse. The find was reported to the archaeological agency Oudheidkundig Bodemonderzoek Zuiderzeegebied (NOP), established a few years previously. The wreck was excavated in two phases in 1949 (28 May - 3 June and 11-24 June).

Fig. 9.1. The excavated wreck, viewed from the stern.

In the intervening week, wooden sheet piling was put in to hold back the continuous influx of water, to allow the groundwater to be pumped out and to stop the sides of the trench and the sections from subsiding. To start with, a trial trench was cut into the side of the drainage ditch to establish the orientation of the wreck. Then longitudinal and transverse trenches were dug across the wreck in order to study the soil profile. The deep position - the highest parts of the wreck lay at a depth of 2 m below the surface - and the nature of the sedimentary layers above and below the wreck suggested that this might well be the oldest wreck found so far in the polder soil, even older than the already excavated cog in lot M107, which according to Modderman dated from the mid-14th century.[2] During the excavation in lot Q75 no finds were recovered that might give an indication of its age. Hence the researchers at the time were dependent on the soil profile, of which some layers elsewhere in the polder had by then been globally dated on the basis of datable objects found within them. After the drawing of the sections and the clearly visible disturbance within the profile caused by the wreck, the dams were removed, and the wreck was further uncovered and cleaned (fig. 9.1). Subsequently the vessel was drawn and photographed.[3]

Six years after these excavations, a Dutch, updated reprint appeared of two articles in English by G.D. van der Heide in *Antiquity and Survival no.2 and 3*. In the second article, on the excavation of shipwrecks in the Zuiderzee area, Van der Heide[4] briefly describes the two ship excavations in the lots M107 and Q75 as follows: "Some vessels of the late Middle Ages have been investigated so far. They are relatively small ships, of which one (Q75) -probably from the XIIth century- was not closely datable because of a lack of datable finds. So here its age had to be determined through analogous dating of soil layers, which was possible because of the well-known medieval sherd layer (viz. here the Vavata shell layer) if around the 13th century and by the second, previously excavated late-medieval ship, which by its stoneware could be identified as having foundered around 1400."[5]

1. Translated by Xandra Bardet.
2. Modderman 1945, 79; the Marknesse cog, chapter 12.
3. The wreck lay with its bow in the watercourse talus at about 10 m from the centre line of boundary ditch Q74/75, oriented SSE-NNW with the bow to SSE. The site lies at 6.5 km southwest of the harbour of Vollenhove and 10 and 6 km from the present mouths of the Zwartewater and the Ganzendiep, respectively (figs 9.2 and 9.3). Ordnance map 21A, coordinates 187.70/ 517.48, with the highest-lying timbers at ca 4 m below NAP. The local depth of the water formerly was ca 2 m (mean low water), Hydrographic map of 1921.
4. Van der Heide 1955, 58-59.
5. M 107, Modderman 1945, the Marknesse cog, chapter 12.

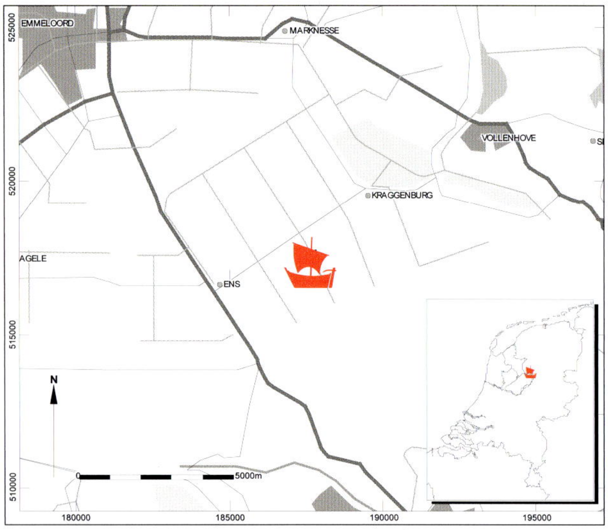

Fig. 9.2. The findspot of the Ens cog.

Fig. 9.3. Ordnance map of the Kingdom of the Netherlands (scale 1:50,000) of 1861, with the location of the Ens cog.

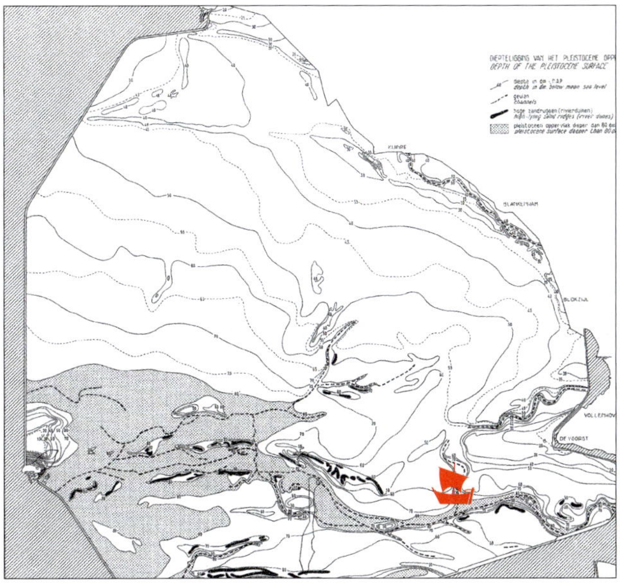

Fig. 9.4a. Depth of the Pleistocene surface (from Wiggers 1955, appendix 2) showing the position of the Ens cog beside the channel of the river Zwartewater/Vecht, immediately north of a sand ridge.

Fig. 9.4b. The Netherlands around 1250: the then Zuiderzee coastline with the location of the Ens cog (after Westerhoff *et all.* 2003, fig. 147).

As a result of the publication in English, the ship in field Q75 internationally received a great deal of attention in the world of nautical archaeology, and played an important part in the theoretical debate on the evolution of the cog. Curiously, the simultaneously mentioned wreck from lot M107, while already characterised as a (small) cog and published, did not share in this international interest. That this *was* the case with the ship to be discussed here was probably due to its possibly 12th-century dating. Crumlin-Pedersen[6] and Ellmers[7] saw the ship as a Waddenzee ship, a link in the transition from the Frisian Waddenzee ship to the great sea-going cogs. Fenwick saw parallels to the Graveney boat.[8]

The typescript of the Preliminary Excavation Report[9] - amusingly to us - records the total number of worker hours spent on the excavation (181.5) and the ensuing wages (155.28 guilders, *i.e.* ca 70 euros, excluding the cost of building the sheet piling, transportation of the motor-driven pump and the like, and the 135% of National Insurance contributions etc.). Under the heading of 'Datable Finds' it mentions that the wreck lies stratigraphically above the 13th-century *Valvata* layer and hence must be older, but it must be younger than the much later deposited Zus layer, which could be dated to the very beginning of the 17th century. In Van der Heide's handwriting, 'older' and 'younger' are deleted, and changed around. The preliminary conclusions of the report state that the little ship must have foundered after the 13th century and before the first quarter of the 17th century. Van der Heide signed off the corrected report on 3 July 1949. Why he mentioned a potential 12th-century date in the reworked Dutch article in *Antiquity and Survival* is unclear. In *Scheepsarcheologie - Scheepsopgravingen in Nederland en elders in de wereld*, with reference to the Q75 wreck we even find "The ship therefore must predate the 12th century"[10] and the museum brochure *Archäologie auf den Meeresboden* reads: "11th-century wreck".[11]

Reinders writes: *"Re-examination of the soil survey showed that the layer of shells (the* Valvata-layer*) was not above but below the shipwreck. So consequently Q 75 should be dated to after the 13th century"*[12] *and "An early 14th-century date corresponds very well with the date of another ship of the same construction, excavated in 1976 at lot N 5 in Flevoland"*.[13]

6. Crumlin-Pedersen 1965 and 1969.
7. Ellmers 1972, 64 and 292.
8. Fenwick 1978, 206.
9. Schokland, 2 July 1949, file NQ 75 RCE, Lelystad.
10. Van der Heide 1974, 421.
11. Van der Heide ca 1974, 17.
12. Reinders 1979, 38.
13. Reinders *et al.* 1980, 7-15; the Dronten-I cog, chapter 11.

9.1.2 Geology and stratigraphy

The Pleistocene surface and overlying deposits
The Pleistocene surface in the southern part of the Noordoostpolder has a strongly varying relief as a result of the east-west course of the river Vecht/Zwartewater (fig. 9.4a). The former riverbed of the main channel lies at least 8 m below NAP (Normal Amsterdam Level), while the Pleistocene surface on either side of the channel rapidly rises to 6 m below NAP or even higher. At the site of the cog, the surface of the Pleistocene sand lies between 8.5 and 9 m below NAP. The wreck lay on the left-hand side of the main channel at about 100 m north of a 200-m-long river dune present in the subsoil. A little further east lies a river dune more than 500 m long. On the Pleistocene (sand/till) surface north of the findspot, at the time of the ship's sinking there most probably still was an area of clay-on-peat land, which extended westward as a peninsula over 3 km long from De Voorst near Vollenhove.[14] The cog lies between 1 and 2 km south of the southwesternmost part of this peninsula. The clay-on-peat land to the south of the old channel around AD 850 still at any rate extended to the present day Zwartemeer.[15] It seems imaginable that north of this, later too, substantial peat islands may have persisted. For instance, overlying the above-mentioned river dunes. This is more or less supported by the mouth of the Zwartewater near the cog findspot, in the unendiked land as indicated on the map of the Netherlands around AD 1250: the Zuiderzee topographical map.[16] The same map (fig. 9.4b) shows that the unendiked land extended southward also south of the peninsula near De Voorst.

Description and dating of the soil profile
The soil profile in the Noordoostpolder differs from that in Zuidelijk and Oostelijk Flevoland. It is marked by distinct stratification, as a result of more even sedimentation. This is the case too at the site of the Ens cog wreck (figs 9.5a and b). In the field drawings of the soil profile, the Zuiderzee deposits are clearly separated. Also the brackish 'Zus' deposit was identified, which in Oostelijk Flevoland has not been recorded. It forms the transition from the Almere deposits below it, which in the Noordoostpolder are known as *Sloef*, *Bandjessloef* or *Humic Sloef*. These are underlain by the earlier-mentioned *Valvata* layer, which overlies a layer of Strongly *Humic Sloef*, below which lies a deposit of Detritus Gyttya. At the base of this lies a thin layer of coarse sand, and below the sand a detritus containing extremely fine sand, a layer of redeposited peat and finally, on top of the Pleistocene, roughly 70 cms of mud and sedge peat. The disturbance of the profile by the wreck is in the lower half of the *Bandjessloef* deposit. Ente[17] writes: "The *Valvata* layer initially was dated to the early 13th century on the basis of sherds reportedly found within it. However, closer investigation has shown this conclusion to be unreliable, because where the *Valvata* layer occurs there are no potsherds and *vice versa*. At best there may be an indirect, fortuitous connection. Gottschalk emphasises the storm flood of 1170. It is tempting to identify this as the one that broke through the peatland barrier between Schokland and Elburg, which then must date the abrupt demise of the Val*vata* molluscs. The Valvata shells were covered over by fluvial sediments from the river IJssel. This date virtually coincides with the presumed increase in activity of the river IJssel."

The Zus layer (the end of the Almere phase) is dated by Van der Heide[18] and Wiggers[19] between ca 1575 and 1625, but almost certainly was formed in the first quarter of the 17th century.[20] If we assume that the *Valvata* layer developed between ca 1175 and 1225, the Almere deposits cover a period of about 400 years. They are about 200 cm thick. At a steady rate of sedimentation, this means about 0.5 cm a year. The top of the profile disturbance by the wreck lies about 80 cm above the *Valvata* layer, which would amount to about 160 years, and imply that the ship was wrecked in the first half of the 14th century. Section 9.8 will return to the dating of the soil profile.

9.1.3 Medieval habitation

There cannot have been any medieval habitation in the immediate vicinity of the wreck, as at the time there was open water[21] or uninhabitable unendiked land east and southeast of Schokland and in the eastern part of the Noordoostpolder. Yet there was habitation at and around the castle of Vollenhove; at Genemuiden; about 6 km west of the wreck site on the island of Schokland;[22] and very probably also on the clay-on-peat ridge between Schokland and Elburg some 10 km southwest of the site of the Ens cog, close to that of the Dronten-I cog.[23]

14. Wiggers 1955, 193, fig. 108.
15. Lenselink & Koopstra 1994, fig. 5.21d; Westerhoff *et al.* 2003, fig. 146.
16. Westerhoff *et al.* 2003, fig. 147.
17. Ente 1973, 146-147.
18. Van der Heide 1955, 6-7.
19. Wiggers 1955, 103-104.
20. The Marknesse cog, chapter 12.
21. Ente 1973, 143-147.
22. Wiggers 1955, fig. 108 and the Nijkerk-II cog, section 6.1.3.
23. Section 11.1.3.

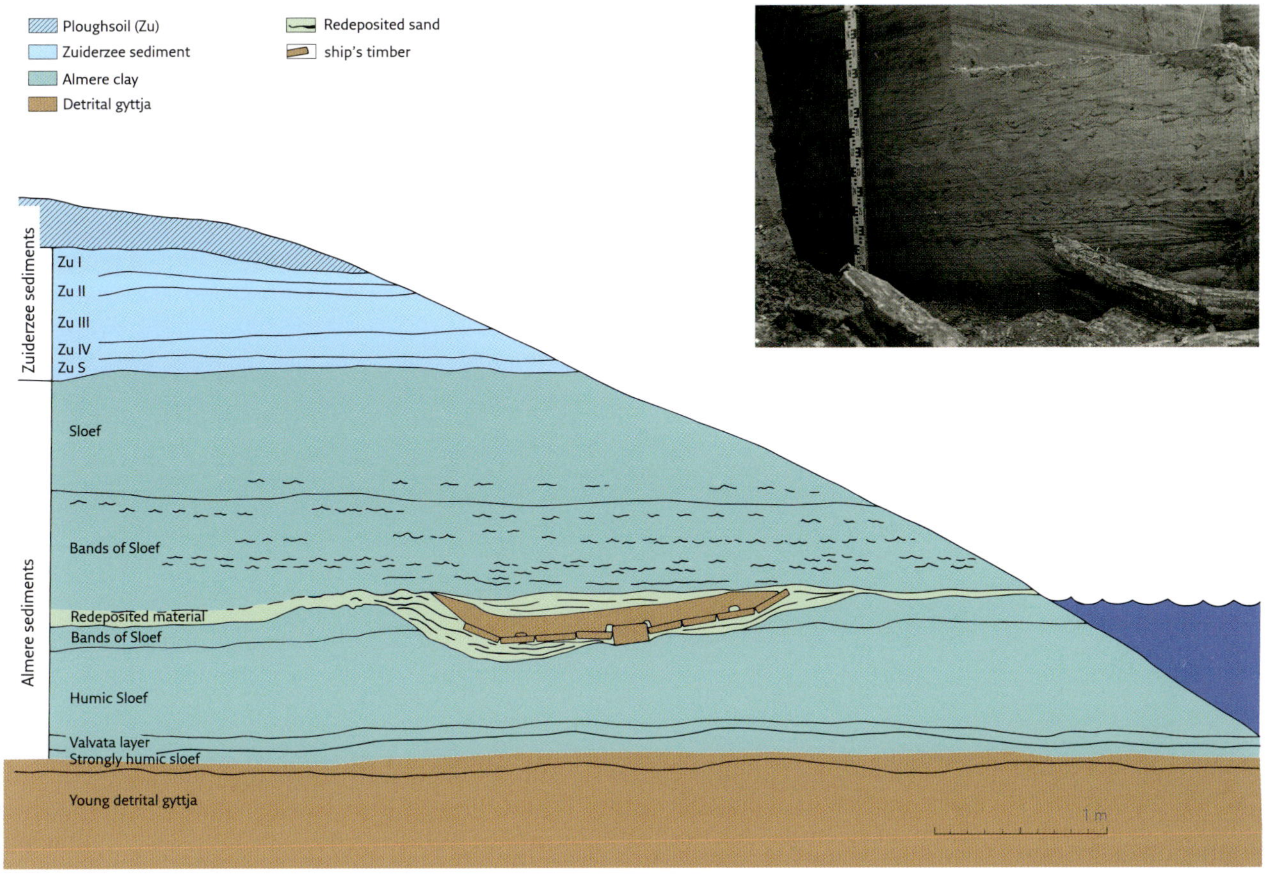

Fig. 9.5a & b. Soil profile with disturbance caused by the Ens cog (partially reconstructed). b. Transverse soil profile with disturbance caused by the wreck.

9.1.4 Condition and wreck formation

Condition

The wreck had sunk almost horizontally into the Almere deposits at a good 2 m below the modern surface. Despite these apparently favourable conservation conditions, the wood according to the excavation report was in a very poor state "... it could be easily cut through with a spade (even the floor timbers)" and "the condition of some of the frames was very poor. They were in part already hollow, which must have occurred even when the craft was still regularly sailing: the cavities were partly filled with mud". The vessel displayed a slight degree of hogging. The underside of the stern hook lay at 2.4 m below the modern surface; the keel plank (amidships) and the underside of the stem hook, at 2.3 and 2.5 m, respectively. Besides, the wreck had slightly listed to starboard. The wreck still comprised the stern- and stemhooks, the keel plank with skegs aft and fore on its underside, most of the bottom (except parts near the stern and stem), some shell plank fragments of the fourth and fifth port and starboard strakes, and (parts of) sixteen frame timbers (Plates 9.I A, B and C and fig. 9.1).

Wreck formation

Owing to the fairly flat shape of the uncovered wreck, the vessel's small dimensions and its relatively modest weight, it probably sank no further than about 50-75 cm into the then soft Almere sediments. Judging by the thick deposit of thin and sometimes undulating layers of sediment overlying the wreck (fig. 9.5b), we may assume that at any rate during and after its sinking, water movement must have been fairly constant here, which at times, intermittently, must have been quite forceful. This is probably an important reason why the ship sank here and why almost nothing survived of the sides, including the attached futtocks and ceiling planks, the stern- and stempost and decking - that is, if at the time they were still present. The total absence of detached timbers suggests that they no longer were there. Taking into account the fact that some of the frame timbers were rotten, hollow and mud-filled, together with the overall poor condition of the wood and a few observed repairs, we may assume that this was an old, written-off, and possibly already incomplete vessel; which during heightened river activity was washed downstream, floating under the surface, finally coming to rest and sinking at the findspot, where it became covered with sediment.

9.2 The ship

In the processing of the data and the vessel's reconstruction of the vessel, good use was made of the previously gained experience in working out the reference wreck (Nijkerk-II). Naturally also inspiration was taken from the other, smaller craft in this study, which this vessel more closely resembles.

9.2.1. Documentation and reconstruction

Documentation

After the detailed drawing of one longitudinal and two transverse soil profiles to a scale of 1:20, a survey drawing from above, a fore-and-aft section across the hooks, the keel plank and the frames, and eight transverse sections were drawn to the same scale. Besides, the stern- and stem hooks and some structural details were drawn separately, the latter not to this scale (Plate 9.I). Of the excavation five black-and-white photos were taken (figs 9.1, 9.5b and 9.9a, b and c). The neatly coloured drawings were probably compiled from measurements, drawings and notes made in the field. During the present study, it has become clear that the drawings in terms of measurements do not match up with each other; also there are (considerable) discrepancies, and that of the preserved hooks which were measured and drawn in 2008, various details had never been recorded, while the excavation report suggests that these were points receiving special attention. The failure to mention treenails, for instance, which were not drawn either, may be explained by inadequate cleaning of the wood (see the excavation photos). Some further examples of inconsistencies are:

- The rake of the stern hook in the drawings is variously 132° and 128°; in reality this must have been about 122°.
- The rake of the stem hook in the drawing is 128°, while in the preserved timber it is found to be 119°.
- The stem hook in the drawings is variously 54 and 57 cm high, while the preserved - and possibly somewhat shrunk - piece measures 60 cm. Its rising part in the drawings is 19 or 20 cm thick while the preserved timber is 16 cm thick, which implies a barely credible shrinkage of 20%.
- Many plank widths in the sections fail to match those in the corresponding places in the top-view survey drawing.
- Van der Heide drew a keel plank 8 cm thick, but in the report put 10 cm; the preserved adjoining horizontal parts of the hooks measure 10 cm.
- According to the description, the scarfs on the keel plank were 25 cm long; according to the drawing and recent measurement of the hooks this is 30 cm.
- Van der Heide writes that the forward skeg was attached to the stem hook and keel plank by four beechwood treenails with oak wedges, drew three, and (later?) wrote into the separate drawing of the stem hook that this too held a treenail. However, it is most likely that the skeg had been attached with five treenails.
- For the attachment of the after skeg just one treenail was drawn, while there must have been at least three.
- In the drawn sections, the carvel-built shell seams all gape 1-2 cm on the outside. The detail sketches, not drawn to scale, of a bottom seam, also clearly indicate that the seam was open throughout the thickness of the planks and filled with moss. Nothing is mentioned about this in the description, or shown in the top-view survey drawing or the photos.
- In a detail sketch, the planks ends at the scarfs are flush, whereas the description explicitly states that the plank ends stood proud of the underside of the bottom.
- The over-all length in the drawing does correspond with that mentioned in the report.

Reconstruction

The vessel fragment, found with its timbers still joined together, yielded sufficient evidence for reconstructing the bottom and the lowermost parts of the sides. The rake of the stem could still be established thanks to the surviving stem hook. This is more difficult for the surviving stern hook, as the rising stern part of the hook, including the gudgeon, had broken away and was no longer present. The rake shown in the reconstruction drawing is the most likely. Analysis of all known cog finds from the Netherlands and Flanders made it possible to achieve a fair reconstruction drawing of the entire hull with the principal dimensions and layout of this small vessel. Plates 9.I and 9.II A show the parts that were actually recovered. The discrepancies and hiatuses in the field drawings and the excavation report have on the basis of today's knowledge and insights been edited out in the reconstruction drawings and the following description. The drawings to scale 1:20 of the sections were manually upscaled to 1:10, the plank widths have been averaged and faired, producing a credible lines drawing. Taking into account the bottom strake widths, assumptions were made about the plank widths in the sides. Lines extended from the rising part of the stem hook, have led to height assessments of the stem- and sternposts, of the gunwale and of the full number of strakes up to the gunwale. The original presence of a mast step, ceiling boards, and an after and fore deck or cuddy may be assumed as virtually implicit. The reconstruction drawings and the scantlings of the non-recovered parts are hypothetical, but probably are a quite close approximation of the original situation and execution.

9.2.2 Construction and layout of the hull

Keel plank/keel beam

The keel including the horizontal parts of the hooks is 8.65 m long. The keel plank itself is 6.5 m long and 10 cm thick; it could almost be called a keel beam. At the scarfs fore and aft it is about 10 cm wide, towards the midships it gradually broadens to a maximum width of 20 cm. The plank is rectangular in section. The flat scarfs joining it with the stern- and stem hook are 30 cm long. Besides the treenails fastening it to the frames, the field drawings show regularly spaced treenails into the keel plank without any obvious connection to another timber. The excavation report mentions nothing on this point, but in the drawing they closely match the described 3 cm thick treenails that attach the skegs. In the preserved frame 4, not only do 2 cm thick oak treenails connect the frame to the keel plank and shell strakes, but also a beechwood treenail of 3 cm diameter was observed above the keel plank and another at the level of the fourth (i.e. D) strake (Plates 9.II B and 9.III A and B). These treenails were not indicated in the field drawings. Whether they remained unnoticed and whether any were present in (any of) the other frames can no longer be established. The most remarkable thing about the keel plank is that it was made of beechwood (*Fagus*), as were the 3 cm thick treenails. This was the first and only time for this wood species to be documented in a shipwreck in the Zuiderzee area. The deviating wood type and the 'pointless' treenails suggest that this was a reused timber, serving to replace the original keel plank which almost certainly had been made of oak. This conjecture is supported by the skegs, to be described below. Another, and possibly more likely explanation for the 'pointless' treenails might be that originally a connecting plank had been attached between the skegs.

Stern hook

The horizontal part of the stern hook is 1.3 m long. Its end at the scarf is about 10 cm wide and its rear end about 4 cm. The horizontal part is somewhat trapezoid in section with the smallest width below, and from its ca 10 cm thick forward end at the scarf with the keel plank, increases to a thickness of 15 cm at the stern and stem. The rising part is similar in section and is 45 cm. Into the top of the flat scarf with the keel plank, three nails were driven to secure the join. On the underside are two short rows of two and three nails with flat, round heads with a 20-mm diameter. Into the sides were cut rabbets 2 cm deep, into which the hood ends of the garboards were spiked: in the horizontal part of the stern hook every 12 cm; in the rising part every 7 cm. Forty cm short of the angle, there is a 3 cm thick beechwood treenail, for fastening the after skeg. Above the scarf to the keel plank is a stich(ed) nail, by which the garboard had been fixed in place (Plate 9.II A and figs 9.6a and b). The (now missing) gudgeon may have been attached both to the stern hook and to the base of the lost sternpost. On either side two nails had been used to secure it. The T-shaped wrought-iron gudgeon had rectangular attachment braces of ca 170x70 mm, with 100 mm long wings which stood proud of the stern by ca 70 mm. Into their ultimate extensions, ca 50 mm wide and converging in a loop, probably an iron ring had been forged, through which the bottom pintle passed (*c.f.* for instance the Spakenburg-I cog, fig. 14.20). The rear of the sternpost as reconstructed has a rake of 122°.

Sternpost

The non-recovered sternpost had been fixed onto the sternhook with a scarf and at first was probably secured to it just with a few nails. The post became thinner towards the top, while the rabbets for the hood ends probably ran obliquely aft at strake F and ended there. The rear surface of the continuation of the sternpost had been mitred.
The sternpost will have been about 2 m long; the total length of the stern including the stern hook, about 2.3 m. Measured perpendicularly at gunwale level, the stern rose about 1.9 m above the keel line.

False sternpost

The vessel never had a false stern.

Rudder

Witness the wrought-iron gudgeon, the vessel had been steered with a stern rudder. The rudder was suspended from the gudgeon below, and at the top almost certainly from a bolt with a ring through which the top pintle passed. Plate 9.II A illustrates what the rudder and its suspension may have looked like.

Stem hook

The stem hook was fully preserved except for its uppermost tip (figs 9.7a and b). The horizontal part is 1.4 m long, at the scarf with the keel plank it is 10 cm thick, towards the rising part it slightly thickens along 25 cm (to 10 cm) and thence slopes upwards to a thickness of 13 cm. From this point, the upper surface continues 70 cm onwards and upwards to the angle of the rising part and there has a thickness of 17 cm. In the sides of the horizontal part rabbets 2 cm deep were cut, into which the garboards had been nailed (at 12 cm intervals, near the rabbets, every 4-5 cm. The rabbets ran on into the rising part at an slanting angle. They held the hood ends of the first two strakes, secured with nails every 4-5 cm. At 50 cm from the outer angle of the stem hook, is a 3 cm thick beechwood treenail for attaching the fore skeg. On the scarf to the keel plank, the stem hook was fastened with three sets of nails driven in from above - rows of two, one and two nails, respectively. They do not pass all the way through the stem hook. From below, the same had been done in two rows, of one and two nails. Into the lower end of the rising part of the stem hook two small holes with a diameter of 1 cm had been drilled. Their significance remains unclear. In the side of the horizontal part, a remnant of a nail can be observed - maybe a stitch(ed) nail for attaching the skeg. Close to the scarf is a stitch(ed) nail by which the garboard was held in place. The front surface of the stem hook (and of the missing stempost) has a rake of 119° with respect to the keel line.

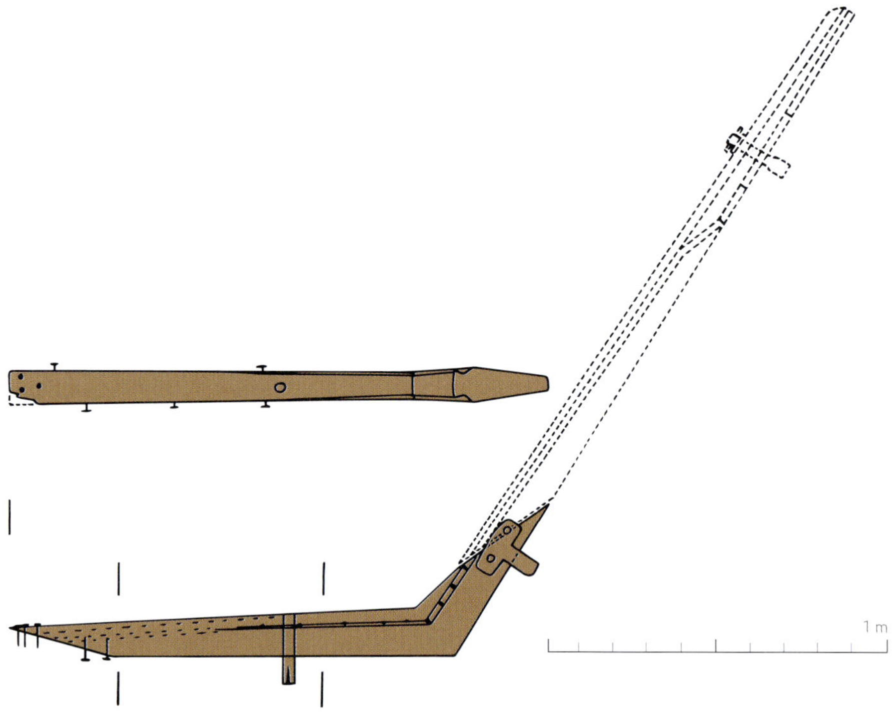

Fig. 9.6a & b. The conserved stern hook. b. (Reconstruction) drawing of the stern construction.

Stempost

The stempost was not recovered. The post had been fastened to the stem hook with a flat scarf and secured with four nails on the inside. The post gradually tapered upwards, and at the top presumably had been no more than 5 or 6 cm thick. The hood ends of the side strakes sat in the rabbets in the stempost probably up to strake F. The subsequent, higher strakes ran on to the purposely mitred front of the stempost. The stempost will have been about 2.35 m long, and combined with the rising part of the stem hook 2.8 m, while the top of the gunwale stood 2.5 m above the keel line, measured perpendicularly.

False stem

There is no evidence of a false stem in this vessel. Given the thin stempost and the probability of the four upper strakes passing alongside it, it seems very likely that the cog originally did have one. Plates 9.II A and 9.III A illustrate what this may have looked like, what its scantlings may have been, how it may have been attached and how it probably extended above the stempost and the gunwale.

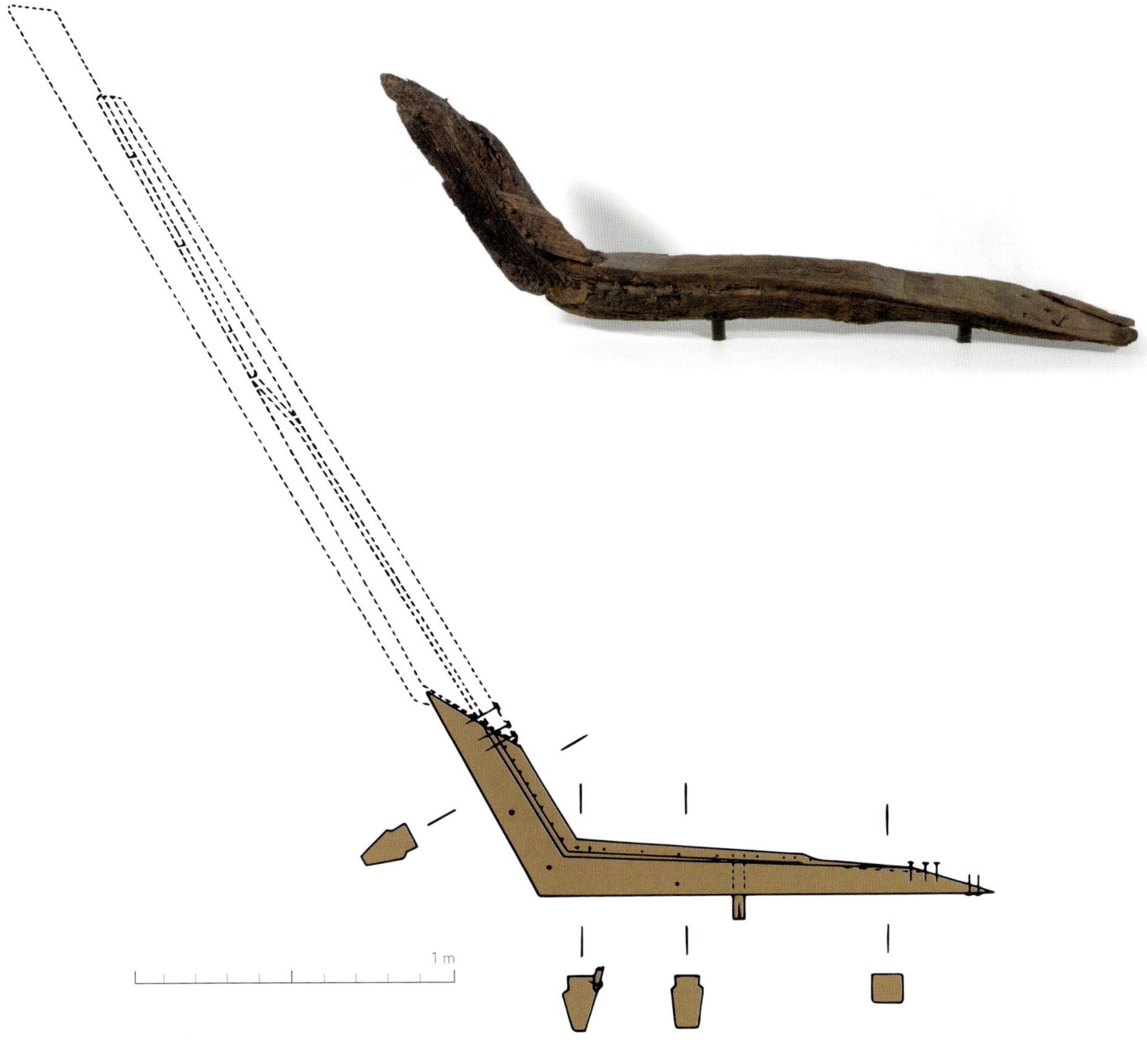

Fig. 9.7a & b. The conserved stem hook. b. (Reconstruction) drawing of the stem construction.

Skegs
Under the keel plank and the hooks, aft and fore an oak skeg had been attached. The after skeg was 2.85 m long, 8 cm thick at the rear and 4 cm at the front; while along its entire length it most probably was a little narrower than the underside of the hooks and keel plank. The lower corners were (severely) abraded. Apart from the above-mentioned thick attachment treenails, the skeg seems to have been fastened to the stern hook at the rear only and not on its fore end. The fore skeg was 3.1 m long, 8 cm thick at the front and 4 cm at the rear. Apart from the five thick treenails, two stitch(ed) nails had been used only on the front and the side. This skeg too had become seriously abraded through prolonged use. The absence of stitch(ed) nails in the skeg ends under the keel plank may be explained by the possibility of the skeg ends initially having been protected by a ca 4 cm thick plank inserted between them. As noted before, this might also explain the 'pointless' thick treenails in the keel plank.

Bottom and sides
The bottom and the sides on either side of the keel plank probably comprised nine strakes (A to I). Each strake consisted of several 2 cm thick planks. As mentioned before, the widths of the planks in the reconstruction have been averaged and faired. Most probably the first four strakes were ca 22 cm wide; the fifth, about 25 cm; and the subsequent four, ca 30 cm. At the stern and stem they are likely to have been a little narrower. The A and B strakes were composed of four planks; the C strakes, most probably of two. The D strake to starboard probably consisted of two, and to port of three planks, while the E to I strakes presumably also comprised two or three planks. The flat scarfs joining the planks have a length of ca 25 cm. The nibs of the overlapping plank ends most probably were ca 0.5 cm thick and on the outside stood proud of the shell. Despite Van der Heide not explicitly mentioning this, the same probably goes for the inside, though as a result of abrasion this may no longer have been visible.

All scarfs were held together by just four nails: two driven in from above and two from below. The nails (fig. 9.8) have round, thin, flat heads, 16-20 mm across. They were driven through the two plank ends, twice beaten sideways and driven back into the wood, *i.e.* double-clenched. A curious feature is the diagonally angle section of the nails with respect to their double-clenched points. Maybe they had been driven in diagonally following the grain of the wood, because of the boatbuilder holding the square-stemmed nail between the fingers and the angle at which he sat or stood while hammering. Judging by the two rows of sintel nail clamps, one above the other, on the sides of both hooks, it appears that the seam between the keel and the garboards had been closed with sintel nail clamped moss caulking both on the inside and outside. Sintel nail clamping on the inside of this seam has not been observed in any of the other Dutch and Flemish cog finds. Also it is clear to see that the sintel lines ran downwards towards the midships. This means that the upper surfaces of the garboards were not flush with the top of the keel plank, but were a few centimeters below it. This might be due to the presumed replacement of the keel plank.

Bottom
The bottom consists of three strakes (A-C) on either side of the keel plank. Van der Heiden in the excavation report writes "Purely clinker-built construction has not been observed on this small vessel, although in the forebody some strakes (planks) did appear to overlap a little". It is indeed remarkable that in the appropriate areas of the planking in the after- and forebody no nails, nail holes or evidence of the like have been observed. Most curious too is the drawing of gaping seams between the planks in the sections. Also in the detail drawing, not drawn to scale, of a bottom seam with moss, moss laths and sintel clamps (Plate 9.1c) it is clearly a somewhat trapezoid gap. Although this cannot be said with absolute certainty, the preserved (dried-out) moss samples and moss-lath fragments seem to indicate that the seams filled with caulking had been triangular in section - taking into account the thickness of the moss lath and the insertion of sintel nail clamps - to a depth of 1.5-2 cm, *i.e.* the full plank thickness. So either the planks abutted on the inside of the shell, or the seam there gaped a few millimeters. The (shrunken) moss lath fragments still are 0.8-1.2 cm wide, so the seam openings on the bottom must have been at least 1.5 cm wide. This could also be inferred from the imprints of the laths on the underside of the sintel nail clamps and from the clamps' dimensions. In sections 9.3 and 9.8 we shall return to the implications of these 'open' seams and the at most very slight overlap of the plank ends at the stern and stem.

For the first two strakes beside the keel plank, four relatively short planks were used. For each of the garboards, two planks of 3.2 m, and two of 2.3 m length. For the B strake to starboard, planks with lengths of 2.5 m, 2.3 m, and two of 2.7 m. To port it is roughly the same. But the C strakes are different. To starboard the aftermost plank is 6 m long, with a forward extension of 4.7 m. The C strake to port is the other way around. Here the builder started aft with a 3.7 m plank, which was extended forward with a plank of more than 7 m. The garboards over their full lengths sit at a slight angle to the keel plank, and towards the stern and stem are twisted so as to be fixed into their rabbets at an angle of almost 90°. The B- and C strakes transversely show a uniform, slight deadrise towards the turn of the bilge. Although we cannot be fully certain, there seems to be evidence of a 'brother-and-sister' pattern, be it that the sequence in one C strake was reversed. In the foremost parts of the A-, B- and C strakes in the bow, no clear overlaps were found, no nails pointing to clinker construction, and no remains or traces of sintel nail clamps on the inside of the shell.

Sides

As for the sides, only midship parts of the D strakes (the bilge strakes) were recovered. And in the forebody, some short pieces of the E strakes. The D/bilge strake to starboard aft probably begins with a plank of over 7 m and is extended forward with a ca 4.5 m plank. To port, the D/bilge strake consisted of three planks, ca 5 m, 2.25 m, and over 4 m long. Both the bilge strakes and the following E strakes had most probably, like the bottom strakes, been built fully carvel-style. No (traces of) nails were found, while the frames did show that the strakes had been attached at different angles, but the frames were not joggled as is usual in clinker construction. However, thanks to the preserved frame 4 it could be established that this had been the case at the top of strake E. Therefore we may assume that the vessel was built clinker-style only from the sixth strake (F) upwards, while the edges of the second to fifth strakes were at most somewhat bevelled aft and fore, but in principle were carvel-built. Besides, it seems quite likely that the F to I strakes ran on to the very rear and front of the stern- and stempost, respectively. The lateral fastenings of the clinker-built strakes will be discussed in sections 9.3 and 9.8.

Although this might seem less plausible, we should yet look to the Kollerup cog as regards the construction of the sides.[24] The craft may have been constructed in clinker fashion only at the transition of strakes E to F, with the planks further up being flush-laid or carvel-built again overlapping (clinker) for just a few meters towards the stern and stem (Plate 9.III D). In the sides too, 'brother-and-sister' patterning is likely to have been applied. The top three strakes in the midships probably sat in a roughly straight, everted line (ca 112°) one above the other. In the stern and bow there was no clear distinction where the bottom transitioned to the sides, the cog had somewhat S-shaped frames. The forebody was a little wider than the afterbody, the sides displaying a distinct sheer, while the gunwale at the bow probably was some 60 cm higher than at the stern.

Sintel (nail) clamped moss caulking

The (triangular to trapezoid, open) carvel seams on the underside of the underwater body were closed with sintel (nail) clamped moss caulking. The double rows of sintel (nail) impressions on the sides of the horizontal parts of the hooks show that the seams of the garboards against the hooks had been caulked both inside and outside. This most probably had also been the case along the keel plank (Plate 9.II C). Around the hood ends of the shell planks on the rising parts of the hooks, no sintel (nail) clamp impressions were observed, now were there any on the inside, on the deadwood-like reinforced parts of the hooks. The sintel (nail) clamps on the horizontal parts of the hooks and along the seams where the garboards abutted on the keel plank, were placed at intervals of 4.5-5.5 cm. This probably applies only for the clamping of the seams between the keel plank or hooks and the garboards, and therefore differs from the following data. In the typescript excavation report there is mention of 10-cm intervals. This was deleted by Van der Heiden and replaced by "6 cm". Therefore 6 cm presumably was the average spacing

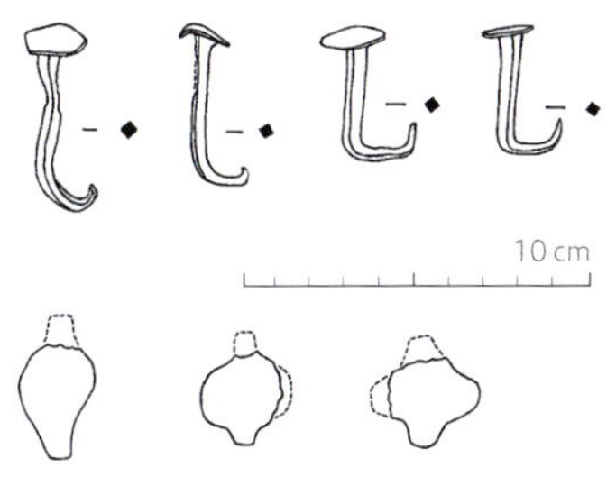

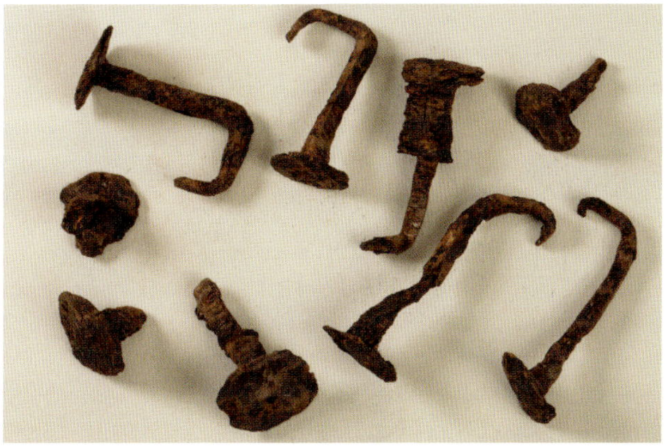

Fig. 9.8a. The double-clenched nails; and the sintel (nail) clamps used in the Ens cog, types C2/D, D and D(1). b. The nails.

24. Andersen 1983, 25; Westphal 1999, 106 Abb. 6.

in the seams of the bottom, and 4.5 - 5.5 cm where the garboards joined the keel plank.

Van der Heide mentions nothing about the seams between the C and D strakes, between the D and E strakes and between the E and F strakes. In the last-named case, it seems likely that in the clinker construction applied there, the seams on the outside and (potentially also) on the inside were provided with sintel (nail) clamped moss caulking. As mentioned before, however, no (impressions of) sintel (nail) clamps were observed during excavation on the inside or outside of the recovered shell. At any rate it is virtually certain that all seams in the carvel-built part of the cog (from the keel plank up to strake E) had been sintel (nail) clamped on the outside. In the vessel three (transitional) types of sintel (nail) clamps were used. They are types C2/D, D and D(1) (fig. 9.8). These types were current in the second to fourth quarter of the 13th century.[25] The applied moss laths are thin strips of oak wood. These strips, 0.8 - 1.2 cm wide, are more or less oval in section, with a greatest thickness of 0.15 to 0.2 cm. Inside the scarfs joining the hooks to the keel plank and in the scarfs joining the strake planks, a thin layer of moss caulking had been applied.

Macroscopic determination and pollen analysis

The investigated caulking sample is of pure peat moss (*Sphagnum*) with a twig of *Erica tetralis*.[26] The sample analysed for pollen contains a great deal of charcoal. Pollen of trees amounted to 48%; herbaceous plants, 52%. Birch 20%, alder and oak 11% each, juniper 3%, hazel, hornbeam, beech, willow and bog myrtle, 1%. Heather 22%, grasses 14%, cereals 7%, arable weeds 5% (*Cruciferae*, mugwort, dock, narrowleaf plantain, black-bindweed). Goosefoot and sea plantain may indicate saline conditions. Local: horsetail 22%, burr-reed and dryopteris fern 1%, *Sphagnum* moss 5%. Indeterminates: 1.3 %.
The moss must have been gathered in oligotrophic peatland. The pollen spectrum resembles that of the caulking from the Nijkerk-II cog (Section 6.2.2), but the relative percentages of the species are very different. This sample must have come from a less open landscape, and a smaller area of peat.[27]

Frames: (V-shaped) floor timbers and futtocks

The rigidity of the ship's hull is determined to a large extent by the frames (figs 9.9a, b, c and d). For the reconstruction (Plates 9.II B and 9.III A and B), it was assumed that there must have been 23 in all (0 to 19 and -1 to -3). In the more-or-less flat part of the bottom, the frames are composed of asymmetrical floor timbers, which are extended on both sides with a futtock up to the top of the gunwale (strake I). In the field drawings (Plates 9.I A and C) it was not always clearly indicated whether the flat scarfs on the floor timbers still were present and intact. From the photos and the nails visible in them for fastening the nib ends of the futtocks it could yet be determined where the scarfs had been in almost all of the recovered frames. In principle the boatbuilder alternatingly placed the curved part of the floor timbers to port and to starboard. Through the differences in length of both the straight bottom part and the curved part up the side, the scarfs to port and starboard viewed from above appear in a staggered pattern (Plate 9.II B). They sit above the B to F strakes. All of them are 10-12 cm thick and wide, and according to the excavation report they were "mostly irregular in shape". The report also states that they were "attached to the keel plank with oak treenails and similarly onto the shell of the ship by one or two treenails to every strake."

Van der Heide's drawing shows frames 1 and 14 treenailed also to the hooks, but this was never the case. From the preserved frame 4 it could be determined that the oak treenails had a diameter of 2 cm. The remark that the planks had been fastened with one or two treenails seems barely plausible. The planks were at least 20 cm wide throughout, for which at least two treenails would be necessary. Such a misobservation may, as remarked before, be due to poor cleaning of the wood, leaving many treenails unnoticed. However, it is not impossible that the observation was correct. Section 9.8 will return to this topic. Aft from frame 3 and forward from frame 12, the floor timbers are V-shaped. Up to and including frame -1 (stern) and 16 (bow), these timbers most probably were extended on both sides with futtocks up to the gunwales. The other frames on the stern and stem comprised an asymmetrical V-shaped floor timber together with a short futtock on one side, or just a crooked timber (the uppermost frames). No V-shaped floor timbers were fastened to the hooks or stern- and stempost. The purpose of the above-mentioned 3 cm thick beechwood treenail in frame 4 above the keel plank was to attach the skeg. On the underside of the (V-shaped) floor timbers, rectangular limber holes had been cut out, about 10 cm wide and 2-3 cm high. In principle, four above the bottom, two above the seams between the garboards and the keel plank, and on either side of these, one above the second or third strake. In the V-shaped floor timbers in the stern and the bow, we see a roughly triangular limber hole on one side. The recovered V-shaped floor timbers did not rest upon the hooks.

25. Vlierman 1996a, Table I and 76, fig. 42.
26. Cappers *et al.* 2000, 578-579, Table 2 and 588, no. 15.
27. Mook-Kamps 1997, 10; see also appendix.

Fig. 9.9a. The frames, the keel plank and shell strakes in the afterbody.

Fig. 9.9b. The frames, the keel plank and shell strakes in the forebody.

Fig. 9.9c. The V-shaped floor timbers 13, 14 and 15, the stem hook and the shell strakes of the underwater body.

Fig. 9.9d. Conserved floor timber.

Mast step

Van der Heide in the excavation report writes: "No mast step was recovered nor any indications of where one might have been. Neither are there any indications of a keelson". In the overview drawing, however, he drew two treenails above the keel plank, in frames 11 and 12 and one treenail plus a hefty nail, standing about 12 cm proud of the frame, in frame 10. There is no doubt that this was a sailing vessel and hence it seems quite likely that the mast stood in a short mast step, which had been fitted with a few (blind?) treenails and a nail onto (and between?) frames 10-12. Curiously, there was no suggestion of a recess for the heel of the mast in a suitably placed frame timber. A long keelson with a broadened part for the mast step seems very unlikely.

Ceiling

The excavation report states that no ceiling boards were encountered, but that there were some fairly high protruding nails in particular places on the floor timbers. These nails, judging by the photos, do not belong to the attachment of ceiling boards, but as mentioned above had served for fastening the nib ends of the futtocks on the floor timbers. The 3 cm thick beechwood treenail in frame 4 above the turn of the bilge (Plates 9.II B and 9.III A) suggests that this was the attachment of a bilge ceiling board. Given the staggered scarfs in the frames and the thin shell planks, it is almost certain that the vessel had relatively heavy bilge ceiling strakes in order to give the ship sufficient stiffness in combination with the frames. In fact, an indispensable structural element. We may assume that these bilge ceiling strakes at any rate were part of an open ceiling.

Very probably, the short mast step had been extended with a plank fore and aft, and on either side of these, between the mast step and the bilge ceiling boards, there also was a ceiling strake. Halfway up the sides there may have been another one, while along the gunwales running from the stern- to stempost another (heavy?) plank would have been fixed (Plates 9.II B and 9.III A and B). Given the light shell planks (just 2 cm thick) and the relatively light futtocks, this top ceiling strake may instead, and indeed more probably, have consisted of a covering inwale on the frame heads, or an inwale triangular in section. Indeed it seems likely, given the small dimensions of the stern- and stemposts, that this timber continued along the inside of the stern- and stempost.[28] The various construction options are represented in plates 9.I.II and III C. A construction like that of the Marknesse cog might be another good alternative.[29]

Deck beams, sail beam, decks and gangways(?)
On the basis of what was recovered, nothing can be said about the construction at deck level. Through comparison with other Dutch finds it does however seem possible to venture a reconstruction. The mast without doubt will have been supported by a mast beam or sail beam. How this was attached to the sides is hard to tell. Most probably this beam was joined to the frame and/or ceiling timbers, and further upwards treenailed to the sides by means of knees. Indeed it seems unlikely that the beam was fastened through the sides with rose bolts - though it cannot be ruled out. Judging by the probable height of the sides and the dimensions of the small ship, it is almost certain that the cog had a foredeck or cuddy and an afterdeck. The planking of the almost certainly 1.8 m long afterdeck and the ca 3.3 m long foredeck would have been supported by probably four and seven transverse beams, respectively. Most probably these beams, large and small, were fastened by oblique treenails to the ceiling strakes and to the frames. For the reconstruction drawings it was assumed that the vessel had no gangways, but these cannot be entirely ruled out. If there indeed were any, they would have run between the fore-and afterdeck at about 70 cm above the top of the floor timbers. Compare the Marknesse and Spakenburg-I cogs on this point.[30]

Pump(?)
No indications of a pump were encountered. The sides of the vessel were about 1.5 m high and any bilge water may have been disposed of by means of a wooden bailing scoop.

Castles
The vessel never had a fore- or aftercastle.

Repairs
In the starboard E strake between frames 9 and 12, there are two aligned small planks, most probably to replace a bad part of this foremost shell plank. Between frames 5 and 6 to port, a plank was nailed on the inside over the seam between the C- and D strakes; while to starboard four nails between frames 2 and 3 on the A- and B strakes suggest that a small plank may have sat there too. Maybe this was an area of wear, where a great deal of bailing had occurred? Already it has been mooted that the beechwood keel plank (actually almost a keel *beam*, as it is five times as thick as the shell planks) was a replacement. The difference in sintel (nail) clamping intervals, together with the moss clamping on both the inside and the outside, supports this suggestion. Some hollow frame timbers, the generally poor quality of the wood, and the worn-down skegs suggest that the boat had reached a considerable age by the time it sank. Indeed it is possible that no skegs were fixed there to begin with. The excavation report does not record any repairs with sintel (nail) clamped moss caulking in (drying-out) cracks in the shell planks.

Wood use
Apart from the remarkably thick beechwood (*Fagus*) keel plank and the thick treenails of the same material, all recovered wood used in the vessel is oak (*Quercus* sp.). The thin planks were probably sawn, and hewn with a broadaxe, but alternatively may have been split from a log. The floor timbers and stern- and stem hook were hewn from naturally grown crooks; the futtocks probably from thick limbs. Also the thinner, 2 cm thick treenails and the moss laths were of oak. In the construction of the shell, it is most likely that the principle of a 'brother-and-sister' pattern was applied.

Fastenings
Apart from the 'pointless', thick treenails in the keel plank, no treenails were observed between the frames that might point to the use of temporary frames. Eleven frames had in principle been fastened by two, 2 cm thick oak treenails to each plank width and by a single one to the keel plank. This means 37 treenails per full frame, the number gradually decreasing towards the bow and stern. In all, a good 800 treenails will have been used for the entire hull and for fastening the crossbeams. It was not established -or recorded- whether these treenails were

28. To this end, compare the small boat K 73/74, Reinders *et al.* 1980, 17-29 and plates 4 and 5.
29. Section 12.2.2.
30. Sections 12.2.2 and 14.2.2.

thicker on the outside of the shell and had somewhat protruding heads, or were simply flush and secured with wedges. The former seems the most likely, but this may have been overlooked owing to abrasion. The skegs (with a plank to connect them?) had been attached with 3 cm thick treenails and some nails. So probably were the bilge ceiling boards. The other ceiling boards were fixed maybe with a few treenails (together with the frames?) combined with some nails.

If there indeed was a covering inwale, this too will have been fastened with treenails. In as far as could be established, few nails were used in the vessel. Just four to fasten every plank scarf. The hood ends of the strakes and the parts of the garboards that were nailed into the rabbets of the hooks displayed rather irregular nail spacing (4-12 cm). For the scarfs in the shell planks and the keel plank, for fixing the strake ends to the stern- and stem hooks and for attaching the futtocks to the floor timbers, overall no more than 300 nails will have been used. If the clinker-built strakes F to I were joined with double-clenched nails, they probably sat at intervals of about 15 cm. Which would require about 6000 nails for this upper part of the shell.

If the attachments had been carried out with treenails 1.5-2 cm thick, intervals of ca 20 cm seem more likely. About 4500 of these would have been needed. The recovered nails were 50-55 mm long, square in diameter with 3-4 mm sides, and had round, flat heads, 16-20 mm across (fig. 9.8). The sintel (nail) clamps probably had an average spacing of 6 cm. In the seams between the keel plank and the garboards they were both on the inside and the outside of the shell; in those between the A and B strakes up to those between the D and E strakes, probably only on the exterior. In the clinker-built construction of the F to I strakes, presumably on the interior only. This means some 200 m of caulking overall, requiring ca 3,300 sintel (nail) clamps. If the clinker seams had sintel (nail) clamped moss caulking on the outside too, and the seams along the ends of the bottom strakes towards the stern and stem as well, there must have been some 1,500 more.

9.3 Construction data

The recovered timbers yielded sufficient evidence to allow reconstruction of the underwater body. Thanks to the preserved hooks, the rake of the stern and stem could be determined. The less than accurate field drawings made this more difficult for the transverse sections. Yet the preserved frame 4, together with the drawn profile dams, made it possible to make reliable estimates of (and to fair) the plank widths and lines. Initially it was assumed that the gaping seams between the strakes in the field drawings were due to careless drawing. However, closer examination has shown that in fact they are likely to be a (approximately) true representation.

9.3.1 Design

The little ship without doubt was built according to rules of thumb familiar to the boatbuilder, and hence suggests that the design must have been a well-established one.

Hull shape

The reconstructed lines drawings (Plates 9.II A, B and D) show a hull with a straight keel and a sheer that towards the bow very probably rose higher than towards the stern, while the sides are unlikely to have been very high. The vessel had straight, inclined stern and stem, whose rake aft and fore did not differ much (122° and 119°, respectively). If the front of the false stem is considered, the difference in rake fore and aft is in fact negligible. On the much younger civic seal of Genemuiden (presumably 1461; fig. 9.10) and on the seal of Kuinre of 1399-1407 (Chapter 13, fig. 13.7) we see cog-like vessels, also with virtually equal rakes of the stern and stem.

Fig. 9.10. The civic seal of Genemuiden (the Netherlands), presumably 1461. After Ewe 1972, 129-152.

The bottom is almost symmetrical in shape, has a longitudinal deadrise, and has garboards affixed at a slight angle and towards the midships a few centimeters below the top of the keel plank. The subsequent bottom strakes slightly rose towards the turn of the bilge at a different, yet slight and constant angle. The bottom is relatively wide amidships, but becomes very sharp aft and fore. Amidships the bilge strakes and the subsequent E and F strakes were affixed at slightly different angles each time. The uppermost three strakes (G-I), very probably were everted in an almost straight line with a rake of ca 112°. Towards the stern and stem the turn of the bilge becomes progressively less distinct and the hull section rounder, in the sharpest-built parts in the bow and stern, slightly S-shaped. The first five strakes almost certainly had been entirely carvel-built, possibly the edges of the abutting strakes had been somewhat bevelled, but there was no suggestion of any true overlap. The four subsequent strakes most probably were clinker-built. The vessel's shape suggests, in part owing to the (later inserted) heavy keel plank and the skegs, that it may have been an easily handled little sailing ship.

Scantlings

There were no traces of temporary frames or any other details relating to the early phases of construction. On the basis of the shape and measurements of the wreck and the reconstructed hull shape, we should be considering the scantlings listed in Table 1.

Layout and tonnage

Little can be said about the layout. Yet it is almost certain that the vessel had an open ceiling, which, on either side of a short mast step extended with a plank aft and fore (not drawn), may have comprised four strakes. The bilge strake most probably was somewhat thicker than the others, and so may the uppermost ceiling strakes. The latter may in fact have been a covering inwale with a square section, or an inwale of triangular section. Yet another possibility is that it had just bilge ceiling strakes with a loose plank-bridge between them, together with a reinforcement of the gunwale as mentioned above. The cog almost certainly had a small deck or cuddy at the bow and stern; to interconnect them, gangways may have been present. At deck level, between the two decks (and any gangways) the vessel most probably was open. Assuming that the 'hold' was the space between the fore- and after decks and that the cargo was stowed no higher than deck level, the following calculation will produce the ship's potential tonnage. The length between the fore and aft decks is 4.6 m, The width of the hold at the foremost edge of the after deck at deck level is 2.4 m, and at the bottom (or the plank-bridge), 1.4 m. Amidships these widths are 2.8 m and 1.8 m; at the after edge of the foredeck, 2.8 m and 2 m, respectively. This means that the average width was 2.2 m. The maximum height of the cargo in the proposed situation would have been about 0.8 m. Thus the capacity of the hold was 4.6x2.2x0.8 m = 8.1 cu m, which amounts to ca 8 gross tons.

Material selection

The Ens cog was entirely built of oak wood. The remarkable beechwood keel beam is most likely to have been a later replacement. The shell strakes, just 2 cm thick, were probably sawn, but may also have been split. The recovered frames were relatively thin and broad and somewhat irregularly shaped. The fastenings of the frames to the shell were executed with 2 cm thick oak treenails;

Table 1. Principal dimensions of the Ens cog

Length of the keel (including stern- and stem hook)	8.7 m
Length of the bottom (between the rising parts of the stern and stem hook)	8.4 m
Greatest width of the bottom (seam between the C and D strakes at frames 7-10)	1.5 m
Overall length at gunwale level	11.2 m
Overall width at gunwale level (at frames 8-9)	3.5 m
Height of the stern (keel line to top gunwale level)	1.9 m
Height of the stem (keel line to top gunwale level)	2.5 m
Height of the sides amidships (at frames 4-9)	1.5 m
Length of the stern	2.2 m
Length of the stem	2.8 m

Length-width ratio of the bottom	5.6:1
Length-width ratio at gunwale level	3.2:1

the attachment of the skegs and probably also the bilge ceiling boards, with 3 cm thick beechwood treenails. The recovered part of the ship contained few iron nails, and as larger wrought-iron elements probably just a gudgeon and a bolt and eye for the upper suspension of the rudder. The iron sintel (nail) clamps are fairly small and irregular in shape; they were used both on the inside and the outside of the seam between the (probably replaced) keel beam. The other flush-laid seams of the shell were caulked only on the outside; the clinker seams were caulked on the inside and/or outside. The caulking material was peat moss (*Sphagnum*).

9.3.2 Construction: techniques and sequence

The typical construction of stern- and stem hook connected by a keel plank, and a relatively wide bottom amidships that narrows sharply towards the stern and stem, was applied also in the Ens cog. Remarkably, not only the usual part of the bottom was carvel-built, as is usual in almost all cogs found to date in the Netherlands, but all three bottom strakes on either side of the keel up to the hooks and stern- and stempost were attached in this fashion. On top of this, the same goes for the subsequent D and E strakes of the sides. All seams between these five strakes slightly gaped on the outside, were triangular or trapezoid, and had been closed from the outside with sintel (nail) clamped moss caulking. The remaining strakes (F to I) most probably were fully clinker-built. The hood ends of the carvel-built planks were set into rabbets in the stern and stem (the F strakes only partially), while the subsequent strakes ran on to the rear and front of the stern- and stempost, and on the bow very probably were protected by a false stem.

Construction phase 1: keel plank/keel beam, stern- and stem hook, stern- and stempost, temporary frames

The building of the vessel without doubt started with the joining of the stern- and stem hook to the (original) keel plank, onto which the stern- and stempost were set and propped up. In the scarfs, a thin layer of moss caulking was applied. No evidence of the use of temporary frames was encountered, but some must have been used on the almost flat, carvel-built part of the bottom; alternatively, some definitive floor timbers may have been inserted from the start.

Construction phase 2: bottom strakes, floor timbers 4 to 12, some futtocks and the lowermost two side strakes

To achieve a degree of rigidity, the fully carvel-built bottom strakes would have been provided with at any rate the floor timbers 4 to 12. Floor timber 4 reached up to about the middle of strake F; the others ended lower, so there must have been at least some (temporary?) futtocks in order to affix the carvel-built lowermost two side strakes.

Construction phase 3: Clinker-built strakes of the sides, clamped moss caulking, V-shaped floor timbers and futtocks.

Once the carvel-built underwater body had been constructed and faired and was held in shape with frames, the four strakes to be attached in clinker fashion were added and their seams caulked. After this, the crooks for the aft and forward floor timbers were shaped and put in position, followed by any further futtocks. Should this ship be similar to the Kollerup cog, the sides from strake F upwards may have been built carvel-fashion like the bottom.

Construction phase 4: Ceiling boards, mast step, covering inwale, deck beams, decks/cuddies, gangways(?), pump(?), sintel (nail) clamped moss caulking, false stem, rudder, mast and spar

In this final phase of construction, the ceiling strakes and the mast step were installed. The bilge ceiling boards, which technically were needed for achieving fore and aft rigidity, could not be inserted at an earlier stage, because first the futtocks had to be fitted. A massive ceiling strake, a covering inwale would keep the gunwale in shape. In this phase too the deck beams, the decks and any gangways and pump were put in place. Finally the shell seams of the underwater body and maybe also the sides were sealed with sintel (nail) clamped moss caulking. The false stem was put on the bow, the rudder hung on the stern, and the mast and spar set up with the rigging.

9.3.3 characteristics

The vessel possesses the following features.

Keel plank/keel beam and skegs

The beechwood keel beam has a thickness of 10 cm (five times that of the shell planks). In the first instance, the cog probably had an oak keel plank rather than a keel beam. The beam was 10 cm wide fore and aft, amidships, 20 cm, it was rectangular in cross-section. The keel beam was fastened to the stern- and stem hook with flat scarfs.

Hooks

The stern- and stem hook had been hewn from naturally grown crooks. Both in the horizontal and in the rising part, they are trapezoid in section, with the broadest surface inward. The rising parts end in an oblique surface: the scarfs to the stern- and stempost.

Sternpost, stempost, false stem and rudder
The stern- and stempost were fixed to the oblique tops of the hooks by just a few iron nails. Upwards they tapered, and at the gunwale they were no more than 6 cm thick. The hood ends of the A to F shell strakes ended in rabbets in the sides of the hooks and posts; the higher strakes ran on to the rear of the sternpost and the front of the stempost. The ship had no false stern. However, a false stem at the bow seems desirable to protect the hood ends of the uppermost strakes and to reinforce the thin stempost. The most obvious solution would be that the false stem started a little below the scarf with the stem hook, and presumably projected above the gunwale. The iron gudgeon showed that the vessel had a stern rudder. The rake of the stern and stem was virtually identical.

Carvel-built underwater body
The three bottom strakes and the two first strakes of the sides were fully carvel-built fore to aft. The garboards sit at a slight angle to the keel beam, joined to it a few centimeters below the surface of the keel beam. Both longitudinally and transversely, the bottom has a deadrise throughout. The seams between the carvel-built strakes slightly gape.

Clinker-built upper part of the sides
The top four strakes of the sides very probably were clinker-built. Owing to the absence of the sixth (F) strake and the failure during excavation to observe nail holes or nails in the top edges of the surviving parts of the E strakes of the forebody, there is no certainty about the joins between the F to I strakes. It seems likely that there were double-clenched nails. Yet it cannot be ruled out that the fastenings were made by means of treenails 1.5-2 cm thick, or by a combination of both.

Ceiling and mast step
The vessel presumably had an open ceiling, consisting of four strakes on either side of a short mast step, which probably sat above frames 10-12. The bilge ceiling boards probably were a little thicker than the other ceiling boards, while the upper edge of the gunwales was reinforced with another thicker timber, a covering inwale of triangular section. Just bilge ceiling boards with loose plank bridges and a reinforced gunwale is also quite possible. The mast step and the bilge ceiling boards may have been notched here and there to accommodate the frames, engaging with them.

Caulking method
The seams of the carvel-built underwater body (strakes A to E) were sealed with sintel (nail) clamped moss caulking only on the exterior. The seams either slightly gaped or the planks touched each other just on the inside of the shell. The triangular or trapezoid gaps were entirely stuffed with strands of moss and covered over with thin laths of oak, which at intervals of ca 6 cm were held in place with sintel (nail) clamps (fig. 9.8). The seams between the clinker-built strakes had presumably been caulked on the outside, and maybe on the inside as well.

Decks, crossbeams and gangways
The cog without doubt had decks both fore and aft. The necessary deck beams, large and small, without any doubt were attached to the frames by treenails. There was no suggestion of protruding crossbeams. Still there may have been some iron rosebolts holding crossbeams in place from the outside of the shell. However, a connection of crossbeams to the gunwale by means of standing knees seems more likely. Whether the decks were interconnected by gangways is hard to tell. Judging by the place of the flat scarfs on the floor timbers, and the dimensions of the vessel, it seems rather unlikely that the Ens cog had stringers in the sides at the transition from the carvel- to the clinker-built parts, as were observed in the Kollerup and Kolding cogs.[31] Yet it cannot be ruled out altogether.

Fastenings
The frames are attached to the shell planks by means of 2 cm thick oak treenails. It was not observed or recorded whether the treenails on the outside had somewhat thicker, rounded heads that stood proud of the shell, or whether wedges had been used to secure them. The thicker beechwood treenails with oak wedges were used for fastening the (later added) skegs and possibly a timber joining the skegs together. Such treenails presumably were used also for fastening the bilge ceiling strakes (which too may be later additions or replacements?). The use of iron nails in the hull was very sparing, just four in every plank scarf. How the strakes of the clinker part of the shell were joined together was not established. The use of double-clenched nails seems likely, though fastening with treenails is also a possibility.

9.3.4 Use

Function
The slight dimensions and light execution of the construction suggest that the vessel was intended for passenger transport and light cargoes. No clues point to use

31. Crumlin-Pedersen 1979, 30-31 and fig. 2.12.

as a fishing vessel. The cog was certainly suitable for use on inland waterways and the Almere lake. In view of its small dimensions and light construction its suitability for navigation over greater distances across the Waddenzee is doubtful.

Wear and repair
The (quite severely) abraded skegs, the rotten frame timbers, and the generally poor condition of the wood, the probably replaced keel plank (and maybe the bilge ceiling boards too), and the several repair planks suggest a relatively advanced age of the vessel by the time it finally sank.

9.3.5 Abandonment(?), wreck formation and embedding

The cog's probably considerable age and poor condition make it likely that it was abandoned. The total absence of the higher and/or any detached parts of the vessel, and the way it lay embedded within the Almere deposits suggest that the findspot was not the place where it originally foundered, or maybe had been cast away. At any rate when the river ran high with a much increased flow rate, such a flat and relatively lightly built craft may have been displaced over a considerable distance.

9.4 Model

No scale model of the Ens cog has been made (so far).

9.5 Inventory

In or around the wreck no finds were recovered, apart from an erratic pebble. All nails, sintel (nail) clamps, moss-lath fragments and moss caulking recovered from the wreck itself are mentioned in the list of finds.

9.6 Cargo

No traces of any cargo were found.

9.7 Ballast

The wreck held no ballast.

9.8 Dating

Soil profile
In section 9.2.1 the dating of the soil profile has already been discussed in detail. Close dating of certain sedimentary layers in the former Zuiderzee floor remains a difficult matter and mainly depends on (closely) datable objects found within them, such as the contents of shipwrecks. For instance, the discovery on an Oude Sloef sand deposit of a typologically dated sarcophagus lid of the second half of the 12th century[32] does not necessarily mean that that layer dates from the 12th century. The layer might be (much) older, or the artefact may have ended up there much later. All it means is that the sand layer does not date from after the second half of the 12th century. The first, undisturbed layer deposited over an artefact indicates the *post-quem* date. The *Valvata* layer at any rate dates from after this period. Besides, only in the case of an (presumed) even sedimentation rate can an assessment be made of when a vessel sank. In the southeast of the Noordoostpolder, the occurrence of the *Oude Sloef* or Almere deposits between the *Detritus Gyttja* and the *Valvata* layer is limited to the areas above the Pleistocene channels.[33]

Fig. 9.5a clearly shows that above the disturbance caused by the wreck the *Almere* (or *Bandjes Sloef*) deposits are alternatingly more or less undulating. The strongly humic layered deposit into which the wreck sank displays a more even sedimentary pattern. The vessel or wreck probably ended up here during an episode of (much) increased fluvial activity. Subsequently there was a prolonged period of variable, at times violent sedimentary activity, during which besides the deposition of new layers also old layers may have been cleared away. Dating the sinking of the wreck to within a relatively short period on the basis of the soil profile therefore remains a fraught endeavour; in this case it can only be stated that it happened roughly between ca 1250 and 1350.

The heightened hydraulic activity apparent from the soil profile and presumable date of the Ens cog's final sinking to the Almere lake bottom puts into doubt the possible link (as proposed by Ente[34]) between the storm flood of 1170 and the increased activity of the IJssel and as a possible causative agent in the breaching of the peaty barrier

32. Ente 1973, 146.
33. Ente 1973, 145-146.
34. Ente 1973, 146-147.

between Schokland and Elburg.[35] It seems more likely that the increased activity of the IJssel, which downstream of Kampen sought an exit in a NNW direction mainly through several distributaries, must have occurred at a later date (*viz.* at the time of, and after the sinking of the Ens cog). Maybe the *Bandjes-Sloef* deposit can be linked to this heightened hydraulic activity. Taking into account the presumably quite advanced age of the vessel at the time of its sinking, dating its construction (see below) cannot yield more information than that this hydraulic activity occurred a considerable time after that date.

Sintel (nail) clamps

No objects of any inventory were found in or near the wreck that might allow a closer dating of its sinking. The encountered transitional type (C2/D) and the sintel clamps D and D(1) are dated between the second and the final quarters of the 13th century.[36] Judging by other, well-dated finds,[37] the small and rather variably shaped central parts of the sintels appear to have been common from the first quarter of the 13th century, and especially around the middle of that century.
Recently (end of 2006) the wood of the dugout vessel Meinerswijk III[38] could be dendrochronologically dated. The wood must have been felled (the vessel must have been built) in the first quarter of the 13th century.[39] In this boat, and in the punt-like craft found nearby, the sintel nails of types B and B/C were used. In some places in the dugout boat, clear traces were found that pointed to repair to the sintel nail clamped moss caulking. This can also be inferred from the two types of sintel nail encountered. Besides, various imprints were found where sintel nails had once been.[40]

Taking into account the fairly rapid change in shape of sintel nail to sintel clamps around the 1150-1225 period (type D probably was in general use by ca 1250[41]), it may be stated that the Ens cog was most probably built in the second quarter or around the middle of the 13th century. The general condition of the wreck suggests an age of ca 50 years or more, which would mean that the wreck must have sunk into the floor of the Almere lake towards the end of the 13th century or in the early 14th century.

Dendrochronology

No dendrochronological analysis was performed. During the drawing of the stern- and stem hook and frame 4 in February 2008, it was established that all three pieces might yet allow such analysis, and that there even is some of the bark in the hooks, required for a precise date, and indeed sapwood in the frame. The drilling of samples may be performed almost without damage and be easily repaired. Tree-ring analysis is of crucial importance for identifying the provenance of the timber (the ship), and to see whether the tree rings match the previously assessed date.

Parallels to other 12th/13th-century finds of ships and ship's timbers

A fully carvel-built underwater body is know also from the Nijkerk-I cog (Chapter 8) and of the Danish Kollerup and Kolding cogs.[42] The two Danish ships date from the end of the 12th century. It is very well possible that the clinker construction of the Ens cog was made with thin treenails. Such fastenings were applied in the recovered punt-like boat from Meinerswijk[43] and vessels of unknown type from Hattem, Amsterdam and Rotterdam.[44] Trapezoid, and subsequently triangular, 'open' seams with sintel nail clamped moss caulking between the carvel-built planks, and the use of mixed thin and thick treenails and their irregular application (one or two treenails per plank width), are known mostly from barge-like river-craft of the 10th to 13th centuries. The trapezoid seams were used up to ca 1200 (or maybe a little longer locally). During the latter half of the 12th century also seams appear that are triangular in section.[45]

The drawings of the punt-like vessel from Meinerswijk indicate that part of the bottom seams had been reamed into a triangular profile.[46] The study of the Ens cog prompted a closer look at the preserved moss caulking from that punt-like vessel. It then became clear that its seams were not triangular, but fully trapezoid and open, with a greatest exterior width of ca 1 cm, and 0.1 - 0.3 cm on the inside of the shell. Triangular seams, covering just half or less of a plank's thickness, presumably first appeared around the mid 13th century. Nails with round-

35. See also the Dronten-I cog, section 11.1.
36. Vlierman 1996a, 76-42.
37. Idem, 68-70, 73 and 74.
38. Reinders 1983, 21-35.
39. RCE file, Lelystad.
40. See Reinders 1983, 26 fig. 19.
41. Vlierman 1996a, 51-52.
42. Crumlin-Pedersen, 1979, 31 fig.2.12; Westphal 1999, 104; Hocker 2000, 50-55.
43. Reinders 1983, 15-20 and Appendix 1.
44. Vlierman 1995, 91-104; Vlierman 1996b, 115-119 and appendix 15; Van Holk 2001, 103-117; Vlierman 2002, 133-134.
45. Vlierman 1996a, 26-28, 53 and fig. 11.
46. Reinders 1983, plate 1D.

heads, and especially with thin, flat heads are known so far only from vessels of the 13th century or older. The early-13th-century punt-like boat from Meinerswijk contained not only nails with flat heads, but also specimens with somewhat convex (pyramidal) heads.

9.9 Summary

Ship

The relatively small vessel from lot Q75 in the Noordoostpolder, excavated in the early years of nautical archaeology in the former Zuiderzee area, has for many years been an important find in international theoretical thinking about the evolution of the cog. Its merely 2 cm thick shell planks, its frames just 10-12 cm thick and wide, and its thin stern and stem indicate that it had been a fairly light ship, built of oak. The rake of the stern and stem was nearly identical. Of particular interest was the fully carvel-built underwater body, which until then was known only from the late-12th-century Kollerup cog, the Kolding cog of ca 1250, and from the Nijkerk-I cog of the second quarter of the 13th century. In the first two ships, however, the ends of the seams towards the stern and stem are in fact clinker-built, which may be due to the considerably larger dimensions of these ships and their greater plank thickness. The Kollerup cog has E strakes which at the top are thicker than the rest of the strake. This produces a quasi-clinker join to the F strakes. Although unlikely, we might consider clinker construction also in the equivalent place in the Ens cog, with the subsequent strakes higher up (also) attached clinker-style. However, it is not altogether impossible that the uppermost strakes of the vessel, as in the Kollerup cog, had been built in carvel fashion.

The shell strakes were each fastened to the frames with two oak treenails just 2 cm thick, sometimes with only one. The carvel seams gaped, and had been closed from the outside with sintel (nail) clamped moss caulking. The clinker seams between the strakes mutually had been secured either with double-clenched iron nails or with thin treenails. The seams between these strakes had been caulked on the inside, but may have been caulked on the outside too. At the scarfs, the plank ends stood proud of the shell by about 0.5 cm both on the inside and the outside, and were fastened with just 4 nails in the scarfs. The boat had no false stern, though presumably it did have a false stem on the bow. The topmost four clinker-built strakes ran on to the front and rear of the stern- and stempost. It is fairly certain that the vessel had cuddies aft and fore, and a mast that probably stood in a short mast step that was fixed to the frames. That it carried a square sail would be obvious. Most probably the hull had an open ceiling, which at the top consisted a thick plank, a shelf beam, a covering inwale or an inwale. The bilge ceiling strakes most probably were a bit thicker than the other ceiling boards, and like the mast step, may have been joggled to fit the frames. Instead of ceiling boards, the vessel may have had a loose plank bridge between the bilge ceiling strakes.

The cog was kept on course with a stern rudder. Together with the Nijkerk-I cog, it is the (Dutch) archaeological proof that the stern rudder existed even before the mid 13th century. The curious beechwood keel beam probably was a replacement, while the two oak skegs have not been observed on any other (Dutch) cog find. The ship most probably was used for passenger transport and light cargoes. No evidence was found to suggest fishing activities. The characteristic hooks at either end of a keel beam, and the shape and dimensions of the underwater body indicate that we are dealing with a small cog, suitable for inland waterways and Lake Almere/the Zuiderzee. Whether it was suitable also for undertaking longer voyages, sailing the Waddenzee, is debatable. The garboards were fixed in an unusual way below the top of the keel beam, by seams closed with sintel (nail) clamped moss caulking from above and below, whereas the seams of the rest of the underwater body were caulked only from the outside. This supports the impression that the beechwood keel beam with its oak skegs was a replacement and an alteration to the original situation.

The reconstructed principal dimensions are:

- Length over the stern and stem at gunwale level — 11.2 m
- Largest width at gunwale level — 3.5 m
  Height of the stern (keel line to the top gunwale) — 1.9 m
- Height of the stem (keel line to the top gunwale) — 2.5 m
- Height of the side amidships (keel line to the top gunwale) — 1.5 m

Dating

From the disturbance of the soil profile we can only establish that the vessel came to rest at the findspot roughly between 1250 and 1350. Presumably the wreck, floating under water, had been washed along from elsewhere as a result of increased fluvial activity, and, judging by its condition, repairs and incompleteness, might already have been abandoned or scuppered, at least 50 years old, by the time it sank to the bottom of Lake Almere near the then mouth of the Zwartewater or one of the contemporary IJssel distributaries. The types of sintel (nail) clamps suggest that the cog was built in the second quarter or around the middle of the 13th century. This means that the intensified fluvial activity of the IJssel must have occurred considerably later than Ente[47] assumed. The 'gaping', triangular caulking seams in the carvel underwater body and the slight number of iron nails with flat, round heads also support a construction date in the 13th century. Tree-ring analysis of the preserved hooks and frame may one day reveal whether the construction date based on sintel (nail) clamp typology is correct, and even indicate the provenance of the timber (and the ship).

47. Ente 1973, 146-147.

The Rutten cog[1]

10.1 Find conditions

10.1.1 Discovery and investigation

On 23 October 1952, Mr R. Terpstra from Kamp Rutten reported the discovery of a shipwreck to Mr G.D. van der Heide of the Archeologisch Museum voor het Zuiderzeegebied at Middelbuurt, Schokland. It had been hit upon during the laying of drain 14 in lot A57 in the Noordoostpolder. Terpstra stated that the wreck lay embedded in peat. In his confirmation of the report, dated 1 November 1952, Van der Heide requested Mr A. Vos of the Nederlandsche Heidemaatschappij at Lemmer to leave the wreck as intact as possible, if necessary relocating the drain somewhat. Terpstra was asked to leave the wreck in place until a suitable moment for its excavation.

On 20 September 1982, the Museum voor Scheepsarcheologie at Ketelhaven received a report by telephone of a find of a shipwreck in lot A58 in the Noordoostpolder. The site was visited the very same day. However, nothing was known about such a find in this field. Yet the tenant farmer did remember that two years before, the neighbours, when improving their drainage, had bulldozed up some ship's timbers from lot A57. To clarify matters, a visit was subsequently paid to Mr D. Klijnstra, the tenant farmer of lot A57. In his yard lay a heap of ship's timbers (fig. 10.2), including some eye-catching ones: a near-complete keelson in three pieces, including a broader mast-step area with the cavity to take the heel of the mast, as well as some parts of massive floor timbers and a complete one. This was sufficient evidence to understand that here was a late-medieval ship of considerable dimensions. The 1952-file had not mentioned the nature of the wreck, nor had any exploratory investigation been carried out up till 1982. In consultation with Mr Klijnstra the site could be visited immediately and the wreck staked out. By probing, it was established that shallow-lying timbers were present over a length of 17 m and a maximum width of 5 m. From a small hole dug at the stern or stem, a proto-stoneware sherd was recovered; it was decided that we were probably dealing with the underwater body of a large and heavily built cargo vessel, probably a cog, dating back to at least the 14th century.[2] Soon after, the timbers languishing in the farmyard were transferred to the museum at Ketelhaven.

The excavation took place between 3 June and 21 August 1985. The shallow position of the wreck was a reason for fully excavating it manually. The staff of the department of Nautical Archaeology of the RIJP were during the first week assisted by a group of vocational pupils of the Copernicus comprehensive school at Dronten, supervised by two teachers (fig. 10.5), and later by volunteers of the

Fig. 10.1. General view from the stern with the repositioned frames, keelson and riders.

Fig. 10.2. The ship's timbers ripped from the wreck in 1980.

1. Translated by Xandra Bardet.
2. The wreck lay at an angle with respect to the boundary ditch A57/58, with the bow towards the southeast, and at ca 80 m from it. The stern lay at ca 90 m away, while the distance to the boundary ditch with the farmyard was 143.5 m (figs 10.3 and 10.4); Ordnance map, sheet Rutten 15H, coordinates 176.300/537.400, with the highest elevation at ca 3 m below NAP, Hydrographic map 1921, low water mean. The findspot lies ca 2 km from - and almost exactly south of - the harbour of Lemmer.

Fig. 10.3. The findspot of the Rutten cog in today's Noordoostpolder.

Fig. 10.4. Ordnance map of the Kingdom of the Netherlands, 1861 (scale 1:10,000), with the location of the Rutten cog.

Fig. 10.5. Removal of the topsoil by students of Copernicus comprehensive school at Dronten.

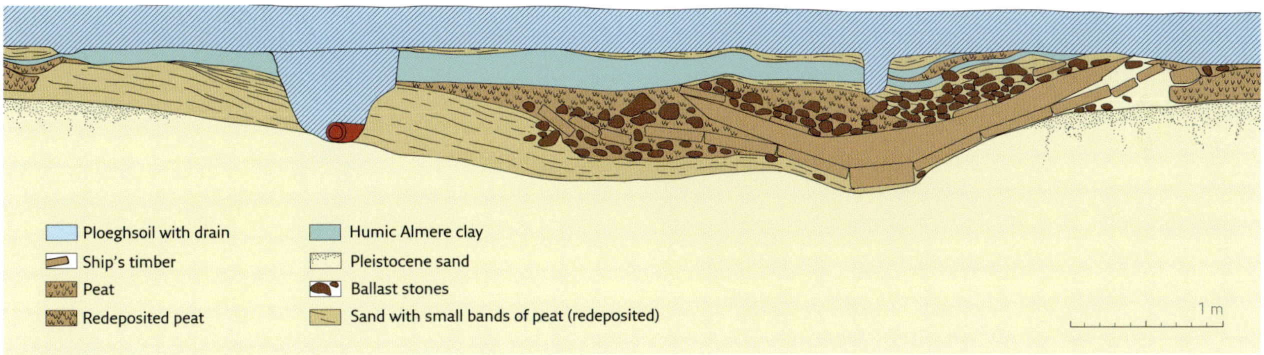

Fig. 10.6. Section showing the soil profile and the wreck with ballast stones in the afterbody.

antiquarian society 'Arent thoe Boecop' from Elburg. Two interns also took part in the field. One of them, A. Vos, wrote an internship report.[3] Initially one fore-and-aft and two transverse (profile) dams were left standing over the wreck, and the topsoil was dug away around them. The transverse sections were taken down further and drawn (fig. 10.6 and Plates 10.I B and C). The sections were analysed by R. Koopstra of the Department of Pedology of the RIJP and explained during an excursion of the staff of this department (figs 10.7a and b).

The fore-and-aft profile dam and the dam across the afterbody were then removed, after which the remarkable variegation of the soil was exposed by rabotage, revealing the part of the ship damaged by the bulldozer in 1980 (figs 10.8 and 10.9). In this phase a start was made on drawing the overview drawing (Plates 10.I A and 10.II A). After the soil disturbance caused by the wreck and the surrounding area had been documented in drawings and photographs, the levels overlying the wreck were taken down right to the timbers (figs 10.10a and b and Plate

3. Vos 1987.

Fig. 10.7a. Mr R. Koopstra of the pedological department of the RIJP, explaining the soil profile to his colleagues.

Fig. 10.7b. The author lecturing on the wreck.

Fig. 10.8. General view from the bow, during the first phase of rabotage, the removal of the longitudinal (profile) dam and the clearing of the part of the wreck damaged by the bulldozer in 1980.

Fig. 10.9. General view of the excavation showing the patchy pattern of sediment: among the dark peat, the pale sand disturbed by the sinking ship. Note the timber fragments left by the bulldozer.

Fig. 10.10a. General view of the excavation during the uncovering of the ship's timbers and the ballast stones.

Fig. 10.10b. The uncovering of the wreck, viewed from the bow.

10.II B), and the overview drawing was completed. Along one long side of the excavated area, pump pipes were installed, allowing the excavation to be kept dry during the fieldwork, to pump up water to keep the timbers wet and to clean them, and to fill the excavation up with water over the weekends. During the investigations, those parts that were not being worked on would be covered with wet jute sackcloth and plastic sheeting to keep the exposed wood from drying out.

After removing the ballast stones and detached construction elements and cleaning, photo's of the situation of the whole wreck were taken (figs 10.11a and b). Subsequently the frame parts, keelson and riders removed in 1980 were put back in their original places as much as possible (figs 10.1 and 10.12). After this, the longitudinal and transverse sections were measured in and drawn (Plates 6.II C and D) and from the top of a cherry-picker, photogrammetric recordings were made by the Meetkundige Dienst, Delft. After the frames were removed, the shell of the underwater body was documented and finally the keel plank and hooks. All ship's parts were transferred to Ketelhaven. Various finds were recovered from the wreck; they are described in section 10.5. The field research drew a great deal of attention from the regional and national press and the NCRV channel featured the excavations in its tv program *Achter het Nieuws* on 12 July 1985. As a result of all the media attention, the excavation was regularly visited by enthusiasts. The Glavimans Symposium 1985 on nautical archaeology in the Netherlands, and the 1985 Annual Report of the Nautical Archaeology subdivision of the RIJP both briefly gave attention to the excavation of the Rutten cog bottom.[4] The stoneware finds were discussed in an overview of stoneware recovered from excavated shipwrecks in the Netherlands.[5]

10.1.2 Geology and stratigraphy

The top view of the excavated area and the cross-sections (Plates 10.I A, B and C) show that around the wreck still parts of the Holocene peat layer had survived. The ship had sunk right through this soft peat and worked itself into the underlying Pleistocene sand to a depth of some 50 cm. It was not until after (most of) the timbers above the underwater body had gone that it was possible for regular sedimentation to occur inside the wreck remnant. On the bottom of the vessel this was mainly redeposited peat, alternating with sandy layers containing thin peaty bands. This material derived from the immediate vicinity. At a higher level, a layer of Almere clay was deposited in some parts. On the southwest side of the wreck it is clear to see that the remaining areas of peat and the sand/clay deposits were disturbed in the cavity created by the sinking wreck, and in some parts were destroyed altogether. These gaps in the peat and in the deposit of Almere sediments became filled with sandy Zuiderzee sediment, interspersed with seashells. This may mean that most of the peat on the west side of the wreck was not eroded until the 17th century or later. The layers of Almere sediment between the surviving 'islands' of peat on the northwest side of the wreck

Fig. 10.11a. General view from the bow: the underwater body and the area of the frames ripped out by the bulldozer in 1980.

Fig. 10.11b. The same, viewed from the stern.

4. Oosting 1987b, 57-63.
5. Vlierman 2010, 192-2.4.

Fig. 10.12. General view from the stern with the repositioned frames, keelson and riders.

revealed a distinct pattern of (strong) currents from the south. Owing to the nearby coast, the sediment sequence is incomplete, and hence provides no closer dating evidence than that the ship must have sunk before ca 1600.

10.1.3 The distribution of land and water; medieval occupation

On the distribution of land and water in the medieval Noordoostpolder area, Wiggers writes that in the 9th century land extended at least 2 to 3 km south of the present town of Lemmer.[6] The area west of this extended even more than a kilometre further southwards, and about 2 km south of Lemmer the coastline ran southwards to the present town of Emmeloord. The findspot of the cog lies in the area that in the 9th century and probably well into the 12th century still was dry land. The vessel presumably became stranded close to the mouth of the peatland channel on the bank of which lay medieval Lemmer (*Lenna*). Medieval potsherds have been found in some places in this original land. Such was the case at ca 1.5 to 2 km west of the cog findspot; in Section K, 5.5 to 6.5 km southeast of it; at ca 8 km south of the wreck; and a little northeast of the modern village of Espel (fig. 10.13). The clay-on-peat land probably was inhabitable only from the 9th century onwards. The erosion of the inhabited landscape

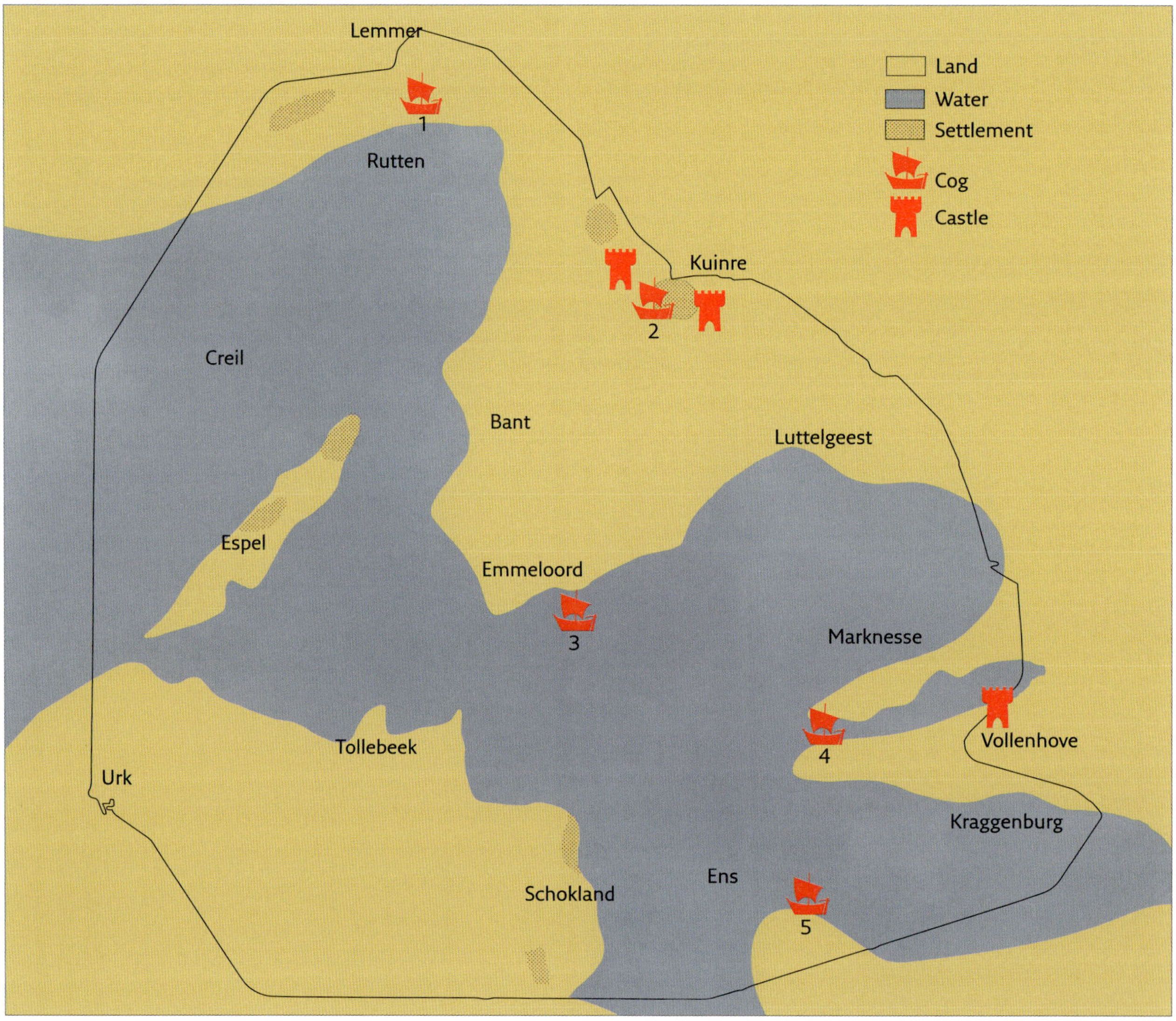

Fig. 10.13. The rough distribution of land and water in the 9th to the latter half of the 12th century, the findspots of late-medieval potsherds, the castles of Kuinre and Vollenhove, and the locations of the Rutten cog (1) stranded ca 1300, the Kuinre cog (2) abandoned ca 1400, the Marknesse cog (3) foundered ca 1375, the Kraggenburg cog (4) foundered ca 1320-1335, and the Ens cog (5) washed downstream ca 1275-1325.

6. Wiggers 1955, 191-195 and fig. 108.

in the Noordoostpolder area locally started even before the turn of the 13th century. It is unclear what was the exact nature of the sherds recovered west of Lemmer (Section A). In the K Section it was mainly 15th- to 16th-century material; around (the castles of) Kuinre, 12th-15th-century material; and from Section C around Espel come 11th- to 13th-century sherds. Similar sherds turned up on the northeast side of Schokland, while mainly 13th-/14th-century material was found on the southwest side of this island. The relatively small number of finds suggests that these findspots do not represent (substantial) settlements. The closest medieval settlement on record was Lemmer. Lemmer is mentioned in documentary sources as early as 1228 and 1236. The adjacent part of the Almere/Zuiderzee was well navigable, which in due course was to make Lemmer a major port for inland shipping and transport across the Zuiderzee.[7]

10.1.4 Condition and wreck formation

The ship's timbers found stacked beside a farmyard in 1982 had been lying exposed for two years and as a result had dried out in an uncontrolled way. Thanks to the excellent quality of the wood, the surface of most timbers had been little affected. Some timbers had become warped to some extent. The greatest damage to the pieces had been inflicted by the bulldozer. Despite its slight depth below the surface, the remainder of the wreck still lying *in situ* also was in fine condition.

The presence of a still intact layer of Holocene peat to the southeast and east side of the wreck, and 'islands' of peat around the southwest to north sides suggest that this undisturbed deposit at the time of the ship's foundering must have been continuous and higher. In other words, the sea was shallower locally and the ship must initially have floated and scraped across the peaty bottom before becoming stuck. Therefore, it seems very likely that this vessel did not come to grief by sinking, but that it became stranded on the then shore or closely inshore.

Judging by the reconstructed height (section 10.2 and Plate 10.III A) and the depth below ordnance datum (ca 3 m -NAP), we may assume that initially the wreck emerged some metres above the water. Its solid construction and massive timbers mean that it would not have been easily ripped apart by the waves. Hence it seems quite likely that all accessible timbers, both above and below the water, would have been recovered by the local population, including the inhabitants of nearby *Lenna* (Lemmer). Most timbers would have done very well for structural timber; other pieces might be gathered for fuel. This would explain the absence of virtually all the ship's parts above the turn of the bilge. This robbing probably occurred quite soon after the stranding, after which the remainder was embedded in, and covered over by Almere sediments. Subsequently the adjacent peatland will have gradually eroded. The wreck then remained unaffected until its discovery and eventual excavation.

10.2 The ship

In describing and drawing the vessel's reconstruction, the data of the reference wreck and those of the other large cogs served as a source of information and comparison.

10.2.1 Documentation and reconstruction

Documentation

It was mentioned above that a survey drawing was made of the situation *in situ* (Plates 10.I A and 10.II A). Also sections across the afterbody and forebody were drawn (fig. 10.6 and Plates 10.I B and C). After the frame timbers removed in 1980 had been replaced into their approximate original positions, one fore-and-aft section and fourteen transverse sections were measured in and drawn. For the latter a graduated pole was used (fig. 10.14). The longitudinal section was measured in with the aid of a horizontal tape measure and a surveyor's level and rod. On the basis of the survey drawing (Plate 10.I A) various sketches were made, showing the pattern of frames and keelson, the shell and various details. In the internship report by A. Vos the observed phenomena and features were described. All parts of the wreck were numbered and after the field research were transferred to the workshop at Ketelhaven, where in 1986 they were drawn to scale 1:10. In the course of the excavation numerous black-and-white photos, colour slides and photogrammetric recordings were made.

Reconstruction

The virtual absence of any parts higher than the bilge strakes renders it impossible to achieve a full hull reconstruction. The overview of the situation found *in situ* and the cross-sections drawn in the field (Plates 10.I A and 10.II C and D) show how the damaged underwater body was uncovered. Several frame parts found detached above the wreck could not be replaced into the wreck in the field, owing to warping; hence they are omitted from e.g. fig. 10.12. For the reconstruction on paper, first a number of field sections (the more complete ones) were drawn,

7. Van der Aa 1846, Vol.7, 181 f.f.

followed by the top view with all recovered and replaced parts of the frames. The longitudinal section drawn in the field was complemented with the recovered frame timbers and the shell planks of the reconstructed port side (Plates 10.III C, 10.II B and 10.I C). Thanks to especially the recovery of the complete lowest port futtock of frame 23 (j in Plates 10.I A and 10.II A), insight could be gained into the strake pattern, and the widths of the planks above the bilge strake up to a height of about 1.8 m above the keel line, and the angle of the everted side could be established. The oblique cross-cut head of the futtock suggests that it had supported a protruding crossbeam. In the Doel-I cog, (Chapter 20, figs 20.23a and b) a similar futtock/crossbeam construction was observed.

The offset on the inboard face of the futtock at the level of strake E suggests that here it supported some transverse connection. To gain an impression of the possible dimensions and lines of the ship, a hypothetical reconstruction of the cross-section at frame 23 up to the gunwale (Plate 10.III C) was made, based on comparable constructions observed in other large cogs. As a starting point it was assumed that the head of the mentioned futtock indeed supported a protruding crossbeam. The height of this beam with respect to the keel line is lower than that in other large cogs. Judging by the length of the keel plus hooks, the rake of the stern and stem, and the width of the hull at frame 23, we are without any doubt dealing with a vessel surpassing the Nijkerk-II, Doel-I and Bremen cogs in size.

Taking into account its massive construction, the vessel must have been higher than any of these as well. Hence it seems very likely that the Rutten cog had two tiers of protruding crossbeams. The hypothetical transverse connection at the level of strake E, mentioned above, may well have supported a floor. Fragments of loose planks of conifer wood (including K in Plates 10.I A and 10.II A) most probably were part of this floor. The hypothetical section at frame 23 shows what the construction may have looked like. The top view (Plate 10.II B) is complemented with the presumable number of side strakes indicated in section 23. Also a side elevation is drawn of the starboard side with stern and stem, and the proposed double tier of protruding crossbeams and the locations of the transverse supports of the presumed floor.

Fig. 10.14. Interns J. Cupido and A. Vos measuring with the graduated rod and drawing the sections.

10.2.2 Construction and layout of the hull

Keel plank

The keel plank is 7.12 m long and provided with two lipped flat scarfs with lengths of ca 50 cm (aft) and ca 60 cm (fore) which join it to the hooks. The overall length of the keel plank plus hooks was ca 15.6 m (Plate 10.II C). At its after end the keel plank is 42 cm wide, midway, 50 cm, and at its forward end, 40 cm. Its thickness is 15 cm aft and 16 cm forward. Its section is virtually rectangular over most of its length, but slightly trapezoid towards the stern and stem, with the smaller side below. At the scarfs, the keel plank is fastened to the stern- and stem hook with four rows of nails (two of four nails and two of three); in both cases, rows driven from below alternating with rows driven in from above. The points of the nails were double-clenched, i.e. beaten sideways and driven back into the wood.

Stern hook

The horizontal part of the stern hook (including the scarf) will have been ca 5.6 m long. The after end had been damaged by the laying of the drainage tube in 1952, which made it impossible to establish its precise length (fig. 10.15 and Plate 10.II C). Its width at the scarf is 42 cm and at its after end ca 16 cm. In section, the horizontal part from the scarf to its after end has a progressively narrowing trapezoid shape with its narrowest face below. Aft from the scarf, the thickness increases from 15 to 18 cm over a length of 3.85 m. This is followed by an upward-curving line until at 4.05 m from the scarf it reaches a thickness of 30 cm. The next part has a straight upper contour that continues to the base of the rising part, where the hook is 40 cm high or thick. In the sides of this

Fig. 10.15. The stern hook, part of the keel plank, the lipped flat scarf, and the sintel nail clamped moss caulking in the seam between the hook and the garboard.

high horizontal part of the hook, 5-cm-deep rabbets were cut, into which the garboards had been nailed. The rising part of the hook is missing, so it cannot be established how high it extended above the keel line, but probably this was between 1 and 1.5 m. The rabbets for fixing the hood ends of the garboards and the two subsequent strakes most probably continued into the rising part of the stern hook.

Sternpost

There is no doubt that on the stern hook stood a sternpost connected to it by a flat scarf. Nothing was recovered of this post. Its rake presumably was ca 117°.

False stern

The cog almost certainly had a false stern. Plates 10.II C and 10.III A show what the stern construction as a whole may have looked like. On the stern hook, sternpost and false stern a number of gudgeons had been fastened for suspending the rudder. Nothing was recovered of the false stern and the gudgeons. Some fragments of corroded iron might have belonged to the latter.

Fig. 10.16a & b The starboard face of the stem hook, with the garboard and behind it the bottom strakes of the port side and the repairs to the stem hook.

Rudder

The Rutten cog was without any doubt steered with a stern rudder.

Stem hook

The horizontal part of the stem hook, including the scarf, has a length of 4.05 m. Part of the rising part had survived, while also some of the detached timbers could be fitted into the reconstruction. Exactly how high this part of the hook extended above the keel line could not be ascertained - as was the case with the stern hook. Probably this was about 1.7 m (Plate 10.II C). On the front of the underside of the stem hook, a piece ca 12 cm thick had in antiquity been cut away over a length of ca 0.9 m and replaced by an elongated-triangular piece that was fastened with nails to the stem hook with a flat scarf and two iron strips. Originally the keel line of the intact stem hook almost certainly was perfectly straight, until it became damaged. As it was replaced, it was taken into account that the ship would be landing on, for instance, sandy shores, and, witness the iron strips, the keel line was given a slight rise (figs 10.16a and b, Plate 10.II C). At the base of the rising part of the hook and in the false stem too some degraded parts had been hewn out and replaced by new inserts which were fastened with nails. We may assume that in the course of the ship's working life the bow was seriously damaged and subsequently repaired on at least one occasion.

Stempost

On the stem hook, and connected to it by a flat and slenting scarf, there must have stood a stempost. Nothing was recovered of this timber. In the reconstruction the stem has a rake of 119°.

False stem

The cog had a false stem as well as a false stern. A small part of it was uncovered *in situ*. As mentioned earlier in discussing the stem hook, the lowermost part of the false stem had been repaired with some inserts. Plate 10.III A shows what the bow construction as a whole may have looked like.

Bottom and sides

The hull on either side of the keel plank was most probably built up out of sixteen strakes (A to P). For the bottom, planks were used with a width of 50 cm or more. In the sides, in as far as could be established, the planks were narrower. For the reconstruction it was assumed that the shell planks through which the crossbeams protruded may have been somewhat thicker and wider than the other planks of the sides. The planks composing each strake were joined with long sloping flat scarfs (1.10 to 1.35 m). The nib ends of the overlapping planks, 4 to 5 cm thick, stood proud of the shell both on the inboard and outboard sides (fig. 10.17). As a result, the overlapping planks in most places had a joint thickness of 10-12 cm. Generally they were fastened together with four rows of nails, sometimes three. Two rows of five or six, and two rows of four or five nails. Alternatingly, rows of five/six, and of four/five nails were driven in from the inside and from the outside. The points of the nails emerging through the planks would be twice beaten sideways and driven back into the wood (double-clenching).

Bottom

The bottom consists of the keel plank and four strakes (A to D) on either side of it. The garboards (A strakes) were fully preserved. Parts of the B to D strakes towards the bow and stern are missing (figs 10.18a and b). The first three strakes are composed of two planks. The port D strake at the stern includes a short third plank (ca 1.2 m long). The planks are arranged according to a 'brother-and-sister pattern', with scarfs in similar positions in equivalent strakes to port and starboard. The garboards

Fig. 10.17. General view of the bottom of the Rutten cog, with the plank ends standing proud at the scarfs. Viewed from the bow.

Fig. 10.18a. General view of the bottom in the afterbody. In the foreground, the scarf joining the stern hook and the keel plank, and the scarfs in the garboards (A strakes).

Fig. 10.18b. General view of the bottom in the forebody with the rising part of the stem hook and the scarf joining the stem hook to the keel plank.

are flush-laid against the keel plank. On the seams with the B strakes they are slightly bevelled, which makes them look as if they are joined clinker-fashion. The lower edge of the B strakes was similarly bevelled. These bevelled edges were not nailed together, and no sintel clamped moss caulking was put onto the seam. The B, C and D strakes are all joined carvel-style, be it that from about frame 8 towards the stern and from frame 22 forward they were wider and were joined clinker-style to the previous strake. Planks up to 10 m long were used for the bottom strakes. Indeed the maximum width of the B strake is over 60 cm, while the thickness of the garboards in some parts is 8 cm. The other planks are on the whole 5 to 7 cm thick. The hood ends were nailed into the rabbets in the hooks, while the clinker-built joins were fastened with double-clenched nails. Apart from this, the connections between shell planks consisted only of the nailed scarfs and the V-shaped floor timbers treenailed to the bottom planks.

Fig. 10.19. Sintel (nail) clamped moss caulking in the clinker seam between the starboard D and E strakes amidships.

Sides

Of the sides, just a few fragments of the bilge or E strakes were recovered. The other strakes up to the gunwales were entirely missing. Thanks to the presence of a few futtocks, especially the lowermost port futtock of frame 23, something can be said about the width of the strakes. In the case of the E to H strakes, this is 35, 26, 23 and 40 cm, respectively. For the hypothetical reconstruction of frame 23 it was assumed that these varying plank widths prevailed right up to the gunwale. The only scarf encountered in the fragments of starboard strake E suggests that this strake consisted of three planks. Judging by the overall length of the F to P strakes, it may be assumed that these too each comprised at least three planks. Above strake D, all planks were joined entirely by the clinker method from stern to stem. In every case, the land in the seams probably was ca 6 cm wide. In the reconstructed section of frame 23, the angle of the side above the lowermost crossbeam was ca 104°.

Caulking method and the shape of the carvel seams; sintel clamped moss caulking

The plank edges of the flush-laid bottom strakes were partially bevelled, which on the underside created a seam of triangular section ca 2 cm wide and 3 cm deep. These seams were filled from the outside with strands of moss and covered with moss laths of split oakwood about 2 cm wide, 0.5 cm thick and cut to an oval section. These laths in their turn were fastened with wrought-iron sintel (nail) clamps sintel nail clamped moss caulking, fig. 10.19). On the inside of the shell these planks abutted on each other. Wherever sintel nail clamped moss caulking was applied on the inboard side of the clinker seams, the outside/

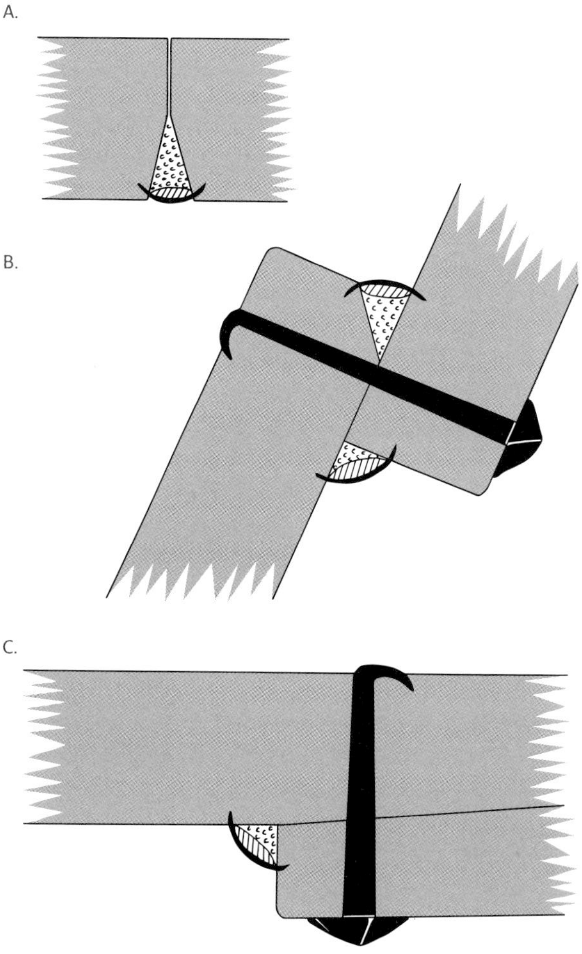

Fig. 10.20. The application of sintel (nail) clamped moss caulking in the seams of triangular section in the carvel-built bottom (A); and between the clinker-built strakes of the sides, inboard and outboard (B); and at the plank ends in the scarfs of the shell strakes (C).

upper edge of the lower strake had been previously bevelled, thus creating a seam of triangular section between this plank and the plank to be fastened above it. These seams are about 2 cm wide and 2.5 cm deep. On the outboard side, the seams of the clinker-joined planks were also closed with sintel nail clamped moss caulking. Here, by contrast, no triangular gap had been made, but the strands of moss were pushed into the angle between the overlapping, 4-5 cm thick plank and the plank below it, and secured with moss laths and sintel clamps. The plank ends at the scarfs are closed in the same manner on the outside (fig. 10.20). Furthermore, sintel nail clamped moss caulking was applied to some cracks in the planks to repair them from the outside (Plate 10.III B). The sintel nail clamps used in the ship are of the type C1/2/(D) and D(1). These shapes were in use during the second and third quarters of the 13th century.[8]

Macroscopic determination and pollen analysis
For caulking, pure *Sphagnum* moss was used in the Rutten cog. The examined pollen sample was found to comprise 60% of tree pollen and 40% of herbaceous plants. The moss must have been gathered in eutrophic surroundings - probably a disturbed area with a small peatbog, arable land and oak woodland with beech and hornbeam (for a detailed description of the findings, see Appendix.[9]

Frames: floor timbers, V-shaped floor timbers and futtocks
The hull is held in shape and reinforced principally by the frames (fig.10.21 and Plates 10.II B and 10.III C). Between the angles of the hooks aft and fore, there are thirty of them (0 to 29). The precise number of frames in the hull cannot be ascertained; probably there were five more fitted to the sternpost (-1 to -5) and seven to the stempost (30 to 36), which would bring the total up to 42. Remarkably frame 26 has no floor timber, just short of the thicker part of the stem hook. Treenails in the shell strakes indicate that there had been futtocks up the sides here. A similar phenomenon was observed in the Nijkerk-II cog, where the same thing was seen in the afterbody as well as the forebody (Plate 6.V A). Section 10.2.2 will return to this feature. With the exception of the floor timbers of frames 1 and 2, those of frames 0 to 29 were all encountered in the excavation. They are massive timbers 22 to 30 cm wide and high or thick. In most of them, the scarfs joining them to the lower futtocks are in roughly equivalent places on the port and starboard sides (frames 3 to 10, 17, 19 and 20 to 29). Of frames 11, 12, 13 and 15 this cannot be said with certainty, because parts of the frames are missing. The floor timbers of frames 14, 16 and 18 (are the only ones that?) run on to above strake F on the starboard side, while to port they end above strake D.
The undersides of the frames had limber holes cut into them. The sharpest V- shaped floor timbers in the stern and stem had one above either one of the seams between the hook and the garboard. The somewhat more widely-curved floor timbers had two: one above each of the seams between the hook/keel plank and garboards. Then follow floor timbers with four limber holes: above the same seams and above the centre of the garboards. The longest floor timbers had six limber holes: two above the seams between the keel plank and the garboards, two above the B strakes and two above the C strakes. Not much can be said about the futtocks: the port futtock of frame 23 was the only one to be recovered intact. It had been joined to the floor timber with a flat scarf and had a straight-cut, oblique head (Plate 10.III C). The lower futtocks beneath the other protruding crossbeams were probably executed in a similar manner. The remainder without doubt ended in a flat sloping scarf, by which they were joined to another futtock above them. It seems likely that the frames between the crossbeams had three tiers of futtocks on both sides. In some places in the waist of the ship, maybe just two.

The near-horizontal scarfs of the V-shaped floor timber of frame 25 suggest that a transverse beam (DV8, Plate 10.III A) may have overlain it, which on one side ended in a knee that constituted a futtock, while on the opposite side a futtock was placed onto this beam - a construction also encountered in the Nijkerk-I cog and the Doel-I cog (Chapter 8, figs 8.14a and 8.35a and b; and Chapter 20, figs 20.20a and b). The head of V-shaped floor timber 25, like the notch in the lower futtock of frame 23, sits at 1 m above the keel line. In the other places where transverse

Fig. 10.21. The floor timbers in the underwater body, viewed from the stern.

8. Vlierman 1996a, 74-36.
9. Mook-Kamps, 1997-6; Cappers *et al.* 2000.

connections are indicated, a similar construction may have existed. On the lower tier of protruding crossbeams instead of futtocks most probably stood knees, which also supported the upper tier of crossbeams. The knees on top of the latter maintained the shape of the bulwarks. The (V-shaped) floor timbers and futtocks were fastened to the shell with generally two treenails (3 to 3.5 cm thick) per plank width. Whether the treenails in the knees had the same diameter is uncertain; they may well have been somewhat thicker. The (V-shaped) floor timbers of frame 2 up to the top of the stern, and those from frame 27 up to the top of the stem were not treenailed to the hooks or the stern-/stempost.

Keelson with mast-step section

Above the horizontal parts of the hooks and the keel plank, a keelson sits on top of the floor timbers, from frame 5 to frame 23 (figs 10.1 and 10.21 and Plates 10.II B and C). It is an important part of the construction for the rigidity of the bottom. Aft, the keelson is 10 cm wide and abaft frame 6 it is 20 cm high. Over a length of 5.55 m the width gradually increases until above the forward face of frame 16 it is 20 cm wide; here it has a height of 23 cm. Immediately forward of frame 16 the wider (48 cm) integrated mast-step section begins, which extends forward over a length of 1.75 m up to the after face of frame 20. The mast-step section is 30 cm high and with chamfered corners merges with the keelson fore and aft. The forward end of the keelson at frame 20 is 23 cm wide and over a length of 1.8 m runs on to above frame 23, gradually narrowing to 12 cm. The height of the forward part is 22 cm throughout. Where required, the underside of the keelson has notches up to 4 cm deep, to engage with the frames. The overall length of the keelson is 9.1m.

Above frames 17 and 18, the top of the broader mast-step section has a rectangular recess cut into it, measuring 78x27 cm. The foremost 64 cm was hewn out to a depth of 13 cm, the remaining length is just 7-8 cm deep. In the bottom of the deepest part is a roughly carved cross with a drainage hole at the centre (fig. 10.22). The 64x27x13 cm hole is the actual cavity that held the mast. The significance of the shallower part is unclear. Immediately against the after edge of this part of the hole, a rectangular recess of 6x12x7 cm had been cut in the middle of the keelson, right above frame 17. This had probably

Fig. 10.22. The mast-step section of the keelson with the hole for the mast heel, the riders on frames 17 to 19, and the forebody with ceiling strakes and frames.

Fig. 10.23. The damaged broad mast-step section of the keelson, and the riders on frames 17 (left), 18 and 19. The middle buttress timber (or rider) has an oblique rabbet at the end above the ceiling strake in the foreground. Viewed from starboard.

held a short wooden upright that supported the protruding crossbeam DB3 midway. Immediately abaft this recess was another rectangular cavity, of 6x8x1.5 cm. Maybe this had held a post that supported the upper transverse beam DB 8. Above frame 15 was a similar hole in the keelson. This latter pair of small holes may instead, and maybe more probably, be linked to the horizontal treenails in the after face of the port rider on frame 18, and to the oblique rabbet in this rider at the turn of the bilge. A similar rabbet had probably also been cut into the starboard rider (fig. 10.23 and Plates 10.II.B and III C). Section 10.3.1 will return to this topic. The keelson was fastened with a single treenail per frame; the broader mast-step section with two (3 to 3.5 cm thick). The treenails run from the top of the keelson to the bottom of the keel plank. Only the foremost treenail is blind and driven into frame 23 obliquely. The keelson was ripped into four pieces and hence badly damaged. The small foremost piece still laid more or less *in situ*.

Riders

To take the lateral pressure of the mast with the sail, the mast-step area of the keelson was buttressed by putting six riders on the floor timbers of frames 17, 18 and 19, three on either side. They support the broad mast-step section and on the underside are joggled in two places to engage with the ceiling strakes running across the floor timbers. These buttress timbers ran on up to the (missing) bilge ceiling boards. They are held in place by treenails 3 to 3.5 cm thick. Some of the treenails run right through to the outside of the shell, while others run blind into the floor timbers. The horizontal treenails in the rider on frame 18 to port and the oblique rabbets in the tops of both riders above this frame have been discussed above.

Ceiling

The ship had an open ceiling. Two ceiling strakes (A and B) were placed on either side of the keelson. They ran from the stern, probably at or abaft frame 3, to somewhere on or near frame 25 in the bow. They are boards 22-38 cm wide and 3-5 cm thick. The tops of the B ceiling strakes are in some areas shaped by adze into a slightly roof-shaped surface. The B strake to starboard has a flat sloping scarf at its after end and hence must have consisted of two boards. The starboard A strake ended at frame 23 (Plate 10.II B). Amidships, the ceiling boards were damaged by the drain-cutter and bulldozer, and large pieces were missing (figs 10.11 and 10.21). The vessel definitely had bilge ceiling boards as well. These would have sat above the seam connecting shell strakes D and E. Given the large notches in the underside of the outer ends of some of the riders, it appears that these ceiling boards were thicker than the others and that their upper surfaces were clearly adze-fashioned into a slightly roof shape. Along the sides, the vessel is sure to have had some further ceiling strakes, but nothing definite can be said about their exact location and number. The ceiling strakes played an important part in the longitudinal rigidity of the ship. The boards were fastened with a few iron nails and with treenails with diameters of 3 to 3.5 cm.

Crossbeams

Already it has been noted that the ship must have had crossbeams protruding through the shell, and that these almost certainly were placed in two rows (Plate 10.III A). The lower row is likely to have comprised four crossbeams. They probably were above frames 5, 11, 17 and 23, with their upper surfaces at ca 1.7 to 1.9 m above the keel line, respectively. They protruded through strakes H (DB 1) and I (DB 2, 3 and 4). The upper row probably consisted of six beams (DB 5, 6, 7, 8, 9 and 10). These beams had probably been inserted above frames 0, 5, 11, abaft or above 17, 23 and 29. Their tops were at a level of 3.25 to 3.55 m above the keel line, and they protruded through shell strakes K/L (DB 5), L/M (DB 6), M (DB 7, 8 and 9) and K/L (DB 10). The forward sides of the protruding heads of these beams might be protected by beam-head guides. Given the massiveness of the ship's construction, the beams will have had a cross-section with sides of about 25-30 cm. The hypothetical reconstruction of frame 23 (Plate 6.III C) shows what the crossbeam constructions may further have looked like. Their spacing (ca 3 m) was derived from other cog finds.

Transverse connections
The notch on the inboard face of the futtock of frame 23 suggests that it supported some transverse connection (Plate 10.III C). The top of this transverse connection or beam was at ca 1 m above the keel line. At the same level were possibly a further six such beams, spaced along the length of the underwater body, apart from this one at frame 23. They might have sat above frames 3, 7, 11, 15, 20 and 25: transverse connections DV 1 to 7 (Plate 10.III A). If indeed there were such beams at the same level, this could point to a floor immediately above the ballast. Sections 10.3.1 and 10.3.4 will return to this and its possible function. There may have been two further transverse connections at the same level as the top of the lower row of protruding crossbeams. One in the afterbody above or close to frame 0 (DV8), and one in the forebody above or close to frame 29 (DV9). These would then have served to locally reinforce the hull. Crossbeams protruding through the shell are less desirable in such areas. Presumably these transverse connections consisted of a knee-shaped timber, of which the upright arm formed a kind of futtock. This part may also have been placed against the inboard or lateral face of a futtock, and in the latter case fastened to the shell. To the other end of the timber a separate knee was fastened. For the execution of such a transverse construction, see also those of the Nijkerk-I cog in chapter 8 (figs 8.14, 8.15 and 8.35; Plates 8.IV A and C).

Stringers
Judging by what has been found in other cogs, it seems very likely that this vessel too had stringers between the top row of protruding-crossbeam constructions, to ensure fore-and-aft rigidity at deck level.

Deck(s)
Of course the question of whether the ship had one or more decks can only be answered speculatively. Above, the possibility of a floor at ca 1 m above the keel line was already mentioned. In the redeposited sediment in the bottom of the hull, five fragments of conifer-wood planking were found (D, E, I, K and L; Plates 10.I A and 10.II A and fig. 10.24). They are 5 cm thick planks of which one side is worn smooth and the other side still clearly displays adze marks (the undersides). Three of them bear scratched marks (so-called *parings*) while the ends are bevelled on the underside. Plank K has a notch which presumably fitted around a futtock. The underside of this plank is partly bevelled towards this notch, while nail holes can be seen along the long side. The plank has a number of blind drilled holes (8 mm across), and at the broken end is a treenail hole with a diameter of 3 mm.

These clearly are floor or deck planks which lay almost loose on the supporting beams and could be easily lifted. The so-called *parings* (Dutch mar.) would facilitate swiftly putting them back in place. A continuous deck may have lain atop the uppermost transverse constructions (Plate 10.III C, frame 23) or else an afterdeck and foredeck interconnected by broad, heavy planks running from crossbeam to crossbeam. These gangplanks may actually have been gangways along the sides, fitted to the shell and around the frames. Still it cannot be ruled out that the conifer planks came from the (upper) deck rather than a floor in the underwater body.

Windlasses
Windlasses will certainly have been present to hoist or lower the heavy sail and yard. Also it seems quite likely that on a large ship like the Rutten cog a windlass was required to operate the anchor.

Pumps
The vessel without any doubt had one or more pumps for discharging rainwater and any bilge water. Nothing has been found of a pump barrel or handle. Areas of abrasion on the horizontal parts of the hooks - between frames 2 and 3 on the stern hook and at frame 26 on the stem hook - suggest that pumps may have been installed there (Plates 10.II.C and 10.III B).

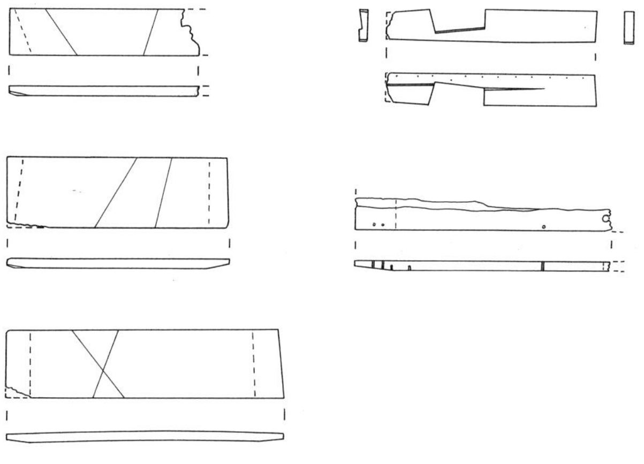

Fig. 10.24. The five floor- or deck planks of conifer-wood, some of them with scratched-in parings, and one with a notch that most probably fitted around a futtock.

Castles
It cannot be established whether the vessel had any (one or two) castles, as no evidence of such has been found in the underwater body. Given the vessel's possible function (Section 10.3.4), it does however seem plausible that there may have been one.

Repairs
Above (in Section 10.2.2) it was already observed that the outside of the bow construction was severely damaged on at least one occasion during the ship's active life, and subsequently repaired. Also, in several places cracks in the shell planks of the underwater body have been recorded that were closed with sintel nail clamped moss caulking on the outside. These repairs were most probably performed even at the end of the construction process. The same may also go for the small repair plank on the outside of the port garboard in the afterbody (Plate 10.III B). To starboard, a small repair plank had been nailed to the inside of strake C at frame 26. Further, a number of small pieces of wood have been noted as repairs in shell seams and to plug knots.

Wood use, fastenings and traces of processing: wood selection and treatment
With the exception of some pieces of conifer planking from the proposed floor in the underwater body, all recovered wooden parts and timbers are of oak (*Quercus* spec.). The planks were probably sawn. Unfortunately (with the exception of A1 bb and D1 bb) the drawing of the planks to scale 1:10 did not record how the tree-ring pattern manifested itself in the plank aspects. The few planks whose grain was sketched suggest that they were rift-sawn. The same can be said of the ceiling boards. The straight and lightly curved floor timbers were probably hewn from naturally curved trunks; the V-shaped floor timbers, from a forked trunk or a trunk and limb. The frames are on the whole neatly square-hewn. It is unclear whether the futtocks gradually narrowed upwards. Presumably this was not significantly the case.

Fastenings
The treenails are of oak wood (*Quercus* spec.) and have a diameter of 3 to 3.5 cm. They are not quite round in section, but cut slightly polygonal. A substantial number of them were examined and were found to have seven to nine faces. In some of them small wedges were observed. The nail-hole plugs (Section 10.3.2), not yet discussed so far, and the split-wood moss laths were also made of oak. In principle, two treenails were used per plank width. This means that in the hypothetically reconstructed hull some 2800 treenails were required for fixing the frames into the shell. This still excludes the crossbeam constructions, transverse connections, any castles and other treenail fastenings. A rough overall estimate of ca 3500 seems quite plausible.

Among the iron nails two formats were distinguished. There are no fully preserved nails. The thickness of the planks and the shanks of numerous double-clinched nails were measured in the field. On the whole, nails were used with a length of ca 140 mm, a shank of square section with faces of 10-15 mm width, and convex to pyramidal heads 35-40 mm across (fig. 10.27-10). The other nails were ca 70 mm long, also square in section (6x6 mm), and with flat heads 1.5 mm thick and ca 20 mm across. An impression of the method of application of the nails and their spacing could be gained only from the overlaps of the B, C and D strakes in the stern and bow, and from part of the clinker-style joining of strakes D and E. The spacing was rather irregular; in some areas there was evidence of (pairs of?) unusually closely set nails. Such was the case also in the fully clinker-built 16th-century cargo-/warship from lot U34 in Oostelijk Flevoland.[10]

An average nail spacing of ca 20 cm in clinker constructions seems reasonable for estimating the required number of nails. The hull contains about 550 m of nailed shell planking, which amounts to 2750 nails. For the scarfs, a further ca 1500 nails would be needed. For fixing the hood ends into the hooks and stern- and stempost, the recovered fragments show that in general two rows of nails would be used; if this continued right up to the top strake, this would account for another ca 350 nails. Hence the overall number of nails used in the hull of this ship would have been at least 4600. This total does not include the number of nails used in the decks and further constructions. The ship would have featured at least 1250 m of seams closed with sintel nail clamped moss caulking (shell seams on the inboard and outboard side, the inboard joins along hooks and stern and stempost, outboard around the heads of protruding crossbeams, and cracks on the outside of shell planks). With an average spacing of 10 cm, this would mean that some 12,500 sintel nail clamps were used on the hull.

Traces of treatment
In describing the bottom planks in the field it was established that in various places in the underwater body black patches occurred, which were most clearly visible beneath the frames when these were removed. At the time thoughts went to scorch marks resulting from the burning that was done to bend the planks. Yet this seems not very likely, as large parts of the planks featuring black

10. File and drawings Ou 34, RCE-Lelystad.

areas had required little if any bending before being fitted into the hull. Another suggested possibility is that the planks were tarred prior to the insertion of the frames. No closer investigation was carried out. The latter option does seem the more obvious, and subsequently was observed in two further instances with even greater clarity. In the Doel-I cog (Chapter 20) and on the shell planks of the above-mentioned 16th-century and fully clinker-built large cargo-/warship from lot U34 in Oostelijk Flevoland.[11] In section 10.3.4 the possible explanation of tarring and its reason will be further discussed.

10.3 Construction data

Unfortunately it is only about the underwater body that anything can be said with certainty. The only well-preserved and some more poorly preserved futtocks offer an insight into the widths of the clinker-built lowermost side strakes, the shape and angle of the sides, and the likely position of a protruding crossbeam. Also the height with respect to the keel line of the lowermost crossbeam could thus be established, while a notch in the inboard side of the futtock suggests that this held a transverse beam supporting a floor. Despite the limited data, yet an attempt has been made to arrive at a hypothetical reconstruction of the side and top views of the hull, and the most completely recovered frame (23) was reconstructed into a cross-section. On the basis of comparison with other large cogs, it may be assumed that the reconstructed length, width and height dimensions and the rake of the stern- and stempost are not too far off the actual ones.

10.3.1 Design

The way in which the underwater body was put together reveals that the construction of such a large and massively built vessel was carried out according to rules of thumb that the shipwrights of those days had long been familiar with. The thickness of the shell strakes is ca 1.5 times that of most other large cogs. Given the reconstructed principal dimensions and the protruding crossbeams probably at 1.8 m above the keel line, it seems likely that the ship had such crossbeams at two levels. The ballast and the probable floor overlying it further suggest that these-features were part of the vessel's original design.

Lines

Despite its size, the ship most probably had elegant lines. Neither lengthwise nor transversely did the bottom have any dead flat. This means that the bottom strakes from the midship point gradually rose towards the stern and stem. The same was the case between the keel plank and the sides. As a result, there are no two identical sections across the hull. Between frames 13 and 19, the A-, B- and C strakes lie side by side in an almost straight line, which slightly rises towards the turn of the bilge on either side. The D (or bilge) strakes, also carvel-built, are fastened to the C strakes at a somewhat steeper angle, after which the lowermost three side strakes (E, F, G) were fixed clinker-wise at progressively steeper angles. Together they create a flowing transition from the bottom to the sides. As they continue to do towards the stern and stem. Probably the afterbody had a somewhat greater deadrise (was more sharply built) than the forebody, with the after frames slightly S-curved. From strakes H/I on, the sides probably continued upwards in a fairly straight, everted line (Plate 10.III C). The reconstructed top view (Plate 10.III B) shows a bottom sharply converging towards the bow and stern, the afterbody being somewhat more slender, the shape of an elongated half walnut. At gunwale level, the afterbody is 10 to 20 cm narrower at similar distances from the stern and bow. The cog has a straight keel and almost certainly also had a straight stern- and stempost. In the reconstruction (Plate 10.III A) the rake of the stern is 117° and of the stem 119°. The sides from strake H upwards (at frame 23) probably had an angle of 104°. The bottom was fully carvel-built over a length of 7 to 8 m, with the B strakes apparently entirely carvel-built alongside the garboards (A strakes).

Measurements and layout

In the underwater body no marks were found that might relate to measurement during construction. With respect to the stern and stem at gunwale level, the mast stood 3 m forward of the midship point on the bottom and measured along the keel plank and hooks at about 1/3 of the length from the bow. It seems quite plausible that a floor had existed between frames 3 and 25 and at about 1 m above the keel line and closely above the ballast. Judging by a piece of planking of conifer wood (K in Plates 10.I A and 10.II A), it would appear that this floor was fitted around the frames at the sides. Besides, it is quite likely that an area roughly between frames 15 and 17 had no floor and that a fireplace was installed here on the ballast stones. A scorched area on shell strake D to starboard might be evidence of a minor fire-tending mishap (Plate 10.III B). In the upper surface of the keelson, above frames 15 and 17, square holes were cut which most probably had held square stanchions. Some blind treenails were identified in the after face of the rider above frame 18 to port, while in both riders overlying this frame an oblique rabbet can be seen in the top of the outer end.

11. File and drawings Ou 34, RCE-Lelystad.

Table 1. Principal dimensions of the Rutten cog *

Length of the keel (including stern- and stem hook)	15.60 m
Length of the bottom (between the rising part of the stern- and stem hook)	15.00 m
Greatest width of the bottom (seams between strakes D and E at frames 15 to 17)	4.75 m
Overall length at gunwale level	22.40 m
Overall width at gunwale level (at frame 15)	9.50 m
Height of the stern (keel line to top gunwale level)	5.05 m
Height of the stern (keel line to top gunwale level)	6.60 m
Height of the side amidships (keel line to top gunwale level)	4.50 m
Length of the stern	6.25 m
Length of the stem	7.70 m

Length-width ratio of the bottom	3.1:1
Length-width ratio at gunwale level	2.4:1

* *All measurements rounded to the nearest 5 cm.*

Rabbets and treenails may be traces of a transverse partition. The small square holes in the keelson may mark the position of 'door jambs' or access to some enclosed floorspace. It may be that in the bow (on transverse connection DV8 and transverse beam DB1) or in the stern (on transverse beam DB4 and transverse connection DV9) floors or intermediate decks were present. A continuous floor from fore to aft at this level seems unlikely, given the slight headroom (0.7 to 1.05 m) above the floor overlying the ballast. Yet it does seem probable that on the upper row of crossbeam constructions a fully continuous deck, or a foredeck and afterdeck were present (Plate 10.III C, frame 23). The headroom between the floor and the deck in the reconstruction amounts to ca 2.4 m.

Tonnage

If the vessel transported cargo, then the space above the supposed floor in the underwater body and beneath the deck resting upon the top row of protruding crossbeams would have been the proper place for it. The spaces aft and forward of this area (on any middle decks) would of course have been needed for the storage of various ship's necessities and personal belongings, and as sleeping quarters for the crew. The floor of the hold would have extended from frame 3 to frame 25 (11.25 m); the average width from side to side was ca 6.5 m, while the height was about 2.4 m. This amounts to a cargo space of 11.25x6.5x2.4 m = 175.5 m^3, which is the equivalent of ca 175 gross tons.

Selection of materials

The hull, treenails and moss laths of the Rutten cog were made of oak wood. The floor in the underwater body, probably of conifer wood. Nothing is known about the decks. The large fastenings such as gudgeons and pintles, the nails and the sintel nail clamps were of wrought iron. Moss had been used for caulking. The thick, cable-laid rope (Section 10.5.2) was made of hemp.

10.3.2 Construction: techniques and sequence

Notable about the keel plank and the hooks is that they were not joined with flat scarfs but with flat, sloping and lipped scarfs, and that the horizontal parts of the hooks are unusually long (stern hook 5.60 m and stem hook 4.05 m), while the keel plank is just 7.12 m long. The mast stood in a mast step integrated in a broader part of the keelson, as has been found in other large cogs as well. The transverse rigidity of the ship's shape and construction is achieved by the frames, the protruding crossbeams and associated constructions on top of them, as well as the transverse connections. The fore-and-aft rigidity is achieved by the keelson, the ceiling strakes and without any doubt also by stringers at deck level. The shape of the everted bulwarks was supported by massive knees on the transverse beams.

About the construction (sequence) of the underwater body we can state that the keel plank and the hooks were naturally first joined up and that the stern- and stempost were mounted on the rising parts of the hooks. The bottom strakes, generally composed of two long planks were subsequently put in place strake by strake, after first being bent into the desired shape. First of all, the garboard planks in the afterbody. Their hood ends were nailed into the rabbets in the stern hook, while the remaining part extending to the midship section was supported and held against the keel plank with stakes. Previously, the forward end of the plank had been chamfered to form the scarf with the foremost plank. Across the seam between the garboard and the keel plank some laths were nailed to keep this shell plank in place. Subsequently the forward planks of the garboard strakes were bent into the proper shape and the ends chamfered to fit to the sloping ends of the after planks. This plank was fastened to the stem hook in the same way, while at the scarf the two planks were nailed together with a few rows of nails, driven both from the inside outwards and the other way around. The subsequent strakes of the bottom were put on in a similar way. This was simpler, because aft and fore the planks overlapped, and hence along a few meters could be nailed in clinker fashion to the previously mounted strakes. Therefore in these areas far fewer or no temporary laths were needed to keep the planks in position.

After the entire bottom was put together in this way, the shape of the hull was given some extra consolidation by means of (short) temporary frames. The positions of these are revealed by the treenails in the planking encountered between the frames. They are found especially near the plank scarfs between frames 16 to 21, between the floor timbers of frames 2 and 3, aft and forward of frame 6, forward of frame 9, between frames 11 and 12, under floor timber 13 and abaft frame 23. Then the floor timbers of frames 3 to 25 would have been shaped and inserted one by one. Simultaneously the temporary connecting laths were removed and the nail holes in the planks were plugged with wooden pegs (nailhole plugs). Also the temporary frames were removed once the adjacent definitive (V-shaped) floor timbers had been fitted. After the floor timbers had been inserted, the keelson and the ceiling strakes were installed above the bottom. Immediately after being burned into a curved shape, the clinker planks of the sides were nailed to the lower, previously fitted strake. Then the end of the plank would be chamfered and fashioned into the scarf joining it to the next plank. As with the bottom strakes, fitting the planks of the sides started at the stern.

Once the level of the lower row of protruding crossbeams was reached, sintel (nail) clamped moss caulking was applied to the shell seams on the inboard side; then the lower futtocks would have been shaped and inserted. Subsequently the ceiling strakes were nailed and treenailed to the futtocks. At this point, the four lowermost protruding crossbeams were placed into previously shaped gaps in the shell strakes. This was followed by further shell strakes, caulking, upper futtocks, and ceiling strakes of the shell up to the level of the upper row of six protruding crossbeams. The same sequence was followed for the bulwarks. The final stage was that of the deck or the decks in the stern and bow, with gangway planks or gangways along the sides. Towards the end of construction, the vessel will have been sealed externally with sintel nail clamped moss caulking. The steadily growing hull of course was well supported with struts. The V-shaped floor timbers, futtocks and transverse connections in the stern (abaft frame 3) and those in the bow (forward of frame 25) very probably were not installed until the intervening part of the ship had been completed at least up to the top row of protruding crossbeams.

10.3.3 Characteristics

The Rutten cog has the following characteristics.

Hull
A large and heavily built vessel with a fine sheer and very probably two rows of protruding crossbeams.

Keel plank
A relatively short keel plank, 15-16 cm thick (i.e. twice or three times the thickness of the shell planks), fastened to the hooks with lipped flat scarfs. Fore and aft its width is about 40 cm. Towards the midships, it gradually broadens to a maximum of 50 cm.

Stern-and stem hook
The hooks have unusually long horizontal arms. The rising parts probably were 1 to 1.7 m high. The garboards and the hood ends of the other bottom strakes were spiked into rabbets in the side faces of the hooks, while the strakes of the sides presumably ran on the forward and rear face of the stern- and stempost, respectively.

Sternpost and stempost, false stern and stem, rudder
Little can be said about the sternpost and stempost, false stern and stem, or the rudder. The ship almost certainly had a false stern and stem, and was steered with a stern rudder.

Carvel- and clinker-built bottom
The bottom of the vessel was built out of the keel plank and four strakes on either side of it. The first three were placed in an almost straight line rising towards the turn

of the bilge; the D strakes, at a slightly greater angle. Both longitudinally and transversely the bottom has no dead flats. The midship section is carvel-built over a length of ca 7 m. Towards the stern and stem, the strakes (with the exception of the seam between the A- and B strakes) merge into a clinker-built construction. The bottom is mostly built from two (broad) planks per strake; these were joined by unusually long (1.10 to 1.35 m) sloping flat scarfs.

Clinker-built sides

From the C- strakes upwards, the bilge strakes and subsequent side strakes in a flowing line continue into the almost straight, everted sides. In the somewhat more slender underwater afterbody, some of the frames may have been slightly S-curved. The overlap (the land) of the clinker construction was about 6 cm, with the planks nailed together at intervals of ca 20 cm. The nails had been double-clenched (the points twice beaten sideways and driven back into the wood on the inboard side). The same goes for the nails securing the plank scarfs.

Caulking method

The carvel seams of the bottom are closed with sintel (nail) clamped moss caulking only on the outside. To this end, the abutting sides of the planks, 6 to 8 cm thick, had been chamfered to a dept of about 3 cm, creating a groove triangular in section and ca 2 cm wide on the seam. The clinker-built sides also had caulked seams of triangular section on the inboard side, while on the outboard side the sintel nail clamped moss caulking had been applied in the angle between the overlapping strakes. The plank ends at the scarfs were externally sealed in the same way. The sintel nail clamping of the moss laths inboard was done before the futtocks were inserted, which is compelling evidence that the shell-first method of construction was followed.

Ceiling, keelson/mast step and repairs

The cog had an open ceiling, and a long keelson with an integrated, broader mast-step section. This broader part was laterally supported by six riders, fixed to the floor timbers on either side of it. Patches set into the base and front of the stem hook and the false stem are indicative of major damage and repair during the active life of the ship.

Crossbeams, transverse constructions, deck(s) and gangways

The vessel probably had crossbeams protruding through the shell at two levels: four below and six above. Besides these, the underwater body probably contained nine further transverse connections: two in the aft and forward at the level of the lowermost crossbeams, and seven at about 1 m above the keel line. The latter probably served to support a floor of conifer wood overlying the ballast. Possibly the ship had one or more between-decks resting on the uppermost transverse constructions and on the aftmost and foremost crossbeams; while the higher crossbeam constructions carried a main deck - or decks aft and fore interconnected by gangway planks or gangways running along the sides.

Fastenings

In the hull, treenails 3 to 3.5 cm thick were used for fastening the frames to the shell. For the crossbeam constructions, possibly somewhat thicker ones. For the clinker-built sides, wrought-iron nails with convex to pyramidal heads were used.

10.3.4 Use

Little can be said with certainty about the use and the function of this cog. Yet a few remarkable features, which in other large cogs were observed also or indeed never before, do suggest a direction in which we might think. In the first place, there is the large amount of ballast which sat mainly abaft the mast in the midship section and the afterbody, and the floor that seems to have been laid directly above it. Also the probably dual row of protruding crossbeams, the heavy construction of the ship and the apparent tarring of the inboard side of the shell are features that this ship may have served a purpose other than transporting cargoes of merchandise. This may be the underwater body of a military vessel, known as a *heerkogge*.

The tarring of the (underwater) shell of the 16th-century merchant vessel/man-of-war in lot U34 in Oostelijk Flevoland may be related to the curious floor installed in that ship. At about 30 cm above the ceiling boards a floor of numerous thin spruce stems had been placed (a kind of load-bed), running from side to side at the turn of the bilge. They were supported by short posts set upon the ceiling strakes. Across the stems lay hurdling, covered by a layer of sand (fig. 27.2b). Most probably a 'pavement' of erratic stones had lain embedded in the top of this layer of sand, while a substantial number of large tiles suggests that this floor abaft the main mast had held an integrated fireplace of sizeable dimensions. The continuous floor, measuring ca 15x6 m, very probably closed off the underwater body; this might encourage decay of the timbers, which people had attempted to counteract by tarring the wood. Of the Doel-I cog, recovered upside-down (Chapter 20), we (as yet) do not know whether it may have held a similar floor. It would appear that shortly before it sank, the interior of this vessel had undergone an elaborate tarring treatment. Whether this cog too might have had a military function cannot be ascertained, but it certainly should not be ruled out.

10.3.5 Wreck formation and embedding

The cog probably became stranded on the then coast or just offshore, close to the presumed mouth of the peatland channel on which was situated the medieval settlement of *Lenna* (now Lemmer). The accessible parts of the wreck may have been dismantled by the inhabitants of the village, reclaiming timber and firewood. Subsequently the remainder of the wreck sank through the peat layer and into the Pleistocene sand, and was probably quite soon filled and covered over with sedimentary material from the immediate vicinity, later locality with Almere clay. Until its discovery, the wreck lay sealed by a layer of Zuiderzee deposits. These same sediments were also encountered close to the west side of the wreck, in 'hollows' which - possibly after ca 1600 - were caused by wave action and currents in both the underwater original peat deposit, and in parts of the depression created by the wreck and filled with Almere sedimentary layers.

10.4 Model

No model was built of the Rutten cog.

10.5 Inventory

10.5.1 Distribution of the finds

Few finds were recovered in and around the wreck. Of the original inventory only non-valuable fragments were found, which is hardly surprising given the assumption that the ship became stranded on or just off the then coastline and subsequently dismantled. Anything of any value would have been removed from the wreck, in the first instance by the crew and later by the ship-breakers. The numbers in the find list (Plate 10.I A) are the find numbers allocated in the field. No definitive accession numbers have (so far) been assigned to the various finds. For the sake of convenience, in this report the described and illustrated finds have been given fictitious accession numbers. These fictitious accession numbers are used in plate 10.II A, in the text, in the captions and in the tables. The items marked with an asterisk* in the list of items and in the tables have not been preserved.

10.5.2 Ship's equipment and inventory

Ship's equipment

The most notable find associated with the wreck and in this category is a length of rope over 12 m long, which lay beneath the forebody to starboard, with one broken-off end close beside it (Plate 10.I A and figs 10.25a and b and 10.26). In the shipwrecks in the former Zuiderzee area, sizeable lengths of rope are not often encountered; in late-medieval vessels, never more than just small fragments. Stassen contributed a description of two pieces.[12] One had been freeze-dried and then treated with E-2250; the second, with Risinii-oil.[13] The cordage was laid in three phases. In the first, the fibers were spun into yarns. In this process they were S-twisted. Which is different from later periods: yarns of natural fiber (in this case

Table 2. Distribution of items

Category	Forebody	Hold	Afterbody
1. Ship's equipment	hawser-laid rope, nails, sintel clamps, rivet, wooden knob	rope, nails, sintel clamps	rope, nails, sintel clamps
2. Operational equipment	-	-	-
3. Administration and navigation	-	-	-
4. Tools	-	-	-
5. Household items			-
6. Cooking, eating and drinking	trivet, stoneware jug, spoon	bricks, globular pots, dish blue/ grey	stoneware jug
7. Victuals	sticks, bones, hazelnuts	stave bowls, splitwood box?, sticks, bones, hazelnuts	sticks, bones, hazelnuts
8. Personal belongings/weaponry	shoe fragment?	sheath decoration?	-

12. Stassen 1992, appendix 5.
13. Wevers, 1990.

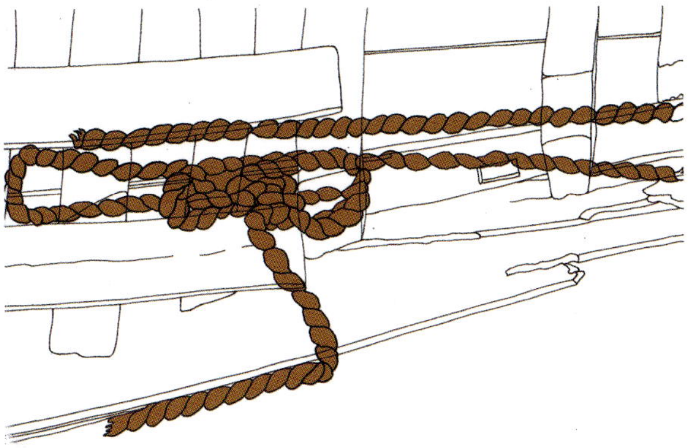

Fig. 10.25 a. The cable-laid rope in situ. b. The cable-laid rope beneath the forebody of the wreck.

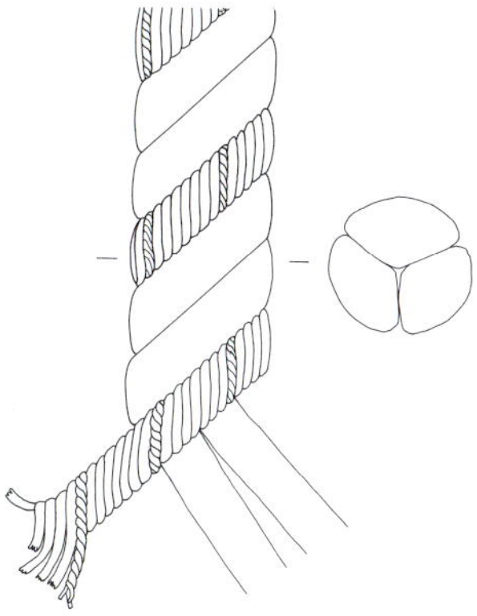

Fig. 10.26a. Fragments of the cable-laid rope after application of different impregnation methods. b. Detail: the structure of the cable-laid rope. After Stassen 1992.

hemp, *Cannabis sativa*) have been Z-twisted from the second half of the 16th century until today. The Z-twisting of the fibers is necessary to employ the natural twist of the bast fiber cells of which a well-prepared fiber mainly consists. In the second phase the yarns were S-twisted into strands, each strand in this case comprising 20 to 22 yarns. In the third and final phase, three strands were Z-twisted to produce the desired cordage.

The fragments described by Stassen show various sizes in the thickness of yarns, strands and finished rope. Other parts of the rope, not examined by Stassen, display yet different thicknesses. On the one hand, this is attributable to the condition of the rope as recovered from the excavation; on the other, the cause may lie in the different methods of conservation. Inconsistent production methods may in part be responsible too. The rope has (or had) a diameter of 6.5 to 6.8 cm, the yarns have a diameter of 3 cm. The coarseness of the fibers indicates that the yarns were made without the use of a rope-jack, and that rope-making technology still was fairly basic. According to the 19th-century rope classification of the Royal Dutch Navy, this cordage would be classified as hawser-laid rope of the *Gijn* type. Such rope would be used for the running rigging and for lifting heavy loads.[14] Stassen states that the rope found with the Rutten cog is best classified as 'three-phase cordage'. The earliest evidence for the use of hemp as a raw material for certain parts of ships dates from Viking days. In the Oseberg ship (ca AD 800), fragments of fabric were found for which hemp had served as the raw material. Analysis of cordage from the former medieval harbour of Schleswig showed 80.4% to be made from bast fibers of trees, 4.9% from animal fibers such as wool and horsehair, and 14.7% from twigs.[15]

14. Fremery, 1878.
15. Körber-Grohne 1989, 11-121.

In the wreck also fragments of three other types of rope were found (fig. 10.27-8, -9, -10). the thinnest is 1 cm thick, consisting of three twisted strands of five spun yarns each. The second is 1.6 cm thick and also consists of three strands of five yarns each, while the third has a diameter of 3 cm and is composed of four strands of five yarns. In the third and final phase of rope-walking the strands were Z-twisted. The twist in the first and second phases was not established, but is likely also to have been a twist. It is unclear what purpose the thinnest and thickest of the three rope types served. However, we do know in the case of the 1.6 cm thick rope, which offers an interesting insight into a particular activity on board. In various places on the bottom between frames 3 and 25, between the frames and in the limber holes on both sides of the keel plank, pieces of such rope were encountered, running fore-and-aft through the limber holes. At frame 25, at the end of a length of it, there was a grip. Without any doubt this was a limber-rope to be periodically pulled to and fro in order to clear the limber holes of dirt and debris, so any bilge- and rainwater could freely drain to the place(s) from where it would be pumped out.

In the afterbody on the keel plank at frame 3, a small piece of wood was found, 8.5 cm long and ca 2 cm thick (a grip), with a groove at one end that held a piece of rope 1.6 cm thick (fig. 10.27-13). At another frame in the afterbody lay a second wooden grip (fig. 10.27-14). In one of the limber holes of frame 19 a small, complete rivet was found (fig. 10.27-15). A curious find, since no rivets were used in the construction of cogs. The two wooden grips and the rivet may have been attached to the string that was pulled to and fro through the limber holes. Such limber-ropes (with grips) running through the limber holes may feature also in the Nijkerk-I cog (Section 8.5.2). In the Rutten cog its significance may be explained by the (permanent) presence of the ballast stones (Section 10.7) and the floor. These meant that the spaces between the frames and the limber holes were not accessible for maintenance. A globular wooden knob with a flat base (fig. 10.27-17) may have been the top of a flagstaff or the like.

Nails of two types were found in the wreck. For the clinker strakes nails were used that were ca 140 mm long with a convex, faceted or pyramidal head (diameter 35-40 mm and 8-10 mm high; fig. 10.27a-11). Those of the other type (fig. 10.27a-12) are smaller (ca 70 mm long) and

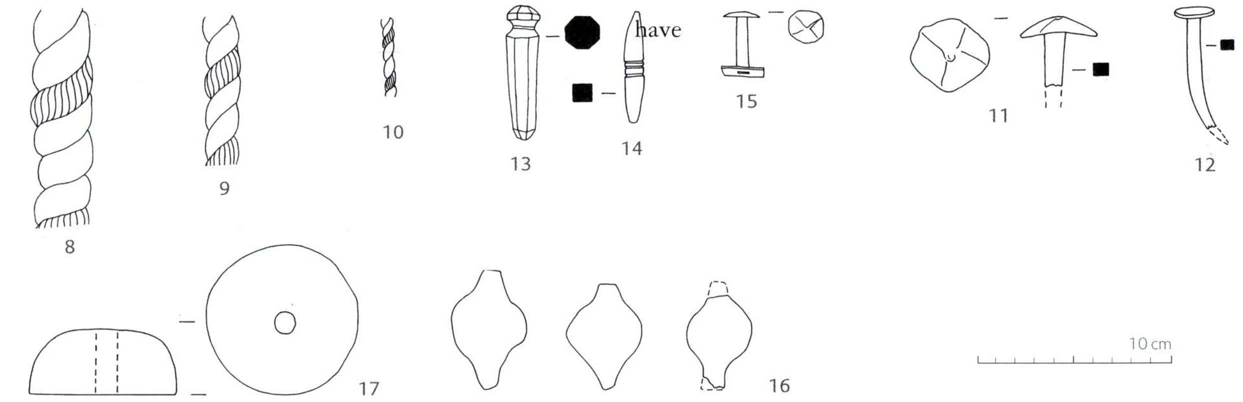

Fig. 10.27a. Three different thicknesses of cordage (8, 9 10); nails (11, 12); two wooden grips (13, 14); rivet (15); sintel nail clamps (16); wooden flagstaff knob(?).

Fig. 10.27b. Fragments of the rope that was pulled to and fro to clear the limber holes.

Table 3. Ship's equipment

Category	Item	Number	Findspot
1.1 Ground tackle	-	-	-
1.2 Sails	knob of flagstaff?	17	afterbody
1.3 Cordage	cable-laid	7	under forebody, sb
1.4 Blocks	-	-	-
1.5 Windlass	-	-	-
1.6 Pump, drain	-	-	-
1.7 Navigation aids	rope, button, rivet	8, 9,10, 13, 14 and 15	on the bottom
1.8 Boats	-	-	-
1.9 Spare parts	-	-	-

flat, round heads (1.5-2 mm thick and ca 20 mm across). The sintel nail clamps have a central piece that in length varies between 28 and 40 mm (fig. 10.27-16). The nails and sintel nail clamps likely to have been part of the hull have been omitted from Table 3.

Working equipment
No items were found in the wreck that might be reckoned to this category.

Documents and navigational equipment
No items were recovered that might have been used for administration or navigation.

Tools
No tools were recovered.

Domestic utensils
No domestic utensils were found with the wreck.

Cooking, eating and drinking; fireplace
Scattered among the ballast stones, principally in the midship and after sections, a considerable number of brick fragments were encountered (Plate 10.I A). One brick had survived intact; it measured 28x13x8 cm. The bricks have yellowish-white inclusions (fig. 10.28). Whether the bricks belonged to a fireplace - there must have been one aboard - cannot be ascertained. No traces of burning or soot were observed. That they were intended as ballast seems unlikely. Maybe there had been a hearth platform of large bricks on top of the ballast or integrated in the floor. The prominent inclusions and the aspect of the bricks closely resemble those of the bricks from which the *Danewerk* near Hüsby in Schleswig had been built. This defensive wall on the southern border of the Danish realm was built by King Waldemar the Great (reigned 1157-1182); the format of the bricks is 26.5x11.5/13x8 cm. The format of the large bricks in the oldest part of the Sankt Petri cathedral at Schleswig is 28x14x8 cm (fig. 10.29).[16] The then new building material, brick, was used especially in the north wing.

Fig. 10.28. Bricks. Fragments with yellowish-white inclusions. Found among the ballast; possibly part of a fireplace.

Fig. 10.29. Two bricks from the oldest part of the Sankt Petri cathedral at Schleswig (ca 1180-1200) and two fragments of bricks from the Danevirk at Hüsby (Schleswig) (second half of 12th century).

16. Gift from Dr V. Vogel, Archäologische Landesmuseum der Christian-Albrechts-Universität, Schleswig.

This part of the cathedral was started in the early decades of the 12th century, and completed between ca 1180 and 1200. The brick type with inclusions was applied also in the 12th-13th-century churches along the Baltic coast between Schleswig and Rügen.[17] This brick format, be it with yellowish-white swirls, also occurs in the western Netherlands (Holland) in the second half of the 13th century[18] Below the keel plank in the afterbody, a small rod of iron was found, which may be a foot of a trivet (19). Below and inside the wreck some pieces of (fen-)peat (28) were encountered. A complete block of fuel peat measures 27x11x6 cm.

Cooking-utensils

Among and under the ballast stones in the hold, various sherds of *Kugeltöpfe* were found - mostly wall sherds and small sherds of pots of the usual size, as well as rim/neck fragments of two large specimens (figs 10.30-4 and -5 and 10.31). Their rim diameter is ca 26 cm. The fragments are sooty, clearly having belonged to cooking pots. Some of the other sherds had lain in open fire. Judging by the fabrics, the sherds belonged to at least ten different vessels of varying sizes. Probably these all were pots that broke well before the ship foundered, and no longer were part of the ship's inventory.

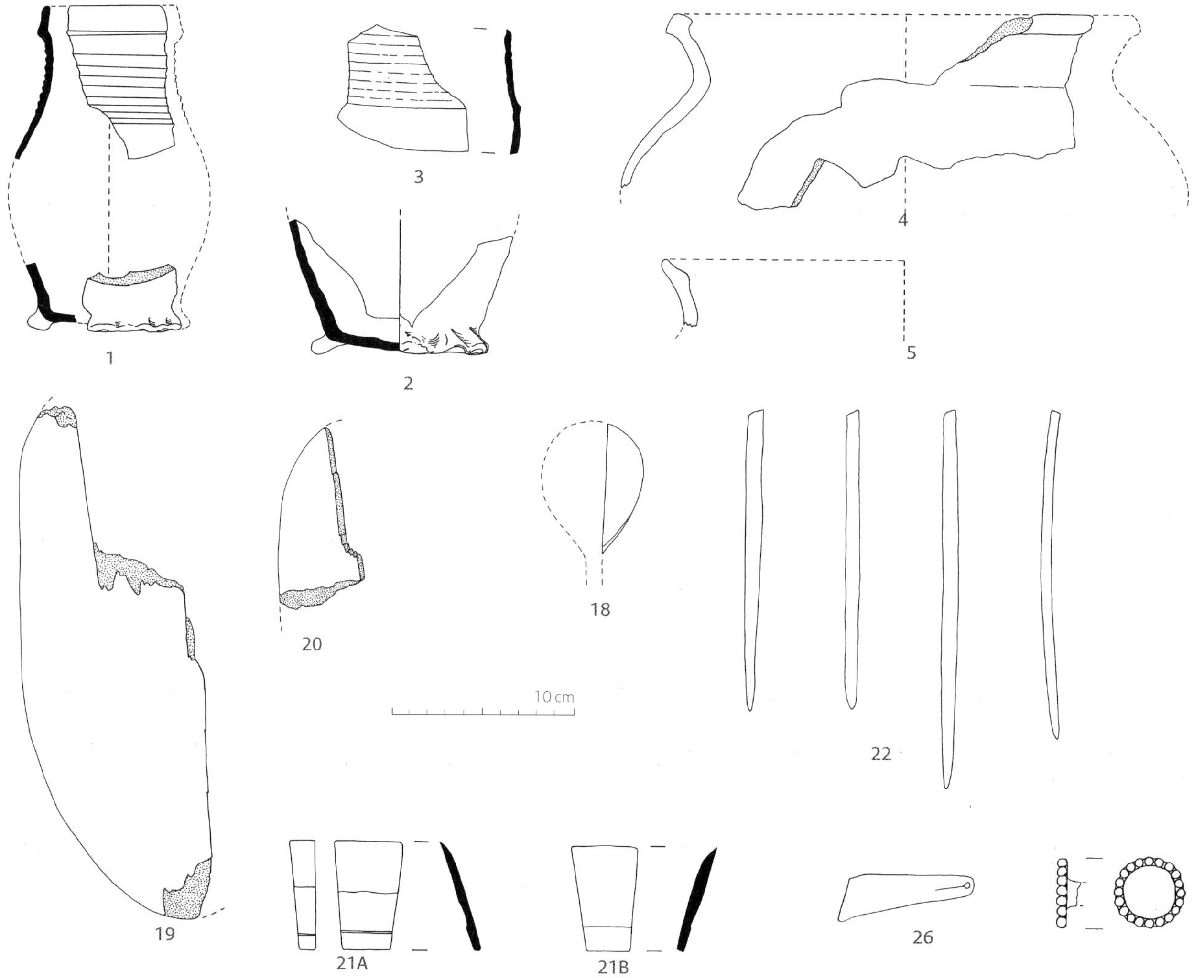

Fig. 10.30. *Kugeltöpfe* (4 and 5); proto-/near-stoneware jugs (1, 2, 3); spoon fragment (18); splitwood boxes? (19, 20); staves of two stave bowls (21 A and B); "cog sticks" (22); pewter(?) button(?) (25) and leather shoe(?) fragment (26).

17. Author's observation, 2015.
18. H.L. Janssen, written comm., October 2014.

Fig. 10.31. Sherds of proto-stoneware.

Eating- and drinking gear

Under a frame in the forebody, part of the bowl of a wooden tablespoon was found (fig. 10.30-18). Sherds of stoneware jugs were found mostly below the ballast in the midship and after sections. They include collared rims, sherds of neck and body, and base rims with pinched feet. Some are salt-glazed and others have an iron-rich slip. The surface and fabric of the various pieces show that most belong in the category of so-called proto-stoneware, and some possibly the near-stoneware that succeeded it. Many of the pieces match the forms and types of pottery known from the late-13th-century fill of the immunity boundary ditch of St Paul's Abbey in Utrecht.[19] Some larger (fitting) sherds will be described here. The small jug (figs 10.30-1 and 10.31-1) is of a slender, pear-shaped model with a collared rim and a rilled neck, and has a pinched foot. Its fabric and surface are a dirty yellow. It can (still) be regarded as proto-stoneware, but is closer to near-stoneware. It was found in the stern.

The neck/wall sherd (figs 10.30-2 and 10.31-2) has rilling on the neck, has a hard, grey fabric and yellowish stripes and spots on the interior. This probably is what is known as 'near-stoneware'. The same goes for the base fragment with a pinched foot of a third jug (figs 10.30-3 and 10.31-3). It is of an orange fabric with a grey core, and has grey-brown saltglaze on the exterior. The sherds of jugs 2 and 3 were found in the forebody. One small base sherd of a dish of blue-grey Elmpt ware (6) was recovered.

Victuals

An uncommon find is the near-complete skeleton of a sheep or goat (figs 10.32a and b). It lay to starboard, partially on and partially below the bottom. The animal may have been on board as livestock, and would have drowned. However, small bones of the tail and hooves are missing, so it may instead have been a skinned carcass. Furthermore a considerable number of bones of cattle and a few pig bones were recovered. These are bones with chopmarks, indicative of butchering or consumption waste. Also various fish bones were collected (24). The full analysis of the bones from the Rutten cog is presented in the Appendix. As parts of food containers may be mentioned three staves of two different stave bowls (figs 10.30-21 a

Table 4. Cooking, eating and drinking

Category	Item	Number	Findspot
6.1 Fireplace	bricks	-	among and on ballast
6.2 Fire-tending tools	trivet?	19	under the afterbody
6.3 Fuel	peat	28	beside and in the wreck
6.4 Cooking equipment	*Kugeltöpfe*	4, 5	among ballast
6.5 Eating- and drinking gear	jugs, blue-grey dish	1, 2, 3, 6	after- and forebody

Table 5. Victuals

Category	Item	Number	Findspot
7.1 Fresh-water supply	-	-	-
7.2 Supply of beverages	-	-	-
7.3 Livestock	sheep/goat skeleton	23	beside the hold, sb
7.4 Food provisions	split-wood boxes, sticks, bones, hazelnut shells	19, 20, 22, 24, 29	hold, stern, on the bottom

19. Smit, 1987.

Fig. 10.32 a. The skull of the near-complete sheep or goat skeleton in situ.

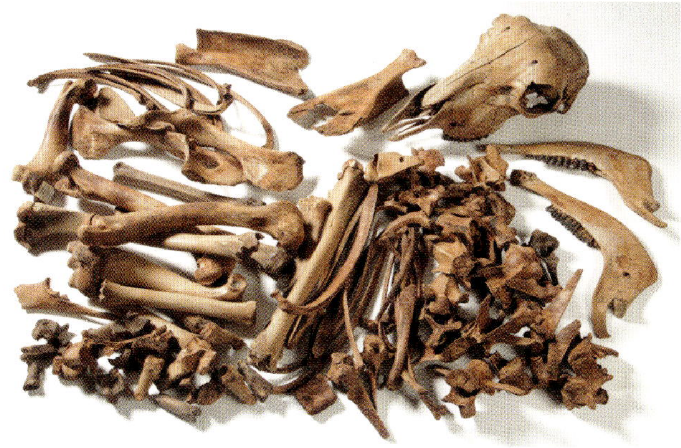

Fig. 10.32b. Skeletal parts of the sheep or goat.

and b). They are staves of the familiar late-medieval low wooden bowls (cf. those from the Nijkerk-II cog, fig. 6.65-90). The two fragments of oval wooden bases or lids might be of split-wood boxes (figs 10.30-19 and -20), and may possibly have served as containers for storing and transporting foodstuffs. The same goes for the large number of sticks (fig. 10.30-22), which probably were part of small, oval woven baskets with a wooden base, which probably held sweet chestnuts or nuts (see the Nijkerk-II cog, section 6.5.3 and figs 6.65-93, 6.66 and 6.67). Hazelnut shells (29) were found scattered about the bottom of the wreck.

Personal belongings/weaponry

Just three small items may be put under this heading. A small, daisy-shaped pewter(?) button(?) 8 mm across and 1 mm thick, with a smooth centre and a beaded rim, and at the back, part of a rod or loop for attachment (fig. 10.30-25). It may have belonged to a knife sheath, shoe or garment. The second is a small piece of leather with a buttonhole (fig. 10.30-26), maybe cut from an old shoe. The third is a small coin (27). The coin is badly corroded, so it can no longer be identified. Nor did X-ray photography offer any clarity.

10.6 Cargo and merchandise

There were no finds pointing to cargo or merchandise.

10.7 Ballast

Mainly in the hold and the afterbody lay a sizeable batch of ballast stones (Plates 10.I A and 10.II A and fig. 10.33). Interestingly, over half of them consist of Scandinavian flint. The other part are of crystalline rocks. The flints almost certainly were gathered on one of the limestone coasts of the Baltic Sea, e.g. on the Danish island of Møn. The total weight of the stones collected during the excavation is 3750 kg. Including the part removed by the bulldozer in 1980, the original total presumably amounted to at least 5000-6000 kg. It is the only instance of a (large) cog being found upon excavation to have carried a substantial and probably permanent weight in ballast. Among the rocks some small fragments of amber were identified, which also points to the stones having been gathered from a beach.

Table 6. Personal belongings and weaponry

Category	Item	Number	Findspot
8.1 Garments	-	-	-
8.2 Footwear	shoe fragment	26	forebody
8.3 Sewing kit	-	-	-
8.4 Tools/weapons	button?	25	hold
8.5 Cash	small coin	27	forebody
8.6 Miscellaneous	-	-	-

Fig. 10.33 The ballast to starboard amidships.

10.8 Dating

Soil profile

The soil profile, so close to the then coastline, offers no other dating evidence for the sinking of the Rutten cog than that it must have occurred before ca 1600.

Finds

The *Kugeltopf* sherds may be labelled 13th-century. The recovered proto-stoneware and near-stoneware sherds provide the closest dating evidence for the loss of the ship. Janssen[20] states that proto-stoneware was especially prevalent during the final quarter of the 13th century and particularly between ca 1290 and 1310, and that it dropped out of use in the second quarter of the 14th century. Shortly after 1300, a fabric known as 'near-stoneware' hit the market. The bulk of the stoneware sherds, belonging to about 25 different and especially small pots which were found among and below the ballast stones, can be reckoned to the proto-stoneware. The few larger sherds from the stern and bow may be classified as proto-/near-stoneware. This means that the proto-stoneware sherds ended up among and under the ballast stones in the final quarter of the 13th century. The proto-/near-stoneware sherds belong to small pots that may have broken shortly before or during the calamity. The vessel most probably became stranded shortly before or around 1300. A small piece of a human *fibula* (30), which lay among the frames in the forebody, a curious find, may be the last remaining relic of a drowned crew member.

Sintel clamps and nails

The sintel nail clamps securing the cog's caulking are of types C1/2(D) and D1, which were in use during the second and third quarters of the 13th century.[21] The large nails with convex, flattened or pyramidal heads were primarily shaped to suit the size of the ship and yield no dating evidence. The smaller, flat-headed nails may be considered a 13th-century version.

Radiocarbon analysis

The radiocarbon laboratory of the University of Groningen dated two timber samples of the Rutten cog.[22] The first sample (GrN-20004) was of an air-dried moss lath which may have come from the heartwood of a thick trunk. The second sample (GrN-20344) was taken from the youngest wood of a plank belonging to the bottom or ceiling. The dates were 1140 ± 30 BP and 959 ± 26 BP, respectively.

Dendrochronology

'RING' examined four timber samples, which could be dated among each other and averaged into an object chronology, ka 5 all.[23] The outcome is AD 1249 + 20 ± 6 years, which amounts to a felling date between 1263 and 1275. The full report by 'RING' is presented in the Appendix. Three of the four samples could be dated with the reference chronology of Poland; the fourth, with that of Lower Saxony (Niedersachsen). The building date of the ship must therefore have been between ca 1275 and 1285. The condition of the recovered timber suggests no great age when the vessel was lost; 10 to 20 years seems quite plausible.

10.9 Summary

Ship

The characteristic hooks, joined by a keel plank, the carvel-built bottom whose strakes for and aft merge into a clinker-built construction, the sintel nail clamped moss caulking and the keelson with its integrated mast-step

20. Janssen 1983, 175.
21. Vlierman 1996a, 74-36.
22. J.N. Lanting, written comm. 16 June 1995, GrN-20004 and GrN-20344.
23. RING-Report April 1994, Hanraets en Jansma.

section, these even at its surprise discovery in 1982 made it clear that the wreck in lot A57 in the Noordoostpolder probably was the underwater body of a large, massively built cog. This idea was confirmed by the excavation in 1985; and a few futtocks came to light which provided evidence about the lower part of the sides. On the basis of the gathered evidence the Rutten cog could be reconstructed with great confidence up to a height of 1.8 m. Comparison with other cogs allowed a hypothetical but reliable, full reconstruction of the most complete frame recovered (frame 23). The oblique, straight head of the futtock of this frame suggests that it supported a protruding crossbeam. The latter's location is some 80 cm lower than its counterparts in the Nijkerk-I and -II cogs; ca 100 cm lower than in the Doel-I cog; and ca 20 cm lower than in the Bremen cog.

Inside the wreck, mainly amidships and in the afterbody, a mass of ballast stones was encountered, weighing 5000-6000 kg. About half of them are Scandinavian flint nodules. Overlying this ballast and probably at about a meter above the keel line there seems to have been a floor that may have extended between frames 3 and 25 and from side to side. Integrated into this floor may have been a large hearth built of bricks. The ballast and floor probably were part of the ship's design from the very start. The relatively low-set protruding crossbeams, the floor immediately overlying the ballast, the heavy construction of the vessel and the likely dimensions of the hull mean that the ship most probably had two rows of protruding crossbeams: four below and six above. Probably the ship had transverse constructions fore and aft, at the level of the lower row, that did not protrude through the shell, and which may have supported between-decks.

The reconstructed hull shows a ship with elegant lines with no dead flat (the hull planking gradually rises from the midships towards the sides and the stern and stem, resulting in different hull sections throughout) and whose bilge and subsequent strakes make a flowing transition to the everted, straight sides. The rake of the stern and stem and of the straight parts of the sides is ca 117°, 119° and 104°, respectively. The bottom is carvel-built over a length of 7 to 8 metres. In the case of cargo transportation the space between frames 3 and 25 could constitute the hold. This would have a capacity of ca 175 gross tons. However, the ship is more likely to have served as a military vessel, a *heerkogge*. Its principal (reconstructed) dimensions are:

- Length over the stern and stem at gunwale height — 22.40 m
- Largest width at gunwale level — 9.50 m
- Height of the stern (keel line to the top gunwale) — 5.05 m
- Height of the stem (keel line to the top gunwale) — 6.60 m
- Height of the side amidships (keel line to the top gunwale) — 4.50 m

The ship did not founder at sea, but probably became stranded on a (submerged) peaty shore, close to the mouth of the peatland channel beside which medieval Lemmer was situated. The actual absence of any items of inventory and equipment, and the absence of any timbers higher than the turn of the bilge suggest that everything of value had been taken from the ship, after which the wreck was dismantled for construction timber and fuel. The remainder became filled with sediment and was eventually covered by Zuiderzee deposits. The only two notable finds were a piece of thick hemp rope ca 12 m long and the near-complete skeleton of a sheep or goat.

Dating

The dendrochronological analysis produced a felling date of the timber between AD 1263 and 1275. This suggests that the ship was built between ca 1275 and 1285. The wood was imported from Poland and Lower Saxony (Niedersachsen). The type of sintel nail clamp was current in the second and third quarters of the 13th century. The proto-stoneware sherds found among the ballast must be of jugs that broke during the active life of the ship; the proto-/near-stoneware sherds must be of jugs that presumably broke shortly before or during the ship's stranding. They offer the closest dating for the calamity, which probably took place shortly before or around 1300. The provenance of the timber, the ballast of Scandinavian flint, and possibly also the transverse constructions with upright knees suggest that the vessel originated and was built at a location on or close to the Baltic coast of Germany or Poland.

Impression of the sailing Dronten-I cog.

The Dronten-I cog[1]

11

11.1 Find conditions

11.1.1 Discovery and investigation

In October 1973, deep ploughing of the soil in lot N5 in Eastern Flevoland revealed ship remains. The find was reported to the Museum for Ship Archaeology of the IJsselmeerpolders Development Authority (RIJP), established nearby since April 1971, and an exploratory investigation took place on the 29th of October. Apart from the frame timbers and planks which had been ripped up by the deep ploughing machinery and found on the surface, a trial trench showed that the seams of the hull planks were closed off with sintel clamped moss caulking. Furthermore, probing showed that over a length of ca 15 m, a width of ca. 3 m, and at a depth of 0,7-0,8 m under ground level a shipwreck was situated. The excavation, salvage of the finds and the documentation of the wreck took place between 2 August and 4 September 1976. When the wreck was excavated it was found to be very incomplete, and that a large amount of frames, planks, the stern hook and the V-shaped timber with the mast-step had been torn from their original position. A few detached parts of the construction and finds were found *in situ*. Wherever possible the timbers were placed in their original positions and the wreck was measured, drawn and photographed (fig. 11.1 and the Plates 11.I A, B, C en D).[2] On the 12th of August the *Meetkundige Dienst* (Delft) made photogrammetric recordings from a helicopter of the Royal Airforce.

Fig. 11.1. General view of the wreck excavation. Viewed from the bow towards the stern.

1. Translated by Jeroen Vermeersch, edited by Xandra Bardet.
2. The wreck lay with its stem in NNE direction at 100 m of the Ketelweg and at 136 m from the boundary ditch of N5/6. The find location is situated approximately against and on top of the eroded eastern bank of the NW-SE orientated peat barrier between Schokland and Elburg, at 1.5 to 2 km from the present-day mouth of the IJssel, the so-called Keteldiep, and at 9 km from Kampen (figs 11.2 and 11.3), Ordnance map page 21C Roggebotsluis; coordinates 181.11/509.12 with its highest parts at ca. 2.3 m -NAP, Hydrographical chart 1921, average low-tide level.

Fig. 11.2. The findspot of the Dronten-I cog.

Fig. 11.3. '*Topographical and Ordnance Map of the Kingdom of the Netherlands*' (scale 1:50,000) of 1861 with the location of the Dronten-I cog.

The wreck and part of its inventory were (albeit concisely) published, the ceramics were treated in the *Corpus voor Middeleeuws Aardewerk uit gesloten vondstcomplexen in Nederland en Vlaanderen* (CMA) and in an overview of stoneware from excavated medieval shipwrecks in the Netherlands.[3] At the time of the study of the excavation data for the report of 1980, little was known and published about archaeological data in regard to ship construction in the late-medieval period. The report therefore only refers to similarities to the Ens cog (Chapter 9), and remarks that the dimensions of the ship deviate from those of the Bremen cog. Apart from this, it was debated whether the vessel had a spar sail or spritsail.[4] Reconstruction of the cog went no further than stating that the vessel had six or seven strakes on either side of the keel. Next to all the other described cog finds in this study, and with the current state of knowledge and insights, it is meaningful to revisit the wreck and - as far as is justified based on the present data and in comparison to the other ships - to reconstruct it. The finds will be discussed in more detail. The locations of the finds are shown with their inventory numbers in plate 11.IA. The list of finds indicates the objects that the numbers refer to.

11.1.2 Geology and stratigraphy

During the excavation initially a baulk was retained amidships in order to study the stratigraphy. This seemed only to be partly possible as a result of the soil having been turned up. The data were recorded and described by R. Koopstra of the Pedological Department of the RIJP as follows: "The ship sank in the Almere phase, where the redeposited layer ends at the interface between the clay-rich Ala and the humus-poor Alc deposits. The date of the Ala layer is therefore of importance in order to determine the period when the ship sank. Figure 11.4 shows the supplemented and partly reconstructed profile with the disturbance by the shipwreck. The top of the present peat is situated at 1.3 m below ground level, the Pleistocene sand at 3 m."[5] At ca 200 m north, ca 400 m east and at ca 1500 m west of the findspot of the wreck, the top of the peat is situated at 0.8-1 m below ground level. The depression in between had been filled with a humus-rich clay deposit. The wreck lay in the upper part of the humic-sandy deposit which was described by the pedologists of the RIJP as a so-called 'foredune'. With regard to the Noordoostpolder, Wiggers assumed that the SLoef Ib layer - in Oostelijk Flevoland referred to as Ala - had been deposited in the second half of the 15th century and the early 16th century.[6] The redeposited layer ends at the base of layer Ala; hence we can assume that the ship sank before the second half of the 15th century. Since the ship is situated along the eastern (i.e. leeward) side of a NW-SE oriented peat barrier, it cannot be excluded that a part of the Alc deposit had been washed away, which suggests an earlier date for the loss of the ship. Also the sharp transition between the Ala and Alc layers (geolog-

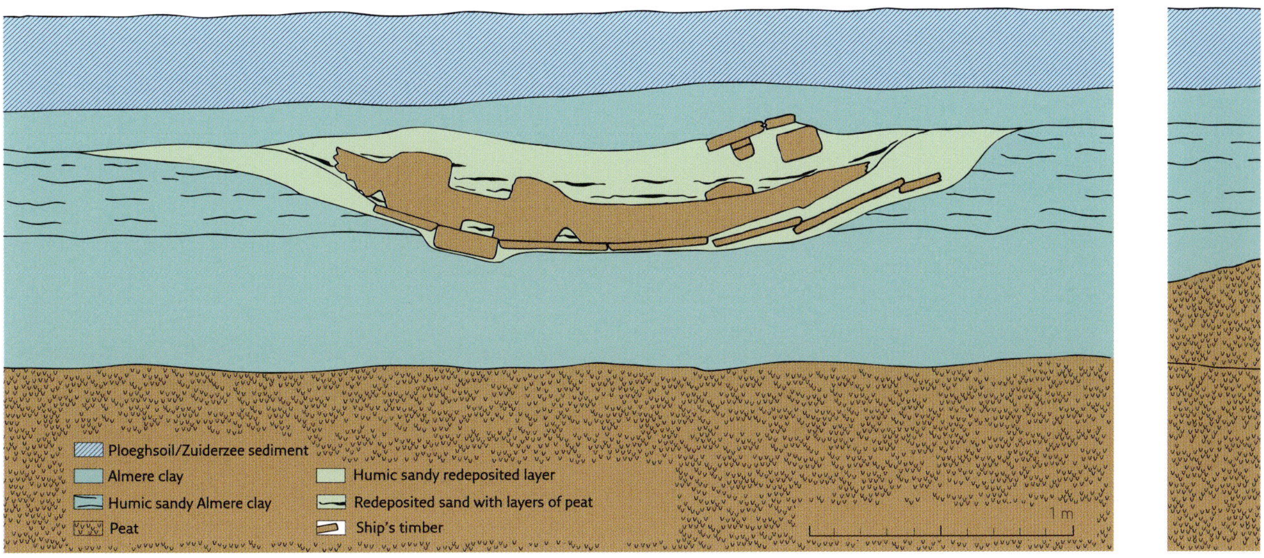

Fig. 11.4. Soil profile with the disturbance caused by the Dronten-I cog (partly reconstructed).

3. Reinders *et al.* 1980; Vlierman 1983 and Vlierman 2010, 192.
4. Reinders *et al.* 1980, 15 en 50.
5. In Reinders *et al.* 1980, 7-8.
6. Wiggers 1955, 88 ff.

ical hiatus) and the completely sandy redeposited layer support this possibility.

the removal of the last remnants of the peat barrier that emerged from the water, if any still did so at that time.

11.1.3 Medieval habitation

In the course of the 1980s and the early '90s, I asked several farmers along the Ketelweg, also on lots N4, 5 and 6, to save any sherds and bones found as they worked their land. This land was situated on the suspected western and eastern margins of the above-mentioned NW-SE-oriented peat barrier. Some sherds date from the Carolingian period, some others from the 13th and 14th centuries. The same phenomenon seemed to occur at some other sites towards Elburg, but at the time we did not manage to find more such evidence. But on a lot on the Oudebosweg, situated *in situ* and at a depth of 0.5 m, a row of large, 14th-century bricks were found (possibly a foundation with the width of one stone). The owner reported that in the course of the years a large quantity of these bricks had been recovered at that same spot, whereby the bricks were all arranged next to each other and were stacked in two or three layers. There were no traces of mortar. During a visit, a 14th-/15th-century redware sherd was found nearby.

The farmer had ignored these finds or just threw them away. The presence of Carolingian and late-medieval sherds along the edges of the original clay-on-peat land had also been observed in the direct surroundings of the Nijkerk-II cog, where most probably this is related to waste from the Carolingian and 13th- and 14th-century rural settlement of Ark (Chapter 6, section 6.1.3).
It is possible that at some locations on the peat barrier between Schokland and Elburg there had been habitation, maybe even into the 14th century. Similarly some domestic waste from the occupation surface on the clay-on-peat land would have ended up in the water through erosion. It seems plausible that the peat barrier had been transected by some watercourses, for example old extensions of the Zuiderdiep and the Reeve. These in particular are places where habitation could have been located (fig. 11.5). In 2012, at a distance of about 1800 m southwest of the site with the bricks, a remnant of a gully was found at which traces of prehistoric habitation were found (the extension of the Reeve?).[7] This is a more or less comparable situation to what was found in the close proximity of the Nijkerk-II cog. Brunnepe and Kampen lie ca. 8 to 9 km away as the crow flies. The 13th- and 14th-century habitation on the southwest point of Schokland is at 6 km from the site with the bricks. The storms of 1367 which resulted in the flooding of the Oosterwoldse Polder and Kamperveense Polder[8] will have contributed to

11.1.4 Condition and wreck formation

The wreck was located at a depth of ca 1 m below the surface and sank almost completely horizontally into the humus-rich clay of the Almere deposits, which had formed directly on top of the peat. The vast majority of the wreck above this sediment was enclosed by a humus-poor clay deposit with a great amount of sandy layers. The filling of the wreck consisted of redeposited sand, while the redeposited sediment around the wreck was humus-rich to sandy. During and directly after sinking, the vessel would have been shaken back and forth quite violently, given the fact that most pots were broken. The excavated inventory and the personal possessions and weapons suggest that the vessel must have sunk against, or in close proximity to the peat barrier or (submerged) shore and that the vessel was abandoned head-over-heels. The findspot of the wreck was located not far from the probable then mouth of the Zuiderdiep, the south-westernmost branch of the IJssel delta, which at that time was still open. After the wreck had settled in - what was then - the sea bed and had sunk partially into it, it probably quite quickly filled with a layer of humic-sandy sediments 10-30 cm thick. After this, the above-mentioned redeposited sand came in. The source of that sand is not quite clear. Most probably it was transported by the river IJssel.

Considerable time after its foundering, possibly dozens of years later, it is probable that the wreck - the upper sides of which must already have been severely damaged - was washed free and broken apart by the waves and/or current. This may also have been caused by ice floes. In this situation the surviving parts of the sides were broken off. The part of the sheer strake, with the channel for attaching the rigging to hold up the mast still preserved on it, must have broken off at an early stage and ended up inside the ship, since the side planks situated lower in the hull have all disappeared. The planks, futtocks, ceiling planks, the decks, beams of stern and stem and transverse connections and even the stern hook all were ripped out of construction. The stern hook was situated in situ on top of the finds in the afterbody, along with some V-shaped frames. Remarkable is the detached find of a strongly deformed gudgeon (figs 11.6a en b) of the rudder suspension, which had originally been attached to the stern hook. It most probably got ripped off from the stern hook together with the rudder and sternpost, and subsequently fell off the pintle/rudder to starboard of the afterbody. The rudder

7. D. Velthuizen, personal communication, January 2013.
8. Gottschalk 1971, 388-390.

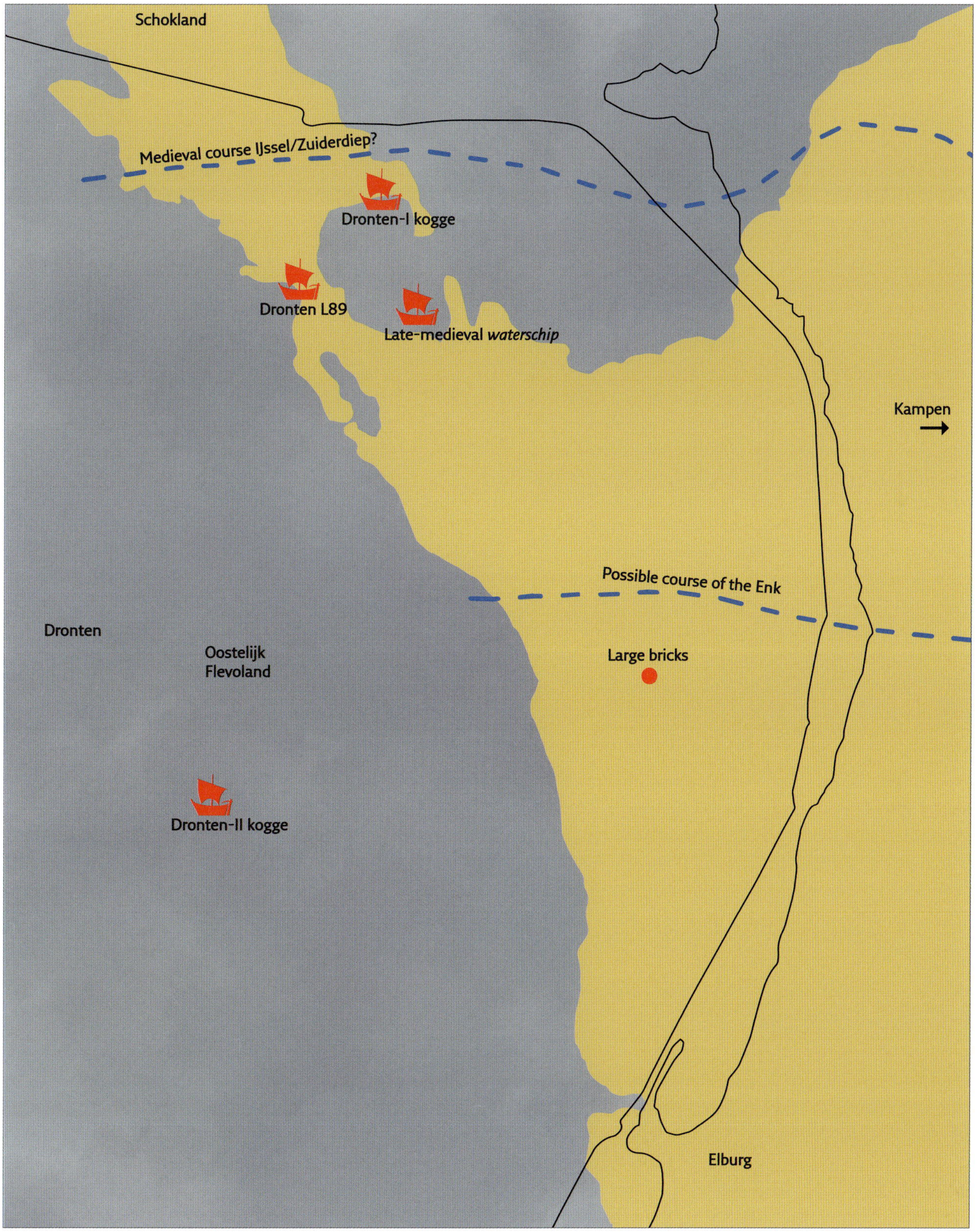

Fig. 11.5. The remnants of the peat barrier in the subsoil between Schokland and Elburg, with the findspots of the Dronten-I cog (1), a late-medieval *waterschip*(2), The Dronten L89 wreck (3), the Dronten-II cog (4), the stack of large bricks (5) and the likely course of some creeks or streams.

Fig. 11.6a & b. The detached, distorted gudgeon.

has not been found. A similar scenario played out with the transverse connections at deck level. Two iron rose-bolts were found next to the wreck at spots almost exactly where they used to pass through the hull to strengthen the connection of the sail beam (Plate 11.I A). The sail beam was lost, and also no remnants were found of the hull planks through which they had passed.

During the excavation, not much more of the ship was found than part of the bottom to port, with plank fragments of the four consecutive strakes of the sides of the forebody, the keel plank with the two (stern and stem) hooks, part of the straight and V-shaped floor timbers, of which one had the recess for the mast heel, a fragment of a bilge ceiling plank and a part of the top ceiling plank or inwale. It seems plausible that the storm period of 1367 can be linked to this destruction of the wreck. After the site settled down and the remaining hull and loose timbers were covered with the upper Almere-clay sedimentary deposit it remained undisturbed until the deep ploughing of the soil started. The wood, especially the thinner planking, in the excavation did not seem to be in a very good state. This was the consequence of the shallow position (higher than the drainage) in the humus-rich/sandy sediment in which it had lain. Moreover, as a consequence of the hot summers of 1975 and 1976, and especially also because of the deep ploughing of the soil, the wood in the soil dried out in a relatively short time.

11.2 The ship

11.2.1 Documentation and reconstruction

Documentation

During the excavation both the situation of the wreck and the location of the detached timbers were noted in a few sketches (figs 11.7a, b, c and Plate 11.I A). After the wreck was cleaned the loose parts were put back in position inasfar as this was still possible, and a lengthwise section along the hooks and keel plank was made, together with 14 cross-sections over the frames, and all were drawn to scale 1:20. This was done by the Department of Ship Archaeology in the usual way: by extending a main tape over the hooks and keel plank and at right angles to this the tapes over the frames, which were levelled above the wreck and fastened to ranging poles (Plates 11.I C and D). Based on the photogrammetric images, an overview of the site was made by the *Meetkundige Dienst* of Delft. During the fieldwork black-and-white photos and slides were made. In 2008 the preserved stern- and stem hooks and the V-shaped floor timber with its recess for the heel of the mast were drawn to a scale of 1:10 in preparation for this study.

Reconstruction

In the plates of the excavation report of 1980, the reconstruction drawings based on the available data were restricted to a side view with the recovered parts, and one cross-section at frame 19 was reconstructed, just up to the six strakes which were attested *in situ*. Apart from these, a few construction details of timbers and of sintel-clamped moss caulking were drawn.[9] As mentioned before, it seems justified to continue the reconstruction on the basis of comparison with the other cogs presented in this study. Some loose ship timbers were now placed in their probable original position and the ship's hull was

9. Reinders *et al.* 1980, Plate 2.

Fig. 11.7a. Finds *in situ*: Sherds of a small *kugeltopf* and a near-stoneware jar, a shoe sole, a knife with its handle and some fragments of large bricks belonging to the fireplace.

Fig. 11.7b. Sherds of a large near-stoneware jar.

Fig. 11.7c. Small near-stoneware jug, fragments of large bricks, and the detached, displaced stern hook *in situ* overlying the afterbody.

made complete in order to get a reconstruction as close as possible to the full original hull form. In doing so it was assumed that the vessel had an afterdeck and a foredeck and most probably also gangways. The resulting ship lines, measurements and layout can be seen as hypothetical, but without doubt are close to the original.

11.2.2 Construction and layout of the hull

Keel plank

The stern- and stem hooks with the keel plank from aft to fore measure 12.4 m, the keel plank itself is 9.1 m including both flat scarfs of 40 cm (aft) and 53 cm, respectively. The top of the keel plank near the scarfs with the hooks is 20 cm wide; in the middle, 28 cm. Its bottom surface at the same locations is 2 cm narrower. The cross section is therefore somewhat trapezoidal. The thickness is about 9-10 cm. The stern hook is nailed to the keel plank from the inside out with two nails situated side by side; the stem hook, with two rows of three and two nails, respectively. From the outside (of the ship) there is in both cases a single row of two nails. The nails do not pass right through the scarfs and therefore are not double-clenched.

Stern hook

The horizontal part of the stern hook (fig. 11.8) has a length of 2 m. At the 20-cm-wide after end of the scarf with the keel plank, its thickness is 10 cm, and over a length of 85 cm it slightly thickens to 11 cm towards the stern. Then follows a 10 cm long part till a thickness of 18 cm is reached. At the transition from the horizontal to its vertical part, the hook has a thickness of 25 cm and a width of 13 cm. The after face of the stern hook goes up perpendicularly for 20 cm and then rakes aft. This face of the stern hook and sternpost is straight and has a rake of 118°. The highest point of the stern hook is 73 cm above the keel line. The sloping flat scarf that connects it to the sternpost runs down from the outer face towards the inner face. The greatest thickness of the rising part directly below the scarf is 23 cm, the width on the inner side is 13 cm. Both the horizontal and the rising arms are trapezoidal in cross-section and are 6 cm in width.

The sternpost was connected with two nails onto the inner side of the stern hook. Through the middle of the scarf the post was joined to the stern hook by a diagonal treenail (Ø 2.5 cm). From the outside, two nails had been driven in diagonally through the upper tip of the stern hook into the sternpost. On the side, the scarf joint was strengthened with a strip of iron plating 200 mm long and 3 mm thick, which was probably fastened onto the upper part of the stern hook and the sternpost with four 25-mm nails with flat heads (Ø 15 mm). In the side of the horizontal part and continuing onto the vertical part, a ca 1.5 cm deep rebate had been cut into the side into which the end of the A strake was nailed. Sintel clamped moss caulking was present in the seams between the garboard strake and the stern hook, both on the inside and outside. It was remarkable that the sintel clamps were applied at irregular distances: on the inner side at 3.5-6.5 cm and at the outside at 6-8.5 cm

The lowest gudgeon was situated 45 cm above the keel line. It is made of wrought iron, has two approx. 40 mm wide and 5 mm thick wings or braces, which are cruciform and fastened to the stern, both connected to it with six nails. Beyond the stern, the two braces come together. To the inside of this loop a ring had been forged. The loop is 10 mm thick at the back, the forged ring 7 mm. The top edges are worn down on both sides of the loop at the level of the ring. This can be a sign of a long use of the ship (Section 11.5.2 and figs 11.8 and 11.19-21).

Sternpost

The sternpost was not found. Figure 11.8 shows the fully reconstructed stern hook and sternpost. The top of the stern will have stood about 2.7 m above the keel line. It seems reasonable to assume that at least the hood end of the second strake, but most probably also the third strake, was fixed into the rebate of the sternpost. The other

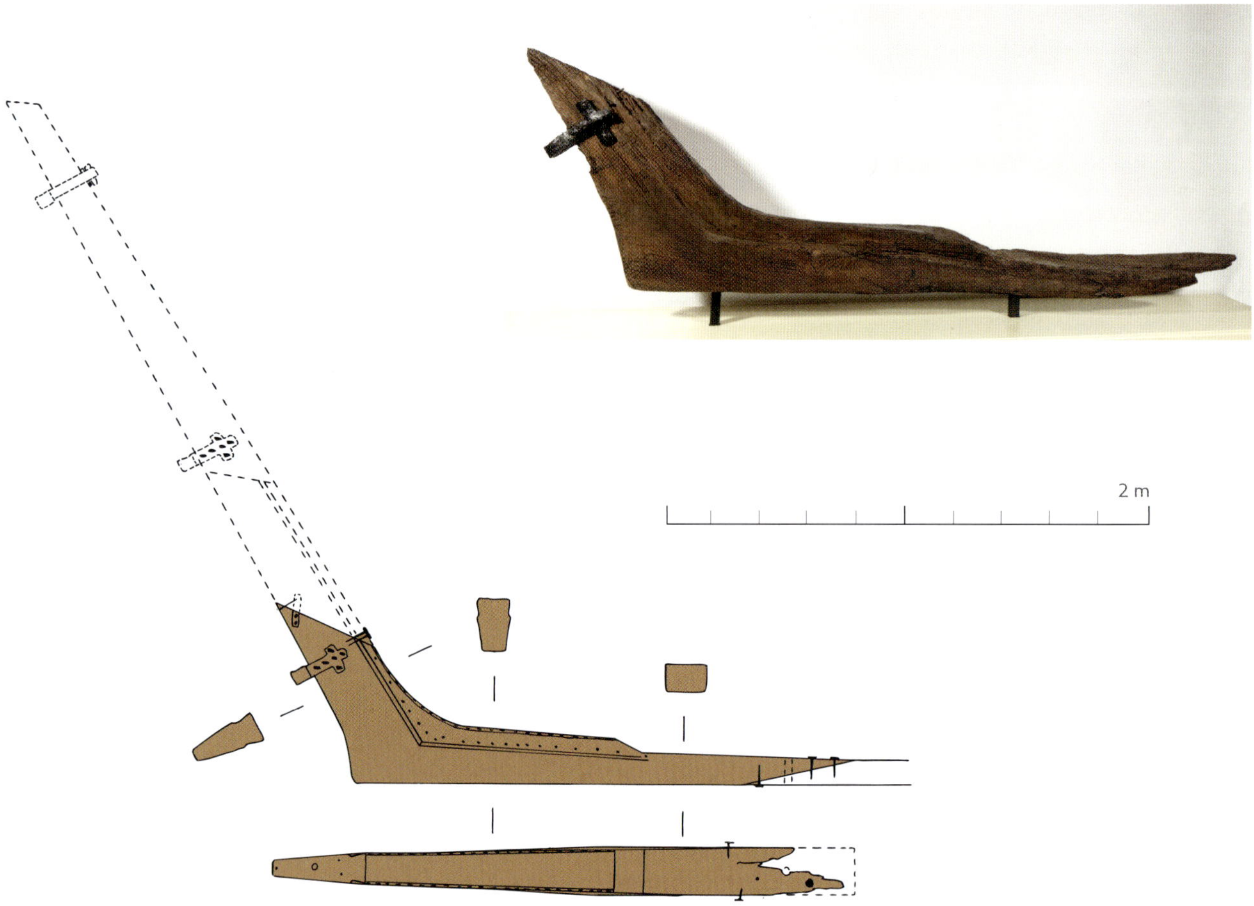

Fig. 11.8. Stern hook.

strakes of the side of the ship most probably ended against the double-bevelled after face of the sternpost. The sternpost was thick enough to have had rebates all the way up, accommodating the ends of all the strakes, but - on the basis of comparison with other cogs - this seems less probable.

False stern

The ship did not have a false stern.

Rudder

As mentioned earlier, the rudder was not found. The gudgeon in any case indicates that the vessel was sailed with a stern rudder. What the rudder would have looked like is depicted in plates 11.II A and 11.III A.

Stem hook

The massive stem hook (fig. 11.9) has a horizontal length of 2.25 m on the keel line. At the 20-cm-wide scarf to the keel plank its thickness is 10 cm, but this gradually increases over a length of 68 cm up to a thickness of 17 cm and then the top surface continues diagonally upward over a length of 17 cm to bring the thickness to 26 cm. The top surface further rises diagonally until the horizontal arm smoothly transitions into the rising arm, onto which the stempost is connected. Here a thickness of 31 cm is reached. The top of the horizontal arm of the hook gradually becomes narrower, and at the beginning of the rising arm measures 16 cm (excl. the 3 cm each of the two rebates). The highest point of the stem hook is 105 cm above the keel line and is 14 cm wide (not counting

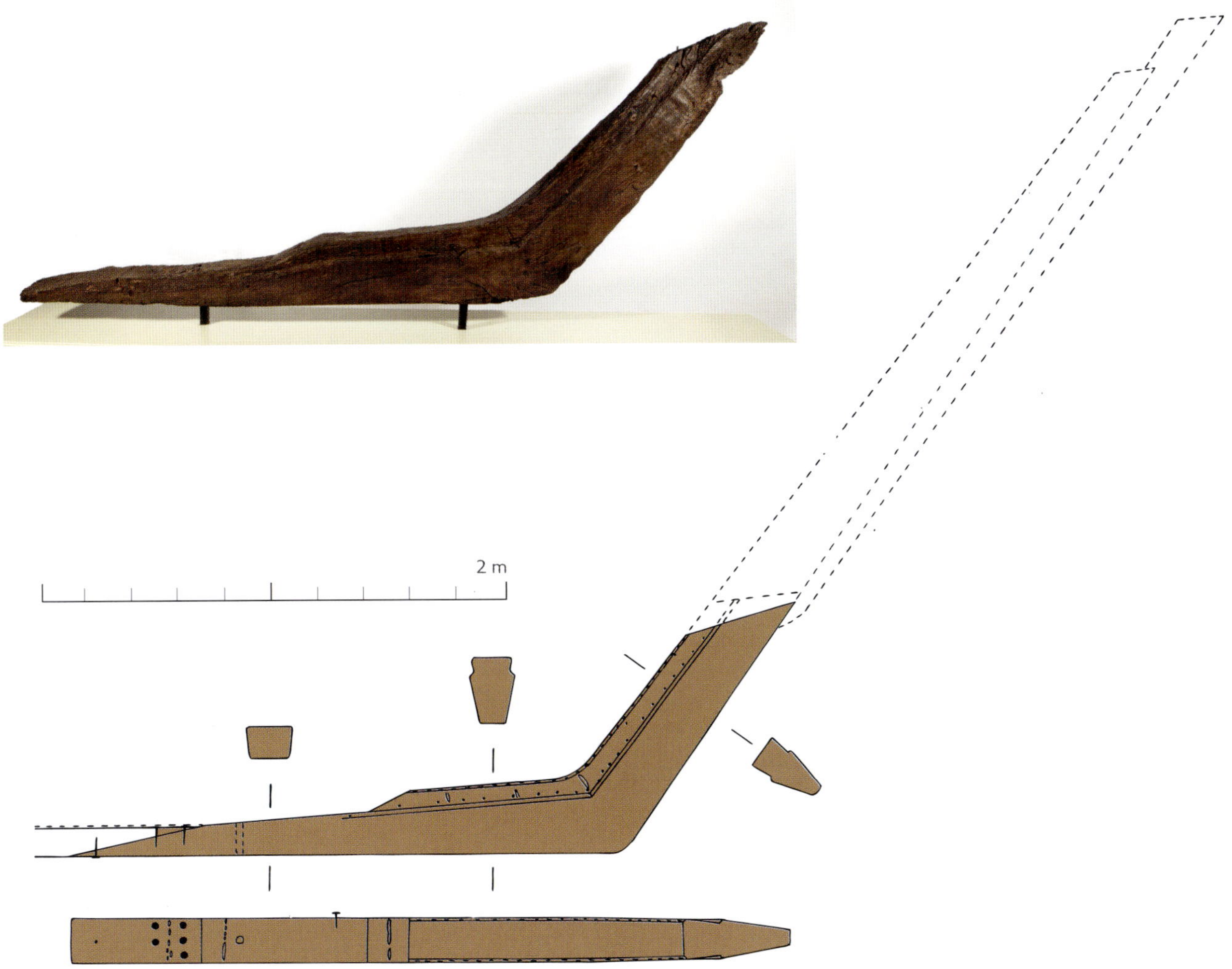

Fig. 11.9. Stem hook.

the two rebates with a depth of 2.5 cm). The thickness of the stempost at the scarf is 26 cm. The horizontal and rising arms of the hook run aft and forward both with a trapezoidal cross-section; and at the front measure 5 cm in width. On the sides of the stem hook, along the top edges of the thick horizontal part and on the rising arm, a continuous rebate is present with a width of 9 cm and a depth of 2.5-3 cm. Into it were fastened the hood ends of the garboard strake (A) and the second bottom plank (B). In the horizontal part of the hook's rebate are three vertical grooves. These are probably the deepest remnants of a mark applied during the ship's construction, to indicate to what depth the rebate was to be cut.

The keel plank was connected to the stem hook with two rows of (three and two) nails. The stem hook was connected to the keel plank with one row of two nails driven from the bottom. The seams between the stem hook and the bottom planks were closed with sintel clamped moss caulking both on the inside and outside. On the inside the sintel clamps were placed at irregular intervals of 5.5 to 6.5 cm and on the outside at 10-15 cm. The outer face of the stem hook (and the stempost) is at a rake of 122° to the keel. As was the case with the sternpost, the measurements were such that here too it might be thought that the plank ends were fixed into one continuous rebate. But here too, this is not very likely, since remains of tar have been seen inside both the hooks.

Stempost

The stempost has not been found. The reconstructed stem hook and stempost (fig. 11.9) show that the stempost, measured at gunwale level, rose to 3.2 m above the keel line. Even though there is no direct evidence for it, as the third and fourth port side strakes were not completely preserved up to the stem, it does seem likely that at any rate the third strake (C) too was nailed into the rebate of the stempost. The following strakes most probably ran to the front of the stempost. The exact shape of the stempost cannot be ascertained. Based on other cog finds it is plausible that it had an even width going upward or even became somewhat broader, and that the forward face from the bottom upwards had a low roof-shaped (double-bevelled) profile, onto which the planks' hood ends were nailed. If there were a continuous rebate right to the top, then the vessel would not have needed a false stem and the front of the stempost would have been flat.

False stem

If, as has been presumed, there was a roof-shaped forward face of the stempost, then the hood ends of the planks from strake D upward would have needed protection with a (thin) false stem. Plates 11.II A and 11.III A show how it could have been constructed.

Bottom and sides

At both sides of the keel plank the vessel most probably consisted out of nine strakes (A to I). On the basis of the recovered planks it may be assumed that the first, second and fourth to eighth strakes had a width of 40-45 cm. The third strake (bilge strake) and the ninth strake (sheer strake) were 35 cm in width. Each strake consisted of 3-4 thick oak planks. On some planks it was observed that the lapstrake overlaps (land) near the stern and stem were narrower than amidships, so during the construction quite straight planks with an overall similar width would have been used. As a result of this, with the bending of the planks a smaller overlap was created towards the stern and stem. The ship's carpenter made an effort to make all overlaps equally. The planks of a strake were connected to each other, from fore to aft, by means of flat scarfs. The nib ends of each overlapping plank were 1-1.5 cm thick and protruded into and outside the hull (fig. 11.10). The plank ends in the scarfs were secured with two rows of four nails. One row from the inside out, the other from the outside inward. It was not recorded whether the nails protruded through the shell and whether they were double-clenched. In the available pictures of the bottom of the vessel this cannot be seen. But given the scarfs joining the keel plank to the stern- and stem hooks, it seems that this was not the case.

Bottom

The bottom on either side of the keel plank and the hooks consists of two strakes (A and B). The strakes at port side are complete in the forebody, while in the afterbody parts of the connection to the stern were missing. On the starboard side only two pieces of the garboard strakes (A) and a fragment of the second strake were found. The garboard strake consists of two planks ca 6.6 m each and the B strake is composed of three planks: aft and fore, ca 5.8 m long; the plank in between, ca 1.95 m. The flat scarfs joining the planks are ca 40 cm long. The planks have a width of ca 45 cm (Plates 11.II B and 11.III B). Of the starboard side this cannot be said with certainty, but assuming similar plank lengths as in the fore and after part of the ship at port side, we can see a symmetrical composition of the bottom according to the so-called 'brother and sister' pattern. The two hood ends of the garboard strake were both nailed into the rebates of the stern- and stem hooks. During the ship's construction, the first plank had been held amidships (just forward of frame 12) against the keel plank with a ca 30 cm long, 5 cm wide and high lath or temporary frame, which was secured across the seam on both planks. And 95 cm forward, the same had been done with a lath of 20 cm (fig. 11.10). The after plank could subsequently be fitted to the scarf of the forward plank, and aft of frame 8 this too was sintel clamped to the keel plank and nailed with a 20 cm long construction timber. These are in fact 'temporary frames', which the ship's carpenter decided not to remove after the definitive frames were put in. In the same way it was done for the second strake. Here a single lath aft of frame 15 was sufficient to hold the A- and B strakes together. Also between the B and bilge (C) strakes in three different positions such laths had been placed during its construction. These 'auxiliary frames' were necessary to keep the carvel-built planks together during the process of construction.

In the keel plank (between frames 10 and 11) only one treenail without any obvious use was found, by which at the start of the construction for example a plank or temporary frame may have been fixed over the full width of the bottom of the ship. At the bilges, the bottom part was presumably also supported by posts. Over a length of about one metre, the garboard strakes were nailed into the rebates of the stern and stem hooks. Also the planks' hood ends were fixed in this way. The nails are irregularly spaced: in the horizontal part of the stem hook at 5 to 15 cm, in the rising part of the stem hook between 4 and 8 cm. Aft, the spacing varied between 4 and 10 cm. The top of the keel plank and the bottom planks are joined amidships in carvel fashion, and between the bilges and frames 9 and 15 form an almost completely flush surface. Aft of frame 9 and forward of frame 15, both strakes gradually turn towards the stern and stem (fig. 11.11), to which the ends of the upper strakes are nailed vertically. At 3.1 m from the stern and 2.8 m from the stem, the flush construction turns into a lapstrake or clinker-built construction. In order to realize this, the B plank has been left wider, whereby an overlap could be realized. The chamfered lapstrake overlaps (the lands) are connected to each other with nails that alternately at the bottom part of the overlap are nailed from the outside inward, while somewhat higher on the overlap they were nailed from the inside out. The nails protruded through the planks, were twice bent and thereafter the tip was driven back into the wood. For the method of double-clenching, see section 6.2.2 and fig. 6.42. The way and extent of caulking the bottom of the ship will be discussed in section 11.2.2.

Sides

Parts of the bilge strake (C) and the fourth strake (D) were present only on the port side amidships and in the forebody of the ship. Of the fifth and the sixth strakes (E and F) only some pieces between frames 18 and 24 were present, while loose inside the wreck a 3.5 m long piece of the uppermost strake (sheer strake, strake I) was found. The bilge strake consisted of three planks, of which the aft- and foremost were ca 5.2 m long with a plank of ca 3.7 m in between. The width of the planks was about 35 cm. Of the second side strake (D) it could be observed that about amidships (under frame 15) a scarf was situated. The forward plank was thought to have been over 6 m in length. Probably it in fact consisted of three planks, of which the middle one must have had a length of 3 m. The strake was 45 cm wide, at the stem approximately 40 cm. Amidships, the bilge strake was carvel-built and placed against the outer bottom plank (B) at an angle of ca 148°. In the after- and forebody the connection was lapstrake.

Fig. 11.10. Part of the bottom amidships with the scarf nibs standing proud, and the small planks nailed across the seams in a few places.

Fig. 11.11. The transition from carvel-built to clinker-built bottom construction near the mast step.

This fourth strake (D) was nailed onto the bilge plank (C) at a similar angle between frames 12 to 16. All consecutive planks were placed at an increasingly obtuse angle and were completely lapstrake from aft to fore in the way described above, in the section describing the bottom.

The width of the land was about 4 cm near the stern and stem, while amidships it was 9-10 cm. Aft of frame 12 and forward of frame 15, the transition from bottom to sides became increasingly flowing and from frame 19 is almost indistinguishable. The sheer strake to port had a width of 35-40 cm. Just aft of the mast, on the outside of the sheer strake and at 8-10 cm below the top of the side, an irregular piece of wood was attached of about 1 m long and 7 cm wide and thick. This small timber on the side against the hull had six vertical openings ca 2.5 by 2.5 cm (fig. 11.12 and Plates 11.II A and 11.III B; compare with the Marknesse cog section 12.2.2 and the figs 12.8, 12.13 and Plates 12.I A and 12.II B). It is one of the channels to which the stays of the mast were attached. The channel was fixed onto the sheer strake with three treenails (Ø 2.5 cm) and two iron nails.

Fig. 11.12. The port channel on the outboard side of the detached sheer strake *in situ*.

Sintel clamped moss caulking and the shape of the carvel seams

The edges of the flush bottom planks are chamfered over almost the entire thickness. Thus reversed V-shaped seams on the bottom with a width of 1.5– 2 cm and with sides of 3.5-4 cm were created. At the seam between the garboard strake and the keel only the strake had been chamfered. The seams were filled with strands of moss which had been twisted into a single strip. These were made into an S-twist and covered with laths of oak, which in cross-section were cut to an oval shape. These ca 1.5 cm broad and 0.5 cm thick moss laths were secured to both sides of the plank seams by iron sintel clamps. On the inside, the planks sat tightly together and no caulking could be seen. The lapstrake or clinker-built planks were caulked on the inboard surface in a similar fashion. As a preparation, the upper/outer edge of the plank below was chamfered. After the next plank had been joined onto the lower one the caulking was put into and above the created V-shaped seams. The lapstrake parts of the bottom and bilge strakes were not treated like this on the outside because the sharply chamfered lapstrake planks were too thin to hammer sintel clamps into. On the other strakes, from fore to aft completely lapstrake, the seams are filled with sintel-clamped moss caulking both on the in- and outside (figs 11.13 and 11.14). The ends of the shell planks were not sintel-clamped, but in the seams and on the scarfs plenty of moss had been applied.

As mentioned, the seams between the plank ends on the stern and stem are caulked and sintel-clamped on the in- and outside surface. The sintel clamps in general are situated at irregular intervals and continue on the inside behind the frames. From this it can be deduced that the shell-first con-

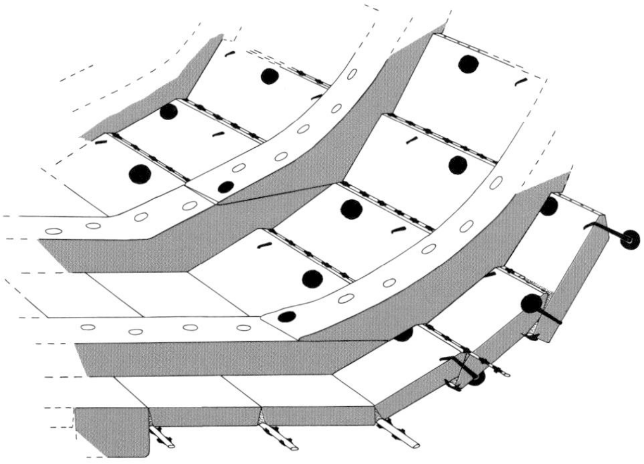

Fig. 11.13. The different shell seams with sintel clamped moss caulking, and the way in which the clinker shell strakes are nailed.

struction method had been used. The sintel clamps in the ship belong to type D2, with a central part 35-40 mm long. This type was in use roughly between 1275 and 1350. More recent research shows that the types used in the ship date from around 1300 (figs 11.19-25, -47 and stern part).[10] The planks were not drawn in detail as later became customary, and it was not recorded how many centimetres apart the sintel clamps have been applied. It was noted in 2008 that on the preserved stern and stem hooks the individual distances could vary. In section 11.2.2 the numbers of sintels used will be discussed.

Determination and pollen research

The moss caulking was analysed and apparently it was pure peat moss (*Sphagnum*) that had been used. The sample that was analysed contained charcoal and the proportions of trees and herbs were 34 and 66 %, respectively. The moss must have been collected in an oligotrophic area. There must have been a large bog and little agriculture in the area (for a detailed description of the pollen, see Appendix).[11]

Frames: straight and V-shaped floor timbers and futtocks

The constructed hull is held in shape and strengthened principally by the frames (figs 11.15a and b). In the reconstruction (Plates 11.III A and B) it has been assumed that the hull had 32 frames (0 to 28 and -1 up to and including -4). Most of the frames were missing. Only (parts of) the V-shaped floor timbers 2, 3, 4, 5 and 6 in the afterbody were present, V-shaped floor timber 5 is complete. Of the frames 9 and 15, situated amidships, only the port side of the floor timbers survived. The frames 12, 14, 15 and 17 had a complete futtock on the port side. Of the V-shaped floor timber with a rebate for the mast (frame 18), the part on port side and the part on the starboard garboard strake were present. Of frames 19 to 25 in the forebody the V-shaped floor timbers had survived, and of frames 19, 22 and 23 the first futtocks as well. The V-shaped floor timbers, the floor timbers and the futtocks were made from irregular pieces of oak and had widths that varied between ca 10 and 20 cm. V-shaped floor timber 18, with the rebate for the mast, was wider and will be discussed below.

The straight floor timbers were about 15 cm thick, the V-shaped floor timbers above the keel plank about 20 cm, and the futtocks 10-12 cm. The V-shaped floor timbers 4 up to and including 19 all had an inverted-U-shaped limber hole 10 cm wide and ca 5 cm high. Frame 6 to 18 also had such limber holes above the outer bottom strakes. In the V-shaped floor timbers 4 and 5 there were triangular limber holes above the garboard strake. In the V-shaped floor timbers 2 and 3 this was also the case, but only above the stern hook and the keel plank. The triangular limber

Fig. 11.14. Sintel clamps and moss laths or split withies in the seam between the port garboard and the second strake, beside the stern hook.

Fig. 11.15a & b. General view from the stern to the stem. b. A closer view of the forebody.

10. Vlierman 1996a, 77; Vlierman in preparation a.
11. Cappers *et al.* 2000; Mook-Kamps 1997, 9.

holes above the garboard strakes are also visible in V-shaped floor timbers 18 up to and including 20; while the subsequent ones (21 to 24) only have limber holes above the keel plank and the stem hook. Even though little was recovered, we can be fairly high confident that both the V-shaped and straight floor timbers in principle have one end just above, or at the upper edge of the bilge strake, while on the opposite side they end a bit higher up. This pattern alternated between the port and starboard sides. The scarf joining them to the lowest futtock was oblique or almost horizontal. The hull planks and the keel plank were in principle connected to the frames by two treenails (diameter 2.5 cm) per plank width, sometimes by only one. On the outside of the shell, these treenails were secured with wedges.

During the drawing of the field sections (Plate 11.I D) it was observed that hull planks had been fixed to the frames with treenails and nails as well. This was also the case with the scarfs of the futtocks that fitted onto the (V-shaped) floor timbers. The V-shaped floor timbers 3 and 21 were the first and last, respectively, to be joined to the stern- or stem hook with treenails. Other cog finds suggest that this was not the case on the other V-shaped floor timbers at the stern- and stem hooks and -posts. These frames were attached only to the shell planks. In the middle of the top surface of V-shaped floor timber 19 is a 45 cm wide and 3 cm deep flat rebate, with three blind treenails in it (fig. 11.16 and the Plates 11.I D and 11.III C). Also on the middle of V-shaped floor timbers 20 to 23, relatively horizontal surfaces can be seen. Only the lowest futtocks 12, 19 and 22 are somewhat undamaged. They all three had an oblique top, the scarf meeting the second futtock. The upper part of futtock 12 reached almost to the middle of the fifth strake (E). The futtocks of frames 19 and 22 reached up to about the middle of strake seven (G).

Fig. 11.16 The frames in the forebody; in the foreground, frame 18 with the mortise for the heel of the mast; beyond it, frame 19 with the shallow rabbet; while further on, more-or-less flat areas can be seen at the centre of (V-shaped) floor timbers 20 to 22.

Mast step

The mast step of the Dronten-I cog is the heavy, V-shaped floor timber 18. It is made out of an irregular piece of oak trunk, of which in some places the dull edge forms the outside of the timber while other sides were straightened (figs 11.17a, b and c). The greatest width of the V-shaped floor timber is 34 cm, the largest thickness is ca 25 cm. Above the keel plank, on the central and widest part (65 cm) of the timber, the recess for the mast step had been

Fig. 11.17a View from the forebody; immediately beyond the surveyor's rod in the wreck is floor timber 18 with its recess for the heel of the mast.

Fig. 11.17b and c. The detached floor timber of frame 18. c. Floor timber 18 with the recess to take the heel of the mast, detail.

hewn out. The wide central part of the timber slightly tapers to both sides of the V-shaped floor timber that measures 16 cm. The rebate measures about 24 x 22 cm and is about 10 cm deep. In the starboard forward corner of the recess, the side of the rebate does not measure 24 but 22 cm, and at the forward part not 22 but 17 cm. The meaning of this is unclear. Possibly the recess once was smaller (17 by 17 cm) and later was adapted for another, somewhat thicker mast. The partly hewn-off treenail in the corner of the recess to starboard could be explained in this way. The bottom of the recess has a drilled hole (Ø 3 cm) to drain off any water, which empties above the limber hole. The V-shaped floor timber was fastened – in contrast to the other frames - with 3-cm-thick treenails rather than 2.5-cm-thick ones.

Ceiling

Some fragments of one ceiling strake have been found (fig. 11.18). It is the bilge ceiling strake, which presumably ran between frame 2 and the forward edge of frame 23. It is a ca 24 cm wide plank with an irregularly adzed double-bevelled upper surface, of which the sides are 3 cm thick and the greatest thickness is 5-6 cm. It was connected onto each frame with one or two treenails. Also some nail holes were seen with which the plank originally was held in place for the carpenters to drill the holes and to insert the treenails. At some frames the plank had been put on top of the lowest futtock, indicating that the latter must have been put in place first. The rebate in the upper part of V-shaped floor timber 19, situated above the keel plank, in which three blind treenail holes were seen, and also the more-or-less flat areas on V-shaped floor timbers 20 to 23 suggest that the top of the forward part of V-shaped floor timber 18 supported a 3-cm-thick plank (with a nail hole) which at its after end was 45 cm wide and 25 cm wide at its forward end.

The lowest futtock of frame 19, at the level of the bottom part of strake F holds two treenails (Plate 11.I D). One of them could be related to a ceiling plank at that spot. Further evidence is lacking. It is certain that there was an open ceiling structure inside the ship. Between the

Fig. 11.18. View from the forebody with the port bilge ceiling board covering the turn of the bilge.

plank above the keel plank and the bilge ceiling planks in the forebody, almost certainly was no other ceiling plank. In V-shaped floor timber 5, a nail was seen in the middle above the starboard garboard strake. Probably between the after side of this frame 5 and the after side of V-shaped floor timber 18, and between the two bilge ceiling planks, a boarded load-bed or underfoot ceiling had been present. Probably the plank above the keel plank started aft of frame 1 (Plates 11.III A, B and C). The side of the ship was at the top strengthened on the inside with a 14 cm wide and 6 cm thick ceiling plank, which can be identified as an inwale. On the under/outboard side it was chamfered, and fitted onto the chamfered tops of the futtocks and to the sheer strake with treenails. In the after- and forebody, the inwale was extended with curved pieces which ended at the first and last frame, or possibly at the very stern and stem. A straight piece of inwale was found inside the wreck. It seems unlikely that more ceiling planks were present on the sides.

Transverse connections, deck beams, mast beam, decks, gangway, windlass and pump

The only indications for transverse connections in the ship's hull are two rosebolts (Section 11.5.2 and figs 11.19-22 and -23 and 11.20), a treenail in strake E just forward of frame 19, and two treenails located beside each other in strake C forward of frame 23, both on the port side. The two rosebolts were part of the constructive connection of the mast beam. The wooden treenails in strakes C and E unmistakably served to join the transverse connections to the shell of the ship. Despite these few clues, judging by other cogs it can be indicated with near certainty what all this must have looked like (Plate 11.III B and C).

Transverse connections

The treenails that were found in the hull on the port side in the forebody also turn up in comparable places in the Nijkerk-I cog (Section 8.2.2 and Plate 8.IV A). These are treenails used to fasten the sides to transverse connections consisting of a beam branching into a standing knee on one side, while at the other end of the beam a counterpart knee was fastened onto it. Or it could be a cross beam running between - and fixed with treenails onto - the inboard surfaces of the sides, with standing knees set onto them at both ends. These transverse connections were meant to stiffen the shape of the vessel fore and aft in the underwater body. For the reconstruction (Plate 11.III A) it has been assumed that the vessel had two such transverse connections both fore and aft (DV 1 to DV 4). A crossbeam has been added onto the standing knees on the transverse connections DV 2 and DV 3, onto which the deck planks could rest. On the deck, knees were placed to which the upper part of the side was connected and which were joined to this deck beam also with treenails. These treenails presumably were somewhat thicker (Ø 3 cm) than the treenails used to connect the frames onto the hull planks. These probably thicker treenails are shown in plates 11.III A, B and C.

Deck beams

Apart from the two deck beams on DV 2 and DV 3, the probable after- and foredeck should have support in various places. Most probably this was realised at frames -1, 1, 5, 18, 19, 22 and 24, while also amidships at frame 12 above the hold, a beam was placed which rested on the frames, lying on the flat tops of the futtocks (Plate 11.III C). At frames 1, 5, 12, 18 and 24 more such beams were probably placed on the ends of the shortened futtocks, at the level of strake G or H. On the deck planking, on top of these beams knees were set which held the top of the shell in place, and together with the beams, deck planks and strakes they strengthened the hull form at deck level. Probably the afterdeck beam rested in rebates in the futtocks of frame -1. Together, the horizontal parts of the knees formed a horizontal deck beam over the entire width of the deck. They were joined by a sloping flat scarf. This was probably also the case above frame 5. The knees on top of frames 1, 12, 18, 19, 22 and 24 probably had shorter horizontal arms (Plates 11.III A, B and C).

Sail beam

The sail beam (above frame 18), as said above, rested most probably also on a shortened futtock. It was heavier than the other beams. The beam most probably had a semi-circular rebate cut into it, into which the mast was partially locked. An additional reinforcement of the beam-sides fastening was realised by means of the two T-shaped rosebolts, which were passed through the sides and nailed to the rear of the sail beam (Plates 11.II A and 11.III C and figs 11.22-22 and -23 and 11.23). It seems likely that such rosebolts were fastened also to the forward face of the transverse beam above frame 5.

Decks

The ship almost certainly had an after- and foredeck. The afterdeck was probably about 4 m long and ended on the beam at the forward edge of frame 5. The foredeck started at the after face of the sail beam (frame 18) and extended all the way to the stempost, some 4.6 m. In the reconstruction (Plate 11.III B) it has been assumed that the planking of both decks ran lengthwise and that the outer planks were fitted around the frames and to the hull.

Hatches have been drawn both in the afterdeck and in the foredeck. The planks most probably were fixed onto the deck beams with nails, but maybe with treenails instead. Whether the seams in the decks had been caulked with sintel clamped moss caulking, as has been attested on the supposed small cog from Hattem, or with thin laths could not be established.[12]

Gangways
Given the reconstructed ship's height it seems self-evident that there were connections along the sides between the after- and foredeck at a level about 1.75 m above the keel line. Even though it is uncertain whether they were thick planks fitted with rebates into the deck beams 5, 12 and sail beam 18 (as in the Nijkerk-I and Doel-I cogs), the reconstruction drawing shows the gangways along the frames and sides of the ship (Plate 11.III B en C). There might have been a coaming bordering the gangways on the inside (see the Marknesse cog Section 12.2.2 and Plates 12.II A, B and C).

Windlass
Most probably there was a windlass on the forward edge of the afterdeck with which the sprit (sail) could be hoisted or lowered.

Pump
The limber holes in the floor timbers and V-shaped floor timbers show that rainwater, bilge water spray and leaked water was to be guided towards a certain part of the ship where it could be pumped out. The handle that was found probably came from the pump stick; the pump tube has not been found. It probably was situated on the keel plank.

Castles
There were no clues pointing to the presence of a castle. It also does not seem likely that there were any.

Repairs
Between frames 16 and 17 and also between frames 18 and 19, there were repair planks nailed onto strake D, with a thickness of 0.5 and 1 cm, respectively. If there was any sintel-clamped moss caulking in the cracks on the outside of the hull planks, this could not be observed or documented.

Timber use and precessing
All ship parts that were found are made of oak *(Quercus* sp.). So were the treenails of the connections, which were wedged into the planks. Even though at the time of the excavation it was not looked at specifically, it can be assumed that the planks were sawn, and the timbers were squared by axe and/or adze. The frames were hewn from irregularly grown parts of a trunk or from thick limbs and roughly squared. The floor timbers and V-shaped floor timbers were generally made from heavier pieces than the futtocks. The hooks were cut from trunk-branch sections and finished with a squaring axe and adze. Of the stem hook it can be said that in relation to the dimensions of the vessel and in comparison to the stem hooks of other cogs it had large dimensions. The recovered ceiling plank on the inside had been roughly adzed into a double-bevelled 'roof shape'. The reconstructed top view of the hull suggests a construction in a 'brother-and-sister pattern'.

Fastenings
The planks of the overlapping parts were connected onto each other with wrought iron nails 80-100 mm long (figs 11.19 and 11.20 C) with almost flat, round heads with a diameter of 25-30 mm. It was not recorded in detail at what spacing the nails were put in. On the preserved stern- and stem hooks, the hood ends of the garboard strakes were nailed at irregular intervals (6-15 cm). It does not seem plausible that this irregular pattern was repeated on the lapstrake or clinker-built planks. An irregular distance between 10 and 16 cm, thus with an average distance of 13 cm, seems more obvious. Very striking is the fact that the nails were alternately driven in from the outside inward and vice versa, whereby those hammered from the outside were positioned 1-2 cm lower. In total, about 215 m of lapstrake connection had been nailed. Including the fastenings of the hood ends on the stern- and stem hooks and-posts, and the probable (longer) nails used for the decks and ceiling planks, the total amounts to 1800 to 2000 iron nails for the entire ship.

The frames are connected onto the shell planks with oak treenails (Ø 2.5 cm). In principle two treenails per plank width, sometimes one. Also the ceiling planks were connected onto the frames with one or two treenails. The V-shaped floor timber with the rebate for the mast was fixed with 3 cm thick treenails, while this was most probably also the case between the transverse connections and the knees, and between the knees and the deck beams. In total about 1400 to 1500 treenails were used in the ship. Apart from the gudgeons, rudder fittings and the rosebolts, few other heavy iron connections will have been used in the ship's hull.

The clinker-built connections were sintel caulked with moss on the inside of the shell, the carvel-built seams on the outside, while the clinker-built parts of the bottom did not have external sintel clamps. The clinker-built connections of the side strakes were on both sides sealed with

12. Vlierman 1996b, 115-119.

sintel clamped moss caulking. This also goes for the seams joining the stem- and stern hooks to the stern- and stem-posts and to the hood ends of the strakes. In total about 465 m of seams had sintel clamps. In section 11.2.2 it was mentioned that the spacing of the sintels has not been checked in the field. On the seam between the garboard strake and the stern hook the interval was seen to be about 6-8.5 cm. Based on comparison with other cogs, this seems a plausible spacing for the ship as a whole. At an average distance of 7 cm, about 6600 sintel clamps would have been used on the cog.

11.3 Construction data

Despite the fact that not very much remained of the cog during the excavation, it had been shown that on the basis of the recovered structural elements and comparison to other cogs, it has been possible to make a reliable reconstruction. Naturally the after- and foredeck are connected by two gangways. This is an assumption, so the depicted execution may be regarded as hypothetical. This also goes for the height of the depicted hull sides and the number of strakes (nine) on either side of the keel plank.

11.3.1 Design

The ship was built according to rules of thumb that go far back. The main dimensions were most probably determined by the available timber. Whether the remarkably heavy stem hook related to a particular function of the ship and whether the strakes passed along the stern- and stem-posts or ended in a rebate higher up on the post could not be said with certainty. On the basis of what is seen on the other cogs, the latter seems less plausible. The mast which is situated quite far forward in the publication of 1980 must be the main reason to suspect the use of a sprit sail.

Lines

The ship had fair lines and a near-symmetrical top view. Aft, the vessel is somewhat narrower than the forebody, while the underwater body slightly rises to the stern. The cog has a straight keel, a straight raked stern and stem, and a medium sheer. The sternpost is 0.5 m shorter than the stempost. They rake at angles to the keel line of 118° and 122°, respectively. The underwater body is almost symmetrical. Amidships the bilges are positioned at a relatively sharp angle to the outer bottom planks. The bottom is composed of a keel plank and (only) two strakes on either side of it. Towards the stern and bow, the bottom planks twist towards the stern and stem. Between the straight floor timbers 9 and 15, the bottom planks, together with the keel plank, form a flat bottom. Apart from the sharp bilges in this midships part of the ship, the other cross-sections show a flowing, curved shape.
In the after- and forebody only a few frames have a faint S-form. The position of the rosebolt-heads determined the position of the strake through which they would go. Assuming that the ship indeed had the reconstructed nine strakes means that the upper two strakes would be positioned at a steep angle. A validation for this can be deduced from the position of the channel securing the stay of the mast to the outside of the top strake. With a more

Table 1. Principal dimensions of the Dronten-I cog

Length of the keel plank (including the stern- and stem hook)	12.40 m
Length of the bottom (between the rising part of the stern- and stem hooks)	11.85 m
Greatest width of the bottom (seam between the B- and C strake at frames 9-15)	2.05 m
Overall length at gunwale level	15.80 m
Overall width at gunwale level (at frames 12 to 16)	5.20 m
Height of the stern (keel line to gunwale level)	2.70 m
Height of the stem (keel line to gunwale level)	3.20 m
Height of the side amidships	2.30 m
Length of the stern	3.10 m
Length of the stem	4.00 m

Length-width ratio of the bottom	5.8:1
Length-width ratio at gunwale level	3.0:1

All measurements rounded to the nearest 5 cm.

everted top strake, the ropes of the stay would rub against the gunwale. The bottom part of the ship has been built in carvel fashion over a length of 7.5 m. In the afterbody and forebody, the planks of the clinker-built parts of the bottom were strongly chamfered on the overlap, in such a way that from the outside these seams looked as if they were constructed in carvel fashion throughout. Only the nails would have shown that this was not the case.

Measurements and layout
On the recovered hull parts no marks were found that could be related to measurements during the ship's construction. The mast is placed in the forebody at 1/3 of the length (at deck level) between the stern and stem. If there was a separate after- or fore-cabin we cannot know. It seems fair to assume that between frames 5 and 18 and the bilge strakes a ceiling or floor was present.

Tonnage
If there was cargo then the most probable hold was situated before the afterdeck, aft of the sail beam and continuing under the gangways. At frame 5 the beam is 4.1 m and the load bed is about 2 m wide. Amidships (frame 12) the widths are ca 4.6 and 2.3 m, respectively; aft of the sail beam, 4.2 and 2 m. This means an average width of 3.2 m. The length of the hold was about 5.7 m, while the headroom between the load bed and the underside of the gangways was about 1.4 m. The content of the hold is therefore 5.7 m x 3.2 m x 1.4 m = 26.5 cu m or 26 gross tonnes of capacity.

Selection of materials
The Dronten-I cog was most probably entirely made of oak wood. The heavy gudgeons, the rudder fittings, the rosebolts and the nails and sintel clamps were made of wrought iron. The caulking - as usual - was moss.

11.3.2 Construction: techniques and sequence

A keel plank with stern= and stem hook are standard features in cog construction, as they are in this vessel. Quite remarkable, as has been stated before, is the strikingly massive stem hook, while the mast step has been placed in a V-shaped floor timber. Among the other discussed cogs, the latter is only seen in the Marknesse cog (in the primary situation, Chapter 12). The structural transverse strength is provided by the frames, the transverse connections below in the after- and forebody, and the deck beams positioned on or near the futtock heads. On the decks, knees were fixed on top of some deck beams, which held the upper parts of the hull in place. The upper part of the hull was also strengthened with a thick ceiling plank or inwale, which on the inside was connected onto the frames. The building phases will not be discussed in detail.

Remarkable are the short laths which are positioned across the seams of the keel plank and the A strakes, the seams between the A and B strakes, and those between the B and C strakes. They sit where in the early stage of construction the plank(ending)s in the carvel-built part of the bottom needed to be held in place while the hood ends were being nailed into the stern- or stem hook. The shipbuilder seems not to have found it necessary to remove these laths when the floor timbers were installed. The two or three planks of each strake of the sides of the ship could relatively easily be nailed to the already realised part of the ship's hull whereby probably initially just the nails from the outside were hammered in, after which further nails were driven in from the inside of the shell. The strakes of the sides may have been constructed up to the sheer strake with some auxiliary frames before the futtocks and deck beams were installed. During the construction, the hull was supported externally with struts.

11.3.3 Characteristics

The Dronten I-cog has the following characteristics.

Keel plank
A keel plank that has about 2.5 - 3 times the thickness of the shell planks, aft and fore 20 cm wide and gradually widening towards the middle where it becomes 28 cm.

Stern- or stem hook
Aft and fore the hooks have flat scarfs for their connection to the keel plank and to the posts. Both hooks have rebates which ran upward up to or into the stern- and stempost to take the plank ends.

Stern- and stempost, false stern and stem, and rudder
The stern- and stemposts were quite thick even in the upper parts. Thus they could have had rebates for the plank endings that ran up to the top. Though in view of other cogs this does not seem probable. If there were continuous rebates then the vessel would not have had a false stem. The ship was steered with a stern rudder.

Carvel-built and clinker-built bottom
The ship's bottom had a quite large carvel-built area though it was fairly narrow. The overlaps in the clinker-built parts were heavily chamfered, especially on the outside, where the lands are very thin.

Clinker-built sides
In cross-section the ship has fairly rounded lines going from the keel plank to the sheer strake. The bilges are angular only at the flat part of the bottom. For the shell, fairly straight planks were used which has as a result overlaps of up to 9-10 cm while at the stern and stem they are no more than 4 cm. The strakes are fixed with double-clenched nails of which the tips were driven back into the wood. The nails were alternatingly hammered from the outside inward and from the inside out. Moreover those hammered from the outside were situated about 1-2 cm lower than the others, hence more or less staggered. At the scarfs joining the planks forming a strake and joining the hooks to the keel plank, the nails were most probably not driven right through the wood.

The caulking method and the shape of the carvel seams
The seams of the clinker-built planks are closed with sintel clamped moss caulking both on the inside and outside. Only the overlapping parts of the bottom strakes are not caulked on the outside because the chamfered overlaps were too thin and would crack when the sintel clamps were hammered in. The bottom part of the ship is caulked from the outside whereby the planks were first chamfered on both edges over the full thickness of the planks which resulted in a triangular seam. There are no repairs with moss caulking on the outside of the hull planks, or at least none have not been seen in the field. Also the seams between the hooks and posts and the hood ends are caulked with sintel clamped moss caulking both on the inside and outside. The sintel clamped moss caulking had been put in place before the frames and futtocks were put in (shell-first method).

Ceiling, mast step and repairs
The vessel had an open ceiling, consisting of three ceiling strakes on either side of the keel plank and of which the first (bilge) ceiling strakes and the upper ones (inwales) were adzed into a double-bevelled, roof-shaped profile. Between the bilge ceiling strakes a ceiling or load-bed could have been laid, while before the mast a tapered plank sat on top of the V-shaped floor timber above the keel plank. Somewhat similar to this plank in the forebody, the middle plank of the cargo floor could have ran further aft.

Transverse connections, deck beams, gangways and decks
Both in the afterbody and forebody at a modest height above the keel line the underwater body was additionally held in shape by two transverse connecting beams which were fixed to the frames and the sides. The cog most certainly had an afterdeck and foredeck, supported by deck beams and gangways enabling the crew to walk the length of the ship.

Fastenings
The frames are fastened to the hull planks with 2.5 cm thick treenails secured with wedges. The treenails attaching the V-shaped floor timber with the rebate for the mast heel, and probably also those in the knees of the transverse connections and the knees on deck, were thicker (3 cm). The construction of the sail beam, possibly also of the beam at the forward edge of the afterdeck, was strengthened with two rosebolts which were inserted from the outside through the hull planks and nailed to the sail beam.

11.3.4 Use

Function
From the findings nothing specific can be said about the function of the vessel. The heavy stem hook could suggest that it sailed more on the Zuiderzee than on rivers and other inland waters. The hold and the tonnage seem to be suitable for bulk and maybe quite voluminous types of cargo. The two lance heads (Section 11.5.2) should probably be seen as the crew's personal weaponry rather than as weapons of a military nature, although this cannot be entirely ruled out.

11.3.5 Wreck formation and embedding

The cog most probably went down on the eastern side of the then presumably still more-or-less intact peat barrier between the isle of Schokland and Elburg. Possibly this was not far from the former mouth(s) of the IJssel. In the first instance, the mast and parts of the upper hull (and decks?) were washed away, after which the wreck would quite soon almost completely have been covered by sediment. Later the wreck was washed free again and further parts disintegrated or drifted away. This can definitely be related to the heavy storms in the area in 1367, while the breaking-off of the stern may also have been a result of heavy ice-drifting.

11.4 Model

No model has been built of the Dronten-I cog.

11.5 Inventory

11.5.1 Distribution of the finds

With the exception of some shoe fragments and the two rosebolts all finds were recovered in the afterbody (Plate 11.I A). Most were found on the bottom of the wreck among the detached V-shaped timbers and the stern hook. The others were found in the turned soil, but most probably originate also from the afterbody. It probably concerns a (near-) complete inventory, the most complete so far from a cog from the former Zuiderzee. This also supports the idea that the vessel sank rapidly and must have been abandoned by the crew in a hurry. In the finds list and in the table, an asterisk (*) marks objects that have been lost.

11.5.2 Ship's equipment and inventory

Ship's equipment

The detached and much deformed lowermost gudgeon has been brought back to its original shape and has been reattached to the stern hook (figs 11.6, 11.8 and 11.19-21). It was fastened with thick, short nails of which the heads were forged into a rectangular shape. The wings or brackets are cruciform while into the loop that stands out from the stern, a ring was forged. The opening for the pintle has a diameter of 54 mm. A fragment of an iron strip (fig. 11.19-24) may be a broken-off piece from the stern hook. During the excavation in the turned soil some nails were found. Eight pieces were 80-100 mm long and had quite flat, round heads. These are examples of the double-clenched nails whose tips were driven back into the wood for joining the clinker-built planks (figs 11.19-26, -57, -58, -59, -60 and -63).

Four nails, three of which have the same size and one longer specimen, (originally) had four small protuberances on the bottom of the head. This shows that the heads were forged on a so-called nail-iron that had some additional holes (figs 11.19-26, -27, -28, -58 and fig. 11.20 C). In figs 11.19-25, -47 and 11.20 C the type of sintel clamp

Table 2. Distribution of items

Category	Forebody	Hold	Afterbody
1. Ship's equipment	-	rosebolts, hilt (grip/handle)	gudgeon, nails, iron plate
2. Operational equipment	-	-	-
3. Administration and navigation	-	-	-
4. Tools	-	axe, adze, caulking iron, awl, grind-/ whetstone	-
5. Household items	-	-	-
6. Cooking, eating and drinking	-	-	stone ware jug, *Kugeltopf*, *Grapen*, sherds, fire tongs, peat*, sample of loam, tile, bricks, flint
7. Victuals	-	bones	bones
8. Personal belongings/weaponry	knives, spearheads	-	-

Table 3. Ship's equipment

Category	Item	Number	Findspot
1.1 Ground tackle	-	-	-
1.2 Sails	-	-	-
1.3 Cordage	-	-	-
1.4 Blocks	-	-	-
1.5 Windlass	-	-	-
1.6 Pump, drain	hilt/handle	55*	hold
1.7 Navigation aids	-	-	-
1.8 Boats	-	-	-
1.9 Spare parts	-	-	-

(D2) is shown. The two unusual T-shaped iron pieces are the rosebolts already mentioned in section 11.2.2. They strengthened the connection between the sail beam and the sides of the ship (figs 11.19-22 and -23, 11.20a and b, and Plates 11.II A and 11.III C). In the excavation report of the Almere-I cog, the bolts were compared with younger examples from other wrecks in the Zuiderzee area and an evolving form was demonstrated.[13] The wooden handle (fig. 11.19-55) probably originates from the pump. The artefacts that belong to the ship's structure are not included in the table below.

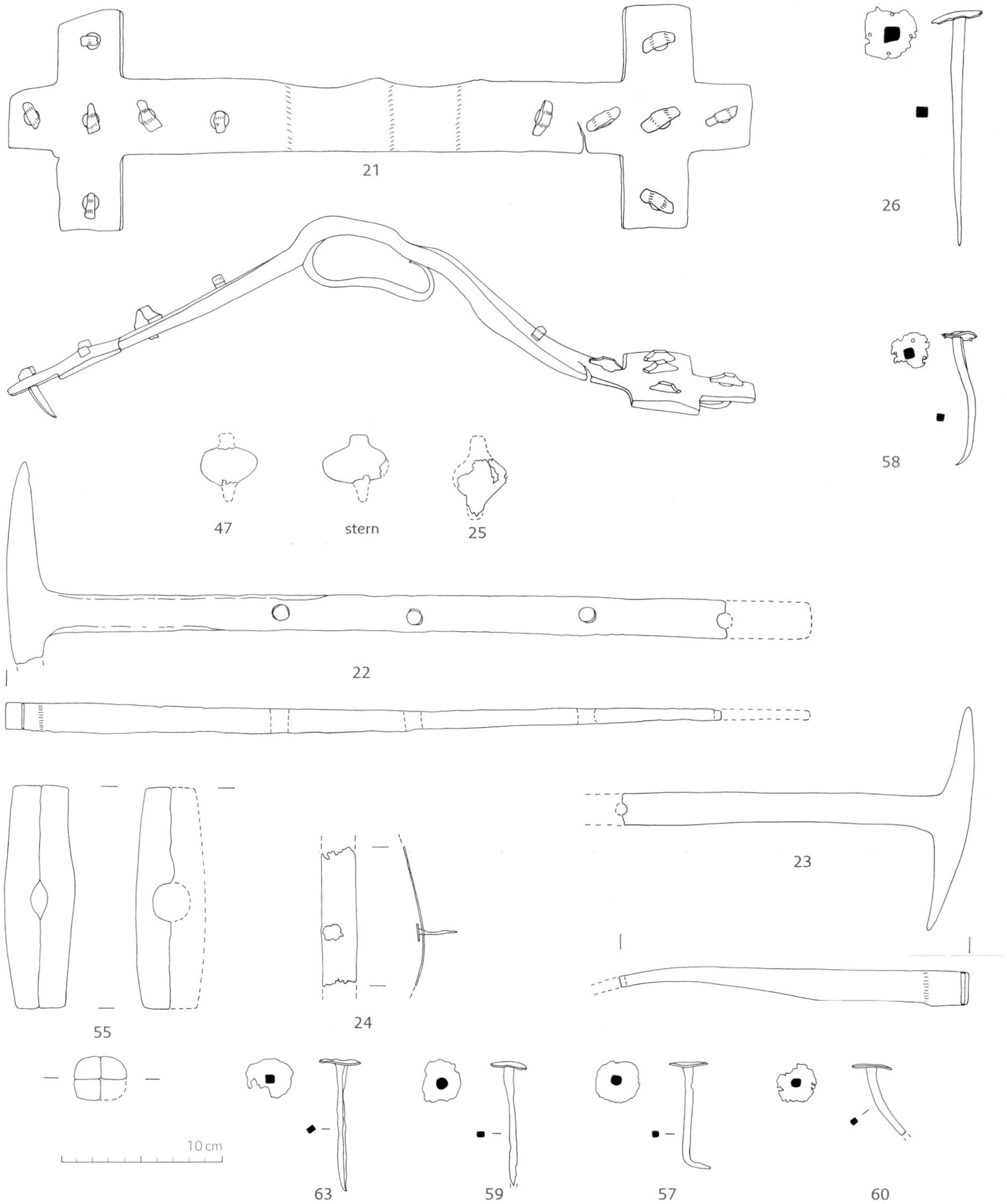

Fig. 11.19. Ship's equipment.

13. Vlierman in: Hocker & Vlierman 1996, 80-81; chapter 16.

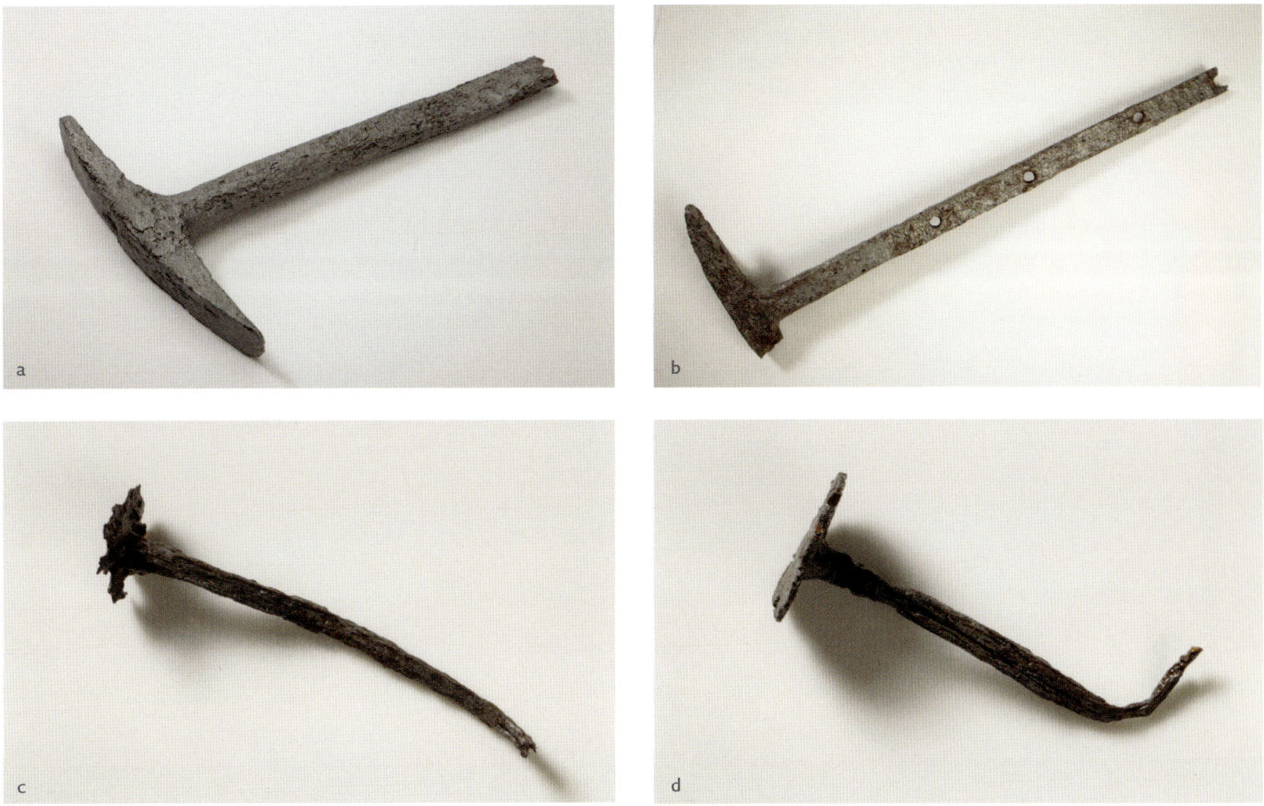

Fig. 11.20. Ship's equipment: a and b Rosebolts, c and d Nails.

Working equipment

In the wreck no objects could be found that fit into this category.

Administration and navigation

Also from this category no objects were found.

Tools

Some tools were found in the wreck.[14] The heavy (1455 gr), almost symmetrical and the left-handed squaring axe (figs 11.21-17 and 11.22) is the only specimen of this lumberjack type to have been found in the Zuiderzee area.[15] The axe is 260 mm long, has a slightly curved edge of 120 mm and a hammer-like butt. The adze (figs 11.21-18 and 11.23) has a gently curved top surface, a relatively thin blade, a rectangular eye with downward extensions on either side. The hammer-like butt has a more-or-less octagonal cross-section and traces of wear on the edges. This shows the intensive use of the adze. The caulking iron (figs 11.21-19 and 11.24) has a straight shape with an octagonal cross-section and a thin blade. It has been compared with other late- and post-medieval examples, from which an evolution in shape could be seen.[16] The bone marlinespike (figs 11.21-39 en 11.25) was made from a beam or tine of a red deer antler. The point is very worn and polished by intensive use. The two whet- or grindstones (figs 1.21-29 and -30, and 11.26a and b) of layered quartz phyllite are almost standard in late-me-

Table 4. Tools

Category	Item	Number	Findspot
4.1 Carpentry	axe	17	afterbody
4.2 Caulking and maintenance	adze, caulking iron	18, 19	afterbody
4.3 Splicing, sailmaking	marlinespike	39	hold
4.4 Miscellaneous	grind-/whetstone	29, 30	hold

14. Reinders *et al.* 1980, 13-14 and appendix 3.
15. Vlierman 1985, 44-46; Vlierman 1992, 15; Vlierman in: Hocker & Vlierman 1996, 76-77.
16. Vlierman in: Hocker & Vlierman 1996, 82-85; section 16.9.

dieval vessels and settlements. In the Middle Ages there must have been an intensive trade with Scandinavia, from where they came. The long and thin example (29) has three flat, smooth sides. The fourth side shows two concave grinding surfaces.

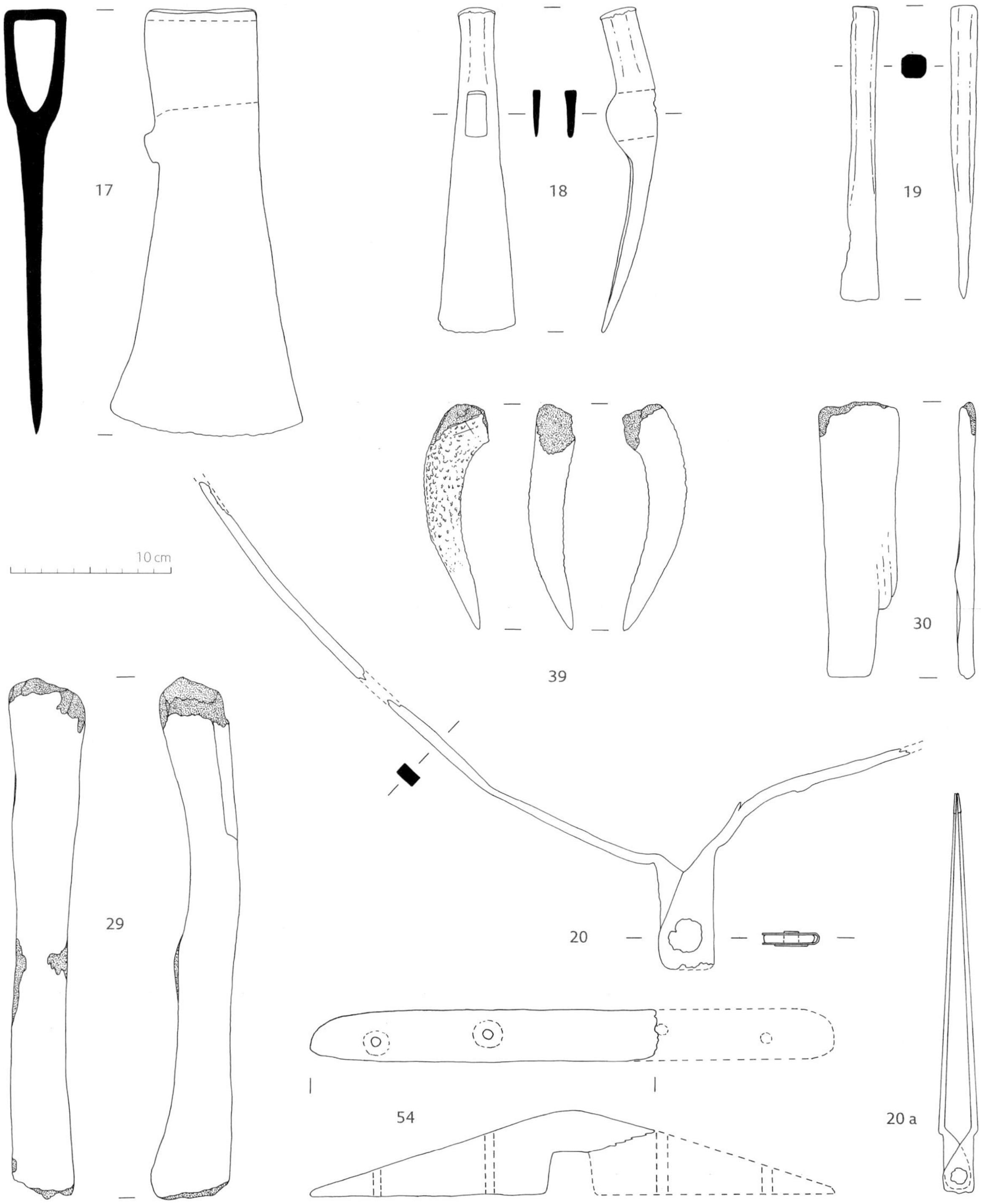

Fig. 11.21. Tools.

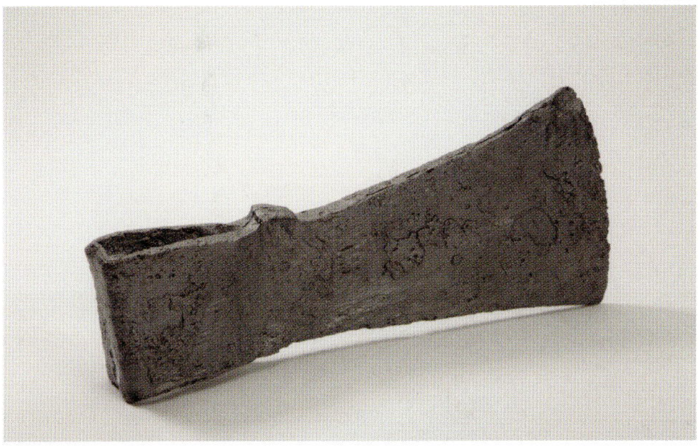
Fig. 11.22. Axe.

Fig. 11.23. Hammer-adze.

Fig. 11.24. Caulking iron.

Fig. 11.25. Marlinespike.

Fig. 11.26a. Grindstone or whetstone.

Fig. 11.26b. Small grindstone or whetstone.

Fig. 11.26c. The two pieces of the fire tongs.

Household items

In the wreck no objects were found that belong in this category.

Cooking, eating and drinking

The ceramics of the Dronten-I cog offer an interesting insight into the kind of ceramics that were in use in late-medieval households (fig. 11.27), in this case aboard a ship. One may assume that the households of ordinary people in the towns and villages were not different in those days.17 It is one of the few assemblages, probably the only one, of a closed and closely dated ceramic find, in which early stoneware, a redware tripod pipkin (Grapen) and a globular Kugeltopf with so-called Besenstrich (brushed) decoration are represented.

Fireplace

In the afterbody a small tile, one complete brick and five fragments of bricks were found (fig. 11.28). The complete brick measures 29x13.8x6.5 cm. On the bricks and on the keel plank in the afterbody some coarse-grained material, presumably loam, was found (50). Also soot could be seen on the tiles. It therefore seems likely that these are remains of the filling of a low wooden box, the ship's fireplace (see also the box from the Nijkerk-I cog, Section 8.5.2. and figs 8.12, 8.50 and 8.51). Of the box itself nothing was found, the planks probably washed away. Such boxes generally are ca 85 by 85 cm and about 20 cm high. A wooden clamp (fig. 11.21-54) probably was attached to the box. We cannot be certain whether the bricks and the tiles together formed its floor; their

Fig. 11.27. The restored pottery from the Dronten-I cog.

Table 5. Cooking, eating and drinking

Category	Item	Number	Findspot
6.1 Fireplace	tile, bricks, sample of loam?, clamp	31, 32- 37, 50, 54*	afterbody
6.2 Fire-tending tools	fire tongs, flint	20, 38	afterbody
6.3 Fuel	peat	57*	hold/afterbody
6.4 Cooking equipment	tripod pipkin(s) *Kugeltopf*	4, 9, 6, 8	afterbody
6.5 Eating and drinking ware	small jug	3	afterbody

17. Vlierman 2010, 201.

combined surface is not more than ca 45x40 cm. Into the loamy fill of the box a smaller floor could have been laid. Indeed the hearth floor could have been completely made of loam, while the bricks were then used to support a pot over the fire (no iron trivet has been found). Furthermore some pieces of cracked flint were (38) found, next to iron fire tongs (figs 11.21-20 en 11.26c) and some chunks of peat (57). Given the findspots, it seems plausible that the fireplace was situated just forward of the afterdeck.

Fig. 11.28. Fragment of a large brick from the fireplace.

Cooking utensils

The cooking ware[18] consisted of a small redware tripod pipkin (*Grapen*) and a small globular *Kugeltopf*. The tripod pipkin has one handle, stands on three feet and has an ornamental rib on the shoulder. On the shoulder there is lead glaze and overall also spots and splashes of glaze on the inside and outside and here and there some reduction green. A sherd of a second specimen probably comes from a small pot that was used on board long before the foundering (figs 11.29-4 and -9 and 11.30). The small *Kugeltopf* with brushed decoration completes the cooking ware found in the wreck. The sherds of probably two other vessels originate from cooking pots that had broken before the wrecking (figs 11.29-5, -6 and -8a and b and 11.31). On the outside were traces of soot, and on the inside cooking residues. Fragment 6, of a neck part, comes from a quite large vessel and is made of grey ware. This pot, which probably had been used on board, may have been a storage jar. No copper kettle or bronze *Grapen* for cooking (larger amounts of) food was found.

Drinking ware

The early stoneware jug (figs 11.29-3 and 11.32) is the only find of a drinking or serving jug. The content is about 1 litre.[19] It most probably was made in Siegburg.

Victuals

The two large stoneware jugs (figs 11.27, 11.29-1 and -2, 11.33 and 11.34) both had a content of about 6 litres and therefore must have been used to store and transport drinking water.[20] One (1) is made of a hard fabric but is not fully sintered. The sherd has a yellowish-grey colour and both on the outside and inside locally has a brown iron-rich slip. It is most probably a product from the area west of Cologne, maybe from Langerwehe or Brühl. The other example (2) was probably made in Siegburg. Both jugs can be classified as 'early stoneware'. The rim sherd of blue-grey ware (fig. 11.29-7) is presumably a storage pot or bowl which had been used on board. A bone of a pig and eight bones of cattle (40-42) derive from meat that once was eaten on board. The bones are discussed in the appendix.

Personal belongings and weaponry

The leather fragments from at least two different shoes (figs 11.35-43, -43a en 1.36) are probably pieces of discarded footwear. They were part of at least one high shoe, which was closed with leather buttons, and the other one possibly of a low model. They match shoe type 25 and possibly a form similar to types 13 and 15 of the ROB overview by O. Goubitz of 20 August 1977.

Table 6. Victuals

Category	Item	Nummbr	Findspot
7.1 Fresh-water supply	Jar	1, 2	afterbody
7.2 Supply of beverages	-	-	
7.3 Lifestock	-	-	
7.4 Food provisions	sherd, bones	7, 40-42	afterbody, hold, foreship

18. Vlierman 1983, 13-25.
19. Vlierman 2010, 192.
20. Vlierman 1983, 13-25: Vlierman 2010, 192.

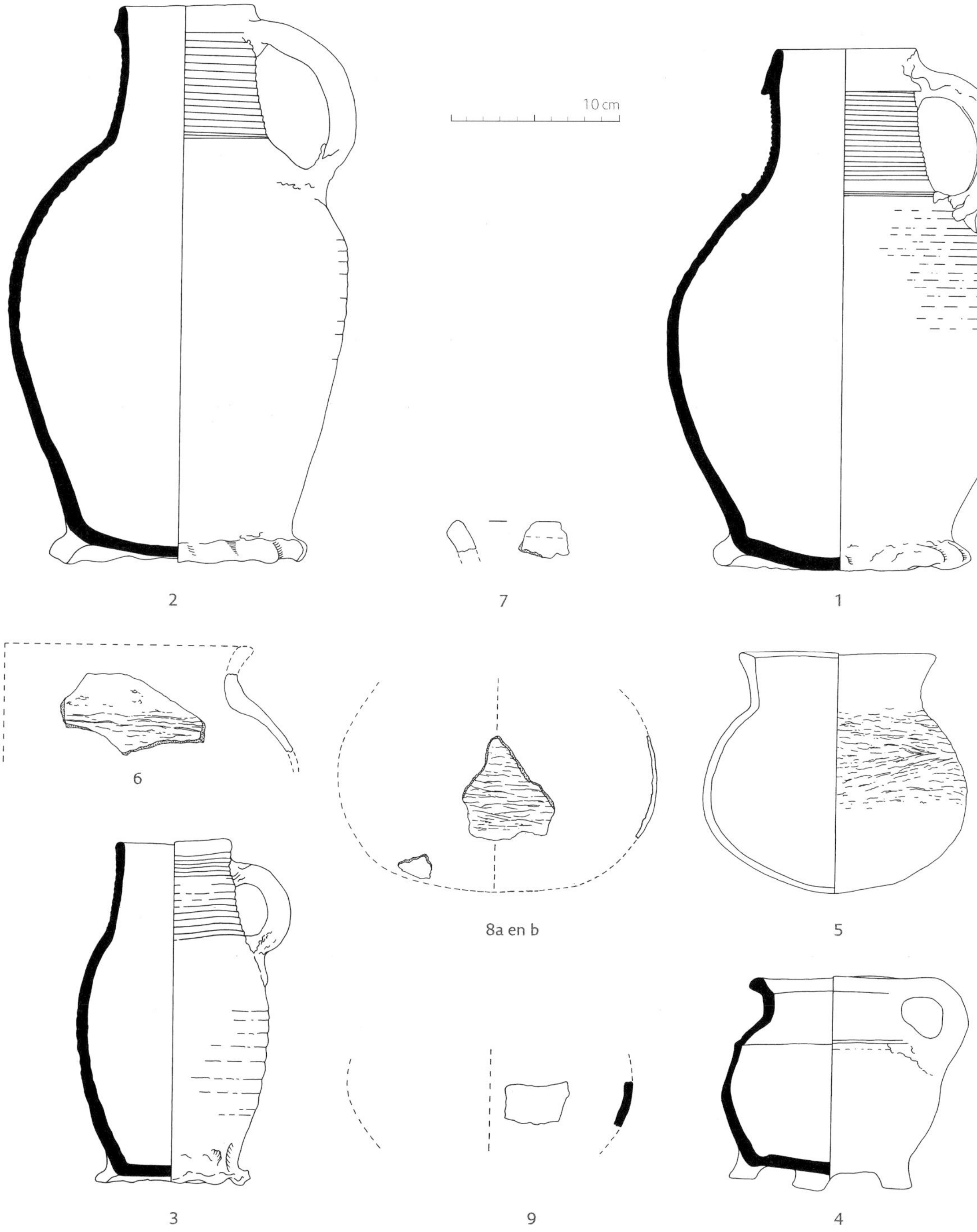

Fig. 11.29. Pottery from the Dronten-I cog.

Fig. 11.30. Redware pipkin (*Grapen*).

Fig. 11.31. Small *Kugeltopf* with brushed surface finish.

Fig. 11.32. Drinking jar of near-stoneware.

Fig. 11.33. Storage jar of Langerwehe or Brühl ware.

Fig. 11.34. Near-stoneware storage jar from Siegburg.

Two remarkable finds are the iron spearheads (figs 11.37-10 and -11, 11.38 and 11.39). This kind of weapon has never before been found in a shipwreck of the former Zuiderzee. The largest (10) has quite a flat point and in cross-section has been forged to a lozenge shape. The socket has an opening running right through it and in cross-section has a decagonal shape. The other has a thicker and shorter point and an octagonal socket. No shafts or parts of them have been found. Maybe before or during the sinking of the ship they came loose and floated away. In section 11.3.4 it already proposed that they were personal weaponry. Of course it cannot be excluded that some soldiers were on board to protect the crew in the often turbulent times. It is also known that the Lords of Kuinre and their cronies executed raids on the Zuiderzee and especially in the area where the Dronten-I cog sank (see the Kuinre cog Section 13.1.3).

Five knives were found in the shipwreck (figs 11.37-12 to -16 and 11.40 to 11.44). The different shapes and dimensions of the four examples (12, 14, 15 and 16) suggest that they belonged to a set of knives that were maybe hunting knives. Often knives would be held in a large leather sheath, together with a dagger. No sheath was found.

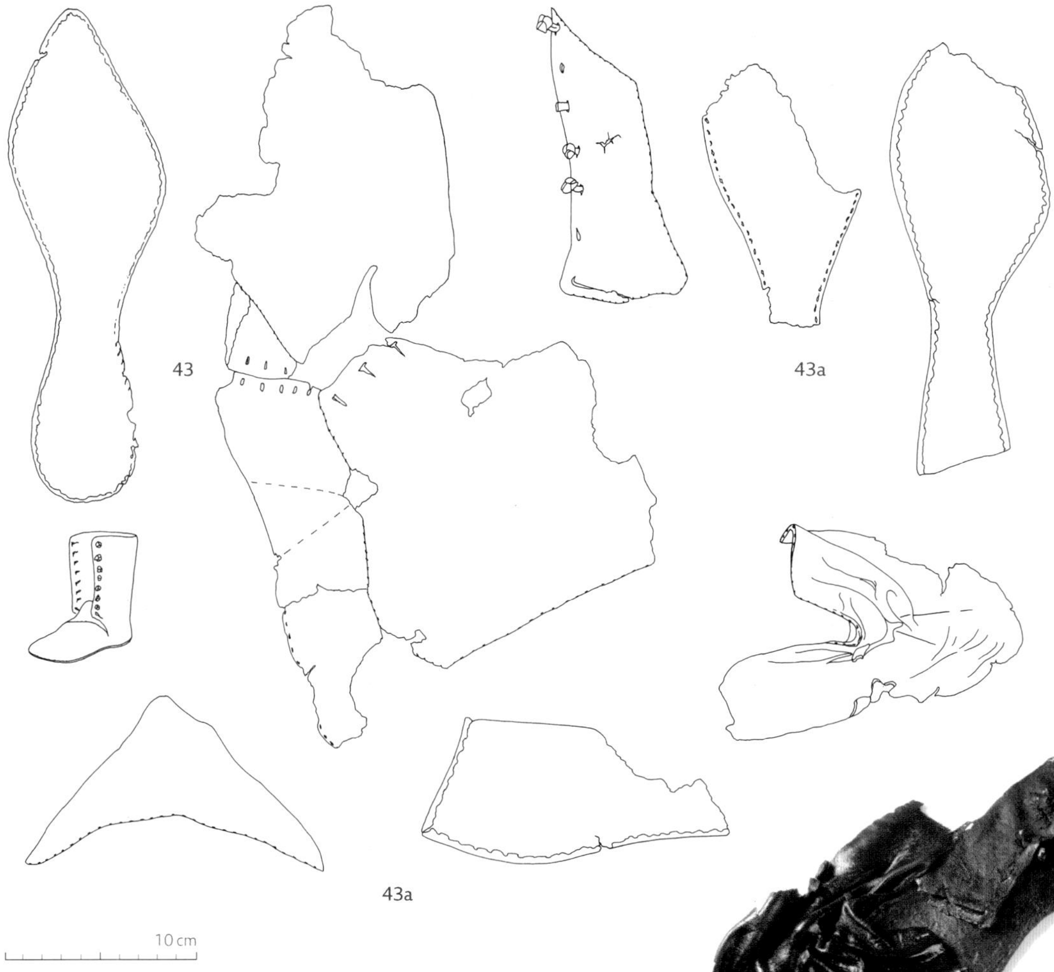

Fig. 11.35. Personal equipment/footwear.

Fig. 11.36. Footwear. Some fragments of discarded shoes.

Knife 12 has a wooden handle that has iron end plates at both ends. The back is slightly curved. Knife 14 has a broad, thin blade with a sharply curved back and a cutting edge positioned diagonally to the handle. The front of the wooden handle has an iron ferrule. Knife number 15 has a curved back and at the blade end of the handle an ovoid plate which has a straight upper edge and a pointy end towards the cutting edge. Possibly two or three thin silver plates had been attached behind it. Knife 16 is straight, and has two almost rectangular copper-alloy plates at the beginning of the hilt. All five knives have a single cutting edge.

Knife number 13 may actually be a small dagger. The knife still has parts of its (possibly walnut) root wood handle with an octagonal cross-section and maybe a somewhat protruding iron guard. The blade is marked with some letters (HC) and five stars. These letters and figures are inlaid with silver. The meaning of the letters is unknown. There may have been yet another letter or star, but that part of the blade had corroded away.

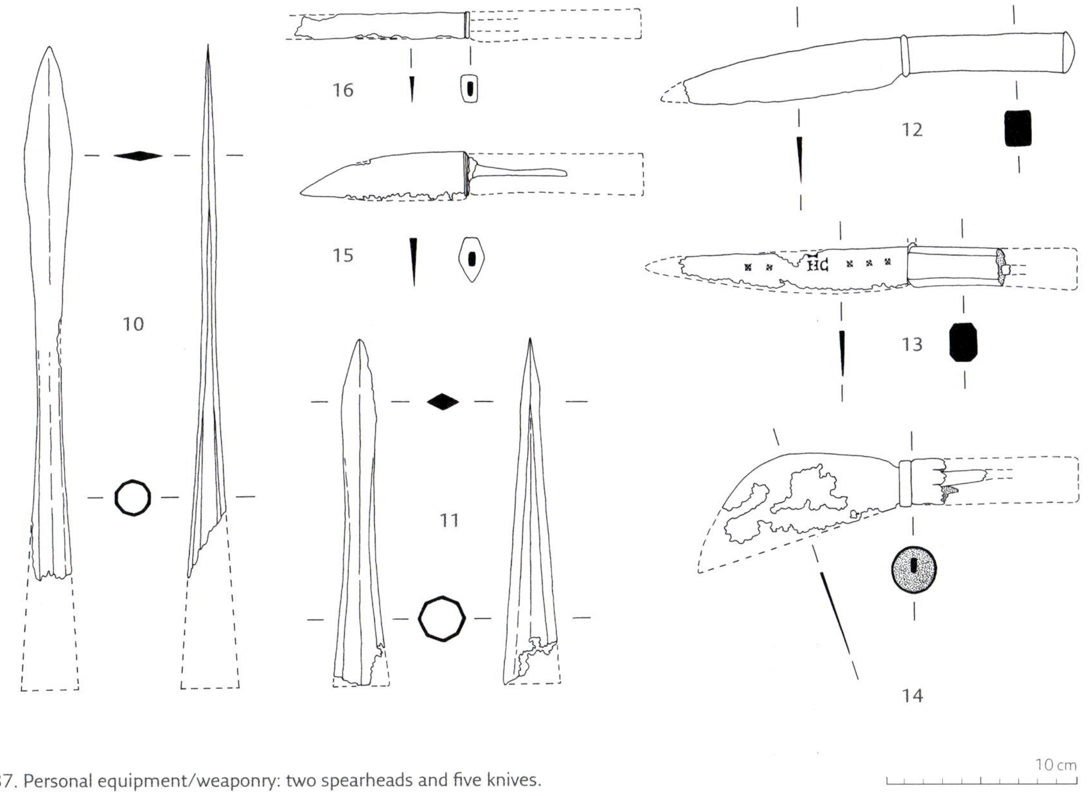

Fig. 11.37. Personal equipment/weaponry: two spearheads and five knives.

Table 7. Personal belongings and weaponry

Category	Item	Number	Findspot
8.1 Garments	-	-	-
8.2 Footwear	high and low shoe	43, 43A	hold
8.3 Sewing kit	-	-	-
8.4 Tools/weapons	spearheads, knives	10, 11, 12 t/m 16	afterbody, hold
8.5 Cash	-	-	-
8.6 Miscellaneous	-	-	-

Fig. 11.38. The larger spearhead.

Fig. 11.39. The smaller spearhead.

Fig. 11.40. Knife with wooden handle (12).

Fig. 11.41. Knife or small dagger (13) with rootwood (walnut?) handle and inlaid with silver initials ('HC') and 'stars'.

Fig. 11.42. Knife (14) with part of its wooden handle and a broad, thin blade with a strongly curved back.

Fig. 11.43. Knife with a curved back (15) with copper and silver plates at the transition from blade to handle.

Fig. 11.44. Small, straight knife (16) with two near-rectangular brass/copper plates at the transition from blade to handle.

11.6 Cargo and merchandize

In the wreck nothing has been found that can point to transported cargo or traded goods.

11.7 Ballast

In the upper forebody and to starboard next to the stem hook five large and three small cobbles were found (Plate 11.1 A and fig. 11.45). It seems plausible that the stones belong in the ship and functioned as ballast. The rocks have not been identified.

Fig. 11.45. View of the forebody with the ballast stones *in situ*.

11.8 Dating

Soil profile

The redeposited layer ends at the base of the Almere-clay deposit Ala, sedimented in the second half of the 15th- or the early 16th century. Given the location of the findspot on the eastern, leeward side of the peat barrier running north-west-southeast between Schokland and Elburg, soil scientists cannot rule out that part of the Alc deposit had been washed away; which - together with other evidence - results in very probably an earlier date of the foundering. Also the sharp interface between the Ala and Alc (geological hiatus) and the redeposited layer which consists completely of sand, supports this assumption.[21] The stratigraphy therefore fails to offer a reliable, close date for the foundering.

Finds

The ceramics offer the best information in regard to the time of the cog's foundering. The small *Kugeltopf* has 13th-century characteristics, which also are found in the first half of the 14th century. The main characteristic is the brushed decoration. *Kugeltopf* vessels with such a decoration are typical of the 13th century. Around 1290 they are in general use, though they also appear regularly in contexts of the first half of the 14th century.[22] The small redware tripod pipkin with one handle and a characteristic ridge just below it is typical of the first half of the 14th century.[23] The same goes for the partially-sintered (early stoneware) jug with iron slip, parallels of which, often decorated with roller stamps on the collared rim and the shoulder, are known from the Limburg pottery centre Schinveld, but above all from Langerwehe and Brühl.[24] The jar (2) and the small jug (3) are characterised by the early-stoneware firing, which generally precedes the introduction of fully sintered stoneware in the years 1320-1330.[25] The other objects found provide no dates. The complete ceramic assemblage, and hence the time of the cog's foundering, can be dated to the first half of the 14th century, with the 1320-1330 period the most likely.[26]

Sintel clamps, nails and the shape of carvel seams

The sintel clamps which were used to secure the caulking into the plank seams are of type D2. This type was in use in the period of ca 1275-1350. Recent research has clarified that the clamps used in the Dronten-I cog especially were in use around 1300.[27] The carvel-built seams of the bottom of the ship have a triangular section which comprises the entire thickness of the planks. Such seams are mainly known from vessels which date before ca 1250. From this time onwards also triangular seam openings occur that comprise only half of the plank's thickness. The carvel seams of the Dronten-I cog still have a 13th-century shape and seem to be a transitional form between seam

21. Koopstra in: Reinders *et al.* 1980, 6-8.
22. Renaud 1976, 37; Janssen 1983, 156-157; Vlierman 1983, 13-14; Janssen & Nijhof 2000, personal communication.
23. Janssen 1983, 158.
24. Bruyn 1962-63, fig. 70-73; Hurst 1977, 226-227, type I.
25. Janssen 1983, 173-175.
26. Vlierman 1983, 14.
27. Vlierman 1996a, 77; Vlierman in preparation a; see also section 29.2.

types B and C in the overview of variations.[28] The same goes for the use of nails with flat heads.[29] On the basis of this evidence, the ship was most probably built at the end of the 13th century or around 1300.

Dendrochronology

The wood of the Dronten-I cog has not (yet) been dated with dendrochronology. The preserved stern- and stem hooks and the V-shaped floor timber with the rebate for the mast are suitable for this, though. In some parts the timbers still consist of the full width of the trunk from which they were made, and also the sapwood is still present.

11.9 Summary

In 1980, the elaboration of the excavation records of the wreck (the Dronten-I cog) found in lot N5 in Eastern Flevoland resulted in the second excavation report in the series 'Flevoberichten' of the IJsselmeerpolders Development Authority (RIJP). The then still very limited (published) knowledge of late-medieval ship finds and the far from completely preserved vessel limited the possibility of a full reconstruction of the hull. After the elaboration of the other cog finds presented in this study it is now possible to do this. Moreover the report of 1980 presented only limited information on the associated finds. The present publication, in a standardized and complete form, gives a better insight into the inventory.

Ship

The vessel does not have a strong sheer; has a straight, raked stern and stem of 118° and 122° respectively; and quite rounded cross-sections. Amidships it has a carvel-built bottom, which consists of only four strakes and a keel plank, and there is a sharp turn of the bilge. Only a few frames in the after- and forebody have a faint S-shaped profile. Viewed from above the ship has an almost symmetrical outline, though the forebody is slightly (10 cm) wider. The bottom is completely symmetrical, mainly carvel-built and between the bilges and frames 9 to 15 completely flat. Towards the stern and bow the strakes twist and turn into clinker-built construction. The planks composing each strake are fastened to each other with 40-cm-long scarfs. The ends of the overlapping planks are 1.5 cm thick and stick out into and outside the shell. The nails used for the scarf connections did not protrude through the planks. The nails of the clinker-built constructions are nailed through and their tips are twice bent over and driven back into the wood. Remarkable are the nails which are driven in alternatingly inside out and vice versa, and differ in level along the overlap depending on their direction. This manner of nailing is known to me only from some ship remains from the province of Brabant and from Antwerp (Sections 26.6 and 26.7).

Another notable feature are the short laths placed perpendicularly across the seams of the underwater body. These are so-called auxiliary or temporary frames, used during the construction of the bottom of the ship and which the ship builder did not bother to remove after the placement of the straight and V-shaped floor timbers. The cog was steered by a stern rudder. The mast was positioned in a rebate in the V-shaped floor timber of frame 18 and was held in place on the mast step with 6 stays. The sail beam, and possibly also the foremost beam of the afterdeck, was made extra secure with two rosebolts which passed through the shell from the outside and were nailed to the beam. The inside of the ship's hull, on both sides of the keel, was longitudinally stiffened with three ceiling strakes. The ship had an open ceiling construction and possibly a load-bed. The lowest ceiling strake (the bilge strake) and the upper ceiling strake (inwale) were roughly adzed into a double-bevelled cross-section.

The ship most probably had an after- and foredeck which were connected by gangways. The sharp parts of the underwater body of the ship were probably strengthened on the inside with two transverse connections, composed of a beam with knees on it at both ends, or a knee-like beam onto which at the other end a knee was placed. These constructions were joined to the sides and to the frames with 3-cm-thick treenails. The decks were lying on beams of which the ends probably were supported by the flat tops of futtocks. The upper parts of the hull were held in position by knees that were fixed with treenails onto the deck beams. The ship had a hold between the after- and foredeck with a content of 26.5 cu m or about 26 gross tonnes cargo capacity. No cargo was found. Some cobbles in the forebody could have been ballast.

The most important reconstructed dimensions are:

- Length over the stern and stem at gunwale level — 15.8 m
- Largest width at gunwale level — 5.2 m
- Height of the stern (keel line to the top gunwale) — 2.7 m
- Height of the stem (keel line to the top gunwale) — 2.3 m
- Height of the side amidships (keel line to the top gunwale) — 3.2 m

28. Vlierman 1996a, 27 fig.11; Nijkerk-I cog fig. 8.24.
29. Ens cog, section 9.8.

Fig. 11.46. Impression of a part of the inventory of the Dronten-I cog.

Inventory

The probably near-complete inventory and the fact that most of the broken ceramics and other items were spread about the afterbody suggest that the 'living quarters' were situated there, and that the vessel had been hastily abandoned. The cog had a hearth: almost certainly a wooden box filled with loam and maybe some pieces of large bricks. The latter could have been used as a trivet on a loamy hearth base. The early stoneware, the *Kugeltopf* with *Besenstrich* decoration and the redware tripod pipkin (*Grapen*) offer a remarkable and unique insight into the number and types of vessels that were in use at any time in a small (onboard) household, possibly the only or at any rate one of the few late-medieval archaeological examples. To operate a small cog like the Dronten-I, a crew of 3-4 persons would have been sufficient. The two spearheads are also noteworthy, as these never before or since have been found in the area of the former Zuiderzee. Presumably they were part of the crew's personal armament. Yet it cannot be excluded that we are dealing with military weaponry belonging to one or a few soldiers: in the area where the cog was found the rogue Lords of Kuinre were operating. The knife or small dagger with the silver-inlaid letters ('HC') and four other knives which could belong to a hunting set are also striking artefacts. The initials might prompt one to link them to the Lords of Kuinre, but no other clues for this were found. Unfortunately the knife's state of conservation was rather poor.

Dating

The time of the cog's foundering cannot be detected from the disturbance of the stratigraphy of the soil. A geological hiatus and the location of the site on or against the peat barrier does not provide a better date than the second half of the 15th- or the early 16th century. The ceramics give the best dating evidence. The small *Kugeltopf* with brushed *Besenstrich* decoration has 13th-century characteristics but also occurs in contexts of the first half of the 14th century. The redware tripod pipkin (*Grapen*), with its typical ridge on the shoulder dates from the first half of the 14th century. The early stoneware jars and jug were produced before the fully sintered stoneware was introduced in 1320-1330. On the strength of the pottery, the foundering of the vessel can be dated to the first half of the 14th century, probably in the period 1320-1330. The sintel clamps that were used in the ship, the nails with flat heads and the triangular-section-seams in the carvel-built bottom which comprise the entire thickness of the planks, do suggest a construction date in the last quarter of the 13th century or around 1300. Dendrochronology has not been applied, but may yet supply more clarity about the date and also identify the origin of the timbers (or even the ship).

The Marknesse cog[1]

12.1 Find conditions

12.1.1 Discovery and investigation

After the Northeast Polder was reclaimed from the sea in 1941, several shipwrecks were discovered during the cultivation of the former seabed. Of the most of these wrecks which were discovered during the war years little has been documented or preserved. Change came however after the discovery, excavation and publication of a late medieval wreck on lot M107, here described as the Marknesse cog (figs 12.1, 12.2 and 12.5a and b).[2] The wreck probably already was discovered in 1943 and reported in the beginning of February 1944.[3] On the 15th of February 1944 under the direction of P.J.R. Modderman a start was made with the excavation. J.P.G. Kloppenburg and D. Huismans[4] were present to study the wreck from a constructional point of view, while from the 17th of March R. Woudstra and J. Lanting[5] documented the wreck archaeologically. The investigation was executed in three phases (15-22 February; 27-31 March and 9-11 May).

Fig. 12.1. Overview of the front half of the ship with the load-bed under foot planks, the gangway and the inwale on the port side.

1. Translated by Jeroen Vermeersch, edited by Xandra Bardet.
2. Modderman, 1945.
3. The wreck is situated at ca 100 m from the ditch along the Lindeweg and at ca 130 m from the ditch along the Marknesseweg and was originally lying in northwest-southeast direction with the bow to the southeast. The find spot is situated at a distance of ca 8 km from Kuinre, ca 10 km from Vollenhove and ca 6km from the northern tip of Schokland. Topographical map 21A coordinates 183.400/525.010 (old coordinates 183.42/524.83, presumably the original location; figs 12.3 and 12.4) with the highest parts at ca 0.4 m below the ground level (ca 3.5 m -NAP). The former water depth on that location was ca 3.1 m. Hydrographical map 1921, average low water level.
4. Scientific assistant and model builder of the Dutch Open Air Museum (NOm) in Arnhem.
5. Draughtsman and field technician respectively of the Biological Archaeological Institute of the University of Groningen.

Fig. 12.2. Overview of the fore and midship section during the excavation on 16th February 1944, after the cargo of bricks had already been removed. Seen from port side in the stern.

Fig. 12.3. Find spot of the Marknesse cog.

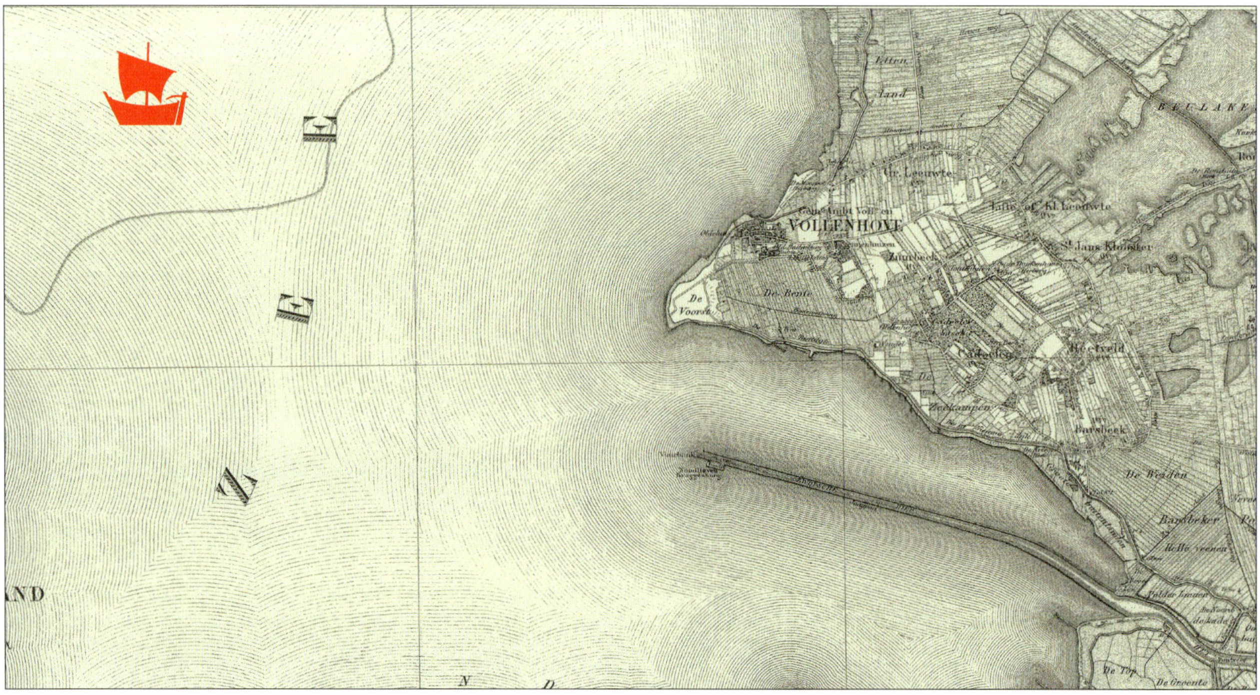

Fig. 12.4. Topographic and military map of the Kingdom of The Netherlands (scale 1: 50.000) from 1861 with the location of the Marknesse cog.

It soon became clear that this ship find from both a historical-nautical constructional and stratigraphical point of view was by far the most important wreck found thus far in the Northeast Polder. This is the reason why from the start measures were taken in order to conserve the ship for the long term. It was assigned to the Dutch Open Air Museum (NOm). [6] In March and April A.E. van Giffen[7] met with R. Lunsingh Scheurleer[8] and W. Voorbeytel Cannenburg[9] to discuss the conservation of the ship. According to the latter conservation was justified, based on the stoneware jug dated around 1400 by J.G.N. Renaud. In the same period contact was made with G. de Vries Lentsch Jr. of the Amsterdam Ship wharf in order to lift the wreck, to transport it and to conserve it. The company already had gathered the necessary experience with the lifting and conservation of the famous ship from Utrecht in 1930.[10]

The absence of an adequate conservation method and the insecure times were the main reasons to bury the wreck deeper in the sand in order to secure it for future lifting and conservation. After the cargo of bricks was taken out and the ship was cleaned (figs 12.5a and b and 12.6), several drawings were made. Initially it was the plan to take the ship apart on the 17th of April 1944. Later it was decided to bury the wreck into a previously dug hole in the soil, located somewhat to the west of the find spot, which was done on the 23rd of June.

Modderman published the wreck as part of his doctoral dissertation and categorised the vessel as a cog.[11] As a result of the *in situ* preservation of some particular wrecks in Flevoland since 1975,[12] in the eighties a growing need arose to investigate the preservation of the meanwhile 40 years buried ship. The exact location of the reburial could unfortunately not be found. Only in December 2005 the wreck was located by D. Velthuizen.[13] From 11-22 August 2008 in the framework of the International Fieldschool for Maritime Archaeology Flevoland (IFMAF) an investigation took place. During the research the aft, mid and fore sections were excavated. Apart from the documentation by the students of the situation *in situ* also samples for dendrochronology were taken. During this post-excavation research it was possible -in the framework of this study and with the knowledge and insights of today- to study the excavated parts and to document and to supplement the various (missing) details.

Apart from the described importance from a historical-nautical constructional and stratigraphical point of view it can be attested that with the research which was executed during the war and also the swift publication of this late medieval cog, actually the basis was laid for the regular ship archaeological research in the Netherlands.[14] Conversations with professor Modderman in 2001, in addition to enthusiastic support for the intended plan to re-examine the wreck in a study on all cog-finds in the Netherlands, provided additional and mostly new information about ship archaeological research during the Second World War. This also led to a visit to the documentation archive of the Dutch Open Air Museum (NOm).

Fig. 12.5a. Overview of the excavated wreck, seen from the bow.

Fig. 12.5b. The same, seen from the stern forward.

6. Modderman 1945, 75.
7. Director BAI.
8. Director Allard Pierson museum, Amsterdam.
9. Director Dutch Maritime Museum/Nederlands Scheepvaartmuseum, Amsterdam.
10. Vlek, 1987.
11. Modderman 1945, 81.
12. Vlierman 1997a, 18-19.
13. Nieuw Land Erfgoedcentrum (NLE), with the assistance of some members of the Archaeological Working Group (AWN) (Van Holk *et al.* 2008, 365-366).
14. Vlierman 1985b, 47.

A beautiful collection of black and white photos of the excavation was found here, many of which are shown for the first time in this study. The photos have now been transferred by the NOm to the archives of the National Service for Cultural Heritage, Lelystad (RCE).

12.1.2 Geology and stratigraphy

The pleistocene surface and covering layers
The stratigraphy around the wreck hardly had been affected by agriculture. At the time of the archaeological research the profile above the wreck unfortunately was already disturbed. The disturbance in the profile caused by the wreck can be seen clearly on three depicted cross-sections by Modderman[15] (fig. 12.6). The geological knowledge about the sediments of the former Zuiderzee area has obviously increased since 1945. Figure 12.7 shows a reconstructed cross-section over the wreck with the load of bricks and the soil profile (disrupted by the wreck) at the location of frame 9. With the current insights, the profile is reinterpreted below.

Description and dating of the soil profile
The wreck rests with the keel plank and stern and stem hooks on the sandy pleistocene surface. At the time of the excavation, this surface was located at a depth of 1.8 m below surface. On the sand there is a layer of Holocene peat of about 70 cm thick. An approximately 15 cm thick layer of redeposited peat or *Detritus Gyttja* was deposited on it. This layer then gradually changes into an approximately 45 cm thick layer of freshwater clay (Almere deposits, referred to in the Northeast Polder as Sloef), in which, according to Modderman,[16] numerous thin layers of redeposited peat occur, particularly at the bottom. Modderman further states that "deeper only freshwater shells of *Valvata, Pisidium, Unico, Sphaerium, Bythinia et al.* occur." On the drawn profiles, at the bottom of this Almere deposit, a clear, approximately 15 cm thick layer is indicated in which these shells would occur. A well-recognizable, approximately 5 cm thick sandy layer with marine shells (*Myas)* has been deposited on the Almere deposits. Above that is an approximately 45 cm thick Zuiderzee clay deposit.

Both the Almere and Zuiderzee deposits are not further subdivided. A photograph of the port side of the midship section (Fig. 12.8) clearly shows from top to bottom that the Zuiderzee clay shows some stratification at the top, but that it looks fairly homogeneous in the bottom 30 cm (sticky clay). The clearly visible light-coloured layer underneath is the sandy layer with marine shells, the Zu IV layer. Modderman[17] speaks of a brackish water transition layer directly below, in which very small *Cardium* shells occur. This is the Zus layer. This is not recognizable in the photo and is not indicated on the profile drawings. The Almere or Sloef deposit below does not show, as described by Modderman, a particularly strong layering with redeposited peat bands in the lower part, but an evenly distributed layer pattern of clay, redeposited peat and sand bands over the entire thickness. At the base of this is a somewhat dark layer with here and there thin sand bands/layers. This is most likely the layer with freshwater shells (*Valvata et al.).* Below it is a dark deposit down to the pleistocene sand. The top of this deposit is approximately 15 cm thick, formed by the *Detritus* deposit with the peat below it.

The oldest Zuiderzee-deposit -the layer Zus- has been dated by Van der Heide and Wiggers between ca. 1575 en ca 1625,[18] while the research of a shipwreck on Lot O99 (ca 3 km south of the find spot of the Marknesse-cog) suggests that the formation of this layer at that location must have happened in the first quarter of the 17th century. During the research on the *beurtschip* B71 in Lelystad in 1980 only small *Cardium* shells were found in the thin disturbance of the profile created by the ship and against the ship's hull. That ship has been dated by dendrochronology just before 1600 (dossier RCE, Lelystad), while, based on the dated objects found on board, the ship must have foundered between ca 1620 and 1625. The shell layer Zu-IV was situated over the wreck. In Eastern Flevoland the brackish transition layer Zus has never been attested by soil specialists of the National Service for the IJsselmeer Polders. If the *Cardium* shells in the washed layer of the ship from Lelystad are a sign that we are dealing with a brakish Zus-layer -which is very likely- then the ship foundered in this phase and it is the most southern attested occurence. In regard to the date of the foundering of the Marknesse-cog it need not mean that the ship sank quite some time before 1600. Wiggers assumed that the formation of each Sloef-layer took about 75 years and therefore came up with the following dates for the Northeast Polder:[19]

Sl(oef)	II c	-second half 13th century
Sl.	II b	-14th century
Sl.	II a	-first half 15th century
Sl.	I b	-second half 15 century and early 16th century
Sl.	I a	-second half 16th century

15. Modderman 1945, afb. 28a, b and c.
16. Idem., 75.
17. Modderman 1945, 75.
18. Respectively Van der Heide 1955, 6-7 and Wiggers 1955, 103-104.
19. Wiggers 1955, 87-88.

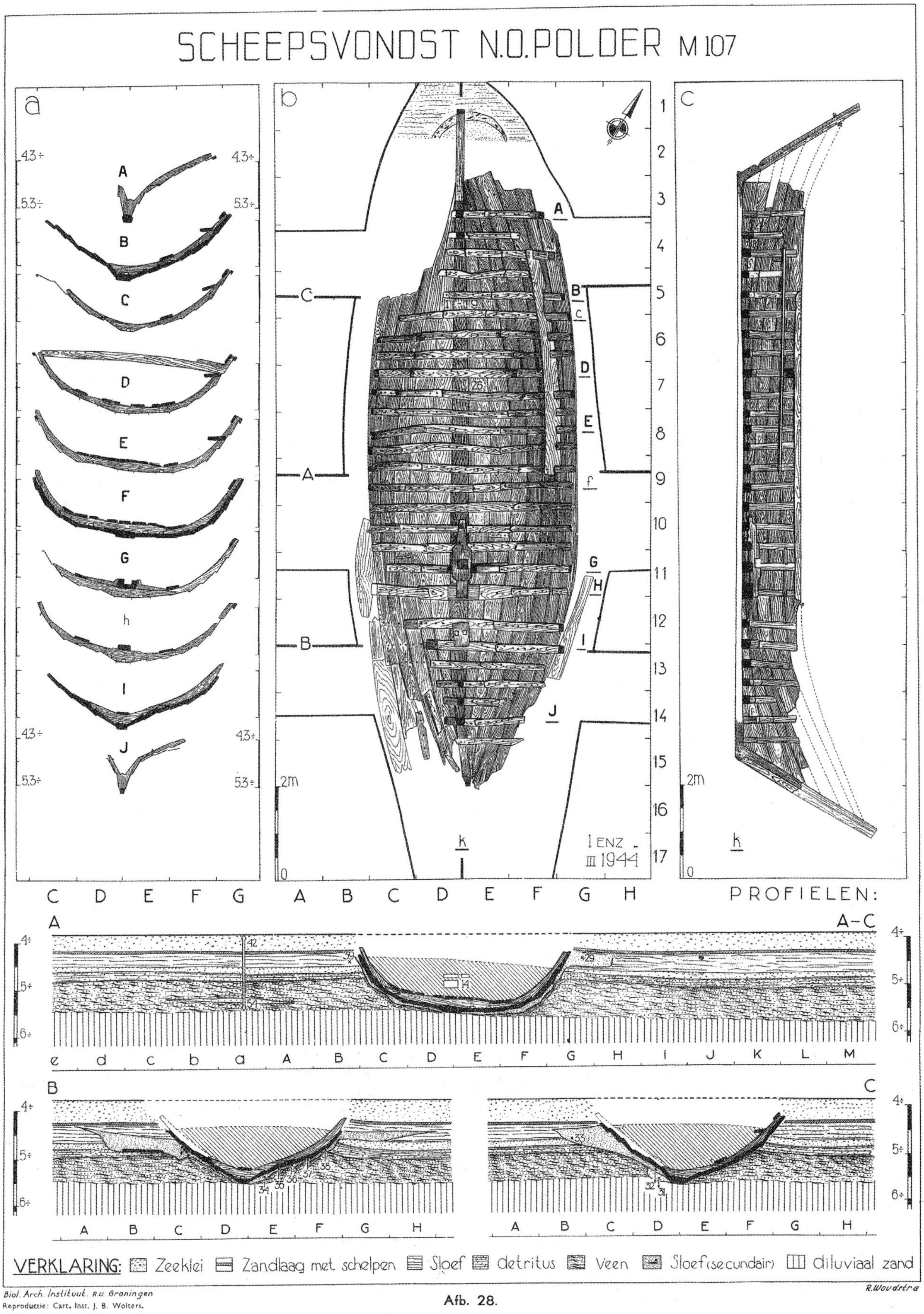

Fig. 12.6. Cross-sections, top view, inner view of the port side and three cross-sections of the wreck with soil profile. *Copy Modderman, 1945 afb. 28.*

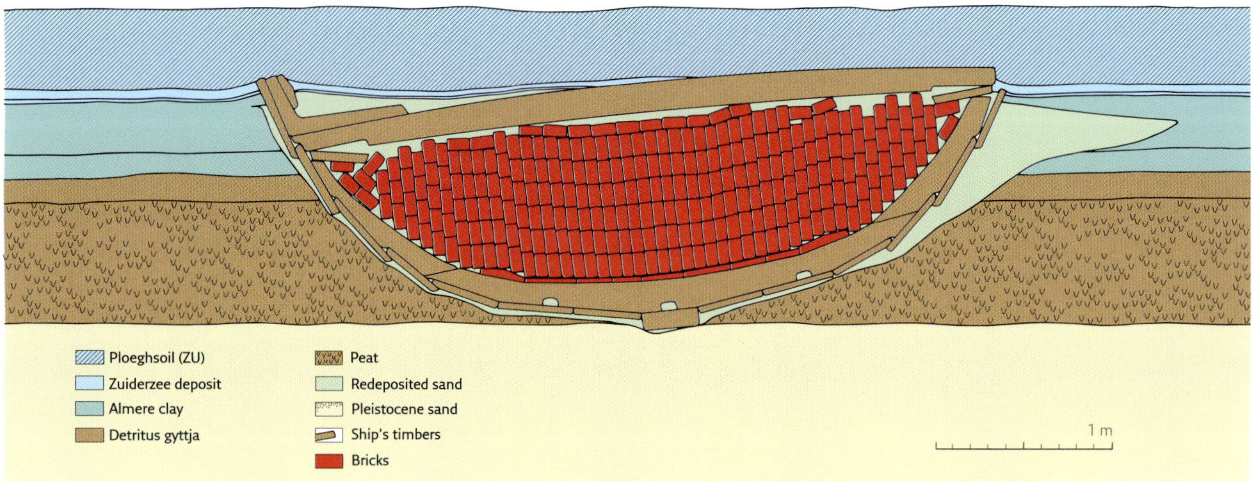

Fig. 12.7. Reconstructed cross-section of the soil profile at the location of frame 9 and the possible stack of the cargo of bricks.

Fig. 12.8. The port side of the wreck with the channel and the soil profile. On the background the labour camp and the farm.

Modderman did not yet know this subdivision. The younger Sloef layers (Sl. I b en Sl. II a) on the Soil and profiles map of the Northeast Polder are missing in the vicinity of lot M107.[20] A geological hiatus in the upper Almere-deposits is therefore very plausible. Wiggers dates the Valvata-layer in the beginning of the 13th century.[21] Ente supposes that its formation took place in the last quarter of the 12th century, presumably in relation to the heightened activity of the IJssel river.[22] Assuming that the Valvata layer was formed between ca 1175-1225, it would mean that the wrecking of the Marknesse cog happened a long time after that. At a steady rate of sedimental deposit during the sedimentation phase of the Valvata-layer and the Almere-layers this would take about 400 years. Therefore a steady rate of sedimentation is not plausible given the presence of peat on this location[23] and the fact that some younger Sloef-layers are absent in the neighbouring lots.

The total package is ca. 45 cm thick, including the Valvata-layer of about 15 cm. The upper part of the redeposited layer sits at about 10-15 cm under the Zu-layer. This would mean that the vessel must have foundered about 130 years (ca 400:3) before the final silting over, hence in the third or fourth quarter of the 15th century. This does not coincide with the date of the stoneware jug (section 12.1.1; ca 1400). If locally none of only partially any Sl. II a and Sl. I b layers were deposited, then an earlier date based on the stratigraphy seems possible. The merely 35 cm Almere deposit and the aforementioned strong, alternating layering implies almost continuously changing conditions throughout the period of sedimentation which at one time a sandy layer is being deposited and at another time a layer of clay or washed peat. Also during this same process layers will have been washed away completely.

12.1.3 Medieval habitation

The distances from the location of the ship to the coast, as specified in section 12.1.1, are also the distances to the nearest settlements/harbours (Kuinre, Vollenhove and Schokland). It is almost certain, however, that the unendiked(?) clay-on-peat land stretched for miles south of Kuinre.[24] According to Wiggers, around AD 1000 the land extended to approximately the current north-eastern periphery of Emmeloord, while sherds were found about 2.5 km north of the cog site, suggesting a settlement on

20. RIJP 1953, page 12.
21. Wiggers 1955, 87.
22. Ente 1973, 146-147.
23. Wiggers 1955, 46 fig. 18.
24. See also the Kuinre-cog section 13.1.2.

the spot.[25] Wiggers writes that perhaps the historical name Vene or Veenhuizen can be associated with this. The youngest fragments date from around 1350-1400.[26] If this situation still existed during the cog wrecking, then the ship sank just a few kilometres south of the coast.

12.1.4 Condition, movement and wreck formation

Condition and displacement
The wreck sank into the Almere-deposit and was completely covered with a layer of Almere- and Zuiderzee-sediments. In the pictures from 1944 the wood seemed to have been in good condition. During the later investigation in 2008 it has been attested that in the last 60 years not that much appeared to have changed. The method of burying the wreck below the groundwater level, or at least in an oxygen-free environment seems to be a very good way to preserve the wreck for a longer time. The strength of the construction must have been very good during the excavation and relocation. The vessel had been moved a few metres sideways and reburied deeper. The highest positioned parts are now situated ca 1.5 m below ground water level. With steel cables and wooden framing on the inside during the displacement it had retained its shape, while in various places supporting wood has been attached to prevent any damage.

The ship parts found detached around and in the wreck were placed in the wreck after it was moved and covered with straw, after which the ship's hull was filled up with sand both around and inside the vessel. The whole was covered with a mixture of Almere and Zuiderzee deposits. The wreck consists of the keel plank, the stern and stem hook, the lower parts of the stem post and false stem, twenty-seven frames or parts thereof, a mast step, bilge ceiling strakes, a floor ceiling and gangways. The port side is still largely preserved up to the gunwale (seven strakes). Aft and fore, however, parts of the hull strakes and frames are missing, some of which have been found in or next to the wreck. Of the starboard shell, the gunwale was completely missing, while in particular the aft 3 m of the top five strakes was not found. The stern- and stempost, a sail beam and two cross beams at deck height, one of which is *in situ*, have also been found. Some repairs and the overall impression (wear on almost all visible parts) suggest that the ship was wrecked a relatively high age.

Wreck information
The wreck originally lay almost upright with the keel plank and stern- and stem hook on the pleistocene sand and with a slight list over port side. As a result, this side has been almost completely preserved, but sagged due to the load of bricks. Because of this heavy, bulky load, the ship must have sunk fairly quickly into the soft and only 45 cm thick Almere deposit. The underlying *Detritus* and peat deposit will also have offered little resistance. The latter can be deduced from the only thin redeposited layer next to the hull of the underwater body. The ship has, as it were, squeezed into the seabed. Some time after, larger disturbances in the soil profile arose around the aft and forebody of the ship in the higher part of the Almere package then present. This was due to the loosening of both the stern- and stempost, the gunwale on the starboard side, a few cross beams and the decks. Part of this then detached wood floated away and has not been found. All of this contributed to the vessel being preserved in good quality and relatively complete.

12.2 The ship

12.2.1 Documentation and reconstruction

Documentation
Despite the short period of field research and the season in which it was conducted, the wreck was extensively documented, certainly for that time. An overview drawing, ten cross-sections, various profile drawings and drawings of some separately found parts, such as the stem post and some deck beams and the upper/forward hull strake on the port side, to which a piece of an inwale was attached. Later, based on the top view and the cross-sections, a longitudinal section with an inside view of the port side was drawn. All drawings are on a 1:20 scale.

Reconstruction
Modderman[27] shows the reconstruction of the port side and writes about a keel beam (in one piece) of 13.4 m long. According to the published drawing, however, this is 12.4 m (presumably a typo). In reality, the total keel length (derived from the top view of the field drawing including the lengths of the horizontal parts of the stern/stem hooks) appears to be 11.8 m. Determining the overall keel length is not that strange, because the ship was not taken apart. At the Ens cog it was first established in

25. Wiggers 1955, 193-194 and fig. 108.
26. In the terrestrial collection of the Nieuw Land Erfgoedcentrum in Lelystad only a few finds of this area are kept, which does not support the assumptions by Wiggers.
27. Modderman 1945, 79 and fig. 28 C; fig. 12.6.

1949 that such a vessel has a 'short' keel plank, extended at both bevelled ends with a stern- and stem hook. It was established in 2008 that this was also the case with the Marknesse cog. Modderman assessed the thickness of the keel plank at 15 cm, probably judging by the outside view at the stern and stem. Here, however, are the thicker, horizontal parts of the stern/stem hook. The actual keel plank is probably about 10 cm thick.

The largest width across the hull determined by Modderman is approximately 4.5 m, while the depth from the keel to the top of the gunwale must have been 1.25 m. The actual width is 4.65 m (see below), while the midship depth was not 1.25 m but 1.5 m. The cross-sections of the field drawings have been manually enlarged to scale 1:10 and reconstructed, from which a reconstruction of the interior view of the port side, of the exterior view on the starboard side and a top view of the presumed deck plan with gangways has been drawn to the same scale (Plates 12 .I A and B and 12.II A and B). Quite a lot of the vessel had been preserved. Moreover, it was still in good shape. This was not the case in the aft and forebody of the ship. The short deck beam found in the forebody, a sail beam, the cross beam found *in situ* above the frame 9 and the complete stern- and stempost made it possible to make a justifiable reconstruction of the entire ship's hull, reflecting the original situation or in any case approaching it very closely. For the interpretation of the field drawings, the photos taken during the excavation in 1944 proved invaluable.

12.2.2 Construction and layout of the hull

Keel plank

The keel plank, together with the keel parts of the hooks, measures 11.8 m. The length and thickness of the keel plank itself could not be determined precisely in 2008. The scarf to the stem hook starts on the inside just before frame 22, 1.9 m from the stem. The location of the scarf with the stern hook was not excavated in 2008. Most likely the scarf on the inside starts at the front of frame 4. Assuming the sloping flat scarfs have a length of approx. 50 cm long, this means that the keel plank must be approx. 8.5 m long. The thickness will be around 10 cm, so that one can almost call it a keel beam. The width at the scarfs connecting the hooks is aft 26, fore 28 cm. From the scarfs, the width towards the center gradually increases until a maximum width of 38 cm is reached. The keel plank probably has a rectangular cross-section. The connection to the stem hook is made on the inside with two rows of four nails. It seems obvious that this was also done from below. Most likely the nails did not protrude from the wood.

Stern hook

The length of the keel section of the stern hook could not be determined, but will be approximately 1.9 m. The width of the scarf to the keel plank is 26 cm, and 14 cm at the start of the sternpost. The connection between the stern hook and the keel plank with nails is probably the same as that of the keel plank with the stem hook. The horizontal arm is approximately 10 cm thick and is trapezoidal in section with the narrowest side below. Under frame 2, its top surface runs diagonally upward to the aft and measures around 17 cm thick. This runs up to 19 cm towards the stern. From there, the rising arm runs up 18 cm obliquely backwards to the scarf with the sternpost. The rising arm is 23 cm thick, while the width of the horizontal arm (14 cm) slightly increases upwards. The rising part is also trapezoidal in cross-section and measures 4 cm on the outside at the bottom. Here is a protruding heel which protects the rudder. The highest (reconstructed) point of the rising arm is 65 cm from the keel line. A 9 cm wide and approximately 3 cm deep rebate has been cut on both sides in both the thickest horizontal and rising part, into which the end of the garboard strake is nailed.

According to the field drawing of the sternpost, it was only fixed from the outside to the top of the stern hook with three nails arranged one above the other. A 3 cm thick wooden treenail (analogous to the stem construction) may have been located at the location of the scarf, which was arranged behind the gudgeon by the broken tip of the stern hook. A wrought iron gudgeon had been attached at about the middle of the scarf between the stern hook and sternpost, the lower suspension of the stern rudder (Plate 12.I A). The gudgeon was detached from the hull with the upper part of the stern hook and was only documented with a photograph (figs12.9a and b). The wings of the gudgeon flare slightly outward due to the shape of the stern hook and bend further outwards because they were nailed onto the hull. The wings are cruciform and each attached with 11 nails. The upper edge of the ring in which the pintle hung is somewhat pointed at the rear in the middle. Thanks to the finding of the gudgeon in February 2009 (see section 12.5.1), the correct dimensions could also be recorded. The reconstructed aft line of the stern is raked 117° to the keel line.

Sternpost

The sternpost, found separately, was still complete and has a maximum length of 2.8 m (fig. 12.10). From the keel line, the sternpost, including the stern hook arm, measures 3.2 m and measures 2.8 m in height. The post is 18 cm wide, while towards the top it gets narrower and at the top only measures 11 cm. The top four hull strakes

Fig. 12.9a & b. The broken off top part of the stern hook with the wrought iron gudgeon, placed in a wooden wheelbarrow, February 16, 1944. b. The loose gudgeon.

ended against each other on the roof-shaped chamfered aft surface of the post. The bottom ones most probably ended all three in rebates, which were arranged in both the stern hook and the sternpost. This is not clearly documented for this post. At 58 cm from the top an iron bolt about 3 cm thick protrudes through the sternpost, with a forged ring on the outside for the suspension of the upper pintle. Furthermore, a hole with a diameter of approximately 2 cm can be seen. Possibly this allowed the helm to be secured with a rope.

Rudder

The gudgeon and the bolt with ring in the sternpost indicate that the cog was kept on course with a stern rudder. Nothing has been found of the rudder itself. Plate 12.I A shows what it might have looked like.

Stem hook

The stem hook is still completely intact in the wreck. The horizontal arm is 2.4 m long, is 28 cm wide at the scarf with the keel plank and 15 cm at the bottom of the rising arm. The horizontal arm is approx. 10 cm thick, its top surface rises just in front of frame 25 and from there approx. 70 cm in one line up to the beginning of the rising arm, where it is 18 cm thick. The sloping flat scarf between the rising arm of the hook and the stempost starts immediately above the thicker horizontal arm. The rising arm is 37 cm thick at the bottom and also forms a toe of 12 to 24 cm thick on which the false stem rests. A hole with a diameter of 3 cm is provided in the toe (fig. 12.12a). Probably meant to hold a rope with which the vessel could be pulled up a slipway. The highest point of the stem hook (the scarf) is situated at 89 cm above the keel line, measured vertically. Along the upper edges of the horizontal arm, 8 cm wide and 3 cm deep rebates are cut into the thick part, which run into the rising arm. The garboard strakes are nailed into this. The stem hook is increasingly trapezoidal in cross-section towards the fore part and measures 5 cm at the front.

Stempost

The lower part of the stempost is connected to the stem hook with a 3 cm thick wooden treenail. It is approximately 1.6 m long, 18 cm wide and 21 cm thick on the inside, approx. 20 cm at the top. The photo (Fig. 12.12b) shows that the bottom three strakes have been nailed into the rebate, while from the fourth strake the planks ran through to the front of the stem. For this purpose, the beam was chamfered roof-shaped at the front. The detached, broken upper part measures 16 cm thick at the top. The piece shows a bevelled part 15 cm wide on the sides at the bottom, at the top this is only 2 cm, these are the slanting sides against which the ends of the upper four hull strakes were nailed. The plank ends of the starboard side probably ran past the plank ends of the port side. In addition to the aforementioned wooden treenail, the post was attached to the stem hook with four nails. The reconstructed stempost is 3.6 m long, together with the stem hook it measures 3.8 m from the keel line, while the top (also the top of the gunwale) rises perpendicularly 3.25 m above the keel line.

Cutwater?/false stem

A 1.4 m long and 15-17 cm wide piece of wood is attached to the 'toe' of the stem hook. The thickness is 5 cm. The piece is chamfered at the top and fixed at the top and bottom with one nail, while three blind treenails (diameter 2.5 cm) have also been used to connect it to the stem hook and stempost. In the condition found, it looks like a cutwater that does not really fit in front of the stempost in terms of cross-section. The four upper hull planks in front of the stempost end in a fragile construction. It therefore seems logical that these were protected with an false stem. Above the part described here, therefore, there should have been a part that possibly protruded above the stempost and the gunwale as shown in Plates 12.I A and 12.II A.

In the upper part of the stempost a nail or treenail is indicated on the field drawing at the fifth strake, while a second one may have been located on the broken/missing part at the height of the centre of the fourth strake. However, this does not seem to be strong enough in terms of construction. It is usual for such a false stem to have a thick iron bolt at the top, while usually also several blind treenails are used. No treenails or nail marks were seen on the upper part of the stempost. It seems that at the time the cog was wrecked the situation was as it was encountered in the excavation, that is, with a somewhat clumsily constructed cutwater. It is, however, possible that the vessel originally had a false stempost and then in one piece. Support for this possibility can be found in

Fig. 12.11. Exterior view of the toe on the stem hook with a hole to attach a rope in order to pull the cog on a slope. Above and next to it you can see the bottom part of the cutwater(?)/external stem post and the connection of the garboard strake on port side.

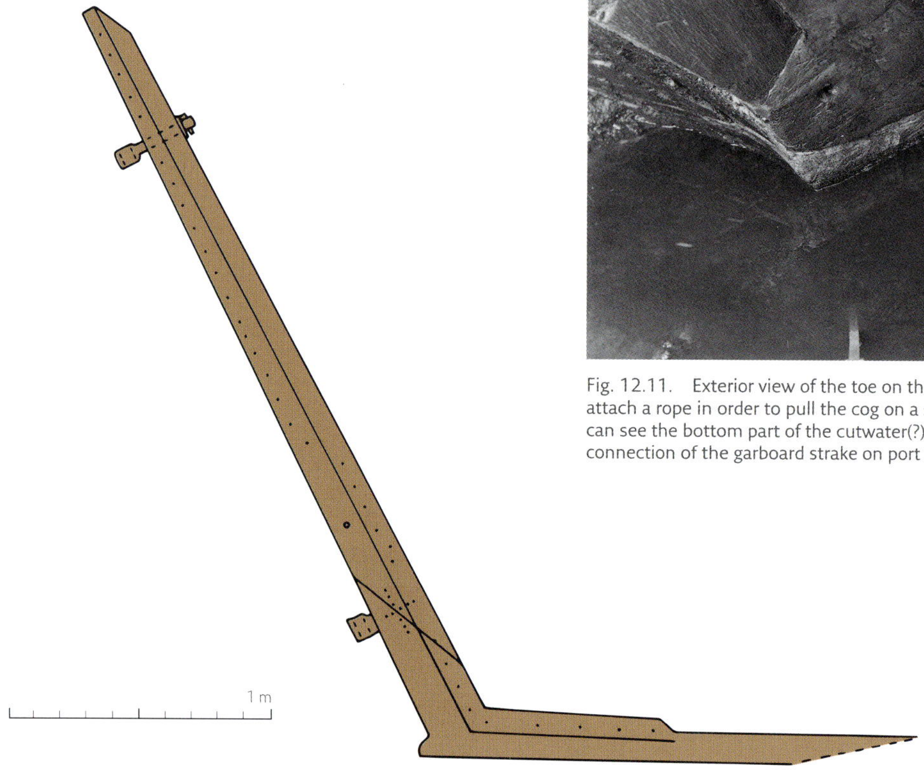

Fig. 12.10. The reconstructed stern construction.

the top section of the stempost, which was carefully investigated in 2008. From a qualitative point of view that looked considerably better than most other ship's timbers. Perhaps there had been a replacement of (the upper part of) the stempost, whereby instead of re-fitting a false stem a cutwater-like construction has been chosen (Fig. 12.12a and b). The front line of the reconstructed false stem is at a rake of 121°.

Bottom and sides

The bottom and sides are made up of seven strakes (A to G) on either side of the keel plank. The planks of the strakes A to F are 3-4 cm thick, the top (G) measures 4 at the bottom and 5 cm at the top. Modderman writes that the garboard strakes are made from one plank and most of the others from two.[28] Strangely enough, the garboard strakes or A-strakes do indeed seem to consist of one plank. However, it cannot be excluded that the scarfs and the protruding nails are situated under the frames and therefore not observed. The B strakes are composed of two, the C strakes are composed of three planks, the D strakes are probably two, the E strakes are probably four, the F strakes are three, while the upper and thicker G strakes are made out of two planks. The connecting scarfs between the planks are sloping flat scarfs and 50-55 cm long, while the overlapping plank ends are approx. 1.5 cm thick both inside and outside and protrude from the hull on the outside. The plank ends are in principle attached to each other with four nails nailed from the inside out and with four from the outside in. Judging by the corresponding locations of the hull scarfs on the starboard and port side, the so-called 'brother and sister pattern' seems to have been applied during construction. Modderman also writes that the hull planks are lapstrake and that the seams are caulked with moss.[29] This moss is held in place by round caulking plates (sintels), which have a point on either side with which they are stuck in the wood. These sintels are arranged close together. It escaped Modderman

Fig. 12.12b. The port side at the stem. The rebated stermpost into which the ends of the first three bottom/hull strakes are nailed. At the top of the sternpost you can still just see the rebate, from which it could be deduced that the fourth strake continued to the fore part of the stermpost. You can also see that the bottom part of the false stem/the cutwater(?) is complete, March 30, 1944.

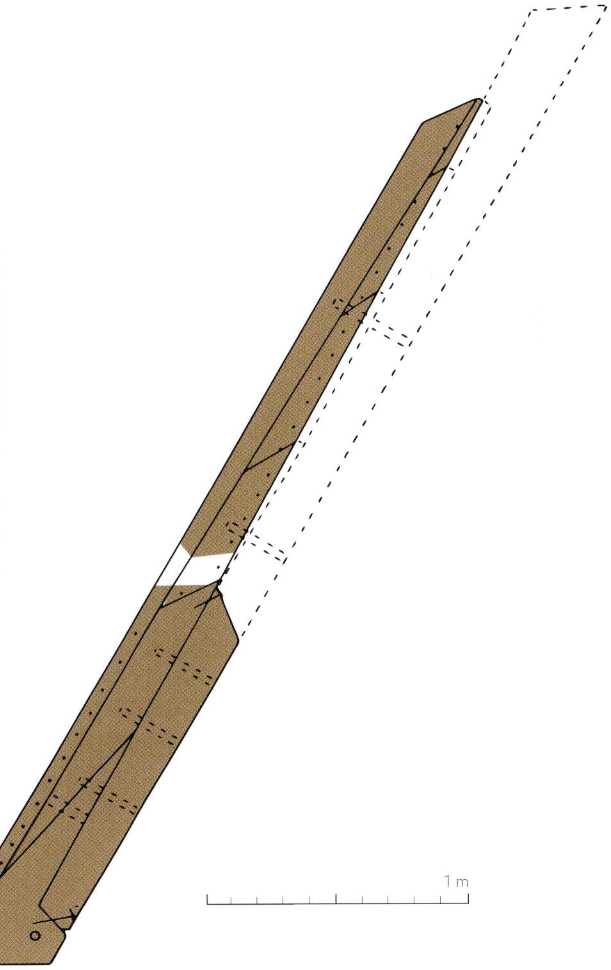

Fig. 12.12a. The reconstructed stem construction.

28. Modderman 1945, 79.
29. Idem.

that the bottom seams are not lapstrake but carvel-built, since they have not been caulked on the inside, and that wooden moss laths had been placed between the moss caulking and the iron caulking plates (sintel clamps) on the lapstrake seams .

Bottom

The bottom of the hull consists of three strakes on both sides of the keel plank (A to C). The str akes are carvel built between frames 9 and 17. This also applies to the A-strakes or garboard strakes against the keel plank, which moreover are situated with their top surfaces at the same height. In this part of the ship, the A and B strakes run slightly obliquely upwards towards the bilges in an approximately constant line, the C strakes are situated against this at a slight angle. Aft and fore the three strakes together form a more fluid form. From frame 9 to the stern and from frame 17 to the stem the planks are clinker built. The overlapping parts, the lands, are chamfered on both planks. These overlaps have been left wider at the outside of the B and C strakes to form the land. The hood ends of the planks have been made slightly thinner to the stern and stem, and are again (almost) carvel-built and are nailed into the rebates in the hooks and posts. The A strakes are made of a 11.8-12 m long plank (or two planks) with a maximum width of approx. 50 cm. The B strakes from two planks, the after two of which are approx. 7.5 m long and the fore one approx. 5.5 m long. The largest width is approx. 40 cm. The C strakes are composed of three planks, aft approx. 2.6 m, the middle one approx. 6.3 m and fore approx. 5 m long. These planks are slightly less than 40 cm wide. Although this could not be established, it may be assumed that the carvel seams are closed on the outside with moss, moss laths and sintels (sintel clamped moss caulking). All the bottom strakes twist towards the posts and are nailed to them in the rebates.

Starboard side

Of the starboard side, most of the strakes D and E and parts of F are still present. However, they are missing over the aft approx. 3 m, while the excavation drawings show that a number of planks have been found next to the forebody, which were probably originally part of the fore part of the hull side. Plate 12.I A indicates what has been found with certainty. On the stern- and stemposts a number of rebates (figs 12.10 and 12.12a) indicate the widths of the various strakes. Therefore there are sufficient data to be able to reconstruct a very reliable plan of the strakes (Plates 12.I A and 12.II B and C). All strakes of the side are completely joined together from sternpost to stempost in lapstrake fashion with an overlap or land of 6-7 cm. Although there is no 100% certainty due to the lack of the aft ones, it seems that strake D is composed of two planks, aft 11.8-12 m long, fore approx. 2.7 m. The E-strake is presumably made of four planks, which have a length of approx. 2.7, 6.5, 3.1 and 3.8 m from aft to fore; however, it may have been three planks. Here too it applies that any scarf has not been attested because they may be under a frame. However, the available data are insufficient to be able to say this with certainty. Only a few parts of the F strake were found.

Port side

The port side is better preserved and almost entirely present. In particular the separately found forward G=plank, with the piece of a (covering) inwale still present in construction, is of great importance for the interpretation of the construction of the entire upper edge of the ship. The plank distribution and dimensions are the same as those on starboard side. The plank widths vary from 38-45 cm. Although this cannot be said with certainty, it can be assumed that the F strake aft and fore had two short planks, approx. 2 and 3.2 m long and a long plank of approx. 11.2 m in between. The thicker G strake at the top consists of two planks, at the aft 8.3, at the foreward part 8.2 m long. These planks have a maximum width of approx. 50 cm. On the outside of this strake there is a wooden channel at the location of the frames 15 to 17, with four openings in order to secure the rigging[30] (figs 12.8, 12.13 and Plates 12.I A and 12.II B).

Caulking method: sintel clamped moss caulking

The hull seams are made waterproof with sintel clamped moss caulking.[31] In the case of the clincker-built planks,

Fig. 12.13. The channel on the port side against the gunwale. At the rear of the channel you can clearly see the strake ends protruding from the hull on both sides near the scarf and the construction of the hull amidships, 30 March 1944.

30. Modderman 1945, 80.
31. Vlierman 1996a.

the bottom plank at the top on the outside is chamfered before the next plank has been placed above it. This creates a triangular opening that is generally approx. 3 cm deep and approx. 1.5 cm wide at the top. In this opening, strands of moss were hammered in, which were twisted into a single 'yarn'. To keep this moss caulking in the seam, oak wood laths with a more or less oval cross-section were laid (1-1.7 cm wide and a maximum of 0.5 cm thick). The length of these moss laths could not be determined. Fig. 12.14a and b show a photo from 1944 and one from 2008 of sintel clamped moss caulking *in situ*. Fig. 12.14c shows the construction and the sintel clamp type. The centre parts of the sintels vary between 28 and 35 mm in length. They are arranged irregularly apart between 4 and 6 cm, measured from centre to centre. The clincker-built seams are only sintel clamped on the inside. This also applies to the seams between the plank ends and the hooks/posts.

A paper thin layer of moss is applied to the plank scarfs. The plank-ends at the scarfs are not finished with sintel clamped moss caulking. Although this has not been established, it is almost certain that the short carvel-built seams between the A and B, between the B and C strakes and the seams between the keel plank and the garboard strakes on the underside are made watertight. The plank edges were therefore chamfered in advance, resulting in a more or less equilateral, triangular seam. The sintel type D used in the Marknesse-cog was applied between about 1250-ca 1375.[32] Some sintels still appear to be similar to the preceding C2 type. This is discussed again in section 12.8. A total of approx. 210 m of seams is provided with sintel clamped moss caulking. Compared to other late medieval ship finds, a striking amount of tar residues have been observed on caulked seams. This may indicate regular maintenance, but perhaps even more to keep the relatively old vessel as watertight as possible. Quite a lot of tar marks have also been observed on the inside of the planking.

Determination and pollen research

A few samples of moss were taken in 2008 for identification and pollen research. At the time of writing this study, the results were not yet known. However, there is at least a mention of peat moss (*Sphagnum* spec.).

Fig. 12.14a. Sintel clamped moss caulking in the lapstrake hull seam. At a loose lying part of the hull.

Fig. 12.14b. *In situ* in between strakes E and F and in between frames 16 and 17 at starboard side, 10 August 2008.

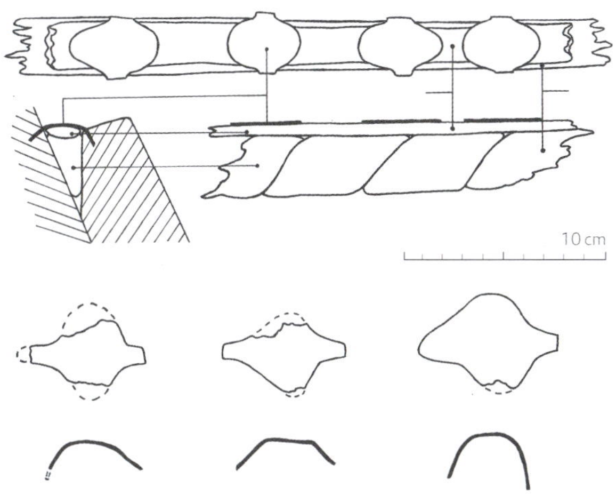

Fig. 12.14c Reconstruction of the sintel clamped moss caulking in a clincker-built hull, the applied moss strands and the used sintel type (D).

32. Vlierman 1996a, 58 tabel II; Before ca. 1300 the sintel type D are generally smaller than the examples from this ship (Vlierman, in preparation a).

Frames
The hull was initially kept in shape by the frames. Modderman[33] only writes about the (parts of) 27 frames. The reconstruction (Plates 12.II A and B) shows that there were 32. In this study they are numbered 1 to 29 from the first frame fore of the stern till the upper one against the stempost. Above the stern from below 0, -1 and -2. The frames 1 to 26 are composed of three pieces; a V-shaped or a floor timber on the bottom, which are extended on the sides with a futtock. The scarfs between these parts are staggered relative to the adjacent frames and are located above the A-, B- or C strakes. Frame 10 is the only one to deviate, since it is made up of four pieces. The frames 0 and 27 are formed of two pieces, one part of which forms the V-shaped floor timber above the stern and a futtock, the other only a futtock. The frames -1, -2, 28 and 29 were probably made from one piece of wood. The floor timber, the V-shaped floor timber and the aft and fore two frames are, in principle, slightly wider and thicker than the futtocks. Floor timber 20 deviates completely from this and is dealt with in the following section. Most floor timbers and V-shaped floor timbers are approx. 14 cm thick and 14-22 cm wide. The aft and fore V-shaped floor timbers above the keel section of the hooks are up to 35 cm thick. The parts of the frames 25 and 26 above the keel are not flat, such as in the aft, but are cut pyramidal. A kind of decoration, of which it is not clear why that was done precisely there, in the foreship.

The futtocks 4, 9 and 20 do not end at the top of the side (strake G), but approximately at the level of the centre of strake F. This is also the case with frame 25, only the top is located there at the middle of strake E. These ends of the futtocks, finished horizontally at the top, were intended to fixate the gangways (Section 12.2.2). The frames -1, -2, 5 to 8, 10 to 19, 28 and 29 have a straight finished top at the top of the side of the hull. The heads of the frames 0 to 4 and 21 to 27 are probably all obliquely finished and ran just below the top of strake G and were inserted in the underside of the covering inwale. This is described in more detail in section 12.2.2. In the frames 7 to 21 above the keel plank and above the A or B strakes three inverted U-shaped, sawn/chopped limber holes can be seen for draining any leakage water or rainwater to a certain point. They are approx. 10 cm wide and 4 cm high. Such holes are also located at the bottom in frames 3 to 6 and 22 to 24, but only above the keel. It is not known whether such limber holes were present in the other V-shaped floor timbers. Probably above the horizontal parts of the hooks, but then most likely triangular in shape and arranged against one of the garboard strakes. The scarfs between the floor timbers and V-shaped floor timbers and futtocks are slanted and relatively short (15-35 cm).

Mast step(s) and a casing for a lowerable mast
Notable in the drawings and photos of the excavation of 1944 (figs 12.5a and b, 12.15a and b and 12.16) were the deviating V-shaped floor timber 20 and the curious short plank with two rectangular holes above the V-shaped floor timbers 21 and 22. The short mast step on frames 17 up to and including 20, with the hole for the mast heel aft of and above frame 19, was the obvious place of the mast at the time of the ship disaster. This mast step with a thicker central part in which a rectangular hole (dimensions 23x22 and 11 cm deep) was cut at the location of the frames 17 and 18, rests on frame 19 and appears to lie in a cut-out part of frame 20. The V-shaped floor timber 20 has a strikingly flat and raised middle part with chamfered edges, measures approx. 50 cm at the top and approx. 60 cm long at the base. The width of this part of the V- shaped floor timber is 33 cm, while the thickness measures 24 cm. The ends of the V-shaped floor timber at the bilges are again equal in thickness to the other frames, i.e. approx. 14 cm.

A check in 2008 confirmed the suspicion that a hole had been cut into the V-shaped floor timber under the front of the mast step (the primary location of the mast) (Plates 12.I B and 12.II A and B). The (thus secondary) mast step is fixed on the frames 18 and 19 with four treenails, two each on frames 17 and 20. A limber hole was cut into the bottom of the mast step. In order to place the mast step on frame 20, the 'rear wall' of the primary hole for the mast was (partially) cut away. The hole probably had the same dimensions as that in the mast step.

Modderman writes about the plank with the two holes which was nailed onto and partly into frames 21 and 22, of which the function is not clear.[34] A possible explanation is given on Plates 12.II A, B and C. The holes were almost certainly intended to place vertical posts. The outer edges of these posts are approx. 37 cm apart. The assumption is that from the posts to close behind the mast (the end of the thickest part of the mast track) a panelling in the form of a casing was made that protected a lowerable mast against bulk or stacked loads in the hold. However, the two holes and the location of the plank are the only concrete indications for such a construction. Something similar is also assumed with the Spakenburg I-cog, where there are also two successive positions of the mast, albeit there at virtually the same place (see section 14.2.2). A floor timber with two holes was found

33. Modderman 1945, 79.
34. Modderman 1945, 80.

Fig. 12.15 . Mast step(s) and the little plank with two holes as part of the formwork around the mast. a. The secondary mast step with two supporting chocks on either side (riders), horizontal frame 20 with the primary hole for the mast under the fore side of the mast step and the little plank with two holes for vertical stanchions which may have served as a part of the construction of a formwork for the lowerable mast. Seen from port side. b. Idem, seen from the forward starboard side, 30 March 1944.

in the Kraggenburg cog (figs 22.2 and 22.6a and b). The lowerable mast and the two sail beams are discussed later in this section.

Ceiling

Under the cargo of bricks lay a loading floor (figs 12.2, 12.16, 12.17 and the Plates 12.II A and B). It consisted of a number of loose oak and pinewood planks. They lay over the frames 7 up to and including 19. The floor, actually the floor ceiling, is bordered on both sides by a 25-32 cm wide bilge ceiling plank. These ceiling planks are slightly thicker than the loose parts of the loading floor, bevelled irregularly in a roof top shape and about 6 cm thick in the middle and about 4 cm on the edges. On the port side the bilge ceiling strake starts slightly before frame 5 and ends at the front of frame 22. The bilge ceiling strake on starboard starts at frame 7 and also ends at 22. The bilge ceiling strakes have extensions both aft and at the front. This could be deduced from the treenails of the fastenings still present, which protrude from the frames (fig. 12.18 and Plate 12.II A). These composite bilge ceiling strakes probably ran from frames 1 to 27 and formed an important part of the longitudinal strength of the ship. They are located above the second and third strakes of the shell, where a large part of the scarfs are also located in the frames. These 'weaknesses' have therefore been reinforced by the bilge ceiling strakes.

During the excavation, ceiling(?) planks of 80-90 cm long and 17 and 26 cm wide, respectively, were found on either side above these bilge ceiling planks. They lay loosely against the futtocks of the frames 14 to 17. It is not clear whether the strakes were extended backwards and forwards and/or whether one or more ceiling planks were also arranged above them. The gangways still present *in situ* at the start of the excavation plus the cargo, which was probably stacked from side to side, suggests that there was no other ceiling strake attached to the vessel between the bilge ceiling strake and the gangways.

Inwale

Along the top edge of the hull on the port side, an approx. 16 cm wide and 4-6 cm thick plank had been placed against the inside of the frame heads 5 up to and including 20. By analogy with the ceiling planks, this plank was also somewhat roof-shaped or chamfered on the

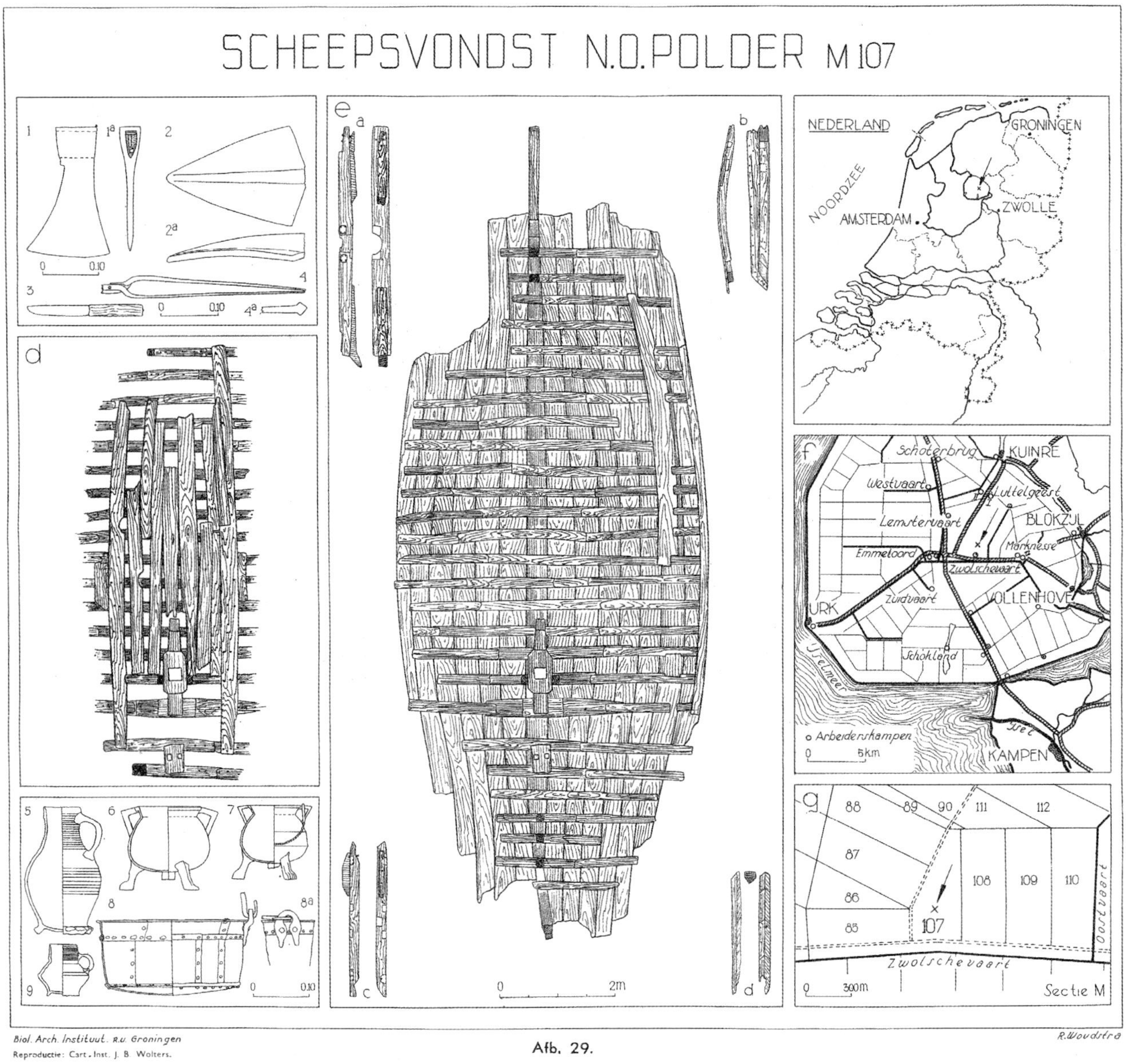

Fig. 12.16. Ship's discovery N.O. Polder M 107. Modderman 1945, fig. 29. D. Bilge ceiling planks, loose floor ceiling (cargo floor), mast step and little plank with holes; E. top view, plank outlines and loose timbers; a. sail beam, b. upper hull plank with stringer or wale, c. small cross beam from the bow and d. the top part of the stempost. And the most important finds.

lower half (Plate 12.I B). Whether we should speak of an upper ceiling or inwale here is not really clear. The latter is probably preferred. The thickness and design indicate that these planks were an important part for the reinforcement and the longitudinal stiffness of the upper hull. An important finding in this context is the front and top portside plank found at or in the forebody with a piece of an inwale still attached (fig. 12.16, E b), which rested on the top of futtocks 28 and 29 and in which the bevelled heads of futtocks 26 and 27 were inserted. The inwale has an oblique scarf on the aft and foreward is attached onto the stem. It seems very likely that this piece of inwale was extended in the same way to frame 20. A similar construction, presumably made of one piece, can also be assumed in the stern, where it was fitted between the sternpost and frame 5. The ends of the inwales are connected to each other on the inside of the posts by a more or less triangular block of wood which is partially inserted into the inner edges and fixed with a treenail onto it. The bilge ceiling planks, as well as the upper ceiling(?) planks and the covering inwales aft and fore are fixed with treenails. Nails were also used in some places, such as at the scarfs.

Fig. 12.17. The ceiling or cargo floor and the bilge ceiling strakes, seen from the forward port side, after the cargo of bricks was removed, February 18, 1944.

Fig. 12.18. Overview of the wreck with bilge ceiling strakes, seen from the forward starboard side. You can clearly see six protruding treenails on port side above frames 22 to 27, with which the forward (missing) bilge ceiling plank was secured. At the aft one treenail of the aft bilge ceiling plank can be seen.

Fig. 12.19. Overview from aft to fore ship with on the left part of the gangway that starts above frame 3 and rests on frames 4 and 9. On starboard side you can see the original straight top of frame 20 and on the port side the gap between the futtock and the top part of the frame between which the missing part of the gangway was attached, March 30 1944.

Gangways, crossbeams and deck beams

It has already been noted that frame numbers 4, 9, 20 and 25 do not extend to the top of the hull, but that they have been left shorter in order to support the gangways (Plates 12.II A, B and C and fig. 12.19). These approx. 10.4 m long, almost 40 cm wide and 5 cm thick gangways were notched to fit around the other frames, from frames 3 up to and including 20 and were placed onto the frame heads and against the inside of the hull. The gangways are fixed with treenails (diameter 3 cm) penetrating the frames and the hull strakes. The treenails are smoothly finished on the inside of the gangways. A very tight construction that, together with the -yet to be discussed- crossbeams, that were attached to it, formed a very important part of the structural and form strength at deck level, both longitudinally and transversely.

The photo (fig. 12.19) shows that the inside edge of the gangway at the frames 3 and 4 is finished vertically (like a seam in the afterdeck), the rest is chamfered. This is common if a coaming was attached to it. However, no nail remnants have been observed or recorded. Also no traces on the sail beam have been found that might confirm the presence of a coaming. The treenails of the gangway connections to the side are, as said, smoothly finished. Any coaming was therefore not put in place at the same time, which is also not very obvious. In any case, it seems very likely that the vessel had a coaming between frames 4 and 9 and (initially) between 9 and 20, which probably did not extend above the crossbeams.

During the excavation, a crossbeam was found *in situ* above frame 9, from which it could be deduced how the transverse structures were assembled. The beam rests on the gangways with a notch in the bottom, so that the bottom of the beam is almost the same level as the bottom of the gangway. The beam runs up to the inside of the hull. A knee is placed on the ends of the beam, the ascending part of which acts as the upper part of the local frame (fig. 12.20). Knees, crossbeam and gangways are interconnected with a number of treenails. The standing knee is also fixed to the upper hull strake with a few treenails. Above frame 4, such a crossbeam was almost certainly also provided, fixed in the same way, but presumably extra heavy (a second beam on top?). Amidships on this beam, very probably a heavy windlass construction for hoisting and lowering the sail was located. There was also a crossbeam against the aft of frame 26, which was also fastened with knees against the side of the hull and to the gangways. This beam was found in the forebody (fig. 12.16 E c and Plate 12.II A and B). Most likely there were above the frames -1, 1, against the rear of 22, above 23 and on frame 28 also crossbeams or beams that supported the after and fore decks.

Primary and secondary sail beam (with knees)

The detached found sail beam with rebate for the mast (figs 2.16E a and 12.21) was initially assumed to belong between frames 18 and 19. During the reconstruction, however, it became clear that this certainly could not have been the case, since the beam proved to be too short for that. It fits exactly above the primary mast step (frame 20; Plates 12.I B and 12.II A and B). It is striking that the rebate for the half cross-section of the mast is not at the fore, but at the aft. Confirmation for this is the complete knee found on one end of the beam, while a broken one is on the other side. A short piece of a timber has been found on the port side against the upper side of the hull. This can be seen on the relevant field section (fig. 12.6 A h, Plate 12.I B, and vaguely on an excavation photo fig. 12.19). The broken knee may therefore have been on this side. However, it does not seem to be entirely impossible that this was originally different: not only this sail beam with a broken knee, but also the deviating frame pattern (Plate 12. II B) suggests that a calamity occurred during the functional period of the primary position of the mast, and made it necessary to adapt the structure.

It is possible that at this time the opportunity was made to make the mast lowerable and to move it backwards. The foredeck could be maintained in this way, including the primary sail beam. An approx. 18 cm high, 3 cm wide and 7 cm deep rebate is chopped into the upper part of frames 18 and 19 on the port side, respectively (fig. 12.22). The secondary sail beam was fitted in this, which was undoubtedly also attached to the sides with

Fig. 12.20. The joint of a knee of the crossbeam above frame 9 on the outer shell and on the gangway at the port side. The bevelled interior of the gangway, where a coaming was originally placed against it, and the somewhat roof-shaped top ceiling plank or inboard, are also clearly visible, March 30, 1944.

Fig. 12.21. The loosely found primary sail beam and a part of the short V-shaped floor timber (frame 29?). At the top a part of the unloaded cargo of bricks can be seen, 30 March 1944.

knees. A 3 cm diameter diagonal pin, arranged obliquely from the inside to the outside and from the top down, which is fixed at approx. 10 cm above the seam between the strakes F and G and a few centimetres for frame 18, confirms the latter. Nothing has been found of this sail beam.

From the distance between the two rebates in the frames, it can be established that the width of the beam must have been approx. 33 cm and the thickness approximately 18 cm. The beam rested on the gangways, possibly was slightly notched to engage with the gangways and fixed onto them, together with the knees, with treenails. Given the location of the beam in relation to the position of the mast in the secondary mast step, it can be assumed that the gap for the half mast thickness was facing forward. Presumably the mast, which could now be lowered, was held in place with an iron bar that was attached to the front. Fig. 12.23 shows the possible primary situation of the mast in the floor timber 20 (a), of the secondary situation in the mast step on the frames 17 to 20 (b) and the mast in the lowered position, enclosed by its casing (c).

Fig. 12.22. The upper part of frame 18 with a rebate on its fore side, in which, together with the rebate in the aft of frame 19, the secondary mast beam was fitted. The meaning of the small rebate above it is unclear. The two treenail holes are from the connection to the inwale on top of the ceiling plank.

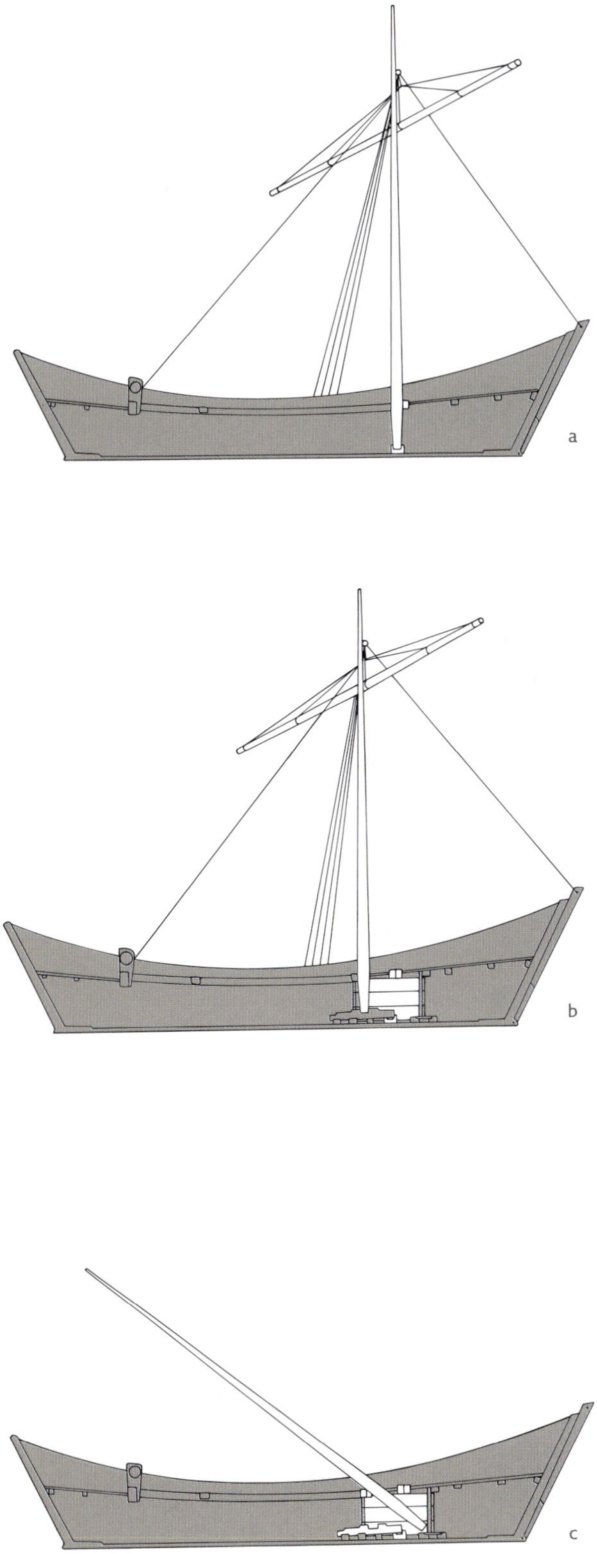

Fig. 12.23. The location of the (lowerable) mast. a. The primary situation of the mast in floor timber 20. b. The secondary situation of the mast in the mast step on frames 17 up to and including 20. c. The mast lowered, within the formwork.

Stringers and decks

In the detached deck- and sail beams different rebates can be seen, which indicate that the vessel as part of the foredeck has had two longitudinal beams or stringers between the primary sail beam and the crossbeam on frame 26 and one between the last beam and the stempost. The first two were positioned (centre to centre) 50 cm apart and on either side of the centre line of the ship; the forward stringer was exactly above it. The stringers had an inverted T-shape. The deck planks that were almost certainly laid transversely lay in the rebates that were created in this way. At the gangway, they most likely rested on a plank that had been placed alongside it, on the rebates in the foreward crossbeam. Immediately in front of the primary sail beam there was/were probably one (or two) hatch opening(s) (Plate 12.II B) and the space between the two stringers may have been covered with a plank. It is not clear what the deck situation was between the two sail beams. Perhaps the space between the beams and the gangway was open. The aft deck ran from the sternpost to the front of frame 4. Most likely the planks of this were arranged lengthwise and, based on what was found on other late-medieval ships in the IJsselmeerpolders, made of irregularly shaped planks that fitted against the inner side of the hull. The deck situations, the layout and the dimensions of the planks shown in plate 12.II B are of course (largely) hypothetical and should serve no purpose other than to offer as complete as possible an impression of the possible original situation.

Pump

No indication of (the location of) a pump was found in the vessel or this was not noticed during the excavation (abraded planks at a suitable location). However, the limber holes in the underside of the floor timbers and V-shaped floor timbers indicate that the water on board was intended to run to a specific location. The most obvious is a (fixed) pump tube just in front of frame 2 or behind frame 25 on the horizontal arm of the stern- and/or stem hook.

Castles

The vessel has not had an aft or fore castle.

Repairs

No repairs were found in the shell planks during the excavation. Modderman makes no mention of any, and nothing is visible on the published field drawings. Among the preserved objects, however, are two small planks covered with textile, which were undoubtedly placed somewhere on the inside or outside of the hull. They are discussed in section 12.5.1. During the follow-up investigation in 2008, it was established at the seam in strake E near the frames 2 and 3 that the seam is very wide and filled with a lot of moss caulking, while a plank appeared to have been placed against the outside. However, this may also have been a reinforcement plank used in the relocation of the wreck. As mentioned earlier, in 2008 the wreck showed a remarkable number of worn edges on the planks and wood, which can in fact only be the result of long-term use. The loose loading floor/ceiling also consisted of a number of unusual and partly incomplete planks. The most noticeable thing about the vessel is the different framing pattern at the initial location of the mast (the futtocks of frame 20 had undoubtedly been heavier at first), maintaining of the original sail beam, the instalment of a new one aft of it, while the mast was placed in a 'new' mast step above the back of frame 19.

The execution of the cutwater or lower part of the false stern also suggests that this is a case of a repair rather than the original construction. In 2008, only a crack sealed with sintel clamped moss caulking on the inside of a plank in the stern indicated a repair. However, it may well be that there are more, that were not seen or recorded as a result of inadequate cleaning. Such repairs, which are often made on the outside of the bottom during the construction of the ship, are very unlikely to have been noticed while moving the vessel in 1944. However, many tar residues have been seen, especially on the lapstrake, sintel clamped plank seams. This indicates regular maintenance and/or that people did their utmost best to keep the probably old ship watertight.

Wood use and processing

With the exception of a single plank of pine as part of the ceiling or loading floor, all parts of the ship are made of oak (*Quercus sp.*). Further, hardly anything could be recorded about the use of wood. However, the planks are most likely tangentially and quarter-sawn and edged with an axe. It was observed in 2008 on various planks and timbers that by no means all sapwood was removed. The plank pattern on either side of the keel plank suggest the 'brother and sister pattern', that is to say, two planks sawn from a tree side by side, at comparable locations on the starboard and port side. The floor timbers, V-shaped floor timbers and futtocks are undoubtedly sawn and hewn from trunks, forks and thick branches. The hooks from forks/trunk-branch parts. The planks in the aft and fore part of the underwater body were almost certainly heated to be bent into the desired shape.

Fastenings
In principle, the vessel has oak treenails with a diameter of 2.5 cm. At a few places it could be established that for the connections of the gangways and at the knees on the crossbeams, treenails with a diameter of 3 cm were used as connection to the ship's hull. A 3 cm thick treenail is also used for the connection of the stempost to the stem hook. In principle, two treenails per plank width are used to connect to the frames; while the frames are fixed to the keel plank, depending on the width of the plank, with one or two treenails. The bilge ceiling strakes and the inwales are also connected with treenails. In total around 950 treenails of 2.5 cm diameter will be used, and 50-60 pieces of 3 cm. A 3 cm thick treenail has also been found in the top of frame 26 on the port side. We may suppose that here and also on the starboard side, for instance a bollard was attached, of which nothing was found. This could very well have been the case at frame 1 in the stern. In some places it could be determined that the treenails on the outside of the hull seemed to have a slightly thicker end and also protrude slightly. Due to the observed overall wear on the wood, it is not possible to say with certainty whether the treenails actually had a thicker end and protruded prominently outside the hull. No wedges have been observed in the treenails, this may indicate that the thicker, slightly protruding convex heads indeed were the case.

75-85 mm long and 5 mm thick wrought iron nails with a round flat, sometimes slightly convex head, the diameter of which varies between 22 and 30 mm (fig. 12.26), were used for fixing the clinker-built shell planks to each other. The spacing of the nails is generally 17.5-18 cm, sometimes 16 cm. A total of approx. 175 m of lapstrake seams has been nailed. An average spacing of 17.5 cm means that approximately 1000 nails have been used for this. The hood ends of the hull strakes are fixed with 5 to 7 nails on the stern- and stempost and with 28 on the stern- and stem hook, in total about 200 nails. For the at least twenty scarfs in the strakes two rows of four nails were used; on the scarfs between the keel plank and both the hooks, two rows of four nails from the inside to the outside were used and vice versa. For securing the toes of the futtocks on the floor timbers and V-shaped floor timbers, generally two nails were used. A total of around 1500 nails therefore were incorporated into the hull. This number does not take into account the possible nails in the scarfs of the bilge strakes and the deck planks. As regards the caulking method it has already been indicated that around 210 m of seam is caulked with sintel clamped moss caulking in the cog. The sintels used for this are spaced at an average distance of 5.5 cm. This means that around 4000 sintel clamps have been used. Apart from the gudgeon and the heavy bolt with ring in the sternpost, no large iron structural elements were found. Naturally, there were also rudder fittings and a heavy bracket was presumably attached to the secondary sail beam for holding the lowerable mast in place. It also seems obvious that in the possible original construction of the stempost and false stem, a heavy bolt was fitted through both posts at the top.

12.3 Construction data

The completeness of the ship's hull as found and the individual structural components found therein in addition made it possible to draw a reliable reconstruction, which at least comes very close to the original form and execution. Only the layout and composition of the decks are an assumption.

12.3.1 Design
No design drawings of late medieval vessels are known. They were built according to pre-existing rules of thumb by the ship's carpenter, while the main dimensions were most likely partly determined by the available timber. The Marknesse-cog has a beautifully lined hull and a relatively high rising forebody.

Form
On the line or plank plan (Plates 12.IA and 12.II C) it can be seen that the cog has a straight keel, sides with a considerable sheer and with the top of the hull at the stern being half a meter lower than at the stem. The hull amidships is even 1.1 m lower than in the stern. The latter is raked at an angle of 117° in relation to the keel line, the false stern at an angle of 121.° The underwater body has an almost symmetrical, bay leaf-like shape and is very sharply built at the bottom of the stern and stem. Because of the sides that are inclined, the vessel widens towards the top and has a slightly elongated, almost oval shape at the height of the sheer strakes. At approximately the same distance from the stem, the forebody is only 10 cm wider than the afterbody. The stern is built slightly narrower than the bow. The cross-sections (Plate 8.II B) show that, in fact, only a few frames have a light S-shape in the forebody. In general, the hull is 'round' almost everywhere up to the top, so nowhere is a sharp transition from bottom to side. The slope of the sides can therefore not really be

determined. Strake F is positioned amidship at an angle of 115° with respect to the keel line, whereas that is approximately 106° for the G strake arranged above it. The virtually nowhere 'dead flat' bottom is carvel-built over a length of just 3.5 m. Further to the aft and to the bow and in all other strakes of the sides, the construction is built in a clinker built fashion.

Dimensions and layout

No marks were found or recorded that could have been an indication of dimensions during construction. The length of the keel between the beginning of both stern- and stempost is 3x the length of the stempost. The length of the sternpost corresponds to the distance from the beginning of the sternpost to the centre of the mast in floor timber 20. It is not possible to say whether there is a relationship between these corresponding dimensions. The original after end of the gangway on the port side and the original place of attachment of the crossbeam found detached in the forebody make it clear that the gangways ran from frame 3 to frame 27 and were therefore integrated into the aft- and fore deck. The aft deck was arranged between the sternpost and the forward side of frame 4, the fore deck initially ran from frame 20 to the stempost. Whether and how the space between the primary and secondary sail beams and the gangways was filled in has not really become clear. It was also not possible to determine whether the hold under the fore- and/or afterdeck was closed off by a wooden bulkhead from side to side; it is not very likely.

Tonnage

Although the floor ceiling/loading floor starts at frame 7 and ends at frame 19, it seems likely that the loading space was located between the fore of the aft deck (forward of frame 4) up to under the fore deck at frame 21 and under the gangways. At frame 4 the width at deck height is about 3.1 m and on the bottom about 1.5 m, amidship it is respectively about 3.9 m and 2.5 m, while with frame 21 this is about 3.4 m and 2 m. Converted, this amounts to an average width of 2.75 m. The length of the hold measures 7.5 m, while the height between the floor ceiling and the gangways is approx. 0.9 m. The capacity of the loading space is therefore 7.5x2.75x0.9 = approx. 20.6 m^3. The space within the formwork around the lowerable mast measures approx. 1.6 x 0.5 x 0.9 m = 0.7 m^3. The net loading space is therefore approx. 20 m^3 or approx. 20 gross tonnes (see also Section 8.6).

Material selection

Analogous to the other cogs, the Marknesse cog is also made entirely of oak. Only one or a few loose planks of the loading floor are made of pine. The gudgeon found and the upper suspension of the rudder, the rudder fitting which was not found, the presumable bracket for sealing the lowerable mast and a possible bolt through the stempost and false stem are probably the only heavy, wrought-iron structural elements of the ship. Furthermore, only wrought iron nails and sintels are used. The sail was probably made of wool (Section 12.5.1).

Table 1. Principal dimensions of the Marknesse cog

Length of the keel (including stern- and stem hook)	11.80 m
Length of the bottom (between the rising part of the stern- and stem hook)	11.30 m
Greatest width of the bottom (seam between C and D strake at frame 14)	2.85 m
Olverall length at gunwale level	14.80 m
Overall width at gunwale level (at the frames 14-17)	4.65 m
Height of the stern (keel line to top gunwale level)	2.75 m
Height of the stem (keel line to top gunwale level)	3.25 m
Height of the side amidship (keel line to top gunwale level)	1.60 m
Lenght of the stern	3.20 m
Lenght of the stem	3.75 m
Height of the hold	0.85 m

Length-width ratio of the bottom	3.1:1
Length-width ratio at gunwale level	3.2:1

** All measurements are rounded to the nearest 5 cm*

12.3.2 Construction: techniques and sequence

The standard keel plank found in cogs, extended at both ends with a hook, has also been used in the Marknesse cog. Remarkable is the protruding cutwater on the bow, on which (originally) the false stem rested. In addition, the more than 10 m long and 5 cm thick gangway planks stand out, which are structurally of great importance for the longitudinal strength of the hull at deck level. The structural strength in the width is formed by the cross-beams attached to the gangways, while the knees fastened to them and the inner sides against and on the frames provide firm, form-retaining sides. The probable construction phases are not discussed in detail here. See the description of the Spakenburg-I cog (Section 14.3.2). It can be noted, however, that the garboard strakes, possibly made from one plank, approx. 12 m long, cannot easily have been burned into their likely shape. It is more obvious that they consist of two planks (Section 12.4). As the carvel-built part of the bottom (A to C strakes) is no more than 3.5 m in length, only a few (short) auxiliary floor timbers and supports would have been needed during the construction of the bottom.

12.3.3 Characteristics

The Marknesse cog has the following characteristics.

Keel plank
A keel plank that probably has about 2.5 x the thickness of the shell planks, which is aft 26, amidships 38 and in the forebody 28 cm wide. One can almost call it a keel beam.

Stern- and stem hook
The keel plank is extended aft and fore with a hook. The stem hook has a toe, with a hole for attaching a rope with which the ship could be pulled up onto a slipway and on which the false stem rests. The stern hook has a heel as protection for the rudder.

Stern- and stempost, false stem and rudder
The posts are placed on the hooks with an oblique scarf. On the stem the vessel has a cutwater, and perhaps (originally) a false stempost protruding above the gunwale. The cog was kept on course with a stern rudder.

Clinker-built and carvel-built bottom
The ship has a remarkably small carvel-built part in the bottom, the other plank connections in the bottom are constructed in clinker-built fashion.

Clinker-built sides
In cross-section, the vessel from the keel to the sheer strake is almost completely built 'rounded'. The sides are constructed entirely clinker-built. The planks are attached to each other with nails, hammered through the wood, bent twice and driven back into the wood (double-clenched nails). This is also done with the scarfs in the strakes. In the case of the scarfs between the keel plank and the hooks, the nails were most likely not hammered right through the wood.

Caulking method
The seams between the planks that have been constructed in a clincker-built fashion are only provided with sintel clamped moss caulking on the inside face, the carvel-built seams only on the outside. Hardly any repairs have been seen with sintel clamped moss caulking.

Ceiling, mast steps, riders, casing and repairs
There is only a small loading bed or floor ceiling made up of loose planks. A ceiling in the sides does not seem very likely (except for two short, loose-lying planks). The ship initially carried a mast, which stood in a hole in a floor timber. Later on, a mast step was fitted to the aft of a floor timber, in which the mast that was made lowerable at the time it was placed, while there was probably a casing around it related to the transport of bulk loads. The broad part of the mast step is supported on both sides of the mast by riders to absorb the sideways pressure. The situation around the first position of the mast suggests a calamity. The resulting damage was probably repaired by maintaining the old sail beam, possibly turned around;
;by installing a new sail beam and a new mast step and replacing the initially presumably heavier futtocks of frame 20 with slightly lighter ones. Furthermore, some small repair planks were found. The ship shows remarkable wear and has many tar residues.

Gangways, crossbeams and decks
The Marknesse cog has gangway planks over 10 m long, that extend into the decks at the aft and fore. The gangways rest on four frames, which are shorter than the other frames. The gangways are fixed to the sides through the frame-heads with 3 cm thick treenails. Five (initially four) crossbeams rest on the gangways and together with the knees fitted on them form a tight connection at deck level to the sides of the hull. There is almost certainly a hatch opening in the foredeck, maybe even two. The deck planks of the foredeck are arranged transversely, in the aft probably in a longitudinal direction.

Fastenings
The frames are attached to the hull with 2.5 cm thick treenails. The treenails probably protruded slightly beyond the hull and they had a slightly thicker, rounded head. As a result of wear, they now protrude a few millimetres from the hull. The planks are connected onto each other with double-clenched nails. The moss caulking in the seams is covered with laths of oak and secured with sintels (sintel clamped moss caulking).

12.3.4 Use

Function
The relatively heavy keel plank, the relatively sturdy stern and stem, the construction of heavy, long gangway planks with crossbeams on them and the open hold indicate a function as a cargo ship for the transport of, in particular, bulk loads, such as the found bricks and remains of a previously transported cargo of shells (Section 12.6). A confinement of the hold with transverse bulkheads seems logical for the transport of bulk cargo, but has not been found, while a casing around a lowerable mast is also very desirable. Making the mast likely to be lowerable points to the possibility that, for example, the vessel could enter a city more easily. In the first instance, one should consider an inland navigation vessel that also sails the Zuiderzee. The strikingly high rising stem of the ship and the once transported cargo of marine shells (Section 12.6) suggest that the vessel probably also went to the Waddenzee and/or Noordzee coast.

12.3.5 Wreck formation and embedding

The possibly not so seaworthy, at least relatively old vessel and the large load of bricks (Section 12.6) have probably (partly) been the reason for the wrecking and sinking rather rapidly into the 1.5 m thick, soft Almere clay and peat. The ship with contents then 'stopped' fairly quickly, after which the minor disturbances of the bottom profile were filled up and there could again be 'normal' sedimentation around and above the wreck. The undamaged presence of a large part of the port side and the detached and also undamaged bow and crossbeams mean that the wreck must initially have been completely covered with Almere deposits. After that, it will at some point have protruded again above the seabed. During this situation the stern and stem, crossbeams and decks must have come loose, part of them drifted away. The consequences of this were that at that time in particular the aft and forward of the cog and the bottom profiles around these parts of the ship, were disrupted considerably. Almere sedimentation layers above the disturbances indicate that the 'normal', heavily layered deposition process has been used for quite some time. This Almere package was then probably (partly) eroded again. This can be deduced from notes in a Modderman notebook on February 15, 1944 "The stones (of the cargo) were sometimes peculiarly eroded because another stone was drilled into it" and "the layer of shells (Zu IV) on the boat (seen myself)." The latter means that the highest parts of the wreck in the beginning of the 17th century (temporarily) stuck out above the seabed. The boat hook (fig. 12.27-16) was also drilled in a brick and found stuck in it. All this means that the Almere package above the wreck must have been almost completely washed away at some point and that the top of the cargo of bricks has been exposed for a long time to occasionally quite powerful streaming water, where the flow direction may have been different. Otherwise the stones would not have been drilled into each other. This hypothesis is supported by the (probable) hiatus already established in section 12.1.2 young Sloef deposits in the immediate vicinity. The Zu-IV shell layer is then possibly one of the first deposited layers that ran undisturbed over the wreck.

12.4 Model

In a letter to the director of the BAI of 13 November 1945, D. Huismans offered to make a model on a scale of 1:10 based on the available field drawings and the notes he made himself. The model would be made from 'best aged oak'. Huismans further writes in the letter that he has no 'time or desire' to make and apply all 3000 iron clamps (sintels), but by way of example, a part of them (around 200 pieces) will suffice. Until recently, it was not clear whether the assignment to build the model was ever given. In 2005, a half model (1: 200 scale) of the Marknesse cog made by Huismans turned up in the Netherlands Open Air Museum. It was in a deplorable condition and people wondered what to do with it.[35] Together with the photos of the excavation, the model was handed over to the RCE (Lelystad) in November 2005, where it was subsequently restored (fig. 12.24).

The Northern Maritime Museum in Groningen also contains a model of the cog from lot M 107, made by W. Pul.[36] The half model by Huismans only shows the reconstructed starboard side. It is remarkable that the garboard strake has a scarf exactly in the middle (based on an observation by Huismans?), that only part of the A and B strakes is carvel-built and the B strake aft and fore at one point is becoming wider and has been constructed in clin-

35. Van Holk *et al.* 2008, 361.
36. Van Holk 2008, 360 fig. 5.

ker-built fashion to the stern and stem. Huismans built the C strake, just like the following strakes, completely clinker-built. He also gives his own (incorrect) interpretation of the stern/stem/keel construction; stern- and stempost that stand on a keel beam and partly cover its cross-cut ends. The keel is converted 12.3 m long and 19 cm thick. The length over the stern and stem is 15.6 m, the largest width over the sides is 5.2 m, the height of the sides at the stern 2.4 m, amidship 1.6 m and at the stem 2.45 m. It is also noticeable that Huismans has added an extra channel on the outside of the sheer strake, which certainly never was there.

The model in the Northern Maritime Museum is undoubtedly (also) made from the field drawings. This model shows the entire found vessel, in which the separately found sail beam is mounted above the mast step and the missing parts of the hull in the aft and forebody have been reconstructed. The number of frames installed in these parts of the ship does not correspond to the original situation, while the gangway planks end at frame 19 (fig. 12.25). In any case, both models deviate considerably from the reconstruction presented in this study. It therefore seems worthwhile to build a third, revised model.

Fig. 12.24. The half model (scale 1:200) by D. Huismans from 1946.

Fig. 12.25. The model of the Marknesse cog by W. Pul. *Northern Maritime Museum / Van Holk)*

12.5 Inventory

12.5.1 Distribution of the finds

In section 12.1.1 the stoneware jug found in the wreck has already been mentioned. Modderman[37] describes all finds and depicts the most important ones.[38] A number of objects appeared to be no longer in the collection during a re-inventory in the 1970s and 1980s. They are marked with an * in the find list. From the larger of the two bronze cooking pots (*grapen*) (6), since the beginning of the seventies only a cast is present, while the copper kettle lacks the loops and the iron handle and the small bronze *grape* lacks the iron handle. Thanks to the photo of the gudgeon found in 1944, it was possible to find the missing gudgeon of the Marknesse cog in February 2009. As a result of the lack of an inventory number, the gudgeon had been housed in a collection of medieval objects that were suitable for the display. The locations of the finds cannot be shown on an overview drawing because they have not been recorded. Modderman only mentions that some utensils were found in the afterbody.[39] During the follow-up investigation in 2008, a few nails, sintels, samples of moss caulking and wood samples were collected. A nail and some sintels are shown (figs 12.14c and 12.26).

12.5.2 Ship's equipment and inventory

Ship's equipment

A wrought iron gudgeon was attached at the location of the scarf between the stern hook and the sternpost (fig. 12.9, Plate 12.I A and fig. 12.26-8). The cross-shaped wings that extended to the hull planks were, measured from the aft, respectively, about 250 and 240 mm long, 55-60 mm wide and about 5 mm thick. The part projecting outside the stern gradually becomes thicker and measures 12 mm at the rear. In the aft part, where the two wings come together in a round shape, an approximately 5 mm thick ring with an inner diameter of 40 mm is forged on the inside. Both wings are fixed on the stern with eleven nails on each side. The nail holes have a diameter of approx. 10 mm. The approx. 13 mm deep areas of wear in the upper edge of the ring on which the rudder fittings rested indicate that the rudder hung almost as much on the starboard side as on the port side during the various boat trips. Also it means that there must have been long-term use. The field drawing indicates that at the top of the detached sternpost a bolt stuck out with an eye, the upper suspension point of the rudder. Detailed dimensions of this cannot be given, but it is almost certainly a bolt of 30-40 mm thick. The sternpost was not seen in 2008. It seems unlikely that the bolt was removed from the stern during the excavation. The bolt (9) mentioned among the finds is probably a different one.

As far as could be ascertained, wrought iron nails 75-85 mm long, 5 mm thick and with a flat, sometimes slightly convex head (diameter 22-30 mm; fig. 12.26-10-1) were used in the vessel. Possibly the nails in the gudgeon and at the attachment of the stern/stem hook and posts were slightly heavier in design. In figures 12.14c and 12.26-11-1 to -5, the sintels used in the cog are shown (type D and the model similar to transitional type C2). Only a small and crumbled piece of rope has been preserved (21). It is very likely that the rope is about 2.5-3 cm thick and is made of three strands, which in turn are made up of three strands, made of undetermined material. It is no longer possible to say whether the rope from the channel on port side was of the same thickness. The boat hook with two equal prongs (figs: 12.26-16 and 12.28c) is a tool that can in fact be expected as standard on board a ship. This type has so far hardly been found in wrecks

Table 2. Distribution of items

Category	Forebody	Hold	Afterbody
1. Ship's equipment	bolt, anchor blade	nail, sintels, reparation planks, caulking/ sail, rope fragments, boat hook	gudgeon
2. Oprational equipment	-	-	forked piece of tool
3. Administration and navigation	-	-	-
4. Tools	-	-	fire tongs, ax, whetstone
5. Household items	-	-	-
6. Cooking, eating and drinking	-	-	kettle, 2 grapes, stoneware jug, drinking jug, shards, peat
7. Victuals	-	-	-
8. Personal equipment/weaponry	-	-	knife, shoe fragment

37. Modderman 1945, 76-79.
38. Modderman 1945, Fig. 29; fig. 8.16.
39. Modderman 1945, 75.

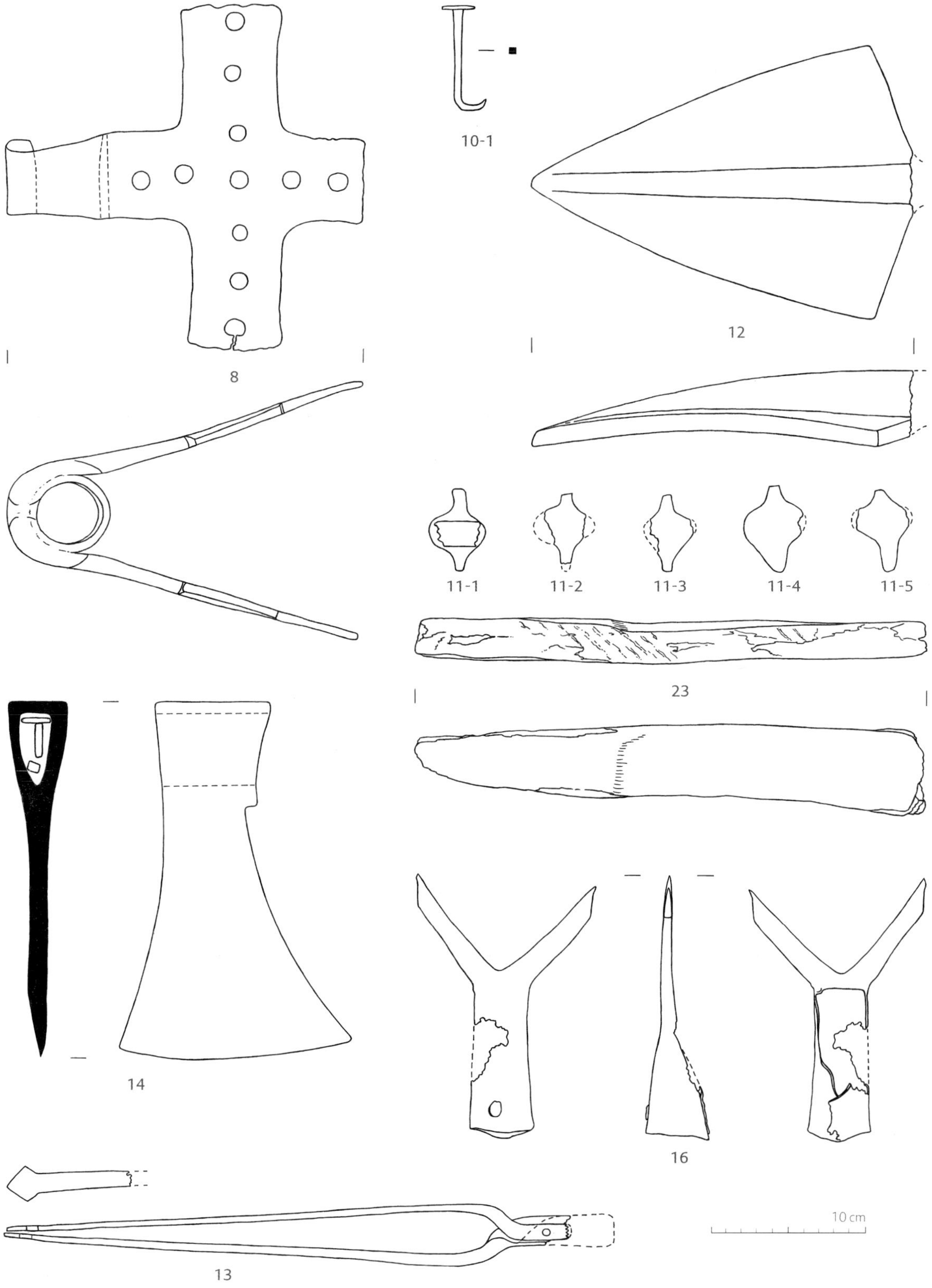

Fig. 12.26. Ship equipment and tools. Gudgeon (8), nail (10-1), sintel type D (11-1 up to and including -5), boat hook (16), anchor blade (12), fire tongs (13), axe (14) and a grind- or whetstone (23).

in the former Zuiderzee area. A few specimens of this model were found in the early medieval harbour area of Dorestad,[40] and among the finds of the Damrak in Amsterdam a number of specimens were also collected during the sinking of the caissons for the new North/South metro line at the Nieuwe Brug site.[41]

Caulking made of pieces of woollen sailcloth

Remarkable are the two oak repair planks with a piece of textile as a caulking and a similar piece without a plank (18a and b, -19a and b and 20, figs 12.27a, b and c). The exact location of the planks on the inside or outside of the hull was not recorded in 1944. The larger plank is 42.5 cm long, 13 cm wide and 0.8 cm thick. It is chamfered on a short side over a length of 1 cm, on the other side about 5 cm. On one long side two round holes with a diameter of 1 cm can be seen, while there are twenty-two nail holes/nail remnants in the plank, of which the disappeared heads had a diameter of 10-15 mm. The locations and the different thicknesses of the nails suggest that they must have been applied on at least two different occasions. On the side where the nails come out, there are still scraps of moss present in some places, from which it can be deduced that initially (and as usual) a layer of moss was applied between the plank and the ship's shell. On the same side the plank also shows light traces of fire. Most likely the moss was later replaced by the piece of textile. The round holes in the plank have to do with earlier use, for example as part of a barrel. The smaller plank measures 32.5 x 9.5 x 0.5 cm and had been fixed with just six smaller nails. The holes in the piece of textile fit exactly with the nails in the plank.

Fig. 12.27a & b. Repair planks (18) and the associated piece of textile caulking (woollen tarpaulin); b. Repair plank with corresponding piece of sail (19). c. Piece of sail (20). You can clearly see the weave (four-twined twill) of the woollen weave.

Table 3. Ship's equipment

Category	Item	Number	Findspot
1.1 Ground tackle	anchor (blade)	12*	?
1.2 Sails	woolen sails (as caulking)	18b, 19b and 20	?
1.3 Cordage	line	21, 22*	at rest and ?
1.4 Blocks	-	-	-
1.5 Windlass	-	-	-
1.6 Pump, drain	-	-	-
1.7 Navigation aids	boat hook	16	hold
1.8 Boats	-	-	-
1.9 Spare parts	-	-	-

40. Van Es & Verwers 2009, 234-237, fig. 194 A up untill and including D.
41. Sintels in het Damrak. Vlierman, in preperation a.

However, the most striking are the pieces of textile. It almost certainly concerns three fragments of a larger part, most likely a discarded sail. The coarse woven, dark brown woollen fabric is made on a wide hand loom and woven in four-tier twill. It is woven from S and Z twisted, irregular, single threads. The warp or warp threads are 1-2 mm thick, the weft threads 2-5 mm. The sail/pieces of fabric have been treated with a tar product, the smell of which still is strong even after more than 600 years. The fabric has a weight of almost 1300 grams per m^2. Based on comparison with more recent woollen sails, Crumlin-Pedersen arrives at a weight of about 1 kg per m^2 (for a sail of 60-130 m^2) if it is tarred to make it water-repellent and as windproof as possible.[42] The Marknesse-cog undoubtedly had a sail of at least 60 m^2.

Operational equipment

The forked iron object with two prongs and a ferrule around the handle (15) is probably to be regarded as a tool in the operational management. The piece is no longer in the collection.

Administration and navigation

No objects were found in the cog that could be attributed to this category of objects.

Tools

The lack of tools for caulking and maintaining the ship and for rope splicing is striking. The carpentry tool is only represented by an axe of the standard axe type, right hand side axe (figs 12.26-14 and 12.28a).[43] The sharp, straight front and below the neck protruding front and the curved and sloping back line of the eye may probably be considered a 'Dutch' design. The object drawing by Modderman (fig. 12.16) shows that a part of the handle with three wedges was still present in the house. The grinding or whetstone (figs 12.26-24 and 12.28b) is 32.5 cm long and used in particular on one side. The narrow end probably served as a handle, since the remaining part was left concave (by a right-handed user?). Vague oblique hewn marks can be seen on the sides. Probably the result of processing to give the stone a handier and smoother shape. The rock is a finely layered quartz phyllite. Such grinding/whetstones have been found in various medieval settlements and shipwrecks. The origin of many of these stones are quarries at Eidsborg in Telemark, Norway.[44]

Table 4. Operational equipment

Category	Item	Number	Findspot
2.1 (Un)loading gear	fork	15*	aft?
2.2 Additional equipment	-	-	-

Table 5. Tools

Category	Item	Number	Findspot
4.1 Carpentry	ax	14	aft
4.2 Caulking and maintenance	-	-	-
4.3 Splicing, sailmaking	-	-	-
4.4 Miscellaneous	grind-/whetstone	24	aaft

42. Crumlin-Pedersen 1997, 190.
43. Vlierman 1985b, 47-49.
44. Kars 1995, 190.

Fig. 12.28. a. The axe (14); b. The grind- or whetstone (24); c. The boat hook (16).

Household items

No objects were found that could be considered household items.

Cooking, eating and drinking: fireplace

One of the preserved bricks (fig. 12.31, right) is a waster. Moreover, the stone is fired harder than the other specimens and shows slight soot marks on one side. There are no indications for the presence of a fixed fireplace in the form of a square, flat box with sand and paved with stones or tiles on top of it. It therefore seems conceivable that a fire was stoked on board on a few bricks. A hearth can be further deduced from the iron fire tongs (fig. 12.26-13) and from a few pieces of a hard peat (32). Based on comparison with the Dronten-I cog fire tongs (Section 11.5.6), it seems likely that the largely absent handle must have been approximately 70 mm long and that the narrow leg handle portion was hinged between a flat U-shaped part of the other leg. The pieces of peat are layered and come from a block of so-called dredged peat, size 13.5x10.5 cm. The thickness could not be determined, but probably would have been 6-8 cm.

Cooking utensils

The cookware found consists of a copper kettle and two small bronze cooking pots or *grapen*. The copper, cylindrical kettle (figs 12.29-7 and 12.30a) had two vertical, iron loops, one of which still has a fragment of one of the two iron rivets. The initially still largely preserved iron handle is also missing. The shape and dimensions of the loops and the handle fragment are taken from Modderman's drawing. The loops were riveted to the wall just below the rim by two holes 2 cm apart. Copper kettles usually also have copper loops. The kettle has a diameter of 260 mm and a height of 130 mm. The capacity is more than 5 litres. The slightly convex bottom was beaten out of one piece, has an approx. 6 mm wide, flange-like recess along the outer edge and an upright part of 8 mm. On the inside on top and against it, an approx. 90 mm high strip was placed, which is fixed on the flange in the bottom with twenty-six rivets. The rivets are made from rolled strips of copper plate and stick out approx. 3 mm from the bottom. This lower piece of the side wall is composed of two large and a smaller plate, which overlap and are also riveted together.

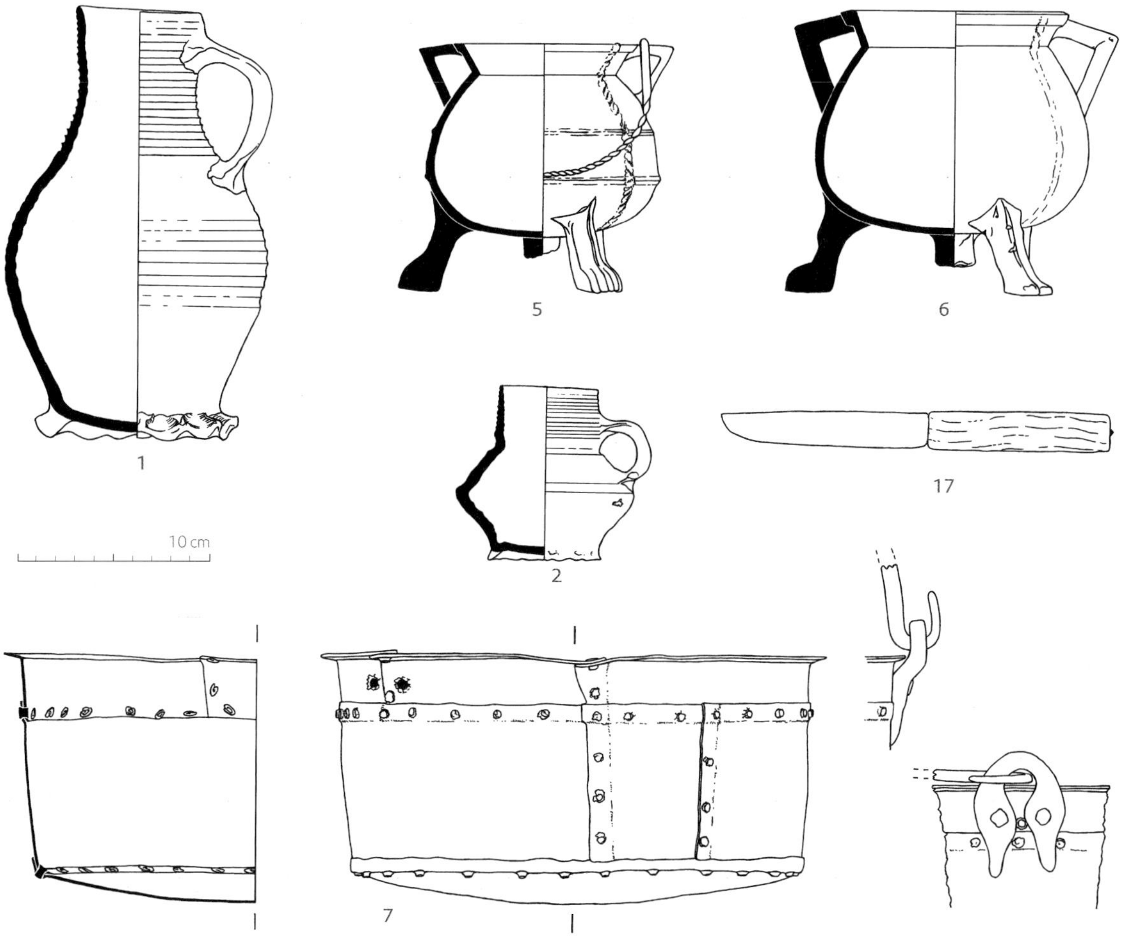

Fig. 12.29. Eating and drinking utensils and personal belongings. Copper kettle (7), bronze *Grapen* (5 and 6), stoneware jug (1), drinking jug (2) and knife (17).

The top of this lower part of the side wall is 10 mm high, 1 mm hammered outwards, a flange against which the upper part of the wall is riveted with thirty-two rivets. The upper edge of this wall part is everted horizontally and 10 mm wide. It is also composed of two large and one small copper plate. The kettle has been cleaned, so that it is no longer possible to determine whether there was a soot deposit on it at its recovery, which is very likely. The slightly convex bottom is made consciously so that the kettle can be placed firmly on a trivet (not found) or on some stones over the fire.

Two small, bronze cooking pots (*grapen*) were also found in the stern. The small one (figs 12.29-5 and 12.30 b) is pear-shaped and 130 mm high, stands on three relatively high (40 mm) claw-shaped feet with five fingers and has an everted neck. At the bottom there is a (broken) casting plug of 11 mm long with a diameter of 20 mm. On each side of the plug are extended, triangular ridges. On one side, the ridge ends against a foot, on the other side between two feet. At right angles to these ridge(s) is the weld between the two halves of the pot, which is finished on the inside, but hardly finished on the outside. On the outside there is in fact a very clear, rough ridge with several cut marks. The *grape* has two ridges around the belly as decoration, while the upper edge of the neck is slightly thickened on the outside. The pot has two triangular, vertical handles to which initially a torsioned, wrought-iron handle was attached. The capacity of the *grape* is more than 0.5 litres. The pot has been cleaned and therefore no longer has any soot traces, which were on it, as is evidenced by a photo taken shortly after the excavation.

The casting of the slightly larger bronze cooking pot (figs 12.29-6 and 12.30c) shows slightly different details. The pot also has a pear shape, is 150 mm high and has three claw-shaped feet 40 mm high. However, the claws are three-lobed, as is also found in red earthenware cooking pots. On the underside there is a 17 mm high and 23 mm diameter casting plug in the middle, while two short, triangular ridges can be seen against this and in line with each other, which are 8 mm high at the casting plug. At right angles to these ridge(s) is the weld, which is completely concealed on the inside and made fairly smooth on the outside. Lacking the small decorative ribs on the belly, the pot does have a slightly thickened rim on the outside of the everted neck. The *grape* has two opposing, triangular, vertical handles connecting the rim and the shoulder. The original pot had no bucket handle when it was found, most likely it had soot traces. The capacity of the pot is more than 1 litre.

a

b

c

d

e

Fig. 12.30. Eating and drinking utensils. a. Copper kettle (7); b. Bronze grape (5); c. Bronze *Grapen* (6) replica; d. Siegburg stoneware jug (1); e. Drinking jug from the Meuse area (2).

Drinking ware

The drinking ware consists of a stoneware jug and a small drinking jug.[45] The yellow-gray stoneware jug (figs 12.29-1 and 12.30d) is 23 cm high, has a slightly convex, sagging bottom and a pinched foot ring, a biconical belly, with approx. 0.5 cm wide rills, a smooth shoulder and sharp, up to 0.4 cm wide rills on the more or less cylindrical neck. A vertical strap handle is placed on the shoulder and against the neck. The jug is made of 'real' stoneware and may be regarded as a typical product from Siegburg, which is widely found in late medieval contexts. The capacity is 1.3 lifters. The jug shows no signs of use. Some of the parts of the foot ring protruding downwards are not worn, even though the jug is wobbly. All this suggests that it must have been practically new at the time of the ship's demise. A sherd (3) has also been found in the ship, but it has not been preserved. Modderman[46] describes the sherd as "A non-profiled rim sherd of yellow-gray stoneware with a brown glaze patch". We are probably dealing with a neck sherd of a drinking jug. There is another sherd in the records (4). However, Modderman does not write anything about this. It cannot be excluded that it is the same sherd (3).

Less common is the drinking jug (figs 12.29-2 and 12.30 e). The only 9.5 cm high jug has a slightly wavy, pinched footring, a biconical belly with a very sharp transition from the lower half to the shoulder and a cylindrical neck. On the lower half and at the beginning of the shoulder there is some 'coarse' rilling. There is also a sharp transition from the shoulder to the neck with two fine 0.8 cm wide rills. In addition, the neck only has fine (0.2 cm wide) rills and a smooth rim. A vertical strap handle is placed between the beginning of the shoulder and the middle of the neck. The small jug has a yellowish glaze on the outside with green (copper oxide) spots and speckles. The ceramic is dirty yellowish white. The jug has a lot in similar with a baby bottle of the same material and the same spotted glaze, in the Van Beuningen collection.[47] It is most likely from a production centre in the Meuse valley. Wear is particularly evident on one side of the neck. Part of it may possibly be the result of long-term use. The two abraded spots at the sharp transition to the shoulder are undoubtedly the result of sliding back and forth on the ship's bottom. The edge of the footring shows damage and wear and tear. The container has a capacity of 0.2 litre.

Victuals

No traces of victuals were found.

Personal equipment/weaponry

The only two objects from the ship that can be counted as personal equipment are a knife (fig. 12.29-17) and a piece of upper leather of a shoe (23). Neither of them is in the collection anymore. Thanks to the description and the drawing in Modderman's publication[48] (fig. 12.16), we know what the knife looked like. "A knife with a cylindrical wooden handle, through which the tang of the blade runs all the way ". Unfortunately nothing can be said about the type of shoe. Personal weaponry, to which we cannot count this household knife, has not been found.

Table 6. Cooking, eating and drinking

Category	Item	Number	Findspot
6.1 Fireplace	brick	28	aft?
6.2 Fire-tending tools	fire tongs	13*	aft
6.3 Fuel	peat	32	aft
6.4 Cooking equipment	2 pipkin, kettle	5, 6*, 7	aft
6.5 Eating and drinking gear	jug, small jug, sherd	1, 2, 3*, 4*	aft

Table 7. Personal belongings and weaponry

Category	Item	Number	Findspot
8.1 Garments	-	-	-
8.2 Footwear	upper leather	23*	?
8.3 Sewing kit	-	-	-
8.4 Tools, weapons	knife	17*	aft
8.5 Cash	-	-	-
8.6 Miscellaneous	-	-	-

45. Vlierman 2010, 193 fig. 8.
46. Modderman 1945, 78.
47. Renaud 1973, 21 and fig. 14 nr. 1.
48. Modderman 1945, 76.

12.6 Cargo and merchandise

The Marknesse cog was loaded with around 5000 bricks according to an 'accurate estimate'.[49] The size (approx. 28x13.5x6.5 cm) indicates that we are dealing with large medieval bricks. At the top of the photo (Fig. 12.21), part of the stones removed from the wreck at an early stage of the excavation can be seen. Fig. 12.31 shows some preserved examples. In 1944 it was not recorded how the stones were stacked in the ship. After the reconstruction of the ship, the most obvious way of stacking was drawn in the hold above the load bed (fig. 12.7 and Plate 12.II A). From frame 7 to the buttress timbers against the mast step (frame 19), 17 rows of bricks can be placed, from side to side and to the bottom of the gangways about two hundred and sixty pieces per row. In total around 4420 pieces. Perhaps three rows were placed on either side of the supposed casing around the mast and up to the primary sail beam. Here 6x100 stones could have been placed, which brings the total to more than 5000 pieces. The weight per stone is around 4 kg. This means that the cog transported 20,000 kg or 20 tonnes of bricks.

Interesting is Modderman's observation "that a layer of shells was found under the loose floor between the floor timbers, which absolutely did not match the stratigraphical soil conditions described above. They are subtruncate surf clams (*Mactra, Spisula* or *subtruncata, Da C.*), with the big shells (*Mactra stultorum, L.*), now the most common shell on our beaches and in the Waddenzee."[50] Modderman rightly concludes that they must be from a previously transported load and presumably were intended to be used to make lime for brick mortar. Whether the "five roof tile fragments with a thin layer of green lead glaze in the cavity" (30) may also be regarded as a remnant of a previously transported load is not certain, but very likely. At the question where was the ship with the load of bricks was heading, because of the small distance, Kuinre castle comes to mind (Section 13.1.3). The bricks found there have a different size. This of course does not mean that the cargo cannot have been destined for the castle. However, another destination is also possible.

8.7 Ballast

The ship probably never had a specific cargo intended as ballast.

Fig. 12.31. Two large bricks from the cargo.

8.8 Dating

Soil profile

In section 12.1.2 the (reconstructed) soil profile with the disturbance by the ship is described. The importance of the ship's discovery for the dating of the various sedimentation layers during the Almere/Sloef period, outlined by Modderman, certainly applied in the 1940s to 1960s. From the beginning of the 1970s, renewed attention was paid to the (disruptions of) soil profiles during the growing number of ship's excavations from that time (sometimes 5 or 6 per year). In conversations with the soil scientists at the National Service for the IJsselmeer Polders (RIJP), it became increasingly clear that the various sedimentation layers did not take place in a period defined by them, as Wiggers assumes for the Sloef deposits among other things, but that the issue is more complex. Where and when, for example, can we speak of an *evenly* sedimentation pattern and how should that be established? The bottom of a certain layer can only be dated generally, because, for example, it is the first undisturbed layer to cover a closely dated (*post quem*) shipwreck.

The brackish Zus layer in the period around 1575-1625[51] is in fact the only layer that can be reliably placed in a relatively short, datable period. It can be inferred from the explanation in section 12.1.2 that the definitive salinization south of approximately the Medemblik-Stavoren line and the estimated duration of that process almost certainly did not start before 1600.[52] This means that the brackish Zus layer must have been formed in the first quarter

49. Modderman 1945, 80.
50. Modderman 1945, 81.
51. Wiggers 1955, 103-104 and Van der Heide 1955, 6-7.
52. Vlierman 2006, 151.

of the 17th century. Together with the established hiatus in the Almere deposits on site, this means that, about the time of the sinking of the Marknesse cog, the soil profile established by Modderman can hardly provide any other information than that it occurred a considerable time after the formation of the Valvata layer and also some time before the deposition of the Zus layer (i.e. after around 1200 and before 1600). If the Sloef II a- and the Sloef I b-layer are indeed missing locally (for the most part), then with the necessary caution, the ship may have been wrecked in the late-14th- or early-15th-century.

Finds
On the basis of a comparison of the Siegburg stoneware jug (1) with the then known finds (dated by coins), Modderman dates the finds to the third quarter of the 14th century.[53] It is remarkable that this dating since 1945 has hardly been better specified. Janssen, in "The dating and typology of the earliest Siegburg stoneware in the Netherlands" came up with a typological dating scheme based on historically datable excavated sites and jugs with coins.[54] The 'real' stoneware jug from the cog corresponds to model II, which Janssen on the basis of historical information dates to the period 1360-1375 and based on coin finds to the period 1365-1385. In excavated sites, Janssen reckons 10 years before the last deposit date, while for coin finds it is set at 10 years after the coin year of the youngest coin. Given the lack of any wear and tear, thus probably a virtually new jug, it can be assumed that the Marknesse cog most probably perished around 1375.

No more can be said about the drinking jug than that it is 14th century and possibly slightly older than the Siegburg jug. The bronze cooking pots and the copper kettle can hardly be dated, or not to a relatively short period of time. The cylindrical, riveted kettle type was found almost unchanged in shipwrecks perishing between ca 1300 and 1650, while, still used as a kettle, they were occasionally found in 18th-century wrecks.[55] Bronze cooking pots are common in late medieval shipwrecks, but also in ships from the 16th-19th centuries. It remains very difficult, just as with the kettles, to classify these *grapen* into 'short' periods of use based on typological characteristics. The discovery of a bronze *grapen* only indicates that the type in question was on board when the ship went down. The date of the foundering is always based on other objects. Due to its great durability, the *grapen* can therefore be (much) older than the rest of the inventory. The other objects are not datable.

Sintel clamps
Waterproofing of the hull seams was done with sintel clamped moss caulking of sintel type D. They are irregular and have different shapes. Some appear to be similar to the preceding type C2, while others correspond to variants D1 and D2. Sintel type D is in use in the period around 1250-1375.[56] Type D is the earliest sintel model, before that there were sintel nails (with a less wide forged middle part, so more like a staple). Well-known finds of type D for the 13th century mainly are small sintels. It has already been stated several times in the previous parts of this study that, given the wear and repairs, the vessel is very likely to have been relatively old at the time of its demise. Judging by the presence of a few sintels that still seem to resemble the sintel nail type C2, a construction date was assumed in the third quarter of the 13th century. This would mean that the ship was about 100 years old when it sank. This appears to be very old, despite the established wear. It is more likely, for example, that the ship was built in an (isolated) shipbuilding region, where the development of the shape of the sintel took place more slowly, in other words where the shape of the sintel was maintained by the local blacksmith for a long time.[57] If this is the case, the construction date may be younger. Dendrochronological research can bring more clarity to this. For the time being, the combined observations point to a date of construction in the first half of the 14th century, while the second quarter seems the most likely.

Dendrochronological examination
During the writing of the chapter on the Marknesse cog, the results of the tree-ring analysis were not yet available.

12.9 Summary

Ship
The small cog excavated by Modderman in 1944 in lot M 107 in the Noordoostpolder and published in 1945 has been an important incentive for the start of regular nautical archaeology in the Netherlands. The vessel has a considerable sheer with a fairly high rising forebody and striking 'round' cross-sections (no sharp transition from bottom to sides). Only a few frames in the forebody have a light S-shape, while the stern is more peaked. The upper edge of the hull has an almost oval shape that tapers towards the stem and stern and is only about 10 cm wider in the bow compared to the stern. The ship has the hooks

53. Modderman 1945, 77-78.
54. Janssen 1988, 311-333 and tabel I.
55. Vlierman 1992a, 55.
56. Vlierman 1996a, 52-53 en 58 tabel II.
57. For this clues have been found among the sintel finds in the Damrak in Amsterdam; Vlierman, in preperation a.

characteristic of cogs with a relatively thick keel plank in the middle, fitted between them. The stem hook on its horizontal part has a forward projecting 'toe' with a hole for attaching a rope. This allowed the ship to be pulled up a slipway. On the toe a cutwater-like lower part of the stem is attached. The sternpost is straight and is at an angle of 117°, the straight stempost has an angle of 121°. The cog was steered with a stern rudder. The angle of the sides cannot accurately be indicated due to the round shape.

Remarkable is the only 3.5 m long part of the bottom that is carvel-built. The rest of the strakes are built in lapstrake fashion. The lower three (bottom) strakes end in rebates in the hooks and posts of the stern and stem. The upper four extend to the aft and fore of the stern- and stemposts. In the bow section the vulnerable hood ends were very likely (originally) protected by an false stem. The plank ends in the sloping flat scarfs protruded about 1.5 cm outside the hull both on the inside and the outside. The lapstrake connection of the individual planks was done with double-clenched nails. The connection of the frames with the hull is done with treenails, of which the slightly thicker, convex heads initially protruded slightly outside the hull. Due to wear and tear, as a result of long-term use, this is no longer visible. The lapstrake seams are only sealed on the inside with sintel clamped moss caulking. There is no certainty about the carvel-built seams of the bottom, but there is almost certainly sintel clamped moss caulking applied from the outside.

The longitudinal stiffness of the hull is determined on the inside on both sides by thick, roof-shaped bilge ceiling planks and by over 10 m long and 5 cm thick gangways, which rested on a few short futtocks and are horizontally connected with 3 cm thick wooden treenails through the frames onto the sides. The transverse strength of the ship is of course primarily formed by the frames. At deck level, the shape of the ship is fixed by a number of crossbeams, which rested on the gangways and are treenailed to each other and to the upper parts of the hull with standing knees. The strength of the upper sides is also formed by roof-shaped, thick planks or inwales, extending at both ends towards the stern and stem, that partly rested on some heads of futtocks and in other places with slanting heads of futtocks that were let in to the underside of the inwale.
The separately found sail beam is related to the primary location of the mast (in frame 20). Presumably as a result of a calamity that occurred, a mast step was then provided on the frames 17 up to and including 20, in which the mast came to stand above the aft of frame 19. On that occasion a new sail beam was placed at deck height, which was not found. The old sail beam had been retained in its original position, though possibly turned around. The backward placement probably also has to do with the decision to install a lowerable mast, and the plank with two holes above the frames 21 and 22 formed the basis for two vertical posts, on which the forward part and sides of a casing around the mast were attached. In the hold there was a load bed of old, partly worn or broken oak and pine planks lying loose on the horizontal frames and between the bilge ceiling planks. The ship is also entirely made of oak. The lapstrake seams are only waterproofed on the inside, the carvel seams only on the outside, with sintel clamped moss caulking. The ship showed a lot of wear and tar residue. At frames 15 to 17 there is a channel against the outside of the sheer strake. The holes in it held four lines of rigging to hold the mast in place.

The most important reconstructed dimensions are:

• Length over the stern and stem at gunwale height	14.80 m
• Largest width at gunwale level	4.65 m
• Height of the stern (keel line to the top gunwale)	2.75 m
• Height of the stem (keel line to the top gunwale)	1.60 m
• Height of the side amidships (keel line to the top gunwale)	3.25 m

The robust and characteristic design of the stern/stem constructions, the thick and wide keel plank, the partly carvel-built and clinker-built hull, the long and relatively thick gangways and the large hold openings indicate that it was a ‘small cog’, a cargo ship particularly intended for transporting bulk, stacked and relatively bulky objects such as bricks, building timber, etc. The ca 5000 bricks (weighing around 20 tons) and remains of a previously transported cargo of marine shells and possibly also a cargo of roof tiles confirm this, while the cargo of marine shells also indicates that the ship (regularly) sailed on the Waddenzee or along the Noordzee coast. The remarkably high forebody suggests that the ship may have been built specifically for visiting those waters. The ship had probably a woven, woollen sail, pieces of which were later used as a seal between repair planks and the hull. It may have been a square sail, while later with the fitting of a lowerable mast may have been changed to a spritsail.

Equipment and inventory

The Marknesse cog had a modest amount of tools and household objects on board. The crew cooked in a copper kettle and in two small bronze cooking pots on a peat fire, which was probably laid on some bricks. The fire was tended with fire tongs. A Siegburg stoneware jug and a small drinking jug with green spotted lead glaze from the Meuse valley were found as drinking utensils. An axe, a grinding- or whetstone, a forked tool and a boat hook are the only tools found. The leather upper of an (old?) shoe and a knife are the objects that can be attributed to the personal belongings category. The significance of carrying a broken anchor fluke on board is not clear.

Dating

The most important evidence for the time of the ship's demise can be derived from the Siegburg stoneware jug. The jug was almost certainly as good as new at that time. The type was produced according to Janssen[58] between approximately 1360-1385. The shipwreck can therefore be set at around 1375. Interpretation according to current insights into the soil profile disturbed by the wreck cannot be determined more accurately than some time after about 1200 and well before about 1600. If there was any geological gap (lack of the Sl II a and Sl. I b), then, with the necessary reservations, it is possible to date the sinking on the basis of the soil profile in the late 14th or early 15th century. This is reasonably close to the date of the stoneware jug.

The (brackish) Zus-layer characterized by small *Cardium* shells in the soil profile above the cog is similar to the shells found in a layer against the hull of a *beurtschip* (B 71 Lelystad), while the Zu-IV shell layer, situated directly over it, was deposited on top of the ship. The ship was built around 1600 and must have sunk around 1625 given the dated artefacts. Soil scientists have not yet established the Zus-layer in Eastern Flevoland. The final salinization of the Zuiderzee, south of the Medemblik-Stavoren line, took place over a short period - during the first quarter of the 17th century. The sintels used in the Marknesse cog, together with the other observations, suggest a construction date in the first half of the 14th century, probably the second quarter. Results of the dendrochronological analysis that are not yet known will probably provide more certainty about this. After the manuscript was closed, the results became known.[59] The dendrochronological research resulted in a felling date of AD 1339 with the reference chronology of NLTWEN03 and of NLNOOR1B. The vessel was probably built in the second quarter of the 14th century in the (northeastern) Netherlands.

Storage, conservation, exhibition and rebuilding the cog

Re-burial of the cog in an anoxic environment in 1944 is an excellent method for preserving important ship finds for a longer period of time in order to postpone the salvage etc. to a moment in the future. The quality of the construction and the wood did not appear to have deteriorated significantly by 2008, as far as this could be established, over the sixty years.

The decision taken in 1944 to exhibit the ship in the future can therefore still be carried out for practical and technical reasons and still seems very worthwhile. In addition to this cog marking the actual start of regular ship archaeology research in the Netherlands, even after the hundreds of shipwrecks dug up in the IJsselmeer polders since the Second World War, it is still the best preserved, most complete and oldest medieval cargo ship for inland waterways and the Zuiderzee, ever found in the Netherlands. Interest in a reconstructed model is minimal, while reconstructing the full-size ship would be a special counterpart to the existing replica of the sea-going, and also 14th-century Nijkerk-II cog, the Kampen cog.

58. Janssen 1988, 311-333.
59. Van Holk 2010, 138; For the complete results see appendix.

The Kuinre cog[1]

13.1 Find conditions

13.1.1 Discovery and investigation

Even before the Noordoostpolder became fully dry in 1942, drains and ditches were dug in lots M131, 132 and 133, immediately south of Kuinre, as part of the reclamation effort. The people of Kuinre followed the proceedings with great interest, not just because of the recently dry, highest part of the newly drained land but also, and especially, out of curiosity after any finds and remains of the castle and the church that were said once to have stood there. As early as May 1941, a small collection of potsherds and bones was accumulating at the town hall of Kuinre, where any finds were reported. In July of that year, the Biologisch-Archeologisch Instituut of the University of Groningen was charged with the archaeological investigation of the former Zuiderzee sea floor.

As a ditch was dug between the future lots M131 and M133, in 1942 a cluster of driven-in piles was cut through. It was found to be the location of the first castle of Kuinre.[2] The same year, a series of test pits were excavated in these fields. A good year later the castle, situated mostly in lot M133, was excavated.[3] About 600 m northeast of this fortification, in lot R77, Van der Heide uncovered the remains of a second castle.[4] Reconstructions on paper by Modderman and Van der Heide in 1948 prompted the decision for the first castle to be marked out in the landscape. Both castle sites have been listed archaeological monuments since 1978. October 1999 saw further archaeological research in and near the Kuinderbos woodland, in the same lots M131, 132 and 133 and in R77.[5]

Meanwhile, between 1952 and 1957 Mr R. Kuiper had been collecting prehistoric flint tools and chips in lot M132. Levelling and drainage works in 1982 allowed investigation and documentation of soil profiles, while also further flint material was found at findspots previously identified.[6] The area in question clearly had a long habitation history, albeit not an uninterrupted one.

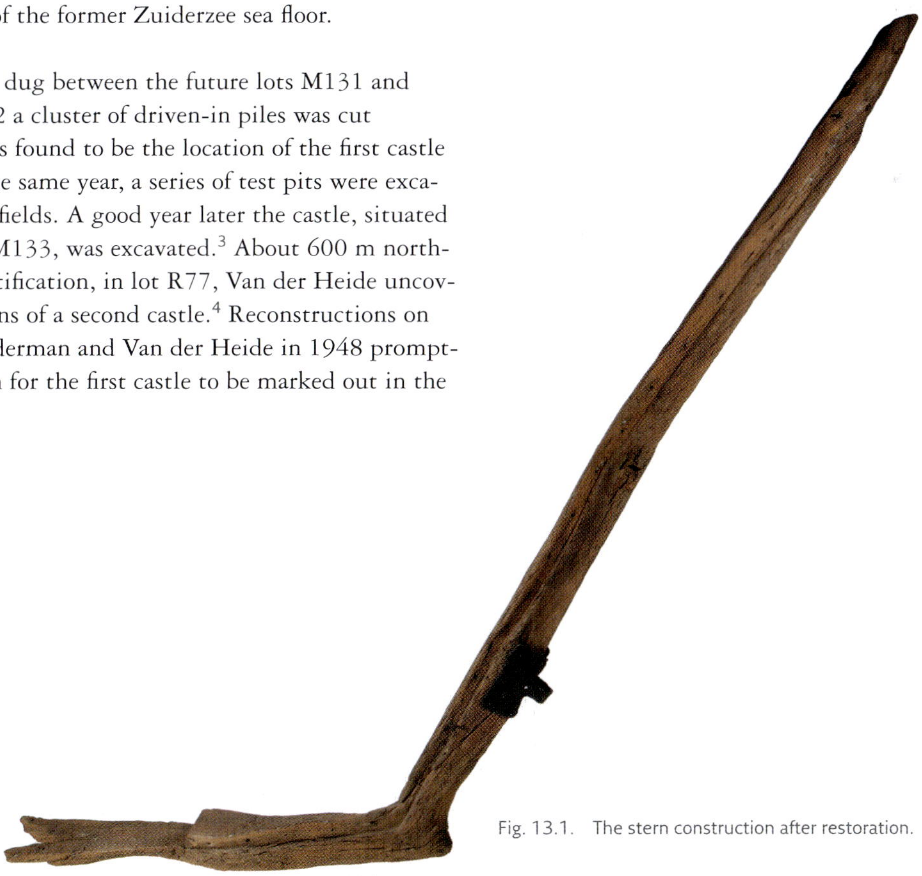

Fig. 13.1. The stern construction after restoration.

1. Translated by Xandra Bardet.
2. Van Doesburg & De Boer 2001, 17.
3. Modderman 1945, 30.
4. Van der Heide 1951, 18-27.
5. Van Doesburg & De Boer 2001, 13 and 22
6. Vlierman 1985c.

During the search activities and test excavations, also various shipwreck remains were encountered. For instance, in 1942 part of a logboat was found to be incorporated in a shore revetment near Kuinre. The wood was radiocarbon-dated to 1450 ± 50 BP, which amounts to AD 550-650.[7] The report by P.J.R. Modderman of September 1943 reads: 'In lot M131 close to the Kuindervaart canal a boat was excavated, which however could not be dated owing to a lack of associated finds'.[8] In the same file is a copy of a note by Modderman, which goes as follows:

> *"Section M 131*
> *A distorted, small, open vessel*
> *Finds: none*
> *Soil: peat*
> *Greatest depth: 1.50 m below the surface"*

In the course of the present study, two black-and-white photos of a ship excavation were noted in the photo archive of RCE-Lelystad. Both are marked "N.O.P. (Noordoostpolder) A ship excavation near Kuinre 1943." Whether the two notes and photographs refer to the same excavation is not explicitly stated, but it is highly likely. The nautical-archaeology file of lot M133 contains the following two remarks by Van der Heide:

> *"January/February 1949. Received a verbal report concerning a shipwreck. In section M, lot 133. Only sternpost recovered. Spoil had been processed. No longer possible to tell whether excavation had previously taken place. No further investigation. Marked as an object of minor significance. 22-2-1955";*
> *and:"Ship find M 133 report (copy)*
> *Coordinates 185.55/533.00, Ordnance map sheet 16C*
> *Investigation of shipwreck in lot M131 I; at stern."*
> *Any further information on the shipwreck and/or the stern construction is lacking.*

In the 1980s I made a fresh inventory of the collection of ship's timbers at the NISA. In this process, the stern hook of a cog was identified. The documentation of these timbers, which at times proved difficult to trace, initially prompted the assumption that this stern hook must derive from field R1 at Kuinre.[9] Subsequently the sternpost was found that fitted to it and it transpired that the complete stern construction had been recorded by Van der Heide in a notebook under number Z1950/I-96 as a stray find from lot M133, with the following description: *"Stern of a vessel with a high, narrow, upright (stern) post"*. It almost certainly was the stern construction mentioned above (fig. 13.1). Which means that the construction must have been found ca 100 m off the Hopweg road, between the two castles and on the western bank of the onetime Kuinder stream (figs 13.2a and b). By stating in his note *"... M133 I; at stern"*, Van der Heide probably meant to say that apart from the stern also a wreck had been found, a second find of ship's timbers on this lot.

In late 1969, the still bare disused grange at Ketelhaven (near Dronten) where in April 1971 the Museum of Nautical Archaeology was to be opened, held the full set of frames and a long keelson with a broader mast-step section. The dried-out timbers came from a not very large or heavily built vessel most probably found/excavated in the Noordoostpolder, which had a carvel-built bottom and clinker-built sides. As the New Year approached, the newly appointed future manager of the museum thought it a good idea to clear out the dried-out old 'junk', and had the timbers carried outside to feed a bonfire. In January 1970 it transpired that it had been Van der Heide's plan to display the set of frames in the future exhibition as a typical example of a late-medieval vessel. The timbers lacked accession numbers and it was never discussed what ship find this actually was.

In the context of the present study, the author has fruitlessly sought to discover a file mentioning a wreck to which the burnt timbers might be attributed. Despite Modderman's mention of a (potentially separate) ship find in field M131, it is imaginable and indeed quite likely that all mentioned finds of ship's timbers (in M131 and M133) came from one and the same wreck, which is illustrated in figures 13.7a and b. In fact, the boundary between the two lots was not very clear because of the early discovery of the first castle (the boundary ditch was not extended through the castle site up to the Kuindervaart canal). Presumably plans already existed in 1942 to plant the Kuinderbos wood, which rendered a precise division of lots less important. The parts of lots M131 and M133 covered by the Kuinderbos wood at any rate were not leased out. In lot R1, on the east side of the Kuinderweg road opposite M132, two late-medieval punt-like boats were found (in the former Kuinder riverbed?).[10]

13.1.2 Geology and stratigraphy

The Pleistocene surface and covering layers

The Pleistocene surface immediately south of Kuinre shows a highly variable relief (ca 1.5 m below NAP at Castle I and ca 4 m below NAP at 3 km from the village). In the northernmost part of M132 lies a ridge running

7. Vlierman 2002, 126.
8. Note by G.D. van der Heide, 25 February 1955, file M 131 NOP.
9. Reinders 1985a, 15 fig. 6.
10. Vlierman 1996b, 66.

southwest-northeast, with a spur to the southeast. Both ridges are clear to see in the aerial photo of 9-4-1989.[11] This ridge, which rises to ca 2.4 m below NAP, carried traces of habitation going back to the Neolithic or the Bronze Age. In 1982, sections and soil profiles were investigated; in the far northeast of the field, the course (or an oxbow lake or a meander) of the Kuinder was identified, its riverbed lying at >5 m below NAP.[12] This depth was established also in the further course of the Kuinder across the Noordoostpolder and towards Blokzijl.[13] In the deeper valleys of the Pleistocene sand lies Holocene peat; while the old Kuinder course partly became silted up and further filled with peaty deposits. Initially clay was deposited on the peat during the Flevomeer phase or the early Almere phase.

The castles and the settlement were built on this clay-on-peat land. Into the 17th century the settlement lay within a fort that succeeded the second castle. By the 18th century Kuinre had moved to its current location.[14] The clay-on-peat cover, which came up to ca 0.5 m below NAP, in the Middle Ages probably extended to roughly the northeast edge of modern Emmeloord, but since then was progressively eroded away by the waves. Just thin patches of it have survived along the coast.[15] During the Almere and the Zuiderzee phases, a sandy ridge was deposited off the coast of Kuinre and towards Blokzijl, the so-called Kuinrezand.[16] The ordnance map of 1861 (fig. 13.3) shows the highest part of it, the Wellerzand. Nowadays the highest parts of the Pleistocene deposits, the remains of the peat, the Kuinrezand and any Zuiderzee clay, have, as a result of compaction of the surface layers, been unrecognisably incorporated into the topsoil.

About 4.5 km west of Kuinre, in what now is the Noordoostpolder, medieval dike remnants running roughly north-south have been observed, up to about 2 km off the dike of the Binnenpolder inland of Kuinre and the Fries Buitendijksveld.[17] It seems very likely that this pattern of dikes on the clay-on-peat land initially extended in a southeasterly direction. Wiggers[18] suggests the same and mentions potsherds of a settlement (Vene or Veenhuizen?) at about 6 km southwest of Kuinre, the youngest of which date from the second half of the 14th century. This peatland with its clay cover was eroded away by the encroaching waves mainly from the early 13th century onwards.[19]

13.1.3 (Prehistoric and) medieval habitation

Above, the two castles of Kuinre were already mentioned. Excavations by Modderman,[20] Van der Heide,[21] Vlierman,[22] Van Doesburg & De Boer[23] together with the substantial number of surface finds in the area have demonstrated that these parts were inhabited during the Neolithic, the Bronze Age and the Iron Age. After the area was overgrown with peat, and the clay was deposited on it, it became habitable again in the Middle Ages.

The following was taken from the introduction to the report on the supplementary archaeological research ('AAO') by Van Doesburg & De Boer.[24] The castles of Kuinre belonged to the Lords of Kuinre, a notorious lineage of ministerials. The name Kuinre is first recorded in 1118 as *swechus juxta Cunre,* though this may just refer to grassland beside the river Kuinder and not as yet to any settlement.[25] In 1165 Bishop Godfried donated a piece of land near Kuinre to the Frisians of Lammerbroek. This donation in all probability was the first spur to the Frisian colonisation that took place in the second half of the 12th century and gave rise to the settlement of Kuinre. Documentary sources from the 12th to the early 15th century refer to the Lords of Kuinre and the existence of a 'stronghold' at Kuinre. Around 1197, the sources mention Heynric die Crane, probably an episcopal ministerial, who from his fortress attacked the Frisians. That same year, a counterattack by Willem, Count of Friesland, saw his castle razed to the ground. A few years later, the Count of Holland and the Bishop of Utrecht reconciled, and the Lord of Kuinre was reinstated in his rights and estates. In 1331 Jan van Kuinre was enfeoffed with among others the seigniory of Kuinre, which included the *'alde bergh'* (the old castle) at Kuinre. This enfeoffment also included the strategically located island of Urk and part of the island of Schokland. This gave the Lords of Kuinre a great position of power, because the waterborne trade routes to and from the mouth of the river IJssel (and the rivers Zwartewater and Vecht) passed along their territory.

11. Foto Atlas van Flevoland, no. 14.
12. Vlierman 1985c, 11-13 and figs 2 and 3.
13. Wiggers 1955, appendix 2.
14. Van Doesburg & De Boer 2001, 10.
15. Wiggers 1955, 96 and fig. 41.
16. Wiggers 1955, fig. 66.
17. Wiggers 1955, fig. 104.
18. Wiggers 1955, 193-194 and fig. 108.
19. Wiggers 1955, 99 and fig. 41; see also section 12.1.3 on the Marknesse cog.
20. Modderman 1945, 30-55.
21. Van der Heide 1951, 18-27.
22. Vlierman 1985c.
23. Van Doesburg & De Boer, 2001.
24. Van Doesburg & De Boer, 2001, 9-10.
25. Formsma, 1934.

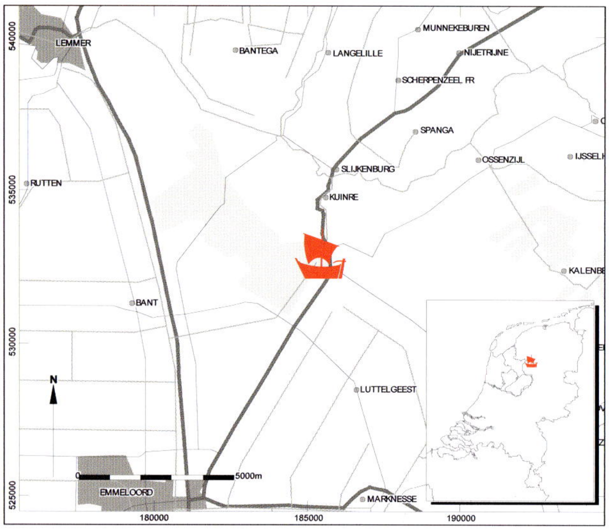

Fig. 13.2a. Modern map (scale 1:100.000)? with the findspot.

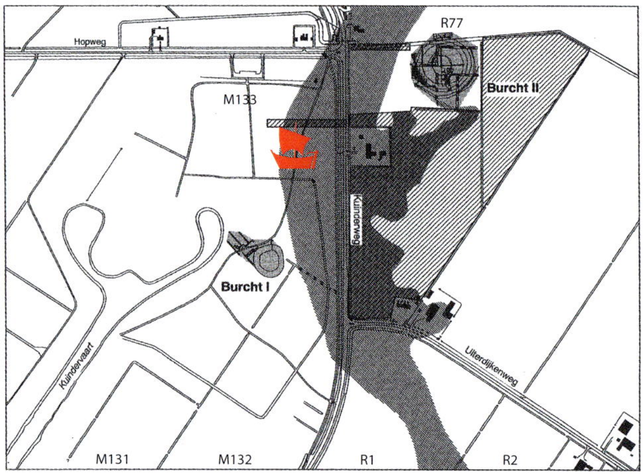

Fig. 13.2b. The location of the two castles (*Burchten*), the Kuinder riverbed and the findspot of the cog stern. After Van Doesburg & de Boer 2001.

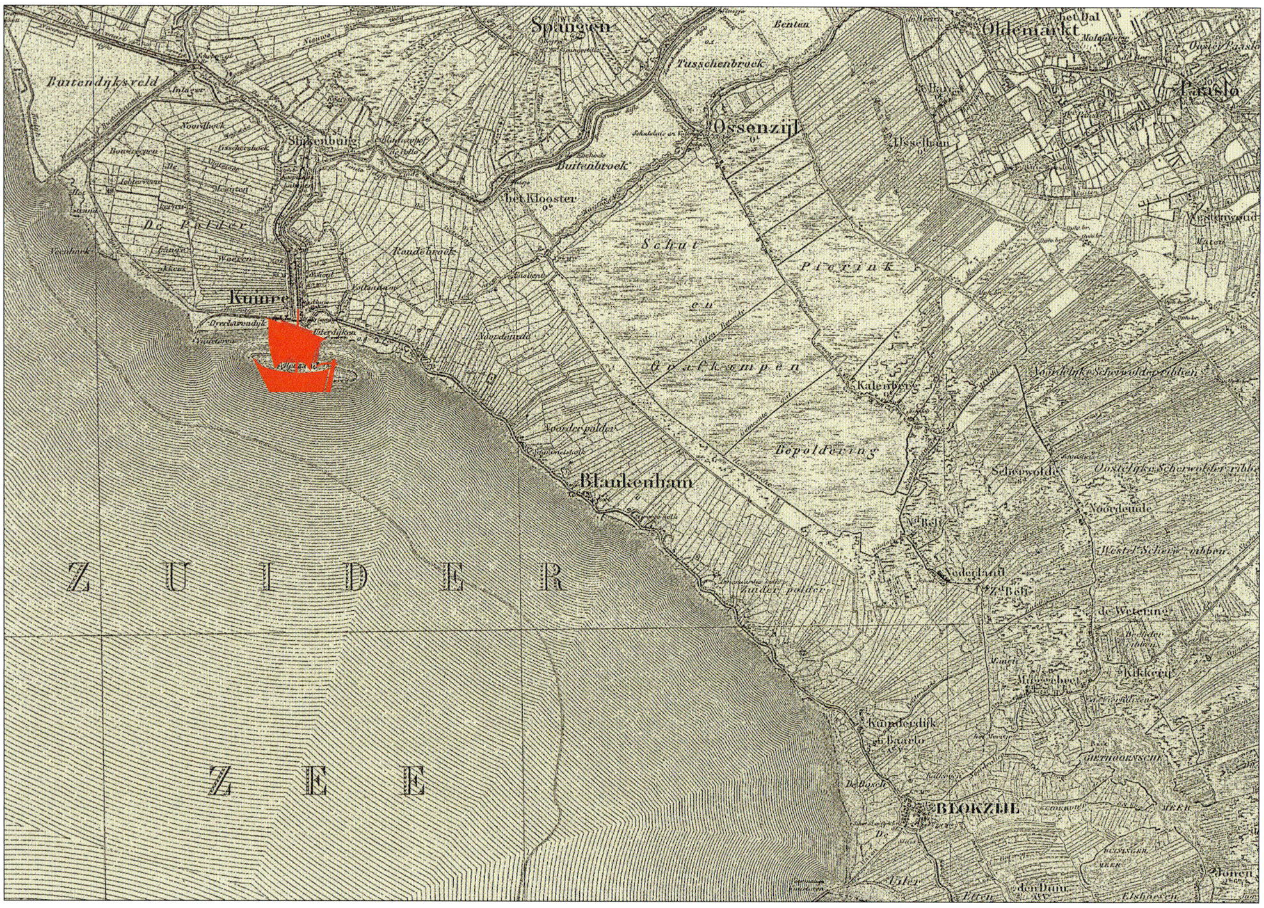

Fig. 13.3. Ordnance map of the Kingdom of the Netherlands, 1861 (scale 1:50.000), showing the location of the Kuinre cog.

They used Schokland as a base for their raids around the Zuiderzee. In the 14th century there were repeated complaints from the councils of the Hanseatic towns of Zwolle, Deventer, Kampen, Staveren, Hamburg and Gdansk about their behavior. The Lords issued coinage in the late 13th and the 14th century, both imitations of existing coinage of sovereign princes and their own coins, mentioning in the circumscription the name of the Lord of Kuinre and/or the place of issue (Kuinre or Emmeloord). In 1396 the Count of Holland landed with a substantial fleet at Kuinre in order to subdue these Frisians.

Given the strategic location of the castle, the cities on the river IJssel (Kampen, Zwolle and Deventer) had a direct interest in the acquisition of the castle by Bishop Frederik van Blankenheim in 1407. This is evident among other things from the fact that Deventer and Kampen in 1416 wanted to call a meeting because they found that the bishop of Utrecht was not fulfilling his obligations with respect to the castle's maintenance. A committee went to Kuinre *om te besien, dat slot op een andere stede te zetten* 'to see whether the castle might be relocated.' The row was eventually settled in 1418 when the Bishop promised that certain funds would be ringfenced for maintaining the *tymmeringe* (structure) at Kuinre. In 1421, Bishop Frederik van Blankenheim wrote: *...van onse slote Cuynre dat wij van nijes ut de gronde doen maken hebben.* '... of our castle at *Cuynre* that we have had raised anew from the ground out of nothing'. On the basis of this passage it is assumed that the stronghold of the Lords of Kuinre had been pulled down and replaced by a new one.[26] Throughout the 15th century the historical sources speak of damage to the castle complex and erosion of the land resulting from storms. The flooding relates to the encroachment of the Zuiderzee in a northeasterly direction. The castle lost its strategic value in 1528, when the Bishop of Utrecht laid down his secular power. Consequently, the castle was pulled down between 1531 and 1536.[27]

13.1.4 Condition and wreck formation

Immediately after the area was drained, the ship's timbers lying in the former seabed would still have been in excellent condition. Especially, as Modderman writes, when it lay embedded in the peat (or in the peaty sediment of the Kuinder riverbed). Despite the absence of suitable impregnation methods for waterlogged archaeological wood during and after World War II, the hook and sternpost are in first-rate condition. Nothing was recorded about the precise find circumstances of the stern construction. About the wreck it is recorded that it lay up to 1.5 m deep in the peat (or the peaty sediment of the Kuinder). Because, as will be explained in section 13.8, this was part of a ship of the second half of the 14th century, it may be assumed with great certainty that the stern (the wreck) lay sunk in the river Kuinder or in a 'harbour basin' leading off it. In 1421 there was mention of a newly built castle, while the site of the first castle still was far from being seriously threatened by the sea - merely by flooding during severe storms. The most plausible explanation is that we are dealing with a discarded vessel, which may have been torn apart by the sea's occasional incursions.

13.2 The ship

13.2.1 Documentation and reconstruction

In the context of the present study, the stern construction was drawn to scale 1:10 in November 2007. The (slight) distortions resulting from the uncontrolled drying-out process in the 1940s and '50s were not recorded in this drawing. So in fact we should speak of a reconstruction drawing. Reconstruction of the entire vessel of course is impossible from just a stern construction and two photographs. On the basis of comparison with the other cogs presented in this study, the preserved stern does allow an informed guess concerning the approximate principal dimensions, while the photos offer an impression of the kind of execution and layout. Such a complete cog stern may prove a key to the global interpretation of the vessel as a whole.

13.2.2 Construction and layout of the hull

Stern hook

From the tip of the heel, which served to protect the rudder, to the end of the scarf on the upper face of the keel plank, the stern hook is 1.3 m long. The flat scarf is 30 cm long. The keel plank was fastened to the hook with two rows of two nails from above and with three nails from below (figs 13.1 and 13.4). The end of the scarf is a good 13 cm wide, while the width at the after end of the horizontal part is just 5 cm. Witness the forward part of the hook, the keel plank was rectangular in section and 7 cm thick. Towards the stern, the thickness of the hook remains constant over a length of 56 cm before it rises to a thickness of 13 cm in a somewhat sloping and stepped line. The section of this thicker part is trapezoid with the broadest side uppermost, gradually becoming sharper towards the stern. The upper line of this after part is also

26. Modderman 1945, 38.
27. Modderman 1945, 35-38.

56 cm long and from this point rises a few centimetres. Here is the lower end of the flat scarf that joins the hook to the sternpost. The highest point of the rising part of the hook - measured perpendicularly - is 50 cm above the keel line. This rising part of the hook is 16 cm thick and also trapezoid in section.

At the thick horizontal keel part of the hook, oblique rabbets can be observed on both sides, about 3 cm deep, which then at an angle continue into the rising piece of the sternpost. Into this rabbet the garboards had been nailed. Above the scarf joining the hook to the keel plank, the garboards were secured with two nails onto the horizontal part. A remarkable feature is the line of sintel clamped moss caulking running downwards and towards the midships on the outside of the seam joining the garboards to the horizontal part of the hook (and the keel plank). It would seem that the top of the keel plank was not, as usual, flush with the top of the garboards, but stood proud of it by up to several centimetres. A similar phenomenon was observed in the Ens cog (Section 9.2.2). In the horizontal keel part of the hook no treenails were found for fastening floor timbers. However, a diagonal treenail (2 cm diameter) did fasten the sternpost to the stern hook.

Sternpost

On the flat scarf of the stern hook stands the complete sternpost. Including the stern hook, the stern from the keel line to its highest point is a good 2.5 m long; the sternpost itself, a good 2.4 m. Measured perpendicularly, the stern (the top of the gunwale) rose 2.35 m above the keel line. Its rake with respect to the keel line is 112°. The sternpost is trapezoid in section, with the broadest face (11-12 cm, including the rabbets) facing inboard. Upwards, the width of the sternpost gradually increases a little. At the lowermost cross-section it is 5 cm, at the second cross-section it is 6 cm. On the forward face, rabbets were cut into either side of it for fastening the hood ends of the bottom and side strakes. These rabbets are extensions of the rabbets in the stern hook. At the base they are ca 3 cm deep. At 1.85 m from the keel line, the rear of the rabbet in an oblique line runs on to the rear of the sternpost, where the rabbet is ca 4 cm deep. Upward

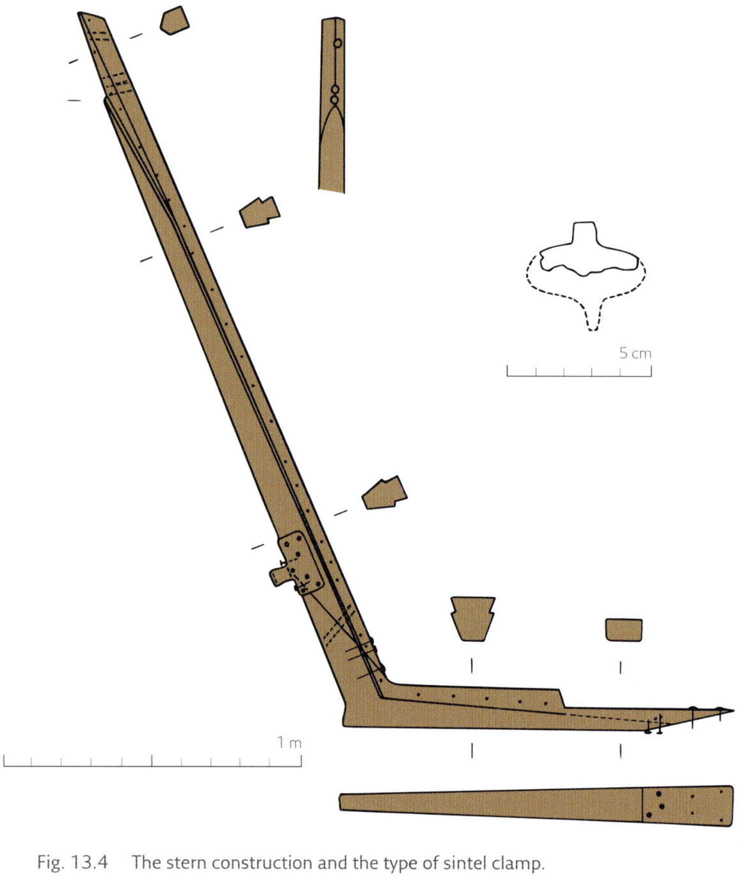

Fig. 13.4 The stern construction and the type of sintel clamp.

from the point where this oblique line ends, the rear face of the sternpost is mitred. The hood ends of the uppermost strakes thus did not end in rabbets, but were nailed side-by-side onto the rear of the sternpost. The difference in width of the two sides of the mitred rear face reveals that the crosscut hood ends of the starboard strakes were covered by the hood ends of the port strakes (fig. 13.4). The very top of the sternpost is 7 cm wide and 8.5 cm thick. Apart from the above-mentioned treenail through the scarf, the sternpost is fastened to the stern hook on its lowermost tip by three 10-mm-thick nails. Their heads have a diameter of 25-30 mm. At 51 and 58 cm from the keel line, nails were observed in the back side. The lower passes through the topmost end of the stern hook. The upper sits in the stern hook immediately above the gudgeon. What function it served there, remains unclear.

The wrought-iron gudgeon (fig. 13.5) is fastened to the lower part of the sternpost. The part of the gudgeon projecting aft of the stern, with the ring to hold the lower pintle of the rudder, sits at the level of the scarf between the hook and the sternpost. The rectangular brackets (210x90x5 mm) are fastened with 7 nails on both sides. The nail holes have a diameter of 12 mm; their heads are 20 mm across. Into the looped end a ring was forged. The opening for the pintle is 45 mm wide. At 24 and 28 cm from the top end of the sternpost are two drilled holes with a 3-cm diameter, which overlap. Their position and dimension suggest that they once held an iron bolt with an eye: the top suspension point of the stern rudder. The presence of two holes can only point to replacement and repositioning of the rudder fittings. One of the holes would have been plugged on that occasion with a treenail. Into the top of the sternpost a treenail hole was drilled with a diameter of 2.5 cm. Without doubt there

Fig. 13.5. The wrought-iron gudgeon on the thin sternpost.

would have been an inwale inboard along the top of both sides, which ended at the stern and stem. To structurally reinforce their end fastenings at the thin sternpost, they were connected to the sternpost by a triangular timber, which was treenailed both to the inwales and to the stern- or stempost (see the Marknesse cog Section 12.2.2 and Plates 12.I B and 12.II A). Also an inwale may have been fastened to the uppermost strakes inboard, of which the triangular piece was a part.[28] The mitred rear face of the upper half of the sternpost shows that at gunwale level the vessel rapidly widened towards the midships, while the rabbets at the lower end show that the sides converged very sharply there.

Bottom and sides

Of the bottom and the sides nothing has survived. Yet something can be said about them. Judging by the nail pattern in the rabbets, it may be assumed that the vessel from the keel plank to the gunwale numbered ten strakes (A to J). The first, the garboard, was about 35 cm wide at the stern; the subsequent four, ca 25 cm; followed by four of 20-23 cm width; while the uppermost strake too presumably will have been ca 25 cm wide (fig. 13.6). Whether the keel plank widened towards the midships and from there narrowed towards the bow, is unclear; yet is likely that it did. About the widths of the strakes amidships of course nothing is known; though 30-35 cm up to 40 cm at most seems reasonable, while the planks probably were 3-4 cm thick (on the basis of half the thickness of the keel plank, and the depth of the rabbets). Assuming that the stern was part of the shipwreck in both photographs (figs 13.7a and b), these do provide some information on the distortion of the hull, on its position in the soil and on its construction.

The principal reason for believing that the photos probably show a cog-like vessel is the heavy crossbeam with a knee on it that held the bulwark in shape. Immediately in front of the figure with a peaked cap and walking stick a second knee is visible and also part of a crossbeam beneath it. The heavy crossbeam (the sail beam?) presumably has a rabbet on the underside where it met the ship's side. As in the Marknesse cog (Chapter 12) this may have rested on a heavy gangway - maybe the straight, narrow timber that below it can be seen longitudinally along the side. The curved futtock poking up in the foreground shows that the side had broken at the turn of the bilge and snapped sideways. The straight head of the futtock suggests that it supported the gangway, while the gangway probably was notched to fit around the frames visible beside them, fitting onto the side. Presumably this is the port side and

28. See Reinders *et al.* 1980, 17-29 fig. 10 and 11 and plate 5.

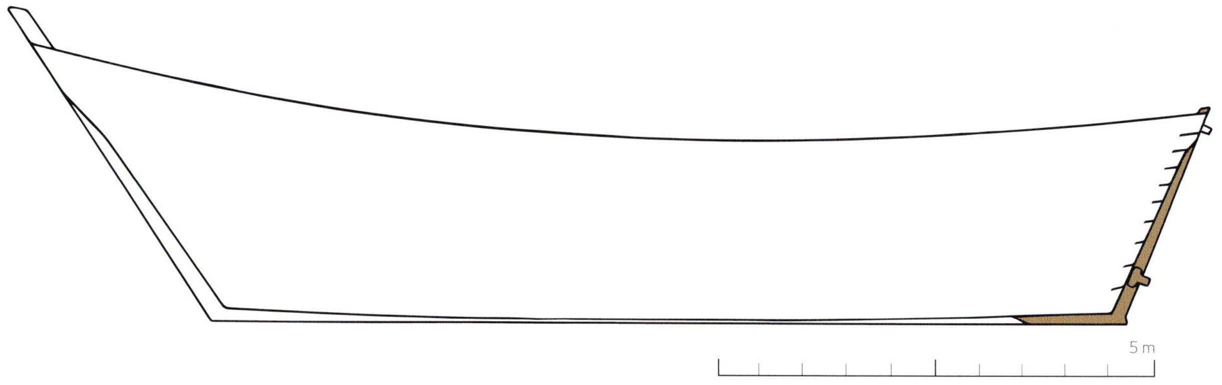

Fig. 13.6 Hypothetical reconstruction of the lines and presumable dimensions of the Kuinre cog.

we are looking at the after face of the (sail) beam. The other knee and deck beam must have been part of the foredeck (fig. 13.7a).

The other photo (fig. 13.7b) shows the clinker-built construction of the port side. Judging by the visible plank widths and the few frames which appear to be undamaged right up to the top, it may be assumed that the sides consisted of at least five, but more probably of six or seven (narrow) strakes. The straight-looking futtock heads probably also mark the height of the gunwale. Further the photo reveals that the afterbody lay deeper in the soil than the forebody, that the vessel was listing to port, and that despite the distortion and the probably detached stern, it was found in a fairly complete state. The picture also reveals that amidships it was a relatively wide ship with an almost oval, walnut-shaped outline at the gunwale, quite sharply converging at the stern and stem. It seems that in the afterbody there are another crossbeam and a (displaced) plank of the afterdeck.

Sintel clamped moss caulking

On either side of the sternpost and the hook and on the outside of the hull, sintel clamped moss caulking has been observed in the seam between the hood ends of the strakes and the rabbet, up to about halfway up the stern length, and between the keel plank and the garboard. The sintel clamps used here (fig. 13.4) have an oval centre 35-40 mm long, and they had been applied at intervals of 4,5 to 5,5 cm intervals. Vlierman[29] classifies them as type E, which was used between the late 14th century and ca 1475. Section 13.8 will return to this subject.

Fig. 13.7a. General view of the excavation at Kuinre, 1943. Probably the port side and the sail beam with a knee.

29. Vlierman 1996a, 53 and 82-58.

13.3 Construction data

For evidence about the construction of the ship, we are confined to the limited evidence offered by the oak stern construction and the two photos. This simultaneous study of all cog finds from the Low Countries, however, makes it possible on the strength of similarities to draw tentative conclusions about the cog's possible dimensions and the interpretation of particular details. The rake of the stern is 112°. In general, the rake of the stem in cogs is about 10° greater than that of the stern, suggesting 122°. The stern hook and the sternpost are relatively light in execution. The width and the thickness of the hook's horizontal part (13 and 7 cm, respectively) also indicate that we are dealing with a relatively lightly built and hence probably not very large vessel. For the length of the keel, measured from the heel of the stern to the forward end of the stem hook, one may reckon four times that of the stern, measured from the keel line. For the Kuinre cog this would amount to a length of about 10 m. With the above estimate of the rake of the stern and stem, the overall length of the hull at gunwale level would have been about 13 m (fig. 13.6), while the greatest width of the bottom and at the gunwale may have been about 3 m and 5.5 m, respectively, if we reckon with the proportions of large cogs; in comparison with small cogs this would be about 3 and 4.5-5 m. At the stern, the gunwale sat about 2.35 m above the keel line, amidships this would have been about 2 m, and at the bow almost 3 m.

Fig. 13.7b. General view of the starboard side in the excavation at Kuinre. 1943.

The bottom and the sides on either side of the keel plank were as good as certainly built out of ten strakes, with planks presumably 3 to 4 cm thick and 30-35 to 40 cm wide. The uppermost strakes (J) ran on to the rear of the stern and ended against their counterparts across the mitred rear of the stern. For the second strakes from the top (I) this was also the case, but only for their upper half. The lower half of the plank fitted into the oblique rabbet. The bow was probably constructed in a similar fashion. The cog did not have a false stern, nor did it probably have a false stem. Eventually the stem may have had a short false stem or a plank to act as a fender protecting the hood ends of the uppermost two strakes. The vessel presumably did have an inwale along the inside or top of the gunwale and probably crossbeams that ran from side to side and were supported by gangways. The bulwark would have been kept in shape by standing knees fixed to the crossbeams and gangways. The vessel almost certainly had a foredeck and an afterdeck. The few frames visible in photo 13.7b show that the wrunghead scarfs, joining the floor timbers to the futtocks, were staggered. The photos show no bilge deals or a mast step. The cog was kept on course by a stern rudder, and - given its dimensions - was probably suitable for sailing inland waterways and the Zuiderzee.

13.4 Model

No scale model was made of the stern construction.

13.5 Inventory

In the fields around castles I and II many (mostly surface) finds have been collected, dating from the period of medieval occupation of the castles, the settlement, and the ship. Yet none of the finds can be directly related to this cog. A notable item, however, is the find of a so-called 'cog stick'[30] in the fill of moat 7 of the first castle.[31]

30. See the Nijkerk-II cog, section 6.5.3.
31. Van Doesburg & De Boer 2001, 54 fig. 29.

13.6 Cargo and merchandise

Naturally nothing can be said about any cargo or merchandise.

13.7 Ballast

No ballast was recorded.

13.8 Dating

Section 13.2.2 mentioned that the sintel clamps used in this ship were of type E, which was used between ca 1375 and 1475. Continued research however has shown that in fact we are dealing with a transitional form between types D and E, which most probably first appeared in the mid-14th century. Also the spacing of the sintel clamps and the presence of a sintel clamp-caulked seam on the outside of the stern (along the hood ends of the strakes) sooner point to a construction date of the cog (early) in the second half of the 14th century. So far, no dendrochronological analysis has been performed on the stern construction, which would be crucial for confirming this dating. The horizontal keel part of the hook probably offers the best possibilities for sampling. The double hole for the rudder suspension suggests repair and possibly replacement, while the findspot too indicates that we may be dealing with the remains of an old, discarded ship. It is not unlikely that such discarding occurred around 1400. The question whether it was a (pillaged?) vessel of the Lords of Kuinre, of the Frisian fleet or maybe one of the ships of the Count of Holland will forever remain unresolved.

13.9 Summary

At the very cradle (1943) of nautical archaeology in the Noordoostpolder (the former Zuiderzee area), a small, distorted vessel was excavated between the sites of the two successive castles of Kuinre, on the west bank or in a bend of the small river Kuinder which later silted up. At the time, dating was impossible owing to a lack of associated finds. In 1949 the complete stern construction which probably belonged to this vessel was recovered and preserved. If, as seems likely, the frames and keelson that were burnt in 1969 belonged to the same vessel, they too would have been stored in 1949. In World War II and shortly after, the newly drained *polder* yielded up a variety of ship's timbers in this vicinity, which were poorly documented. Given the hard times, this is not surprising; and such old timbers, once dried, would have made a welcome source of fuel. The dimensions of the stern and the two photos indicate that this was a not very heavily built vessel, the gunwale at the stern rising 2.35 m above the keel line. Comparison with other cog finds from the Zuiderzee area suggests that the ship would have been ca 13 m long from stern to stem and numbered ten (narrow) strakes on either side of the keel plank.

The structural rigidity in the upper part of the hull was probably largely ensured by gangways and crossbeams supported by them. The ship almost certainly had an afterdeck and a foredeck. The 'light' execution is also an indication for its use. It is more likely to have been a ship for transporting passengers or relatively light and not too bulky cargoes than one intended for heavy loads such as bricks (cf. the Marknesse cog, Chapter 12, and the Spakenburg-I cog, Chapter 14). This also suggests that it was a craft for inland waterways and the Zuiderzee. The civic seal of Kuinre (fig. 13.8) probably shows a small cog. Maybe this (discarded) shipwreck is one of the vessels with which the infamous Lords of Kuinre carried out their piratical raids across the Zuiderzee. Yet it may equally be a vessel of the Frisian fleet, one of the ships of the Count of Holland or a ship of any other provenance. The spacing and the type of the sintel clamps used in caulking the shell seams point to a construction date in the second half or the final quarter of the 14th century. Tree-ring analysis may confirm this date and tell us whether the timber (or the ship) came from the Netherlands or elsewhere.

Fig. 13.8 The civic seal of Kuinre. Presumably 1399; certainly used in 1407. After Ewe 1972, 145-88.

The Spakenburg-I cog[1]

14.1 Find conditions

14.1.1 Discovery and investigation

During the investigation into a clinker-built *waterschip*[2] (a type that was built between roughly mid-14th century to mid-16th century for fishing and transport of living fish, Section 27.4.) on 29th October 1975 on lot Nz 42 in the southern part of the province of Flevoland, it was reported that a second wreck (Nz 42-II) had been discovered during drainage at the back of the plot (figs 14.1, 14.2 en 14.3). The site of the new discovery was measured with prods, from which it could be deduced that wood was present over a length of about 11.5 m and a width of about 3.5 m. The top of the bottom of the ship lay at about 0.5 m below the surface. During the measuring and the further investigation of the drainage trench it also transpired that a load of bricks had been stored in the middle of the ship, some of which had been destroyed by the draining machine. No other finds were discovered on this occasion, which meant that a (global) dating of the ship was not possible. As further investigation was deemed necessary, the wreckage site was marked with poles. The Agricultural Department of the IJsselmeerpolders Development Authority was requested to reserve the wreck for that purpose, and was also asked that the location be spared further land treatment, as the parts of the ship were situated quite near the top soil.[3]

Fig. 14.1. Overview from the sternpost after the severely damaged ceiling planks (with the exception of the bilge ceiling and the mast step) had been removed.

1. Translated by Stefanie Hoss.
2. Pedersen, 1996.
3. The shipwreck lay 22 m from the axis of the lot ditch of lot Nz 42/43 and 40 m from the axis of the drainage ditch between the Eemmeerdijk and the lot. The vessel lay northeast-southwest, with its bow pointing southwest, about 0,8 km northeast of the current harbour entrance of Spakenburg. Topographical Map 32 B, coordinates 155.467/474.680, with the highest lying parts at ca 25 cm -MV/below the surface (ca 2.4 m -NAP). The former water depth at that point had been approx. 1.6 m, Hydrographic Map 1921, average low tide/mean low water.

Fig. 14.2. The find location of the Spakenburg-I cog.

In the autumn of 1978, the creation of recreation park Eemhof on lots Nz 42, Nz 43 and Nz 44 was announced. Several large ponds were to be constructed for the park and the excavated soil would be used to raise the areas between the ponds, on which the recreational homes were to be built. In view of this new use, the vessels had to be excavated as soon as possible. In addition to the aforementioned *waterschip* and the wreck discussed here, this included two other vessels, namely a second clinker-built *waterschip*[4] and a fourth vessel on lot Nz 43. All that was known about the latter wreck was that it seemed to be a small vessel, whose peservation - given that the top of the Pleistocene sand layer appeared quite high in the profile – seemed to be not very good. However, during the excavation of this wreck it emerged that this ship was a relatively complete, cog-like vessel.[5] In 1979, the permanent staff of the Department of Maritime Archaeology excavated the cog-like vessel and the two *waterschepen*. In view of this overloaded excavation

Fig. 14.3. Ordnance map of the Kingdom of the Netherlands, 1861 (scale 1:50.000), showing the location of the Spakenburg-I cog and Hulkestein Castle.

4. Nz 43/44 Zuidelijk Flevoland, dossier RCE.
5. Van de Moortel,1991; the Spakenburg-II cog, chapter 15.

schedule and the time pressure, it was decided that a group of young people, members of the Dutch Youth Association for the Study of History (N.J.B.G.) and some French guests would excavate the wreck with the bricks, supervised by H.R. Reinders[6] and K. Vlierman (fig. 14.4a and b). Quite soon after the start of the excavation, it became clear that in this case too, it was very likely a vessel with the characteristics of a cog. The excavation and field documentation were carried out between 30th July and 25th August 1979. A rectangular excavation trench was set up above the wreck to be excavated, with two profile dams left standing. These crossed over the width of the ship and sat at even intervals over the length of the wreck. The resulting smaller trenches were subsequently excavated using hand tools. The parts of the vessel lying separately above the wreck and the load of bricks were measured and drawn in situ (Plate 14.I A) and then removed. After the data for the soil profile had been recorded (fig. 14.5), the two dams were also removed.

Fig. 14.4a. Overview of the excavation work and the measuring and drawing in situ of the pile of bricks as seen from the stern.
b. The load of brick at midships, the afterbody and the tent camp of the N.J.B.G. youths.

6. The then head of the department of maritime archaeology.

Fig. 14.5. Reconstructed soil profile in relation to frame 21 and the overall position of the wreck in relation to the Pleistocene surface.

After this, the wreck was cleaned and photographed and the badly damaged, deformed and broken ceiling was drawn in situ, or rather, the original plank shapes were recorded (figs 14.6a, b and c and Plate 14.I B). With the exception of the mast step and the two bilge ceiling strakes, the fragments of the ceiling planks were then removed, and the wreck was cleaned further (figs 14.1 and 14.7). Photogrammetric images of this situation were taken on 8th August from a Royal Dutch Air Force helicopter by the Geometric Service in Delft, followed by a second series of images of the shell on 22th August (fig. 14.8). The locations of the finds discovered during cleaning are recorded in Plate 14.I.A. The numbers shown in the circles correspond with the inventory numbers in the find list. The finds are dealt with in section 14.5. A longitudinal section across the keel plank, the mast track and the stern and bow stems and a large number of cross-sections have been measured and drawn (Plates 14.I C and D)

14.1.2 Geology and stratigraphy

The Pleistocene surface and covering deposits

The soil profile around and above the wreck was barely affected; only the upper approx. 25 cm had been slightly ploughed. In this southernmost part of Flevoland, the soil structure is very irregular due to the strongly varying height of the Pleistocene sand layer (0.25->1.5 m below ground level; see also section 6.1.2). On top of this sand layer, in which the A and B horizons are still largely present[7], remnants of the Holocene peat layer are still present in the deeper parts and in the vicinity of the cog up to approx. 0.8 m below the surface. The wreck was located on the eastern side of a depression in the Pleistocene surface that is more than 1 km wide (depth at the wreck up to ca 1.5 m below ground level), which rapidly (within 1 km) deepened to > 2 m towards the north. The site is located approximately 4.5 km east-south-east from the mouth of the river Eem. The depression was filled with Almere deposits (Ala) above the peat, with a layer of Zuiderzee/IJsselmeer clay (fig. 14.5) on top of that.[8] Up to 1.25 m below the the surface, no peat was found in the vicinity of the wreck. It therefore seems quite possible that no peat had formed in at least parts of the approx. 1.5 m deep depression, or that the peat had already disappeared before the ship sank here.

It is very likely that the depression in the Pleistocene surface indicates the access route into the harbour of Spakenburg/Bunschoten. A number of higher peaks of the Pleistocene sand layer which are up to approx. 0.5 m below the surface high are located approximately 350 m to the east and 1.3 km west of the site, lined up approx. in a north-southerly direction. It is possible that these peaks formed the shape of the coastline of the clay-on-peat landscape (outside of the dikes?). The inlet or peat channel in this landscape that almost certainly continued to the Spakenburg/Bunschoten harbour was probably not very wide at the location where the wreck was found. Although the soil profile does not provide any indication for this, it seems quite conceivable that dikes were situated along this peat channel. During heavy storms, holes could have been made in these dikes, which one tried to close. Later, the sea may yet have swallowed up these dikes, together with the land behind them. The Almere-layer above and around the wreck could not be classified any further and almost certainly shows several hiatuses. The bottom profile can therefore provide no other information than that the ship sank before the salinization of the Zuiderzee (after about 1600).

7. Koning & Ente 1986, explanation for page III of the soil code and profile map of Zuidelijk Flevoland.
8. Koning & Ente 1986, appendices I, II and map III.

Fig. 14.6. The severely affected and deformed ceiling, a. Seen from aft to the forebody; b. The part in the afterbody, seen from starboard; c. The part in the forebody part seen from starboard.

Fig. 14.7. The cleaned area with the floor timbers, the bilge ceiling strakes and the mast step, seen from starboard aft.

Fig. 14.8. Photogrammetric image taken by the Geometrical Department from Delft, from a Royal Dutch Air Force helicopter. The wreck with frames, bilge ceiling strakes and mast step.

14.1.3 Medieval settlement

The location of the wreck is only 800 m from the current harbour of Spakenburg. However, the most important medieval town in the vicinity was Bunschoten, a little to the south. As a result of a dispute in the Prince-Bishopric of Utrecht between two bishops appointed at the same time at the beginning of the 15th century, Bunschoten was set on fire in 1428 and the walls, gates and towers were demolished.[9] At the time, Bunschoten was situated in a strategically not insignificant location close to the mouth of the River Eem, on navigable waters with a harbour (Spakenburg), and only a short distance from the border with Guelders as well as only 3.5 km from Hulkestein Castle, built by the Duke of Guelders in 1427.[10]

The cog was found in an extension of the water channel that ran from the Zuiderzee to Bunschoten, which leads to the conclusion that it probably sailed on these waters during the period that Bunschoten was a city (see section 14.8). In the Late Middle Ages (and later), the southern coast of the Zuiderzee was regularly attacked by storms and the water cleared away pieces of land. Old Naarden, for example, which was destroyed and abandoned around 1350-1355, was situated in the present peripheral lake northeast of Naarden[11], while it is also known that the dikes in the coastal area east of the Spakenburg-I cog were regularly moved inland.[12] It seems very likely, that this could also have been the case north of Spakenburg and that the wreck might have had something to do with it (see below).

14.1.4 Condition and wreck formation

Condition

The clay package above and next to the wreck was torn during the hot, dry summers of 1975 and 1976. As a

9. Van der Aa 1840, 822-823.
10. See Nijkerk II cog, § 6.3.5.
11. Bakker 2004, 6.
12. See Nijkerk II cog § 6.1.3.

result, the parts of the ship which were found separately and located the highest, namely the upper parts of the frame and the ceiling planks, had been severely damaged, dried out, shrunk and torn. Despite its location at about 100 m from the Eemmeerdijk, there was no seepage water present high in the profile, because the Pleistocene surface is located at about 1.5 m below ground level. The ship had sunk almost horizontally into the Almere clay layer. The keel plank and stern- and stem hook showed slight 'hogging', so that the bottom of the aft stern hook was at approx. 1.2 m below the surface, the stem hook at approx. 1 m, while the depth of the keel plank was approx. 0.95 m amidships.

As a result of the horizontal sinking of the vessel, the shape of the underwater hull and the presence of (hard) sandy layers in the clay package, the vessel had not sunk to the Pleistocene sand, but 'floated' in the Almere-clay layer. The wreck, which had slightly sagged in width, consisted of the entire bottom with the keel plank and the stern- and stem hooks, the lower parts of the stern- and stemposts, the false sternpost, as well as the lower part of the false stempost and parts of the sides. The floor- and V-shaped timbers were present on the bottem, and some of the frames still had their futtocks. In addition, the complete ceiling, including the bilge ceiling strakes, the mast step and some remnants of the ceiling in the sides were also still present. Laying on top of the forebody was the broken off front part of the starboard shell with a separate, complete futtock and some other parts of the shell strakes of both the starboard- and port sides. Parts of an aft deck or aft cuddy were discovered above the afterbody of the ship as well as a complete starboard futtock (see Plate 14.I A).

Formation of the wreck

It has not become clear whether the vessel sank during a storm, for example, or whether, given the relatively large number of repairs and the almost non-existence of other finds, it had become old and redundant and was abandoned in an inlet adjacent to the waterway. Noteworthy is the broken off, large fluke of an anchor stuck into the starboard bottom approx. amidships from below and the load of unused bricks (Sections 14.5.1 and 14.6). These are the only clues that suggest that the ship was used as a sort of caisson to help close a hole in a dike. This can be done by hanging an anchor from another ship under the cog and then hauling it up. The anchor was forced through the hull of the ship, which filled with water and sank. The anchor may have broken during this action. The sea may have subsequently washed away the dike[13] and the land, causing the wreck to partially protrude above the new seabed. Ships sailing over the wreck, fishing activities and floating ice will have caused most of these protruding parts of the ship to disintegrate and drift away over time. However, another scenario is that the location was chosen as anchorage long after the sinking of the vessel discussed here and (part of) an anchor was lost during normal anchoring. Such an action would also have contributed to the wreck formation process (figs 14.9a and b and Plate 14.I A-14).

Fig. 14.9a. The broken off anchor blade, which stuck to the underside of the bottom at midships starboard.

Fig. 14.9b. The same view after the frames had been removed.

13. For instance during the 1552 flood, see Van der Aa 1840, 823.

14.2 The ship

During the analysis of the field data, the results of the reconstructed base wreck (see Nijkerk-II, Chapter 6) were used. In addition, the other, smaller cogs were also examined in this study.

14.2.1 Documentation and reconstruction

Documentation
The wreckage was measured by hand in situ and drawn to a scale of 1:20. Top views were made of the parts found separately and of the load of bricks, as well as the distribution of the finds, of the ceiling and bilge ceiling and a detailed overview of the mountings of the first mast step. Furthermore, a longitudinal section was made over the keel plank and the stern and stem, fourteen cross sections over the frames and one over the load of bricks. In addition, an overview of the shell was drawn, with on it a sketch of the structure of the wood, in connection with the clearly visible 'brother and sister' pattern. In addition, sketches were made of several frames to indicate from which part of a tree they had been taken (Section 14.2.2).

The Geometric Department made stereophotogrammetric recordings in two phases, while the various excavation phases were recorded in slides and black-and-white photographs. In 2007 the constructions of the stern and stem hooks, which are standing in the open depot of the RCE-Lelystad, were drawn in detail on a scale of 1:10 for this research report (figs 14.10a and b and 14.11a and b). The shell strakes are numbered from the keel plank to the inwale and from aft to fore (strakes A to G). The frames are numbered 0 to 29 (from the foot of the stern to the V-shaped floor timber against the stem and -1 to -3 (from the lower to the upper floor timbers against the stern).

Reconstruction
The bottom with keel plank and stern- and stem hooks, the remnants of planks from the sides and the frames found in construction together provided reliable information for the reconstruction of the underwater hull and the lower parts of the sides. Thanks to a happy coincidence, the upper part of the starboard side had broken off at the bow. It was lying above the forebody of the ship, together with a complete, clinker-built futtock (24). Together with the complete 2 futtock of the port side of the afterbody, which had also been found separately, this provided very reliable information about the height of the sides of the fore- and afterbody of the vessel and the total number and width of the strakes on both sides of the keel plank.

The remains of the aft deck, together with the deck beam found in the afterbody, provided reliable information about its construction, its position and its height in relation to the keel plank. Lathes on the ceiling, square holes in the 22 frame and grooves in the mast step are indications of a construction around and mainly in front of the mast, while wear marks in the top of the keel plank between the 22 and 23 frames make it likely that the tube of a pump was located here during the final phase of the ship's life.

Comparison with some vessels made it possible to make a realistic reconstruction of the decking, the transverse connections and their construction and of the inwales of the ship (Plates 14.II and 14.III).[14] Taking into account differences in dimensions due to shrinkage, drought, expansion of the wood due to moisture, deformations plus the possibility of measurement errors and the rounding off of measurements when drawing to scale 1:20, the plank lines of the top and side views were brought into line and can be assumed to vary no more than a few centimetres, if at all from the original sizes. All main sizes mentioned are rounded off to 10 cm.

14.2.2 Construction and layout of the hull

Keel plank
The overall length of the keel plank, together with the keel parts of the stern- and stem hooks (from the rear of the stern hook to the front of the false stem at the bow), is 13 m. The keel plank itself is 10 m long and approximately 10 cm thick. It is 20 cm wide aft and 23 cm wide fore, while it is 30 cm amidships (Plate 14. II B). The keel plank is almost rectangular in cross-section and has rounded (worn?) corners at the bottom. The keel plank has 20 cm long sloping flat scarfs as connection to the stern- and stem hooks, which run from front to back and from top to bottom. Both ends are attached to the hooks with nails. A row of five nails going from the inside out and two rows of two nails from below are located on the sloping flat scarf that is the connection to the stern hook. Two rows of three nails have been used for this purpose on the scarf connecting the keel plank with the stem hook. The nails are slightly convex and have heads that were forged into four planes (diameter 15 and 25 mm). None of these nails is protruding through the keel plank or stern and stem hooks.

At 6 m from the rear of the stern hook, a treenail with a diameter of 2.5 cm was placed just off the centre of the keel plank. Two more treenails were placed at 6.5 m, one with the same diameter, the other measuring 4 cm. The thinner of the two is located 6.5 m from the rear of the aft

14. The comparisons used are the Marknesse cog, Chapter 12 and Hocker & Vlierman 1996; also see the Almere-I cog, chapter 16.

stern hook, the wider of the two treenails is located 6.5 m from the front of the false stem fore (Plate 14. II B). All three treenails are located between the frames. The two thin treenails were probably used during the construction of the vessel to determine the approximate centre, but must in any case and almost certainly indicate the location of the temporary frames during construction. The thick treenail was more likely a plug for the hole where, during cleaning at the shipyard (rinsing out cargo residues such as brick grit), the water was drained out of the ship. The sides of the keel plank do not have a rabbet for attaching the garboards, these A strakes were placed cold against them and were not attached to the keel plank. Between the frames 21 and 22, 22 and 23 and between 23 and 24, wear marks could be seen in the top of the keel plank. These are indications that (at one point) the tube of a pump had been placed here.

The stern hook

The length of the stern hook (from the back of the heel, which acted as a protection for the front of the rudder) to the front of the sloping flat scarf connection to the keel plank is 1.35 m (figs 14.10a and b). The end of the sloping flat scarf is 20 cm wide and measures 7 cm at the back. The keel section is slightly trapezoidal in cross-section and has the same thickness or height (10 cm) at the point of connection with the keel plank. Some 50 cm from the flat scarf, it suddenly increases in steps to a height of 17 cm. This line then continues for almost 60 cm to the rear and then goes upwards and backwards in a chamfered line towards the flat scarf with the stern. The highest point protrudes 70 cm (measured at right angles) above the keel line. The bow also has a trapezoidal cross section, is at the widest on the inside and measures 8 cm at the back. The sides of the keel part of the hook are cut into the wider rear part up to 2 cm beyond the beginning of the elevation, in order to have enough wood at this critical point for a watertight connection of the garboards, which were then nailed into a rabbet, twisted to almost vertical. The rear line of the hook/stern beam sits at an angle of 119° to the keel line. The rear end appears to be slightly worn at the bottom over a length of about 20 cm.

Fig. 14.10a. The stern hook with the fragments of the sternpost and the false stern with the iron gudgeon.

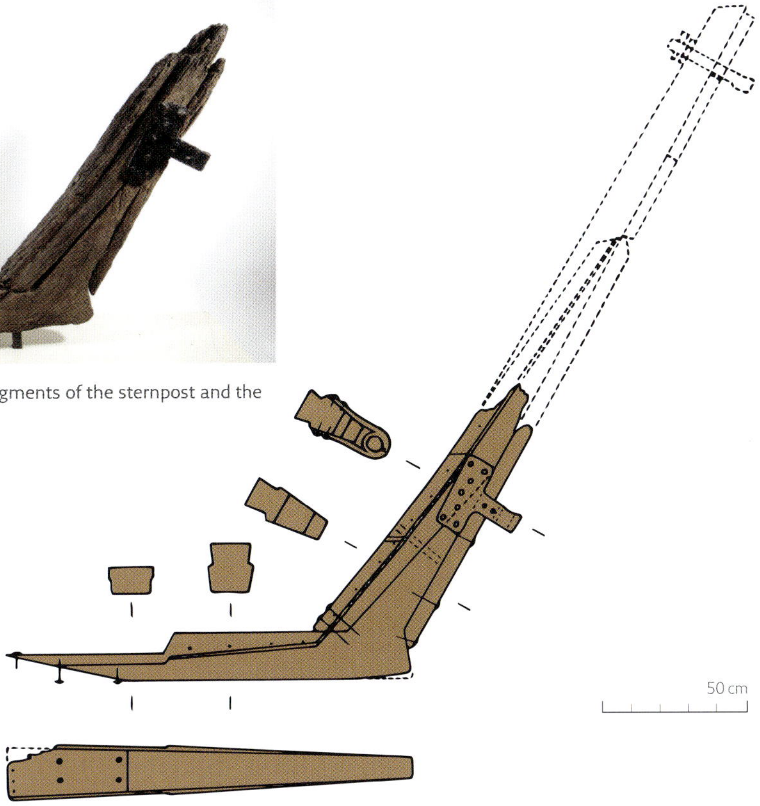

Fig. 14.10b. The sternpost knee construction with the gudgeon.

Sternpost

The inside of the sternpost was placed on and against the stern hook. It is attached to the stern part of the hook with a treenail (diameter 2.5 cm) and with three nails through the lower part plus one through the upper point of the stern hook. It is likely that only the A strakes or garboards were nailed to the front of the beam in a rabbet that was cut out stepped and slightly bevelled and is approximately 3 cm deep. Before the ends of the other strakes had been fixed to them, the rabbets had been smooth towards the top. The iron gudgeon (the lowest suspension point of the rudder) was located with the lowest, aftmost point at 50 cm above the keel line and was positioned at right angles to the rear line of the stern beam. The flat, T-shaped 'wings' are approx. 7 mm thick and gradually become thicker (10 mm) towards the 60 mm high hanging ring. The almost rectangular plates of the wings (260x100 mm) are fixed with seven nails on both sides onto the sternpost and stern hook. The wings were originally also attached to the false stern with two bolts or nails. A 15 mm thick ring was forged inside the curved end of the gudgeon (with a thickness of 10 mm). The ring had an opening of 55 mm diameter, which formed the hole for the lowest rudder pin. Only 1.1 m was recovered from the sternpost, which originally had of course been longer and almost certainly consisted of a single piece of wood.

False stern

Above the heel of the stern hook, a part of the false stern, about 85 cm long, was preserved. It was attached to the back of the hook with three nails under the gudgeon. The beam is slightly trapezoidal in diameter (7 and 8 cm wide), only 5 cm thick and only partially fills the space between the stern hook/-post and the gudgeon's ring. There are holes in the wings of the gudgeon between the ring in which the rudder was suspended and the back of the false stem (diameter 14 mm), through which the gudgeon had originally been attached to the false stem with two nails or a bolt. However, in the situation found during the excavation, these holes were located behind the false stem. We can conclude that the replaced false stern was a clumsy repair, which was too thin. The vessel must have originally had an false stern with a thickness of 7-8 cm.

Rudder

The iron gudgeon on the stern indicates that the vessel had a stern rudder. Neither the rudder nor the upper suspension point was found during the excavation. Plate 14.II A shows what the rudder and the suspension most likely looked like.

Stem hook

The stem hook was also completely preserved (figs 14.11a and b). The length of the sloping flat scarf with the keel plank to the front is 1,8 m. The scarf had a width of 23 cm, and on the front at the bottom 8 cm. The keel section is 10 cm thick, but – just like the aft stern hook – some 50 cm from the flat scarf, it suddenly increases in steps to a height of 17 cm. From this point onwards, the line then runs bevelled upwards and forwards over 25 cm, where the flat scarf with the stem starts. The (broken off) top, measured at right angles to the keel line, had a height of 77 cm above it. The line from the front of the hook (and from the missing sternpost) sits at an angle of 122° to the keel line. The keel section is increasingly trapezoidal in cross-section towards the front. The attachment of the garboards is similar to that on the stern hook. In the sides, 3 cm deep rabbets were cut out on the inside of the ship to attach the ends of the garboards. These rabbets ran through into the stempost, which was not found. The stem hook is attached to the keel plank with two rows of three nails: one row from the inside to the outside, the other the other way round. The nails were not driven completely through the wood. On the scarf, the stempost was attached to the hook with a treenail (ø 3 cm). It was no longer possible to determine whether this treenail had passed (blind?) through to the false stem, but it is unlikely.

False stem

Only the lower 85 cm are left of the false stem. On the ship's side, the piece fits onto the front of the trapezium-shaped stem hook. It is 4 cm wide on the front and the highest part still preserved measures 2.5 cm (due to shrinkage). The thickness at the bottom is more than 16 cm, while the thickness at the highest preserved part seems to be about 15 cm. This must mean that the false stem gradually becomes thinner towards the top. The underside of the stem gradually increases towards the front, perhaps it was made this way or perhaps this is the result of wear and tear. The false stem was attached to the lower part of the stem hook with two nails, nailed in at an angle. An iron plate each was attached to both sides, both of which were 230 mm long, about 100 mm wide at the ends and 40 mm wide in the middle. They are each fastened with six nails to the sides of the stem hook and the false stem. It is not known how the attachment was further up. It is very likely that the stempost and the false stem were joined together with a thick iron bolt at the top (Plate 14.III A). Such a connection has been found in other cases.

Bottom of the ship and sides

The A strakes or garboards consist of two planks, the other strakes probably were all originally made from three pieces. It seems that the C- and D strakes on starboard had four planks, but this probably was a repair (Section *repairs*). The planks of each strake are connected with sloping flat scarfs of 15-20 cm in length and generally fastened with two rows of five nails (sometimes only three or four nails); one row from the inside to the outside, the other row the other way around. The points of the nail had been bent over twice and hammered back into the wood.[15] The plank ends were about 1 cm thick and protruded out of the shell at the scarfs, both on the inside and on the outside. The surface was still almost completely present, with more and more missing from the sides upwards. The parts that were found are indicated in the plates 14.II A and B.

Thanks to the presence of the broken off piece of the starboard shell found at the bow, and a futtock associated with it, as well as a futtock from the port side of the afterbody found separately, it was possible to establish that the vessel had seven (A to G) strakes from the keel plank to the gunwale. It could also be determined that the top three strakes did not end in a rabbet of the stempost, but rather stuck out along it. This was probably the case in the back as well. These plank ends were protected at the front by the false stem. The bottom gradually merges into the sides, a 'real' turn of the bilge is not easy to indicate and can only be seen amidships because of the somewhat angular placement of the C- and D strakes there. This transition gets less and less clear towards the afterbody and the forebody and the hull has a rounded shape in cross-section here, which turns into a slightly S-shaped form towards the stern and stem (Plate 14.III A). The top of the garboards and the B-courses form a more or less flat line rising to the sides, while the next two strakes were placed at a stepped angle against them. At amidships, the vessel has more or less straight, slightly sloping sides with an angle of approximately 112°. The top line of the keel plank is more or less equal to the top line of the garboards.

Fig. 14.11a. The restored stempost construction.

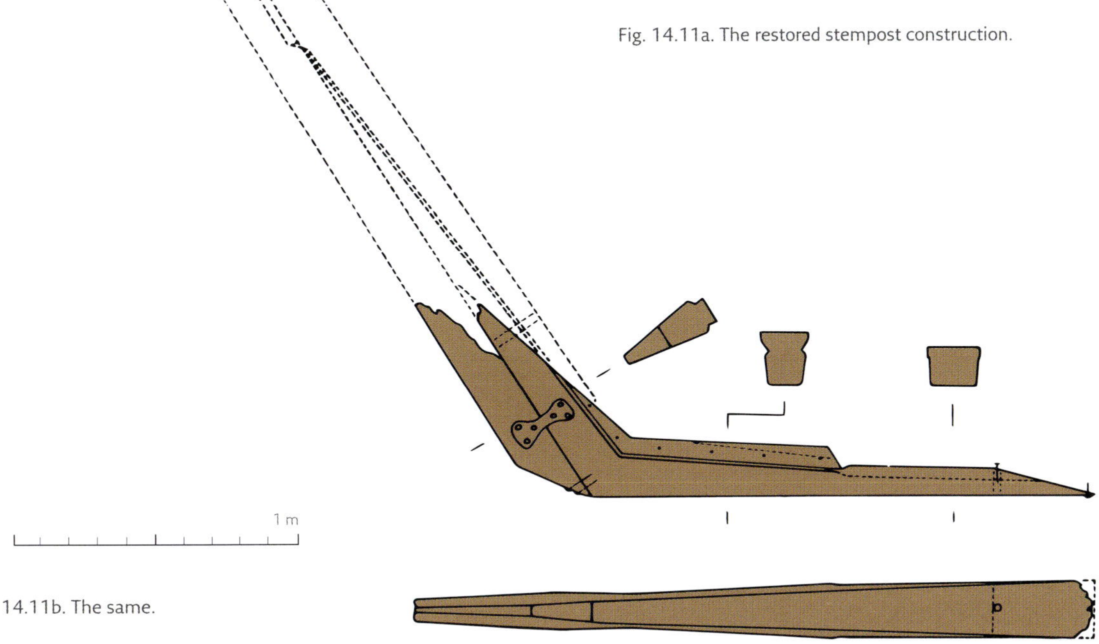

Fig. 14.11b. The same.

15. See also the Nijkerk-II cog, fig. 6.42.

Bottom of the ship

The bottom on either side of the keel plank is composed of three strakes (A, B and C). The A- and B strakes are in the carvel-built part (as mentioned before), in a line going up to the bilge, the C strakes were placed at an angle to them. The bottom has no dead flat in the length and is clinker-built towards aft and fore (from V-shaped floor timbers 5 and 20 respectively). Planks of approx. 50 cm wide and 4 to 5 cm thick were used for the A- and B strakes, which gradually become narrower towards the aft and forebody (approx. 45 and 40 cm respectively at the cross-sections over V-shaped floor timbers 1 and 25). The C strakes measure approx. 45 cm amidships, and also become narrower and are about 18 cm wide at the cross sections mentioned above. The scarfs of the strakes were staggered and there is a very visible 'brother and sister' pattern (Section 14.2.2).

The carvel-built planks are not connected to each other, but remain in place through the connection with treenails on the frames. The ends of the garboards are nailed into the rabbets of the hooks, the B- and C strakes with the ends in the rabbets of the stern- and stemposts, while the clinker-built parts of the three strakes are nailed together every 12-18 cm in the manner described above. The C-strakes are connected at the top to the D strakes completely clinker-built. The A strakes are composed of a plank of about 6 m in length (aft) and 6.8 m in length (fore). The three planks of the B strakes are 3.6, 4.8 and 5.3 m long from aft respectively, those of the C strakes 4.1, 5.2 and 5 m. This is different on the starboard side. There, the B- and C strakes are shorter in the middle (4.3 and 5.4 m) and have been extended by sections of 1.9 and 2.1 m.[16] The seams of the scarfs are all slanted from top to bottom and from front to back; a logical construction, based on the obvious direction in which the vessel is sailing. From the places where the B- and C strakes turn into the clinker-built afterbody and forebody of the ship, the planks are 3-6 cm wider because of the overlap or land. Based on the depth of the rabbets in the stem and stern hooks and the sternpost (approx. 3 cm), the planks become slightly thinner towards the stern- and stemposts and it is almost certain that they end up almost carvel-built in the rabbets. No bilge laths were found.

Starboard side

From the starboard side only some parts of the lowest (D-) strake were connected to the bottom. A broken off piece of the shell had been found above the forebody, which had originally been joined against the bow. Next to it was a fragment of more than 5 m long that consisted of parts of the F- and G strakes still connected to each other (Plate 14.II A and B). Thanks to the front part of the shell and the two separately found futtocks of 2 (port side) and frame 24, the width of the various strakes could be determined on these cross sections and thus also the height of the shell at that location. The D strake was 35-38 cm wide, while the E strake was quite a bit wider > 50 cm. The upper two strakes (F and G) were about 40 cm wide, and had probably been a little narrower towards the stern and stem. The staggered pattern of the scarfs probably continued upwards. All the planks of the shell were attached to each other from back to front clinker-fashion with nails that were applied every 12-18 cm and in the same manner as with the scarfs. The dimensions of the nails used are also the same as for the scarfs. The ends of the D strakes are the only ones that were nailed into the rabbets in the stern- and stemposts, just like the bottom strakes. The top three strakes protruded past the stern- and stemposts and were attached to the most likely triangularly chamfered back of the sternpost and to the sides of the stempost, respectively. The overlap of the planks attached to each other in clinker fashion was 4-6 cm.

Port side

From the port side, only parts of the D strakes remained preserved, the forebody was found separately above the foreship, and some parts of the E strakes (Plates 14.II A and B).

Sintel clamped moss caulking

The planks of the shell attached to each other clinker fashion were made watertight on the inside by means of sintel clamped moss caulking. The seams of the carvel-built bottom were only caulked from the outside. The seams between the C- and D strakes were the first to be completely caulked from the inside. The connecting seam between the C- and D strakes had also been caulked from the outside in some places, such as on the starboard side over a length of approx. 3 m (amidships and in the rear half of it) and in the forebody on the port side over a length of 4.4 m and 0.5 m respectively. This is almost certainly all sintel clamped moss caulking, which was applied as repair, not during construction. The clamping of the seams mentioned first was carried out before the frames were placed in the ship, while the seams of the bottom were most likely caulked shortly before the launch. The seams between the stern- and stem hooks and the stern- and stemposts and the shell were also caulked from the inside, insofar this could still be determined. This is not the case at the ends of the plank scarfs in the strakes, but there was a thin layer of moss ('paper') between the two parts lying on top of each other.

On the outside, the seams of the planks of the sides attached to each other in clinker fashion were also closed here and there with sintel clamped moss caulking. As mentioned above, these are repairs. As far as observed, there was no sintel clamped moss caulking used with the repair planks used

16. For an explanation see the section *repairs*.

from the outside (Plate 14.II B), but only moss. Iron sintel clamps of type D2/E were used to fix the moss and the lathes put on top of it; this type was in use during the last quarter of the 14th and the first quarter of the 15th century.[17] The oval middle part of the sintel clamps (figs 14.20-46 and -47 and 14.21c) is about 40-50 mm long. They generally were used every 7.5 cm (sometimes every 6 cm) centre to centre. During construction, approx. 180 m of plank seams were caulked, which corresponds to approx. 2500 sintel clamps for the entire hull. This does not include the approx. 20 or more metres of external repairs. The moss lathes were made of oak approx. 1.5 cm wide and 2 to 3 mm thick and more or less oval in diameter, with sides of about 1 mm thick.

Identification and pollen analysis

The caulking consists of a single species of moss; peat moss (*Sphagnum*). No further subdivision of the genus was possible. Of the three samples examined, one originates from a bottom seam (47) and one between a repair board and the shell (49). The third sample (46) consisted of a combination of *Sphagnum* with other plants.[18] Peat moss was usually collected in wet, muddy or peaty environments, e.g. a heath, marsh, ditch or places where reed grew.[19] *Sphagnum* grows abundantly, both in upland bogs as well as in nutrient-poor (oligotrophic) parts of the peat in the lowlands of the Netherlands.

It seems logical to assume that for both practical and economic reasons the moss was collected in an area that was not far from the construction site of the ship/shipyard. Pollen analysis of the last sample (*Sphagnum* and other plants) from the seam between the stem hook and the garboards shows that the sample contains little pollen and a lot of charcoal; 42% trees, 58% herbs. Alder 16%, oak 9%, hazel 6%, birch 5%, hornbeam 2%, beech 1%, willow 1%, elm 0.5% and pine 2%. Heather 18%, grasses 11%, grasses of the *Cyperus genus* 3%, cereals 17% (buckwheat 0.3%), field weeds 2% (lamb's tongue and sorrel). Local: low percentages of wood fern and bracken and 15% of peat moss. Indeterminates: 1.7%. This means that the moss was collected in an area with nutrient-poor to nutrient-rich conditions; the beech and hornbeam will have been on richer ground. A cultivated area bordering a bog seems quite possible. It is possible that the cultivation was with buckwheat.[20] But the tiny percentage of buckwheat suggests that this is more likely to be more like a 'stray' crop.

Straight and V-shaped floor timbers and futtocks

The bottom and sides were kept in shape with frames (fig. 14.12). The ship originally had 33 frames; 0-29 and -1 to -3 (Plates 14.III A and B). In the more or less flat part of the bottom of the vessel, these frames were made up of asymmetrical floor timbers, extended on one or both sides with a futtock. Towards aft and fore, these are asymmetrical V-shaped floor timbers, most of which also form the frames up to the top of one of the sides, while on the other side this was done with a futtock. Some of the upright sections of the floor timbers and futtocks probably did not extend beyond the top of the E strake, about which more information can be found in section 14.2.2. The frames 0, -1, -2, 27, 28 and 29 probably stopped under an inwale. The frames 1, 2, 3, 23, 24 and 25 may have been recessed into the bottom of the inwale. The frames -2 and -3 and 26, 28 and 29 probably only consisted of a V-shaped floor timber. The scarfs between the two kinds of floor timbers are relatively short and are chamfered slightly upwards towards the shell.

Frame 22 deviates from this; the V-shaped floor timber has a chamfered flat scarf on both sides, above the C and D-strakes respectively, which was fixed from top to bot-

Fig. 14.12. The frames and floor timbers (with remains of the futtocks) in the forebody, seen from starboard midships.

Fig. 14.13. The limber holes in the floor timbers in the afterbody.

17. Vlierman 1996a, 79.
18. Cappers *et al.* 2000, 578-579 en 588 nr. 27a, b and c.
19. Cappers *et al.* 2000, 584.
20. Mook-Kamps 1997, sample 10, see also appendix.

tom at the back instead of at the front of the frames. Both futtocks were thus fixed behind it. This is a very unusual construction, which suggests that these pieces were replaced, so we are looking at a repair.[21] The scarfs between the two kinds of floor timbers and the futtocks are staggered in height in relation to the frames. They are located above the A-, B-, C- and D strakes (Plates 14.III B and C). The frames are 16-25 cm wide and generally 14 cm thick or high, some a bit lower, while the V-shaped floor timbers in the foreship are higher (23 and 25 cm). As far as could be determined, the futtocks turn less broad and thicker in an upward direction. A small board was nailed above the keel plank on top of some V-shaped floor timbers to support the ceiling planks. Both kinds of floor timbers had drain holes on the underside, which were in the shape of an inverted U, 5-6 cm high and 10-12 cm wide. There generally were two or three such holes in each floor timber (one above the keel plank and one above each A- or B strake), and sometimes two (Plate 14.III C and fig. 14.13). The aftmost and foremost floor timbers had limber holes on the starboard or port side of the hooks and stern- and stemposts, sometimes on both sides; these limberholes are generally triangular in shape.

Mast step(s) and the construction for a removable mast

A mast step was installed above the keel board (frame 16-21; Plates 14.I B and 14.III A, B and C and figs 14.14a, b and c). It consisted of a block of wood of 2.25 m length with a maximum width of 70 cm at the front and 40 cm at the back. At 85 cm from the back, the sides stepped inwards 4 cm wide as seen from the front. At the back, the thickness was 3.5 cm and at 20 cm from the front 15 cm, while from this point to the front, the block was chamfered down to a thickness of 3.5 cm. From the point where the piece was stepped on both sides to the front, the sides were also chamfered. Above and in front of frame 19, there was a square hole of 28x28 cm and 6 cm deep cut into the top; the hole for the foot of the mast. There was no drain hole drilled in the hole for draining off water. There was also no crosscut in the bottom, something found quite often in Flevoland shipwrecks. [22] However, a small cross was made on the top of the gudgeon's ring (Section 14.5.1 and fig. 14.20). The exact meaning of these crosses is not known. It could be a dedication and/ or a protection for the ship from disaster. In this context, a coin placed under the base of the mast is also often mentioned. Among the hundreds of wrecks excavated in the IJsselmeer region, no coin was ever found in the hole for the mast. Whether or not the application of such crosses indicates a regional origin cannot be said either.

Grooves that are about 3 cm wide and deep were cut in the block from the back of the hole for the mast towards the front. They seem to lie a little further apart at the front than at the mast (35 and 28 cm). The mast step was fixed to the floor timbers 16 to 20 with two treenails (diameter 3 cm) each. Slats with a width of 3 cm and rounded on the outside were nailed to the ceiling in front of the mast step in line with the grooves. They continue the line of the grooves in the mast step as it were, until the end of the ceiling at the beginning of frame 22. Before the ceiling, two square holes of 6 x 6 cm and 4.5 x 5 cm respectively were cut into the same frame. The grooves, slats and cut-in holes are clear indications of a more or less rectangular formwork around and in front of the mast, of which the planks were nailed at the front against poles placed vertically in the cut-in holes (Plates 14.I B and 14.III A and B). This type of formwork is customary for masts that could be struck and is intended to prevent cargo from ending up here, which would result in the mast being unable to be struck at any moment. It is very likely that the formwork continued upwards up to the crossbeams.

After removing the mast step and cleaning the frames underneath, it appeared that a 34 cm wide and 2 cm deep section had been cut out at the mast hole in the mast step in the top of frame 19. The hole had slightly chamfered sides and was not exactly above the keel plank, but slightly towards the port side. Frame 20 also had two square holes, which corresponded in terms of location and dimensions and also were in line with the other two in frame 22. In addition to the treenails for connecting the frames to the bottom planks and the treenails for securing the mast track, other treenails were identified in the frames 16 to 20. One in frame 16, three in 17, five in 18, two in 19 and four in 20. This indicates that the vessel must previously have had a different mast step. The number and position of the pins suggests that the block was approximately 1.8 m long and at least 50 cm wide while, judging from the one pin on this side, the rear section probably was narrower. It is possible that the hole for the foot of the mast ran through this mast step and that the mast was in the recess in the top of frame 19 (Plate 14.II C). The two square cut holes in 20 suggest that the formwork for lowering the mast may initially have started before the mast.

Ceiling or loading floor

The vessel has had a completely closed ceiling or loading floor. Between the two 5 cm thick and approx. 35 cm wide bilge ceiling strakes, which gradually narrows towards the rear to approx. 20 cm towards, a mosaic of planks and boards existed which together formed the last

21. Also see the section on the Nijkerk-II cog, section 6.3.2.
22. Also see the section on the Nijkerk-II cog, fig. 6.26.

Fig. 14.14a. The mast step with the hole for the mast heel.

Fig. 14.14b. The mast step after the ceiling had been removed, seen from starboard.

Fig. 14.14c. The mast step with the hole for the mast heel and the grooves for the formwork, seen from the fore portside.

ceiling (Plates 14, I A and B and Plate 14, III, B and C). These planks and boards varied in thickness, 2, 2.5, 3 and 3.5 cm. It was abundantly clear that many of these boards were structural elements that were installed during repairs which, judging from the differences in the quality and conservation status of the wood, were certainly not all carried out at the same time. The planks were nailed to the floor timbers in more or less random places. Some planks were laid over each other by means of a chamfered side or had a scarf of a few centimetres in the longitudinal direction. It seems that one (sometimes) tried to make the seams as (water)tight as possible. The condition of most of the ceiling planks turned out to be very bad during the excavation, as mentioned before (fig. 10.6). The loading floor started a few centimetres from the rear of frame 4 and ended a few centimetres in front of frame 22. The thicker mast step, protruding above the floor, was completely enclosed by the planks. The bilge ceiling strakes were attached to each frame and the hull plank underneath it with two wooden pins (ø 2.5 cm), in some rare cases with a single one. In addition to being part of the ceiling, the boards also formed an essential part of the ship's longitudinal bracing due to the presence of a large number of scarfs in the frames. Four lathes were nailed to the ceiling planks in front of the mast step, which held the planks of the mast's formwork in place. At the back, some of the planks were slightly shorter. The spaces between these ceiling planks and the bulkhead that may have been present there were partially filled with short pieces of round wood that were nailed to the floor timber.

Ceiling/inwale

During the excavation, some 3 cm thick plank fragments were also found above the bilge ceiling strakes. The vessel therefore not only had a closed ceiling, but was also equipped with a closed bilge ceiling in the sides up to a certain height. This probably consisted of two strakes, of which the lower ran, like the ceiling planks, from frame 4 to 22, while the upper one most likely ran from frame 2 to 25. Along the top of the sides, the vessel undoubtedly must have had a plank on the inside, or a slightly thicker inwale, perhaps a thicker board. This could be deduced from the shape of the heads of the futtocks of 2 and 24. However, it seems very possible or perhaps even probable that instead of planks from (approximately) these frames towards the stern and stem, inboard boards had been fitted, in which - analogous to the cog of Marknesse[23] - the frame heads were held (Plates 14.III A, B and C).

Transverse connections and sail beam (with knees) and gangways

The vessel has undoubtedly had crossbeams for the cross bracing and for the support of the decks. With the exception of a deck beam under the aft deck (near frame 0), no transverse- or crossbeams were found. It is very likely that the ship had frames in a number of places that ended straight at the top of the E-strakes (frames 4, 11, 19 and 22). It seems very likely that a relatively thick gangway (approx. 8-10 cm) with a maximum width of approx. 40 cm (Plates 14.III A, B and C) rested on the heads of these frames on both sides. Judging from the construction found in the Marknesse cog,[24] we may assume that the three aft crossbeams rested on the gangways. They were made slightly thinner at the bottom and were attached with treenails. The beams did not protrude through the shell. Also, the ends were not fastened with so-called rose bolts, as was found with the Dronten-I cog and with the Almere-I cog.[25] Such rose bolts would undoubtedly have been found during the excavation.

The beams were almost certainly attached to the hull with knees, which stood on top of them and also formed the extension of the frame to the top of the edge of the side. In the case of the Marknesse cog, this was done with approximately 3 cm thick treenails. The rear and front knees (on frames 4, 19 and 22) were probably not placed directly on the crossbeam, but as elsewhere, on the deck planks. Between the crossbeams above frames 4 and 11 and between frames 11 and 19, a coaming was probably nailed to the inside of the gangways. The third crossbeam (above frame 19) was the so-called sail beam. In analogy to the Markenesse cog, it almost certainly was heavier than the other one and probably had a semi-circular hole at the front, into which the aft half of the mast cross-section fell. In such a construction, the mast was usually held in place with an iron bracket.

Stringers and decks

Above the afterbody were several (parts of) planks of the aft deck and the only remaining deck beam (the straight beam that runs almost perpendicularly under the deck planks in the middle of the photo, fig. 14.15). During the reconstruction it became clear at what height and above and against which frame (0) the deck beam had been attached with wooden pins placed at an angle. The recesses for enclosing frames in the chamfered side of some planks indicate that the deck ran from side to side (Plates 14.II C and 14.III B). It is likely that similar deck beams were placed at the frames -1 and 1, while the front of the rear deck planks ended on the first crossbeam above frame 4. The planks had a thickness of 2.5 cm, were irregular in shape and lay lengthwise in the ship. The seams were covered with small planks of 1 cm thickness, which were fastened on both sides of the seam with small nails. It could not been established whether there was any caulking underneath the planks. As mentioned earlier, the knees on this crossbeam were probably placed on the deck planks. It is not certain whether the aft deck had a hatch opening to the aft, but it is likely. The most obvious place for this is above the keel plank between the beams above frames 1 and 3 (Plates 14.III A, B and C).

From the aft deck, the crew could walk via the gangways to the foredeck. Based on the recesses in the sail beam and the crossbeam, which were found separately in the Marknesse cog[26], it seems very likely that stringers (on which the deck planks lay) had also been installed on this ship between the third and fourth crossbeams and between the fourth crossbeam and the stem beam. The most likely solution is that these planks were fastened athwart the ship to the frames, on the stringers and on the beam clamps. The deck was undoubtedly fitted with a hatch opening. As shown in plate 14.III.B, this opening may have been between the third and fourth transverse beam, or even on both starboard and port sides. In this situation, however, the hatches rise above the cargo hold. An access to the front bulkhead is more likely to have been located in front of the fourth crossbeam. It seems logical that the hatch would have been fitted above the keel plank, just as

Fig. 14.15. The remains of the afterdeck above the afterbody.

23. See chapter 12.
24. Modderman 1945, figs 30-32 and 36; chapter 12.
25. Chapter 7, Hocker & Vlierman 1996, 33 and 52 and 80-81; chapter 16.
26. Modderman 1945, fig. 29 E a and b; chapter 12, section 12.2.2.

it was in the aft deck. In that case, there could not have been just one stringer towards the stem, there must have been two, between which the hatch opening was located.

Pump

There is no evidence of a pump. There was a clear wear spot in the top of the keel plank in front of the fourth crossbeam, between frames 22 and 23, which would be an indication of the location of a pump tube. Noteworthy are the similar wear spots between frames 20 and 21 and between frames 21 and 22. Perhaps the pump initially stood on one of these places or it was possible to insert a loose pump tube into the opening for lowering the mast inside the formwork.

Castles

No indications were found that the vessel had an aft or fore castle.

Repairs

Several repairs were found on and in the vessel. The false stem aft that was too thin and the two short shell planks on starboard in the C- and D strakes between the frames 5 and 10 in the afterbody are some examples. The ends of the latter transitioned almost smoothly into the original planks at the scarfs (figs 14.16a to c). This does not correspond to the original scarfs. Here, the ends of the boards protrude about 1 cm above the shell. Nails hammered in from the outside through the original bottom planks and into frame 11 suggest that these ends were nailed down during the removal and replacement of bad parts. The local application of sintel clamped moss caulking on the outside at the clinker-built shell strakes, the 'patchwork' of planks of the ceiling and a former mast step preceding the current one were already mentioned above.

In addition, repair boards with a layer of moss underneath were nailed in place in various spots both on the inside and on the outside of the shell. Another type of repair, possibly carried out at the same time, are the planks with chamfered ends, which the ship's carpenter nailed from the outside into the B- and C strakes on the starboard side of the afterbody and almost amidships in the D strake on the port side. The bad part of the planks in question had been removed from the shell beforehand. The seams between the planks and the original shell planks were only made watertight with moss. These different repairs, which very likely were carried out at different times, in addition to the high number of repairs in total suggests that the vessel had been used intensively and certainly was no longer young when it sank. No cracks in the shell planks were recorded that were sealed with sintel clamped moss caulking.

Wood use and processing

All parts of the ship excavated (including the treenails and moss lathes) had been made from oak (*Quercus sp.*). The planks had been sawn with the grain or quarter sawn, with the sides contoured with an axe and the sapwood removed. Immediately after cleaning the skin, it was no-

Fig. 14.16a. Repairs. Small planks in the afterbody.

Fig. 14.16b. Repair. A small plank above a hole in the bottom in the midship section.

Fig. 14.16c. The sloping flat scarfs in the C- and D strake, which do not protrude out of the shell, at starboard in the afterbody and the nails hammered in from the outside in frame 11.

ticeable that the underwater hull (and probably also the sides) had been built with care and that the composition and dimensions on both sides of the keel plank were remarkably symmetrical. This was the cause why in this wreck, for the first time during a ship's excavation in Flevoland, more specific attention was paid to the use of the wood and to the use of the so-called 'brother and sister' pattern (fig. 14.17). This was mainly prompted by the wood structure of several planks, which displayed almost the same pattern in corresponding places on the starboard and port sides of the shell. Nowadays it would have been possible to determine whether they were indeed from the same tree and from how many trunks the planks of the underwater hull were made with the help of dendrochronological research and research of the wood grain patterns. However, at the time of the excavation dendrochronological research was still in its infancy in the Netherlands.[27]

In the case of the wrecks in the IJsselmeer Region, a number of samples were collected per ship only from 1975 onwards, so that such research could be carried out in the future if necessary. In 1985, this was done for a number of shipwrecks.[28] The floor timbers with rising frame sections were cut from a trunk or the section of a trunk or branch. Figure 14.18 shows a number of floor timbers and V-shaped floor timbers of the ship and the part of the tree from which they were made, based on the established wood structure. The futtocks or the upper part of the floor timbers (if they ran through to the inwale), are slightly less thick and wide at the top than the floor timbers i.e. the parts in the underwater hull. The futtocks were cut from thick branches or thin trunks, the hooks and other knees from parts of trunks and branches, as well as the V-shaped floor timbers, which could also be made from crooked trunks, while the crossbeams would have to be taken from a single trunk. Whether there was already a cultivation of trees being made to grow with curved or forked pieces is not known. It is likely that the desired shapes were still selected from natural grown trees in the forest at that time.

Fasteners

The two thin treenails between the floor timbers of frames 12 and 13 and between frames 13 and 14 are the only indications of the use of construction frames at the beginning of the construction of the ship. Generally, all frames were connected to the hull with two treenails (Ø 2.5 cm) per plank width, fitted slightly staggered. The connection of the frames with the keel plank was made with just one blind treenail each. The bilge ceiling planks were pegged to the shell plank with two treenails (occasionally with just a single pin) at the same time as the frames. From frame 1 to the aft and from frame 25 to the fore, the V-shaped floor timbers were not attached to the stern- and stem hooks and beams. This means that for frames 1-25, at least twenty-nine treenails per frame were used, a little less to the fore and aft. The gangways and the transverse- and deck beams, as well as the knees placed on them were probably also connected in this manner, but possibly with 3 cm thick treenails, so that 900-950 treenails will have to have been made for the whole vessel.

The clinker-built joint between each of the planks of the strakes, the connections of the scarfs and the fastening of the gudgeon were made with nails. The nails were 100-110 mm long and had a convex head, forged in four planes (pyramid-shaped) with a diameter of 25-35 mm (figs 14.19, 14.20 and 14.21a). All nails were driven through the

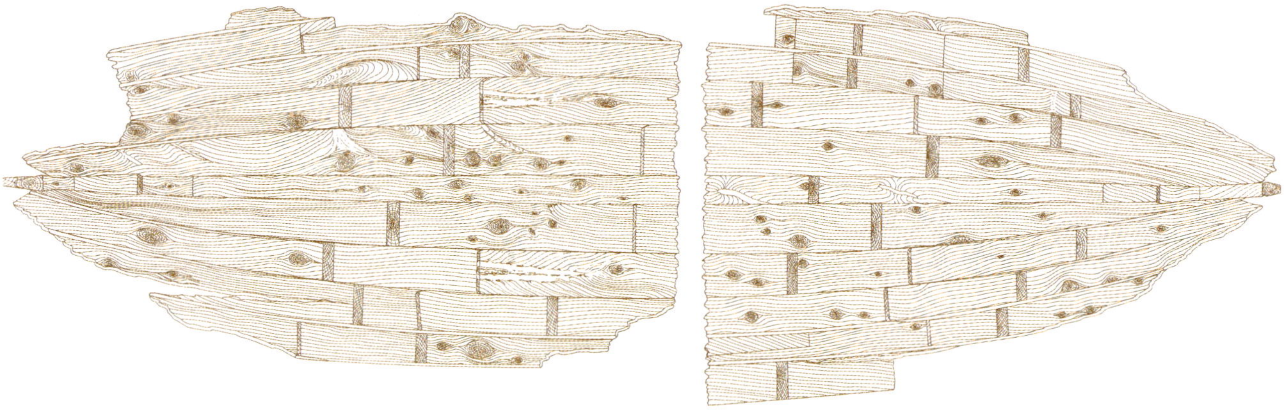

Fig. 14.17. The symmetrical plan of the planking of the shell, showing the sketched structure of the wood, the cross-sections of the planks indicating that they have been sawn from the same tree and placed in a similar place on either side of the keel plank ('brother and sister' pattern).

27. Vlierman 1983, 4.
28. Van Holk 1986 en 1987.

Fig. 14.18. Some frames with the sketched structure of the wood (knots) and the probable places in the trees from which they have been cut.

overlap of two planks, bent, driven further, bent again and hammered back into the wood, so called double clenching (*ijzeren naaiwerk* in Dutch). For the way of applying the nails see the Nijkerk-II cog.[29] The nails were applied every 12-18 cm. Together with the nails used at the scarfs and with an average distance of 15 cm, this results in 1250 nails for the entire hull. The smaller nails for the ceiling, bilge and decks are not included in this calculation. Approximately 2500 wrought-iron sintel clamps were used during construction to attach the moss lathes and the moss caulking. The iron gudgeon, the iron pieces which had been attached flat against the stem hook/false stem and the upper rudder suspension (which was not discovered during the excavation), the rudder fittings, the iron bracket used to secure the mast to the sail beam and the bolt, which most probably protruded through the stem hook/false stem at the top, were probably the only large(r) connecting elements of the vessel fashioned from (wrought) iron.

14.3 Construction data

All the excavated parts and details of the ship were discussed above, as well as the reconstruction that could be deduced from them. Examination of all the data yielded sufficient information to be able to speculate on the design, the (order of) construction, the layout, use, abandonment and the formation of the wreckage and its embedding in the sea bottom.

14.3.1 Design

Medieval design drawings are not known; they probably did not exist or were only brief sketches with some of the most important measurements. The vessels were built in accordance with the rules of thumb known to the shipbuilders. The dimensions of the available wood probably in part determined the main dimensions of the ship. This must also have been the case with the Spakenburg-I cog. The (reconstructed) ship looks as if it had originally been a vessel of beautiful lines and skilful build.

Shape

The plan of the lines, or rather, of the planks (Plates 14.II A and 14.III D) shows a hull with a straight keel, little sheer in the not very high sides and a straight, raking stern and stem. The stern was at an angle of 119° to the keel line, the stem at an angle of 122°. The ship had a stern rudder. The underwater hull had an almost symmetrical shape like a laurel leaf, was very sharp at the bottom of the stern and stem, got wider and wider upwards due to the sloping sides and had a slightly elongated, almost oval shape at gunwhale height. The forebody was only 10 cm wider than the afterbody. The forebody was relatively fully built, while the afterbody had a more streamlined shape and rising stern.
In the bow and in the afterbody, the frames have a weak S-shape, amidships an U-shape with sides inclined outwards (112°). An angular rounded bilge can only be seen amidships. The planks of the carvel-built bottom, which did not have a dead flat anywhere, were increasingly twisted towards the stern and stem. At certain points they transformed into a clinker-built construction and were attached at an angle of (almost) 90° to the stern and stem constructions. The D- and next three strakes of the sides were constructed completely clinker-built. The shape of the underwater hull suggests that it must have been a good sailing vessel.

Dimensions

Two thin treenails in the keel plank may have been used to determine the mid-point during the start of the construction of the ship. The rearmost is 6.5 m from the front of the false stem and the foremost at an equal distance from the back of the stern hook. This length almost corresponds to twice the length of the reconstructed stem hook construction (approx. 3.2 m). The aft stern construction was 2.6 m long. Both stern- and stemposts will have protruded slightly above the hull, while the false stem may have been slightly longer. Thanks to the relatively large number of cross-sections drawn in the field (Plate 14.I D), a very reliable reconstruction of the ship could be made up to strake D. The shape of the two futtocks of frames 2 and 24 found separately from the ship made it possible to determine the position of the sides, the total number and width of the strakes and the height of the hull. The only deck beam found in the afterbody provided important information about the shape and dimensions of the ship at this place and the height of the aft deck. In general, the reconstruction presented in this study provides a realistic and reliable representation of the main hull shape dimensions, the supposed deck plan and the layout.

Classification and tonnage

The fully closed loading floor or ceiling and the closed bilge ceiling in the sides indicate that the vessel will have regularly transported loose cargo, and perhaps was even built for it (Plates 14.I B and 14.III A). The indications of a formwork around a mast that could be struck support this guess, while the ending of the loading floor at a few centimetres from the outside of the floor timbers suggests that vertical and transverse bulkheads may have been installed here, which closed off the whole. The entire, almost 8.4 m long hold (from frame 4 to 22), was - measured between the tops of the bilge ceiling - 2.4 m wide at both the rear

29. Nijkerk-II cog, section 6.2.2 and fig. 6.42.

Table 1. Principal dimensions of the Spakenburg-I cog

Length of the keel (including stern- and stem hook)	13.0 m
Length of the bottom (between the rising stem- and stem hook)	11.8 m
Greatest width of the bottom (at frames 13-14)	3.1 m
Overall length at gunwale level	15.2 m
Overall width at gunwale level (at frames 13-14)	4.8 m
Height of the stern (keel line to top gunwale level)	2.2 m
Height of the stem (keel line to top gunwale level)	2.7 m
Height of the side amidships (keel line to top gunwale level at frame 11)	1.7 m
Length of the stern	2.6 m
Length of the stem	3.2 m
Height of the hold	0.9 m

Length-width ratio of the bottom	3.8:1
Length-width ratio at gunwale level	3.2:1

All measurements are rounded to the nearest 10 cm.

and front and 2.8 m wide in the middle. The height to the bottom of the gangway was 0.9 m, while the widths at this height and at the aforementioned locations were 3.2, 3.4 and 3.8 m, respectively. This corresponds to an average width of approximately 3 m. The rough content of the hold is therefore 8,4x3x0,9 = 22,7 m³. The formwork around the mast was approximately 1.8 m long, 0.3 m wide and 0.9 m high = approximately 0.5 m³. The net loading space was thus about 22 m³ or more than 20 gross tons. The spaces under the decks aft and fore were undoubtedly intended for the storage of spare materials for the ship, such as (anchor) ropes and sails, but also for (ship's) tools and personal belongings, other parts of the ship's inventory and possibly even for sleeping. If cooking ever took place aboard the ship, it will probably have taken place in the hold.

Choice of materials

The Spakenburg-I cog was built entirely of oak, in analogy with the other (small) cogs discussed in this study, with the remaining parts and fasteners used also showing an almost identical picture. Nothing was found of the mast and the spar or sprit. They were probably made of Scots pine (*Pinus sylvestris*). The shell was meticulously constructed in the so-called 'brother and sister' pattern, while the frames were also regular in shape and size.

14.3.2 Construction: techniques and sequence

The characteristic construction of the keel plank and stern- and stem hooks in the large cogs (see previous chapters) was also used in the smaller versions of this type of ship. As the upper edge of the sides was more than half as low as that of the large cogs (Nijkerk-II cog 4.3 m, Spakenburg-I cog 1.7 m and the length 20 m and 15.2 m respectively), it does not appear to have been necessary for the construction to connect the sloping sides by crossbeams protruding through the shell. In the case of the Spakenburg-I cog, it is very likely that very heavy gangways were chosen instead, which connected the aft deck to the foredeck and on which three of the four transverse beams that ran from shell to shell were fastened. The gangways with crossbeams lay on short frames, finished straight at the top. Knees were placed on the crossbeams (and the deck planks attached to them), which provided the connection between the beams, the gangways and the upper parts of the hull's sides. They were connected with (thicker?) treenails.

The ends of the upper shell strakes of the large cogs, which extend to the front of the stern and stem beams as previously noted, were also used in the smaller vessels (in the case of the vessel described here, the top three). The thicker bilge ceiling planks and probably also the upper ceiling planks in the sides almost certainly formed an important part of the ship's longitudinal strenghtening. The upper ceiling planks, which are presumed to have been thicker, probably were extended aft and fore with inwales which had been placed along the top of the hull's edge. They reinforced the upper edge of the hull (see the Marknesse cog, Chapter 12). The construction only started after it was decided which type of ship to build and for what kind of transport it would be used. It is possible to deduce from the ship that the construction was carried out in a number of phases according to a well-known, traditional pattern and in an expert manner.

Construction phase 1: keel plank, stern- and stem hooks, stern- and stemposts and temporary floor timbers
After being sawn, cut and adzed into shape, the keel plank and stern- and stem hooks were probably the first to be assembled on a wooden frame. Subsequently, the stern- and stemposts were placed and braced. The shipbuilders probably only cut the rabbets into the stern- and stem hooks and beams in order to attach the ends of the bottom strakes into them when that strake was being installed. Two thin treenails in the keel plank are the only indicators of the temporary attachment of building frames. They seem to have been sufficient for the temporary attachment of the ends of the two planks of each A strakes and were then removed. The chamfered scarfs between the keel plank and the stern- and stem hooks were provided with a thin layer of moss (paper) before they were nailed together.

Construction phase 2: bottom strakes
The A strakes were the first to be formed and installed. As mentioned above, the temporary use of two building frames in the middle of the ship seems to have been sufficient to be able to do this without too many problems, because the ends of the pre-formed planks were nailed against and into the rabbets of the stern- and stem hooks. The other ends could easily be clamped against the underside of the building frame in the middle of the ship with a prop or support (a thin post). Hence, just two frames were sufficient during the start of the construction of the underwater hull. Next, the three planks of the B strakes were placed in the same manner, from aft to fore. The use of two frames also seems to have been sufficient for these planks. The aft planks are rather short and are largely nailed to the garboards in clinker-built fashion. The second plank was nailed to the scarf of the first plank and could be clamped with a prop amidships to the aft building frame. The third and foremost plank was fitted just like the first, aftmost plank. The planks of the C strakes followed in the same manner. In order to make the transition from carvel to clinker-built, the matching parts of both planks were chamfered. Each time before the following plank was placed against the previous strake, the outer edge of the lower plank was chamfered at the parts to be attached to each other clinker-built fashion. In the case of the carvel seams, the boards touching each other were partially chamfered on the outside. In both cases, a triangular opening was created with sides of approximately 2 cm, intended for the insertion of the sintel clamped moss caulking.

Construction phase 3: (V-shaped) floor timbers above the carvel-built part of the bottom and the D strakes
During the third construction phase, the floor timbers and the V-shaped floor timbers were placed on the carvel-built part of the bottom (frames 6 to 20). Some of them continued on one side above the D-course and could only be cut into the right shape and installed after this strake had been placed as well. It is therefore quite possible, that initially a part of these fifteen floor timbers was cut to shape and placed, and only then the D strakes were cut and after that, the floor timbers that ran up to the D-strakes.

Construction phase 4: shell strakes E to G and the application of sintel clamped moss caulking
As soon as the underwater hull was sufficiently solid in shape, the planks of the E- to G strakes could be fitted. This was followed by waterproofing the seams of the clinker-built shell strakes between the stern and stem hooks/beams and the plank ends of the shell strakes on the inside of the ship.

Construction phase 5: Futtocks of the frames in the carvel-built part of the ship, V-shaped floor timbers and futtocks in the after- and forebody, the bilge ceiling, the upper ceiling strake extended with an inwale and the mast step
In order to give the ship's hull sufficient strength, the futtocks of frames 6 to 20 were attached to the sides in this phase. It is likely that some temporary connections were also made to determine the desired width of the ship and to consolidate this during the further construction. It is also quite possible that the gangways, with the crossbeams lying on them, were already installed in this phase. This would mean that no temporary cross-connections were required. Subsequently, the fashioning and placing of the V-shaped floor timbers and futtocks took place in the afterbody and forbody (respectively behind and in front of the ends of the gangways), and the fitting of the two bilge planks and ceiling or inboard planks along the upper edges of the sides. These were probably extended on both sides towards the stern and stem with an inwale, partially resting on the frame heads and in which a number of them had been sunken. All frames plus the bilge ceiling were then pegged to the shell, generally with two treenails per plank width. The mast step, which had been fashioned by then, was then fixed to the frames with blind treenails.

Construction phase 6: gangways with coamings, crossbeams, deck beams, decks, knees on the crossbeams, windlasses and bollards
The rising timber parts or futtocks of frames 4, 11, 19 and 22 were levelled at the top and did not run further

than about the top of strake E. The heavy gangway planks were placed on the first three of these frames and enclosed the frames located between them. The three crossbeams were inserted at the bottom across the width of the gangway, so that they partially subsided in between. The knees were then placed on the crossbeams, which formed the connection with the upper sides. Knees, beams and gangways were pegged together. Between the crossbeams 1 and 2 and between 2 and 3 (the sail beam) a so-called coaming was placed against the sides of the gangways. It is not known whether there were hatches over the resulting openings in the holds. No indications of this have ever been found in late medieval ships. The earliest example dates from around 1600.[30] In this ship, notches were found for the first time on the coaming, on which the hatches had been placed.

In the after- and forebody, deck beams were installed, which were fixed between and against the frames with treenails. In addition, the foredeck most probably had stringers, on which transverse deck planks were attached. On the shell sides, these planks must have been placed on a shelf clamp. The planks of the aft deck lay in the longitudinal direction of the ship. No indications were found, but it seems plausible that the vessel must have had a windlass for hoisting the sail and for lowering and erecting the mast and the spar or sprit. The windlass was most likely attached to the first crossbeam at the front of the aft deck (Plates14.III A, B and C). Another windlass for the anchor on the foredeck also seems desirable. The vessel undoubtedly also had bollards. These may have been formed by futtocks that protruded above the sides.

Construction phase 7: Ceiling, belaying rack, pump, rudder, mast, rigging, spar or sprit and sail and the sintel clamping of the seams on the underside of the bottom
During the final phase of the construction of the vessel before launching, the ceiling, transverse bulkheads and the formwork for the mast step were placed and the seams of the bottom were fitted with sintel clamps on the outside. Some of these operations may also have been carried out while the ship was in the water. The installation of the mast and rudder will probably have been carried out during the latter situation. In this final phase of the construction, other fittings such as for example the belaying rack, and all kinds of smaller aids for operating the ship and securing the rigging were also fitted. The belaying rack could also be integrated into the inwale along the upper side of the vessel in the after- and forebody of the ship, but then had to be installed during construction phase 5. This last phase saw the pump being placed and the hatches covering the openings of the spaces under the aft- and foredeck. Finally, the spar or the sprit and the sail were fitted to the mast and the rigging was finished.

14.3.3 Characteristics

The data recorded and subsequently analysed showed the following elements to have been characteristic for the Spakenburg-I cog.

Keel plank
The vessel has a keel plank, which is more than twice the thickness of the shell strakes. Aft, the plank is 20 cm wide, and fore 23 cm, while from both ends it gradually widens to 30 cm towards the centre. The keel plank is almost rectangular in cross-section and is fitted between the keel parts of the stern and stem hooks by sloping flat scarfs like a connecting beam.

Hooks
The stern- and stem hooks were cut from a naturally grown form of a trunk/branch of a tree and gradually become less broad from the top to the bottom and from the inside to the outside (trapezium-shaped in cross-section). They form the aft and fore parts respectively of the keel plank and the lower parts of the stern- and stemposts. The tops of the stern and stem sections are chamfered, the sloping scarfs with the stern- and stemposts

Stern- and stemposts, false stern- and stemposts and stern rudder
The stern- and stemposts were fitted to the hooks with nails and with a single treenail. The sternpost (and the sloping flat scarf to the hook) starts directly above the thickened keel part of the knee, becomes slightly less thick upwards and has grooves up to and including the D strake for securing the planks ends. The top three strakes continued and were nailed to the triangular back of the beam. The stern originally had a 7-8 cm thick false sternpost, which ran from the heel at the bottom to the E strake. The scarf between the stem hook and the stempost at the front started slightly above the thickened keel section. The stem also slightly increased upwards and was therefore slightly less thick in the side view. The same probably applied to the false stempost which presumably protruded from the keel line to well above the collar and the stem. The fore shell strakes up to and including strake D also ended in rabbets in the sides of the stem, while the strakes above them ran along the chamfered sides to the front. The ends of these planks were protected by the false stem, which was placed against them. Aft, the upper three strakes probably ended up with the ends against each other on the back of the stern or the ends of the planks ran

30. Shipwreck B 71 Oostelijk Flevoland, file at RCE-Lelystad.

against those of the other side (starboard and port side). The ship was equipped with a stern rudder.

Carvel- and clinker-built bottom

A striking feature of the bottom was the clearly visible 'brother and sister' pattern. The bottom had no dead flat lengthwise and ran a little upwards from the first two strakes next to the keel plank in a stretched line towards the bilges. The third strake was set carvel fashion at a slight angle to this. At a certain point in the after- and forebody (at frames 5 and 20), the carvel fitted B- and C-strakes widen towards aft and fore and become clinker-built. During the assembly of the underwater hull, not all floor timbers could be placed above the carvel built part, because it was decided to fit the scarf to the futtock of some frames above the bottom, and for others above the D-strake. This staggered application of the scarfs improved the structural strength in the bilges, but it also meant that the D-strakes had to be fitted first, before the structurally important bilge ceilings strakes could be fitted after the (last) futtocks had been attached. The construction of the ship seems not to have been a 'bottom first' construction, but rather a transitional form between 'bottom first' and 'shell first'.

Loading floor/ceiling, mast step, casing and repairs

The vessel had a completely closed ceiling or loading floor, with a short mast step integrated into it, and was bounded in the transition from the bottom to the sides by thicker bilge ceiling strakes, which also fulfilled an important bracing function in the construction in connection with the scarfs in the frames on both sides. Above the bilge ceiling, the frames were concealed by two touching ceiling planks, while the rear and front of the cargo hold were probably closed with bulkheads. There was a wooden casing around and in front of the mast, so that the mast could be lowered at all times. Such a casing was in fact intended to prevent a loose or sliding load from lying in front of the mast. It can be deduced from the cut section and extra treenails in the frames under the mast step that the ship originally had had a different mast step. The mosaic of planks and small planks of the ceiling indicate that it has been repaired several times. A large number of repairs on and in the shell and the replacement of the false stern support this impression.

Clinker-built sides and double clenching

The D-strake was also installed at a slight angle to the C-strake. From the top of the D-strake upwards, the planks of the strakes from the stern to the stem were completely clinker-built and amidships the planks were attached to each other in an almost outstretched line, raked outwards. This was carried out with nails, bent over twice and driven back into the wood, called double clenching (*ijzeren naaiwerk* in Dutch). This construction method makes it possible to install the planks without any frames strake for strake. The shell strakes were probably fitted in one phase up to the top. The V-shaped floor timbers and futtocks in the after- and forebody could only be placed after the seams in the clinker-built part of the underwater hull had been fitted with sintel clamps.

Caulking method

The seams between the parts of the shell strakes that were attached in clinker-built fashion to each other were only provided with sintel clamped moss caulking on the inside of the ship. The seams between the carvel/built parts of the bottom were caulked from the outside shortly before launching. The sintel clamped moss caulking in some parts of the seams on the outside of the clinker-built part of the shell is almost certainly the result of repairs.

Gangways, crossbeams and decks

The vessel most likely had thick gangways with three of the four transverse beams placed on them. This, together with the knees placed on top of the crossbeams, made for strong cross bracing. The crossbeams were not connected to the shell with rose bolts, but with knees, which also formed the connection with the upper sides of the ship and kept it in shape. The (thicker?) upper ceiling strake, directly below the top of the shell, formed an important part of the upper deck construction of the sides and the longitudinal bracing. These strakes had probably been made heavier aft and fore (like an inboard) and they also formed the belaying rack at certain points. It was not possible to determine whether the vessel was fitted with a gunwale. However, based on the types of cargo that were probably transported, this seems likely. As can be seen in the plates, the gunwale may very well have been open at the top between frames 4 and 22. The gangways formed the connections between the aft and foredecks, while the coaming attached to them would not have been as high. The decks probably contained access hatches for the rear and fore holds.

Fasteners

Oak treenails were used to attach the frames to the shell planks. It was not recorded whether wedges had been struck into the treenails. They were hammered from the outside to the inside into pre-drilled holes. It is possible that hicker treenails were used by means of the crossbeam constructions with the gangways and the upper shell. The clinker-built shell planks were fastened with wrought-iron nails, the ceiling and deck planks with smaller ones.

The moss caulking and the moss lathes attached above it were secured in or above the seams with wrought iron sintel clamps. Relatively few heavy wrought-iron connecting elements were used in the ship: the rudder fittings, the suspension points for the rudder at the stern and possibly a bracket for securing the mast in place. Blocks or rigging were not excavated, with the exception of a piece of rope and an iron ring, the mast and the spar or sprit.

14.3.4 Use

Function

The construction and execution of the ship make it absolutely clear that we are dealing with a cargo ship for the transport of - at least also - loose cargo. The load of bricks found at the excavation is an example of this, despite the fact that in this case, it may have had a different purpose. The two bricks and the grit of other bricks found in the ship, are in any case proof of a cargo transported previously. The two large openings between the three rear crossbeams also indicate the desirability of being able to relatively easily load and unload large and possibly cumbersome objects or parts (e.g. construction timber and piece loads). The elongated shape, the relatively low sides and the mast that can be struck make it clear that the ship was an inland vessel, a freight vessel designed to bring her cargo into the cities or up to the walls of a city or a castle. The maximum draught in loaded condition would have been about 1.25 metres.

In determining the length and width of the ship (15.2 and 4.8 m), the builder may not have taken into account (or may not have needed to take into account) the common sizes of the main Dutch locks at the time. It was not until 1536 that Charles V allowed the construction of a new lock on the east side of Gouda. This was wider (almost 16 feet) than the Gouda Inner Lock and many other locks.[31] If these measures were expressed in Amsterdam feet, the new lock would have been about 4.5 m wide, if in Rhineland feet, it would have been about 5 m wide. It seems that at the time of the launching of the cog, the Dutch locks could not be passed with this ship. This information and the location of the wreck suggest that it was a ship built especially for the coastal area of the Zuiderzee and the adjacent waters that were suitable. The ship may have had a relationship with Bunschoten, which perhaps even was the home port and the location of the shipyard where the cog was built.

Wear and tear and repairs

The rounded corners on the outside of the keel plank and stern and stem hooks may be the result of prolonged use. The traces indicating that the mast step discovered was not the first, the many times repaired loading floor, the repairs on and of the shell and the replacement of the aft false stem, the futtock of frame 22 and the short inset planks in the B- and C strakes on starboard also suggest a relatively high age at the time when the ship sank permanently.

14.3.5 Dismantling and sunk on purpose

The high age of the vessel and its general condition, which may have been considered to no longer be seaworthy, suggest that the vessel was abandoned and, for example, sunk on purpose in an inlet next to the shipping lane or perhaps even used as a 'plug' to close a hole in a dike. In order for the vessel to sink, a hole in the bottom may have been pulled with an anchor, during which action the anchor blade was broken off and left behind. The (valuable) load of unused bricks could thus be explained (combating such a calamity) and has then served to effectively sink the vessel and keep it in place.

14.3.6 Wreck forming and embedding

The ship was probably sunk completely on purpose after all the objects and any parts of the inventory that were considered to be of some value had been removed beforehand. The ship sank almost upright into the soft, humorous Almere-deposit, as a result of which there is hardly a layer of washed-out sediment underneath the wreck. This supports the idea that the vessel may have sunk into a hole already partially filled with eroded material. The disturbances found higher in the profile may have occurred later as a result of the disturbance of the highest parts of the ship when the wreck protruded above the newly created seabed, as a result of the eventual clearing of the land by the sea.

31. Smits 1995, 468.

14.4 Model

A scale model of the Spakenburg-I cog has not (yet) been made.

14.5 Inventory

14.5.1 Find distribution

It has already been indicated in sections 14.1 and 14.3.6 that hardly any finds were discovered in the wreckage and that the ship was most likely abandoned and sunk on purpose. Moreover, the objects that were discovered were fragmentary and clearly no longer of value. All objects have been included in the find list, while the locations of the finds are indicated in plate 14.I A.

14.5.2 Ship's equipment and inventory

Ship's equipment

An iron gudgeon was attached to the stern construction (figs 14.19 and 14.20). On the upper edge of the ring on which the rudder was resting, there was a small crosscut exactly in the middle of the back of the ring, perhaps a sign of consecration or protection against harm. On the starboard and port sides, this upper edge shows clear signs of wear and tear, indicating long use. Different sizes of nails were used in the ship (figs 14.20-8, -9-1 to 5, -10-13 and 14, -10-20 to 22, -10-24, -11-1 and 2 and -12-1 and 2 and 14.21a). The heavy iron nail (8) was found unattached in the stern and most likely came from the cross-beam construction or another construction in that area (windlass?). The nail was inserted through 7.5 cm thick wood and had been bent at right angles and hammered in. The five iron nails under 9 were found scattered throughout the ship. Some of them are straight and may, due to their length, have served, for example, as a connection of a futtock to a floor timber. The two bent specimens may have been used to attach the D-strakes to the head of a futtock. The six iron nails under 10 are examples of so-called doubled clenched nails, which were used for the connection of the shell planks to each other and the scarfs in the strakes. The iron nails (11), which are approximately 80 mm long, were used to fasten the ceiling planks to the frames. This also applies to the longest nail 12. The other, small nail was probably used to fix the 1 cm thick planks, which were placed over the seams in the aft deck. The sintel clamps used to seal the seams in the shell are of type D2/E (figs 14.20-46 and -47 and 14.21b). The broken iron ring (figs 14.20-7 and 14.21a) may have been part of the rigging.

Fragments of a well-preserved, woollen fabric were preserved in the fore hold, lying on and against the stem hook. The largest piece measures about 40x50 cm, the second-largest about 15x15 cm and the third is a piece

Fig. 14.19. The gudgeon on the stern post construction.

Table 2. Distribution of items

Category	Forebody	Hold	Afterbody	
1. Ship's equipment	sail	-	ring	
2. Operational equipment	-	-	-	
3. Administration and navigation	-	-	glass of a compass	
4. Tools	heather broom	-		
5. Household items	-	-	-	-
6. Cooking, eating and drinking	-	stoneware juglets	-	
7. Victuals	barrel fragment, bone, walnut shells	bones, walnut shells	bones	
8. Personal equipment/weaponry	shoes	-	shoes	

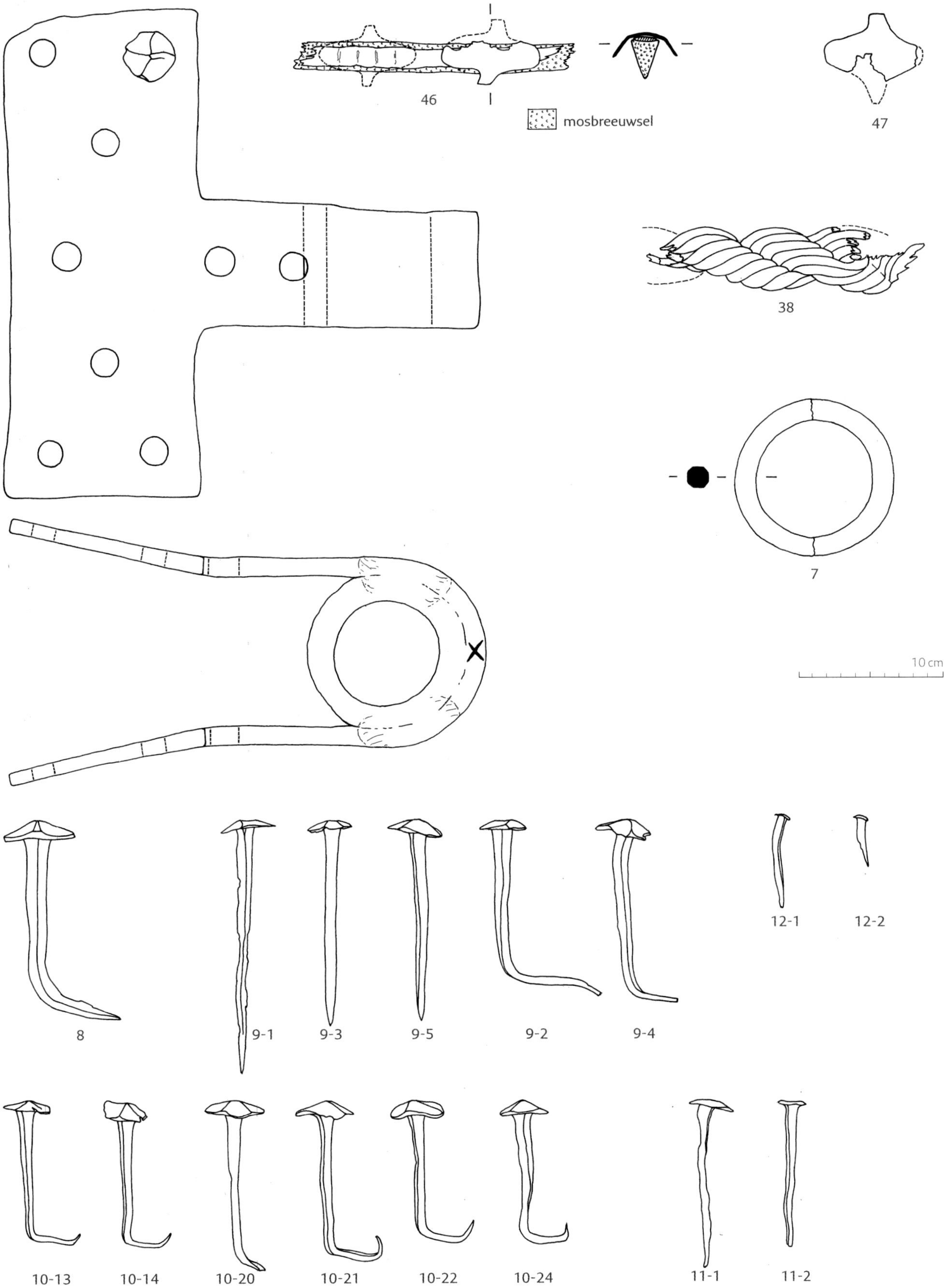

Fig. 14.20. Ship's operational equipment. Gudgeon, ring (7), nails (8, 9-1 - 5, 10-13 and -14, 10-20 - 22, 10-24, 11-1 and -2 en 12-1 and -2), sintel clamps (46 and 47) and rope (38).

Table 3. Ship's equipment

Category	Item	Number	Findspot
1.1 Ground tackle	-	-	-
1.2 Sails	sail	15 and 16	forebody
1.3 Cordage	ring, rope	7, 38	forebody
1.4 Blocks	-	-	-
1.5 Windlass	-	-	-
1.6 Pump, drain	-	-	-
1.7 Navigation aids	-	-	-
1.8 Boats	-	-	-
1.9 Spare parts	-	-	-

Fig. 14.21a. Ring.

b. Bent nail.

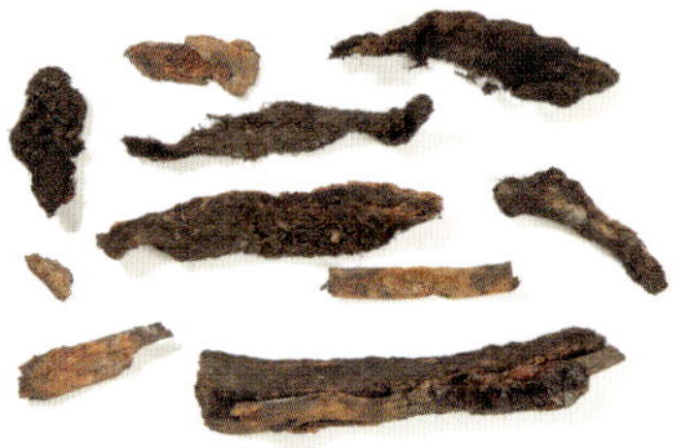

c. Several pieces of moss caulking, oak moss lath and moss lath with a sintel clamp (right) and the print of another sintel clamp (left).

Fig. 14.22a. The woven pieces of woolen sail (15 en 16).

b. Reinforcement of the selvedge of the sail.

with a thickened edge and some loose threads (figs 14.22a and b). On the large piece, a small part of the thickened edge or selvedge is still present. These are fragments of a sail. Two coarse, S-twisted, spun yarns, which are twisted together to a Z-twisted yarn of 3 to 5 mm length, were used to make a sail woven in flat or plain weave. It is very likely that these were strips woven on a relatively wide loom and then sewn together. A thicker thread was sewn through the selvedge with whip stitching in order to reinforce the outer edge of the sail. The strips were probably also joined to each other with whip stitching. The yarn used for this purpose was twisted from two yarns, each 3-5 mm thick, as described above. The sail weighed approximately 700-750 grams per m^2. There were no visual traces of tar, but it was undoubtedly dipped in some kind of organic tar, judging by the still vaguely perceptible smell. This might be confirmed by research. The sail may therefore have originally been heavier. Crumlin-Pedersen writes that more recent woollen sails used a fabric of approximately 500 grams per m^2 for a small sail of 20 m^2, and for a sail of 60-130 m^2 needed approximately double the amount.[32] The Spakenburg-I cog probably had a sail of about 7 m wide and 9 m high (approx. 65 m^2), which is quite similar to the data from Crumlin-Pedersen.

A small piece of rope was found between the floor timbers in the forebody. It was made of vegetable fibres. The approximately 3.5 cm thick piece of rope (fig. 14.20-38) probably came from the rigging. It was twisted out of two cable strands, which in turn were composed of six twisted yarns.

The broken blade of a large anchor (figs 14.9a and b) which was discovered stuck into the underside of the bottom certainly was not part of the cog's equipment, but must have been lost by another vessel. For example, during the earlier described possible process of using the ship as a 'plug'.

Operational equipment

Nothing has been found in the ship that can be considered part of the operational equipment.

Administration and navigation

On the starboard side under the ceiling, traces of writing chalk have been found between several floor timbers (41). Pieces of chalk were used to mark parts of the cargo. A quarter of a round window glass that was 2 mm thick (figs 14.23-4 and 14.24) almost certainly comes from a compass, presumably a so-called 'botervloot' ('butter dish') compass. Such compasses were turned out of one piece of wood or were composed of a round, flat wooden base around which a relatively low and thick wall of staves was made. A few completely preserved specimens have been found in a fish transport ship, a so-called 'ventjager' (transporting the fresh fish caught by other ships to the land and fresh food to the ships), near Swifterbant.[33] They date from about 1700. The glass from the Spakenburg-I cog may, in view of its location (in the stern between the rafters and under the aft deck), be regarded as part of the original inventory of the cog. It is the earliest example so far of the use of a compass on board a cargo ship that sailed the Zuiderzee.

Tools

In order to be able to repair the ship, there are almost always some tools on board. It is therefore striking that they are completely missing in this ship, and all the more reason to think of the ship as having been abandoned, with everything of any value having been taken away before it was sunk on purpose. The only maintenance tool found was a 30-40 cm long heather broom (19) that lay above the stem hook. It was of very poor quality and has not been preserved.

Household items

On board a (small) late-medieval cargo ship, no objects such as chairs, cages, lighting, etc. are to be expected as household goods. In the Spakenburg-I cog, no indications of such finds were discovered.

Table 4. Administration and navigation

Category	Item	Number	Findspot
3.1 Ship's papers	-	-	-
3.2 Stationery	chalk	41	hold
3.3 Financial/Commercial matters	-	-	-
3.4 Orientation and dead reckoning	compass glass	4	afterbody
3.5 Timekeeping	-	-	-
3.6 Signalling and distress signals	-	-	-

Table 5. Tools

Category	Item	Number	Findspot
4.1 Carpentry	-	-	-
4.2 Caulking and maintenance	heather broom*	19	forebody
4.3 Splicing, sailmaking	-	-	-
4.4 Miscellaneous	-	-	-

32. Crumlin-Pedersen 1997,190.
33. Ship wreck H 41 Eastern Flevoland; Vlierman 1991, 102.

Cooking, eating and drinking
The two fragments of flat roof tile of red pottery (figs 14.23-5 and 6 and 14.24d and e) show traces of soot. On the latter there are also remains of mortar. They may have been part of a fireplace (perhaps in secondary use). They originally probably were parts of unglazed roof tiles of about 22-27 cm long and 1.5 cm thick. Such roof tiles were used in the Netherlands until the beginning of the 16th century. Flat roof tiles were only found in the west of the Netherlands, in Flanders, northern France and England.[34] In the east of the Netherlands, in Westphalia and the Rhine area, the common roof tiles were of the nun-and-monk pattern. The province of Utrecht lies just on the edge of the flat roof tile area; in Amersfoort and Harderwijk, both nun-and-monk roof tiles and flat tiles were named as an alternative to a slate roof, but east of these places, there are (almost) only nun-and-monk roof tiles. In Utrecht, the usual roof tile type is 1.5 to 2 cm thick and has a surface area of approx. 26x16 cm.

The fact that at one point fire had probably been made on board (for cooking) can be deduced from the nine (fragments of) peat sods (fig. 14.24h). These were somewhat layered sods from low peat bogs with occasional inclusions of fairly coarse plant remains. They were discovered lying near the stem hook and starboard next to the mast step. A piece of white quartz (Rhine or Meuse gravel) and the chunk of dark flint (probably a northern glacial erratic), displaying dents on almost all protruding edges (figs 14.23-53 and 14.24f and g), were most likely used to strike fire. They were lying between the floor timbers in the afterbody and on the ceiling behind the load of bricks, respectively.

No cooking or eating utensils were found. A sherd of Siegburg stoneware and a sherd of stoneware with a brown iron slip (Langerwehe) on the outside come from two small drinking jugs. (figs 14.23-1 and -2 and 14.24a and b). The Siegburg and Langerwehe sherds both dated to the 14th century, The Siegburg sherd,[35] which may be

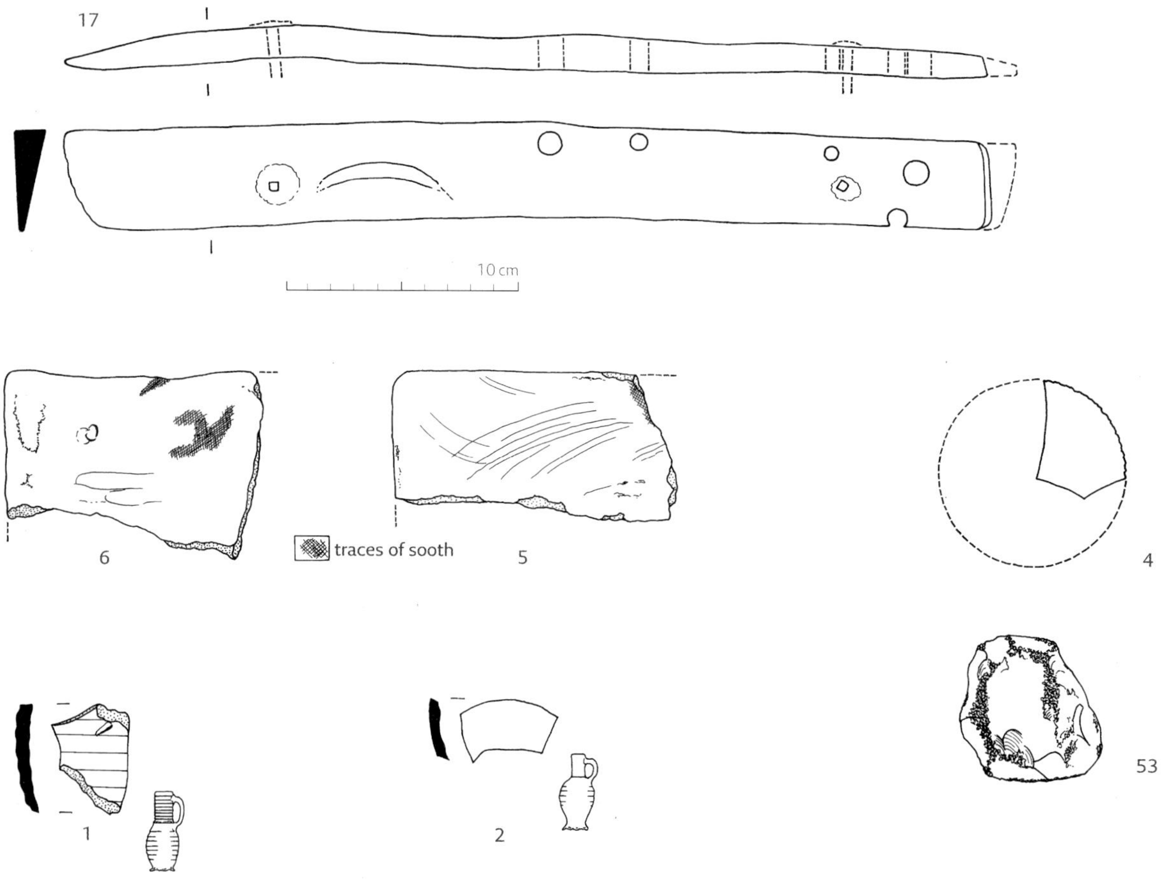

Fig. 14.23. Compass glass (4), roof tile fragments (5 and 6), two stoneware sherds (1 and 2) and fire strikers (53) and a fragment of the bottom of a barrel (17).

34. Kip 1990, 92,93 and fig. 54.
35. Type IV/V, Janssen 1988, 326 table 1.

Fig. 14.24a. Sherd of stoneware (1);

Fig. 14.24b. Sherd of stoneware (2).;

Fig. 14.24c. Fragment compass glass (4);

Fig. 14.24d. Fragment roof tile (5);

Fig. 14.24e. Fragment roof tile (6);

Fig. 14.24f. Piece of quartz (used as flint for fire striking?);

Fig. 14.24g. Piece of flint (for fire striking);

Fig. 14.24h. Peat.

Table 8. Cooking, eating and drinking

Category	Item	Number	Findspot
6.1 Fireplace	roof tiles, quartz, flint	5 and 6, 52 and 53	near mast step, afterbody
6.2 Fire-tending tools	-	-	-
6.3 Fuel	peat sods	39-1-9	forebody/near mast step
6.4 Cooking equipment	-	-	-
6.5 Eating and drinking gear	stoneware jugs	1 and 2	amidships

dated slightly earlier, was located on the port side wedged under floor timber at the bottom of the bilge. The other sherd was found between the floor timbers and under the bilge ceiling on the starboard side amidships. These jugs must have once formed part of the inventory, but possibly not at the same time. A third sherd, which certainly did not belong to the ship, was a red earthenware sherd (3) found in piled up soil on the Zu IV shell layer next to the ship. It is a rim shard of a square bowl, dating to the (late) 17th-18th century, which probably settled in a small depression above the wreck on the seabed.

Victuals

A small oak plank with a rounded and slightly tapered exterior has five holes, a residual nail and a nail hole and, as a result of impregnation, a barely perceptible (part of a) burnt-in mark. It was originally part of a barrel bottom. The diameter of the barrel was approximately 39-40 cm. The largest holes (diameter 9-10 mm) were probably meant for a plug, but this is less certain of the two smaller ones (diameter 6 mm, see figs 14.23-17 and 14.25a). This small part of a discarded barrel was discovered to have been found in secondary use, nailed to the ceiling against the inside of the bilge ceiling above the floor timbers 9 and 10. It had been cut until the section was triangular, so as not to protrude too much above the ceiling. If it originally had had more than the current thickness of 1.5 cm, it would have protruded above the bilge ceiling and may have served, for example, as a support for a short ladder that had been placed against the coaming in order to climb in and out of the hold. Walnuts and hazelnuts were once eaten on board, as evidenced by the nutshells

Table 7. Victuals

Category	Item	Number	Findspot
7.1 Fresh-water supply	barrel bottom fragment	17	hold
7.2 Supply of beverages	-	-	-
7.3 Livestock	rat's scull	36	next to afterbody
7.4 Food provisions	walnut and hazelnut shells, bones	37, 26 - 35	forebody, amidships, and under mast step

Fig. 14.25a. Fragment of the bottom of a barrel (17).

Fig. 14.25b. Walnuts and hazel nut (37).

found (fig. 10.25b). Ten bones were found scattered in the wreckage, proving that meat had been consumed on board. The skull of a rat was found on the port side next to the afterbody. It was discovered in the Almere-deposition and may therefore have been from a rat that lived on board. The bones were determined and examined by F. Laarman (see appendix).

Personal equipment/weaponry

No personal equipment or weaponry is to be expected in an abandoned ship, at most some items that were broken or scrap, which might have been kept as suitable for a secondary use. Parts of discarded footwear, frequent finds in shipwrecks in Flevoland, were also found in the Spakenburg-I cog. Cuts from the parts of the discarded footwear that were still in good shape could be used for repairs, for example. Numbers 20, 21, 22 and 24 refer to the upper parts of four shoes. Of the first, the complete upper part was preserved, which was largely worn out along the connection with the sole. O. Goubitz was kind enough to impregnate and reconstruct the shoe; a new sole and a new stitching edge were added (fig.14.26a and 14.27-20). In the upper of number 21 there is still a piece of the lace.

Only little has been preserved of the other three shoes and the pieces are in a poor state of conservation. However, it is clear from the remains that these were two different types of high shoes reaching up to the ankle, both with a single sole attached from the outside (without a heel) sewn to the upper using a stitching edge. Maybe the remains represent two pairs. One type was closed with laces on the side (figs 14.26a and 14.27-20), the other pair (figs 14.26b and 21) was closed above the instep of the foot. Goubitz dates both models as typical for the 14th century, perhaps until the beginning of the 15th century.[36] The design is similar to the types 50 and 60 from the excavations of the Rijksdienst voor het Oudheidkundig Bodemonderzoek (ROB) in Dordrecht, Heer Heyman Suysstraat, a closed find complex of 218 shoes and many fragments. Type 50 was worn in the period of ca 1350-1425, type 60 ca 1380-1450. The complex must have been dumped around 1400 and covered with earth.[37]

In the Spakenburg-I cog, a baldy tanned rectangular cut piece of leather with a rough hole inserted at the end was also excavated and a crumpled piece of leather (figs 14.26b and 14.27-25 and 23), which may have been cut out of a boot, shoe or other larger object at one time or another.

Fig. 14.26a. Reconstructed and restored ankle shoe or high shoe with side closure (20).

Fig. 14.26b. Upper leather of a shoe (21).

Fig. 14.26c. Two pieces of leather that have been put on top of each other (25).

36. Goubitz, oral communication, february 1980.
37. Goubitz *et al.* 2001, 111-119.

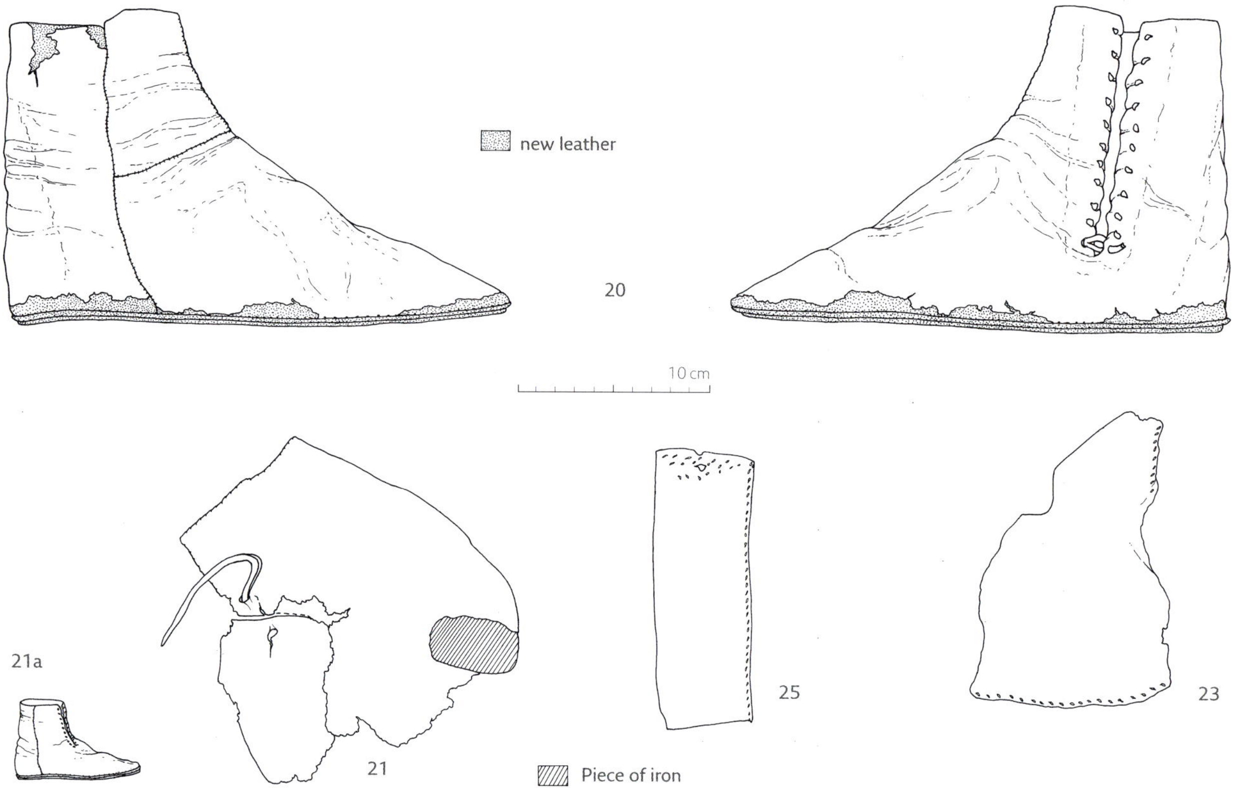

Fig. 14.27. Personal equipment/weaponry. Reconstructed ankle shoe or high shoe (20, fragment of the upper leather of a shoe (21 and reconstruction 21a) and two pieces of leather (23 and 25).

Table 8. Personal belongings and weaponry

Category	Item	Number	Findspot
8.1 Garments	-	-	-
8.2 Footwear	2 pairs of shoes, patches from shoes?	20, 21, 22 and 24 23 and 25	after- and forebody afterbody
8.3 Sewing kit	-	-	-
8.4 Tools, weapons	-	-	-
8.5 Cash	-	-	-
8.6 Miscellaneous	-	-	-

14.6 Cargo and merchandise

In section 14.1.1 it has already been stated that there was a load of bricks in the hold approximately amidships (figs 14.4a and b and Plates 14.I A and 14.III A). If the assumption is correct that the load had been placed there in order to effect the sinking and holding in place of the abandoned vessel, then these bricks did not form part of a cargo. However, it is almost certain that the vessel did (regularly) transport such loads, as evidenced by the pieces of brick found under the ceiling between the floor timbers next to the mast step (54). The two handmade bricks (fig.14.28), which were also found under the ceiling between the floor timbers, next to and behind the mast step, are other examples of this. Their dimensions are approx. 21x10,5 x 4,5 to 5 cm. The brick pieces differ in ware from each other and from the load of bricks in the hold. The brick pieces are of a bright red and relatively 'soft' ware, the two hand-formed bricks of a greyish red to yellowish and relatively hard ware, with a ringing sound when struck. The bricks of the batch in the hold are red and of a hardness that lies roughly in between the previous two. Of the approximately 900 stones excavated in the hold during the excavation, 100 were measured,

Fig. 14.28. One of the bricks.

demonstrating that they are very irregular in shape and size. Lengths were 22, 22.5, 22.7, 22.9, 23, 23.2, 23.6 and 23.8 cm, widths varied between 10.1 and 11.7 cm and heights or thicknesses between 4.9 and 5.9 cm. Many stones still had an imprinted edge of the mould in which they were formed, demonstrating that these were new, unused stones.

14.7 Ballast

The batch of bricks described above certainly did not function as ballast. The ship probably never had any (permanent) ballast on board, as was regularly found in the wrecks of fishing vessels in Flevoland.

14.8 Dating

Soil profile

Because of the disturbance of the soil profile, the date of the final sinking of the ship was put by R. Koopstra of the Soil Science Department of the RIJP between ca 1550-1600.[38] The reason for this was that the wreck had been covered by the Zu IV shell layer, which was deposited around 1600 or a little later in this area[39] and 'floated' at the top of the Almere-clay package, which cannot be further classified. The height of the Pleistocene sand and the occasionally still present Holocene remains of peat, or in other words the relatively shallow depth of the water and the fact that at the time of the sinking of the ship there was probably a sailing route of at most 0.5-1 km width through the then still present low (and perhaps diked) clay-on-peat landscape, probably explains how it is possible that the Almere-sedimentation package was not completely deposited so shortly below or near the coast; there are gaps that cannot be determined exactly. It is therefore not possible to date the ship more accurately on the basis of these data than into the period as has been shown above. As the following will show, the soil profile is even unable to at least give a good global date of the ship's demise.

Finds

O. Goubitz dates the types of discarded shoes found in the wreck as typical for the 14th century, but occasionally still found in the early 15th century. The two sherds of stoneware are the only other objects that can provide some information about the date of the ship's sinking. The Siegburg sherd is of a type of jug and ware that is considered to have been a common commodity in the third quarter of the 14th century. The jug with the brown iron slip dates from the same or a slightly later period. The only thing that can be said about the pieces of roof tile that may have been reused in a firebox is that such tiles were used to cover roofs in the (Western) Netherlands for a number of centuries and up until the beginning of the 16th century. The fragments of roof tiles can therefore make no other contribution to the dating of the ship's demise than that this probably took place before the beginning of the 16th century. The stoneware fragments and the shoes provide the best dating information.

Sintel clamps

The sintel clamps used in the ship, of type 2D/E, were in use in the fourth quarter of the 14th and the first quarter of the 15th century.[40] Since the excavation, more and more precisely dated sintel clamp finds were published. Based on these, the type used in the Spakenburg-I cog may be assumed to most probably date into the late 14th century.

Dendrochronological research

No dendrochronological research of the wooden parts was carried out. During the drawing of the stern- and stem constructions in 2007, it was discovered that it would be possible to do dendrochronological research on both stem hooks (figs 14.10a and b). A roughly estimated 100 year-rings were counted in the thickest parts of the keel parts, while the (outer) finishing ring may still be present (in the stem hook). In the case of the same place at the stern hook, this is roughly estimated about 50-80 year-rings. The drilling of samples could be carried out in places that are not conspicuous and covered almost without any damage.

38. Koopstra, oral communication, august 2nd, 1979.
39. Vlierman 2006, 151.
40. Vlierman 1996a, 79-50.

14.9 Summary

Ship

The ship, built entirely of oak, had a beautiful hull shape and some of the building elements that are characteristic of cogs; stern- and stem hook constructions connected by a keel plank and a carvel-built bottom, which at aft and fore turns into a clinker-built construction. In addition, the clinker-built shell planks were joined together with so-called double clenching and the ends of the boards protrude about 1 cm from the shell at the scarfs. Due to the relatively low sides, the transverse strenghtening of the ship did not involve the use of crossbeams protruding through the hull, but rather the use of beams that remained inside the ship's shell. Three of the - most likely originally four - transverse beams were pegged to the gangways, which lay on frames that did not fully extend upwards; the fourth was located under the foredeck. The connection to the sides was formed by standing knees, placed on the crossbeams and fixed to the upper side with treenails. The vessel had an afterdeck and a foredeck, connected by gangways in the sides, a mast that could be struck, probably with a sprit sail woven of wool, and was equipped with a stern rudder. The line plan or rather the plank plan shows a relatively low-sided hull with a straight keel, little sheer and a straight, raking stern and stem. At the rear, the stern line is at an angle of 119°, in front under 122° with respect to the keel line.

Amidships, the top three strakes are attached to each other in an almost straight line at an angle of 112° The bilge is rounded at an angle, but rounded further to the rear and front. The frames in the afterbody and forebody are weakly S-shaped, while the forebody is fuller than the raked stern. At the fore, the ship is approx. 10 cm wider than at the comparable location aft. The horizontal plane of the sides therefore shows an almost oval shape, tapering towards the stern and stem. The upper three shell strakes continued to the rear and front of the stern- and stemposts. At the fore, the ends of these strakes were protected by a false stem, at the aft, they ended up against each other on the roof shaped back of the stern. The part of the hull visible above water was clinker-built. The vessel was expertly built according to the traditional cog building tradition and in a way that may best be described as a combination of 'bottom first' and 'shell first'. The 'brother and sister' pattern of the shell strakes used during construction is remarkably symmetrical.

The clinker-built shell seams are only sintel clamped on the inside, the bottom seams in the carvel-built part only on the outside. What is striking is the complete absence of dry cracks in the outside of the skin, which would have to have been sealed with sintel clamped moss caulking. This indicates an effective and fast construction. The manner of sawing and of placing the hull planks will undoubtedly have contributed to the prevention of these cracks. The use of treenails to attach the floor timbers and the V-shaped floor timbers on the keel plank and the fact that the nails on the scarfs between the keel plank and the stern- and stem hooks did not penetrate the wood, indicates that the shipbuilder acted in a very well-considered manner. This considerably reduced the loss of strength of the ship's backbone.

The most important, reconstructed main dimensions are:

- Length over the stern and stem at gunwale height — 15.2 m
- Largest width at gunwale height — 4.8 m
- Height of the stern (keel line to the top gunwale) — 2.2 m
- Height of the stem (keel line to the top gunwale) — 2.7 m
- Height of the side amidship (keel line to the top gunwale) — 1.7 m

The completely closed ceiling or loading floor and the closed ceiling in the sides, together with the bulkheads that may have closed off the cargo hold aft and fore, indicate that the ship must have been intended for the carriage of loose cargo and (bulky) piece loads. This is supported by the formwork around the mast (which could be struck). The two large hold openings between the gangways are another indication of this. It is unknown whether hatch covers were used to close the hold above these openings at the time. Until now, it seems that this was not the case in the period in question and with this type of vessel. The load of bricks and the indications of the earlier transport of similar loads are examples for the possible cargo carried by the vessel. The ship, which is not very large, appears to be relatively heavily built in terms of stern and stem construction and the thickness of the planks used. This fact supports the suspicion that the vessel may have been designed specifically or mainly for the transport of heavy loads/building materials.

Based on the typical characteristics, the shape of the hull and the design, it can be assumed with great certainty that this is a small cog for the transporting of cargo on inland waterways and the Zuiderzee. The mast that could be struck and the relatively shallow draught of the hull demonstrate that the ship was able to sail into or close to a city; for example until near a building site of houses, city walls or a castle. The width (4.8 m) of the vessel indicates that the cog probably could not make use of the Gouda Inner Lock and many other Dutch locks during its sailing period. The repairs to the vessel show that the vessel achieved a relatively high age (approx. 30-50 years), while the find location of the vessel suggests that the ship was abandoned and sunk on purpose either in an inlet next to the approach route to Bunschoten/Spakenburg or perhaps even used as a 'caisson' in a dike in which, for example, a hole had been dug during a heavy storm. It also seems not inconceivable that the vessel had a relationship with Bunschoten (home port?), and perhaps even was built there.

Equipment and inventory

Only a few discarded parts of objects were found, which were once part of the inventory and the personal belongings of the crew. The remnants of the woven woollen sail and the compass glass may be regarded as unusual finds. The roof tile fragments probably come from the Sticht of Holland.

Dating

The two sherds of stoneware belong to jugs that may be considered common goods during the second half of the 14th century. The footwear types were in vogue in the last quarter of that century or at the beginning of the 15th century. The sintel clamps used in the hull were probably made in the last quarter of the 14th century, which roughly indicates the construction period of the ship. Dendrochronological research of the stern and stem construction can probably provide more certainty in this respect. The shoes and sherds are from old, discarded and broken objects. It can therefore be assumed with considerable certainty that the vessel sank in the first half of the 15th century, most likely in the period around 1410-1435. The estimated lifetime of the cog falls within the period that Bunschoten was a walled city (1383-1428).

Impression of the reconstructed Spakenburg-II cog with four blocks and the vertical beams for a construction.

The Spakenburg-II cog[1]

15

15.1 Find conditions

15.1.1 Discovery and investigation

On 19 July 1971, a survey was carried out on lot Nz 43 in the southern part of Zuidelijk Flevoland. A few days earlier, a shipwreck had been discovered during the clearing of the drainage trenches. After checking the position with prods and examining the exposed pieces of wood, it was assumed that this was the bottom of a rather disjointed wreck of a clinker-built vessel. The area in which the wood was recorded covered approximately 10x4.5 m, while the wood lay no deeper than 0.5-0.6 m below the present surface and in the Pleistocene sand. On the basis of these data it seemed plausible at the time that these were the remains of a vessel probably dating to the 16th/17th century, which did not need to be reserved for systematic research, but which were interesting enough for further investigation.[2]

In the autumn of 1978 it became known that recreation park Eemhof would be constructed on lots Nz 42, Nz 43 and Nz 44. In view of the intended purpose for these lots, a *watership* on lot Nz 42, a second one near the boundary ditch between lots Nz 43 and Nz 44[3], a second wreck on lot Nz 42 (the Spakenburg-I cog, Chapter 14) and the wreck to be discussed here had to be excavated as soon as possible. Based on its shallow location in the Pleistocene

Fig. 15.1. Overview of the excavation with the frame parts found separately repositioned in their original position and the light coloured sand filling of the hole caused by the ship in the Pleistocene surface.

1. Translated by Stefanie Hoss.
2. The shipwreck lay between the 45th and 46th ditch (560 m from the centre of the ditch Nz 43/Nz 46), 185.5 m from the Nz 42/Nz 43 ditch. It was located with its stem to the west-southwest, about 1.2 km northeast of the current harbour entrance of Spakenburg (figs 15.2 and 15.3) and only 400 m east of the site of the Spakenburg-I cog (chapter 14). Topographical Map 32 B, with the highest lying parts at ca. 25 cm below the surface (ca. 2.2 m -NAP). Locally, the former water depth had been approx. 1.6 m Hydrographic Map 1921, average low tide.
3. Pedersen, 1996; file Nz 43/44, RCE-Lelystad.

Fig. 15.2. The find location of the Spakenburg-II cog.

Fig. 15.3. Ordnance map of the Kingdom of the Netherlands, 1861 (scale 1:50.000), showing the location of the Spakenburg-I and the Spakenburg-II cog (right) and Hulkestein Castle.

sand and other data from the initial inspection, it was assumed that the wreck of lot NZ 43 could be excavated, documented and cleared fairly quickly. However, at the beginning of the excavation it became clear that this was a relatively complete vessel with cog-like characteristics.

In 1979, the permanent staff of the Department of Nautical Archaeology of the RIJP excavated the wreck between 27 April and 31 May (fig. 15.1), and the timbers were then transferred to the rinsing floor at Ketelhaven so they could be drawn in due course. Initially, a rectangular trench was staked out around the wreckage and a baulk was retained in order to study the disturbance of the soil profile. The northern trench was later extended to accommodate the detached ship's timbers lying next to the wreck and to record a disturbance caused by the ship in the top of the Pleistocene surface. An overview drawing was made of the excavated situation and of the parts of the ship lying above and beside the hull. The locations of the finds are also indicated on this drawing (Plate 15.I). The finds will be discussed in section 15.5 and mentioned in the find list.

In 1985, Aleydis van de Moortel was looking for a nautical-archaeology research project to obtain a degree from Texas A & M University, USA. H.R. Reinders, head of the Department of Nautical Archaeology of the RIJP, drew her attention to the as yet unpublished file of the cog-like vessel in lot Nz 43. In 1986/87 Aleydis completed the report in Texas. The result has been published in the ' *Flevoberichten*' series of the RIJP.[4]

New insights resulting from research on the other cogs and additional information, such as dendrochronological results, indicated that it would be useful to re-examine the vessel and investigate it in a standardized way together with all other cog finds from the Low Countries.
For the presentation in this study, the overview drawing of the situation as excavated and the reconstruction drawings are sufficient. The available (field) data have been used for this purpose.

15.1.2 Geology and stratigraphy

The Pleistocene surface and covering deposits
During the 1971 investigation, the soil profile around and above the wreckage was found to have been barely disturbed. Only the top 25 cm had been lightly ploughed. The southern part of Zuidelijk Flevoland has a very irregularly structured Pleistocene sand surface, which greatly varies in height (0.25->1.5 m below the present surface). On top of this sand, in which the A and B horizons are still largely present, remnants of the Holocene peat layer are still present in the deeper parts.[5] The wreck lay at the highest point (0.3 m below the present surface) of a sand ridge, directly on the eastern margin of a more than 1 km wide depression in the Pleistocene surface. At a good 100 m to the west, the depression was already 1.5 m deep and within 1 km quickly deepened northwards to > 2 m.

The site is located approximately 4.5 km east-southeast of the mouth of the river Eem. In view of the disturbance of the Pleistocene sand surface (fig. 15.4), it seems likely that the ship, which had probably already sunk in the vicinity, was washed against the sandy outcrop and became stranded against it. It then worked its way into the sand. On both the Pleistocene sand next to the wreck and the redeposited sand with thin bands of humic Almere-clay mixed with clay found inside of the ship, there were still

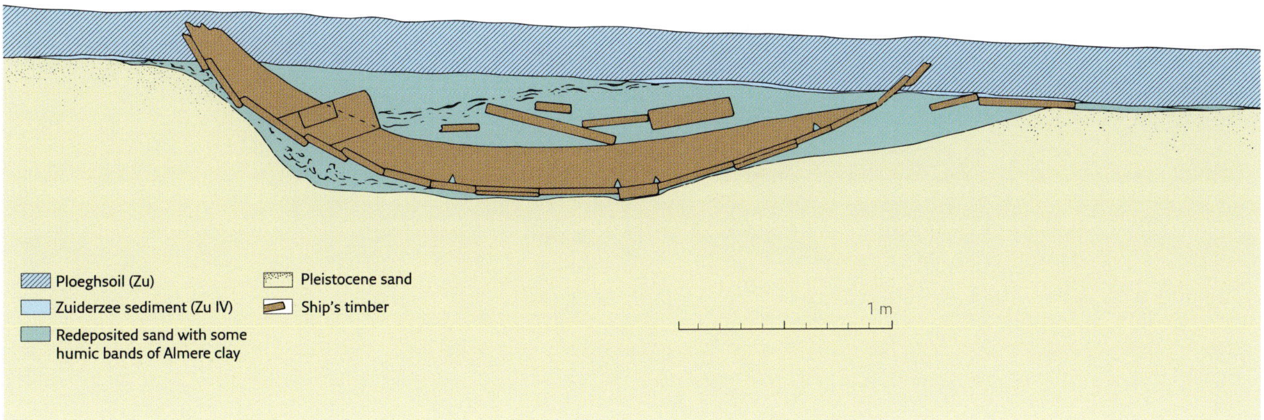

Fig. 15.4. Reconstructed soil profile at frame 13/14 with the disturbance in the Pleistocene surface by the ship.

4. Van de Moortel, 1991.
5. Koning & Ente 1986, explanation of sheet III of the soil and profile map of ZuidelijkFlevoland.

some remains of the lowermost Zuiderzee deposit (Zu-IV-shell layer). Above it was the topsoil consisting entirely of Zuiderzee clay. There was no question of a 10 cm thick Almere deposit as reported by Van de Moortel.[6] The soil profile does not provide any closer information for the time of the sinking of the cog than that this must have taken place (well) before ca 1600.

15.1.3 Prehistoric and medieval habitation

In the Pleistocene sandy outcrop on which the ship was stuck, two prehistoric hearths and some pieces of worked flint were found during the investigation. They may date from the Mesolithic or Neolithic periods. The nearest medieval settlement was Bunschoten, with Oud-Naarden a little further to the west.

15.1.4 Condition and wreck formation

Condition

Despite the shallow location of the wreck, the parts that still were more or less joined together were of a reasonable to good quality. The detached stem and some thin shell planks lay within the redeposited sand or on the Pleistocene surface. They were covered with Zuiderzee deposit/topsoil. They were in poor condition (fig. 15.5) and could only be measured and drawn on site.

Wreck formation

It was clearly visible in the planed bottom of the trench on the sandy outcrop that the ship initially must have drifted up the sandy outcrop almost from due north with the stern first, initially, before becoming stuck. The vessel then rocked in the water that was about 1.5 m deep on the spot at that time and created a hole in the sand. Presumably after some time, the cog again shifted some 4-6 m southward, while (at the same time) the forebody turned more to the southwest. The ship finally remained lying with a slight list to port.

The cog probably foundered near what then was the southern coast of the Zuiderzee, which was most likely further north than the current shoreline (see also Sections 14.1.2 and 14.1.4 of the Spakenburg-I cog). The vessel was unable to work itself into the hard sand any further and therefore, from the moment of sinking, it had protruded from the seabed for the most part. The disintegra-

Fig. 15.5. Overview of the badly preserved shell planks at the portside forebody.

6. Van de Moortel 1991, p. 17-18 and fig. 6.

tion of the ship will therefore have started fairly quickly. It is striking that the forward half of the I-strakes with pieces of the inwale was still present, while the planks of the four strakes below had (largely) disappeared. This probably has to do with the parts of the connecting transverse deck beams that were also found, which may have remained joined to the I-strake planks for some time before disintegrating. The first parts that came loose probably drifted away on the water. Given the shallow depth of the water and the proximity of the coast it can also not be ruled out that the highest parts, such as the decks or cuddies, were demolished and carried off by salvagers. Other parts first became partially loose and ended up beside the hull, where they were eventually covered with sediment.

The keel plank, stem hook and stern hook of the ship were still present and a large part of the stempost. Of the starboard side almost the entire first three strakes (A to C) remained, strakes D and E for more than three quarters and half of the F strakes. Of strake G only the middle plank (of three) was still present. The H-strake was missing, while the forward half of the I-strake was still present. On the port side, the first four strakes were fully preserved. Of the next three (E, F and G) the planks were missing at the stern and stem. Of the H-strake, the middle two (out of a total of four) planks remained. As with the starboard side, the I-strakes on the port side were still present along the forward half (Plates 15.II A and B). Fragments of the inwale were still attached to these I-planks. Of the frames, all of the (V-shaped) floor timbers were discovered on the bottom, as well as a number of the futtocks on the sides. The V-shaped floor timber (A 26, Plate 15.I) which connected both sides above the stem post as frame 21 was also found. Furthermore, some planks of a small floor amidships, a transverse bulkhead, pieces of several transverse connecting beams, a chock with a hole cut in to take the heel of a mast or a stanchion against the port side and a few knees that once were placed on the transverse beams or a cuddy.

15.2 The ship

15.2.1 Documentation and reconstruction

Documentation

In addition to the overview drawing of all parts of the ship found *in situ*, supplemented by the locations of the finds discovered (Plate 15.I), the parts found separately have been returned to their original places as far as possible (figs 15.1 and 15.6). One section on the longitudinal axis and seven cross-sectional drawings (all to a scale 1:20) were drawn of the ship in this state. Photographs and slides were taken and the Geodesic Service at Delft made photogrammetric recordings from a Royal Dutch Air Force helicopter. During the excavation, the film-maker of the RIJP shot several takes for a film about the activities of the IJsselmeerpolders Development Authority and for a possible documentary/film about ship's archaeology in the IJsselmeerpolders. In 1983, all parts stored at Ketelhaven were drawn to a scale 1:10.

Reconstruction

The hull, together with the loose timbers found within and besides it provided sufficient information to largely reconstruct the ship. Van de Moortel made cardboard replicas of the parts drawn 1:10 and used them to build a model. The highest surviving piece of shell plank (strake I) on the port side in the forebody with a piece of inwale on the inboard side made it clear, together with the straight and horizontally finished heads of three futtocks (frames 9, 10 and 11) on the port side, that the I-strake formed the top of the gunwale (Plate 15.I).[7] The corresponding part of the starboard I strake was recovered as well. This gunwale formed also the top for J. van der Zee's construction of the oak scale model (see section 15.4).
Van de Moortel mentions some knees (A 1 and A 20, Palte 15.I). However, the crooked piece of a frame A 26, which she also interpreted as a knee, actually is the aforementioned frame 21. Van de Moortel assumes, on the basis of their shapes, that the two knees had been used somewhere amidships. On other cog-like ships, such knees are attached to the deck beams. In larger cogs, knee-shaped bulkheads are usually put in such places, to which the upper strakes of the sides are treenailed.[8]

Fig. 15.6. The repositioned frame parts, the chock with a hole against the port side and the remaining planks of the ceiling/loading floor.

7. Van de Moortel 1991, figs 62 and 67.
8. Van de Moortel 1991, p. 77-79.

15.2.2 Construction and layout of the hull

Keel plank

The keelplank is 6.5 m long and has a width of 19 cm at its forward end and 15 cm at the after end of the scarf joining it to the stern hook. At the beginning of the scarfs, the keel plank is 5-5,5 cm thick, in the middle 5,5-6 cm. Sloping flat scarfs between the keel plank and the stern- and stem hooks, running from fore to aft (in the direction of travel) is common for cogs. The scarf to the stem hook is 30 cm long; and that to the stern hook 20 cm. However, the latter does not run as described above, but the other way around. This will be followed up in the description of the hooks and the section dealing with *repairs*. The keel plank is fastened to the hooks with just a few nails.

The total length of the keel plank plus the horizontal parts of the hooks attached to it is ca 9.3 m. Van de Moortel assumes that the keel plank with the horizontal parts of the hooks never formed a straight line, but that the keel line rose at the stern and stem. She considers it very likely that the vessel was deliberately built in this way.[9] However, the field data show that the keel plank has an irregular, straight bottom line, the front part of which is slightly bent downward. The adjoining stem hook continues in the same downward-curved line. The bottom line of the horizontal part of the stern hook is also curved and runs down towards the end. The keel plank and stern hook thus sag towards the stem and stern; this deformation is found regularly in shipwrecks (in the IJsselmeer polders) and is known as 'hogging'; it is the result of the wreck formation process: currents whirl around the stern and stem of a ship that has already partially subsided into the seabed, which often cause the sediment to wash away. Because of their weight, the stern and/or stem often slowly subside. Depending on the lines of the ship, the strength of the hull and the seabed structure, this deformation can be small or even considerable. Even a heavy load may not suffice to counteract this.[10] The Spakenburg-II cog was undoubtedly built with a straight keel line. Between frames 13 and 14, the top of the keel plank was abraded, very likely a sump directly forward of the floor.

Stern hook and skeg

The horizontal part of the stern hook has a length of 1.76 m. The tip of the scarf with the keel plank has broken off, so that the original length of the stern hook will have been ca 1.85 m. During the drawing of the ship in 1983, the top of the part of the keel plank scarf was damaged. According to a note, there was a slight hollowing between frames 3 and 4 (a second sump?). It is not clear why the scarf was placed the wrong way around. The most obvious explanation is that it was a replacement. This will be discussed in the section on *repairs*. The thickness is 8 cm near frame 3, increasing to a thickness of 14 cm towards the base of the rising stern part. The width of the horizontal part is 15 cm near the scarf and 8 cm at the base of the rising part. The back of the rising part is 4-5 cm wide. Both the horizontal and the rising part are trapezoidal in cross-section, with the narrow side on the outside. The stern/rising part protrudes 28 cm above the keel line at the rear and has a sloping flat scarf of 30 cm long, which starts at the top of the back and runs down to the horizontal part. In the very top of the rising part is a nail and just below halfway down the scarf, a treenail (Ø 2.3 cm), both of them fastening the hook to the sternpost. At the top in the sides of the horizontal part, rabbets have been cut that start at about 40 cm aft of frame 3 on 0 and are at the rising part 2 cm deep and 4 cm wide. The hood ends of the garboards or A strakes were nailed into these. The rear line of the stern has a rake of about 128°.

A skeg was nailed under the stern hook. The piece was 1.16 m long and protruded 5 cm at the back. The skeg had the same shape in section as the horizontal part of the stern hook, was 5 cm thick at the back and 2 cm at the front. The protruding rear end prevented floating debris from getting caught between the stern hook and the rudder.

The sternpost

Van der Moortel describes the timber A7, which was found next to the hull (Plate 15.1) as 'probably a piece of the sternpost'.[11] According to the list of timbers found loose next to the hull, the piece, which is ca 1.5 m long is a 'thin piece of planking (possible fragment of a shell strake)'. The piece was not drawn true to scale in 1983. Of the stern, nothing was found. The vessel did not have a false stern.

A sternpost had originally been placed on the sloping flat scarf of the stern hook, whose shape concurred with the shape proposed by Van de Moortel. This beam was about 8 cm wide on the inner side, 4-5 cm at the back and had a thickness of about 12 cm. The lower point rested in a groove that was V-shaped and 1-2 cm deep which had been cut into the keel part of the stern hook. A rabbet of 3 cm wide and 2 cm deep had probably been hacked in the sides along the front, into which the ends of the shell strakes of the forbody had been nailed. For the execution, see the piece of the stem that has been found.

9. Van de Moortel 1991, p. 36.
10. Wrecks with cargoes of bricks have been found several times in the IJsselmeer polders. Despite these heavy loads, there was sometimes hogging of more than 0,5 m.
11. Van de Moortel 1991, 43-45 en fig. 37.

Rudder

On the stern hook there are no indications of a gudgeon. The lower gudgeon had probably been attached to the stern beam just above the sloping flat scarf. Despite the lack of concrete indications, it can be assumed that the vessel had a stern rudder.

Stem hook

The horizontal part of the stem hook measures approx. 0.95 m. The front ends in a vertical line and forms a small lace piece, as it were a protrusion on the fore side of the stempost, which must have added to the vessel's stability and sailing qualities. A triangular piece of approx. 35 cm long was inserted at the bottom of the sloping flat scarf, a repair that considerably weakened the connection between the keel plank and the stem hook. In order to overcome this at least to some extent, a small board and a filler were placed on the top of the keel plank and the stem hook. They were clamped under the V-shaped floor timbers of the frames 18 and 19 (fig. 15.7 and Plate 15.III A). This construction will be discussed again in the chapter on *repairs*. At the flat scarf, the keel is 19 cm wide and 7 cm thick, while it is 12 cm wide and 14 cm high or thick at the beginning of the stern section. The forward face has a rake of approx. 132° and rises approx. 50 cm above the keel line. The stem section has a sloping flat scarf of 60 cm long, running down to the keel section. Similar to the stern hook, rabbets of 0-2 cm depth and approx. 3 cm width were hacked into the sides of the horizontal section. These rabbets continue at an acute angle into the stem section and in the stempost. Approximately halfway through the scarf, there is a treenail for the attachment of the stempost. At the highest point, which is broken off now, there probably was a nail.

Stempost

With the lower part more or less in its original position, the stem was folded forward (fig. 15.8 and Plate 15.I). During the excavation, this piece still measured approx. 2.25 m in length. During the drawing in 1983, there was still about 1.4 m left. The missing part was of very poor quality and could not be retrieved. The inside of the beam is 12 cm wide, the outside 4 cm. The thickness is 15 cm. On the inside of the beam, rabbets had been hacked into the sides, which were 4 cm wide and 2 cm deep. The plank ends of the hood ends had been nailed into them. A hole (Ø 3 cm) was drilled through the stem at 90 cm from the lowest point of the scarf. This made it possible to fix a rope to the ship in order to pull it onto a slope. The lower 1.2 m was almost straight. The remainder of the stempost was of poor quality and lay in the ground slightly bent forward. Van de Moortel assumed that the resulting curved shape of the stem had been the original shape.[12] Despite the fact that some seals with cogs do show such a curved stem, it is far from certain that this also was the case for the Spakenburg-II cog. Rather, this seems to be a result of the deformation of the piece through warping and the poor state of preservation in which it was found. The oak scale model by Van der Zee (Section 15.4) is executed with a straight stem. This is also assumed in this study (Plates 15.II and 15.IV A). The vessel did not have any false stern either.

Fig. 15.7. Stem hook and the sharply narrowing forebody.

Fig. 15.8. Overview from forebody to afterbody. In the foreground the stempost fallen forward.

12. Van der Moortel 1991, 39-45

Fig. 15.9. Overview of the preserved shell. Seen from forebody to afterbody.

Bottom and sides

The bottom seems to be composed of five strakes on either side of the keel plank, placed carvel-fashion against each other; these merge into a clinker-built construction aft and fore. However, the ends of the planks are chamfered over a length of some tens of centimetres, again placed more or less carvel-fashion against each other and smoothly nailed into the rabbets in the stern post (Plates 15.III B and 15.IV A). The bottom has no dead flat, either in the length or in the width, but in cross-section runs smoothly in a slightly curved line to the turn of the bilges and the sides. The turn of the bilge is thus 'round' on both sides, with only a slight kink to be observed between the third and fourth strakes between the frames 8-12. If this is the actual turn of the bilge, then the bottom is composed of six strakes plus the keel plank, with the lower two strakes of the sides built carvel-fashion.

Bottom

The garboards or A strakes and the B strakes on the port side are composed of three planks. A long middle one in the carvel-built part and a shorter one in the clinker-built part afterbody and forebody. The C strakes are composed from five short planks, while the D- and E strakes were each composed of four or five planks. The plank scarfs are chamfered and 25-35 cm long. The ends of the planks are 0.5-1 cm thick and protrude from the skin both outside and inside (fig. 15.9). The thickness of the different planks (and even within a single plank) varies between 2.5 and 3.5 cm, while the planks are just 2 cm thick at their ends, which sit in the rabbets of the stern- an stempost. The middle planks of the garboards are about 40 cm wide and become narrower towards the stern and stem. The other strakes of the bottom vary in width between about 20 and 30 cm. The middle planks of the A strakes are about 4.85 m long, all the others are shorter.

Sides

The F strakes were the first of the side strakes to be completely clinker-built. But it is very likely that in the last 10 to 20 cm, these also turned into carvel and smoothly sat in the rabbets. Due to the absence of parts of, and even entire planks, it is not possible to indicate the number of planks per strake with any certainty. The number of planks for the F strakes is at least four, possibly five. For the G strakes the number is at least three, but quite probably four planks. The H strakes also probably had four planks, while the I strakes very likely only had two. The F- and H strakes are quite narrow (approx. 25 cm), while the G- and I strakes were approx. 40 and 43 cm wide, respectively. The overlap (the width of the land) of the clinker built planks is 4 cm, and in some places even 5 cm. These planks were fasted with nails, which had been bent over twice and driven back into the wood, generally, every 15-16 cm. However, in several places the distances between the nails are very irregular and vary between 7 and 30 cm. Whether the closely spaced nails were caused by introducing extra nails (e.g. repairing leakage of the seam) cannot be determined on the basis of the recorded data, but seems to be quite possible. The planks are attached to the frames with treenails.

The few loose knees or parts of knees found separate from the hull may be assumed to have been attached to transverse beams on a deck or cuddy.

Sintel clamped moss caulking

The seams of the carvel- built part of the bottom have been made watertight on the outside with sintel clamped moss caulking. The moss laths used for this were made of oak and 1-1.5 cm wide. They have an elliptical diameter and a thickness of 0.5 cm. The sintel clamps are of the transitional type D2/E and have a central section of approx. 35 mm long. This type was in use during the period 1350-1425.[13] The seams of the clinker-built parts of the vessel are sintel clamped on the inside, and in some places also on the outside. The latter were very likely repairs, just like the nails in the clinker-built part mentioned above.

Van de Moortel writes that insufficient information has been found to reconstruct the shape of the sintel clamps. But actually, this information has been found (figs 15.10a and b). The sintel clamps were applied room-and-space

Fig. 15.10a. Clamped moss caulking in the seams between the clinker-built shell strakes on the inside; b. Sintel clamp.

13. Vlierman 1996a, 79-48.

every 5.5-6 cm and, according to her, overlapped each other by about half if they would have had a mid-section of 50-60 mm.[14] However, that is not the case at all. The moss lath was visible over a length of approximately 2-2.5 cm between the clamps. On the outside, cracks had been filled in various places with sintel clamped moss caulking and in some short parts of the clinker-built seams. These cracks most likely arose during the construction of the vessel and the pieces of clamped moss caulking in the clinker seams on the outside are probably repairs. A total of 200-250 m of seams and cracks were sintel clamped.

Identification and pollen analysis

One species of moss has been used for the caulking, peat moss (*Sphagnum spec.*): trees 21%, herbs 79%. The pollen found with the moss indicate that the moss was collected somewhere on the richer ground of a river valley with tree growth. There was also some arable farming of perhaps rye and buckwheat in the vicinity.[15]

Floor timbers, V-shaped floor timbers and futtocks

The transverse stiffness of the hull was mainly guaranteed by the frames. Of these, twenty sat above the keel plank and the horizontal parts of the stern and stem hook (frame 1 to 20), two against the stern (0 and -1) and three against the bow (21-23), a total of twenty-five pieces. At the floor timbers above the bottom, the scarfs with the futtocks were not consistently fixed alternately to the starboard and port sides. The scarfs of the floor timbers at the utmost rear and front are placed above the second to fourth shell strake on both sides of the keel plank, while they are at the fourth to sixth shell strake for the other floor timbers. Frame 14 is made up of two croked timbers. Particularly striking are the eight more or less straight floor timbers of the frames 6 to 13, on which the floor was placed. They are 14-20 cm wide. The frames in the stern are slightly more irregular in shape, but have similar widths. Remarkable are the first four frames (14 to 17) in the foreship, directly in front of the floor just mentioned. They vary in width between 20 and 30 cm. The ones at the front (frames 18 to 20), are again comparable in width to those of the afterbody. The V-shaped floor timber against the afterbody (frame 21) was about 20 cm wide. Striking is the larger spacing between the frames 13 and 14 in the sides, but also the difference in spacing between the frames of the afterbody and the foreship in comparison to the spacing ´between the frames under the floor. In addition, a number of futtocks were found that were thicker at the top and that partly extended to the upper edge of the inwale or even higher. These fasted the inwale to the shell, so to speak. In the reconstruction, they were placed near the frames 0, 5, 18 and 22 (Plates 15.III A and B and 15.IV A, frame 5).

Mast step?

Up to now, small cogs – a group into which the Spakenburg-II may be included – have been found with short mast steps, which were attached to the frames above the keel plank. Another manner was to fashion a hole for the mast's heel into either a heavy floor timber or a V-shaped floor timber. No indications of this were found on this vessel.

On the port side there was a heavy chock of 100 cm long, 47 cm wide and 20 cm thick against the lower two strakes (F and G) of the side. One of the inner sides of the chock is straight. On the side of ship's shell, the chock is notched at the frames 13 and 14. The part in between joins up to the shell. The chock is attached to the frames and the shell strakes with four treenails (Ø 3-3.5 cm). In the upper part, a square hole of approx. 17x17 cm and 9 cm deep has been cut in the part between the two frames. A limber hole had been drilled in the bottom of the hole (Ø 2.5 cm). A U-shaped piece of iron plate was attached around three sides of the top of the hole (fig. 15.12a). Noteworthy is a concave wear spot (Ø 8-10 cm) in the top of the chock, directly in front of the square hole with the fitting in the corner against the back of frame 14 and the inside of the shell (fig. 15.12b).

On the starboard side, the shell planks at the frames 13 and 14 were only present up to and including strake E. Next to the afterbody, a piece of plank lay on the starboard side, which had a treenail hole with a diameter of 3-3.5 cm, presumably from strake G. It seems quite possible that a similar chock was attached to the starboard side. More or less at the same location, a similar piece of iron plate was discovered there as had been found on the chock on the port side.

Van de Moortel rightly notes that such chocks have not been discovered earlier and that their function is uncertain. According to her, the most likely explanation is that the chocks were used in order to fit heavy vertical beams in them, for some kind of hoisting construction. Two lifting devices independent of each other would exert enormous pressure on one of the sides; in addition, the shape of the ship is absolutely not suitable for this. She also discusses the possibilities of a tripod (twin mast) and a tabernacle mast.[16]

At frame 5 on the port side, a mortise with a diameter of 3-3.5 cm was noticed in strake G. In a fragment of a separately found piece of shell plank, a mortise with the same diameter was discovered. With all due caution we

14. Van de Moortel 1991, 68.
15. Mook-Kamps 1997, sample 11, see appendix.
16. Van de Moortel 1991, 81-85.

Fig. 15.11. Overview of the wreck with repositioned frame parts and the light-coloured filling and 'route the ship took' (the light stripe in the top left corner of the excavation pit).

Fig. 15.12a. The chock with the hole with iron mounts on three sides. Seen from the front.

Fig. 15.12b. The same from above.

can speculate whether the vessel did not have four such chocks instead of two (Plates 11.III.A and B). However, the construction may also have been different. We will return to this and to the possible meaning and function of the chocks in section 15.3.4.

Work floor/ceiling and the transverse bulkhead(s)
Three ceiling planks were found in situ in the wreck (Plate 15.I and figs 15.6 and 15.8). The exact length of two of them could not be determined because they had broken off at the rear end and the missing pieces were no longer available. Their width was approximately 40 cm, while the thickness varied between 2.5 and 3 cm. The third board was 2.4 m long, 32 cm wide and 1.5 cm thick. The planks were fastened to the frames along the edges with nails. It is very likely that the ship had a work floor between a few centimetres from the rear of frame 6 and a few centimetres from the front of frame 13. The floor had been approx. 2.75 m long and probably approx. 2 m wide at the rear and approx. 2.3 m wide at the front (Plates 15.III B and 15.IV A).

Loose and partly above the work floor lay a 40 cm high and 4 cm thick wooden bulkhead with a straight top and a curved lower line (A 32, Plate 15.I). One end tapered to a point, but was broken off. The other side ended more or less straight and had a height of about 18 cm. A part was broken off along the top and there was a nail hole at the edge. Initially it seemed obvious that the bulkhead had been on the front of frame 13, directly against the front of the work floor. But the shape of the top of frame 13 is flatter than the shape of the bulkhead. It does however fit exactly on frame 14, joins against the chock on the port side and ends on the starboard side against the frame. The latter means that the possible chock on the starboard side did not continue to this place (as it does on the other side), but that it possibly ended shortly before it. Nor were any treenails with a diameter of 3 cm found in the starboard crooks of frame 14, which may imply confirmation of the above (Plates 15.III B and 15.IV A). It does not seem unthinkable that the stern also had a transverse bulkhead. If the construction of the forebody and afterbody were more or less the same, this bulkhead would have stood at frame 5. We have already observed that a mortise with a diameter of 3 cm was discovered in strake G on the port side at the point of frame 5. This may indicate the attachment of a chock.

Ceiling and inwales
Except for the planks of the work floor or ceiling, no indications for ceiling planks were found in the ship either in the forebody or in the afterbody of the sides.

A reinforcement of the top of the ship's hull with an inwale could be established. On the port side, between the strakes 8 to 15, a part of the I-strake with the inwale was still connected to the hull. The part that went up to the stem lay broken next to the hull. From the starboard side inwale, two pieces were found next to the hull that represent approximately the same part of the inwale of the I-strake plus the front part of the inwale (Plate 15.I, hull 1 and 2 port side, hull 1 and 2 sb and A 41). The front inwale part of the starboard side is 2.65 m long, 8 cm wide at the top and 6 cm wide at the bottom and has a height of 14 cm. The height and width of the inwale gradually decrease towards the stern to 7 and 3 cm, respectively. At the other end the piece has a flat scarf of 40 cm length. The part of the inwale amidships is 14-15 cm high and has a width at the top that increases from 8 cm to 9,5 cm at the centre of the ship's side. The pieces of inwale are attached to the I-strake with thin treenails (Ø 2 cm), in many cases set in pairs. Between frame 20 and 21, there is a treenail with a diameter of 3 cm both on the starboard and port sides of the ship. They may be an indication of a knee which had been placed between the frames on the front cuddy. In the port fragment of the inwale, four other thick treenail fragments were observed between frame 21 and the bow. Perhaps these were the attachments for an extra piece of wood (fig. 15.13a) as part of, for example, an anchor windlass?

Tranverse beams and decks
A beam was found on the port side next to the stern (A 37, Plate 15.I). One end was complete and finished at an angle and has an oblique notch at the bottom. This end had been attached with an oblique pin (Ø 3 cm). The other end was broken off. The beam was 17 cm high and 13 cm wide and fits on the top of the bulkhead and the chock on the port side and was probably attached to frame 14 and hull plank G with the treenail. The missing starboard side probably continued in a knee shape to the top of the hull. On the port side, a futtock had probably been placed at the end of the beam, which had been secured with 3 cm thick treenails. This may be the short futtock that was discovered laying next loosely on the chock between the frames 13 and 14 during the initial phase of the excavation (fig. 15.14 and Plates 15.I [A 43] and 15.IV A).

Four further pieces of crossbeams or deck beams were found above the foreship (A 23, A 25, A 27 and A 31; Plate 15.I). They were all broken off on one side and A 25 on two sides, so that the original lengths cannot be determined. Those ends that were complete are all chamfered. In the chamfered part of A 23, there was a treenail hole with a diameter of 3 cm and there were three nails in

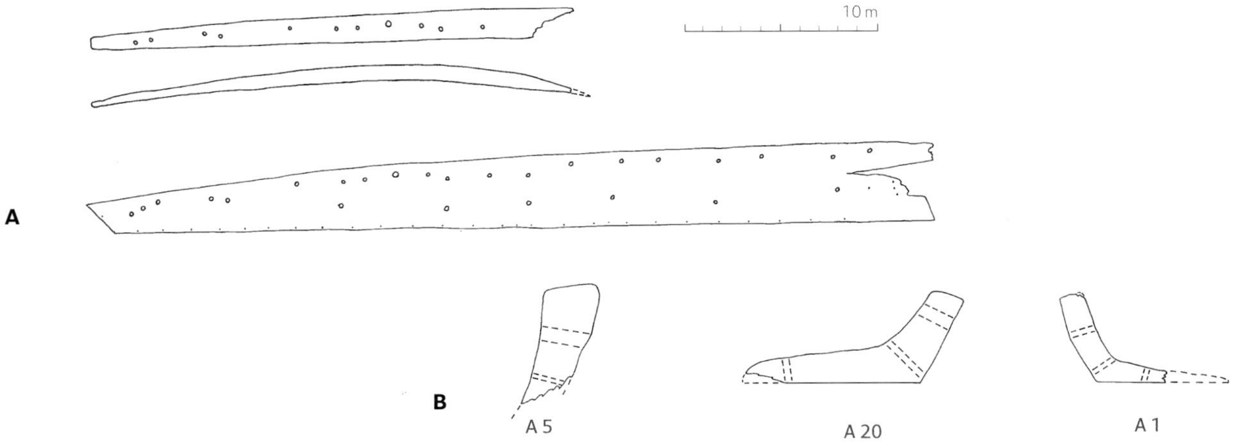

Fig. 15.13a. Drawing of a piece of the I-strake with inwale from the starboard side of the forebody; b. Drawing of (deck)knees A 1, A 5 and A 20.

the top at 90, 105 and 115 cm from the chamfered end. The cross beam A 25, had two nails and A 27 had a treenail hole of 2.3 cm diameter in the chamfered part and a treenail in the top at 10 cm from there. A remark in the drawing of this piece mentions that 'there were probably planks against the top of this side'. A 31 was unfortunately not drawn on a scale of 1:10. The width and height of the beams A 23, A 25 and A 27 are approx. 10x10 cm, that of A 31 was approx. 17x12 cm. It seems very plausible that these were small deck beams from the forebody, which were attached to the inside of the frames. This may have been equal in height to the top of the cross beam above frame 14, but more likely slightly higher (Plates 15.III A and 15.IV A).

If we look at the shape of the hull in this area, it would be possible to walk on the V-shaped floor timbers of frame 15 and the spaces between frames 13 and 14, between 14 and 15 and between 15 and 16. This is hardly possible in the front part of the vessel. It therefore seems very plausible that there was a small deck or cuddy of about 3.25 m long from frame 16 to the stem. The situation is similar in the afterbody: while it may have been possible to walk on the V-shaped floor timbers of frame 4, and between frames 3 and 4 as well as between 4 and 5 and between 5 and 6, this is not the case more aft. From frame 3 to the stern, the vessel would have had a cuddy of about 1.75 m in length. We will return to this in the part on *Function*.

In addition to the pieces of transverse beams or deck beams, some small, upright knees or pieces thereof have been found (A1, A 5 and A 20, Plate 15.I and fig. 15.13 b). They probably stood on a transverse beam or deck. In the top of A 5, there is a hole with a diameter of 8 cm. This is also the case in the top of A 20. These pieces could have been the heads of bitts, or something similar. This means that the part with the hole must have at least protruded above the board. A 5 was broken off at the bottom. A 20 measures 31 cm below the hole to the deck or cross-beam where the knee was attached. A 1 was fastened to the gunwale and to the deck or a beam with a few 3 cm thick treenails. This was also the case with A 20, but some of the treenails only had a diameter of 2.5 cm.
The treenails in the horizontal parts of the knees indicate that the deck planks or cross-beams to which A 20 was attached were not discovered and that it's upper edge could not have been more than approximately 30 cm below the upper edge of the inwale. If the knee had been at the level of the beam above frame 14, the holes with a diameter of 8 cm would have gone on through the I-strakes. However, this does not seem to be very plausible. The vessel

Fig. 15.14. Overview amidships. A short futtock is standing detached on the chock against the port side shell, between frames 13 and 14.

probably had a (heavy) transverse beam/rider beam in the forebody, onto which the A 20 knee was attached. In the afterbody, the situation was probably different and there were more or less separate bitts (Plate 15.IV A, frame 5).

Pump

The wear spot on the keel plank between the frames 13 and 14 is the only indication that a pump tube may have been installed here. In view of the low height of the inwale in this part of the vessel (approx. 1 m) and the wide space between the frames, it seems more likely, however, that leakage and rainwater was taken out of the vessel with a bailer. This will also have been the case between frames 3 and 4.

Castles

There is certainly no question that this vessel ever had a castle or castles.

Repairs

A number of repairs were already mentioned during the treatment of the keel plank and the stern and stem hook, such as the reverse scarf at the rear and a triangular piece inserted at the stem hook. It seems likely that the stem hook was replaced. This can be done relatively easily, because the horizontal part was only connected to a frame with a single treenail and for the rest just with a few nails with which the garboards were fixed in the rabbets. This suspicion is further reinforced by the fact that the stern is less sharply tapered. Most cogs have a sharper rise at stern than at the front. At the scarf between the stem hook and the keel plank, a triangular piece was inserted from below. This weak repair was partly offset by clamping two pieces of wood under the two frames above. Between frames 9 and 10 on the starboard side, a repair board was nailed to the inside of the garboard (fig. 15.15). In addition, the extra nails in some areas of the seams have been mentioned as possible indications for the repair of leaking seams. This suggests that the vessel was quite old when it sank. This suspicion is reinforced by the irregularly worn hull planks and the remnants of tar on the inside. The sintel clamped cracks on the outside of some of the hull planks are likely to be repairs that were carried out shortly before the vessel was launched.

Wood use and processing

All parts of the vessel found are made from oak (*Quercus sp.*). The wide planks of the H- and I-strakes were plain-sawn, while all the others were quarter sawn, with the sides straightend by hewing with an axe. Sapwood was still present on a single piece. The plank plan shows a striking symmetry on the starboard and port sides. So the built clearly follows the 'brother and sister' pattern. What is also striking is the use of straight, even planks and futtocks under and next to the work floor, while the frames in the forebody are a little heavier than those at the stern and the frames in the after- and forebody have a more capricious shape. The single floor timber with a rising frame section on one side (two crooks) and the stem hooks were cut from the parts of trunks/branches of trees. The keel plank, stern- and stemposts and the (heavy) floor timbers were made from trunks, which probably also applies to most of the futtocks.

Fasteners

Treenails with a diameter of approx. 2.5 cm have been used to attach the frames to the hull planks, some of which were clamped with wedges. The inwale was attached to the upper hull planks with thin treenails (Ø 2 cm). The chock(s) and rising(deck) knees and the transverse beams were attached with 3-3.5 cm thick treenails. The keel plank and the narrow ca 2 m long ends of the bottom and hull strakes at the stems are attached to a frame with one treenail, while the rest of the planks is attached with two treenails per plank width. They are placed in a straight line with respect to each other, sometimes slightly staggered. The frames -1, 0, 1, 2, 20, 21, 22 and 23 were most likely not attached to the stern- and stem hooks and beams. Approx. 40 treenails (Ø 2.5 cm) were used for the frames above the stems, approximately 110 for the narrow plank ends at the stems and approx. 560 for securing the other frames, thus approx. 710 pieces in total. To secure the inwale with 2 cm thick treenails, 150-170 pieces were used. If the vessel had four chocks and four (heavy) transverse beams with rising knees and eight deck beams, then at least 50 treenails with a diameter of 3-3.5 cm were used for this.

Fig. 15.15. Repair plank on the bottom between frames 9 and 10 at starboard next to the keel plank.

The clinker-built seams were fixed with nails. These are bent twice on the inside and driven back into the wood with the tip. Generally, the nails were applied every 16 cm, room-and-space. In most cases, six nails were used for the plank scarfs, three from the inside to the outside and three the other way around. A total of approximately 150 m of clinker-build seam was nailed. This amounts to about 950 nails. The hull had almost sixty scarfs, which means another 350 nails, so we are speaking of a total of about 1300 nails. Smaller nails were probably used for the work floor and the cuddies. These must have been a few hundred.

The clamps were applied every 5.5-6 cm. A total of 150-200 m seam was fixed with clamps. This means that about 2500-2750 sintel clamps were used. The only heavy pieces of iron which were found are the remnants of the bolt, which probably came from the head of the heavy knee A 5 and pieces of U-shaped plate from the chocks. In addition, the heavy iron fasteners of the vessel consisted of gudgeons and rudder fittings.

15.3 Construction data

In the previous, the wreck was described as it had been discovered, the components were named and the reconstruction that could be derived from them was described. Despite the fact that this led to a reliable hull reconstruction, a number of questions remain unanswered with regard to the deck level and its layout. On the basis of all the data collected, the reconstruction drawings show one version of how it might have looked like. In the section *Function*, we will sketch what may have stood in the chock(s).

15.3.1 Design

The Spakenburg-II cog is a relatively small vessel with a low board. Significant are the even, straight and 15-17 cm wide frames under the work floor while the frames in the aft and fore are more irregular in shape and the frames in the forebody are heavier. It seems as if the frames 6 to 13 were deliberately shaped in this way and were intended specifically for this part of the bottom, while the shipbuilder was less concerned about the other frames. However, it does seem that a deliberate choice was also made for the heavier frames 14 to 17 in the forebody and that the rising parts of frame 14 in the sides were made to bend forward in order to create more space for the chock and the stanchion to be placed in it.

Shape

The plank plan, the side and top views and the cross-sections show a ship's hull with a straight keel, little sheer and with a flattening edge in the forebody towards the stem. At 3 m from frame 11 to the rear and to the front, the vessel is 10 cm narrower at the rear than at the front, measured along the inwale. However, when compared with the naked eye from above, the ship's fore and aft halves seem almost symmetrical. This is partly due to the small difference in the fall of the stern and stem (129° aft and 132° at the fore). The cross-sections show that the bottom turns quite smoothly into the sides and that there is no sharp bend anywhere. This 'round' shape continues in the aft and the frames get a very weak S-shape from the fifth frame towards the stern. In the fore section this is the case from frame 15, while from frame 17 there is an increasingly pronounced S-shape. Noteworthy are the five strakes on both sides of the keel plank, which were placed carvel-built next to each other. At the stern and stem, the planks turn into clinker-built constructions within only 2.5-3.0 m (respectively aft and fore). From strake F onwards, all subsequent strakes from stern to stem were clinker-built.

Dimensions

Between the front of frame 9 and a small repair board in the starboard garboard, two parallel grooves of 7 mm wide and 4 mm deep were identified; a V-shaped groove of similar dimensions was discovered in the repair board. It is not clear whether these are marks which were related to the hull dimensions during construction. However they are located in the middle (half the length) of the work floor. More difficult to explain is the zig-zag scratch in the same plank at frame 13 (Plate 15.II B).

Layout and tonnage

The work floor and the bulkhead are indications for the layout of the vessel. The bulkhead fits exactly on frame 14 and joins perfectly against the inside of the chock on the port side. The connecting beam (A 37) found loose next to the hull may also have been used there. The reconstructed height of the top of this beam (approx. 65 cm above the bottom) does not seem to complicate the draining of water from rain and leakage. At the back of the work floor at frame 5, the situation may have been similar. The probable bail spot in this area is located between frames 3 and 4. The spaces of approx. 1.5 m long from the work floor in the stern and forebody are possibly suitable for standing and walking.

To the fore and aft of the work floor, it must have been much more difficult to stand or walk. It therefore seems very likely that the ship had an aft- and forward deck. In the back, this would have been necessary in order to have a space to stand and operate the rudder. Several fragments of deck beams, which have been found in the forebody indicate that there must have been a deck here as well. This reduction of the work floor and the careful manner in which this structure and the underlying frames had been constructed, allow the deduction that the vessel was used for activities other than the stowage of cargo. In this context, it does not seem to make much sense to talk about the cargo capacity of the vessel.

Choice of materials

All excavated parts of the Spakenburg-II cog were made of oak (*Quercus spec*). Approx. two-thirds of the hull planks were sawn into quarters, the other were plain-sawn. The large amount of sapwood near the keel plank and the stern- and stem hooks was surprising. The grain pattern of the planks was not recorded, as in the case of the Spakenburg-I cog (Chapter 14, fig. 14.17). However, the fact that the scarfs were applied in almost the same places both at the starboard and port sides make it more than likely that they had been laid in the 'brother-and-sister' pattern.

15.3.2 Construction; techniques and sequence

The relatively small and low Spakenburg-II cog was built in several phases. In the first phase, the keel plank with the stern and stem hooks and posts were build-up. The (generally short) planks of the strakes A to E were installed next, always starting from aft. The nail in the keel plank and the small wooden nail wedges in the adjacent bottom planks make it likely that a construction or auxiliary frame had been placed between the (future) frames 4 and 5, against which at least the aft planks of the A-, B-, C- and D strakes had been fixed. Wooden nail wedges had been fixed in the A- and B strakes between the frames 14 to 16, which formed the temporary fastening for a construction frame for the middle planks. Where necessary, the planks of the other bottom strakes were probably temporarily fasted by short planks (Plate 15.II B).

Then the floor timbers and V-shaped floor timbers of the frames 5 to 16 were adjusted and the planks were fixed to them with treenails. The planks of the following strakes of the sides were nailed clinker-build fashion from stern to stem. This was possible without any real temporary fastenings. After the upper (I)- strakes had been fitted and the clinker-build seams had been provided with sintel clamped moss caulking on the inside, the futtocks of the frames 5 to 16 could be fitted and fastened. It is likely that the aft and fore frames were fitted and fastened next. In order to reinforce the upper edge of the hull's sides, a sectioned inwale was placed on the levelled tops of the frames. Near the stern and stem, the starboard and port side inwales were probably connected with each other via V-shaped pieces of wood

In the final phase of the construction, the heavy chocks, the transverse beams, the work floor and the transverse bulkhead or bulkheads, the aft and fore decks and the bollard-like construction elements with the 8 cm large holes were installed. The rudder and the vertical stanchion(s) in the chock(s) were separate elements and it seems likely that the stanchion(s) were not always standing in the chock(s). We will look at that again in the part on *function*.

Table I. Principal dimensions of the Spakenburg-II cog

Length of the keel (including the stern and stem hook)	9,.0 m
Length of the bottom (between the rising part of the stern and stem hook)	9.05 m
Greatest width of the bottom	2.80 m
Overall length at gunwale level	12.40 m
Overall width measured at gunwale level (at frame 12)	4.30 m
Height of the stern (keel line to top gunwale level)	1.70 m
Height of the stem (keel line to top gunwale level)	2.05 m
Height of the side amidships (keel line to top gunwale level at frame 12)	1.25 m
Length of the stern	2.25 m
Length of the stem	3.15 m
Height of the hold	1.05 m

Length-width ratio of the bottom	3.2:1
Length-width ratio at gunwale level	2.9:1

* *All measurements are rounded to the nearest 5cm.*

15.3.3 Characteristics

Keel plank

The Spakenburg-II cog has a keel plank that is about 6 cm thick, about twice the thickness of the hull planks. The sloping scarf between the stern hook and the keel plank runs from back to front and from top to bottom, which is precisely opposite to what is usually found, which makes it likely that the stern hook was replaced at some point.

Stern and stem hook

The stern hook is longer than the stem hook, which may also be due to a replacement. The upright parts of the stern and stem are not high; aft approx. 30 cm, and at the fore approx. 50 cm. Both have a sloping flat scarf connecting them to the stern- and stempost. The sternpost is positioned in a notch in the horizontal part of the stern hook, which is only a few centimetres deep.

Stern- and stempost and rudder

The vessel only had a stern- and stempost, no false stern and stem. Both posts almost certainly had rabbets in the sides, right up to the top. The ends of the planks had been nailed into these and then finished to be smooth. The ship undoubtedly had a stern rudder that was hung in a gudgeon at the bottom and into a bolt with an eye at the top.

Carvel-built and clinker-built bottom and clinker-built sides

The bottom, consisting of six or ten strakes plus the keel plank, and the strakes of the sides were built according to the 'brother and sister' pattern. In general, relatively short planks were used, which meant that not too many temporarely frames were needed during the construction. The ends of the planks are all nailed smoothly into rabbets in the stern- and stempost. Most of the frames were made up from a flat and a V-shaped floor timber with two futtocks in the sides. Frame 14 actually consists of two crooks, with the scarfs on the starboard side above the A-, B- and C strakes. The surface of the bottom has no dead flat, either in length or in width, but smoothly turns into the sides, with hardly any visible transition from the bottom to the sides.

Ceiling or work floor, inwale and chock(s)

The vessel had a clearly defined and relatively small ceiling, which may better be described as a work floor. It is noteworthy that ceiling planks seem to have been absent from the after- and forebody and along the sides. The inboard along the upper edge is in fact the only reinforcement of the longitudinal strenghtening in the vessel. It is placed on the levelled tops of the frames and fixed to the upper hull strake with treenails. The unusual chock with its square hole, which was excavated set against the hull on the port side probably had a counterpart on the starboard side. In addition, it does not seem inconceivable that a similar set of chocks was also present in the afterbody.

Caulking method

The clinker-built seams were only caulked with sintel clamped moss caulking on the inside; the flat carvel-built seams on the outside. The sintel clamped cracks on the outside of some hull planks almost certainly date from the end of the construction period, shortly before the ship was launched.

Decks, transverse beams and bollards?

The vessel undoubtedly had a deck both aft and fore, supported by thin beams that were fastened against the inside of the frames. Between the chocks and on them, the ship had a transverse beam, while underneath there was at least one bulkhead on the forebody. The knee and the head of a bollard (?), both with a hole of 8 cm, indicate a specific function.

Fasteners

The frames were attached to the hull planks with treenails (Ø 2.5 cm). The chocks and some transverse beams and (deck) knees were connected to the inwale with 3-3.5 cm thick treenails. Wrought-iron nails were used for the clinker-built seams, for the scarfs, for the decks and for the work floor.

15.3.4 Use

Function

A number of construction elements and the layout (as far as it could be reconstructed), the low sides, the chock(s) and the bollards or the rising knee parts of a transverse connection make it clear that we are most likely not dealing with a cargo ship. Van de Moortel has already remarked that the chock discovered on the ship most probably served as a location for a vertical beam for some kind of hoisting construction. I agree with this hypothesis. Hardly anything is known about workboats with one or more hoisting structures, certainly from the Medieval period.

A working boat from the Danube in Austria can help us to get an idea what kind of hoisting structure would have been possible. Figure 15.16 is a model in the Schiffahrtsmuseum in Spitz on the Danube, built after an 18th century description. This working boat, which consists of two vessels connected by beams, was used to remove large stones from the riverbed in order to keep the river navigable. Such a workboat could have been used to place dolphins or wooden posts as bridge pillars and to help salvage sunken vessels and large objects. Bearing in mind that the Spakenburg-I cog was found just a few hundred metres away, and the hypothesis that this ship may have served to clamp down a broken dike, it does not seem inconceivable that the Spakenburg-II cog may indeed have been a work barge (one of two vessels?) comparable to the boat in the Schiffahrtsmuseum in Spitz and been used for similar work.

Wear and tear and repairs

The keel plank was clearly subject to wear and tear on the underside at the scarfs. This hypothesis is supported by the stern hook, which most likely had been replaced, as well as the replaced piece near the scarf between the keel plank and the stem hook. The irregular thicknesses - sometimes within a single plank – indicate a high amount of wear and a relatively high age of the vessel when it sank. The extra nails in some parts of the clinker-built seams and the plank on the starboard garboard also indicate repairs during the period of use of the cog.

15.3.5 Wreck formation and embedding

The vessel may have sunk during a calamity (and bad weather?) and very likely subsequently drifted under water and ran aground at the sand head/find spot. It then drifted a little further and finally worked its way partly into the sand head. Part of the wreckage remained above the ground, and some of these parts of the hull became detached and drifted away. Given the shallow depth at the location and the proximity to the coast, it is also quite likely that parts of the ship were demolished and taken away. The wreck was filled with washed Pleistocene sand and sand interspersed with thin bands of humic Almere-clays. The whole was then covered with Zuiderzee deposits, which were largely absorbed in the topsoil.

15.4 Model

The model builder of the Department of Maritime Archaeology built a number of test models from cardboard and aircraft plywood of the Nijkerk-I cog (Section 8.4.1). Van de Moortel built a cardboard model of the Spakenburg-II cog.

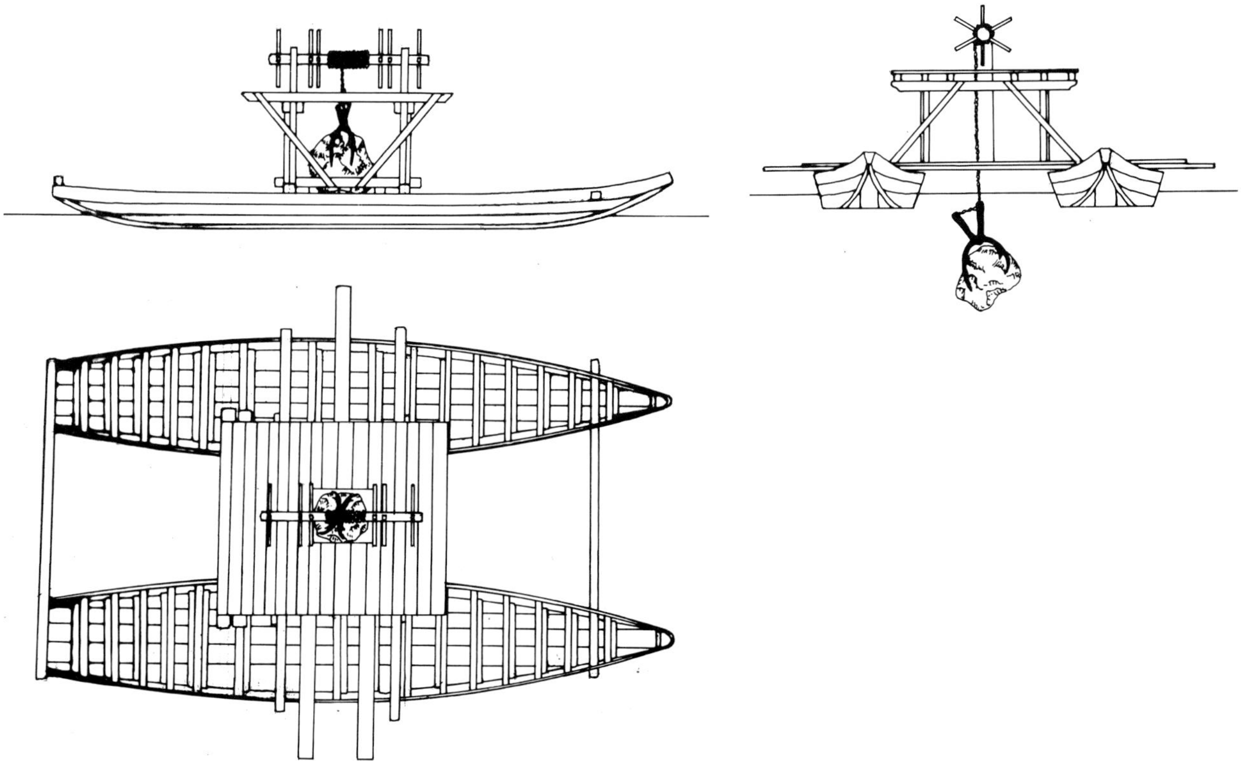

Fig. 15.16. Reconstruction model of an 18th century work barge on the Danube in Austria, which was used to lift heavy stones from the river bottom. Schiffsfahrtmuseum Spitz, Austria.

Fig. 15.17. The reconstruction model at scale 1:10 by J. van der Zee.

The results were not deemed satisfactory by the researchers of the Department of Maritime Archaeology and were the cause to build new test models of the Nijkerk-I from oak wood. Later, an oak model of the Spakenburg-II was also built (fig. 15.17). However, the results of this study provide reason enough to build another new 1:10 scale model of the Spakenburg-II cog.

15.5 Inventory

15.5.1 Distribution of finds

Only a small number of artefacts were found in and around the wreck of the Spakenburg-II cog. Because of their precise find location (not in the infill of the wreck), some of them can even be regarded as not belonging to the vessel. The artefacts are described in the find list. The artefacts that can be included in the ship's original inventory with a reasonable degree of certainty are listed with their inventory number at their find location in plate 15.I. They are briefly described below. Furthermore, it does not seem useful to list the small number of artefacts in tables. In principle, it can be stated that there were in fact no artefacts belonging to the inventory in or near the wreckage. If they did originally exist, they probably were washed away or salvaged.

15.5.2 Ship's equipment and inventory

In and around the wreckage, a curved and flat piece of iron (6) and fragments of a few iron bolts (7 and 8) were excavated in the sand that had been moved by the ship. The first may be a similar piece to the one found on the port side chock. The bolt fragments almost certainly come from the heads of rising knees of transverse beams or bollards. Some pieces fitted to piece A 5, which had been found separately. Furthermore, some treenails fragments with and without arks or wedges, nail and sintel clamp fragments, tar samples and sintel clamped moss caulking were collected.

A piece of relatively densely woven textile with tar residues may come from a sail (fig. 15.18).

Three sherds of red pottery (1) may have been in the ship when it sank. The other sherds found are (much) older or younger than the period in which the ship must have sunk, while the find locations also indicate that they cannot be considered part of the ship's inventory. The same applies to the heel of a shoe and for at least one (16) of the three bones of cattle found.[17] Some fragments of brick may have had a relationship with the vessel, but the others, plus two oval and flat limestone net weights did not. The latter are probably from a net lost on the wreck site.

Fig. 15.18. Piece of woolen and tarred fabric of a sail.

17. See appendix.

15.6 Cargo and merchandise

Irrelevant.

15.7 Ballast

No ballast was found in the wreck. It is uncertain whether the two boulders found on the starboard and port side between the frames 6-8 (Plate 15.I) belonged to the ship.

15.8 Dating

Soil profile

The soil profile provided no other dating than that the vessel must have sunk (well) before ca 1600.

14C and dendrochronology

Until the 14C and dendrochronology studies in 1994, the dating of the ship was no more precise than 'late medieval'. The 14C study by the University of Groningen produced a date of felling of the wood used between 1300 and 1400 AD.[18] The dendrochronological examination of the year rings by RING (using the reference chronology of Netherlands/Westphalia) resulted in a felling date of 1402-1414 AD.[19]

Sintel clamp typology

The sintel clamps used were of type D2/E, which was in use between ca 1350-1425.[20]

All in all, we can conclude that the vessel was almost certainly built in the first quarter of the 15th century. Probably continuing to the top of the rabbets in the sides of the sternposts, no sintel clamp caulking the seams on the outside and the iron bolt remnants in the bollards/upright knee parts support a dating in the late 14th/(early) 15th century.

15.9 Summary

Ship

The relatively small and low-hulled Spakenburg-II cog has an aft and fore section that looks almost symmetrical from above. From the side, the vessel has a light sheer, with the front section flattened at the top of the hull. The ship has the keel plank with stern and stem hook that is characteristic of cogs. The stern- and stempost beams had rabbets in the sides up to the top, into which the ends of the hull planks had been nailed. The plank hoods and stern- and stempost were attached to each other and smoothly finished.

The cross-sections show a remarkably 'round' shape, a carvel-built bottom consisting of six or ten strakes plus the keel plank and clinker-built sides. The aft and fore planks were nailed together in clinker-fashion along a few meters. The seams were provided with sintel clamped moss caulking from the inside, while the carvel-built seams were caulked from the outside. Only the front section clearly shows S-shaped frames. The straight frames at the work floor are noteworthy, as are the more capriciously formed ones in the afterbody and the forebody, while they were also heavier in the forebody.

The work floor was mainly located in the aft half of the ship. Furthermore, the vessel had no ceilings in the afterbody, the forebody or in the sides. The edge of the ship was reinforced with an inwale, lying on the levelled tops of the frames and attached to the upper hull strakes with thin wooden treenails. The cross strenghtening was initially formed by the frames, but also by a number of transverse beams and deck beams, which were attached to the frames and the hull on the inside with diagonally applied treenails.

Very striking is the heavy chock attached to the port side against the frames 13 and 14. It seems very likely that there was a similar one at the same location on the starboard side. It is also quite likely that the ship had another pair of these chocks behind the work floor. A function as mast step or steps does not seem very plausible. In the hole in the chocks a vertical stanchion or beams were probably placed as part of some kind of hoisting construction.

The hull shape of the ship is quite unsuitable for hoisting, as the vessel would probably tip over at the slightest action. However, hoisting would be possible in the case of two vessels that were fasted to each other at hull height with space between them. In other ships, heavy beams were used for this purpose, which spanned both the two vessels and the intervening space. Another possibility is a hoisting system for placing poles along the banks of a

18. With two standard deviations (so 95% certainty). Written report J.N. Lanting, Department of Archaeology Rijksuniversiteit Groningen, 22 March 1994.
19. For the complete rapport see appendix.
20. Vlierman 1996a, 79-48.

river or the coast of the sea, with the heavy beams lying over the ship's sides and up to the bank or coast. The 8 cm wide holes in the bollard and the rising knee were probably intended to fix the heavy beams over the sides and keep them in place. However, there are not many indications to support this hypothesis. It is quite possible that the ship was used with a completely different (unknown) function. What has become abundantly clear with regard to the ship's function, however, is that the Spakenburg-II cog must have been some kind of workboat and not a cargo ship or fishing vessel.

The wear spot next to the hole in the chock on the port side was perhaps caused by the underside of a sprit, which was probably attached to the vertical stanchion together with a sail. However, the vessel had a deck both aft and fore, and at least one, but more likely two transverse bulkheads.

The reconstructed principal dimensions are:

- Length over the stern and stem at gunwale height — 12.40 m
- Largest width at gunwale level — 4.30 m
- Height of the stern (keel line to the top gunwale) — 1.70 m
- Height of the stem (keel line to the top gunwale) — 2.05 m
- Height of the side amidship (keel line to the top gunwale) — 1.25 m

Equipment and inventory

There were almost no artefacts found that can be attributed to the equipment or the inventory of the ship. A small piece of woven woollen fabric may have originated from a sail.

Dating

The dendrochronological research resulted in a felling date of 1402-1414 AD (with the reference chronology of Netherland/Westphalia). The vessel was probably built in the first quarter of the 15th century. It seems very likely that the small cog was probably built in the (northeastern) Netherlands. The wear and tear, the residues of tar and the repairs all indicate that the ship was quite old when it sank, presumably either in the second quarter or around the mid-15th century.

The Almere-I cog[1]

16.1 Find conditions

16.1.1 Discovery and investigation

On 2 April 1986, the construction of a sewage system for a new housing development at Almere in Zuidelijk Flevoland brought to light ship's timbers, a potsherd and some human bones (figs 16.2 and 16.3).[2] The find was reported, and the very same day NISA employees started an evaluating investigation. This revealed that the find probably was a fairly complete late-medieval vessel. Subsequently the wreck was excavated between 3 and 23 April. This was possible also thanks to the contractor of the sewerage works, the firm Van Gelder, who made available a pump to keep the excavated area dry (fig. 16.4) and supplied an excavator on request.

Time pressure for excavating the ship, recovering any finds and documenting the wreck was considerable: the wreck lay at a crucial point in the sewerage system, which had to be operational as soon as possible (figs 16.5a and b). For this reason it was decided at any rate to do a survey drawing (scale 1:20) in which the position of all (loose) timbers and the findspots of all items were to be marked (Plates 16.I and 16.II A and B). The usual longitudinal and transverse sections could not be measured in and drawn, owing to the limited time available and the harsh weather conditions (figs 16.1 and 16.6a, b and c). On the basis of this survey drawing, all parts were numbered, lifted and transferred to the Museum voor Scheepsarcheologie at Ketelhaven. There the timbers were stored under water in specially excavated basins. In the autumn of 1987 all parts were transferred to the rinsing floor in the museum's workshop.

June 1988 saw the arrival of Fred Hocker and three other graduate students from Nautical Archaeology Program at Texas A & M University (USA) for a period of seven weeks as part of a collaborative programme between Texas A &

Fig. 16.1. General view of the excavation.

1. Translated by Xandra Bardet.
2. The wreck lay in future housing development no 13 of Almere-Stad, Phase 1, 2E West. It lay almost horizontally, oriented NW-SE (bow-stern), at ca 17 km east of the contemporary harbour of Amsterdam (Damrak). Ordnance map sheet 26A, Oostvaardersdiep; coordinates 142.55/478.60. The former depth of the water locally was ca 2.7 m (Hydrographic map 1921, mean low-water level).

Fig. 16.2. The find spot of the Almere-I cog.

M University and our own organisation (NISA), to clean and draw all the ship's timbers.

In the course of the summer all parts of the ship could thus be documented and a provisional reconstruction was started. The principal parts of the bow (stem hook, stem post and frames) were reassembled on the rinsing floor, also in order to display them to the participants of the Fifth International Symposium on Boat and Ship Archaeology (ISBSA 5) (fig. 16.7). In 1991 Fred Hocker presented the first results of the investigation in his PhD thesis *The development of a bottom-based shipbuilding tradition in Northwestern Europe and the New World* (unpublished). Vlierman in 1992 and 1993[3] published three articles on items from the ship's inventory, and two further articles, together with A.M. Koldeweij, about the remarkable clapper from the wreck.[4] The excavation report appeared in 1996 in the series *Flevoberichten* published by Rijkswaterstaat Directie IJsselmeergebied.[5]

Fig. 16.3. Ordnance map of the Kingdom of the Netherlands (scale 1:50,000) of 1861, showing the location of the Almere-I cog.

3. Vlierman 1992b, 1992c and 1993.
4. Vlierman & Koldeweij 1992-1993, 323-363 and Vlierman & Koldeweij 1996-1997, 155-162.
5. Hocker & Vlierman 1996.

The format and subdivision of that excavation report served as a model for the standard format applied to all excavation reports in the present study of all cog finds in the Low Countries. Because of new insights resulting from this study, and for the sake of completeness, this wreck is included here and treated again. Part of Hocker's construction drawings have been elaborated/revised and redrawn in accordance with the standard format of the plates to the present publication. The text of the 1996 excavation report has been retained with some adaptations and abridgements; some passages have been omitted.

16.1.2 Geology and stratigraphy

At discovery, so much of the overlying and surrounding soil profile had been stripped away that useful evidence about any disturbance caused by the wreck could no longer be extracted. The recovered items at any rate showed that this certainly was a ship of 15th-century antiquity and hence that the wreck must have lain entirely within the Almere clay deposit. It was overlain by a complete Zuiderzee-clay deposit. Locally the top of the Pleistocene deposits lies at ca 5.5 to 6 m below the surface.

16.1.3 Medieval habitation

The wreck site lies closest to the town of Muiden at the mouth of the river Vecht (at ca 10 km). The contemporary harbour of Amsterdam (the Damrak) lay at a distance of 17 km.

16.1.4 Condition and wreck formation

Condition

Until its discovery, the bottom of the vessel was still complete. Over a length of ca 5 m, the sides of the midship section had survived up to the gunwales. At the stern and bow, parts of the uppermost two strakes were missing. The stern- and stem hooks, as well as the lower part of the

Fig. 16.4. The excavation in progress.

Fig. 16.5a. General view of the uncovered wreck, viewed from the bow.

Fig. 16.5b. The same, viewed from the stern.

stempost were still part of the original construction. The upper part of the stempost had broken off and lay beside the forebody. The sternpost was missing. On the bottom, all frame parts (straight and V-shaped floor timbers) still were present. At the place of the lost sternpost just one, while against the stempost two crooked floor timbers had survived. A third lay dislodged beside the forebody. A few of the midships futtocks had survived all the way up to the gunwales. In the after- and forebody these frame parts were largely absent. Some lay loose within the wreck, others had disappeared. The mast step was still in its place, and large parts of the underfoot ceiling and ceiling boards were found in their original position. A few planks lay loose and had drifted away from their position. Also some detached gangway planks and parts of an covering inwale were encountered. The condition of the wood was fair to good. During the mechanical digging for the sewer works, a strip over 1 m wide had been cut athwartships out of the afterbody. The dislodged fragments were no longer retrievable (Plates 16.I and 16.II A and B, and figs 16.6a to c).

Wreck formation

The vessel sank into the sea floor virtually upright, and presumably emerged from the waves by about a metre for some time. The wreck eventually was probably damaged by fishermen sailing across it with their nets. The erratic stones at the stern and around the bow probably were net weights that were lost when the nets became entangled with the submerged wreck (Plates 16.I and 16.II B). Alternatively, the wreck was damaged by ice floes, but given the depth of the water (over 2.5 m deep this seems less

Fig. 16.6a. The ceiling, viewed from the stern. b. General view after removal of the detached timbers. c. The mast step and remains of the ceiling.

Fig. 16.7. The composite bow and frames of the forebody in the workshop of the Museum voor Scheepsarcheologie at Ketelhaven.

likely. In the wreck-formation process, the sternpost was torn from the hull, pulling away with it parts of the upper shell strakes in the afterbody. The stempost broke, similarly ripping away parts of the upper shell strakes of the forebody. The planking of the after and fore decks became detached and floated off. This also goes for the deck beams and the heavy transverse beams. These probably became detached and drifted away when the sides collapsed outward once the connecting constructions at the stern and stem had given way. The by now waterlogged, massive gangway planks and other structural elements sank into the hull (Plates 16.I and 16.II A), where they became covered by redeposited Almere sediments. Subsequently fresh layers of Almere sediment were laid down over the wreck, followed after ca 1600 by a deposit of Zuiderzee clay.

16.2 The ship

16.2.1 Documentation and reconstruction

Documentation

Documentation in the field comprised the above-mentioned survey drawing together with photos and colour slides of the various stages of the excavation. In 1988 the lifted timbers were drawn to a scale of 1:10 and close-up pictures were taken of structural details and the like.

Reconstruction

With the data from the drawn timbers, Hocker compiled a reconstruction on paper.[6] Just nine frames from the midships section were amenable to a reasonably reliable hull-shape reconstruction of one or both sides. The components of these frames (floor timbers and futtocks) were damaged, incomplete or possibly (to some extent) distorted, owing to the fact that the hull had been ripped apart or had collapsed, become splayed and had broken just above the turn of the bilge. Hocker worked on the assumption that the cross sections of the ca 5 m long midships section had similar dimensions throughout; in other words, that the midships had a constant width across the gunwales. However, subsequent research on all recovered ships in the cog-building tradition in the Low Countries has shown that such was not the case in any of the vessels built before ca 1550. It can be stated that the widths across the gunwales, at similar distances from the stern or stem, in the forebody are always at least 10-20 cm greater than in the afterbody. Even in the almost 16 m long boat with cog-like features from lot K73/74 (section 26.1), where just under half of the length is built with hard chines to connect the flat bottom with the everted sides, the widths at the gunwales at 3 m from the stern and stem differ by 15 to 20 cm. A uniform width of the hold does not appear as a common feature of Dutch inland cargo craft until the 18th century, probably having evolved in the 17th century.

Hocker's reconstructed sections a, f and g[7] probably give a good rendition of the original lines and dimensions of these parts of the hull. In the other cross-sections, the uppermost shell strake appears more or less turned inward. Comparison with other late-medieval vessels makes this quite unlikely. The rear surface of the top futtocks is usually vertical with respect to the keel line or even inclines outward a little. This means that the sides, up to the uppermost shell strake, were everted (Plates 16.IV C and D). Hocker also assumed that the base line of the keel plank and hooks rose towards the stern and bow by about 5 and 7.5 cm, respectively. Yet we now know the vessel to have had a straight keel line. The present study of all wrecks in the cog-building tradition has demonstrated that the pattern of frames in the after- and forebody of the Almere-I ship must have differed from the reconstruction of Hocker (Plate 16.IV A and C). The length-width ratio across the gunwales is a little greater than it is in the seagoing cogs.

16.2.2 Construction and layout of the hull

Keel plank

The keel plank is 9.94 m long and connected to the stern and stem hooks by sloping flat scarfs at either end. Together they are ca 12.90 m long. The scarfs are (stern) 40 cm and (bow) 47 cm long. The lands in the scarfs are joined with two rows of double-clenched nails: one row

6. Hocker & Vlierman 1996, enclosures 1 to 5.
7. Hocker & Vlierman 1996, enclosure 3.

driven in from outside, the other from within. Aft, the keel plank is 28 cm wide, forward 25 cm, and towards the midships it gradually widens to 32 cm. Throughout its length, the plank is 5.5 to 6.2 cm thick. The underside is unevenly abraded; it is thought that the keel plank may have been ca 7 cm thick originally. It is rectangular in section.

Sternhook

The horizontal part of the sternhook is 1.44 m long; and at the scarf joining it to the keel plank, 28 cm wide. Towards the rear, it gradually narrows. At the base of the rising part, the horizontal part still is ca 20 cm wide (including the two rabbets for attaching the lowest bottom strakes). At the end of the scarf the horizontal part is 7-8 cm thick; at the base of the rising part, 20 cm. On the rear is a toe to protect the rudder and to stop debris from fouling the rudder by getting caught between it and the stern. The rising part of the stern hook with its sloping flat scarf stands 53 cm above the keel line. In the flat scarf are two treenails (3.5 cm across): the principal fastenings connecting stern hook and sternpost. Besides, small holes showed that the sternpost had been attached also with nails, towards the ends of the scarf. Abrasion on the rear surface and nail holes in the sides of the rising part of the hook mark the location of the lowermost gudgeon.

The rabbets to take the hood ends of the bottom strakes reveal that the ship must have had a sternpost that was fairly thick at its base (ca 40 cm). The timber was not, as is usual, joined to the stern hook just by a flat scarf, but had also over a good 20 cm rested on the topof the thick horizontal part of the stern hook, in a way similar to the forward construction.

Sternpost

The sternpost was no longer present, but almost certainly was shaped similarly to the stempost. The ship probably did not have a false stern. The upper part of the post had probably held a forged bolt with an eye: the top suspension device of the rudder. The rake of the sternpost (rear face) will have been about 120°.

Rudder

The stern rudder was not recovered.

Stem hook

The stem hook had been fully preserved. The horizontal part is 2.11 cm long, and at the scarf joining it to the keel plank is 25 cm wide. As in the stern hook, the horizontal part narrows towards the angle, and at the base of the rising part of the stempost is a good 20 cm wide (including

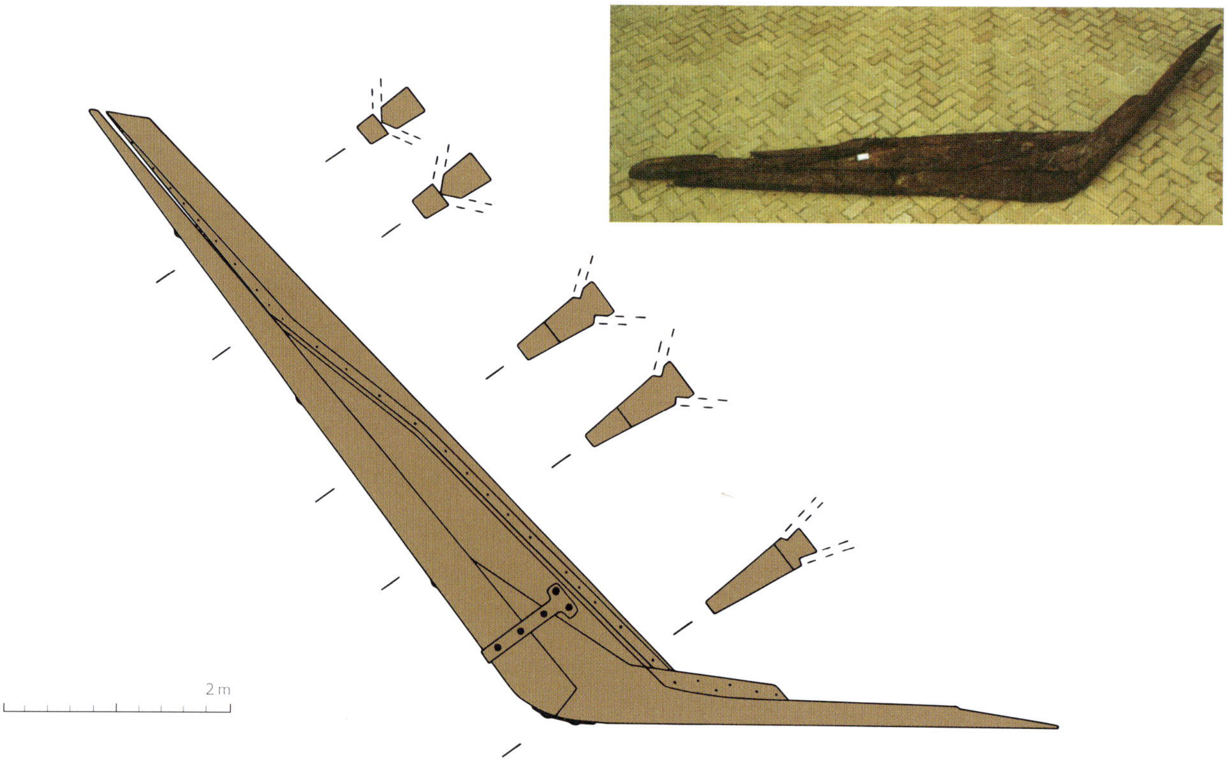

Fig. 16.8a. The complete bow construction. b. The bow construction composed of the stem hook, the stempost and the false stem.

the rabbets for taking the hood ends of the first bottom strakes). Its thickness at the scarf is almost 10 cm. Midway along, its upper surface 'jumps', and at the base of the stempost the horizontal part of the hook is 20 cm thick. The rising part of the stem hook stands 68 cm above the keel line. On the front, it has a projecting 'toe', analogous to the toe on the stern hook. The slanting top of this toe serves to support the false stem nailed to the stempost above it. The rabbets in the sides of the horizontal arm of the hook obliquely run on into the sides of the stempost. Both the horizontal and the rising arms of the hook are trapezoid in section, with the narrow side outward.

Stempost

The stempost like its after counterpart rested not just on the sloping flat scarf of the stem hook, but also, in part, on the top of the horizontal part of the stem hook. The stempost at the top of the scarf is 30 cm thick and tapers upwards: at the tip, just 16 cm. In the sides, towards the rear it contained rabbets into which the hood ends of the bottom and side strakes were nailed. The bottom 50 cm of these rabbets run parallel to the rear of the stempost. The line of the subsequent metre at the top ends at ca 7 cm from the rear surface. The next 70 cm from there runs obliquely upwards towards the front of the stempost. From this point upwards the hood ends of the sides continue towards the centre line of the mitred front of the stempost, onto which they are nailed fast.

False stem

Against the front of the rising part of the stem hook and the stempost, a false stem was fastened. This fully covers the forward hood ends of the side strakes.

The composite bow construction is 3.4 m long and, measured perpendicularly, rises ca 2.7 m above the keel line. Its rake is ca 127°. As in the stern, so the stempost too was fastened to the stem hook with two treenails (3.5 cm across) and below these with a nail. The false stem is fastened to the stempost by three long, forged nails, and to the stem hook by two nails. At the lower end, the construction was reinforced with a forged iron plate ca 60 mm wide, which was nailed to the sides of the construction. (figs 16.8a and b and Plate 16.III A).

Bottom and sides

The bottom is composed of three strakes on either side of the keel plank. The top of the garboards or A strakes is flush with the top of the keel plank. Amidships, the first two strakes rise ca 5 cm towards the turn of the bilge. The third strake is still carvel-built but adjoins strake B at a slightly greater angle and at its outer edge rises about 10 cm above the keel line. Longitudinally too, the bottom is dead flat at no point. Both aft and fore, the carvel seams between the A- and B strakes and between the B- and C-strakes merge into a clinker-built construction (Plate 16.IIIB).

The next two strakes (D and E) form the turn of the bilge. They are both fully clinker-joined to the C-strake and to each other, at a sharper angle. The remaining three strakes

Fig. 16.9a. Overview of the bottom with the plank scarfs of the 'brother-and-sister' pattern of construction.

Fig. 16.9b. Detail: some of the plank ends standing proud of the bottom in the scarfs amidships.

Fig. 16.9c. Keel plank, stem hook, stempost and shell strakes in the forebody.

of the sides too are nailed together clinker-style from stern to stem. The overlapping lands of the clinker joins are 5-6 cm wide. The space between the double-clenched nails varies between 14 and 18 cm.

The planks of the bottom, of the turn of the bilge and of the lowermost two side strakes were abraded, decayed, and repaired in several places; they had probably been 4.5 cm thick to start with. The top strake of the sides was 5.3-5.7 cm thick, barely abraded and of far better quality than the other strakes. The planks of the bottom had a maximum width of 48 cm, at the bilge this was 50 cm, while the planks of the sides were up to 55 cm wide. In all, the vessel had eight shell strakes (A to H) on either side of the keel plank.

The shell was put together according to the brother-and-sister pattern. This is evident from the corresponding positions of scarfs in the port and starboard strakes (figs 16.9a to c). The A strakes are composed of two planks; the B- and C strakes, of three. The bilge strakes D and E are made up of two planks, and the three strakes of the sides (F, G and H) consist of three planks on either side. The flat sloping scarfs joining the planks are ca 30-40 cm long. At the scarfs, the plank ends are 1-2 cm thick, proud of the inside and outside of the shell. The scarfed plank ends are secured together with two rows of five nails each: one row driven in from the outside, and the other from within. The nail points are twice beaten sideways and driven back into the wood (double-clenched nails, Plate 16.III B).

Sintel clamped moss caulking

The seams of the shell are closed with sintel clamped moss caulking. The carvel seams of the bottom as usual are caulked on the outside. The clinker seams, both on the inside and the outside. The moss was twisted into single, short strands (ca 10 cm long), driven into the seams, and covered over with laths of split oak (moss laths) which in their turn were secured with sintel clamps of the type D2/E (fig. 16.10). This type was used in the period ca 1375-1450.[8] The caulking of clinker-built seams on the outside of the shell was no longer customary from the mid 14th century onwards. Hocker suggests that it may have been applied to this ship retrospectively to combat leakage. Also various pairs of small holes were observed which suggest that the sintel clamps were locally replaced or that lengths of seam were re-caulked.[9] In several places, cracks in the wood had been closed with sintel clamped moss caulking on the outside. Probably this had (in part) been done even during construction.

Moss determination and pollen analysis

For caulking, a single species of moss (*Sphagnum* spec.) was used. The encountered pollen suggests a growing environment of the moss in an area of open vegetation. For its analysis and conclusions, see Appendix.[10]

Frames: floor timbers and futtocks

The frame pattern is regular and above the bottom comprises 24 frames. The very aftmost and foremost frames were not treenailed to the hooks; just to the sides. Nor were any frames fastened to the stern- or stempost. Because the hood ends of the upper two strakes are missing both at the stern and stem, it could not be established how many frames had been present in the stern and bow. Hocker assumes three each for stern and bow. However, four would seem more plausible. This would raise the total to 32 (-3 to 0, 1-24 and 25-28). Most of the frames are composed of a straight or crooked floor timber extended by a futtock at either end. The scarfs between the floor timbers and the futtocks are alternatingly above the

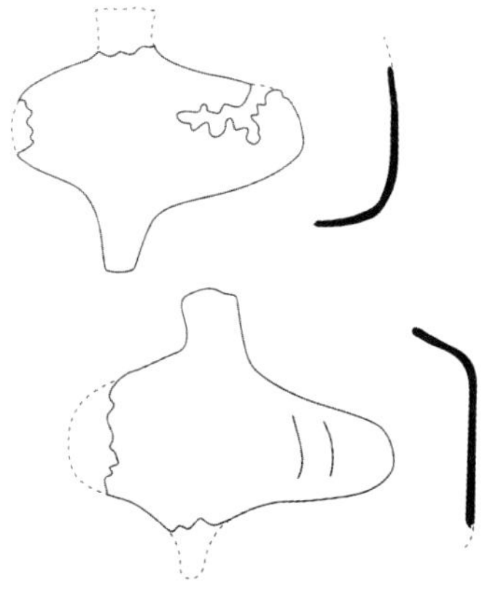

Fig. 16.10. The type of sintel clamp applied in the Almere-I cog: D2/E.

Fig. 16.11. The pattern of frames, viewed from the stern.

8. Vlierman, 1996a and fig. 51; section 29.2.
9. Hocker & Vlierman 1996, 29.
10. Cappers *et al.* 2000, 577-590; Mook-Kamps 1997, Almere W13.

second or third bottom strake and at the turn of the bilge or a little higher up. In the stern and bow, the crooked floor timbers cover most of the height of the sides. The V-shaped floor timbers of frames 0 and 25 most probably did not rest on the hooks or stern-/stempost (Plates 16.IV A, B and C, and fig. 16.11). Limber holes were cut into the underside of the floor timbers to allow leakage and rainwater to flow towards the pump.

Mast step

Viewed from the side, the mast step consisted of a block of wood. It was 1.77 m long, with a somewhat convex upper surface, a maximum width of 34 cm, and a thickness of 23 cm. In its upper surface, exactly between the underlying frames 17 and 18, a cavity for the heel of the mast had been cut (ca 28x14 cm and ca 11 cm deep). On the rear side of the hole was a rabbet ca 6 cm wide and 5 cm deep. This had probably held a plank, the upper end of which was fixed to the sail beam. On the underside the mast step is joggled, to engage with frames 17 to 20. The block does not sit exactly above the keel plank, but about 5 cm to starboard (Plate 16.IV A, B and C, and figs 16.12a and b). The treenails observed in the block revealed that it had previously been part of at least one other construction. Indeed it seems likely that the block may originally have been part of another ship. The position of the mast step suggests that the ship may have carried a sprit sail, while the vertical plank aft of the mast may have been a guide for lowering and raising a lowerable mast. Still it seems possible that the vessel originally carried a square sail.

Ceiling

The vessel has a fully closed underfoot ceiling of 3 cm thick deal boards. The planks of the sides were also 3 cm thick, but of oak. The situation as recovered made it clear that the underfoot ceiling planks ended forward in an irregular fashion, above, or close to, frame 21. Aft this could no longer be established, because this part of the wreck had been cut through and lost at the time of discovery. Still it seems quite likely that the underfoot ceiling and the ceiling planks along the sides ran from frame 4 and ended forward at frame 21: a hold with a length of ca 7.5 m. The port side still held a piece of the top ceiling strake in its original position. This oak plank was thicker than the others (4.4 cm). Its forward end was fastened to frame 23. It is quite likely that the other end started at frame 1. This heavy ceiling strake/shelf clamp without a doubt also served as a stringer, an important longitudinal reinforcement of the sides of the vessel (Plates 16.I and 16.IV B).

At the beginning and end of the underfoot ceiling, Hocker drew transverse bulkheads to separate off below-deck accommodation fore and aft. A few remains of nails on the back of the floor timber of frame 21 might support this idea. Given the slight headroom (ca 1.2 m) between the frames and the decks, this does not seem very plausible, though. In the Spakenburg-I cog (Chapter 14, Plate 14.III A) too, partitions were drawn in these places. In the case of both ships, the reality of such partitions seems doubtful.

Like the Marknesse cog (Chapter 12), this ship is an early example of the late-medieval use of cheaper deal planks for the underfoot ceiling. Depending on the nature of the ship's cargoes, its underfoot ceiling - or individual planks in it - might need replacing from time to time.

Inwale

On the highest surviving part of the starboard side, a length of inwale or covering inwale had survived, resting on or attached to the heads of frames 12 to 16. To port lay a further, detached fragment, ca 4.5 m long (Plates 16.I

Fig. 16.12a. The mast step and the remains of the ceiling, viewed from starboard.

Fig. 16.12b. The same, viewed from port, after removal of the ceiling boards and cleaning of the timbers.

and 16.II A). The pieces were 7-8 cm wide and 15 cm high. The oblique notches in their outward-facing side engaged with the also oblique futtock heads. In smaller boats, such reinforcements of the gunwales tend to be roughly square in section (ca 8x8 cm), and lie across the squared-off heads of the frames. They are fastened to the top shell planks and the frames (see the Dronten-II cog, Chapter 17; and the boat from lot K73/74, Section 26.1). There we are dealing with covering inwales. In the case of more massive timbers, as here in the Almere-I, we may speak of a proper inwale.

The inwale on either side of the ship probably consisted of three parts, joined by sloping flat scarfs. In the Marknesse cog (Chapter 12), the central piece consisted of a heavy uppermost ceiling board and a true inwale only fore and aft. Without a doubt, their ends met the (inner face of the) stern- and stemposts, to which they were fastened. Aft and fore, there almost certainly was a small crook connecting the ends of the starboard and port inwales. In the bow, this crook and the one below it supported a rectangular plank (a good 50 cm long and wide and 6-16 cm thick), which was treenailed to them. A ca 8 cm deep notch had been cut into its 16-cm-thick top edge. Presumably this was intended to support the mast and/or sprit with the sail when these were lowered (Plates 16.IV A, B and C). Various treenail and nail holes were observed in the top of the covering inwale fragments. Hocker assumes that the top of the inwale (with the upper edge of the top shell strake) was fully covered by a broad plank (a gunwale proper). Whether there was any question of such, we can no longer tell. Given the many repairs in and on the hull, these traces may in fact reflect the attachments of short repair planks. So far, no true gunwales pre-dating the 17th century have been identified in the former Zuiderzee area.

(Deck) beams, decks and gangways

The head of frame 17, immediately aft of the mast, had been cut square, a short distance above the transition from strake F to strake G. On it, it without doubt a fairly massive sail beam had rested. Nothing was recovered of the beam itself. Yet to starboard, a curiously shaped rosebolt was found: a flat rod of forged iron with holes for attachment, and at its end a large, thin, somewhat convex head. The bolt had been inserted through the shell from the outside and had been nailed to the beam (Plates 16.III A and 16.IV C, and figs 16.13a and b). Its counterpart to port probably was lost together with the sail beam. A little forward of frame 7 both to port and starboard the shanks of similar rosebolts passed through the G strakes. The connecting beam, now also lost, most probably rested upon the gangway planks with notches on its underside. The rosebolts served as additional reinforcements of the connections between the beam and the sides and in order to hold the everted sides in place. The function of the rosebolts is similar to that of wall anchors in buildings. In section 16.9 they will be compared with rose bolts from other shipwrecks in the Zuiderzee area.

The square top of port frame 23 at strake G level suggests that a connecting beam lay on top of this as well. Apart from some beam and knee fragments[11] no further parts of connecting or deck beams were recovered. Comparing this with the Marknesse cog, for instance (Chapter 12), which in many respects resembles the Almere-I, it may be assumed that the vessel besides the beams on frames 7, 17 and 23, did have a further set of connecting and deck beams.

The notched thick plank up against the floor timber that connects the port and starboard inwales at the bow, and the vertical guide plank aft of the mast heel suggest that this vessel had a lowerable mast. If this was the case, the ship must have had longitudinal stringers at deck level on both sides which ran at any rate from the sail beam to the connecting beam above frame 23. On the front of this connecting beam a single stringer may have extended up to the stempost. In this situation, a relatively short mast (in the case of a sprit sail) would have been lowerable between the stringers without too much trouble, and be laid forward into the notched plank at the bow. The connecting beam above frame 23 would then not be an obstacle. The stringers may have had a section in the form of an inverted T; the rabbets might then have held planks to keep the deck closed when sailing. Short deck planks cover the space between the stringers and the sides. Alternatively, and maybe more plausibly in the case of this vessel, there may have been relatively heavy stringers, rectangular in section, connected to the sides by also quite substantial deck knees that were fastened to the inwales, the frame and the uppermost strakes of the sides.[12] The foredeck almost certainly extended from the sail beam to the stempost. On either side of the mast there may have been hatches giving access to the hold and the fore-cabin. The afterdeck presumably also had a hatchway.

Hocker's reconstruction indicates that the beam with rosebolts above frame 7 was the first transverse construc-

11. Hocker in: Hocker & Vlierman 1996, 34.
12. By way of example, see section 27.4, fig. 27.4. The early-16th-century *waterschip* from lot K 84 in Oostelijk Flevoland had two masssive stringers in the forebody, which were connected to the gunwales by knees. This vessel also had massive inwales, triangular in section, running from stern to stem. Probably transverse deck planks were placed between the knees.

tion aft of the sailbeam. Compared with other cogs, the distance between these two beams (ca 4.5 m) is rather large. In most cases, it is no more than ca 3.75 m. The distance from the aftmost beam to the the stern is also ca 4.5 m. This seems a rather long afterdeck for a vessel of this size. Despite there being no concrete evidence, it is more likely that at the spot (above frame 4) that Hocker indicated in 'Enclosure 3' there also was a heavy transverse connection. This beam then probably also formed the front of the afterdeck. On the front and on the beam there may well have been a windlass for lifting heavy items into and out of the hold and for lowering the mast. The planking of the afterdeck probably lay longitudinally. In a few places, deck knees will have formed the connection between the deck and the bulwark. Inside the wreck lay two gangway planks (Plates 16.I and 16.II A; figs 16.5a and b). The foremost ends of these were fully intact; the after ends had broken off. The starboard plank still measured almost 9 m. The planks were up to 30 cm wide and were 7.5 cm thick. This length means that both fore and aft they were an integral part of the decks. Besides, they were important structural parts ensuring the transverse and longitudinal rigidity of the ship at deck level. These thick gangway planks were treenailed to the futtocks and the sides. Between the futtocks, they were nailed to the sides. They probably ran from frames 3 (aft) to 23 (fore). Their inboard edge had been bevelled. Most probably a coaming had been nailed against this edge. However, no trace of such was observed.

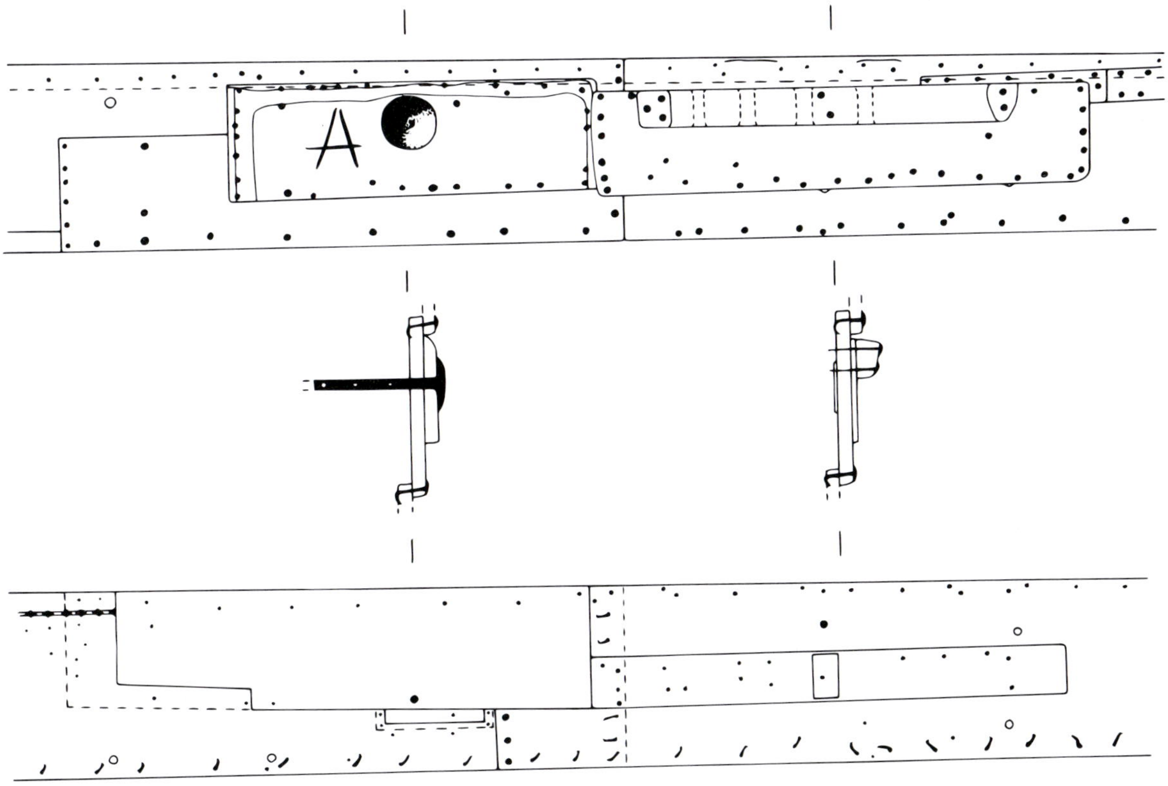

Fig. 16.13a. Repairs, and the rosebolt to port holding the sail beam (frame 17). On the top of strake G, immediately against the underside of strake H (not drawn) the support for the rigging of the mast can be seen. Beside the head of the rosebolt is a scratched-in (maker's) mark.

Fig. 16.13b. The head of the rosebolt in the repair plank and the (maker's) mark. The piece of the shell plank of strake G was placed upside-down. The edge with the nail heads should be below.

Pump
No traces of wear were found on the bottom that might indicate the location of a pump barrel. The ship certainly had a pump; a suitable spot would have been between frames 2 and 3 (Plate 16.IV B).

Repairs
The vessel stands out for its numerous, and occasionally quite remarkable repairs. Not all of them are shown in the drawings. The most notable are the repair planks on both the inside and outside of the lowermost port bottom strakes at the stern (Plates 16.III A and 16.IV D). In order to insert repair planks from within, the ship's carpenter must have temporarily removed, or replaced, some of the frame timbers. Besides, the various repairs around the rosebolt are quite eye-catching (figs 16.13a and b). The treenails in the mast step that had no function in the construction as recovered, suggest that the block had earlier served a purpose in another place or manner, or may have come from a different vessel altogether. Given the vessel's presumable date of construction (section 16.8), the externally sintel clamped shell seams maybe were thus treated to remedy a leak: sintel clamped caulking seams from the outside began to be an obsolete practice in the second quarter of the 14th century, and at any rate by the middle of that century. Besides these, other (double) sintel clamp imprints have been identified that are certain to represent repairs. (At least part of) the cracks closed with sintel clamped moss caulking on the outside of the shell planks are likely to have been repairs made during the working life of the ship. Almost certainly there would have been some cracks that were caulked even during construction. The uppermost strake of the sides appears to have been of decidely better-quality timber than the other strakes, and may be a later replacement. These observations, together with the abraded exterior of the shell and hooks, imply that the ship must have been of considerable age by the time it sank.

Wood selection and processing
All parts of the hull are of oak (*Quercus* sp.); the boards of the underfoot ceiling are of pine. The planks were sawn. It was not established whether they were rift- or quarter-sawn timber. The former seems quite likely, as the planks were fairly broad. The timbers of the frames, the stern- and stemposts and crossbeams were made from trunks and heavy limbs. The stern and stem hooks were hewn from trunk-branch crooks.

Fastenings
For the connections between the shell strakes and the frames, treenails 2.5 cm in diameter were used. Generally two per plank width; in some cases just one. Some 1000 treenails in all. Whether in the transverse and knee fastenings at deck level thicker treenails were used, is unknown, but plausible. In some treenails, pyramidal wedges were observed.

In the clinker planking, iron nails are set at 14-18 cm intervals. Seams totalling 175 metres, which means a requirement of about 1100 nails. For spiking the hood ends into the hooks and stern- and stemposts, and for joining the shell planks in the 26 scarfs, a further ca 400 nails were needed. At least 400 to 500 shorter nails were used in fastening the deck planks and the ceiling boards.

The moss laths of the sintel clamped moss caulking were fixed with forged sintel clamps set at intervals of 5 to 6 cm. The carvel seams of the bottom had been sintel-clamped on the outside only; the clinker sides, both inside and out; overall, some 400 m of shell seams. At an average spacing of 5.5 cm, this means that for the ship some 7500 sintel clamps were required. This still excludes the sintel clamps on the cracks in the shell planks and possibly also the deck seams.

The gudgeons (not recovered), and the rudder fittings, the plate holding together the bow construction, and the four rosebolts probably were the only heavy forged fastenings in the cog.

16.3 Construction data

16.3.1 Design

Design, lines and measurement system
So much had survived of the oak hull that a reliable reconstruction of the ship could be made. Most probably there was no question of a preliminary design in late-medieval shipbuilding. Shipwrights built their ships 'by sight'. The lines were determined by well-established and traditional rules of thumb, artisanal knowledge and expertise. In the hull no marks were observed that might point to particular measurements. Beside the head of the rosebolt on the repair plank there is a scratched-in mark (figs 16.13a and b). Its meaning is unclear. Barely any evidence had survived about the transverse connections and deck layout. For a reconstruction of these (Plates 16.IV A, B and C) however, we could easily turn to the cogs of Marknesse and Spakenburg I (Chapters 12 and 14). It is especially the data from the former that provided important evidence. Apart from some minor differences in detail,

Table 1. The reconstructed principal dimensions of the Almere-I cog

Length of the keel (Including the stern- and stem hook)	ca 12.75 m
Length of the bottom (between the rising part of the stern- and stem hook)	11.60 m
Greatest width of the bottom (at frames 14-18)	ca 3.20 m
Overall length at gunwale level	ca 16.25 m
Greatest width at gunwale level (at frames 14-18)	ca 4.80 m
Height of the stern (keel line to top gunwale)	2.60 m
Height of the stem (keel line to top gunwale)	2.65 m
Height of the side amidships (keel line to top gunwale)	ca 1.90 m
Length of the stern	ca 3.00 m
Length of the stem	3.40 m
Height of the hold	1.05 m

Length-width ratio of the bottom	3.6:1
Length-width ratio at gunwale level	3.4:1

the descriptions relating to the build, the lines, and the construction phases and sequence of these ships will also apply to the Almere-I, and for this reason will be omitted here.

Plugged-up treenail holes in the keel plank and in the first two bottom strakes on either side of it indicate the places where at the start of construction temporary frames had been installed to hold the carvel-laid planks in position. Such holes were observed between frames 6 and 7, between 11 and 12, between 15 and 16 and between frames 21 and 22. The treenails were of course removed before the permanent frames were fitted.

Layout and tonnage

The underfoot ceiling actually indicates a tripartition within the vessel. With or without transverse bulkheads, there is at any rate a (cargo) hold ca 7.5 m long. The sharply built after-/and forebody are poorly accessible, and in fact suitable only for e.g. stowing tools, (spare) cordage, sail and any personal possessions. The nature of some of the finds (see section 16.5) suggests that there may have been one or a few passengers during the ship's final voyage. The foremost 1.5 m of the ceiling was under the foredeck: a suitable place for passengers to shelter during bad weather. The after part of the floor, between frames 4 and 7, may have been the crew's quarters. The items recovered there support this idea. No fireplace was found, and it would appear that when the ship foundered none was present (any more). A wooden box paved with tiles or bricks to serve as a fireplace may have stood on the floor here, or, more probably, forward of frame 7. If initially the purpose was to transport bulky cargo on the floor such as bricks, timber or heavy crates and barrels, then capacity was available even under the gangways and over the full length of the ceiling. This means a hold of ca 7.5 x 3.7 x 1.1 m = ca 30 cu m, that is a maximum of ca 30 gross tons.

16.3.2 Characteristics

The following typical features of the Almere-I cog are of particular structural interest:

Keel plank

A keel plank that originally was up to twice as thick as the shell planks. Aft and fore, the plank was 28 and 25 cm wide, respectively; towards the middle gradually increasing in width to 32 cm. Its section was rectangular.

Stern and stem hooks

The hooks were fastened to either end of the keel plank by means of flat scarfs. Aft and fore, they had a so-called heel and toe. These served to protect the rudder and to prevent debris from accumulating between the rudder and the stern; and forward, to support and protect the lower end of the false stem.

Stern- and stempost, false stem and rudder

The stern- and stempost too were joined to the hooks by sloping flat scarfs. Remarkable was their attachment by a pair of thick treenails. Also notable was the thickness of the stern- and stempost at their lower ends. They did not, as usual, rest only upon the flat scarfs on the rising arms, but also on the horizontal parts of the hooks. Above the projecting toe of the stem hook and against the stempost, a false stem had been affixed. This did not extend any higher than the stempost or the gunwale. A support had been fastened to the inside of the stempost, onto which the mast could be lowered. The A- to D strakes had been spiked into the rabbets on the sides of the stern- and stempost; strakes E to H, onto the stempost's mitred front; while these hood ends were covered and protected by the false stem. The same construction probably also applied to the stern, be it without a false stem. The cog was fitted with a stern rudder.

Clinker-built and carvel-laid bottom
The ship had a carvel-laid bottom that was about 8.5 m long. The six, fairly wide bottom strakes merged into a clinker-built construction both aft and fore. The bottom, both longitudinally and transversely, was dead flat at no point.

Clinker-built sides
Amidships, the vessel had fairly steep sides. The transition from the bottom to the sides is rounded in every part; nowhere does it have an angular turn of the bilge. This results from the way the outer bottom strakes (C), the bilge strakes (D) and the lower side strakes (E) were attached at ever-changing angles. From strake C upwards all strake planks were joined together in clinker fashion, with double-clenched nails. The hull was built from planks 48-55 cm wide, according to the so-called 'brother-and-sister' pattern. The shell planks were attached to the frames by means of treenails.

Caulking method
The carvel seams had been sintel-clamped on the outside; the clinker seams of the sides, both inside and out. Judging by the presumable construction date (Section 16.8), it seems quite possible that (a major part of) the clinker seams were sintel clamped on the outside by way of repairs to an old, leaky ship. An alternative possibility that cannot be ruled out is that the vessel came from an isolated shipbuilding area, where the traditional way of exterior sintel clamping was still adhered to, despite having been abandoned elsewhere by the mid-14th century. Another possibility is that the ship was extremely old by the time it sank.

Mast step
The vessel had a short mast step with a rabbet to hold a vertical guide plank aft of the mast. Which may be an indication for a mast that could be lowered forward upon the support fixed to the stempost. The mast step must at one time have been attached in another way or in another place, maybe in another ship. The position of the mast, the guide plank and the support on the stempost suggest the use of a spritsail and a short, lowerable mast.

Ceiling and repairs
The Almere-I cog had a fully closed underfoot ceiling, which sideways continued almost up to the transverse connections. The uppermost ceiling board, the shelf clamp or beam clamp, was thicker than the other boards. The floor boards were of pine, the others of oak wood. During its sailing life, the ship was repaired on multiple occasions and in various parts. Indeed it is not impossible that the uppermost strakes of the sides were at some point replaced.

Gangways, transverse connections/deck beams and decking
The heavy and long gangway planks were part of the decks both aft and fore. The ship almost certainly had four fairly massive transverse connecting beams, of which in any case the middle two (the sail beam and the beam above frame 7) had extra strong fastenings to the sides with forged iron rosebolts, which had been inserted from outside the shell, and on the inside were nailed to the beams. The vessel probably had a foredeck ca 5.5 m long and an afterdeck 3.5 m long. Presumably a windlass stood against and on the front of the afterdeck.

16.3.3 Use

Function
The heavy stern- and stem constructions, the long, thick gangway planks, the fully closed ceiling and the fairly large opening(s) allowing large items of cargo to be hoisted into and out of the hold suggest that we are dealing with a good example of a small freighter cog. For example, for carrying timber, bricks, floor tiles, peat and heavy crates and barrels. Actually the old ship may no longer have been sufficiently seaworthy for this purpose, but still served for example to carry passengers. It appears to be a typical cog for inland transport on the (large) rivers and the then Zuiderzee inlet, and is comparable to the ship depicted on the wooden barrel vaulting of the Oude Kerk church in Amsterdam, dating to ca 1400 (fig. 16.14). Keeping in mind the remarks in section 14.3.4 on the Spakenburg-I cog, one might wonder if this ship also might not be too wide to pass through many of the inland locks in the western Netherlands. It was not until 1536 that Charles V granted permission for a new lock to be built east of Gouda, that was 16 feet wide, wider than the existing Goudse Binnensluis and many other locks. If these were 'Amsterdam feet', the new lock was ca 4.5 m wide; if they were 'Rijnland feet', its width was ca 5 m.[13]

13. Smits 1995, 468.

Fig. 16.14. A sailing small cog with a spritsail. Painting in the vaulting of the Oude Kerk church in Amsterdam, dating from ca 1400.

16.3.4 Wreck formation and embedding

The possibly no longer seaworthy ship may have foundered in a storm, and subsequently sank fairly rapidly into the soft Almere-clay sediments of the Zuiderzee sea floor to about halfway up the hull. Timbers protruding from the sea floor would have been dislodged by fishermen's nets and washed away. Other waterlogged and detached parts remained in place, lying above, beside and within the wreck. In due course the wreck became filled up with and covered by Almere sediments, and from the early 17th century onwards a full Zuiderzee deposit developed on top.

16.4 Model

The excavation report of 1996 describes how the scale model (1:10) was built.[14] Here a brief description will suffice.

For building the model, the survey drawing (Plate 16.1) and the various photos and colour slides taken during the excavation were consulted. Because no sections of the hull could be drawn in the field, the modelmaker resorted to using the scale drawings of the recovered and numbered parts prepared by Hocker and his fellow students, but especially Hocker's reconstruction drawings of the vessel.[15] The model was built on the basis of these reconstruction drawings; while the scale drawings of the various timbers served to copy the relevant components as realistically as possible (fig 16.15).

It is not a research model. In other words, the model was not made to establish how the original vessel would have been built and shaped, but as a three-dimensional version of the reconstruction on paper. Supported by a template

Fig. 16.15. The model scale 1:10. The brown coloured parts are the recovered timbers

14. J. van der Zee & D. van den Broek in: Hocker & Vlierman 1996, 41-49.
15. Hocker in: Hocker & Vlierman 1996, Enclosures 1-5.

(a sort of mould) the ship's hull was composed plank by plank after the keel plank and stern- and stem constructions had been set up. After this, the template for the hull's shape was removed and the frames were fashioned and inserted. Further almost the entire ceiling, a piece of gangway plus the hypothetical coaming to port, a full and a partial crossbeam plus deck knee were put in place, as well as the support on the stempost on which the lowered mast or the sprit and the sail could rest. For the sake of transparency, the recovered parts are marked, as in the plates, by their brown colour, in contrast to those not recovered.

16.5 Inventory

16.5.1 Distribution of the finds

In the excavation report of 1996, the items recovered from the wreck are classified according to the system developed to compare the inventories of small to medium-sized vessels of the Zuiderzee.[16] Here they are grouped under eight headings, a classification designed to suit the composition of inventories of late-medieval vessels.[17]

The descriptions of the items in the above-mentioned excavation report need no adjustments and together with the relevant illustrations have been adopted integrally. The inventory numbers in the list of objects (Plate 16.II B) return in the text, in the illustrations and in the findspots around the wreck (Plate 16.II B). The rosebolts and the block are not included in Table 2, as they were part of the ship and the rigging.

16.5.2 Ship's equipment and inventory

Ship's equipment

Three rosebolts (29, 30 and 31; figs 16.18a and b), bolts with a large, somewhat convex, round head (ca 140 mm across), served as extra reinforcements of the fastening of two transverse beams to the sides, above frames 7 and 17 (Section 16.2.2 and Plates 16.III A and 16.IV C). There must have been four such bolts. The heads of the rosebolts were on the outside of the G strakes (figs 16.13a and b). The first 100 mm of the bolt below the head is roughly round in section (25 mm across). This was the part passing through the actual shell. It continued as a forged, flat strip with some (three or more) holes in it. The overall length of the bolts could not be established, as all of them had broken and the end parts must have drifted off with the beams to which they were attached. These flat ends had been fastened with thick nails to the after surfaces of the beams. In section 16.9 these rosebolts are compared with parallels from the Zuiderzee area, including those from the Dronten-I cog. Three flat iron fragments (32) come from the bow construction (Plate 16.III A and fig. 16.8a). They are parts of an iron band that had been fitted around the front and the sides of the false stem and fastened to the sides of the stem hook and stempost. This fitting provided extra strength to the connection between false stem, stem hook and stempost. For the same reason, another flat piece of iron had been nailed across the seam between the 'toe' of the stem hook and the false stem. Inventory number 38 comprises ten nails of different sizes. Numbers 40 (figs 16.18-38 and 16.18-40), 41, 42 and 83 represent nine sintel clamps of the D2/E type, which was in common use between ca 1375 and 1450.[18]

A tapering oak plug (47) may have served to close a hole in the bottom. Such holes might be drilled to drain the hull when the vessel was being overhauled for mainte-

Table 2. Distribution of items

Category	Forebody	Hold	Afterbody
1. Ship's equipment	-	rope, nails, sintel clamps	-
2. Operational equipment	spade	ring	tool handle
3. Administration and navigation	-	writing chalk	-
4. Tools	adze, saw handle	marlinespike, grease, flint, (iron?) slag	axe
5. Household items	-	-	-
6. Cooking, eating and drinking	-	cooking pots, tripod, bowls	knife, jar
7. Victuals	-	jar, mussel shell	jars, bones
8. Personal belongings, weaponry	shoe, sheath	spur, clasp(?), clapper, belt(s), bag	dagger, coins

16. Reinders, 1985b.
17. Vlierman, 1992b; Vlierman 1993; section 24.4.2.
18 Vlierman 1996a, 80-51.

Part of the inventory of the Almere-I cog.

nance. No such hole was found; there may have been one in the destroyed part of the afterbody. Or it may have been a stopper for a different purpose altogether.

The single-sheave block (figs 16.18-44 and 16.19) was found amidships underneath the splayed port side. It still held remains of rope from the hitch. Probably the block was part of the rigging. The solid, rectangular, 32.5 cm long body of the block is of ash wood (*Fraxinus* spec.). The block was made from the heartwood of a tree. It is roughly square in section with chamfered corners. Inside an opening ca 20 cm long and 4.7 cm wide, it holds a pearwood (*Pirus* spec.) sheave (13.3 cm across). The sheave has a groove in its rim, into which a rope could be laid. Immediately above the boxwood (*Buxus* spec.) axle, the block was diagonally drilled through from two sides towards the top. Within it were remains of rope from the hitch that had run around the underside of the block, also in a groove. At the top there had been a rope by which the block was attached to *e.g.* the mast or the rigging.[19] Woodworm holes and wear both to the block and the sheave suggest that it was very old by the time the ship foundered. A replica of the block was made and tested for effectiveness and tensile strength. Some further replicas were incorporated in the rigging of the Kampen Cog.

Presumably it is a brace block with remains of the brace hanger or pendant, which according to Van Beylen[20] hung immediately below the yard of a square-rigged ship. It served for reeving the halyard by which the yard was adjusted. Alternatively, the old block may have been used for *e.g.* adjusting the sprit, if this ship was rigged with a spritsail. A fragment of rope (62) with a diameter of ca 1.8 cm, and the nails (34 to 37, and 39) have been classed with the spare parts of the ship's equipment. Yet they (or some of them) may have been part of the ship itself. Of the four nails constituting no 37 it should be noted that they have four small nubs on the underside of their relatively large

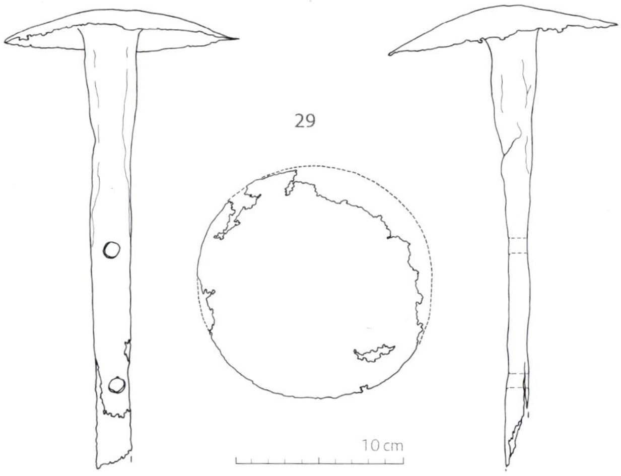

Fig. 16.16. Rosebolt, inventory number 29.

Fig. 16.17. The three rosebolts (29, 30, 31).

Table 3. Ship's equipment

Category	Item	Number	Findspot
1.1 Ground tackle	-	-	-
1.2 Sails	-	-	-
1.3 Cordage	line	62	hold
1.4 Blocks	brace block	44	beneath port side
1.5 Windlass	-	-	-
1.6 Pump, drain	-	-	-
1.7 Navigation aids	-	-	-
1.8 Boats	-	-	-
1.9 Spare parts	nails	34-37, 39	hold

19. Van Beylen *et al.*1972, part 1, 233.
20. Van Beylen, 1985,40.

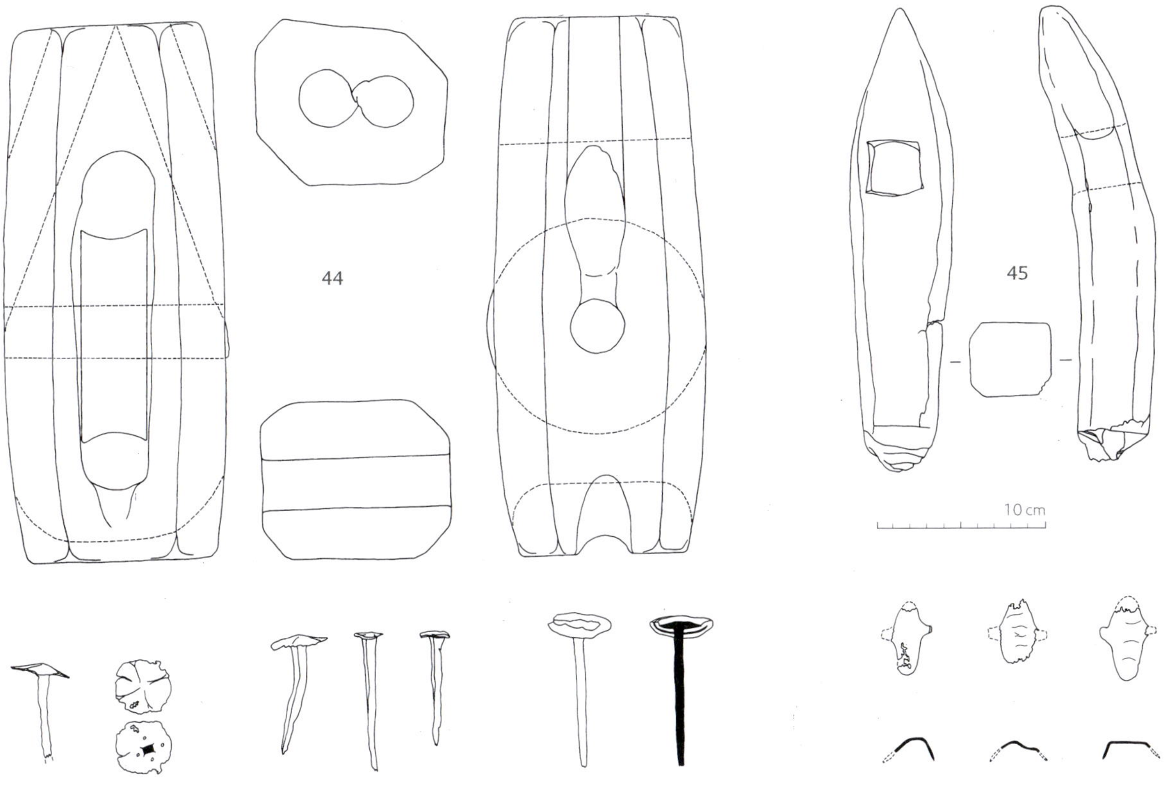

Fig. 16.18. Brace block (44), piece of a tool or anchor (45), nails (37, 38, 39) and sintel clamps (40).

heads (fig. 16.18-37). This means that they were shaped on a moulded nailheader. Under the head of nail 39 there was a two-layered fragment of leather, presumably serving to close a leaky (nail-)hole (fig. 16.18-39).

Operational equipment

Just one item can possibly be attributed to the category of operational equipment: the iron spade with a fragment of its ashwood (*Fraxinus*) handle (figs 16.20 and 16.22-20). The tool, with a cross-section like that of a spade, has a round handle case with a nail hole, and was forged from a single piece. The front of the blade is almost flat; the rear has a protruding ridge extending from the handle case. The sides and the cutting edge were so thin that they had entirely rusted away, so the original shape cannot be fully ascertained. Yet it is likely that the sides sharply converged into a point. Its shape is unsuited to shovelling bulk cargo such as grain, and differs from specimens found earlier in the Zuiderzee area. The model does match that of (later) peat-cutting spades, and is virtually identical to the spades once used in Wales for constructing and maintaining roadside earth banks.[21]

The keyhole-shaped iron ferrule (fig. 16.22-33) also was put under the heading of equipment for loading and unloading. Such ferrules were used for *e.g.* reinforcing the connection of the business end of a tool to its handle.

Fig. 16.19. Brace block (44).

An unusually shaped and hewn-off piece of oak wood (fig 16.18-45) may be the end of a handle of some tool. It could also be part of an anchor. It is slightly curved and square to oblong in section. The corners are chamfered; making its section in fact octagonal. The top end tapers to a point. The broadest, pointed end contains a near-square hole. On the underside of the hole the edges are somewhat

21. Examples and information: National Museum of Wales/Welsh Folk Museum, St. Fagans.

Table 4. Operational equipment

Category	Item	Number	Findspot
2.1 (Un-)loading gear	spade, ferrule tool handle?	20, 33 45	forebody, hold, afterbody
2.2 Additional equipment	-	-	-

chamfered. No traces of wear are evident around the hole. If the hole served to suspend this object, it certainly was not frequently used. A second and maybe more obvious possibility is that it was part of some device, such as an anchor. In Ship 1 from the harbour of Kalmar[22] an anchor was found that was composed of an erratic boulder enclosed by two forked wooden crotches (see also the Nijkerk-II cog, Section 6.5.3 and fig. 6.53a-131). If the piece indeed derives from such an anchor, it should of course be classified with Ground tackle in table 3.1.1. Finally, the piece may have been an as yet unknown ship's timber.

Administration and navigation

Items relating to administratiion are hardly to be expected in late-medieval vessels. An irregularly shaped piece of chalk of 6x5 cm (55) shows some small abraded surfaces. This may be due to its use for marking particular parts of the cargo. Writing chalk has been found on other occasions aboard shipwrecks in the Zuiderzee area. The three coins will be discussed under Personal equipment (below) and Dating (Section 16.8). No navigation devices were encountered.

Tools

The category of tools matches what is found in other late-medieval ships in the former Zuiderzee area. It consists of the most essential toolkit required for repairs and maintenance of the ship, the rigging and the sail. The recovered carpenter's tools are an axe, an adze, and the handle of a saw. The axe with a large part of its oak handle (figs 16.21 and 16.22-21) is of the standard type: a left-hand broadaxe.[23] The handle remnant is 44 cm long. In section 16.9 this axe type is compared with some other axes from ships in the current study.

A fine and well-preserved adze (figs 16.22-22 and 16.23b) is the second carpentry tool. It has the usual curved shape and a hammerlike butt end. On the underside (*i.e.* handle side) the blade is somewhat mitred in section and has a 15 cm long, sharp cutting edge. The hammer part towards the case becomes a little narrower and is forged into an octagonal shape. The burrs on the edge of the hammer surface suggest regular use. The manner in which the adze is fixed to the handle is a curious one: in contrast to all other specimens from ships found so far in the Zuiderzee area, the adze does not have a closed handle case that projects below the blade, but two more than 120 mm long, straight, flat flanges, with a width of 17 to 21 mm. The handle is set tight between these flanges. The adze itself has a rectangular cavity into which the very top of the handle is fitted. A rivet passed through the ends of the flanges and the handle, holding the handle in place. A small part of the ashwood (*Fraxinus* spec.) handle was preserved. The blacksmith's mark, a little cross, was stamped into the blade. An almost identical specimen was found during the earth-moving works at the Damrak site for the new Noord-Zuidlijn of the Amsterdam metro (find no 0524-001). Otherwise no domestic or foreign finds of this type are known to the author.

Table 5. Administration and navigation

Category	Item	Number	Findspot
3.1 Ship's papers	-	-	-
3.2 Stationery	chalk	55	hold
3.3 Financial/Commercial matters	-	-	-
3.4 Orientation and dead reckoning	-	-	-
3.5 Timekeeping	-	-	-
3.6 Signalling and distress signals	-	-	-

22. Åkerlund 1951, drawing 5b.
23. Vlierman 1985b, 33-34.

Fig. 16.20. (Peatcutting) spade (20).

Fig. 16.21. Axe with handle (21).

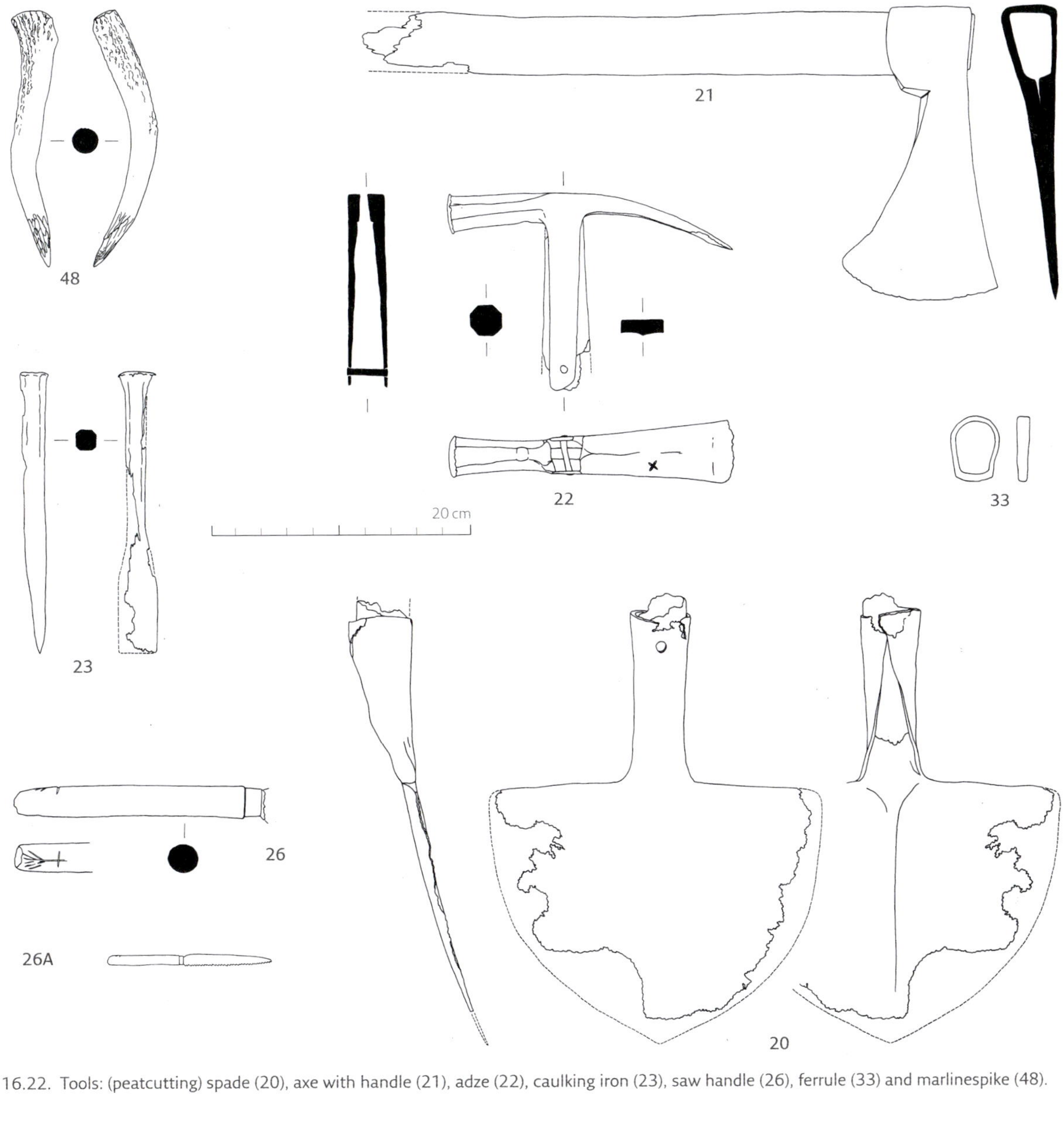

Fig. 16.22. Tools: (peatcutting) spade (20), axe with handle (21), adze (22), caulking iron (23), saw handle (26), ferrule (33) and marlinespike (48).

Fig. 16.23a. Marlinespike (48).

Fig. 16.23b. Adze with fragment of the handle (22).

The beechwood handle (*Fagus* spec., fig. 16.22-26) probably belonged to a handsaw. It has a diameter of 2 cm and is 17.5 cm long. Near one end, it has a scratched-in, stylised tree(?). At the other end is an iron ring that reinforced the connection with the blade. The first part of the blade is still present. Goodman describes various medieval saws and reproduces a drawing from the *Bedford Book of Hours* (in the British Museum) of an early-15th-century model with a curved blade.[24] One of the two 17th-century handsaws from the Nordiska Museet in Stockholm shows a marked resemblance to the saw from the Almere-I wreck (fig. 16.22-26a). Such handsaws were in general use among ship's carpenters.[25]

Handsaws, or parts thereof, have also been recovered from a large flat-bottomed cargo ship (B55, Oostelijk Flevoland), which probably foundered around 1500[26], and in the *waterschip* from lot W10 in the same *polder*,[27] which went down in the second half of the 16th century. The handle of the former is missing, though the ring survives and has a diameter of 30 mm. Of the latter, the handle is 53.2(!) cm long, has a diameter of 3-4 cm and in shape resembles the handle from the Almere-I cog. The blades of the two saws differed in shape; the Almere-I blade at any rate was narrower.

The toolkit for caulking and maintenance is represented only by a caulking iron (fig. 16.22-23). The tool, 203 mm long, has only partially survived, but still allows reconstruction of its original shape. The shank is octagonal in section and 135 mm long. Towards the blade, its circumference slightly decreases. The narrow blade has a constant width right to the working edge and measures 30 mm. The burrs on the butt surface indicate that the caulking iron had been regularly used (for repairs). In section 16.9 it will be further discussed in terms of typological development.

The marlinespike made from a tine of red-deer antler (figs 16.22-48 and 16.23a) is worn and polished as a result of frequent use. The tip had cracked lengthwise and was given a fresh point with a knife. Marlinespikes are used in splicing rope. The use of antler marlinespikes has continued even after the introduction of iron ones. One of the last surviving enterprises in the Netherlands where rope-walking is still done in the traditional way, the firm of Deetman's at Elburg, actually prefers the use of antler marlinespikes, because they will do less damage to cordage made from softer fibres.[28]

The lump of grease (fig. 16.24) has the shape of the inside of a small crock (a stoneware jug?). In some areas, the surface shows imprints of textile, which might indicate that the grease had been wrapped in a rag of sailcloth or the like. In the top, around a cavity, several narrow, sharply tapering grooves can be seen, probably left by a needle. Aboard sailing ships there usually was a horn or pot containing grease in which sail needles would be kept, to stop them from rusting. The shape of the lump of grease does not match any of the recovered crockery.

A splinter of grey-black flint (57) and a small piece of (iron?) slag (59) have been classified as Miscellaneous.

Fig. 16.24. Lump of grease (60).

24. Goodman 1964, 122-132.
25. Saw reconstructed after Goodman 1964.
26. Reinders *et al.* 1980, 40-41 and Appendix 11.
27. Reinders *et al.* 1978, 26.
28. Personal comm., J.C. Deetman, 1988.

Table 6. Tools

Category	Item	Number	Findspot
4.1 Carpentry	ax adze saw handle	21 22 26	afterbody forebody forebody
4.2 Caulking and maintenance	caulking iron	23	forebody
4.3 Splicing, sailmaking	marlinespike, lump of grease	48 60	hold hold
4.4 Miscellaneous	flint, iron(?) slag	57, 59	hold

Household items

It is as good as certain that small cargo vessels in the late Middle Ages lacked any separate workfloor or living quarters for the crew. Objects belonging in the category of household items are barely to be expected, nor were any found.

Cooking, eating and drinking

Fireplace

A little more can be said about this category. Eight small pieces of peat (61) indicate that - at some point - there was a fireplace on board. However, nothing was recovered of any fireplace: the usual loose wooden box paved with tiles or bricks. Nor was there any sign of tongs, or of flint for striking fire. Also the very meagre few sherds of redware pots (see below) suggest that at the time of the the ship's demise there was no (longer a) fireplace, so there were no on-board cooking facilities.

Cooking ware

The following potsherds could be attributed to cooking ware. Some rim, wall and base sherds of a redware cooking pot (fig. 16.25-7). A large fragment of the base of a similar pot (8). Three stick handles of redware skillets (figs 16.25-13, -14 and -15). A number of small base and wall sherds of a tripod or cooking pot (17) and some grey (burnt?) sherds of a redware pot (18). Most of these sherds bear traces of soot, with the exception of handles 13 and 14. Handle 15 shows abrasion. A small sherd, which strongly resembles *Kugeltopf* ware (19) is also abraded, which may account for the loss of any traces of soot.

It is hard to distinguish between cooking ware and tableware aboard late-medieval ships. Traces of soot point to the former, but may also have resulted from reheating left-overs. Bruijn discusses the link between the shape and the function of simple earthenware items for domestic use.[29] He believes the fragments of the three crocks with similar rims (5, 6 and 9) should be classified as cooking pots. The sherds of pot 5 allow a partial reconstruction (fig. 16.25-5). The rim profile and the flowing outline of the shoulder are similar to those of a one-handled cooking pot.[30] The base displays a thick blob of glaze. The inside of the pot had just a few small spots of glaze. The same goes for the outside and the rim. The handle is flattened, however which according to Bruijn is typical of two-handled cooking pots.[31] Remarkably, the sherds of pot 5 lack any trace of soot.

Tableware

A rim and a few base sherds (fig. 16.25-12) come from a bowl without a handle. Its profile and dimensions are similar to a specimen illustrated by Bruijn.[32] The sherds (fig. 16.25-10) and comparable rim fragments (11) can also be attributed to a bowl. Yet the wall sherd does suggest a greater rim diameter, more like that of a stick-handled bowl, the so-called *kromsteert* ('curved tail').[33] The flowing transition from wall to relatively wide rim is different; therefore a bowl with a stick handle seems unlikely. Further features of interest are the claw feet and the absence of soot. On the inside the vessels were probably fully lead-glazed; on the outside, only partially. The profile and the dimensions match those of a find from Leiden. Bitter surmises that these forms might have served as 'dinner plates'.[34] If indeed these bowls were 'tableware', which seems quite plausible, these would be the earliest known examples of 'plates' aboard a late-medieval ship in the Zuiderzee area. The absence of a soot deposit and the wide rim is why they have been classified not as cooking ware but as tableware. The pottery from the

29. Bruijn 1979, 57-104.
30. Bruijn 1979, 61 and 63 and no 3 in figs 18 and 21.
31. Bruijn 1979, 69 and fig. 83.
32. Bruijn 1979, 84 fig. 41 no 6.
33. Bruijn 1979, 81.
34. Bitter 1989, 117-118, fig. 16, type 32.

Almere-I cog is described in detail in part 7 of the *Corpus Middeleeuws Aardewerk* (*CMA*).[35]

The knife (figs 16.27-25 and 16.29) is considered with the tableware, though maybe it should belong under the heading of 'Personal equipment'. The 20 cm long knife has a straight back, which runs from the handle to just short of the tip of the blade. The handle was made from two strips of (beech?) wood. Fitted to either side of the flat tang, they are fastened with small rivets. The end of the handle is protected by a metal plate or cap attached to the tang. The knife is similar to specimens from Amsterdam. These have been dated to the late-14th/early 15th century and the second quarter of the 15th century.[36] In the knives

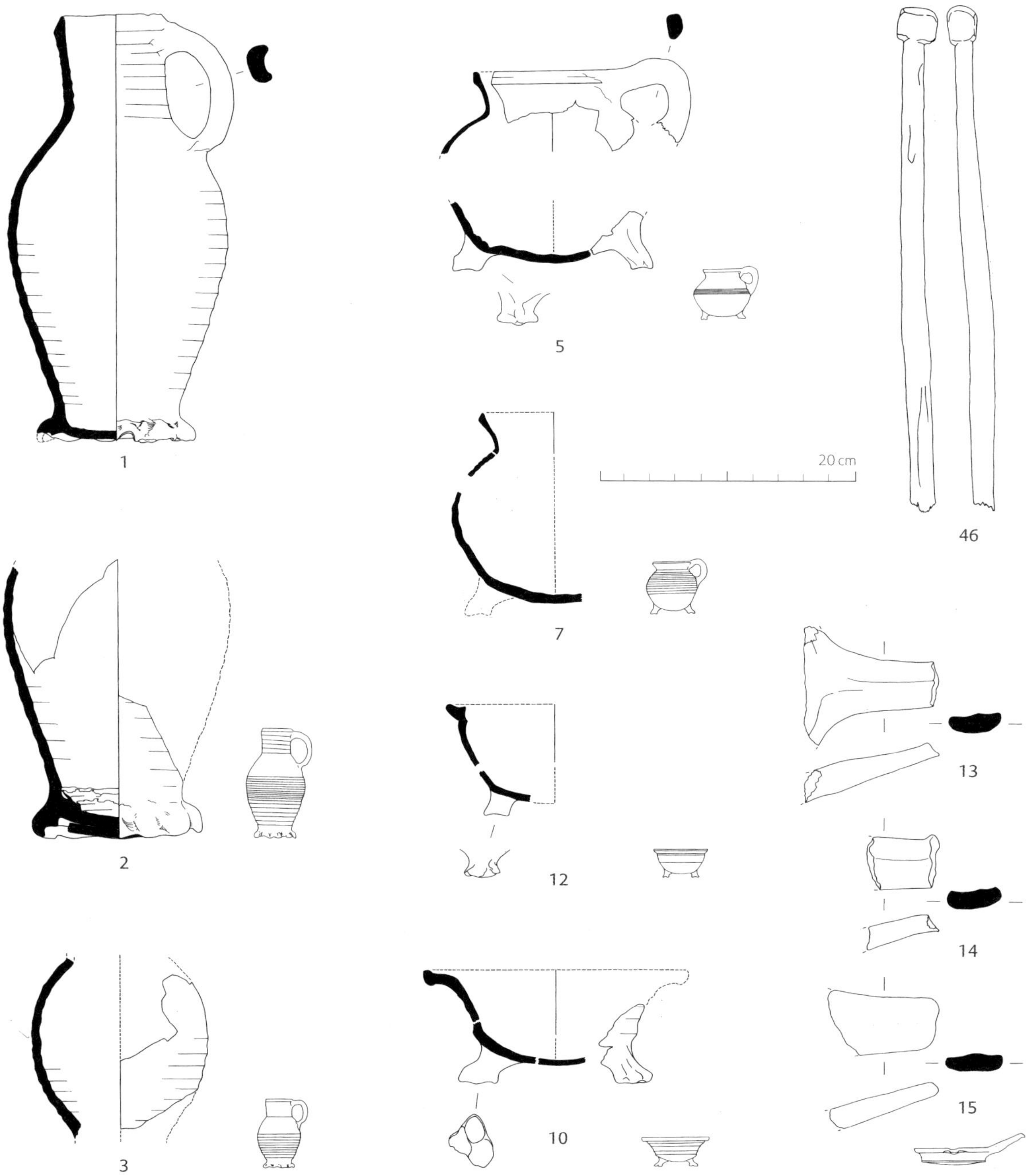

Fig. 16.25. Cooking, eating and drinking: stoneware jugs (1, 2, 3), cooking pots (5, 7), bowls (10 and 12), skillets (13, 14, 15) and spurtle (46).

35. Vlierman, 1992c.
36. Baart *et al.* 1977, 325-326 and 328-329, numbers 619 en 621.

from Amsterdam, however, the cap at the end of the handle is rounded. The same type of knife has also been found aboard the earlier-mentioned cargo ship from lot B 55 in Oostelijk Flevoland.[37]

Drinking ware

The stoneware sherd (4), maybe from the neck of a Siegburg funnel beaker, definitely belongs to drinking ware. As no other sherds were found, we may assume that the beaker had broken well before the shipwreck, by which time the other sherds had all been thrown overboard. The same probably also goes for the handles of skillets 13, 14 and 15, and the top part of stoneware jug 3.

Miscellaneous

A roughly worked, broken stick of alder wood (*Alnus* sp.) still is 37.5 cm long and a little over 1.5 cm thick. It has a knob - rounded rectangular (and abraded?) - at one end (fig. 16.25-46). It has been placed in the category Miscellaneous. Its context offers no clues as to the function of this broken implement. Despite a large gap in time, the piece closely resembles an oakwood parallel, over 30 cm long and ca 1.8 cm thick, exhibited at the Museum für Volkenkunde at Osnabrück (Germany). There the object is described as a *Quirl*, a stick for stirring, churning or beating (scrambled eggs). It was made by a Polish prisoner-of-war in World War II and served as a kitchen implement. In Scotland such a stick is known as a *spurtle* or a *theavel*, for stirring porridge

Victuals

On the assumption that the vessel made no major voyages, not much in the way of provisions would be expected on board. Distances on the Zuiderzee are modest, and there were many potential ports of call for purchasing food en route on a (near-)daily basis. No wooden barrels or parts of barrels were recovered, but any present may have

Fig. 16.26. Stoneware jar (1).

been washed away. But two stoneware jars were found in the afterbody (figs 16.25-1 and -2 and 16.26). Number 1 is complete. Of number 2, just sherds of its lower part were recovered. This jar was probably broken by the digger cutting the sewer trench. The missing parts must have been removed with the extracted soil, and despite a search were not retrieved. It is likely that we are dealing with two near-identical jars, most probably produced in the well-known ceramic centre of Langerwehe, east of Aachen. Their exterior had been dipped in a brown, iron-rich slip, which was covered with salt glaze. The inside

Table 7. Cooking, eating and drinking

Category	Item	Number	Findspot
5.1 Fireplace	-	-	-
5.2 Fire-tending tools	-	-	-
5.3 Fuel	peat	61	forebody
5.4 Cooking equipment	cooking pots, skillets tripod?	5-8 and 18 13, 14 and 15 17	forebody, hold hold hold
5.5 Eating and drinking ware	bowls, Siegburg jug	4, 10, 11 and 12	hold, afterbody
5.6 Miscellaneous	stirring stick	46	hold

37. Reinders *et al.* 1980, 41 and enclosure 11.

of the neck of the complete specimen showed drips of the brown slip. An imprint and a purply hue at the greatest circumference and remains of a fused-on rim of another crock on the outside of the base indicate that number 1 had been stacked inside a kiln for firing. A sherd from another stoneware jar is fused onto the base of number 2. These imperfections suggest that we are dealing with a potter's seconds (or thirds).[38] In various places on the outside of these jars areas of wear can be observed. Taking into account the hardness of stoneware, it is likely that these jars had seen prolonged service by the time the ship sank. Stephan shows a near-identical jar.[39] Bruijn depicts a similar model, with a somewhat narrower neck.[40]

The fragments of a third, smaller stoneware jar (fig. 16.25-3) have a yellowish colour throughout. The small jar probably had its provenance in the Rhine-Meuse region. The distribution of the sherds and the black deposit on the top of the broken edge suggests that it may have been the bottom half of a jar that was reused, *e.g.* for paint. The missing base sherds probably were lost with the spoil from the sewer-digging. The outside bears an orange-brown slip and a light salt-glaze.

It is hard to tell whether these were storage jars or vessels to drink from. Indeed neither excludes the other. The content of jars 1 and 2 is about 2.5 litres, which is appropriate for a modest store of fresh water or another beverage. The rim at any rate is suitable for drinking from.

Bones

A number of bone fragments were found in the afterbody. Three were of cattle: a coarsely broken piece of a left leg (*tibia*, 51), a fragment of a right *radius* (53) and a burnt piece of a rib (*costa*, 52). A rib fragment with butchering marks (*costa*, 50) is of a goat or sheep. The bones indicate that occasionally meat was carried on board. The mussel shell (*Mytilus edulis*, 58) too may be regarded as provisions. In any case no mussels occurred in the Zuiderzee when the ship sank.

Personal equipment/weaponry

This final category comprises objects classified as personal possessions and weaponry. No remains of clothing were found. Footwear is represented by the almost complete upper of an ankle boot fastened with a lace on the instep (figs 16.27-63 and 16.28). The model is similar to type 8 from 's Hertogenbosch.[41] The vamp had been cut off immediately above the seam connecting it to the sole. The upper was no longer part of a shoe when the ship went down, but was kept as repair material. In the lace holes there were remnants of three leather laces (shown beside the vamp in fig. 16.27).

Of a dagger (figs 16.27-24 and 16.29), only the blade was recovered. Its shape and profile however do allow comparison. The blade originally was at least 230 mm long and at its base 26 mm wide. It has a single cutting edge and a straight, even, ca 4 mm wide back. One side of the blade is forged flat, but the other was fashioned with a slight ridge. The tip has corroded away. Yet at the end of the preserved part it is clear that the blade there becomes a thicker, awl-like spike. The blade is similar to the somewhat thinner specimen from wreck B 55 in Oostelijk Flevoland.[42] The root-wood dagger hilt from that ship has survived. The shape of the hilt accounts for the name 'kidney dagger' or 'ballock dagger' (fig. 16.27-24a). Such daggers were very common from the 14th- to the 16th-century and feature in many 15th- and 16th-century images. A famous example is the painting *The Whore-hopper or Awl-bearer* by Hieronymus Bosch, dating from ca 1500 (fig. 16.30). Part of a decorated leather sheath (fig. 16.27-64) may belong with the dagger. Yet the two were not found in close proximity.

Table 8. Victuals

Category	Item	Number	Findspot
6.1 Fresh-water supply	-	-	-
6.2 Supply of beverages	jars	1, 2 and 3(?)	afterbody, hold
6.3 Livestock	-	-	-
6.4 Food provisions	bones, mussel shell	50-53 58	afterbody afterbody
6.5 Miscellaneous	-	-	-

38. Vlierman in collaboration with Kleij 1990, IX.
39. Stephan 1980, 111 fig. 8.13.
40. Bruijn 1979, 129 and 130 no. 11.
41. Goubitz 1983, 277-279.
42. Reinders *et al.* 1980, 39 and Appendix 11.

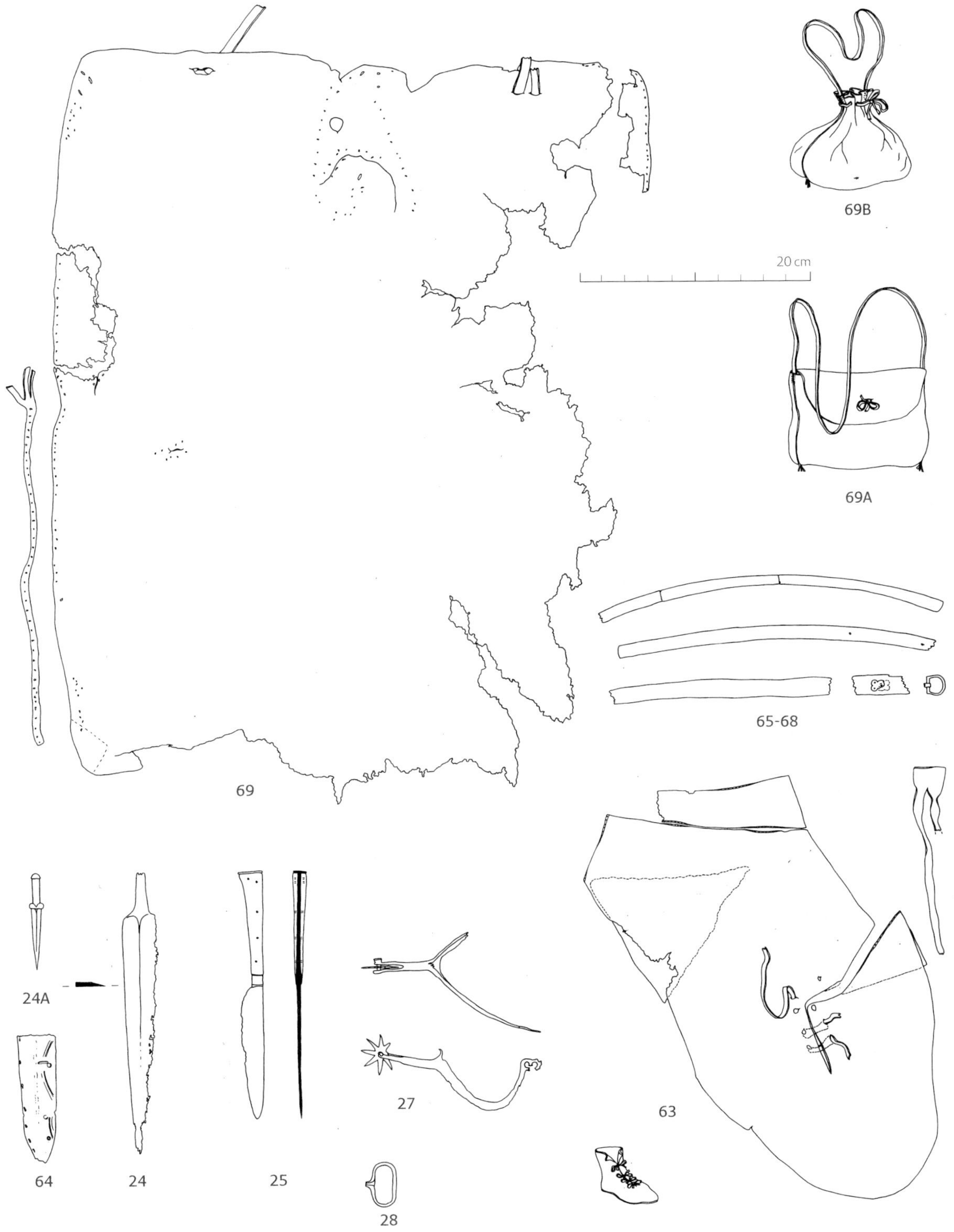

Fig. 16.27. Personal equipment and weaponry: kidney dagger (24), knife (25), spur (27), ferrule (28), shoe (63), sheath (64), belt fragments (65 to 68), pouch and shoulder bag (69, 69 A and B).

Fig. 16.28. Shoe (63).

Fig. 16.29. Kidney dagger (24) and knife (25).

Judging by their small number, the three silver coins (fig. 16.33, 70-72) may be considered pocket cash. They were found beside the sternhook, where the heel of the missing sternpost should have been. After cleaning, they were identified at the Rijksmuseum Koninklijk Penningkabinet. The coins are discussed under Dating (Section 16.8).

The final items in the category of Personal equipment are: a spur, a buckle(?), a wooden clapper and a number of leather fragments which together with a poorly preserved piece of leather once belonged to a pouch, knapsack or shoulder bag.

The iron spur (fig. 16.27-27) has a complete and a broken terminal. At the back, where the two terminals meet, there is a forged iron shank holding a small rowel. The front of the complete terminal has a flat, rectangular end with two holes in it. These holes served for attaching a strap by which the spur could be fastened across the instep and under the sole. A forged protuberance on the top rim at the back would have protected the rider's Achilles tendon. The rowel is an eight-pointed wheel. Being of a very common 15th-century type, the spur is very similar to one from the Zwanenburgwal/Moddermanstraat in Amsterdam.[43] The purpose of the small iron item (fig. 16.27-28) has remained unclear. It was found beside the spur and was recorded as a 'buckle(?)'. Maybe the two were in some way connected.

The wooden clapper (figs 16.32, 16.39 and 16.41) is the finest and most remarkable artefact to have been found on board. Initially it was thought that the three round clapper blades were part of the leather bag or pouch (see below), as they were found among the leather remains. Only after conservation did it become clear that they belonged with a figurine. The clapper and its possible significance, function and provenance are described at the end of this chapter.

Fragments of a poorly preserved and incomplete rectangle of leather (fig. 16.27-69) are the remnants of a bag or pouch. The piece is 59 cm long and 48 cm wide. A distinct fold indicates that the fragment had originally been folded in two, so it must have been a rectangular type of bag, ca 28 cm high and 48 cm wide. A folded-over trimming had been stitched into the well-preserved side

Fig. 16.30. The whore-hopper or The awl-bearer by Hieronymus Bosch, ca 1500.

43. Baart *et al.* 1977, 436.

Fig. 16.31. Two peasants and standing peasant with shoulder bag. Pieter Breughel the Elder.

Fig. 16.33. The three silver coins (70, 71, 72).

Fig. 16.32. The clapper (43).

Table 9. Personal belongings and weaponry

Category	Item	Number	Findspot
7.1 Garments	-	-	-
7.2 Footwear	ankle boot	63	forebody
7.3 Sewing kit	-	-	-
7.4 Tools, weaponry	kidney dagger sheath	24 64	afterbody forebody
7.5 Cash	coins	70-72	afterbody
7.6 Miscellaneous	spur, buckle(?), clapper, strap fragments, bag/pouch	27 28 43 65-68 69	hold hold hold hold hold

seams. The trimming emerged ca 4 cm beneath the bottom of the bag and had been cut into three strips to form a kind of fringe. The trimming of the other side too was recovered, be it without such a fringe. The upper end of the trimming was folded over and with its stitchholes fits onto the upper edge of the bag. The inside, where the two halves match up, shows a triangular impression with stitch-holes, probably the place where the carrying strap had been attached with a patch. The best-preserved part of the roughly cut-off upper edge shows a lip ca 5.5 cm wide, midway along, formed by cutting away V-shaped pieces ca 1.5 cm deep on two sides. In and below the lip, four relatively large and several smaller stitch-holes can be seen. To the right and left of the lip, the upper edge has been cut off somewhat convexly. Roughly midway along each of these 'convexities', and 1.5 - 2 cm below the top edge, a close-set pair of holes, ca 0.7 cm in diameter, are seen. One pair of holes still held the remnants of a split leather strap. In the other pair were the remnants of a similar lace. By these drawstrings the bag could be pulled shut like a pouch. Below the lip is a round hole, ca 1.2 cm across, with a pointed end on one side. Below it is a curved cut, forming about three-quarters of a circle 5 cm in diameter. Stitch-holes along this line and in a roughly triangular shape around the cut and the hole indicate that various repairs were carried out. In another area of the bag, stitch-holes are seen around a tear.

A folded rectangle stitched together at the sides is not an obvious basis for a pouch. Hence it seems likely that the bag originally was a rectangular shoulder bag with a flap, as shown for instance in the drawings by Pieter Breughel the Elder: *Two peasants* and *Peasant, standing* (fig. 16.31).[44] Arguments for this are the roughly cut-off top and the absence of stitchholes along the top, which originally must have been neatly edged. The old shoulder bag was most probably transformed into a pouch or knapsack. The two pairs of holes with straps below the cut-off top edge would have served to draw together the top of the pouch. The function of a crumpled, triangular scrap of leather, found detached near the pouch remains, is unclear. Its edge had been finished with overhand stitching. Maybe it was part of the original flap by which the top of the bag had been closed? No comparable Dutch finds are known.[45] Figures 16.27-69a and b show the presumable models of the original bag and pouch. The likely link between the bag/pouch and clapper will be discussed at the end of this chapter. Fragments of various thin and narrow leather straps and a small buckle (figs 16.27-65 to -68) may have been part of the bag as well. Though some may have belonged to the spur, or served another function altogether.

16.6 Human bones

Just a few of the over 350 shipwrecks excavated so far in the former Zuiderzee area have yielded human remains. At the discovery of the present wreck, a number of bones were recovered that presumably come from the afterbody. Despite intensive searches in the sewer-digging spoil, no further parts of the skeleton were found. The nine bones (49 A-I) were studied by Ms E.G. van Vliet, who supplied the following description:[46]

- *'Tibia*, right, with vivianite around both ends
- *Tibia*, left, with vivianite around both ends
- *Fibula*, right, with deformation of parts of the shank (as a result of infection or injury?)
- *Radius*
- *Ulna*; the proximal end is missing
- *Os metacarpale III?*

44. Besançon ca 1950, figures 69 and 70. The drawings are undated.
45. O. Goubitz, written comm., 1989.
46. Ms E.G. van Vliet (Kampen), written comm., 17 November 1989.

- rib
- rib
- Small fragment, possibly part of a *scapula*.

There is insufficient material for reconstructing a reliable picture of the individual in question. The deformations to the shank of the right-hand fibula may be indicative of an affliction. Specialised pathology may yet provide answers about the nature of any disease and the person's age. This definitely was an adult.'

At the end of this chapter the results of palaeopathological research by G.J.R. Maat; the small fragment was found to be animal bone.

16.7 Cargo and merchandise

Some finds have so far not been discussed. They are a fragment of a roofing tile (56) and about twenty fairly large erratic stones. Some of the stones were found inside the wreck. The rooftile fragment is of a curved model and presumably was left from a transported cargo. The erratics have maximum dimensions between 15 and 30 cm. Most were found in the upper layers of sediment at the stern and bow. Probably they have no direct link to the ship. They may well have been net weights, lost on the wreck by fishermen. A small erratic stone (54), a so-called *windkanter* (a stone eroded by the wind into an often ridged shape) was found inside the ship. Many such quartzites of southern/central European origin and eroded into *windkanters* may be found in the Veluwe region, east of the site.

16.8 Dating

Globally dating the sinking of the Almere-I cog by the disturbance that it caused to the soil profile was no longer possible, as the overlying soil had already been dug away. Yet soon upon discovery a stoneware sherd and the sintel clamps told us that we were almost certainly dealing with a 15th-century vessel.

A number of artefacts provide further evidence for dating the ship's foundering, the three silver coins being the most obvious. These are:

- a *groot*, Guelders, Willem I (Arnhem), ca 1390
- a *nije penning*, Guelders, Reinald IV (Arnhem), ca 1420
- a ¼ *brede plak*, Utrecht, Frederik van Blankenheim, ca 1395-1420

The *terminus post quem* of the ship's sinking lies around 1420. The circulation of coinage was quite brief, owing to strong inflation; as a result, we can put the *terminus ante quem* around 1430-1440. The absence of coins of the Netherlands under the Burgundians, issued in large numbers after 1434, supports this hypothesis. However, three coins are insufficient for any statistical interpretation. That individual coins might circulate for decades is demonstrated by the *plak* issued by Frederik van Blankenheim († 1423) in the shipwreck K 73/74 in Oostelijk Flevoland, which as its final date has 1457.[47]

As for the coins of Utrecht and Guelders, just a few have been found after 1425, and virtually none after 1435, in any of the various (large) Dutch coin assemblages with closing dates between 1430 and 1440. On the basis of this evidence, the presumed period of the ship's foundering is more likely to be early than late in the second quarter of the 15th century.[48]

This date is supported by the traditional and parallel dating of the stoneware jars (1 and 2) of Langerwehe ware. Jars with these characteristic ridged rim profiles were produced around the turn of the 15th century. The use-wear traces and scratches on the jars suggest that they had been in use for some considerable time. The various redware sherds are closely comparable to, among others, the typologically parallel wasters from a kiln find in Utrecht, an assemblage dated to ca 1400.[49] The sherds included no forms that could be fairly definitely dated to the second half of the 15th century.[50]

The shoe type can be roughly dated to 1350-1475,[51] but according to current insights, the basic form of this type was rarely worn any longer after ca 1400.[52]

47. Van Gelder, Jaarboek Munten en Penningen 1971/1972; A. Pol, written comm., 2 December 1987.
48. Vlierman 1992c, IV.
49. Bruijn 1979, 22.
50. Vlierman 1992c, V.
51. Goubitz 1983, 278.
52. Goubitz, written comm., 1992.

The axe, the knife, the dagger and the spur may on the basis of comparison be dated to the 15th century – with a preference for the first half of that century, and no later than the middle of that century.

The dating of the coins, the traditional and typological dating of the pottery and the other (global) datings indicate that the cog sank in the first half of the 15th century, with 1450 as the ultimate limit. The calamity almost certainly occurred (early) in the second quarter of the 15th century.[53]

Tree-ring analysis, without a clear result, was performed by RING at Amersfoort (see Appendix). The condition of the wood and the attrition, the repairs and the overall impression of the ship suggest that the vessel was at least 15 to 30 years old by the time it was lost. A construction date around ca 1400, probably before rather than after the turn of the century, is the most likely. Still it cannot be ruled out that the ship was even older than this.

16.9 Comparison and typological evolution of the rosebolts, the axe and the caulking iron

During the working-out of the data relating to the equipment and the inventory for the excavation report of 1996, it became clear that there might be some typological evolution with respect to some of the items. Already some publications in this vein had appeared.[54]

The rosebolts

In sections 16.2.2 and 16.5.2 three rosebolts (29, 30 and 31; figs 16.13a and b, 16.16 and 16.17) were discussed, which played an important part in securing two transverse constructions.

The remarkable shape and dimensions of their heads (diameter ca 140 mm!) had not earlier been encountered in ships in the former Zuiderzee region. However, their function and construction were previously known. In the 17th-century *modderschouw* (mud-barges) of Amsterdam, rosebolts were found with square heads of 45x45 mm. Their flat bolt shanks, provided with nubs, were fastened

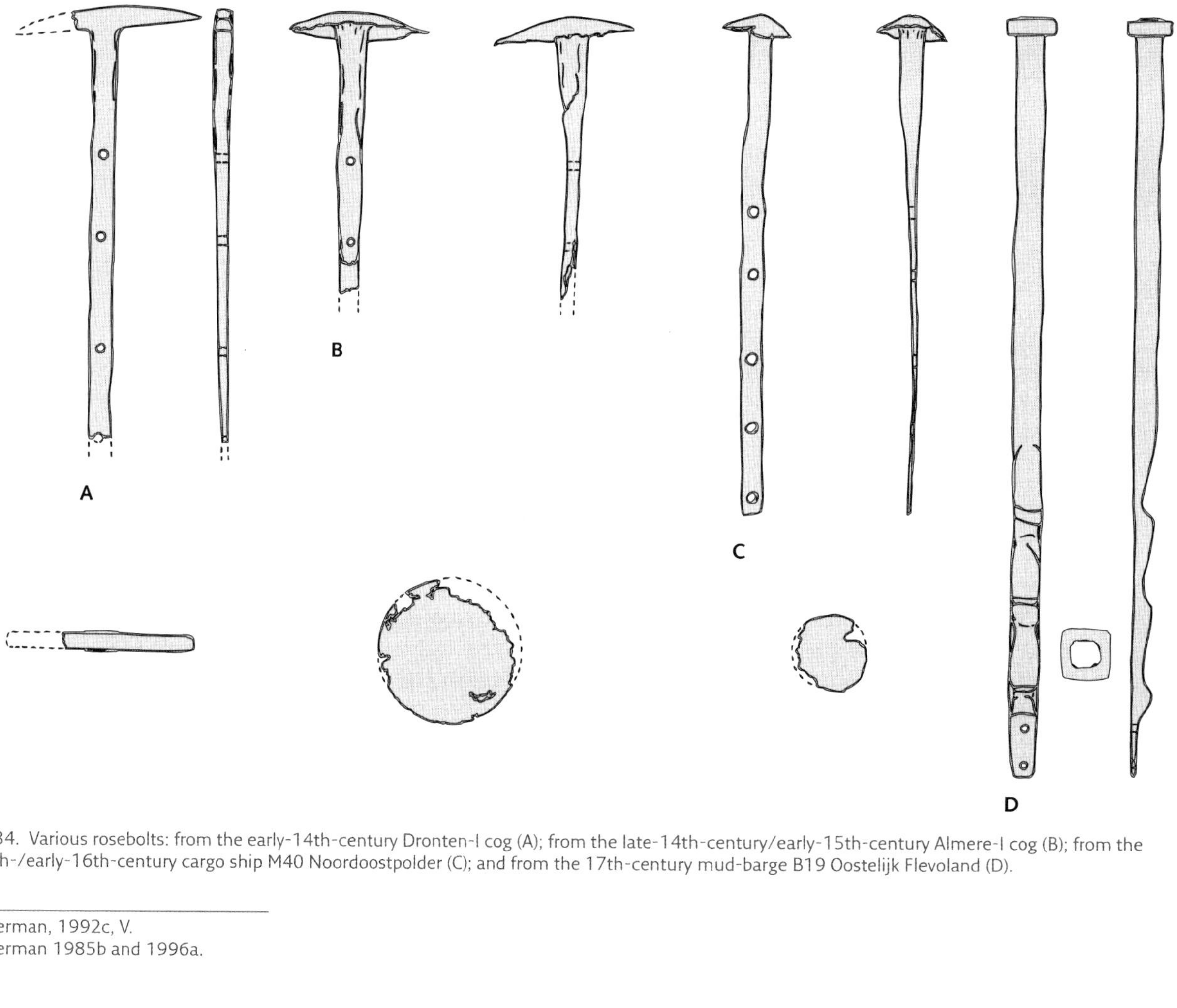

Fig. 16.34. Various rosebolts: from the early-14th-century Dronten-I cog (A); from the late-14th-century/early-15th-century Almere-I cog (B); from the late-15th-/early-16th-century cargo ship M40 Noordoostpolder (C); and from the 17th-century mud-barge B19 Oostelijk Flevoland (D).

53. Vlierman, 1992c, V.
54. Vlierman 1985b and 1996a.

with iron clamps.[55] The rosebolts from a 16th-century cargo ship display a greater resemblance to those from the Almere-I cog.[56] These bolts have a round head with a diameter of ca 75 mm and a flat bolt shank with five (nail) holes in it. The Dronten-I cog, sunk in the first or second quarter of the 14th century,[57] contained two T-shaped items of iron that most probably served the same purpose. Figure 16.34 shows the various types of rosebolt. Despite the small range of four types, there are clear indications of typological evolution.

Such bolts are known mostly from medium-sized ships with an open hold and an after- and foredeck or cuddy. They serve the same function as wall anchors in buildings: to counteract outward pressure on a wall or the sides of a ship. They are in fact comparable to the heavy protruding crossbeams in cogs, and appear on some late-medieval civic seals as well.[58] The T-shaped model may possibly (also) be associated with Small Cogs. The three round knobs on the shell of the ship depicted on the earliest seal of Harderwijk (second half of the 13th century, fig. 16.35)[59] suggest rosebolts similar to those in the Almere-I cog.

Fig. 16.35. The earliest civic seal of Harderwijk (second half of the 13th century). After Ewe 1972.

The axe

The axe from the cog (figs 16.21 and 16.22-21) is comparable to the axes from the Marknesse cog (Chapter 12, section 12.5.2, figs 12.26 and 12.28a) and the axe from the boat with cog features from lot K 73/74 in Oostelijk Flevoland. These vessels date from the second half of the 14th, and the mid-15th century, respectively.[60] In comparing the axes from late-medieval vessels.[61] It was established that the protruding front below the neck, combined with a somewhat convex curved underline of the handle case, might be considered a 'Dutch' design feature. The axe from the Almere-I is different. The downward-curved underline of the case is known from English models.[62]

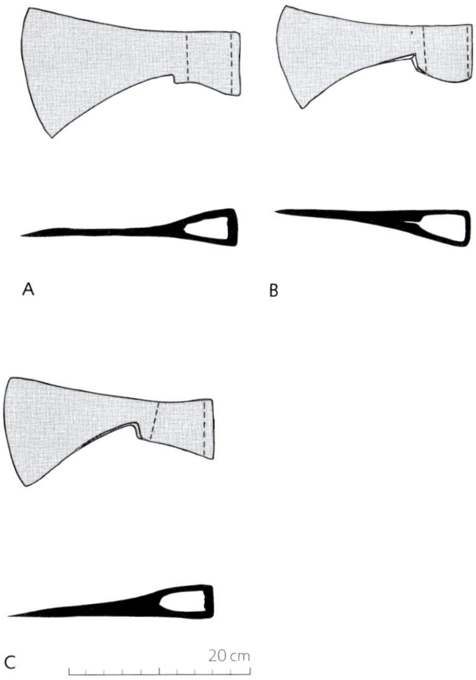

Fig. 16.36. Axes: from the Marknesse cog, second half of the 14th century (A); from the Almere-I cog, first quarter of the 15th century (B); and from the boat with cog features K73/74 Oostelijk Flevoland, mid 15th century (C).

55. Reinders *et al.* 1984, shipwrecks B 19 and Mz 6; shipwreck Az 87 II, Zuidelijk Flevoland, file RCE-Lelystad.
56. Wreck M 40 from the Noordoostpolder, file RCE-Lelystad.
57. Reinders *et al.* 1980, 15; Chapter 11, sections 11.2.2, 11.5.2 and 11.8, and figs 11.19-22 and -23 en 11.20a en b.
58. Ewe 1972.
59. Ewe 1972, 63 and 132 (catalogue no 62).
60. Modderman 1945, 79 and fig. 1; and the Marknesse cog, see section 12.5.2 and fig. 12.28a; and Reinders *et al.* 1980, 28 and plate 7, respectively.
61. Vlierman 1985b
62. Idem, 34-35

The caulking iron

The caulking iron (fig.16.22-23) was compared with specimens from two late-medieval cargo vessels.[63] Figures 16.37a, b and c show the three late-medieval caulking irons and illustrate the variation in blade shapes. The forged handles are straight and octagonal in section. When we compare some 16th- and early-17th-century caulking irons , it is striking that one from a mid-16th-century carvel-built *waterschip* (W10, Oostelijk Flevoland) has a curved blade, which is the case also with one of the two caulking irons from a small cargo vessel with a carvel-laid bottom and clinker-built sides.[64] Their short handles are round in section and usually have flat, round butts. The blades are wider too. [65]

The reason for this change in shape can be explained by a change in building tradition. The late-medieval (partly) clinker-built ships on the whole have relatively wide, triangular seams between the overlapping planks, which were filled with moss. These seams are easily accessible with short, straight caulking irons. Carvel-built ships lack such wide seams. The caulking method had to co-evolve. For filling the straight, narrow seams with caulking without damaging the edges of the planks, a thinner and broader blade was required. For inserting and jamming the caulking, carvel construction requires an overhand method (fig. 16.38). To do this effectively, the blade of the caulking iron should be slightly curved. The round, flat butts that we do not see in older specimens, were without doubt designed to stop the mallet from slipping and striking the wood beside the seams. Also they helped to avoid the hands being hit. In this context, Witsen at the time appropriately remarked:[66] *"When the planks are cut obliquely upwards, the ship is easy to caulk."*

The caulking irons in the toolkit found aboard the late-16th-century cargo vessel SO1 in the Waddenzee[67] show that large ships in those days carried long and short, narrow and broad, and straight and curved caulking irons.[68] The same range is still found in 19th- and 20th-century toolkits of shipyard carpenters.

The generally wide seams in the river craft on the German river Main too were filled with moss caulking. It was in-

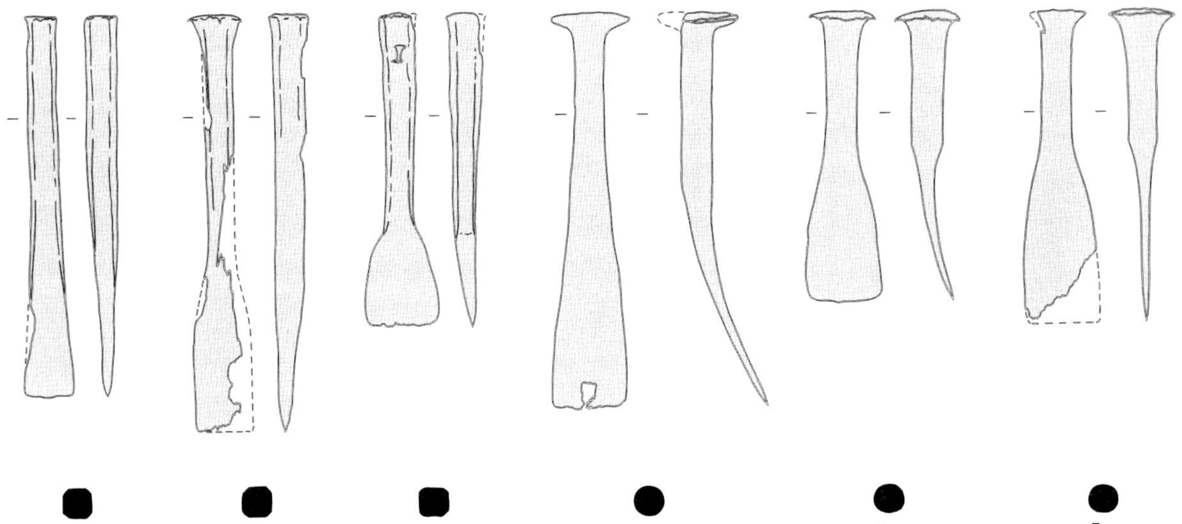

Fig. 16.37. Various caulking irons: from the Dronten-I cog, early 14th century (A); from the Almere-I cog, first quarter of the 15th century (B); and from the boat K73/74, mid 15th century (C); from the *waterschip* W10 Oostelijk Flevoland, second half of the 16th century (D); and from the small cargo ship, G116 Oostelijk Flevoland, early 17th century (E and F).

63. From lots N 5 and K 73/74 in Oostelijk Flevoland, Reinders *et al.* 1980, 14 and 25, plates 3 and 7; and the Dronten-1 cog, see section11.5.2, figs 11.21-18 and 11.24.
64. Reinders *et al.* 1978, 41 (find no. 45) and G 116, Oostelijk Flevoland, file at RCE-Lelystad, respectively.
65. Vlierman 1992b, 15.
66. Witsen 1671, 48a.
67. Vos 2012, 26 and fig. 1.4.
68. Depot of ships' inventories, RCE-Lelystad.

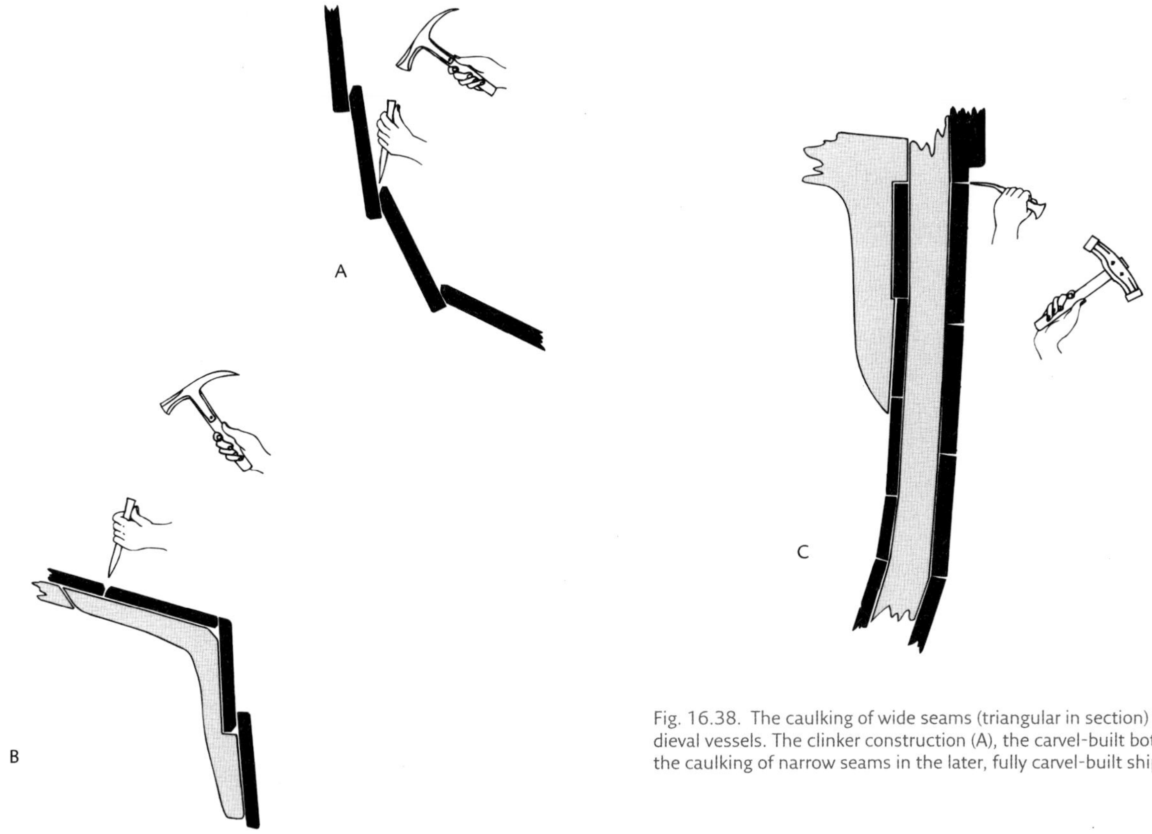

Fig. 16.38. The caulking of wide seams (triangular in section) in late-medieval vessels. The clinker construction (A), the carvel-built bottom (B) and the caulking of narrow seams in the later, fully carvel-built ships (C).

serted by means of wooden tools, known by the names of *Stopholz*, *Triebel*, *Mooshölle* and *Holzkeil*. The *Klopfholz* or wooden mallet was used for jamming the caulking tight.[69] In the Bremen cog, an oakwood object was found that Keweloh interpreted as a *Kalfatholz*.[70]

Keweloh refers to his article on the *Triebel* from the Main region, the *Keidel* or *Mooskeidel*, as the tool was called in the Upper Rhine region, and the *Schöppel* along the Danube. Further he notes that this kind of tool was known only in the southern part of Germany. He also mentions similarly fashioned wooden objects belonging to a sailmaker's toolkit, but states that in the case of the Bremen cog it should most probably be regarded as a *Kalfatholz*.

At the start of the building of the Kiel cog, the caulking was inserted with a *Kalfatkeil*; but fairly soon the builders switched to the 'modern' caulking iron. Baikowski explains the change as follows: the mixture of cow hair, moss and pitch could not be jammed tight with a caulking iron by contrast, the modern material with which the Kiel cog was caulked was suitable.

Like its original model (the Nijkerk-II cog), the replica, the Kampen cog, was caulked with moss (*Sphagnum*). In emulation of the original manner of caulking in the Kiel cog, all the moss was driven into the seams with a broad wooden 'chisel' and an iron mallet. The business end of the 'chisel' was 10 cm wide and 1 cm thick. Some experimenting showed this method of caulking to be quite effective. The wooden caulking tool or *Keilholz* can be simply made from a piece of oak wood, and has a reasonable use life. For caulking the seams of the Kampen cog, four of these were made. After some experimentation it transpired that the moss caulking on the Kampen cog could equally have been inserted with a regular caulking iron without damaging the moss or the wood.[71]

In the over 200 shipwrecks in the former Zuiderzee region whose caulking method was investigated, only moss was employed to fill the seams, no oakum. In the cogs, pure moss was used in combination with moss laths and sintel clamps (clamped moss caulking). Cowhair had been used only in some fully clinker-built and riveted ships.[72] The Doel-II cog so far is the only exception. Generally just one or two caulking irons will be found in the standard on-

69. Hanke 1995, 279-280.
70. Keweloh 1983, 7-10.
71. Personal comm. C. Sars and A. van den Heuvel (Sars Houtbouw), 1996.
72. Vlierman 1996a, sections 1.2 and 1.3.

board toolkit.[73] No wooden caulking tool was ever found in the over 350 excavated wrecks in the Zuiderzee region.

16.10 Summary

Ship

The keel plank and the hooks together make up the 'spine' typical of cogs. The stern- and stemposts are straight and have a rake of ca 120° (aft) and ca 127° (bow). These posts are thicker at their lower end than is usual in the older cogs, and not only rest upon the sloping flat scarfs of the stern- and stem hooks, but also on the hooks' horizontal arms. The ends of the strakes of the bottom are fixed into rabbets in the sides of the stern- and stemposts. The side strakes end on the mitred rear surface of the sternpost and the mitred front surface of the stemhook, respectively. Besides, the hood ends of the side strakes on the bow are protected by a false stem, nailed across them and onto the front of the stempost and stem hook. The height of the stern and stem above the keel line differs little, which together with the high sides amidships gives the vessel a gentle sheer. The top view of the ship shows a somewhat more elongated outline than that of the seagoing small cogs, and at similar distances from the stern and stem is ca 20 cm wider fore than aft.

For the hull relatively wide, and up to ca 7 m long oak planks were used. Most of the strakes are composed of three planks; the remainder, of two. They are arranged according to the so-called 'brother-and-sister' pattern. During construction, four temporary frames had been used to keep the carvel-laid bottom planks together for the time being. In the stern and bow, the carvel construction of the bottom merges into a clinker-built one. The strakes of the sides are joined in clinker fashion. The seams are closed with sintel clamped moss caulking. The ship has no sharp transition from the bottom to the sides, but a rounded turn of the bilge. This is achieved by the changing angles at which the outer bottom strakes and the lowermost two side strakes are joined together. Some frames in the stern and the bow have a slight S curve. The hull numbers eight strakes on either side of the keel plank.

The cog had a closed underfoot ceiling of pine boards running from frame 4 to frame 21. The short mast step was fully integrated into the underfoot ceiling. The adjacent ceiling on the sides too was fully closed, but was made of oak; the uppermost ceiling strake was thicker than the others and ran from frame 1 to frame 23. It was a shelf clamp and at the same time functioned as a stringer to augment the longitudinal rigidity of the sides. The vessel had at least two massive transverse connections: one forward of, and against frame 7, and the other resting on the truncated futtocks of frame 17 (the sail beam). Their connections to the sides were additionally strengthened by forged-iron rosebolts with exceptional heads, ca 140(!) mm across, of a model not previously encountered (in shipwrecks in the former Zuiderzee area. Besides, there must have been further transverse connections in the ship. At deck level, there were gangways made of massive planks, ca 30 cm wide and over 9 m long. They ran on into, and were integrated in the after- and foredecks. The gangplanks were fitted around the futtocks and against the shell. Presumably a coaming was attached along their inboard edge.

A small number of futtocks preserved up to the gunwale had been chamfered at the top and inboard side. These sloping futtock heads engaged with notches in the rear/outside face of the few inwale fragments that were also recovered. The vessel most probably had an inwale almost rectangular in section along the inboard side of the upper strakes of the sides. It is likely to have consisted of three parts on the starboard and port side, which continued right up to the stern- and stemposts. The ends of the inwales were joined together at the stern and stem by a small V-shaped timber. Against the inboard side of the V-shaped timber in the bow, at the top and almost vertically, a short, thick plank had been treenailed with a large notch in its top edge. Most probably this plank was intended to hold the lowered mast and/or sprit with the sail.

Nothing can be said with certainty about any decks or cuddies. Yet it seems very likely that the vessel had an afterdeck ca 3.5 m long and a foredeck ca 5.5 m long. The front of the afterdeck probably carried a windlass. The deck planks of the afterdeck presumably were longitudinally placed; those of the foredeck would have been short, and placed transversely between the sides and stringers which ran forward on either side of the mast. The stringers may have been held in place by 'short' or 'half' deck knees (between the stringers and the sides) that were fastened both onto the stringers and against the sides (inwale, futtocks and top shell strake). The foremost part of the foredeck also may have held deck timbers or knees running from side to side. The ship may have carried a short, lowerable mast and a spritsail.

In 1988, Hocker, together with some other graduates of Texas A&M University (USA), drew to scale 1:10 the wreck's timbers salvaged in the rescue excavation, and reconstructed the ship on paper. A scale model was built

73. Vlierman 1996a, 1992b, 14-15 and table 2.

on the basis of this reconstruction. New and evolving insights, resulting especially from the analysis in the present study of all known cog finds in the Low Countries, made it desirable and worthwhile to reconsider this wreck too. Another reason was the intention at the outset of this study to draw, describe, analyse and publish in a monograph *all* cog finds from the Netherlands in a standardised way.

The reconstructed principal dimensions of the Almere-I cog are:

- Length over the stern an stem at gunwale height — ca 16.25 m
- Largest width at gunwale level — ca 4.80 m
- Height of the stern (keel line to the top gunwale) — ca 2.60 m
- Height of the stem (keel line to the top gunwale) — ca 2.65 m
- Height of the side amidships (keel line to the top gunwale) — ca 1.90 m

Equipment and inventory
A few sherds of a stoneware jar and a number of human bones were the first finds at the discovery of the wreck. They came from the part of the afterbody that was destroyed by the mechanical digger. Despite searches in the spoil, it proved impossible to complete the skeleton or the jar.

In the course of the excavations, the various findspots of the items were recorded in a survey drawing (Plates 16.I and 16.II B). The items are listed under 83 inventory numbers and discussed in the text under various headings. They are classified according to a system designed for the inventories of late-medieval vessels. This classification yielded interesting evidence of, among other things, multifunctional use - and hence smaller numbers of items per ship compared to inventories of 17th-century and later ships. In section 16.9 some recovered items and fastenings are compared with specimens from other ships in the former Zuiderzee area, that display evolution in their form.

The ship's equipment is considered to include the above-mentioned three (of the originally at least four) rosebolts. Similar striking, large globular heads appear to feature on the side of the cog on the earliest, 13th-century civic seal of Harderwijk. The recovered brace block is the oldest known example from a shipwreck in the Zuiderzee area. A curious, hewn-off piece of oak wood may have been part of an anchor. The operational equipment is represented by a (peatcutting?) spade, which in shape resembles spades that once were used in Wales for constructing and maintaining earthen banks along roads.

Among the tools generally present aboard late-medieval ships are an axe with a handle, an unusually shaped adze, a caulking iron, a hand saw and a marlinespike. Certain details in the design of the axe and the adze appear to be 'not quite Dutch'. The tools bear a closer resemblance to English specimens. A remarkable find is a lump of grease in which almost certainly impressions of a sail needle are visible.

Curious too is the absence of a hearth or any evidence of one. Also fire-tending implements and cooking ware are lacking. The only evidence that probably some fire-tending and cooking took place on board at some time, are a few chunks of peat and redware sherds of cooking pots with soot on them. Two jars of Langerwehe stoneware were carried on board, probably containing a store of water or some beverage. They each had a content of 2.5 litres. Remarkable are the redware sherds of some bowls with claw feet. They had no handle and may have been early precursors of dinner plates. At any rate they are the earliest examples of such tableware aboard ships in the former Zuiderzee area.

There are a few exceptional items that can be reckoned to people's personal equipment and weaponry. These are the spur, but above all the unique birchwood clapper and the presumably closely associated leather bag or pouch, items which have received a good deal of attention and will continue to do so. With our current understanding, it is thought that we may be dealing with the clapper and beggar's pouch of a possibly affluent, 'healed'/misdiagnosed leper. The possible affluence is inferred from his person's presence aboard a ship, as well as the remarkable recovery of the rider's spur.

The three silver coins produced the most precise evidence for dating the foundering of the ship. This was supported by the global (and/or parallel conventional) dating of the pottery, the spur, the kidney dagger and the shoe, to yield a (by archaeological standards) close dating in the 15th century, most probably early in the second quarter of that century.

The lines, dimensions and construction of the vessel suggest that the Almere-I must have been a Small Cog for transporting heavy cargo in bulk (such as building materials) across the Zuiderzee and (major) inland waterways. Whether its itinerary included the 'inland route' between the Zuiderzee and the river Scheldt, is uncertain. It was not until 1536 that Charles V granted permission to build a new lock east of Gouda, through which ships 16 feet wide might pass. Like many other locks in Holland (*i.e.* the western Netherlands), this Goudse Binnensluis previously was too narrow. The many repairs, the wear on the hull, the possible replacement of the uppermost side strakes and maybe even the sintel clamped moss caulking on the outside of the clinker seams, which was no longer customary even in the second half of the 14th century, suggest that the ship was quite elderly by the time it foundered. The dendrochronological analysis produced no result, so no certainty can be had on this point. It is imaginable that the vessel was not even seaworthy any longer, and was only rarely used. This might explain the absence of a hearth and galley equipment. It may be far-fetched, but in conclusion might we not conjure up a wealthy, misdiagnosed leper with his obligatory attributes (clapper and beggar's pouch) who with his horse had been refused passage on other ships?

Human bone material by G.J.R. Maat

Palaeopathological analysis of the eight human bones from the early-15th-century cog Almere-I produced interesting results. The skeletal remains were found to be of a male, 20-40 years of age, who is likely to have drowned when the ship foundered. Closer examination showed that his stature was ca 1.69 m, and that in his growing years he suffered at least seven episodes of ill health that left radiologically detectable scarring in his shinbones. The man moreover had extensive bone abnormalities resulting from scurvy. In the course of healing, the associated bleeds at the attachments of tendons and muscles, as well as around minor bone cracks under the membrane had been replaced by bone tissue. The locations of these bone abnormalities are very characteristic of this condition, which results from chronic vitamin-C deficiency.

Formerly scurvy would affect the population of Scandinavia and the Low Countries in early spring on an almost yearly basis. Its cause was a lack of this vitamin in late winter, when its principal source, fresh fruit and vegetables, was not yet available.

The identified lesions are not characteristic of leprosy, although it cannot be ruled out that the man was a leper when he died. Of course, after the manifestation of the typical skin lesions caused by leprosy, such as sores and lumps, it may take years for the disease to affect the bone through bacterial invasion.

A remarkable clapper: significance, function and origin

A.M. Koldeweij & K. Vlierman

The presumably almost complete inventory of the Almere-I cog includes a remarkable wooden clapper (section 16.5, figs 16.32, 16.39 and 16.41). This clapper gave rise to two articles in the journal *Antiek*.[74] In the first article, an elaborate description of the object was presented and questions were raised regarding its origin: could it be a leper's clapper and what might be the significance of the figurine that forms the clapper's handle. Without these questions being answered it was then assumed that the clapper was most unlikely to be a leper's, and that it was unclear who or what the female figure represented. As regards the latter, some suggestions were put forward, based on various interpretations of the inscription on the clapper: the words '*hilgi-nin*'.

Meanwhile, in response to written replies and debates sparked by the hypotheses in the first article, some new ideas were formulated and it seemed that the 'clapper problem' could be resolved.[75] The reactions and new data were published in the second article in *Antiek*. They also prompted a request for palaeopathological analysis of the human bones, which also were recovered from the wreck (section 16.6 and after this).[76] The two articles are summarised below. The reader is referred to the full articles in *Antiek* for more detailed information about medieval images of lepers with clappers, about miracle stories and leprosy wonder cures, and about the tradition of people with infectious diseases warning with clappers, which is a good deal older than is generally assumed.

The figurine represents a slender female figure, in a long, pleated mantle. The mantle is closed at the front with buttons(?) and around the waist is gathered by a belt. The hair is tightly combed with a central parting, and braided into a single plait at the back, which hangs down to the waist. The young woman wears a braided headband (fig. 16.39). At her wrists, on the front, we can see that she wears an undergarment, the cuffs of which are visible. Her left hand rests upon the hilt of a sword hanging somewhat diagonally. In her right hand she holds a flat, round disc with a six-pointed star at the centre. The hem of the mantle hangs down to the sockle, hiding the feet. Letters are engraved on the front and back of the sockle, the sides of which are decorated with crossing lines. Beside the plait on the back is the remnant of what may have been an iron suspension(?) eye.

The two round, 'necked', movable clapper blades are carved on the outside. The fixed, third clapper blade in between is undecorated. The front blade displays a seven-pointed star, constructed out of three arrow-shaped motifs. The star actually consists of a single uninterrupted ribbon. Its seven-sided heart is decorated with a seven-petalled flower. The interspaces are filled in with small and large trefoils. Around the star's circumference a wavy ribbon is woven through its points.

The blade on the rear of the clapper shows a star with nine points, constructed from three interlaced equilateral triangles. As on the front, a wavy ribbon is woven through the points of the star. Carved at the centre is a large flower with two tiers of seven petals around its heart.
On the narrow necks of the movable clapper blades, two different owner's or maker's marks are engraved (figs 16.39 and 16.41). The left edge of the front clapper blade bears a small button. On the other five sides of the blades are the bases of broken-off buttons, presumably ornamental features. They may (also) have served to tie the movable parts together. Apart from the five broken knobs, the softwood (birchwood, *Betula* sp.) clapper displays no use-wear or abrasion. Hence we may assume that the object was no more than one or two years old when the ship foundered in the early 15th century.[77] The clapper appears to be unique in the Netherlands (and beyond), and raises questions about its function, iconography and provenance.

74. Vlierman & Koldeweij, 1992-1993 and 1996-1997.
75. With thanks to A.M. Hagemeijer (Broekland), S.M.C. Ebbinge-Wubben (Utrecht) and P. le Blanc (Stichting voor de inventarissen van Kerkelijke Kunstcollecties in Nederland, Utrecht).
76. Maat, 1996.
77. Vlierman, 1992b; and section16.8.

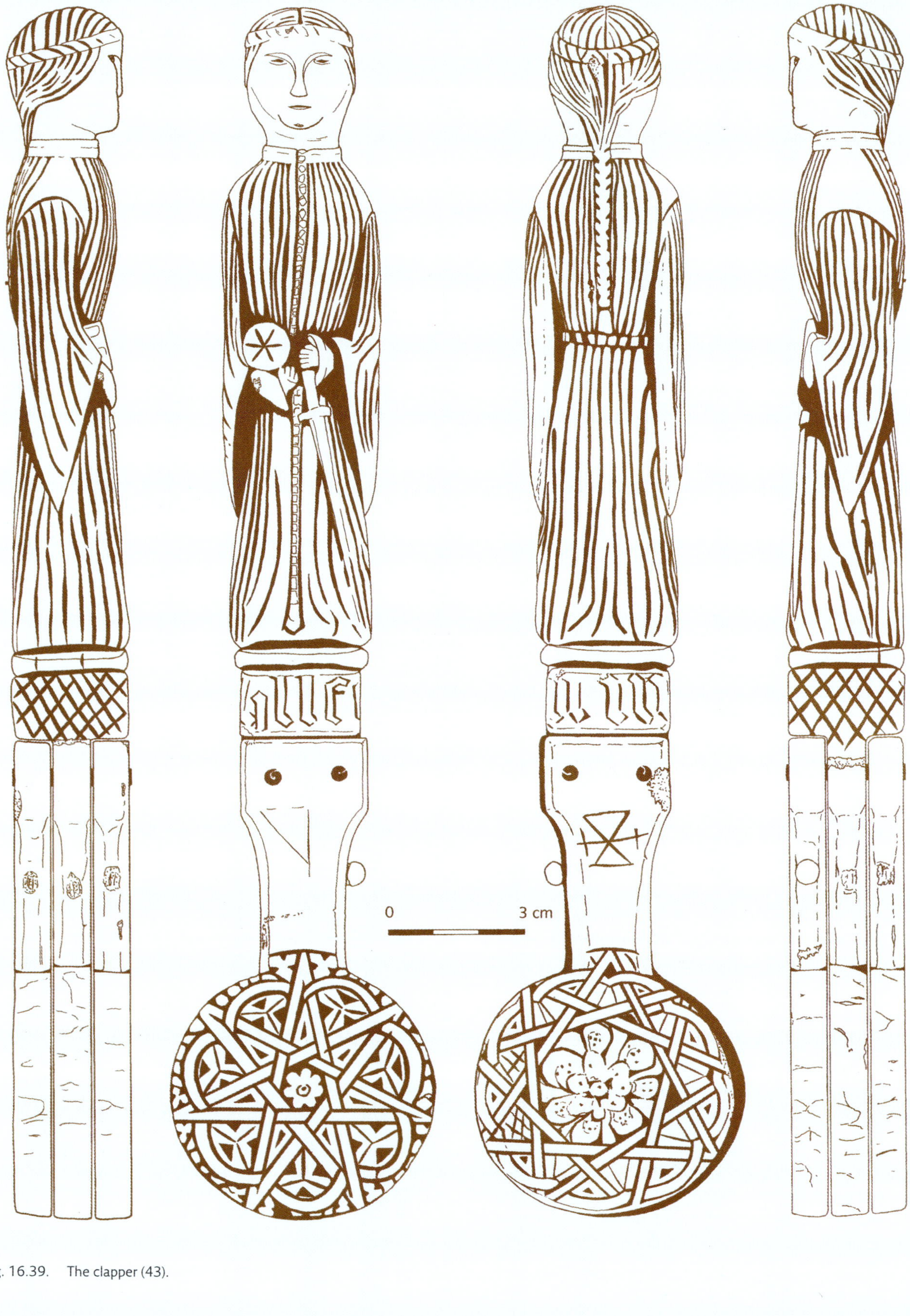

Fig. 16.39. The clapper (43).

Initially the clapper was thought to have belonged to a leper. The more evidence about this much-feared disease in the Late Middle Ages was gathered, the less likely this scenario began to appear. The reasons for this are quoted from a publication by Leyten.[78] 'When the disease became evident to people, patients were obliged to report to the authorities, including the surgeon, who would examine them and make a pronouncement. If the disease was confirmed (at the so-called '*schouwing*' or visitation), patients would be handed a letter stating the diagnosis, after which they were to abide by the regulations governing the behaviour of lepers in their town of residence and its surroundings. The afflicted were also given a clapper to warn those in the vicinity about the sickness. Upon a leper's death, all his or her possessions, including the clapper, would be destroyed or otherwise disposed of.'

It seems hardly likely that anyone showing symptoms of leprosy and carrying a clapper should be admitted aboard a ship, where close physical proximity is virtually inevitable. A second reason why this seemed unlikely to be a leper's clapper it its elaborate decoration. Would anyone banned from all social intercourse have been afforded such a fine clapper? Lepers' clappers normally were quite humdrum objects, at best adorned with the city's coat of arms and/or a date. Figure 16.40 shows an example from Haarlem with the date 1738 and the initials LP, together with the associated note stating the surgeon's verdict (the *Schouwbriefje* or *Vuijlbriefje*), which belonged to "A Spinster about 30 Years of age by the name of Wijntje Gerrits of Cülenburg'. Both objects are kept at the Museum Elisabeth Weeshuis at Culemborg.

Besides such simple clappers with mere initials, dates or city arms, a few other decorated examples are known. Meulemeester discusses a leper's clapper from the Museum Onze-Lieve-Vrouwe ter Potterie in Bruges (Belgium), and another decorated one from St. Walburga's church in the same city. Remarkable about these two clappers from Bruges are the trapezoid clapper blades. This is also the shape of the clappers in drawings and paintings that Meulemeester reproduced with his article.[79] We shall list them here too: *The diabolical leper* by an anonymous 16th-century artist;[80] an etching by Rembrandt (1606-1669) depicting a leper;[81] and a *grisaille* painting showing a blind beggar with his sick, leprous wife, carrying their child on his back, by A. Pieterszoon van de Venne (1589-1662).[82] The trapezoid model is known from other medieval paintings as well.

The 18th-century specimen from Culemborg has round-oval clapper blades, while another in Museum De Lakenhal in Leiden is rectangular with rounded corners. The latter is said to have belonged to Quirijn Janszoon van Leyden, 1572.[83] The clapper bears the city arms of Haarlem on one of the moving clapper blades, and the letters LP in a circle. On the other blade are the figures 1188. Apart from the shape of the clapper blades, the clapper is comparable to the one from Culemborg, while the dating too probably differs little. The second digit 1 most probably is a 7, making the date 1788. This also tallies with the overall impression of the clapper and the condition of the wood.

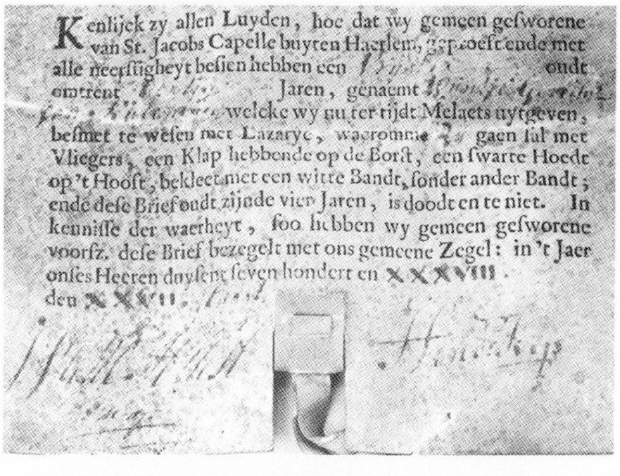

Kenlijck zy allen Luyden, hoe dat wy gemeen gesworene van St. Jacobs Capelle buyten Haerlem, geproefd ende met alle neerstigheyt besien hebben een [illegible] oudt omtrent [illegible] Jaren, genaemt [illegible] welcke wy nu ter tijdt Melaets uytgeven, besmet te wesen met Lazarye, waeromme [illegible] gaen sal met Vliegers, een Klap hebbende op de Borst, een swarte Hoedt op 't Hooft, bekleet met een witte Bandt, sonder ander Bandt; ende dese Brief oudt zijnde vier Jaren, is doodt en te niet. In kennisse der waerheyt, soo hebben wy gemeen gesworene voorsz. dese Brief bezegelt met ons gemeene Zegel: in 't Jaer onses Heeren duysent seven hondert en XXXVIII. den XXVII. [illegible]

Fig. 16.40. The clapper and the Schouw- or Vuijlbriefje (visitation certificate) of Wijntje Gerrits van Cülenburg (1738).

78. N. Leyten, Koninklijk Instituut voor de Tropen, Amsterdam, written comm. 1989.
79. Meulemeester, 1987-1988, 393-398.
80. Amsterdam, Rijksprentenkabinet, inv. no. OB 3080.
81. Amsterdam, Rijksprentenkabinet, inv. no. B 400/B. 171 111.
82. Rotterdam, Museum Boymans-Van Beuningen, inv. no. 2119.
83. Overvoorde 1924, 62-63 no. 740.

Significance, function and provenance

There are few if any similarities between the clapper from the Almere-I and the later specimens discussed above and those in the more familiar medieval images. Nothing seemed to indicate that this item ever was used as a leper's clapper, and the context of the find - a cog - made this even less likely. Nowadays clappers are known only as children's toys, but until quite recently they served to produce sound in both religious and secular settings.[84] This variation in secular signalling instruments - maybe for instance the town crier or other, comparable events or procedures - usage in Christian liturgy instead of the church bells in the final three days of Holy Week - or in folk religion.

The context - the cog - fails to contribute to any further, 'functional' explanation of the clapper. Nor does its specific form bring us any closer to a solution, in particular because it has no parallel, neither typologically nor chronologically. Our sole remaining source of evidence is the decoration on the clapper itself. The designs on the round clapper blades are made by the incision technique. This may be of interest in terms of folk art, but does not contribute to a more specific explanation. Of greater relevance without any doubt is the handle, with its figurine and inscriptions, but their meaning too is not straightforward. In the first article in *Antiek* an attempt was made anyway to interpret the meaning of the handle. Yet it should be noted that this was done with the utmost caution and that the conclusion was highly tentatative.

As described above, the clapper's handle is worked into the figure of a young woman wearing a long plait down her back. With her left hand she touches the hilt of a sword that hangs from the belt around her waist. In her right hand she holds a round, hard-to-define object. This valiant lady stands on a sockle inscribed with Gothic lettering. The text, incised in imitation calligraphy, is difficult to transcribe, but appears to read: *hilgi* (front) - *nin* (rear; fig. 16.41). Read the fourth letter as a capital in Humanist script and in the middle omit the other letters. It may be that the woodcarver - who presumably was illiterate - copied some example to the best of his ability, putting the (Gothic) Blackletter 'd' in reverse, as often happened.[85] If this is the case here, we should read the inscription as *hildi - nin*. Which immediately prompts the next question: what is the meaning of this word or these words? Some solutions have already been suggested, but unfortunately they failed to produce a satisfactory answer. The suffix *nin* may be a diminutive of *hilgi*, or more prob-

Fig. 16.41. The socle with the inscriptions 'HILGI' (front) and 'NIN' (rear), and the two owner's or maker's marks; the one surviving button and the fastenings of the movable clapper blades (rolled pieces of leather).

84. Zeller 1902, 214-215; Adrian & Mielke 1903, 436-438; Andree 1910, 250-264; Perkmann 1932, K. 1443.
85. Koldeweij 1987, 305-317.

ably *hildi*. Variations on *hilte*, *hil* or *helte*, meaning hilt or handle;[86] the name of the Valkyrie Hilda, Hilde or Hille, who features in Norse mythology;[87] or *Hillebille*, which folklorists tell us is the onomatopoeic name of a signalling instrument from Central Europe.[88]

The response to the first article, the ensuing debate and some new evidence prompted a revision of the intial, tentative conclusion that it could not truly have been a leper's clapper.

The female figure displays two distinct attributes: a sword in her left hand and a round object in her right. The latter should not, as was suggested rhetorically in 1993, be interpreted as a clapper, but as a wheel. In fact, the young woman can be identified as St Catherine of Alexandria: *Hilgi nin*, Holy/Saint Nina - Nina being short for Catharina.[89]

St Catherine was one of the most popular saints in the Late Middle Ages; she was venerated in various capacities. In southern Germany she was the particular patron saint of lepers. In late-medieval hospitals the chapel or the altar would often be dedicated to St Catherine.[90] Leprosariums were dedicated to St Catherine in the German towns of Esslingen (Baden) and Fulda (Hessen), both on the river Neckar, and in the Swedish town of Skännige. St Catherine's Hospital at Esslingen was founded in 1232 and lies within the town's ramparts north of the old city church.[91]

At Fulda, St Catherine's leprosarium was built outside the defences in 1272.[92] In the Netherlands, the leprosarium at Kampen was under the patronage of the Holy Virgin and Saints Catherine and Magdalen. It was built c 1380 beside the regular hospital.[93] This does not mean that the meaning can be inverted and the role of St Catherine as the patron saint of the lepers is overestimated. Among the great number of patronages (over 35) in the medieval bishopric of Utrecht, St Catherine had been selected for very different reasons, which explains this.[94]

Late-medieval illustrations of individuals with clappers are fairly numerous, but at the same time limited: lepers can be recognised by their clappers, with which they warned other people to keep their distance. Infection was avoided by isolating the afflicted. A miniature in a mid-14th-century manuscript, the *Miroir Historical* by Vincent de Beauvais, for example, shows two lepers being denied access at the city gate. In the same vein are two near-identical copies - dated ca 1400 - of the *Riègle des soeurs de l'hôpital Notre Dame* at Tournai, which clearly states that the sisters were absolutely forbidden to allow lepers or suspected lepers into their hospital, not even for a single night.[95]

'Se il y aucuns dont on puisse auoir souspechon qu'ils soient infect ou entechie de aucune maniere de meselerie, on ne les doibt en aucune maniere recepvoir, non mie {même} pour l'espace de une seule nuit, {mais gui ils} demeurent hors des herberges de israhel!
'If there are any whom one might suspect to be infected or marked by any kind of leprosy, one must not in any way receive them, not even for the duration of a single night, but they remain outside the Hostels of Israel.'

According to ancient German laws, lepers were to be banished altogether. They were deemed to be 'as dead', *Tamquam mortui*. In any case it was impossible to fully achieve this in the Late Middle Ages, nor, for various reasons ranging from religious to financial ones, was it desirable.[96]

A book illumination of the mid 14th century in which a clapper is shown as a leper's attribute offers a serious explanation for the presence of the clapper in the cog (fig. 16.42). It deals with, among other things, the miraculous healing of a leper. Considering this marvel story with today's knowledge, we might conclude that this was yet another case of misdiagnosis giving rise to a miracle. Whatever the truth of the matter, such an explanation is irrelevant to the function of the pious story in its historical context. The miracle story illustrates a precious missal that was copied and illuminated in Paris around 1350 for the Abbey of Saint-Denis.[97]

X-ray photographs of the recovered human tibias (figs 16.44 and 16.45) show small scars indicating that the man whose shinbones they once were, had suffered at least seven episodes of ill health. There also was a serious form of bone abnormality that can be linked to healed scurvy.

86. Stallaert 1890, 594.
87. Hoffmann-Krayer 1932, k.1.
88. Adree 1985, 103-107; Weinhold & Hoops 1895, 327-329; Schwartz 1896, 444-445; Weinhold 1898, 347; Zaretsky 1905, 93-94; Jacoby 1932, k.2 and Perkmann 1932, k.2-3.
89. Torsy, 1976; Engelen & Marx 1993, 28.
90. Frutaz 1961, kol. 60-62.
91. Scholkmann 1992, 451-463.
92. Ehbrecht 1989, kol. 1022-1023.
93. Ketting 1992, 148.
94. See also: Kok 1958, 91-113.
95. See: Smeyers 1993, 191-196, cat. 61 and 62.
96. Seiler 1992, 469-471.
97. London, Victoria and Albert Museum, ms. 1396-1891.

Fig. 16.42. Image of the miracle story of the healing of a leprous beggar with clapper, from Missal from St. Denis.

Fig. 16.43. Image on a Siechen Ordnung from Nürnberg (Germany), showing two lepers with a clapper and beggar's pouch.

The location of the deformation is typical of this disease. Bone alterations are not typical of leprosy, but the dead man may yet have been suffering from this affliction when he died. Once the characteristic symptoms of leprosy such as boils and sores started to erupt on the skin, it might be years before bacterial infection of the bone occurred. Also it should be remembered that the bones did not necessarily belong to the owner of the clapper. There must have been more people aboard.

If indeed there was a person aboard with a visible (or maybe cured) affliction and a clapper to make his approach known among the people, then this need not be too surprising. Misdiagnosis must have been quite common in those days.[98] He may have been registered as suffering from some form of the incurable leprosy, which had miraculously healed owing to supernatural intervention. Such cures, which came to be labelled as miracles, were not unique. Like the two that were referred to in the first article in *Antiek*, in the *Miracle Book of Our Lady of 's Hertogenbosch*, and the above-mentioned cure in the Abbey church of Saint-Denis. Besides, lepers would travel, and not just into town for their assessment, but also to centres of health care or pilgrimage. Regulations might be ignored, for instance if a leprous traveller paid ready money. Not everyone was equally scared of lepers.[99]

A new regulation, issued in ca 1484, a so-called *Sichkobel,* of one of the hospitals on the four main roads outside Nürnberg (Germany) on its front shows a pair of lepers, one well recognisable by his raised clapper, the other by a clapper tied to his right wrist. Both have their beggar's pouches hanging from their shoulders (figs 16.43).[100] The beggar's pouch, which like the clapper serves as a common and iconic attribute in paintings featuring lepers condemned to beg, prompt such ideas.

Initially it was assumed that the three blades of the clapper were part of the poorly preserved leather bag (for its description, see section 16.5). Once the wooden parts had been cleaned and impregnated, it transpired that the undecorated blade had broken off and actually was the middle blade, which fitted to the sockle of the figurine. Only then was it clear that the artefact was a clapper. Further study prompted the conclusion that there must have been a link between the bag/pouch and the clapper. Now it is assumed that the clapper was contained in the bag during the calamity, when it broke. A bag or pouch

98. Writen comm. A.M. Hagemeijer, 1993.
99. Written comm. S.M.C. Ebbinge-Wubben, 1993.
100. Boockman 1986, 250.

is an unusual find aboard ships of the Zuiderzee. With the exception of some leather (salt-?) bags or pouches from an early-17th-century *beurtschip* (ferry), so far[101] no other leather bags or pouches have been found in the over 350 excavated shipwrecks. Besides, no comparable finds from the Netherlands are known.[102] After the fragments were impregnated, attempts were made to make a reconstruction of its original shape. Figures 16.27-69a and b show two different models. The receptacle probably started as a rectangular (shoulder) bag with a flap (A). Later the flap was cut off and two pairs of holes were cut along the top edge. Through these holes a strap/lace was strung which could be drawn tight. The old, worn-out shoulder bag was thereby transformed into a pouch (B), which might be suspended from the belt.

It does not seem too far-fetched to assume that we are here dealing with what - besides the clapper - was the usual attribute of a (cured?) leper: a beggar's pouch. If the clapper belonged to a man of means, he might have been allowed aboard, for a financial consideration, to make a crossing, maybe together with his horse.

101. Shipwreck B 71 Oostelijk Flevoland, file and depot RCE-Lelystad.
102. O. Goubitz, written comm. 1989.

Healed scurvy, a Dutch example from the fifteenth century. Paleopathological analyses of some human bones

G.J.R. Maat in collaboration with J.H. Lens, A.A.H. van Immerseel and H.A. Lonnée, Afdeling Anatomie en Embryologie, Universiteit van Leiden)

Introduction

In 1986, a small cog was excavated near the town of Almere by the Department of Ship Archaeology of the Department of Transport, Public Works and Water Management,[103] since 12 April 1995 known as the Netherlands Institute of Ship- and underwaterArchaeology (NISA). The site was situated in the IJsselmeerpolders, an area of low land reclaimed from the former North Sea inlet, the Zuiderzee. The vessel had sunk in the first half of the fifteenth century. Besides the inventory of the ship, which included a wooden clapper used by lepers to make their presence known, some human bones were found (49 A-H, figs 16.44 and 16.45a). The latter were presented for palaeopathological analyses.

Material and methods

The human bones comprised eight more or less complete bones: a right and left *tibia* and a right *fibula*, a right *radius* with corresponding *ulna*, a right *metacarpal* III, a left rib IX and a right rib X. All bone material had changed in colour to an even dark brown. The texture was well-preserved.

The stature (standing height in life) was calculated from the long bones in order of preference according to the smallest standard error of estimate.[104]

X-rays were made using an Enraf-Nonius (type Elinos 90/40; 90 kV; 3.3 mamp; exposures 0.3 and 0.5 sec.; focal-film distance 150 cm). The age of the Harris's lines, the so-called growth-arrest lines, which are seen as radiopaque transverse scars in the distal tibia, was calculated according to Maat.[105]

Results

In general, the bones had robust proportions. All epiphyses, in particular the ossification sites near the ends of the long bones, had fused. All joints were unscathed. No signs of osteo-arthritis, a degenerative joint disease, were found. Not even the start of marginal lipping along edges of joints could be detected.

The stature, calculated from the average maximal length of both tibias (36.0 cm) was 169.05 ± 4.00 cm.

A total of seven Harris's lines were recorded in the X-rays of the distal tibias. The lines had developed at the age of 9.2, 12, 12.7, 13.2, 14, 14.5 and 15 years. Two of them were strong lines. They were formed at the ages of 9.2 and 12.7 years.

Gross anatomical inspection of the bones revealed the presence of widespread uniform pathological bone changes. Especially the long bones of the lower extremity were most severely affected. Subperiosteal bone appositions were found at many places, but in particular near attachments of muscles and fibrous joints (figs 16.44 and 16.45a). These appositions were situated on the original periostical surface. The X-rays showed that the underlying cortex was not involved in the process. In many places, the build-up of new bone tissue had enclosed blood vessels and tendons which originally ran over the outer surface of the bones (fig. 16.45b). From inspection of the *tibias* it became clear that we were dealing with symmetrical changes. Exemplary were the appositions at the soleal line for the popliteus muscle, at the inter-osseous margin for the inter-osseous membrane, at the *fibular* notch for the ligaments of the distal *tibio-fibular* joint, and at the malleolar groove for the flexor retinaculum. In addition, extensive patches were seen all over the diaphysis, while smaller ones were found on the metaphysis. The *fibula*,

103. Vlierman & Koldeweij 1992-1993.
104. Trotter & Gleser 1952 and 1958.
105. Maat 1984a.

known for its numerous attachment sites for muscles, membranes and ligaments, was almost completely covered with new bone tissue which had enclosed many blood vessels over long trajectories. The bones of the right forearm also showed a distinct apposition on the radius at the attachment of the quadrate pronator muscle.

With the exception of the distal joint surface of the right ulna, which showed a little bony elevation, all joint surfaces were smooth.

Discussion

The correlation between the bones clearly indicated that they had originally belonged to one individual. In addition, from the archaeological context in which the human remains were found, it could be concluded that the person had most likely died at the time of wrecking of this small cargo vessel, *viz.* in the second quarter of the fifteenth century (Section 16.8). The robustness of the bones, the fusion of the epiphysis, and the intactness of the joints indicated that we were dealing with an adult man 20-40 years of age when he died.

The calculated height of our man of approx. 169 cm seems to be small when compared with the average standing height of 182 cm for Dutch adult men in 1980.[106] But when compared with the height of Dutch men living before the beginning of the industrial revolution with its positive growth shift, his height was not at all unusual. During the seventeenth and eighteenth centuries, adult male stature ranged from 166.0 to 166.7 cm.[107] Adult conscripts from 1825 also had an average height of 169 cm.[108]

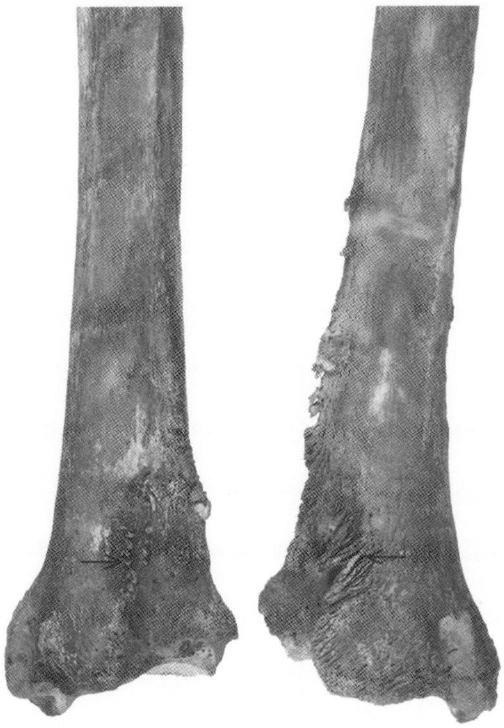

Fig. 16.44. The left and right *tibias*.

Fig. 16.45a. The two corresponding *tibias*. b. The left *fibula*.

106. Roede & Van Wieringen 1985.
107. Maat 1984b.
108. Oppers 1966.

The finding of seven growth-arrest lines indicated that our man had suffered at least seven health disorders, since these scars tend to obliterate during life due to the normal remodelling of bone tissue. Apparently, disease or malnutrition had temporarily retarded his growth several times. The disorders at the age of 9.2 and 12.7 years had caused distinct scars. In comparison: Dutch whalers from the seventeenth and eighteenth centuries, who were known to have experienced unfavourable growth conditions, showed an average of 2.4 lines per individual.[109]

Paleopathologically, our man represents a very interesting case. His bone changes match extremely well with known data on skeletal pathology due to healed scurvy.[110] Scurvy is the result of a vitamin C deficiency. The vitamin is abundantly present in fresh fruit and vegetables. But since the common diet in those days lacked these foodstuffs during the winter period, scurvy occurred almost yearly in early spring as an endemic disease among the population of the Low Countries[111] and the Nordic countries.[112] Sailors in particular were susceptible to this disease, as their diet largely consisted of preserved foods without any vitamin C. Indeed, from the above-mentioned findings on Harris's lines we know that the man had regularly been in a poor state of health. Lack of vitamin C produces inferior collagen, the latter being an important body-building protein. As a result, minor traumas easily lead to severe bleeds in and around the joints, and to endofractures with subperiosteal haematomas. During the healing phase these haematomas may reabsorb, or may ossify as bone appositions. Apparently, the latter process had dominated in our case. Bone changes were found at all known predilection spots, *i.e.* at the attachment spots of muscles and fibrous bones, and at the metaphyses. Because of the physical strain resulting from body weight, the bones of the legs are always far more seriously affected than those of the arms. In this case, infectious diseases and tumour growth could be excluded as probable causes, since no lytic or eroding bone changes were found.

The slight thickening found on the distal ulnar joint surface was due to osteochondritis dissecans. In this disease, a small piece of articular cartilage is extruded into the joint cavity. It may adhere at any place on the articular surface. It would not have led to any medical consequences here. The disease is most common in young men.

109. Maat 1984a.
110. Maat 1984c.
111. Hirsch 1860.
112. Holck 1986.

COGS, SMALL COGS AND BOATS

Cogghen, Kleene Cogghen ende Schuten

Quoted from a letter by a German merchant in Bruges addressing the entire Hanseatic League in 1401

SPA
UITGEVERS

Made possible by:

Ms. H.C. van der Velde-Doyer

Ms. B. Espeet

Mr. P.B. Zwiers

Stichting professor van Winter fonds

Stichting Hanzefonds

P.J.R. Modderman Stichting

provincie Overijssel

Provincie Flevoland

Stichting voor de Middeleeuwse Archeologie

Rijksdienst voor het Cultureel Erfgoed
Ministerie van Onderwijs, Cultuur en Wetenschap

Ruud van Beek Stichting

Stichting Promotie Archeologisch Onderzoek SPAO

Stadsarchief Kampen

Kampen Archeologie

NISS

BurodeBrug

Prins Bernhard Cultuurfonds

Stichting Promotie Archeologie SPA

Vaderlandsch Fonds ter aanmoediging van 's-lands zeedienst

Admiraal van Kinsbergen Fonds

Vereeniging de Prins Hendrik Stichting

Directie der Oostersche Handel en Reederijen

Colophon:

©2021 SPA uitgevers in coöperation with Karel Vlierman

ISBN:	978-90-8932-048-3
Author:	Karel Vlierman
Editors:	Hemmy Clevis, Hans Janssen and Jos Koldeweij
Translators:	Xandra Bardet, Hemmy Clevis, Stefanie Hoss, Christine Jefferis, Joy Phillips, Sasja van der Vaart-Verschoof, Jeroen Vermeersch
Graphic Design:	Hidde Heikamp
Lay-out:	Hidde Heikamp
Plates:	Coen Bouwstra, Hidde Heikamp
Drawings:	All drawings and impressions are made by Karel Vlierman
Photographs:	Unless otherwise by Karel Vlierman and the Rijksdienst Cultureel Erfgoed
Print:	Huvé Graphics

SPA UITGEVERS

SPA uitgevers
Assendorperstraat 174-4
8012 CE Zwolle
The Netherlands
info@spa-uitgevers.nl
www.spa-uitgevers.nl

COGS, SMALL COGS AND BOATS

The thirteenth- until sixteenth- century Dutch and Flemish archaeological finds from the Hanseatic shipbuilding tradition seen in a broader perspective

Karel Vlierman

Overview of the Doel-I cog seen from the front with the visible stem hook, the keel plank, the stern hook, the planks of the bottom and the overlapping planks of the sides and the hole made by the crane. At the starboard side the colleagues of NISA.

Contents

IV Results, analysis, summary and conclusions

Fig. 17.1. General view from the stern, showing the erratic stones, the rudder, and in the background the boundary ditch M61/62.

The Dronten-II cog[1]

17.1 Find conditions

17.1.1 Discovery and investigation

During the cutting of the ditch between lots M61 and M62 in Oostelijk Flevoland in the late summer of 1959, the diggers hit upon a shipwreck. The find was reported by a surveyor of the Department of Pedology of the Directie Wieringermeer (Noordoostpolderwerken). On 26 October 1959 an inspection was carried out, during which it was established that at the bottom of the ditch lay the bow or stern of a small, narrow vessel built in the clinker fashion. Ship's timbers had already been removed from the eastern side of the cutting. In the opposite side a few shell strakes were visible, from which it could be inferred that the sides of the vessel must have been fairly high in relation to the narrowness of the craft. No finds came to light. The age of the wreck could not be determined and it was decided to preserve it for closer investigation.

In the context of new policy regulations regarding wreck protection and prioritisation of excavation, a substantial number of ship finds that had been inspected and briefly investigated in the 1970s and early '80s were once more evaluated by the Subdepartment of Nautical Archaeology of the RIJP. The findings were recorded in a standardised report format, which included information about the wreck's precise location, any protective measures required, dimensions and depth of the wreck, its state of conservation and completeness, soil conditions, the likely vessel type, the presumable age and its archaeological significance.

The re-examination of the wreck in lot M61 took place on 4 and 5 October 1982. This confirmed that it was a lightly built vessel, of which ca 6.5 m survived. It lay with a slight list to port and in the trial trench was found to have a port side surviving right up to the top. On the heads of the slender futtocks there still was part of a covering inwale, which revealed that the craft had had a keel plank and three shell strakes on either side of it. The stern was badly damaged and broken, but still fairly complete, while - quite exceptionally - a large part of the stern rudder lay aft of the hull. The greatest width of the hull recorded then was ca 2 m across the gunwales, while the highest and the deepest timbers of the wreck lay at ca 0.4 and 1.4 m below the surface, respectively.

In the afterbody lay a layer of small erratic stones, and some more were found at the bottom of the vessel near the bottom of the ditch. Higher up against the stern some small planks were observed that might have been part of a cuddy. It was concluded that this probably was a special type of boat not earlier encountered in the IJsselmeer polders. The sintel clamps, the soil profile and the overall impression suggested a 15th-century date, most probably one in the latter half of that century. Priority for excavation was granted because the overlying clay deposit was displaying drought cracks up to 5 cm wide right down to the Pleistocene sand (owing to the hot summers of 1975 and 1976), while the entire sediment layer in which the wreck lay was bone-dry, in part as a result of its location beside the ditch. Cleaning the eastern face of the cutting revealed that the soil profile there was undisturbed. This meant that the bow did not extend into lot M62, but had ended in the ditch cutting. *In-situ* protection of the wreck would be impossible.[2]

The little wreck was excavated between 19 June and 11 July 1995 (fig. 17.1), documented in the field, taken apart and transferred to the workshop of NISA at Ketelhaven. There the pieces were drawn to scale 1:10 in January 1996. The excavation was succinctly described in NISA Yearbook no 2.[3] Apart from the erratics mentioned above, a few fragments of red brick and four half bricks, there were no associated finds.

17.1.2 Geology and stratigraphy

In the course of the 1982 inspection, R. Koopstra of the Department of Pedology of the RIJP examined and recorded the soil profile (figs 17.4 and 17.5).[4] On the Pleistocene sand lies a 10-cm layer of redeposited Pleis-

1. Translated by Xandra Bardet.
2. The wreck lay at an angle to the boundary ditch M61/62 with the stern to the northwest (332°) and 30 m from the ditch between fields M56 and M61/62 (figs 17.2 and 17.3). Ordnance map sheet 20 H Dronten, coordinates 178.366/501.063, with the highest-lying timbers at ca 2.7 m below NAP, Hydrographic map 1921, mean low water. The site lies ca 4 km west of the peaty barrier between Schokland and Elburg that was eroded away in the late Middle Ages (see fig. 11.5, Dronten-I cog).
3. Schreurs *et al.* 1995, 36-38.
4. R. Koopstra, written report of 13 December 1982.

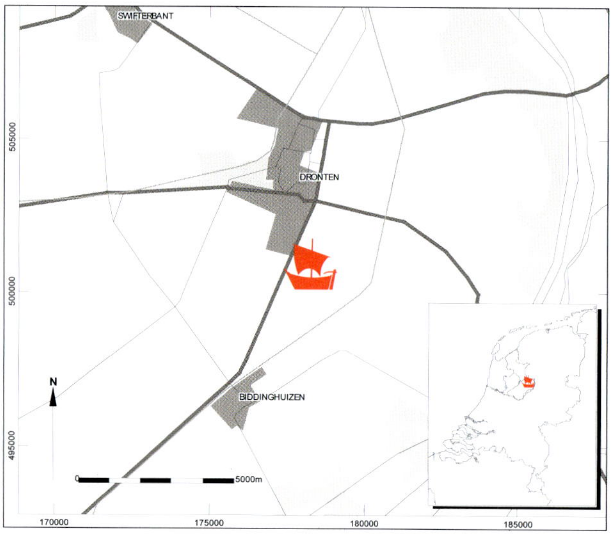

Fig. 17.2. The findspot of the Dronten-II cog.

Fig. 17.3. Ordnance map of the Kingdom of the Netherlands (scale 1:5,000) of 1861, showing the location of the Dronten-II cog.

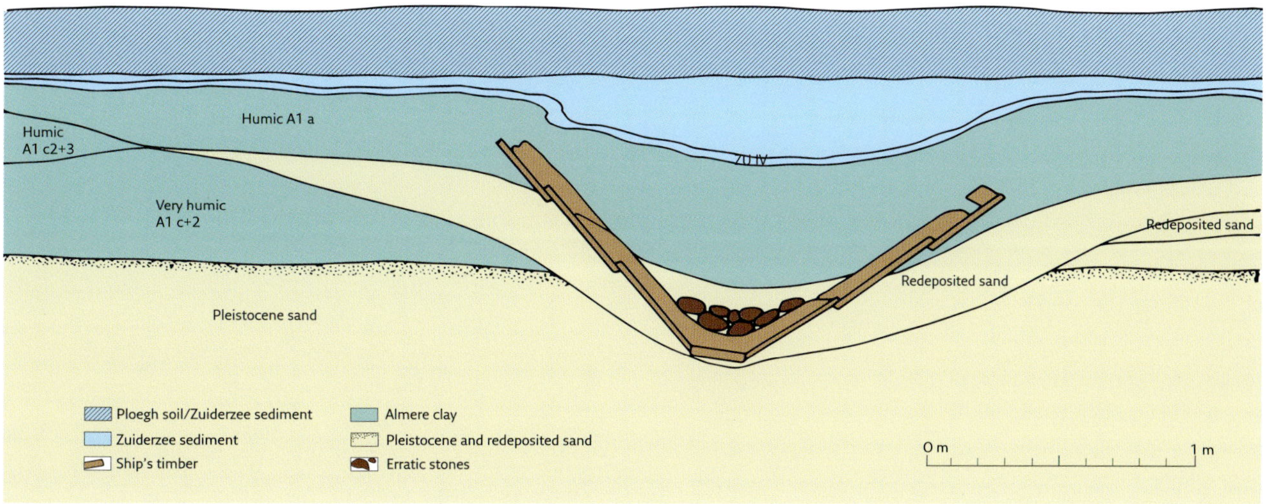

Fig. 17.4. Soil profile disturbed by the Dronten-II cog (partially reconstructed).

tocene sand. This is overlain by a ca 30-cm layer that can be reckoned to the Almere deposit Alc2+3, the lower half of which consists of strongly humic clay, which may be a young detritus-gyttja (Flevomeer deposit). On top of the Alc2+3 deposit lies a layer of Almere clay (Ala), ca 15 cm thick. This was covered by a distinctive sandy layer with seashells (Zu IV), followed by younger Zuiderzee deposits and the ploughsoil. The disturbance of the soil profile by the ship was clear to see in a sharply ascending line on both sides of the boat, ending roughly halfway up the Ala deposit. Further it is remarkable that the redeposited sediment and the infill of the cavity in the sea floor caused by the wreck displayed two layers. First a 20 cm layer of redeposited sand petering out to both sides, which also lay inside the wreck, and above it a fill of Almere clay (Ala).

Wiggers globally dates the layer of Sl(oef) Ib (named Ala in Flevoland) to the second half of the 15th century and the early 16th century.[5] In the case of gradual sedimentation this would put the foundering of the vessel in the final quarter of the 15th century. The redeposited Pleistocene sand most probably comes from eroded higher-lying Pleistocene outcrops immediately south and west of the site. Over the past decade, various prehistoric artefacts have been gathered from the surface of the fields in the close vicinity, which have come to the surface as a result of mechanical turning-up. These probably are implements deriving from a settlement on the sandy outcrops. Ca 2 km north of the site, the bottom of a channel in the Pleistocene sand was identified during the cutting of the ditch along the Dronterweg road in the late 1950s. In 2012 a similar thing occurred in lot P 35 on the Oudebosweg road.[6]

17.1.3 Medieval habitation

The nearest late-medieval town, Elburg, lies at 8.5 km from the site. The hamlet Kamper-Nieuwstad, on the Kamper dike between Elburg and Kampen at (the mouth of) the Reve stream, lies ca 9.5 km away.

Fig. 17.5. Section across the afterbody, viewed from starboard; in the foreground, the covering inwale fragment resting on the frame heads.

5. Wiggers 1955, 88.
6. D.E.P. Veldhuizen, pers. comm. January 2013.

17.1.4 Condition and wreck formation

The base of the stern of the wreck lay at a depth of ca 1.5 m below the surface, the keel was slightly hogged and at the M61/62 boundary ditch lay with its keel plank at a depth of 1.25 m. After sinking through the Almere clay deposit, it had worked itself into the redeposited and primary Pleistocene sand by another 0.6 m, causing a disturbance to the soil profile that at the top was at least 6 m wide (while itself being a good 2 m wide). Currents and the 'loose' sand must initially have made the wreck move about quite a lot as it sank, after which at least 20 cm of redeposited sand settled around and inside the wreck. From this time on the vessel probably lay motionless, allowing the cavity in the sea floor to fill up with Almere clay, which eventually fully covered the wreck. The upper part of the Ala deposits, and the Zuiderzee deposits cover the site without interruption.

As noted previously in section 17.1.1, the conservation of the timbers was not very good. Even at the inspection in 1982, the generally thin planks and light frames were found to be considerably affected. The dry soil conditions would without doubt make for their rapid further decay. The excavation in 1995 proved this assessment to be correct, the uppermost 20-25 cm of the wood observed in 1982 by then having completely disintegrated, and the remainder of the wood being in considerably worse condition. Apart from this deterioration we can now, after completing the investigation, conclude that the quality of the wood/the vessel even when it sank most probably was less than optimal, and that the vessel had undergone various repairs. The overall impression of the light frames, in as far as could be established, was that they were worn down through prolonged use. The presence of a batch of erratic stones in the afterbody and the impression that there may have been some in the bow as well, suggests that they may have been put there to deliberately sink the boat. The report on the initial inspection of 26 October 1959 makes no mention of timbers of the bow or erratic stones, which indicates that they had already been removed with the spoil from the ditch.

17.2 The ship

17.2.1 Documentation and reconstruction

Documentation

After the wreck was excavated and cleaned (fig. 17.1), a two-part survey drawing was made to scale 1:20 by means of a 3-D pantograph. The cross-sections were drawn to scale 1:10 with the aid of a 2D-Eichstaedt pantograph; the longitudinal section, with a graduated device constructed by RIJP at the request of NISA. After the drawing procedure, general views and detail photos were taken of the wreck, as well as video recordings. Finally the frames and strakes were numbered, lifted and wrapped in clingfilm to prevent further desiccation. Then the pieces were transferred to Ketelhaven, where they were drawn to scale 1:10. Of the top views and cross sections drawn with the pantographs, fair copies were made on graph paper. Unfortunately it was not marked on the field drawings which parts of the frames could be preserved, so a considerable part of the timbers drawn to scale 1:10 at Ketelhaven could no longer provide reliable evidence for the purpose of reconstruction. This was a consequence of the poor condition of the wood, causing parts of it to disappear in the course of lifting and subsequent washing.

Regrettably during the excavation and the dismantling of the boat no notes were kept on, for instance, the location of any scarfs in the frames, the number of treenails per plank width, the spacing and number of nails in the clinker joins, or the spacing of sintel clamps. Only occasional remarks in the drawings, e.g. of the keel plank, "the edge is full of sintel clamps" and "close-set row of sintel clamps with moss lath", are documented observations regarding such details. When a few samples were taken of the sintel clamped moss caulking, which could be investigated as part of this study in 2012, it was found that the sintel clamps in the "close-set row" actually overlapped (figs 17.9-3a and 4a). In the contribution to the Annual Report some construction details do get mentioned. Also it should be noted that the place of the frames in the longitudinal section drawing (measured in with the graduated device) in several places deviates from the top view drawn with the aid of the 3-D pantograph (Plates 17.I A and B). Which of the drawings is correct, is impossible to establish. For the reconstruction it has been assumed that the data from the longitudinal section are the more reliable.

Reconstruction

The constructions and details that *were* drawn and recorded, together with the evidence gathered during the 1982 inspection did provide enough evidence to compile a reliable and complete reconstruction of the construction, the lines and the dimensions of the vessel. It had a fine sheer, which rises a little more forward than aft, and almost certainly had a cuddy both aft and fore; and above the mast step was a thwart, which held up the mast which passed through a hole in it (Plates 17.III A and B). There are some remnants of nails in the top of the V-shaped floor timbers of frames 5,

6, 7 and 8; yet, judging by the shape of the cross-sections drawn at these spots, it does not seem very likely that the boat had a ceiling between the after cuddy and the thwart. This point will be returned to in section 17.2.2.

17.2.2 Construction and layout of the hull

Keel plank

The keel plank and both hooks together were ca 7.5 m long (Plates 17.II A and B). The keel plank itself is 5.5. m long, including the hooked scarfs at either end, which join the keel plank to the hooks. The after scarf is 38 cm long, while the forward scarf too would have been about 40 cm long. The keel plank at its after end is ca 20 cm wide; its width at the fore end would have been about 30 cm. From the hooks (the stern- and stem hook) the keel plank gradually broadens to a width of 52 cm amidships. Its thickness varies between 3 and 4 cm and originally will have been 4 cm or a little over. The keel plank is fastened to the stern hook with two rows of three, and one row of two nails. The three-nail rows were driven in from the inside outwards; the remainder, the other way around. The nail points protruding through the overlapping timbers were beaten sideways and driven back into the wood (double-clenched).

Stern hook

The horizontal part of the hook, including the heel at the rudder, has a length of 1.32 m (Plate 17.III A and fig. 17.6). At the scarf joining it to the keel plank it is 20 cm wide, and at the foot of the rising part, 7 cm. The end of the horizontal part at the scarf is 2 cm thick and stands proud of the keel plank. From there the upper surface gradually rises and below frame 1 it jumps 1 cm in height. The thickness of the horizontal part at the base of the rising arm is 15 cm. The rising part of the hook ends 40 cm above the keel line, measured perpendicularly. It is joined to the sternpost by a flat scarf sloping inward. The outer face of the stern has a rake of 124° with respect to the keel line. The rising arm is just 3 cm wide at the back. Including the hook, the stern was 1.5 m long. On the rear, it has a triangular heel measuring ca 10x10 cm at its base, which served to protect the rudder. Both arms of the hook taper into a trapezoid cross-section downwards and towards the rear. In the sides are hewn rabbets ca 3 cm deep into which the garboards were spiked.

Fig. 17.6. The stern hook.

Sternpost
On the flat scarf of the stern hook stood a sternpost, fastened with two sturdy nails passing through a rectangular iron plate ca 20 mm thick. The rabbets of the stern hook continued up into the sides of the sternpost, to hold the hood ends of the garboards and the shell strakes. The inboard face of the sternpost is 5 cm wide, the rear face, 3 cm. Across the scarf joining the sternpost and hook, on both sides, the T-shaped braces of a wrought-iron gudgeon were nailed, of which the after ends meet at the back of the stern, with an opening (ca 25 mm across) for holding the lowermost pintle (fig. 17.9, 1). At the top, a wrought-iron bolt passes through the head of the sternpost, at the after end of which is a round, forged extension also with a hole 25 mm across, to take the uppermost pintle. On the inboard end of the bolt there probably had been a flat ring and a pin to secure the bolt. The stern hook with the sternpost, measured perpendicularly, rose 1.25 m above the keel line (Plate 17.III A).

False stern
The vessel had no false stern.

Rudder
The rudder is a remarkable find. A rudder is only rarely found with the wrecks in the IJsselmeer polders. Like the sternpost, the rudder was badly damaged and incomplete. Most probably it had been secured with a length of rope passing through a hole (2.5 cm diameter) in the rudder and presumably a pair of holes of the same diameter in the uppermost strakes on either side of the sternpost. The rudder remnant is a plank ca 20 cm wide and 3 cm thick. Most probably a second plank was fastened to the rear of this plank, also 20 cm wide, creating a 40-cm-wide rudder blade. The two parts were joined by wrought-iron fittings. The rudder originally was ca 1.9 m long (figs 17.1 and 17.7 and Plates 17.II A and 17.III A). The pintle(s) had a length of ca 200 mm, and a diameter of ca 18 mm (fig. 17.9, 2).

Stem hook, stempost and false stem
Nothing was recovered of the stem hook and stempost. Most probably they were executed in a more massive format than the stern, and like the stern provided with rabbets into which the hood ends of the garboards and strakes were fitted. The overall length of the rising part of the hook and the stempost probably was 1.9 m, while the head of the stempost measured perpendicularly to the keel line would have stood 1.5 m high. The rake with respect to the keel line will have been ca 128°. Lacking a false stern, the vessel would have had no false stem either.

Bottom and sides
On either side of the keel plank the vessel was constructed from three strakes (A to C, fig. 17.8). Adjacent to the flat keel plank were the garboards (the A strakes) with a slight deadrise, which towards the stern and stem were twisted so the hood ends could be nailed into the rabbets in the hooks and stern- and stempost. The keel plank and garboards together formed the bottom of this small craft. The subsequent strakes of the sides (B) were nailed to them in clinker fashion, as were the topmost strakes (C) to the B strakes. The B strakes sit at a somewhat steeper angle to the garboards than the garboards to the keel plank. To this end, the upper and outer sides of the garboards at frames 5 to 9 had been bevelled to a thickness of just

Fig. 17.7. The uncovered rudder with the sternpost fragment.

Fig. 17.8. The hull with frames and the erratic stones in the afterbody, viewed from the boundary ditch.

1-1.5 cm. Towards the stern and bow these planks did retain their full thickness of 2.5 to 3 cm. The hood ends of the B-strake planks had been fastened at about the same angle as those of the garboards. This also goes for the C strakes, be it that aft of frame 0 and forward of frame 12 they were somewhat more everted.

The keel plank consisted of a single timber. The B- and C strakes comprised two planks each, which were joined by flat scarfs ca 35 cm long. The plank ends at the scarfs were about 1 cm thick and (originally) stood proud of the inboard and outboard surface of the shell. The scarfs in the garboards occupy the same position to port and starboard. The after planks are 4.35 m long and the forward ones 4 m, while the strakes are ca 40 cm wide and ca 35 cm at the stern and stem. In the next (B) strake the after plank to starboard is 3.95 m long and to the port-side probably ca 5.8 m. The forward planks of these B-strakes had been 5 and 3 m long, respectively. Their width was 40-45 cm; at the stern and stem, ca 30-35 cm. The uppermost strakes (C) had the scarfs probably in equivalent places. In this strake the after planks were 5.5 m long and the forward planks 4.3 m. In the afterbody and midship section, the planks of the C strakes were ca 35 cm wide, but in the forebody they probably gradually widened (to ca 40 cm). The thickness of the planks ranged between just 2 and 2.5-3 cm, as a result of the age of the vessel, wear and tear, and further degradation of the timber due to its desiccation in the soil. It seems likely that the planks originally had been ca 3 cm thick. The overlap (the land) of the clinker-built strakes is 5-6 cm wide. At the scarfs, the planks of the strakes are joined with two rows of three or four nails: one row driven in from the inside and the other from the outside. The nails were double-clenched, the points protruding from the planks beaten sideways twice and driven back into the timber.

Caulking method

In the scarfs joining the keel plank and the hooks, and in the scarfs between the planks of the six strakes, a thin layer of moss caulking had been inserted. On the lands of the clinker-built strakes no moss caulking was observed. Probably there never had been any. The seams between the stern hook and the garboards slightly gaped in V-fashion and had been closed with moss, moss laths and sintel clamps (sintel clamped moss caulking). The same was observed on the outside of the stern hook and sternpost. Most probably this caulking continued right up to the top of the stern, while we may assume that the same was the case on the stem. Besides, the seams between the keel plank and the two garboards were closed on the underside with sintel clamped moss caulking, with actually overlapping sintel clamps. No such overlaps occurred on the hooks and the stern- and stempost. It appears that there they were fixed with interspaces. On the outside of the B- and C strakes to starboard two cracks in the planks were closed up with sintel clamped moss caulking. On the seams at the stern (and stem), sintel clamps 55-60 mm long were used; smaller ones (35-40 mm) on the seams between the keel plank and the garboards and on the cracks within the planks.

Whether this difference in sintel clamps is indicative of different (far apart in time?) episodes of application, is impossible to tell. The smaller specimens can be reckoned to type D2; the larger, to type E1 (figs 17.9-3a and 8b). If indeed these were different types that were used for instance during construction and during a subsequent episode of repairs or additional caulking, then the larger (typologically younger) ones must have been employed later. However, this seems unlikely, given the areas in the ship where they were encountered. No detailed investigation could be performed, so real certainty cannot be obtained. Therefore we here assume that for the seams at the stern and stem larger specimens were used; and on the seams between the thin planks, the smaller ones. Therefore it is more appropriate to speak of transitional types D2/E1. The dating of these sintel clamps can be put around 1375/1450.[7] The moss laths were made of split oak wood, 1.5 cm wide and cut to an oval cross-section, with a maximum thickness of 0.5 cm. No pollen analysis was performed on the moss caulking. The species of moss is peat moss (*Sphagnum*).

Frames

The stiffness of the craft was primarily provided by the frames. They are quite light. On the bottom, these timbers are 7-9 cm wide and ca 7 cm thick. Upwards, their width is quite constant, but their thickness reduces to ca 6 cm. Insofar as can be established, these small frames consist of two parts, joined with flat sloping scarfs and in most cases with a nail. No pattern was observed of the scarfs to starboard alternating with scarfs to port. The positions of the scarfs appear to be quite random. Frame 3 probably had a vertical flat scarf above the seam between the keel plank and the port garboard. Frame 6 consisted of three timbers, the two shortest ones on the port side.

The aftmost and foremost two frames in the stern and forebody probably were just V-shaped timbers. The vessel numbered a total of 17 frames (-2 and -1 in the stern and 0-14 towards the stem). Forward of and against frame 8 was a short floor timber with the hole for the heel of the mast (frame 8a; Plates 17.III A, B and C and figs 17.10a

7. Vlierman 1996a, 81.

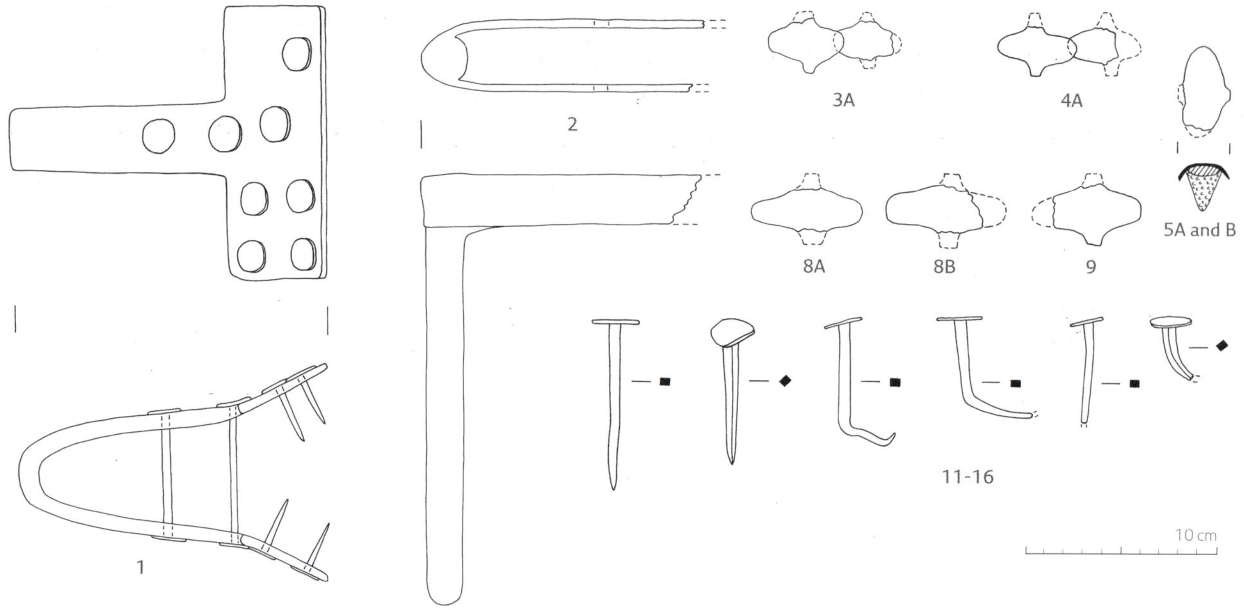

Fig. 17.9. The gudgeon, the rudder fittings, the different nail types and the sintel clamp types D2 and E1 used in this vessel.

Fig. 17.10a. General view of the wreck with its frames, viewed from the boundary ditch; in the foreground, the short floor timber with the hole for the heel of the mast, and two erratic stones.

Fig. 17.10b. General view of the midship section with its frames, viewed from starboard; to the right, the short floor timber with a hole for the heel of the mast, and two erratic stones.

and b). Into the lower faces of the various floor timbers were cut limber holes about 5 cm wide and 2 cm high. Frames 3 to 9 had three each, one above the keel plank and one above each of the garboards. In the V-shaped timber of frame 2 there were two: above the seams between the keel plank and the garboards. The same probably went for frame 10. No limber holes were observed in frames -2, -1, 0, 1 and 11-14. Probably none were needed, as the boat most probably had cuddies both fore and aft.

Covering inwale

A fortunate coincidence during the inspection in 1982 was the observation of a small piece of covering inwale, still part of the construction, on the frame heads against the inboard face of the topmost shell strake of the port side. This covering inwale reinforced the entire gunwale; it was 6 cm wide and 5 cm thick, and probably ran on along the inboard side of the stern and stem. It was an essential part of the construction for the topside planking, which like the other strakes was just 3 cm thick (fig. 17.5 and Plates 17.III A, B and C).

Mast step

There was no such thing as a proper mast step in this boat. Against the forward face of frame 8, above the keel plank and the adjoining garboards, was a small floor timber (8a), ca 65 cm long and 15 cm wide, into which a rectangular hole (10x7 cm) had been hewn above the middle of the keel plank. The part of this timber aft of the hole was about 8 cm thick; forward of the hole, 6 to 7 cm. Both ends of the timber were roughly cut or worn down.

Ceiling or plank bridge?

Despite the presence of several nail remnants in the top of some floor timbers, is seems rather unlikely, judging by the shape of the cross sections, that the boat had a ceiling between the after cuddy and the thwart and maybe also forward of this. Removable plank bridges, for instance like those in 20th-century punts (fig. 17.11), are a potential alternative (see also the miniatures in section 29.4.4, figs 29.5a and b). However, no archaeological evidence of plank bridges has ever come to light.

After and fore cuddy

In the course of the 1982 inspection, some plank fragments were observed high up against the stern, which might be attributed to a cuddy. By the time the wreck was excavated in 1985, they had disintegrated. Yet it seems quite certain that there had been a cuddy, given the sharp, narrow shape of the stern which allowed no proper footing for the crew. Yet no definite evidence was uncovered, as a result of the absence and/or decay of the vessel's timbers that might have provided such clues. Most probably a cuddy, 1.55 m long, was constructed between the stern and the front of frame 2, composed of longitudinal planks that laterally were fitted to the shell and around the frames. The cuddy probably sloped slightly towards the midships, and was supported by V-shaped timber -1 and three transverse beams fixed to the vertical parts of frames 0, 1 and 2. The top surface of the cuddy at its forward edge sat about 45 cm above the keel line.
In the bow the boat probably had a similar cuddy of comparable length, starting at the after face of frame 10 (Plates 17.III A and B). The presence of these two cuddies is to some extent corroborated by the absence of limber holes in the V-shaped timbers situated below them. The boat with cog features from lot K73/74 in Oostelijk Flevoland served as a model for the reconstruction suggested here.[8] The soil profile disturbed by this wreck, and the coins and stoneware jugs found within it put its foundering in (the beginning of) the second half of the 15th century.[9] The sintel clamps suggest that it was built in the final quarter of the 14th, or the first half of the 15th century.[10] The general condition of the boat indicates that it most probably dated from the second quarter of the 15th century. The small cargo vessel from lot K73/74 will be briefly discussed in section 26.1.

Fig. 17.11. Giethoorn punts, the one on the left with an after cuddy; note the thwart and duckboards. After Berk 1984, 37.

8. Reinders *et al*. 1980, 17-29, plates 4 A, C and D, and 5 C and figs 10 and 11; section 26.1, fig. 26.1.
9. Vlierman 1983, 6.
10. Vlierman 1996a, 80.

Thwart
In the second starboard strake (B) both aft and forward of frame 8 various nail remnants were observed and recorded. Most probably these were attachments points for two 'futtocks' that ran from the outer edges of the A strakes or garboards and continued up to the top of the B strakes (Plates 17.III A, B and C). These 'futtock' heads and the top edge of the B strakes without any doubt supported a thwart, ca 37 cm wide and 4-5 cm thick, presumably fastened to the futtocks and shell with oblique treenails. Midway along the thwart there would have been a round hole for holding a mast upright and in position.

Repairs
Already mentioned above were the two cracks in the starboard B and C strakes, closed up on the outside with sintel clamped moss caulking. Given the remains of nails, the after part of the crack in the C strake between frames 3 and 4 on the inboard side probably had a thin plank patch. Similar evidence was seen in other areas of the starboard and port shell. During the preliminary investigation in 1982 there also were small planks both in the port C strake and on the seam joining the keel plank and the starboard garboard between frames 2 and 3. Whether layers of moss had been placed between these planks and the shell was not recorded, but it is quite likely. The small planks must represent repairs made during the working life of the craft. Closing up cracks externally may even have been done during the construction phase (Plates 17.II A and B and 17.III A and B).

Timber selection and processing
The vessel was built entirely of oak wood (*Quercus*). The hooks were hewn from naturally formed trunk-limb sections. So probably were the curved floor timbers and frame members. The planks, in as far as could be established, appear to have been rift-sawn. Of what wood the treenails were made was not recorded. The frame timbers were fairly regular and squared. The reconstructed planking plan suggests a shell construction according to the so-called 'brother-and-sister' pattern, despite the longer B strake planks having on one side been fastened forward, and aft on the other (Plate 17.II B).

Fastenings
The fastenings of the keel and shell planks to the frames were made by means of small treenails (ø2 cm). In as far as could be established, two treenails per plank width. Also some treenails of 1.5 cm diameter were observed. The covering inwale was fastened to the top shell strake with treenails 2 cm thick. Wedges were observed in the top of some of the treenails. Whether (in some parts) also connections between the covering inwale and the frames were present could not be verified, but seems likely. The clinker construction between adjacent shell strakes, in the scarfs within the strakes, and of the fastening of the shell strakes to the hooks and stern- and stempost were executed with iron nails. Nails were observed also in the scarfs within the frames. The spacing of the nails in the clinker construction could not be verified throughout. Most probably they had been driven in, from the outside, every ca 15 cm. Their protruding points were, double-clenched: twice beaten sideways and thus driven back into the timber. The nails for joining the hooks to the stern- and stempost were of a larger size. Small nails were used for affixing the repair planks. Fig. 17.9 shows the different nails. In the sintel clamped moss caulking, forged sintel clamps of two different sizes were used. Most probably, the only massive forged-iron fastenings in the craft were the gudgeon and bolt with ring, and the rudder fittings with the pintles.

17.3 Construction data

Already it was noted that the remains of the craft were sufficient and were adequately recorded to allow a reliable reconstruction of its build, its lines and dimensions. On the one hand this revealed that it incorporated several features typical of the cog-building tradition, while on the other there are elements that are quite alien to it; nor do the lines and dimensions of the craft fit the description of a cog. In section 4.1 it was stated that in the present study all vessels with (several of the) cog-specific structural features will be termed cogs. In certain cases we might modify this to 'cog-like', but this option was rejected. In the case of the present vessel - the Dronten-II cog - the lines, the dimensions and some structural details are notably different from all other cog finds treated in this study. Yet is it a vessel built according to the traditional cog-building method. The points on which it diverges from the other cogs are attributable to differences in size, function and construction period.

17.3.1 Design

The vessel was clearly built according to a sequence and to lines familiar to the shipwright. For a 'cog', it had a notably strong rake of the stern, a fore and aft cuddy and a thwart that with its hole supported the mast at almost 1/3 of the craft's overall length. The sharp aft and forbody, as well as the cross-sections in between, suggest that it was a sailing boat that handled well (Plates 17.III C and D).

Most probably it carried a spritsail. No evidence of rowing was found, but it should definitely have been possible to row this boat.

Lines

The boat has a fine sheer, which rises a little higher forward than aft. The top view also shows a sleek and slender planking pattern with relatively thin frames spaced some 50 to 70 cm apart. It has a laurel-leaf shape sharply converging towards the stern and stem. The stern is 25 cm lower than the stem. The stern has a rake of 124° with respect to the keel line; the stem would have had a rake of ca 128°. The angle of the top strake at frame 8 is about 122°.

Scantlings and layout

This craft is the smallest known specimen to display a number of typical cog features. Its length over all is 9.45 m, its greatest width is 2.35 m and amidships it is just 67 cm high. It had an after and a fore cuddy 1.55 cm long, and below them storage space for a few personal belongings, tools or other utensils. Between the after cuddy and the thwart and forward of the thwart were two open spaces, 3.4 and 0.9 m long, respectively. In section 17.2.2 it was established that those parts of the boat most probably had no ceiling, but possibly loose plank bridges instead. It was an open boat with a short mast set a little less than 1/3 of the length from the bow, which stood in a small mast step and probably passed through a hole in a broad thwart.

Tonnage

If the boat was intended for carrying cargo, the spaces between the cuddies were available for this. Their combined length was 4.75 m. Cargo would not have been stacked any higher than the level of the cuddies (ca 0.4 m). The widths of the bottom and halfway up the B strakes at frame 2 are 0.25 and 1 m, respectively; at frame 6 these dimensions are 0.5 and 1.5 m; and at frame 10, 0.5 and 1.3 m. The average width thus is ca 0.7 m. 4.75 x 0.7 x 0.4 = ca 1.3 cu m. Hence the boat had a maximum capacity of 1 to 1.5 gross tons.

Material selection

The Dronten-II cog was built entirely of oak, though the wood species of the treenails was not investigated. The gudgeon, the bolt with ring for holding the topmost pintle, the rudder fittings, the various nails and the sintel clamps are of forged iron. The caulking material was *sphagnum* moss.

Table 1. Principal dimensions of the Dronten-II cog

Length of the keel (including the stern- and stem hook)	7.50 m
Length of the bottom (between the rising part of the stern- and stem hook)	7.10 m
Greatest width of the bottom (at the frames 6 and 8)	1.35 m
Overall length at gunwales level	9.45 m
Overall width at gunwale level (between frames 6 and 8)	2.35 m
Height of the stern (keel line to top gunwale level)	1.25 m
Height of the stem (keel line to top gunwale level)	1.50 m
Height of the side amidships (keel line to top gunwale level	0.65 m
Length of the stern	1.50 m
Length of the stem	1.90 m

Length-width ratio of the bottom	5.25:1
Length-width ratio at gunwale level	4.00:1

All measurements rounded to the nearest 5 cm.

17.3.2 Construction: techniques and sequence

In the - partial - vessel as excavated a number of small, apparently non-functional treenails 1.5 and 2 cm thick were observed in the keel and shell planks (Plates 17.II B and 17.III A and B). In the keel plank between frames 3 and 4, and forward and aft of frame 7, in the after starboard garboard plank immediately aft of frame 4 and between frames 5 and 6, and in the after garboard plank to port, aft of frame 3 and between frames 5 and 6. Probably there had been some also between frames 9 and 10 in the forebody. Without any doubt they relate to the attachment of temporary construction frames during the positioning of the garboards or A strakes at an early stage of the vessel's construction.

The boat features the hooks and keel plank which are characteristic of cogs. What is different is that the rabbets for the hood ends continue up to the top of the stern- and stempost. The flush-laid bottom planks against the keel plank, joined only by means of the frames, and the clinker construction of the sides with double-clenched nails too are typical of cogs. A further feature that we see also on other cogs is that of the plank ends standing proud of the shell (internally and externally) at the scarfs in the strakes. The relatively thin frames, the continuous covering inwale running along the gunwale and the thwart are remarkable features. So are the absence of sintel clamped moss caulking or indeed of any caulking on the overlapping lands of the clinker shell strakes. Some of these features most probably are directly linked to the dimensions and function of the vessel, and to reinforcing the gunwale construction and protecting the frame heads. Besides, the period in which the boat was built will certainly also play a part.

Construction sequence

After positioning and joining the keel plank, hooks and stern- and stempost, first the after planks of the A strakes were put in place with the aid of temporary frames affixed between the (future) frames 2 and 3, 3 and 4, and 5 and 6. The hood ends of the planks were fitted into the rabbets on the stern hook and sternpost. Then the same was done with the forward planks of the A strakes. To hold these in position, temporary frames were placed forward and aft of (the future) frame 7. Presumably there also were one or two temporary frames between (the future) frames 9 and 10. After this, the planks of the B strakes and then the C strakes could be fashioned and in a similar sequence nailed immediately to the garboards and the B strakes, respectively. The strakes of the sides were arranged according to a 'brother and sister' pattern, in which generally two planks sawn side-by-side from a tree are used in equivalent places to port and starboard in the hull. Tree-ring analysis (section 17.8) revealed that the two investigated planks of the garboards came from the same tree. Although this was not established, we may assume that all of the planks were cut from the same tree. The greatest width of the keel plank (52 cm) indicates that this must have been a substantial trunk; one of the samples displayed an erratic tree-ring pattern, which suggests that the tree was a thick and gnarled one. After the insertion and fastening (with treenails) of the permanent frames, the temporary ones could be removed. The covering inwale and the cuddies were the last to be added. The sintel clamped moss caulking between the shell planks, the hooks and the stern- and stempost on the inboard side must have been applied before the permanent frames were put in. The exterior caulking of seams between the keel plank, hooks and stern- and stempost was not performed until the vessel was wholly completed.

17.3.3 Characteristics

The Dronten-II cog has the following features:

Keel plank

A keel plank of less than twice the thickness of the shell strakes (4-4.5 cm), with a width of 20 cm aft and 30 cm forward, and gradually broadening to 52 cm in the midship section.

Hooks

The hooks were joined to the keel plank by lipped flat sloping scarfs. In their sides are rabbets for fixing the hood ends of the garboards.

Stern- and stempost and rudder

The stern- and stempost have rabbets running to the very top for fixing the hood ends. The posts are joined to the top of the hooks with flat sloping scarfs. The boat had a stern rudder with a blade probably composed of two planks and suspended by two pintles. It is most likely that the rudder was secured to the hull by a length of line running through a hole in the rudder and presumably also holes in the upper shell strakes near the stern.

Carvel-built bottom

The boat has a bottom consisting of just the keel plank and two garboard strakes, which in every part have some deadrise towards the turn of the bilge. Towards the stern and stem, the curve becomes ever steeper and the planks are twisted until they end almost vertically at the stern- and stempost.

Clinker-built sides
Amidships, the lowermost strakes of the sides (B) were nailed to the garboards at a slight angle to the garboards' bevelled outer/upper edge. Towards the stern and bow these strakes were twisted almost parallel to the garboards or A strakes. The upper strakes (C) are similarly everted along almost their entire length. Only in the very last metre fore and aft were they a little more everted.

Ceiling/plank bridges, mast step and repairs
The craft most probably had no (permanent) ceiling, but maybe removable plank bridges. The mast step is fashioned in the form of a short floor timber immediately forward of frame 8. During its working life the vessel underwent various repairs.

Cuddies and transverse beams
In the after- and forebody were cuddies of planks about 3 cm thick, fitted to the sides around the frames. Probably they were supported by three light transverse beams that were fastened to the inner face of the frames and by a V-shaped floor timber in the aft and fore above the stern- and stempost.

Fastenings
The shell planks are joined to the frames by 2 cm thick treenails. The treenails were secured in the planks by small wedges. This also goes for the fastening of the covering inwale to the top shell strake. For the clinker connections, the scarfs and the repair planks, wrought-iron nails of various sizes were used. The moss caulking and the moss laths were held in place with wrought-iron sintel clamps.

17.3.4 Use

The lines of the boat, the slightly higher sheer at the bow and the covering inwale suggest that the boat was built (also) for open water (the Zuiderzee) and not just for inland waterways along rivers, canals and lakes. In the latter case there would have been less likelihood and need of an covering inwale, and a punt-like model would have sufficed (compare, for instance, the 15th-century punt from Hellendoorn).[11] Transportation of passengers and light, easily handled cargo on a regional route seems the ship's most likely function, though its use as a fishing vessel should not be ruled out. However, no evidence of such was found. The small cargo vessel presumably built in the late 15th or early 16th century, found at the drowned village of Nieuwlande in the Oosterschelde estuary, is yet another interesting vessel for comparison.[12]

17.3.5 Wreck formation and embedding

The erratic stones, which we may safely assume were not intended as ballast, the wear and tear and the repairs to the boat, and possibly also the location of the site, are clues suggesting that the vessel was purposely abandoned and sunk. The relatively thin (just 30 cm) and soft uppermost Almere clay deposit, which formed the sea floor at the time of sinking, and the lines of the boat meant that it could quite easily and presumably also quite rapidly work its way through the clay deposit and up to 60 cm deep into both the redeposited and the primary Pleistocene sand. The first sedimentary layer around and inside the boat in the disturbed cavity will have settled there quite soon after its sinking. The hogging is a result of the currents around the stern and stem, which left them hanging unsupported in the cavity, weighing down the keel ends. After the wreck settled into the sandy deposits the vessel was probably fully covered over by Almere clay, followed by the Zuiderzee deposits.

17.4 Model

No model of the Dronten-II cog has (as yet) been built.

17.5 Inventory

In or around the wreck no implements were found.

17.6 Cargo and merchandise

No traces of cargo or merchandise were found.

17.7 Ballast

The recovered erratic stones (figs 17.1, 17.5, 17.10 and 17.12) almost certainly did not serve as ballast, but were placed inside the vessel to sink it. In all, there are 135, which were not closely determined as to rock type. A rough sorting produced the following classification in terms of rocks and numbers:

granite	73
gabbro	4
porphyry	8

11. Vlierman 1996b, 53-67.
12. Vlierman 1996b, 33-52; section 26.10, fig. 26.5.

Fig. 17.12. The erratic stones in the afterbody, viewed from port.

granite-porphyry	1
quartzite	32
sandstone	7
Skolithos linearis	1
limestone	1
flint	2
quartz	1
total	130

Among the stones lay (large and small fragments of) four half, red bricks and part of a misfired brick. A remarkable item is a single large stone with a weight of 33 kg. The overall weight of the erratics and bricks in the afterbody is about 90 kg. Possibly a similar batch lay in the bow. The nature and composition of the erratic assemblage suggests that it may well have been gathered in an area like the nearby Veluwe or on the Land van Vollenhove, some 12 km Northeast of the site.

17.8 Dating

Soil profile

The lower and upper boundaries of the redeposited sediment occur roughly halfway up the Almere Ala deposit. This layer was deposited in the second half of the 15th, and the early 16th century. Assuming a constant rate of sedimentation, this would imply that the craft sank in the final quarter of the 15th century.

Finds

No (datable) finds were encountered.

Sintel clamps

The sintel-clamps applied in this vessel are of the (transitional) type D2/E, which were used in the final quarter of the 14th century and the first half of the 15th century. Their close spacing and indeed eventual overlapping was known so far only from shipwrecks dating from the 15th century and the first half of the 16th century.[13] Leaving (large parts of) the clinker seams of the shell devoid of sintel clamped moss caulking was known so far only from a few vessels of the 1425-1525 period, such as the ear-

13. Vlierman 1996a, 28.

lier-mentioned small cargo vessel from lot K 73/74, the punt from Hellendoorn and the small cargo ship from the Oosterschelde estuary at Nieuwlande.

Dendrochronology
Samples were sawn from the keel plank and from the after planks of the garboards to port and starboard. In the analysis by RING the measurements of the latter two samples were averaged and produced the best match with DLS 9 (i.e. Mittel- Westniedersächsiches Tiefland) with a date of AD 1155-1276, and a Gleichlaufigkeit of 68.2. The youngest ring dates from AD 1276 + at least 20 ± 6. A felling date after AD 1296 + 6. No sapwood boundary was observed, so an unknown number of heartwood rings may be missing. Hence the actual date of felling was an unknown number of years after AD 1296 ± 6 (see Appendix).[14]

17.9 Summary

Vessel
The boat from lot M61 in Oostelijk Flevoland, the Dronten-II cog, has a fine sheer, rising somewhat towards the bow, and gunwales forming a laurel-leaf outline, sharply converging towards the stern and stem. The straight stern- and stempost have a rake of 124° and 128°, respectively. The transverse sections of the hull are sharp towards the stern and stem, but more rounded amidships. Special features are the rabbets for fixing the hood ends in the sides of the stern - and stempost, which run up the full length of these posts, and the survival of the rudder.
Amidships the garboards or A strakes slightly rise towards the somewhat angular bilges. The bottom comprises just the keel plank and the two garboards. The latter were joined carvel-style to the keel plank with the seams closed externally with sintel clamped moss caulking. The B and C strakes were fastened to the garboards and interconnected in clinker fashion with double-clenched nails, their tips driven back into the wood. Thin layers of moss caulking had been put within the scarfs joining the keel plank to the hooks, joining the hooks to the stern- and stempost, and joining the planks that formed a strake. No caulking was present on the lands by which the overlapping clinker strakes were fixed to each other, nor was any sintel clamped moss caulking found in the upper edges of these seams. The keel plank had a thickness of 4 to 4.5 cm, which is less than twice that of the shell planks. The boat had frames 7-9 cm wide and 7 cm thick, which at the top were ca 6 cm thick. On the inside of the gunwale, a 6x5 cm covering inwale was inserted, resting on the frame heads. In the aft and the forebody it had cuddies fitted to the sides. If it engaged in transportation of (non-bulk) freight, loads would probably not have exceeded 1 to 1.5 tons. The craft is likely to have been well suited for carrying passengers with hand luggage. No evidence was found for the boat's use as a fishing vessel, though this is not impossible. There had been a mast, supported by the hole in the small floor timber forward of frame 8 and by a hole in the thwart. It probably carried a spritsail, and presumably could be rowed as well. Judging by the shape of the hull, it must have been easy to handle as a sailing boat, suited to navigating rivers and lakes and, in fair weather, also the Zuiderzee. Presumably it was at home in the region (the coast of Guelders and the environs of Kampen). Its wear and tear and repairs, as well as the recovered erratic boulders, suggest that it was an old, written-off craft that was deliberately weighted and sunk.

The principal dimensions are:

• Length over the stern and stem at gunwale height	9.45 m
• Largest width at gunwale level	2.35 m
• Height of the stern (keel line to the top gunwale)	1.25 m
• Height of the stem (keel line to the top gunwale)	1.50 m
• Height of the side amidships (keel line to the top gunwale)	0.65 m

Inventory
No utensils of any kind were encountered.

Dating
The dendrochronology of the bottom planks produced a date that amounted to an unknown number of years after AD 1296 ± 6. The tree from which the planks were sawn was a fairly thick one (with a diameter of at least ca 70 cm). A tree 150-200 years old would be a likely candidate. About 120 tree rings were counted in the samples. This would imply a felling date around 1350 or later. It was concluded that the vessel probably was an old, written-off craft: a working life of 30 to 50 years would be quite plausible. The D2 type of sintel clamp as employed here fits in the latter 14th century, type E1 was most probably used especially from ca 1400 and in the first half of the 15th century, but locally may have remained in use somewhat longer.[15] The predominant absence of sintel clamped moss caulking in the seams, the overlapping sintel clamps where they were used, and the

14. Jansma & Hanraets, written report, August 1996.
15. Vlierman, in prep. a; and section 29.2 in the present publication.

continuous rabbets in the stern- and stempost too suggest a date from the mid 15th century on. Hence it seems that the craft was built around the mid 15th century or indeed more likely in the second half of the 15th century and that it was sunk at the end of that century or in the early 16th century. This would fit with the disturbance of the soil profile recorded during the inspection in 1982, and the presumable date of sinking inferred at the time. Quite remarkable is the correspondence in side elevation of the Dronten-II cog to the small vessels depicted on the civic seals of Kuinre and Genemuiden, dating from ca 1400 and the mid 15th century, respectively (figs 17.13a and b). Indeed the Kuinre seal clearly shows a vessel that numbers three strakes on either side of the keel plank, the other, four strakes.

Fig. 17.13a. The civic seal of Kuinre of 1399 (probably); 1407 (certain). After Ewe 1972, 145-88.

Fig. 17.13b. The civic seal of Genemuiden of the mid 15th century (presumably 1461). After Ewe 1972, 129-52.

18

The Swifterbant cog[1]

18.1 Find conditions

18.1.1 Discovery and investigation

In June 1962, a shipwreck was struck during drainage works on lot G 77 in Eastern Flevoland. Shortly afterwards, the find was the subject of a short assessing survey in the drainage trench. A small copper kettle with a pouring rim and iron handle was found in the wreck during the survey. This early type of kettle was supposed to date from the 16th century and the excavator Van der Heide therefore assumed that the ship dated to the same period.[2] The dimensions, the type and the completeness or other information on the ship were not recorded in 1962. In the period 13th-25th June 2001, an extensive assessing excavation was carried out. After the circumference of the wreck had been determined by probing, a trench was dug across the entire width of the ship amidships and smaller trenches were dug at the stern en stem.[3] It could be established that this was the rather flat and carvel-built bottom of a vessel, the planks of which twist towards the stern and stem and are clinker-built there. The seams of the clinker-built part of the structure have sintel clamped moss caulking on both the inside and the outside. The sintel clamps used are of type D. The objects and bones found inside the wreck are dealt with in section 18.5 and in the find list and appendix. During the assessing excavation, some wood samples were taken for dendrochronological research. The presence of the stern and stem hooks and the construction characteristics mentioned lead to the assumption that the vessel was a cog. The wreck is still at the site of its discovery and is awaiting full excavation.

18.1.2 Geology and stratigraphy

The cog is located in the area where several sites of the Swifterbant culture have been found. On top of the river dunes/gullies landscape in the subsoil, there is peat or displaced peat (peat which had been taken from elsewhere through flooding and deposited here which then sedimented onto the sea floor), on which a package of Almere- and Zuiderzee deposits have been found.[4] The lower Almere deposit consists of clay with displaced peat or Detritus bands. On top of this, there was a sandy layer with also displaced peat or Detritus bands, the wreck has sunk into this (fig. 18.3). In the disturbance created by the wreck and inside the wreck, displaced sand was deposited first, interspersed by a layer of Almere-clay. On top of this lay a thicker layer of Almere clay, which was then covered with a shell-rich (*Mya arenaria*) Zuiderzee deposit of displaced sand. On top of that lay Zuiderzee and IJsselmeer deposits. These can no longer be discerned, however, because they have been completely included in the topsoil. Amidships, the profile was largely disturbed, probably as a result of digging the drainage trench in 1962.

In which Almere deposit the ship has sunk could not be determined. However, it can be said that the Almere deposit was probably higher or that the vessel did not sink entirely into the seabed at the time. The parts of the ship above the turn of the bilge had already disappeared completely before the sandy layer of sand gapers (*Mya arenaria*) was deposited undisturbed over the wreckage. It is very likely that this part of the Zuiderzee had turbulent currents at various times and that, in addition to the decay of the cog, the wreck may have been covered with sediment and then largely washed free again several times. There may have been a wide current channel running north-south at the site, which may explain the strikingly sandy sedimentation package. As to dating the sinking of the ship on the basis of the soil profile, the only thing that can be said is than this probably took place before about 1600.

18.1.3 Medieval habitation

At the time of the shipwreck Urk was the nearest medieval village/island (approx. 10 km).

1. Translated by Stefanie Hoss.
2. Report assessing survey ship G 77 Oostelijk Flevoland, 22 June 1962. RCE-Lelystad. In a letter dated 20 April 1967, Van der Heide writes about a probable date for the ship in the 15th or 16th century.
3. The wreck is located north-northeast/south-southwest facing north-northeast (320°) with the prow at 140 m from the Vuursteenweg and 100 m from the ditch G 76/77. The site is located about 10 km south of Urk (Fig. 18.1 and 18.2). Topographical map, page 20 G Lelystad; co-ordinates: stern 169140.094/509570.598, stem 169149.643/509581.224, with the highest parts at ca 3,8 m -NAP, Hydrographical Map 1921, medium low tide.
4. Ente *et al.* 1986, 45-91 and figures 2.2 and 2.7.

Fig. 18.1. The site where the Swifterbant cog was discovered.

18.1.4 Condition and wreck formation

Condition

The stern is situated the deepest (ca. 1.2 m below surface or ground level), the highest part of wood found was only 40 cm deep. During the assessing excavation, the conservation quality of the wood was good to moderate. The soil structure in the immediate vicinity mainly consists of a sandy package with displaced peat and clay bands. The topsoil consists of a sandy clay package. The three drainage pipes across the wreck plus the disturbance and the excavation trenches from 2001 have disrupted the soil structure. This means that conservation conditions have deteriorated since then, with the result that the quality of the wood continues to deteriorate at an increased rate.

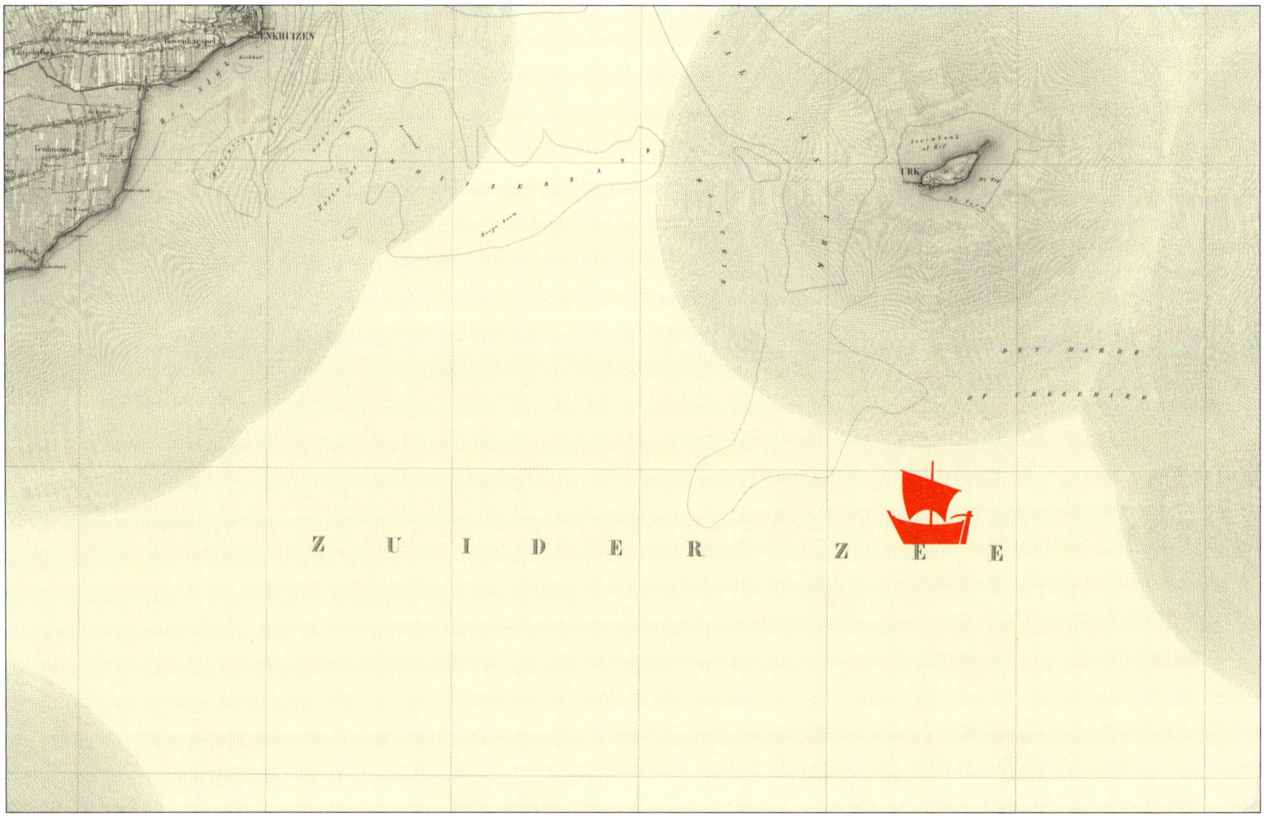

Fig. 18.2. Topographical and military map of the Kindom of the Netherlands (scale 1:50.000) from 1861 with the location of the Swifterbant cog.

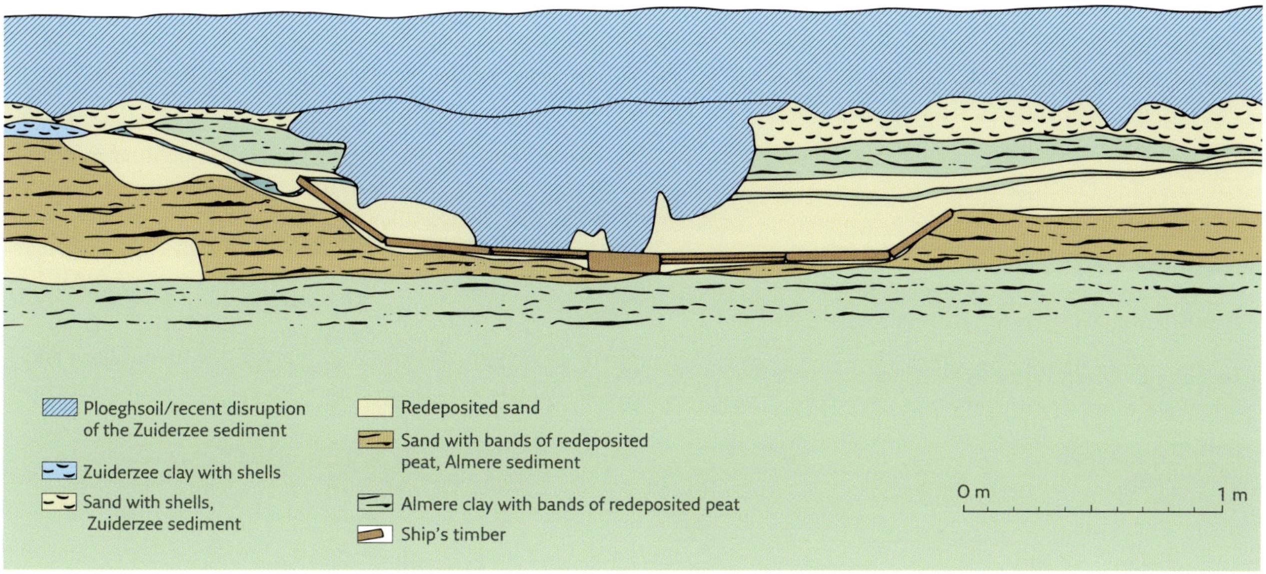

Fig. 18.3. Soil profiel with the disturbance by the Swifterbant cog (partly augmented).

Wreck formation

The ship may have sunk during a storm on open water. The island of Urk was probably larger at the time than it is now, but was certainly at a distance of 8-9 km from the wreck site. How deep the Zuiderzee was at that time cannot be said with exactitude. What is certain is that the ship sank with the keel to at least 4 m below the average low tide line of the time. This means that the entire hull immediately disappeared completely under water. The objects excavated from the inventory of the ship supports this notion. It is likely that the crew had to leave the ship in haste or died in the disaster. It has already been said above that the vessel may not have sunk completely into the seabed, but that parts may have stuck out above the seabed. These parts disappeared, after which the remaining wreck was eventually covered with sediment. It seems plausible that the disappearance of the sides and other parts that were higher on the ships took place relatively shortly after the shipwreck.

18.2 The ship

18.2.1 Documentation

The excavated parts of the ship and the disturbance of the soil profile were measured, drawn and photographed. A top view was drawn of the excavation trenches above the stern, amidships and the fore (Plate 18.I). In addition, a cross section of the disturbed soil profile and three cross sections across the aftership, amidships and the foreship were drawn.

18.2.2. Construction and layout of the hull

Only fairly small parts of the wreck were excavated during the assessing excavation in 2001, so that we can only report on these with any certainty. Based on comparisons with the other ship finds in this study, and in particular with the wreck of the Dronten I cog (Chapter 11), a brief discussion is given of what the unexcavated parts may look like.

Keel plank

The vessel has a keel plank whose full length could not be determined in the assessing excavation. Only 37 centimeters remained of the damaged scarf with the stern hook. The plank was 22 cm wide at the scarf and 27 cm amidships. The thickness is 7 cm amidships and 9 cm at the aft scarf.

Stern- and stem hook

The stern and stem hooks (figs 18.4a and b and 18.5a and b) were severely damaged and pulled apart by the draining machine, and some pieces have disappeared. The excavated pieces have been left in situ and are awaiting further study and documentation. Approximately 85 cm remain of the stern hook's keel section, which is still connected to the keel plank. Of the section extending further aft, a piece of ca 85 cm in length was found in the drainage ditch. The piece was 15 cm wide at a short distance from the top of the sternpost section and 13 cm wide at the lower part, it has a thickness of 22 cm. The stern hook gradually increased from the start of the scarf with the keel plank to a thickness of 17 cm and then presumably jumped to a greater thickness of 22 cm. A severely damaged piece from the stern section that was ca 35 cm long was also found in the drainage trench. The keel part of the stern hook was at least 1.6 m long. The keel part of the stem hook was broken off by the draining machine and partly set vertical. The front part plus the stem part are missing. The original length of the stem hook could not be determined. The rabbets for securing the garboards are ca 10 cm high and 3 cm deep in both hooks.

Stern- and stempost

Nothing was found of the stern and stempost. It also seems very unlikely that parts of them are in the unexcavated remains of the ship.

Fig. 18.4a. The afterbody, showing the keel plank and the destroyed stern hook with the drainage put above and (partly) through it. Seen from starboard.

Fig. 18.4b. The same, as seen from the stern hook. The nail heads visible on the floor boards are from the connection scarf to the keel plank.

Fig. 18.5a. The forebody with the keel part of the stem hook, severely damaged and partly pulled upright by the machine putting in the drainage. Seen from port.

Fig. 18.5b. The same, as seen from the stempost.

Rudder
The vessel undoubtedly had a stern rudder. Nothing has been found of it either.

Bottom of the ship
The bottom of the ship consists of the keel plank plus the stern and stem hooks and on either side of these two strakes (A and B). Amidships, both strakes are placed carvel next to each other in an almost straight line and run slightly up to the turn of the bilge. Although not the whole bottom of the ship has been excavated, it may be assumed that the strakes were composed of two planks and that – at a similar distance from the stern- and stemposts - the bottom was 4-8 cm narrower at the rear than the front. In the excavation trench across the midship, the ends of the aft planks of the A strakes or garboards may be seen (Plate 18.I and fig. 18.6a and b). On the other side of the floor timber, the nails of the scarf in the C strakes or bilge strakes can be seen on the port side. The two rows of nails of two and three nails respectively in the B strake on port side amidships are from a repair board on the outside. The full length of the strakes has not been determined. The length of the bottom (keel plank plus the keel parts of the stern and stem hooks until the beginning of the rising parts of the stern- and stemposts) will have been ca 11 m. The width of the A strakes amidships is 46-47 cm, while it is 40 cm aft and 36 cm to the fore. The B strakes are 37-39 cm wide amidships, 35 cm wide aft and 37 cm wide to the fore. The planks are 3 cm thick, while the ends of the planks are about 1.5 cm thick at the scarfs and protrude above and below the shell. The length of the scarfs was not documented. The planks were fastened to each other with a few rows of nails, hammering them in from the inside to the outside and vice versa. Whether the nail points pierced through the skin, were bent twice and driven back into the wood was also not documented. The wreck lists slightly over the port side (Plate 18.I).

Fig. 18.6a. Overview of the trench across amidships with two floor timbers and the ends of the aftermost planks of the garbaords or A strakes. On the right in the foreground above the turn of the bilge a small lath (auxiliary frame). Samples were sawn out of the turn of the bilge in the foreground and from the floor timber on the right. Seen from starboard

Sides
From the sides, only some large pieces of the bilge or C strakes have been found. The width of this strake amidships is 39 cm and the thickness is 3 cm. On the port side, parts of the bilge strakes may be seen aft and to the fore, they are missing on the starboard side (fig. 18.6a). The bilge strakes have been placed carvel against the B strakes. The strakes above that are missing.

Sintel clamped moss caulking
The seams between the B- and C strakes are open on the outside, creating a triangular seam that is 2-2.5 cm wide, probably covering all or almost all of the thickness (3 cm) of the boards. In this seam, strands of peat moss (*Sphagnum spec.*) were hammered, covered with moss lathes, which in

Fig. 18.6b. Detail. The triangular small lath (auxiliary frame) on the turn of the bilge.

turn were fastened with sintel clamps. This probably was also done in the other carvel seams of the bottom. The seams of the clinker-built planks have sintel clamped moss caulking on both the inside and the outside. During construction, the planks were chamfered to create triangular openings of 2.5x2.5 cm before being set against each other. The moss hammerd into these seams was secured with moss lathes, which were 2 cm wide and 0.5 cm thick, and fastened with sintel clamps. On the inside, the seams between the garboards and the stern and stem hooks were also fitted with sintel clamped moss caulking. Whether this was also the case on the outside has not been documented, but is probable. Fissures in the ends of the aft garboards were also closed in this manner. The sintel clamps were placed every 8-9 cm centre to centre and are of type D2 (fig. 18.7 and 18.8). This type was in use in the period ca 1275-1350. Recent research suggests that the model used for the Swifterbant cog was mainly in use around 1300.[5]

Frames: floor timbers, V-shaped floor timbers and futtocks

In the small excavation trench at the stern, a small V-shaped timber was documented sitting just before the beginning of the scarf with the stern hook. The next two V-shaped timbers were missing. The distance between the V-shaped timbers was (centre to centre) ca. 50 cm. In the trench across amidships, two floor timbers could be seen, a third protruded out of the wall of the trench towards the fore of the ship. There also was a V-shaped timber noted at the bow. The distance between the aftmost and foremost V-shaped timber is ca. 10 m. This means that there were at least twenty, but most probably twenty-two frames on the bottom of the ship between the beginnings of the stern- and stemposts. They were extended with futtocks, but of these nothing was found, with the exception of possibly two pieces on the starboard side amidships. The floor timbers were 16-19 cm wide and 17 cm thick and they were neatly contoured. The wooden frames were generally attached to the shell planks with two treenails (Ø 3 cm) per plank width. The treenails are positioned in a barely staggered pattern. A triangular limber hole was noted on the underside of the aftmost V-shaped timber on the port side. Three limber holes with an inverted U-shape were noted in the foremost floor timber amidships above the keel plank and the B-strakes. The V-shaped timber in the forepart also had an inverted U-shaped limber hole on the starboard side.

Mast step

Unfortunately the mast step was not documented during the assessing excavation. It can be assumed that the hole for the stepping of the mast was carved out in a wide and possibly thicker floor timber (compare the Dronten-I cog, section 11.2.2 and figs 11.16, 11.17a, b and c and 11.18).

Fig. 18.7. Detail. Sintel clamps of the D2 type on a lath in the clinker-built seam between the A- and B-strakes in the afterbody. Seen from the ship.

Ceiling

It is not possible to state with certainty whether the vessel had ceiling strakes. However, it seems very likely that this must have been the case at least above the turn of the bilge, in order to compensate for the 'weak' scarfs between the floor timbers and the lower futtocks. It also seems logical that there were a few strakes of an open ceiling higher up in the sides.

Traverse connections, deck beams, sail beam, decks, gangways, windlass and pump

Obviously nothing can be said with certainty about the structural parts situated above the bottom of the vessel or a windlass or a pump. However, it seems reasonable to refer to the hypothetical reconstruction of the Dronten-I cog (Chapter 11 and Plates 11.II A and B and 11.III A to D) in order to get a reliable impression of what the vessel might have looked like.

Castles

No indications of a castle have been found.

Repairs

Apart from the two fissures in the aft garboard which were sealed with sintel clamped moss caulking and the nails of a repair board on the outside of the B strake on the port side amidships, no repairs were found in the excavated sections of the ship.

Wooduse

All parts of the ship excavated were made of oak (*Quercus spec.*). The opposing scarfs in the planks of the bottom of the ship on starboard and port side and the fissures in them, which were in the same places and were sealed with sintel clamped moss caulking, suggest that the con-

5. Vlierman 1996a, 77; Vlierman in prep. a.

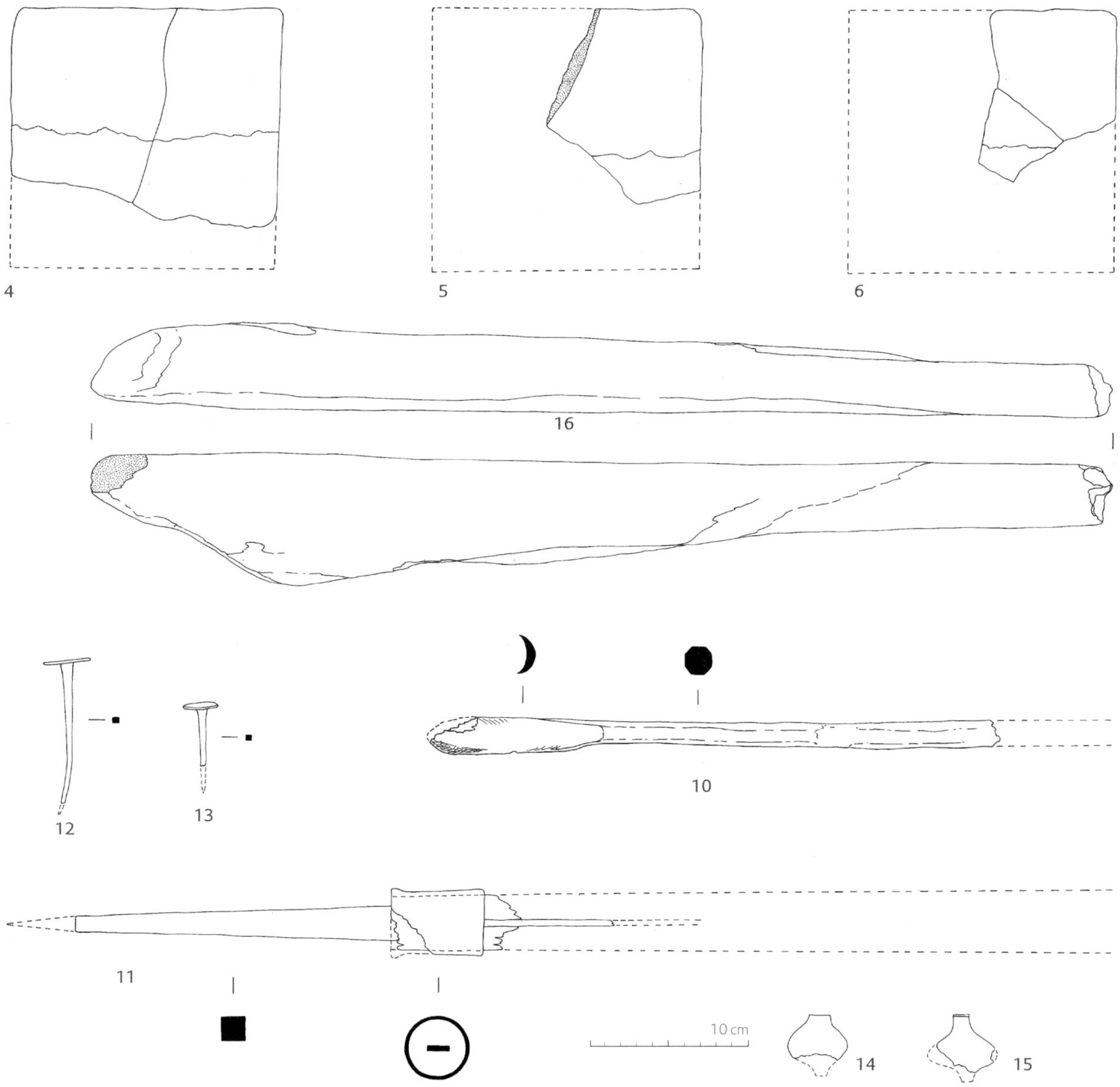

Fig. 18.8. Ship's equiment, fireplace, tools and weapons. Tiles (4, 5 and 6), spoon drill (10), lance point (11), nails (12 and 13), sintel clamps (14 and 15) and whetstone (16).

struction must have taken place according to the so-called 'brother and sister pattern'. As far as could be seen, the frames were neatly contoured, while it is likely that the planks were sawn.

Fasteners

The frames were attached to the shell planks with 3 cm thick oaken treenails, while the clinker-built joints were fastened with wrought iron nails ca. 100 mm long. They have been twice turned at 90° degrees and then driven back into the wood on the inside. The nails were fitted every 17 to 18 cm (measured centre to centre). It has not been documented whether there were any nails driven in from the inside to the outside, as in the Dronten I cog. The nails had flat, round heads (Ø 30 mm). A smaller type of nail, which was ca 60 mm long and had a flat head of 23 mm in diameter, was also used in the vessel. The moss caulking and moss lathes were fixed with wrought iron sintel clamps of type D2. The central parts of the clamps were 40-45 mm long. Nothing has been found of the heavier wrought iron gudgeons and the rudder fittings.

Table 1. Estimated principal dimensions of the Swifterbant cog

Length of the keel (including the stern- and stem hook)	ca 11.5 m
Length of the bottom (between the rising part of the stern- and stem hook)	ca 11.0 m
Greatest width of the bottom	ca 2.0 m
Overall length at gunwale level	ca 15.0 m
Overall width at gunwale level	ca 5.0 m
Height of the stern at gunwale level	ca 2.5 m
Height of the stem at gunwale level	ca 3.0 m
Height of the side amidships	ca 2.0 m

Length-width ratio of the bottom	ca 5.5:1
Length-width ratio at gunwale level	ca 3.0:1

All measurements were rounded off to 10 cm

18.3 Construction data

Despite the fact that the wreck was not (yet) completely excavated and not everything could thus be documented, some things may be said about the vessel with the help of comparison. The ship had a carvel-built bottom, consisting of a keel plank and two strakes on either side. The carvel construction turned into a clinker-built fastening of the strakes in the after- and forebody. The vessel had a stern and a stem hook, both connected to the keel plank by scarfs. The seams were sealed with sintel clamped moss caulking, at the clinker-built parts these were both on the inside and the outside. The C- or bilge strakes sat at an angle against the carvel-built part of the ship, with the cross- sections becoming continuously rounder towards aft and fore. Both the construction and the dimensions are very similar to the Dronten-I cog (Chapter 11), which also dates to the same period (Section 11.8). It is likely that the Swifterbant cog was slightly smaller. The triangular 'slats' found at right angles to the bilges are auxiliary frames used during construction to clamp the bilge strakes against the outer strakes of the bottom. They were fastened with two nails (figs 18.6a and b) and not removed after placing the floor timbers. Similar slats were also found in the Dronten-I cog (fig. 11.10 and Plate 11.II B).

Characteristics
The Swifterbant cog has the characteristic keel plank of a cog: approximately twice as thick as the shell planks and narrower aft and to the fore (ca 20 cm) than it is in the middle (27 cm). The keel plank was connected to the stern and stem hooks with sloping scarfs. The ship had a relatively narrow bottom (ca 2 m) and upright, everted bilge strakes. The transition of the bottom strakes from a carvel-built construction amidships to a clinker-built construction in the after- and forebody could also be documented. The (almost completely missing) sides must undoubtedly have been entirely clinker-built. The seams of the clinker-built construction were provided with sintel clamped moss caulking on both the inside and the outside. In the relatively small vessel, remarkably thick treenails were used (Ø 3 cm). The wrought iron nails have flat heads.

Function
Brick grit was found during the excavation, possibly originating with a load. The most likely explanation is that this was a small cog intended for the transport of goods and people. Judging from the inventory excavated (see below Section 18.5), it seems very likely that cooking took place on board, from which we can deduce with some caution that the vessel probably (also) made voyages of several days.

18.4 Model

No scale model was made of the Swifterbant cog.

18.5 Inventory

18.5.1 Distribution of the finds

The objects found during the excavation in 2001 were mainly located in the afterbody. The copper kettle found in 1962 was found amidships while digging the drainage trench. The kettle could not be retrieved in the collection

Table 2. Distribution of finds

Category	Forebody	Hold	Afterbody
1. Ship's equipment	sintel	-	nails, sintels
2. Operational equipment	tar	-	-
3. Administration and navigation		-	blackboard chalk
4. Tools	-	-	spoon drill, grindstone?, whetstone
5. Household items	-	-	-
6. Copoking, eating and drinking		kettle*	floor tiles, brick, pipkin, beaker, jug, flint*
7. Victuals	-	-	storage jar, bones
8. Personal belongings/weaponry	lance	-	spindle whorl

of finds in 2013. Although it is not certain whether the objects collected to date comprise the entire inventory, they are described here in accordance with the finds from the other cogs and presented in tables. The objects marked with * in the find list and in the tables could not be retrieved in the find collection in 2013.

18.5.2 Ship's equipment and inventory

Ship's equipment

Two types of nails were found in the wreck (fig. 18.8-12 and -13). They had thin flat heads and a diameter of 30 and 22 mm respectively. The larger model was used for the clinker-built parts. The sintel clamps used in the vessel were of type D2 (fig. 18.8-14 and -15). A small L-shaped wooden clamp (23) may originally have been nailed somewhere against the hull. No other objects belonging to the hull and ship's equipment were found. The objects that belong to the hull were not listed in an extra table.

Operating equipment

No objects were found in the wreck that could be classified in this category.

Administration and navigation

Twenty-two bigger and smaller chunks of blackboard chalk were found in the afterbody. It is possible that the chalk once came aboard in one large piece. It might then possibly have come from Møn in Denmark or Rügen in northern Germany, for example. It could not have been collected in the Netherlands as an erratic, because it is far too soft for that. The containers for cargo, such as barrels, were marked with this blackboard chalk. Blackboard chalk has been found several times in shipwrecks in the former Zuiderzee region.

Fig. 18.9. The spoon drill (10).

Tabel 3. Administration and navigation

Category	Item	Number	Findspot
3.1 Ship's papers	-	-	-
3.2 Stationery	blackboard chalk	18	afterbody
3.3 Financial/Commercial matters	-	-	-
3.4 Orientation and dead reckoning	-	-	-
3.5 Signalling and distress signals	-	-	-

Fig. 18.10. The quartz phyllite whetstone (16).

Fig. 18.11. Fragments of three floor tiles (4, 5 en 6).

Tabel 4. Tools

Category	Item	Number	Findspot
4.1 Carpentry	spoon drill	10	afterbody
4.2 Caulking and maintenance	-	-	-
4.3 Splicing, sailmaking	-	-	-
4.4 Miscellaneous	whetstone, grinding stone, tar	16, 17, 28	afterbody. afterbody, forebody

Tools

During the 2001 excavation, a part of a wrought iron spoon drill was found (figs 18.8-10 and 18.9). The upper part of the shaft and the handle are missing. The diameter of the auger part (the spoon) is 25 mm. During the construction of the vessel, a different auger was used, as the diameter of the wooden pins is 30 mm. A whetstone of quartz phyllite almost 65 cm long was found in the afterbody (figs 18.8-16 and 18.10). A more or less rectangular piece of coarse-grained sandstone (17) may have been intended as a grinding stone. However, traces of use can not or hardly be seen. A large chunk of tar (28) in the forebody under the ballast stones may have been intended for maintenance.

Household goods

Household good are not to be expected on board of a not very large late medieval vessel and they were not found.

Cooking, eating and drinking: fireplace

It seems very likely that the vessel had a fireplace in the afterbody. Although no soot traces were found, it may be assumed that the fragments of three floor tiles (figs 18.8-4, -5 and -6 and fig. 18.11) were part of this fireplace. This probably also applies to the half of a brick (19). The floor tiles are of a striking type and have never been found in a wreck excavated in Flevoland before. They have an unusual size (probably 17x17x1.7 cm) and are covered for more than half with a streaky white engobe, with a yellow lead glaze on top of it (4 and 5), while the other half is covered with a dark brown/green lead glaze (6). Another and perhaps more likely possibility is that these are not floor tiles, but roof tiles. Generally, these (Burgundian) roof tiles were only found in the west of the Netherlands, in Flanders, Northern France and England. In Westphalia and the Rhineland, the common roof tiles were of the arched upper and lower pattern, also called 'monks and nuns'. In 2014, a 14th century fireplace was found on the grounds of the former Sint Geertruidengasthuis in Kampen.[6] The fireplace was composed of vertically placed pieces of similar roof tiles. In Kampen, the complete roof tiles measured 25x16x1.5 cm and were finished on the top for over 1/3 of the length with a transparent lead glaze, covering the part of the tile that would have remained visible on the roof.[7] It seems plausible that the tiles (together with others) were placed in a square wooden box filled with sand as a firebox. Similar fireboxes have been found in the Nijkerk-I cog (Section 8.5.2 and figs 8.9, 8.12, 8.50, 8.51 and 8.52) and two *waterschepen* (Section 29.3.1 and figs 29.7a, b and c). According to the report by G. Wilgerson, flint (30) was found in the afterbody, however, this could no longer be found in the find collection in 2013.[8]

6. Also see section 28.2.
7. Oral communication M. Klomp, Archeologie Gemeente Zwolle, November 2015. Indications for the production of roof tiles were found in the 'Bemuurde Weerd' suburb of medieval Utrecht. Also see the Spakenburg-I cog, section 10.5.1 and fig. 10.24F.
8. Report Og 77, RCE, Lelystad, 17 juli 2001.

Fig. 18.12. The bronze pipkin (9), the brick fragment (19) and to the left of the V-shaped floor timber the whetstone (16) with the spoon drill next to it (10) in situ in the afterbody. Seen from starboard.

Fig. 18.13. The bronze pipkin (9).

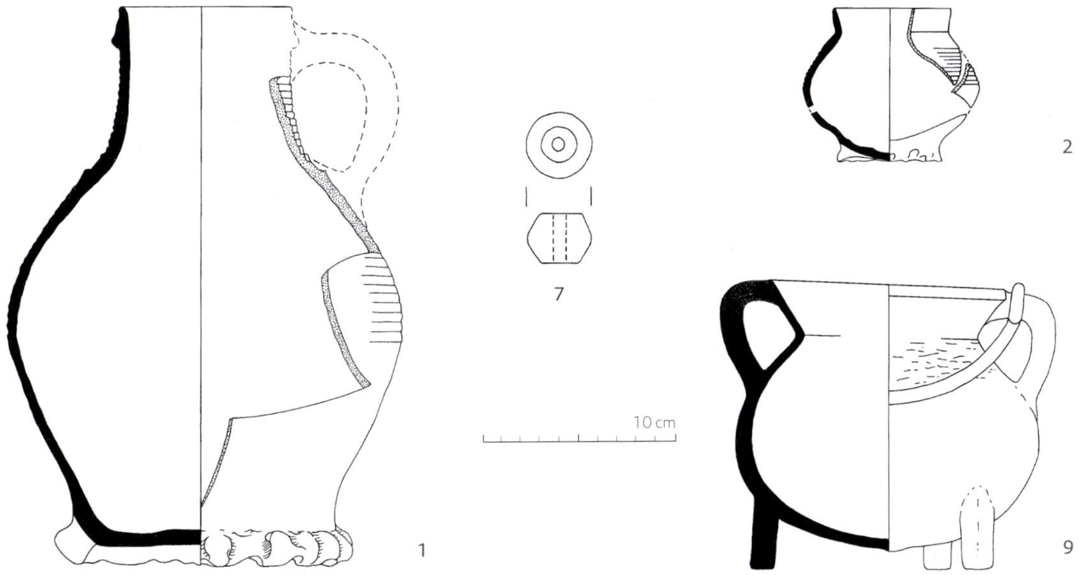

Fig. 18.14. The near-stoneware jug (1), the near-stoneware beaker (2), the bronze pipkin (9) and the spinning whorl (7).

Cooking utensils

The copper kettle was undoubtedly an important piece of cooking-ware (8). The kettle could not be retrieved in the find collection in 2013, so we will have to make do with the description by Van der Heide. According to him, it was a small model with a pouring rim and an iron handle. Based on other kettles from shipwrecks in the former Zuiderzee area, this information can be supplemented with some certaincy with the information that it probably was a cylindrical model, composed of copper plates riveted together with rivets of rolled-up scraps of copper plate and two vertical handles of copper or iron. This is a well-known type that was regularly found in shipwrecks dating from ca 1300-1650 and even in 18th century vessels (for example the Marknesse cog, chapter 12, figs 12.29-7 and 12.30a).[9] The second object used for cooking is the small bronze pot or grape with an iron handle (figs 18.12, 18.13 and 18.14-9). The pot was almost spherical and had three almost vertical legs with a rounded triangular cross-section. The outside had soot deposits and the inside food deposits. The grape has not yet been cleaned. It looks as if the shoulder of the grape has a '*Besenstrich*'-like surface. Bronze cooking pots also occur regularly in late-medieval ships and are still documented in inventories from the 17th and 18th centuries (see the Marknesse cog Chapter 12, figs 12.29-5 and -6 and 12.30b).

9. Vlierman 1992a, 50-59.

Drinking utensils
The small spherical pot/beaker of near-stoneware (figs 18.14-2 and 18.15-2) is from a well-known type. Bruijn depicts four 14th century (period V) examples from Schinveld and Reineking-Von Bock a 14th-century Rhineland specimen.[10] Seven sherds of near-stoneware with a brown-grey salt glaze are from a slender drinking jug (fig. 18.15-3). It seems likely that the jug was already broken (long) before the shipping-disaster.

Victuals
The (sherds of the) large near-stoneware jug (figs 18.14-1 and 18.15-1) may be regarded as a container for storing and transporting fluids, for instance drinking water. The jug is directly comparable to jug 1 from the Dronten-I cog (Section 11.5.2 and figs 11.29-1 and 11.33) and belongs to a common type with a characteristic collared rim. This jug, the near-stoneware beaker and the other jug belong to the production period preceding the introduction of the fully sintered stoneware in the years 1320-1330.[11] Four animal bones were found in the wreck, two of which could be determined to have come from a duck, one is a cattle bone and the fourth a sheep/goat bone. The latter seems to be the only bone that may have been part of the meat stock on board. The cattle bone is a piece of skull with hornbeam that may have been washed in, while the duck bones probably should not be considered part of the ship either. The bones are treated in the appendix.

Fig. 18.15. The near-stoneware jug (1), the near-stoneware beaker (2) and the sherds of the near-stoneware drinking jug (3).

Table 5. Cooking, eating and drinking

Category	Item	Number	Findspot
6.1 Fireplace	tiles, brick	4, 5, 6, 19	afterbody
6.2 Fire-tending tools	flint?	30*	afterbody
6.3 Fuel	-	-	-
6.4 Cooking equipment	kettle, *Grapen*	8*, 9	amidships,
6.5 Tableware	beaker, drinking jug	2, 3	afterbody

Table 6. Victuals

Category	Item	Number	Findspot
7.1 Fresh-water supply	jug	1	afterbody
7.2 Supply of beverage	-	-	-
7.3 Livestock	-	-	-
7.4 Food provisions	bone	22	afterbody

10. Bruijn 1962-63, fig. 89; Reineking-Von Bock 1976, 139-102.
11. Janssen 1983, 173-175.

Personal belongings and weaponry

The piece of iron composed with wood (figs 18.8-11 and 18.17) most probably was part of a lance. The wrought iron tip, whose top had broken off, was still 200 mm long at the time of the excavation. Originally it must have been ca 240-250 mm long. The lance tip plus the ring has now a weight of about 600 grams. The tip is square in cross-section, measuring 10x10 mm at the broken top and 22x22 mm at the lower end on top of the wrought iron ring that sat around the end of the shaft. The ring was 58 mm in diameter and had a round cross-section with a wall thickness of 2-3 mm, at the top 5 mm. The lance tip had a narrowed tang with which it had been inserted into the top of the now almost completely missing shaft. The broken end of the tang is 9 mm wide. During the excavation, the lance still had a part of the shaft of about 15 cm length. However, the conservation condition turned out to be so poor that only some pieces of wood in the ring were preserved. The shaft had a diameter of 3.5 cm at the top, possibly widening to 4 to 5 cm downwards. The type of wood has not (yet) been determined. It is remarkable that in the inventory of the Dronten-I cog, mentioned several times already, also contained two lance tips. The Dronten-I and the Swifterbant cog are the only late-medieval vessels with such weapons on board found in the Zuiderzee region. Another curious find in the wreck of the Swifterbant cog is the spindle whorl (figs 18.14-7 and 18.16). It was found on the pipe in the drainage trench above the stern hook. Although it has been listed here as part of the ship's inventory, the chances are even that it was an object washed into the wreck, and originating from a medieval settlement.

18.6 Cargo and merchandise

According to G. Wilgerson's internship report, concentrations of shells and brick dust were found throughout the wreck and listed as remnants of possible cargoes.[12] If the shells collected from these concentrations are represented by sample 25, these are beach gapers (*Mya arenaria*), which originate from the Zuiderzee deposit, so they cannot have been cargo. The brick dust was not included in the finds taken from the ship, but may very well represent the residue from cargo. However, another possibilty is that these are the remains of further bricks from the fireplace.

Fig. 18.16. The spinning whorl (7).

Fig. 18.17. (Broken off) lance point with tang and ring.

Table 7. Personal belongings and weaponry

Category	Item	Number	Findspot
8.1 Garments	-	-	-
8.2 Footwear	-	-	-
8.3 Sewing kit	spindle whorl	7	afterbody
8.4 Tools/weapons	lance	11	forebody
8.5 Cash	-	-	-
8.6 Miscellaneouss	-	-	-

12. Dossier Og 77, RCE, Lelystad, 17 juli 2001.

18.7 Ballast

Twenty-one fist-sized rocks (erratics) have been found in the forebody above the stem hook (fig. 18.18). The stones were not determined by a specialist, but a rough determination yields the following data:

Granite	11
Pegmatite	1
Red sandstone	1
Sandstone	5
Quartzite (southern)	1
Granite porphyry	1
Porphyry	1
total	21

The total weight of the stones is ca 25 kg. This is hardly enough to work as ballast. The group of erratics may very well have been collected in, for example, the Land of Vollenhove (with the exception of the southern quartzite), but also have been possible on the Veluwe. They also may not have been collected in one place.

18.8 Dating

Soil profile

The seabed profile disturbed by the ship does not provide any information other than that the sinking of the ship must have taken place before ca 1600.

Finds

The near-stoneware (the jug, the beaker and drinking jug) provide the best information for the time of the sinking of the cog. The vessel must have sunk in the first half of the 14th century; the period may probably be further narrowed to the period ca 1320-1330, when the production of near-stoneware was given up in favour of fully sintered stoneware.

Sintel clamps, nails and the shape of the seams of the carvel-built part

The sintel clamps used in the vessel are of type D2, which was used in the period 1275-1350. Recent research suggests that this type, and in particular this size, was in

Fig. 18.18. The forebody with the floor timbers, the ballast stones and the drainage. Seen from above and from amidships.

use around 1300. It has been established in other vessels that nails with flat, thin heads like the ones found in the Swifterbant cog occur mainly in 13th century vessels. The seams of carvel-built shells which are composed of two planks that have an inverted V-shaped opening over (almost) the entire thickness of the board are also typical of 13th century vessels and are particularly well known from vessels dating from before ca 1250. The bilge seams of the Swifterbant cog are triangular over the full thickness of the planks, this was not documented during the excavation for the other seams of the carvel-built part of the hull. It is likely that – as in the Dronten-I cog - we are dealing with a transition between the seam shapes B and C here (see seam shape variants).[13] Sintel clamps, nails and the shape of the seams suggest that the vessel was most likely built at the end of the 13th century or around 1300.

Dendrochronology
The samples sawn for dendrochronology were examined by RING and could be dated with the reference curves NLWestphalia, NLHist-3 and NSS 7 (Niedersächsischer Küstenraum). Their fell dates are ca AD 1305, AD 1301 ± 8 and after AD 1300 ± 6, respectivley (for the full report see Appendix). A construction date between 1305-1310 for the Swifterbant cog therefore seems the most likely, while the construction site may have been in the northeastern Netherlands. A period of use of the vessel of 15-25 years seems perfectly normal.

18.9 Summary

Ship
A complete description of the wreck is not possible and would make little sense at the moment. It must first be fully excavated and documented. However, it can be established that this is the wreck of a small cog, the carvel-built bottom of which consisted of a keel plank with stern and stem hooks and two strakes on either side of the keel plank, that the carvel-built part amidships turns into clinker-built parts aft and fore and that the bilge or C strakes were placed at an angle against the B strakes. The frames were placed approximately every 50 cm. The treenails have an unusually large diameter of 3 cm, as they usually do not exceed 2.5 cm in diameter for vessels of this size. The flat-headed nails are also striking. Similarities with the analysis of the Dronten-I cog in this study (Chapter 11) provide sufficient comparisons for the Swifterbant cog to envisage the latter as a slightly smaller, but otherwise probably comparable vessel. Excavation of this vessel, preferably as soon as possible, may ensure that the remaining information does not get lost, such as the type of mast step and whether the nails were nailed into the shell both from the outside in and vice versa. The site's conservation conditions suggest that there will be hardly anything left of the wreck within a decade or so.

The most important main dimensions (estimated) are:

- Length over the stern and stem at gunwale height — ca 15.0 m
- Largest width at gunwale level — ca 5.0 m
- Height of the stern (keel line to the top gunwale) — ca 2.5 m
- Height of the stem (keel line to the top gunwale) — ca 3.0 m
- Height of the side amidships (keel line to the top gunwale) — ca 2.0 m

Inventory
The objects found so far are undoubtedly part of an inventory that originally was more extensive. There certainly was a fireplace, which means that cooking on board must have been possible. The small copper kettle and the small bronze grape suggest a small crew, which fits well with the dimensions of the ship. The (floor?) tiles excavated are of a type not previously found in the Zuiderzee area. In Kampen a fireplace was excavated in 2014 with comparable examples dating to the same period. In addition, the lance tip may be regarded as an unusual find.

Dating
The period in which the cog sank cannot be reconstructed with the help of the disturbed soil profile. However, thanks to the near-stoneware found, this is possible. The ship must have sunken in the first half of the 14th century, and the fact that the near-stoneware vessels on board belong to the last generation to be made before it was replaced by fully sintered stoneware help us to narrow the period to ca 1320-1330. This indicates that the vessel almost certainly sank around 1325. The type of sintel clamp used was common around 1300, while flat nail heads and inverted V-shaped seams of carvel-built hull parts which run along (almost) the whole thickness of the planks generally date to the 13th century. Based on this information, a construction date at the end of the 13th century or around 1300 seems most likely.

Dendrochronological research confirms this. The construction date probably lies between 1305-1310 AD. The reference chronologies with which these dates were obtained also suggest that the vessel may have been built in the

13. Vlierman 1996a, 77; The Nijkerk-I cog, fig. 8.24.

north-eastern Netherlands. The dates of the sinking and the similarities in construction between the Swifterbant and the Dronten-I cog seem to be so close (with the information available), that we can suggest a relationship between the two, with the necessary reservations. However, there is no hard evidence for this. At most, dendrochronology of the stern and stem hooks and the floor timber with the hole for the mast heel of the Dronten-I cog could yield a comparable date and possibly also indicate the same origin region of the wood, the building region of the ship.

The Almere-II cog[1]

19.1 Find conditions

19.1.1 Discovery and investigation

In the ecological zone to be created on the south side of the Oostvaardersplassen, a nature reserve, a shipwreck was uncovered on lot Cz 46 during excavation work on 20 August 1999.[2] Shortly after the report was made, the site was checked and a dislodged piece of sintel clamped moss caulking was taken away. The sintel clamps present on it could not immediately be recognised as a dateable model. It could be either type D or type E, dating between approx. 1275-1350 and approx. 1350-1450 respectively.[3] A dating of the ship before 1500 was therefore certain, while it could not be ruled out that it dated to the 14th century or even earlier. On the basis of sketches and the information provided by colleagues, it seemed that it could be a cog(like) ship. On 24 and 25 August and between 31 August and 3 September 1999 an assessing survey was carried out. While checking the underground with prods, wood was met over a length of more than 14 metres. On the north side, at the part already dug out for a future lake, the wreck was severely damaged and there were dislodged parts of the ship in the vicinity. (figs 19.3a and b). Above the widest part of the wood checked with prods, an excavation trench about 2 m wide was dug, and at what later turned out to be the foreward, an excavation trench of approx. 2x1.75 m (Plate 19.I A). The wreck is still at the site and is awaiting complete excavation.

The most remarkable observation was that the vessel had sunk upside down and almost horizontally into the seabed. The ship has a carvel-built bottom, whose planks turn into a clinker-built construction towards the front. On 31 August, it was established that the seams of the clinker-built construction had been caulked with sintel clamped moss caulking on both the inside and the outside. On the basis of information already known at the time, this meant that the construction of the ship had to date into the first half of the 14th century. From about 1350 onwards, these seams were only caulked on the inside. During the survey, two samples for dendrochronological research were sawn from the ship. These first two samples could not be dated. The discovery of the stem hook and the stempost made it quite clear that the ship was a cog.

19.1.2 Geology and stratigraphy

The site of the Almere-II cog is located above the more than 1 km wide underground river basin of the Eem in the Pleistocene surface, which extends into the North Sea.

Fig.19.1a. Overview of the excavation work for the extension with an ecological zone on the south side of the Oostvaardersplassen with the site of the Almere-II kogge.

Fig.19.1b. The pieces of the cog which had been torn apart from it, lying as they were found after the initial report.

1. Translated by Stefanie Hoss.
2. The wreck lay in a northwestern-southeastern direction (145°), with the bow towards southeast. The ship sank ca 11 km eastsoutheast of Marken (figs 19.1 and 19.2). Topographical map leaf 26A Almere-Buiten; coordinates: bow X: 147790.845, Y: 492788.172, stern X: 147781.892, Y: 492800.065, with the highest lying parts at ca 3.6 m below surface Hydrographic map 1921, average low tide/mean low water.
3. Vlierman 1996a, 52, 53 and 58 table II.

Fig.19.2. The findspot of the Almere-II cog.

This means that the wreck has subsided into a package of Holocene deposits at least 5-7 m thick.[4] The upper 50 cm of the ground consists of Zuiderzee and IJsselmeer deposits, marked at the bottom by a layer of shells (Zu IV). As far as could be determined and taking into account the occurrence of settling, this layer lay undisturbed on top of the wreckage. It can be regarded as an early-17th-century deposition in this part of the former Zuiderzee. The soil profile has not been documented in detail. The visible package under the Zu IV deposit consists of finely stratified Almere-clay deposits. The bottom profile does not provide any other information for dating the sinking of the vessel than that it must have taken place before ca. 1600.

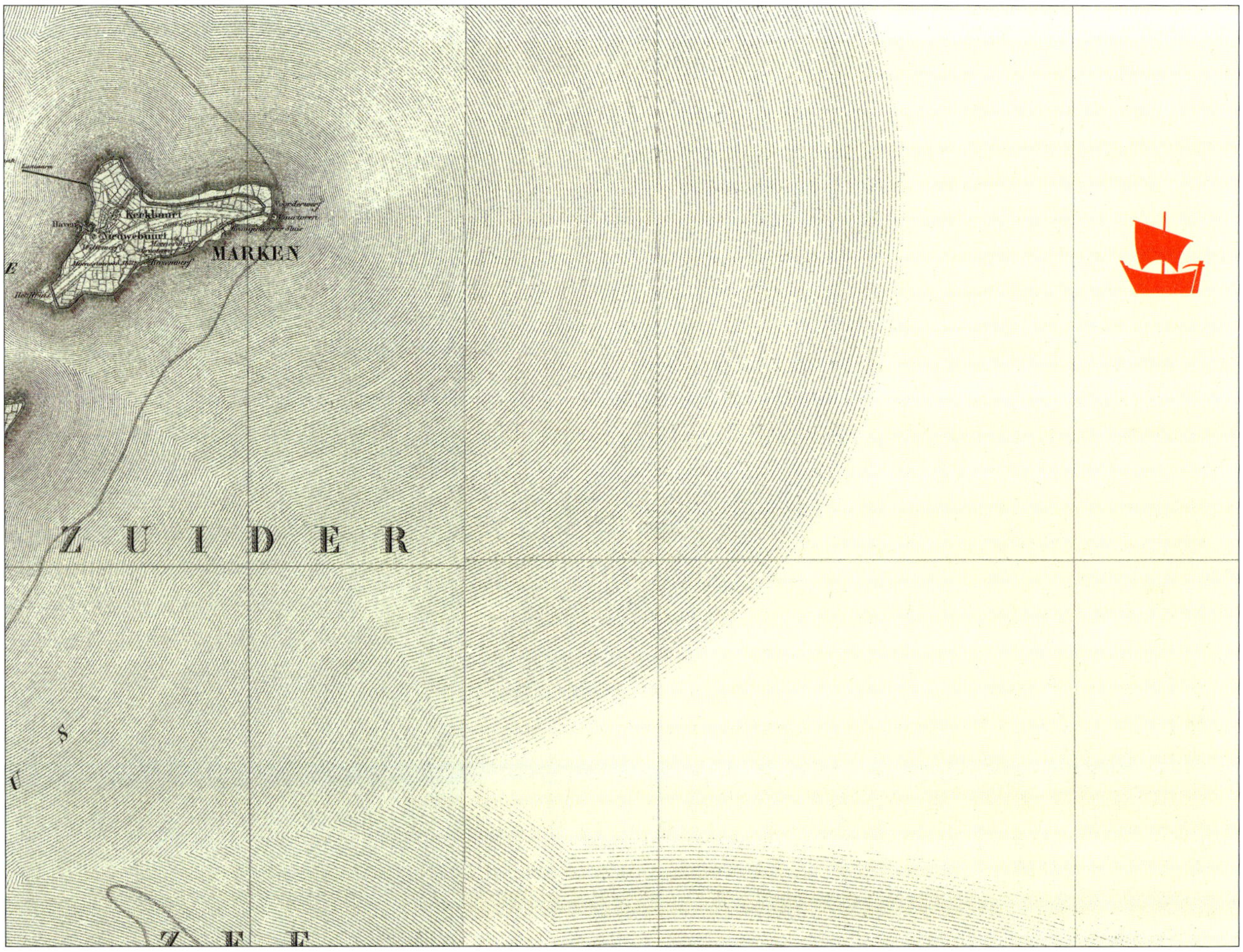

Fig.19.3. Topographical and Military map of the Kingdom of the Netherlands (scale 1: 50.000) from 1861 with the location of the Almere-II cog.

4. Map depth Pleistoceen in m below topsoil 1990. IJsselmeerpolders Development Authority, Scientific department soil science; Lenselink & Menke 1995, leaf 6.

19.1.3 Medieval occupation

The late-medieval village and island of Marken is closest to the place where the cog sank (ca 11 km). The mouth of the river Vecht near Muiden lay at about 17.5 km southwest of the site.

19.1.4 Condition and wreck formation

Condition

Despite the shallow location of the highest parts of the ship, the quality of the wood was good during the survey. This was mainly due to the high ground water level, the fact that the topsoil had barely been worked and the lack of drainage pipes. During the excavation work, the starboard side bottom planks in particular had been pulled out and thrown to the side, and the keel beam was broken and partly moved, as well as a number of frame pieces were pulled out of the ground. Whether these frame parts had initially been more or less in the original construction context could not be verified. The dislodged V-shaped floor timber (Plate 19.I D) probably belongs in the front half of the vessel, about halfway between the foremost floor timber still present amidships and the beginning of the stem. Outside the stempost, which runs diagonally downwards and could be followed with the prods to a depth of ca 2.85 m below surface, wrecked wood was found to a depth of ca 1.3 m below the surface during the survey. Based on the position of the stem and the fact that the seabed originally probably resembled quicksand, it is not inconceivable that other parts of the ship may also still be present at greater depths.

Wreck formation

It is very likely that the cog perished during a heavy storm. The fact that the ship's hull is upside down and lies almost horizontally in the seabed in my opinion indicates that this must have been a violent event. The vessel must have capsized, and the mast with the spar and sail must have gone overboard immediately. Else, the hull would have ended up on the seabed with a list and then subsided. Another possibility is that, because of the flat and relatively wide bottom, the ship would have subsided into the seabed in more or less its normal position. It is rare for a vessel to be found completely upside down. An example is known from the Noordoostpolder of a fully clinker-built and relatively sharp/rounded late-medieval vessel. The most striking example is the Doel-I cog found near Antwerp (Chapter 20). As a result of lying upside down, the keel plank of this vessel sank down in the middle). Plate 19.I B shows that this must also have been the case with the Almere-II cog, before the keel plank was pulled into pieces and partially moved. The 'unnatural' position of the ship will certainly have contributed to the deformation and displacement of certain parts. However, in this manner, the hull could sink relatively quickly into the soft seabed. The stem hook, and possibly also the stern hook, probably protruded above the seabed at least initially, but perhaps also for a longer period of time. It was probably only (much) later, that the stem hook was pulled loose, perhaps due to heavy drift-ice or a trawling net, and ending up next to the ship (Plate 19.1 A).

19.2 The ship

19.2.1 Documentation

The parts excavated during the survey were measured and drawn (Plates 19.I A to E). Photographs were taken of the various situations.

19.2.2 Construction and layout of the hull

Despite the relatively small parts of the wreck that could be seen, quite a lot can be said about the Almere-II cog with relative certainty. Due to the fact that the wreck is upside down, this is obviously not possible with regard to the layout. However, a comparison of the data collected from the Almere-II wreck with those of the other cogs analysed in detail in this study made it possible to make a reliable reconstruction of a complete frame from the front half and from amidships. It also proved possible to reconstruct the front section of the keel plank with the stern hook and the stern post (Plates 19.II A and B), while various construction details and the type of the nails and sintel clamps could also be recorded.

Keel beam

What is particularly striking about this vessel is that it does not have a keel plank, as was found in the majority of the cogs, but has a 12 cm thick keel beam (that is four times the thickness of the hull planks). Approximately 7.5 m of the keel beam was excavated, with a number of pieces severely damaged by the digger (fig. 19.4). The largest width of the keel beam is 20 cm amidships, at the end of the scarf with the fore hook, this is 16 cm. The original total length from fore to aft hook could not be determined.

Stern- and stem hook

The stern hook was not discovered. The stem hook is completely on the south side next to the forebody (Plate 19.I A and figs 19.5a and b). The keel section is 2 m long. At the rear, this part has a sloping flat scarf with a length of 48 cm, which protrudes 2 cm above and below. At 73 cm from the end of the scarf, the upper line rises almost perpendicularly to a thickness of 20 cm. The upper line continues for 90 cm to the corner where the stem section and also the scarf for the connection to the stempost starts, it has a thickness of 25 cm at that point. The front of the keel section is rounded. Whether this is wear and tear or whether it has been intentionally made this way could not be ascertained. The highest point of the front side of the stem section now extends 58 cm above the keel line but is probably damaged. The original height probably was 67 cm. A 3 cm deep rabbet was cut into the sides, in which the garboards were nailed. Details such as the spacing of the nails and whether the seams between the stern hook and the garboards had been caulked on the outside were not recorded. The rabbet for securing the end of the garboard continues into the stem. The width of the hook is 16 cm at de scarf with the keel beam, and 14 cm at the start of the stempost. The cross-section is slightly trapezoidal with the smallest width at the bottom.

Fig.19.5a. The stem hook next to the foreship. In the foreground, the underside of the two floor timbers at the stempost. On the left, the lowest part of the stempost with the sloping flat scarf.

Fig.19.4. The excavated rear part of the broken keel beam.

Fig.19.5b. The forebody of the ship lying upside down, with in the foreground the underside of the two V-shaped floor timbers, with to the left and right of them the second and third bottom strakes, the ends of which are fixed in rabbets in the stempost. To the right of the foreship, the stem hook and between the hook and the foreship probably the lower (A) strake, which had come loose.

Stern and stempost
Nothing was found from the sternpost. The stempost probably is still in the position in which the hull sank into the seabed. It could be determined with the prods that the beam runs downwards to a depth of ca 2.85 m below surface, which means that it is still ca 3.5 m long. As far as could be determined, the beam is 14 cm wide on the inside, trapezoidal towards the front and has a flat front of 7 cm wide. It is certain that the lower three (bottom) strakes end in the rabbets that run through from the hook (fig. 19.6). They were fastened there with nails at every 10 cm (Plate 19.I B and fig. 19.7). The fastening of the beam to the scarf with the hook was also done with nails.

Rudder
Nothing was found of the sternrudder.

Bottom
The bottom of the cog is made up of the keel beam with three strakes on both sides of it (A to C). These strakes were placed carvel-fashion next to each other. Their straight upper line rises amidships ca 15 cm from the keel beam to the turn of the bilge. The angles increases to the fore (and to the aft), halfway up the foreward, it is 26 cm (Plates 19.I D and 19.II A and figs 19.8a and b). The widths of the three strakes amidships are ca 45, 44 and 44 cm (A to C) respectively, the thickness is 3 cm. The scarfs in the strakes are approximately 30 cm long, while the plank ends are 1-1.5 cm thick and protrude from below and above the shell, respectively. The total length from stern to stem can only be determined approximately (Section 19.3). The greatest measured width of the bottom is ca 2.9 m. The bottom was 2.5 m wide at the V-shaped floor timber from the foreward.

Fig.19.6. The foremost pieces of the bottom strakes at starboard, the two floor timbers and the lowest piece of the stempost with scarf.

Fig.19.7. The two floor timbers at the stempost and to the left of them the bottom planks of the starboard side. At the top (the dark piece of wood) you can see the flat scarf of the stempost and on it the (brown) nails of the fastening with the stem hook.

Fig.19.8a. Overview of the excavated underside of the midship section with the bottom strakes to port, the scarfs in the strakes, the sintel clamped moss caulking, the keel beam and the two floor timbers visible to starboard.

Fig.19.8b. The part of the underside bottom at port side with the sintel clamped moss caulking in the seams and in the repaired cracks in the plank ends, the wooden fastening pins and the keel beam.

Fig.19.8c Fragment of a plank with sintel clamped crack on the outside.

Fig.19.9a. The forebody floor timber found separately, with the futtock found separately placed against it. The two pieces do not belong together.

Fig.19.9b. The same.

Sides

The only thing that can be said about the sides is that the bilge strake (D) was placed at a fairly sharp angle (amidships approx. 130°) carvel-fashion against the C-strake of the bottom and that the upper part of the side had an angle of approx. 115°. The turn of the bilge becomes increasingly smooth forward and aft (Plate 19.II A). The futtock found dislodged (Plate 19.II E) indicates that the planks of the clinker-built sides were 35-37 cm wide, plus an overlap of 6 cm = 41-43 cm. Although the futtock does not belong on the dislodged V-shaped floor timber also found (Plate 19.I D), it is depicted on it on both sides, in order to arrive at a hypothetical frame reconstruction (Plate 19.II A and figs 19.9a and b). The slanted top of the futtock makes it absolutely clear that there must have been a second futtock.

Sintel clamped moss caulking

The seams of the bottom planks placed carvel-fashion against the keel beam and against each other as well as the seam between the C- and D strake or turn of the bilge were provided with sintel clamped moss caulking on the outside. In order to be able to do this, the edges of the planks were bevelled at about half the thickness of the planks, creating triangular openings of 1.5 x 1.5 cm. In the case of the clinker-built construction, these triangular seams had approximately the same dimensions, with the caulking carried out both on the inside and on the outside. The sintel clamps have a central section of 45-50 mm long. Peat moss *(Sphagnum* spec.) was used as a caulking. The moss laths are oval in diameter (1.5 x 0.5 cm) and made of oak. The dendrochronological study (Section 19.8 and Appendix), could date the construction of the vessel to a period that includes the end of type D and the beginning of type E. It thus seems very likely that we are dealing with a transitional type.

Frames

Two floor timbers were found in their construction context in the excavation trench opened amidships during the survey (Plates 19.I A and 19.8 A and 19.10). The broken piece of frame for the foremost floor timber may be of a narrower, third piece, but may also be a remnant of a rider. Unfortunately, it was not recorded whether this piece was discovered directly against the bottom (floor timber) or whether it lay deeper, that is on a floor timber. The two complete floor timbers are 13-14 cm thick and 21 and 24 cm wide. At the bottom, there are limber holes above the seams between the keel beam and the garboards and above the middle of the C strakes. The V-shaped floor timber found dislodged from the forebody is 25 cm wide and 12 cm thick. The limber holes in it were located above the seams between the keel beam and the garboards and above the B strake. The two V-shaped floor timbers at the bow are 22 cm wide and were (as usual) not attached to the stern hook. The frames were generally connected to the shell planks with two treenails (Ø 3 cm), sometimes they were attached staggered in relation to each other, in other places more in a straight line. The floor timbers and the V-shaped floor timbers were probably both attached to the keel beam with blind treenails.

Mast step

During the survey, the rear side of a piece of wood was recorded located right before the foremost floor timber amidships. It was located deeper than the frames (the horizontal U-shaped dashed line on the keel beam (Plate 19.I A). On the drawing from above, it is noted as 'end of mast step'. If this was indeed the case, then it must have been a short mast step. In order to visualise this in the reconstruction (Plate 19.II B) the mast step of the Marknesse cog (Chap-

Fig. 19.10. The port side of the midship section with the two complete floor timbers en the limber holes in them.

ter 12) was drawn in a dotted line. However, at the moment it is not possible to say with certainty whether this really was such a short mast step.

Ceiling
Although nothing was been established in this respect, it may be assumed that the Almere-II cog had an open ceiling, and at least (thick) bilge ceiling strakes in order to support the 'weak' scarfs between the floor timbers and the lowest futtocks. These were undoubtedly placed above the D-shell strakes. Above the bottom, in the middle between the mast step and the bilge ceiling, there may also have been a ceiling board. However, this was not seen between the two floor timbers amidships (fig. 19.10). It is therefore quite possible that a more or less loose ceiling was fitted between the two bilge ceiling strakes, in which the mast step had been integrated (for a comparison see the Spakenburg-I cog, Chapter 14). The ship undoubtedly had a number of ceiling strakes on the sides.

Crossbeams, transverse beams, deck beam, sail beam, decks, gangways, windlass and pump
The only available indicators that give some idea of the dimensions of the vessel are the stempost, which protrudes for about 3.85 m at a right angle above the keel line, the futtock, which extends up to ca 2 m above the keel line (Plate 19.II A and B), the maximum width of the bottom and the distance from the probable position of the mast to the base of the bow post. On the basis of a comparison with the other cogs in this study, it is possible to guess at how the construction of the vessel must have been above ca 2 m in height. In my opinion, this could have been done in two ways. One is the use of transverse beams protruding through the shell, the other is the use of transverse beams on the inside of the vessel. Plate 19.II A illustrates both versions, showing the possible construction with crossbeams though the shell on the port side and the other with transverse beams on the inside of the ship on the starboard side. The same applies to the possible position of the deck or decks. There may have been an entire deck from the aft to fore, but another possibility is an aft- and a foredeck, connected by gangways along the sides. The deck or decks were supported by the crossbeams and the deck beams. The ship was undoubtedly equipped with a windlass for hoisting and lowering the yard with the sail and one or two pumps.

Castles
Nothing can be said about the possible presence of one or two castles.

Repairs
The cracks on the underside of the ends of the bottom planks which had been sealed with sintel clamped moss caulking (Plate 19.I A and figs 19.8a and b) are the only repairs found. They were almost certainly carried out at the end of the construction.

Wood use
All the wood found in the cog is oak (*Quercus spec.*). The keel beam is the only one of which it was noted that it is made of half the thickness of a trunk. A clinker-built shell fragment, which was included in the collection of the Cultural Heritage Agency (RCE), shows that the planks were plain-sawn. The frames found are neatly edged.

Fasteners
The frames are fastened with 3 cm thick oak treenails. It was not recorded whether wedges had been struck into the treenails. Whether the pins for attaching the crossbeams had been fixed with thicker pins is uncertain. The mutual fastening between the clinker-built constructions is with 80-100 mm long nails with flat heads (Ø 30 mm) and a shaft thickness of 8x6 mm. They are applied every 15 cm. The moss caulking and the oak moss lathes were fixed every ca 6.5-7.5 cm with sintel clamps of the transition type D/E. The centre sections of the clamps are approx. 50 mm long.

19.3 Construction data

The data recorded during the survey provide only limited information about the original vessel. However, a comparison with the other cogs found in the former Zuiderzee area allows us to make some statements about its probable size and design. The largest measured width of the bottom is ca 2.9 m. The largest reconstructed width across the sides is ca 6.3 m. If there is indeed a short mast step with the position of the mast approximately 4 m behind the beginning of the stempost, then the bottom probably was 11.5-12.5 m long and the length of the keel beam plus both hooks was ca 12.2-13.2 m. The rake of the stempost is ca 120°, the rake of the sternpost will have been slightly less, so that a total length over the stern and stem at gunwale height is probably between 16.5-17.5 m. The straight lines rising to the sides of the bottom, and the relatively sharp angle at which the bilge or D-strake was fitted are very striking (amidships ca 123°). At ca 145° and 155° respectively, this angle is much less sharp in the Dronten-I cog (Plate 11.III C) and the Swifterbant

Table 1. Principal dimensions of the Almere-II cog

Length of the keel (including the stern- and stem hook)	ca 12.2-13.2m
Length of the bottom (between the rising part of the stern- and stem hook)	ca 11.5-12.5m
Greatest width of the bottom	ca 2.9 m
Overall length at gunwale level	ca 16.5-17.5 m
Overall width at gunwale level	ca 6.3 m
Height of the stern (keel line to top gunwale level)	ca 3.7 m?
Height of the stem (keel line to top gunwale level)	ca 3.9-4.0 m
Height of the side amidships	ca 3.2 m

Length-width ratio of the bottom	4.0 à 4.2:1
Length-width ratio at gunwale level	2.6 à 2.8:1

All measurements rounded to the nearest 10 cm

cog (Plate 18.I B). Based on the reconstructed sections of the crossbeams from the foreward and amidships (Plate 19.II A), the vessel (also) had relatively straight, sloping sides. The width of the strakes on the sides is very regular. The slanting top of the dislocated futtock, which had a stepped, flat, vertical part of only 17 cm high on the outside at the top indicates that the sides had at least six strakes from the bilge. However, in relation to the probable length and width of the ship, there must generally have been two futtocks on top of each other above the floor timbers. The height of the sides of the vessel amidships will therefore almost certainly have been more than 3 m. The remaining parts of the stempost indicate that it must have protruded at least 3.85 m above the keel line. All of this demonstrates that we are dealing with either a 'small' large cog with crossbeams protruding through the shell or with a 'large' small cog, whose crossbeams were installed inside the shell (transverse beams).

Characteristics
The Almere-II cog is very unusual in having a keel beam of 4x the thickness of the shell planks on which the frames were most likely blindly pegged. The keel beam is attached to the stern hooks with sloping flat scarfs. The ship has a relatively wide bottom consisting of three strakes, of which the straight upper lines run towards the bilges at an ever-changing angle (the bottom had no dead flat anywhere). The bilge- or D strakes were placed at a rather steep angle against the outer bottom strakes. The ship has the usual carvel-built bottom, which turns into a clinker-built construction to the aft and fore. The ends of the planks most likely ended in rabbets in the stern hooks and sternposts. The planks from the sides probably ended at the front of the stem, which is flat at the front. The ends of these planks were most likely covered and protected by a false stem. The frames were fastened with treenails 3 cm in diameter. This differs from most other excavated cogs, where the treenails were 2.5 cm thick. The seams of the clinker-built parts of the shell were protected by sintel clamped moss caulking on both the inside and the outside. The carvel-built seams were only caulked on the outside, with the triangular openings having a size of about half the thickness of the planks. The nails used in the vessel have flat heads and the sintel clamps used are of the transition type D/E.

Function
Of course, nothing can be said with certainty about the function of the vessel.

19.4 Model

No scale model of the Almere-II cog has been made.

19.5 Inventory

No finds were discovered during the survey that could have been part of the inventory or equipment of the vessel.

19.6 Cargo and merchandise

No remains of a possible cargo were found.

19.7 Ballast

This also applies to the ballast.

19.8 Dating

Soil profile
The soil profile above the wreck demonstrated that the ship must have sunk before the beginning of the 17th century.

Finds
As no finds were discovered, no indication could be obtained from them about the period in which the cog sank.

Sintel clamps and nails
The flat-headed nails are best known from 13th-century cogs. The sintel clamps used on this vessel were difficult to characterize as either type D or E. Thanks to the results of the dendrochronological study (see below), it later became clear that the construction of the vessel dated to a period that included the end of type D and the beginning of type E, making it very likely that the sintel clamps used on the Almere-II cog were of a transitional type.

Dendrochronology
Initially, two samples of the wood were taken for dendrochronological study, one from a dislodged piece of a shell plank and the second from the keel beam. Neither could result in a date. Subsequently, strips were sawn from three of the bottom strakes on the port side. These samples lack the (outer) finishing ring, so that the established data, including the estimate of the missing finishing rings, yielded *post-quem* dates of respectively after AD 1327 ± 6, after AD 1314 ± 6 and after AD 1310 ± 6. The reference chronologies used to obtain the dates are NLHist-2 and NLWestfalen. It is therefore very likely that the cog was built in the second quarter of the 14th century, while the north-eastern part of the Netherlands may be seen as the region where the construction took place. The research report of the dendrochronological study by RING is fully reproduced in the appendix.

19.9 Summary

The wreck of the Almere-II cog was not excavated completely during the 1999 survey and can therefore not be fully described. However, the parts that were exposed and those surveyed with prods provided sufficient information that - in addition to the details observed and recorded – it was possible to give a reliable description of the probable dimensions of the vessel, its shape and construction on the basis of comparison with the other cogs from the former Zuiderzee area. First of all, the vessel lies upside down, which is not often found. The vessel does not have the relatively thin keel plank commonly used with cogs, but a keel beam that is 4 times the thickness of the shell strakes. It is very likely that the vessel had a bottom without a dead flat anywhere, whose three bottom strakes on either side of the keel beam run up in a straight line to the bilges. The bilge or D strakes were placed carvel-fashion against the outermost bottom strake. They are at a fairly sharp angle amidships, while the sides probably have a fairly straight and outwardly inclined form of clinker-built strakes up to the top of the sides. The carvel-built bottom, which changes in the usual way to a clinker-built construction towards aft and fore, the use of hooks and a raked stem demonstrate that this vessel is indeed a cog. The reconstructed lengths, widths and heights indicate that the vessel can be described as a medium-sized variant.

The main estimated dimensions are:

- Length over the stern and stem at gunwale height — ca 16.50-17.50 m
- Largest width at gunwale level — ca 6.30 m
- Height of the stern (keel line to the top gunwale) — ca 3.70 m?
- Height of the stem (keel line to the top gunwale) — ca 3.90-4.0 m
- Height of the side amidships (keel line to the top gunwale) — ca 3.20 m

On the basis of the available information, it is not possible to say with certainty whether this vessel was a 'small' large cog, with crossbeams protruding through the shell or a 'large' small cog with crossbeams fastened to the inside of the hull. The sintel clamping of the seams both on the inside and on the outside indicates that the vessel must have been built before ca 1350. At first, it could not be determined whether the type of sintel clamp used on the vessel could was type D or E. However, the results of the dendrochronological study showed that the vessel was most probably built in the second quarter of the 14th century, which indicates that the sintel clamps must have belonged to a transitional type between D and E. The construction site of the cog is likely to have been in the northeastern Netherlands. No finds that could have belonged to the inventory or equipment of the vessel were found in the wreckage. The location of the vessel in/under a partly artificial island in a swampy part of the Oostvaardersplassen probably provided fairly good conditions for its conservation. From the point of view of increasing the knowledge about the cog, it would be particularly intriguing to determine whether the vessel is a 'small' large cog or a 'large' small cog and whether the sides and the afterbody are still present in deeper layers.

Karel Vlierman November 2019

The Doel-I cog[1]

20.1 Find conditions

20.1.1 Discovery, salvage and investigation

During the excavation works for the construction of the quay walls for the western tidal container dock (known as the Deurganckdok) near the town of Doel in Flanders (Belgium), a shipwreck was discovered in 2000. The Archaeological Department of Waasland (ADW), which was consulting on the excavations, had already found and measured a fragment of a ship's timber in early September. A short time later, an upside-down wreck was found 100 m towards the Scheldt. While digging a drainage ditch in the construction site, a significant chunk had been removed from the intact hull of the ship, which represented approximately 20% of the total hull volume (fig. 20.2 and Plate 20.I). This damage could have been avoided, or at least limited. The silted-up channel in which the shipwreck was found is known as the 'Deurganck' or 'Grote Gat', and served as a drainage channel for the southern part of the Doel polderlands until about 1980 (fig. 20.3 and 20.4). The current course (which predates 2000) returns to a late medieval predecessor, the 'Tonnekin', which was listed in 1257 as the border between the parishes of Kieldrecht and Kallo.[2]

Due to the complete lack of knowledge and expertise on nautical archaeology in Flanders, the Netherlands Institute for Ship and Underwater Archaeology (NISA), part of the Dutch National Service for Archaeological Heritage (ROB), was asked by the ADW to come and take a look and offer their advice on how to handle the find. During their first visit on 28 September 2000, A.F.L. van Holk and R. Oosting, staff members of the NISA, recommended excavating the wreck – very likely a cog – to free it fully from the soil. In the first two weeks of October, the ADW did so using a crane. By that time, the ROB had sent a letter stating that the find should be considered to have European significance, making it absolutely necessary to ensure comprehensive documentation and conservation *ex situ*.[3] The conclusion was shared by nautical archaeologists from some twenty-five countries when I announced the find and showed several slides of the shipwreck during the 9th International Symposium on Boat and Ship Archaeology (ISBSA 9) in Venice on 7 December 2000. The NISA advisory report was sent to the ADW in October. The Institute for Archaeological Heritage (IAP) notified the Flemish Government, in this case represented by Minister L. Sauwens (Flemish Minister of Local Authorities, Spatial Planning, Housing, Monuments and Landscapes), who immediately visited the findspot; his Ministry then decided to preserve the cog.

During my first visit to the site on 16 October, I was asked by R. van Hove, director of the ADW, the body responsible for the excavation, with the approval of the then director of the IAP (Professor G. de Boe), to consult on the salvage, to oversee the process, and to assemble the required documentation of the field survey *in situ*. The compiled data could be used and published in the present study, which was already underway at that time. On that occasion, I offered the ADW my services in writing the nautical archaeology technical investigation report, if they were interested. The ROB enabled me to produce the report requested by the ADW, in the interests of the find and my research.

Fig. 20.1. Artist's impression of the Doel-I cog from the Deurganck dock; and the civic seal of Elbing, ca 1350.

Fig. 20.2. Overview of the excavated port side of the afterbody with stern hook, crossbeams protruding through the shell with beam-head guides, and the damage to the ship's hull caused by the digger, 28 September 2000.

1. Translated by Joy Phillips.
2. Van Hove 2005, 51,
3. Letter from the Dutch National Service for Archaeological Heritage (Rijksdienst voor het Oudheidkundig Bodemonderzoek, ROB), reference NISA 000148, dated 6 October 2000.

Fig. 20.3. M.G.I. Topographical map of Belgium, map sheet 7/6, ca 1975 with the findspots of the cog and the submerged late medieval settlement. Based on Van Hove 2005, 50.

The shape of the ship's hull, the carvel-built bottom, the clinker-built sides, the protruding crossbeams with their beam-head guides, the presence of a stern hook and stem hook, a straight stempost, and the sintel clamped moss caulking on the shell seams clearly confirm that this was definitely a large cog (probably Hanseatic). The marks left on the shell from the butterfly clamps (all iron nails and sintel clamps had dissolved completely over time from exposure to salt water and groundwater) had already clearly shown that the vessel had almost certainly been built before ca 1350.[4] The fully excavated shell of the vessel was sprayed clean several times using a steam cleaner to remove the rusty, caked, somewhat cemented layer of sand and clay with plant and shell fragments. The shell was then described, and sketches, photos and slides were taken of the various surveys and details. Working with Ms A.E.M. Hanraets from RING, the most suitable locations on the hull for taking tree-ring samples with a saw were marked, in coordination with the disassembly plan.

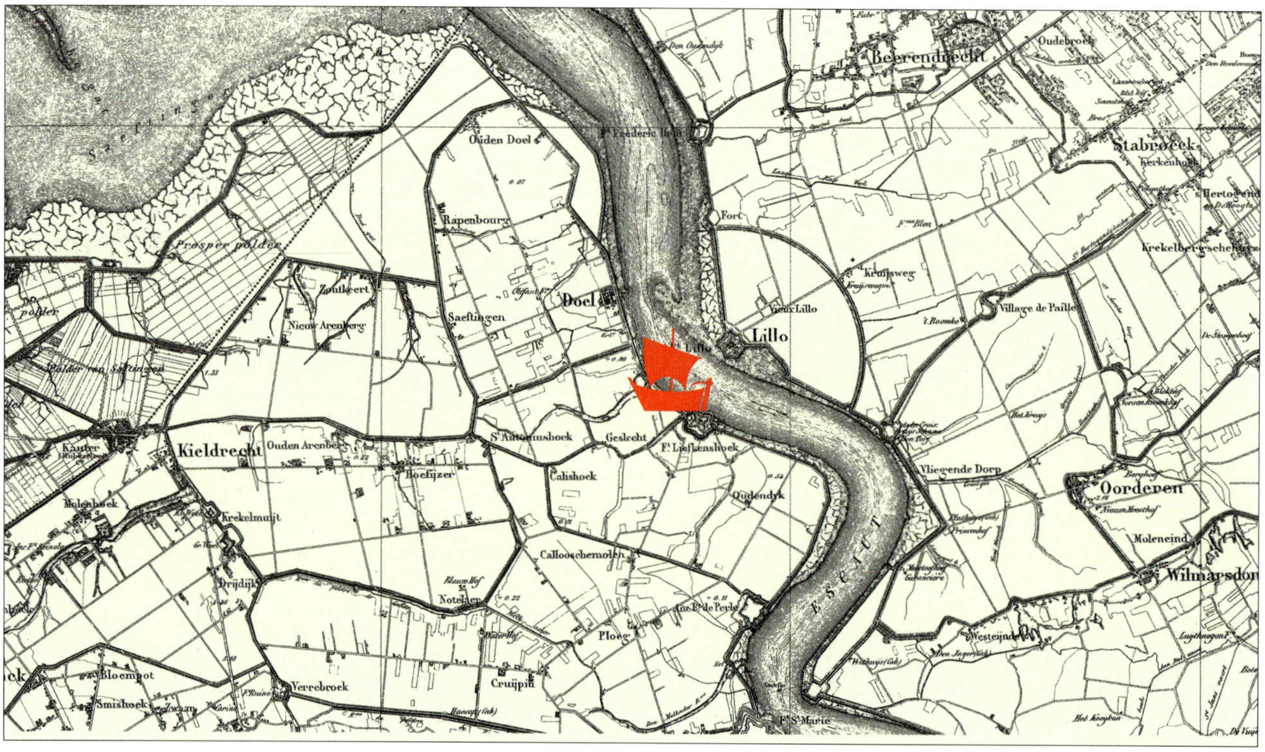

Fig. 20.4. Topographical and Military Map of the Kingdom of the Netherlands (scale 1:50,000) from 1861 with the location of the Doel-I and Doel-II cogs.

4. Vlierman 1996a, 58 table II, type D2.

During the period that the various types of work were taking place and the ship sections were being salvaged, the shipwreck was covered in ground-cover membrane whenever and wherever no work was being done. The ground-cover membrane was regularly sprayed with water to keep the wood from drying out. The pieces that were removed and salvaged were placed on rigid wooden pallets built to size in advanced, wrapped in ground-cover membrane, and then covered in plastic sheeting. During the salvage period, the pallets were stored in close proximity to the site, and the packaged wood was kept wet (fig. 20.5). By 15 December 2000, in accordance with the arrangements made with the Project Management of the new dock (Deurganckdok), the entire shipwreck had been dismantled and removed. From January 2001 on, all the pieces of the cog were stored under water, resting on 115 pallets of 4.5x1m in 30 iron containers of 5x2.5x1m.

Thanks to the highly professional approach and dedication of the ADW with support from the IAP, making grateful use of the years of expertise contributed by the NISA, it was possible to complete this enormous undertaking on schedule, despite the limited time frame and generally poor weather conditions. The discovery of the cog attracted widespread media interest in Belgium and internationally. At the express request of the Minister, the findspot was opened to the public on 22 October 2000. Public interest was massive; an estimated 12,000 visitors came to the site. In addition to the written records and the sketches, photos and slides that were taken, the position of the hull was recorded using a 3D scanner (fig. 20.6).[5]

After the documentation, it was time to start disassembling the shell planks. One incidental benefit, which saved a great deal of time and (otherwise unavoidable) damage to the wood, was the fact that all the iron fastenings had dissolved. The only resistance that might be encountered while taking the shipwreck apart was the wooden treenails, which held the shell planks to the frames. However, given their condition, they were not anticipated to cause any significant problems. That proved to be the case during the implementation (fig. 20.7). Once the sediment had been removed from the spaces between the exposed frames (fig. 20.8), the overall pattern of the frames was documented. Because the shipwreck had been upside-down, the clinker-built strakes of the sides, which had been added last when the ship was being built, were now lowermost. As a result, it was necessary to start disassembling from below, working upwards, and from stem to stern, due to the overlapping scarfs. The tilted starboard side (the wreck had keeled over) was removed first, followed by the port side, and finally the planks of the bottom. The heavy hooks, the keel plank, the stempost and all the floor timbers, the keelson and the large futtocks were lifted from the wreck with a crane on 29 November; the heavy crossbeams were removed a week later. The sediment from the hull was scraped down with shovels to salvage any loose pieces or objects of any other nature undamaged.

Fig. 20.5. Overview of the excavation works by the Deurganckdok with the wrecked cog by the red container and the white wagon. The ship parts that were already salvaged and packaged on pallets are in the foreground.

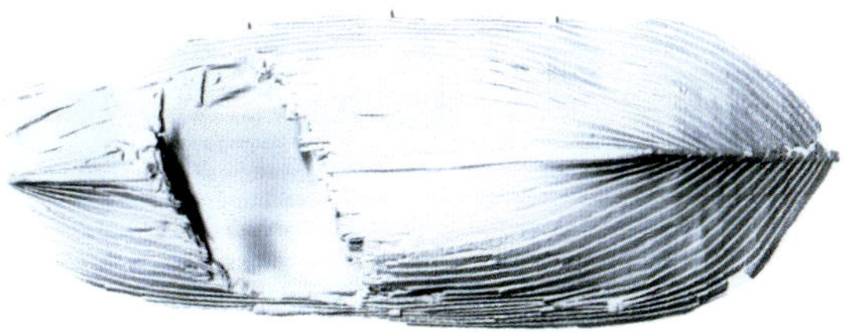

Fig. 20.6. A milled scan model of the Doel-I cog, showing the carvel-built bottom, the keel plank, stem hook and stern hook, the damage caused by the digger, the clinker-built shell strakes of the sides, and the crossbeams with beam-head guides protruding through the shell. Based on Van Hove 2005, 54.

5. 3D-Solutions & Technics, Temse/Bornem.

Fig. 20.7. Manual removal of the numbered shell planks on the starboard side.

Fig. 20.8. Removal of sediment between the frames on the port side.

On 14 December 2000, a study day entitled 'The Doel Cog immortalised: possible applications of 3D scanning in archaeology' was held in the Flemish town of Sint-Niklaas. Most of the draft investigation report on the cog had already been written in the evenings during the consultation on the disassembly and salvage. In the media, 'armchair experts' objected to piecemeal salvage; critical comments swelled to near-mythic proportions (e.g. about the conservation status of the cog).[6] Opinions were expressed that the ship's hull could perhaps be lifted out intact. However, this was impossible, due to the absence of any iron nails and the poor quality of the treenails holding the planks in place, many of which had already broken under pressure from the hull. Effectively, the shell strakes were lying more or less loosely on the frames.

Starting in January 2001, the ADW consulted with the Flemish Government and various local authorities to find a suitable location in the Waasland area, and to find funding for a building and for the entire subsequent process involving the Doel cog, which would take many years. As the curator and the authority that held title to the find, the ADW wanted me to lead the process of documenting the parts of the cog, in the capacity of consultant and researcher; to familiarise the employees who would be working on the project with the fundamental principles of nautical archaeology; and to work with the ADW staff to publish the archaeological/technical investigation report.

Upon request, a draft master plan and budget were drawn up for a new warehouse set up to accommodate the work, where all 33 containers with cogwood could be stored.[7] In this warehouse, which could for example be built at Fort Liefkenshoek by the Scheldt, just a short distance from the location where the -later- two shipwrecks (see Chapter 21) were found, the parts could be cleaned and documented, the wood could be conserved, and the ship could eventually be reassembled/reconstructed and put on display. The plan included an option for keeping the public updated on the progress of these activities from the outset. Financial approval from the then Flemish Minister of Public Works, Energy, the Environment and Nature (Kris Peeters) to have the Maritime Access division of the Waterways and Maritime Affairs Administration (AWZ) provide funding for Phase I of the next steps for the two cogs in 2005 meant that progress was (finally) being made after the salvage. This ensured that the scientific evidence and marine engineering details were secured. 24 months of full funding were allocated to this phase of the project, to be spread out over a period of 2–3 years.[8]

However, archaeological policy in Flanders proved less stable than expected. The Flemish Government (specifically the IAP, by then known as the Flemish Heritage Institute (VIOE)), decided that this work could not be done by the ADW in conjunction with a private individual, and that there should in any case be an accredited institution overseeing the undertaking. An American university was contacted and the funding already allocated for the implementation of Phase I was redirected to commission an action plan. The plan did not differ in any essential way from the proposal I had already submitted in the draft master plan.

In the end, the containers of cogwood stood outside exposed to the elements for 10 years. The wood deteriorated in quality, with the likely consequence of a much more expensive conservation process, should it ever come to that.[9] Moreover, R. van Hove, director of the ADW and the driving force behind the cog project, grew weary of the constant obstruction and relinquished the management and scientific rights on behalf of the ADW. In the second half of 2010, others in Antwerp started on documentation of the ship sections as well as the conservation study. Remarkably, the responsible authorities and the persons designated to do this work have to date made no effort whatsoever to obtain access to the field data I recorded, which I believe is essential for the elaboration, reconstruction and description of the cogs. They only have the numbered pieces of wood. Rudiger van Hove will not be distressed by this oversight, since he sadly passed away on 15 August 2011; a great loss to Flemish archaeology, and in particular for the Waasland region.[10]

20.1.2 Geology and stratigraphy

During the excavation works in the Deurganckdok, observations were carried out and recorded that could refine and supplement the overall impression of the geomorphology and paleo-landscape of the lower Scheldt region. Chronologically, this involves a period from 65,000–25,000 years ago (the last phase of the Weichsel Glacial

6. Van Hove 2005, 51.
7. Vlierman, February 2003.
8. Van Hove 2005, 50-69.
9. Vlierman, 21 September 2006; lecture during the 2nd International Colloquium on Maritime and Fluvial Archaeology in the southern North Sea area. 21-23 September 2006, Bruges. The cog finds at Doel. Book of Abstracts. Ostend, 2006, 29-32.
10. A joint publication by the ADW staff and myself is no longer likely to happen, unfortunately. The field data I collected would be more than enough to write a detailed investigation and salvage report. However, in the present and the next chapters, I merely provide a number of illustrations, a preliminary reconstruction of the lines, and a concise text on the field surveys in situ and what I can say 'off the cuff' regarding those matters and the two cog wrecks.

Stage) until the Late Middle Ages.[11] The shipwreck was found upside-down in the late medieval channel known as the Tonnekin. Figure 20.9 shows the cog in the foreground looking like a beached whale. Opposite from the dock section that still needed to be excavated, the soil profile can be seen, exhibiting alternating dark and light sections (with the Doel-II cog within it). The top layer of sand had been artificially levelled-up. The fill was probably partly deposited by a catastrophic event. The cog was found at approx. 6-7 m below sea level.

20.1.3 Prehistoric and medieval habitation

Various traces of prehistoric and medieval habitation were found in the subsoil at the construction site for the new dock (Deurganckdok). The key details from a short summary of the various finds will suffice here. It concerns a number of significant Stone Age sites, late medieval field boundaries, and a submerged settlement. We can reference the remains of an Upper Palaeolithic community of hunter-gatherers from the Federmesser industry (dated around ca 12,000 BP), two concentrations of flint artefacts from early Mesolithic hunter-gatherers dated at 8830 and 8860 BP respectively, late Mesolithic early Neolithic habitation which shows ties to the Swifterbant culture of the western Netherlands from ca 5600-5400 BP, and a site with finds showing similarities to the Michelsberg culture of central Belgium. These finds can be dated ca 5100 BP.

Late medieval spatial features have been recorded dating back to the second half of the 12th century to the early 13th century and from the late 13th and early 14th century. Rows of closely packed alder branches may indicate dike structures from the 13th and 14th century. About 700 m south of the town of Doel and approximately 400 m north of the Tonnekin (Deurganck), a settlement was found that was probably formed in 1407, and was lost during the St. Elizabeth's flood of 1424. The settlement itself is unknown, but clearly served some purpose related to water, considering its location along the Scheldt (fig. 20.3) and the orientation of the parcels along a (navigable) canal about 20 m wide. Despite the fact that a great deal of evidence was lost at this site, it has yielded a wealth of data on dates for early-15th-century ceramics, since it

Fig. 20.9. The overturned Doel-I cog, the soil profile as observed in the part of the Deurganckdok being excavated, with the nuclear power plant in Doel along the Scheldt River in the background. The Doel-II cog was slightly to the left below the power plant, in the light-coloured dip in the ground.

11. Van Hove 2005, 59.

was deposited in the period from 1407-1424. The pieces from the flood layer offer a perfect impression of the range of pottery used in 1424.[12] A bit farther away, there were various villages in the area around the time that the cog sank. The late medieval city of Antwerp was about 17 km upstream on the Scheldt.

20.1.4 Condition and wreck formation

Condition

The 'unnatural' position of the cog caused significant deformation of the ship's hull, although that may not be immediately apparent at first glance. A straight keel line should be envisioned from the front tip of the bottom of the stem hook (at the rightmost corner of the container in fig. 20.10) to the back of the stern hook (the corner protruding farthest behind the person in the green jacket). This means that the keel plank has sagged 0.5-1 m amidships (hogging). The wavy shell strakes, which seem very narrow in some places, already indicated in an early stage of the field investigation, along with the lack of connecting nails, that they had also collapsed into each other and warped. After removing the shell strakes, this did indeed prove to be the case, at least in part. Moreover, even the futtocks of the frames had shifted over and along each other (fig. 20.11). All this meant that no solid structures were left, not even in parts of the vessel.

The seemingly intact hull was therefore solely due to the mound of sand on which the shipwreck rested, once the vessel had been fully exposed. The wood also proved uneven in quality across the wreck, and some planks were cracked or broken due to the deformation of the hull. The wooden treenails that attached the shell planks to the frames had lost structural integrity and had already broken off in many places. While cleaning the hull, it had already become apparent that the outside of the shell must initially have been liberally covered in tar, more than has ever been seen in the hundreds of shipwrecks in the IJsselmeer polders. After disassembling the shell planks, it became clear that there was an even thicker layer

Fig. 20.10. Overview of the starboard shell of the wreck with the bow in the foreground, the stem hook protruding high above the sagging keel plank with the lower section of the false stem, the crossbeams with beam-head guides protruding through the shell, and the 'wavy' sagging shell strakes.

12. Van Hove 2005, 57 and 58.

Fig. 20.11. The starboard side amidships after the shell planks of the side were disassembled and the frame sections were numbered. To the right, the head of a crossbeam protruding through the shell. Futtocks shifted away from each other can clearly be seen. On the third frame from the right, the floor timber has slid more than two plank widths over the bottom futtock.

on the inside of the shell, and the frames had also been treated in some way. The odour of tar was still distinctly present. Further investigation will have to show exactly what substance(s) or tar had been used. Besides the recent damage due to the crane, the top four strakes to starboard are almost entirely missing, as well as the top seven strakes on the afterbody. Nearly all the strakes on the port side of the forward half of the ship are still fully intact. On the afterbody, the top five strakes are missing, as well as pieces of the sixth through thirteenth strake near the stern plus most of the sternpost, the entire false stern and the rudder.

Wreck formation

The fact that the shipwreck was upside-down obviously immediately raises questions. Such a situation has only been observed a few times in the IJsselmeer polders, among the nearly 450 registered shipwrecks discovered to date. In all cases, it concerned hull shapes that were more or less comparable, and all of them were also late medieval wrecks. The Almere-II cog (Chapter 19) dating from the same period is one example. The dimensions of the Doel-I cog, and particularly its height, raise some questions about how such a vessel could have turned turtle. It is almost impossible to imagine any other explanation than that the ship was most likely not carrying any cargo and/or ballast. The cog may have been tipped over fairly easily by a wall of water in the late medieval branch of the Scheldt (the Tonnekin) during a high storm tide, due to its shallow draught and high sides rising well above the waterline. Even if there were ballast stones on board, this may still have happened, and possibly even sooner. If the stones shifted, the uneven ballast may have even made it easier for the cog to capsize. The massive crossbeams and the knees placed above them (the latter were not found), may also have contributed to the (further) capsizing of the cog due to the weight they added in the upper section of the ship. It could not and cannot be established whether the cog capsized at or near the findspot. It may also have happened somewhere upstream or downstream from that location.

The shipwreck was found in the northern section of the Grote Gat (fig. 20.3 and 20.4) and was probably situated in a curve of the late medieval channel. This surmise is supported by the soil structure and the presumably 'stable' anchoring of the top of the stempost in an area where the channel grew shallower, which probably held the wreck in place at the location where it was found. The pressure caused the stem hook to slide along the lower end of the stempost. The ship was orientated nearly north-south, with the stem facing south. The hull was probably intact when the disaster occurred. The wear and tear observed in various areas and the almost complete lack of tar on the outside of the shell, especially on the starboard side, makes it likely that this deterioration was primarily caused by the abrasion of the sediment-heavy, seashell-laden waters of the Scheldt as the tide came in and out over a long period of time – probably decades. The afterbody, which was orientated towards the deeper part of the channel, would in part have regularly been free of the channel bottom during this period. As a result, fish could swim freely under and inside the wreck. A number of flatfish skeletons were found in the enclosed hull. During the wreck formation process, the first parts to come loose and wash away were probably the false stern and false stem. After that, pieces of the shell planks on the afterbody would have gradually been lost, as well as the futtocks and floor timbers attached to those planks, up to and including the two most aft situated protruding crossbeams, the knees on all crossbeams, and most of the sternpost. Any interior timbers and decks would also have been washed away.

20.2 The ship

20.2.1 Documentation

Before starting disassembly, the outside of the shipwreck was first described. Sketches, photos and slides were taken of the various structures, peculiarities and details. The entire hull was measured with a 3D scanner to record the field survey as it was found. The drawings and milled scale models to be produced from these scans could help in the reconstruction of the vessel.

20.2.2 Construction and layout of the hull

The exceptional position of the Doel-I cog made it possible to examine the entire exterior of the shipwreck, a situation that almost never occurs. In the IJsselmeer polders, nearly all shipwrecks are (by sheer necessity) excavated, and then measured and sketched on the inside. It has only been possible to examine part of the exterior of a few ships, and only if they had been keeled over. The full exterior of a cog with all its details had not been seen previously, and was exceptionally valuable to offer a clear impression of what these striking ships would have looked like. This was the direct reason for depicting the sintel clamped moss caulking as realistically as possible on the drawings of all the cogs in the plates to this study. The position of the Doel-I cog was why it was only possible to state anything about the frames with any certainty after the shell planks had been removed; the presence of crossbeams could be deduced from the ends protruding through the shell.

Keel plank

The keel plank was attached to the stern hook and stem hook with nibbed scarfs. The collective length from stern to stem was approx. 14 m. The keel plank was 23 cm wide and 12 cm thick aft, and 32 cm wide and 14 cm thick to the fore. Despite the thickness of 12-14 cm, it can still be considered a keel plank rather than a keel beam, since it is only 2.5-3x thicker than the shell planks.

Fig. 20.12. The frames of the afterbody (port side) with the stern hook and rabbet for attaching the first two bottom strakes and the rest of the (cracked) stern post.

Stern hook

The stern hook (fig. 20.12) broke off at the scarf with the keel plank when the drainage ditch was being dug. It was 4.1 m long from aft to fore. Its width was 23 cm at the scarf with the keel plank and 14 cm aft. There was a triangular heel protruding about 10 cm along part of the width of the afterbody, which was once connected to the false stern. The rising stern had a back line at an angle of 118° compared to the keel line and was approx. 1.1 m long from there. Rabbets were hooked into the sides and nailed to the ends of the A and B strakes and the lower 5 cm of the C strakes. The remaining parts of the final strakes continued until the back of the sternpost. The stern section had a flat scarf connecting it to the sternpost.

Sternpost

The sternpost was attached to the stern hook with a long sloping flat scarf. The bottom rested on a triangle cut out of the stern section. The rest of the post, approx. 1 m long, was broken off at the top and there were cracks at multiple locations along the length from the pressure (fig. 20.12). The post was 25 cm wide on the inside, 12 cm on the outside at the top of the scarf with the sternpost and 4 cm wide 55 cm above that.

False stern

The false stern is missing, so it is not possible to state anything about its precise shape and dimensions. Similarly, there is no data on the number of gudgeons for the suspension of the rudder. For the possible shape and execution of the entire stern construction with the rudder, see the Nijkerk-II cog (Chapter 6, Plate 6.V A).

Rudder

The rudder was not found. Similar to all other cogs, it can be assumed that the vessel had a stern rudder.

Stem hook

The keel section of the stem hook was 3 m long from the scarf with the keel plank to the bow. Its width was 32 cm at the scarf and 15 cm at the front. The keel section, like the keel plank, had a rectangular cross-section; the lower edges were significantly worn. The forward 50 cm of the bottom had been worn down, with the wear increasing to as much as 10 cm towards the bow. This was probably the result of beaching the ship many times, as was also found in the Nijkerk-II cog (Plates 6.V A and B). The stem section must have been 1.3 m at the front, measured from the keel line. The rake was 124°. Wide rabbets had been cut into the sides, with the first two bottom strakes nailed into them (fig. 20.13).

Fig. 20.13. The port view of the stem hook on the V-shaped timbers and the shifted and broken lower section of the stempost.

Fig. 20.14. Overview of the frame pattern from forward to aft, with the stem hook and stempost still in situ.

Stempost

The stempost was placed against the back of the stem section of the stem hook with a sloping flat scarf about 1.5 m and attached to it with a wooden treenail (Ø 2.5 cm) plus a number of nails. The pressure pushed the stem hook approximately 35 cm over the thin lower end of the stempost, causing the lower part of the stempost to bend and break. The stempost was approx. 6.8 m long. The width was 23 cm right above the scarf with the stem hook, 27 cm in the middle, and 25 cm about 40 cm from the top of the gunwale. From there, the back line of the post ran forward and upward, and was 18 cm thick at the top. The post was 29 cm wide on the inside. The front had been chamfered at the corners. At the bottom, this had been done across a thickness of 10 cm (the extensions of the rabbets in the stern hook). Towards the top, the front became increasingly roof-shaped and the diagonal ends became lower until it was nearly flat at the top (fig. 20.14). The ends of the shell strakes were fastened to this. The total length of the stempost plus stem hook, measured from the keel line, was approx. 7.3 m. In the lower half, there were in any case one, possibly two (completely rusted away) iron bolts (Ø 30 mm) and a wooden treenail (Ø 2.5 cm). In the upper half, there were two wooden treenails (Ø 4 cm) and an iron bolt (Ø 30 mm), which would have held the false stem in place.

False stem

Towards the bow, against the stem hook, the lower section of the false stem was still present. This is due to the fact that the section, 1.15 m long and 18 cm thick, had originally been nailed onto the stem hook and the garboards or A-strakes with two iron fittings. These fittings, like all the other iron, are no longer present. However, the marks left on the wood show that the iron plate was approx. 60 mm wide. The T-shaped ends were 260 mm high at the top and 330 mm high at the bottom, and 80 mm wide. They were attached with nails. The bottom of the false stem started about 14 cm above the keel line. The width was 14 cm at the front, 15 cm at the bottom against the stem hook, and 16 cm at the top. The top of the lower piece of the false stern had been cut at an oblique angle, for the scarf with the top part of the (missing) false stern. It is unclear whether the piece that was found is a repair – in other words, if the false stem was initially constructed from a single piece, which seems very plausible.

Bottom and sides

The cog had eighteen strakes (A through R) on either side of the keel plank and stern and stem structures. Figure 20.15 and plate 20.I show what was found of those. The strakes on the bottom were not straight, but irregular in width (31-55 cm). The strakes on the sides were remarkably narrow (34-40 cm); on the visible part of the clinker-built sections of the bottom strakes that can be seen from the outside, another 8-16 cm must be deducted for the overlap, and 7-9 cm must be deducted from the side strakes. The strakes also varied in thickness, even within the same strake (3.5-6.5 cm). In general, the thickness of the bottom strakes varied between 5 and 6 cm, while the thickness of the side strakes varied between 4 and 5 cm. The scarf joints between the strakes were 55-85 cm long.

Bottom of the ship

The bottom of the ship consisted of the keel plank with the stern- and stem hooks, with four strakes (A through D) on either side of that. In the carvel-built section, the bottom was not flat at any point, and the first three strakes (A through C) ran in one straight line from the keel plank to the turn of the bilge. The fourth strake (D) was also carvel-built against it, but at an angle. Forward and aft of that, the

Fig. 20.15. The Doel-I cog in the Deurganckdok during the visiting day for the public on 22 october 2000. Photo ADW..

strakes transitioned from carvel-built to clinker-built. The ends of the A- and B strakes were nailed into rabbets in the stern- and stem hooks; this only applied to the lower part of the C-strakes. The rest continued diagonally until the middle of the outside of the stern- and stempost. The strakes on the bottom were composite of two planks. They were disparate in shape and had been fitted together like puzzle pieces. The strake ends at the scarfs protruded from both sides of the shell, inside and outside. The ends were about 4 cm thick. The wide overlap of the planking in the clinker-built sections (7-16 cm) was worth noting.

Port side

The port side had been preserved best. The forebody of the ship was almost entirely present up to the top of the gunwale (Plate 20.I). To the extent that it can be stated with certainty, the shell strakes E through R were composite, made from two or three boards. As already stated, the overlaps of the planking were surprisingly wide (7-9 cm), and as much as 13-16 cm for the Q and R strakes. At the stem, the planks ended on the roof-shaped front of the stempost. In principle, this applied to the upper twelve strakes, positioned in rows of two, alternating between starboard and port. Where the planks ran from port to the starboard side of the stempost, the corresponding planks from starboard were placed over them and continued to the port side of the stempost. The overlapping ends of the planks were chamfered and were no thicker than 0.5 cm at the tips. This striking construction has not been seen before. The structure was designed to create a front surface that was as flat as possible in order to attach the false stem on it. Based on the marks on the plank ends, the false stem was approximately 30 cm wide (fig. 20.16). The more or less trapezoidal opening in the upper two shell strakes near the stem post (between frames 34 and 35) may have been where a gangway was extended. It may also have been intended for throwing out and retrieving the anchor (fig. 20.17). A similar opening was also found in a more or less comparable location on the Nijkerk-II cog (Plates 6.V A and B).

Three heads of the protruding crossbeams could be seen in the M strakes, and another could be seen in the forebody through the N strake. In the afterbody, a fourth beam very probably protruded through the M strake, and a second through the N strake. Between the two forward crossbeams to port, there was a rectangular hole about 14x11 cm in strake L (between frames 27 and 28). It was presumably where the end of a drainage channel opened to the outside, to remove the pump water from the ship. For comparison, see the possible execution on the Nijkerk-II cog (Section 6.2.2 and Plates 6.IV A and B [cross-section of frame 2] and Plate 6.V A). It is surmised that the location where the drainage channel previously penetrated the shell was between frames 26 and 27.

Starboard side

The upper two strakes (Q and R) are almost completely gone on the starboard side. When the shipwreck capsized, they probably snapped off the frame pegs and were washed away. Beyond that, it is sufficient to refer to the details provided above on the port shell. There was no indication of a drainage hole for pump water on the starboard shell.

Caulking method: sintel clamped moss caulking

Because the vessel was upside-down, it was possible for the first time to see a clear (and nearly complete) overview of how and where the seams of a cog are caulked on the outside with sintel clamped moss caulking. On most of the spaces between the bottom strakes, about 0.5 m of the aft clinker-built connections were not sintel clamped, and about 1.5 m of the forward connections. The carvel spaces along the bottom were approx. 3 cm wide at the lowest

Fig. 20.16. The front of the ship with the striking, overlapping plank ends on the starboard and port side. The mark left by the missing false stem can clearly be seen (no brown tar residues). The lump of rust slightly above the middle is one of the iron bolts used to attach the false stem.

Fig. 20.17. Overview of the port side with the damage caused by the crane, the heads of the crossbeams protruding through the shell, two of them with beam-head guides, and the trapezoidal opening in the gunwale of the forebody.

point. The adjoining planks had been partly chamfered, creating a triangular opening. These openings were filled with strands of moss caulking, covered with oval-cut oak moss lathes (2.5x0.8 cm) and held in place with sintel clamps, type D2/(E?). Starting from the clinker-built transitions between strakes E and F until strakes J and K on the outside of the starboard side, the transitions between the strakes were not sintel clamped, except for a few points near the stern- and stempost. On the port side, this was the case up to and including the transition between strakes I and J. The rest of the spaces, up to and including the transition between strakes Q and R, had been completely caulked with sintel clamped moss caulking, except for a few points near the stern- and stempost. Clamped moss caulking was also used around the heads of the crossbeams protruding through the shell. No caulking was used on the cross-cut plank ends at the scarfs. This also applies to the cross-cut plank ends that were held in place in the rabbets in the stern- and stem hook. To the extent that there was opportunity to do so during the salvage, the loosened shell planks on the inside were examined (fig. 20.18). It is probable that all the spaces between the clinker-built planking were sintel clamped on the inside. A layer of peat moss (*Sphagnum spec.*) had been applied between the overlapping planks at the scarfs.

Fig. 20.18. The inside of a recently disassembled shell plank with the imprints of two futtocks, oak moss lathes and moss caulking. The location of the sintel clamped moss caulking also indicates the width of the overlap (right).

Frames

The frames provide the basic structure to hold the hull in shape. Along the bottom of the ship, that structure is provided by the floor and the V-shaped floor timbers and planks, which rise higher along the sides, in principle alternating between starboard and port. Futtocks were attached to them with flat sloping scarfs. Some are only attached to one, while a lower and upper futtock are attached to other frames (figs 20.19a, b and c). The lower part of frames 1 and 3 in the afterbody and frames 26 and 28 in the forebody had an unusual construction. At frame 1, there was a small symmetrical floor timber on which a second timber was placed, which was higher on the port site (figs 20.20a and c). At the V-shaped frames 3, 26 and 28, two naturally curved pieces of wood had been stacked on the similarly small and symmetrical V-shaped floor timbers, each of which had a horizontal beam section and curved upwards to starboard and port respectively as futtocks (figs 20.20b and c). In principle, all frames were fastened to the shell planks with two wooden treenails (Ø 2.8-3.5 cm) per plank width.

Keelson with mast step section

An essential element in large cogs is the keelson with an integrated wider, thicker section at the mast: effectively the mast step. In principle, this always consists of a single piece. The keelson ran from the back of frame 5 to the front of frame 26 and was approx. 9.75 m long in total. The forward piece, approx. 5.5 m long, was still in its place; the rest, about 3.5 m, had been scooped away by the crane and was no longer present. The remaining aft section had moved. The bottom of the keelson fitted over the frames. To allow that to happen, a notch about approx. 3 cm had been cut into each frame. The parts that slide between the frames had been hollowed out on the lower end (and at the corners), creating 'archways' between the frames. The back end of the keelson was 19.5 cm wide, widening to 30 cm just behind the mast step section. The thickness showed a similar progression: 15 cm at the back and 20 cm above frame 16.

Fig. 20.19a. Overview of the framework on the port side with the stem-post and stem hook.

Fig. 20.19b. The framework of the afterbody viewed from the stern, with in the foreground the lower V-shaped floor timbers on the stern hook and the broken sternpost. The ends of two planks from the open ceiling can also be seen.

Fig. 20.19c. The bottom of the floor timbers amidships and in the forebody, with the limber holes to let water drain out. The transition from carvel-built to clinker-built construction of the strakes along the floor timbers can be seen in the background.

Fig. 20.20a. The V-shaped floor timber construction of frame 1 in the afterbody.

Fig. 20.20b. The timber/curved wood construction of frame 3. The protruding ceiling boards and the sediment layers in the soil profile of the deposits filling the hull.

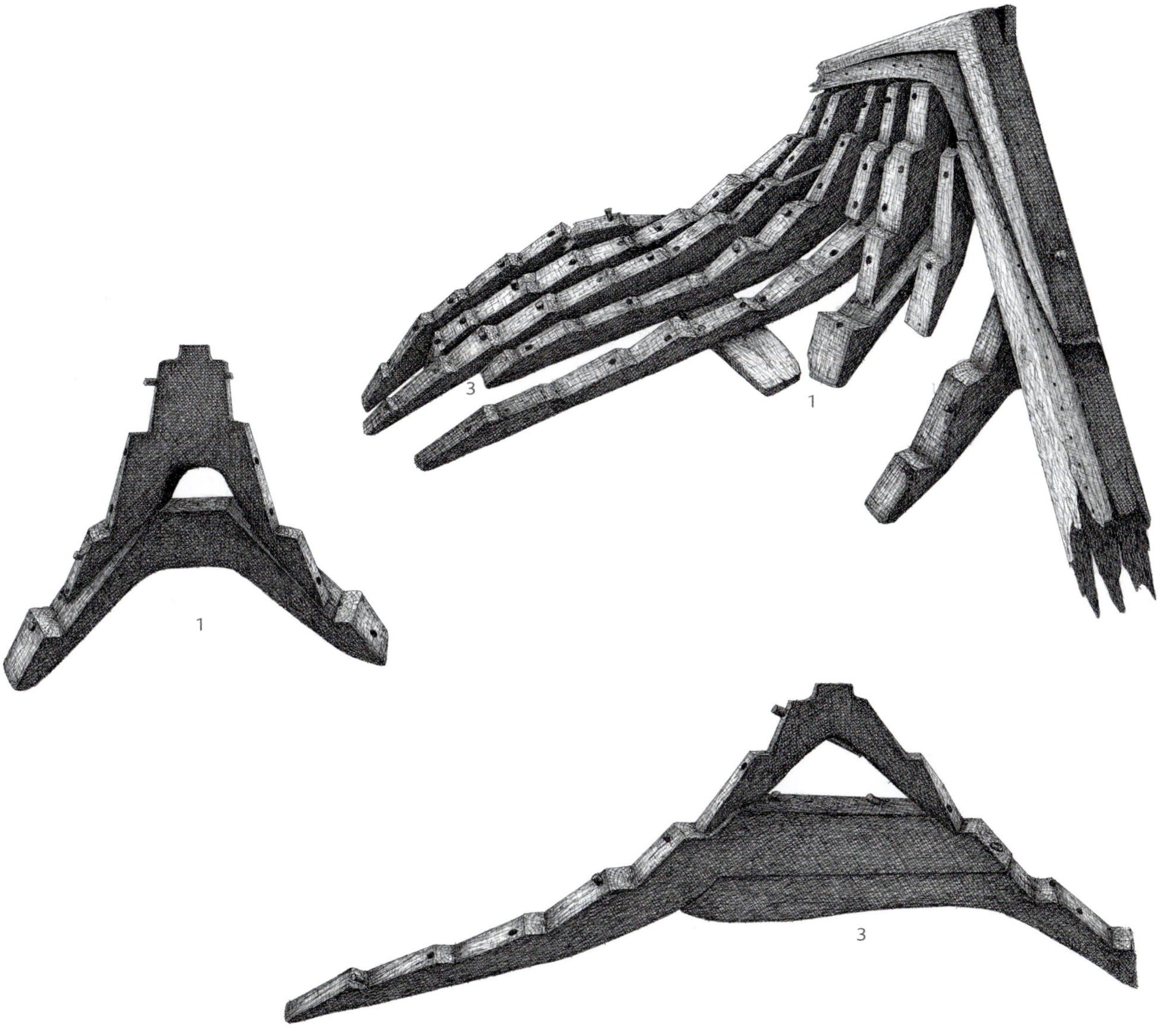

Fig. 20.20.c. The V-shaped floor timber constructions of frames 1 and 3 and their position among the frames in the afterbody.

The mast step section was located between frames 16 and 21. It was approx. 1.75 long and 26-27 cm thick. The width was 37-39 cm. Above frames 17 and 18, a hole had been cut in the top which was 56.5 cm long, 24 cm wide and 11 cm deep (fig. 20.21). In the middle of the 'bottom' of the mast cup, a drainage hole had been drilled (Ø 3 cm), with an X cut over it to allow runoff and rainwater from the mast to drain into the hole. There were two small, flat surfaces on the 'bottom' of the mast cup, forward and aft; the aft surface was outlined by a groove that had been cut into the surface. The section with the X was 33 cm long and probably indicates where the mast would have been. The significance of the two flat surfaces is unclear. The forward part of the keelson was 32 cm wide in front of the mast step section, narrowing to 20 cm at the bow. The thickness was 23 cm (20 cm above the frames). The upper edges of the aft and forward section of the keelson had been chamfered. Rectangular holes had been cut in the top above the back of frame 22 and above frame 26. The first was 9.5x8 cm, 1 cm deep on the mast side and 3.5-4 cm on the forward side. The hole above frame 26 was 7.5x6cm and 4 cm deep. A vertical stile or post was probably placed in the final hole to support crossbeam 5. During the salvage, a (compressed) post was found between frames 25 and 26.

Riders

The wider mast step section of the keelson was secured on both sides by three riders. These 'extra timbers' take on the sideward pressure from the lower side of the mast. These riders were 1.4 m long, 13-5 cm wide and 19-22 cm thick, and the tops were nearly even with the top of the mast step section. They had notches on the underside so they would fit over the A- and B ceiling strakes. It was not yet possible to establish whether the riders had been fastened to the shell at the same time as the underlying floor timbers, or had been added later and e.g. attached to the timbers with blind treenails.

Ceiling

Ceiling strakes were placed on the frames, not tightly fitted against each other, but at intervals of approximately one plank in width: an 'open ceiling'. Only a few fragments of nine ceiling strakes were still present in the afterbody; pieces of fifteen were found in the forebody. The pieces in between had been scooped out by the crane, along with a piece of the keelson and part of the shell and futtocks, and were no longer present. Due to the position of the ship and the salvage situation, it was hardly possible to determine whether or not the planks were still in their original places. Field survey sketches were made. The planks varied in width between 30 and 39 cm, and were 4 or 5 cm thick. It would not be possible to place them back into their original location until the reconstruction of the ship. For now, it seems plausible that the

Fig. 20.21. Part of the broader mast step section of the keelson with the rectangular hole for the heel of the mast. There is an X cut into the 'bottom' and a drainage hole for letting out the rainwater runoff from the mast.

Fig. 20.22. The exposed crossbeam 3. Viewed from port to starboard.

cog had seven ceiling strakes on either side between the keelson and the crossbeams. There would have been at least one above the crossbeams, and it seems plausible that there would have been two, possibly with a third above it, or a (covering) inwale.

Crossbeams with knees

The crossbeams protruding through the M and N strakes were already mentioned in the description of the port side of the shell. These were the most important transverse connections that formed the crossbeam and determined the strength at deck height, in combination with the rising knees that were originally placed on top of them, ensuring that the everted sides did not actually fall outwards. The six beams and other pieces were numbered from aft to fore, following the practices used for the other ships in this study. Beams 1 and 2 were missing. Crossbeam 3 was right underneath the piece of the ship that was scooped out by the crane (fig. 20.22). It was the only crossbeam for which there was sufficient time available during the salvage process to record all the details and dimensions. At the top, between the rabbets in which the M strakes fell, the beam was 6.12 m long: the width of the ship measured from the interior of the shell. The beam was 29x29 cm. A rabbet 12x6.5 wide and 1 cm deep had been cut away on the top and bottom at the centre of the beam. This rabbet was most likely intended to hold a post that would prevent the beam from sagging. The bottom of the post was probably placed in a hole in the top of the keelson. Wooden treenails (Ø 3.8 cm) were secured in the beam. They were used to fix the heavy knees that were placed on top.

Fig. 20.23a. The front of crossbeam 3 with the head protruding through the shell, the rabbet in which the M strake fell, the triangular notch in the upper side (= lower side) for the frame head that the beam rested on and the notch for a gangway board on the lower side (= upper side).

Figure 20.23a and b show the beam head of crossbeam 3 on the port side. Figure 20.23a shows the beam head from the front, with the rabbet for the M strake to the left. At the bottom (= upper side) there was a notch for a gangway plank and at the top (= lower side) there was a triangular notch for the frame head on which the beam rested. Figure 20.23b shows the lower side of the beam head with the rabbet for the M strake in the side, the rabbet for the upper side of the L strake, and next to it the slanted opening for the frame head on which the beam rested. Beside it was the rabbet for the ceiling strake. Figure 20.2 shows the crossbeam head and the beam-head guide fixed over it on the outside of the shell. Crossbeams 4 (fig. 20.24) and 5 had broken in half. Only the ends of crossbeam 6 with the heads protruding through the shell were found; the rest had vanished. The knees that had originally been on the crossbeams were not found. Crossbeams 3 through 6 were situated above frames 10, 17, 26 and 28 respectively. Beams 1 and 2 were very probably placed above frames 0 and 3.

Fig. 20.23b. Idem. The underside with (from left to right) the rabbet for the upper side of the ceiling strake, the triangular notch for the frame head, the notch for the L strake of the shell, the notch in the side for the M strake, and the rounded beam head.

Fig. 20.24 The broken crossbeam 4 in the layered sedimentation in the ship's hull. Viewed from aft. Photo by ADW.

Beam-head guides
Beam-head guides or fenders were placed against the fronts of the crossbeam heads (fig. 20.25). This was intended to prevent the vessel from hooking onto a quay or post, for instance while tying up the ship, with all the consequences that would entail. The type of wood used for the guides had not yet been determined, but it may have been ash (*Fraxinus spec.*). Only one guide was still present with the Nijkerk-II cog, found separately. With the Doel-I cog, there were five guides and a fragment of a sixth. They were found against the fronts of crossbeams 3, 5 and 6 to starboard and port respectively. They differed in length and were triangular in cross-section, with a rounded side towards the outside. From the crossbeams, the beam-head guides sloped forward to a point.

Stringers, deck(s) and windlass(es)
Nothing was found of any structure on and above the crossbeams, other than the frame sections and shell strakes. However, it seems very plausible that the vessel had fore and aft beams on the horizontal sections of the knees with an intermediate piece inserted in between (for a possible execution, see the Nijkerk-II cog, Plates 6.IV A and B and 6.V A), which would have provided structural support for the vessel, both longitudinal and at deck height. It is impossible to state anything about a deck or decks. It is worth noting the recesses in the tops of the crossbeams close to the sides of the ship. Gangway planks 4-5 cm thick may have laid here, if there was an aft deck and a forward deck. However, that would mean that those planks were at the same height as the tops of the crossbeams, although they would have been expected to be at deck height, i.e. on the horizontal sections of the knees. Obviously, the ship would have had a windlass to hoist and lower the spar with the sail. There may also have been a windlass or capstan on board to operate an anchor.

Pump(s) and drainage channel
The description of the port shell mentions a small rectangular hole, which may have been for the end of a drainage channel for pump water. The hole was located between frames 27 and 28, just under crossbeams 5 and 6. Confirmation of the possible location for a pump shaft can be deduced from the larger interval (40 cm) between frames 27 and 28, above the keel section of the stem hook. The intervals between the other frames varied between 21 and 29 cm. There was a comparable situation above the stern hook, where the interval between frames 3 and 4 was 52 cm. The vessel presumably had a pump in two locations. (For a possible execution of pump shafts and drainage channels, see the Nijkerk-II cog, Plates 6.IV B and 6.V A.)

Fig. 20.25. The crossbeams 5 and 6 with the beam-head guides attached in front of them, the roof-shaped adze-hewn shell planks, the gouged ornamental grooves with rust concretions from the nail heads and moss lathes from the clamped moss caulking in cracks in the planks. At the top right, several rounded heads of wooden treenails can be seen protruding slightly above the shell. The brown spots are tar residue.

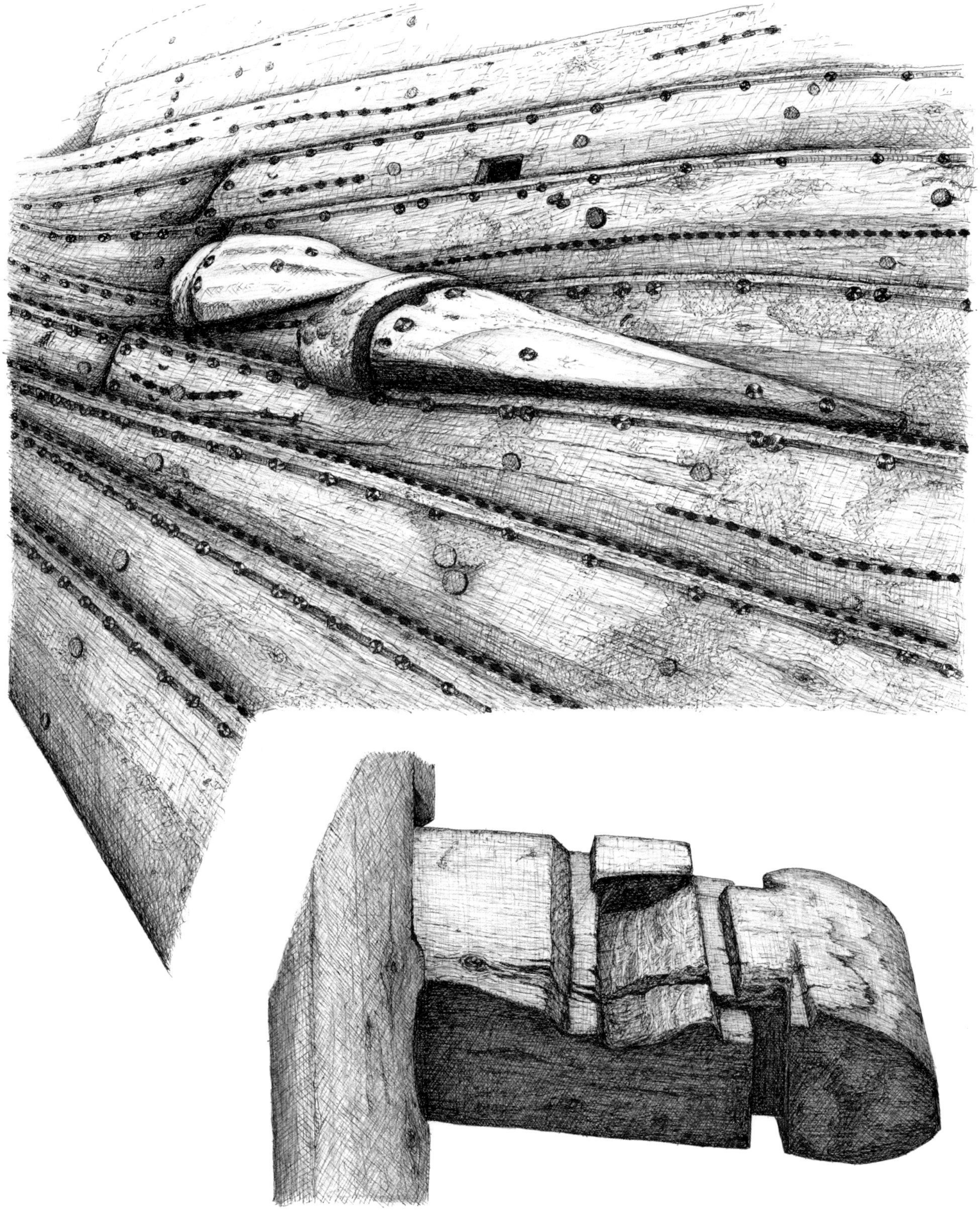

The Doel-I cog. Part of the port shell with two heads of protruding crossbeams with guides. Above them is a rectangular hole through which most probably the drain for pumped-up bilge water emptied. The nail heads of the clinker-built side are seen in the gouged ornamental grooves. Sintel clamps on the seams and cracks. The underside of a protruding crossbeam head with rabbets for the shell planks, the ceiling boards and the (bevelled) frame head.

Castles
No evidence was found during the salvage that would indicate the presence of an aftercastle and/or forecastle. Such indications may have been found while the elements of the ship were being cleaned. It seems probable that the Doel-I cog would have at least had an aftercastle.

Repairs
Repairs were detected at various places on the shell of the vessel. These repairs can be grouped into two categories. One category was most probably done at the end of the ship's construction (using sintel clamped moss caulking to close cracks in the planking that had formed as the planks dried). The other category was presumably done while the ship was sailing. These generally involved small repairs, such as fitting new pieces of wood into knots in a plank; in some locations, small boards were nailed onto the ends of the planks at the overlapping scarfs. The lower piece of the false stem that was found against the stem hook may also have been a repair. It will not be clear until after cleaning to what extent e.g. repairs using small planks were performed on the inside of the shell. Overall, the hull appears to have been several decades old when the ship sank. More repairs seem likely.

Wood use and woodworking, decorations and fastenings
The cog was made of oak (*Quercus spec.*). Most of the tree-nails were as well, although a few were made of willow (*Salix spec.*).[13] The moss lathes were cut from oak boards. The larger elements of the ship, such as the keel plank, the stern- and stempost, the false stern and stem, the crossbeams, the stringers and the larger floor timbers, were cut from straight tree trunks; the elements were cut into rough shape and then adze-hewn. The V-shaped floor timbers and the stern- and stem hooks were cut from curved tree trunks or trunk/branch sections, contoured, and then smoothed with an adze. The planks were sawed with the grain from long, straight tree trunks. The outside of the vessel, particularly the shell strakes from the turn of the bilge upward, were all adze-hewn in more or less a roof shape. As a result, the higher planks attached there against the top were slightly more vertically oriented. A very remarkable feature, not seen previously in cogs (and other shipwrecks) in the Netherlands, is the gouged grooves in the lower edge of the clinker-built strakes. The nails of the lower fastenings of the strakes were placed in this decorative hull element (figs 20.25, 20.26 and 20.27). With its narrow strakes, accentuated even more by the decorative grooves and the six beam heads protruding through the shell and their beam-head guides, and the sheer of its construction, the cog must have looked graceful.

Fig. 20.26. The port side of the forebody with the heads of the crossbeams 5 and 6, the roof-shaped adze-hewn shell strakes and the gouged ornamental grooves.

Fig. 20.27. Ms A.E.M. Hanraets (RING) and K. Vlierman on the ship's hull near the stem hook, searching for the most suitable plank sections for tree-ring analysis. The warping and shifting of the 'loose' shell strakes of the starboard shell are clearly visible. The ornamental gouged grooves can also be observed clearly.

13. J.-P. Van Roeyen, verbal announcement, 10 October 2002.

Fastenings

The planks and frames were connected to each other with wooden treenails. They were 2.8-3.5 cm in diameter. Some treenails seem to be (or have been) slightly wider towards the outside, and have a slightly curved head rising somewhat from the shell (fig. 20.25, top right). Amidships, there were in principle two treenails per plank width. Aft and forward, where the planks were narrower, there was only one. The treenails in the crossbeams for the connection to the knees and the connection between the knees and the shell were 3.8 cm in diameter. It has not yet been determined whether some treenails on the inside had arches or wedges. Several thick iron bolts (30 mm) and wooden treenails 4 cm thick were used to connect the false stem to the stem post. It is interesting to note that iron bolts were used here, but not for the crossbeam constructions.

The clinker-built shell planks were nailed on. These iron nails have dissolved completely, but the marks and holes indicate that the heads would have been approx. 30 mm in diameter and that the shafts below the heads were 8x8 mm thick. It cannot be determined with any certainty how long the nails were and whether they were bent and hammered back into the planks until the various parts are cleaned. It is interesting to observe that the nails were used in pairs in some places. The interval between the nails varies widely due to the varying overlap in the planking. The interval is 16-22 cm, average 19 cm on the clinker-built transitions between the A/B and F/G strakes; between 13 and 18 cm, average 15.5 cm at the G/H and I/J strakes; between 17 and 24 cm, average 20.5 cm (plus one extra nail at each frame) at the J/K and L/M strakes; and between 16 and 24 cm, average 19.5 cm at the M/N and Q/R strakes. In total, approximately 650 m of clinker-built overlap was nailed into place. Taking all this into account, it means that at least 4400 nails were used for the ship's shell. In total, it is estimated that sintel clamped moss caulking was applied to approximately 1660 m of seams (and repaired cracks) on the inside and outside of the shell. The average distance between the sintel clamps, centre to centre, is 7.5 cm. In total, at least 22,000 sintel clamps were used.

20.3 Construction data

Based on the above, the observations outlined above can only be partially interpreted. Definitive conclusions will only be possible after the vessel has been reconstructed based on the sections that were measured and sketched. Reconstruction 'on paper' would be possible, but there is a very good chance that it would not be easy. This may seem to be an acceptable cross-section at this point, despite warping of the various parts of a specific frame – but it may later become apparent that the frame does not produce fair plank lines with the similarly reconstructed cross-sections of the frames aft and forward, which may have warped in different directions. In principle, therefore, the vessel will have to be reconstructed from the keel plank, similar to the Bremen cog, in order to restore the original hull shape. Despite these limitations, it is possible to make some preliminary statements about the design, construction, use and wreck formation based on a comparison with the other large cogs in this study.

20.3.1 Design

Similarities to the other large cogs – such as the hull shape and the characteristic stem hook, stern hook and keel plank, a carvel-built bottom that transitions into a clinker-built construction aft and forward, clinker-built sides, crossbeams protruding through the shell, seams sealed with sintel clamped moss caulking – in combination with the period in which the ship was most likely built (second half of the 14th century, see section 20.8) clearly show that we are dealing with a type of vessel that has been well known for some time now, as well as the way in which such ships were built.

Lines

Although the vessel has not yet been reconstructed, so no precise dimensions and lines are available for the various cross-sections, the photos and notes that were taken provide sufficient data to offer an approximate description of the probable lines of the hull. The Doel-I cog has a striking hull with high sides, a straight keel, and a plumb stern and stem. It is difficult to determine the exact sheer at this time. The sheer was most probably higher at the stem than at the stern, with the gunwale approx. 1 m higher at the stem than at the stern. The midship section of the gunwale was probably about 1.6 m lower than at the stern. The underwater body curves more sharply aft than forward. The carvel-built section of the bottom is nearly flat, hardly curving at all from the keel plank to the turn of the bilge. The more or less rounded transition from the bottom to the sides is formed by three strakes, attached to each other at an increasingly sharper angle. Other than that, the sides were probably nearly straight, curving outwards at an angle of approx. 108° amidships. Aft and forward, the clinker-built sections of the bottom strakes transition more smoothly into the sides; the frames in the afterbody presumably are slightly S-shaped. When viewed from above, the gunwale was very probably shaped like an elongated half-walnut, taking on a pointed shape towards the stern and stem. The ship was probably 10-20 cm narrower across the gunwale towards the aft than towards the bow, at a comparable distance from the stern and stem. Due to the narrow strakes along the sides and the ornamental grooved seams along the bottom edge of the side strakes, accentuated even more by the nails in the clinker-built planking over the strakes and the six beam ends with beam-head guides protruding through the shell, the ship probably looked more elegant than it actually was.

Measurements

In the Late Middle Ages, 'feet' were a frequent unit of measurement. Converted into centimetres, a foot varied between regions and cities. Since it is never definitively known which city a vessel comes from, and the dimensions within the shipwrecks may vary due to shrinkage or expansion of the wood, the measurements have been provided in metres and centimetres for this cog, and for all the other cogs in this study. It is not possible to state much about the measurements used in constructing the Doel-I cog at this time. Until the pieces have been cleaned, it will not be clear whether there are any markings related to the ship's construction. It does seem fairly certainly by now that the primary measurements of a wooden ship were determined in part by the availability of timber lengths. If an order was placed for a ship 20 m long, it may very well have been delivered in a length of 19.8 m, but could also have ended up being 20.15 m long.

It is not possible to determine the precise primary measurements of the vessel at this time. It will not be possible to state the exact dimensions until the frames and plank lines have been reconstructed, and eventually the entire hull. The rake of the stern and stem and the angle of the sides have already been mentioned (stern hook 118°, stem hook 124° and the sides amidships 108°), but these measurements should also be considered approximations for the time being. Based on the data available now, we can assume estimates for the primary measurements as provided in Table I.

Internal structure

It is not possible to say much about the internal structure (yet). The vessel may have had probably transverse bulkheads here and there. Any notches on the side of the frame or other indications of fastenings may indicate whether there may have e.g. been a floor above the ceiling boards of the bottom.

Tonnage

Obviously, it is also impossible to say anything about the cargo capacity with any certainty at this point. For instance, if there was a cargo hold between frames 3 (crossbeam 2) and 26 (crossbeam 5), the sides of the ship and the bottom of the deck, it could have had a total capacity of 10x5x3 m = 150 m³, or 150 gross tonnes.

Table 1. Estimated dimensions of the Doel-I cog

Length of the keel (including the stern- and stem hook)	ca 14.0 m
Length of the bottom (between the rising part of the stern- and stem hook)	ca 13.3 m
Greatest width of the bottom (at frame 17)	ca 3.1 m
Overall length at gunwale level	ca 20.0-21.0 m
Overall width at gunwale level (at frame 17)	ca 7.0 m
Height of the stern (keel line to top gunwale level)	ca 5.3- 5.4 m
Height of the stem (keel line to top gunwale level)	ca 6.1- 6.2 m
Height of the side amidships (keel line to top gunwale level)	ca 4.4- 4.5 m
Length of the stern	ca 6.0- 6.2 m
Length of the stem	ca 7.3 m
Length of crossbeam 1 (to outside of shell at the level of frame 1)	
Length of crossbeam 2 (to outside of shell at the level of frame 3)	ca ?
Length of crossbeam 3 (to outside of shell at the level of frame 10)	6.2 m
Length of crossbeam 4 (to outside of shell at the level of frame 17)	ca 6.3 m
Length of crossbeam 5 (to outside of shell at the level of frame 26)	ca 6.1 m
Length of crossbeam 6 (to outside of shell at the level of frame 28)	ca ?

Length-width ratio of the bottom	ca 4.3:1
Length-width ratio at gunwale level	ca 2.85-3.0:1

All measurements are rounded to the nearest 10 cm

Materials
To the extent that the materials could be established during the salvage, the ship was built from oak (*Quercus spec.*). The various elements were made from tree trunks and trunk/branch sections, depending on the size and form. The planks were sawn and adze-hewn into rough roof-like shapes on the outside after they were attached to the hull. The frames were generally neatly contoured and smoothed with an adze. The parts of the shell strakes visible from the outside were strikingly narrow (24-32 cm). The wooden treenails were generally made of oak, some (with the rounded heads?) made of willow (*Salix spec.*), and the fenders were made from another, different type of wood that has not yet been determined. Only on the stem post, it was determined that the thick wooden treenails were supplemented by thick (Ø 30 mm) wrought-iron bolts that were used to attach the (missing) false stem and pieces of T-shaped iron plating that were used to hold the bottom section of the false stem in place. Obviously, the stern rudder was suspended from the rudder fittings with wrought-iron pintle fittings in the gudgeons on the stern. Thick wooden treenails were used to attach the (missing) knees to the crossbeams and onto the shell. The clinker-built planking was nailed on with iron nails, which were almost certainly folded over twice on the inside and then pounded back into the wood. The moss caulking and the moss lathes were held in place with wrought-iron sintel clamps.

20.3.2 Construction: techniques and sequence
For an indication of the techniques used to build the ship and the sequence of construction, it is sufficient to refer to what has already been written on this subject regarding the Nijkerk-II cog (Section 6.3.2). The construction data would have been almost identical, differing only in minor details.

20.3.3 Characteristics
The characteristics of the ship are also very similar to the Nijkerk-II cog. In principle, they are nearly identical. However, one striking phenomenon on the Doel-I cog should be noted regarding the caulking method used here. The clinker-built seams between strakes E/F through to the seams between strakes J/K were not caulked with sintel clamped moss caulking on the outside of the starboard side. On the port side, this was the case for the seams between strakes E/F through I/J. It is very likely that they had already discovered that there was not much point in doing so, since the iron sintel clamps (which were only 1-2 mm thick) would corrode and dissolve fairly quickly in saltwater. As has already been established, from about the mid 14th century on, the seams of the cogs (except for the carvel-built seams of the bottom) were no longer sintel clamped on the outside of the shell; sintel clamped moss caulking was only applied on the seams on the inside of the ship from that point on. The Doel-I cog and the Marknesse cog are the two earliest examples of this evolution in caulking methods known to date.

Another very striking characteristic of the ship is the ornamental grooves added along the lower edge of the clinker-built strakes (figs 20.26, 20.27 and 20.28), which have also been nailed. A third remarkable structure is the trapezoidal opening in the upper two shell strakes to the port side of the forebody. In terms of shape and location, it is directly comparable to the opening found in the Nijkerk-II cog. The possible correspondence to a drainage channel for pump water is also striking. Finally, we can note the remarkable timber constructions of the frames 1, 3, 26 and 28 in the underwater body (figs 20.20a, b and c).

20.3.4 Use

Function
When looking at a vessel like the Doel-I cog, it is initially almost an automatic reflex to think of a Hanseatic cargo ship. If that is the case, we are presumably looking at a medium to large ship of that time, with a load capacity of at least 150 gross tonnes, more or less. The intensive application of tar on both the inside and the outside of the ship is very unusual and may be due to the age of the vessel when it sank. In that case, the tar may have been applied to an old, possibly deteriorating and leaking ship in an attempt to keep it sailing for longer. Alternatively, there are several examples where there was confirmed or presumable use of intensive tar application on the inside of a ship. The most striking example is the 16th-century armed trading vessel/warship from lot U 34 in Eastern Flevoland.[14] In this vessel, part of a remarkable floor was found, just above the ceiling boards on the bottom, which probably completely closed off the underwater body from side to side. This is a situation that should preferably be avoided in a wooden ship, and may have been the reason for the extra-intensive application of tar. Traces of interior tarring were also found in the Rutten cog, which also seems to have had a floor in the underwater body. With some caution, it would be possible to contemplate some other function for the Doel-I cog than cargo transport. It may have been a relatively old trading ship that was converted for use for specific military purposes.

14. File and drawings, Ou 34, RCE-Lelystad; the Rutten cog, sections 10.2.2 and 10.3.4.

Wear and repairs
The forward 50 cm of the bottom of the stem hook is quite badly worn, probably the result of beaching many times on a sandy shore, for example. The limited number of repairs that were observed on the outside and carried out while the ship was sailing may increase once the inside of the planking is cleaned. Looking at the overall impression of the condition and wear and tear of the ship's hull, it can in any case be presumed that the vessel was several decades old when it sank.

20.3.5 Wreck formation and embedding

It can be presumed that the Doel-I cog capsized during a catastrophic, tsunami-like floor wave, turned turtle and sank. Whether this occurred in the medieval channel known as the Tonnekin is impossible to say. The ship may also have capsized in the Scheldt River, upstream or downstream from the findspot, and then carried into the Tonnekin by an ebb or flood tide. It is presumed that the afterbody was regularly free of the channel bottom initially, after which the higher sections, of the afterbody in particular, slowly but surely came loose and drifted away. Eventually, the wreck embedded itself in the channel bottom and settled to a fixed position, after which sedimentary deposits filled and covered what remained.

20.4 Model

No (reconstructed) model of the Doel-I cog has been built.

20.5 Inventory

Other than a significant number of what is known as 'cog sticks' (see the Nijkerk-I and -II cogs and the Rutten cog, sections 8.5.2, 6.5.3 and 10.5.2), nothing was found in or near the wreck.

20.6 Cargo and merchandise

No merchandise or any other cargo was found in or near the wreck.

20.7 Ballast

There were no traces of any ballast either.

20.8 Dating

Sintel typology
Based on the marks left by the type of sintel clamps that were used, it could be established with some certainty on 16 October 2000 that the vessel was almost certainly built in the first half of the 14th century, and must in any case have been built before ca 1350. The sintel clamps were model D2, which was used between ca 1275-1350, while a few marks seem more indicative of the successive type E sintel clamps.[15] Dendrochronological analysis then clearly showed that this was indeed a transition type D2/E (or the first steps towards that type) and that it was already being seen (in certain areas?) in the second half of the 14th century.

Dendrochronology
On 7 November, in consultation with Ms. A.E.M. Hanraets (RING), the places were marked on the ship's hull where tree-ring samples should be taken (fig. 20.27). During

Fig. 20.28. The tree-ring sample sawed from frame S8 on the starboard side (object/code kka 17.1, felling date AFTER 1296 AD ± 6; scale 1:1).

15. Vlierman 1996a, 58 Table II.

the study day in Sint-Niklaas on 14 December 2000, the results of the study were announced. Of the eighteen samples, fourteen could be dated with reference chronologies from the Niedersachsen regions, subregions Südniedersächsisches Bergland, Nordwestdeutsche Bergland Schwelle, and Ostniedersächsisches Tiefland; two could be dated with the reference chronology NLWestfalen and one with that of Ost-Friesland. The samples can be dated most closely with the various chronologies from Niedersachsen. It is therefore likely that the wood came from that area. The trees from which the samples were taken were felled during the summer or winter, but before the trees sprouted new leaves in the subsequent year, i.e. between summer AD 1325 and spring AD 1326. This felling date is generally possible for all the dating samples. For the full report from RING, see appendix. Figure 20.28 shows the tree-ring sample/pattern of frame 8 starboard. It is very probable that the ship was built between 1325 and 1335. These dendrochronological data were used by the Wood Technology lab at the University of Ghent, together with the data from the Doel-II cog, to investigate where the wood from both ships came from.[16]

Soil profile

The soil profile cannot offer any conclusive evidence about when the vessel sank.

Finds

None of the ship's inventory was found in or near the wreck; some inventory is usually found with the shipwrecks in the former Zuiderzee region, often providing an accurate indication of when the ship went down. The general impression of the ship's condition (wear, repairs) as such indicates that the vessel would have been at least several decades old when it sank. We can therefore assume that the Doel-I cog sank in the third quarter of the 14th century. Historic sources tell us that the worst storm in the Netherlands in the 14th century took place in 1362.[17]

20.9 Summary

Ship

The striking structural elements – keel plank, stern hook, stem hook, a carvel-built bottom that transitions into a clinker-built structure towards the afterbody and forebody, clinker-built high sides, plumb stern and stem rising steeply, crossbeams protruding through the shell, and shell seams sealed with clamped moss caulking – clearly indicated not long after the find was made that the Doel I shipwreck was a large cog, most likely Hanseatic. In terms of lines and dimensions, it is directly comparable to the Bremen cog (fig. 1.1a and b) and the Nijkerk-II cog (Plates 6.V A and B). Built from oak wood, the vessel was put together with wooden treenails that were 2.8-3.8 cm thick. Some treenails have a slightly curved head on the outside of the ship, rising somewhat from the shell. Viewed from the side, the cog very likely had a sheer rising towards the stem and the shape of the hull is not particularly exciting. The strikingly narrow strakes (24-32 cm) along which the lower edges have been embellished with grooves, in which the nails that fastened the clinker-built planking were placed, collectively create a compact pattern of closely packed lines. As a result, it is very probable that the vessel had a fairly elegant appearance. This was probably accentuated further by the six heads of the crossbeams protruding through the shell, with their beam-head guides, and the sintel clamped seams.

A very remarkable detail is the lack of sintel clamped moss caulking on the outside, in the seams of the clinker-built strakes along part of the hull, which would have been just below and above the water line in loaded and unloaded state. It had already been discovered that the thin (1-2 mm) wrought-iron sintel clamps corroded and dissolved fairly quickly due to the constantly changing conditions – dry vs. wet with saltwater – so there was not much point in using them. It was already known that ships from the Hanseatic region stopped using sintel clamped moss caulking on the clinker-built strakes on the outside of the shell from ca 1350 on. The Doel-I cog and the dendrochronological dating of the ship clearly show that this development had already started in the second quarter of the 14th century.

The ornamental grooves along the seams have been seen on vessels from the Baltic region. The striking construction of some floor timbers and the first transverse connections in the afterbody and forebody (frames 1, 3, 26 and 28) also offer indications that the vessel may have been built in Niedersachsen, possibly somewhere along the Baltic coast. The tree-ring analysis clearly showed that the majority of the wood from the sampled ship parts came from the sub-regions of Niedersachsen.

Another noteworthy feature was the extent to which the ship had been tarred, both on the outside and on the inside. In comparison to several other finds, it could be cautiously considered that the ship may have had a different function than transporting trade goods – for example a ship several decades old that was converted for use for military purposes. The varying intervals between the nails in the clinker-built strakes is also worth noting. The aver-

16. For that report, see section 21.8 and appendix.
17. Buisman 1995, 633.

age intervals vary from 15.5, 19, 19.5 and 20.5 to 22 cm. Whether or not this was a deliberate choice, for example with the aim of using fewer nails, is unclear. It may have been that different people nailed the clinker-built strakes into place in different parts of the ship. It is also possible that a few strakes were nailed into place at a time, with (significant) time intervals in between work periods. Although it would not be possible to say anything definite about the dimensions and exact shape of the cog until after reconstruction, the field documentation does provide sufficient data to offer approximate indications. The most important preliminary and estimated primary measurements are:

- Length of the stern and stem at gunwale height — ca 20.0-21.0 m
- Largest width at gunwale level — ca 7.0 m
- Height of the stern (keel line to the top gunwale) — ca 5.3-5.4 m
- Height of the stem (keel line to the top gunwale) — ca 6.1-6.2 m
- Height of the side amidships (keel line to the top gunwale) — ca 4.4-4.5 m

It is very likely that the ship capsized during a tsunami-like storm tide. Whether this occurred in the medieval channel known as the Tonnekin is uncertain. The ship may also have come loose in the Scheldt River, somewhere upstream or downstream from the findspot, eventually turning turtle in the mouth of the Tonnekin and becoming covered in layers of sediment.

Dating

Shortly after the find, it was possible to establish, based on the marks left by the sintel clamps, that the vessel must in any case have been built before ca 1350. Dendrochronological analysis confirmed it. The wood of the Doel-I cog was felled during summer of AD 1325, or in any case before the trees sprouted new leaves in the spring of AD 1326. It seems very plausible that the ship was built somewhere in the second quarter of the 14th century, and more precisely probably between ca 1325 and 1335. There is less certainty about when it sank. Based on the repairs identified so far and the overall condition of the ship's hull at the time of the disaster, it can be assumed that the vessel must have been at least several decades old at that time. Therefore, it is very likely that the shipwreck occurred in the third quarter of the 14th century. There are historic sources that refer to a storm flood as described. The worst storm in the Netherlands in the 14th century took place in 1362.

The Doel-II cog[1]

21.1 Find conditions

21.1.1 Discovery, salvage and investigation

Following the spectacular find and salvage of the Doel-I cog (Chapter 20) in the autumn of 2000, the bottom of a second cog was found in July 2002 in the Deurganckdok near Doel in Flanders (Belgium), only 50 m from the findspot (fig. 21.1). This shipwreck was also upside-down and, like the first wreck, situated about 6-7 m below sea level on the bottom of the Tonnekin channel, which was noted in the mid-13th century as the border between the parishes of Kieldrecht and Kallo[2] (see figs 20.3 and 20.4 of the Doel I-cog).

Based on the experience gained from the Doel-I cog, the employees of the Archaeological Department of Waasland (ADW) in Sint-Niklaas excavated most of the wreck within the same month to get a clear impression of the size and nature of the find. It was then covered in soil again due to archaeological research elsewhere in the Waasland region. In August-October 2002, the wreck was uncovered again (fig. 21.2) and documented. K. Vlierman visited the site on 20 August and 10 September to examine the visible sections, to record data on the shell, to describe the sintel caulking and frames, etc., to take photos and slides, and to advise the ADW on what to do with the remains of the ship. Although the wreck consisted only of most of the bottom structure of a cog, the wreckage did provide supplementary and previously unknown information, and would definitely be worth keeping and preserving for scientific research and exhibition. Like the first cog at the Doel site, the pieces of the ship were salvaged and stored in containers under water.

21.1.2 Geology and stratigraphy

The geological and stratigraphical details were identical to the data on the Doel-I cog; see section 20.1.2 for further information.

21.1.3 Prehistoric and medieval habitation

Further details on this aspect are also available in the previous chapter on the Doel-I cog.

The straight and V-shaped floor timbers in their original positions beneath the stem hook, after removal of the timbers torn apart and detached by the mechanical digger.

21.1.4 Condition and wreck formation

Condition

There was little left of the original ship: only the forward section of the keel plank/stem hook and most of the three bottom strakes (A, B and C) on the starboard side, except for the forward ends, which were missing. Only the forward A-plank or garboard and several pieces of the forward B-plank were still present on the port side. It is likely that the port side of the ship's bottom was about as complete as the starboard side before the wreck was found. Until the digger disturbed the site, the planks that were still present were held in place by the floor timbers and beams that they were attached to, which were in turn attached to the keelson and the A-strake of the ceiling to starboard. The six riders, three on each side of the wider mast step section

Fig. 21.1. Overview of the bottom of the Doel-II cog, seen from fore to aft.

1. Translated by Joy Phillips.
2. Van Hove 2005, 51.

of the keelson, that had been affixed on the floor timbers were still present. The quality of the wood was reasonable to good during the field investigation.

Wreck formation
It is impossible to provide a casual answer as to how and why only the bottom of such a ship was found. The most obvious explanation would be that it was the bottom of a vessel that had already been mostly destroyed, which sank in the Tonnekin during a catastrophic disaster similar to the storm in which the Doel-I cog turned turtle, perhaps in even the same incident, and was then covered in sediments. It is also impossible to exclude the possibility that the cog was initially mostly complete, but that the hull broke along the scarfs between the floor timbers and beams and the lower futtocks, during the disaster and the subsequent wreck formation process, after which the sides and so on floated away in fragments. In most of the cogs included in this study, the transition from bottom to sides was relatively weak due to these scarfs in the frames above the turn of the bilge. Striking examples include the shipwrecks of the Nijkerk-I cog and the Nijkerk-II cog, the sides of which were found beside the underwater body. The stem of the Nijkerk-II cog had in fact folded forward entirely, taking large parts of the starboard and port sides along with it. The reason why the sides of these cogs were still present is because the wrecks probably fell apart slowly in a relatively calm environment (against a peated coastline and in a 'harbour basin' respectively, see Plate 8.I and fig. 8.1 as well as Plate 6.I and fig. 6.1 respectively). Due to the current of the Scheldt River and the tidal influences, the Doel cogs were in an entirely different situation.

21.2 The ship

21.2.1 Documentation

The field survey was measured and drawn by employees of the ADW. After the shell planks were removed, the same was done to the frames and keelson as well as the ceiling strake (Plates 21.I A and B). Photos and slides were taken of the various surveys and details. In addition, various particulars were described and the (V-shaped) floor timbers and beams etc. were sketched.

21.2.2 Construction and layout of the hull

In this case as well, the exceptional upside-down position meant that the outside and bottom of the side could be observed more closely in the field than the inside. With the exception of the outer edge of the C strake to starboard in the afterbody, where the overlap with the D strake could be seen, the other planks of the shell as found were fully carvel-built, only connected to the frames with wooden treenails.

Stern hook and stem hook
Very remarkable, and never before seen in a cog, there was no keel plank that was connected to the stern hook and stem hook with sloping flat scarfs. The Doel-II cog only has two hooks with a long keel plank section; the stem hook from the scarf with the stern hook to the front is 8.4 m long. It varies in width. From the end of the sloping flat nibbed scarf (approx. 0.9 m long with a 2.5 cm notch), the width was 36 cm at the bottom; the widest part was 40 cm wide, about the level of the mast, while the width at the start of the worn section at the front was 21 cm. The keel plank section was more or less trapezoidal in cross-section, with the widest side facing upwards. At the stated wider points at the scarf and mast, that width was 37.5 and 41.5 cm respectively. The keel plank section was 14.5 cm thick at the front and 15 cm thick at the scarf. In the front 1–1.5 m, the stem hook has a higher (knee) section that transitions upwards into the (missing) stem section at a fairly sharp angle.

The bottom and front of the hook have been worn down considerably (to 17 cm) over a length of approx. 45 cm. Wide rabbets have been cut into the sides of the thick section for attaching the ends of the garboards (figs 21.2a and b). At the dump site for the soil excavated by the digger, it was still possible to save a piece of the keel plank section with nibbed scarf from the stern hook. It was 3.15 m, including the scarf which was 0.9 m long. At the scarf, it was 36 cm wide and 15 cm thick; at the broken-off end, it was 31 cm and 16.5 cm respectively. The total recovered length of the keel (= the stern- and stem hook) was approx. 9.65 m. The total length from the back of the stern hook to the front of the stem hook would have been approx. 13 m.

Stern- and stempost, false stern, false stem and rudder
Obviously, it is not possible to state anything with certainty about the sternpost, stempost, any false stern or stem, or the rudder. It was not possible to determine from the stem hook how steep the rake of the stempost would have been. With the Doel-I cog as an example, a rake of approx. 124° seems entirely plausible; approx. 118° is the most likely figure for the stern.

Bottom of the ship
The bottom of the ship consisted of the stern- and stem hook with three strakes on either side (A, B and C). The strakes, where they were still present, were built from two planks, the longest of which was the forward-most of the A strakes. The longest plank at the B- and C strakes to

Fig. 21.2a. The starboard side of the stem hook with the worn front.

Fig. 21.2b. The front of the bow hook on the port side with on the left the onset of the upright and missing part of the stem post, the higher keel section and the wide groove in which the garboard was nailed.

starboard was the aft and forward plank, respectively. The planks were even in width: 45-50 cm amidships at the A- and B strakes, 30-35 cm at the C strakes. The planks were slightly narrower towards the stern and stem. The A strakes were approx. 6-7 cm thick on the side of the stern and stern hooks/keel sections, and approx. 5.5 cm on the side of the B strakes. The adjoining B strake is of the same thickness there, and 5 cm thick on the side of the C strake. The C strake has a similar gradient from 5 cm to 4.5 cm. No transition between carvel-built and clinker-built planking was observed between the A- and B- or B- and C strakes on the pieces that were found. The planks were fully flush and carvel-built from aft to fore. The C strake to starboard was chamfered towards the afterbody over a length of nearly 3 m and a width of approx. 10 cm (the overlap) to attach the overlapping section of the D strake.

The overlap was attached with nails applied at irregular intervals, with the impression at several spots that they had been pounded into the strake in pairs. The planks of the strakes were attached to each other with 50-70 cm long sloping flat scarfs. The plank ends were 1-2 cm thick and protruded above and below the shell respectively. Both strake ends were attached to the scarfs with two rows of three to five nails: one row from outside to inside, the other vice versa. The nails were folded over twice and pounded back into the wood. Although no exact cross-sections were measured and sketched in the field, with sketches only noting the dimensions of the various frame sections, it can be stated with fairly great certainty that the bottom of the ship was not flat at any point along the ship's length or width; at most, it had a short section in the middle (1-2 m) that would have been nearly horizontal in length and width. At its widest point, between frames 17-19, the bottom was about 3 m wide; the length between the rising stern and stem sections was approx. 11.8-12.2 m.

Sides

No part of the sides survived. Based on the sections of the (V-shaped) floor timbers that were found, it can be stated that the D strakes were carvel-built against the C strakes at an angle, and that the E-strakes were the first that were very likely entirely clinker-built from the sternpost to the stempost. That was undoubtedly also the case for all strakes above that point.

Sintel clamped moss caulking

The carvel-built seams between the stern- and stem hook and the garboards or A-strakes were triangular across nearly the entire thickness of the bottom strake (6-7 cm) and were approx. 1.7 cm wide at the bottom. The other carvel-built seams were triangular in shape along approximately half of the plank thicknesses. A very remarkable feature that had not been seen previously: the seams were filled not only with moss caulking, but with a combination of moss and animal hair. In the seam between the stern- and stemposts and the garboards, a single strand of twisted thread made from animal hair was applied from the outside. A single strand of twisted fibre made from moss was presumably also used in the seam (fig. 21.3). Neither of the materials have been identified yet. The composite caulking was tarred (the scent was still clearly present), pressed into the seams with moss lathes, and clamped into place using type D2/E sintel clamps, which were placed 6.5-7.5 cm apart, centre to centre (fig. 21.4). Sintel clamped moss caulking was also used inside the ship on the seams between the high part of the stem hook by the bow and the garboards.

The same applied to the butting at the scarfs of the bottom planks on the outside and around the repair planks against the underside of the bottom to starboard. Several repairs were also seen. Two small planks in the B- and C

Fig. 21.3. Detail. The side of the keel section of the stem hook. The lowest, dark strip is the section of the keel protruding under the bottom. The light ochre-brown strip above it is the 'moss yarn'. The lowest, hollow part of this was filled with the 'yarn' of animal hair, which was laid on top of the keel.

Fig. 21.4. Detail. The keel section of the stem hook (top) and a moss lath with the print of a clamp above the seam between the stem hook and the garboard.

strake in the starboard half of the bottom were fitted into place and sintel clamped all around. A third small plank was added against the bottom of the A strake on the port side, with sintel clamped moss caulking all around the edges. In the afterbody, two cracks in the B strake on the starboard side were closed using this method. It cannot be stated whether sintel clamped moss caulking was or was not used on the clinker-built seams on the outside of the ship's sides. A layer of moss caulking had been applied in the rabbets of the stern hook and on the scarfs between the stem and stern hook and the planks of the bottom. The sintel clamps used here had a middle section of approx. 45 mm long.

Frames

Of the frames, twenty (pieces of) V-shaped floor timbers were found; the parts from the forebody on the starboard side were still where they were originally located in the structure (Plate 21.I B). The port sections were gone as a result of the excavation works of the Deurganckdok. The sections of the frames in the afterbody had been moved, broken and/or destroyed entirely by the digger, but the aftmost four floor timbers were still almost entirely complete. The floor timbers and beams were 20-27 cm wide, varying in thickness between 12-15.5 cm. The sections of the floor timbers above the back parts of the stern hook were 18-25 cm thick (figs 21.5a and b). The starboard side of floor timber 8 was the highest one that remained intact and continued to about the top of strake E. The D- and E strakes were approx. 30 cm wide at that spot. The other frames had a scarf with the bottom futtocks, to the extent that they were still present, which extended to the top of strake D at most. There were rectangular limber holes on the bottom of the floor timbers and beams: at frame 8 above the seams between the hook and the garboards, above the seams between the A- and B strakes, and above the seams between the B- and C strakes. In the starboard sections of the floor timbers for frames 17 and 18, pieces had been replaced. We will come back to this in section 21.2.2.

Some distance from the site where the cog was found, a futtock was found in August 2002; it had cracked lengthwise and the bottom was flat and oblique. The outside had been hewn in a stepped line, indicating a top strake that would have been slightly more to the inside. From bottom to top, the strake widths were 30, 26, 22, 28, 25.5 and 30 cm respectively. A diagonal hole had been bored through the bottom strake for a treenail about 3.5 cm in diameter. The futtock was 1.68 m long and 14 cm thick (Plate 21.I D). It is very likely that it was a futtock that was on a crossbeam protruding through the shell between the shell and the outside of the rising section of a heavy deck beam knee (see the Nijkerk-II cog, Plate 6.IV B). The futtock could have come from this ship, but it could also have been from the Doel-I cog. The gunwale above the crossbeams of the Doel-I cog was approx. 1.7 m high. The futtock is most likely to have come from the Doel-I cog (from crossbeams 1 or 2). In any case, the narrow strakes placed one beside the other were definitely placed there.

Keelson with mast step section

Besides the stern hook and stem hook, the keelson was an important part of the longitudinal rigidity. The keelson was 9.8 m long and situated above frames 0 through 21, which meant that it started strikingly close to the stern (Plate 21.I B). The aft section was 5.5 m long, 13 cm wide and 11 cm thick towards the stern, and 25 cm wide and 19 cm thick at the start of the wider mast step section. The mast step section flared diagonally into a section 37 cm wide and 28 cm thick, which was approx. 2.15 m

Fig. 21.5a. The rear four surviving (V-shaped) floor timbers, as they were found. To the right, the starboard side of the bottom strakes, in the foreground part of the keelson is visible.

Fig. 21.5b. An overview of the floor timbers still in their original position under the stem hook. The pieces pulled loose and broken by the excavator were already removed.

long. The forward section was approx. 2.15 m long; it was 25.5 cm wide and 16 cm thick right in front of the mast step section, and 15 and 11 cm respectively at the bow. At the frames, the bottom had been notched to a depth of approx. 3 cm, depending on the height of the frame, and the remaining pieces fell in between them. The corners of these pieces had been chamfered in a slight curve here, creating the appearance of small archways. A rectangular hole was cut into the top of the mast step section for the heel of the mast. The hole was above frames 14-15. A water drainage hole had been bored into the 'bottom' of the hole, emptying out in between the frames.

Riders

The mast step section received extra support on each side from three riders placed on floor timbers 14, 15 and 16. They were notched at the bottom so they could cover the A ceiling strakes entirely and partially cover the ends, with an oblique surface where they were fastened on a part of the probably roof-shaped finished tops of the bilge ceiling (figs 21.6a and b).

Ceiling

The ship had an open ceiling, of which the majority of the A strake and a fragment of the bilge ceiling (B strake) on the starboard side were still present. The A strake (still) consisted of two planks that were put on top of each other with a sloping flat scarf measuring 58 cm in length. The ends are 1.5 cm thick and were projected above the other plank. The planks have a thickness of 4.5 to 5.5 cm. The B-strake fragment is 4 to 4.5 cm on the side.

Crossbeams with knees, beam-head guides, stringers, deck(s), windlasses and pump(s)

The ship parts protruding above the bottom of the ship were not found on the site. Judging by the measurements on the bottom of the ship and the long keelson with the mast step section, one can assume that the Doel-II cog was a mid- to large-sized ship, which almost certainly had shell-protruding crossbeams with knees, beam-head guides, stringers, a deck or multiple decks, windlasses, and one or more pumps.

Castles

It is impossible to offer any certainty as to whether the ship had one or two castles, but it certainly seems quite possible that it would have had an aftercastle.

Fig. 21.6a. The underside of the keelson with notches for the frames. The three riders on the starboard side with the notches for the A ceiling strakes and the chamfered notches at the ends with which they rested on a section of the roof-shaped top of the bilge ceiling-strake.

Fig. 21.6b. The same, with the A ceiling strake (with scarf) and the fragment of the bilge ceiling in situ.

Repairs

The description of the sintel clamped moss caulking also mentions the small repair planks in and on the underside of the bottom of the ship (fig. 21.7). These are almost certainly repairs that were made when the ship was still sailing. It is surely no coincidence that frames 17 and 18 had also been repaired (Plate 21.I A, B and C). The ship had to have been severely damaged in that spot and then repaired. A point of interest is the two (new) wedge-shaped and, judging by their shape, were presumably applied by the carpenter without removing the ceiling strake. In other words, the pieces were inserted by sliding them in under the ceiling strakes. The two tears in the B strake of the bottom of the ship that were patched with sintel clamped moss caulking were almost certainly repaired at the end of the build. It seems likely that the ship was fairly old (several decades?) at the moment it sank fully. The extent to which the underside/front of the hook was worn down is impossible to say.

Woodworking

The pieces of the ship that survived were made of oak (*Quercus spec.*). The planks were sawed and contoured by a squaring axe and/or an adze; the frames were probably hewn from tree trunks, contoured, and smoothed with an adze. The stern- and stem hooks were taken from long trunk sections with a large side branch.

Fastenings

The frame sections were in principle fastened with two oak treenails (Ø 2.9-3.7 cm, most between 3.2 and 3.5 cm) on each plank width. With the exception of 21, the keelson was only fastened with one treenail on the frames and keel sections of the stern and stem hooks. Frames 13 to 16 had two treenails placed in alternating positions. Some treenails had a slightly curved head on the outside of the ship, rising somewhat from the shell (fig. 21.8). Other had wide wedges and a dowel was seen at one treenail. The wrought-iron nails that were used for the clinker-built fastenings were completely dissolved by the salt (ground) water. The impressions made by the nails indicated that they had heads measuring about 30 mm in diameter, while the nail holes indicated a rectangular shaft measuring a section of 5x6 mm. The only piece of bottom strake with those visible marks had had nails that were placed at irregular intervals, varying from 12 to 18 cm. The wrought-iron sintel clamps were also dissolved entirely. The mid-sections were about 45 mm long. The riders were presumably only attached after the ribs and ceiling strakes were attached to the shell with treenails, and may have been attached only to the floor timbers, and partly with blind treenails.

Fig. 21.7. One of the repair planks on the underside of the bottom, sealed all around with sintel clamped moss caulking.

Fig. 21.8. Detail. Two treenails with a convex head that protrude from the shell.

21.3 Construction data

Obviously, the parts of the ship that were found offer insufficient data to fully reconstruct or describe the vessel. It is only possible to make definite statements regarding the bottom of the ship, along with the stern and stem hooks, the keelson and riders, and the ceiling strakes. Viewed collectively, however, that does constitute enough information to compare the shipwreck with the other cogs in this study and offer a reliable impression of which type of ship it was and what the dimensions, shape and presumable structure would have been.

21.3.1 Design

The bottom of the ship as found at the site displays enough characteristic elements to state with certainty that it was a vessel in line with the cog-building tradition. The six strakes of the bottom (three on either side of the stern and stem hooks/keel plank sections) were most likely entirely carvel-built. There is a chamfered overlap on the C strake on the starboard side, which indicates that the aft (and presumably also forward) sections of the D strakes were the first that were clinker-built. It is very likely that all the strakes from stern to stem were clinker-built, starting from the transition between the D- and E strakes.

Lines

The bottom of the ship follows the lines characteristic of cogs, with hardly any straight sections, gradually transitioning into a steep afterbody and forebody. The underwater body probably curved more sharply towards the stern. The transition from bottom to sides consisted in any case of two, but probably three strakes (D through F), placed snugly against each other at an increasingly steep angle. That transition would have been more fluid towards the stern and stem. The vessel undoubtedly had a plumb stern and stem and most likely also had everted sides. When viewed from above, the ship was very probably shaped like a half-walnut, with the afterbody being a bit narrower than the forebody. The lines and top view were comparable to the Bremen cog, the Nijkerk-II cog and the Doel-II cog (see figures 1.1a and b and Plates 6.V A and B and 20.I).

Measurements of internal structure and tonnage

It goes without saying that it is not possible to offer very definitive statements about the measurements. Based on the dimensions of the bottom, the long keelson and the length from the back of the stern hook to the front of the stem hook (the keel line), in combination with the presumable rake of the stern and stem and the angle of the sides in the midship section (these are assumed based on the data from the Doel-I cog) and the presumable heights of the gunwale at the stern and stem and the midship section, it is amply apparent that we have a midsized to large cog constructed with a number of crossbeams protruding through the shell. We can assume estimates for the primary measurements as provided in Table 1.

Obviously, it is not possible to offer any statements about the internal structure of the ship. There is a notch on the front of the floor timber at frame 16, on the starboard side, measuring 34 cm in length and approx. 3 cm in depth, with a rabbet of 4x4 cm near the turn of the bilge. Parts of a transverse bulkhead may have been located there. However, these features are located under the ceiling, so they certainly would not have been in place when the ship was sinking. The total cargo capacity of the ship was almost certainly more than 100 gross tonnes.

Table 1. Principal dimensions of the Doel II-cog

Length of the keel (including the stern- and stem hook)	ca 13.0 m
Length of the bottom (between the rising part of the stern- and stem hook)	ca 12.0-12.3 m
Greatest width of the bottom	ca 3.0 m
Overall length at gunwale level	ca 18.9-19.6 m
Overall width at gunwale level (at the frames 17-19)	ca 6.5- 6.7 m
Height of the stern (keel line to top gunwale level)	ca 4.5- 5.0 m
Height of the stem (keel line to top gunwale level)	ca 5.5- 6.0 m
Height of the side amidships (keel line to top gunwale level)	ca 3.8- 4.1 m

Length-width ratio of the bottom	ca 3.9-4.1:1
Length-width ratio at gunwale level	ca 2.9:1

All measurements are rounded to the nearest 10 cm

Materials

The parts of the Doel-II cog that were found were all made out of oak (*Quercus spec.*). The planks were sawed with the grain or quarter-sawn. The stern- and stem hooks/keel sections, floor timbers and beams were cut from tree trunks and trunk/branch sections, contoured and then smoothed with an adze. The wooden treenails and moss lathes were made out of oak as well. Some treenails had a slightly curved head on the outside of the ship, rising somewhat from the shell, while others had a wide wedge. A dowel was seen at one treenail. The nails that were used had round heads measuring about 30 mm in diameter and a rectangular shaft. The caulking material in the seams was made out of two singular strands of entwined 'thread,' one made from moss and the other from animal hair. They were pounded into place side by side, held in place with oak wood splints (moss lathes) and fastened with wrought-iron sintel clamps.

21.3.2. Construction: techniques and sequence

For these aspects, please refer to what has been written about this on the Nijkerk-II cog (Section 6.3.2).

21.3.3. Characteristics

The characteristics of the bottom of the Doel-II cog are the completely carvel-built construction, a keel that consisted solely of a stern hook and a stem hook with long keel part sections, which had been fastened to each other with a flat nibbed scarf, and the caulking material in the seam which was made not from the usual moss, but from two strands or 'threads' of moss and animal hair respectively. Regarding the lines and construction of the upper parts of the ship's hull, please refer to the Bremen, Nijkerk-II and Doel-I cog.

21.3.4 Use

Function

Obviously, it is not possible to offer any statements about the usage of the ship. It can be assumed that it was used as a trading vessel, as the most likely option.

Wear and repairs

The repair planks against the underside of the bottom were undoubtedly applied when the ship was sailing.

21.3.5 Wreck formation and embedding

The upside-down bottom could have come from a ship that had been wrecked previously, but it is also entirely possible that the ship's hull broke apart at the turn of the bilge, after which the sides etc. floated away. These events must have taken place during a catastrophic storm surge, possibly the same that capsized the Doel-I cog. The bottom structure of the ship was covered by sediment after definitively sinking at the findspot in the Tonnekin.

21.4 Model

No scale model was made of the Doel-II cog.

21.5 Inventory

There were no objects found near the wreck.

21.6 Cargo and merchandise

No cargo or merchandise was found.

21.7 Ballast

No ballast was found either.

21.8 Dating

Sintel typology and a carvel-built bottom

The imprints of the sintel clamps that were fully dissolved show that the sintel clamps were sintel type D2/E, conclusively dating the build back to the first half of the 14th century. A complete carvel-built ship's bottom, as seen in cogs in this study, is only known in ships built before ca 1300.

Dendrochronology

Seven samples from the Doel-II cog were tested at the Laboratory of Wood Technology (UGhent-Woodlab) in Ghent, Belgium.[3] One sample still had 13 sapwood rings. Afterwards, the individual growth rings were compared and a middle curve *Cog1m* could be calculated. A proper match was found using chronological referential information from the Baltic area (what is now Poland and eastern Pomerania/North Poland). The final date determined for dendro-code Cog1m is 1326, logged *after AD* 1328 +9/-6. For the complete report, which also covers the area of origin for the wood from the Doel-I and Doel-II cogs, see appendix. The vessel was almost certainly built in the second quarter of the 14th century.

21.9 Summary

Ship

Even though no more than a large part of the bottom of the ship was found, it can be assumed that it concerns the remaining wreckage of a mid- or large-sized cog, judging by the lines of the ship and the various parts, which included a long keelson. What is remarkable is the completely carvel-built shell of the bottom of the ship, the keel that was comprised only of a stern hook and a stem hook, rather than a stern- and stem hook with a keel plank in between, and the use of 'threads' of moss and animal hair as caulking for the sintel clamped moss caulking in the seams of the bottom. The two repair planks that were inserted, 2 m and 2.5 m long, and the repairs made on frames 17 and 18 suggest substantial damage to the ship, which was sustained during the use of the ship and then repaired. With the measurements of the bottom of the ship, using the angles measured on the Doel-I cog at which the sternpost and stermpost and the sides curve outwards relative to the keel line, it can be assumed that the Doel-II cog had the preliminary estimated measurements displayed in:

- Length over the stern and stem at gunwale height — ca 18.9-19.6 m
- Largest width at gunwale level — ca 6.5-6.7 m
- Height of the stern (keel line to the top gunwale) — ca 4.5-5.0 m
- Height of the stem (keel line to the top gunwale) — ca 5.5-6.0 m
- Height of the side amidships (keel line to the top gunwale) — ca 3.8-4.1 m

Dating

During the visit on 20 August 2002, the impressions made by the sintel clamps already suggested that the cog must have been built during the first half of the 14th century. Tree ring testing (dendrochronological analysis) confirmed this. The wood that was analysed originated from eastern Pomerania, North Poland and Poland itself. The felling date of the dated trees with which the ship parts were made dates back to *after AD* 1328 +9/-6, which seems to confirm that it was built somewhere in the second quarter of the 14th century. Indications of a date for the disaster/when the remaining wreckage sank can only be extrapolated from the repairs that were found. Estimating an age of several decades certainly seems acceptable. Therefore, it is very likely that the shipwreck occurred in the third quarter of the 14th century. The use of animal hair as caulking was predominantly seen in the Baltic region. Together with the origin of the wood that was used, that could mean that the ship might have been built in Baltic coastal region of northern Germany/Poland.

3. Haneca, 20 March 2003, report on dendrochronological research: Cog ships at Deurganckdok.

The Kraggenburg cog[1]

22

22.1 Find conditions

22.1.1 Discovery and investigation

In May 2007, ship's timbers were found during excavation work for water features at the then still future Brennels Buiten experience centre near Kraggenburg in the Noordoostpolder. The find was reported via Nieuw Land Erfgoedcentrum to the Rijksdienst voor Archeologie, Cultuurlandschap en Monumenten (RACM), location Lelystad. RACM staff visited the site on 23 and 24 May. On 15 June it was reported that more ship's timbers had been discovered, but that this time the timbers were not loose pieces, but a fragment of a ship with the timbers still more or less in place. The nature of the ship's parts found (a stern hook, relatively heavy parts of the frame, a part of the widened mast step of a keelson, a number of floor timbers and a part of the bottom whose timbers were still connected, (fig. 22.1) and that of a number of the finds, including sintel clamps and fragments of near-stoneware, led to the assumption that the wreckage had to be that of a cog, and quite an old one at that. During the inspection in May of 2007, a dendrochronology sample had been sawn out of the ship's timber. The pieces of ship's timbers found unattached to the remains of the ship had been transferred to the RACM workshop in Lelystad. As part of the programme of the International Field School for Maritime Archaeology Flevoland (IFMAF), the wreck was studied in an exploratory manner in July 2010.[2] It was subsequently re-covered with soil pending a full investigation.

Fig. 22.1. Overview of the wreck as seen from starboard aft. Open day for the public on 17 July 2007.

22.1.2 Geology and stratigraphy

The Pleistocene surface and the cover

The presence of the Vecht/Zwartewater river basin running from east to west and the original 'De Voorst' peninsula determine the structure of the soil in the south-eastern part of the Noordoostpolder. The former river bottom of the main stream is > 8 m -NAP, while the Pleistocene surface is rapidly rising on both sides of the stream. At the site of the cog, north of the north-western tip of the 'peninsula' still present in the underground, the top of the Pleistocene sand lies at approx. 1.2 m below surface. A Pleistocene stream channel lies directly south of the site, probably the most south-western course of the Steenwijker Aa/the Steenwijkerdiep, which had run along the moraine about 1 km north of Vollenhove and probably bent southwards somewhat west from the find spot of the cog. From there it had run into the main stream of the Vecht/Het Zwartewater. At the location where the cog was found, this channel bottom is located at approx. 6 m -NAP (fig. 22.4). It is not possible to say how far west the coast was at the time of the sinking of the cog. Before the sea flood of 1170, the western border of the country would have stretched far above the island of Ens (Schokland).[3]

Description and dating of the soil profile

The topsoil is approx. 30 cm thick. Underneath is a sandy Almere-clay deposit. This deposit is somewhat layered at the top, with at the bottom some more humus bands than higher up. Under the topsoil, shell remains of the ZU IV layer were found. The Zuiderzee layer cannot have been thicker than 25-30 cm, while the top of the Almere deposit has been ploughed. The wreck lies on the Pleistocene sand(cover). During the wreck formation process, the ship is likely to have caused a 'pit' in the sand. The origi-

1. Translated by Stefanie Hoss
2. Blok & Van den Hoek, 2017. The scattered parts of the wreck lay and lie on lot T 25, 171-176 m from the shoring of the ship's route Zwolse Vaart and 6.5-8 m from the centre of lot T24/25. The site is located approximately 3.8 km west of the port of Vollenhove (figs 22.2 and 22.3). Topographical Map 21 A Ens, (global) coordinates 188.570/521.960, with the highest parts at ca. 1,2 m below surface (between ca. 3,75-4,3 m -NAP). The former water depth at the site was approx. 2.4 m, Hydrographical Map 1921, medium low tide.
3 Van der Aa 1848, 816.

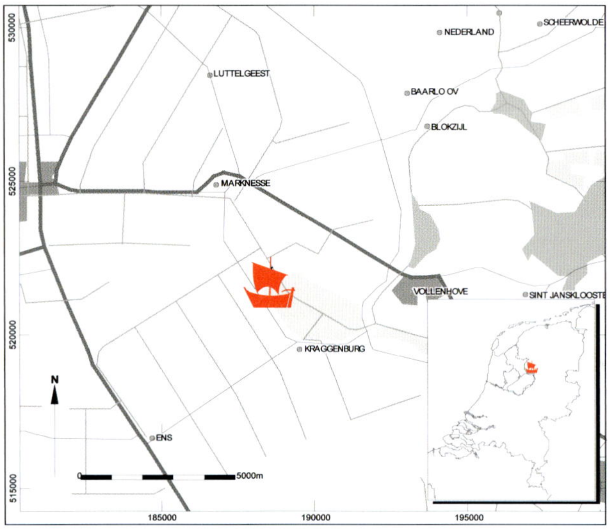

Fig. 22.2. Findspot of the Kraggenburg cog.

Fig. 22.3. Topographical and Military Map of the Kingdom of the Netherlands (scale 1:50.000) from 1861 with the location of the Kraggenburg cog and Vollenhove castle.

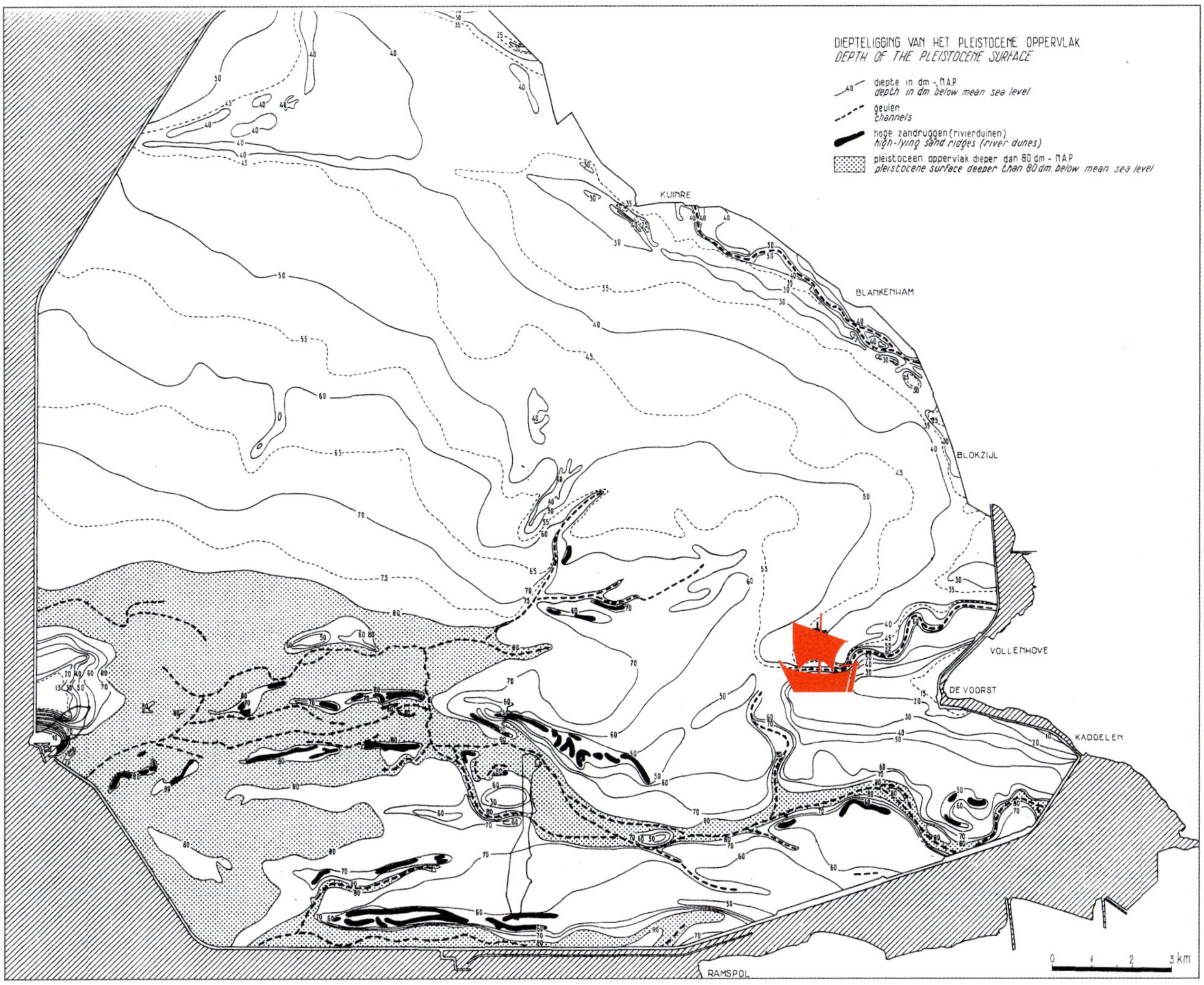

Fig. 22.4. The location of the Kraggenburg cog slightly to the north of the stream channel (Steenwijker Aa/Steenwijkerdiep) in the Pleistocene surface and the depth of it (after Wiggers 1955, appendix 2).

nal top may thus have been slightly higher than the depth of approximately 1.2 m -below surface recorded with the wreck. During the 2007 survey, a hard peat/bog iron layer was found in the recently dug drainage trench, in which the wreck had been discovered (and damaged), at approx. 3.5 m below surface. This could indicate the bottom of a stream channel, which was covered by sand. The soil profile cannot provide any dates relating to the sinking of the ship other than that it must have taken place (some time) before approx. 1600.

22.1.3 Medieval habitation

The medieval settlement closest to the find spot of the wreck was Vollenhove. The completely disappeared castle of Vollenhove, which had been located directly south of the church at the harbour, had been founded in 1168 by Godefried van Rhenen to keep the neighbouring Frisians in check. In 1354, Vollenhove received city rights from Jan van Arkel, the bishop of Utrecht. In the first half of the 14th century, the castle and the city had experienced turbulent times: the castle had been besieged in 1309 by the independent peasants of nearby Stellingwerf. The then bishop of Utrecht, Guido van Avesnes, had come to the relief of Vollenhove with a considerable army in ships.[4]

22.1.4 Condition and wreck formation

Condition

The ship's fragment (most of the bottom of the ship with a number of floor timbers still more or less in place) lies almost horizontally in the sand. The approximately 50 cm wide drainage trench responsible for the discovery of the ship was dug at an angle through the starboard half of the bottom (fig. 22.5). During this, the shell planks plus the

4. Van der Aa 1848, 815-822.

parts of the frame above them and the keelson with riders were severely damaged and moved. The detached stern hook was found about 50 m away from the ship's fragment; a loose frame was even found about 300 m south of it. Prior to the excavation work for the water features, the wreck probably consisted of almost the entire bottom of the ship with at least one stern hook and the keel plank, the first two strakes on either side of the keel plus large parts of the third bottom strakes. Next to the stern part of the ship lies a part of the clinker-built shell of the starboard bottom. There were at least ten floor timbers on the bottom, while the pieces of keelson and riders were probably still attached to the frames. A few V-shaped floor timbers were found near the ship's fragment. It is not clear whether they were moved as a result of the recent excavation work or whether they were already washed out of the ship's hull during the wreckage process. The same may apply to the stern hook and the frame found loose, while they may also have been pushed off the bottom as a result of the movement of ice. The quality of the wood varies between very poor and reasonably good.

Wreck formation

The vessel probably got stuck on the shallows with a south-westerly wind (storm?) and worked its way through the thin Almere clay package and into the underlying sand(cover) fairly quickly. It is also likely that the elements of the vessel became detached quite soon after the stranding, with the sides at the turn of the bilge breaking up, collapsing and drifting away. In most cogs, as here, the bilges are quite vulnerable because the scarfs between the floor timbers and the first futtocks are usually located at the lower two or three strakes of the sides (Plates 6.I and 6.II A and B of the Nijkerk-II cog). The remaining bottom with frame and keelson was then covered with Almere-sedimentation layers and finally with a thin package of Zuiderzee deposits.

Fig. 22.5. View from rear to front, with on the right the drainage trench slanting through the starboard part of the bottom and on the right separately found ship parts and washed-up timber. Open day for the public on 17 July 2007.

22.2 The ship

Initially, no more could be said about the ship's fragment and its various parts than what could be seen through observations at the wreck on 17 July 2010 during the 'open day' for the public at the excavation site plus observations on the pieces found previously that had been taken to the RACM in Lelystad in 2007. The report of the IFMAF exploratory study day in 2010 was published in 2017. A number of dimensions and details of the parts and the ship's fragment were derived from this and incorporated into the present text in mid-2018. However, the following is mainly based on the similarities and differences that could be identified in this study thanks to the writing up of the reports of all the other cogs. All the ship's timbers that were excavated are of oak (*Quercus spec.*).

The bottom and the sides

It can be said with certainty that the bottom consisted of the keel, the stern and stem hook and three strakes on either side of them; also that the ship had been carvel-built, turning into lapstrake or clinker-built construction both aft and fore. On the port side of the afterbody, only a few fragments of shell planks from the side were found. Other parts of the sides (shell planks, frames and ceiling boards) were not discovered. The preserved part of the underwater hull is about 9 m long. The largest width of the bottom is 2.5-2.7 m.

Keel

The keel has been preserved at a length of 6.1 m and is striking in its thickness and width. At frame 4, the width is 11 and the thickness 13.5 cm, at frame 10 respectively 15 and 15 cm, while at the scarf with the stem hook, the keel is 10 cm wide and 13 cm thick. The thickness of the bottom planks varies between 3 and 4 cm. These measurements mean that the keel is more of a keel beam than a keel plank. The latter is common for most of the cogs in this study, with a thickness of 2.5 to 3 times the thickness of the planks almost being the rule. To the fore, the keel beam has a sloping flat scarf (the connection to the stem hook). The aft is broken off, so that the full length of the keel (including the stern hook) cannot be determined. Between frames 6 and 7 there are three or four nails that were driven into the wood from the inside to the outside, while between frames 9 and 10, three nail tips were discovered that had been inserted from the outside and then bent and hammered down, plus a nail hole. They lead to the assumption that the keel beam was composed of two parts connected by a sloping flat scarf of approx. 1.5 m.

Stern- and stem hooks

The stem hook found loose had been broken in two parts. The horizontal (keel) part includes the sloping scarf with the keel of approx. 1.3 m long. The chamfered stern part of the hook is about 1.25 m long and protrudes up to about 1.1 m above the keel line. The front line of the stem hook has a rake of approx. 122°. In the sides of the stem hook, rabbets have been cut in order to attach the first ca 70 cm of the garboards (in the keel section) with nails. In the stem section, there are rabbets for attaching the plank ends of the strakes with nails (Fig. 22.6). At the top there is a sloping lip scarf forming the connection to the stempost (which has not been found). The stern hook and sternpost were also not found.

Strakes

The garboards or A strakes are ca 35 cm wide, the B- and C strakes are 40-43 cm wide and 3-4 cm thick. The garboards were almost certainly composed of two planks, while the B- and C strakes were made up of at least three planks. The planks of each strake had been joined to each other with sloping flat scarfs of ca 40 cm long and connected with nails. The overlapping board ends protrude ca 1-2 cm out of the sheathing. The clinker-built joints were made with iron nails with a length of 60-100 mm, which have relatively flat heads with a diameter of 28-30 mm.

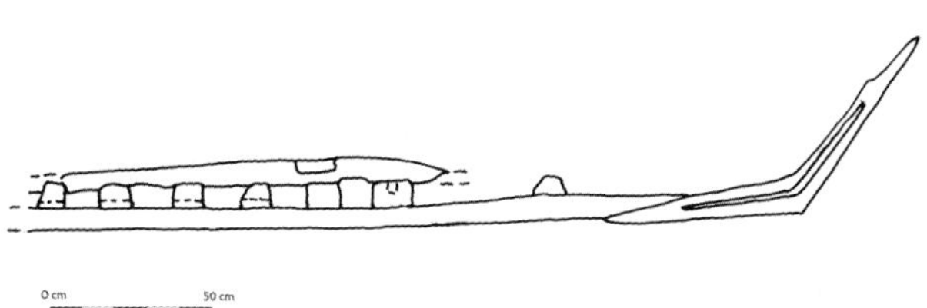

Fig. 22.6. Reconstruction of the front section of the keel beam with stem hook. Drawing after: IFMAF Blok & Van den Hoek 2017, fig. 3.10.

Caulking method: sintel clamped moss caulking

The carvel seams of the bottom were sealed on the outside with sintel clamped moss caulking. The clinker-built part was caulked at least on the inside, but probably, judging from the date of the wreck (Section 22.8), most likely also on the outside. The latter could not be determined with certainty, because the wreck was not dismantled and the seamed parts are in fact no longer available. It is striking that the carvel seams of the bottom on the inside have sintel clamps applied every 25 cm or so and that it appears that the seams, which have an inverse V-shape, cover all, or most of the thickness of the plank.[5] This has only been observed once before, in a 13th-century cog.[6] The sintel clamps that were used are of the D2 type, which was in use between ca 1275-1350.[7]

(V-shaped) floor timbers

Indications for at least 19 frames could be established. In addition to the wreckage, some parts of the frame were found loose. The floor timbers are 14-21 cm wide and 13-17 cm thick. Three of them are the narrowest; they sit about 25-30 cm apart and have inverted trapezoidal limber holes on the underside. Two of the limber holes are above the seams between the keel and the garboards and one above the second and third strake on either side of the keel. The floor timbers are fastened to the keel with one wooden pin and connected to the shell planks with two pins (Ø 2.5-3 cm) per plank width. Nail holes were present on the sloping scarfs found on the floor timbers (the scarfs joining them to the futtocks that had originally been on top of them). These nails had been hammered through the thin ends of the futtocks instead of a wooden pin. Next to the wreck, at least one piece of a futtock was found. On the stepped outside, flat surfaces of the plank widths of five clinker-built strakes of the shell can be seen. The height of the flat surfaces/width of the planks varies between 25 and 32 cm. The V-shaped floor timbers in the fore part of the bottom are not connected (with treenails) to the keel beam and stem hook. Two square holes were cut in the upper part of the foremost floor timber/frame of the bottom that was still present (figs 22.6 and 22.7a and b). They will be discussed again later in connection with the keelson/mast step.

Keelson/mast step

The beam found in 2007, which had been broken into pieces and had the hole for the foot or heel of the mast was broken off at one end. The other end, the front end, was probably 'cut off' at an angle by a machine during excavation work. The almost 2.5 m long piece could be reconstructed back into its original position during the exploratory study in July 2010. The almost rectangu-

5. Blok & Van den Hoek 2017, 24.
6. Nijkerk-I cog, Chapter 8, Section 8.2.2.
7. Vlierman 1996a, 58 table II.

Fig. 22.7a. The front part of the bottom, seen from starboard, with on the right the floor timber with the two rectangular holes.

Fig. 22.7b. Overview of the bottom with on top left the recent disturbance of the drainage trench. In the middle of the front, the floor timber with the two rectangular holes and the light coloured narrow keel beam underneath. Next to/under the bottom planks in the foreground are pieces of (light brown) washed-up wood.

lar shape of the thickest part with the hole (dimensions 30x12.5 and 8.5 cm deep), the mast step section, seems to have been part of a long type of keelson. It is widest at the hole and gets both narrower and less high towards aft and fore. The underside of the keelson has cut-outs to fit over the frames. Instead of a long keelson there may have been a shorter one, similar to the one found in the Marknesse cog (Section 12.2.2 and figs 12.15a and b). Nevertheless, it is remarkable that in the Kraggenburg cog, two rectangular holes were cut in the front of the floor timber on the bottom, in which vertical posts must have stood. The foremost part of the mast step/keelson had been attached in between the holes at the frame (figs 22.7a and b). The slightly dried up upper part (showing cracks) in the flat part of the left-most frame at port side is probably where one of the riders was attached, while the part above the keel, on which the mast step section probably rested, is slightly higher than the rest. In the Marknesse cog, two of these holes have also been found at probably almost the same place in the ship (figs 12.15a and b). However, they had not been made in a frame, but in a short, thick plank, which had been recessed into the top of two frames and attached. In the case of the Marknesse cog, it is assumed that the vertical posts formed part of a box-shaped construction protecting a removable mast and kept it free of the load in the hold (Section 12.2.2, figs 12.23a and b and Plates 12.II A, B and C). Remarkable are the two rows of treenail holes in the step for the mast, one of which is located above the frame. Usually one treenail-sized hole is used to discharge the rainwater that runs down from the mast. The other holes may be related to an earlier use of the piece of wood.

Repairs

On the bottom, a number of repair planks were found above bad spots in the planks. They have been attached with nails. Some of them are partly below the frames, from which it can be deduced that they were probably already installed during construction.

22.3 Reconstruction

The above observations and data are insufficient to make a definitive analysis and reconstruction of the vessel. However, it is certain that these are the remains of a cog. It is not possible to say whether it is a large cog (with a longitudinal keelson, integrated mast step section and protruding crossbeams) or a smaller, seagoing cog. The latter, however, seems the most plausible.

22.4 Model

No scale model of the Kraggenburg cog has been made.

22.5 Inventory

In and near the wreck a number of objects or parts thereof were found that were part of the ship's inventory and equipment.

22.5.1 Distribution of the finds

The locations of the finds in 2007 have not been measured in. During the investigation in 2010, the X and Y coordinates in relation to the main ribbon were determined, the Z coordinates in relation to NAP. Due to the lack of sufficiently correct data, the tables simply state the finds from the category in question. Whether they were found in the aft, middle or foreship or outside is not taken into account here. The numbers of the objects shown in the text, the find list and in the tables are the field find numbers. With the stoneware and earthenware sherds, only the lowest find number is mentioned. In the list of objects, the other find numbers of the sherds belonging to the same jug or pot are given in brackets behind the description.

22.5.2 Ship's equipment and inventory

Ship's equipment

A few pieces of rope are the only finds that can be placed in the category of ship's equipment.

Operational equipment

There were no items found in this category.

Administration and navigation

Nothing from this category was found in the wreck.

Tools

A number of tools were found in and near the wreck. An axe with handle and a carpenter's adze represent the carpentry tools. A marlinspike belongs to the craft of rope splitting and sail making, while a grindstone should be classified under miscellaneous. The axe (Fig. 22.8-32) is of the standard axe type with an almost symmetrical

Table 1. Distribution of items

Category	
1. Ship's equipment	line
2. Operational equipment	-
3. Administration and navigation	-
4. Tools	axe with handle, carpenter's adze, marlinspike, grindstone
5. Household items	-
6. Cooking, eating and drinking	bricks, fire tongs or trivet, flint, peat, coal, sherds of *Kugeltopf* (globular) pots and stoneware
7. Victuals	sherds of blue-grey and stoneware, bones
8. Personal equipment/weaponry	-

Table 2. Ship's equipment

Category	Item	Number	Findspot
1.1 Ground-tackle	-	-	-
1.2 Sails	-	-	-
1.3 Cordage	line	3*,9*	-
1.4 Blocks	-	-	-
1.5 Windlass	-	-	-
1.6 Pump, drain	-	-	-
1.7 Navigation aids	-	-	-
1.8 Boats	-	-	-
1.9 Spare parts	-	-	-

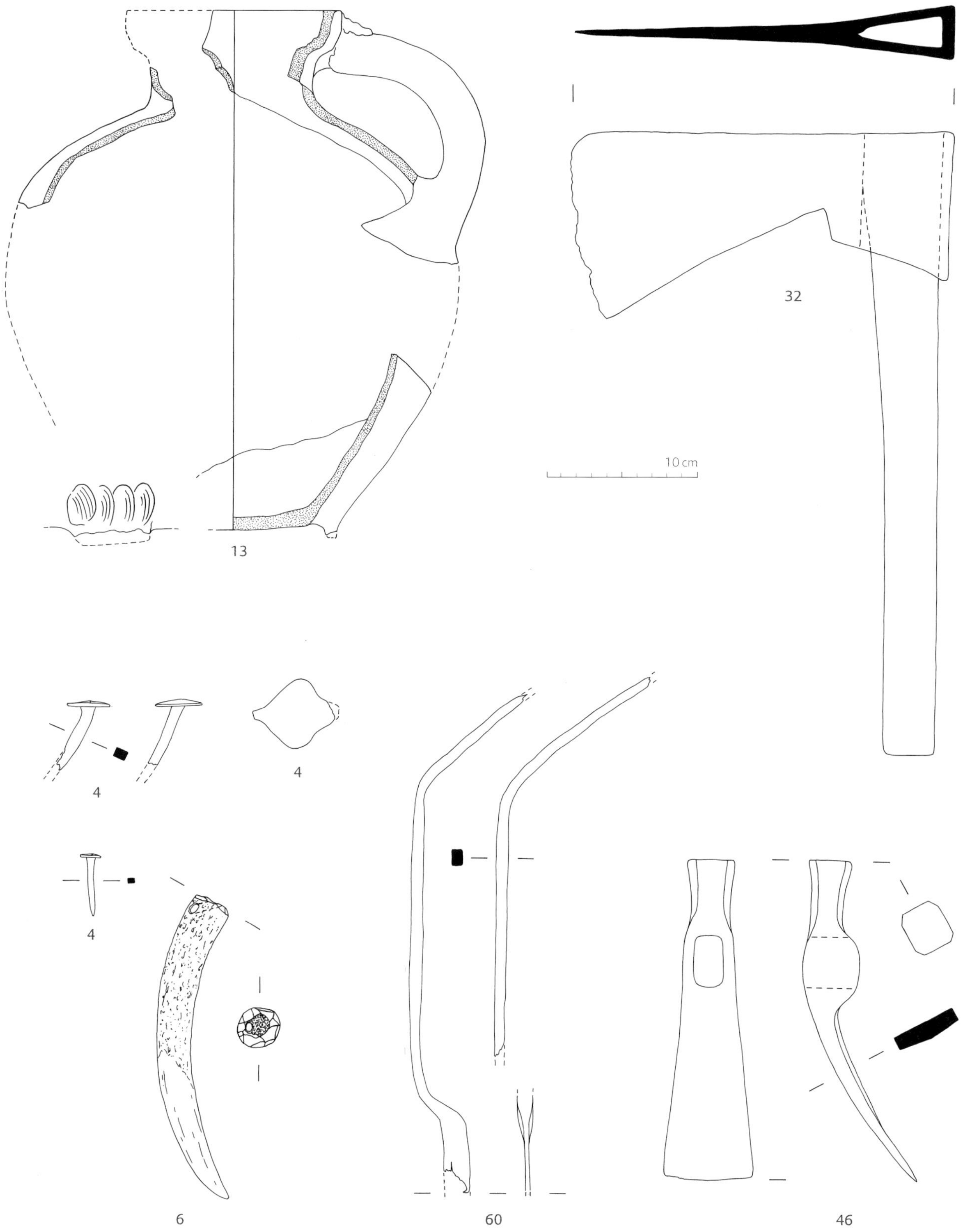

Fig. 22.8. The axe with handle (32), the adze hammer (46), the split pin (6), the fire tongs (60), the blue-grey ware jug (13), a sintel clamp (4) and nails (4).

cross-section.[8] The cutting edge is damaged so that it cannot be said with certainty that it was a symmetrical axe. It may have been sharpened as a left-handed broad axe. The elongated triangular blade has a fairly straight top line that continues to the eye. The lower line of the leaf is also almost straight and runs diagonally up to the eye. The front of the eye runs diagonally backwards to the also sloping and slightly hollow lower line. The shape of the axe is very similar to the axe from the Nijkerk-II cog.[9] The handle of the axe is 33 cm long and ca. 3 cm wide, ca. 4 cm at the top near the eye. The handle has an oval section and was broken into three pieces. The type of wood has not yet been determined.

The (ship) carpenter's adze (Fig. 22.8-46) is the other piece of carpentry equipment. The blade is relatively strongly curved and has an edge of 55 mm wide. The hammer part is forged octagonal, while the two sides flanking the eye are extended downwards, as it were, in order to better attach the adze to the handle (which was not found). The adze is very similar to the example from the Dronten-I cog.[10] The third tool is a split pin or marlinspike, made from a cut-off antler point of a red deer (fig. 22.8-6). The curved point is worn smooth by splitting. Higher up, clear wear and tear from prolonged use can be seen. At the top, a small hole has been drilled at an angle, through which a string or leather strap could be attached, for example, to hang the marlinspike on a belt. The fourth object in the tool category is a rounded/worn-off piece of red sandstone (41). The stone has almost certainly been used as a grinding stone. There may be traces of soot or fire on it, which cannot be explained very well.

Household items

There were no household items found.

Cooking, eating and drinking: fireplace

A true fireplace was not discovered. However, a number of pieces of red brick (very large and thick bricks) have been found, some with soot or fire marks (14, 18, 21, 26, 33 and 47). Two pieces (26 and 33) are almost complete and measure 28x14x6.5 cm and 27.5x13.5x6.5 cm, respectively. They were almost certainly part of a fireplace consisting of a flat wooden box.[11] In the Kraggenburg cog, a similar box was probably filled with sand with the bricks side by side (a floor) on top. The fact that the crew had used fire and/or cooked on board can also be deduced from the remains of fire tongs (fig. 22.8-60), whose pieces are incomplete and bent. The fire tongs from the Dronten-I cog (fig. 11.20-20 and 20a) and the Marknesse cog (fig. 12.26-13) are comparable. The two pieces of peat (42 and 53) and a piece of coal (22) also point in that direction. Furthermore, a number of lumps and chunks of flint (1, 2, 4 and 11) have been found, which may have been intended to make fire.

Cooking ware

The small sherds of *Kugeltöpfe* (13, 30, 34 and 56) are the only sherds found that can be attributed to one or more cooking pots.

Tableware

Two wall sherds of near-stoneware (25 and 58) may have come from a (drinking) jug. They have a greyish yellow sherd and yellow stripes on the outside.

Table 3. Tools

Category	Item	Number	Findspot
4.1 Carpentry	axe with handle, carpenter's adze	32, 46	-
4.2 Caulking/maintenance	-	-	-
4.3 Splicing/sailmaking	marlinspike	6	-
4.4 Miscellaneous	grinding stone	41	

Table 4. Cooking, eating and drinking

Category	Item	Number	Findspot
6.1 Fireplace	bricks	14, 18, 21, 26, 33, 47	-
6.2 Fire-tending tools	fire tongs, flint	60, 1*, 2*, 4, 11*	-
6.3 Fuel	peat, coal	42, 53, 22	-
6.4 Cooking ware	*Kugeltopf* sherds	13, 30, 34, 56	-
6.5 Eating and drinking gear	sherds proto-stoneware	25, 58	-

8. Vlierman 1985b, 34.
9. Chapter 6, section 6.5.2, figs 6.49 and 6.50 A en B; Vlierman 1985b, 46-48, fig. A 3.
10. Chapter 11, section 11.5.2, figs 11.21-11.23; Reinders *et al.* 1980, 14 and Appendix 3.
11. See the Nijkerk-I cog, chapter 8, section 8.5.2 and figs 8.50, 8.51 and 8.52.

Victuals
Sherds of at least three large near-stoneware jars have been found (12, 39 and 54). The first has a pinched foot and a finely ribbed neck, of which the upper rim was smooth. The outside is brown-grey, the inside greyish yellow. A second large near-stoneware jar (39) has a dark grey sherd. The outside is greyish brown, while the inside is striped greyish orange. Presumably the upper rim had a collar neck. The jar is probably directly comparable to jar 2 from the Dronten-I cog.[12] The sherds of the third jar (54) are orange on the outside (discoloured?). There seems to be soot on the inside. It is possible that the sherds come from a jar that had been broken long before the sinking of the ship, of which a part had had a second function. The sherds of a fourth jar that may have served as a beverage container are made of blue-grey pottery. The jar has a more or less flat bottom with (four) squeezed-out shell feet, an inverted pear-shaped body, a short narrow neck with a collar edge and a sausage ear (figs 22.8-13 and 22.9). The near-stoneware jars undoubtedly come from Langerwehe/the surroundings of Cologne, the blue-grey jug from Elmpt. The bone fragments (18 and 49), which are unidentifiable, undoubtedly originate from meat that was once consumed on board.

Personal equipment/weaponry
No objects were found that could be classified in this category.

22.6 Cargo and merchandise

No cargo or merchandise was found in or near the wreck.

22.7 Ballast

No ballast was found in or near the ship either.

Fig. 22.9. Two sherds of a blue-grey ware jar, adze hammer, split pen and a piece from fire tongs.

22.8 Dating

Soil profile
The location of the cog is close to the coast, where it was rather shallow due to the nearby high ascending Pleistocene surface. As a result, only a relatively thin Almere- and Zuiderzee sedimentation package was able to deposit there. The homogeneous character of the Almere sedimentation does not allow for subdivision. Therefore, the soil profile does not provide any other data with respect to the dating of the sinking of the cog, other than that this must have taken place (a considerable time) before ca 1600.

Finds
The finds can help dating the sinking of the ship. *'Kugeltöpfe'* globular pots were in general use at the end of the 13th and beginning of the 14th century. The sherds of near-stoneware of three storage jars and a drinking jug provide the most accurate date for the ship's sinking. Near-stoneware precedes fully sintered stoneware, which was introduced in the years 1320-1330 and became common very quickly.[13] It therefore seems very plausible that the ship sank in the early 14th century or shortly afterwards, as sherds of stoneware would almost certainly have been found if the ship had sunk any later. The similarities between the axe and the adze from this cog to the exam-

Table 5. Victuals

Category	Item	Number	Findspot
7.1 Fresh water supply	proto-stoneware, blue-grey jug	12, 39, 54, 13	-
7.2 Supply of beverages	-	-	-
7.3 Livestock	-	-	-
7.4 Food provisions	boden	18, 49	-

12. Chapter 11, section 11.5.2 en figs 11.29-2 en fig. 11.34.
13. Janssen 1983, 173-175.

ples found on the Nijkerk-II and Dronten-I cogs respectively, plus the pottery found in the latter wrecks support this date: the Nijkerk-II cog is dated to around 1350 and the Dronten-I cog to the first half of the 14th century (with a preference for the transition between the first and second quarter). Both of these wrecks, even though their finds are so similar to the Kraggenburg cog, can of course not provide hard data for the dating of the sinking of the Kraggenburg cog, but they at least support the date given by the presence of near-stoneware and the absence of true stoneware.

Sintel clamps and nails
The sintel clamps used to fix the moss battens and the moss caulking are of the D2 type, which was in use between ca 1275-1350. Recent research suggests that this model in particular (fig. 22.8-4) was in use around and after ca 1300.[14] The shape of the carvel seams (triangular over the entire or almost the entire thickness of the planks), and the presence of sintel clamps applied every 25 cm on the inside above the seams are techniques known from the 13th-century Nijkerk-I cog and older vessels. The nails used have rather flat forged heads (fig. 22.8-4) which generally also indicates a date in the 13th century.

Dendrochronology
Seven samples of the wreckage were taken in 2010 for dendrochronology. Three were taken from bottom planks and four from parts of the frame. The three bottom planks and one of the frames could be dated with the reference chronologies NLNOOR 1E, NLNOOR 10 and NLHOOR 01, that is wood used in the Northwestern Netherlands and in the city of Hoorn, the Netherlands. The most recent felling date is AD 1310 ± 5. So it is very likely that the vessel was built between ca 1310-1320. The full report of the dendrochronological study is given in the appendix.

22.9 Summary

Ship
The wreck and the parts found loose on lot T 25 in the Noordoostpolder were undoubtedly the almost complete bottom of a cog before the disruption during the excavation work for the water features. The ship's hull probably broke up relatively soon after the probable foundering on the hard seabed and not far from the coast. The stern- and stemposts, the sides and the deck(s) then drifted away. The objects from the inventory found in the ship indicate that the ship sank due to a catastrophe and was not sunk intentionally as, for example, a ship at the end of its useful period. Little can be said about the ship itself. What is certain is that the bottom on either side of the keel consists of three strakes and that there was probably a keel beam instead of the more usual keelplank, and that this keel beam was made of two parts connected by a scarf. It is uncertain whether the mast step section found with the hole for the heel of the mast was part of a long keelson or of a short one, such as was found in the Marknesse cog (Chapter 12). In other words, it remains unclear whether we are dealing with a large cog (with a long keelson and crossbeams protruding through the shell) or with a smaller seagoing cog. Two small rectangular holes that were observed in the foremost surviving frame suggest a similar construction as found in the Marknesse cog.

Inventory
Large and thick bricks with burn marks, parts of a fire tongs, peat, coal, flints and some sherds of *'Kugeltopf'* clearly indicate that the ship had a fireplace and that cooking was done on board. The sherds of three large near-stoneware jars and of a blue-grey jar may have come from objects intended for storing beverages. A few sherds are from a near-stoneware drinking jug. An axe with a handle, a carpenter's adze, a deer horn marlinspike and a grindstone are the tools found in the wreck.

Dating
The date of the foundering of the ship cannot be seen (even approximately) from the disturbance of the soil profile. The incomplete sediment pattern so close to the coast and the homogeneous composition of the Almere package provide no information other than that the vessel must have perished well before ca 1600. The near-stoneware sherds found in the wreck provide the most accurate dating of the sinking of the ship. Near-stoneware was in general use until the introduction of the fully sintered stoneware in the period ca 1320-1330. The foundering must have occurred in the first half of the 14th century, or more precisely, probably in the period ca 1320-1335. Dendrochronological research has made it clear that the youngest felling date of the dated wood is 1310 ± 5. This means that the vessel was most probably built in the period 1310-1320, most likely in the north (west) of the Netherlands. The type of sintel clamps used on the caulking on the inside and the shape of the seams of the bottom planks (which have an inverted V-shape that probably extended over the entire thickness of the planks) are only known from vessels dating to the 13th century or earlier. The construction of the Kraggenburg cog probably represents a transitional phase that may have taken place earlier in other areas.

14. Vlierman 1996a, 58 table II; Vlierman, in preparation a; Section 29.2.

The Kampen (IJssel) cog[1]

23.1 Find conditions

23.1.1 Discovery, salvage, field evaluation and investigation

At the request of the Directorate-General for Public Works and Water Management (Rijkswaterstaat), Programme Directorate for Room for the River (RWS-PDR), between 15 and 31 October 2012 the Archeologisch Diensten Centrum (ADC Maritime), working with Baars CIPRO, performed a field evaluation of a cog-like shipwreck in de Lower IJssel near Kampen (figs 23.1a and b and 23.2).[2] The wreck was found in 2011 during preliminary archaeological research that took place in the context of activities to dredge the summer bed. Fourteen objects were found outside of the cog-like wreck, including two smaller shipwrecks and a number of structural elements that almost certainly belonged to the cog. The two smaller wrecks were a punt-like boat and a barge-like vessel.

As of January 2015, the ADC ArcheoProjects, represented by W.B. Waldus and K. Vlierman, agreed to lend Vlierman's expertise as a senior advisor specialising in ship archaeology, and specifically his knowledge of late medieval cogs, to the Isalacogghe consortium.

The overall goal of the project was to safely, professionally and methodically document, salvage and preserve the archaeological remains from the findspot of the IJssel cog, in accordance with the tender documents provided by the Directorate-General of Public Works and Water Management (Rijkswaterstaat) and in mutual consultation with the archaeological and non-archaeological stakeholders involved in this project.

The boat and the barge

On 29 September 2015, object 14 which had initially been typed as a punt was salvaged by the Isalacogghe team. Based on the information from the 2012 field evaluation, it had already been established that the visible part of the wreckage strongly resembled a vessel that had been excavated by RIJP in 1971 on lot K 73/74, approx. 5 km west of Dronten.[3] Immediately after the salvage, it was evident that the similarities were striking, and that the near-complete vessel from Flevoland would be able to provide the missing information for the incomplete shipwreck taken from the IJssel, making it possible to achieve a full reconstruction of the IJssel boat on paper. An offer was made to measure the salvaged wreck by hand, draw the wreck and use 3D recordings of the frames to make reconstruction drawings. The set of drawings was completed at the end of 2015. After that, I contributed to the report on the wreck.[4] Section 26.1 describes the boat on lot K 73/74 in Eastern Flevoland, while section 26.2 describes the vessel near the IJssel cog. Section 26.3 describes the barge fragment salvaged on 22 October 2015 that was found beside and partly under the forebody of the cog. Section 26.4 focuses on the (original) function of the boats and the barge.

The cog

During the visit to the diving vessel on 23 October 2012, it was possible to work in direct contact with a diver to advise on where to look for sintel clamps from the caulking and identify which accessible parts of the wreckage would be suitable for dendrochronological analysis. At that time, it was not possible to say anything about the date of the ship with any certainty. A sample of the sintel clamped moss caulking taken immediately after that clearly showed that the clamps used here were type D2/E. This model was in use from the late 14th century to the mid-15th century.[5] The cog was most probably built in the first half of the 15th century. The Exploratory Field Survey, Underwater Evaluation[6] states that the stem was found on the south side of the wreck, which was oriented from south-southwest to north-northeast. However, the shape of the contour line, the windlass cylinder and beams in appendix 27 of ADC report 3300 suggest that this must have been the afterbody of the vessel. During the dig in 2015, it was proven that it was in fact the afterbody. In some cases, the correct position can be determined at an early stage by looking at the direction of one or several sloping flat scarfs in the shell strakes. Other tentative impressions and conclusions in response to the ADC report, completed at the end of 2014 and provided

1. Translated by Joy Phillips.
2. Waldus *et al.* 2012.
3. Reinders *et al.* 1980, 17-29, 43-46 and Appendices 3, 4 and 5; Vlierman 1983, 3-11; idem 2010, 193-194.
4. Waldus *et al.* 2018, Chapter 12.
5. Vlierman 1996a, 53.
6. Inventariserend Veldonderzoek Onderwater-Waarderend ; Waldus *et al.* 2012, 31.

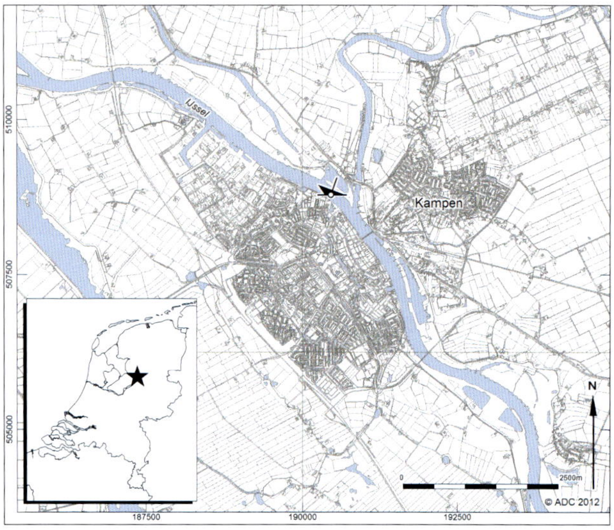

Fig. 23.1a. Ordnance map of Kampen and surroundings, with the cog's findspot in the river IJssel (ADC Maritime).

Fig. 23.1b. The river IJssel at Kampen with the wreck location.

Fig. 23.2. Topographic and Military Map of the Kingdom of the Netherlands (scale 1:50,000) of 1859 with the location of the Kampen (IJssel) cog marked on it.

in Chapter 23 of the present manuscript, have not been taken into consideration. They will be covered later on, in the definitive description.

On 10 February 2016, the shipwreck of the Kampen (IJssel) cog (also known as the IJssel cog) resurfaced (fig. 23.3).[7] Two days later, it was finally possible to enter the wreck, which had been laid on a pontoon tied to the IJsselkade, near Van Heutzplein in Kampen (fig. 23.4). It was the first opportunity to explore the inside of the vessel and see the complexity of the wreck due to warping and the detached structural elements still present inside the shipwreck. Also what this meant for the documentation and reconstruction of the vessel. The researchers assumed at the time that it would be possible based on the planned documentation to create 3D reconstructions of the hull using 3D photography and special computer software.

The agreement of January 2015 provided for several days of consulting during salvage and documentation. Providing some of the documentation and making the reconstruction drawings of the boat had already changed this advisory task. During documentation of the loose elements found in and around the cog wreck during the excavation, which had already been salvaged and transported to Lelystad, I was asked to join the research team.

Based on decades of experience in documenting shipwrecks in the province of Flevoland and reconstructing all the other cogs found in the Low Countries, it quickly became clear to me after further investigation of the shipwreck that the 3D reconstruction of the IJssel cog would probably only be possible if 2D reconstruction drawings were made first, based on supplementary data collected manually. Neither the 3D nor the 2D reconstructions had been taken into account in the agreements with the commissioning client (Rijkswaterstaat). Based on those agreements, subsidy applications had been submitted to various granting authorities.[8] In the summer and early autumn of 2016, there was enough opportunity to collect the minimum data deemed necessary, based on the assumption that it would be possible to produce 2D reconstruction drawings. It was also agreed at that time that all the research data, photographs and drawings of the shipwreck and the finds, and the 2D reconstruction drawings, as well as the monograph on the IJssel cog to be produced by ADC, would also be allowed to be used in this overview of all cog finds from the Low Countries. Reconstruction was able to start in January 2017. During a study trip on 29-30 June 2017, the drawings that had been made up to that point, at a scale of 1:10, were displayed in the German Maritime Museum (*Deutsches Schiffarhtsmuseum*, DSM) in Bremerhaven.[9] The full set of drawings was completed on 30 November 2017. The method used for the 2D reconstruction of the cog and the boat is described in a separate section of the ADC monograph[10] and in sections 23.2.1 and 26.2. This information is vitally important for planning future projects, which could ultimately consider creating a 3D reconstruction of the vessel for the purpose of producing lines plans of the ships and performing all sorts of calculations regarding water currents, sailing characteristics and so on.

Fig. 23.3. The cog being lifted from the river IJssel, 10 February 2016 (ADC Maritime).

Fig. 23.4. The cog being placed on the pontoon in the IJssel in front of the Van Heutzplain, 12 February 2016 (ADC Maritime).

7. For detailed information on the shipwrecks, I made grateful use of ADC Monograph 24 (Waldus *et al.* 2018). With regards to the principles, organization and technical aspects of the salvage and the 3D photogrammetry etc., please see Chapters 3, 4, 6, 7, 8, and 9 of the monograph.
8. Grants were given by the Netherlands Organisation for Scientific Research (NWO), the municipality of Kampen, the province of Overijssel and the Samenwerkende Maritieme Fondsen (an alliance of six historical maritime funds in the Netherlands). The project will be published in the International Journal of Nautical Archaeology (IJNA).
9. During the Fourth Project Workshop from the North Sea to the Norwegian Sea - Interdisciplinary studies on the Hanseatic League, FROUG 2017.
10. Vlierman & Waldus. In: Waldus *et al.* 2018, Chapter 15.

23.1.2 Geology and stratigraphy[11]

The middle and lower IJssel Valley during the Pleistocene/Holocene eras

Based on research in the middle and lower IJssel Valley, it can be confirmed that the IJssel formed a locally functioning river valley during the transition from the Pleistocene to the Holocene. Strikingly, the lower course of the IJssel contains various meanders that are comparable to the Overijsselse Vecht River in terms of morphology, depth and date. The Lower IJssel may have functionally been part of the river Vecht in the Early Holocene. Based on 14C research, it has been confirmed that the Lower IJssel started to drain water from the Rhine as early as the 9th century. Archaeological finds dating from this period that have been found in the top layer of oxidized peat support this date estimate. It was not until the Early Middle Ages that a definitive breakthrough would have been achieved near Zutphen, enabling the Middle and Lower IJssel to be connected to the Rhine (again). To date, it is not yet certain exactly when the delta near Kampen formed. This process probably started around the Late Middle Ages, when the growing influence of the sea in the IJssel estuaries led to a decrease in water discharge. The increasing influence of the sea and the increased influx of sediment via the IJssel River eventually led to the complete silting up of the course of the IJssel and its estuaries. At the beginning of the 17th century, the IJssel River had become nearly unnavigable due to its shallow waters. In 1694, the IJssel had silted up entirely near the junction with the Rhine, so no water was coming in from the Rhine.

The drilling data that is available, combined with official data on the Current Dutch Elevation (AHN) and depth measurements of the bottom of the IJssel, make it possible to create a schematic cross-section of the river valley approximately 1 km west of the Eilandbrug, a bridge in Kampen (fig. 23.5). There is a layer of peat about 1 m thick over the Pleistocene sand. On top of that, on the north side, there is a packed layer nearly 5 m thick alternating between clay and fine sand. On the south side, this packed layer consists of coarse sand mixed with gravel, with clay on top. The bed of the river is in the Pleistocene sand.

'Medieval water management theory'

Starting from the second series of exploratory investigations at the findspot in 2012, it became clear that the well-preserved 15th-century shipwrecks were not the only extraordinary aspect; the context of the maritime landscape also warranted further investigation. The discovery of three shipwrecks at this location in the IJssel suggested that these vessels were had deliberately sunk here, presumably with the intent of influencing the river's course. Compiling data to test this 'medieval water management theory' played a central role in the dig of 2015. With this in mind, the first step was to map out the ground layers around the shipwrecks by taking a series of soil samples. Perpendicular to the IJssel, a total of five profiles were sampled from the bottom of the IJssel spanning a distance of approximately 1 kilometre. These data were combined with highly detailed data on the topography of the river bottom that was recorded using a survey vessel. The data from these 'multibeam' recordings make it possible to

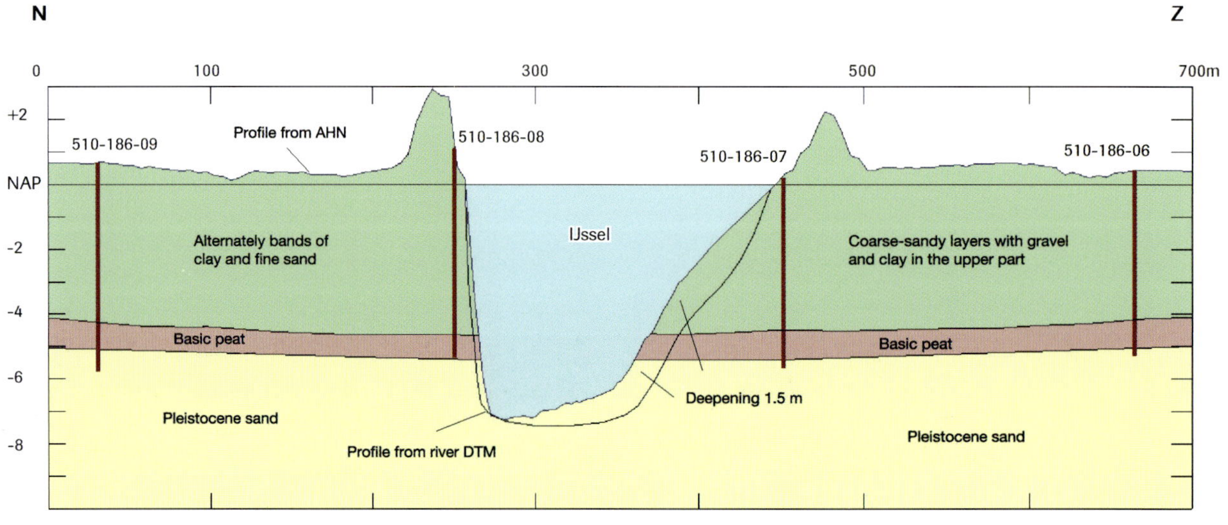

Fig. 23.5. North-south profile of the IJssel valley at ca 1 km west of the Eilandbrug (bridge) (ADC Maritime).

11. Much of the information in this section is derived from Waldus *et al.* 2012, 8-13; Vlierman & Waldus 2017, 84-89 and from Van Dinter. In: Waldus *et al.* 2018, 119-133.

Fig. 23.6. Relief map of Kampen combined with multibeam data of the IJssel riverbed (blue) and the identified fossil channels (broken lines) (ADC Maritime).

meticulously map out the shipwrecks, the ripple marks of the IJssel, and other objects in the riverbed with an accuracy of 3 to 5 cm. When all this information is combined with detailed topographical data on the land, it produces an overview that makes it possible to analyse old soil structures across large areas.

Geological research has shown that the morphology of the bottom of the IJssel is determined by the presence of a layer of highly compacted peat (the Basal Peat Bed or Nieuwkoop Formation). This peat bed was formed from 4000 BC on. Carbon dating has indicated that the top of the peat bed dates back to 2450-2138 BC. The layer extends across large parts of the IJssel river bottom and is covered by a thin layer of sand that forms small ripple lines in the current. It is interesting to note that clear intersections of the peat layer can be observed at a number of locations (fig. 23.6). These are an arch-shaped intersection at the findspot of the IJssel cog and a channel-shaped intersection oriented towards the Noorderdiep. These are zones where meandering channels of the IJssel estuaries have slowly etched their routes into the peat bed. The connection to the Noorderdiep is clearly evident, but it is less straightforward to reconstruct the extension of the channel at the findspot of the IJssel cog, since topographical details are partly camouflaged by the buildings of Kampen. However, it is possible to reconstruct a channel oriented in a westerly direction. It is conceivable that the current Pleviersingel canal and the ditch along Sint Nicolaasdijk in Kampen mark the remaining traces of an IJssel channel that no longer exists. For now, we will refer to this channel as the Brunneperdiep.[12]

When the position of the shipwrecks is analysed in relation to this reconstructed channel, it can be established that the IJssel cog and the barge were perpendicular to the current and the boat was nestled against the inside of the bend. The question then arises as to whether this would have closed off the entire channel, or whether the shipwrecks were sunk in order to serve as a groyne or head. It is also possible that there was a temporary intervention in which the vessels served to stem the current, providing shelter to dam up the Brunneperdiep. Based on the position of the shipwrecks in figure 23.6, the second or third option seems likely. A riverbank reinforcement of split beechwood trunks was found along the boat. Dendrochronological analysis indicates that they were felled in the winter of 1451/52. It can be assumed that beechwood was used in the construction shortly after the tree was felled. Another interesting fact was discovered there: the same tree trunks used to protect the riverbanks were also found in the boat. They were located in the midship section along with a number of small spars, clearly preserved as the remains of the cargo. This fact makes it plausible that these trees were felled around the time when the vessel sank. This may also apply to the other two shipwrecks. In both cases, a minor catastrophe may have occurred. In the first situation, the boat sank while the riverbank protection was being put in place. In the second situation, there may have been a planned intervention in water levels that ended in the loss of vessels and equipment. It is known that Kampen struggled with silting in the IJssel and its estuaries, especially in the 15th century.

On this topic, Van der Aa[13] wrote in the mid-19th century: *"In the year 1475 it was concluded that navigation of the IJssel had become somewhat cumbersome due to sandbanks that had formed in front of the estuaries of the river, gradually creating a growing obstruction that prevented loaded vessels from entering the inland waters. To eliminate these hindrances, signif-*

12. The paleogeographical reconstruction of the IJssel delta in the 14th century by J. Dirkx, P. Hommel and J. Vervloet, Kampereiland. Een wereld op de grens van zout en zoet [Kampen Island: A world on the boundary between saltwater and freshwater] (Utrecht 1996) suggests reconstructing the Zuiddiep at the location of this channel. In this reconstruction, the channel curves in a northern direction back towards the IJssel channel. Maps from the 17th century and later also show a 'Zuyderdiep' in the IJssel estuary, which is confusing. For this reason, and because the topographical data makes it very plausible that the channel continued westward until the Zuiderzee, the channel will be referred to as the Brunneperdiep in the context of the investigation of the IJssel cog.
13. Van der Aa 1845, 278.

icant works were undertaken and major costs incurred to close off the Zuyderdiep and other discharges of water from the Regtediep. And so the Zuiderdiep, behind Brunnepe, 33 rods wide (approx. 120 m) and more than 2 fathoms deep (approx. 3m), was dammed in 1479. Another branch, running between the Greente and the Zuiderwaard towards the sea, received similar measures, while efforts were made to tame the current by riverbank works along the Raas, including building or lengthening dykes. In the year 1540, a Koningsberg man by the name of Pieter van Hoerne, who had already deepened that port most proficiently, dredged the right side of the IJssel estuary to a sufficient depth, thus preventing further silting up. This work does not appear to have withstood the test of time, while new improvements were later made on the waterways draining the IJssel."

It is possible that the work being done to dam or narrow the Brunneperdiep was already in full swing in a nearly dry IJssel riverbed when the river's flow unexpectedly spiked again. Such biblical deluge scenarios can be presumed based on a note in the Kampen city chronicles from 1479, describing that *"swair water van boven"* (which translates as 'heavy water from upstream') caused a breach of the *Suderdiep* (Brunnerperdiep) dam.[14] It can be assumed that the term *'swair water'* in that context refers to a strong and sudden increase in the amount of water flowing from the IJssel. It is unlikely that the reference from 1479 can be linked to the boat and the riverbank reinforcement. It is possible, however, that there were multiple attempts to dam the Brunneperdiep. However, the position of the cog and the aak perpendicular to the river in this specific position and the way in which the cog and the aak remained intact suggest an intentional intervention, and therefore still support the water management theory. In the event of a disaster, the shipwrecks would not have been left where they lay, or there would most likely have been visible traces of breakage or dismantling, indicating an attempt to remove as much of the shipwrecks as possible and to salvage valuable ship parts.[15] It can be stated that the intervention did not have a significant impact on the IJssel; the map made by Jacob van Deventer in 1560 (fig 23.7) shows no traces of their presence, and the findspot can be reconstructed precisely where it is now: underwater and in the middle of the IJssel. It did not pose problems until centuries later. The shipwrecks were an obstacle to ships and fishermen, as is shown by the hundreds of netting weights and four grapnels that were found in the wreck during the excavation.

Fig. 23.7. Bird's-eye view of Kampen by Van Deventer, ca 1560. The findspot of the cog is marked in the river IJssel (ADC Maritime).

23.1.3 Prehistoric and medieval habitation

During the excavation of the cog in 2015, several prehistoric deer antlers and flint tools were found on what would originally have been the south bank of the channel (the Brunnerperdiep). These finds will not be discussed in more detail here.[16]

Archaeological finds such as earthenware pottery, footwear and sintel clamps indicate that the settlement of Kampen was probably founded in the second half of the 12th century.[17] Blockmans writes: "From the second half of the 12th century on, a number of towns arose at trading junctions for long-distance inland shipping. Dordrecht, Kampen, Zwolle and Muiden were founded in the Maas, IJssel and Vecht estuaries; Zutphen and Arnhem were strategically situated at a river fork and a confluence. Utrecht's commercial importance declined from the late 13th century on, since the connections along the inner-city waterways proved less advantageous than for example the connections along the IJssel."[18]

It can be assumed that interventions in the delta landscape date back as far as the very earliest periods in which the delta was forming. Very few excavations have taken place in the outskirts of Kampen thus far. The scarce information that is available indicates that the area was reclaimed from the 9th-10th century on. Habitation in the area was on

14. Kampen City Chronicles, online publication: http:/kroniekvankampen.huygens.knaw.nl/home/, accessed on 5 March 2017.
15. Nothing has been found that supports these scenarios, although it is conceivable that very rapid silting could have occurred, rendering it impossible to conduct salvage operations. It is known that the IJssel delta has moved several metres a year at localised points.
16. Amkreutz & Brattinga and Raczynski-Henk. In: Waldus *et al.* 2018, Chapters 16 and 17 respectively.
17. Van der Heide 1952, 144; Van Nie & Smit 1997, 67-75; Barwasser 1997, 85-89; Vlierman 1997e, 113-121.
18. Blockmans 2010, 76.

terps and high donks (sand dunes overlaid by peat or clay). My own observations in the 1990s confirmed habitation in the 11th and 12th century on a donk slightly north of De Enk (near the new route of the N50 motorway) and at a number of locations in the clay-topped peatlands between Kamper-Nieuwstad and Elburg. Kampen was granted city rights in 1236.[19] There are reasons to assume that Kampen was founded as a trade settlement by enterprising merchants from the Rhineland.[20]

The city is strategically situated on a historically significant shipping route between the Rhineland and the Zuiderzee area. In 1928, German historian R. Häpke depicted the IJssel River – along with the cities of Deventer, Kampen, Zwolle, Zutphen and Doesburg – as a 'harbour channel' between the Rhine and the Zuiderzee, with strong ties to the Germanic hinterland. Häpke pictured it as a triangle, with Kampen at the apex and Maastricht and Dortmund at the base. The important study by F. Petri in 1961 confirmed and built on Häpke's insights and is still generally accepted today. According to Häpke, the strong ties to the German hinterland were characteristic of the IJssel and Zuiderzee regions. Petri's substantiated thesis suggests that the economic regions of the German Lower Rhine and the Dutch IJssel were inextricably linked.[21]

23.1.4 Condition and wreck formation

Condition

During the field evaluation back in 2012, it was discovered that the ship sections protruding above the riverbed and the pieces of oak wood exposed later in the testing wells were well preserved. The documented top view of the exposed sections and the reconstructed cross-section at the level of crossbeam 3[22] (fig. 23.8) suggest that the ship's hull was presumably reasonably intact up to about deck height. The shipwreck was tilted to starboard, which is why the starboard side was deeper underwater and the crossbeams protruding from the shell were still present up to about halfway along the gunwale in some places. During the excavation, it became apparent that this was also the case about midship height. However, the aft and forward sections of this side are missing in a tiered pattern from about halfway up the stem and stern to the preserved part of the midship section higher up. Separate elements were found beside the forebody to starboard that presumably came from the missing section of the side at that point. The bottom half of the ship's hull was fully present, but warped. The afterbody, including the bottom half of the stern construction, was completely twisted to starboard and badly damaged. The structures of the port side were still present from aft to forward, until about one strake below the lowest row of the crossbeams protruding through the shell. All but two of these were incomplete, still hanging from the starboard shell, or detached and lying loose in and beside the ship's hull, along with other structural elements.

A remarkable find was the discovery in 2012 of part of the rigging with putting irons.[23] During the excavation in 2015, the other part of the rigging with putting irons and deadeyes was found in the structure against the outside of the starboard shell. Another remarkable feature was the presence of riders against the inside of the starboard shell as well as loose inside the shipwreck. These beams 2-3 m long were vertically fastened to the ceiling strakes against the top half of the sides. There was also an absolutely spectacular find of the remains of a domed brick oven with a fireplace beside it on the deck, of the starboard side in the afterbody. It is the oldest medieval example to date of a permanent fireplace in a vessel from the Low Countries.

Wreck formation

As stated previously, the Zuiderdiep had a depth of approx. 3 m; the depth of the channel near the cog would not have been much different. This means that the cog, which sank at a right angle to the direction of the current (deliberately or otherwise), would have initially protruded at least 2 m above the water line. If it served as a caisson,

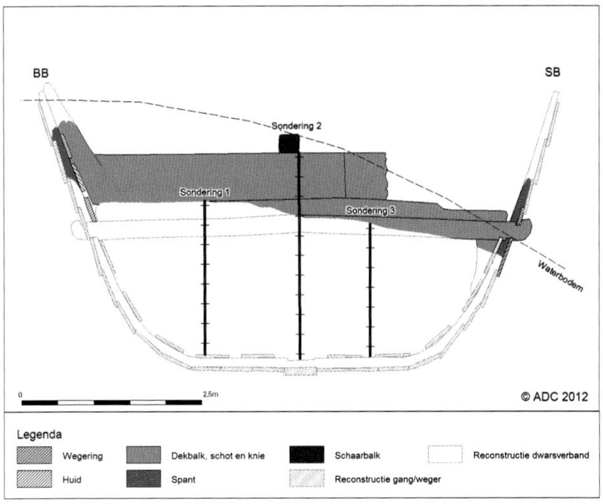

Fig. 23.8. Reconstructed cross-section along crossbeam 3. After Waldus *et al.* 2012, fig. 26 (ADC Maritime).

19. Stadsarchief Kampen 2019.
20. Kossmann 1990, 119.
21. Looper 2010, 110; idem. 2017, 19.
22. Waldus *et al.* 2012, Appendix 27 and fig. 26.
23. Waldus *et al.* 2018, fig. 33. Object 11.

then the ship's hull probably filled up with earth/clay. Boulders would have been more likely, but only a few were found. The raised pattern of the hard Pleistocene channel bed undoubtedly affected the shape of the ship's hull from the start. Over time, the hull would have slowly moulded itself to the shape of the riverbed and banks. After that, the changing current strength of the IJssel would undoubtedly have caused the starboard side of the vessel, which was pointing upstream, to be undermined regularly, gradually tilting to starboard. Moreover, the new situation that emerged as a result would have changed again as new sediment was deposited, ultimately causing the vessel to list heavily to the side. Over the centuries that followed, the upper strakes of the port shell gradually disappeared (probably also as a result of ice drift), and the heavy crossbeam constructions on that side were unsupported and gradually drooped downwards. The various pieces separated and floated away or ended up next to or inside the wreck. The weight of the crossbeam constructions suspended from the starboard shell caused that side of the shipwreck to become increasingly warped and damaged as well. During the excavation of the hull, it was not only the weight of the crossbeam constructions that put pressure on the starboard hull, but also the vast weight of the entire starboard side. Considering the chosen salvage method, it was therefore obvious that the hull structure – which was centuries old, warped, and had deteriorated over time – would slowly continue collapsing and warping during the work to excavate and lift out the shipwreck, especially the entire starboard side and the afterbody.

23.2 The ship

23.2.1 Documentation methods

In 1981, the Ship Archaeology subdivision of the RIJP decided for the purposes of documenting the Nijkerk-I cog to supplement the customary field sketches (top view and cross-sections) with drawings at scale 1:10 of all the structural elements, to be drawn after disassembly of the shipwreck.[24] The drawings show various perspectives of each frame section, precisely indicating the details and including the parts that are presumably missing, marked in dotted lines. Besides the drawings of the keel, stern and stem, the frames and any crossbeams or deck beams, the 1:10 drawings of the *plankuitslagen*, drawings showing details of dowel/treenail holes and repairs) are in principle indispensable in creating reconstruction drawings, building models and/or building a full-sized replica.[25] These reconstruction drawings will also be indispensable for any efforts to rebuild/reconstruct the wreck of the boat and the IJssel cog, restoring the warped sections found within the structure and the separate fragments found in and around the wrecks by returning them to their original shape and inserting them at the correct locations.

In investigating a complex three-dimensional object like a shipwreck, reconstruction drawings are the best way to synthesise the numerous field sketches in perspective, the assumptions regarding the original positions of the ship parts that were found, and the discussions about the ship's construction during the course of the work. For these reasons, the separation between documentation (securing the archaeological information) and reconstruction is artificial in nature. All those involved were under the impression that it must be possible to create a reconstruction of the shipwrecks, considering that they were largely intact; as already indicated in section 23.1.1, options were explored for how to realise the reconstruction. This chapter focuses on the two-dimensional reconstruction of the cog. For more information about the boat, please see section 26.2.[26] Besides the drawings based on the 3D imagery (such as Plates 23.I, 23.II and 23.III), several structures were measured by hand and drawn: a side elevation of the stern and stem construction, a top view of the underwater body, and the accessible parts of three cross-sections (spanning frames 11, 20 and 28). These structural drawings include the measurements and details that were not shown and/or visible on the 3D drawings. In addition, the information was also recorded in a written format. Subsequently, the first attempts were made to reconstruct the lower parts of the cross-sections in their original form and to supplement them upwards from there. The two intact crossbeams above frames 20 and 28 were essential to determining the correct shape and width of the hull at that point. The shape of the outside of the vertical riders that were fastened against the inside of the starboard shell determined the correct position and shape of the sides, especially the parts above the crossbeams. This combination of methods was used to arrive at the final results

Other than the data and find surveys recorded underwater, the shipwrecks were documented on dry land. On the one hand, the goal was to produce the most accurate possible 2D and 3D documentation and description. On the other hand, as already determined during the investigation and data processing on other cogs,[27] it is not possible to effectively complete the documentation process and the reconstruction of a ship as separate processes. In the case

24. See Chapter 6, figures 6.13, 6.14 and 6.15.
25. Vlierman 1997a, 19-21, idem 1997f, 60 and idem, sections 6.2.1 and 8.2.1.
26. A reconstruction of the barge was not included in the scope of the project, but could in principle still be done.
27. See elsewhere in this publication.

of the finds from the IJssel, and in particular the cog, this separation would be even more problematic because the shipwreck was salvaged in its entirety. For example, it was only possible to take measurements, etc. on a few locations on a continuous shell strake running from stern to stem.

The composite top view of the wrecked cog (fig. 23.9 and Plates 23.I, 23.II and 23.III), which was compiled from thousands of 3D recordings using a special computer programme, yields beautiful and dimensionally accurate snapshots of the current situation. However, it does not yield enough information to reconstruct the entire ship's hull. The same goes for 3D images of detached structural elements. To make substantiated reconstruction drawings of the hull shape, especially in the field survey of the IJssel cog, it was necessary to rely to largely on the 'old-fashioned' methods: manually collecting data and drawing ship parts, structures and details. When analysing a piece of wood, 3D images will record the precise outlines and such details as treenails, etc., but cannot identify what parts of the piece have been worn away or broken off, what is missing and where the wood is warped. Similarly, 3D imagery cannot be used to determine how the piece was fastened in place, or what it was connected to. In other words, when documenting a shipwreck, it is not enough to simply record the elements that are found; simultaneous investigation/interpretation is also needed to place the recorded data in the right perspective. Knowledge and experience are vital. Another issue with obtaining measurements and dimensions is the fact that waterlogged archaeological wood shrinks as it dries (even after chemical impregnation with e.g. polyethylene glycol [PEG]). Lengthwise, wood shrinkage is usually limited to 0.1-0.2%. However, planks and boards tend to shrink by 3-5% in width, and sometimes losing as much as >10% of their width. During the documentation of these two shipwrecks the measurements were recorded shortly after the salvage, based on the original dimensions in a waterlogged state.

The structural integrity of the shipwreck offered the option to research a medieval ship structure in its entirety, in a coherent context. At the same time, essential elements of the shipwreck were not accessible. These were the elements that were covered by supporting structures and the structural elements that were hidden behind frames and ceilings. For example, the precise measurements of the shell planks could only be determined in a disassembled state. Obviously, temporarily disassembling the shipwreck was not an option. Many of the measurements were therefore taken manually, using elaborate measures.

At the same time, in accordance with the descriptions of the other cogs, the reconstructions include the information collected from all phases of the investigation: the preliminary investigation, the underwater documentation and observations from other shipwrecks of cog-like vessels. The reconstruction process will be described from the keel plank to the inside wooden structures. The section ends with an assessment of the situation at the end of the reconstruction

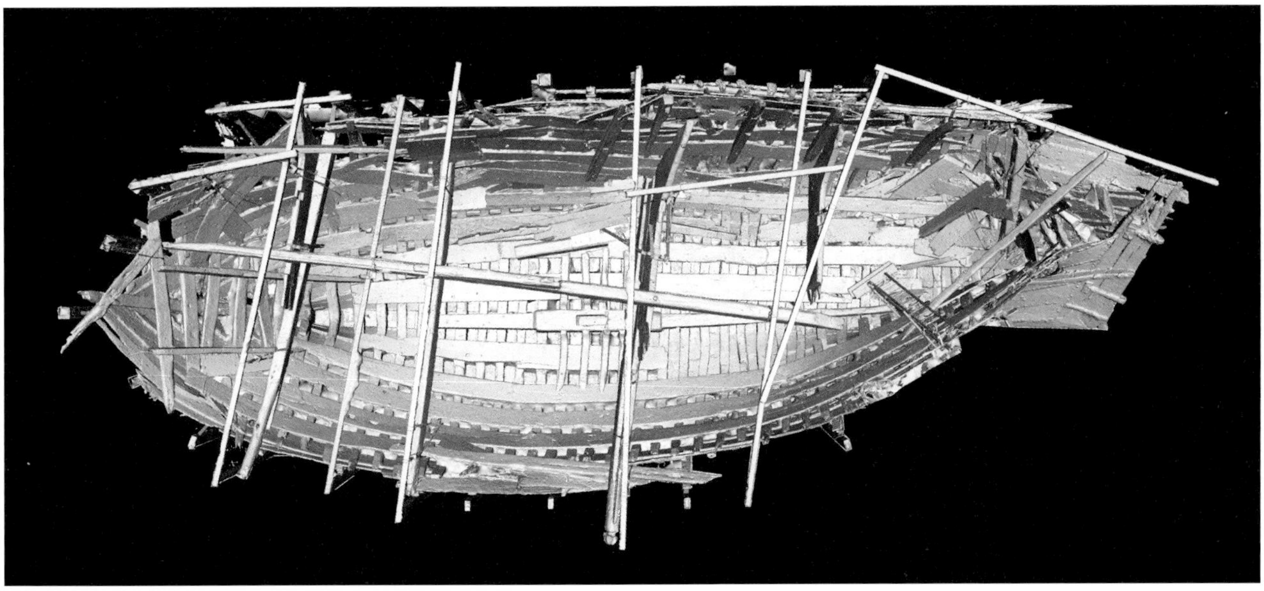

Fig. 23.9. Composite 3D image of the IJssel cog in the conservation workshop at Lelystad, summer 2016.

process (in November 2017): the elements of the IJssel cog that could not be reconstructed based on archaeological and iconographical data, of which the investigators working on the project were able to form an impression.

Field survey of the wreck of the IJssel cog and the parts salvaged separately since spring 2016.

Since 19 February 2016, the shipwreck has been installed in the conservation warehouse in Lelystad; by necessity, extensive supports have been added. As a result, the outside of the starboard and port sides is only partially accessible for examination and documentation. On the inside of the hull, it is possible to closely examine the bottom of the ship and the port side from the stem construction, together with the keelson on that structure and its riders and ceiling strakes, all the way to the start of the stern (Plate 23.I). These ship sections are still in reasonable shape, structurally speaking, until shortly below the crossbeams protruding through the shell. This is the case in the forebody up to sections of strake I, and in the afterbody up to parts of strake G. On the starboard side, remnants can still be seen of the shell strakes of the general midship section until the second-highest strake (Q), and several futtock heads still protrude from them; the uppermost strake of the gunwale (R) would have been fastened onto these. From this 'midship section' of the starboard shell on, the shell strakes are missing in a tiered pattern in the afterbody and forebody. Remnants of strake I are the highest shell strake still present at stern and stem. The stempost that collapsed forward, plus the detached false stem found beside the shipwreck, were still present till strake I (approx. 4.25 m above the keel line). The stern construction remained intact up to strake H (approx 3.25 above the keel line).

The inside of the starboard side is completely or partly inaccessible for examination and documentation. The entire afterbody, abaft and upwards from the start of the sternpost on the keel, is completely inaccessible. The reasons are:

- The entire starboard side is still mostly intact due to the fact that most of the heavy crossbeams protruding through the shell and the connecting beams fastened in between and the rising knees are/were still attached to the shell, while this was not the case on the port side.
- The enormous weight of these 'loose-hanging' crossbeam constructions, the condition of the almost 600-year-old frames, shell strakes and ceiling strakes, and the fastenings predominantly consisting of wooden treenails is what caused the collapse of the entire shell construction on the starboard side. The height amidships in the current situation is approx. 4.25 m above the keel line. The current reconstruction shows that this must originally have been 5.85 m (Plates 23.V and 23.VIII).
- The loose ceiling strakes and buttresses/vertical riders on the starboard side that were now strewn about inside the wreckage make it impossible to view the frames and shell (Plate 23.II). It is also impossible at this time to determine where the various pieces were originally attached.
- The absence of the connection between the crossbeam constructions and the port side hull, the fact that the afterbody lay in the IJssel above the deeper and older channel bed and its rising flank, means that the starboard shell from about the middle of the ship, including the port shell from about frame 1, was keeled over and collapsed to starboard, twisted, and essentially almost entirely distorted and warped.
- The remains of the barge were found on the starboard side of the vessel, extending beyond and partially under the forebody.

During the exploratory investigation and field evaluation and in preparation for salvaging the hull, all the ship parts that were found as separate elements and/or had become detached from the ship, and had been found adjacent to, above and inside the shipwreck, were measured at the findspots, numbered and salvaged.

The separate structural elements that had been stored in containers underwater were cleaned, documented and sketched in summer 2016. 3D recordings were made of the elements that were sketched and described, including detailed information, important for the reconstruction of the ship's hull. Plank fragments that were salvaged separately and can only be traced back to the source during the final phase, if at all, have not yet been documented in detail. Separate attention was given to the parts of a loading bed of branches, twigs and straw in the forward part of the hold and to planks and upright posts of inside panelling in the afterbody. The deck section on the starboard side between crossbeams 1 and 2, with the remains of a brick oven and fireplace on top, had already been documented while the vessel was still submerged.

23.2.2 Construction, reconstruction and layout of the hull

The team started manually measuring and drawing the stern and the shell strakes attached to the port side of the stern. The same method was used for the stem construction. Measurements were taken on the outside to determine the exact length of the keel plank together with the horizontal sections of the stern- and stem hooks from stern to stem (16.5 m). The seams of the sloping flat scarfs between the rising parts of the stern- and stem hooks and the sternpost and stempost placed on top of them deviate nearly 2 cm at the front of the stern and at the back of the stem. This means that the sternpost and stempost are sagging abaft and forward, respectively. The inaccessibility of the underside of the stern and stem constructions on the inside of the ship made it difficult to determine what the correct position would have been. Similarly, the correct shape of the stem hook could also be reconstructed only with great difficulty.

Keel plank
The keel plank measured 10.60 m in length and has two sloping, flat and nibbed scarfs for the connection to the stern and stem hooks. The aft scarf is 0.68 m long, while the forward scarf is 1.20 , staggered by 2 and 3 cm respectively. The keel plank is thicker on the forward end (15 cm) than abaft (10 cm). The width varies from 29 cm at the aft scarf, to 39 cm at a distance of 2 m forward of the mast, to 35 cm at the forward scarf. The keel plank is rectangular in cross-section, but gradually takes on a somewhat trapezoidal shape towards the stern- and stem hooks, with the narrowest width facing downwards. The connection between the keel plank and stern- and stem hooks is fastened into place with nails, probably two rows at each scarf: one row from the outside in, the other from the inside out. It was not possible to determine if this was in fact the case and if the nails were pounded through both parts, double-clenched and driven into the wood (probably not, i.e. fastened with blind nails) It also was not possible to determine with any certainty how many nails were used per row – probably 4 on the aft scarf and 5 on the forward scarf. An extra wooden treenail in frame 5 secured the aft scarf, providing an additional point of connection. The forward scarf probably had the same at frames 28, 29 and 30, but this cannot be checked at this time.

Stern hook
The keel section of the stern hook is 3.10 m long. The 68 cm nibbed sloping, flat scarf with the keel plank is located on the front. The top protrudes about 2 cm above the keel plank. At the end of the scarf on the underside, the keel section is 13 cm thick. The top gradually increases to a thickness of 16 cm over a distance of 1.38 m from the start, then angles diagonally up and abaft for a length of 7 cm, and is 23 cm thick at that point. From that point on, just before frame 2, the top line continues for another 1.24 m until reaching a thickness of 30 cm. The line then continues at a sharp angle, moving obliquely up and abaft for 83 cm (the scarf with the sternpost). This point is 90 cm above the keel line. The top of the stern hook is 29 cm wide at the scarf, and 22 cm wide at the start of the sternpost. The cross-section is trapezoidal in shape along the entire length, with the narrowest side facing downwards, and gradually grows sharper towards the stern. The back of the stern construction is 7 cm wide. The aft section of the stern hook is locked into place between the lowermost part of the sternpost and the false stern (fig. 23.10a and b and Plate 23.VIII). The back of the stern hook is also diagonal, forming the scarf on which the false stern rests.

Sternpost
The sternpost is no longer intact. Although the afterbody is inaccessible, it was possible to determine that the beam would have had an elongated triangular shape from a side perspective. The underside of the sternpost is standing on the keel section of the stern hook across a length of 25 cm, and the sternpost is resting against the sloping flat scarf. At 90 cm above the keel line (the highest point of the stern hook), the post is 40 cm thick. On top, about 4.90 m above the keel line, it was probably no more than 2 cm thick. The inside of the sternpost was probably about 22 cm wide along the entire length. A rabbet approx. 2 cm wide and 3 cm deep was cut into the sides of the post, continuing at a sharp angle into the sides of the stern hook. This rabbet continues until 2.55 m above the keel line and then takes a sharp turn towards the back of the false stern. The ends of shell strakes A through E were nailed into these rabbets.

False stern
A beam 7.30 m long was placed against the diagonally heeling back line of the sternpost and stern hook, with the false stern behind it. The beam is 43 cm wide at its widest point at the bottom, and 15 cm wide at the top. From strake F on, the beam is slightly mitred on the aft side. An L-shaped piece was inserted on the back of the underside, about 2 cm thick at the top and forming a hook at the bottom. This heel was there to prevent flotsam and jetsam from becoming entangled between the stern and the rudder. The sternpost, stern hook and false stern were connected to each other by means of a U-shaped wrought-iron plate approx. 10 mm thick and 80 mm wide. The piece encloses around the back and is fastened into place with six thick nails or rivets. The thin part of the heeled piece of wood was also fastened to the false stern with two nails at the top. The main stern and false stern were very probably fastened together by five iron bolts that were 30 mm thick: three with a flat head on the outside that do not protrude into the inside, and two that do protrude. On top of that, a flat iron ring was fastened against the

stern and clamped into place with an iron cotter. There probably was a fifth in the top of the false stern, including a ring and a cotter. The top three protruding treenails probably had a large eye on the back where the pintles for the rudder would hang. Two gudgeons were added in the lower half of the stern construction, with the pintles inserted through them. The gudgeons were forged into a T-shape on the front and made from iron plating 80-90 mm wide and 10 mm thick. They were attached with nails. It is not inconceivable that the central suspension of the rudder would have been a gudgeon rather than a bolt 30 mm thick.

Rudder

The stern rudder was sadly not found. Relying on iconographic images and an archaeological find at an early 16th-century *waterschip* in Flevoland, a possible indication of how the stern rudder might have looked is provided on the reconstruction drawings of the Nijkerk-II cog (Plates 6.V A and B).

Stem hook

The keel section of the stem hook measures 3.97 m in length. On the back is a 1.20 m sloping, flat and nibbed scarf with the keel plank. The scarf jumps 3 cm higher and 3 cm lower than the keel plank. The top is even with the top of the keel plank and is 15 cm thick at that spot. The top line gradually increases to a thickness of 21 cm over a distance of 1.43 m from the scarf, then angles diagonally up and forward for a length of 5 cm, and is 25 cm thick at that point. From that point on, the top line con-

Fig. 23.10a. The port side of the afterbody with stern construction. The timbers in the background are the shell planks of the broken starboard side.

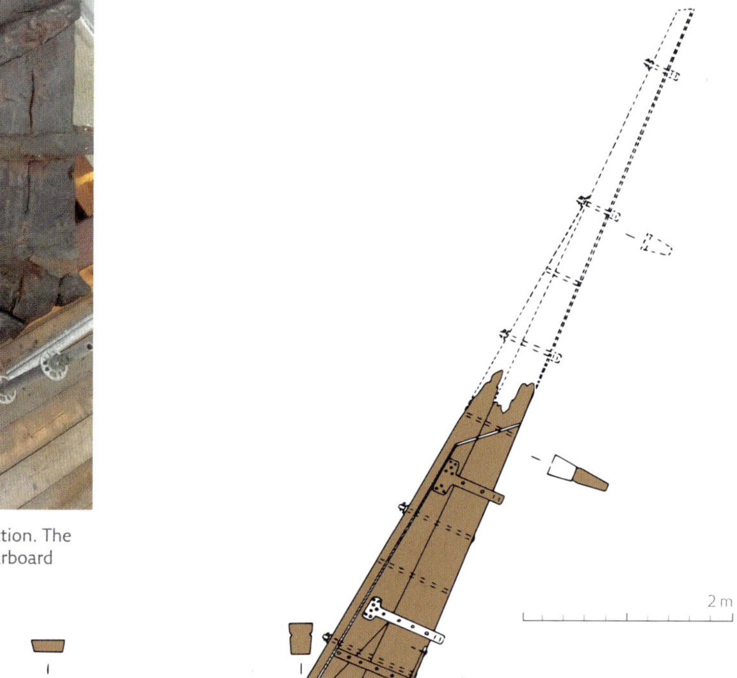

Fig. 23.10b. Reconstruction of the stern construction.

tinues upward for another 1.20 m until reaching a thickness of 45 cm, then continuing up and forward at a fairly sharp angle for a length of 64 cm. In essence, this is the scarf with the stempost and also the highest point of the stem hook (75 cm above the keel line). There is a 'stop-water' or short treenail (Ø 3 cm) rammed into the sharp angle. These wooden treenails were inserted to prevent water from seeping in through the seam. They are primarily found in (much) younger ships and, to the best of my knowledge, have not previously been seen in a late-medieval vessel. On the front of the stem hook, there is a diagonally protruding element, just as was found in the stern construction. The lowest section of the false stem rests on top of that. The top of the stem hook is 35 cm wide at the scarf and 23 cm wide at the beginning of the stempost.

The cross-sections of the hook are trapezoidal in shape, growing sharper towards the stem, with the narrow side down. It is 14 cm wide at the front (the lowest section of the false stem). The entirety of the stem construction is displayed in figure 23.11 and in Plate 23.VIII.

Stempost

Similar to the stern construction (hook, post and false stern), the stem construction consists of a stem hook, stempost and false stem. The entire construction was preserved up to approx. 4 m above the keel line. The stempost was almost surely composed of two pieces. The lowest piece rests on the slightly angled top of the stem hook and has a maximum thickness of 43 cm. The top part of the post was probably approx. 6.60 m long, 4 cm thick at the top, and extended until the top of strake K. That puts it well above strake I, on which the lowest row of crossbeams rests. It is 22 cm wide. The top piece of the stempost probably started about 4 m above the keel line, was approx. 2 cm thick at the bottom, was probably approx. 5.45 m long, and was 18 cm thick at the top. It would almost certainly have been 22 cm wide along the entire length. A rabbet 2-3 cm wide and 3 cm deep has been provided on the sides of the stempost, extending at the top until the bottom of strake F. From that point on, the rabbet continues slightly upwards to the front of the stempost, where (similar to the stern construction) a stop-water has been placed in the seam between the stempost and false stem. The rabbet continues approx. 1.20 m along the bottom in the keel section of the stem hook. The front

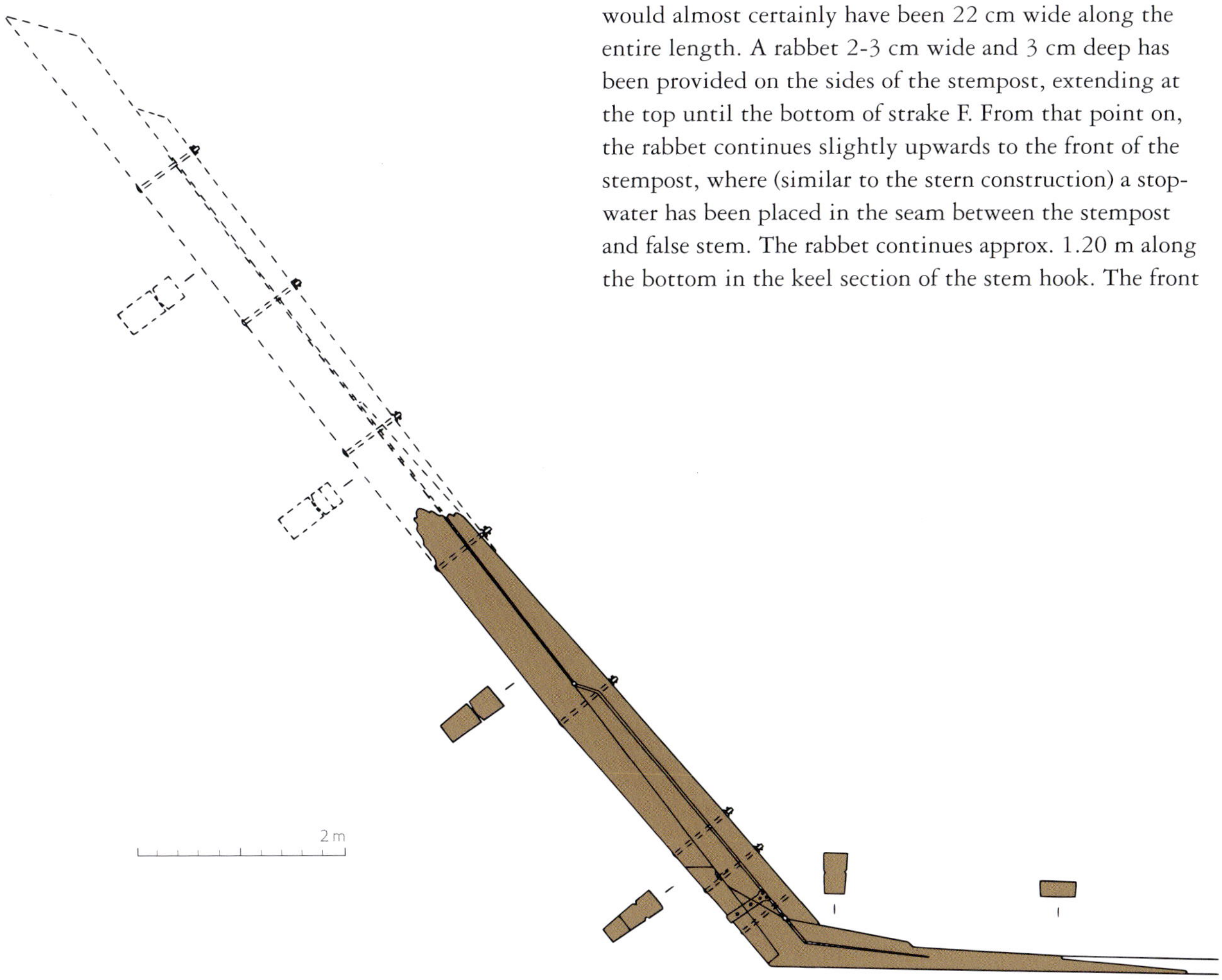

Fig. 23.11. Reconstruction of the stem construction.

of the stempost is slightly mitred from strake E towards the stem, so the ends of shell strakes F through Q could be nailed to it. Between frames 35 and 36, a short board 4 cm thick was nailed against the inside of the stempost. It is not clear what the purpose of this board would have been.

False stem

The false stem of the cog is comprised of two parts. The lowest section is well over 1.20 m long and almost horizontally finished on the top. The piece is 14x14 cm at the bottom and 20 cm thick and 14 cm wide at the top. The top section of the false stem was most likely more than 10 m long, gradually widening towards the top (from 35 cm to approx. 4 m above the keel line). This means that the false stem was probably approx. 45 cm thick towards the top. The width was probably the same along the entire length. In the preserved part of the stem construction, it has been confirmed that there were four iron bolts 30 mm thick. They formed the connection between the stempost (stem hook) and the false stem. The lowest did not protrude into the inside, but the other three did, and had a flat iron ring and a cotter. The bolts on the front have convex, somewhat rounded thin heads. It is highly probable that the stem sections were originally fastened together with eight of these bolts. About 60 cm above the keel line, the lowest section had an extra fastening: it was connected to the stem hook and stempost with an iron plate 10 mm thick and 80 mm wide. The U-shaped plate wrapped around the front and was nailed in place. The back of the plate was widened as it was forged. The top of the lowest piece of false stem was fastened against the stempost with a skew nail on both sides. The ends of strakes F through Q that were nailed to the front of the stempost were covered by the false stem.

The shell planks of the bottom and sides

It was possible, using all of our wits and acrobatic skills, to measure most of the underwater body (the keel plank, stern- and stem hooks, bottom strakes and the start of the sides) on the inside of the shipwreck. Using the collected data, an overview of the bottom of the ship was drawn by hand, piece by piece; these drawings also showed the keelson and riders, the ceiling strakes and the visible parts of the frames. Using the 3D recordings, it was possible to conduct a partial check of the observed measurements and intervals between the frames and the dimensions of the specified parts and the entire overview. A few minor, negligible differences were observed.

To reconstruct the sides as well as possible, the widths of the shell strakes on the port side were measured at various locations on the outside, distributed across the length of the ship. The lengths of strakes A through G were also measured as accurately as possible along the lower edge of the strake fitted above it, and from stern to stem. This final measurement was used to help determine the true lengths of the strakes, obtained by tracing lines from the reconstructed cross-sections, checking them and determining whether they were correct. On various occasions, the cross-sections had to be adjusted – especially in the bows and in similar parts of the afterbody – to achieve an acceptable and substantiated plan of the lines and planking. The differences detected between the reconstructed lines of the strakes and the measured lines of the strakes varied 10 and 20 cm. The researchers felt that this was an acceptable and negligible difference. Similar discrepancies were detected and resolved on various occasions while creating the reconstruction drawings of all the other cogs from the Low Countries.[28] To illustrate: the open seam observed between the stern- and stem hooks and the stern- and stempost (approx. 2 cm) has major consequences for the reconstruction of the rake of the stern and stem, over 7 m and over 10 m long respectively. That is not even taking into account any warping that was not observed or determined in the present and missing parts of the stern- and stempost.

The lowest strakes (A through E) are a little wider than normal at the stern and stem connections compared to several metres away. The widths of strakes I through R, which are completely missing on the port side, were measured along the outside of the starboard side wherever possible. Nothing can be said with any certainty about the width of these strakes across the aft 5 m and forward 8 m, approximately. During the reconstruction, it was assumed that shell strakes J through R would follow the widths recorded amidships, narrowing slightly towards the stern and stem. The ends of strakes A through E were nailed into rabbets in the stern- and stem hooks, sternpost and stempost. With the exception of the R strakes, which probably ended some distance from the stern and stem, all strakes from strake F onward continue until the back of the sternpost and are nailed to its sides. In the forebody, the plank ends of these strakes were nailed against the (partly) mitred front surface of the stempost and then fully covered by the false stem.

Timbers known as wales have been fastened against the outside of shell strakes N and P, against the underside of the clinker-built strakes above that point. So these are not the wooden wales that were common in (much) later shipbuilding, but strakes about 10-13 cm thick and approx. 20 and 15 cm wide or high strakes respectively, which were applied to the outside of the shell strake. A similar construction was

28. The present manuscript.

Fig. 23.12a. The wreck of a large cargo vessel or warship of the first half of the 16th century in lot U34, Oostelijk Flevoland. Clear to see on the shell planks of the collapsed part of the port forebody (right) are the three rubbing strakes fastened with iron bolts and treenails.

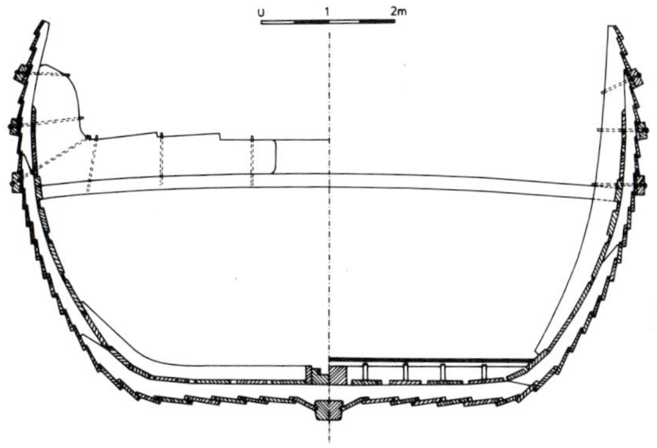

Fig. 23.12b. Reconstructed cross-section of the U34 wreck. To the left, the crossbeam-and-knee construction and a rider on the spar/open ceiling. To the right, a vertical rider, three rubbing strakes and the load-bed of branches, twigs and straw, resting on short pillars (after Reinders & Oosting 1989, Abb. 5).

observed earlier on the 16th century transport vessel or warship at lot U34 in Oostelijk Flevoland (figs 23.12a and b).[29]

The final aspect of the reconstruction of the hull concerns the positions of the plank seams in the strakes and the repairs detected that used small planks, as well as the cracks that were sealed with sintel clamped moss caulking. These are shown on the reconstruction drawings of the port side (Plates 23.IV and 23.V). Fig. 23.13 shows a piece of shell on the port side near the stern.

Fig. 23.13. Cracks on the outside of a shell strake near the stern, closed with sintel clamped moss caulking (ADC Maritime).

Caulking method: sintel clamped moss caulking

Before the lowest sections of the frames in the ship's hull were put into place, the seams of the clinker-built construction were first waterproofed on the inside with sintel clamped moss caulking. The clinker-built seams, the carvel-built seams and the cracks in the planks on the outside were only waterproofed using this method shortly before the launch. This type of caulking is characteristic of the cog-building tradition.[30] The carvel seams of the bottom were only sealed with sintel clamped moss caulking on the outside. The sintel clamp type D2/E used in the construction of the IJssel cog was in use in ca 1375-1450. Many of the sintel clamps observed already seem to be more similar to the next type. This makes it very probable that the vessel was built after ca 1400. In some repairs, distinctive types of the subsequent model E1 were confirmed, which was probably already in general use around ca 1450. The connecting seams between the stern and stem hooks, the sternpost and stempost, and the ends of the shell strakes were also sealed with sintel clamped moss caulking on both the inside and the outside of the hull. No trace of such caulking was found on the plank ends of the scarfs and around the crossbeams protruding through the shell. As far as could be ascertained (Plate 23.V), the outside of the shell shows cracks in various places that were sealed with sintel clamped moss caulking. The lengthwise cracks found in the wood were probably caused by the planks shrinking and drying during the construction of the vessel, which would have already been sealed shortly before the launch. Sintel clamped moss caulking was also used to waterproof the edges around some repairs in the shell. A layer of moss was placed under the small repair planks nailed onto the shell and between the scarfs in the shell strakes.

29. Reinders & Oosting 1989, 106-122, abb. 5.
30. Vlierman 1996a; idem. the present manuscript, section 29.4.1.

Determination and pollen research of the moss caulking
Dozens of caulking samples have been collected from the cog. Six samples were examined (203 through 208).[31] One sample from the construction phase (204), the others from repairs. In three samples, the pollen were well preserved (204 through 206). The caulking consists of one species of moss (*Sphagnum spec.*). The appendix shows the observations of the pollen and the pollen diagram.

Three samples of moss caulking (155, 163 and 215) were examined at Naturalis.[32] Besides the *Sphagnum* moss already found in the aforementioned examination, a single stem of *Calliergon richardsonii (Mitt.)* was found in sample no. 155. This species is not native to the Netherlands; it is an Arctic-Boreal species found in the northern hemisphere, widespread in Fennoscandinavia (including Iceland and Spitsbergen) and just reaching Denmark in the south. This could indicate that a one-time minor repair took place somewhere in the Scandinavian region. The other *Sphagnum* species found in this sample and the two others have a distribution area that includes the Netherlands. The pollen spectrum analysis also showed no indications of a Scandinavian origin. Consequently, that single stem is an interesting find that adds a dash of detail to the biology of the IJssel cog. Besides moss remains, twelve of the samples taken show remains of heather as well. In most cases, it was found to be common heather (*Calluna vulgaris*), although cross-leaved heath (*Erica tetralix*) was found in a couple samples.

Frames: (V-shaped) floor timbers and futtocks
Where possible, the scarfs between the floor timbers and futtocks were fastened to the port side. They offer an impression of the alternating scarf pattern (Plates 23.VI and 23.VII). The scarfs indicated on other pieces that were not observed or are missing have been estimated. They were derived from scarf patterns in other analysed cogs[33] and only indicated here to show that these frames must have been composed of various elements. There were 36 frames above the keel plank and stern- and stem hooks, starting at the base of the sternpost with 0 and ending above the start of the stempost with no. 35. Seven more frames are still present above the stempost (36 through 42). They must originally have continued until frame 52. This final frame is based on a comparison to the Bremen cog, and was possible an inverted reinforcement, that would have been applied along the top of strake Q and along behind the stern. Nothing can be said with any certainty about the frame construction above the sternpost. Judging by the frame pattern in the forward section, there were probably 14 of them (numbers -1 through -14). The top frame would probably have been similar to the one on the stempost. In total, that means the vessel would originally have had 67 frames. With the exception of the top two, it is very probable that most of the frame heads ended nearly even with the top of the gunwale (strake R). This was presumably different on the aft and forward ends because of the superstructure that was almost certainly present. The V-shaped timber of frame 6 has a nearly straight scarf above the seam between the stern hook and the garboard (strake A) on the port side. Above the stern hook, this V-shaped timber has a cleat on it which is nailed to the A ceiling strake (fig. 23.14). The exact purpose of this has not been clearly identified.

The frames are understated compared to the measurements of the vessel and other, older cogs. The (V-shaped) floor timbers are 18-22 cm wide and approx. 15 cm thick/high. Some are a bit wider or higher towards the aft and forward ends. It was not possible to show (in all locations) with any certainty whether the futtocks placed on top had been lightened, as is customary. Some are 16 cm wide on top, but most are 10 to 12 cm thick at the top. Since the shipwreck was not disassembled, it cannot be stated with certainty whether temporary frames were used during the construction of the underwater body (nailpegs in the boards of the ship's bottom). However, this seems highly likely. The floor timbers of frames 5 and 33 were the first and last that were fastened to the keel plank and stern- and stem hooks with wooden treenails. The other frames were not connected to the rest of the stern- and stem hooks, sternpost and stempost, only to the shell planks. This is one of the characteristic features of the cog-building tradition.

Keelson with mast step section
On the inside of the vessel, the hull is reinforced lengthwise using what is known as a 'keelson ' that was placed on the frames and above the keel plank (fig. 23.15). The beam was 9.20 m long and can be divided into a forward, middle and aft section. The aft section starts right after frame 7 and is 14 cm wide at that spot. At the beginning of the mast step (above frame 18; amidships), the width is 29 cm, while the average height is 20 cm. The section with the mast step is approx. 2.50 m long, 40-43 cm wide and 28-30 cm thick. A rectangular hole was cut into the top (60x19 cm and 11 cm deep) in which the foot of the mast was placed. A drainage hole was drilled in the underside to let out any water. The forward section of the keelson measures 30 to 24 cm (at the front) and is also approx. 20 cm thick. The keelson ends above frame 30. In principle, the underside of the keelson has been notched several at the locations of the frames, causing it to 'sink' between the frames. These parts were chamfered in a curve on the sides.

31. Moolhuizen & Van Ash. In: Waldus *et al.* 2018, section 24.5.1.
32. Kruijer & Leusink. In: Waldus *et al.* 2018, section 24.5.1. See this section for more detailed information on the two studies on moss caulking.
33. The present manuscript, the Plates.

Fig. 23.14. Chock on the floor timber of frame 6, fastened onto the A ceiling board (ADC Maritime).

Ceilings

Besides the keelson, the ceiling strakes on both sides also play an important role in the longitudinal strength and rigidity of the vessel. The vessel had at least twelve ceiling strakes on each side, which together created what is known as an open ceiling. The exact location and dimensions of the lowest five or six strakes can be determined with (reasonable) certainty. However, it cannot be stated where they started in the afterbody. A 'pattern' was chosen for the reconstruction that was derived from the still present ends of the ceiling strakes in the forebody (Plates 23.VII and 23.VIII). Nothing can be stated with certainty regarding the correct placement of the ceiling strakes on top of the crossbeams protruding through the shell (Plate 23.II). This can only be confirmed if a complete reconstruction is done of the ship's hull. They are indicated on the reconstruction drawings to provide the most comprehensive overview of the ship's structure possible. The ceiling strakes are generally 40 to 50 cm wide and approx. 5 cm thick. The B ceiling strakes are 8 cm thick.

Riders

To redistribute the kinetic forces that come through the mast, spar and sail while sailing, spreading them as much as possible through the rest of the hull, the mast step section of the keelson above frames 20 through 23 is supported from the side by riders, eight knee-shaped pieces of wood pegged to the floor timber which enclose the A ceiling strakes and end against or on top of the thick B ceiling strakes (fig. 23.16).

Vertical riders

Thick beams known as buttresses or vertical riders about 2-3 m long and 16-20 cm thick were distributed as evenly as possible along the length of the ship, placed vertically against the sides. The shape of the vertical riders of the starboard shell found *in situ* and those that could be placed later were essential in order to reconstruct the lines

Fig. 23.15. Keelson with a broadened and heightened part for the mast step, with the hole for the heel of the mast; the riders, the ceiling, the frames in the bow, and the two side stems.

of the top part of the starboard side. They were intended to reinforce the top 3 m of the sides. Several of these vertical riders were found in the shipwreck, some of which had presumably originally been attached to the port side. The inside view (Plate 23.VIII) shows – with as much substantiation as possible – how and where they were probably attached. Only after a complete reconstruction of the starboard side, including the ceiling, would it be possible to use the nailholes to determine the exact positions of these vertical riders. In the 16th-century freighter or warship from lot U34 from Flevoland, already mentioned previously, similar riders were found against the inside of the ceiling boards (figs 23.12b and 23.17).

This type of vertical rider was also found in certain locations on the outside of the ship's sides. They had a similar function. A fragment of such a vertical rider was found at frame 19, before the forwardmost deadeye for the rigging of the mast (Plate 23.V). It seems plausible that this buttress/vertical rider went from strake K to the underside of strake R; in addition to reinforcing the gunwale, it also served to protect the rigging. There was a heavy iron bolt protruding through the upper end of the preserved top

Fig. 23.16. The mast-step area with the port riders on top of the frames and the A ceiling board.

Fig. 23.17. The vertical riders and shelf clamb on the inside of the collapsed port shell of the large, 16th-century cargo vessel or warship from lot U34 in Oostelijk Flevoland, after removal of some of the shell planks.

Fig. 23.18. The standing starboard knee (8) of crossbeam construction 8, found beside the afterbody (ADC Maritime).

of the starboard knee above crossbeam 7 (at the location of frame 39), extending so far from the surface that a vertical rider was probably attached there as well. A similarly long, protruding heavy bolt was also found in the detached heavy crossbeam knee (8) found by the port side of the afterbody (fig. 23.18 and Plate 23.VI B). It seems very likely that similar vertical riders were added at multiple spots all along the outside, distributed across the sides. Several vertical riders are marked on the side elevation (Plate 23.V) in the locations where they were presumably fastened.

Side-stems

In the bow of the forebody, 'side-stems' were placed on either side of the stem and about 1.5 m apart centre to centre, on the inside over the frames (fig. 23.9 and Plates 23.VI F, 23.VII and 23.VIII). The underside of the side-stems starts between frames 36 and 37. The missing top half probably ended at the same height as the top of the stempost. The side-stems are 13 cm wide and 20 cm thick or high. The underside is notched, letting them drop several centimetres and locking them into place between the frames. They were fastened to the frames with wooden treenails. At frames 39 and 42, holes (Ø 4 cm) were drilled athwartships through the beams; their purpose is uncertain. The function of the side-stems was to reinforce the bow, much like as the vertical-riders and decking-edge timbers against the sides (fig. 23.19).

Crossbeam constructions

The protruding crossbeams and crossbeam fragments that were found, both still in the ship's hull and detached, as well as the knees with connecting beams, and the interpolated beams in between, are very important to the transverse structural strength of the hull at deck height and to fasten and retain the everted gunwales above that. These constructions also served as attachment points for the superstructure in the forebody and afterbody of the ship, the windlass and the mast shaft. (It is probable that there would only have been a possible vertical conducting board behind the mast). Although only five crossbeam constructions were found *in situ*, it can be assumed that there were seven or more, possibly even eight in total. These crossbeam constructions vary in shape and appearance; it was possible to reconstruct them either with certainty or hypothetically, depending on the parts that were recovered (Plates 23.VI A through G). The crossbeams were placed at various levels. The lowest level found *in situ* (at strake I) has been numbered from aft to forward (1 through 5). At a higher level, again starting from the back, crossbeam constructions 6 and 7 were next, and crossbeam 8 was above that (over no. 1). Crossbeam no. 8 was reconstructed based on a detached knee (8), while crossbeams 7 and 8 are hypothetical but can be substantiated in the light of the complete reconstruction.

Fig. 23.19. The forebody of the Bremen Cog with frames, side stems and the foremost crossbeams.

The rising knee sections above crossbeam no. 1 were originally higher. Crossbeams 1 and 2 were probably slightly curved. The filled bulkheads on top are trapezoidal when viewed from the side, so that the top of the knees that were fastened on top on them were (nearly) horizontal. The other crossbeams were probably straight. They all protruded approx. 20 cm from the shell. Beam-head guides were nailed against the front of the beam heads and fastened against the shell. They were intended to prevent that the ship from e.g. getting hooked by a pole while mooring. They may have also served as water guides when the ship was heavily loaded and the beam heads were under the surface of the water. Without the fenders or beam-head guides, the beam heads would have caused drag and slowed down the vessel. Such beam-head guides have been found previously on large late-medieval shipwrecks[34] and are visible on the ship's hulls that were portrayed on some contemporary city seals. Against the outside of the shell on the afterbody, a nearly horizontal beam was attached on both sides, measuring 2 m long and 9-12 cm thick, and at the back 14 cm wide. The beams are located 1.30-1.40 m above the keel line against strake C. They were almost certainly intended to create/build up extra water pressure on the rudder at the sharply angled stern for maximum efficiency in steering.

Crossbeam construction 1 (Plate 23.VI B)

The starboard half of crossbeam construction 1 was found. This consists of over half of a slightly curved beam, a trapezoidal intermediate piece and a heavy rising knee.

34. For example in the Nijkerk-II cog and the Doel-I cog, in sections 6.2.2 and 20.2.2 respectively.

Only the top part of the rising section (438) of the port knee was found. The intact starboard construction was lifted out of the wreckage before salvage and is now set up beside the shipwreck in the conservation warehouse. The beam is approx. 22 cm high and 25 cm wide. It rests on the futtocks of frame 3, on the top of shell strake I, and protrudes through strake J to the outside. The protruding head has been rounded off. The intermediate piece is approx. 10 cm high and 13 cm wide amidships. Against the side, it reaches a height of 28 cm and abuts the ceiling. The entire intermediate piece consisted of two of these parts. The rising knees rest on those intermediate pieces. The horizontal sections are approx. 50 cm high, while the rising sections are approx. 1.50 m high. Originally, they were almost certainly higher (approx. 2.20 m); the heads have been sawed off. In the outside of the rising section, a recess was provided to attach a 'bench' (404) made from wood similar to decking-edge timber. There was presumably a similar recess in the missing top section for the 'wooden bench' (or decking-edge timber) (448) that was found detached. The underside of the starboard knee is about 2.30 m, including a 35 cm nibbed scarf. Assuming that the lower missing port knee was also 2.30 m long, there must have been a connecting piece in between measuring 2.20 m in length, including two nibbed scarfs of 35 cm. The three sections were fastened together with wooden treenails 3 cm thick placed through the nibbed scarfs. The beam, intermediate pieces and knees were fastened together with thick (Ø 4 cm) wooden treenails. The rising sections of the knees are fastened to the sides with three diagonally positioned wooden treenails of the same diameter. There is a round iron bolt 30 mm thick through the top of the knee, with a ring and cotter on the inside, and a thick head on the outside of the lowest wale. Two wooden treenails (Ø approx. 3 cm) are placed one above the other on the front side of the beam and the knee. This was probably to fasten an upright post that continued until the underside of the aftercastle. This upright post also probably functioned as one of the cheeks for the large windlass. There are recesses in the upper front of the horizontal sections of the knees in which the aft ends of long fore and aft beams rest, positioned on both sides of the vessel's centreline. Between these spaces and the sides, recesses have been provided for shorter fore and aft beams.

Crossbeam construction 2 (Plate 23.VI C)

Almost all the elements of crossbeam construction 2 were found. The starboard side of the beam is intact. Much of the port side is missing. The beam is slightly curved and rests on the futtocks of frame 11 and on the top of shell strake I. The head protrudes through strake J and is rounded off. The beam is 26 cm high and wide. There would have been two trapezoidal intermediate pieces on the beam, although the port-side piece is missing. They were connected by a 35 cm nibbed scarf. The section to starboard is about 18 cm high above the ship's centreline, and 35 cm at the side. It is 13 cm wide and abuts the ceiling. Heavy rising knees rest on the intermediate pieces on either side. The starboard knee is still in the shipwreck. The other (250) and the connecting piece (258) were also hoisted out of the shipwreck before the salvage. Both knees were 2.65 m on the underside. The horizontal sections are 64 cm high and 13 cm wide. The rising knee sections were originally approx. 1.70 m in height. The top of the starboard knee is damaged, and a piece of the port knee was chopped or sawed off. The connecting piece has not remained fully intact towards the top; the piece is 1.85 m long and was fastened to each knee with a 30 cm nibbed scarf and five wooden treenails (Ø 3 cm). There are two recesses in the top on either side of the ship's centreline. The long fore and aft beams rested in these recesses. On the starboard and port side, rabbets were cut on both front and back for the shorter fore and aft beams. The beam, the intermediate pieces, the knees and the connecting piece are fastened together with three thick wooden treenails (Ø 4 cm). The rising knee sections are connected to the sides with two diagonally inserted and similar wooden treenails. The heads of the knee sections are fastened with a round iron bolt 30 mm thick with a ring and cotter on the inside, and a thick head on the outside of the lowest wale. There is a treenail hole in the knee at 1.5 m from the ship's centreline, which was probably the attachment point of an upright post for the aftercastle.

Crossbeam construction 3 (Plate 23.VI D)

With the exception of the missing knee on the port side, crossbeam construction 3 is essentially intact. The beam, the intermediate piece consisting of two parts, and the starboard knee are still in the wreckage. The port ends of the beam and the intermediate piece are damaged. However, the protruding head of the beam is still present on the port side, making it possible to determine the exact width between the sides. The beam is almost straight, measuring approx. 30 cm high and wide amidships. The intermediate piece is also straight, 25 cm high and 14 cm wide. The beam rests on the futtocks of frame 20 and on the top of shell strake I. The head protrudes through shell strake J. The starboard knee rests on the intermediate piece, with a scarf about 25 cm long on the starboard side held in place with one treenail. The underside is 3.70 m long. Above the centreline, there is a 65 cm hooked scarf which is the connection to the port knee. No connecting piece was used between these two knees. They were probably fastened together on the scarf with six wooden treenails (Ø 3 cm). The horizontal section of the knee is approx. 65 cm in height, while the rising knee section is 1.65 m. The beam, intermediate pieces and knees were fastened together with three wooden treenails (Ø 4 cm). The rising sections of the knees are fastened to the sides with two treenails. Inserted through the top of the knee, there is a round iron bolt 30 mm thick, with a ring and cotter on the inside, and a thick head on the outside of the lowest wale. The two long fore and aft stringers were not resting in recesses, but on top of the knees. They were attached to the crossbeam construction with iron bolts 30 mm thick, with a ring and a cotter on the bottom. Small rabbets in the front and back of the top section show the locations of the shorter fore and aft stringers. On the front of the underside, above the centreline, there is a rabbet approx. 30 cm wide and 5 cm deep with two nail holes in it. It was most likely the location of a guiding board for the mast placed at an oblique angle, which stood behind the hole for the foot of the mast in a diagonally cut groove in the top of the mast step. The plank was not found, as far as could be ascertained. Two elongated triangular rabbets were cut into the front and underside of the beam to starboard. The significance of these is unclear. There were no visible nail holes in the rabbets.

Crossbeam construction 4 (Plate 23.VI E)
The port-side half of the intermediate piece of crossbeam construction 4 was not found. The beam is almost entirely intact, although the port head is slightly damaged. The beam is approx. 28 cm high and wide in the middle and 25 cm high at the sides. The beam rests on the futtocks of frame 28 and on top of shell strake I, and the head protrudes through strake J. The intermediate piece to starboard (235) that was found is 33 cm high and 13 cm wide. On the starboard side of the midship section, there is an approx. 30 cm nibbed scarf held in place with four treenails. There are two knees (236 and 252) on the intermediate piece, which was originally composed of two sections. The beam is still hanging in the shipwreck, while the knees and the middle section were removed from the shipwreck before the salvage. The horizontal sections of the knees are approx. 62 cm high and 13 cm wide. The knee sections are approx. 1.70 m high. Both knees are attached to the intermediate piece and the beam with two wooden treenails (Ø 4 cm). The rising sections of each knee are connected by a single horizontal wooden treenail and by an iron bolt 30 mm thick that has a ring and cotter on the inside, and a thick head on the outside of the lowest wale. The two fore and aft stringers in the middle are resting on the top of the knees. They were fastened on the horizontal parts of the knees, together with the diagonal beams, fitted with blind thick iron nails.

Crossbeam construction 5 (Plate 23.VI F)
Crossbeam construction 5 consists only of a beam with two knees on it, i.e. no intermediate piece and connecting piece. The nearly intact beam and the starboard knee are still hanging in the wreck. The port side knee was not found. Both knees have a short rising section. The top and inside of that section forms a sloping hooked scarf on which the top pieces were placed. This piece is missing on the starboard side. The top piece of the port side is still hanging in the forebody. The straight crossbeam rests on the futtocks of frame 35 and on the upper edge of shell strake I. The beam head protrudes through strake J to the outside. The beam is 25 cm high and 23 cm wide. At the base, the starboard knee is about 3.10 m long, 48 cm high and 12 cm wide. The piece has a 60 cm hooked scarf amidships, which forms the connection to the port knee. This hooked scarf does not continue to the beam. The end of the port knee was presumably less high. The knees were attached to each other with three wooden treenails (Ø 3cm). There is a recess for the long fore and aft beam in the top of the knee. There is a rabbet for the short fore and aft beams halfway long the side on the front and back. The highest point of the rising knee section is about 1.25 cm above the beam, and approx. 1.35 m on the port side. Together with the piece placed on it, the knee sections protruded approx. 2.55 m above the beam. The connection between each knee section and the beam consisted only of two wooden treenails (Ø 4 cm). At the rising knee sections, the connection was a wooden treenail inserted diagonally through the hooked scarf and two iron bolts (Ø 30 mm) with ring and cotter on the inside and thick heads on the outside of the two wales attached on top of each other. Recesses were provided on the outside of the rising knee sections to attach two side boards placed vertically one above the other.

Crossbeam construction 6 (Plate 23.VI A)
A knee hanging in the afterbody is still fastened to part of an intermediate piece and part of the crossbeam. The port knee (116) was removed from the shipwreck shortly before the salvage. Together, they make up crossbeam construction 6. The straight crossbeam rests on the futtocks of frame 0 and on top of shell strake K. The heads of the beam protruded outwards through shell strake L. The beam is 20 cm high and 22 cm wide. The intermediate piece, which remained intact and originally consisted of a single beam, is also straight, measuring 18 cm high and wide. The knees rest on either side on the intermediate piece. The underside of both is approx. 2.20 m long. The tops of the horizontal sections taper gradually towards the middle, and are 35 cm high and 12 cm wide at the bend in the rising knee sections, and 25 cm high and 12 cm wide at the start of the flat scarfs that would have joined to a connecting piece that was not found. The rising knee sections are approx. 1.20 m in height. On the starboard half of the

crossbeam construction, it could be determined that the knee, the intermediate piece and the beam had been fastened together with two wooden treenails (Ø 3cm), and three treenails against the starboard shell. A recess for enclosing the decking-edge timber (448) was provided in the outside of the rising section. A lath was nailed against the top edge on the front of the knees. On each side of the centreline, a wooden treenail can be seen in the beam, probably the fastening for an upright post to support the aftercastle or poop.

Crossbeam construction 7 (Plate 23.VI G)

Beside the forebody on the port side, the rising section of a knee was found, still fastened to a futtock and a piece of wale. The knee section is the only piece found of crossbeam construction 7. Two vertical nail holes (Ø 4 cm) were founded in the lowest heavy decking-edge timber in the forebody. Most likely treenails providing the connection between a crossbeam, an intermediate piece and a knee. The nail holes in the timber are near frame 39, about halfway between crossbeam construction 5 and the stem. A logical spot, from a structural perspective. In this reconstruction, the crossbeam rested on the futtock of frame 39 and on the upper edge of shell strake K. A recess for fastening a (top) side board was provided in the outside of the knee section. A small square hole was made through the head, oriented lengthwise in relation to the ship, and a square rabbet was cut into the inside. The top part of the knee was connected to the side with two iron bolts (Ø 30 mm) with a ring and a cotter on the inside and thick heads on the outside. The length of the bolts suggests that the heads were on the outside of a vertical rider.

Crossbeam construction 8 (Plate 23.VI B)

A heavy rising knee (8) was found next to the afterbody on the port side during the exploratory research in 2012, and was later salvaged. The outside of the rising knee section is different from all the other knees in terms of the shape and position of the outside line, and therefore cannot have been one of the missing knees from crossbeam constructions 1, 3, 5 or 7. The findspot between/at the level of crossbeam constructions 1 and 2 suggests it would have been above crossbeam construction 1. The base is approx. 2.55 m long and 50 cm high and has a rising knee section of approx. 1.35 m. The piece is 12 cm wide and has a nibbed scarf of approx. 35 cm. The knee almost certainly stood on a trapezoidal intermediate piece and was connected to that piece and to the crossbeam below it by means of two wooden treenails (Ø 4 cm). In the base of the rising section, it would have been fastened to the side with a treenail. Through the head, it was fastened with a wrought-iron bolt 30 mm thick which has a ring and cotter on the inside. The length of the bolt protruding on the outside suggests that it was inserted through a vertical rider. A recess for fastening a side board is visible on the outside of the knee. In the knee, treenails were added to attach upright posts for the superstructure/aftercastle or poop.

Reconstruction of the cross-sections

While manually recording the top view of the ship's bottom with keelson and ceiling strakes, the course of the bottom strakes and lowest shell strakes was measured as much as possible at the location of frames 0, 3, 11, 20, 28 and 35. This would provide the basis for later reconstructions of cross-sections with crossbeam constructions of the hull. The manually collected data was supplemented using cross-sections made at more or less the same locations based on 3D photos. The cross-sections of the 3D recordings were especially important to determine the lines of the port-side hull, which was still present and in more or less the original shape, and to detect details such as the widths of the ceiling boards on the vulnerable hull, which was difficult or impossible to access. Based on the information available, a preliminary top view of the ship's bottom and an initial attempt at a cross-section at frame 20 were drawn and discussed with the other members of the research team. At that time, agreements were made as to the minimum additional data that would need to be collected, in order to be able to make a hull reconstruction.

The crossbeams determine the width of the vessel and – based on the outside of the knees – the angle and lines of the port side. Crossbeam 3 (at frame 20) is the only one that was fully preserved and also is at more or less the original location. Due to the recesses in the beam ends provided for the shell strakes that slotted into them, it was possible to measure precisely how wide the ship would have been at that spot (6.76 m, measured on the inside of the shell and on top of the beam). Crossbeam 4 at frame 28 is nearly intact. The width of the ship could also be measured from that beam (6.68 m). The other beams were not found, or were no longer intact. Using the notches on the underside of the long stringer beam at the starboard side, which was still fastened to crossbeam 3, it was possible to determine exactly where crossbeams 2, 3, 4 and 5 were originally placed in relation to one another. What could not be determined exactly due to the deformation of the shell at that spot was where crossbeam 3 protruded through the shell on the starboard side. This may have been above, abaft of or slightly forward of frame 20. Frame 20 with crossbeam 3 was chosen to be used as the starting point for the reconstruction drawings; it is assumed that the futtock of the frame was attached vertically to the shell and that the crossbeam was on top of that. Using that as a starting point, the cross-sections across crossbeams 4 and 5 and forward and then across crossbeam 2 were gradually compiled, and the course of the planking lines in the top view was determined, along with the data on the longitudinal cross-section. The drawings had to be adjusted various times, because small discrepancies crept into the collected dimensional data. These differences had to be corrected to achieve a hull reconstruction, which would naturally have had (nearly) flowing plank lines. Ignoring these minor discrepancies would result in growing differences in the plank lines of the ship's hull moving upwards, which could not be substantiated.

Stringers

The two long stringers on either side of the ship's centreline consist of two sections. The sections are fastened together with 57 cm sloping and hooked scarfs and have a combined length of 15.45 m. The aft section of the stringer on the starboard side rested in a rabbet in the top of the starboard knee above crossbeam 1. The longest stretch of the back piece (205) is 3.80 m long, 20 cm wide at the back, and 23 cm high at the start of the hooked scarf. The width is 15 and 18 cm respectively. This piece has rabbets on both sides of the top to insert small transverse beams and planks. The forward section of the long stringer is 27 cm high and 21 cm wide aft, and 13 cm high and 15 cm wide forward. Recesses several centimetres deep were provided on the underside, at crossbeam constructions 2, 3, 4 and 5, which closed around the top of the knees. This is the only information available to determine exactly where the crossbeam constructions were positioned in relation to one another. A rabbet 4 cm deep and wide is visible on the starboard side along the entire top edge. The detached, transverse deck planks were laid in that rabbet. From frame 29 forward, a similar rabbet is also present on the side of the ship's centreline. At the height of frames 24 through 29, a sloping beam rising towards the mast is attached to the stringer. 20 cm high aft and 3 cm high forward, and 15 cm in width along the entire length. A piece of the long stringer from the port side in the shipwreck, with a similar oblique beam on top (251). Judging by the model of the Bremen cog (fig. 23.20), these sloping upwards were probably parts of a 'hatch coaming'.

Short, loose stringers, also resting in rabbets in the heavy knees, were fitted between that starboard stringer and the starboard side, about halfway along and also lengthwise, between crossbeams 1 and 2, between 2 and 3, between 3 and 4, between 4 and 5 and probably also extending between 5 and frame 43 against the stem (also on the port side). The tops of the long and short stringers have rabbets on the edges that are 4x4cm wide and deep (Plates 23.VI A through G). There were no visible nail holes in the rabbets.

Decks

Crossbeams 1 through 5 almost certainly all rested on or in the upper ledge of strake I, enclosed in strake J in rabbets cut into the top and side of the beams. It means that they protrude through the shell differently in relation to each other and to the keel line (along with the sheer). This difference in height was resolved by the intermediate pieces and the knees placed on them. The long fore and aft beams were composed of two pieces each and start at the knees above crossbeam 1 and continue till frame 43. The top line of these fore and aft beams descends about 3 cm from fore to aft in relation to the keel line. On both sides, short deck planks (1.10-1.35 m long) were laid in a transverse direction in the aforementioned 4x4 cm rabbets on the outside of the long fore and aft beams and the innermost rabbets of the short fore and aft beams, and in the rabbets on the outside of the shorter one and the shell. This shows a deck construction that extended from crossbeam 2 (frame 11) all the way to frame 43 against the stem. This was also the case between both long stringers, from crossbeam 4 (above frame 30) until the stem. There was probably a hatch in the space between these stringers, the mast and crossbeam 4, which could be used to access the hold. The aforementioned short beams on the stringers, installed diagonally towards the aft section and increasing in thickness, were probably part of this construction (Plates 23.VII and 23.VIII).

No holes from treenails or iron nails were found anywhere in the rabbets for the ends of the long and short stringers and in the rabbets where the deck planks would have laid. This means, in all probability, that they laid in the rabbets without fastenings. This was already observed dur-

Fig. 23.20. The construction of the hatchway in the 'main deck' between the two long stringers in the model of the Bremen Cog (between the upper coil of rope and the mast).

Fig. 23.21. Impression of the main deck with barrels, the stringers, the oven, the fireplace, the windlass, and the cargo on a between-deck and in the hold.

ing the investigation of the Nijkerk-II cog[35] and was also been detected in the Bremen cog. On both sides of the long stringers, there was a detached deck all on the same level that was approx. 12 m long and 2.50 m wide. The similarly detached deck section between the stringers and the forebody was about 4.5 m long. It is striking that none of these deck planks were found during the excavation.

Another construction was situated between crossbeam constructions 1 and 2 in extension from this deck. The short stringer on the starboard side (194) is also detached. A rabbet was cut in the top of this beam on the port side, as described above, in which rested transverse planks that were irregular in width and extended until the starboard shell. At four places, beams were placed on top of those, oriented in the same direction. On top of that, there was a floor of planks 30-35 cm wide laid lengthwise (figs 23.30a-c). On that floor, a brick oven and fireplace were built. Based on the rabbets present on the other side of this short stringer and in the longer stringer parallel to it, the same or a similar deck construction would have been present between them, and in any case also between both long stringers. This may have also been the case for the port side. Even though all the elements of this section were detached, it is probable that there was a permanent floor in the caboose or cooking area/fireplace.

Crossbeams 6 and 7 were situated on a higher level, above frames 0 and 39 respectively. Crossbeam 6 probably rested on top of strake K, while crossbeam 7 probably protruded through that strake. A beam about 20 cm thick probably rested on the knee of crossbeam 1. It would have supported a small horizontal deck of planks placed lengthwise which rested on the top of crossbeam 6 near the front side. This would be the small deck from which the windlass was operated. On the top of the knees on crossbeam 6 and in a straight line until frame 9, against the sternpost, there was probably a small deck for the helmsman. It is plausible that these planks were also placed lengthwise.

Extra decks or tweendecks?

Based on the ship parts found loose in the ship, a second deck level can still be reconstructed. The most important evidence to support this is a shelf clamp that was against the underside of the crossbeams on the starboard and port side. It was attached to the ceiling board and the frames behind it with wooden treenails. The 11 m specimen on the starboard side was 30 cm wide/high at most, trapezoidal in cross-section with the widest side facing up, and was a maximum of 13 cm wide at that point, and 9 cm wide at the bottom. The height and width gradually tapered abaft and forward from there. The top of the beam clamp was diagonally notched to precisely enclose crossbeams 1 through 5. Similar notches were found in the middle between these notches. A scarf at the front end suggests that the shelf clamp was probably extended. The shelf clamp as it was found spanned the space between just after crossbeam 1 until just before crossbeam 4. The forward missing section presumably continued until well under crossbeam 5 (Plates 23.VI A through F, 23.VII and 23.VIII). A fragment of its counterpart on the port side was also found.

The diagonal rabbets between the crossbeams in the top of the shelf clamp undoubtedly held detachable crossbeams that covered the entire width of the ship. However, none of these beams was found. There may have been a completely removable 'tweendeck' for which the crossbeams also had 4x4 cm rabbets along the upper edges. The planks that fit into this would have been placed lengthwise. The space between this deck and the continuous deck at the height of the fore and aft stringers was approx. 1 m between beams 2 and 5 and approx. 90 cm between beams 1 and 2.

It is conceivable that there were originally even more deck levels. Notches in the buttress/vertical rider against the starboard shell at frame 23 and presumably also at frame 31 (249) suggest that crossbeams could also have been placed there, making it possible to add another detachable deck at that height (approx. 4.75 m above the keel line). It would have been approx. 1 m above the continuous main deck on the fore and aft beams. However, no indications were found that this deck would have continued all the way to deck beam 2 (Plate 23.VIII).

Sideboards

On the starboard side in the afterbody, a curved beam was found which measures approximately 4 m long, 20 cm wide and 12 cm thick (446). The recesses and straight stretches on the diagonal outside enclosed around the frames and against a ceiling board of the ship's side. This is a side support, generally intended to offer a step up from the deck, making it possible to e.g. look out over the gunwale, or to do certain things that require access to that part of the ship. Other recesses on the inside indicate that there must have been a knee at that location. Recesses in various crossbeam knees confirm this. Using these traces, we were able to determine where this decking-edge timber was located: from frame 4 through 6 at the stern. The side support was fastened to the frames with horizontal wooden treenails (Ø 3cm). The rising knee section above crossbeam no. 1 was originally higher. In the reconstruction, the top is between 0.85 m and 1.40 m below the gunwale (the top of strake R; Plates 23.VI A, 23.VII and 23.VIII).

35. Section 6.2.2.

Fig. 23.22. Painting by Hieronymus Bosch (ca 1514), featuring a ship in which sideboards can be seen aft and fore (ADC Maritime).

On the starboard side in the forebody, there is still a sideboard more or less in its original place (476). It is a curved piece of wood about 5 m long, no more than 47 cm wide, and 10 cm thick. Towards the forward end, the beam broke lengthwise. The total length of this side support must have been approx. 5.40 m. The front end probably continued until just before the side-stem. Similar to the side board in the back, the curved outside is chamfered and has been provided with notches for the frames and straight pieces that abutted a ceiling board. A recess on the inside at the aft section fitted around the knee on crossbeam 5. Crossbeam construction 7 was fastened to the decking-edge timber with two wooden treenails (Ø 4 cm). This warrants taking into account that the beam may not have protruded through the shell. Determining the original construction required a great deal of brainpower and calculating (Plates 23.VI G, 23.VII and 23.VIII). The section of the sideboard in front of beam 7 was probably fastened to the frames with four wooden treenails (Ø 3 cm). They were inserted diagonally from the top down. The back section was fastened with five horizontal treenails running through the piece. The side support was approx. 45 cm above the continuous deck and 1.90-2.70 m below the upper edge (Plates 23.VI G and 23.VIII). It cannot therefore have been intended as a step to make it possible to look out over the gunwale. The pieces were presumably primarily intended as reinforcement for the bows. Rabbets in the top/inside edge suggest that small diagonal beams lay in them at those spots. This could indicate the presence of a small deck between the two side supports to port and starboard, crossbeam 7 and the stem. Figure 23.22 shows a detail in a painting by Hieronymus Bosch (approx. 1514) on which sideboards can be seen.

A recess in the outside of the knee on crossbeam 7 (above frame 39) and a small rectangular opening through the top of the knee above crossbeam 5 suggest that there were also sideboards at that height, which probably extended all the way to the sidestems. However, these were not found. In the reconstructed field survey, the back of the side support was approx. 95 cm below the gunwale, while the front was about 2 m below it. It seems very plausible that there was a small deck about 3 m long, which may have had e.g. an anchor windlass on it, located between these timbers and between crossbeam 7 and the stem. This small deck must have been accessible (in the middle) by a stair from the deck that rested on the stringers. A square hole has been cut in the inside of the knee. It may have been intended to attach a railing – or, given the height above the deck, possibly a small intermediate beam – of a fence or gate on the back of the small deck.

'Bench'? and lockercabinet

On the front of crossbeam 6 and continuing to the fireplace, there was a thick beam similar to a sideboard placed against the starboard side, measuring 1.90 m long, 40 cm wide and 10 cm thick. It was at approx. 40 cm above the deck on the fore and aft beams. The forward end seems like a very appropriate place for the cook to sit. On the beam and against the front of the knee on crossbeam 3, there was probably a construction approx. 30 cm wide, 25 cm deep and approx. 20-40 cm high (figs 23.23a and b, in the upper right corner above the gutter). This has every appearance of having been a storage locker/small cabinet with a sloping flap, in which the cook could store his tools of the trade, such as fire strikers, spoons and knives, etc. (Plates 23.VI B, 23.VII and 23.VIII).

Fig. 23.23a&b. 3D underwater images of the oven and fireplace on the 'main deck' in the starboard part of the afterbody. To the right against the knee of crossbeam 1 we see the small locker/cabinet? with sloping flap. Frontal and top views (ADC Maritime).

Pump(s) and drain channel
On the keel section of the stern hook, between frames 3 and 5, there was an open area where the pump shaft would have been. Frame 4 had no V-shaped floor timber, only futtocks. The remains of a woven pump barrel were found at the findspot (fig. 23.24). Also, several fragments were found that were part of the pump rod (fig. 23.25). Plates 23.VI B, 23.VII and 23.VIII show how this pump probably worked: the pumped-up water was caught in a wooden cistern and then drained from the ship using an attached wooden gutter (204). A large part of this gutter, which was covered with a board, was found *in situ* next to the fireplace, right in front of crossbeam construction 1 (fig. 23.26). The gutter was more or less square in cross-section (14 cm wide and 12 cm high) and hewn from a single piece of wood. The bottom section is 7 cm thick, while the sides are 4 cm thick. The feed opening is 7x6 cm. Drain channels (or evidence of their presence) were found in the Nijkerk-I and -II cogs.[36]

A woven basket or small tub was found between frames 35 and 36, at the height of the underside of the stempost, in similar fashion to the sternpost. The ship's structure does not provide any evidence of a pump in this part of the cog.

Windlass(es)
A heavy windlass was found in top of the afterbody. The round spindle (Ø 46 cm; fig. 23.27) is 3.05 m long and has protruding 7 cm gearing with a diameter of 20.5 cm. More or less square holes approx. 14x14 cm have been cut at right angles in three locations. They go straight through the drum. These are the openings in which the levers or bars were inserted by several of the crew to hoist the spar with the sail. The 'windlass cheeks' between which the spindle would have turned were not found. Judging from several thick treenails in crossbeam 1 and the knee on top of that, it is probable that there were vertical posts placed against and on top of the front of the beam construction and fastened somewhere above that (Plates 23.VI B, 23.VII and 23.VIII).

Fig. 23.24. Remains of a wicker basket in which the pump barrel probably stood, close to frame 4 on the stern hook.

A beam almost 3.40 m long (160, fig. 23.28) notched with various recesses was found in the hold between crossbeams 1 and 2. A semi-circular hole with a reconstructed diameter of approx. 13 cm suggests that the spindle of a windlass revolved in there. The tiered rabbet below it suggests that the piece was fastened against a crossbeam construction. The diameter of the axis of the spindle does not correspond to that of the specimen described above. The piece found was probably part of a smaller windlass of about 30 cm in diameter. It is not clear where it was fastened. One possibility would be next to the large windlass above crossbeam 1, on the starboard side.

Shifting boards?/bulk-heads
In the afterbody, directly in front of and against crossbeam construction 1, positioned lengthwise in the cog and on either side of the keel plank, sections of two partitions were found that had collapsed forward (figs 23.29a and b). The partitions consist of several vertical plank scraps nailed between two small posts and on what was probably originally four small crossbeams in between. They are marked as shifting boards, which would have been put in place to prevent the cargo from moving around during transport. Looking at the width of the shifting boards (only approx. 90 cm), however, one wonders what kind of cargo could be held in place with these. These inside shifting boards were naturally only addressed at the very end of the reconstruction of the ship's hull.

Although the shifting boards were slumped forward, evidence was found that suggested that they were nailed at a right angle against the front of crossbeam 1, lengthwise in relation to the ship, and presumably originally continued until the deck on which the oven and fireplace stood. During the underwater investigation, it was established that the shifting boards were on the starboard and port side of the keel plank/stern hook. A notch on the side of the ship's centre in the fore and aft beam to starboard suggests that the forward post of the starboard bulkhead was fastened in that notch. If the shifting boards had been vertical, then they would have been resting on/fastened to the lower half of the C ceiling strakes. However, this could not be confirmed. In any case, this means that the shifting boards protected the vital pump-barrel, reserving a space of about 30-40 cm on all sides. In addition, the wooden pump barrel was always accessible via the ends of the A ceiling strakes. This (reconstructed) situation is

36. Sections 6.2.2 and 8.2.2.

Fig. 23.25. Various wooden parts of the pump.

Fig. 23.26. Top view of the wooden pump drain, U-shaped in section, with a plank cover, to starboard between the fireplace and the knee of crossbeam 1 and on the 'main deck' (in the frame). 3D underwater photo (ADC Maritime).

Fig. 23.27. The large spindle of the windlass, found detached (ADC Maritime).

Fig. 23.28. Part of the cheek of a small windlass (ADC Maritime).

Fig. 23.29a. 3D underwater image of the shifting boards?/bulk-heads on either side of the pump barrel (ADC Maritime).

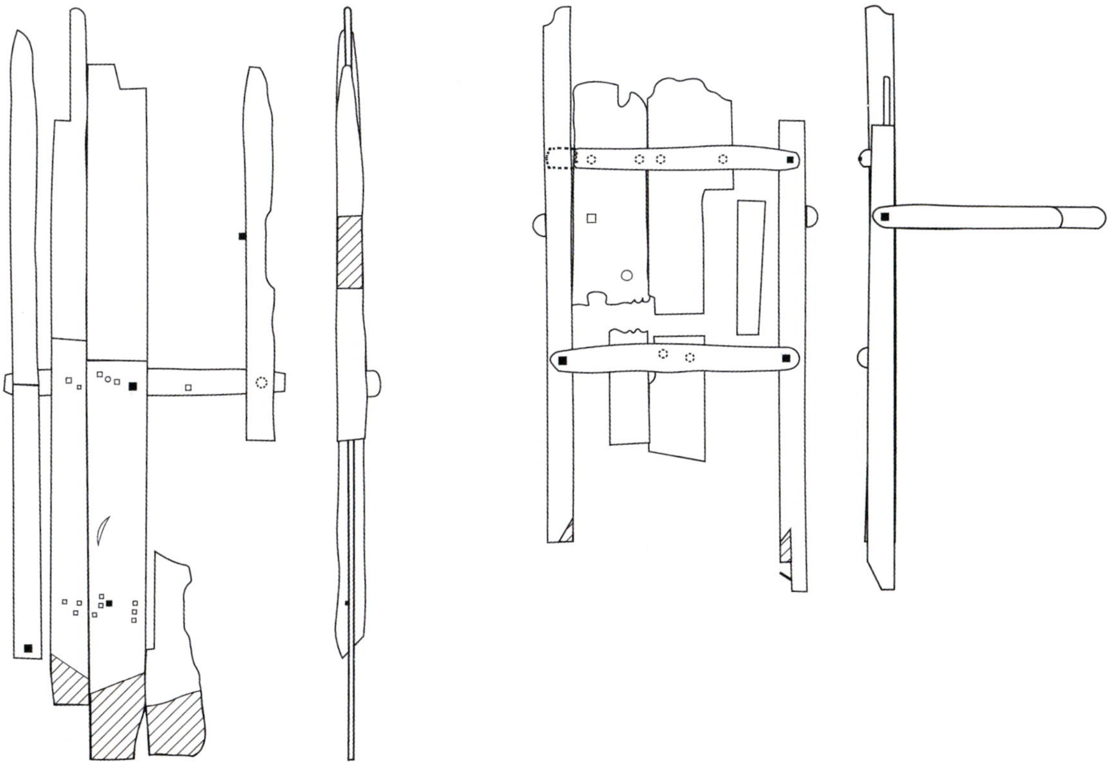

Fig. 23.29b. The remains of the starboard and port shifting boards?/bulk-heads (ADC Maritime).

provided in Plates 23.VI B, 23.VII and 23.VIII. Rabbets were cut into the underside of the B ceiling strakes above frame 3 which are similar in width to the posts of the bulkheads. Notches were found on the inside of the V-shaped timber of frame 5. The interval between the notches and rabbets correspond to the width of the bulk-heads. It was initially thought that the lower ends of the posts of the bulk-heads were fastened in those rabbets. This would mean that the bulk-heads were not positioned vertically and parallel in relation to each other, but together formed a funnel shape, with the bulk-heads about 1.25 m apart at the top and about 0.45 m at the bottom, with the smooth sides facing the pump. In this shape, they would have functioned as a sort of disposal chute. For example, wastewater from the caboose/fireplace surrounding could be poured in, which would then be discharged via the pump and the gutter. However, in terms of vertical positioning, the rabbets and notches in the B ceiling strakes and in frame 5 do not correspond to the rabbet in the fore and aft stringer and the fastening of the vertical post of the bulk-heads against the front of crossbeam 1. It is therefore not possible to provide a definitive and all-explanatory answer.

Oven and fireplace: a 'caboose'?

In the other cogs that were examined, the movable fireplace was covered along with the cooking and baking utensils that were found. In the Kampen (IJssel) cog, in view of the very probably permanent construction on the deck, it seems more appropriate to discuss these elements in the context of the layout and structure of the vessel. The section on *decks* already mentioned the divergent reconstruction of the deck section between crossbeam structures 1 and 2 and that there was a permanent floor in a caboose (figs 23.21 and 23.31a - d). The description of the below indicates that the oven and fireplace were situated under the forward section of the castle deck. This deck between crossbeam constructions 1 and 2 (similar to the Bremen cog)[37] most probably did not span the entire width of the ship, but consisted of two 'promenades', one on each side, with an open section approx. 2.50 m wide between them. Indications were found in the Bremen cog that spaces were created under these 'promenades' by installing (partial) vertical walls (fig. 23.30). It seems very plausible that this was also the case in the IJssel cog. In my opinion, it is doubtful whether we can still consider that a cabose was here. It seems rather that this marks the emergence of the permanent, separate galley.

Figure 23.31a-e consists of three drawings of the deck section between crossbeam constructions 1 (right) and 2, and between the short stringer and the starboard side. The top drawing shows the field survey as it was found. The second drawing shows the planks that were placed lengthwise in the ship, while the third drawing shows the plank floor below that. This plank floor, effectively the 'main' deck, is made out of planks that were placed athwartships in the rabbet in the stringer and nailed against a lath that was placed against the ceiling strake on the starboard side. Visible burn marks on this floor indicate that there were probably fireboxes of some sort on this spot before. There were seven small beams oriented in the same direction. The plank floor on which the brick oven and fireplace was built rests on those beams. These planks do not have any visible burn marks. The spaces between the small beams on the deck were most likely filled with loam mortar. The drain channel (covered by a plank) can be seen on the right on the various drawings.

Fig. 23.30. The remains of the starboard and promenade decks on the model of the Bremen Cog with (partial) vertical walls.

The domed brick oven with a floor that was also partially made from large floor tiles, and the fireplace next to it (figs 23.22a and b and 23.30a, b and c) can also be considered a unique find. 'Cooking and cooking utensils on (inland) ships 1300-1900'[38] covers the fireplaces known from archaeologically examined ships up to that time. In the oldest large cog from the Netherlands (the Nijkerk-I cog)[39], a wooden chest was found measuring approx. 80x80 cm and 20 cm high, filled with loam. A small peat fire would have been lit in the middle of the loam floor for cooking or baking. Such chests were still in use until at least c 1600, and well into the 20th century on fishing vessels on the Zuiderzee. However, from the 14th century on, they were filled with sand which was topped with a light floor of bricks or tiles with chalk mortar.[40] Evidence of permanent fireplaces were found in the shipwrecks of

37. Tanner 2017.
38. Vlierman 1992a, 50-59.
39. See section 8.5.2 and figures 8.9 and 8.12.
40. Chapter 29, figures 29.13a, c and c.

A

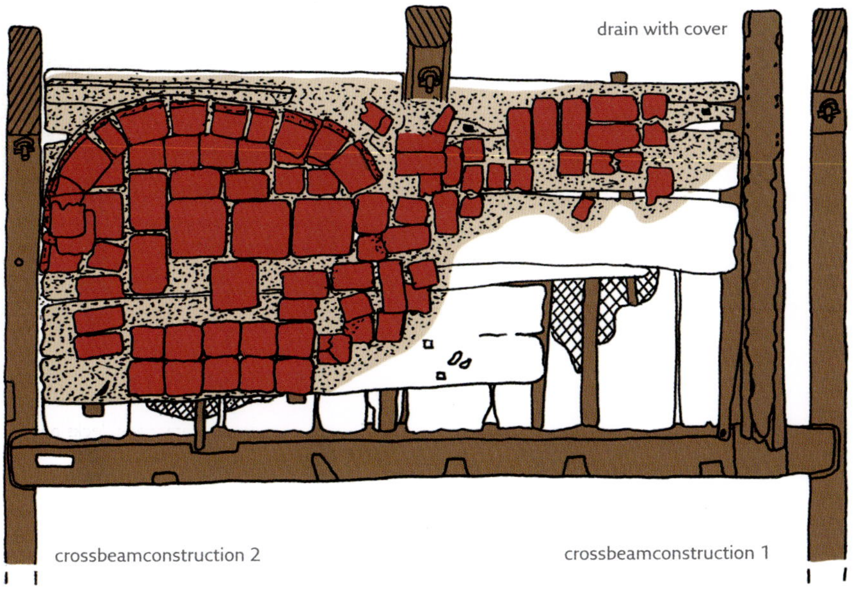

B

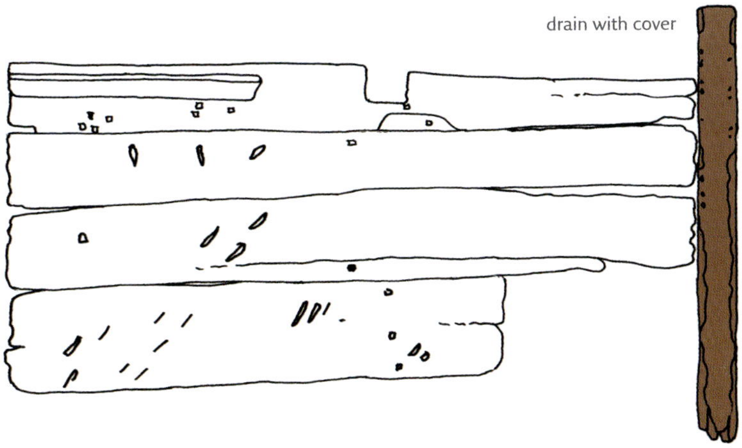

C

D

Fig. 23.31a. The oven and the paved fireplace, top view.To the right and left, the standing knees of crossbeam constructions 1 and 2, and the covered drain close to crossbeam 1. Below is the short stringer lying loose in its rabbets with mortises for fitting in short transverse beams; and to the left, a rectangular hole for a stanchion of the starboard promenade pier of the castle deck (ADC Maritime).

Fig. 23.31b. The plank floor underlying the oven and fireplace (ADC Maritime).

Fig. 23.31c. The original 'main deck' with stringer, drain, short beams and charred patches (ADC Maritime).

Fig. 23.31d. Some floor tiles from the fireplace inside the oven (ADC Maritime).

two small cargo vessels from Flevoland, dating back to the early and mid-16th century respectively.[41] Thus far, the permanent fireplace and oven in the IJssel cog – which in any case date back to the mid-15th century – are the oldest medieval examples in a vessel in the Netherlands that are confirmed archaeological finds. The fire boxes from the older cogs were placed on the ceiling at the bottom of the ship.

At its largest point, the partially reconstructed oven (Plates 23.VII and 23.VIII) had an outside measurement of approx. 1.45 m, a depth of approx. 1.35 m and a height of approx. 90 cm. At least two sizes of bricks were used in the oven (approx. 22x10x4.5 cm and approx. 27x13x6 cm). The unglazed flags (or flagstones) used here are 25x25x5 cm. Below the oven floor was a second floor made out of bricks with the dimensions stated above, and of floor tiles or small flags (or flagstones) measuring 16x16x4 cm. The mortar used in the masonry, the exterior finish and a fire-retardant layer added on the surrounding wood were presumably made from loam, although it was confirmed while the shipwreck was still under water that the mortar was completely gone. Clear traces of it were visible on the underwater photo (figs 23.23a and b). There is no information on how the access point of the oven would have been structured. The oven may have had a presumably straight front with an opening of approx. 50x50 cm covered by a small iron door, but such a door was not found. The opening could also have simply been closed during baking by stacking bricks to block it off. Separated by a low 'wall' built from a few stones, the brick fireplace was situated against the starboard side, between the oven and the drain channel; the fireplace was found partly *in situ*. The brick floor was reconstructed as well as possible using the remains that were found. The floor had a total surface of approx. 1x1 m and the bricks used for the floor were approx. 20x10x5 cm. Loam (see above) was used below the floor and in the mortar joints. A relatively thick layer of this material was applied on the wood around and behind the floor. The double floor of bricks and tiles in the oven and the loam layer under the fireplace were added to prevent the wood below from charring or even catching fire. A peat fire was built in the middle of the floor. There was an iron trivet above this fire, parts of which were found in the wreck (figs 23.32, 23.36 and 23.63).

The superstructure and an opening in the port side

Although detached parts of the superstructure were found in the forebody and afterbody, it is hardly possible to reconstruct it with any substantiation. The remnants do very clearly show that there was a superstructure, however; for this reason, a possible rendition of the superstructure is sketched onto the reconstructions with dotted lines (Plates 23.VI, 23.VII and 23.VIII).

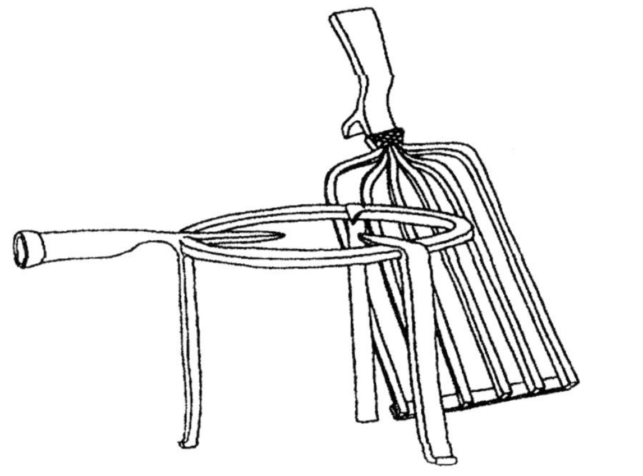

Fig. 23.32. Two types of iron trivet. Drawing after Vlierman 1992a, 53 fig. 36.

The first is a heavy knee (8) that was almost certainly above protruding crossbeam 8. That piece was found beside the afterbody on the port side. The rising knee section from this element of the ship has a strikingly shaped outside curve on the shell side. This is completely different from all the other knees (compare Plates 23.VI A through G). In Plates 23.VI B and 23.VIII, dotted lines have been used to indicate the knee, the possible crossbeam construction and the presumable location in the ship's hull (above crossbeam 1). The iron bolt (Ø 30 mm) that protruded through the head of the knee shows that there had to also have been a vertical rider against the outer shell. A number of nail holes on the knee almost certainly indicate the attachment points for vertical posts pointing upwards. Treenails found in crossbeam construction 1 below also suggest the presence of vertical posts (for the windlass). These posts probably (also) supported the deck of an aftercastle/poop, amongst other structures. Rectangular holes/rabbets observed in the top of the stringers to starboard, between crossbeams 1 and 2, indicate the location of other posts. These suggest that the forward side of the castle deck continued until just before the large windlass, and the deck continued above the fireplace and oven until slightly before crossbeam 2 across a width of 2.50-3.00 m. This was also the case on the port side. In other words, the castle deck/poop was most likely U-shaped, similar to the Bremen cog: a castle deck with two 'promenades' pointing forward.

The two curious knee-shaped pieces of wood (392 and 393, fig. 23.33a) are almost certainly a matching pair. Altogether, the total length is approx. 2.30 m, while the width/height of the long ends is approx. 20 cm, with a thickness of approx. 30 cm. Rabbets with remains of shell

41. Shipwrecks Nz 3 and L 89.

Fig. 23.33a. Two knee-shaped timbers with rabbets (392 and 393), found beside the afterbody; probably they once were joined together (ADC Maritime).

Fig. 23.33b. The larger cog drawing in the mid-15th-century *Digestum Vetus* of Kampen. Municipal Archive, Kampen.

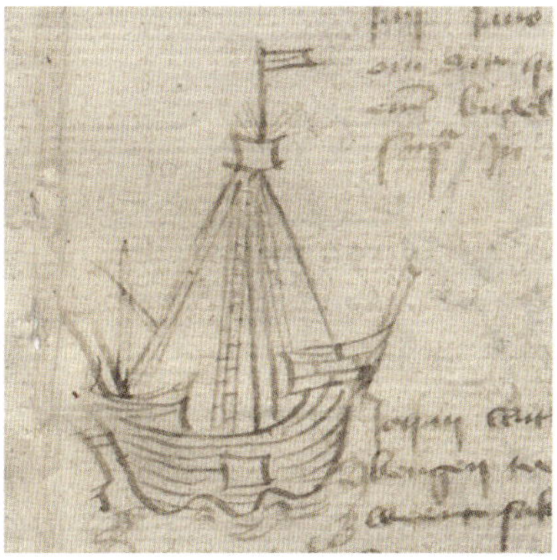

Fig. 23.33c. The smaller cog drawing in the mid-15th-century *Digestum Vetus* of Kampen. Municipal Archive, Kampen.

Fig. 23.33d. IJssel cog, portside. Impression of how timbers 392 and 393 may have framed the top of the opening.

planks can be seen on the outside, as well as the nails with which they were fastened (or the nail holes where they would have been located). Slots 9 cm wide and 7 cm deep were cut into the thickness of the knee-shaped parts. They are situated in the outer half from the middle. A rabbet of approx. 14x7x5 cm was cut into each of the horizontal sections. The pieces/the connected pair was fastened to the framing timbers with wooden treenails 3 cm thick.

It was initially thought that this was part of the trim around an opening in the shell between strake R and the castle deck, for example above the domed oven and fireplace to let out smoke. However, the two pictures of a cog in the *Digestum Vetus* from Kampen (figs 23.33b and c and 23.43) suggest that it is more likely that there was an opening in the port side immediately above and between crossbeams 2 and 3, which protruded through the shell. Judging by where the two pieces were found (next to the port side, at the level of crossbeams 6 and 1) and the picture of a similar opening in the *Krack* by Master WA (fig. 23.44), it is not inconceivable that the opening was between crossbeams 1 and 2. This location may be more logical, considering the rigging of the mast. The shape, function and placement were not yet considered in the monograph on the IJssel cog.[42] The adjusted side elevation (Plate 23.V) shows where this opening could most likely have been placed. The two pieces together probably formed the upper trim around the opening, while a similar construction provided the underside of the opening. The knee-shaped pieces were fastened against the (extra-heavy?) futtocks at that point. Three thick planks (approx. 40 cm wide and 9 cm thick) were stacked on top of each other in the slots to close off the opening. They were clamped into place with two upright posts, 14 cm wide and 7 cm thick, inserted into the rabbets in the top. In turn, the upright posts were probably held in place by two horizontal beams that rested on each side of the opening in clamps that were placed there. Fig. 23.33d shows what the envisioned structure would have looked like from the inside. The underside of the opening was probably about at the same height as the tweendeck below the 'main deck.' By removing the unattached short deck planks and short stringer of the main deck on the port side, between the long stringer and the side of the ship, a work floor could be created from which the crew could take on cargo and stow it in the tweendeck, or offload items of cargo. The cargo hatch was most likely 'protected' abaft and forward on the outside by a vertical rider (Plate 23.V).

A third piece mentioned here is a futtock (265; fig. 23.34). The frame section was found on the port side in the wreckage forward of crossbeam 4. On the reconstruction cross-sections A and F in Plate 23.VI, and in Plates 23.V and 23.VIII, a dotted line indicates the shape and shows where multiple similar futtocks could have been part of the superstructure in the forebody and afterbody. Another part of the superstructure is the slightly curved

Fig. 23.34. The S-curved futtock (265) which most probably reflects the shape of the cross-section of the side between the gunwale (the R strake) and the presumable roof construction of the forebody. A similar shape may also have been applied in the raising of the gunwale up to the after castle deck.

Fig. 23.35. The beam with mortises (170), maybe the foremost beam of a roof above part of the castle deck.

42. Waldus *et al.* 2018.

beam approx. 4.5 m long (170; fig. 23.35) found on the port side of the afterbody. The monograph on the IJssel cog[43] suggests that it may have been the shield on the front of the aftercastle. Further research suggests that it may even be more likely to have been the forward cross-beam of a roof construction above the castle. In that case, beams would have been fastened lengthwise in the wide rabbets on the underside. Upright posts were placed in the square holes of approx. 10x10 cm, forming a kind of lattice from the top until e.g. the castle deck. Similarly, it does not seem impossible that these 'latticed' posts would have continued until the beam that was most likely placed on the knees of crossbeam construction 1. In the latter case, the large windlass would have had to be moved slightly abaft. Fig. 23.36 shows where the beam could have been and what the field survey might have looked like in the first hypothetical scenario.

Fig. 23.36. Sketch showing what the foredeck in front of the large windlass and up to the castle deck may have looked like.

43. Waldus *et al.* 2018, section 14.10, fig. 14.8.

Fig. 23.37a. A purposely rounded piece of wood, possibly part of a mast partner.

Fig. 23.37b. A slightly curved piece of wood with notches at the ends (325); function unclear.

Fig. 23.37c. A wooden block 510/522). The sheave is missing.

Three other detached pieces of wood that were found are also noteworthy: a semi-circular, flat piece of wood (282; fig. 23.37a), a slightly curved piece with recesses on the ends (325; fig. 23.37b) and a belaying post (510/522; fig. 23.37c). The first piece could be part of a mast partners. The slightly curved piece was presumably a guiding beam that may have been attached against the gunwale. However, the opening in the head could also been intended for something to rest in it. The belaying post is clear. The piece 95 cm high was probably originally attached to the (missing) disc lengthwise in the ship on a crossbeam.

Repairs

The cog wreck was salvaged in its entirety and supported in a great many places along the outside. As a result, it is only partly possible to determine where and to what extent there were repairs to the hull. It was possible to obtain a fairly comprehensive overview of the port side, with the exception of the bottom of the ship (Plate 23.V). The repairs can be divided into two types. First, there are the cracks that were sealed with sintel clamped moss caulking. The majority of the cracks probably formed during construction and were caulked shortly before the launch. The second type involves damaged planks repaired in place. At various locations, small planks (generally with butt joints) were inserted and the seam all the way around the repair plank was sealed with sintel clamped moss caulking. The depiction here probably offers an accurate impression of the 'repair pattern' on the rest of the hull. On the inside of the ship, small repair planks were observed in various locations, as well as repairs to sintel clamped moss caulking a few places.

Wood use and fastenings: types of wood and process

The ship's hull was made of oak (*Quercus spec.*). Most of the planks were probably rift-sawn. It could not be confirmed whether the planks also included pieces of sapwood. The keel plank, stern hook and stem hook, sternpost and stempost, false stern and false stem, keelson, the various parts of the crossbeam constructions, the fore and aft beams, the supports of the superstructure and the structural elements of the windlass(es) were made of wood that had been sawn and/or chopped from tree trunks or pieces of trunks/branches. They were all fashioned into the desired form using bearded axes and adzes. The elements of the frames were cut from tree trunks or thick branches and chopped and/or adze-hewn into the desired shape to fit.

Fastenings

The shell planks were fastened to the frames using oakwood treenails (Ø 2.5 cm). The elements of the crossbeam constructions were fastened to the sides of the ship using wooden treenails that were 4 cm in diameter. In addition, thick wrought-iron bolts (Ø 30 mm) were used at crucial connection points. The clinker-built shell strakes were fastened together with wrought-iron nails. The moss in the seams was held in place by moss laths made of oakwood, which were in turn held in place with wrought-iron sintel clamps.

Loading bed of branches, twigs and straw

Remnants were found of a loading bed, more or less between crossbeam constructions 3 and 4 on the starboard side between the mast step and about the middle of ceiling strake C.[44] Remnants of this floor were also spotted abaft of the mast on the starboard side, during the documentation in summer 2016. It seems very plausible

44. Dijkstra in: Waldus *et al.* 2018, Section 14.14.

that this loading bed originally covered more or less the entire bottom of the ship between frames 10 and 33. The top was probably almost even with the top of the mast step section of the keelson. Small beams, thin trunks and thick branches were placed lengthwise on the lower sections of the keelson and between and on ceiling strakes A and B. They were placed in such a way, and sometimes pre-sawn, to create a more or less horizontal ship's bottom with short transverse planks on top of that. Some of the small planks were repurposed from elsewhere. There were narrow planks, branches and narrow tree trunks of alder, oak and pine. Thin branches were laid lengthwise on this 'planking floor', with a layer of the same material on top in a transverse direction. Then two layers of thin branches and twigs, which were also laid approximately at right angles to each other. On top of that, in a similar fashion, were two layers of reeds. This floor (fig. 23.38) was definitely meant to transport the cargo (probably mostly barrels) in safe, dry conditions as much as possible. The composition of the floor allowed for airflow through the spaces between the bottom of the ship, the frames and the ceiling boards. Spraywater and rainwater coming down from above through the gaps between the separate elements of the composite decks could also drain into the spaces between the frames and be discharged through the narrow limber holes and via the pump.

Fig. 23.38. Sketched reconstruction with 'main deck' and 'between decks', the load-bed of branches, twigs and straw and a cargo of barrels.

A somewhat similar piece of loading bed was found behind the mast in the Nijkerk-II cog.[45] Pine was represented by several quartered tree trunks approx. 1 m long and approx. 20 cm in diameter. On top of those narrow tree trunks, there was a piece of what had probably been an old wickerwork basket. The branches were covered in a layer of straw thatching with a layer of coarse sand over it. In the Nijkerk-II cog, it was assumed that it had possibly been a small piece of flooring in a separate room that was added by a fisherman who made their home in the shipwreck rather than it being the loading bed of a cog. However, with the find in the Kampen (IJssel) cog, this theory will have to be revisited. We are most likely dealing with the remains of loading beds characteristic of late-medieval vessels. There are examples from several other shipwrecks in Flevoland.[46] Dijkstra[47] refers to an English source from 1312. The ship *Nicholas of Westminster* was prepared that year to transport 63 barrels of wine. They were not stacked directly on the ceiling, but on a loading bed which was probably made from reeds.

The mast, sail and stays

We will not discuss the rigging and so on of the mast, the mainsail and all other related matters at this point Since the (lowest) deadeyes, putting irons and chainwales or channels are an integral part of the hull, they will be discussed. Several blocks found detached in the wreck will be discussed in section 23.5.2. It is the oldest example to date of how the mast was rigged on a large, late-medieval ship found in the Netherlands through archaeological research.[48] The original nine putting irons (round in cross-section, Ø 30 mm) point somewhat diagonally downwards from the top and from the outside to the inside through the two chainwales on the outside of shell strakes P and N, which are attached to the shell strakes (fig. 23.39 and Plates 23.V and 23.VI C). The completely

Fig. 23.39. Channel with the (lowermost) deadeye, on the starboard side.

45. Section 6.2.2, figures 6.28, 6.29, and 6.30 and plates 6.I and 6.II A.
46. Van Holk 1986, shipwrecks J 137 and O 28 in the Noordoostpolder; Reinders & Oosting, shipwreck U 34 Eastern Flevoland. Annual Report 1986, fig. 3.
47. Dijkstra. In: Waldus *et al.* 2018, Section 14.14.
48. Chainwales for the rigging were found in the small Dronten-I and Marknesse cogs, in section 11.2.2, fig. 11.12 and section 12.2.2, fig. 12.13 respectively.

missing putting iron (third from the back) may have been related to the loading/unloading hatch in the port side. It seems plausible that there was a vertical rider at that location. The deadeyes are the pear-shaped, flat type found previously on multiple occasions and in late medieval fishing vessels in Flevoland. The forwardmost deadeye was located about 1.10 m abaft of the mast, while the aftmost deadeye was at 4.30 m.

23.3 Construction data

Based on all the information, and with no small effort, a reliable set of reconstruction drawings of the hull was produced. As far as possible, the drawings also indicate the layout, providing several suggestions for the superstructure. Based on the reconstruction, it is now possible to say a few things about the lines of the ship, its construction and use, the ship's hull which was probably deliberately sunk as a caisson or discarded, and about the wreck formation and embedding.

23.3.1 Design

The Kampen (IJssel) cog was constructed in accordance with the (cog-building) tradition as had been used by shipbuilders for centuries. Blueprints and building plans were almost certainly not in use at that time, nor would ship-builders necessarily have been familiar with the concept. What is immediately noticeable is that the hull is much higher and therefore also longer than all its 14th century predecessors known to date.

Lines

Plates 23.V and 23.VI H show a ship's hull with a high gunwale. The ship has a straight keel, a straight raking stern and stem, everted sides and a subtle sheer that is higher in the front than in the back. The vessel was held on course with a stern rudder, as evidenced by the gudgeons and other suspension points. The stern has a 114° rake in relation to the keel line, while the stem is at a 130° angle. The bottom of the ship (the keel plank, stern and stem hooks and strakes A through C) is shaped like a bay leaf; at similar distances from the inside of the stern- and stem, it is approx. 60 cm narrower aft than forward. This difference between aft and forward is maintained in the sides, but gradually decreases towards the top. The difference at gunwale height was approx. 20 cm. The underwater body of the vessel angled sharply towards the stern and stem, gradually transitioning from there towards in the middle in a more or less 'round' shape, with a flat bottom amidships curving slightly towards the turn of the bilge. The bottom of the ship is dynamic, without any flat longitudinal surfaces, and angles slightly towards the everted sides (at an angle of approx. 114° amidships). Towards the stern and stem the bottom gradually transitions into the sides. The frames in the back section exhibit a slight S-curve. Strakes A through D are carvel-built. The aft 6.50-7.00 m and the forward 4 m of strakes A, B and C are clinker-built. The strakes of the sides are entirely clinker-built.

Elevated ship's hull

It had already been established during the 2012 visit with the diving vessel that the clinker-built seams of the cog were waterproofed with sintel clamped moss caulking on both the inside and the outside. As far was known at that time, this practice was used till ca 1350. After that point, the clinker-built seams of the outside were no longer sealed with sintel clamped moss caulking. However, the thick iron bolts that were then visible at crucial connections and the type of sintel clamps used would be appropriate to a date later than ca 1400. If we compare the hull of the Nijkerk-II cog[49] (built in the second quarter of the 14th century) with that of the IJssel cog, we see a general correspondence in the overall dimensions and measurements of the heights in relation to the keel line, the sternpost, and the lowest section of the stempost (without the top section and the false stem and stern). The knees on crossbeams 1 through 5 extend until about the middle of the O strake. The O strake is also the top strake of the sides on the Nijkerk-II cog. That vessel is approx. 4.30 m high amidships. The Kampen (IJssel) cog is approx. 4.95 m at the same point. The locations where the crossbeams protrude through the shell are approx. 25 cm higher than the keel line. In its initial approach, the IJssel cog is built almost exactly the same as its predecessor of a nearly hundred years before, but larger. The Bremen cog (dating back to ca 1380) slots somewhere in between the Nijkerk-II cog and the IJssel cog in terms of measurements (and date). Figure 23.40 shows the outlines of the sides of the Nijkerk-II, Bremen and Kampen (IJssel) cog with the heads of the potruding crossbeams at the same scale.

In late summer 2016, it was discovered that long shelf clamps had been placed against the ceiling towards the bottom, against the lowest crossbeams. This may have been done during construction, but also (much) later. The same goes for the vertical riders installed against the ceiling to reinforce the structure of the highest part of the sides. Several samples were taken for dendrochronological analysis to see whether there might be a difference

49. Chapter 6.

enumerated above, it cannot in my opinion be excluded that the cog may have been elevated after all at some point in time, and that a castle resembling a poop deck with a roof structure were presumably added at that time, as well as a roof structure in the fore-part of the ship.

Measurements
It is impossible to state whether temporary frames were used during the start of construction, especially while assembling the carvel-built section of the underwater body, which is very likely. In other cogs, that could in many cases be determined based on small pyramid shaped or conical wedges used to close off the nail holes left by the temporary frames, after they were removed and before the final frames were added. For the same reason, it was similarly impossible to determine whether any markings/indications of e.g. the exact location of the mast were located.

Internal structure and tonnage
The sharply built lowest sections of the afterbody and forebody, from frame 3 abaft and from frame 35 forward respectively, are not particularly suited for placing cargo. It is not inconceivable, however, that there may also have been decks in these sections of the vessel at the level of the shelf clamps forward from the removable tweendeck, at the height of the top of the lowest row of crossbeams protruding through the shell. If they existed, they would have been approx. 3 m in length and been approx. 6 m at the widest point. However, concrete evidence that these decks were present has not (yet) been found.

The section of the ship's hull between crossbeam constructions 1 and 5 and the sides can without a doubt be considered the cargo hold and is 12.50 m long. The height from the ceiling or loading bed to the underside of the 'main deck' (on the fore and aft stringers) is approx. 3 m. No evidence was found that the hold was bordered by e.g. vertical partitions installed athwartships. The largest width below the 'main deck' is approx. 6.50 m at crossbeam 1, approx. 7.00 m at crossbeam 3, and approx. 6.50 m at crossbeam 5. At the lowest point (at the height of the C strakes of the ceiling), these crossbeams are approx. 2.50 m, 5.50 m and 4.00 m respectively. This works out to an average width of approx. 5.10 m. The cargo hold is therefore 12.50x5.10x3.00 m = approx. 191 m^3 in size. If the space between the extra deck above the 'main deck' is added to the calculation, that adds approx. 41 m^3 for the entire section forward of the mast. If the section abaft of the mast to crossbeam construction 2 could also have been equipped with an extra deck, it would have added another 14 m^3 approximately – so an estimated maximum loading capacity of just under 250 m^3 or 250 gross tonnes.

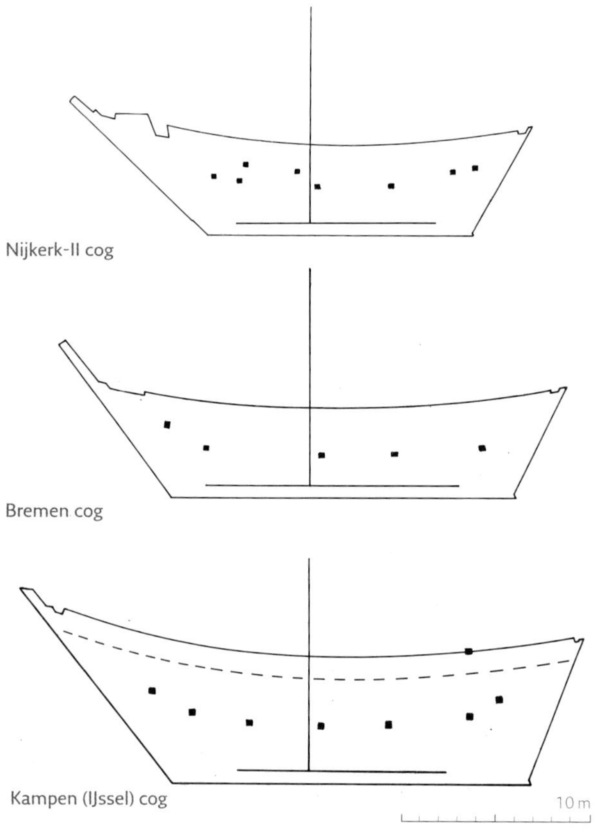

Fig. 23.40. Outlines, drawn to the same scale, of the Nijkerk-II cog, the Bremen Cog, and the Kampen (IJssel) cog. They show the heads of the protruding crossbeams. In a broken line, the probable original height of the side of the Kampen (IJssel) cog.

between the general year of construction and the possible later installation of shelf clamps and vertical riders. The outcome of that examination was not able to show whether there must have been a specific moment when the ship's hull was raised/enlarged. The heads of a number of the rising knees on the crossbeams have been sawn or chopped off. This clearly indicates that structural changes were made at that location. Because of the collapsed starboard shell, it is no longer possible to determine whether – at the top of a possibly initially lower gunwale (top of the O strake) – scarfs can be observed in the frames that would support that. It can be noted that the top three strakes (P, Q and R), especially the first two, are noticeably narrower than all the other strakes. It is also worth noting that the large decking-edge timber in the forebody was put into place at the top of strake O, approx. 1 m abaft of and 1.75 m below the gunwale. It would have been suitable to stand on and perform actions outside the vessel. For the time being, it is assumed that the hull was built in a single, continuous period. Taking into account the structural features

Table 1. Principal dimensions of the Kampen (IJssel)-cog

Length of the keel (including the stern- and stem hook)	16.15 m
Length of the bottom (between the rising part of the stern- and stem hook)	14.65 m
Greatest width of the bottom (at frames 20-24)	3.60 m
Overall length at gunwale level	24.90 m
Overall width at gunwale level (at frames 27-31)	8.50 m
Height of the stern (keel line to top gunwale level)	6.95 m
Height of the stem (keel line to top gunwale level)	5.85 m
Height of the side amidships (at frames 11-28)	8.15 m
Length of the stern	7.20 m
Length of the stem	10.45 m
Height of the hol (at frame 28)	3.45 m

Length-width ratio of the bottom	4.1:1
Length-width ratio at gunwale level	2.9:1

* *All measurements are rounded to the nearest 5 cm.*

As discussed above, there was presumably an 'open caboose', and the space under the protruding parts of the castle deck/ poop between and above crossbeams 1 and 2 presumably had vertical partitions (at least partially). The oven, fireplace and drain channel were to starboard, while there may have been e.g. crew quarters to port. Such quarters may also have been provided on the possible decks in the sharply built afterbody and forebody, in addition to storage for ship's necessities (sailcloth, extra ropes and blocks, etc.).

Materials

The Kampen (IJssel) cog is built entirely of oak (*Quercus spec.*). That includes the treenails that fasten the shell strakes and ceiling strakes to the frames (Ø approx. 2.5 cm) and the thicker treenails (Ø 4 cm) to connect the crossbeam constructions. Assuming that there are two treenails per plank width, at least 5000 of the thin treenails and about 75 thick treenails were used. Besides that, the two long stringers are only fastened in a single place (at crossbeam construction 3), while the heads of the rising knees are fastened to the shell with one thick (Ø 30 mm) wrought-iron bolt. The knee sections on crossbeam construction 5 were fastened with two bolts. Some of the vertical riders were attached at the top with these bolts. Five of these bolts were used in the stern construction, of which three (or two) had a thick ring on the outside for the suspension of the rudder. There are seven of these 30 mm bolts in the stem construction. Almost all of them have thick heads (Ø 50-70 mm and 25 mm high). On the inside, almost all of them have a flat ring against the wood that is held in place with a cotter in a slot in the bolt. Apparently, it was assumed that the bolts had to be removable. There were definitely more than 40 of these heavy, long iron bolts used in the hull. Obviously, it is not possible to make any statements regarding the number of treenails and bolts needed for the superstructure.

Wrought-iron nails with flat, pyramid-shaped heads (Ø 25-30 mm) were used to fasten together the clinker-built planks. The nails were hammered through the overlapping parts of the plank and double-clenched on the inside, with the tip driven back into the wood. They are 120 to 150 mm in length. In some places, they are nailed in pairs: two iron nails at an interval of 6-8 cm, about 10-13 cm from the next pair (fig. 23.41). In most places, however, they were at irregular intervals of 12-15 centre to centre. In total, approx. 780 m was nailed down between the keel plank, the R strakes and against the stern- and stempost through the ends of all strakes. This means that, with an average interval of 13 cm, approx. 6000 nails were needed for the ship's hull. At least ten nails would have been used on each scarf in the keel and shell strakes. The number of scarfs could not be determined. Assuming two scarfs in the keel and in strakes A through D, and three in the other strakes, that brings the total number of scarfs to sixty, for which at least 600 nails would have been used. This brings the total number needed up to approx. 6600 nails. As far as the superstructure is concerned, it is obviously impossible to make any statements regarding the number of nails that were needed. The nail holes that were found which may be related to the inside of planks suggest a fairly limited use of (probably smaller) nails.

All the pieces that were found indicate that the deck elements, including any extra deck levels or tweendecks (fore and aft stringers [with the exception of two thick iron

Fig. 23.41. Detail of the paired nails in the clinker fastenings.

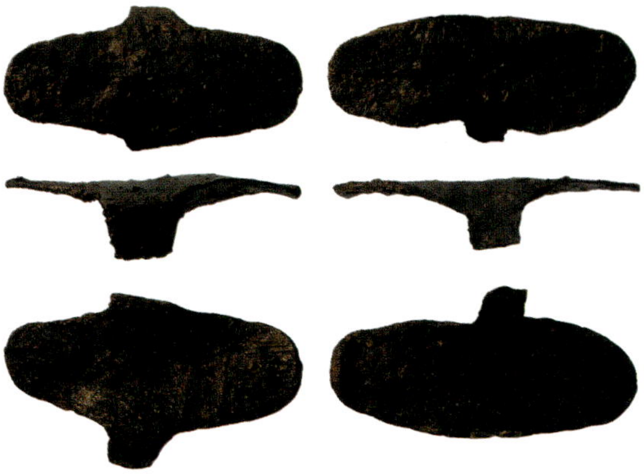

Fig. 23.42. Sintel clamps from the IJssel cog (type D2/E).

bolts] and deck planks), were not fastened down. No deck planks were found. It does not seem inconceivable that the removable deck planks were made not of oak, but of some other wood such as pine. The gudgeons, the fittings of the rudder and stern, and the putting irons have not yet been mentioned; all of these were heavy wrought-iron fastenings.

The clinker-built seams of the ship's hull were waterproofed using sintel clamped moss caulking on the inside as well as the outside. The sintel clamps used are type D2/E (fig. 23.42) and have an oval top with two protruding ears. The oval top is 550-620 mm in length, although some are a bit shorter. They are 1.5-2 mm thick and placed at intervals of approx. 7 cm centre to centre. In total, about 1650 m of seams was sealed with sintel clamped moss caulking between the strakes and between the plank ends and the stern and stem. This means that at least 23,500 sintel clamps were needed. This does not include the sintel clamps used on the caulked cracks and around repairs. In total, at least some 25,000 sintel clamps would have been used in the IJssel cog.

23.3.2 Construction: techniques and sequence

During the construction of the Kampen (IJssel) cog, the keel plank and stern- and stem constructions were obviously the starting point or base. The top half of the stern- and stempost was not found, which means we cannot be completely certain of the reconstructed length and execution of the vessel. However, the version shown in the reconstruction drawings seems to be the most logical in terms of the shape of the sections of the stern- and stempost that were found. If the ship's hull was initially lower (only up to strake O), then the stern- and stempost would have been different at the top and were lengthened later on. On the Nijkerk-II cog, for example, the top section of the sternpost is over 2 m long, and connected with a nibbed scarf.[50]

Construction phase 1: keel plank, stern hook and stem hook, sternpost and stempost, false stern and temporary floor timbers

The keel plank and the stern and stem hooks were first fastened together with sloping flat nibbed scarfs running from forward to aft and from top to bottom. The entire keel construction was probably set atop on reasonably high launching blocks. This was done so that the treenails to connect the shell planks to the frames could be hammered in at a later stage and the sintel clamped moss caulking could be applied to the underside. Although the sternpost and stempost are (in principle) resting on the stern hook and stem hook respectively, without fastenings, they would also have been put into place during this first construction phase for fastening the strake ends of the ship's bottom. It is not inconceivable, and may even be probable, that the sternpost and false stern were first fastened together with wooden treenails and then placed on the stern hook. The same presumably applies to the lowest section of the false stem. In both cases, the different elements were fastened flat against each other with a wooden treenail and a U-shaped piece of wrought iron. The actual false stem, approx. 10 m long, was attached in a (much) later phase of construction. The free-standing stern and stem constructions were braced with thick struts that held them in place while the ship was being built.

Hypothetically, the next step would have been to attach three to five temporary floor timbers to the keel plank, around the level of frames 7 and 29 and in between them. The planks of strakes A through C and perhaps even D were pinned into place against those temporary timbers and frames and nailed down. Small poles were used to hold the planks in place during this process, added as supports underneath.

50. Chapter 6, fig. 6.16b and plate 6.VI A.

Construction phase 2: strakes of the bottom and bilge
The second construction phase consisted primarily of adding the bottom strakes. They started with the aft planks of the A strakes or garboards. The planks were bent into shape using heat (known as 'hot moulding'), and then fitted against the keel plank/stern hook; a flat scarf was added on the forward end. The planks were then nailed into the rabbets in the sides of the stern hook, supported with short poles, and nailed into place against the temporary floor timbers. Then the two forward planks of the garboards were heat-moulded, fitted and fastened into place in the same way. A small layer of (tarred) moss was applied on the scarf between the two planks. The corresponding parts were fastened together with two rows of nails: one row from the outside in, the other from the inside out. In principle, the nails go all the way through both planks, and are then double-clenched: the tips are hammered down and driven back into the wood. On the clinker-built sections of the strakes, the upper and lower edge were chamfered before the strakes were put into place, creating what would become the oblique side of the V-shaped seam where the sintel clamped moss caulking would be applied. The overlapping parts of the clinker-built planks were also fastened together with double-clenched nails. After hot moulding, the planks of the B-, C- and D strakes were finished and attached in a similar fashion. The next construction phase would begin as soon as the bottom of the ship had achieved the desired lines.

Construction phase 3: floor timbers and keelson
In this phase, the final floor timbers from frame 7 through frame 30 were made, fitted into the ship's hull, and attached one by one in their final location with wooden treenails that were hammered inwards from the outside of the shell through pre-drilled holes. After several were attached, they would remove the temporary frame spanning the length, close the nail holes with wooden nail hole plugs, small pyramid or conical wedges, and install the final (V-shaped) floor timber. They would wait with attaching the floor timbers abaft of frame 7 and forward of frame 30. First, the clinker-built sections of the plank seams had to be sealed with sintel clamped moss caulking on the inside.[51]

To provide extra rigidity for the bottom of the ship at that point, the long keelson was then fitted to the tops of the floor timbers and fastened with (blind?) wooden treenails. A solid base for constructing the lowest part of the sides.

Construction phase 4: shell strakes to crossbeam height (top of strake I), caulking of the seams on the inside, lowest futtocks, ceiling strakes to height of crossbeams and riders
In this phase, the nearly straight planks were heat-moulded/prepared strake by strake and clamped into place on the preceding strake already fastened onto the hull. (*Boeitangen* – wooden planking tongs– would have been used for this purpose.) Beforehand, the tops and bottoms of all these planks (strakes E to I) were chamfered in preparation for caulking the seams and provided with flat scarfs. From strake E on, all strakes are fully clinker-built from stern to stem. It is very probable that the shipbuilders added the corresponding strakes on both sides simultaneously. This gradually gave the vessel its shape, ensuring that the overall structure remained in balance. The process of attaching the strakes was repeated until strake I was in place.

Next, the seams on the inside were waterproofed with sintel clamped moss caulking[52] and the carpenter repaired any weak spots in the planks, in most cases by nailing thinner short planks against those spots for support, with a small layer of tarred moss underneath. After the stretch of shell was finished on the inside, the next step was shaping the lowest futtocks, which were then fastened into place (using treenails) with sloping flat scarfs on the floor timbers and against the shell. The heads of the lowest futtocks were squared off at the points where the protruding crossbeams would be placed. The other futtocks were shorter or longer, creating a pattern alternating between the sloping flat scarfs and the top futtocks.

This was followed by the planks of the ceiling strakes between the keelson and the underside of the lowest row of crossbeams protruding through the shell. There is an 'open ceiling': the ceiling strakes are pegged to the frames with treenails with limited gaps between the planks (approx. 10-15 cm). At the thick, wide section of the keelson, the mast step section, supports known as riders were added on both sides between the mast step and until or onto the thicker B-ceiling strakes. They were pegged onto the floor timbers of frames 20 through 23 and enclose the A ceiling strakes.

Construction phase 5: the lowest five crossbeams, shell strakes J and K, crossbeams 6 and 7, shell strakes L through O or R, caulking of the seams, second futtocks (and third futtocks where relevant), the V-shaped floor timbers and the futtocks in the afterbody and forebody
Crossbeams 1 through 5 were situated above frames 3, 11, 20, 28 and 35 respectively, resting on the futtocks and on the upper edge of the I strakes. Square holes were sawn in

51. See image of the Kampen cog during construction, on the cover of the book.
52. Chapter 7, the Kampen cog, figures 7.6 a through e.

the planks of the J strakes, which fit into the slots cut in the tops and sides of the beam heads. This was how the beams were secured in place. After that, the planks for the K strakes were prepared, sawing the holes for crossbeams 6 and 7. Then the process of adding the shell strakes continued until reaching the top strake. If the ship was built in a single continuous construction period, this would have been up to strake R. If the vessel was initially lower, it would have been up to strake O. After this, the seams on the inside were caulked, the second and possible third futtocks were fitted into place, and then the ceiling strakes were added.

Once the (lowest) crossbeams had been put in place and the shell had been built all the way to the top, the lines of the ship's hull would have been rigid and stable, especially between frames 7 and 30. This was the point at which work could continue on the afterbody and the forebody. First, the seams on the inside could be caulked from stern to stem, although some of this caulking may already have been done earlier. The (V-shaped) floor timbers were not attached to the stern and stem hooks or posts, only to the sides. Finally, in this phase, the ceiling strakes were lengthened with planks in the afterbody and forebody.

Construction phase 6: crossbeam knees, intermediate and connecting pieces, side supports and inner stems.
In order to hold the everted bulwarks in place from strakes J through O above crossbeams 1 through 7, and to achieve the intended rigidity of the hull, heavy rising knees were placed on both sides of each crossbeam. A connecting piece was added if necessary, depending on the length of the horizontal sections of the knees. The different elements were connected to each other with (lipped) sloping flat scarfs and fastened with wooden treenails. Depending on the position in relation to the keel line, intermediate pieces were added to the various crossbeams between the beam and the knees, to even out the tops of the knees and create a level deck. The rising knee sections have protrusions on the outside that slot into the sides between the ceiling strakes. The various elements of the crossbeam construction were fastened to each other and then to the sides with wooden treenails 4 cm thick. Moreover, a thick (Ø 30 mm) wrought-iron bolt was inserted through the heads of the knees in most cases.

There is a 'bench' in the afterbody, consisting of a side board of 10 cm thick. It was installed at an appropriate height to have originally served as a side board if the gunwale only continued until strake O. The side support in the afterbody more than 1 m over the 'main deck' was suitable to use in the reconstructed hull lines for actions outside the vessel (depending on how the superstructure was constructed at that point). There is a heavy, wide plank approx. 50 cm above the 'main deck' in the forebody, from frame 32 through frame 44. It would only have been useful for performing actions outside the vessel if the gunwale was no higher than strake O. Judging by the recesses in the outside of the rising knee above crossbeam 7 (frame 39), another heavy plank (or side support was also placed there, higher up. This would have been suited for actions outside the vessel at gunwale height until strake R. Besides this function, the sideboards were definitely also intended to reinforce the structure of the bows and corresponding elements in the afterbody (stringer).

The centre-to-centre intervals between crossbeam constructions 1 and 2, 2 and 3, 3 and 4, and 4 and 5 (approx. 3.40 m, 3.50 m, 3.40 m and 2.50 m respectively) ensured that the hull was strong and rigid between frames 3 and 35. Crossbeam constructions 6 and 7 provided rigidity for the structures in the sections between crossbeam 1 and the stern, and between crossbeam 5 and the stem. Crossbeam constructions 6 and 7 were almost certainly part of the hull (although possibly in a slightly different form), if the ship was not initially higher than strake O.

The question is whether the side-stems had already been fitted in the situation outlined above. The Nijkerk-II cog certainly did not have them.[53] Their purpose was to strengthen the bows.

Construction phase 7: fore and aft beams, 'main deck' and extra deck
On top of the knees on crossbeams 1 through 5, two long fore and aft stringers are placed on either side of the ship's centreline. They are comprised of two parts. A section between crossbeams 1 and 2, and a forward section connected to it with a long hooked scarf. The back section is approx. 3.80 m long, while the forward section is approx. 12.10 m. In the back, the stringer rests in a rabbet cut into the front of the knee. At the location of the four other crossbeam constructions, there were recesses several centimetres deep on the underside of the forward stringer, closed over the top of the knees. The long fore and aft stringers were only fastened to crossbeam construction 3 with a 30 mm thick iron bolt. Other than that, the fore and aft stringers are not fastened. The front end stopped around frame 41, against the stem. Short fore and aft stringers were placed in the knees of crossbeam constructions 1 through 5, about halfway, on either side between the long fore and aft stringers and the sides of the ship. These were also placed without fastenings in rabbets in the aft and forward sides of the knees. All fore and aft stringers have rabbets cut into the top edges, measuring 4 cm wide and deep. No holes from iron nails or wooden treenails were found in the rabbets of the (complete or partial)

53. Plates 6.IV B and 6.V A.

fore and aft stringers. Short deck planks were placed athwartships between the two longer fore and aft stringers and the shorter ones, and between the shorter fore and aft stringers and the ceiling of the sides. None of these planks were found. They were undoubtedly removed from the ship before it was used as a caisson. So the IJssel cog had a 'main deck' that was fully detached, except for two iron bolts, which extended from crossbeam construction 1 all the way to the frames against the stem. There was a hatchway forward from the mast. Between crossbeam construction 1 and crossbeam construction 6, which was placed behind it the first and at a higher level, there were short decks between the top of crossbeam construction 6 and the stern, made of planks placed lengthwise. These decks were respectively intended for providing access to operate the heavy windlass and the helm of the rudder. The sections of the ship between crossbeams 2 and 3, crossbeams 3 and 4, and crossbeams 4 and 5 could be cleared out entirely by removing the deck planks and shorter fore and aft stringers all the way to the loading bed of branches, twigs and straw above the ceiling on the bottom of the ship to load and unload cargo. The deck planks and shorter fore and aft stringers between crossbeam constructions 1 and 2 were also detached and therefore completely removable if needed. The brick oven and fireplace (installed later?) suggest, however, that this probably never happened.

Below that, thick shelf clamps are fastened with treenails against crossbeams 1 through 5 on either side against the ceiling strakes. In the middle between each pair of crossbeams, oblique rabbets have been sawn into the top, in which heavy crossbeams could be laid. It is almost certain that planks were laid on top of that, also lying loose, but lengthwise in relation to the ship. This made it possible to create an extra cargo hold about 1 m high – which would have stayed dry – below the 'main deck'.

Construction phase 8: windlass(es), superstructure on the afterbody and forebody, loading/unloading opening or hatch

Two wide posts/'windlass cheeks' were fastened against the front and above crossbeam construction 1. The heavy windlass, over 3 m long, was between them. This windlass was used to hoist and lower the spar and sail. There almost certainly would have been a smaller windlass, which may have been located to starboard beside the larger one. The small deck for the helmsman could probably have been reached via a small ladder or staircase going up from the small deck for operating the windlass.

It seems plausible that the cog had an aftercastle, with strake O as the gunwale, which could have been accessed from the main deck via one or more sets of stairs.[54] The 'windlass cheeks'/posts and other vertically placed posts would have supported the castle deck. Nothing else can be stated with certainty on this matter. In the reconstructed field survey (up to shell strake R), it is almost certain that the domed oven and fireplace were sheltered, and the separate space in which they stood presumably had partial vertical walls. The rabbets in the short fore and aft stringers in front of the oven confirm the presence of upright posts. There was probably a similar space on the port side. The deck between these spaces (approx. 3x3 m) had to remain open. The pump would have been operated from there, and the lines between the windlass and the mast and spar ran from that point. The IJssel cog most likely had a castle deck/poop (promenade) approx. 3 m wide on the starboard and port side and from beam 1 to 2, with an opening approx. 2.50 m wide in between the two sides. It is impossible to make any statements as to exactly what the sides of the superstructure towards the stern would have looked like based on the very few structural elements that were found. The curious curved pieces (392 and 393) were presumably on the port side between crossbeams 2 and 3. The two pieces together most likely were used as part of the window-frame' around a hatch used for loading and unloading. They were found on the port side next to the afterbody. Figure 23.33d shows what the inside view of the (shut) hatch or loading/unloading opening might have looked like. Figures 23.73a and b show where the hatch presumably would have been located on the port side. If such an opening was added during construction, it probably would have been during construction phase 4.

When the aftercastle/poop deck was being added above crossbeam construction 1, another crossbeam construction was first put into place at about gunwale height (Plates 23.VI B and 23.VIII). In the knee that was found from this construction (8), nail holes can be seen that indicate that upright posts were attached against the knee to support the poop deck. On the reconstructed cross-section (Plate 23.VI B), dotted lines indicate that these posts continued until a possible roof construction above the deck. After the cross-section was drawn, it became evident that the execution of the build could also have been slightly different. Section 23.2.2 surmises that the detached crossbeam that was found (170) may have been the forward beam of this very roof construction. The length of the beam (approx 5.10 m) and the treenails in the ends suggest that it may have been attached to the upright posts at approx. 2.50 m from the centreline of the ship. This would mean that the roof construction would have been over 5 m wide, leaving an uncovered walkway approx. 2 m wide on either side. Figure 23.36 shows what this might have looked like. The S-shaped futtock 265 (fig. 23.34 and Plates 23.V and

54. Such as e.g. on the Nijkerk-II cog, plate 6.V A.

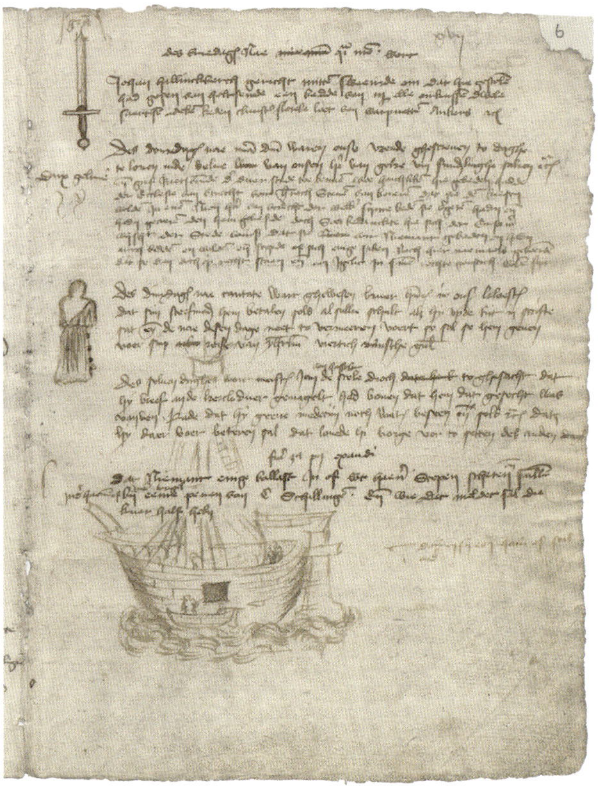

Fig. 23.43. Drawing of a cog in the mid-15th-century *Digestum Vetus* of Kampen. Municipal archive, Kampen.

Fig. 23.44. A kraeck, by 'Master W A', ca 1475. After Assaert 1974, 110.

23.VIII) suggests the possible shape of the section of the sides in the upper deck above strake R in the forebody (presumably also in the afterbody). It would have been a supporting beam of a superstructure above the forebody. More direct indications for this have not (yet) been found. The idea that there might have been a roof construction above the forebody and castle deck was inspired in part by a mid-15th century depiction of a cog in the Kampen *Digestum Vetus* (fig. 23.43) and the illustration of a 'kraeck' by Master WA from 1475 (fig. 23.44). Figure 23.45 shows a model of the *kraeck* based on the copper engraving by Master WA.

Construction phase 9: pump and drain channel, belaying racks, layout of the deck, false stem, beam-head guides, rudder, mast, sintel clamped moss caulking in the seams on the outside, rigging and spar with sail

The finishing touches were applied in this final phase of construction. The pump and drain channel together were a very important system for the vessel. In rainy and stormy weather, the pump had to be operated almost constantly or at least very regularly. Rainwater and any spray and leakage would trickle down between the frames and through the deck seams and end up in the bottom of the ship. The shifting- boards/bulk-heads found beside the location of the pump shaft (near frame 4) were 90 cm wide and smooth on the pump side. The posts and short beams of the shifting boards were on the outsides. They were probably not vertical, but rather placed at an oblique angle against the inside of the long fore and aft stringers, closer to the pump shaft at the lower end. This positioning formed a 'funnel' for dumping wastewater. There was no other way to remove wastewater from the vessel, since the sides were about 2 m high at that point. The shifting boards were certainly also intended to reserve space for the pump shaft. The 'baffle' function proposed in the monograph[55] seems less likely, in view of the limited width of the shifting boards. No evidence was found to suggest multiple shifting boards or partitions. Beam-head guides were fastened against the front of the crossbeams protruding through the shell and against the shell itself. Several became detached and were found next to the shipwreck (fig. 23.46). In this phase, the putting irons with deadeyes would probably also have been installed for the rigging of the mast.

It is hardly possible to state anything else about the deck layout (capstan, belaying posts and/or knights belaying racks, etc.). A single belaying post was mentioned. The rudder was made, the gudgeons and rudder fittings were forged and placed, and the false stem was put into place (although that may already have happened in an earlier

55. Waldus *et al.* 2018.

Fig. 23.45. Model of a kraeck in the National Maritime Museum in Antwerp. After a copperplate engraving by 'Master W A'. After Assaert 1974, 97.

phase of construction). The mast was erected, the rigging was fitted, and the spar was attached. The seams and cracks in the planks on the outside of the ship's hull were waterproofed with sintel clamped moss caulking shortly before the launch. Traces of tar were also observed on the outside of the shell (from maintenance).

Renovation/changes to the cog?

In the preceding sections, it was noted in relation to certain features of the vessel's construction that the cog may originally have been built with a lower gunwale. These notes referenced the roughly corresponding dimensions of the Nijkerk-II cog, dating from the second quarter of the 14th century. Reference can also be made to the Bremen cog, built in the last quarter of that century. The shipwreck of the Bremen cog, which remained more intact, showed many similarities in terms of the layout and set-up of the deck and above that level. In the first stages of construction, the hull of the Kampen (IJssel) cog was built almost exactly the same as these two predecessors. The much higher sides are worth noting. To provide sufficient strength and rigidity to those sides (which are approx. 1 m higher) as part of the hull, vertical riders were placed on the inside and outside, distributed along the length of the vessel. These beams, 2-3 m long, were fas-

Fig. 23.46. One of the detached beam-head guides.

tened vertically against the ceiling and the outside of the shell with treenails. In addition, the top part of the sides was reinforced by two wales. These beams were 10-20 cm thick and high and probably fastened against the N- and P strakes with treenails (more or less) from stern to stem. The tops of these wales abutted the overlapping underside of the O- and Q strakes respectively. These wales were held in place with extra iron bolts 30 mm thick that went all the way through the shell, the futtocks, the ceiling and the heads of the knees on top of the crossbeams, and through the top of some vertical riders. It is not clear at this time whether there were structural renovations to the hull, or if it was raised. However, it is certain that there is a distinct difference between the top three and the rest of the strakes of the hull, and that a number of changes were made in the knees of some of the crossbeam constructions. If the hull was rebuilt or its height was increased, it seems very plausible that the aftercastle/poop deck would also have been modified at that time. In addition, I think it is very likely that the forebody had a canopy, and a part of the poop deck did as well.

23.3.3 Characteristics

Like the other cogs, the IJssel cog has a number of specific characteristics, which correspond to the specifics of the other cogs. However, some may differ and/or supplement the list of general features.

Keel plank

The keel plank is thicker forward than aft, and is three to four times thicker than the shell strakes. The plank together with the horizontal sections of the stern- and stem hooks make up the keel. The plank is the widest amidships and narrows gradually towards the stern and stem. The cross-section is in principle nearly rectangular. Towards the stern and stem, it becomes slightly trapezoidal with the narrowest side underneath. The keel plank is connected to the stern and stem hooks by flat sloping and nibbed scarfs.

Stern hook and stem hook

The stern- and stem hooks, which were sawn out of a singular piece of tree trunk or branch, were fastened to the keel plank with two rows of nails. The rising sections of the stern- and stem hooks are not very high.

Sternpost, stempost, false stern and stem, side-stems

The sternpost and false stern were fastened together with wooden treenails and iron bolts (with an eye for a pintle). The rising part of the stern hook tapers to a point and is enclosed between the stern sections. The lowest part of the false stern had a hooked shape to protect the rudder. The stempost is comprised of two parts, as is the false stem. The two composite beams are fastened together with thick iron bolts. The false stem almost certainly rose well above the gunwale. The stern construction is trapezoidal in cross-section, with the narrow side abaft. The stem construction is probably the same width on the inside and outside. Rabbets were notched into the sides of the stern- and stem constructions, and the ends of shell strakes A through E were nailed into the rabbets. From the F-strakes on, the planks end at the back of the false stern and are nailed to the sides of it. In the forebody, the strakes from the F strakes on end at the mitred front of the stempost and are nailed to it. These nailed plank ends were covered entirely by the thick, wide false stem.

Carvel-built and clinker-built bottom and clinker-built sides

The leaf-shaped bottom of the ship is angled sharply towards the stern and stem. It is composed of the A-, B- and C strakes, and is 60 cm narrower abaft than forward, measured at the same distance from the stern and stem. The bottom strakes were placed side by side in a carvel-built construction, only connected to each other by the floor timbers to which they were pegged with treenails. The garboards are also not connected to the keel plank. The seams between the (twisted) aft and forward sections of the A- and B strakes transition from carvel-built to clinker-built from frame 13 towards the stern and from frame 28 towards the stem. The planks were fastened together along the overlap with double-clenched nails/*naainagels* (folded over and hammered back into the wood). The strakes of the sides are entirely clinker-built.

A long keelson with a mast step section, an open ceiling with thicker bilge ceiling strakes and riders.

The keelson (or inner keel) was primarily intended to reinforce the underwater body. In addition, the thicker, higher section served as the mast step. On the underside, the keelson is in principle notched where the frames would be and falls neatly into place around them. The parts that are inserted between the frames had been chamfered and rounded on the underside. The C strakes or bilge ceiling strakes are thicker than the other ceiling boards. This is because of the scarfs between the floor timbers and the lowest futtocks, which could be said to create a weak construction in the turn of the bilge transitioning from the bottom to the sides. The ceiling strakes that continue all the way to the gunwale, together with the keelson, provide the longitudinal rigidity of the ship's hull on the inside.

Caulking method

All the seams of the clinker-built planks of the hull were waterproofed with sintel clamped moss caulking on both the inside and the outside. This includes the seams between the stern/stem and the ends of the strakes. Cracks were also sealed with sintel clamped moss caulking. The seams around the heads of the crossbeams protruding through the shell were not.

Crossbeam constructions

In any case, the vessel had seven crossbeam constructions, probably eight in total at a later stage. They are different in design and consist of crossbeams protruding through the shell, intermediate pieces and heavy, upright knees on both sides, possibly with a connecting piece. The tops of the lowest five constructions are at nearly the same height from the keel line. The beams rested on the upper edge of the I-strakes. The sixth and seventh beams protrude outward through the J- and K strake respectively. Beam-head guides were fitted against the front of the beam heads and fastened against the outside of the shell. An eighth crossbeam construction was probably above crossbeam 1 at gunwale height.

Decks and superstructures

The stringers and shelf clamps (and parts thereof) that were found definitively confirmed that the (large) cogs had decks that were almost entirely detached and easily removed. This had already been suspected of the Nijkerk-II cog and the Bremen cog. The Kampen (IJssel) cog clearly had a 'main deck', and an extra deck that could be installed approx. 1 m under that on crossbeams placed on shelf clamps. It would presumably also have been possible to add such an easily installed deck above the main deck. The vessel probably had a U-shaped aftercastle or poop deck abaft. A roof construction presumably covered part of that and the forebody.

Vertical riders, wales and horizontal beams on the outside of the aft underwater body

The vertical riders against the inside and outside of the sides and the wales against the N- and P strakes are the earliest known examples of this type of hull reinforcement. As such, they are directly related to the construction of the highest parts of the sides of the IJssel cog. The beams about 2 m long that were attached nearly horizontally and to the side of the underwater body more than 1 m above the keel line beside and on the stern were most likely intended to maintain sufficient pressure on the rudder, for example if the vessel was not carrying a full cargo.

'Caboose'
The domed brick oven and fireplace on the starboard side in the afterbody on the 'main deck' suggest that there was an 'open caboose' in this part of the ship. It is the oldest known example of a permanent fireplace in a cog, which was moreover built on a deck and not in the form of a movable wooden chest containing a fireproof base (of bricks or tiles) situated in the bottom of the ship. It is unique to have an oven on board a late-medieval ship.

Fastenings
The shell strakes were fastened against the frames with wooden treenails in the customary diameter of 2.5 cm. The wooden treenails fastening the different elements of the crossbeam constructions and their connections to the shell were approx. 4 cm thick. In addition, wrought-iron bolts were also used, including for the connections fastening the elements of the stern- and stem constructions. The nails for the clinker-built connections were placed in pairs at certain points, generally relatively close together. In most of the other cogs, they were placed at intervals of approx. 15-16 cm, compared to an average interval of 13 cm on the IJssel cog. The fact that the clinker-built seams were waterproofed with sintel clamped moss caulking on both the inside and the outside can be considered 'outdated' for that time period. After ca 1350, the other cogs from the Low Countries no longer waterproofed the clinker-built seams of the outside with sintel clamped moss caulking. The presence of the rigging of the mast, including thick putting irons and several deadeyes, can be considered a remarkable find. It is the oldest medieval example to date from this region.

Loading bed of branches, twigs and reed
It could also be confirmed for the first time that cogs had loading beds on top of the ceiling strakes. These loading beds were assembled from small beams, logs, twigs and branches, with several layers of reed on top of that. The floor was 20-30 cm thick, designed to safely stow barrels and other items, and also to let the rainwater and spray seeping through the seams in the decks to drain off into the bottom of the ship and the pump, where it could be channelled back off the ship through a gutter.

23.3.4 Use

Function
Besides the lines and dimensions of the vessel, the layout and easily removable deck sections between the crossbeam constructions, and the loading bed, no tangible evidence was found in the IJssel cog to suggest what the ship was used for. However, it is almost certain that the ship was built as a cargo vessel. To increase the cargo load that the ship could transport, and probably to stow the cargo in drier conditions, extra decks could be added about 1 m below and above the 'main deck'; the ship's height had already been raised, either during construction or later. Several large (ballast?) boulders were found in the shipwreck which were almost surely taken aboard in the western Baltic region.

The presence of a domed oven for baking, probably primarily for bread, is striking. The monograph on the cog[56] suggests that this may have been provided for a convoy of cogs heading out on a trade voyage. A few limestone sling balls (fig. 23.47) were found in the wreckage. It does not seem inconceivable that certain voyages might have included soldiers on board in addition to the crew. This would require extra food. In my opinion, this theory is supported by the fact that the oven was built on an original deck section not fastened to the ship, which was moreover reinforced by short beams and planks. It also suggests that the masonry work for the oven was not done during the build, but later. The removable decks that could be added between the main deck and the ceiling may also have been intended to accommodate soldiers during a voyage.

Wear and repairs
Wear and tear was observed on various elements of the ship. The repairs that were found and changes in the hull also suggest that the cog already had a fairly long operational life behind it when it sank. Based on the estimates

Fig. 23.47. The three limestone sling balls.

56. Waldus *et al.* 2018.

from other shipwrecks from Flevoland, the Kampen (IJssel) cog was certainly at least several decades old.

23.3.5 Discarded and used as a caisson

The vessel may no longer have been considered seaworthy, and was almost certainly used as a caisson to block off one of the IJssel's stream channels. Judging by the presumable age of the ship when it met its fate, this suggests that it most probably happened in the third quarter of the 15th century. All usable parts (such as the mast, the spar and all decking, with the exception of the deck planks under the oven and fireplace) would have been removed from the ship for repurposing or to be used as firewood. The hull was then most likely filled with soil, city waste and clay to make it sink and keep it in place.

23.3.6 Wreck formation and embedding

The shape of the IJssel channel in which the ship sank in a transverse position contributed to the hull's warping. Over time, the changes in current and water pressure would have repeatedly washed away the riverbed around the shipwreck, causing it to heel over more and more to starboard. Ice drift would also have played a role in the warping process, particularly in the superstructure and top sections on the port side and the aft and forward section on the starboard side of the hull. The ship's hull would probably occasionally have protruded above the riverbed. In other periods, the shipwreck was mostly covered in sediment. Moreover, it does not seem unlikely that the shipwreck(s) created an erosion channel after settling on the riverbed, and then steadily sank deeper from there.[57]

23.4 Model

There are no tangible plans to build a model at 1:10 scale based on the 2D reconstruction drawings. When reconstructing or recreating the cog again after preserving the shipwreck, it would be strongly advisable to build a scale model at the same time, in order to correctly identify the warped elements and the detached pieces found beside the wreckage and put them correctly into their original locations. That would also make it possible to try out all kinds of variations to achieve more clarity about the execution of the superstructure in the afterbody and forebody.

23.5 Ship's inventory, or objects that were part of the fill or washed in?

23.5.1 Introduction and distribution of the finds

Throughout the investigation, various objects or parts of objects were found above, beside and in the hull of the cog. A number of the finds were related to the location of the shipwreck (buoyage, anchors, dredging anchors and netting weights), while others could have been part of the ship's inventory and equipment. Assuming that the cog was used as a caisson to dam up a stream channel of the IJssel near Kampen, and was filled with soil to facilitate that purpose, it seems very plausible that part of that would have been city waste. As usual, such waste would have contained all sorts of discarded utilitarian objects or fragments of such objects. In addition, the people of Kampen and the inhabitants of towns and villages upstream undoubtedly threw all kinds of rubbish into the IJssel over the centuries since then.[58] The majority of the finds, especially in the top layers of the sediment deposited in the cog, can therefore almost surely not be attributed to the ship. All the objects that were in principle still usable would have been removed from the ship if the vessel was discarded and/or if the vessel was still sufficiently accessible after sinking. Other shipwrecks in Flevoland, which were found to contain (nearly) complete equipment and inventory of tools and other utensils, almost always also contain fragments of objects which often pre-date or are more recent than the objects that were used shortly before the ship perished.[59] The younger sherds are in the top layers of the sediment above the shipwreck and do not belong to the ship. They had washed in. Sherds of a broken pot or jug that ended up amongst the ballast or between the frames, many of which had worn edges, may have been lying there for several years before being covered by sediment. Bones from meat-based meals consumed at various times are nearly a standard feature in shipwrecks. The younger sherds are in the top layers of the sediment above the shipwreck and do not belong to the ship. They had washed in. Sherds of a broken pot or jug that ended up amongst the ballast or between the frames, many of which had worn edges, may have been lying there for several years before being covered by sediment. Bones from meat-based meals consumed at various times are nearly a standard feature in shipwrecks.

In the monograph on the IJssel cog[60], the various authors categorised the found objects by material and discussed them based on specific functions. In the chapters on the other cogs in this publication, the found objects are de-

57. Van Dinter. In: Waldus *et al.* 2018, 132.
58. Schilder 1983, *Digestum Vetus.* 21 August 1456. It is prohibited to dump manure or rubbish, by carts or by barrows, from the bridge in the IJssel on the city side of the marking (the hand) that is on the bridge. Similarly, manure or rubbish may not be dumped on the dyke or alongside the road on the opposite side of the IJssel. All this under penalty of a 100-shilling fine, of which half will be given to the person reporting the violation.
59. Vlierman with cooperation from Van Kleij, 1990.
60. Waldus *et al.* 2018, ADC monograph 24, Chapters 18 through 23, 25 and 26. See the monograph for the list of objects.

scribed according to a predefined set of categories in a specific order. These categories are:

1. Ship's equipment
2. Operational equipment
3. Administration and navigation
4. Tools
5. Household items
6. Cooking, eating and drinking
7. Victuals
8. Personal equipment/weaponry

Given the reasons listed above for why all or most of the found objects may not have been, or probably were not part of the equipment and inventory of the IJssel cog, it was decided not to sort the objects into categories for this shipwreck. The monograph offers a more detailed discussion of many of the objects; in this publication, it is sufficient to summarise and depict certain noteworthy objects (and groups of objects) and their individual dates in relation to the period in which the cog must have been built, sailed the waters, and sank into the IJssel. In addition, based on many years of experience with excavating shipwrecks in Flevoland, an estimate has been of the extent to which the finds inside and near the cog may have (had) a direct relationship with the ship's inventory.

23.5.2 The finds

Stoneware and pottery

Sherds of stoneware and earthenware pottery are almost always found on board in shipwrecks. This was also the case in the IJssel cog. About 157 sherds of utilitarian pottery were gathered, dating back to the Late Middle Ages and the Modern Era. The researcher(s) note that these finds, in particular the small and incomplete objects, should be interpreted with caution. In the lowest, clay-rich layer of the sediment inside the cog, sherds were found that may indicate that the shipwreck was filled with sediment and city waste dug up for this purpose.[61] This assumption would mean that the objects found in

Fig. 23.48. The funnel jug of Siegburg ware (201).

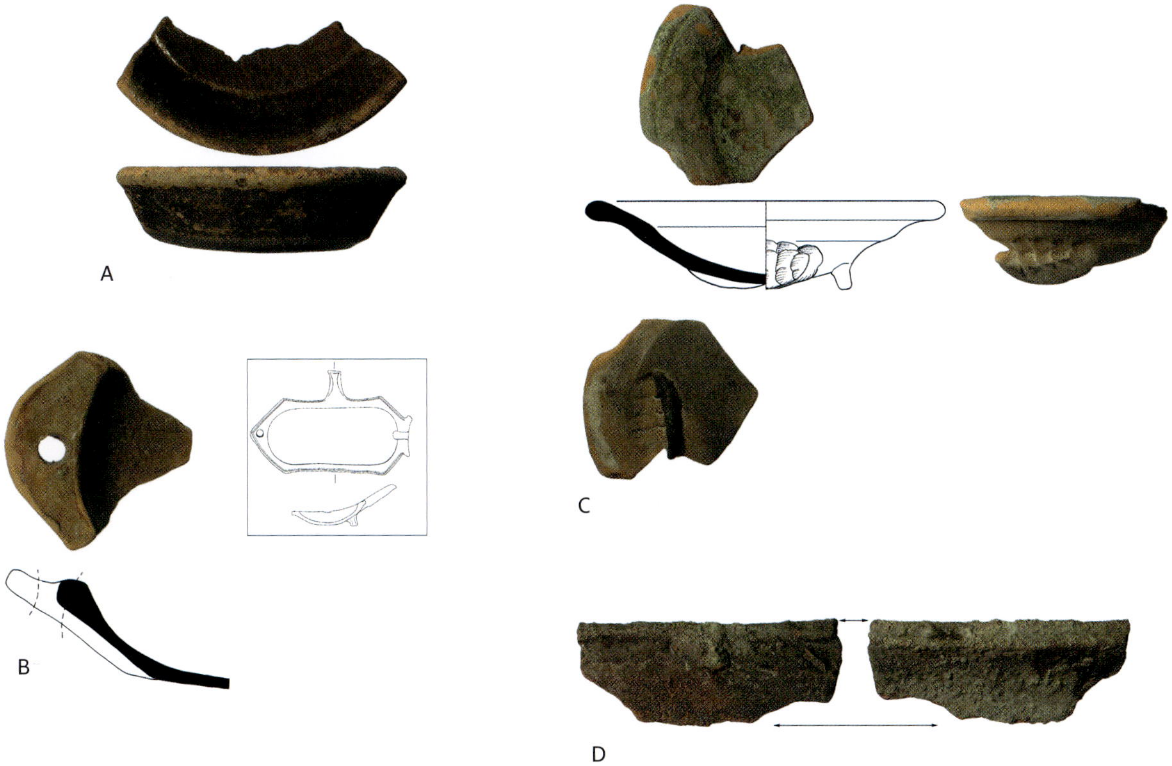

Fig. 23.49. Sherds of redware: a. skillet (151); b. dish (126); c. dripping-pan (97); d. dripping-pan (115).

61. Jaspers. In: Waldus *et al.* 2018, 253-262.

the higher layers of sediment can mostly likely also be considered part of that deliberate fill, or may have washed into the wreck later. The nearly complete Siegburg 'trechterkannetje' or funnel jug (201, fig. 23.48) dates back to ca 1375-1425 according to the Deventer System. Several sherds of red earthenware pottery from a frying pan (151), a plate (126) and two grease catchers (97 and 115) (figs 23.49a-d) date back to 1300-1525, 1375-1525 and twice 1375-1500 respectively. These fragments also came from the fill inside the cog. An intact tripod pipkin made of red earthenware pottery (88) was found with the wreck of the IJssel cog. It had become attached to a grapnel (fig. 23.50). This object dating of this cooking pot, possibly a misfire, is 1375-1500.

One extraordinary find is a near-complete relief made from white earthenware pottery depicting an angel holding up the Veil of Veronica (or 'Manoppello Image') with an imprint of the face of Jesus on it. Object dating: 1375-1525 (17, fig. 23.51). Figure 23.52 shows a painting by the Master of Veronica (c. 1420) depicting Saint Veronica with the veil (*Sudarium*).

Ceramic building material

The domed oven and fireplace in the cog were made of tiles and bricks. Three sizes of tiles and bricks were found.[62] No remains of lime mortar were found on the bricks or tiles. Judging by the traces in the wood under and immediately around the oven and fireplace, it can be assumed that loam mortar was used in the masonry and gradually dissolved and washed away. Loam mortar was probably also used as a fire-retardant layer between the fireplace floor and the deck on which it was placed, and as mortar between the bricks of the floor itself. The Nijkerk-I cog had a sort of fireplace in the form of a wooden chest filled with loam, approx. 20 cm high.[63]

Three kinds of tiles of various sizes were used in the oven floor. The floor consists of two layers. The smaller tiles and several bricks were in the lowest layer. The actual oven floor was composed out of large and medium-sized tiles with bricks around the outer edge (figs 23.31a - d and Plates 23.VII and 23.VIII)). This surface was used for baking. Unglazed and glazed tiles were used. The glazed tiles generally had a dark or black lead glaze, while a couple were dark brown or green, and one tile was slipware with a light green glaze over it.

The researcher[64] extensively discusses the various sizes and fabrics, the possible or probable origin and date of the bricks and tiles. In his conclusion, he asserts that these are new bricks, not re-used specimens. Size, material and other features suggest that they date back to the second half of the 15th century or early 16th century.[65] The tiles that were used correspond to tiles used elsewhere dated to the 15th century. In combination with the date of the bricks,

Fig. 23.50. Grapnel or creep anchor corroded onto a redware *Grapen*.

Fig. 23.51. Whiteware relief of an angel holding up St. Veronica's veil, bearing the image of Christ's face.

Fig. 23.52. Painting by the 'Veronica Master', ca 1420. National Gallery of Art, London.

62. Melkert. In: Waldus *et al.* 2018, 263-268.
63. Section 8.5.2 and figures 8.9, 8.12, 8.50, 8.51 and 8.52.
64. Melkert. In: Waldus *et al.* 2018, 263-268.
65. The large bricks in that shape and size were no longer made after ca 1300, so these must have been re-used from elsewhere. Report by H. Clevis.

Table 2. Bricks from the oven and the fireplace

Description	number	length	width	thickness (in cm)
Large bricks	20	31.5-32	14.5-15	7-7.5
Medium-sized bricks 1	41	23.5-24	11.5-12	5.5
Medium-sized bricks 2	13	21.5-22	10.5-11	5.5
Fragment of different size	1		13	5
Fragment of small brick	1		10	4

Table 3. Floor tiles from the oven floor

Description	number	length	width	thickness (cm)	fabric	surface
Tile, large	4	27	27	5.5	red fired	black glaze
Tile, medium	26	17.5	16.5	4	grey fired	unglazed
Tile, small	16	11.5	11.5	2.5	red fired	mostly black glaze, several dark brown/ green, 1x slip below light green glaze

it can be assumed that all the ceramic building materials can be dated to the mid-15th century and that the domed oven was built at a later stage, not immediately after the construction of the cog was completed.

Natural stone

Nineteen large boulders were found: six in the bottom of the ship, and the other thirteen spread through the wreck. Other than that, about twenty smaller boulders. The nature and composition of the identified assortment of boulders suggests that these may have been ballast stones, which were not collected as loose boulders in the Netherlands, but presumably near the south-east coast of Sweden in the Baltic Sea.[66] However, it should be noted immediately that this conclusion has been drawn on the basis of a small assortment of stones. This also brings up the question whether these were actually ballast in the first place. They may have been left behind from the last time the ship was ballasted.

A few of the smaller stones stand out. Three cannonballs of sandy limestone 6.8-6.9 cm in diameter were found near cross beam 2 on the port side (177), while two others were found in the sandy layer of sediment with netting weights (308 and 309). Given the circumstances of the finds and their locations, they could be considered part of the cog's inventory or attributed to the objects that washed in or were lost on the wreckage (which had been used as netting weights) (fig. 23.47). In terms of dating, they are a very good fit for the 15th century. They are made out of *ledesteen*, sometimes referred to as 'Flemish arduin', a type of stone that was mined in quarries west of Brussels. Cannonballs of this white, sandy limestone were fairly common and have thus far mostly been found in the southern Low Countries and the rivers region.

A small cylinder of Bentheim sandstone with a square hole measuring 2x2 cm (61, fig. 23.53) was revealed while lifting out the cog. This relatively coarse sandstone comes from quarries in Germany, several miles across the border from the Dutch town of Oldenzaal. This sandstone was transported via the (Overijsselse) Vecht to the staple market in Zwolle. Building blocks for large buildings were made from that type of stone; the royal palace on Dam Square in Amsterdam, for example, was built of Bentheim sandstone, but it was also used to make feeding troughs for cattle. Round, flat whetstones and grinding wheels of all sizes were another important product. The find described as a possible 'bearing'[67] was most likely the last remaining usable part of such a grinding wheel. An iron shaft with a crank handle was inserted through the square opening. This type of grinding wheel was used very widely for centuries by artisans, in households and on ships. Generally, the stone was fastened onto a wooden block beside a wooden water trough in which the grindstone was rotated. The piece that was probably no longer usable as a grindstone may have been repurposed as a netting weight.

Another stone is a piece of a large, narrow bit of chlorate-phyllite rubble (50, fig. 23.54). It is an Eidsborg phyllite from a quarry near Eidsburg in southern Norway. This type of rock is very suitable as a grindstone or whet-

66. Melkert. In: Waldus *et al.* 2018, 269-274.
67. Melkert. In: Waldus *et al.* 2018, 272.

61.001

Fig. 23.53. Remnant of a round grindstone with a square hole, made from Bentheim sandstone. (61).

50.001

Fig. 23.54. 'Semi manufactured' grindstone or whetstone of chlorite-phyllite from southern Norway.

stone, and was in use as early as the Viking era; it was in common use throughout the Late Middle Ages. These stones were also found in other shipwrecks and in other cogs discussed in this study.[68] They were probably primarily purchased as rough stone, developing the widely varied grinding surfaces during use.

Eleven fragments of roof slate were found in the bottom of the wreckage of the IJssel cog. These seem to consist of three types of slate, probably originating from the Rhineland and the Maas (or Meuse) area (e.g. Fumay). The findspot, between the frames, suggests that this could possibly be remnants of a cargo that had been transported at some time.[69] It is not very likely that the cog itself would have brought this cargo from the region from which the stones originated.

Objects made of wood, rope fragments and peat

Over a hundred objects made of wood were collected during the excavation. 82 of those were barrel staves and stave fragments, while 12 were sections of barrel lids/bottoms, and 15 were other objects: parts of the pump system, elements of the rigging, tools and eating utensils.

The pump system

As mentioned previously, the remains of a woven tub or basket of willow branches was found at the detached timber of frame 4 in the afterbody (fig. 23.24). Moreover, remnants of such a basket were found in the forebody between frames 34 and 35. There were no other indications of any pump found in the immediate vicinity of that findspot. A number of objects were found beside the open space above the stern hook in the afterbody, which were part of a pump system (fig. 23.25), while on the deck right against the front of crossbeam construction 1 on the starboard side, a wooden drain channel with a lid was found. Section 23.2.2 already discussed this in some detail. The reconstruction drawings in Plates 23.VI B, 23.VII and 23.VIII show what the pump system might have looked like. The rainwater, spray and bilgewater, and presumably also the wastewater from the 'caboose', flowed through the seams between the loose decking was funnelled between the frames into the bottom of the ship, trickling down through limber holes in the underside of the floor timbers to the lowest point in the afterbody. From there, it was pumped out through a wooden tube (a hollowed trunk approx. 20-25 cm in diameter). In later times, the water pumped up through the pump tube would run across the watertight decking to scuppers in the gunwale. However, that was not yet the case when the cog was built. In the IJssel cog, a cistern construction was undoubtedly placed near or around the pump shaft to collect the water. From there, the water would be discharged from the ship via the gutter connected to the bucket. Evidence was found in the Nijkerk-I cog that indicated the presence of a drain channel, while an intact drain channel

68. Melkert. In: Waldus *et al.* 2018, in the Dronten-I cog, Section 11.5.2, figs 11.2 and 11.26 B; in the Marknesse cog, section 12.5.2, figs 12.26 and 12.28 b; and in the Swifterbant cog, section 18.5.2, figs 18.8, 18.10 and 18.12.
69. Melkert. In: Waldus *et al.* 2018, 275-282.

was still present in the Nijkerk-II cog.[70] The pump barrel itself and parts of the cistern construction were not found. It is not entirely clear what the execution of the pump would have looked like above decks, in other words how the water was pumped up. A possible version has been indicated on the reconstruction drawings. In the later inland vessels and fishing vessels of the Zuiderzee, wooden pump shafts were used well into the 19th century. They were found within ship structures on various occasions. There would have been a 'bundle' with a check valve in the bottom to prevent the water from flowing back in. There would have been a small hilt or handle at the top of the pump rod. Operating it would be comparable to a bicycle pump. On the inland vessels, the pump shafts are approx. 2.5-3 m long. The pump shaft of the IJssel cog must have been about 4 m long. Whether there would have been a wooden pump rod of max. 4 cm in diameter seems questionable (probably too fragile). A short wooden rod or an iron rod might be more likely. There would probably have been ropes threaded through the limber holes lengthwise in relation to the ship, which would have been tugged back and forth sometimes to keep the holes clear of obstructions. These ropes would have been knotted or had small wooden sticks attached to them.

Elements of the rigging[71]

Besides the large windlass already mentioned above, a second one that was probably smaller, and a detached belaying post, a number of objects were found that can be attributed to the standing and running rigging. Several should probably be considered spare parts. On the starboard chainwale with putting irons, five of the presumably original nine deadeyes are still present (Plates 23.V and 23.VI C). They are the lowest of a set that are attached to the ring of a putting iron with an eye. They are pear-shaped and flat in appearance (fig. 23.39), and date back to the first half of the 16th century at the earliest. However, they were apparently already in use at the start of the 15th century. Rounder, more bulbous models came into use from ca 1600.

The rigging includes a double fiddle block (90, fig. 23.55.). The block is 49 cm long, 10 cm wide and 8 cm thick. The largest disc is 11 cm in diameter and 3.5 cm thick. The small disc is 9x3 cm. On the edge of the large disc, there is a hole (Ø 4.5 cm) to attach a rope catcher. In the Almere-I cog, which dates back to approximately the same period, a single-disc block was found.[72] It is possible that it was a 'brace block' which was used to operate (reeve) the runner which moved the spar. Furthermore, four loose discs of blocks were found in the IJssel cog, varying in thickness and diameter. They are made out of maple, elm and beechwood (find numbers 156, 182, 296 and 393). Two finds, almost surely parrel beads or stones (132 and 165, fig. 23.56), would have been threaded onto the parrel-strap of the spar to make it easier to raise and lower the sails. The parrel beads allowed the parrel-strap to roll along the mast. Two wooden rope thimbles were also found (188 and 189, fig. 23.57). These rope thimbles would have been reeved in (the end of) a line or rope. The thimbles have a hole through which another rope could be guided. The rope thimbles are also equipped with locking holes to sew them into the loop. They are made of hornbeam and beechwood respectively. Number 189 still had a piece of rope around it. Several other pieces of rope were also found (114, 123

Fig. 23.55. Fiddle block (90).

70. Section 8.2.2 and 6.2.2, figs 6.35 and 6.36a and b, and plates 6.IV B and 6.V A.
71. Waldus. In: Waldus *et al.* 2018, 278-279.
72. Vlierman & Hocker 1996, 53-55 and figs 26 and 27; the present manuscript: section 16.5.2, figs 16.19 and 16.20.

Fig. 23.56. The two wooden beads (kloten). (132 and 165).

Fig. 23.57. Two wooden rope thimbles (188 and 189).

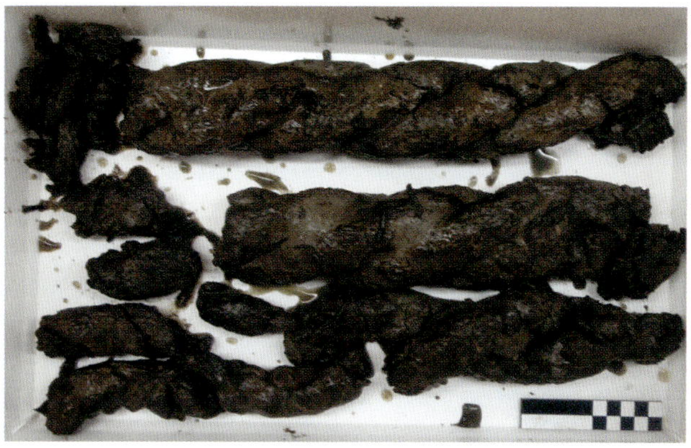

Fig. 23.58. Lengths of rope (114, 123 and 152).

and 152, fig. 23.58). The pieces are approx. 2 cm in diameter, are single-strand, Z-twisted, and made from flax or hemp.

Eating utensils

Fragments of a plate made from ashwood (154, fig. 23.59) with a reconstructed diameter of 27 cm and a maximum height of 6 cm is the only eating utensil that may have possibly belonged to the ship's inventory. Such plates are rare in the hundreds of shipwrecks in Flevoland, although they were almost certainly widely used. When a vessel sank, such lightweight wood items would be the first to float away. They only stay behind in the ship's hull throughout the entire process of sinking and wreck formation if there is a closed space to retain them. It is therefore more likely that the plate was part of the dug-up fill in the hull, or washed into the wreck.

The peat that was found could have been part of the last batch of fuel used on board. This type of fuel is commonly found in late medieval and post-medieval shipwrecks in the Netherlands.

Fig. 23.59. The treen dish of ashwood (154).

Objects made of iron[73]

According to the researchers, the 4 four-armed dredges which were found – three of which were in the cog, while the fourth, as stated above, was attached to the cooking pot grapnel – were too light to have been used as an anchor for the large IJssel cog. The same goes for the small stock anchor with part of the stock still attached (56, fig. 23.60) which was found in the forebody. It is only 1350 mm long. However, as the researchers indicate, very little is known about the dimensions, weight etc. of anchors in relation to historic ships. In the case of the IJssel cog, it can be assumed with reasonable certainty that they were lost by other vessels after becoming entangled in the wreck.

The pike or boathook which was found in the clay layer deposited in the forebody of the shipwreck can be considered part of the ship's equipment. It is a well-known type (190, fig. 23.61a), which until now was only known from shipwrecks from the 16th century and later in the former Zuiderzee area.

The piece of iron chain with links in three different sizes (67, fig. 23.61b) certainly belongs in the cog. A nearly identical chain was attached to the inside of the stem of boat K73/74 from Flevoland, which has also been dated to circa 1425 based on the sintel type used in the construction.[74] The item interpreted by the researchers as a caulking iron/chisel or wedge is a small iron rod (85), which was most likely one of the cotter that was inserted through the ends of the 30 mm connection bolts.

67

Fig. 23.61a. The fragment of chain with three links (67).

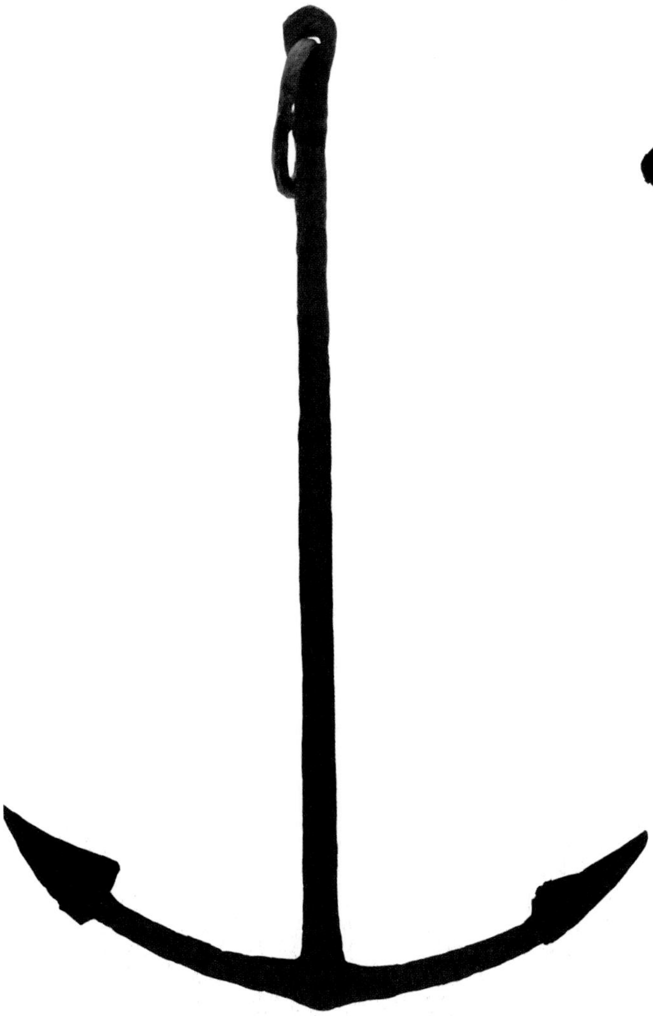

56

Fig. 23.60. The small stock anchor from the forebody (56).

190

Fig. 23.61b. The boathook (190).

73. Langelaar & Waldus. In: Waldus *et al.* 2018, 288-295.
74. Reinders *et al.* 1980, 17-29, 43-46 and plates 3, 4 and 5; Vlierman 1983, 3-11; idem 2010, 193-194; section 26.1 of this publication.

The fragments of two wrought-iron spades from the fore-body (figs 23.62, 104 left and 162 right) are a type that I had only seen before in the Almere-I cog.[75] They could possibly be considered part of the presumed fill used in the hull of the IJssel cog.

The four flat wrought-iron bars (99, fig. 23.63) came from a rectangular iron trivet, a cooking rest or fireplace grate. The original width of the iron cooking rest was 560 mm and it had legs measuring 80 mm in height. Round and rectangular iron rests were found in various shipwrecks in Flevoland (fig. 23.32). The pieces were found next to the oven. The significance of the fifth piece of iron 'lattice' depicted here is unclear.

330

Fig. 23.62. Fragments of two iron spades (104, left, and 162).

404

99.001

Fig. 23.63. The parts of the trivet (99), and the piece of iron framing of unclear function.

75. Hocker & Vlierman 1996, 55 fig. 28 and 29-20; present publication Section 16.5.2, figs 16.21 and 16.22.

Personal possessions[76]

Seven more or less intact pilgrim badges were found in the shipwreck (4, 16, 30, 58.1, 131 and 167, fig. 23.64). All of them date from the 15th century. It cannot be determined whether they belonged to one of the crew members of the cog. They may also have ended up in the wreck by way of the dug-up soil used in the fill. The find numbers 4, 16, 30 and 58 (the final one shown in fig. 23.65) are all badges of the Miracle of Amersfoort. Badge 58 shows the discovery of a clay figurine of the Virgin Mary, which was tossed into the outer canal of Amersfoort in 1444 by a woman named Geertje Arends during her pilgrimage to the Saint Agnes convent. The housemaid Margriet Albert Gijsen discovered the statue under the ice near the Buitenpoort, after she had a vision three times in which she was told to go to this location. The statue was eventually placed in the Chapel of Our Lady in Amersfoort and various miracles have been attributed to it. The Miracle of Amersfoort and the pilgrimage to that location were especially popular with skippers, sailors and others who worked on the waters. The fact that most of the pilgrim badges found here references the Miracle of Amersfoort cannot be a coincidence in the context of the maritime hub of Kampen.

Find no. 131 is a fragment of a badge depicting Saint Quirinus of Neuss. Quirinus was a Roman tribune who is said to have converted to Christianity together with his daughter Balbina. He became a martyr when Christians were persecuted under Hadrian. His remains are kept in the Saint Quirinus Church in Neuss, which is named after him. Various healings are attributed to his relics and cult. Find no. 167 is a badge depicting Saint Adrian of Geraardsbergen (East Flanders). Adrian was a Roman officer who died a martyr's death after refusing to torture Christians who would not worship the Roman gods. After his death, his mortal remains ended up in Geraardsbergen by way of Rome, Spain, Lisbon and Constantinople. Once there, pilgrims would visit him in the St. Adrian monastery because he was said to offer protection from the plague. Find no. 6 from 2012 is a badge depicting the holy sacrament 'Three Hosts' from Wilsnack (Brandenburg, Germany). It shows three blood-spattered, unblemished hosts (the sacramental bread) that a priest named Johannes Cabbuz found in the ruins of the Wilsnack Church, after it had been burned down in 1383 by Heinrich von Bülow's troops. The drops of blood were attributed to Jesus himself, and this place of pilgrimage has been called the Church of the Holy Blood ever since.

167

58

131.001

6

Fig. 23.64. Various pilgrims' badges.

30.001

Fig. 23.65. A pewter text strip reading 'AVE.MA[RIA]' (30).

76. The following text is adopted in its entirety from Langelaar & Waldus. In: Waldus *et al.* 2018, 293-294.

The other personal possessions of a religious nature also included find no. 30 (fig. 23.65). It concerns an inscription made of tin with letters in Gothic style beaten into it. Presumably it says: AVE.MA[RIA]. It had three holes in the flat side and three eyelets on one of the longer sides. It may have been a pendant.

The handle with the start of a blade (92, fig. 23.66) is from a weapon known as a bollock or ballock dagger or kidney knife. The original dagger was probably about 30 cm long; this type was a well-known late-medieval personal weapon. These types of daggers were also found in the 15th-century wrecks of the Almere-I cog, the lot K73/74 boat and the cargo vessel on lot B55, all three located in Flevoland.[77]

A conical, round wrought-iron object (105, fig. 23.67) may have been a buckler. It measures 140 mm in diameter and is 120 mm in height.

The Neurenburg jeton (58, fig. 23.68) could be hung up or nailed to something. Jetons were part of a financial administration system based on tokens or counters and a counting-board, similar to an abacus. The jeton found in the cog dates back to the first quarter of the 16th century and must therefore be counted among the finds that were washed into the wreckage.

Fig. 23.66. Wooden handle of a kidney or ballock dagger (92).

Fig. 23.67. A round, conical object of forged iron (a shield boss?) (92).

Fig. 23.68. A Nuremberg jetton (58).

77. Hocker & Vlierman 1996, 67 68 figs 32 and 33-24/section 16.5.2 of this publication, figs 16.27 and 16.29; Reinders *et al.* 1980, 27 and plates 7-10/section 26.1, fig. 26.7c of this publication, and Reinders *et al.* 1980, 39 and plates 11-12.

Leather finds[78]

During the investigation, remains of leather shoes and other objects were found in various places. With the exception of find no. 108, which was probably a fragment of a set of bellows made out of sheep or goat leather, all other fragments are cow leather.[79] The fragments mainly come from various shoes. A number seem to have been cut up. A striking find is an intact cutting pattern for the side panel of a mid 14th century half-high laced boot. The shoe seems to never have been made or assembled. Only the aforementioned piece of leather from a bellows and the leather on the piston head of the pump can be attributed to the cog with a fairly high degree of certainty. The rest could have been in the dug-up soil and city waste used in the fill, or have washed into the wreck. Besides complete boots and shoes, discarded leather items or bits of leather are almost always found in the shipwrecks in Flevoland. They were used for all kinds of repairs. Some of the fragments could certainly have remained in the cog when it sank.

Animal bone material[80]

The bone material found in the cog was thoroughly interpreted by the researcher. In total, over 17 kg of animal bone material was collected, consisting of 866 fragments. The identified animal species are: cow, sheep/goat, sheep, wild goose/domesticated goose, pig, chicken, horse, cat, dog, beaver and fish. The majority of the bones came from the meaty part of the livestock. The chopping and cutting marks show a clear slaughter technique: the animals were first halved by sawing or chopping straight through the spine from head to tail. The animals were then divided into smaller, standardised pieces. They would have been divided up roughly (probably with an axe), chopping straight through the bones instead of at the joints (fig. 23.69). The pieces of meat were about the same size, probably so they would fit into barrels. Only a couple bones show teeth marks. Judging by the quality of the bones, it seems very plausible that the meat was packed into barrels, and presumably salted. However, it may also have been dried or smoked. There were no burn marks on the bones, so the food was not prepared in combination with fire. It is plausible that some of the barrel staves that were found could be linked to how this meat was stored.

Staves[81]

A hundred barrel parts were found in the wreckage of the IJssel cog. The staves and sectional lids are all made of oak wood. The staves vary in length between 70 and 125 cm. Judging by the lengths, nine different groups of barrels can be identified. Unfortunately, it was not possible to reconstruct a complete barrel. Various markings were found (fig. 23.70). Dendrochronological research has shown that the oak used here primarily originates from the Polish coast, with several from the Maas Valley/Germany, and that the trees used to make the barrel parts must have been felled between 1362 and 1418. Besides the animal bone material found, indicating that the meat may have

Fig. 23.69. Some fragments of animal bones with butchering marks.

78. Blom. In: Waldus *et al.* 2018, 283-287.
79. The finds have been described in extensive detail in Waldus *et al.* 2018.
80. Aalders. In: Waldus *et al.* 2018, 309-325.
81. Schoute. In: Waldus *et al.* 2018, 327-337.

Fig. 23.70. Four parts of barrels with markings.

been stored in several of the barrels, it should in my opinion be assumed, based on the dating and the probable date when the cog sank, that a large part of the (discarded) staves would have been part of the loading bed.

23.6 Cargo and merchandise

The barrels may also have been used to transport other contents besides the meat. Apart from the pieces of slate and the rough piece of grindstone/whetstone made of Eidsborg phyllite, indications were also found that suggest that the cog may once have transported a load of grain.

23.7 Ballast?

The boulders already mentioned in the section on *natural stone* may once have functioned as ballast (or part of the ballast).

23.8 Dating

Initially, based on the type of sintel clamp (D2/E) used for the caulking, it was assumed that the cog must have been built in the first half of the 15th century. This dating was corroborated by the presence of a strikingly large number of heavy iron connection bolts. Such heavy iron bolts are rarely if ever found in the 14th-century cogs discovered thus far. However, there was a contradiction in the use of the sintel clamped moss caulking in the seams on both the inside and outside of the clinker-built shell planks. Until the discovery of the IJssel cog, it had been established based on previous investigation of all cogs from the Low Countries that clamped moss caulking using sintel clamps on the seams of the outside was no longer performed on (large) vessels) after ca 1350.

Dendrochronological analysis of several samples initially yielded a general felling date of after 1399 and showed that wood from the Netherlands, Germany and north-western Poland was used in the construction of the ship. A further twenty-two wood samples were examined in 2015 and 2016.[82] The felling intervals of the eight samples with the latest dates are all between ca 1400 and 1420. A construction date between ca 1415-1420 seems to be the most probable time frame.

No hard evidence was found to determine the year in which the cog sank/was used as a caisson. The object dates of the various finds from in and near the cog, to the extent that they could be dated, do place this event in the first half/middle of the 15th century. The dendrochronological dating of the barrel staves (the youngest dating back to 1418) are essentially the only 'firm' dates for objects that were related to the cog in one way or another. This dating fits amazingly well in the time period when the vessel was built and could be explained by suggesting that they were aboard the cog while it was sailing and were put into the loading bed once the barrels were no longer fit for use.

The repairs on the shell and the general wear and tear on the hull suggest that, in comparison to the hundreds of shipwrecks investigated in the IJsselmeer polders, the date when the IJssel cog sank was somewhere in the third quarter of the 15th century.

Table 4. Dendrochronological dating of the IJssel cog 2012, 2015/2016

Sample no.	end of year	felling interval
M31a	1276	between 1296 and 1308
M31b	1286	between 1296 and 1308
M173	-	-
M186	-	-
M187	1385	after 1393
M188	1372	after 1372
M225	1398	between 1406 and 1421
M226	1412	between 1412 and 1414
M227	1412	between 1412 and 1414
M228	1381	after 1398
M229	1388	after 1398
M230	1414	late summer/winter of 1414/1415
M231	1387	between 1393 and 1408
M232	1384	after 1392
M233	1387	between 1393 and 1408
M234	1347	after 1357
M235	1371	in 1394 or 1395
M236	1386	after 1394
M237	-	-
M238	1418	late summer/winter of 1418/1419
M239	1417	late summer/winter of 1418/1420
M240	1418	late summer/winter of 1418/1421
M241	-	-
M242	-	-

23.9 Summary and comparison with the other, older cogs

Ship

The shape and construction of the IJssel cog: straight stern and stem on stern and stem hooks respectively, with a keel plank in between, a carvel-built bottom with a sharply angled afterbody and forebody, with the ends of the strakes on the bottom slowly transitioning into a clinker-built construction, fully clinker-built sides that have crossbeams protruding through the shell, on which heavy rising knee constructions were placed, and shell seams waterproofed with sintel clamped moss caulking. These are the features characteristic of a 14th-century cog. However, the oakwood vessel has a keel length (including stern- and stem hooks) of 16.15 m and sides that are approx. 1.5 m higher than the previously known predecessors of this type of vessel. In comparison: the Nijkerk-II cog has a keel length of 12.30 m, while the Bremen cog is approx. 15.70 m. These vessels were built in the first and second half of the 14th century respectively. Although dendrochronological analysis did not provide conclusive results to determine whether the ship's hull had been rebuilt/elevated, various indications were found in the construction that leave this option open.

The higher/raised sides were structurally reinforced on the inside and outside of the ship's sides by vertical riders : thick wooden beams 2-3 m long, held in place with wooden treenails and positioned vertically against the ceiling strakes and the outside of the shell (fig. 23.71 and Plate 23.VIII).

82. See appendix.

Fig. 23.71. Two vertical riders placed against the starboard ceiling, and the knee of crossbeam construction 3.

Fig. 23.72. Exterior of the starboard shell with the two wales/rubbing strakes and also a channel with deadeyes and iron fasteners.

In addition, wales were fastened to the outside on the N and P strakes to give these highest parts of the hull strength and rigidity (fig. 23.72). The fastenings connecting the crossbeam constructions and vertical riders to each other and the hull included not only the customary thick wooden treenails, but also wrought-iron bolts 30 mm thick. These iron bolts are very rarely found in the cogs from before ca 1400.

Its principal reconstructed dimensions are:

- Length over the stern and stem at gunwale height — 24.90 m
- Largest width at gunwale level — 8.50 m
- Height of the stern (keel line to the top gunwale) — 6.95 m
- Height of the stem (keel line to the top gunwale) — 8.15 m
- Height of the side amidships (keel line to the top gunwale) — 5.85 m

On the inside, the IJssel cog has a long keelson with a thicker mast step section supported on the sides by four pairs of riders. Between the keelson and the gunwale, the ship had an 'open ceiling' on either side: a total of twenty-four strakes consisting of fairly wide planks fastened onto the frames at fairly small intervals. The keelson and ceiling strakes together provide the longitudinal rigidity of the ship's hull on the inside. The seven (probably later eight) heavy crossbeam constructions reinforced the lateral structure.

In the deck elements still present in the Bremen cog and the Nijkerk-II cog, indications had already been found suggesting that certain deck planks were probably not fastened to the ship. The Kampen (IJssel) cog provided the proof that all deck elements were detached! The confirmation was unanticipated and incredibly wonderful. It means that they apparently journeyed across the seas with loose-lying decks. The 'main deck' on the tops of the crossbeam constructions primarily consisted of two thick, wide fore and aft stringers about 1.25 m apart, on either side of the ship's centreline, which extended from crossbeam 1 to the frames against the stem. The fore and aft stringers were only fastened in one spot (on crossbeam construction 3) with a thick iron bolt. Short fore and aft stringers rested in rabbets in the knees of the crossbeam constructions, about halfway and similarly lengthwise, between the long fore and aft stringers and the sides of the ship. The long and short fore and aft stringers have 4x4 cm rabbets cut into the top on either side, in which short deck planks were laid athwartships. The 3-3.5 m spaces between the crossbeam constructions and the sides could easily be cleared completely to load and unload the ship. In fact, this is comparable to the later inland vessels and modern barges. One would expect that the loose parts of the decks were secured somehow. The only vessel examined to date which contained indications of deck hatches fastened on the coamings against the gangways above deck level was an 18 m long passenger and packet vessel (ferry) that probably perished in the Zuiderzee around 1625.[83] The shipwreck can be viewed at Batavialand in Lelystad.

There were almost certainly two short decks in the afterbody of the IJssel cog that were placed slightly higher and at different levels. The first to operate the windlass, and the second for the helmsman. The planks of these decks were probably placed lengthwise in relation to the ship.

83. Shipwreck B 71 Eastern Flevoland.

What is 'new' to cogs is the addition of a shelf clamp against the underside of the crossbeams on either side, abutting the ceiling. Recesses were cut into the top of this shelf clamp about in the middle between the crossbeams, in which additional crossbeams could be laid. These were removable support beams for 'tweendecks' that could be added between the crossbeam constructions. Their planks must have been approx. 1.70 m in length, and were detached and oriented lengthwise in relation to the ship, between the support beams and the crossbeams. This 'tweendeck' was approx. 1 m under the 'main deck' and provided a dry hold, possibly for transporting special cargo. Sleeping quarters, e.g. for soldiers, is also an option. Evidence was found which suggests that the crew could also install an extra deck about 1 m above the 'main deck'.

The cargo hold below the main deck between crossbeam constructions 1 and 5 and the sides offered approx. 190 m^3 in total cargo space. If the extra, higher deck was installed, then the cargo space could have been raised to nearly 250 m^3 in total, or approx. a gross weight of 250 tonnes. There was a loading/unloading hatch approx. 2 m long and 1.25 m high on the port side. The hatch probably could be closed with three planks 9 cm thick, which were in turn clamped in place by two vertical and two horizontal beams. The hatch was most likely located between crossbeams 2 and 3 and right below the 'main deck' at the spot indicated in the sketches of the *Digestum Vetus* (figs 23.33b and c and Plate 23.V).

Various evidence supporting the presence of a superstructure was found in the afterbody and forebody of the ship. There was presumably a canopy or roof construction above the forebody. There was probably a U-shaped castle deck or poop deck in the afterbody, a shape that was also used in the Bremen cog. There was probably also a canopy or roof construction above a part of that deck. However, the number of elements found from the superstructure were too limited to determine a definitive description. The hypothetical superstructure/roof constructions were derived in part from the mid 15th century drawing of a cog in the Kampen *Digestum Vetus*, and a model based on the illustration of a *kraeck* from 1475 drawn by Master WA (figs 23.36, 23.43, 23.44 and 23.45).

In the Nijkerk-II cog and several other late medieval shipwrecks in Flevoland, indications were found that suggest the presence of loading beds in slightly different forms. It was definitively established in the Kampen (IJssel) cog that such floors would very probably have been added in the cargo hold of many, possibly even most late-medieval cargo vessels. They were intended to be used to properly stow the cargo (primarily wooden barrels). In the Nijkerk-II cog[84], it was assumed that it could have been a 'living floor' added by a fisherman who stored his fishing gear in the shipwreck. If it turned out to have been part of the loading bed on that vessel as well, then the Nijkerk-II cog would have had a cargo hold of approx. 140 m^3 rather than approx. 100 m^3 which means a gross weight of 140 tons. In any case, the loading bed observed in the IJssel cog was installed above the ceiling of the underwater body between the mast and frame 35 (crossbeam construction 5), but probably also behind the mast from frame 5 on (crossbeam construction 1). In addition to stowing the cargo, the loading-bed also had to allow air to circulate to prevent the inside of the underwater body from sealing off and rotting. Several ships (such as the Doel-I cog and the large 16th-century freighter or warship from lot U 34 in Flevoland) was observed to have been generously tarred on the inside of the underwater body to prevent the wood from rotting. The loading bed was composed of short, chopped logs and pieces of twigs and branches varying in thickness. In between, there are often thin planks, such as old barrel staves, with a few layers of reed or straw on these twigs in alternating criss-cross directions. Sometimes a layer of sand was be applied over that; the large cargo vessel of lot U 34 had a floor of boulders. These loading beds are also water-permeable, which is very important because the rainwater and any spray that came in through the deck seams had to drain off to the bottom between the frames. The water would flow through limber holes in the underside of these frames to the deepest point in the afterbody. From there, it would be discharged from the ship with a pump, via a cistern connected to a wooden gutter on the starboard side. Wastewater from the 'caboose' would be diverted to the pump through the 'funnel-shaped' shifting boards. Hosing/discharging wastewater from the 'main deck' would have been impossible, since the gunwales were more than 2 m high.

The presence of a domed oven, probably for baking bread, as well as a permanent fireplace, are the first examples of permanent hearths aboard a late-mediaeval ship excavated in the Netherlands. All the other cogs probably had removable wooden chests filled up with loam or sand, with a floor of bricks or tiles.

The Kampen (IJssel) cog can most likely be considered a representative of one of the last vessels built according to the cog-building tradition. It is a 'key find', which marks the transition from the cog-building tradition that held sway for over three centuries to the later, larger carvel-built vessels with more than one mast and more sails. The lines and measurements of the vessel on the reconstruction

84. See sections 6.3.3 and 6.9 and figs 6.28. 6.29 and 6.30.

drawings can be considered more than 95% accurate. Figures 23.73a and b show impressions of the Kampen (IJssel) cog and what it would have looked like before (a) and after (b) a renovation.

Finds

All sorts of different finds were collected in and around the wreckage of the IJssel cog during the various phases of the investigation. There are only a few that can be directly related to the vessel and its original crew with some certainty. The two iron spades, for example, could have been left behind when the ship's hull was being filled with soil, clay and city waste. Sherds of stoneware and pottery from broken jugs and pots etc. may (if they were found between the frames) also have been part of the cooking, drinking and eating utensils once used on board, which would have been part of the ship's inventory. General object datings can be provided for many of the finds that fit within the period in which the vessel was built, when it sailed, and when it was deposited as a caisson. The majority most likely ended up in the ship's hull along with the city waste used in the fill, or came from waste dumped into the IJssel that ended up near and in the wreck and then became covered with sediment over time. Remnants of leather are (also) common in shipwrecks. The pilgrim badges could have been lost by crew members, but they could also have ended up in the ship along with the city waste used in the fill. The barrel staves that were found and had probably been discarded were presumably mostly incorporated into the original loading bed.

Date, origin and last function of the IJssel cog

It was already possible during the field evaluation in 2012 to determine a general date for the construction of the cog based on the typochronological dating of the sintel clamps in the caulking along the seams. The sintel clamp type D2/E, was used in ca 1375-1450. Together with the striking use of heavy iron bolts, it was clear that the vessel was most likely built in the first half of the 15th century. Preliminary dendrochronological analysis shortly after resulted in a dating after 1399. Dendrochronological analysis of multiple samples in 2015 and 2016 resulted in a definitive felling date for the trees that yielded the wood used to build the ship. Construction most likely took place between ca 1415 and 1420. Judging by the general condition and wear of the cog and the previous repairs, it seems very plausible that the vessel went down in the IJssel in the third quarter of the 15th century.

The wood that was used originates from the Netherlands, Germany (Westphalia) and north-west Poland. The moss used for the caulking (various species of *Sphagnum)* occurs throughout the Netherlands. The twelve caulking samples examined contained moss, but remains of heather were also found. It seems very likely that the moss was collected in the Veluwe region. The presence of a stem from another moss species (*Calliergon richardsonii (Mitt.)* suggests that a caulking seam may have been repaired in Scandinavia. After ca 1300, ship wood was increasingly imported from other areas. It seems very plausible that the IJssel cog would not only have called at Kampen as its home port, but may also have been built there or elsewhere in the IJssel region. The shipyard discussed in section 28.2, located on the Burgel in Kampen, could not have been where the vessel was built. The land in question was given a different use in the second half of the 14th century.

Is the Kampen (IJssel) cog depicted in the Digestum Vetus? A hypothesis

The well-known city chronicle *Digestum Vetus* from Kampen (1448-1478) contains very realistic drawings depicting dozens of diverse objects and people using accurate proportions. The sketches are related to the subjects described for each date.

20 May 1455. Above the large, well-known drawing of the cog (figs 23.33b and 23.43) it says: "No one may, on punishment of a fine of 100 shillings, throw ballast out of their ship in the harbour or the Borgel (Burgel). He who brings the perpetrators to us shall receive half of the fine."

9 August 1455. The smaller cog (fig. 23.33c) says: "Johan Witte sailed to Bergen in Norway for Hollandish merchants. The Aldermen and Council have notified him that if the city of Kampen or its citizens are inconvenienced of suffer losses from this, they shall recover this from Johan."[85]

Johan Witte is mentioned again in 1460 in connection with a collision with a ship of fellow Kampen resident Johan Sellen. Both vessels were probably anchored off the isle of Texel. Witte's ship went adrift, probably as a result of severe weather, and rammed into Sellen's vessel, causing the latter to sink; fourteen men drowned. Witte managed to save his ship, but Sellen presented a damage claim to the court of Kampen against him. Witte was willing to pay for the damages, but Sellen thought the compensation he offered was too low. The aldermen and council then demanded clarity about the cargo and the damage.[86]

The confirmed loading/unloading hatch in the port side of the IJssel cog was not found in archaeological research on

85. Transcriptions by K. Schilder 1983.
86. Jager 2015, 382.

earlier medieval vessels and may be considered a remarkable construction. Such striking elements would usually be emphasised in medieval pictures. This was also the case in the drawings from Kampen. It cannot be a coincidence that both sketches show the same number of strakes between the water line and the gunwale (8), that the hatch is located below the top two shell strakes, and that it covers about three strakes in height. The measurements of the hatch are in proportion to the ship's hulls (including the IJssel cog). It can also be seen on the largest sketch that the opening is above the crossbeams that protrude through the shell/the beam-head guides.

If the Kampen (IJssel) cog is indeed depicted here, then this drawing shows the ship's hull at its presumably original height, three strakes lower. This would mean that the ship built around 1420 was modified/raised *after* August 1455. It could also be assumed that Johan Witte's cog is depicted here, that Kampen suffered losses through his actions, and that the city authorities either confiscated the old, rebuilt vessel, or reached an agreement with Witte for him surrender the vessel as compensation for damages so it could be used as a caisson in the IJssel channel they planned to dam up. In that case, it would have taken place after ca 1460.

Fig. 23.72a. Impression of what the Kampen (IJssel) cog may have looked like before ca 1455. Based on the well-known drawing in the *Digestum Vetus* of Kampen.

b. Impression of what the Kampen (IJssel) cog may have looked like after ca 1455.

innografix

The Enkhuizerzand Cog[1]

24.1 Find conditions

24.1.1 Discovery and investigation

During the deepening of the channel in the IJsselmeer in the northern approach to the Houtribsluizen locks[2], a ship find occurred on 17 March 1999. In accordance with the applicable procedures, the discovery was reported to the Netherlands Institute for Ship and Underwater Archaeology (NISA), part of the Dutch National Service for Archaeological Heritage (ROB). On 22 March, a piece of ship wood was delivered to NISA, where it was recognised as a fragment of a rudder. This offered sufficient grounds to take a look at the other pieces of ship wood that had been pulled apart by the cutter suction dredger, all of which had been taken on board the dredger. This initial viewing took place on 23 March. All the fragments were examined and some thirty pieces were sufficiently recognisable to have them transferred to NISA for further investigation and to saw off suitable samples for dendrochronological analysis. The pieces were significantly fragmented and torn apart, so it was not possible in all cases to confirm which parts of the ship they were from. In any case, there were several parts of the rudder, a piece of a gangway?, smaller pieces of beams and frames (from (V-shaped) floor timbers), a fragment of a clinker-built V-shaped timber and a number of pieces of 3 cm thick shell planking with nails from the clinker-built fastenings and bearing impressions of sintel clamps, several 6 cm thick pieces of a bulkhead?, and a beam fragment with a thick iron bolt

The sintel clamped moss caulking detected in the clinker-built seams and the type of sintel clamp used suggested a dating (well) before ca 1550, while several sherds of red pottery found between the pieces of wood confirmed that dating and suggested that the find might be even older. The nails of the clinker-built fastenings were placed at intervals of approx. 15 cm centre to centre and double-clenched (folded over twice and pounded back into the wood). The sintel clamps were pounded into place over the moss lath in the seams on the inside of the ship at intervals of 4.5-6 cm. A thick layer of tar was found on one of the pieces of a shell plank. Besides the two sherds of red pottery, twenty fragments of red brick (approx. 22.5x10.5x5.5cm in size) were found amongst the wood, as well as a small piece of granite and a small piece of flint, a limestone fishing-net weight, a lead sinker weight (for a fishing-net) and a rib fragment. The wood was exceptionally well preserved.

The subsequent dendrochronological analysis yielded dates for two of the seven samples. One had a felling date of after AD 1456 ± 6, the other of AD 1465 ± 6. They were dated using the reference curve for N1Westphalia C. Germany (Hollstein) and N1Westphalia (Tisje) EU1 (EU project 96) respectively. For the full report, see appendix.

24.2 The ship

Also in March, sonar mapping and diving reconnaissance were commissioned at the location to estimate the condition and size of the ship's remains. This fieldwork took place on 26 March 1999 and was carried out by the ADT/SNAP combination. The preliminary report of the sonar mapping and diving reconnaissance describes an object 25-30 m in length and 15 m wide.[3] Heavy beams (up to 35x35 cm in cross-section) protruded from the sediment along the northern and southern sides. The deepest point at the site was slightly more than 6 m; the top of the remains appeared to be at 4-4.5 m.[4]

24.3 Preliminary construction data

The observations made up to that point clearly indicated that a sizable, solid shipwreck dating from the late medieval period had been found at the location. It could also be assumed from the observations that the remains were still mostly situated close together. A ship at least 20 m in length had to be taken into account. The remains were spread out over an area of 30x15 m. The vessel had

1. Translated by Joy Phillips.
2. Client: Directorate-General for Public Works and Water Management (Rijkswaterstaat), IJsselmeer Regional Directorate, IJsselmeer-Markermeer service district; dredging specification IJA 1773).
3. Sonar scan, Lelystad harbour, 26 March 1999.
4. Below the level of 26 March 1999.

been built in the second half of the 15th century and was at least partly clinker-built in construction. The nailing method (double-clenched by folding over twice and pounding back into the wood) and the sintel clamps used on the clinker-built seams on the inside of the shell were characteristic of the cog-building tradition. The heavy beams protruding along the northern and southern side could probably also be considered cog-like features, although it had not yet been established whether they were crossbeams protruding through the shell. On the other hand, these crossbeams indicated that the vessel was presumably positioned more or less perpendicular in the former sea-bottom, and that the hull seemed to be present, in any case up to the height of the beams/deck?. The difference in depth findings suggests that an area surrounding the shipwreck had been more or less exposed to a height of 1.5-2 m, forming an 'island' in the dredged channel.

Immediately following the discovery, the commissioning client (Rijkswaterstaat – RWS) cordoned off a large area around the findspot to keep the dredging equipment away from the site.[5] On 30 March 1999, a consultation took place between representatives of the RWS IJsselmeer Directorate and the Netherlands Institute for Ship and Underwater Archaeology (NISA), part of the Dutch National Service for Archaeological Heritage (ROB). At that time, it was agreed that ROB/NISA would prepare this advisory report, including alternatives for protection and conservation of the findspot, so that RWS could use the advisory report to develop an action plan for final resolution in accordance with the RWS-ROB agreement.

The 'Advisory Report on Ship Find in Northern Approach to Houtribsluizen Locks' notes the following regarding the observations that had taken place thus far.[6] A large, well-preserved 15th-century (or late-15th-century) ship is sufficiently rare as to make it a find of exceptional scientific, cultural and historical value. Without any doubt, the ship find and the findspot are of national and supranational importance and in every sense worth protecting. It should therefore be handled in accordance with national archaeological policy. The advisory report also describes the management choices, explains the various options and, where possible, provides a provisional, estimated calculation of the costs. Taking all these considerations into account, the urgent advice of ROB/NISA is to examine – in line with general national policy – whether adjustments to the plan and *in situ* protection are among the possibilities. In the event that there are convincing reasons not to do so, ROB/NISA advises to have estimates and quotations prepared for launching the project and disassembling the shipwreck on dry land.

According to W.B. Waldus,[7] the shipwreck was left in place, the channel was rerouted, a drill survey was done by RAAP,[8] a dive inspection was carried out and a multibeam recording was made.

24.4 Model

Building a scale model is not yet possible.

24.5 Inventory

The two sherds of red pottery that were found are very probably from two different small cooking pots. It seems very likely that a presumably near-complete inventory could be found on board the vessel. The twenty red brick fragments may come from a fireplace, but could possibly also be part of the ballast. The rib fragment can be considered part of the victuals, while the limestone fishing-net weight and lead sinker weight were presumably lost by a fisherman whose net caught on the wreck.

24.6 Cargo and merchandise

It has not yet been established whether there was cargo or merchandise on board. There is certainly a chance of that.

24.7 Ballast

Apart from the brick fragments, which may possibly fall into that category, no signs of ballast have been found.

24.8 Dating

The sintel clamps and the two sherds of red pottery initially provided an approximate dating of the ship of (well) before c. 1550. The subsequent dendrochronological

5. Between transverse trench 9625 and 9675.
6. ROB/NISA 30 May 1999, cover letter NISA U991111, T.J. Maarleveld, 4 June 1999.
7. Verbal report, February 2017.
8. Borsboom, 2002. The find is now known as Shipwreck IJH1-Val2: IJsselmeer shipping channel Amsterdam-Lemmer.

analysis clearly showed that one of the samples yielded a felling date of after AD 1456 ± 6, while a second sample indicated a felling date of AD 1465 ± 6. This means that the vessel was presumably built in the third, possibly in the beginning of the fourth quarter of that century. It is not yet possible to make any statements about the year in which the shipwreck occurred, although the exceptionally good quality of the wood seems to indicate that the ship was not all that old when it sank.

24.9 Summary

In March 1999, during the deepening of the channel in the IJsselmeer in the northern approach to the Houtribsluizen locks, workers discovered a shipwreck. The pieces of the shipwreck that had been pulled apart during the discovery were brought aboard the cutter suction dredger. The find was reported to the Netherlands Institute for Ship and Underwater Archaeology (NISA). Shortly afterwards, all the salvaged parts were examined on the dredger and some were then transferred to NISA for further study and to saw off samples for dendrochronological analysis. The results of the analysis, which yielded a construction date for the ship in the second half of the 15th century, made it extremely advisable to arrange immediate sonar mapping and diving reconnaissance to gauge the condition and size of the ship's remains.

Double-clenched nails (folded twice and pounded back into the wood) in clinker-built planks, sintel clamps, and also the presence of heavy crossbeams, plus the size of the ship's remains (25-30 m long and 15 m wide) suggested that we were dealing with a vessel at least 20 m long. It had presumably sunk in a more or less perpendicular position in the bed of the IJsselmeer, and the hull was most likely intact, in any case up to the height of the beams/ deck? The quality of the wood was very good. The nailing method used on the clinker-built shell planks, the sintel clamps and the heavy crossbeams suggested that we were dealing with a large vessel. The dating, the probable type of ship (cog?), the completeness of the find, and the good quality of the wood make it a remarkably rare find. Moreover, the find can be described as having exceptional scientific, cultural and historical value.

Without any doubt, the ship find and the findspot are of national and supranational importance and in every sense worth protecting. It should therefore be handled in accordance with national archaeological policy.

Shipwrecks classified as cogs[1]

Once all the cogs had been worked out and drawn, it was clear that two other wrecks, a field evaluation and an excavated vessel, should really also have been studied during the investigation (Medemblik, Zeebad and G 37, NOP). It was noted previously that this was eventually abandoned for practical reasons. For the sake of completeness, these shipwrecks are briefly discussed in sections 25.1 and 25.2 below.

25.1 Medemblik, 'Zeebad'

In 1982, T.J. Maarleveld[2] investigated a late-medieval wreck in what is known as the *Zeebad* in the IJsselmeer near Medemblik. The wreckage consisted of the nearly completely carvel-built bottom with frames, which transitions aft and forward into a clinker-built construction. The bottom of the ship was approx. 15 m long and had a maximum width of 3.6 m. The ship probably had a long keelson with an integrated (wider?) mast step section, supported on the sides by riders (Plate 25.I). The seams were sealed with sintel clamped moss caulking. During the investigation, it could not be determined whether the vessel had a stem hook connected to a keel (or keel plank). It could be determined that there was probably a false stern. The ship has a stern fastened onto the keel (or keel plank?) abaft, rather than a stern hook with a sternpost fastened onto it. This was the reason why the shipwreck was not directly included in the cog study. Other noteworthy features included are the beams and poles(?) in the forebody, which may indicate a loading bed of branches above the ship's bottom. The large number of boulders (ballast?) in both the afterbody and the forebody are also striking. The bricks found in the wreckage may have come from a fireplace. It seems very likely that the ship was built in the 15th century. It was not possible to substantiate a more precise dating at that time.[3] To the best of my knowledge, no additional data has been collected since the 1980s. The shipwreck will not be discussed in more detail here. The results of the present study on the cogs from the Low Countries clarified that we are almost certainly dealing with the underwater body of a large, seagoing cog here. It would be advisable to determine whether the wreckage is sufficiently protected or whether a full investigation should be carried out to secure the data. Very little is known about large cogs from the 15th century.

25.2 Cargo vessel G 37[4]

In 1955, the wreck of a cargo vessel was excavated on lot G 37 in the Noordoostpolder. The entire underwater body and the nearly complete port side were found. This side had become detached slightly above the turn of the bilge and along the scarfs between the various floor timbers and the futtocks of the hull, and had collapsed sideways and folded inwards over the wreckage. The starboard side was missing (fig. 25.1). The complete stern composed of two sections, with gudgeons for the rudder suspension, was still attached to the overturned port side (Fig. 25.2). Much of the stem was lying loose near the wreck (the forward and top parts were missing). The ship did not have a sternpost attached to a stern hook and a stempost attached to a stem hook, but rather a stern and stem placed on a keel approx. 15 cm thick. This was the most important reason why the shipwreck was initially not included in the cog study. Van der Heide dated the wreckage to the 13th century on the basis of a number of Pingsdorfer sherds collected during the excavation.[5]

On the basis of the field drawings, it can be said that this is a vessel with a keel consisting of two parts, interconnected by a flat scarf. A 'skeg' that was 2.25 m long and 7 cm was nailed to the underside of the scarf, and a small plank 40 cm was nailed on top of it, both intended to reinforce the point where the keel was weakened by the scarf. The composite keel would have been 13.5-14 m

1. Translated by Joy Phillips.
2. Maarleveld 1984, 45-61.
3. Idem, 61.
4. File G 37 NOP, RCE-Lelystad.
5. Van der Heide 1974, 423.

Fig. 25.1. The shipwreck in lot G37 in the Noordoostpolder. Viewed towards the stern. The part of the collapsed starboard shell overlying the mast step has been removed.

Fig. 25.2. The collapsed part of the port side with the stern attached to it. Clear to see are the gudgeons, and the rabbet in the side of the sternpost, into which the hood ends of the side strakes had been fitted.

long. The plumb stern and stem were placed at the ends. The bottom of the ship was carvel-built, transitioning into a clinker-built construction abaft and forward. The plank ends of the bottom and sides were nailed into rabbets in the sides of the stern and stem. Planks up to 60 cm wide were used for the bottom and sides. The seams in the carvel-built section of the bottom were sealed on the outside with sintel clamped moss caulking. On the inside of the ship, these seams were covered with thin laths 5-6 cm wide that were nailed on either side of the contiguous strakes. The undersides of the floor timbers were notched where these laths were placed.[6] The clinker-built seams were clamped with sintel clamps on the inside and outside. At the level of the carvel-built section of the bottom (the hold), the lowest strakes of the sides were nailed into place against the outer strakes of the bottom at a fairly sharp angle, and protrude downward several centimetres at that point. These seams at the turn of the bilge were waterproofed using sintel clamped moss caulking on the inside as well as the outside.

A keelson consisting of three sections, about 9.2 m in total, was fastened onto the frames above the keel. The aft and forward sections were approx. 20 cm wide at the ends and approx. 25 cm wide near the mast step section. The mast step itself was nearly 3 m long, 53 cm wide at the widest point, and 30 cm thick. In some places, the aft section of the keelson was notched on the underside to fit over the frames below it. The mast step section was notched on the underside for all the underlying frames. The section received extra support on each side from two riders placed on the floor timbers. On the forward end, the mast step section continued until the keelson, with about 20 cm space between them abaft. Ceiling strakes were placed at intervals on either side of the keelson, creating an 'open ceiling'. The top ceiling strake against the portside had a cross-section similar to an inwale (10 cm thick at the top, 6 at the bottom, and 18 cm wide). This inwale was at deck or gangway height.

The ship had everted, clinker-built sides. The shell strakes were fastened to the frames with wooden treenails. The slightly convex heads of the treenails protruded from the shell. The strakes were fastened to each other with nails through the overlap at close intervals. At its widest point, near the mast step, the bottom of the ship was approx. 3.7 m wide. The length from stern to stem at gunwale height would have been approx. 18 m. The broadest width of the gunwale was approx. 5-5.5. m, while the side must have been about 2-2.5 m high amidships. The intact stern was approx. 4 m long and was approx. 3.7 m above the keel line, measured at right angles. The incomplete stem was still approx. 2.5 m long. When measured at right angles, it must have originally protruded at least 4 m or more above the keel line. The vessel therefore had quite a sheer. The mast was situated about 1.5 m forward of the centre point in the ship's length. No beams were found from transverse connections. The vessel most likely had an aft and forward deck and possibly gangways. 278 boulders were found in the ship as ballast. Identification took place by W.T. Hellinga.[7] Most of the boulders were heavy, in order to achieve the greatest possible weight with as few stones as possible. The array of stones does not represent a normal association, so nothing can be said about a possible collecting area. It is worth noting that there were 41 limestone boulders among them.

Soil experts[8] visited the site on 21 September 1955 and found that the wreck was <u>in</u> the Sloef IIa sedimentation layer and <u>below</u> Sloef 1b. According to Wiggers, these layers date from the first half of the 15th century and the second half of the 15th/early 16th century respectively.[9]

On the basis of a cursory comparison with the ships covered in more detail in this study, it can be provisionally stated with regard to the shipwreck in lot G 37 that we are very probably dealing with a relatively large cog used on inland waterways. The ship undoubtedly had transverse connections in the form of crossbeams, not protruding through the shell. The clinker-built seams sealed with sintel clamped moss caulking on the inside as well as the outside initially suggest a construction date before ca 1350. Unfortunately, the type of sintel clamp used was not recorded. However, the fully rabbeted stern and stem for fastening the plank ends suggest a later date (late 14th century at the earliest). Together with the data from the soil experts, a 15th-century dating seems most likely for the time being.

6. Rieth, 2010. The same construction was also found on a small cargo vessel from the Canche River, just north of the Somme (France). Those seams were caulked with flax fibres. Dated ca 1425.
7. W.T. Hellinga, stones from ship G 37; file G 37 NOP, RCE-Lelystad.
8. H.H. Wiggers, De Jong and Spanjer.
9. Wiggers 1955, 87-88.

Fig. 26.1. The boat built in the cog-building tradition from lots K73/74 in Oostelijk Flevoland. Viewed towards the stern.

26

Boats and barges with cog-like features[1]

The Dronten-II cog (Chapter 17) still retained various cog-like features, but could more accurately be described as a boat or *Schute* due to its dimensions and length-width ratio.

In addition to the vessels discussed in the previous chapters (6 and 8 through 25), a number of late-medieval shipwrecks were investigated in the former Zuiderzee area, in Noord-Brabant and in Antwerp, in which at least some feature from the cog-building tradition were identified. These shipwrecks were 15th- and 16th-century vessels on lots K 73/74, B 55 and L 89 in Oostelijk Flevoland and objects 3 and 14 beside the Kampen (IJssel) cog. In addition, a number of ships were excavated at the end of the 19th century in Antwerp dating from the 13th or possibly even the 12th century, as well as a find of 13th-century ship wood in 's Hertogenbosch and a (river)barge from the Meuse river near Kessel (in the province of Noord-Brabant). Several late-medieval vessels and river boats from another building tradition are also covered, which had been found across the Netherlands. In order to indicate the regional differences in construction, the section on this group also briefly discusses a boat from the Oosterschelde (Eastern Scheldt) estuary in which no cog features were found. The vessel dates back to ca 1500.

26.1 Boat K 73/74[2]

The striking vessel – 15.8 m long, 2.35 m wide and only 0.75 m high – that was excavated in 1971 approx. 5 km west of Dronten in Eastern Flevoland (fig. 26.1 and Plate 26.I) displays a number of (later) features from the cog-building tradition. The stern and stem (shaped much like stern and stem hooks) were attached on either end of the central strake of the bottom; there were five bottom strakes in total. The boat had a carvel-built bottom that was mostly flat (the outer strakes on both sides of the bottom curved slightly upwards towards the turn of the bilge), connected via bent frames to the slightly everted sides, which consisted of two clinker-built strakes (fig. 26.2). The bottom of the ship transitioned smoothly into a clinker-built construction abaft and forward. The planks of the shell strakes and the parts of the covering inwale were fastened together with nibbed scarfs, rather than the usual flat scarfs in the strakes; the plank ends were 1-2 cm thick and protruded outside the shell. The plank ends of the strakes in the bottom and sides were nailed into rabbets in the sides of the straight raking stern and stem. The overlapping planks of the clinker-built constructions were fastened together with double-clenched nails (folded twice and driven back into the wood). The seams of the carvel-built section of the bottom were sintel clamped from the outside. The clinker-built seams in the afterbody and forebody were sealed on the inside with sintel clamped moss caulking.

A covering inwale was installed on the squared-off frame heads along the inside and from stern to stem, to reinforce the upper edge of the gunwale and maintain its shape. The vessel had an aft deck and a foredeck. Towards the aft side of the foredeck, on either side between the frames, short planks were attached on which a (detached) rowing seat could be laid (fig. 26.3a and b). Vertical holes could be observed at various locations in the largest piece of inwale still intact in the forebody of the vessel. These holes were found at the location of the forward 'mast step' and halfway along the foredeck: two holes close together (Plate 26.I). They were probably used for placing rowlocks or thole pins. A similar hole

Fig. 26.2. Part of the port side shell. Part of the carvel-laid bottom, a sharp turn of the bilge, and the slightly everted clinker-built side with L-shaped timber/hard chine and a covering inwale.

1. Translated by Joy Phillips.
2. Reinders *et al.* 1980, 17-29, 43-46 and plates 3, 4 and 5; Vlierman 1983, 3-11; Vlierman 2010, 193-194.

Fig. 26.3a. The forebody with the small deck, the small planks in front of the deck on which a thwart could be laid, and the length of chain on the inside of the stem.

Fig. 26.3b. The small beams supporting the foredeck, and the frames in the bow.

was found in the covering inwale forward of the aft deck. The aft deck consisted of two sections. The short aft section against the stern was at gunwale height, while the main part of the aft deck was about 40 cm above the keel plank. A small vertical wall was nailed against the forward side of the small aft deck, with a small lockable door in it. It provided a small space, e.g. for storing a few personal possessions.

Two(!) mast steps were found in the low, elongated ship's hull (fig. 26.4). The mast (which was probably short) and the sail and auxiliary sail could be placed in the forebody or nearly amidships, depending on e.g. wind force and the type of water the vessel was sailing on (river, lake, Zuiderzee). There was definitely no permanent mast. The schuyt had no sail beams or other transverse connection in those locations. Similarly, no evidence was found to indicate attachments for rigging. However, the two mast steps could also indicate a different construction, which may even be more obvious. More on this topic in section 26.4, on the (original) function

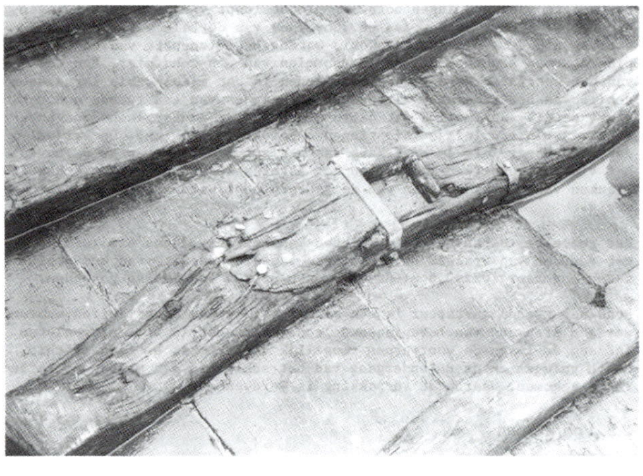

Fig. 26.4. The after mast-step with the sloping scarf in the floor timber, and the reinforcement of the construction by means of two iron strips.

of the two boats and the barge.

Inside the small ship, there were two stoneware jugs, some red earthenware sherds and a sherd of glass. In terms of tools, an axe with handle, two adzes with a handle (and handle fragment), an auger, a pair of pliers and a caulking iron were found. A beautiful kidney dagger was found with remnants of a scabbard, which would also have held a smaller knife. Fragments of a padlock, a hook with an eye and bent end, a hinge (from the door above the aft deck?), pieces of chain, of which the longest part was attached to the inside of the stem (see fig. 26.3a), and some spare nails and sintel clamps. A triangular piece of iron with points would presumably have been fitted into the sole of a shoe, making it easier for the wearer of the shoe to walk on ice without slipping. A wooden scoop, a piece of woven woollen sailcloth, a strip of leather and several bone fragments were the other finds. A number are depicted in figure 26.5.

The eleven silver coins (fig. 26.6) found together with a number of other objects near the aft 'mast step'(see Plate 26.I) may be regarded as pocket money. They were described by Van Gelder.[3] They are:

1	BISHOPRIC OF UTRECHT, Frederik van Blankenheim, plak (Deventer)
2-4	BISHOPRIC OF UTRECHT, David van Bourgondië, stuyver
5	CITY OF KAMPEN, double plak
6	CITY OF KAMPEN, half plak
7	CITY OF KAMPEN, quarter plak
8	CITY OF GRONINGEN, jager 1458
9	BISHOPRIC OF MUNSTER, Johan van de Palz, pfennig
10	FLANDERS, Philips de Goede, braspenning (Namur)
11	FRANCE, Charles VII, plaque (Tournai)

3. Van Gelder 1972, 186-191.

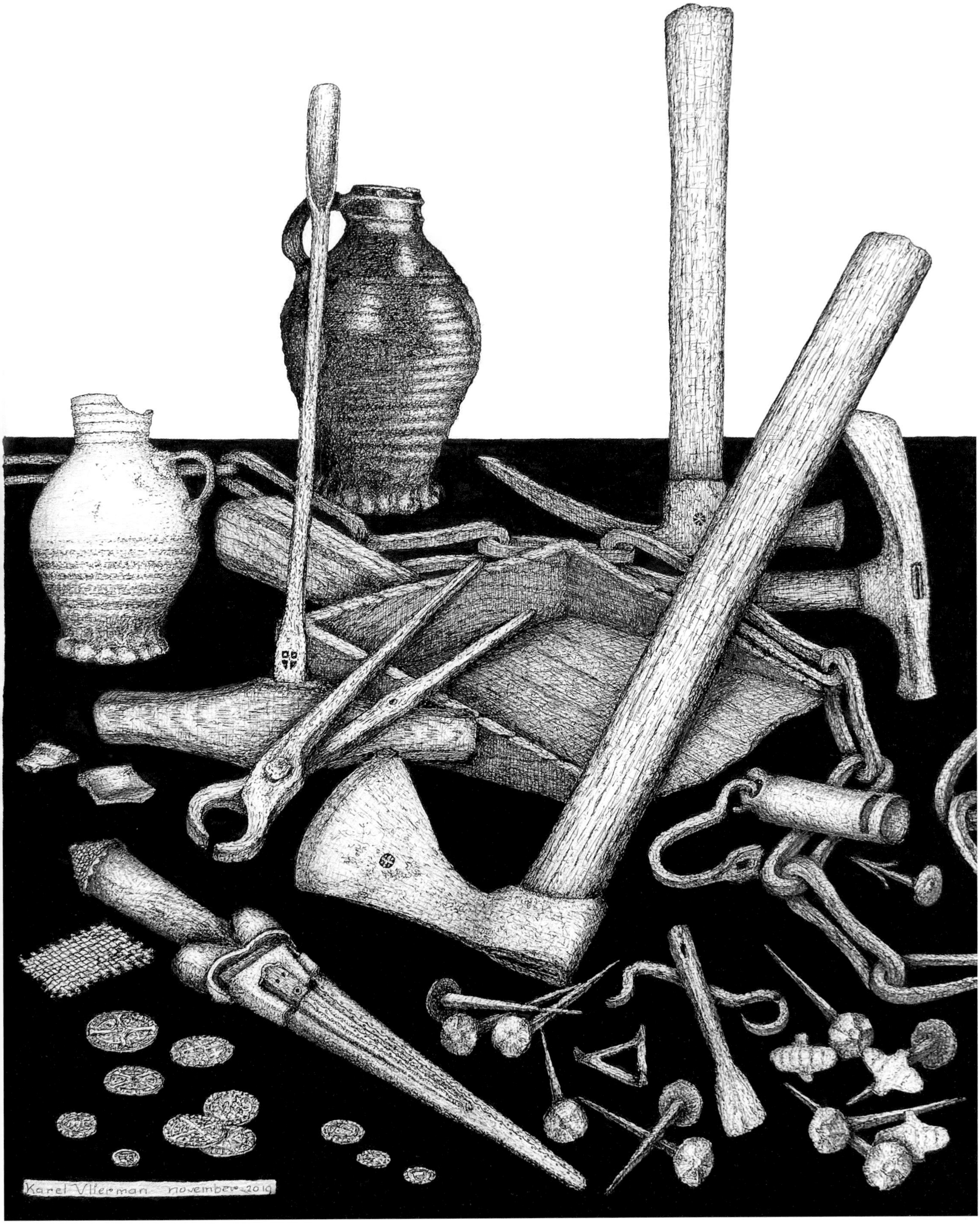

Karel Vlierman november 2019

1

2

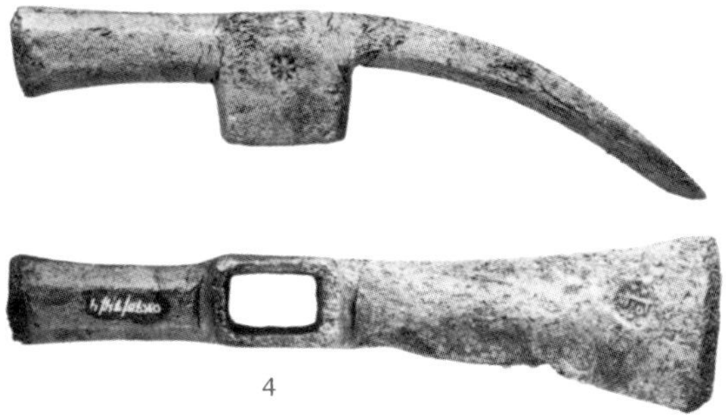

4

5

6

7

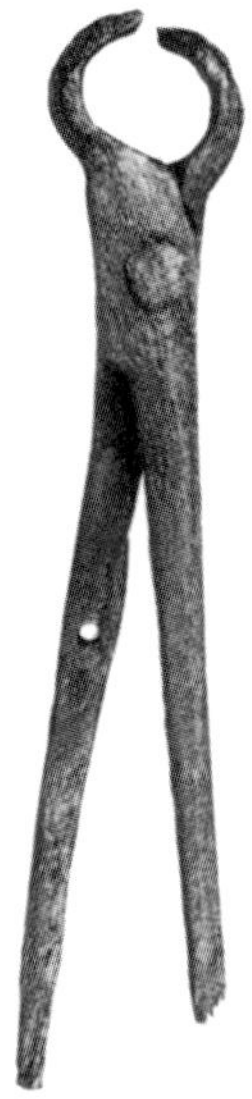

8

9

Fig. 26.5. The finds from the K 73/74 boat. 1-2 The stoneware jugs; 4-9 The tools; 10 and 36 The ballock or kidney dagger with its sheath and the wooden bailing shovel.

36

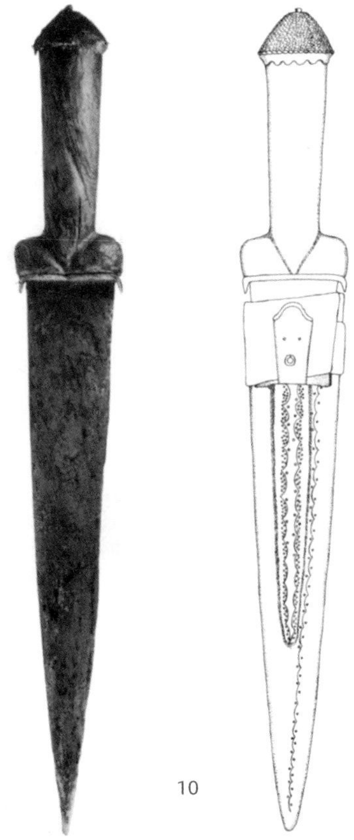

10

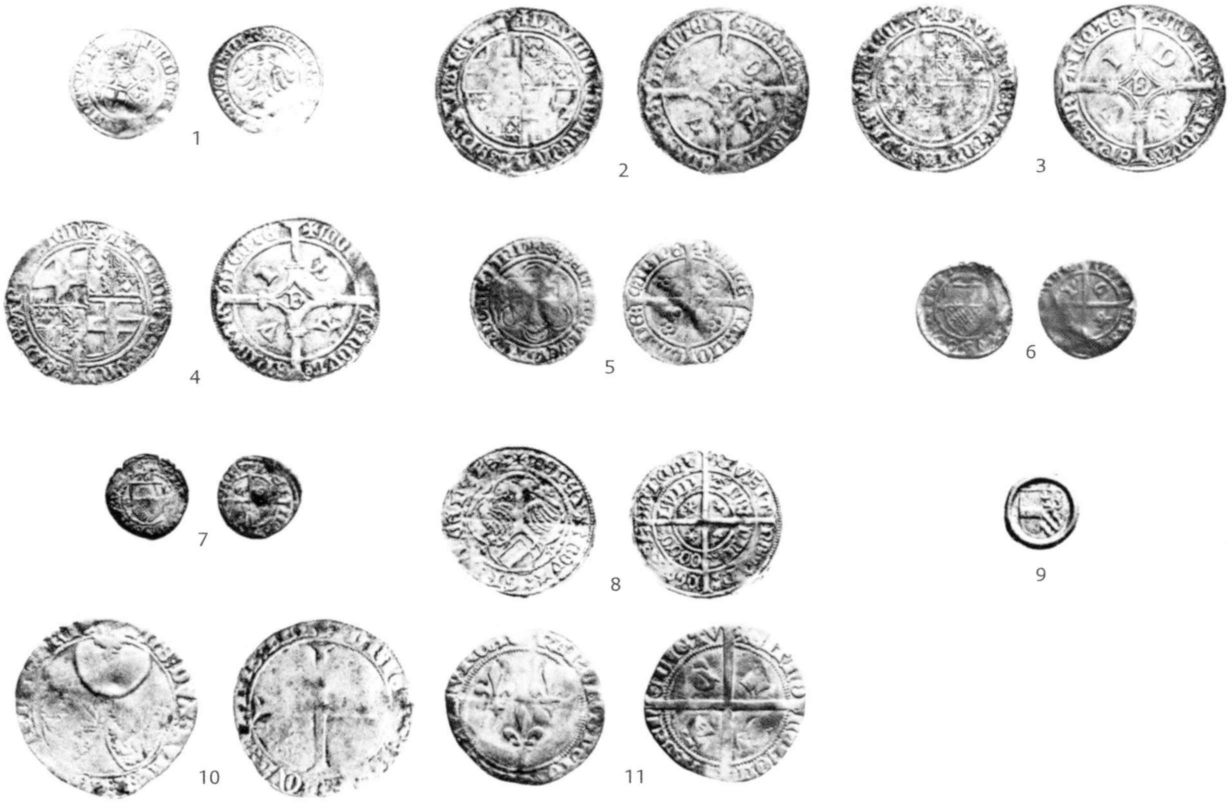

Fig. 26.6. The eleven silver coins from the boat.

From a numismatic perspective, the three coins from Kampen are the most interesting. The find confirms the assumption expressed previously that the city coinage in Kampen must have started as early as the second quarter of the 15th century. Of this oldest series, only the smallest coins had been known thus far, and dated properly due to the finds in Elst and Barneveld. This find adds a new denomination to the series and also dates what has long been known as the largest unit in the same period. On the basis of the coins found, the small ship perished around ca 1460.[4] Based on the type of sintel clamp used (D2/E1) and the observed wear and tear, it may be assumed that the vessel was built between c. 1375 and 1450, with the period around ca 1425 seeming most likely.[5] Dendrochronological analysis in 2001 did not yield a date.[6]

26.2 Boat near the Kampen (IJssel) cog[7]

The remains of two smaller vessels were found with the Kampen (IJssel) cog (see Chapter 23): object 3, a barge-like vessel, which is covered in the next section, and a boat (a punt-like vessel/object 14). Based on the drawing from the field evaluation[8], it was already quite certain to me that the latter vessel had to be similar to the boat from lot K 73/74 in Eastern Flevoland described above. When the shipwreck was lifted out of the IJssel on 29 September 2015 and placed on the archaeological pontoon, this did indeed prove to be the case (figs 26.7a and b). Quite a lot of the boat was still present: about three-quarters of the bottom, over half of the port shell, nearly all the frame sections attached to that, pieces of the covering inwale, several deck-beams and a board from the aft deck plus the entire stern. A very special find was the presence of the rudder, which was still attached to the stern. All the data from the boat combined provided more than enough information, along with the option of comparing it to the vessel on lot K 73/74, to document the wreckage with cog-like features on the pontoon and then make a set of reconstruction drawings (Plates 26.II, 26.III, 26.IV and 26.V).

Documentation and reconstruction

An overview of the exposed parts of the boat had been drawn during the underwater field evaluation in 2012.[9] After the wreckage had been laid on the archaeological pontoon on a bed of sand, on 29 September 2015, the unattached elements that had been removed before the salvage were returned to their original location wherever possible. The nearly complete afterbody and rudder were significantly warped. The same applied to the bottom of the vessel, which had taken on the shape of the riverbed of the IJssel, the channel banks and the peat border over the centuries. Once the bottom had been laid out in its original shape as well as possible in the bed of sand, the preserved parts of the lowest strake on the port side could be placed against the frames again. The pieces of the second, upper strake had to be documented separately. In addition to the planned and executed 3D images (fig. 26.7b), it was necessary to measure and draw a top view by hand, marking all the details relating to the construction, the caulking method, the repairs and, in particular, the warped afterbody. Wherever possible, the afterbody had to be documented as if it were already in its reconstructed shape. In my opinion, it would not have been possible to e.g. try to

Fig. 26.7a. The boat (object 14) close to the IJssel cog being lifted from the river IJssel. 29 September 2015.

Fig. 26.7b. The reassembled wreck of the boat on the archaeologists' pontoon. 30 September 2015

4. Reinders *et al.* 1980, 27.
5. Vlierman 1996a, 80 fig. 52.
6. Vernimmen, RING 2001, item OK 73/74, file RCE-Lelystad.
7. Punter-like, item 14, Waldus *et al.* 2012, 39-40 and appendix 21; Waldus *et al.* 2018, chapter 12.
8. Waldus *et al.* 2012, 39-40 and appendix 21.
9. Waldus *et al.* 2012, appendix 21.

achieve this afterwards based on the 3D recordings. Once the overview had been measured by hand and completed, the frames were removed. The elements of the frames of which reconstruction cross-sections were going to be made were laid on their side and the lines were documented using 3D recordings. Drawings of those elements at a 1:10 scale, along with the manually documented top view at the same scale, were the references for the final reconstruction of the vessel, to the extent that those elements were found. The most intact section of the bottom on the port side offered sufficient tangible evidence to reconstruct the missing forebody with more than 95% certainty, based on the 'smaller sibling' on lot K 73/74 in Eastern Flevoland, which was almost certainly the same age. At the end of the process of disassembling the wreckage, it was also possible to detach the rudder that had been compressed against the port hull, and the grapnel that had rusted onto it. 3D recordings were also made of the rudder.

For a description of the vessel, it is sufficient to refer to the report on the boat from K 73/74[10] and the preceding section. There were sufficiently similar indications and measurements to establish that the length from stern to stem, the greatest width over the gunwale, and the height amidships must have been (almost) the same (15.80 m long, 2.35 m wide and 0.75 cm high amidships). The stern of the vessel from the IJssel was slightly thicker. The missing stem probably would have been as well. A remarkable feature were the recesses about halfway up in frames 4, 23 and 25. They undoubtedly served to lay in (detached) rowing benches (Plates 26.III A, B and D and 26.V A). The two holes in the frames for installing vertical posts ('mast steps') were 2.30 m apart in the K 73/74 vessel, and 1.80 m apart in the boat found in the IJssel. Figure 26.8 shows the hole in frame 18. The forward post was 4.40 m from the stem of the first vessel, and approx. 4.60 m on the other boat. The clinker-built seams in the afterbody and forebody, in the sides and along the stern and stem were sealed on the outside and inside with sintel-clamped moss caulking. The same had been done to the seams in the carvel-built section of the bottom and on repairs to cracks in the boards, but those were sealed from the outside. A number of seams in the bottom had widened quite a lot due to shrinkage, for example after the vessel had been out of the water for some time, and were sealed with moss caulking again so the vessel would be usable once more. Strands of moss were placed in these seams (again) and covered with the customary moss lath, which was fastened into place with sintel clamps. Along the seams that were open widest, small boards were nailed over the seam on the inside (Plates 26.III B and 26.IV B). In one place (between the keel plank and the first strake on the port side in the afterbody), a willow switch was laid on the moss to hold it in place (Plate 26.II F). Other than that, it is striking to note the large number of repairs on the inside and outside of the pieces of the wreck that were found (Fig. 26.9).

Rudder

Folded over against the port side, the rudder was still suspended from the stern (fig. 26.10). Only the top and the tiller were missing. The rudder was assembled from three parts. The actual rudder was made from a board approx. 1.6 m long, measuring approx. 35 cm wide at the top and 55 cm wide at the bottom. An extra piece of rudder blade consisting of two sections was fastened into place on the back. The first section was a board approx. 70 cm long and approx. 35 cm wide, with a second board attached to it measuring approx. 60 cm long and nearly 20 cm wide. In total, the functional rudder blade under water was approx. 60 cm high and 100-120 cm long. The rudder blade was approx. 5 cm thick at the stern and 10 cm aft. The three parts of the

Fig. 26.8. The after of the two 'mast steps' in the boat.

Fig. 26.9. Repair plank on the inside of the lowermost (bilge)strake of the port side shell. The plank continues beyond the L-shaped timber/hard chine.

10. Reinders *et al.* 1980, 17-29, 43-46 and plates 3, 4 and 5.

Fig. 26.10. The rudder with an unusually long blade, still connected to the stern and partially preserved. On the rear of the blade, and extending above it, is the grapnel corroded onto it.

rudder were fastened to each other with two blind treenails and rudder fittings between each part. At the bottom, the rudder was suspended from a treenail in a gudgeon; at the top, it was suspended from a treenail in a ring on a bolt protruding through the stern and secured on the inside with an iron ring and cotter pin. There were two holes in the rudder and one in the stern, through which a rope could be inserted to secure the rudder to the vessel (Plate 26.IV.A).

Finds and cargo

Several finds were collected in and around the wreckage of the boat. About half were in the wreckage. Pilgrim badge (find number 4), one of the badges for the Miracle of Amersfoort, was already mentioned under *personal possessions* in section 23.5.2 on the IJssel cog. Table 1 lists the finds from inside and around the shipwreck.

Table 12.6 of the ADC Monograph on the IJssel cog describes find 11 as a wooden draw-hook. The hook-shaped piece 25 cm long (fig. 26.11) from the afterbody near the stern is assumed[11] to have been a tool much like a pick-hook intended for pulling nets out of the water, for example. However, it was in fact undoubtedly a rowlock or thole pin. The fully intact object was cut from a thick branch that included the base of a side branch. The long end would have been inserted into one of the holes in the covering inwale, i.e. with the hook facing up and the rest pointing down. A loop of rope would have been hung over this 'hook', through which an oar was inserted. The wear marks in the corner between the pin and the hook clearly indicate that it was used frequently. The whetstone (9) may have belonged to a crew member of the boat. It goes without saying that the prehistoric finds have no relation to the vessel. In terms of object dating, some of the other finds do fit within the period in which the boat sailed, but they were most likely washed into the wreck or left lying nearby.

11

Fig. 26.11. The wooden rowlock.

Table 1. Overview of items found in and around the wreckage of the boat

Find no.	Description	Findspot	Dating
1	iron bolt (from cog?)	outside the wreckage	
2	two stoneware jugs	outside the wreckage	
3	piece of whetstone	outside the wreckage	
4	pilgrim badge, tin	in the wreckage	1450-1500
5	book lock	in the wreckage	
6	blade	in the wreckage	prehistoric
7	bullet-shaped object	in the wreckage	
8	scraper and small tool	in the wreckage	prehistoric
9	whetstone	in the wreckage	
10	chunk of iron ore with vivianite		
11	draw-hook, wood	in wreckage near stern	
12	sherds of Siegburg, Langerwehe and proto-stoneware		
13	hook made of antler		prehistoric
14	flint, tools?		prehistoric
15	grapnel	attached to rudder	

11. Waldus *et al.* 2018, 280.

There were split beechwood trunks lying in the boat, mostly oriented lengthwise and in the middle part of the wreck. The smaller trunks that had not been split were lying athwartships between the frames. Split tree trunks were also found around the wreck, which had been used in the riverbank revetments. In total, about two m^3 of wood was found (fig. 26.12). Considering how it was positioned, it would seem to have been part of a cargo. However, that raises the question of why it did not float away. Perhaps the wood was tied together. The fact that the split trunks were found both in the boat and in the riverbank revetments and lying loose around the wreck-side does in any case indicate that the vessel, the cargo and the revetments had a significant connection. It is certainly not inconceivable that riverbank structures were built outwards from the boat, and that a disaster took place that caused the vessel to end up on the bottom of the river.

Fig. 26.12. (Split) roundwood from the boat, presumably cargo.

Fig. 26.13. Two type-E sintel clamps from the boat.

Dating

Immediately after the salvage, it could be determined that like on the cog and the barge, sintel clamp type D2/E (fig. 26.13) had been used.[12] This model was in use from the last quarter of the 14th century to the first half of the 15th century. The most likely typological specification is type E from the transition between the first and second quarter of the 15th century. The rather large number of repairs already mentioned above and the wear and condition of the wood suggest that the boat was relatively old. Sintel clamps used in one of the repairs show a clear similarity with type E1, from which it may be deduced that the repair was carried out somewhere in the middle or third quarter of the 15th century. A total of five dendrochronological samples were analysed.

Samples M13 and M71 provided the most information. The sample from frame 21-BB can be considered a reliable indication of the vessel's construction data; the felling date was ca 1435. The investigated samples from the beechwood poles from the revetments beside where the wreckage lay and from the cargo of the boat provided a felling date: autumn/winter 1451/1452. It can be presumed that the felled wood must have been used fairly soon after the trees were logged. A disaster in the early 1450s that caused the boat to sink and put a stop to the construction of the riverbank revetments seems very plausible. This means that the vessel would have been about fifteen years old at that time. However, the extent of the repairs that used small boards and planks, the repairs done on the moss caulking, and the overall impression of the ship parts (wear and tear) indicate that the vessel was older, or that the boat had seen very intensive (previous) use.

Comparison and interpretation

The vessel(s) were approx. 8 m long and had a nearly flat bottom. The bottom transitioned smoothly into a clinker-built section extending about 3.5 m towards the stern

Table 2. Dendrochronological dating of samples from the boat near the IJssel cog 2015/2016

Sample no.	end of year	felling interval
M13	1431	ca 1435 (1431-1448)
M22	1406	after 1412
M29	-	-
M37	1400	after 1406
M46	-	-
M71	1426	after1432
M79	-	-

12. Vlierman 1996a, 53 and Table II.

and over 4 m toward the stem. The straight stern and stem in the form of stern- and stem hooks had a rake of 129° and 127° respectively. Where the carvel-built part of the bottom transitioned to the sides, a clinker-built section was fastened onto bent frames. The length-width ratio is 6.7:1. The clinker-built seams were waterproofed on the inside, while the carvel-built seams were waterproofed on the outside, using sintel clamped moss caulking. With the exception of the bent frames, all the features of the cog-building tradition are present.[13] If we imagine away the flat-bottomed section in the middle, what remains is a small cog with the length-width ratio of about 3.1:1 so common in those vessels, but then with lower sides. In German, Förster describes the vessel from lot K 73/74 as a *Binnenkogge* or inland cog.[14] Keeping in mind the historical source from ca 1200 that Heinsius references,[15] in which the designation *Kocke* is also mentioned three times in relation to a type of river ship, this classification seems quite possible. It seems better to me, however, to classify the type under the collective term of boats (*Schuten),* which were still built (largely) according to the cog tradition.

The construction, dimensions and proportions do not correspond to the punts from the north-western Overijssel region, as had been built nearly unchanged for centuries, and still remain nearly the same to this day. Berk covers the contemporary punt in extensive detail, identifying the distinction from the punt-like boats based on a definition by N. van den Sigtenhorst, which reads as follows: "A punt is a fairly slim (length-width = approx. 4.2:1) bent-frame vessel of more than 8 m in length, with a straight stern and stem on a usually double, lanceolate bottom, with sides – usually consisting of a single wide strake – forming a rather sharp bilge, with gunwales curving inwards slightly or standing vertically."[16]
What should be added to this definition is that the vessel always has a bottom that is not flat at any point (and gradually curves upwards from the middle towards the stern and stem).

When comparing a flat-bottomed medieval vessel from Meinerswijk near Arnhem (probably from the 13th century), Reinders[17] mentions the similarity with the contemporary punt and gives the only characteristics of this vessel:

- the bottom was not flat at any point;
- the bottom consisted of three strakes;
- the side consisted of a single strake.

Vlierman described a *'bootien'*, probably from the 15th century, the smallest variant of a punt, which was found in the Regge near Hellendoorn.[18] The features of this small boat are:

- the bottom was not flat at any point;
- the bottom was somewhat lanceolate and had two strakes;
- the everted side consisted of one strake that was wider at the stern than in the middle;
- the plumb stern was knee-shaped and fastened onto the bottom.

The length-width ratio of late-medieval and later punt-like and *praam*-like vessels found so far, certainly from Overijssel (and the greater surrounding area), ranged from around 5:1 to well over 6:1. Together with the characteristic structural elements, the boats from lot K73/74 from Flevoland and from the IJssel near Kampen can therefore more accurately be referred to as boats from the cog-building tradition.

26.3 Barge-like shipwreck near the Kampen (IJssel) cog[19]

On the starboard side, in front of and partly under the cog, lay the wreck of an barge-like vessel (item 3; fig. 26.14a). The barge lying diagonally beneath the wreck of the cog did not make it easy to conduct the underwater investigation, nor to lift it from the water. During the deposition process, the part of the small vessel that protruded diagonally above the river bottom broke off and disappeared into the fast-flowing current of the river. During the salvage on 22 October 2015, it was only possible to retrieve a large part of the bottom and frames. The other loose-lying pieces had already been numbered under water and brought up individually. As had been done in September with the boat, the various parts had been reassembled – as well as possible – on a bed of sand on the archaeology pontoon (fig. 26.14b). The photogrammetric underwater recordings were indispensable in this process. 27 parts of the shell and 32 parts of the framework could be reassembled with certainty. 18 parts of the shell and 21 pieces of the frames could not be fitted into the construction with certainty. Working from the reconstructed field survey, overview drawings and five cross-sections were made based on the photogrammetric recordings (Plates 26.VI A, B and C).[20]

13. Vlierman 1996a, section 2.4.
14. Förster 2008, 355. This classification can in my opinion be used for the larger vessel on lot L 89 in Flevoland, which is covered in section 26.5.
15. Heinsius 1956, 72.
16. Berk 1984, 13; working group on Historical Ship-bulding at the IJssel Academy in Kampen.
17. Reinders 1983.
18. Vlierman 1996b, 53-67.
19. Item 3, Waldus *et al.* 2012, 38 and appendix 20; Waldus *et al.* 2018, chapter 13.
20. Waldus *et al.* 2018, chart appendices 7, 8 and 9.

Fig. 26.14a. The position of the barge to starboard under the bow of the cog.

Bottom of the ship

The part of the bottom that remained intact was made from oak and was approx. 11.60 m long, indicating a long, flat-bottomed vessel with a *heve* (the sloping upward part of the bottom) to the back or the front of the ship. The bottom was assembled from four strakes that were 3 cm thick, carvel-built and fastened onto the frames. The bottom strakes were composed of three boards. The scarfs between the three planks of the outer strake on the port side were, highly unusually, butt joints. This suggests that repairs may have been done here. The other scarfs were sloping and flat, all placed from stern to stem angled from top to bottom. This means that the partially preserved *heve* was the front of the ship. At its widest point, the bottom of the ship (at frame 4) was approx. 1.95 m wide. According to the drawing (Plate 26.VI.B), it could have been as much as 2.00-2.05 m at the aftmost point that could be measured. The width at frame 11 was over 1.80 m, about 1.35 m at frame 17, and more than 0.75 m towards the front of the *heve.*

Frames and 'mast step'

During construction, the strakes of the bottom were initially connected by floor timbers spanning most or all of the bottom width. This is the case with frames 4 through 17. The boards were fastened onto those timbers with wooden treenails made from oak and willow (Ø 2.5 cm), two for every board width. In principle, curved trusses (bent frames) were placed against these timbers to fasten the sides against the bottom and keep them in shape. On the starboard side, these curved pieces of wood were placed on the forward side against the timber; on the port side, they were placed against the aft side. In general, the horizontal parts of these curved pieces of wood were 50-80 cm long. They were extended in several places, essentially consisting of two timber sections placed against each other. A wide timber was originally fastened into place at the start of the *heve.*

Fig. 26.14b. The reconstructed part of the barge on the archaeologists' pontoon, 23 October 2015.

Most of the rising sections of the curved pieces of wood had a bend on the outside. That means there was an inward-sloping gunwale. At the end of the *heve*, the sides were almost vertical. Part of a 'mast step' was still present at frames 9 and 10, above the centreline of the vessel. It was fastened onto the frames with several wooden treenails, measuring 103 cm, and was originally approx. 36 cm wide and 7 cm thick. The hole cut into it would probably have been 15x15 cm in size and 5.5 cm deep. The 1.5-cm thick bottom of the hole rested on the two adjacent timber sections of frame 10.

Sides

The sides were about 1 m high and consisted of two clinker-built strakes. The lower strake, which was most likely assembled from two boards that were fastened together with a flat scarf, was 53 cm wide at its widest point and sloped outward slightly. Only one piece of planking from the starboard side remained intact of the upper or second strake or gunwale, which sloped inward slightly. The board still had a maximum width of approx. 30 cm and was damaged along the upper edge. The nails were placed at irregular intervals (10-17 cm), both on the clinker-built fastenings between the two side strakes and at the fastening point of the lowest (bilge) strakes against the bottom. The covering inwale, in which the chamfered heads of the frames were recessed by a maximum of 4 cm along the lower side, was 14-18 cm high. The top of the inwale was level with the upper edge of the gunwale. Together with the overlap of 5 cm at the lower strake, the original strake of the gunwale was almost certainly also approx. 50 cm wide.

Covering inwale

The covering inwale was 6-8 cm wide and attached to the frame heads with wooden treenails. There are vertical holes in the preserved pieces of the starboard and port side, positioned in the same place relative to each other. They may have been used for placing rowlocks or thole pins. Some are mounted in pairs, such as at frame 11, possibly also between frames 4 and 5, and also at 18. A similar hole is also visible between frames 15 and 16. This one has no port-side counterpart. Section 26.4 addresses the function of these holes.

Waterproofing

The carvel-built seams of the bottom are waterproofed on the outside with sintel clamped moss caulking. The clinker-built seams and the bilge seam are caulked on the inside and outside. The edges of the boards were chamfered during the construction of the vessel and before they were fitted into the ship's hull, to create a V-shaped opening along the seam. Singular spun 'threads' of *sphagnum* moss were pounded into place into this opening, held in place with narrow oak wood splints (moss laths), and fastened with wrought-iron sintel clamps.[21] The sintel clamps used here (type E) were applied at 5.5-6 cm intervals heart to heart. A layer of moss was laid on the flat scarfs, and tar residue was found in various places in the wreck.

Wood use and repairs

The pattern of the wood grain, especially of the middle two boards of the bottom, and the resulting cracks show that the boards were sawn from adjacent slices of a tree, folded open, and installed on either side of the bottom. This is known as 'brother and sister pattern'. Repairs were found at various locations in the part of the hull that remained intact. The outermost strake of the bottom on the port side which may have (partially) been replaced has already been mentioned. Other than that, various knots were sealed with moss, as were a number of nailholes that were no longer being used. Cracks in the wood, usually lengthwise, or in any case in line with the wood grain, were carved out in a V-shaped groove and waterproofed with sintel clamped moss caulking. Some cracks were covered with small repair boards, which were nailed into place over a layer of moss. Repair boards were also installed at certain point in one strake, and sintel clamped all around.

Finds

A number of prehistoric flint tools were found in the wreck, as well as seven iron nails and a sintel clamp of a different type than that used in the barge, a plain iron bolt and an iron bolt with an eye, a number of stoneware sherds, a religious badge, and the remains of a large wickerwork eel trap. The nails and bolts may have been connecting elements of the vessel. The bolt with an eye (26) might have been part of the rudder fittings, if there was a stern. The religious badge (17), depicting St Veronica and her veil, was already covered in section 23.5.2 in relation to the cog.

The most striking find in the barge, however, is a wickerwork eel trap (figs 26.15 and 26.16). Fishermen often place eel baskets and fish traps near shipwrecked vessels. The position of the barge, diagonally under the forebody of the cog, with the open side facing upstream and approximately in the middle of the current, undoubtedly meant that some part of the wreck was regularly exposed, protruding out of the river bottom. That would have been an ideal place to put a large eel basket or fish trap (180 cm long and about 120 cm in diameter at the mouth of the trap). A small part of the wickerwork basket was salvaged.

Only one comparable specimen has been found thus far, which was discovered in 1984 while excavating a pond in the Oostvaardersplassen nature reserve.[22] This eel basket or buck (fig. 26.16) had a reconstructed length of approx. 150 cm and an opening or mouth of approx. 90 cm. They cannot therefore be considered comparable to the bell-shaped eel traps used in the 20th century, which are approx. 55 cm long and 33 cm wide. The Oostvaardersplassen trap was found in a deposit of Almere clay. Its location in the clay pack suggests that it dates back to a period well before 1600. A historical document entitled *De Vischtuigen* – the full title of which translates roughly as 'The fishing implements permitted in our fatherland according to the pre-

21. Vlierman 1996a.
22. Vlierman in: 1984 Annual Report of the Nautical Archaeology Subdivision, 10-13 figures 2 and 3; idem 2006, 151-161 and figures 19 and 20.

Fig. 26.15. Photogrammetric image of the eel-trap in the barge, top view.

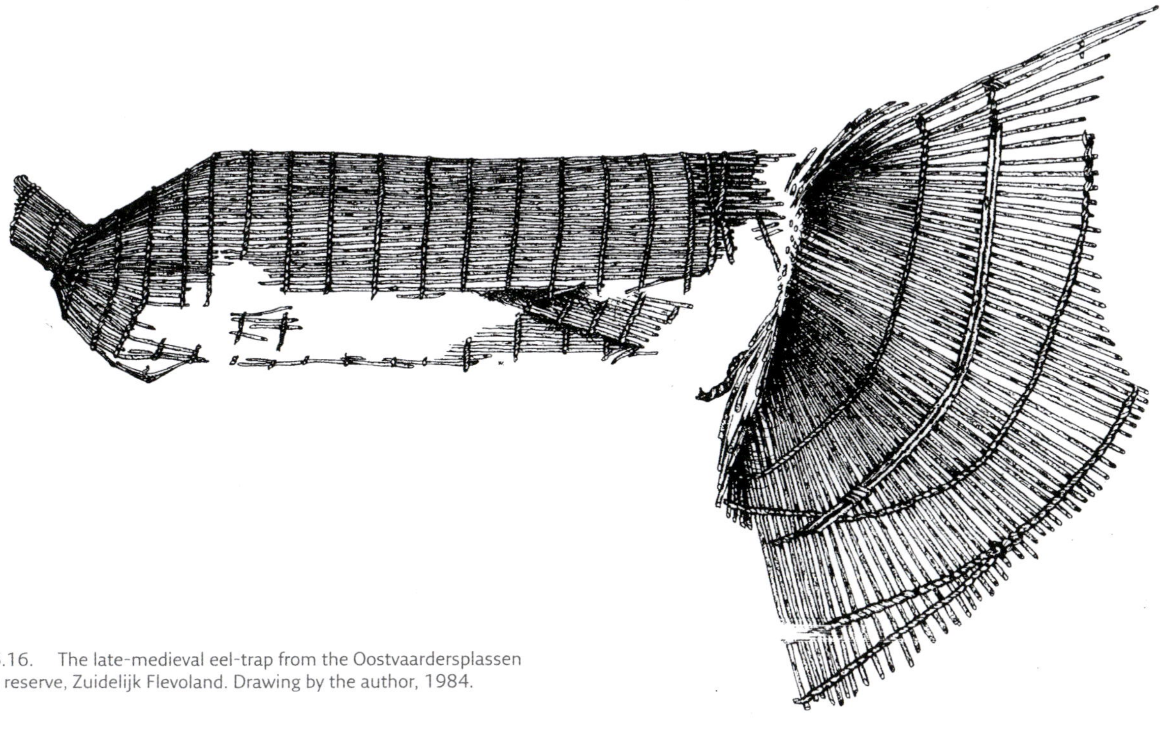

Fig. 26.16. The late-medieval eel-trap from the Oostvaardersplassen nature reserve, Zuidelijk Flevoland. Drawing by the author, 1984.

vailing regulations, ordered by province'[23] – describes an eel trap used in Overijssel and Drenthe that bears a strong resemblance to the specimen found in Flevoland. The wickerwork trap described therein was 170 cm long and had a circumference of 90 cm at its largest point.

Dating

Dendrochronological analysis of a sample of sapwood (44) gave a dating of 1409 ± 1. The felling date is therefore 1410. The wood came from a tree of local origin.[24] The barge was probably built shortly afterwards.

Reconstruction and interpretation

It is not possible to arrive at a complete and fully substantiated reconstruction of the vessel based on the pieces that have remained intact. However, we can add some notes about this find to indicate what we might consider and what could be said about it in comparison to previous finds of late medieval river vessels that could be classified as a barge. Firstly, the original dimensions and lines of the vessel. Obviously, it is impossible to say anything about the missing section with any certainty. The alternating flat scarfs in the planks of the bottom were located in the afterbody between frames 2 and 5. In a symmetrically constructed bottom, which was very common in late medieval vessels, this could mean that the midway point in the length of the vessel would have been approximately around frame 3. In that case, the barge would have been approx. 20 m long. The greatest measured width across the barge-like vessels and the two boat gunwales (at frame 7) is 2.40-2.50 m. That would indicate a length-width ratio of approx. 8:1. In other covered boats above, the ratio was generally between 5 and 7:1. The barge from the IJssel would probably have been about 16-17 m long at most, although it should be noted that a larger brother or sister of the boats was 20-21 m and 4 m wide, with sides nearly 1 m in height.[25] Another detail that is more difficult to explain is that the vessel widened steadily towards the aft end, or stern. The most obvious assumption is that the barge was built symmetrically, so it would also have had a similar *heve* of about 75 cm wide at the aft end. This would mean that the width from approximately frame 4 abaft would not have widened as shown in the drawing (Plate 26.VI B), but would rather have grown narrower. Given the state of preservation of the plank ends, this does not seem inconceivable. Another possibility is that the *heve* in the afterbody was wider than in the forebody. Alternatively, the vessel may have had a stern rather than a *heve* in the afterbody. A good example of the latter is depicted in the foreground of a winter scene by Hendrick Avercamp dating from 1609 (fig. 26.17). There are even two treenails protruding from the covering inwale at the stern, with a strap running between them into the vessel which could be used to steer or scull. The forward end clearly consisted of the flat plate indicating a *heve*. Judging by the height of the person next to the stern, the vessel depicted here must have been just under 10 m long. As an aside, it can be noted that the boat or barge depicted here is very similar in structure to the two boats and the barge, and that Avercamp's highly realistic paintings were probably also made primarily in Northwest-Overijssel, an area in which shipbuilding continued virtually unchanged from the Late Middle Ages well into the twentieth century.

Another striking fact is that the barge was built on bent frames and with an inward-facing gunwale. In *Section 1.4.7 Origin of the contemporary punt from Overijssel*[26], one of the writings is about a *'geboeijde prame'* ' (a small flat-bottomed river vessel with a gunwale) that was built in Zwolle in 1635; this may already have been the case elsewhere ten years earlier. A *'weyschuit'*-like river vessel with gunwales is depicted in the painting 'View of Delft' by Hendrik Cornelisz. Vroom, dated 1615. The painting entitled 'Fishing for Souls' by Adriaan Pietersz. van de Venne, dated 1614, shows four common late-medieval boats with everted sides, as well as one with an inward-facing gunwale.[27] In 1995, the wreck of a small fishing vessel was excavated on lot H 61 in Flevoland, probably a flat-bottomed *'pincke'* or pink. During the modelling process, it became clear that the vessel had an inward-leaning gunwale. The objects found from the inventory suggest that the ship sank in the first quarter of the 17th century.[28] The discovery and dating of the barge near Kampen in the IJssel River clearly shows that the practice of building on bent frames and with an inward-leaning gunwale was already applied in the early 15th century.

26.4 (Original) function of the two boats and the barge

The boat by the IJssel cog (Section 26.2) was almost certainly used in the installation of riverbank revetments in the final phase of its use. The many repairs and signs of wear that were found suggest that the vessel was already

23. Hoek 1899, 5 and 6 ill. 4.
24. Waldus *et al.* 2012, 44.
25. See section 26.5.
26. Vlierman 1996b, 64-67.
27. Vlierman 1996b, fig. 44 and 45.
28. Schreurs *et al.* in: NISA Yearbook 2 1995, 38-42; Vlierman 2006, 159-160.

Fig. 26.17. A barge-like boat in a *Winter Landscape* by Hendrick Avercamp, 1609. To the right, in the covering inwale at the stern or bow, is a pair of rowlocks; to the left, behind the man and woman, the sloping upward part of the bottom *(heve)* of the vessel (Rijksmuseum Amsterdam).

quite old when it sank, or had seen intensive use over a shorter time period. The execution of the boat also suggests that it was probably not built for this use. The better and more intact boat on lot K 73/74 (Section 26.1), which had tools, drinking utensils and various personal possessions on board, indicates that there were several people on board at the time of the presumable disaster: one or two to row and steer the vessel, and possibly a passenger.

With regard to the two holes in several floor timbers in the boats and the single hole in a block of wood on one timber of the barge (although it does not seem inconceivable that there were two in this vessel as well), it would be almost automatic to contemplate holes for the base or heel of a mast. It was common practice, for example, for punt-like and praam-like vessels to then also have a thwart or sail beam against which or through which the mast could be placed and supported. No evidence of a thwart or sail beam was found at the location of the 'mast steps' on the port side and inwale of the K 73/74. Similarly, there was also no rigging to hold a mast in place. If there had been, then several vertical holes in the inwale would have shown signs of wear, or there would have been a fastening (rigging cleat) on the outside of the gunwale. Neither was the case. Unfortunately, the relevant pieces of the sides of the boat and barge by the IJssel cog were not found. In all likelihood, the vessels did not have a 'real' mast and would not in principle have sailed. The vertical holes in the covering inwales placed close together were almost certainly intended for placing rowlocks or thole pins. In the boat from the IJssel River, evidence was found that indicates two removable rowing benches in the forebody and one in the afterbody. It would also have been possible to sit on the foredeck. The boat from Flevoland had a rowing bench that could be placed on short planks directly behind the foredeck. Above the forward side of the aft deck, there were clamps with two holes for rowlocks against the inside of the covering inwale. They may also have served for the placement of a bargepole, which would have been used to push the vessel forward from the shore.

The two holes in the floor timbers may have held poles approx. 2 m high on which a horizontal beam 2.5-3 m long was attached. A sail may have hung over this beam, which would have been attached to the covering inwale on both sides, forming a sloping canopy on either side of the beam, under which e.g. passengers could sit. Judging from the vertically drilled holes in the inwales at the level of the hold, it may also have taken some other form. For example, willow branches about 4-5 cm thick could have been placed in them and bent over the horizontal beam on the poles, fastened into place on them, and curved from gunwale to gunwale. Good examples of the latter can be seen on three late 15th and early 16th-century Flemish illustrations (fig. 26.18, 26.19 and 26.20).[29] On the seal of Kuinre from 1399 (Chapter 13, fig. 13.8) and on the boat on the 14th/15th-century mural in the Ratskeller in Wismar (fig. 26.21), a similar 'canopy' was presumably situated in the forebody. The vertical holes drilled into the covering inwales aft and fore were undoubtedly used to insert rowlocks/thole pins.

Kampen vessels?

The boat and barge from the IJssel were almost certainly directly connected to Kampen. The boat from Flevoland sank definitively approx. 13.5 km southwest of the IJssel delta. It may for instance very easily have floated there, largely under water.[30] The striking resemblance to the boat by the IJssel cog and the presence of coins from that city suggest that this vessel, too, was most probably connected to the Hanseatic city. The three vessels were almost certainly intended to carry people with hand luggage across the IJssel (ferry boats/barges) and/or to transport them within the region. They can probably be considered the predecessors of the tow boat. The two boats were probably tied up at jetties, while passengers could walk directly onto the riverbanks from the barge. The *heve* at the front made it easier to get on and off the boat. It would also have been possible to take along (small) farm animals. Although no evidence was found to indicate its presence, it does not seem inconceivable, especially if the ships were mainly used to transport passengers, that

Fig. 26.18. Boat with a roof. Miniature in a calendar for the month of May, in Book of Hours and Calendar, Bruges or Ghent, ca 1480. British Library. After Flatman 2009, 55.

Fig. 26.19. The same.

29. Flatman 2009, 55 and 115.
30. On lot K 45 near Dronten, a small armed vessel from the second half of the 17th century was excavated. Part of the port-side afterbody was gone as a result of an explosion. Years later, the missing piece was found 25 kilometres to the south, past Harderwijk (Vlierman 1997a, 157-166).

Fig. 26.20. Flemish miniature featuring a roofed boat. The Golf Book, Bruges, ca 1520-1530. British Library. After Flatman 2009, 115.

Fig. 26.21. A roofed boat beside a cog. Wall painting in the Ratskeller, Wismar, 14th/15th century.

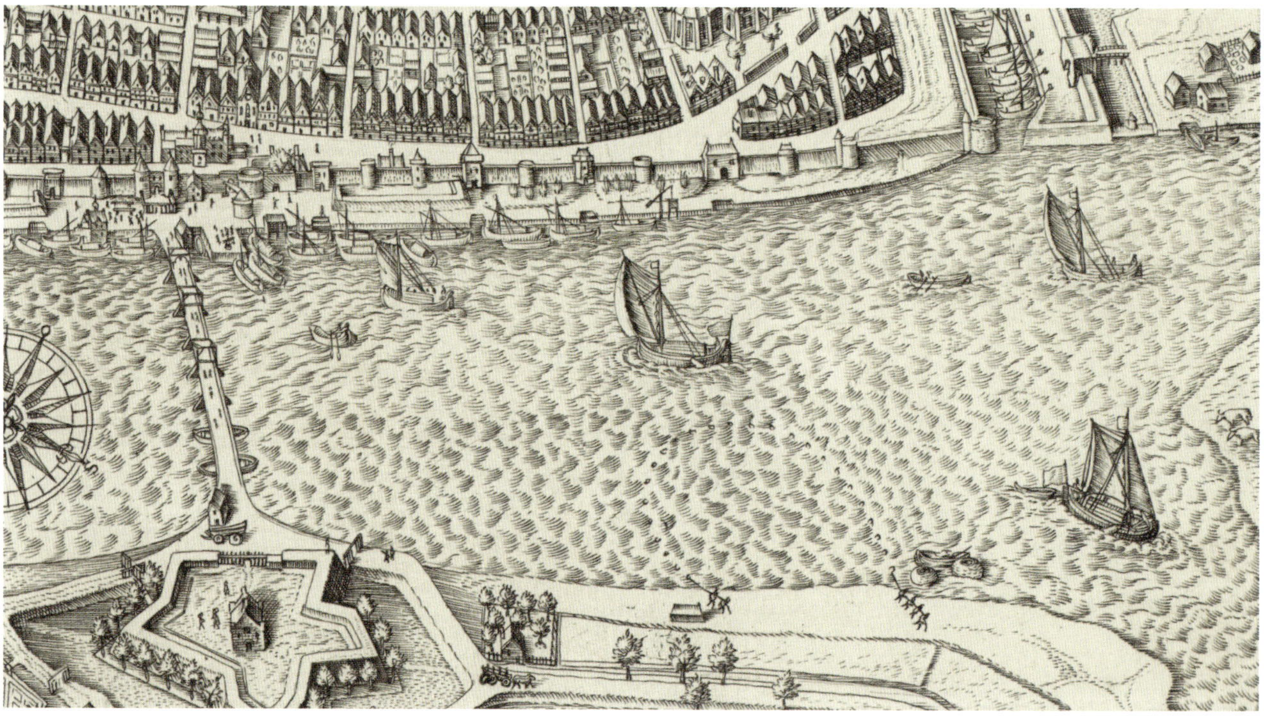

Fig. 26.22. Map of Kampen (detail) by P. Utenwael, 1598. Near the bridge across the IJssel and roughly at the findspot of the IJssel cog, are two boats with two oarsmen in the forebody.

removable wooden decking would have been placed over a large part of the bottom (see Chapter 17, fig. 17.11 and the 15th-century miniatures in Chapter 29, figs 29.4 and 29.5). The map of Kampen by P. Utenwael from 1598 shows two boats with two rowers seated one behind the other at the IJssel bridge, approximately at the findspot of the cog (fig. 26.22). The painting 'View of Kampen from the other side of the IJssel', attributed to Claes Bellequin and dating from 1653-1660[31], even shows a boat with two rowers, a helmsman? and a sloping canopy amidships, sailing near the IJssel bridge. These rowed boats were apparently part of a regular 'ferry service' near Kampen for centuries. It certainly does not seem too daring to assume that the three vessels were built in Kampen, or at least in the Lower IJssel region.

26.5 Large boat/small cog L 89

A third vessel, larger (approx. 20-21 m long, approx. 4 m wide and 0.8-0.9 m high amidships) and similar in construction and hull lines to the boats described in sections 26.1 and 26.2, was excavated north of Dronten (figs 26.23a and b). Earthenware and coins indicate that this vessel perished in the third quarter of the 16th century, while dendrochronological analysis clearly showed that the wood used in its construction must have been felled ca AD 1545 ± 6 in the Central Netherlands and in western and southern Lower Saxony.[32] In contrast to the boats from K 73/74 and from the IJssel near Kampen, this vessel had a keelson approx. 7.5-8 m long made from a board 6-10 cm thick. A square hole had been cut into the keelson, continuing into the underlying frame (27a) to fit the base of the mast. A rectangular notch at the back of this hole probably served for the insertion of a vertically placed board. This was presumably in relation to a removable mast. The vessel undoubtedly had a sail beam, but it was not found. That means that, unlike the two boats, this vessel was clearly a sailing ship. In addition to the hole for the base of the mast, a rider approx. 25-30 cm long and approx. 6 cm high was nailed on both sides of the keelson to help absorb the lateral pressure from the mast. In any case, the vessel had 38 frames, and the tops of the futtocks were finished straight against the sides. A covering inwale was fastened into place on these frame heads. The bottom of the hold area was flat and equipped with bent frames. Triangular battens were fastened in the sharp, almost right-angled turn of the bilge, held in place with wooden treenails into the lower strakes of the sides and nailed onto the outer strakes. Moss caulking was applied between the battens and the bottom/sides. Aft and forward, the carvel-built bottom strakes transitioned into the familiar clinker-built construction of the cogs, and the seams on the inside were sealed with sintel clamped moss caulking. This was also the case on the seams of the clinker-built sides. The plank ends of the flat scarfs were 1-2 cm thick and stood proud of the shell both on the inboard and outboard sides. The plank ends were fastened together on the scarfs with at least five nails per row: one row from the inside out, the other vice versa. The vessel had a straight stern which was placed at the end of the central bottom strake/keel plank and fastened to it with three nails. The wooden fireplace with a floor made of red bricks was located approximately amidships on the port side. The cooking, eating and drinking utensils as well as the tools and personal possessions were in the immediate vicinity of the fireplace.

Fig. 26.23a. General view of the mid-16th-century cargo vessel with cog features in lot L89 in Oostelijk Flevoland, showing frames and keelson. Viewed from the bow. In the afterbody is the break in the ship resulting from its being stranded on the western edge of the submerged peat barrier (see section 11.1.2 and Fig. 11.5): b. The forebody after removal of the frames. Clear to see are the side strakes' transition from carvel to clinker construction, and the hard turn of the bilge with the barely everted sides. Viewed from the bow.

As in the two boats, when we imagine away the elongated midship section (about 1/3 of the total length) and connect the afterbody and forebody to each other, then it is clear

31. Kampen Municipal Museum.
32. File L 89 Eastern Flevoland, RCE-Lelystad; RING report, January 1997.

beyond a doubt that we are looking at a small, shallow cog. This larger, single-masted vessel from lot L 89, with a single mast step and a fireplace, could in my opinion possibly be considered a late model of a small freight cog (*Cleene cogghe*) which would have been 'old-fashioned' in its construction. Förster's[33] classification of the boat K 73/74 as a *Binnenkogge* or inland cog does seem acceptable for this larger sailing vessel. In addition, it is not too daring in my opinion to contemplate that the vessel may have been constructed in Kampen or at least in the Lower IJssel region.

Fig. 26.24. The reconstructed afterbody of the inland cargo vessel B 55-I (Oostelijk Flevoland) with its stern and clinker-built sides, in the workshop of the Museum of Maritime Archaeology at Ketelhaven, near Dronten.

26.6 Cargo vessel B 55-I

A fifth shipwreck find which can be mentioned in the context of the cog-building tradition was excavated in the centre of Lelystad.[34] The vessel had a completely flat, slightly lanceolate and carvel-built bottom against which the lower strakes of the clinker-built sides were secured with wooden treenails. In the sharp transition from bottom to sides, battens were nailed into place on the inside to reinforce the bilge construction. The part of the stern that was still present was partly attached to the end of the central plank of the bottom (fig. 26.24). The seams of the bottom were sealed from the outside with sintel clamped moss caulking, while the clinker-built seams were caulked on the inside. The type of sintel clamp used here was model E, typologically in use between the first and the third quarter of the 15th century.[35] The entire bottom found in situ measured 18.5 m long and approx. 3.5 m wide at the widest point. The largest width is approx. 3/8 from the stem. At the same distance from the stern, the bottom measured 3.25 m wide. If the stern and stem had a similar rake, the vessel would have measured approx. 23 m long from stern to stem, with a width of around 5 m around the gunwales at its widest point. The height amidships would have been about 1.2 m, while the gunwale at the stern would have risen approx. 1.9 m above the bottom. A remarkable find was the presence of the rudder, which lay under the port side of the afterbody. The pewter jug found in the wreck dates from the beginning of the 16th century, possibly ca 1500. The stoneware and earthenware and a kidney dagger that were found are generally dated to the second half of the 15th century. Other exceptional finds include the upper part of a stabbing/striking weapon, a weapon known as a *goedendag* or godendac, and a handsaw. Other than that, a copper kettle with an iron handle, pieces of a bronze cooking pot, an iron padlock, a knife and several pieces of large floor tiles were found. The objects were all lying on and near a small floor in the afterbody (figs 26.25a-c and Plate 26.VII). Dendrochronological analysis was not carried out. The double-clenched nails used for the clinker-built fastening of the planks and the sintel clamped moss caulking are (still) the only characteristic structural elements reminiscent of the cog-building tradition, and the reason why the vessel is mentioned here. It was definitely not a cog. Whether this large vessel should be called a boat is doubtful, to say the least.

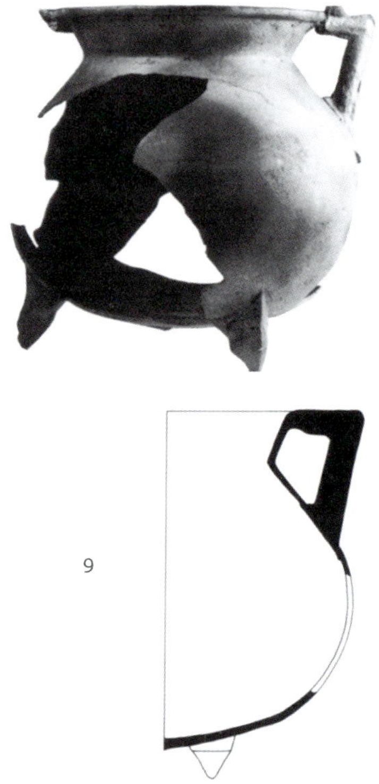

Fig. 26.25a & b. The galley utensils from the B 55-I wreck. The drinking ware and storage jars from the cargo ship.

33. Förster 2008,355.
34. On lot B 55 Eastern Flevoland; Reinders *et al.* 1980, 31-50.
35. Vlierman 1996a, 83.

1

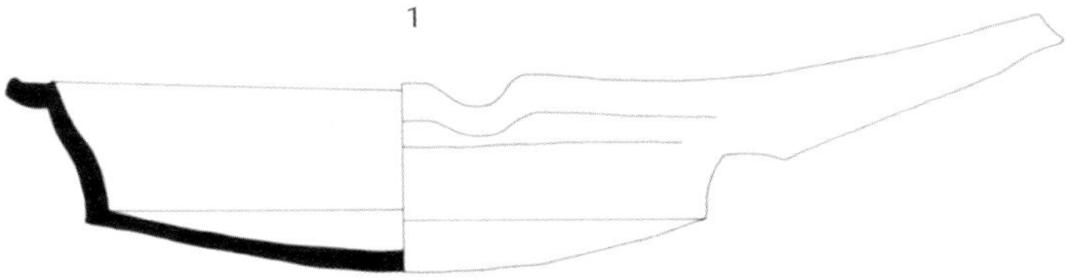

8

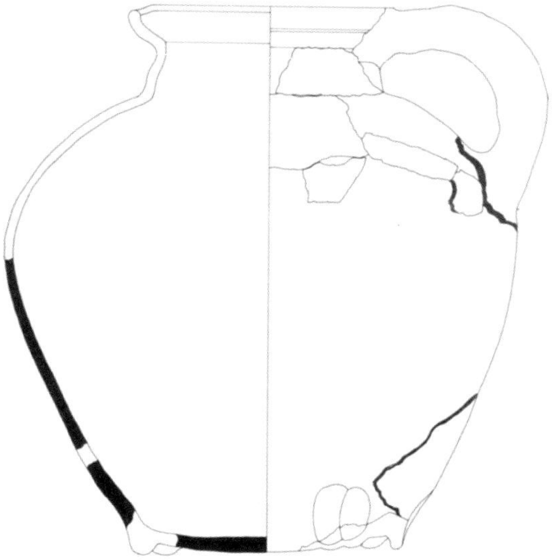

3

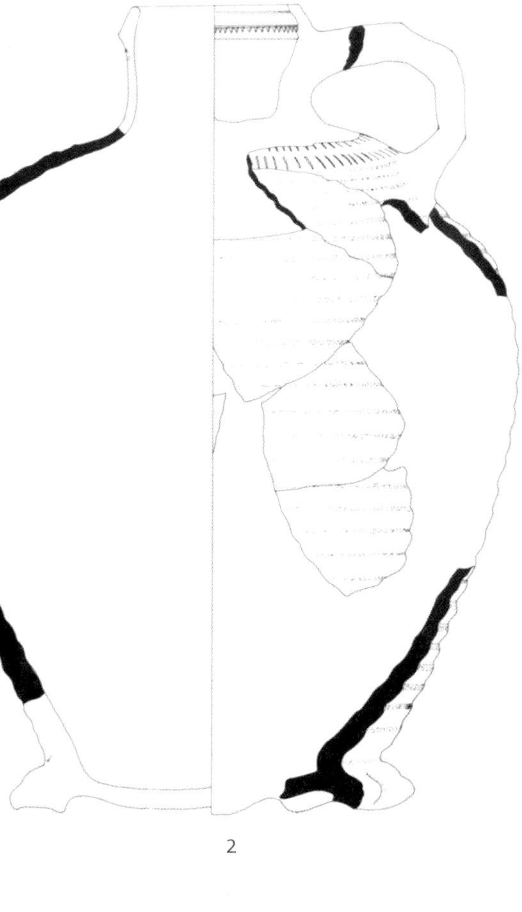

2

Fig. 26.25c. The pewter jug, the weaponry, knife, padlock and handsaw from the B 55-I wreck.

Impression of a part of the inventory of cargo vessel B55-I.

26.7 Shipwrecks from the Lefèbvre Dock in Antwerp, 1884

The publication on finds involving ships and ship wood in Dutch medieval cities[36] mentions a number of river barges of the type known as barge, and barge-like planking boats, as well as ships of an unknown type and several seagoing vessels. The publication was an (initial) inventory of all medieval wrecks of seagoing and inland ships until ca 1300 that had been discovered in the Netherlands up to that time. A remarkable group of various vessels mentioned in the publication was discovered during the excavations for the Lefèbvre Dock in Antwerp in 1884 (fig. 26.26).[37] It was possible to identify two shipwrecks as barge-like planking boats, while two other vessels were of a type that was unknown at that time, and the fifth was a seagoing vessel. It now seems possible to identify the last three vessels as well. Despite the lack of information about the use of stern and stem hooks, other structural details, and steering with a stern rudder or steering oars, it is very likely that the oak-wood shipwrecks 3 and 4 were early examples of boats with cog-like features. Shipwrecks 1 and 2 were barge-like vessels, which also had these features. Shipwreck 5 can now be designated with certainty as a small cog. The five vessels were primarily intended to navigate the inland waterways. The first four could probably be rowed and/or pushed forward with a bargepole, while the fifth could be sailed. It is perhaps not too daring to once again reference the aforementioned *Kocke* among the 13th-century, but more likely 12th-century boats, in relation to river ships in the *Willahelm* dating from ca 1200.[38]

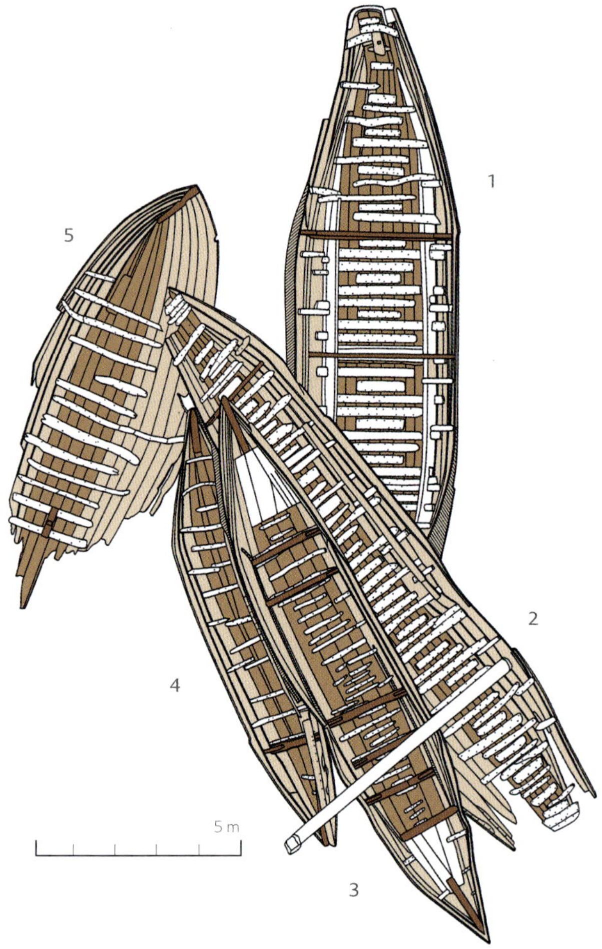

Fig. 26.26. The five shipwrecks found in the Lefebvre dock at Antwerp in 1884. 1 and 2: barge-like plank boats with sharply converging bow and stern; 3 and 4 boats with cog features; 5 a small (seagoing) cog; and a detached crossbeam (from a cog?). After Hasse 1907.

26.7.1 Two barge-like planking boats

*The bar*ge-like planking boats 1 and 2 (fig. 26.26) were the largest of the five vessels; boat 1 was approx. 20 m long, 4 m wide and 1.50 m high, while boat 2 was approx. 15 m long, 3 m wide and 1.20 m high. These boats had a carvel-built bottom sloping and tapering up to the flat plate of the *heve*, which was only approx. 80 cm wide at the ends. They would therefore not have been suitable to transport ox-drawn carts, for example. They had clinker-built sides that were fairly everted (the bottom of the vessels was 2.20 m and 1.00-1.20 m wide respectively), with an abrupt transition from bottom to sides. The carvel-built seams of the bottom were completely open and had been waterproofed with sintel clamped moss caulking. On the underside, the moss was held in the seam with triangular moss laths and sintel nails; on the inside, it was only held in place with sintel nails (type A2, in use in the 2nd/3rd quarter of the 12th century).[39] A striking feature is the construction of the clinker-built sides. A thin layer of moss was applied between the overlapping planks, while the planks were fastened together with nails pounded in alternately from the outside to the inside and vice versa. According to the drawing[40], the nails were driven in blind (not with the tip protruding through the wood), while the text refers to U-shaped double-clenched nails that are folded down and driven back into the wood. Barge number 1 had at least three, but possibly originally five transverse connections resting on gangways and secured with knees to the gunwale. The vessel most likely also had an aft deck and foredeck or platform. Barge 2 had at least two but more likely four

36. Vlierman 2002, 119-148.
37. Hasse 1907; Vlierman 2002, 132.
38. Heinsius 1956, 72; Wolfram van Eschenbach (1197-1210).
39. Vlierman 1996a, 63.
40. Hasse 1907, 9 planche IV figure 5.

transverse connections, similar to barge 1, and most likely also an aft deck and foredeck or platform. Presumably no gangways, but there was a ceiling board in the sides.

26.7.2 Two boats

The Antwerp vessels 3 and 4 (fig. 26.26) are the oldest examples found to date of boats from the Low Countries which have features from the cog-building tradition. Vessel 3 was approx. 13 m long, 2.5 m wide and 1.2 m high amidships, while vessel 4 was approx. 10 m long, 2.2 m wide and 1.0 m high amidships. The overview drawn by Hasse clearly shows that both vessels had a stern and stem (or stern- and stem hooks?), a (largely?) carvel-built bottom, and clinker-built sides. They were also equipped with transverse connections that were fastened to the gunwales with rising knees. On vessel 3, there were as many as five. On vessel 4, there was still one beam present. In vessel 3, a platform or small deck could also be seen both aft and forward. Unfortunately, it is no longer possible to determine whether these ships were steered with a stern rudder or with an oar. The fastening of the clinker-built planking and the execution of the sintel-clamped moss caulking was the same as described above in the barge-like planking boats 1 and 2. The type of sintel clamp used was also A2.[41] Vessels 3 and 4 differ from barges 1 and 2 in their stern and stem.

26.7.3 A small cog

The fifth shipwreck (fig. 26.26) can now definitely be confirmed as a small (seagoing) cog. The small vessel was approx. 13 m long, approx. 5 m wide at the widest point across the gunwale, and approx. 1.5 m high amidships at the gunwale. The keel plank (?) was widest in the middle and tapered towards the stern and stem. It was not determined whether the vessel had a keel plank with stern- and stem hooks, although it does seem very probable. The shape of the stern as drawn suggests that not only the three strakes on either side of the keel, but also the first two strakes of the sides were fastened into place in rabbets in the stern. In the afterbody, the second (B) strake was fastened to the A strake in a clinker-built construction across a length of approx. 2.5 m. The other strakes of the sides presumably ended against the aft side of the sternpost. The vessel still had a short mast step fastened to two frames. The forebody was missing. Hasse dated the wreckage to the 15th century, but it seems more likely that it would have been older.

26.7.4 A crossbeam from a cog?

A beam approx. 6.5 m long and approx. 30 cm wide (and thick) was lying diagonally across shipwrecks 2, 3 and 4 (fig. 26.26). According to the drawing, one end was slightly rounded, while the other may possibly show the head of a crossbeam protruding through the shell with the groove into which a shell plank would have fitted. On the rounded-looking side, the protruding head may have been lost. The dimensions of the beam correspond to those of the shipwrecks known thus far of large, seagoing cogs and the Atlantic *'nef'* depicted on seals. It does not seem too daring to assume that it was indeed such a beam.

26.8 Ship wood in 's-Hertogenbosch

In the archaeological investigation at Postelstraat in 's-Hertogenbosch, it was established in the examination of the wood of an barge-like vessel, probably dating from the mid-13th century, that the outermost bottom strake was fastened onto the lowest strake of the sides with wooden treenails and nails. The clinker-built construction was fastened only with nails. It is striking to note that the nails in the piece of wood that was found were all pounded in from the inside out, folded over and driven back into the wood. The possibility cannot be excluded that other parts of the vessel that were not found may have had nails pounded from the outside inwards.[42] The seam between the outermost bottom strake and the lowest plank of the sides, and the seams between the strakes of the sides, were cut into a V-shaped groove along the inside and sealed with sintel clamped moss caulking. Instead of a barge, we could also contemplate a boat with cog-like features here.

26.9 Barge, Kessel (North Brabant)

In the clinker-built fastenings of the two strakes of the sides in the barge from the Meuse river near Kessel (North Brabant), which presumably can be dated to the 15th century, the nails were pounded in from the outside and vice versa, in identical fashion to the finds from Antwerp. In both cases, they protruded through the two strakes that were fastened together, and the double-clenched nail points had been folded down twice and driven back into the wood. The carvel-built seams of the bottom were sealed from the underside with sintel clamped moss caulking. On the inside, only the bilge seam had been waterproofed with moss, moss laths and sintel clamps.[43]

41. Hasse 1907, 16-17 and planche IV, fig. 4 and 5.
42. Vlierman 1988, 97-101; idem 1996b, 107-110 and appendix 13.
43. Vlierman 1996b, 19-32.

26.10 Parts of (river) vessels from a different building tradition?

In addition to the vessels mentioned above, there are another four 13th-century ship finds from the Netherlands that could actually have been covered in the previous group. They are different, however, primarily due to different fastenings used on the clinker-built constructions. These vessels did not use (double-clenched) nails, but wooden treenails. These are the finds from Meinerswijk, Amsterdam, Hattem and Rotterdam. There are similarities to the river vessels of the type barge and barge-like planking boats.[44] However, it is striking to note that the first three (almost certainly) tapered to points at the ends, and that the plank ends must have been fastened into place against a stern hook/sternpost (see Antwerp 3 and 4). That is not certain with regard to the vessel from Rotterdam. It may also have involved a construction method as seen in Antwerp 1 and 2 (fig. 26.26).

The Meinerswijk 2 find was a punt-like vessel made of oak wood; the entire carvel-built bottom rose to aft and forward, and must very probably have ended in stern- and stemposts or stern- and stem hooks with a short horizontal section. The sides probably consisted of two strakes in a clinker-built construction and fastened to each other and to the bottom with narrow wooden treenails. The seams in the bottom were sealed on the underside, while the seams in the turn of the bilge and the clinker-built seams were sealed on the inside with sintel-clamped moss caulking. The type of sintel nail used (B) corresponds to the type used during construction in the Meinerswijk 3 tree-trunk boat found nearby. Dendrochronological analysis of the wood of this vessel yielded a felling date of AD 1216 ± 5 or later; the reference chronology used was N1 south.[45] A construction date between ca 1200-1250 seems to be the most probable time frame for both vessels. Meinerswijk 3 had open seams on the bottom that were sealed with sintel clamped moss caulking; moss laths and sintel nails were used on the outside, while the inside only used sintel nails. Other than this vessel, this practice is only known from the Nijkerk-I cog (Chapter 8).

In the oak-wood ship pieces found under the walls of the fortifications by Nieuwezijds Kolk/Nieuwendijk in Amsterdam, the clinker-built planks were also fastened with wooden treenails.[46] Only small parts could be viewed. The ship fragment under the northwest tower/northern wall showed that the carvel-built bottom gradually transitioned into a clinker-built construction towards one of the ends of the ship and in the similarly clinker-built sides. Very unusually, the seams of the bottom were sealed with sintel-clamped moss caulking on the inside. In the clinker-built fastenings, the sintel clamped moss caulking was only used on the inside in the corners against the overlapping planks. The planks of the bottom were connected to each other by floor timbers about 40 cm wide and only 6 cm thick at most. They were attached to the shell with wooden treenails. The treenails had slightly thicker heads rising somewhat from the shell. On the inside, they were clamped onto the clinker-built constructions with an arch in the plank. The vessel, which was originally certainly over 18 m long and 4-5 m wide, was built between 1275-1300 and originated from the Central/South German area or the Meuse Valley.

The third vessel in which the clinker-built oak planks were fastened together with wooden treenails was found at Hattem.[47] The ship had a maximum width of approx. 3.5 m. The length is not known, but is at least 13 m. The bottom of the ship is carvel-built. It was unfortunately not possible to determine whether the bottom transitioned into a clinker-built construction towards the stern and stem, but it seems very plausible. The more or less rounded transition from the bottom to the sides consisted of three strakes applied at an increasingly sharper angle. The floor timbers of the bottom were only 3.5 cm thick, while the curved pieces of wood for connecting to the sides were a maximum of 6.5 cm thick. The clinker-built seams were sealed with sintel clamped moss caulking on both inside and outside, while the seams of the bottom were sealed on the underside. The aft deck or foredeck found loose in the vessel tapered to a point. This indicates a sternpost or stempost. The plank seams of the deck were sealed on the top with sintel clamped moss caulking. Based on several finds in the ship and the type of sintel clamp used, the vessel is thought to date from the late 13th or early 14th century. Dendrochronological analysis did not offer any clarity.

The oak-wood ship parts (ship 4) that were used in a lock in the Willemsspoortunnel in Rotterdam came from a vessel with a flat bottom and clinker-built sides.[48] The clinker-built sides were also fastened together with wooden treenails. The sintel clamped moss caulking was applied on the outside along the bottom, and probably only on the inside of the clinker-built sides. Since the pieces only constitute part of the vessel, and other pieces

44. Vlierman 1997, 75-77 and 84-85.
45. Reinders 1983, 15-20, 40-41 and appendix 1; RING report no. 2006025, 11 April 2006.
46. Vlierman 1995, 91-104, figs 43 and 44.
47. Vlierman 1996b, 115-119 and appendix 15.
48. Van Holk 2001, 103-117.

containing key features are missing, it is not possible to reconstruct the dimensions. In any case, it was not a small vessel. We could possibly contemplate something around the dimensions of the vessel from Amsterdam. The few pieces of planking that taper to a point towards one end could indicate that the afterbody or forebody was narrower. However, it is questionable whether there would have been a stern or stem against which the planks ended. Another possibility would be a (strongly) tapered flat plate or *heve* towards the forward end, comparable to the Antwerp 1 and 2 vessels (fig. 26.26). The vessel was probably built in the region between the Meuse, Moselle and Rhine Rivers between 1265 and 1275.

Fig. 26.27. The boat recovered from the Oosterschelde inlet near Nieuwlande/Krabbendijke. Rebuilt in the workshop of the Museum of Maritime Archaeology at Ketelhaven, near Dronten.

26.11 Boat from another building tradition in the Oosterschelde estuary

A vessel similar in size to the boat from lot K 73/74 (and near the IJssel cog) was found near the drowned village of Nieuwlande in the Oosterschelde (Eastern Scheldt) near Krabbendijke.[49] This small vessel had a flat, somewhat lanceolate bottom, with the stern and stem placed largely on the middle strake of the bottom, and sides consisting of two clinker-built strakes. An inwale was fastened into place along the tops of the frame heads. Small vertical planks were placed at various locations in the top of the inwale: supports for a splashboard. About 1/3 of the ship's length measured from the stem, the small ship had a mast step with a sail beam about 25 cm wide above it. Slightly in front of the mast step, the imprint of an iron plate could be seen on the outside of the uppermost shell strake. It may have been from putting irons for the rigging. In contrast to the boats and the barge from Sections 26.1, 26.2 and 26.3, this vessel clearly did sail. There was a thwart approximately 55 cm wide towards the stem (fig. 26.27). The seams were not sintel clamped, but were instead sealed with moss caulking and moss laths, which were fastened into place with small wooden wedges. The vessel very likely dates from the late 15th or early 16th century. Approximately 10.8 m long with a maximum width of approx. 1.8 cm across the sides and a height of approx. 55 cm amidships, the vessel can be classified as a boat. Shipbuilding according to the cog-building tradition was apparently not in evidence in Zeeland ca 1500, or in any case no longer practiced.

49. Vlierman 1996b, 34-52 and appendix 4.

Early-16th-century completely clincker-built *waterschip* with a straight, raked stern and stern. Wisentbos, Dronten 1971.

15th- and 16th-century clincker-built ships[1] 27

A few of the excavated late-medieval shipwrecks in the IJsselmeer polders were certainly not built along the lines of traditional cogs, judging by their structural features. To indicate how specific the cog-building tradition is, the most important differences are summarised in brief sections below. The presumable origin of the ships indicates that it is very likely they were not built in the former Hanseatic area.

27.1 Four fully clinker-built ships

Several cargo vessels

Several 15th or 16th-century cargo vessels were discovered and excavated in the Flevopolders over the years, completely clinker-built from keel to gunwale with a stern and stem attached to it. The caulking in the seams of these vessels consists of animal hair, while the wooden treenails attaching the frames to the outside protrude slightly and have a rounded head. Some frames display a striking construction in the underwater body at the stern and stem. A small (low) V-shaped floor timber does not have the customary futtocks above it on both sides, but a knee-shaped transverse connection which is also the futtock on one side. On the other side, a futtock has been placed on the transverse connection.[2] Also noteworthy: the loading beds of branches and twigs that were found, which had been installed above the ceiling boards. These ships are not very large (presumably up to approx. 17 m long); they probably came from the Baltic Sea region and are examples of the Nordic ship-building tradition.

In contrast, the cogs have a carvel-built bottom which transitions aft and forward into a clinker-built construction and clinker-built sides. The carvel-built sections of the bottom were only attached to the frames with wooden treenails. The clinker-built boards were attached to each other with double-clenched nails. The wooden treenails mostly had a flat head and were flush with the outside of the shell. In the cogs, the keel consists of a short keel plank, extending into knee-shaped stern and stem hooks, which also constitute the lower section of the stern and stem respectively. The construction based on a floor timber with a knee-shaped transverse connection on top, of which the rising section is a futtock, was also found in the Doel-I cog.[3] In that vessel, there were two floor timbers placed on top of each other and two knee-shaped transverse connections placed on top of each other. In the cogs, the shell seams were waterproofed using sintel clamped moss caulking.

Another small cargo vessel that was fully clinker-built and riveted, which had slightly different lines except for the remarkable transverse connections, was found upside down, just like the Doel-I cog. Dendrochronological analysis showed that the test wood samples were indicative of a Dutch and southern English curve and that the felling date must be approx. between 1415-1430.[4] A potentially comparable vessel was found on lot B 36 in the Noordoostpolder.[5]

27.2 A 16th-century variant

A coaster 20 m long and 5 m wide is another example of a fully clinker-built cargo vessel. Here, too, the boards are fastened to each other with rivets. However, the boards of the bottom deviate from this: a number of thin wooden treenails have been used with some distance between them, followed by three rivets. This is a recurring pattern throughout the underwater body. The strakes along the sides are only fastened with rivets. The ship had a sharply straight stern and a curved stem, probably a covered afterbody and forebody, and gangways. Besides the more customary cooking, eating and drinking utensils, a stokehole (a small wooden box) with a floor of beautiful Gothic tiles bearing inscriptions, and tools. Several remarkable cargo samples were also found. There were seventeen bar-

1. Translated by Joy Phillips.
2. Files J 137 and O 28 Noordoostpolder, RCE-Lelystad.
3. Chapter 20, section 20.2.4 and figs 20.20a and b.
4. Shipwreck G 35 Noordoostpolder; RING report October 1999, file RCE-Lelystad.
5. Shipwreck B 36 Noordoostpolder, file RCE-Lelystad.

rels containing fish remains, probably herring, a number of bars of lead, and some grain. The most extraordinary, however, was a number of colf clubs. Coins had already indicated that the ship had presumably perished around 1540. The tested wood samples could be dated by the reference curve NLHist-2 (Central Netherlands) and provided a common felling date between 1531-1533 AD.[6] Perhaps what we have here is a vessel related to the *waterschip* vessels in terms of construction, so we can with some caution perhaps estimate that its origin lies in the Netherlands (possibly the western region).

27.3 A 16th-century large freighter/ warship

More than 95% of the late-medieval vessels excavated in the former Zuiderzee area were no longer than approx. 20 m. The vessel from lot U 34 in eastern Flevoland was about 30-35 m long from stern to stem and presumably had three masts (fig. 27.1). So far, it is the only large vessel discovered that sailed on the Zuiderzee at the same time as the latest large cogs or shortly afterwards. The vessel is briefly discussed here, due to the very striking use of sintel clamped moss caulking on the inside of the underwater body to waterproof the seams, although the ship's hull is fully clinker-built. A second reason is the remarkable loading bed on spars just above the ceiling. In several large cogs (Rutten, Doel-I and Strandbad Medemblik), evidence has been found which may also indicate the presence of a somewhat similar floor, as well as its possible significance in relation to the function of those vessels.

The ship has a keel with rabbets, with the garboards nailed into the rabbets. A straight, raking stern of approx. 7 m is attached to that, with stepped rabbets notched into the sides for the purpose of attaching the ends of the shell planks in the lower half. Viewed from above and from behind, the upper half of the stern has a T-shaped profile, with the ends of the shell strakes nailed to the sides along the outside. The vessel has a curved stem. The bottom of the ship and the sides are fully clinker-built from keel to gunwale. The strakes of the entire underwater body were fastened together in a repeating pattern of three narrow wooden treenails placed at intervals with an arch and a rivet. The strakes of the sides are only fastened to each other with rivets. A thin layer of moss has been applied

Fig. 27.1. Excavation of a large, fully clinker-built warship or cargo vessel (a '*hulk*'?) of the 16th century, viewed from the bow. Uncovered in 1987 in lot U 34, Oostelijk Flevoland. To the right, part of the port side with rubbing strakes that have collapsed inward. Extra riders are present on the ceiling/load bed.

6. Morel 1993, 25-37; RING report December 1997, file M 11 Eastern Flevoland, RCE-Lelystad.

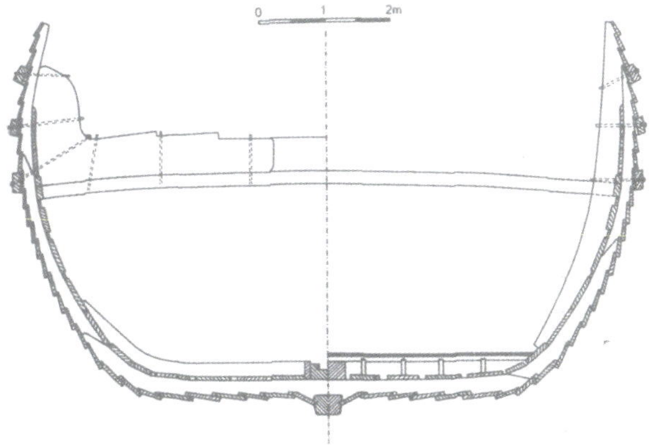

Fig. 27.2. Reconstructed cross-section. On the right is a 5 m long vertical rider. After Reinders & Oosting 1989, *Abb.* 5.

Fig. 27.3. The load-bed with the layer of sand and several erratic cobbles of the floor.

along the overlap, while the seams of the underwater body have been sealed with sintel clamped moss caulking (as already noted in the previous section).

Above the keel on the massive frames, a broad keelson has been attached, with a hole for the heel of the (main) mast about amidship. The ship probably also had a forward and aft mast and was equipped with various square sails. The keelson is supported by riders, not only where the mast would have been, but at various locations athwartships; the riders are attached to the ceiling strakes and the underlying frames. Vertically positioned heavy vertical riders about 5 m long have been placed on the inside of the sides to reinforce the gunwales. On the outside, there are three thick extra strakes (wales) fastened on the strakes of the top half of the sides (figs 27.1 and 27.2). The transverse connection at deck height is provided by heavy crossbeams, with similarly heavy transverse bulkheads/connected knee sections on top of them. The gunwale is attached against the rising parts of these knee sections. The crossbeams or connecting beams no longer protrude through the shell. In other words, this ship was built using another, subsequent method than the cogs.

A loading bed on spars was installed above the ceiling boards, which had been fastened with fairly little space between them. The spars were covered with wickerwork and topped by a packed layer of sand. The spars are supported by short poles resting on the ceiling boards (fig. 27.3). Approximately amidships, a sizable fireplace lined in large tiles was placed on the packed sand. A floor made of boulders was laid behind and in front of this fireplace. The floor plus fireplace covered the entire width of the ship, completely blocking off the ship's bottom below that area.[7] To prevent wood rot, this closed section of the underwater body was generously tarred. Various stones and iron cannonballs were found in the wreckage, as well as a mould for casting bullets, a hinged mould for making lead musket balls, and a hatch for a cannon port(?). The vessel is similar in many ways to the Copper ship' from the Gulf of Gdańsk.[8] Dendrochronological analysis compared with the reference curve EU 8 (region Poland) provided a common felling date between AD 1532-1537.[9] The presumable function of the ship is thought to be an armed freighter. However, it is equally likely that the vessel could have been a warship. It will be difficult to establish certainty on that matter. The same applies to the type of vessel. It is still not clear when we are dealing with the '*hulk*' referenced in historical sources. Reinders & Oosting postulate that research done on archaeological shipwrecks that are already known (the wrecks listed in section 27.1) may offer an answer, and do not exclude the possibility that this type name could also be applicable to the vessel described here.[10]

27.4 Completely clinker-built and completely carvel-built *Hollandse waterschepen*

The *waterschip* was the most characteristic vessel of the Zuiderzee and is mentioned in historical sources as early as 1339 as a merchant vessel intended for the transport of live fish. In the 15th and 16th century, it was (also) used as a

7. Reinders & Oosting 1989, 106-122; idem in Annual Reports 1986 and 1987 of the Nautical Archaeology subdivision.
8. Litwin, 1979 and 1980.
9. Ring report investigation 1994, dated October 2000. File U 34 Eastern Flevoland, RCE-Lelystad.
10. Reinders & Oosting 1989, 118.

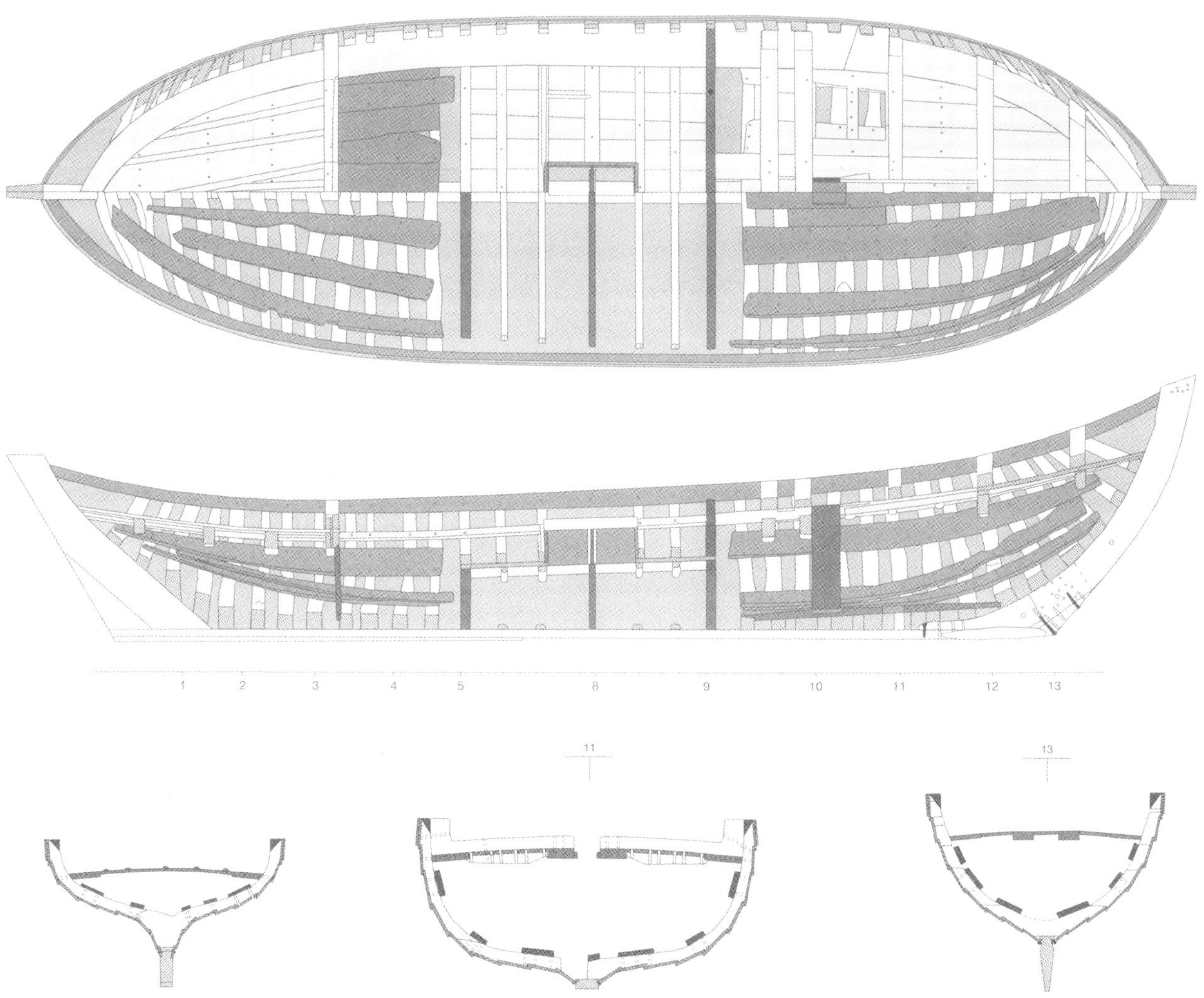

Fig. 27.4. The reconstructed top and interior views, and some cross-sections of the fully clinker-built *waterschip* of the first half of the 16th century, from lot Nz 42 in Zuidelijk Flevoland. After Pedersen 1996, enclosures 3 and 6.

fishing vessel. The fishermen used trawls, sometimes fishing in convoys, and probably provided a significant percentage of the supply of fresh fish, especially in Holland. Most of the shipwrecks that were found were built and perished in the 16th century. There were two types of *waterschepen*. One type was a 16 m long, fully clinker-built ship which was in use until approximately the mid 16th century.[11] The other, a fully carvel-built vessel of about 20 m long,[12] was in use from the mid 16th century until about 1830.[13] The first type sailed the Zuiderzee together with the (latest) cogs. The hull shape, internal structure and layout of the two types hardly changed at all over the centuries.

It seems plausible that the 14th- and 15th-century *waterschepen* would not have looked much different than the *waterschepen* from the first half of the 16th century. The type is similar to the small cogs in many ways in terms of lines and structure: the hull shape, the dimensions, the length-to-width ratio and the construction of the transverse connections with raised knees at deck level, the use of double-clenched nails on the clinker-built fastenings of the shell planks, and the seams on the inside of the ship that were – very remarkably – sealed with sintel clamped moss caulking. The 16th-century *waterschepen* had a keel on which straight, raking sterns were placed. They had curved stems. The fully clinker-built construction and the

11. Pedersen, 1996.
12. Reinders *et al.* 1978.
13. Vlierman 2006, 154-155; the watership was also mentioned as a tow barge as early as 1598 (Crone 1926, 240) and was later used frequently to tow the ship camels (external flotation tanks) used to bring the ships of the Dutch East India Company (VOC) to and from Amsterdam. This type of ship became obsolete as a result of the opening of the North Holland Canal between Amsterdam and Den Helder.

curved stem are the main points of difference compared to the small cogs. It is unfortunately impossible (as yet) to state whether the older (14th- and 15th-century) *waterschepen* would have had stern and stem hooks with posts attached to them, similar to the cogs (fig. 27.4). Extrapolating from the excavation, it was possible to reconstruct a large part of the deck plan and gunwales from another fully clinker-built *waterschip* built in the first half of the 16th century[14] (figs 27.5a-c). This vessel was the first where it was established that the transverse boards of the forward deck were laid between the fore and aft beams, deck girders and sides without fastenings. The *waterschip*

Fig. 27.5a. The wreck of a fully clinker-built *waterschip* in lot Mz 22 Zuidelijk Flevoland with a jumble of detached timbers of the deck and bulwarks.

Fig. 27.5b. General view of the deck plan, reconstructed at the end of the excavation area. Viewed from the afterdeck. In the foreground is the fish-well deck with its trough, and the gangways on either side. At the top of the photo, the half topside crossbeams with standing knee timbers connecting the stringers to the sides of the foredeck.

Fig. 27.5c. The same. The afterdeck with deck beam, topside crossbeam with standing knees, and the bulwark with inwale. In the foreground, part of the fish-well deck and the gangways.

14. File RCE. Shipwreck Mz 22 Southern Flevoland. A reconstruction of the vessel is currently being built at the Batavia shipyard in Lelystad (since 2017).

was said to be of Hollandish origin.[15] Only one excavated example from the early 16th century has a straight, raking stern and stem (fig. 27.6)[16] and thus bears even more resemblance to the cog. The origin of this vessel may possibly be situated closer to the eastern side of the Zuiderzee, for example around the estuary region of the IJssel River. In the mid 16th century, and fully carvel-built *waterschip* from lot W 10 in eastern Flevoland, only the seams of the deck fish-well were still sealed with sintel clamped moss caulking.[17] The type of vessel may have emerged from the same predecessor as the cog. We will return to this in part IV in the conclusions.

Fig. 27.6. The fully clinker-built *waterschip* dating from the first half of the 16th century, recovered in the Wisentbos park near Dronten (lot K 84). It was excavated in 1971 and here is viewed towards the stern. It is the only known *waterschip* with a straight, raked stem as well as stern. Most remarkable was the presence of the complete rudder.

15. Amsterdam and Zuidwoude in Waterland; Pedersen 1996, 93. It is also known as the *Marker waterschip*.
16. Shipwreck K 84-II O.Fl., file RCE, Lelystad.
17. Reinders et al. 1978, 15 and plate 12 i.

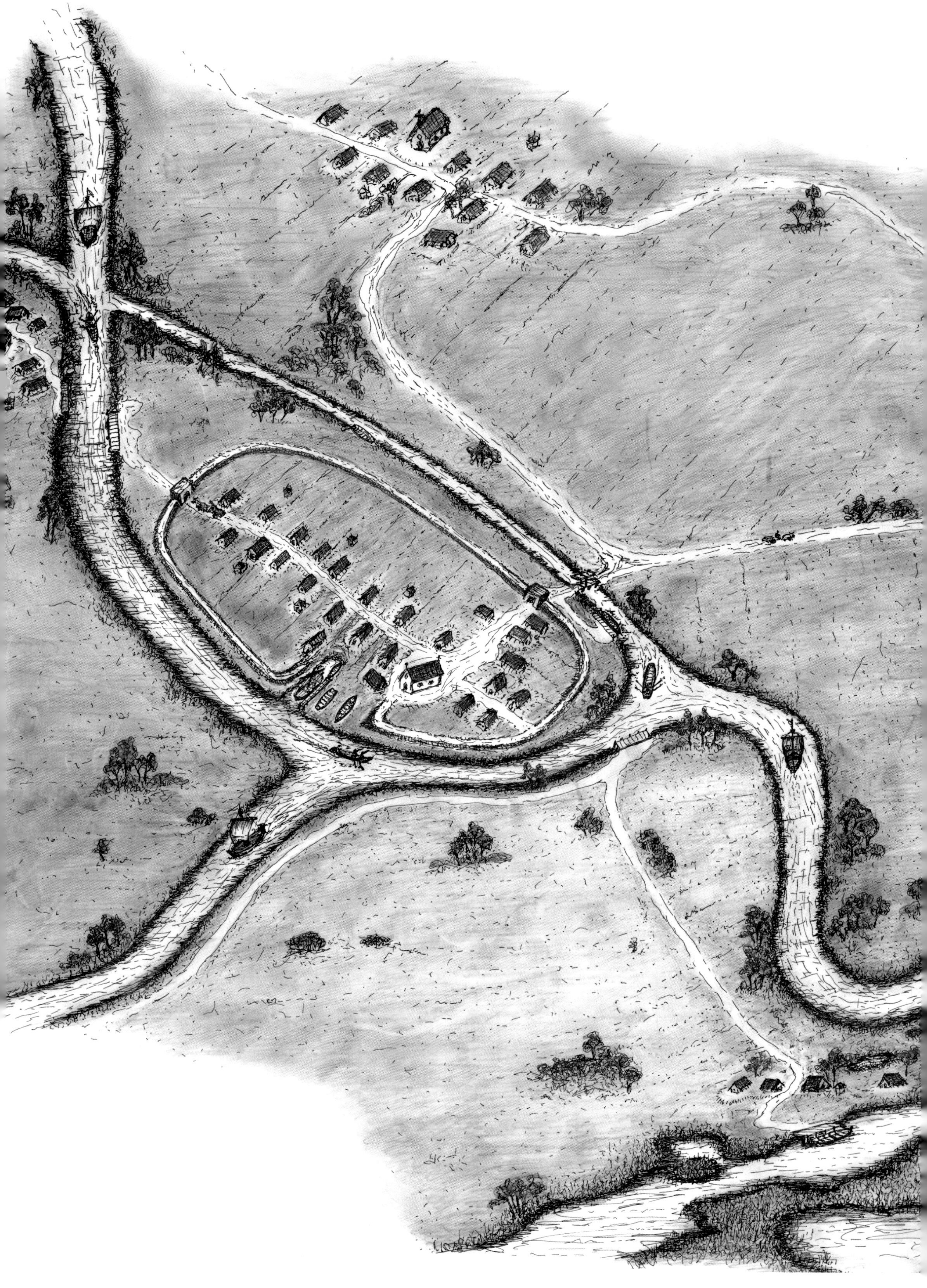

Medieval shipyards[1]

28.1 Possible shipbuilding sites of cogs

Virtually nothing is known about medieval shipyards in Northwest Europe. This is hardly surprising because the archaeological evidence for them is scarce. A gently rising bank along a navigable stretch of water in the vicinity of a settlement, village or town and situated at a geographically strategic point would appear to be a necessity. In addition, equipment such as round posts and a capstan are the first requirements in order to launch a ship of some size and to pull it ashore, as well as sufficient manpower, supports, ropes and tools of course. Figure 28.1 shows a rising bank and a simple shipyard on an inlet of the Bay of Volos near the village of Amaliapolis in Thessalia, Greece. The largest vessel (a kaïki) is, as regards the shape of the hull and its dimensions, directly comparable to the small cogs discussed in this study. On such sandy and stony ground one would not expect to find any objects of value, such as tools, after the shipyard fell into disuse. They would have been taken. At most, unsuitable means of attachment such as broken treenails, wood chippings and possibly some remains of tar or something like it would have been left behind. The latter items are hardly to be expected in the Mediterranean and at appropriate locations by the sea. Only objects that are covered by an enclosing layer of sedimentation (such as clay) can be preserved. The soil composition in the basins of the great rivers in the Low Countries on the North Sea from Flanders to Denmark and that of the Zuiderzee is eminently suitable for this.

Fig. 28.1. A small shipyard at Amaliapolis in Thessaly, Greece.

1. Translated by Christine Jefferis.

In addition to what is now known about the provenance of the wood used to build the ships discussed in this study (tables 1 and 2) and taking into account the fact that ships from after ca 1300 need not have been built in the area where the timber was felled but could also have been built elsewhere from imported wood, it is clear that there is great similarity between the large shipwrecks of Bremen, Nijkerk, Rutten, Doel and Kampen. It may also be stated with a substantial degree of certainty that all of them were most probably built in the core area of the Hanse on the coasts and along the great rivers discharging into the sea there.

Table 1. Cogs, small cogs, boats and barges built in the cogbuilding tradition in the Low Countries, in chronological order of date of felling/latest date established via dendrochronological (D) analysis c.q. building period and areas of provenance of the timber.

Ship (discovered in, land* and excavated in)	Date building period	Provenance of the wood	Reference
Antwerpen 1 (B) (1884)	1275-1300 (1125-1175)		Hasse 1907; Vlierman 2002 and this publication
Antwerpen 2 (B) (1884)	1275-1300 (1125-1175)		Hasse 1907; Vlierman 2002 and this publication
Antwerpen 3 (B)(1884)	1275-1300 (1125-1175)		Hasse 1907; Vlierman 2002 and this publication
Antwerpen 4 (B) (1884)	1275-1300 (1125-1175)		Hasse 1907; Vlierman 2002and this publication
Nijkerk-I (1980/1981)	1225-1250		This publication
Ens (1946/1949)	1225-1275		This publication
Rutten (1952/1985)	1263-1275 (D)	Poland/Lower Saxony	RING Dec. 1993; this publication
Rotterdam 4 (1991)	1263-1264 (D)	Between Meuse, Moselle and Rhine	Van Holk 2001
Amsterdam (1994)	1276-1277 (D)	Central/South Germany/Meuse Valley	Vlierman 2002; RING 1999; Toebosch 2011
Hattem (1978)	1250-1300		Vlierman 1996a and b; Vlierman 2002
Swifterbant (1962/2001)	ca 1305 (D)	Netherlands/ Westphalia/ Lower Saxon coastal area	RING juli 2001; this publication
Dronten-I (1971/1976)	1300-1325		Reinders *et al.* 1980; this publication
Doel-I (B) (2000)	1325/1326 (D)	Lower Saxony	RING Nov./Dec. 2000; this publication
Almere-II (1999)	1321/1333 (D)	Nederland/Westphalia	RING sept. 1999; this publication
Doel-II (B) (2002)	1322-1337 (D)	East Pomerania (North Poland), Poland and Baltic	Haneca, Gent March 2003; this publication
Nijkerk-II (1981/1983)	1335/1336 (D)	Apeldoorn/ East Netherlands/ Westphalia	RING Dec. 1993; this publication
Marknesse (1944)	1339 (D)	North Netherlands	Modderman 1945; Van Holk 2010; this publication; RING 3 November 2008
Kraggenburg (2007/2010)	1310-1320 (D)	Northwest Netherlands	RING 8 December 2010; this publication
Kuinre (1943)	1350-1375		This publication
Spakenburg-I (1975/1979)	1385-1400		This publication
Almere-I (1986)	ca 1400	?	Hocker & Vlierman 1996; RING, Nov. 1996; this publication
Spakenburg-II (1971/1979)	1402-1414 (D)	Netherlands/Westphalia	Van de Moortel 1991; RING June 1994; this publication
Antwerpen 5 (B) (1884)	1400-1425 (13th century?)		Hasse 1907; Vlierman, 2002
Dronten-II (1959/1995)	after 1296 (D) (1425-1475)	Middle-West Lower Saxon lowland	RING August 1996; this publication
Kampen (IJssel) cog (2011/2016)	after 1399 (D)	Netherlands, Germany	Waldus *et al.* 2012; RING 2012063; this publication
Dronten, K 73/74 (1963/1971)	1425-1450		Reinders *et al.* 1980; Vlierman, 1996a, Vlierman 2010 and this publication

Kampen (IJssel) barge (2011/2015)	ca 1410 (D)	Netherlands, local	Waldus *et al.* 2012; RING 2012063; this publication
Kampen (IJssel) boat (2011/2015)	1400-1450		Waldus *et al.* 2012; this publication
Enkhuizerzand (1999/)	ca 1465 (D) (1470-1485)	Netherlands, Westphalia, North Germany	RING March 1999; this publication
Medemblik, Strandbad (1982/)	1400-1500		Maarleveld 1982; this publication
G 37 NOP	13th century 1400-1500		Van der Heide 1974; this publication
Dronten, L 89 (1996)	ca 1545 (D)	Netherlands, West and South Lower Saxony	RING January 1997; this publication

* (B) = Belgium

Table 2. Finds with characteristics of the cogbuilding tradition in Europe, outside the Low Countries, and the date of felling/latest date established via 14 C and dendrochronological (= C en D) analysis c.q. building period and areas of provenance of the wood

Ship (discovered in, country* and excavated in)	Date building period	Area of provenance of the wood	Reference
Kollerup (DK) (1978)	after 1153 (D) (end 12th century) 1150 1150	South Jutland	Andersen 1983; Westphal, 1999; Daly *et al. 2000;* Crumlin-Pedersen 2000
Kolding (DK) (1943/1999-2001)	1188/1189 (D)		Hocker 2000 en Hocker/ Dokkedal 2001 *et al.*
Skagen (DK) (1962/1994)	1193 (D)	West Denmark	Dokkedal 1996
Kuggmarken (S)	1215 (D)	South Denmark	Hocker 2004; Van Holk 2010
Kronholmen (S) (1995)	1225-1250 (C)		Förster 2008
Oskarhamn (S) (1985, 1987)	ca 1250 (D)		Förster 2008
Rostock-Warnemünde, Jachthaven (G) (2004)	1280 (D)		Förster 2008
Helgeandsholmen 2 (S) (1978-1980)	ca 1330/before 1530		Crumlin-Pedersen 2000; Förster 2008
Dars (G) (1977, 2000-2004)	1298-1303/after 1335 (D)		Förster 2008
Lille Kregme (DK) (1982, 1986, 1992)	1358 (D)		Förster 2008
Vejby (DK) (1976)	1372 (D)	Southeast Baltic	Bonde & Jensen,
Bremen (G) (1962)	1380 (D)	Bremen	Lahn, 1992
Skanör (S) (1993-)	1388/1390 (D)	West Baltic	Hörberg, 1995

* (DK) = Denmark, (S) = Sweden, (G) = Germany

Apart from the probable locations of a late-medieval shipyard in Kampen, Bruges[2] and probably in Duisburg that are discussed in the following section, to date I know of no shipyards from the Hanseatic period. Cologne, Neuss, Duisburg, Wesel and Emmerich on the Rhine, Doesburg, Zutphen, Deventer, Hattem and Kampen on the IJssel mentioned above, Zwolle, Hasselt and Genemuiden on the Zwarte Water and the towns of Harderwijk, Elburg, Vollenhove?, Kuinre? and Staveren on the east coast of the Zuiderzee are actually the most obvious locations in the Low Countries of the German Empire where, since the beginning of the 13th century (possibly even earlier in Cologne and Duisburg[3]) the construction of large and small cogs may have taken place. As well as Kampen, Hasselt and Zwolle in particular would appear eminently suitable from a geographical point of view due to their position on waterways easily navigable for larger vessels. In the French/(South) Flemish coastal area and especially in Boulogne, Nieuwpoort and Bruges and surroundings this may also have been the case from the middle of the 12th to into the first half of the 13th century.[4] Antwerp and Ghent may also be included among these possible early building locations. The ships that were almost certainly built along the Flemish coast and along the Lower Rhine were probably not yet entirely similar in construction to the later, 13th- and 14th-century cogs. The oldest name of cog recorded in writing concerns two seagoing Flemish ships that were deployed during the Second Crusade in 1147.[5]

In the Zeeland-Holland area and based on the Vlaardingen seal from ca 1300 and its geographically favourable position at the confluence of the Oude Maas and Lek rivers, this town also appears to be a possible site, while Dordrecht will undoubtedly also have played a part. In the more northerly North Sea region, cogs were probably also built in South Jutland, the surroundings of Ribe, (initially) in the period ca 1150-1250. There is no evidence that this occurred at a later date. It is very likely that in the course of the 13th century this did take place particularly in Bremen on the river Weser and Stade, Buxtehude and Hamburg on the Elbe, where cogbuilding is more to be expected. In the Baltic region the most suitable places for building Baltic cogs were initially Lübeck, Wismar, Rostock, Barth, Stralsund, Gollnow/Stettin, Wollin, Cammin, Danzig and Elbing.

In this respect, it does not appear insignificant that the ships on the mid-13th-century seals of Wismar, on the 13th-century seals of Stralsund, on the two 13th-century seals of Danzig and on the mid-13th-century seal of Elbing are the earliest examples to depict a cog, whereas this is not the case with the Lübeck seals dating from ca 1200, 1250 and 1280 respectively, and two 14th-century seals of nearby Neustadt (in Holstein). Shipbuilding in the latter two towns probably remained under Nordic/Scandinavian influence for longer. This hypothesis is underpinned from a geographical point of view and on the basis of, for example, the trading area of the Hanse merchants on the *Caerte van het Oostland* by Cornelis Anthoniszoon from 1543.[6] The other seals of German Baltic towns mentioned that portray cogs all date from the 14th century or later. Up to now it is still not clear whether the 13th-century seals of Wismar, Stralsund, Danzig and Elbing depict ships that correspond to the finds from the Low Countries described in this study or whether they show Baltic cogs of softwood and completely clinker built. Some of the construction details of the Danish oak cog finds from Kollerup, Kolding and Skagen show typical solutions/choices from the Scandinavian shipbuilding tradition.

Dendrochronology

The provenance of the wood samples was established for sixteen of the twenty-nine shipwrecks from the Low Countries that were built in the cog tradition (tables 1 and 2). In a number of these, especially from recent excavations, the result is based on several samples. In others it is derived from a few, and occasionally only from a single sample. The reliability of these results with regard to all the wood used in the ship concerned is therefore questionable. Preferably at least ca 20 different and suitable samples should be sampled. In addition, there should also be a close examination of the section of the ship hull from which the samples were taken and whether there were any later repairs. Thanks to the availability of a suitable reference curve it is possible under the most favourable circumstances to establish not only the felling date of the wood but also the place (region) where the tree grew. It is, moreover, very probable that vessels predating ca 1300 were built in the region from which the wood came. After this date, wood was increasingly imported, due to the fact that in certain areas (for example, Flanders) timber suitable for building was hardly, or was no longer available.

Despite these restrictions it is notable that the early ships (from before ca 1300) appear to be typical inland/river vessels which must have been built in the Rhine, Meuse and Moselle areas. There is only one case of a large cog. In this

2. Vlierman 1996a, 38-39 en 75.
3. Vlierman 1996a, 36-37, 59, 61, 62 and fig. 18. The finds of sintel clamps indicate that on both sides of the stronghold/Königspfalz shipbuilding probably took place on the banks of the Rhine between the 9th and 13th centuries (Alte Markt and Niederstrasse).
4. Vlierman 1996a, 38-39 and Vlierman in preparation a and b.
5. Bill *et al.*, 1997,158.
6. Henn 2010, 14.

Rutten cog, wood from Poland and Lower Saxony was used. It would therefore appear likely that the ship was built in or originated from somewhere on or near the German/Polish Baltic coast. The large and remarkable quantity of ballast of Baltic flint, possibly brought on board ship shortly after building, underpins this hypothesis. The wood of the Doel-I cog comes from Lower Saxony. This ship has remarkable ornamental grooves made with a gouge on the lands, in which the nails of the overlapping fastenings are driven. These grooves are so far only known in ships built in the completely clinker built and clenched Viking or Scandinavian tradition. From the same tradition are V-shaped floor timbers fastened together fore and aft in the underwater body. These were found in the Doel-I and Nijkerk-I cogs. The latter ship also had several tongue-shaped plank scarves in the sides, also only known from the Viking shipbuilding tradition. A shipbuilding site on the Lower Saxon/South-west Baltic coast (surroundings of Lübeck, Wismar) is quite possible in the case of both ships. The wood from the Doel-II cog comes from North Poland and East Pomerania. The caulking (moss and animal hair) used in the seams makes it likely that it was built on the Polish/German Baltic coast. The use of animal hair as caulking material is typical of Scandinavian and Slavic shipbuilding.

The wood from the remaining cogs dated by dendrochronological analysis originates from the North and East Netherlands, Westphalia and Lower Saxony. These cogs all date from after ca 1300, so that it cannot simply be assumed that they were built in the region from which (most of) the wood originates. That they were all built somewhere in the core area of the Hanse (the IJssel basin and the Zwarte Water/the Vecht and along the lower reaches of the Weser and Elbe) appears, however, most probable.

28.2 Finds in Kampen linked to ships/ shipbuilding

During the archaeological excavation on the Kok site at the beginning of Oudestraat in Kampen in 1991 a number of sintel (nail)s/clamps were found, among other things, which point to the possibility of their manufacture/a smithy and the caulking of vessels in the vicinity.[7] In addition to these finds, several sintel finds are also discussed in the Vlierman 1997[e] article which had been collected since 1950 in Voorstraat 58-60, in the Graafschap, near Burgwalstraat/Boven Nieuwstraat and at the Koornmarkt. These sintel clamps date typologically from the period ca 1150 - ca 1375 and in particular from the 13th and the first half of the 14th century. The article ends with the recommendation to pay special attention to aspects of ship archaeology during future archaeological investigations. This could, for instance, throw more light on the possibility of a shipyard during the development of the Hanse town of Kampen.

In 2002 I received from T.J. Broek in Kampen one sintel nail (sintelnagel), nine sintel clamps and a nail head which he had collected in the 1990s with his metal detector in the meadow on the northwest side of the Oude Trekvaart. This is now a ditch running from (the bastion on) the town canal outside the Cellebroederspoort in the direction of the Zwarten Dijk. On the map of Kampen by P. Utenwael from 1598, the canal on which two small sailing boats can be seen and on which there were standing one or more windmills (fig. 28.2). The small iron objects probably came from the nearby part of Kampen together with other town refuse and were strewn there, a customary way of dealing with refuse in late-medieval towns.[8]

Fig. 28.3 shows the sintel clamps. The elongated sintel nail (fig. 28.3-1) appears unused and is of the A1 type. A date in the 12th century seems very likely. If the object was used and, for instance, came from a river barge from the Rhine area, a 13th-century date would also be possible.[9] The small sintel clamp (fig. 28.3-2) is type D and

Fig. 28.2. Map of Kampen by P. Utenwael, 1598. The box indicates the excavation site in the former precinct of St Gertrude's Hospital and the field bordering on the Oude Trekvaart canal.

7. Vlierman 1997e, 113-121.
8. These objects related to shipbuilding/maintenance were donated to the Archeologische Dienst Zwolle in 2015 in connection with the shipbuilding yard discussed here.
9. See section 29.5.1.

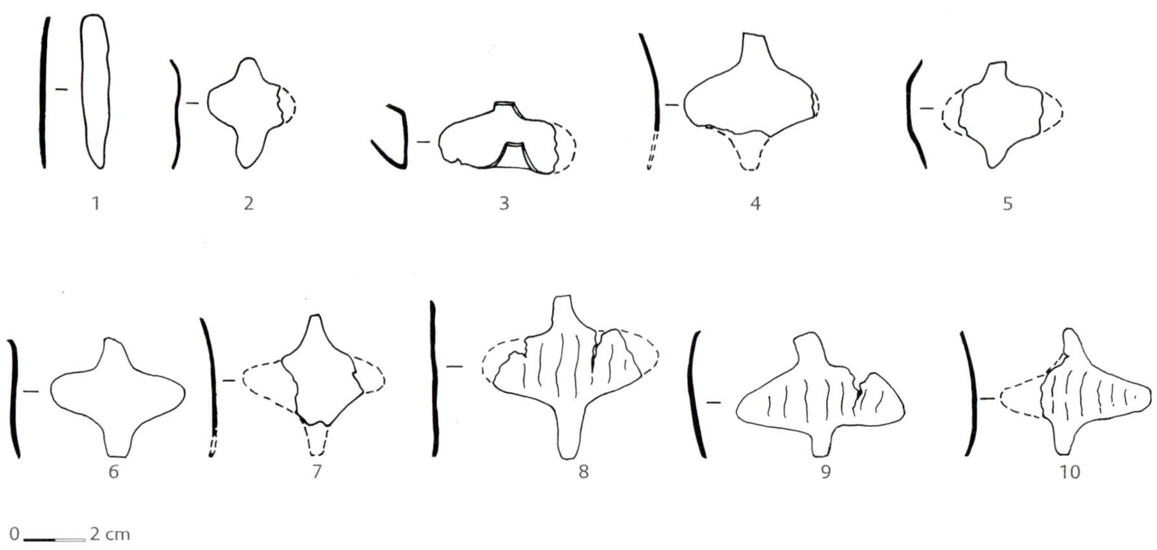

Fig. 28.3. The various types of sintel nails and sintel clamps from the field on the Oude Trekvaart canal outside the Cellebroederspoort gate.

was in use around the middle of the 13th century. Five sintel clamps (fig. 28.3- 3-7) are type D2 with a probable date ca 1300-1375. The remaining three (fig. 28.3- 8-10) are, on the basis of the ribbed traces of smithing, type E(1). These are sintel clamps made of two strips of iron welded together at right angles. They most probably date from the period ca 1375-1450.

In 2011 and 2014 archaeological excavations were carried out on the site of the former Huize Margaretha, situated between the Burgwal, Burgwalstraat, Boven Nieuwstraat and Geerstraat, the site of the former Geertruidengasthuis (hospital) (fig. 28.2).[10] According to historical sources the Gasthuis hospital was founded in 1382 or perhaps even earlier (in 1371). In 1598 the Geertruidengasthuis (hospital) was united with St. Catharinagasthuis (hospital) and came under a joint name. The Gasthuis hospital, especially the characteristic ward in the Boven Nieuwstraat, existed in its medieval capacity until 1897. The area concerned belongs to the oldest part of Kampen. In addition to the foundation remains of the Geertruidengasthuis (hospital) in the Boven Nieuwstraat, remains of the walls of several houses on the Burgwal side were found as well as two fine and well-preserved brick fireplaces. The houses were almost certainly built in the first half of the 14th century. Under one of the wall remains short posts and pieces of ship wood had been used as log foundation (fig. 28.4).[11] Scattered over the site, on and in the old occupation layer in which there were a number of pits filled with manure, close to a hundred small wrought-iron objects were found, among other things. These objects included almost eighty sintel clamps, several nails and clinch-nails and a number of unknown objects. A heavily tarred piece of sailcloth, a small wooden tar barrel, a wooden structure and wood chips are finds that are also connected with ships/ship maintenance/shipbuilding. On 20 June 2014 a visit was paid on request to the excavation to examine the finds.

Fig. 28.4. Ship's timbers on a house-wall foundation of fascine and piles, and part of a windlass construction in the excavation of the Geertruidengasthuis (St Gertrude's Hospital) site in Kampen.

10. The investigation was carried out by the Archeologisch Diensten Centrum (ADC: from 17 October to 25 November 2011 and subsequently by the team Archeologie of the municipality Zwolle (from 28 April to 4 July 2014). The data on the excavation, dendrochronological dating and the historical information were made available by M. Klomp, Archeologische Dienst Zwolle.
11. Dendrochronological analysis of a piece of ship wood produced a felling date of 1337 +/- 5.

28.2.1 Iron objects

Sintel nails and sintel clamps

More than half of the caulking clamps/nails can be classified as sintel nail types B and C, more probably a transition type B/C (fig. 28.5- 1-5), and sintel clamps type D (fig. 28.5-6-9). The remaining specimens as type D2 with transitional form to type E (fig. 28.5- 10-14).[12] Roughly half had been used and a number of these had probably been bent during hammering into the ship's shell, were no longer usable and thrown away. The remaining specimens are 'new' and unused.

Nails

Four large nails were found. Three are complete and partly bent. The fourth is incomplete (fig. 28.6- 1-3). They are nails ca 120-125 mm long with large heads wrought in facets (Ø 32-40 mm). One specimen was bent over twice and may be regarded as a so-called 'naainagel' in Dutch, a double-clenched nail, a method of fastening ('iron sewing' in Dutch) like that used in the cog building tradition to attach planks together through the overlap.[13] The nails were certainly intended for or originated from a vessel of considerable size.

Rivets

Certainly not from a vessel in the cog building tradition are two rivets (fig. 28.6- 4 and 5). They are fastenings from (a) vessel(s) that was completely clinker built in the Nordic (Scandinavian) method. In this case too it was probably a medium size to large vessel. The rivets are complete with rivet plate. This means that they are used specimens. A repair of a ship from another region would appear to be the most plausible explanation for the presence of these fastenings.

Joints(?)

Two small iron objects which are very unusual in design have never been observed before (fig. 28.7- 1 and 2). One specimen is ca 100 mm long, the other ca 55 mm. From the side view they are Z-shaped. The vertical part is on one side ca 1.5 mm thick and toothed. The other side is

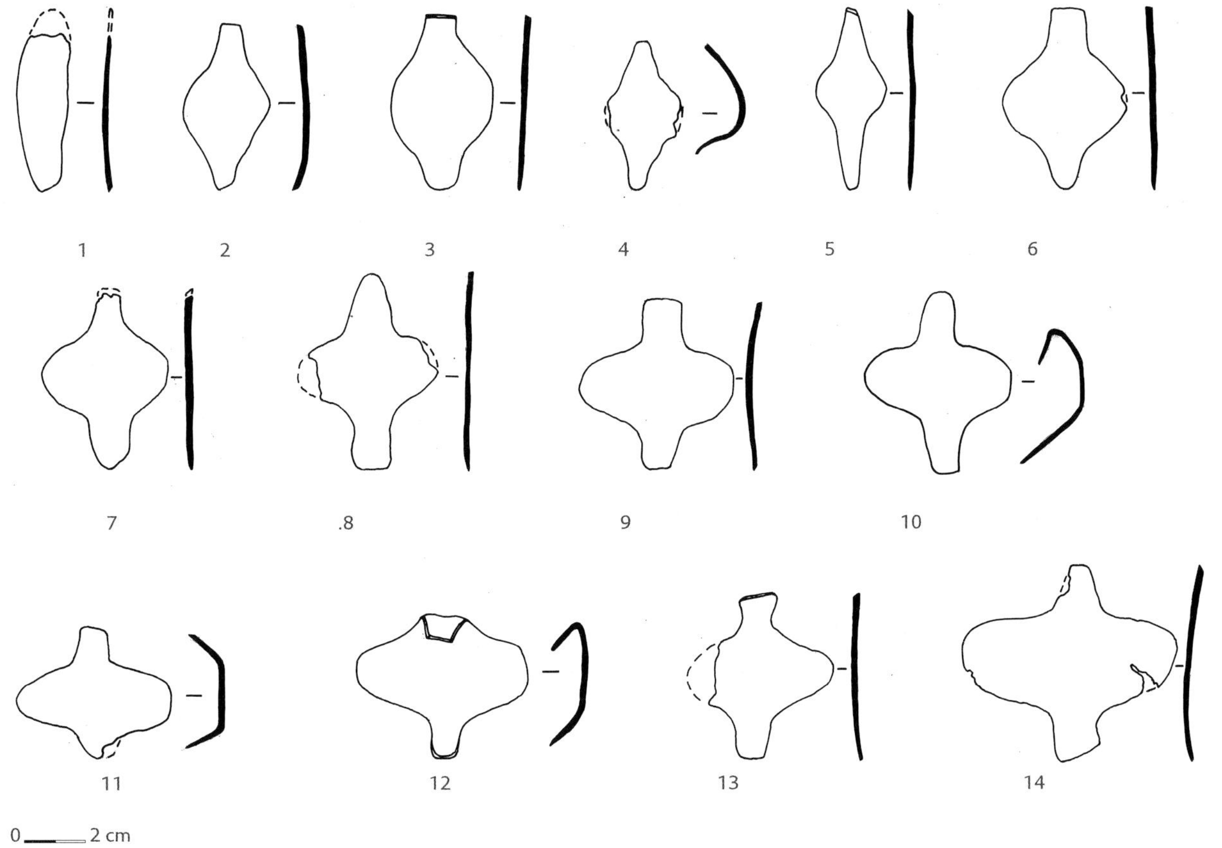

Fig. 28.5. The iron sintel nails and sintel clamps.

12. Vlierman 1996a, 58 table II.
13. For the method of attachment of these double-clench nails (naainagels in Dutch) see chapter 6, the Nijkerk-II cog, fig. 6.42.

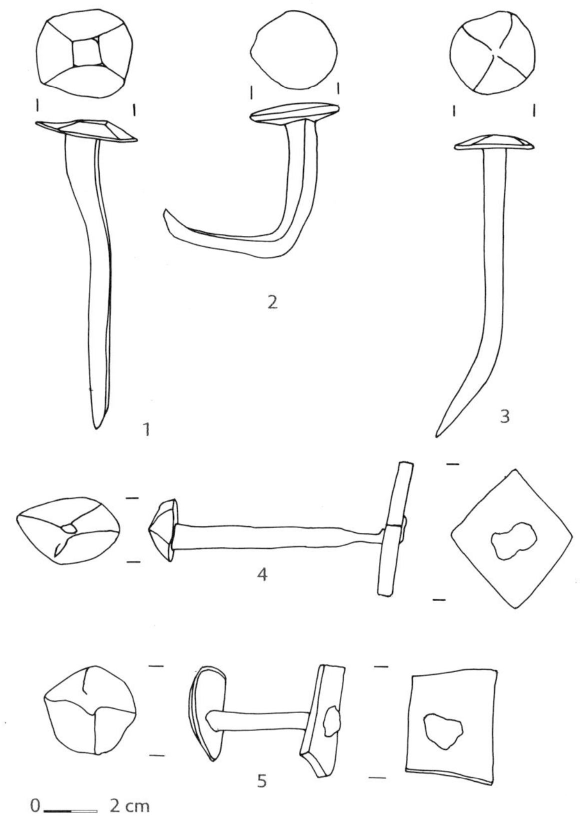

Fig. 28.6. Nails and rivets.

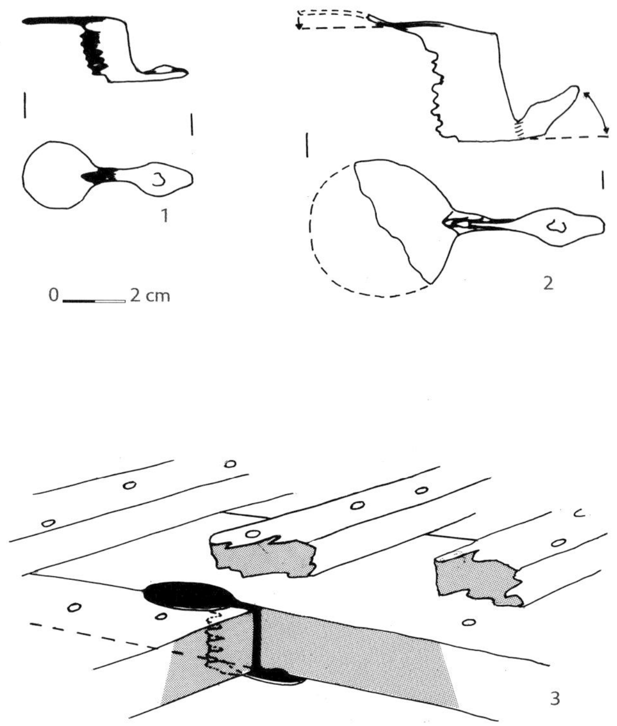

Fig. 28.7. Iron fastenings(?) and their possible application.

3 mm thick. On top or underneath? a 1 mm thick, round and flat piece with a diameter of ca 45 and 22.5 mm respectively has been wrought at right angles to the vertical part. On the other side of the vertical part there are laurel leaf shaped pieces 25 and 18 mm long and 3 and 2 mm thick respectively. In the large specimen this is thickened in the middle to 9 mm, in the smaller one there is a ca 5 mm thick knob.

In shipbuilding, two methods of joining planks were used. Carvel is with planks edge-to-edge and only fastened on the frames. Clinker is with the lower edge of one plank overlapping the other. Nails or rivets are hammered through the overlapping part.

They could perhaps be joints/connecting pieces used to repair a carvel-built bottom of a ship for instance. On various occasions it has been observed during a maritime archaeological excavation that poor parts were sawn out of bottom planks and replaced by a better piece of plank. It must have been difficult to fit a rectangular piece of plank into a bottom plank. It is not inconceivable that these iron joints were used for this purpose. The vertical part was then hammered with the toothed side in the seam between two bottom planks. The round, thin top was on the inside of the ship, under a frame, for instance, and the laurel leaf part on the outside. The top and bottom held the shell planks in place, and it would not have been too difficult to push the repair plank under the laurel leaf part and keep it in place (fig. 28.7-3). The repair plank could subsequently be fixed to the frames with treenails and/or nails.

(Fire) striking irons?

Among the objects found are seven unusually shaped items (fig. 28.8- 1-7). These are ca 130-140 mm long strips of iron 7-10 mm wide, with a 4-10 mm centre part and thinner wrought ends. All the specimens end in a point on one side, the other side is flat and in some cases has a small hole in it. The thinner ends are bent around, giving them a clamplike shape originally. These ends are now bent. In Dordrecht a somewhat similar, though

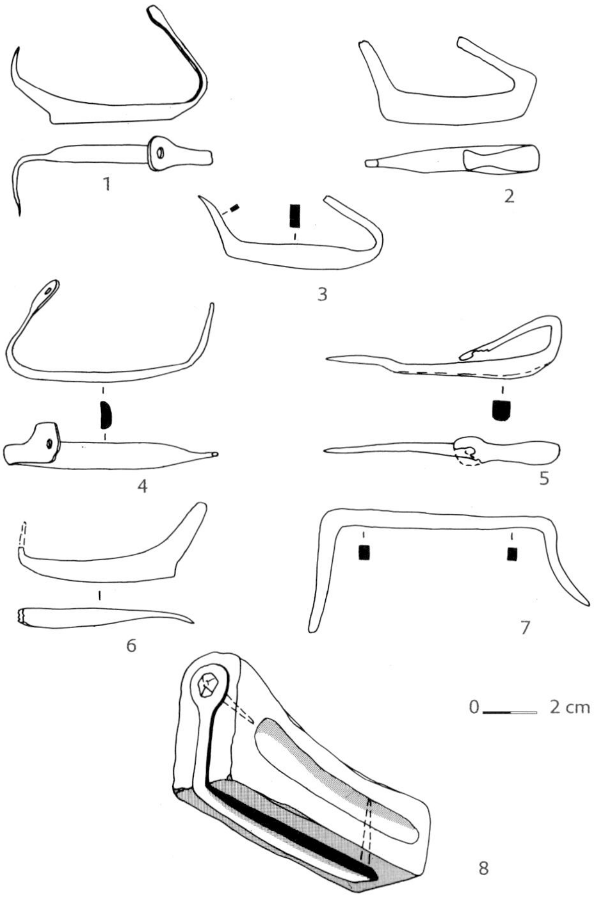

Fig. 28.8. Strike-a-lights and a reconstruction with a wooden handle.

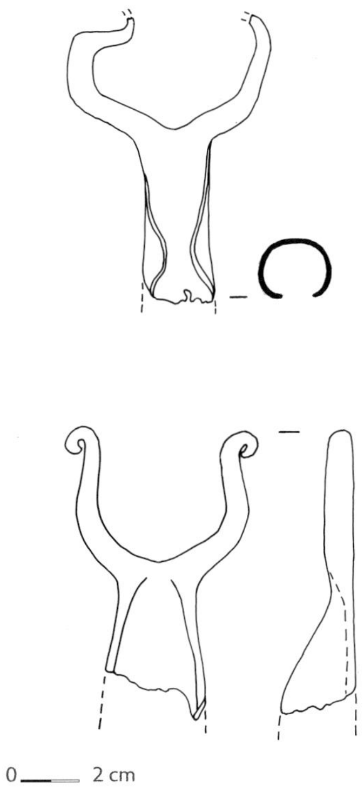

Fig. 28.9. Boat hooks.

smaller specimen was found.[14] At the time it was suggested that this was possible a semi-manufacture for a nail or sintel clamp. However, the flat end with a hole in several of the specimens from Kampen presupposes a completely different function. They were probably both a strike a fire, which were hammered on two sides into a wooden handle or with one side in the wood and with the flat end with the hole nailed against the front (fig. 28.8- 8). Two strikes a fire in the shape of spectacles were found in a 16th-century *waterschip*.[15] Number 7 in fig. 28.8 may possibly have had a different function, for instance, as a clamp to hold a tree trunk during sawing.

Boathooks

Figures 28.9-1 and -2 show two boathooks. They are of a forked type with two teeth, generally used to push a boat clear of the mooring. They are indispensable tools in a shipyard. A more or less similar specimen, though pointed, was found in the Marknesse cog (Section 12.5.2, fig. 12.28c). The actual/true boathook, a shaft with a long square sectioned point and a curved hook is so far only known from 16th-century and younger ships from the IJsselmeer Polders.

28.2.2 Wooden objects and a piece of sailcloth

Ship wood

In pit 5 under wall S421 lay a futtock (wooden beam S425) on a number of pieces of plank lying crosswise in fill 2. It is a ca 1.7 m long piece, 15 cm wide and ca 10 cm thick (fig. 28.10- 1). The outer side is largely stepped. It is a shape that indicates that this is part of a frame from the clinker-built side of a ship. The successive lands show that planks of > ca 43 cm wide had been attached to them. The planks were attached with two wooden treenails (Ø 3 cm) per width of plank. The bottommost part of the futtock was not stepped but flat, while the end had

14. Vlierman 1996a, 31 fig. 15b (right).
15. Reinders *et al.* 1978, plate I. A so-called Burgundian fire-iron is depicted on several of the tiles of the fire box of this wreck; this manuscript chapter 29, section 29.6.1 and figs 29.13a, b and c.

probably been sawn off for the purpose of the secondary use. A plank had probably been attached against the 37 cm long, lowest flat part which fitted flush against the overlapping strake above it.

A similar piece of wood (S425 HW) is a futtock 1.8 m long split lengthwise (fig. 28.10- 2). This piece, that was probably originally also ca 15 cm wide and 10 cm thick suggests that it was in any case attached against five overlapping planks one on top of the other. The top four were > 40-43 cm wide. These planks were also fastened with two wooden treenails 3 cm in diameter per width of plank.

A third frame part (S425 from fill 3) is ca 1.55 m long, 27 cm wide and 14-15 cm thick (fig. 28.10- 3). The curved piece has, from top to bottom, four stepped plank widths of > 35-40 cm, 33, 43 and 33 cm respectively (originally). The lowest part measures ca 22 cm. The plank attached to it lay edge to edge and under a slight angle to the bottommost clinker built plank. It is part of a floor timber that had been attached to the hold side of the fore or afterbody. It was attached through the top and following plank with a single wooden treenail, through the widest strake below this with none, through the bottommost lapstrake planks with two treenails and through the lapstrake plank probably with two treenails (Ø 3 cm) originally. The

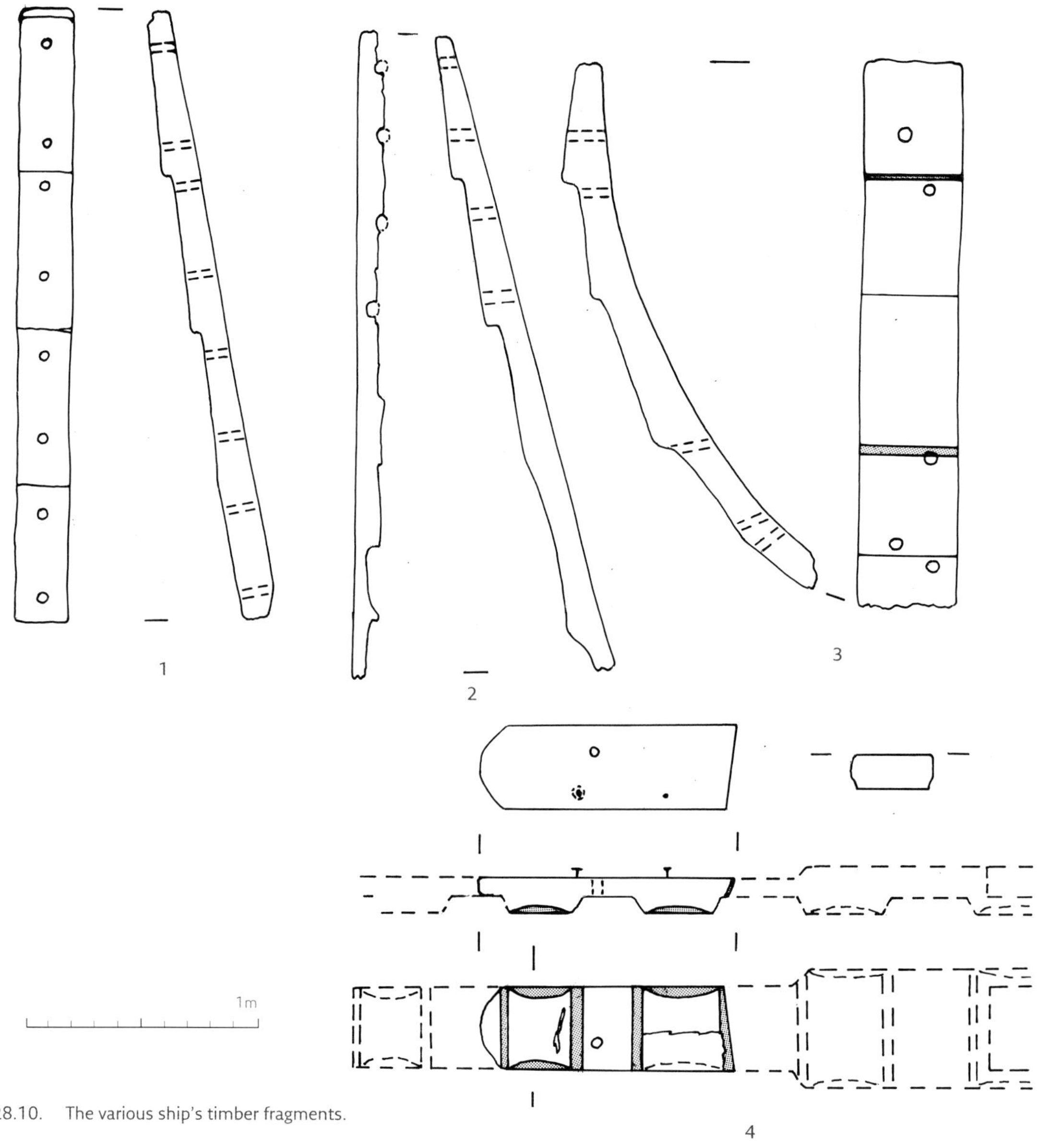

Fig. 28.10. The various ship's timber fragments.

fourth small beam found with nailholes of varying diameters (S425 from fill 1) is almost certainly not ship wood and will not be described further.

The last piece of ship wood was found on the dump. It is a very recognisable piece of a so-called keelson or inner keel (fig. 28.10- 4). A long beam with fore and aft section generally narrower towards the ends, and in between, a wider and generally thicker integrated part in which a hole for the heel or foot of the mast was cut. This beam, sawn off at one end and pointed at the other is 68 cm long, ca 24 cm wide and 10 cm thick. Halfway along it and underneath is a 5 cm deep notch measuring 13.5 cm at the top and 19 cm wide at the bottom which fits over the top of a floor timber. At both ends one can see the oblique/slanting sides for the next floor timbers. The keelson was fastened to the floor timbers with treenails (Ø 2.5 cm). The centre-to-centre distance between the frames is ca 37 cm. The lower edges of the parts between the floor timbers have been hollowed ('archways'). Long keelsons such as this are typical construction features of the large (Hanse) cogs. Examples are the keelsons of the Nijkerk-I cog from the middle of the 13th century and of the Doel-I and -II cogs, both from the first half of the 14th century.[16] The widths and maximum thicknesses of the mast step sections of the Doel-I are 37-39 and 26-27 cm respectively, of the Doel-II 37 and 28 cm respectively and of the Nijkerk-I 50 and ca 25 cm respectively.[17] The fore and aft sections of the keelsons are less wide and thick. It is, therefore, quite possible that this is a piece of keelson from a large cog and probably the front or rear part. The notch for a floor timber which is ca 19 cm wide supports this hypothesis. The centre-to-centre distance between the frames of ca 37 cm is comparatively small. In most cogs this is between 45-55 cm.[18] The beam is probably suitable for dendrochronological analysis.

The design of the frames plus the general date of the profile and other finds confirm that these are certainly ship parts from the cogbuilding tradition. The length and shape of the pieces of futtock and floor timbers presuppose a gunwale height of the vessel(s) from which they originate of approximately 2-2.5 m. This means that they are probably from a small cog or related type. The 3 cm thick treenails are better known from the large ships and especially for fastening/attaching crossbeams.

A windlass

In pit 5-level 1 of the 2014 excavation, part of a wooden structure was found (fig. 28.11, 1a and b). The structure consists of a horizontal beam 1.54 m long and 16 cm wide and thick (S428). At one end the beam is 8 cm thick over 27 cm, while at 27 cm from that end there is a notch on the underside 14 cm wide and 4 cm deep. In the upper side about half the length of the thick part and in the middle of the beam a rabbet has been cut 16x5 cm and 10 cm deep. In this rabbet was the remains of a vertically placed beam 21x18 cm (S419). On one side of the beam

a.

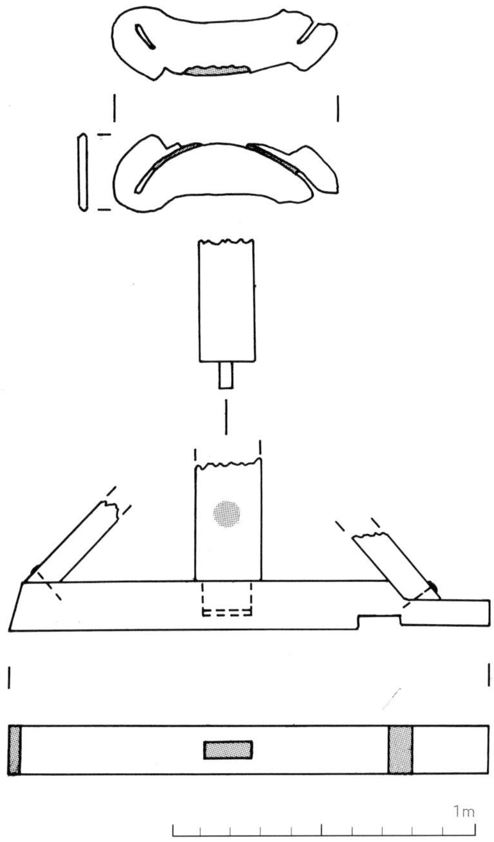

b

Fig. 28.11. The recovered part of an oaken windlass, the fragment of the presumed roof and a reconstruction.

16. Chapter 8, section 8.2.2, chapter 20, section 20.2.2, chapter 21, section 21.2.2 and figs 21.7a and b respectively.
17. Appendix 8.II B.
18. See section 3.2.

there was a circular impression with a diameter of 8 cm in the middle and at ca 20 cm above the top of the horizontal beam. The length of this beam is 46 cm including the 8 cm long part in the rabbet. The top part has rotted away. On the top of the horizontal beam S428 the remaining parts of two small obliquely placed beams 10x10 cm were nailed. They are still 30 cm (S426) and ca 35 cm (S427) long, respectively. The original, complete structure was one of the sides of a windlass.

A remarkable piece of worked wood (S640, fig. 28.11a top) is a 73 cm long and 23 cm wide plank, which has been somewhat rounded. The 3 cm thick plank is rounded along the edges, damaged and worn. On one side a curved groove 1.5 cm deep and 1.5-1.8 cm wide was cut, which (originally) ended ca 10 cm from the ends of the plank. Over the last 18 cm of the groove it changes on both sides into an opening through the entire thickness of the plank. My assumption is that it was one of the sides of a roof structure on top of the capstan. Figure 28.11b shows a reconstruction of the windlass with roof.

A tar-barrell-bucket

In a manure or refuse pit (S415) a wooden stave-built bucket was found (find no. 282, fig. 28.12) . The bucket/barrel which is conical in shape was still copiously covered in tar inside and out. The barrel/bucket is 27.5 cm in height, has a base measuring 34.8 cm in diameter and the top is 28 cm in diameter.

A piece of sailcloth

During my visit to the excavation on June 14, 2014, I saw on the Burgel side a large crumpled up piece of textile drenched in tar lying in a pit in the 14th-century surface layer. Closer examination revealed that this was a comparatively coarsely woven woollen fabric, most probably a fragment of a discarded sail (fig. 28.13). In the Nijkerk-I cog a piece of woven woollen sail ca 60 cm tall and 80 cm wide was discovered (section 8.5.2, fig. 8.46). In the Nijkerk-II cog two fragments ca 70x55 cm and one fragment ca 40x30 cm were found (section 6.5.3, figs 6.55a and b). In several other cogs from this research small pieces of woollen sail were found. The measurements of the piece of sail from Kampen are not yet known, but will certainly be several square metres. The lump approximately one metre long has not yet been unravelled but is, at any rate, a unique find and it deserves a thorough investigation. The large fragments from the Nijkerk-I and -II cogs were woven in a so-called plain weave on a comparatively broad loom. The fabrics consist of S-twisted, coarsely spun threads which are twined together into Z-twisted spun yarns. Several fragments of selvage were also observed. Some pieces have brown and ochre strips in the fabric and are tarred.

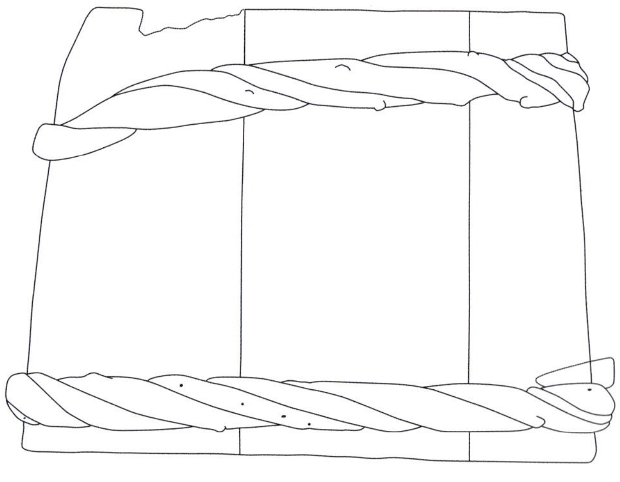

Fig. 28.12. The wooden tar-barrel/-bucket.

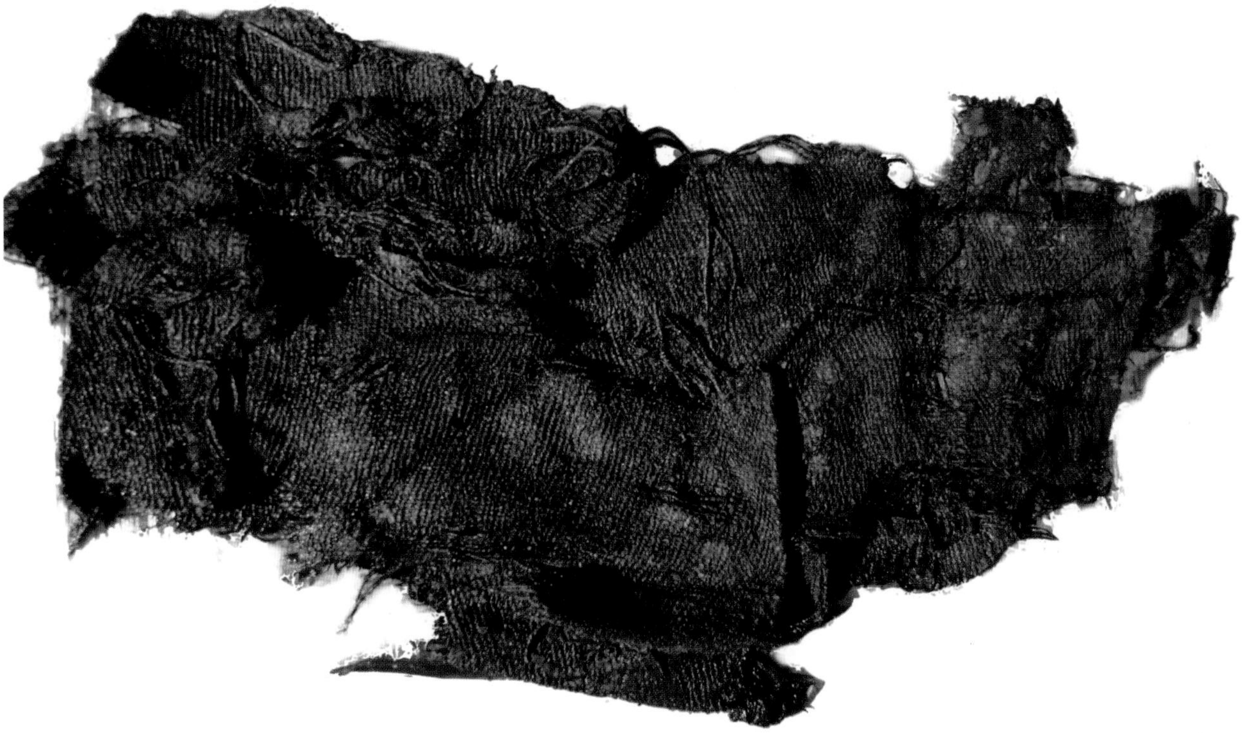

Fig. 28.13. The fragment of sailcloth.

28.2.3 A shipyard?

The objects described above are all directly related to ships, shipbuilding and ship maintenance/repair. The sintel clamps found earlier in and near the oldest part of Kampen also point to this and to the presence of a smithy where the iron objects must have been made. The typo-chronological dates of the sintel clamps suggest that the oldest were made between the end of the 12th and in the first quarter of the 13th century (types B and C). Around 1250, sintel clamp type D was already in use throughout the Hanse region and type D2 probably after ca 1300. Type E develops around the beginning of the last quarter of the 14th century. Approximately half of the caulking clamps found are 'new' and unused. A substantial number of the remaining sintel clamps were probably distorted while being hammered into the ship's hull, unusable and thrown away. On the one hand, the sintel (nail)s are a very likely indication that there must have been a smithy in the vicinity. On the other hand, they were almost certainly thrown away or lost on the site where they were found. In other words, maintenance of ships was at any rate carried out on the spot. However, it appears more likely that there was a shipyard here where ships were built and repaired between the end of the 12th to the third quarter of the 14th century. The windlass with which ships were pulled up on land fits in with this picture, whereas the ship parts discovered in a wall foundation probably came from a large and a small vessel of the cogbuilding tradition which were probably scrapped in the shipyard. Dendrochronological analysis of a piece of ship wood yielded a building date for the vessel to which it belonged of 1337± 5. The discarded strikes a fire indicate that fires were regularly made there (burning the shell strakes). Wood chips in a pit point to woodworking, while the tarred piece of sailcloth, the tar barrel and the boathooks all belong in a shipyard.

Is a shipyard at this site realistic? When this was first suspected, the course of the soil profile was first examined. There is indeed a natural slope down to the Burgel. Unfortunately, during the archaeological excavation of the site no soil borings could be made in the direction of the Burgel to establish whether there was a decline to a natural watercourse. It has long been suspected that at any rate the southeast course of the Burgel between the IJssel and Geerstraat was originally a natural stream channel.[19] In a distant past this might even have been the course of the entire Burgel. In fact, my assumption is that this was indeed the case. Moreover, it seems very likely that a dug (?) branch with a large wide bend at the Geerstraat ran towards the Cellebroederspoort and further in the direction of the Zuiderzee, the previously-mentioned

19. Speet 1986, 12.

Oude Trekvaart (fig. 28.2). The Trekvaart was already mentioned in the 15th century, but then under the name Reeve.[20] If there is an opportunity in the future for excavation/soil borings along this route, it might provide an answer as to the breadth and depth of the channel. In other words, whether (also large) cogs were able to sail from the IJssel to the shipyard on the later site of the Geertruidengasthuis (Geertruiden hospital). It was not possible to establish a clear relation between the foundations discovered of houses/buildings from the first half of the 14th century and the assumed shipyard. The same goes for the exact, original position of the windlass. The absence of sintel clamp types that came into use after ca 1375 corresponds astonishingly well with a new and different function of the site, the building of the gasthuis hospital in 1371 or 1382. Kampen is thus the first Hanse town in the Netherlands where the location of a medieval shipyard has been established. The dates of the construction of the IJssel cog and the boats and barges from Flevoland and near the IJssel cog[21] show that they cannot have been built at this shipyard.

28.2.4 The shipyard on an old course of the IJssel: a hypothesis

After writing the above in 2016 for the publication on the excavation[22], in 2017 I received the results of the investigation into the maritime landscape context surrounding the findspot of the IJssel cog, boat and barge.[23] The contour map of Kampen in the article together with multibeam data on the IJssel bed shows the location of the shipwrecks and the mid-15th-century bank revetments on the inside bend of an IJssel course, which was given the name Brunneperdiep.

Jager[24] discusses in chapter 6 the town 'terp' of Kampen and deals with the origin and development during the Late Middle Ages on the basis of the results of excavations carried out in various parts of the town until 2014. With the help of three small maps he illustrates the topographical stages with the situations of around 1337, 1390 and 1505[25] (fig. 28.14). He begins the chapter with the sentence: "Kampen's oldest past is difficult to pin down." In section 6.2 the remark: "Typical of the Kampen soil is that there are also fossil beds of the IJssel present". In the period 1324-1337 the old town centre expanded northwards. The old town boundary was probably located along the Gasthuisstraat. Between this street and Houtzagersteeg there was possibly a canal which connected the Burgel to the north with the IJssel, so that the urban agglomeration was situated on an island.[26] The combined AHN multibeam map with the reconstructed course of the Brunneperdiep (figs 23.6 and 28.15) shows that the IJssel cog lay on the south side against the bank which in around ca 1450 lay in the middle of the present-day IJssel. This bank bends away upstream in the direction of the Van Heutszplein.

Although at the moment the hard evidence for this is lacking, on the basis of current knowledge and insight (March 2018) it does not seem too bold to assume that the Burgel in the 11th and a large part of the 12th century roughly followed the present course of the Burgel upstream from its beginning at the IJssel/the Graafschap, curved north at the Karpersteeg and subsequently diagonally over the Botervatsteeg, the Van Heutszplein and right over the present course of the IJssel into the Ganzendiep.[27] Roughly at the Van Heutzsplein was probably the branch of the Brunneper- diep. It is even highly probable that the Burgel was not a tributary of the river but the then main stream of the IJssel. At Wilsum it flowed quite close to the 11th-century church, then straight to the north and subsequently as the most northerly route along the Scherenwelle. From there it flowed westsouthwest as far as the later Venedijk, then with a large meander into a part still known as Oude IJssel (immediately south of the Pieter Zandt school), joined the present course of the river, meandered further in a northerly direction and then curved away at the Graafschap/Burgel. In this hypothetical situation the foundation of the settlement of Kampen would not have taken place on the left, but on the right side of the river IJssel, therefore literally in the Oversticht. This would explain the odd position of the Overijssel Hanse town on the Gelderland side of the river (fig. 28.16).

In relation to Speet's summary,[28] I assume that the oldest settlement of Kampen should not be sought around the Sint-Nicolaaskerk, but along the bank of the IJssel. This oldest settlement was characterized by ribbon development on small terps located along the river which grew together

20. Speet 1986, 12.
21. Second quarter 15th century. Chapter 23 and sections 26.1 to 26.4.
22. Vlierman. In Klomp 2018, 257-267.
23. Vlierman & Waldus 2017, 84-89.
24. Jager 2015, 129-152.
25. Jager 2015, 148, figure 56.
26. Jager 2015, 158.
27. The Ganzendiep is described as 'De IJssel' from the present-day IJssel to Grafhorst and beyond (the fork between Ganzendiep and Goot) in the Cadastral Atlas Overijssel 1832 on the Minute Plans of Kampen, IJsselmuiden and Grafhorst and on the Topographical and Military Map of the Kingdom of the Netherlands (scale 1:50.000) from 1859. The further course of the present-day IJssel is referred to as 'Regterdiep'.
28. Speet 1986, 12.

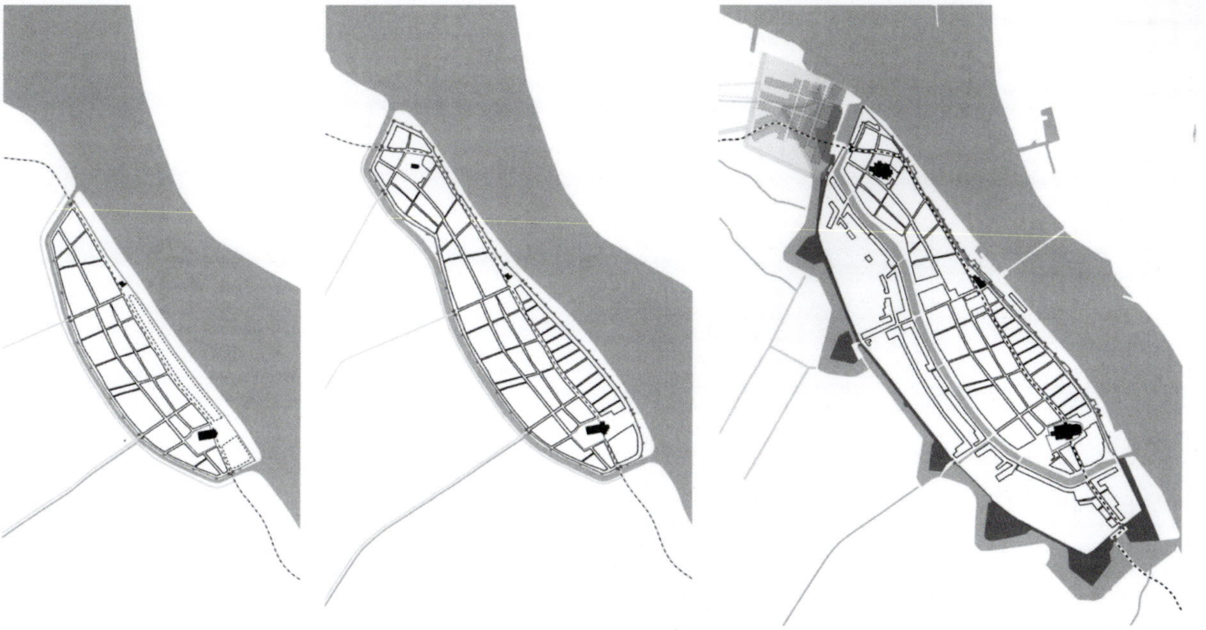

Fig. 28.14. The evolving layout of the city of Kampen: around 1337, 1390 and 1505. After Jager 2015.

Fig. 28.15. The combined AHN and multibeam relief maps with the reconstructed course of the Brunneperdiep stream. Map: ADC.

Fig. 28.16. Presumable geographical setting of Kampen, ca 1175: the IJssel riverbed running along the Burgel, Botervatsteeg and Ganzendiep, the branching of the Reeve at the Geerstraat, and the settlement with its northeastern moat coinciding with today's IJssel course. Projected onto the 'Topographical and Military Map of the Kingdom of the Netherlands' 1859 (1:50,000).
1. The town of Kampen. 1A. the northeastern moat of the settlement coinciding with the current riverbed. 1B the Oude IJssel watercourse. 2. Brunnepe. 2A Brunneperdiep watercourse. 3. hamlet Het Oenen. 4. Zalk. 5. motte castle Buckhorst near Zalk. 6. motte castle Voorst near Westenholte. 7. Wilsum. 8. IJsselmuiden. 9. Grafhorst. 10. Ganzendiep watercourse. 11. Noorderdiep watercourse. 12. Zuiderdiep watercourse. 13. Reeve watercourse. 14. De Enk watercourse. (Map: K. Vlierman 2018)

to form a dike (Oudestraat). The difference being, however, that the IJssel did not follow its present course, but that of the Burgel. The fishing village of Brunnepe, that possibly also existed at that time, did lie on the Gelderland side and on the Brunneperdiep.

The marshy area on the south side of the settlement was bordered on the southwest side by a stream that ran almost as far as the Oude IJssel, the Enk. This flowed into the Zuiderzee past Kamper Nieuwstad. On the east side the stream continued as far as the hamlet. Het Oenen/the present route of the N50. During storm surges the water in this stream ran on into the IJssel, and during high river water levels probably also in the reverse direction. The Enk formed the natural boundary with Gelderland. At the end of the 13th century dikes began to be constructed near Kampen. Around 1302 the St. Nicolaasdijk and the Zwartendijk were in existence and somewhat later the Slaper. In 1345 the Zwartendijk was extended to the Enk and connected via the Slaper and Het Oenen with the Venedijk. In the second half of the 15th century Kampen and Kamperveen decided to build a dike between the Slaper and the Hogeweg, which was completed in 1478. This Enckdijk was storm-swept repeatedly by the Zuiderzee.[29] Before ca 1340, the county of Gelre (Gelderland) could only be reached overland from Kampen by means of a ferry across the Enk at Het Oenen. During the archaeological investigation as part of the Ruimte voor de Rivier (Space for the River) project/construction of the Reevediep, ADC-ArcheoProjecten discovered possible landing stage/jetty structures close to the N50 perhaps connected with this ferry. However, dendrochronological analysis[30] has revealed that these were 16th-17th -century structures. The polder area (Buitendijks and Dronthen) to the west of the Zwartendijk was protected on the sea side in 1597 by the Dronther Zeedijk[31] and was bordered on the southwest side by the Enk. This polder land near Kamper Nieuwstad extended at least 6 km westward in the period from the 12th to the beginning of the 14th century. On the northern part of this peat coast/barrier which ran from Schokland to Elburg, there were possibly several small settlements in the Carolingian period and in the 13th century.[32]

29. Jager 2015, 87 and 92.
30. Van Daalen 2017.
31. Van Dalen 2017, 93.
32. Chapter 11 De Dronten-I cog, section 11.1.3 and fig. 11.5.

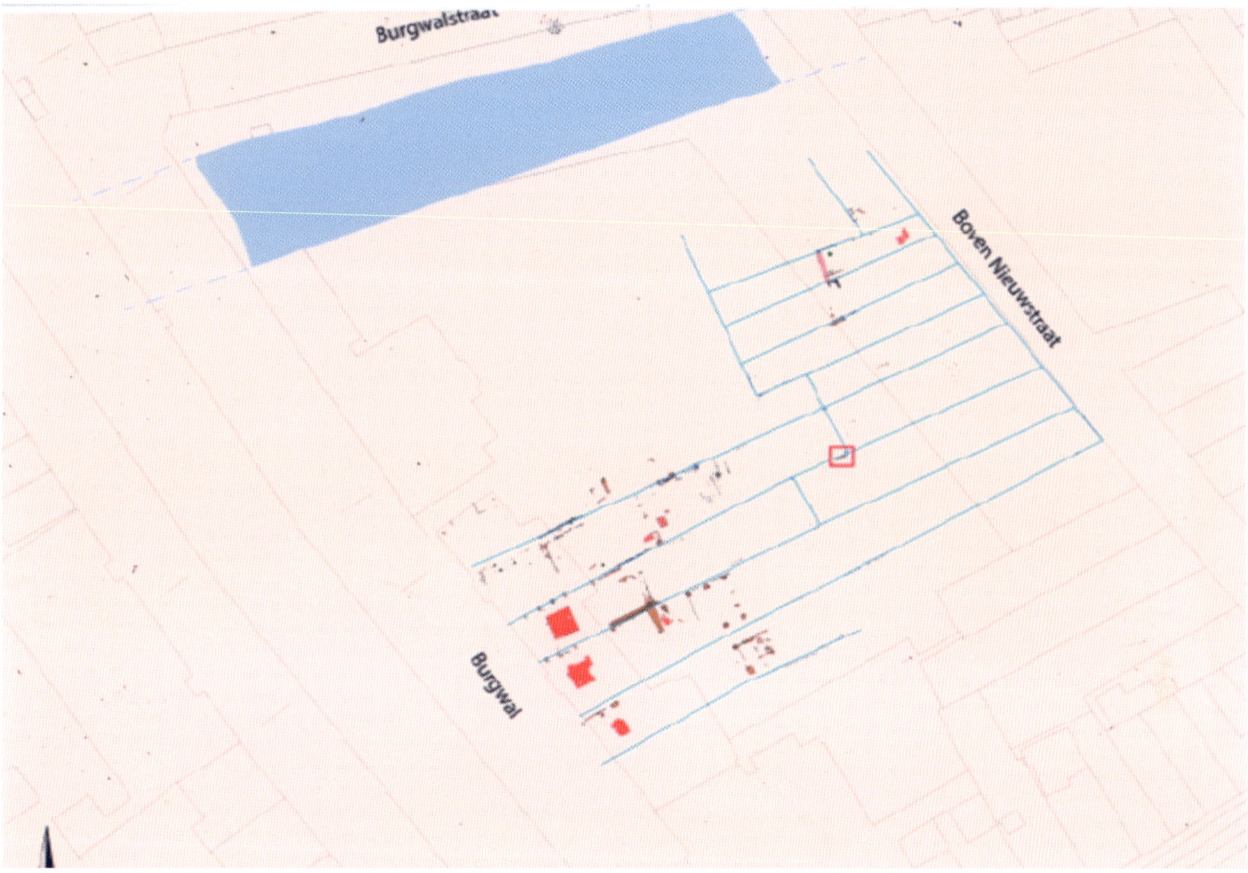

Fig. 28.17. Plan of the excavation site Geertruidengasthuis with the revetted watercourse on its northern boundary. Map: Municipal Archaeological Service of Zwolle.

In section 28.2.3 mention was made of the Oude Trekvaart/Reeve, the waterway that branched off from the Burgel to the southwest in the direction of the Zuiderzee. If the Burgel, as assumed above, was actually the (stream channel of the) IJssel in the 11th and 12th centuries, it seems very probable that the river forked at this spot and that there was originally no dug watercourse to the Zuiderzee, but a natural stream that was easily navigable.[33] Soil borings along these routes could, as has already been remarked, shed more light on this in future.

In 2017 I became aware of the results of the archaeological excavation of 2011 at the north part of the Geertruidengasthuis (hospital).[34] The forged iron finds of that date relating to the shipyard were processed together with the 2014 finds.[35] On the south side of the Burgwalstraat a watercourse c. 14.5 m wide on average was observed parallel to this street (fig. 28.17). In excavation pit 2[36] revetments were found on either side of it. Jager[37] considers the watercourse to be a connecting canal between the Burgel and the IJssel and part of the moated (oldest) settlement. During the archaeological excavation the quay revetments could be traced over a length of ca 70 m (from the Burgel to nearby the Boven Nieuwstraat). On the north side the revetment consisted of a low plank wall that was kept in place with short posts driven into the ground. On the south side a sturdy quay structure had been built consisting of vertically driven posts. These were in pairs consisting of one oak post rectangular in cross section and one round ash post. Along the canal two rows of posts were driven and these were connected breadthwise by heavy oak partitions. The partition beams on the canal side partly reached beyond the beam so that heavy oak planks could be attached to them. The plank structures

33. Schilder 1983, figure 106. Digestum Vetus: Sabbato post Petri ad Vincula (7 August 1456). While the Sheriffs, Councillors and sworn commoners were convening in the Town Hall, six ships full of Utrecht soldiers came sailing up the river Reeve. The sworn commoners then demanded of the Council that the alarm bell be sounded. However, nothing happened. The City of Kampen signed up 150 of the soldiers, while the remainder sailed on to Deventer. It does not mention whether the journey was continued by ship, which seems most probable. This could mean that the voyage over the Reeve from the Zuiderzee via the Burgel to the IJssel was still possible without hindrance (via one or two locks?).
34. Bouma 2012.
35. Vlierman. In: Klomp 2018, 257-267; Vlierman this section.
36. Bouma 2012, 30-31.
37. Jager 2015, 143-144.

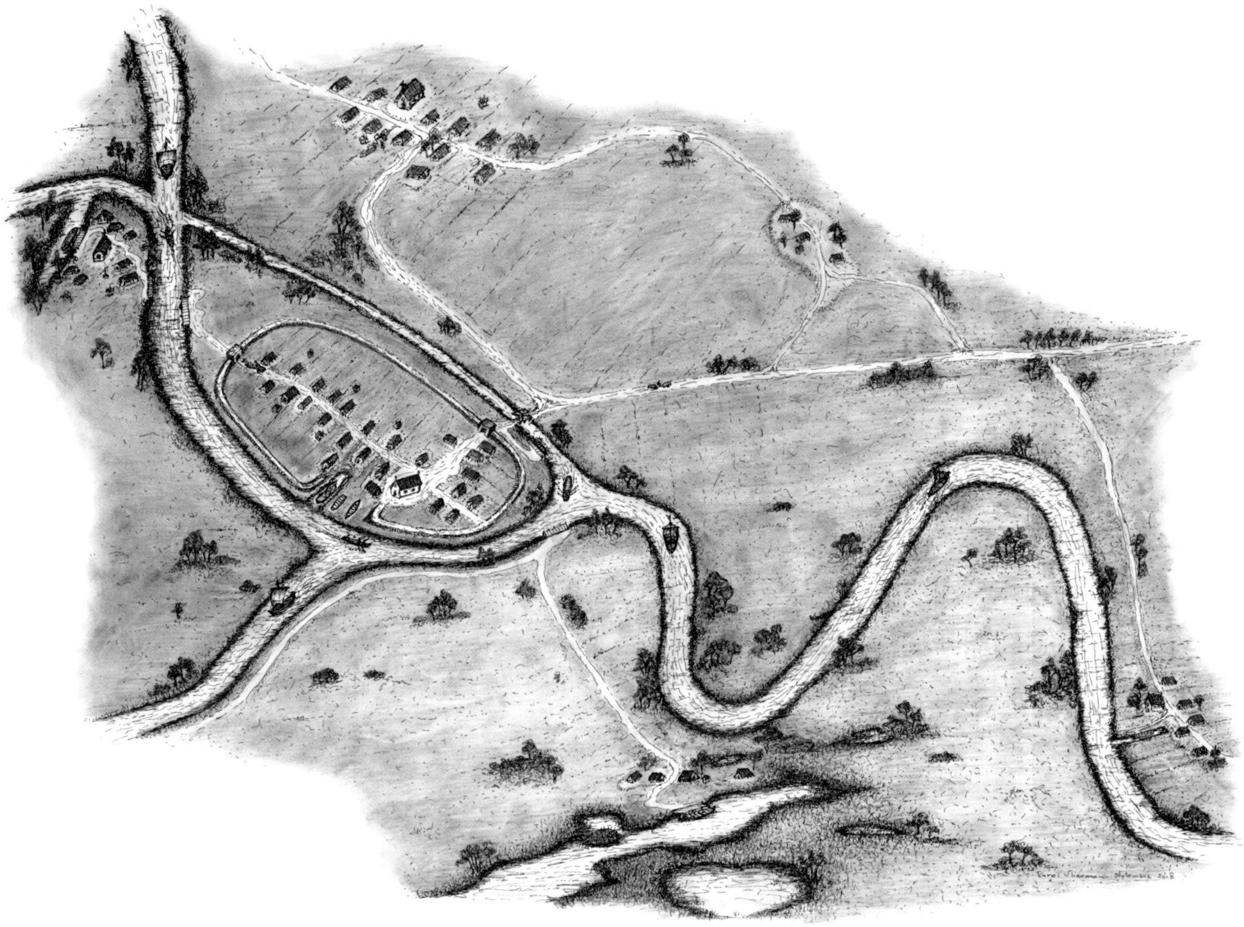

Fig. 28.18. Hypothetical reconstruction of geography of the town of Kampen with the shipyard to the right of the old course of the IJssel ('Burgel'), which follows the Botervatsteeg and the Van Heutzplein before entering the Ganzendiep, and when it passes the Geerstraat splits into the Reeve, ca 1175. Drawing by the author, 2018.

were partly nailed, and this formed a quay where ships could be moored. At the bottom, the revetment consisted of sturdily attached wickerwork. The raised layer of the shipyard site against the top of the revetments contained early-14th-century archaeological finds. At a certain point the canal was filled in. The layers of fill in the canal have been established with certainty as far as 1.60 m-NAP (Normal Amsterdam Level). The bottom of the canal may possibly have been up to ca 2.80 m deep; below this, there is peat.[38]

It is possible, as Jager assumes, that the watercourse formed part of the surrounding moat of the oldest settlement. My view, however, is that it was a widely dug watercourse directly connected to the shipyard, which did not run (much) farther than the present-day Boven Nieuwstraat and that the surrounding moat of the oldest settlement consisted of the part of the IJssel (Burgel) up to and including the Nieuwe Markt and run north of the Gasthuisstraat to the present IJssel.[39] More likely perhaps is that the most northerly surrounding wall lay there and that the enclosing ditch followed the Burgel (IJssel) via the Karpersteeg/the Van Heutszplein/the Ganzendiep and that the northeast part of the canal was a dug watercourse which for example ran roughly in the middle of the present course of the IJssel from the beginning of the Burgel as far as, say, the Van Heutszplein. The Zwolseweg ended at the possible and perhaps built slightly more to the east predecessor of the Koornmarktspoort.

Assuming that the Burgel originally did form the main stream of the IJssel in the second half of the 12th century, it would be natural to expect that at catastrophically high Almere-/Zuiderzee water and/or river water the northeast part of this canal would start to form the beginning of the present course of the IJssel along Kampen. The location of the shipyard, slightly beyond the fork with the Reeve on the bank of the northwardly curving IJssel

38. Bouma 2012, 30-31.

39. Van Winsen & Van Velzen 2012, 24 (map with the shape of Kampen ca 1324).

and outside the oldest surrounding wall was a logical and practical choice.[40] When the settlement/town expanded in a northerly direction the canal next to the shipyard was kept on until the site was given a new function (Gasthuis hospital). The established length, breadth and depth of the watercourse was in the 13th and 14th centuries very probably suitable for mooring even (several comparatively) large cogs (fig. 28.18). Apart from cogs, the shipyard on the Burgel/IJssel in Kampen appears to have been a place where inland vessels and barges were built and maintained.

28.2.5 The sailcloth[41]

In section 28.2.2 the discovery of a crumpled bundle of tarred sailcloth was mentioned and illustrated (fig. 28.13). It was found in context 49, a dungpit. Location: pit 1, find-numbers S78 and 482. The dungpit lay under pit S58. The context date is 1325-1375.[42] During the excavation in 2014 it was initially assumed that it was perhaps a comparatively recently buried piece of linoleum and an attempt was made to remove it from the excavation level. As a result the bundle was damaged and the cloth which was in lengths of fabric partly rolled or folded up was torn and frayed in several places. Shortly after its recovery it was established that it must in all probability have been an old and discarded piece of cloth with, in some places, the lengths of fabric, which had originally been sewn together with light-coloured thread, still in place. It was later observed that remains of the sewing thread were visible along almost all the other parts of the seams and selvages.

In October 2018 the municipality of Kampen made it possible for the bundle to be disentangled, cleaned and for the piece of sailcloth to be reconstructed as far as possible. Archeologie Zwolle arranged the facilities necessary, while garden centre H. Palland in Kampen made available a suitable cleaning table measuring 3x1x0.8 m. On this table the approximately 650-year-old hard bundle could be soaked in water and the surfaces which had stuck together ('petrified') carefully separated and folded out. The

Fig. 28.19. Sailcloth. a. An 'unfolded' fragment of the woollen fabric prior to further cleaning. b. A piece of sailcloth being brushed clean. c. Rinsing down the brushed fabric fragment. d. View of the cleaned bundles. Municipal Archaeological Service, Zwolle.

40. Compare the topography of Wilsum and the watercourse connecting it with the river IJssel. (figs 28.16 en 28.18).
41. Identification fabrics, weaving techniques and similar textile finds Sandra Y. Comis.
42. Klomp 2018, 99.

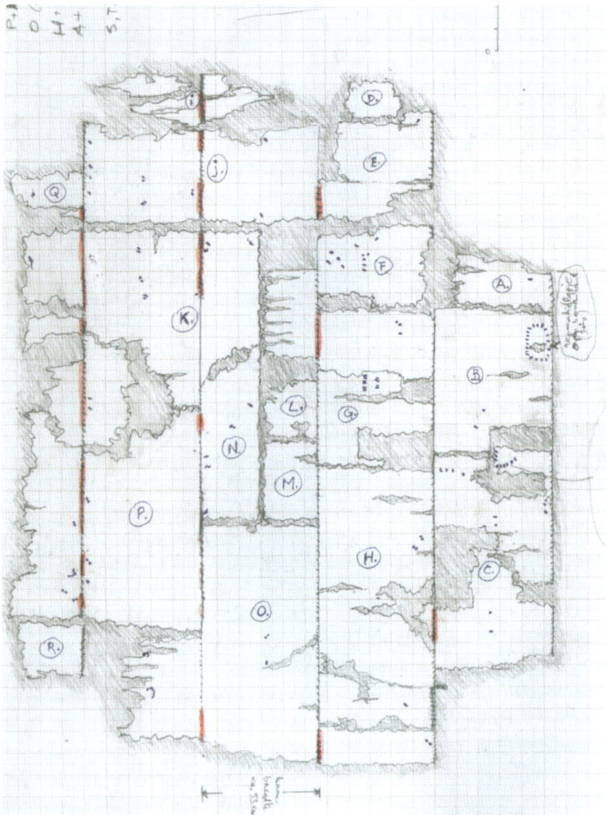

Fig. 28.20a & b. The composite piece of woollen sailcloth and the working drawings of the composite fragments of the woollen sailcloth. (Photo Archaeological Service Zwolle).

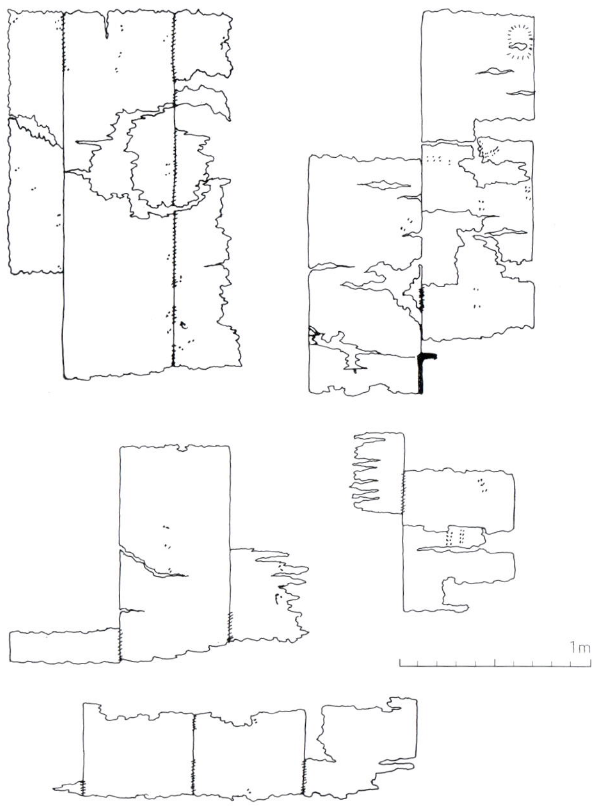

Fig. 28.21. The pieces of sailcloth that were still partially stitched together.

fragments which gradually came free were then mostly brushed clean on both sides (not all the tar remains were removed [figs 28.19a-d]). It quickly became clear that this was not only a brown and rather coarse woollen fabric, but that in several places there were also pieces of a light-coloured finer fabric in between. The fragments of textile belonging together were transferred in a wet condition and in the order that they had been in the bundle to the quarters of Archeologie Zwolle. On a large improvised table the fragments were subsequently laid out and flattened as far as possible, with the pieces still attached to each other forming the starting-point The lengths of fabric (so-called kleden in old Dutch) are ca 53 cm wide. The remaining loose parts of stripes were then laid against the other fragments on the basis of the shape and the selvages. This was done in order to gain an overview of all the fragments and of the total surface area. When laying out the fragments it was assumed that the outer side of the layers of cloth in the bundle was still the same side of the sailcloth. For further investigation of the fabric and for making reference notes, the various fragments in the working drawing were given a letter code (A to R). The fragments that are still attached together have one letter, with the exception of N/P.

In the case of one fragment that was still sewn together (J) it could be established with certainty that the piece had originally consisted of at least four stripes. In three fragments (K, N/P and O) of three. One fragment of a stripe appears to have been cut through lengthwise and roughly in the middle, while another length of fabric has a frayed edge and is still ca 30-35 cm in width. The latter fragment led us to assume that the sailcloth possibly originally consisted of at least five lengths of fabric. The drawing and the photograph show the composite fragment (fig. 28.20a and b). Despite the fact that the various fragments could not be fitted together with certainty, it gives a good impression of what it might have looked like. The assembly shown measures a maximum of ca 3.20x2.50 m (ca 8 m^2). Based on the nature of the 1.5-2 mm-thick fabric plus the fact that the fragments were found together in the ball/lump, it may be assumed with great certainty that this was originally one single piece of sail. Figure 28.21 shows the fragments that certainly belong/are sewn together.

All the fragments of the woollen lengths of fabric are frayed on top and bottom, so the seam was not originally finished. It is therefore not possible to establish an exact original length. If the original piece of sailcloth consisted of five lengths of fabric it would have had a width of ca 2.65 m. In addition to the remains of the light-coloured thread with which the lengths of fabric were sewn together with an overhand stitch and single thread (fig. 28.22), approximately 1 cm long pieces of thread can be seen in various places on the right or top side of the cloth. Both ends of this thread have been sewn through the cloth and in some places are still knotted together on the wrong

Fig. 28.22. Two lengths of sailcloth sewn together with an overcast stitch.

Fig. 28.23. Two lengths of stitched through linen yarn, one with a knot.

Fig. 28.24. A worn area in the woollen fabric and the stitched trough linen yarn with which the repair patch (now lost) had been sewn on. Also the cut-off(?) thread along the selvedge/the thread attaching it to the adjoining length (see fig. 28.25).

Fig, 28.25a. The still stitched-on fragment of cable.

Fig. 28.25b. The hemmed edge with a double linen yarn and at right angles to it, the protruding hemmed edge of an adjacent length (?).

Fig. 28.26a & b. The two largest pieces of linen sailcloth and another part with a seam (drawing), and a photo of the larger piece.

or reverse side (fig. 28.23). In all probability the piece of linen sailcloth which will be described below was attached to the underside of the woollen cloth in this way. Unfortunately this could not be established with certainty, nor could a clear pattern of attachment. The sewn attachments are found scattered over the entire woollen cloth. In some places in the fragments B, C, F and G linen attachment threads of minor repair fragments can be observed. These (missing) repair fragments were most probably sewn over worn patches in the woollen fabric (fig. 28.24). Approximately in the middle of the length of fabric on the right (C) the end of a ca 1 cm thick bundle of threads has been sewn on the top side with a double thread (fig. 28.25a). The rest of the bundle is missing. A loose piece of a woollen bundle of thread was also found between the textile fragments. At the lower end of fragment H a ca 20 cm long edge can be seen which has been hemmed with a linen thread (fig. 28.25b). There is also a hemmed fragment which possibly served at right angles to the selvage as a finish of a(n) (outer?) length of fabric originally ending there. In general, the overhand stitches are comparatively irregular, particularly so in the case of the repair fragments originally sewn on.

Between the parts of the woollen cloth that were crumpled up together were various fragments of a thin (1 mm), finely woven and light-coloured sailcloth (S to V. The other fragments are collected under the letter W). It did not prove possible to fit the fragments together again but they undoubtedly all belonged to a single piece of sail. As

Fig. 28.27a. The hemmed edge of the linen sail with the sewn-in rope.

b. Fragment of the linen sail with the overlapping hems of two adjacent lengths joined by a hem stitch on two sides.

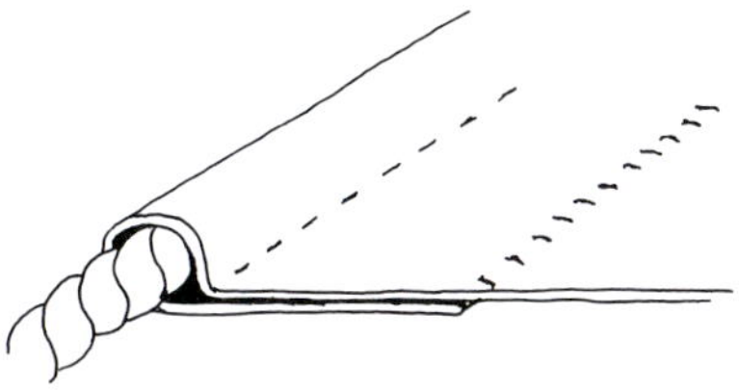

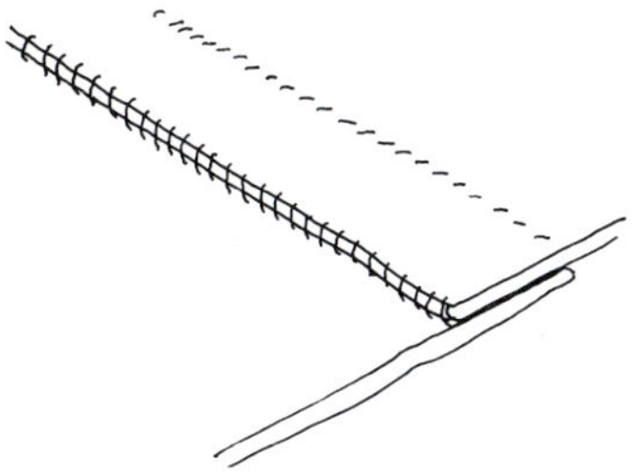

Fig. 28.27c. Construction of a and b.

mentioned above, it could not be established with certainty that it had been sewn to the reverse side of the woollen sail. However, the stitching sewn in the woollen cloth makes it almost certain that this must have been the case.

The two largest pieces measure ca 90x45 cm and ca 68x58 cm respectively (fig. 28.26a and b). The piece first mentioned (S) has one frayed edge. The other side of the strip still has the selvage. This seam is folded ca 6 cm and attached to the reverse of the sail with an overhand (sailmakers) stitch. At the folding line of the folded part there is a 3-mm-thick cord on the inside. The cord is secured with a thread which has been tacked (interval between stitches 3-5 mm) through the double cloth. Such pieces of cord or rope (a so-called leech) are sewn in or on the outer edge of a sail to reinforce it and to prevent it from flapping. The strip including the seamed edge is still ca 53 cm wide. The other piece (V) is frayed on all sides, no selvages can be seen. This means that the length(s) of fabric must have been at least ca 60 cm wide. There are some fragments in which the join between the two lengths of fabric is present. The selvages overlap each other 0.5 cm and are sewn together on both sides with a hem or sailmakers stitch (figs 28.27a-c). The stitches are sewn every 5 to 6 mm. Compared with the sewing of the woollen fabric, the join here is much finer. In places where this join between two stripes has been observed it was not clear whether this was a piece of sailcloth made up of for instance two (or more) stripes of full width put together or that it was a single stripe with the remnant of a second. Based on the preserved fragments it could in the latter case have been roughly one piece of sailcloth ca 2.5 m long and at least ca 70 cm wide. It is not inconceivable that the piece of cloth was originally larger, (almost) as large, for example, as the woollen sailcloth, or that several pieces had been used which were not found.

The heavily tarred,[43] stiff sailcloth could hardly be rolled up or folded into a manageable package which could be deposited in the refuse pit. It was probably cut up into pieces (largely along the seams). On the basis of the hard, homogeneous mass of initially melted tar found in some of the doubly folded fragments, it appears very likely that they were transported warm, in other words straight from the tar or tanning kettle to the refuse pit, crumpled up and covered with a layer of soil.

43. The type of tar has not yet been identified. Sails and nets were (later?) also tanned. Tannin was originally a product of oak bark which was also used in leather tanning. Later, cachou was used, sap from a tree of the same name. The sails and nets were boiled in a tanning kettle to extend their durability (Sopers 1974, 26).

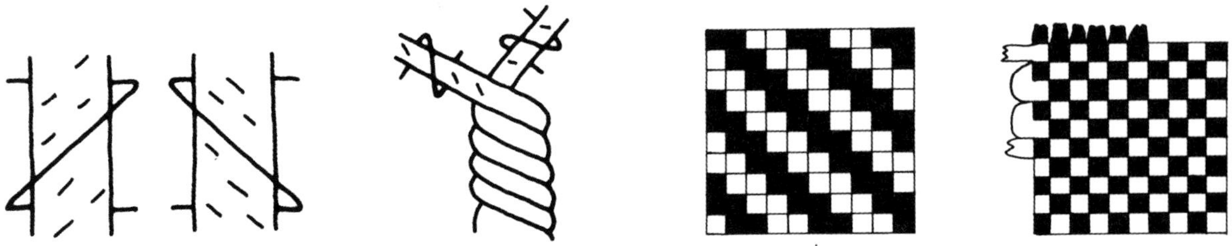

a b c

Fig. 28a. On the left a S-Z twine; on the right a S (ZZ) twine. b. fourshed twill; the warp is black coloured, the weft is white. c. plain- or tabby weave.

Woollen cloth and weaving technique

The whole package eventually proved to consist of over 40 fragments of a brown woollen fabric. For this fabric Z and S-twisted threads were used; this means that the direction of the fibres twisted together corresponds to the oblique line in the capital letters Z or S (fig. 28.28a). This fabric consists of two systems of threads crossing each other at right angles. This is called the warp, consisting of threads running lengthwise in the fabric and the weft with threads running breadthwise. The woollen fabric is woven in four shed 2/2 twill weave in which the weft thread is passed over two and under two warp threads. In every row the weft thread skips one warp so that the binding points form a diagonal line in the fabric (fig. 28.28b).

In the warp direction there are 10 to 12 thin Z-spun threads per cm. It is remarkable that in most of the fragments there are 6 to 8 thicker S-twisted weft threads per cm, so that this fabric belongs to the category of medium-fine textiles. On closer inspection, however, there proved to be Z-twisted weft threads present in four of the fragments. Initially it was thought that there must have been two different fabrics here, but in fragment J it was clear that S-twisted weft threads were first used for weaving and at a certain stage Z-twisted weft threads were used. The fragments A, E, F and two strips of O have Z-twisted weft threads.

Most fragments have two selvages, the places where the weft changed direction during weaving. As a result, it was possible to establish the width of the fabric as being 52-54 cm. This varying width may be due to the weaving, or to tar remains still more or less present. At any rate, the average width of the runners was ca 53 cm. This indicates the use of a loom that could be it operated by a single person. The 0.5-cm-wide selvages each consist of three double warp threads in four shed 2/2 twill weave.

In various places the woven strips are sewn together with a 2-mm-thin S-plied linen thread, formed by twisting two Z-spun threads together in an S direction. The fabrics were laid on top of each other with the right sides facing and sewn along the edges with large overhand stitches. After folding the seam open a 0.8-cm-wide strip with stitching was visible on the underside. The distance between the sewing holes is 4-7 mm.

In fragment P there is a small cross seam. The cloth is folded back over a length of 30 cm. There is no stitching present. Both the thickening of the edge of the fabric and the stitching of fragment H indicate that there is a seam here of at least 18 cm in length. The cloth is folded back 0.6-0.8 cm and secured both on the upper and underside with linen thread overhand stitches. Where this seam ends there is a bundle of threads bound with linen sewing thread at right angles to the seam. This is possibly the remains of another fragment of fabric.

In various parts of the cloth the linen thread remains mentioned above are present at places where a cord (?) or repair fragment was attached on one side. Because these are no longer present, only projecting threads can be seen on the upper side. In fragment B the thread remains point to the original presence of an oval repair fragment 21 cm long and 10 cm wide. This repair fragment is now missing. On the underside of fragment C there is a 1.2-cm-thick bundle of linen threads secured with linen thread. This bundle consists of at least three cables.

Among all the fabrics a woollen cord, without any tar remains, was also found loose. This 22-cm-long and 1.1-cm-thick cord consists of ca 10 S-twisted threads which are eventually plied together in a Z direction. This possibly served to attach the cloth to something.

Linen cloth and weaving technique
Among the fragments of woollen cloth there were about twenty fragments of a linen fabric with measurements varying from 6x7 to 58x68 cm. The latter size means that the width of the cloth was at least 58 cm. For this fabric 15 Z- and 15 Z-twisted warp and weft threads were used. These textile finds have a linen or plain weave in which each weft thread goes over and then under a warp thread (fig. 28.28c). On the long sides of some fragments there is a selvage. No particular measures were taken to reinforce this selvage. In five fragments there is a seam. This is formed by laying two strips with selvages on top of each other over a distance of 0.5 cm and attaching them both on the upper and underside with a linen thread and overhand stitches. This thread is 0.3-0.5 mm thick and consists of two Z-spun threads twined in an S direction. The distance between the sewing holes is 0.5 cm (figs 28.27b and c).

Along one fragment (S) There is a 6 cm wide and (still) almost 40 cm long seam made by folding the cloth with selvage and securing it with overhand stitching. There is a 3 mm thick linen cord present in the fold. This cord (leech) consists of three S(Z,Z)-plied threads which are later twisted together in a Z direction.

Colours/dyes and state of preservation
Until the end of the 19th century textiles were dyed with natural/vegetable dyes that are not colourfast. Moreover, these dyes dissolve fairly quickly in acid soils, after which the fabric takes on the brown colour of the soil. As a result of being treated with tar probably more than once, the woollen fabric now has a dark-brown to black colour.[44] The linen cloth was originally white to ash-grey, but is now light-brown due to the same treatment. The state of preservation of the various fragments of sailcloth is good. However, the woollen fabrics are now fragile since most of the tar has washed out and the natural lanolin has gone. Consequently these fragments must be stored flat and not folded,[45] otherwise the fabric will eventually break. It may be considered exceptional that the linen fabric has been preserved. Normally speaking, it is mainly the woollen and silk fabrics made from animal fibres that are preserved in the predominantly wet and acid soils of the Netherlands. Vegetable fabrics on the other hand decay quickly due to bacteria and fungi. The most important factor influencing the condition of both types of material/fabrics is the plentiful and probably repeated treatment with tar.

Similar textile finds from Rotterdam and Zwolle
The discovery of the numerous woollen fragments with typical seams proved to have parallels from Rotterdam, which, though they date from the fifteenth century, are virtually identical.[46] During the Markthal excavation in Rotterdam, four very large and five smaller fragments of woollen fabric were discovered, which, like the finds from Kampen, are woven in a four shed 2/2 twill weave with 10 thin Z-twisted and 4 thick S-twisted weft threads per cm. These fragments are also very large and soaked in tar. The largest fragment is 105 cm long and 67 cm wide. One of the fragments has a narrow selvage made up of thinner warp threads, and the warp threads there are doubled. On the underside of many fragments the remains of a similar type of stitching are still visible. Although the sewing thread has disappeared the impressions left by the linen (?) thread can clearly be seen. These finds were discovered in the vicinity of the harbour and at the time it was suggested that they were parts of a tent-cloth, covering or sail.[47] In one fragment a 4-6-mm-thick woollen Z(S,S)-twisted thread still projects.

The linen finds, on the other hand, are comparable to various finds from different findspots. For example from Zwolle, where during the Praubstraat excavation four linen fabrics from the fourteenth and fifteenth centuries were found, two of which also had 5 Z-x 15 Z-twisted warp and weft threads.[48]

The presence of textile with tar is also known from various findspots and sites where, or near where, there were activities connected with shipbuilding and maintenance.

In section 28.2.2 it was mentioned that in the Nijkerk-I and -II cogs several sizeable fragments (ca 80x60 cm) of tarred woollen sailcloth were found. Smaller fragments were discovered in the Marknesse and Spakenburg-I cogs and in the boat of lot K 73/74.[49] They are described in the relevant sections.

Original function and reconstruction
Despite the fact that there is no certainty on this point, it does not seem very likely that it is a piece of sail with which a cog, for example, or other vessel was propelled.

44. There may (in addition) also be a natural colour brown. In the Nijkerk-I cog (section 8.5.2) a fragment of tarred woollen sailcloth was found in which there are light-coloured strips with 6-8 warp threads alternating with wider brown strips.
45. All the textile fragments were placed in a specially constructed wooden chest between acid free paper and in December 2018 they were stored in the archaeological depot in Kampen.
46. Comis 2013, 418 and 423-425.
47. Comis 2013, 418.
48. Vons-Comis 1987, 43-45.
49. Sections 8.5.2, 6.5.3, 12.5.2, 14.5.2 and 26.1, respectively.

Fig. 28.29. Flemish miniature with a covered boat. Photo The Golf Book Bruges ca 1520-1530. British Library/Flatman 2009, 115.

The piece of linen cloth almost certainly was part of a sail and probably that of not too large a ship. Up to now it is the oldest known piece of late-medieval (ship's) sail made of this material. The most probable explanation is that, in the case of the Kampen find, these are fragments of the largest part of a woollen sail used to cover a tent-like covering of a vessel. Because the comparatively coarse and probably old fabric would constantly have lost particles of the tar layer in a dry condition and in wind and was possibly not entirely waterproof either, the finely woven linen sailcloth was attached to (part of) the underside and the whole thing then tarred.

In section 26.4 the probable function of several late-medieval (ferry) boats from Flevoland and from the IJssel at Kampen is discussed and the assumption is made that a number of the round holes made vertically in the covering inwales must have been intended for holding, for example, curved willow branches as part of the structure/support of a covering or tent-cloth. The remarkable two 'mast steps' in the boats fixed behind each other were almost certainly not meant to hold masts, but upright posts which were connected with a horizontal beam or beams over which sailcloth was laid to form a kind of medieval party tent. One of the miniatures from ca 1520-1530 depicted in section 26.4 shows a covering of this type with a sail partly rolled up or folded back (fig. 28.29).

The boats and barge from Flevoland and near the Kampen (IJssel) cog were built in the first half of the 15th century. The design/construction of the vessels indicates that this type must already have existed for a long time. The two 13th- or possibly even 12th-century barges from Antwerp[50] corroborate this. The two 'mast steps' are ca 1.80 m apart in the Kampen boat, and 2.30 m apart in the boat from Flevoland. If these vessels had been equipped with a tent/cover as described above, it would have consisted of length of fabrics running over the width of the boat. In the case of the sail from the shipyard at Kampen that would mean that it must have been made of 4 to 5 lengths of fabric. Based on the persons depicted in the miniature, the length of the covering would appear to be ca 2 m, while the headroom under the sailcloth would be 1.50-1.70 m. The barges are ca 2.30 m wide over the gunwales. If there was a similar sail, a length of 3-4 m would have been sufficient to cover the entire tent/cover as far as the gunwales. The sailcloth may have been knotted to a horizontal beam with the piece of woollen cable still attached roughly in the middle of length of fabric B/C. There must also have been a similar cable on the other side as well as at the seams of the upper and underside in order to attach the cloth to the vertical supports on top of the gunwale. One or two horizontal beams would probably have been required to carry the heavy, tarred sail weighing ca 15-25 kg. The sailcloth may also have had toggle-like fastenings. Figure 28.30 gives an impression of what all this may have looked like. The old, discarded, and originally probably about 8-10 m^2 in size tent cloth consisting of woollen and linen fabrics is in all probability the largest piece of late-medieval textile discovered during an archaeological excavation in the Netherlands (Europe) to date.

50. Section 26.7.2, vessels 3 and 4.

Fig. 28.30. Artist's impression of the sail on a 'Kampen' boat.

Results, analysis, summary and conclusions

IV

29 Results and analysis[1]

This chapter presents the findings as known by late 2014, when the manuscript was completed. The information about the Kampen (IJssel)-cog and the boat and barge found with it had been taken from the report on the preliminary investigation of October 2012.[2] This also goes for the data and measurements in tables 1 and 5 (in sections 29.3.1 and 29.3.3, respectively). The subsequently revised data and measurements appear in the updated chapter 23 and in sections 26.2 and 26.3.

29.1 Construction methods of (seagoing) vessels in North-western Europe in the Late Middle Ages

The investigation of shipwrecks in the Netherlands with the construction characteristics of the Bremen cog yielded the surprising conclusion that nearly all shipwrecks from ca 1250-1450 found so far in the Low Countries can be considered part of a characteristic construction method that was established by researchers as early as 1985. They were mainly found in the former Zuiderzee region. Most of them are small cogs and boats. It can in my opinion be assumed that these craft, as well as several of the large seagoing vessels that were found, were most likely built in the region(s) along the eastern coastline of the Zuiderzee and the river basin of the IJssel. In the case of the other large cogs, the lower reaches of the major rivers that flow along the Lower Saxon North Sea coast should primarily be considered. The other Dutch finds were built according to the Nordic or Viking shipbuilding tradition, or originate from it. They are probably all from regions outside the Netherlands. In North-western Europe between ca 1000-1450, judging by the archaeological finds known to date, seagoing ships were almost certainly built according to two methods.

1. The Nordic ship-building tradition. This method of construction used for the Viking ships was already known in the Early Middle Ages and was almost certainly (later) applied along the coasts of what was once Frisia, on the British Isles and along the Flemish/French Channel and Atlantic coast. The construction method has in principle been used for wooden vessels in Scandinavia since that time.
2. The cog-building tradition. It almost certainly originated in the second half of the 12th century. The last (large) cogs were probably no longer navigating by 1500. The construction hardly evolved at all until ca 1400.

The most striking difference between the vessels from the two building traditions is the hull shapes. A vessel type that was (initially) low and symmetrically built in top view, with curved (and high) rising sterns and stems, was characteristic in the Nordic tradition. The cog-building tradition featured a relatively high ship with a straight and raking stern and stem, which had a top view where the afterbody was less wide than the forebody at a similar distance from the stern and stem. Other features of the Nordic tradition included the completely clinker-built construction, with the planks fastened to each other with rivets, and the overlaps in the first group were pre-treated with caulking made from animal hair. The cogs had a largely carvel-built bottom and clinker-built sides. The clinker-built planks were fastened together with double-clenched nails (folded twice and driven back into the wood). After that, but before the frames were put in place, the seams of the clinker-built planking were made watertight with sintel clamped moss caulking on the inside of the vessel. The seams on the outside of the ship's hull were made watertight in the same way shortly before the launch.

A small number of carvel-built shipwrecks of seagoing vessels have been found, dating from the (late) Roman period and the 7th century and originating from Great Britain, Flanders and France. It is impossible to say whether this method of construction continued in Western Europe until the Late Middle Ages. I am not aware of any archaeological or historical evidence to support that at this time.

In addition to construction methods for seagoing vessels, construction methods for watercraft to navigate rivers and other inland waterways have existed on the European continent for centuries: dugout canoes, log-boats with gunwales, river barges of the type known as barge, and barge-like planking boats. As early as the Early Middle Ages, moss was used to make the seams watertight in these vessels. It was held in place with thin wooden splints or laths, which in turn were fastened with iron clamps (sintel [nail] clamps), a technique known as sintel clamped moss caulking.

1. Translated by Joy Phillips and Xandra Bardet.
2. Waldus *et al.* 2012.

29.2 Other late medieval vessels

The shipwrecks from the Nordic tradition that have been discovered so far were almost certainly not built in the region of what is now the Netherlands, according to my best judgement. The early vessels (before ca 1200) may have been built in the coastal regions of what was once Frisia.[3] Other than that, the regions where these entirely clinker- built vessels were constructed can be found in Scandinavia, the Baltic region, the British Isles and along the Flemish/French coast of the Channel and Atlantic Ocean. The late-medieval *waterschip* – entirely clinker-built and, just like the cogs, with the planks fastened together by double-clenched nails (folded twice and driven back into the wood) and the seams sealed with sintel clamped moss caulking – was probably built in Holland (somewhere near Amsterdam).

Historical sources indicate that there were many different names in use for types of late medieval ships, both seagoing vessels and craft that navigated the inland lakes and waterways. To date, none of these names can be linked with any certainty to the finds in Western Europe, which now number in the dozens. The cog seems to be the only exception.

29.3 The groups of vessels in the cog-building tradition

Even when the first find was published (the Marknesse cog, Chapter 12), Modderman already noted that it involved a small type of cargo vessel.[4] Written sources mention cogs, small cogs and boats.[5] A 'new' element in this study is that the similar, characteristic features of the construction method and components of the various vessels are essentially identical, and that almost all vessels found to date from the period in question and a specific region were built according to this method.[6] As such, they provide the archaeological and structural evidence that there were indeed large and small cogs. It is very likely that the other, smaller inland vessels that had some or all of these features would have been collectively referred to as boats. It has also become clear that there is no definitive cog as a type of ship, but rather as a building tradition according to which these vessels were constructed, although the craft themselves varied widely in function and dimensions. It seems obvious that the large, seagoing cog must have been a prominent and inspiring vessel within that diverse range of cogs, from the perspective of the inhabitants of north-western Europe at that time. Even now, when the world's largest container ship sails into the port of Rotterdam, its arrival is considered big news. One could call the large cogs the first 'container ships' of the Late Middle Ages, or the Hanseatic period. Figure 29.1 shows the contours of all the vessels covered here, depicted at the same scale.

29.3.1 *COGGHEN*: large (Hanseatic) cogs

Among the known shipwrecks with cog features, some stand out because of their length and height. In contrast to e.g. other types of ships from the same period (insofar as this could be determined thus far), the amidships height of the sides of the 14th/15th century vessels is particularly remarkable (4 m or more). This height is also linked to crossbeams that protrude through the shell. The length from stern to stem is 20 m or more, while the length-width ratio across the gunwales does not exceed 3.2:1. The Nijkerk-I cog, which was most probably built towards the middle of the 13th century, was over 18 m in length and had a gunwale height of almost 3.5 m. The dimensions used for such vessels may have been characteristic of that period and/or in a specific region. In any case, it certainly seems acceptable to assume that the vessels with cog-like features, crossbeams protruding through the shell, the specified dimensions, and a length-width ratio over the gunwales of 3.2:1 or less, are the *COGGHEN* mentioned in the written sources: the large, seagoing (Hanseatic) cogs.

Unfortunately, there have not (yet) been any archaeological finds of 12th/13th-century vessels with crossbeams protruding through the shell along the English and French/Flemish coasts of the Channel – the vessel (referred to as a *nef*), that is probably depicted on various contemporary city seals. As far as the dimensions of these ships are concerned, it is therefore impossible to say anything with certainty, but I do not think they would have been much different from those of the large cogs.

In the Netherlands, Germany and Flanders, five or six shipwrecks would qualify for designation as a 'Hanseatic cog'. There are another six shipwrecks from the Netherlands, Denmark and Flanders which were less intact, which could almost certainly also be given that designation. They are shown in Tables 1 and 2, together with the (approximate) main dimensions to the extent that they are known or have been approximated. The probable periods in which the vessels were built are also stated.

3. Vlierman 2002, 142 and idem 2004, 28.
4. Modderman 1945, 81.
5. Luns 1985, 36.
6. Most probably the region roughly situated between the Danish southern border (Jutland) and the north-western German coastline of the North Sea and Baltic Sea in the north, from Flanders to the Lower Rhine area in the south, and in between the eastern coast of the Zuiderzee and the IJssel river basin.

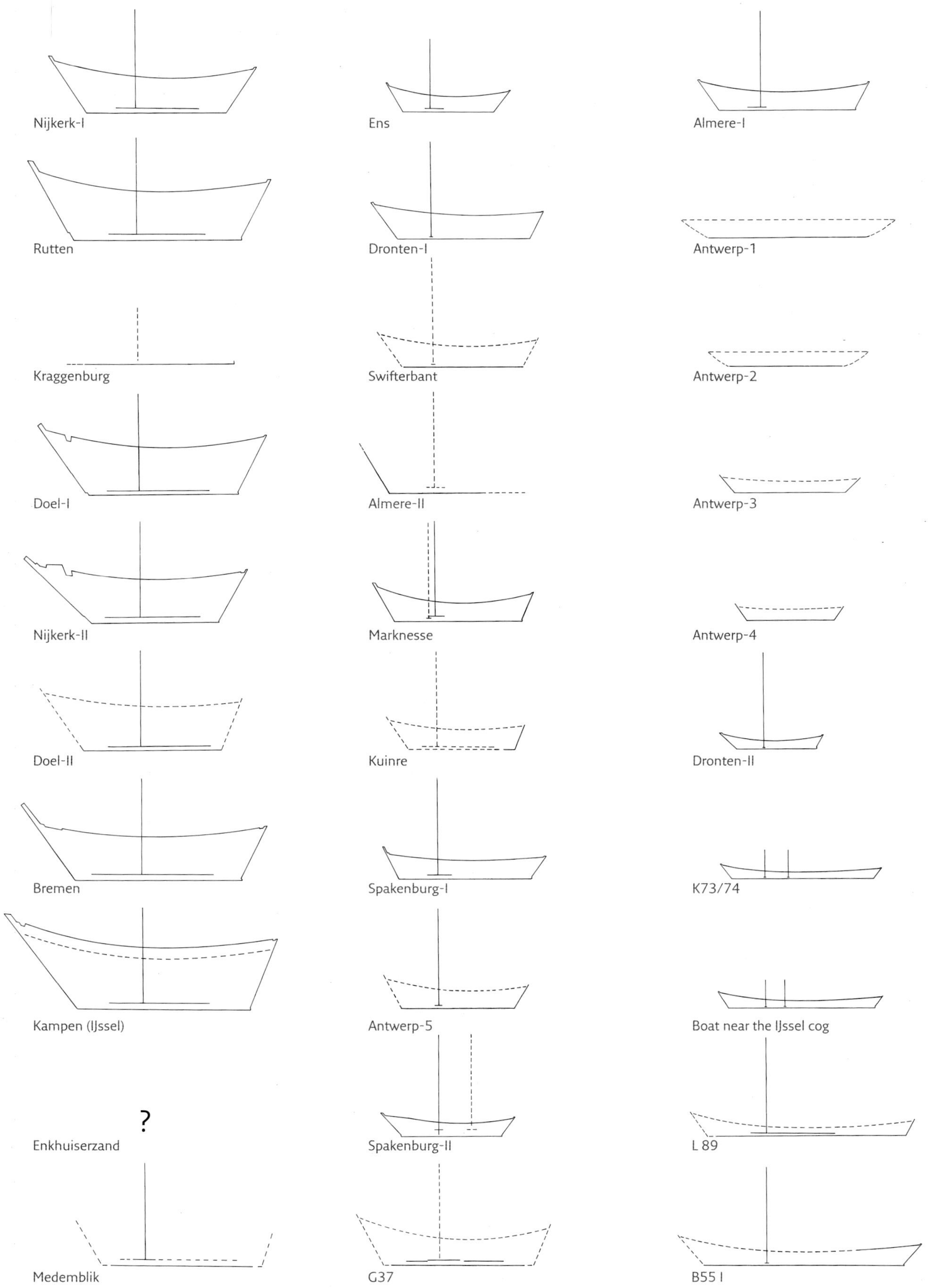

Fig. 29.1. Overview of all discussed vessels in the cog-building tradition: hull contours with the keelson/mast step and location of the mast (scale 1:100).

The dimensions of the Doel-I cog, the Kampen (IJssel) cog and the Enkhuizerzand cog are (rough) estimates. They cannot be determined precisely yet; the wreckage from Doel-I was upside-down and had sagged, collapsed and warped in the middle, while the data on the Kampen (IJssel) cog are not yet available, and the Enkhuizerzand cog has not yet been excavated.[7] It will be necessary to wait for the reconstructions of the ships' hulls. A striking feature in the five 14th and 15th-century cogs is the seemingly near-identical lengths from stern to stem, the heights of the sides, and the length-width ratio across the gunwales. Moreover, measured at gunwale height, they are all 10-20 cm wider in the forebody than in the afterbody, at similar distances from the stern and stem. The width of the Nijkerk-II cog is distinctive. It is very likely that this strikingly larger width is the result of the presumed renovation/adaptation of the ship in relation to a possible later function, which may have been military in nature. It is also worth noting that the older Nijkerk-I cog is one step smaller than the other five.

The first four cogs in the table, but also the Kolding, Rutten, Kraggenburg, Doel-II and Vejby cogs, had a long keelson with an integrated wider mast step section. It is a long, heavy beam that was mounted over the frames and above and over almost the entire length of the keel plank. The keelson is also referred to as the inner keel and is primarily intended to add longitudinal rigidity to the structure of the underwater body. In vessels with cog-like features, this long keelson has only been established with certainty in the large ships. It has not yet been confirmed in the Enkerhuizerzand cog. However, this type of keelson with mast step section is not reserved to large cogs, nor to the cog-building tradition. In fact, we already see it in the large river vessels dating from Roman times (such as the Zwammerdam 2 and the Woerden 7), but also in the mid-16th century (for example in the Mary Rose, the flagship of Henry VIII, and in the large, armed and completely clinker-built 'hulk' on lot U 34 in Eastern Flevoland)[8] and in later vessels.

Regarding the find at Kraggenburg, it can be noted that hardly anything can be said about the dimensions of the vessel. The wreckage comprises most of the ship's bottom with about ten frames (or partial frames), a large part of – very unusually – a keel beam (in two parts), the stem hook and a part of the keelson with mast step section. Sintel clamps on the carvel seams on the inside, and presumably carvel seams that (nearly) covered the entire thickness of the planks in a reverse V-shape, are only seen on a few older vessels. The dimensions of the large cogs that were not found fully intact have been estimated as accurately as possible based on the available data and by means of comparison to the first four vessels listed in table 1. The ship from Kollerup,[9] which was built around the mid-12th century, was – due in part to the lack of crossbeams protruding through the shell – a completely different 'large' vessel with cog-like features, with a lower construction. The reconstructed comparable dimensions are approx. 20.0 m long, approx. 6.0 m wide and approx. 2.2 m high respectively. The ship had a recess in a floor timber for the base of the mast. The shipwrecks in the IJssel before Kampen, on the sand flats of the Enkhuizer- zand and in the Zeebad in the IJsselmeer near Medemblik (Chapters 23 and 24 and Section 25.1) should be investigated further in order to obtain more certainty about the constructions and dating, etc.

The Kampen (IJssel) cog

With regard to the Kampen (IJssel) cog, it should be noted that the information provided for this purpose has been derived from the published data in the report of the field evaluation by Waldus *et al.* from 2012 and the preliminary conclusions I derived from that. The descriptions of all the other cogs in this study as well as the analysis and conclusions on these vessels were ready by the end of 2014. They have remained unchanged. The actual dimensions, construction and reconstruction etc. of the Kampen (IJssel) cog as they are now known have been incorporated into chapter 23 and the accompanying plates. The (divergent) conclusions drawn from this wreckage with regard to e.g. changing construc-

Table 1. Approximate dimensions of the large (Hanseatic) cogs that were most completely recovered

Ship	Length from stern to stem	Greatest width	Height amidships	Construction dating
Nijkerk-I	18.3 m	6.9 m	3.4 m	1225 - 1250
Doel-I (B)	ca 20.0 - 21.0 m	ca 7.0 - 7.5 m	ca 4.2 - 4.7 m	1325 - 1330
Nijkerk-II	20.0 m	8.7 m	4.3 m	1325 - 1350
Bremen G)	21.0 m	7.8 m	4.2 m	1380
Kampen (IJssel)	ca 21.0 m	ca 7.0-7.5 m	ca 4.0 m	1400 - 1450
Enkhuizer zand	> 20.0 m	?	?	1470 - 1485

7. The Kampen (IJssel) cog was salvaged in early 2016.
8. Rule 1982, 14; Vlierman 1997, 26 fig. 17.
9. Crumlin-Pedersen 2000, 234, 238 and 239; Daly, Eriksen & Englert 2000, 61.

tions and the development of the large seagoing cog in the 15th century are provided in the summary of chapter 23.

Certain characteristic building elements and constructions in the case of a few large cogs may possibly indicate a *Heerkogge* or war-cog or a trading vessel that also had a military function, while clear differences (in details) can be indicated between the 13th-, 14th- and 15th-century vessels. For instance, rather than thick wooden treenails, thick iron bolts were increasingly used in the beam constructions, and the extent to which the shell seams were sintel clamped began to decrease in the second quarter of the 14th century. The seams of the clinker-built sides were initially made watertight using sintel clamped moss caulking on the inside as well as the outside. After that, they were no long sintel clamped on the outside. From ca 1400 onwards, the ends of the planks were probably also gradually nailed into rabbets in the sides of thicker and wider stern- and stemposts rather than ending against each other on the mitred fore or aft side of the stern or stem. In the course of the (second half of the) 15th century, the construction consisting of keel plank/stern and stern- and stem hooks/sternpost and stempost was replaced by a keel with a stern- and stempost placed on it.

29.3.2 *CLEENE COGGHEN*: small cogs

The smaller seagoing cogs

Ewe depicts a large number of seals which depict the characteristic hull shape of the cog (Section 4.1), but do not show any beams protruding through the shell.[10] For that reason, it can be assumed that we are probably dealing with smaller, seagoing cogs in those cases. The oldest example is the seal of Staveren (first half of the 13th century). The second seal (from the second half of the 14th century) also lacks this information about the construction, although it does very explicitly indicate the overlapping planks. The seals of Stralsund from the 13th century, the second half of the 13th century, ca 1300 and 1329 respectively do not have any protruding beams, although many other details were depicted. This also applies to the oldest seal of Elbing (1242), while the cog on the second seal from this city from ca 1350 does depict protruding crossbeams. The seals of Wismar from ca 1250 and ca 1350, the seal of Vlaardingen (ca 1300), the seals of Danzig from the second half of the 13th century, ca 1300 and ca 1400, the seal of Gollnow (the younger, secret seal dating from the 14th century), the seal of Kiel (14th century), the seal of Kuinre ca 1400, the seal of Genemuiden presumably from the first half or middle of the 15th century, and the oldest seal of Nykøbing (15th century) also belong to this group of vessels. In particular, the seals of Gollnow, Kuinre, Genemuiden and Nykøbing seem to depict vessels that are not very large. For that reason, it is quite possible that we are not dealing with small seagoing vessels, but perhaps rather with small cogs for inland lakes and waterways and in the Dutch situation specifically the Zuiderzee and the waters of South Holland and Zeeland.

Four or five shipwrecks from Flevoland can probably be counted among the small seagoing cogs. These vessels, like the large cogs described above, have a length-width ratio of 3.2:1 or less across the gunwales, do not have protruding crossbeams, and have sides approximately half as high as those of the large cogs. In these ships, the beams are transverse connections that are fastened into place on the inside of the sides. The shipwreck of the Dronten-I cog has provided the most information to date. The Spakenburg-II cog seems to be fairly complete. Van de Moortel[11] assumes that the inwale which was fastened against the upper edge of the highest shell strake found was also the upper edge of the relatively low gunwale. The model made by J. van der Zee from oak on a scale of 1:10 (fig. 15.17) also made that assumption. Re-examining and reconstructing the ship on paper in the context of this study provided additional information about a possible function as a working vessel for bringing up sunken objects or e.g. pile-driving into the beds of lakes and waterways. Table 3 shows the approximate dimensions of the reconstruction by Van de Moortel. The approximate dimensions of the reconstruction in this study are shown in brackets behind those figures.[12]

Table 2. Approximate dimensions of six large (Hanseatic) cogs that were less than intact upon discovery

Ship	Length across stern to stem	Greatest width	Height amidships	Construction dating
Kolding (DK)	ca 18.5-19.5 m	?	?	1190-1210
Rutten	22.4 m	9.5 m	4.5 m	1275-1285
Kraggenburg	?	?	?	1310-1320
Doel-II (B)	ca 18.9-19.6 m	ca 6.5-6.7 m	ca 3.8-4.1 m	1330-1350
Vejby (DK)	ca 16.0-17.0 m	?	?	ca 1375
Medemblik	? (bottom ca 15 m)	? (bottom ca 3.6 m)	?	15th century

10. Ewe 1972.
11. Van de Moortel 1991.
12. See chapter 11 and plates 11.I to 11.IV.

Only limited data is available for the other three vessels from Flevoland, because they have not (yet) been fully excavated and documented. Vessel 5 of the vessels found in 1884 in the Lefèbvre Dock in Antwerp[13] can very probably, judging from the shape of the drawing of the top view (figs 3.2 and 26.26), also be considered part of this group. The dating assumed by Hasse on the basis of the soil profile (first quarter of the 15th century) seems less acceptable: according to the drawing, the ship was situated partly below vessel 2, which he dated to the beginning of the 13th century. The sintel nails used in vessel 2 suggest that the vessel may be even older than that (second or third quarter of the 12th century).[14] It therefore seems quite possible that vessel 5 was older than that.

In contrast to the large cogs, the smaller cogs do not have a long keelson. The Dronten-I cog had a recess in a floor timber for the base of the mast. This cannot (yet) be said of the Swifterbant cog, but it is most likely the case with this ship as well. The Antwerp 5 had a short mast step that was fastened into place on two floor timbers or planks. The Almere-II cog may have had a short mast step attached to the frames, but this is not certain. Due to the estimated gunwale heights amidships and at the stem (2.5-3.0 and approx. 4.0 m respectively), it can therefore not entirely be ruled out that this shipwreck was a large or mid-sized cog. Even less can be said with certainty about the Kuinre cog. Judging by the shape, dimensions and height of the stern and the approximate dimensions derived from that, the vessel can be counted among the smaller seagoing cogs. If the burned frames and keelson did indeed come from this ship, which seems very likely, then this would be the only smaller vessel so far that had an integrated mast step in a long keelson. The Spakenburg-II cog had two (possibly originally four) chocks placed against the sides, in which the poles of a hoisting construction were probably placed. The approximate main dimensions of the Dronten-I cog have been reconstructed. The dimensions shown in Table 3 for the other five vessels are (provisional) assumptions.

Like the long keelson in the large cogs, the recess for the base of the mast in a floor timber or plank was not limited to the cog-building tradition and the Hanseatic period, but seen as early as Roman times.[15]

The vessel that can be defined as a mid-sized cog that was suitable for sailing on the inland lakes and waterways as well as the Zuiderzee and the Waddenzee is the presumably 15th-century wreck from lot G 37 in the Noordoostpolder (Section 25.2). However, the excavation data for this shipwreck must first be fully worked out in order to obtain more certainty about this. It differs from the other, older cogs in the use of a keel with a stern and stem placed on it instead of a keel plank that is extended aft and forward with stern and stem hooks. The stern and the mast step have remained intact and may possibly be able to provide more certainty about the dating by means of dendrochronological analysis. For the time being, the shipwreck has been classified among the small, seagoing cogs.

The small cogs for inland waterways and the Zuiderzee

As already noted above, the vessels on the 14th and 15th century seals of Gollnow, Kuinre, Genemuiden and Nykøbing may be more suitable to be interpreted as small cogs for inland waters and the Zuiderzee. In the case of Gollnow and Nykøbing, the Zuiderzee should, of course, be replaced by the Oderhaff and River Oder respectively, and the waters between the Danish islands of Lolland and Falster. Without any doubt, the vaulted painting on the south side of the Oude Kerk in Amsterdam (fig. 29.2) depicts such a late 14th-century inland cog. Two open sailing ships can be seen on the beautiful miniature from the *Apocalyps in Dietsche*, which is discussed in section 29.3.5 below. These are also clearly examples of the small cogs referred to here, which were loaded with goods transshipped from seagoing vessels and transported inland. Dutch shipwreck finds that belong to this group of inland cogs are the Ens cog, the Marknesse cog, the Spakenburg-I cog and the Almere-I cog.

Table 3. Approximate dimensions of the six or seven small seagoing cogs

Ship	Length from stern to stem	Greatest width	Height amidships	Construction dating
Dronten-I	15.8 m	5.2 m	2.1 m	1300-1325
Swifterbant	ca 15.0 m	ca 5.0 m	ca 2.0 m	1300-1325
Almere-II	16.5-17.5 m	ca 6.3 m	ca 3.2 m	1325-1350
Kuinre	ca 13.0 m	ca 5.0 m	ca 1.8 m	1350-1375
Antwerp 5 (B)	11.5 m-13.0 m	4.0-4.5 m	1.5-1.8 m	1400-1425 (13th century?)
Spakenburg-II	11.75 m (12.3 m)	4.3 m (4.5-4.7 m)	1.2 m (1.9 m)	1415-1430
G 37 NOP	ca 18 m	5-5.5 m	2-2.5 m	15th century

13. Hasse 1907.
14. Vlierman 1996a, 63-13; Vlierman 2002, 132.
15. Vlierman in preparation b; for example the Bruges boat and Mainz 3.

Fig. 29.2. Small cog with spritsail. Vault painting in the Oude Kerk church, Amsterdam, ca 1400. Photo: Oude Kerk, Amsterdam.

The approximate width of the Almere-I cog shown in the table between brackets is the largest width across the gunwales as I have now assumed it. All other vessels analysed and reconstructed in this study were found to have a slightly larger width in the forebody than at a similar position in the afterbody. The reconstruction shown by Hocker[16] has the lines that were customary in cargo ships for inland navigation in the Netherlands in the 18th-20th century.

29.3.3 *SCHUTEN*: boats with cog-like features

Chapter 26 briefly covers a number of small vessels (schuyts). They characteristically have at least some features of the cog-building tradition. This chapter also includes several excavated vessels where corresponding characteristics have also been established. However, as far as their dimensions are concerned, they would certainly not be included in that category. The types of these vessels are not known, but they should be mentioned in this study. The *Schuten* or boats include the early examples from Antwerp (several 13th or possibly even 12th-century barges and boats for inland navigation), 13th-century ship wood from 's Hertogenbosch, a river-barge from Kessel (Noord Brabant) dating from ca 1500, and several vessels dating from the 15th and even 16th century from Flevoland and near Kampen in the IJssel. Unfortunately, it is not possible to say anything with certainty about the finds from Antwerp with regard to construction, steering and other cog-like features, because the relevant data were not recognised, not sufficiently determined and/or not sufficiently recorded during the discovery of the vessels and their description. This is fully understandable, since archaeological knowledge about medieval ships and similar finds was still almost non-existent at the end of the 19th century. In fact, we can conclude that Hasse, in publishing his work[17] and despite the limitations mentioned above, has produced a valuable document about vessels from the beginning of the Late Middle Ages/the cog-building tradition in the Netherlands. Vessels 1 through 4 are in any case the earliest examples from the Low Countries where it is possible to speak with certainty of constructions and structural details that are part of the cog-building tradition. The small seagoing cog (number 5), which can be designated with certainty, also indicates that vessels from this building tradition were sailing in Antwerp waters during the period in question. The remarkable custom of applying nails to the clinker-built constructions from the inside as well as the outside, which was also determined in the find of 13th-century ship wood from 's-Hertogenbosch and in the Kessel barge from ca 1500, suggests that it can cautiously be concluded that this may have been the custom at that time and in that part of the Low Countries. This method of fastening the planks together was also found in the Dronten-I cog.

The use of double-clenched nails (folded down twice from the inside and the outside, with the nail tips pounded back into the wood) was already observed in the large river vessels from Roman times. On the few known carvel-built finds of Celtic/Roman seagoing vessels (Section 3.2), it was

Table 4. Approximate dimensions of the four small cogs for inland waterways and the Zuiderzee

Ship	Length from stern to stem	Greatest width	Height amidships	Construction dating
Ens	11.2 m	3.5 m	1.5 m	1225-1275
Marknesse	14.8 m	4.65 m	1.6 m	1340-1350
Spakenburg-I	15.2 m	4.8 m	1.7 m	1385-1400
Almere-I	15.2 m	4.2 m (ca 4.5 m)	2.0 m	ca 1400

16. Hocker & Vlierman 1996, plate 3.
17. Hasse 1907.

only done from the outside. On the similarly carvel-built vessel Port Berteau II,[18] probably built shortly after 600 (still after Late Antiquity), the planks are fastened onto the frames not with the mortise-and-tenon joints used in the Roman vessels in the Mediterranean, nor the clenched nails of the seagoing Celtic/Roman vessels, but with wooden treenails. The plank ends were nailed against the sides of the stern and stem. According to the reconstruction by Westphal, the part of the stem that was found curves along the underside into the keel and is fastened on top of it (derived from the Roman finds). However, the connections between the keel and the stern and stem were not found, so there is no certainty about that. The vessel was 14.6 m long, 4.8 m wide and 1.4 m high and equipped with cross-beams protruding through the shell. The seams were sealed with caulking made from plant material. In the construction of the sides, in addition to the usual planks, several thicker, rounded specimens were also used. This can also be seen in the 11th-century log-boats with gunwales found at Van Hoornekade and the Waterstraat in Utrecht[19] and in the river barge at the final findspot.

29.3.4 13th-century river vessels

It seems very plausible, based on dendrochronological research, that the finds of the river vessels Meinerswijk 2, Amsterdam and Rotterdam 4[20] are examples of a shipbuilding tradition from the region comprising the major river basins of the Rhine, Moselle and Meuse. This tradition may have extended in the Rhine basin all the way to the upper reaches of the IJssel river. Unfortunately, it was not possible to determine the dendrochronological dating and provenance of the wood from the ship find in Hattem.[21] What the vessels in question have in common is that the clinker-built shell planks were fastened with small wooden treenails, in contrast to the cog-building practice of using double-clenched nails, which were folded down twice and pounded back into the wood. In particular, the gradual transition from carvel-built bottom to clinker-built sides and the sintel clamped moss caulking used in the vessels from Amsterdam and Hattem show a clear resemblance to the cog-building tradition, or perhaps rather show evidence of its influence. In any case, the (late) 13th-century age of the vessels indicates that the construction method using treenails may have persisted for a long time in these regions. It is striking to note the difference between the nails in the clinker-built fastenings of the Antwerp vessels and the few other ship wood finds from the south-western Netherlands dating from the same period or even earlier. Although some construction details and the caulking method do seem to indicate some kinship with the cog-building tradition, the vessels therefore cannot be considered cogs.

29.3.5 A miniature displaying the full spectrum of cogs and boats

In the *Apocalypse in Dietsche*, presumably made between ca 1403 and 1409, twenty-three very beautiful and colourful miniatures are depicted.[22] Five of them depict vessels. One of these (fig. 29.3) shows five different types of vessels, which can undoubtedly be interpreted as a large seagoing vessel (a cog?), two small cogs and boats which were most probably built in line with the cog-building tradition. First, the large cargo vessel with a single mast and a crow's nest. Towards the stern, we see a stern rudder and a castle. The hull is not very realistically depicted. The shell strakes rise very sharply towards the stern and stem, which is why it would initially be logical to consider e.g. a hulk. However,

Table 5. Approximate dimensions of barges and boats with at least some features of the cog-building tradition

Ship	Length from stern to stem	Greatest width	Height amidships	Construction dating
Antwerp 1 (B)	ca 20.0 m	ca 4.0 m	ca 1.5 m	1125-1175 (1275-1300)
Antwerp 2 (B)	ca 15.0 m	ca 3.0 m	ca 1.2 m	1125-1175 1275-1300)
Antwerp 3 (B)	ca 13.0 m	ca 2.5 m	ca 1.2 m	1125-1175 (1275-1300)
Antwerp 4 (B)	ca 9.5 m	ca 2.0-2.5 m	1.2-1.5 m	1125-1175 (1275-1300)
's-Hertogenbosch	? (fragment)	?	?	ca 1250
Dronten-II	9.0 m	2.5 m	0.9 m	after 1296 AD, (1425-1475)
Dronten (K 73/74)	15,8 m	2.4 m	0.8 m	ca 1425
Kampen (IJssel, object 14)	>11.25 m (ca 15-16 m)	ca 2.0 m	ca 0.8 m	1400-1450
Lelystad B 55	ca 23.0 m	ca 5.0 m	ca 1.2 m	1450-1500
Kessel (N.Br.)	ca 12.5 m	ca 2.1 m	ca 0.7 m	ca1450-1550
Dronten (L 89)	ca 20.0-21.0 m	ca 4.0 m	ca 0.8-01.9 m	ca 1550

18. Westphal 2004, 9-21.
19. Van de Moortel 2011, 90-95 and fig. 10.
20. Respectively: Reinders 1983, 15-20 and 40-41; Vlierman 1995, 91-104 and Van Holk 2001, 103-117.
21. Vlierman 1996b, 115-119.
22. De Hommel-Steenbakkers 2001.

Fig. 29.3. The miniature Folio 21r: 15th-century miniature in Apocalips in Dietsche 18. Chliché Bibliothèque nationale de France, Paris. From De Hommel-Steenbakkens 2001.

given the dating of the miniature and the ship's single mast, it is more likely that this vessel may have been intended to be a cog. It is possible that the miniature painter had never seen a large cog before, or never taken a close look at one. A dinghy is attached to the ship by a rope. Between the seagoing ship sailing away and the walled city lies a vessel that can undoubtedly be described as a small cog, as they are now known from archaeological finds. This ship has in fact been faithfully reproduced. It has a single mast, a rolled-up mainsail, a stern rudder with tiller and a stempost protruding above the gunwale and a (covering) inwale. It is also clear to see that a false stem has been placed against the stempost. A cargo consisting of various items (barrels, sacks and a chest) is displayed in the open vessel.

Next to the ship, there is a small boat with a platform at either end and a transverse connection amidships. A slightly smaller vessel is depicted behind/above this small cog. It also has a single mast and a small mainsail, and the person seated in the afterbody is rowing the little ship. The oars rest between two dowels in the inwale. The person, like all the other figures in the miniature, is depicted facing the viewer. He is rowing backwards, although the little ship is sailing forwards. The miniature painter apparently thought this contradiction was less relevant, but did attach importance to the fact that the vessel could be both sailed and rowed. Three sacks containing goods are lying amidships in the vessel. The little ship has a lot in common with the small cog. Both of the vessels could be considered boats.

This impression of several vessels in action tells a comprehensive tale about the most important mode of freight transport in north-western Europe in the Late Middle Ages. A seagoing vessel known as a *Cogghe* or cog would retrieve goods from distant countries. Using *Cleene cogghen* or small cogs, these goods were then transported from the seaports further inland (via the Zuiderzee, the Scheldt estuary and the major rivers), and then transferred to the even smaller cogs and *Schuten* or boats, which sailed, rowed and poled the goods to locations inaccessible to the other vessels.

29.3.6 Small cogs in the *Digestum Vetus* from 1456

The Kampen city chronicle known as the *Digestum Vetus* also contains, among other things, two lesser-known drawings by the city archivist depicting six small vessels (two of which are most probably cogs) as well as a large cog and a smaller vessel (figs 29.4a and b).[23] The last vessel, according to the accompanying text, may be a *boeijer* or a *pleijt*, both of which were shallow-draft coastal vessels used in that region. They are highlighted here because four of the vessels shown clearly depict a curved canopy over the hold. From the miniature described above and from archaeological finds, until I became aware of the drawings in the Kampen city chronicle in 2018, I had suspected that the small cogs were probably open.

29.4 Characteristics, construction sequence and catalogue

This section analyses and discusses the results of the technical investigation of the shipwrecks and indicates the extent to which the results correspond to or deviate from the information known and published up to circa 2000. In addition, the section once again covers the features of the cog-building tradition and additional characteristics, and the most important data and details are presented in a catalogue of all finds.

Reinders discussed the characteristic structures of the nine shipwrecked cogs from the IJsselmeer polders that were known in 1984.[24] He successively covered the keel plank, the stern- and stem hooks, the sternpost and stempost, the shell strakes of the bottom and sides, the fastenings of the shell strakes of the bottom and sides onto the stern and stem, the fastenings of the overlapping planks, the caulking method, the scarfs, the construction of the mast step and the crossbeams (protruding through the shell). One of the nine finds he covered, the stern hook from lot R 1 in the Noordoostpolder, was later completed with the fully preserved sternpost. When the sternpost was found, it soon became clear that the entire construction of the afterbody had been found not on lot R 1, but on lot M 133 nearby. During this study, it was discovered that the stern construction was probably part of a cog that had already been found in 1943, of which only two photographs have survived (Chapter 13, the Kuinre cog). The typical maritime names and descriptions are explained in the glossary and in figure 1.4.

29.4.1 Characteristics

A slightly different approach to the constructions and ship parts has been chosen here, in which the descriptions by Reinders are reiterated. They are supplemented with the data on the other cogs from the Low Countries that are now known. For each construction element, a detailed description is provided as to what this supplementary information has yielded and whether or not there is confirmation of the previous analyses. The characteristics listed in section 2.4 could be determined on the basis of the Bremen cog and several Dutch and Danish finds. They have been in general

23. Schilder 1983, text and figs 106 and 112. For text on 106, see footnote to section 28.2.4.
24. Reinders 1985a.

Fig. 29.4a. Six vessels (small cogs) on the river Reeve near Kampen. City statute book Digestum Vetus, 7 August 1456. Municipal Archive, Kampen. Schilder 1983, drawing 106.

b. A large cog and a boeijer or pleijt. City statute book Digestum Vetus, 21 August 1456. Municipal Archive, Kampen. Schilder 1983, drawing 112.

use since around 2000, at least for the large sea-going cogs. They are repeated here. These characteristics, or at least the majority of them, are found in all excavated vessels in the cog-building tradition. The characteristics are:

- A keel plank with stern and stem hooks and a plumb sternpost and stempost.
- An underwater body that is relatively slender aft and forward and relatively flat and wide amidships.
- A carvel-built bottom which transitions aft and forward into a clinker-built construction.
- The planks of the carvel-built part of the bottom are not fastened to each other, but connected by means of the frames.
- Clinker-built and everted sides.
- The clinker-built planks were fastened together with double-clenched nails (folded twice and driven back into the wood).
- The plank seams of the shell were sealed with sintel clamped moss caulking.
- Beams that protrude through the shell with heavy knee sections/bulkheads placed on top of them which provide the connections at deck height and above.

All structural elements, the caulking method and the lines of the ship are discussed in the following sections:

1. The bottom: the keel plank with stern and stem hooks, sternpost and stempost, and the bottom strakes
2. The sides
3. The caulking method
4. The frames, mast step, ceiling boards and inwale
5. The crossbeams, and/or transverse connections
6. The decks and castles
7. The rudder, mast and sail and the ropes and blocks
8. The lines of the ship and the length-width ratios

1 The bottom: the keel plank with stern- and stem hooks, sternpost and stempost, and the bottom strakes

Keel planks

The bottom is a composite construction consisting of the keel plank with stern- and stem hooks and two or more bottom strakes. Reinders had determined that all finds from the former Zuiderzee area had a keel plank that varied in width between 18 and 34 cm and had a thickness or height of 7-10 cm. They deviated from vessels dating from the Viking era, and the vessels from the following period up to ca 1200 from Denmark and England, and the vessels from the Slavic ship-building tradition. All these vessels had a T-shaped keel in cross-section, with the first bottom strakes nailed to the underside of the protruding parts of the top of the keel. With the cog finds from the

Low Countries that are now known, it has become clear that the variation was even greater than previously presumed. The keel plank width amidships varied between 20 and 52 cm. Most were elliptical with straight ends (the scarfs with the stern and stem hooks). At those scarfs, the planks were generally wider on the forward end than on the aft end, varying between 10 and 42 cm; the width was sometimes the same forward and aft. A few cogs had a straight keel plank that was slightly wider forward than aft. The thickness of the keel planks on all these ships varied between 4 and 16 cm. On the Kraggenburg cog, the cross-section is nearly square (13x15 cm). The small Ens cog could almost be said to have a keel beam. In addition, skegs and a shim had been inserted under the stern- and stem hooks and in between, elevating the keel even higher.

In cross-section, the keel planks were almost rectangular, usually slightly trapezoidal, with the widest side being at the top/inside of the vessel. Amidships, the difference in width between the top and bottom did not exceed a few centimetres. As the plank narrowed towards the stern and stem, the trapezoidal cross-section tapered steadily, continuing smoothly into the keel sections of the stern and stem hooks. The sides of the keel planks were smooth. These strakes were not fastened into place on the keel plank. On the one hand, the variation in the execution of the keel planks is related to the dimensions of the ship in question; on the other hand, it may also have been a design used at a certain shipyard. The term 'keel plank' has been chosen if the plank thickness is no more than 3 x the thickness of the shell planks of the ship. In most cases, the keel plank was short. It was extended aft and forward by the horizontal section of the stern- and stem hooks. Together they formed the total length of the keel.

Stern- and stem hooks

The most eye-catching and remarkable structural elements of cogs are the stern- and stem hooks. The cog builders most likely did not have the courage or know-how to make a keel with a sternpost and stempost on the ends: a construction that would create a seam between the two structural elements at two very crucial and vulnerable points. This was certainly not preferable if we assume, for example, that the ships (regularly) landed along a coast or were beached on the mud flats. It is almost certain that this was a reason, in the earliest vessels, to opt for a (strong) naturally grown knee-shaped piece of wood for this transition from keel to stern and stem, with one section comprising the extension of the keel plank and the other comprising the lower part of the rising sternpost and stempost. The stern- and stem hooks can be compared to the connecting pieces between the keel and the stern and stem in vessels from the Scandinavian or Nordic shipbuilding tradition.[25] The construction was probably very satisfactory, because it was maintained until well into the 15th century.

The stern- and stem hooks were made from a section of a tree consisting of a trunk and connected branch, where the trunk was used for the keel plank section. The keel plank sections of the stern- and stem hooks were in principle – usually depending on the dimensions of the vessel – short or long, with the section for the stem hook being longer than the section for the stern hook. The stem hooks were generally also more robust than the stern hooks, sometimes remarkably heavy (Dronten-I cog). Stern- and stem hooks were also used with a long keel plank section. In that case, the stern- and stem hooks were fastened to each other with a sloping flat scarf about amidships, i.e. without a short keel plank in between (Doel-II cog). The stern- and stem hooks were attached to the keel plank with flat sloping scarfs. The scarfs started at the top towards the stem and ran sloping downwards and abaft (moving in the direction that the vessel would sail). The overlapping sloping parts of the stern hook and stem hook and the keel plank were fastened together with at least two rows of nails, and sometimes four rows in larger vessels. The number of nails per row also varied. If there were two rows of nails at a scarf, one row was driven upwards into the wood from below, and the other row was done in the other direction. The protruding nail tips were in principle folded down twice and driven back into the wood (double-clenched nails). Sometimes they were driven in blind (i.e. not protruding through the wood). The cross-section of both the keel section and the stern- and stem section of the stern- and stem hooks was trapezoidal, just like the keel planks. This trapezoidal shape tapered steadily downward towards the aft and forward ends of the keel section and the stern- and stem sections. The aft end of the small cogs was often no wider than approx. 6 cm. Rabbets were notched into the sides of the stern- and stem hooks, in both the horizontal section and the stern- and stem sections, and the ends of the bottom strakes were nailed into the rabbets. In the keel plank section, this was not done along the entire length. Sometimes the end of the outermost bottom strake was (also) nailed onto the sternpost and/or stempost.

The point of connection with the sternpost and stempost was also a sloping flat scarf. The lower end of the scarf generally started at or just above the keel section and ran sloping upwards towards the aft and forward side of the stern and stem respectively. The knee shape of the stern- and stem hook was still in use in the early 15th century, but was no longer part of the keel plank. It has been established in smaller vessels from that time that the construction was the full sternpost and stempost with a knee-shaped lower sec-

25. Crumlin-Pedersen 1967, 99.

tion that was placed on the middle bottom strake and generally protruded partly beyond the end of the plank on the forward end to protect the cross-cut end of that plank. The keel section of the sternpost and stempost gradually became shorter and shorter. In the course of the 15th century, the stern- and stem hooks completely disappeared, coinciding with the placement of the sternpost and stempost directly on the keel. Sometimes the sternpost and stempost were also fastened into place partly on top of the keel and partly against one of the sides of the keel with a flat nibbed scarf. In both cases, the knee joint was then replaced by a filler piece connecting the keel to the stem and sometimes to the stern, known as the deadwood. There are also examples known where the keel (or keel plank) was left thicker at the ends, forming what could be referred to as the keel plank section of the stern and stem hooks. The sternpost and stempost were then placed on this thick end.[26]

Sternposts and stemposts

The sternpost and stempost were placed on the sloping flat scarfs of the stern- and stem hooks. They were generally only fastened to each other with a few nails through the thin end of the scarf of the stern hook, stem hook, sternpost and/or stempost. Sometimes these fastenings were supplemented by a wooden treenail. The sternpost and stempost were straight and trapezoidal in cross-section, with the widest side on the inside of the ship. They generally tapered towards the top. The width usually remained the same. In a few cases, the sternpost and/or stempost also widened towards the top. The thickness of the upright section of the stern- and stem hooks is usually also the thickness of the base of the sternpost and stempost. However, sometimes the sternpost and/or stempost was thicker, not only resting on the scarf but also on the (usually thicker) end of the keel section of the stern hook and/or stem hook respectively. The overall impression is that this was particularly common in inland cogs that may have been specially built for heavy transport, e.g. building materials. All cogs have straight, raked stern- and stem constructions, in which the stem would in principle have a greater rake. The difference between stern and stem was sometimes minimal. In 13th- and 14th-century vessels, the rabbets for securing the ends of the shell planks did not, in principle, continue into the sternpost and stempost. Only the ends of the bottom strakes were nailed into a rabbet in the stern- and stem hooks. The plank ends of the strakes of the sides continued past the mitred, chamfered aft or forward side of the sternpost and stempost and were nailed against them. A striking construction was used in the stem of the Doel-I cog. The planks end in a zigzagging two-by-two pattern, alternating between the starboard and port side of the front of the stempost. It was only towards the end of the 14th century that the cogs began showing a sternpost and stempost and a stern and stem, with the ends of the planks increasingly being secured in a rabbet towards the top until all the plank ends were eventually fastened in this way.

False sterns and stems

The vulnerable plank ends of the sides, which were nailed against the front of the stemposts, were in principle always protected with a false stem. In a few cases, the vessel also had a false stern; this usually only covered the lower half of the stern. In the case of the Nijkerk-II cog, the false stem was narrower at the bottom than at the top. Together with the stempost, where the opposite was the case, this means that the composite thickness was therefore almost the same everywhere.

Bottom strakes

The description provided by Reinders states that the bottom of the vessels consisted of a varying number of strakes in a carvel-built construction. These carvel-built elements were not fastened to each other, nor to the keel plank; rather, they were only connected by the frames against which they were fastened into place with wooden treenails. The number of strakes depended on the dimensions of the vessel and the widths of the planks that were used. The planks were usually relatively straight and about the same width along the entire length. The thickness varied between 2 and 8 cm, also depending on the size of the ship. When fitting the planks into the hull, each plank was generally cut to (narrower) size towards the stern and stem. Sometimes the construction involved using the suitable wood to the fullest possible extent, so the planks of the bottom strakes were cut to fit together like puzzle pieces. In the Rutten cog, bottom strakes were used in widths of 60 cm or more.

The cross-sections per ship show that the shape of the bottom varied considerably amidships: from horizontal or near-horizontal, rising slightly towards the turn of the bilge, and then transitioning into the sides at an angle, to a more 'curved' bottom rising gradually and unobtrusively transitioning into the sides from the keel plank. In principle, the top of the keel plank was always even with the top of the adjacent garboards. The cross-sections changed aft and forward as a result of the underwater body which tapered towards the stern and stem. In order to obtain the desired lines, the planks were heat-moulded so they could be twisted until the parts near the stern and stem had reached an angle of almost 90° relative to the bottom of the ship. The bottom strakes were made from multiple planks fastened together by means of sloping flat scarfs. These scarfs were generally 30-45 cm long. In the smallest vessel (the Ens cog), they were no longer than 25 cm; in the largest (the Rutten cog), some scarfs are 1.1-1.35 m long. The nib ends of the overlapping

26. For example in the early-16th-century *waterschip* K 84 II from Eastern Flevoland. See section 27.4.

planks at the scarfs, 4 to 5 cm thick, stood proud of the shell both on the inboard and outboard sides. The overlapping parts were fastened together with between two and four rows of nails, and nailed into place. The sloping flat scarfs were angled from top to bottom and from forward to aft.

Transition from carvel-built to clinker-built construction of the bottom strakes

In the 13th-century cogs, at least the first two bottom strakes from stern to stem were in principle completely carvel-built. In the case of the Nijkerk-I cog, that was even the case in the first seven strakes, although the edges of the planks were chamfered and placed on top of each other in the aft and forward meters at the stern and stem. This was where the bilge strakes and the first two strakes of the sides were also fastened into place in the ship's hull in a carvel-built construction. In the 14th-century cogs, the bottom planks at the stern and stem were nailed together in a clinker-built construction along a length of approx. 2-3.5 m starting from the second strake (B strake). For this purpose, these parts of the planks were left wider at the bottom before being fitted to the hull. The wider piece (the overlap) was chamfered along the inside to make it thinner. The same was done to the top edge on the outside of the board that was already attached. The two diagonal surfaces were then clamped together and fastened to each other with double-clenched nails (folded twice and driven back into the wood). This transition from carvel-built to clinker-built construction between two successive shell strakes is very remarkable, and has only been seen in the cog-building tradition. The construction may therefore also be considered a characteristic feature of late-13th to 15th-century cogs. The planks in the transition from the bottom to the sides (the turn of the bilge) were, with the exception of the parts at the stern and stem as mentioned before, carvel-laid against the outer strakes of the bottom. The connection to the first strake of the sides is completely clinker-built. The combination of carvel-built bottom and clinker-built sides differs from the Scandinavian and English vessels. These ships were completely clinker-built, and the planks (of the earliest vessels) were split. The planks of the cogs were sawn.[27]

2 The sides

Depending on the type of vessel – (large) cog or small cog – the sides were either high or low, completely clinker-built, and everted. Sometimes the final pieces at the stern and stem were sanded smooth to ensure that the overlapping plank end did not become damaged. The sides of the small Almere-I cog, which was probably built at the end of the 14th century, deviated from this trend. At the level of the hold, the vessel's sides were nearly vertical, although they were less steep than Hocker's reconstruction indicates. Hocker also assumes that the width of the hold would remain the same throughout, which is not the case.[28] This is one of the earliest examples known thus far of emerging changes in shipbuilding practices (within the cog-building tradition) that started around that time, and in this case involving an inland vessel. The well-known 18th to 20th-century inland vessels in particular had sides that remained equidistant in width and were built vertically (creating a cargo hold that was larger and more practical for loading). Until now, I had assumed that this change had not started until the 17th century.

In general, the planks of the sides in the large cogs were regular in shape and less wide than the planks of the bottom. Some vessels were built using relatively narrow planks (e.g. the Doel-I cog, where the planks of the sides were 34-40 cm wide, including the overlap of 6-18 cm). The decorative grooves gouged into the lower edge of the shell strakes of this ship are remarkable. The planks of the shell strakes were fastened together with sloping flat scarfs, similar to the bottom strakes. The plank ends of the scarfs, which were 1-2 cm thick and sometimes as much as 4 cm thick, also protruded from the shell. The several striking longue-shaped plank ends of scarfs on the outside of the sides of the Nijkerk-I cog are decorative elements and, like the decorative grooves on the Doel-I cog, only known from the Viking ships from the Baltic region. This may offer a clue as to the region in which the ships were built. The clinker-built shell planks on the cogs were fastened to each other with double-clenched nails, which were folded down twice and driven back into the wood. The interval between the nails was usually 15-16 cm. The nails were hammered in from the outside. It has been established on a few shipwrecks and ship timbers that the nails were driven into the wood alternately from the outside to the inside and vice versa (e.g. the vessels from Antwerp and the Dronten-I cog). The Antwerp finds as well as finds from 's-Hertogenbosch and Kessel (North Brabant) suggest a construction method that was applied in the south-western Netherlands and in Flanders. As already discussed in relation to the sternposts and stemposts, the plank ends of the strakes of the sides were nailed into place against the mitred aft or forward side of the sternpost and stempost respectively. From the end of the 14th century, they were increasingly fastened into place in the rabbets in the sides of the stern- and stempost.

3 The caulking method

Previous researchers[29] have given attention to the method

27. Crumlin-Pedersen 1983, 7 and 17.
28. Hocker in: Hocker & Vlierman 1996; chapter 16.
29. Crumlin-Pedersen 1969, Ellmers 1972 and 1975, Sopers 1974, Arnold 1977, Cederlund 1978, Reinders 1979 and 1984.

used for caulking various vessels from different building traditions. Extensive and lengthy research has shown that the shell seams of all cogs (from the Low Countries) in the Hanseatic period were caulked in a characteristic manner.[30] The method was carried out with moss, moss laths and sintel clamps, and was almost certainly derived from the method used for centuries in Western Europe, even before the cog-building tradition emerged, to make the shell seams of river vessels watertight. The seams of inland and seagoing vessels from other shipbuilding traditions from the same period were made watertight by other means, specifically with animal hair drenched in tar that was clamped between the overlapping planks. There are a few late-12th-century and 13th-century ship finds from the Nordic/Viking shipbuilding tradition, in which small, leaking sections of shell seams were repaired with sintel clamped moss caulking. Sometimes a mix of different methods was used in the first half of the 16th century, when the use of sintel clamped moss caulking gradually disappeared. On a few occasions, a cog was found to have had part of the moss caulking between the keel plank and the garboards replaced by animal hair, a material mainly used in the Scandinavian shipbuilding traditions. This suggests that the ship may have been built (or repaired?; Doel-II) in the Baltic region. The seams on the inside of the shell were sintel-clamped before the frames were put in place. This offers evidence that the ship was built according to the 'shell first' method. The seams on the outside of the ship's hull were made watertight with sintel clamped moss caulking only shortly before the launch.

4 The frames, mast step, ceiling boards and inwale

Frames

There are a few things that can be said about the frames. First, with the exception of the top timbers against the sternpost and stempost, the frames are always composed from two or more parts, and the parts are connected by sloping flat scarfs. In the small cogs and boats, the different frame sections were generally approximately equal in thickness and width; sometimes a few frame sections near the mast and at the stern and stem were somewhat more robust. In the large cogs, the floor timbers on the bottom of the ship were generally noticeably thicker than the futtocks against the lower part of the sides, and the subsequent (second) futtocks were often slightly less robust. In addition, the (V-shaped) floor timbers on the bottom received more attention than the futtocks (generally not made from sapwood, and the pieces were generally square-hewn). A number of floor timbers were generally somewhat thicker near the mast. The futtocks were often not square-hewn, or only partly. Futtocks regularly had not had their sapwood removed, and the bark was sometimes still present. Irregularly shaped pieces of wood were also sometimes used. In general, the floor timbers were 15-25 cm wide and 10-15 cm thick. The V-shaped floor timbers were often somewhat thicker. Obviously, the dimensions were also tailored to the size of the ship. The V-shaped floor timbers in the afterbody and the forebody were <u>not</u> attached on the stern- and stem hooks and stern- and stempost with wooden treenails. This was probably intended to avoid weakening them. The frames in the cogs from the Low Countries were placed at intervals of approx. 50 cm centre to centre. In the vessels from the Scandinavian shipbuilding traditions, this interval was more than 60 cm, and usually around 80-90 cm. The frames of these vessels are also usually less thick.

Mast step

It was already established in the 1980s that the construction of the mast step was quite different in the cogs. There was no increase in the variety of types resulting from the later finds from the Low Countries. It is now clear that the large cogs, which had crossbeams protruding through the shell, always had a long keelson with an integrated wider and thicker mast step section, which had a rectangular recess for placing the mast. The underside of the keelson is usually recessed at the level of the frames, enclosing the top of the frames by several centimetres. This type of keelson is not restricted to the cog-building tradition and was in fact already seen as early as Roman times. The Kuinre cog could possibly be an exception with regard to the sole presence of a long keelson in large vessels. It is in fact not certain that the frames and the long keelson seen in Ketelhaven in 1970 belonged to the small vessel depicted in two photos from 1943, and/or belonged to the stern construction that had remained intact.

The mast step sections were supported on both sides by riders, short timbers that rested on the frames and continued until the bilge ceiling strakes. They were intended to absorb some of the lateral pressure of the mast. In the Nijkerk-II, Rutten, Doel-I and -II cogs and probably also in the cog from Medemblik Zeebad, there were three riders on either side. The Nijkerk-I cog and the Kampen (IJssel) cog had two and four respectively. The number of riders in the Kraggenburg and Enkhuizerzand cogs is not yet known. It is also not yet clear whether the Kraggenburg cog definitely had a long keelson. The pieces broken off on both sides near the mast step do not offer conclusive evidence. There may also have been a short keelson with a widened mast step section, as was found in the Marknesse cog. In that last vessel, it was also the secondary mast step, sup-

30. With sintel clamped moss caulking, Vlierman 1996a.

ported by a single rider on either side. Initially, the mast on this ship was in a recess in the top of a thick, wide floor timber. The Dronten-I cog also had a recess for the base of the mast in a thick and wide (V-shaped) floor timber. In the other small cogs, the mast step is in principle always formed by a short, rectangular beam with a square recess that was fastened into place on several frames. Some cogs had a knee-shaped mast support on the keelson behind the mast, or a thick plank placed in the aft side of the recess for the base of the mast and fastened into place at the top against the sail beam. The chocks against the sides of the Spakenburg-II cog should most probably not be regarded as a mast step, but were rather intended for the installation of posts for a hoisting device.

Ceiling

The large cogs have an 'open ceiling': planks laid lengthwise in the ship, usually with a plank-width gap between the planks, fastened into place against the frames on the inside of the ship. The ceiling was usually composed from a number of planks connected to each other with sloping flat scarfs that together formed a ceiling strake. Thicker planks were almost always used at the turn of the bilge, frequently with recesses cut into the underside where the frames would be. The remaining pieces fell between the frames. The main purpose of the ceiling strakes was to reinforce the longitudinal rigidity of the hull. In principle, the small cogs always had thick planks in the bilge ceiling. Often, judging from the traces left, the tops were mitred with an adze. This has also been observed in some large cogs. The thick planks of the bilge ceiling were located above the relatively weak scarfs between the floor timbers and the lowest futtocks of the frames. In addition to the bilge ceiling, the small cogs generally also had one or several ceiling strakes along the sides, and often had a continuous ceiling of planks above the nearly horizontal part of the bottom between the bilge ceiling and the after- and forward deck. The short mast steps were enclosed between the two. It has been observed that the assortment of thick, thin and damaged planks was sometimes disorganised, which indicates that they had to be replaced on a fairly regular basis. They were often made not from oak, but from cheaper conifer wood, and usually only fastened into place with a few nails. Evidence was found in some large vessels that suggested the presence of an extra loading floor installed about or up to 30-60 cm above the open ceiling on the bottom of the ship.

Inwale? on large cogs

In the large cogs, it has not been determined to date how the top of the gunwales was finished on the inside. For the time being, I am assuming that this would have been a somewhat thicker ceiling board.

Inwale in small cogs and schuyts

It is possible to say more about this aspect in relation to the small cogs. In several 14th-century vessels, it was possible to establish that they had a ceiling board along the top of the straight part of the sides. Along the top of the bow curving towards the stem and the corresponding parts in the afterbody, these ceiling strakes continued into inwales. The inwales and the ceiling strake were connected with sloping flat scarfs. In cross-section, the curved pieces were triangular in shape with uneven sides. They were attached to the chamfered frame heads and abutted against the inside of the shell strake at the top. There are also examples of more trapezoidal pieces of inwale , which had slanted rabbets cut into the outside, enclosing the slanting heads of the frames. There are also vessels in which the sections of the triangular or trapezoidal inwales were installed (in pieces) from stern to stem. In addition to providing a practical finishing touch to the gunwales, the inwales also provided an important reinforcement and added longitudinal rigidity to the gunwale. They have also been found on multiple occasions on the large, 16th-century fishing vessels of the Zuiderzee, known as *waterschepen.*

In the even smaller boats, the gunwale was often reinforced by a smaller reinforcing timber which was referred as an covering inwale. Usually square or rectangular in cross-section, these covering inwales were placed along the inside of the boat, binding together the squared-off heads of the frames. The pieces were connected to each other with sloping flat scarfs and pegged onto the top shell strake with wooden treenails. Several blind treenails were usually also driven into the heads of the frames from above. There are also examples of these square inwales that had slanted rabbets cut into them from the outside, enclosing slanting frame heads.

5 The crossbeams and/or transverse connections

Crossbeams on the large cogs

As has already been indicated several times, the large, high seagoing cogs probably always had crossbeams protruding through the shell. These beams were square and approx. 25-30 cm in cross-section, and had a rabbet all the way around at the point where they intersected with the clinker-built shell. The top of a shell strake was enclosed in the underside of these rabbets. In the sides and in the top of the beams, the rabbet would enclose the strake installed above it, in which a square opening has been cut to fit beforehand. The beams usually rested on the top of squared-off or slanted futtock heads. The beams protruded from the shell, standing proud by about 15 cm, and were protected on the front by guides known as beam-head guides. These crossbeams, together with the stringers, were essential for the transverse and longitudinal strength of the vessel at deck height and above. Together with the close-fitting construction of the underwater body, they combined with the frames to ensure the lines of the entire ship's hull would retain their shape. The entire, robust construction at deck height almost gives the impression that this is where the actual strength of the hull was located, with a watertight tub 'suspended' beneath it.

Heavy knee bulkheads were placed on the beams on the inside, which were usually assembled from beams approx. 25-30 cm high and approx. 15 cm thick, which were placed on the crossbeams protruding through the shell. The rising knee sections were then placed on top of those beams. This raised beam and knee bulkheads usually consisted of two sections that were fastened to each other with a sloping flat scarf approximately amidships. Sometimes there are three parts, for example knee bulkheads to starboard and port with a intermediate piece or connecting piece in between. This construction rose approx. 80-100 cm above the crossbeam (Kampen [IJssel] cog). The rising knee section came up to approximately gunwale height. On the outside of these knees, the sections above the deck and about 1 m sections of the everted sides were fastened into place with futtocks/frames. Recesses were cut into the upper edge of the horizontal sections of the knee bulkheads for placing the stringers lengthwise in the ship.

Transverse connections on the large cogs

Besides the crossbeams protruding through the shell, crossbeams were sometimes also used – particularly in the narrower parts of the afterbody and forebody – that did not protrude through the shell. In this study, they are referred to as transverse connections. The Nijkerk-I cog is the best example. In principle, it is a crossbeam that is cut at an angle on one end and fastened into place against the shell, with a rising knee attached to it that also serves as a futtock. On the other side, the beam is knee-shaped. Sometimes two of these knee-shaped beams were placed on top of each other, with the branch sections forming the futtocks. As stated, they were fastened into place against the hull, and sometimes also against the nearest frame lengthwise in the ship. There are also examples in which the beam plus knee sections were pegged with wooden treenails against the inside of a frame on both sides. In the Doel-I cog, this construction was used for several sharply V-shaped floor timbers in the lower part of the afterbody and forebody, with the beam section laid on the squared-off tops of the floor timbers and with the rising knee sections forming the lower part of the futtocks. This construction is also known from the Kalmar- I find (Sweden) and from the Vejby cog (Denmark). On the basis of a number of small details and (presumed) provenance of the wood in these ships, I believe that this is a typical construction method from the Baltic region/the Nordic shipbuilding tradition. A large windlass was generally fastened into place on and against the crossbeam at the forward side of the after deck. Handspikes could be inserted into the windlass to hoist or lower the sail and yard. On some ships, it was probably possible to stow the lowered sail plus the yard between the stringers under the forward deck. In the early cogs (see the 13th-century Nijkerk-I), the lowered yard plus the sail were probably placed on a forked wooden support near the stem, as depicted on some seals.

Transverse connections/deck beams in the small cogs and boats

In principle, the small cogs and boats have at least two transverse connections at deck height towards the front of the after deck and towards the aft side of the forward deck. In addition, in the small cogs, we see a third beam amidships across the hold. Under the after deck and forward deck, there were obviously also a number of deck beams, and sometimes two or four stringers placed lengthwise in the ship in the forward deck. Short knees (approx. 50 cm high) were placed on either side on the (three) transverse connections, against which the upper gunwale was fastened into place. These two knees often had long, horizontal sections and were connected to each other with a sloping flat scarf approximately amidships. Besides the beam that provided the transverse connection, there was also a timber known as a deck girder. The ends of the deck planks were usually enclosed between the beam and the deck girder. If there was a third transverse beam above the hold, it rested on both gangways and was also connected to the upper gunwales by rising knees. The beam plus deck rider or thick beam at the back of the forward deck was often also the sail beam, with a semi-circular recess in the beam to place and secure the mast against it. Just as in the large cogs, a windlass was usually placed against and on the transverse connection at the start of the after deck; this windlass could be used to hoist or lower a sail and yard or sprit.

In addition to securing the transverse beams against the sides and the frames with wooden treenails, there was sometimes also an additional connection to reinforce these important structural elements for the transverse connection. Fastenings known as 'rosebolts' were used on the Dronten-I and Almere-I cogs. Similar to wall anchors, these iron fastenings had a hammer-shaped or curved head and were inserted through the shell from the outside and secured with thick nails on the aft side of the transverse connecting beam.

6 Decks and castles

Decks on the large cogs

The planks of the after deck and forward deck were not nailed to the stringers and the deck beams attached to them in a transverse direction. Thick gangway planks were fitted between the after deck and forward against the sides, enclosing the frames and continuing to the inside of the shell. In vessels from the early 14th century and before, these may have been straight planks that ran from crossbeam to crossbeam at a distance of a few centimetres from the inside of the frames and rested in recesses in these beams. Based on what we know so far, it seems very plausible that the large cogs were open between the after deck and forward deck until the early 14th century. Only then would there have been a gradual transition towards completely closed decks (with hatches?).

A number of late medieval shipwrecks from the IJsselmeer polders provided some insight into the composition of the decks. In addition to the aforementioned stringers and deck beams supporting the decks, it appears that the large and small cogs had planks on the aft deck which were installed lengthwise, enclosing the frames in the sides and continuing until the shell. The planks were fastened to the beams with wooden treenails and/or nails. Sometimes the deck seams were sealed with sintel clamped moss caulking. There are also examples of thin, narrow planks that were placed over the deck seams and nailed into place on the deck planks on both sides of the seams.

As stated, in the large cogs as well as the small cogs, thick gangway planks up to approx. 50 cm wide were placed between the after deck and forward deck at deck height and against the sides. These are often long planks that extend into the decks. Only a few small cogs were found to have a coaming about 20 cm high on the inside of these gangways. This plank placed slightly inwards between the deck girders prevented people from falling too easily into the hold. The coaming was later used as the base for the deck hatches. It is not yet possible to say whether the large cogs also had coamings around any possible deck hatches.

For the time being, it appears that the large cogs had transverse planks in the forward deck running from the stringers to the starboard and port shell. They were probably supported by short beams placed lengthwise at about that halfway point, running from crossbeam to crossbeam against the frames. Deck riders were (sometimes?) placed above the crossbeams, with the deck planks between them. The late-medieval finds to date suggest that these short transverse deck planks lay almost completely unattached between the stringers and the shell (secured with only a few nails). This may have been so they could easily be removed for loading and unloading? In the reconstruction drawings of the Nijkerk-II cog and the replica ship (the Kampen Cog), it has been assumed that the ship had a continuous deck from the aftmost crossbeam to the forebody, with transverse deck planks resting on short beams. There is no certainty about this. The vessel had a slightly lower, smaller deck of planks placed lengthwise running from the aftmost crossbeam to the stern.

Decks and cuddies on the small cogs and boats

The after deck and forward deck in the small cogs and boats were supported on the hold end by fairly thick transverse connections/beams; in the small cogs, the forward deck was often supported by the sail beam. These vessels probably also always had at least two stringers, sometimes four, and a thick deck beam about halfway between the sail beam and the stem. In addition, the deck planks were supported by thinner, short deck beams that were usually fastened into place against the inside or sides of the frames with wooden treenails and/or iron nails. The planks of the forward deck were also placed in a transverse direction, as described above for the large cogs. Presumably there were hatchways in the forward deck beside the mast. In the after deck, the planks were probably placed lengthwise in relation to the ship. Above some of the transverse connection, knees approx. 50 cm high were fastened on the deck for attaching the upper gunwales and transverse connection beams to the sides. Slightly thicker wooden treenails were often used for fastening the shell and the frames. The after decks had planks placed lengthwise and presumably also a hatchway. The cuddies of the boats, which were generally even lower, most likely did not have hatchways. The spaces under the cuddies could not have been used for much more than storing any tools or personal possessions while under sail.

Castles

Although many seals depicting cogs also show that they had castles, there is hardly any evidence of such structures in the archaeological finds. Only the Nijkerk-II cog seems to have had a separate castle both aft and forward. The Bremen cog had an aftercastle that was more like a poop deck. Parts of what may have been a poop deck or aftercastle were found beside the Kampen (IJssel) cog.

7 The rudder, mast and sail and the ropes and blocks

The rudder

All vessels known to date that were built according to the cog-building tradition were kept on course with a stern rudder. It is very likely that this was also one of the characteristics of this shipbuilding tradition. 12th- and 13th-century ships from other shipbuilding traditions were kept on course with steering oars, as far as can be determined thus far. Ships depicted on some 13th-century English seals and the vessels on the baptismal fonts from Zedelgem (Belgium) and from the cathedral of Winchester (GB), both from the 12th century, probably show the ship type known as *nef*, which was Norman/built according to the Nordic tradition and innovated with a stern rudder. The Skagen cog (Denmark) built ca 1200 and the cogs from the Netherlands (Nijkerk-I and Ens) built around the mid 13th century are the oldest archaeological evidence to date that cogs were fitted with a stern rudder. There is no certainty as to whether or not the vessels from the Lefebvredok in Antwerp, dating from the 13th century or possibly even before, were kept on course with a stern rudder.

The mast and the spar or sprit

Masts, spars, sprits and rudders have almost never been found in the cog finds to date. The only exceptions to this are the rudder of the probably early-15th-century boat from the cog-building tradition, recently found next to the Kampen (IJssel) cog, and of the Dronten-II cog. With regard to how the rudders were implemented on cogs, we must therefore rely on the late-medieval depictions and on the rudders of the early-16th-century *waterschepen* that were found on several occasions. Evidence has been found in a few (small) cogs that suggests the presence of a removable mast that was often placed slightly closer to the stem. In those cases, it is also possible that these vessels were (for the first time?) running a spritsail. Rudders were often secured at the stern with a piece of rope. However, after some time, this rope would have come loose or snapped, after which the rudder could be lifted out of the gudgeons by the current, for example, and then drifted away. As the polders were cultivated, detached post-medieval ship's rudders were regularly found in the fields.

Sails

Pieces of coarse, woven woollen fabrics were found in several 13th- and 14th-century cogs. These were almost certainly pieces of sails. Some pieces show vertical and alternating light and dark strips in the weave of the fabric. During an excavation in Kampen in 2014, a unique and very large piece of a woollen sail was found at the presumable site of a 14th-century shipyard. The woollen sail remnants were all tarred to keep them intact for longer and improve their windproof properties.

The ropes and the blocks

Pieces of rope from the rigging etc. have only been sporadically found in and around the cogs. A nice disc block was found in the Almere-I cog. The block dating from ca 1400 is the oldest found in the Netherlands to date. In the Kampen (IJssel) cog, the entire rigging and several of the deadeyes for the lines before the mast were observed for the first time. A large piece of a heavy (anchor?) rope was found under the Rutten cog. The rope was made of hemp. Remnants of rope were found in the limber holes of the frames in some of the cogs. These ropes ran through the limber holes lengthwise in the ship. These limber-ropes could be periodically pulled to and fro in order to clear the limber holes of any dirt and debris, so any bilgewater and rainwater could drain to the place from where it would be pumped out. The water was pumped from there to approximately deck height and then via a wooden gutter to a scupper in one side of the shell (see the Nijkerk-II). The pump(s) were in principle placed in the deepest part of the vessel, usually above the keel plank, stern hook or stem hook in the afterbody and/or forebody.

8 The lines of the ship and the length-width ratios

The lines

Both the large and small seagoing cogs had a bottom shaped like a laurel leaf from the top view, tapering sharply towards the stern and stem and slightly narrower aft than forward. Amidships, the bottom was nearly horizontal in both length and width, rising slightly to the turn of the bilge. At the point of the hold, the bottom gradually transitioned into the everted sides, or the side was fastened into place against the outermost plank of the bottom at an angle. Towards the stern and stem, the bottom gradually transitioned into the sides and the frames (aft) were often S-shaped. Both large and small seagoing cogs were also shaped like a laurel leaf at gunwale height, with the forebody always being approx. 10-20 cm wider than the afterbody. The gunwales curved towards the stem at the bows, and curved towards the stern in the corresponding pieces in the afterbody. Cogs always had a straight stern and stem, with the stem showing a sharper rake than the stern. In small(er) vessels, we often see an almost identical rake in the stern and stem. Depending on the size and possibly also the function or range of operation of the vessel in question, there was a sharply rising sheer from stern to stem. Sometimes the gunwale was fairly flat. The sheer almost always rose higher towards the stem than in the afterbody.

Length-width ratios

Reinders discussed the dimensions of the bottom of the currently known cog shipwrecks in chapter 3. He was only able to say something about the lengths and greatest widths of the stern and stem and the gunwales about a few vessels.[31] Now, with the larger number of finds and thanks to the reconstruction drawings of most of the ships, it is possible to say more on this matter. Moreover, more clarity is now available about the positions of the masts in relation to the length of the ships at gunwale height (fig. 29.1). Further on in the catalogue, the entry for each ship lists the particular details of the specific vessel, the lengths and widths of the bottom and from stern to stem and across the gunwales, and the length-width ratio. In the large and small seagoing cogs, the length-width ratio is always 3.2:1 or slightly less. The straight stern and stem and the length-to-width ratio from stern to stem and across the gunwales have never previously been spoken of as features of the cog-building tradition. Both can probably be considered as such.

The hulls of the cogs from the Low Countries were made of oak. As an exception, the keel plank on the Ens cog was replaced by one made of beech. Other than that, only a few pinewood ceiling planks have been found.

The presence of certain constructions and/or some of the above-mentioned characteristics on other vessels, which differed from the cogs in terms of the dimensions and length-width ratio, made it clear that these vessels were in principle built in the same way and can therefore be considered part of the same building tradition. The differences compared to the 'true' cogs involve, on the one hand, the type, dimensions, length-width ratio and probable functions. On the other hand, they are examples of developments over time. These vessels classified in the group of boats are among the oldest (section 26.6) and youngest (Chapter 17 and Sections 26.1, 26.2 and 26.3) examples of the cog-building tradition in the Low Countries. The lengths and widths of the bottom and from stern to stem and across the gunwales, and the length-width ratio derived from those dimensions, are listed in the catalogue. There are also ship finds that have a few cog-like features which must be categorised as a different type of ship and/or part of a different building tradition. They may possibly have been influenced by the cog-building tradition.

Ship reconstructions on paper. Possibilities, limitations and recommendations

One of the (primary) aims of this study was to investigate the possibilities and limitations of reconstructing the excavated shipwrecks on paper and producing standard (reconstruction) drawings of all the vessels covered here. The possibilities, limitations and recommendations are:

- The reconstruction drawings on how the ships were constructed should preferably be on a scale of 1:10 in order to obtain a reliable plan of the planks and to be able to display details accurately.
- While the drawings were being made, it was determined that, in order to obtain an acceptable reconstruction of the lines, the ship's planking lines at actual size should be within a margin of 2-5 cm, taking into account any shrinkage or expansion of the wood. If this turns out to be impossible, then the starting points (the reconstructed cross-sections) will not be accurate.
- Before it is possible to write anything meaningful about a ship find, all reconstruction drawings must be satisfactorily completed. If this takes place in reverse order, the text will undoubtedly need to be updated after the drawings are completed.
- A scale model should preferably be built. However, life-size reconstruction (experimental archaeology) is much to be preferred in order to arrive at a reliable replica of the lines of the hull. When making a model to scale, the smaller planks are easier to force in a certain direction than when using full-sized oak planks 4-5 cm thick. In other words, a model built to scale 1:20 may look very good, but it may possibly not have the right lines or dimensions. In order to address this issue to some extent, it was decided at an early stage of the modelling process at the Netherlands Institute for Ship and Underwater Archaeology (NISA) to switch from scale 1:20 to scale 1:10 and to work in oak. Test models of aircraft plywood or cardboard produce unreliable results. However, in most cases it is not realistic to assume that a ship can be rebuilt 1:1, due to financial considerations, among other things.

29.4.2 Sequence of the work and construction phases

The research done on the various shipwrecks, the model-building and reconstruction of the Nijkerk-II (the Kampen Cog) clearly showed that the large and small cogs were in principle built according to a predefined sequence and that the building process took place in phases. The sequence of work is described in the construction phases outlined below.

Phase 1

Setting up the keel plank and the stern and stem hooks and fastening them together. It is likely that the sternpost and stempost were placed on that immediately or later, depending on the dimensions of the ship to be built and the height of the rising sections of the stern- and stem-posts and stern- and stem hooks. This entire assembly was put together on structures known as piles, in which the stern- and stem constructions were braced with poles. A thin layer of moss caulking was applied to the flat scarfs between the various elements.

31. Reinders 1985a, 23-25.

Phase 2
Also depending on the dimensions of the future vessel, several or many temporary timbers were attached to the keel plank for short-term support. These were evenly distributed the part of the bottom that would later be carvel-built. In this phase, no temporary frames were used in the narrow afterbody and forebody. After that, the planks for the garboards and the subsequent bottom strakes were put into place, starting from the stern and working forward. The first board was pre-shaped (heat-moulded), cut to fit, provided with a sloping flat scarf for the connection to the next plank to be added, and nailed into the rabbet in the stern hook and against the underside of the temporary timbers. Judging from the nail hole plugs that were found, in some situations there must also have been short planks to temporarily connect two planks at a certain location, for example. On two vessels, several small temporary frames were left in place between the frames (Dronten-I and Swifterbant). Depending on the length/height of parts of the temporary timbers that continued into the sides, the angled bilge strakes were also put into place in this phase of the construction of the ship's bottom. The aft and forward planks of the second and subsequent bottom strakes were left wider in the sections of the bottom that were twisted into a nearly perpendicular angle, thus overlapping the strakes that had been placed before/below that, and nailed into place with double-clenched nails. In the early 12th/13th century ships, the twisted ends up to the stern hooks placed against each other in a carvel-built construction. However, the edges resting on each other were sometimes bevelled and fastened to each other with a single nail. The bottom of the ship, assembled and 'suspended' only from a few temporary frames, and nailed against the stern- and stem hooks, was supported on the underside by short poles. Between the successive planks of a strake, a layer of moss caulking was applied to the scarfs, after which the plank ends were nailed to each other.

Phase 3
After this, the (V-shaped) floor timbers and planks above the carvel-built part of the bottom were marked off, cut to fit and fastened into place on the bottom strakes with wooden treenails. If necessary, the temporary frames were replaced by definitive frames immediately or later. After removing the temporary frames and before placing the final frame sections into place, the nail holes were closed with small wooden pegs known as nail hole plugs. To reinforce the rigidity of the entire assembly, a long keelson with integrated wider mast step was placed over the floor timbers and planks and above the keel plank on the larger vessels. It is likely that ceiling boards were already put into place above the bottom. In the smaller vessels, a short mast step was placed, or the recess for the heel of the mast was cut into a wider floor timber or V-shaped floor timber. After sintel clamped moss caulking was applied to the clinker-built seams between the bottom strakes in the afterbody and forebody and between the garboards and the stern- and stem hooks, the V-shaped floor timbers in these parts of the ship were put into place and fastened with wooden treenails. The V-shaped floor timbers were not fastened with wooden treenails to the stern- and stem hooks and sternpost and stempost, only to the shell planks on either side.

Phase 4
Then , the planks of the strakes of the sides were shaped, cut to fit, provided with sloping flat scarfs, and nailed into place from aft to forward, placing each plank in a fully clinker-built construction against the preceding strake. In the large vessels, this 'shell first' method was used to build up to the place where the crossbeams protruding through the shell were meant to go. In the smaller vessels, this method continued up to the top of the gunwale. A thin layer of moss caulking, also known as moss paper, was applied between the overlapping plank ends at the scarfs. After that, the seams on the inside were provided with sintel clamped moss caulking and then the (lowest) futtocks were placed. In the large ships, at this stage of construction these futtocks were probably only placed between the future crossbeams that marked the aft and forward end of the hold. They would have waited to install the frames that were still missing in the afterbody and forebody until the crossbeams protruding through the shell had been installed and the top strakes of the sides had been constructed. This was important, because deformation of the hull due to the weight of the afterbody and forebody and the stern and stem and all the frames was entirely conceivable until the crossbeams protruding through the shell were put into place.

Phase 5
In the large vessels, the crossbeams protruding through the shell were put into place in this construction phase. In the small ships, this would very likely have been the transverse connections that rested on shorter futtocks inside the gunwales and the heavy gangways which had already been placed on top. These transverse connections were fastened on the gangways and against the sides with wooden treenails. In the large vessels, the strakes were then assembled in the top part of the sides, which were usually more than 1 m high. The clinker-built seams of these strakes were then sintel clamped on the inside. Then the (second) futtocks were placed. The remaining futtocks and the floor timbers against the sternpost and stempost of the sharply tapered afterbody and forebody were also

put into place in this phase. The vessel was then provided with ceiling strakes above the turn of the bilge and in the sides to provide rigidity for the entire structure and to reinforce the longitudinal rigidity. Above the turn of the bilge, these ceiling strakes were generally thicker, and sometimes notched on the underside so they would fall into place over the frames. In the smaller vessels, the short mast step was also placed in this phase (or earlier) and the gunwale was reinforced with an inwale.

Phase 6

The small vessels were then fitted with the after and forward decks; in many cases, the heavy gangway planks were integrated into these decks. In the large ships, heavy knee constructions were assembled on the crossbeams against which the upper parts of the everted sides were fastened. In many cases, stringers were probably installed lengthwise in the ship, above the middle of the forward deck. Together with the crossbeam constructions, these stringers ensured the longitudinal and transverse strength of the ship at deck height.

Phase 7

In the final construction phase of the large vessels, the remaining parts of the decks, the castles (if any), the deck layout, the false stern and stem, the rudder, and the mast and the rigging were added. Before the ship was launched, the shell seams were made watertight with sintel clamped moss caulking on the outside. Until the second quarter of the 14th century, all the seams were sintel clamped on the outside. From that time on, in any case from ca 1350 on, this practice was limited to the carvel seams of the bottom, and sintel clamped moss caulking was only used on the seams on the inside. After the Kampen (IJssel) cog was salvaged on 10 February 2016, it could immediately be determined that the seams on the outside had been made watertight entirely using sintel clamped moss caulking. This is contrary to the time frame stated above, since the provisionally assumed dating indicated that the ship had been built in the first half of the 15th century. How this should be interpreted is not yet clear at this time. The ship may be older than was assumed in February 2016, or it is possible, for example, that the degree of caulking used in the IJssel region did not evolve for some time, whereas the practice did continue evolving in other regions. Further research may be able to answer this question. In vessels built in the first half of the 15th century, we also see the practice of sintel clamping on the inside of the ship gradually diminishing; sometimes the method was only used on clinker-built seams in the sharp afterbody and forebody. In other vessels, such as the striking, entirely clinker-built large fishing vessel of the Zuiderzee, the *waterschip*, this method of caulking remained fully in use on the inside until well into the first half of the 16th century. In a carvel-built watership built ca 1550, only the seams in the fish-well cover were still sealed with sintel clamped moss caulking.

29.4.3 Catalogue: Inventory of structural data and phenomena per ship

This section discusses the exceptional features and constructions of the ships covered in this study, one ship at a time. These are standardised summaries of the construction and sailing history of the vessels. The most striking finds of the ship's equipment, inventory and personal equipment found in the wrecks are also listed. The data is displayed schematically and listed point by point per ship. The standard summary is structured as follows:

Name of the ship and the chapter/section that covers the ship. Any previously published excavation report is referenced between brackets.

- Type of vessel, date of construction/foundering and possible construction region;
- Circumstances of the find, surrounding soil, wreck formation and preservation;
- Relation to the late-medieval surroundings of the findspot/ship;
- Dimensions and proportions of the reconstructed ship:
  A. Length of the keel plank (including stern and stem hooks)
  B. Length of the bottom (from the start of the rising stern to the start of the rising stem)
  C. Broadest width of the bottom
  D. Length from stern to stem at gunwale height
  E. Broadest width across the gunwales
  F. Height of the side amidships
  G. Length-width ratio of the bottom
  H. Length-width ratio from stern to stem and across the gunwales
- Construction method used for the bottom and the sides;
- Sintel clamped moss caulking;
- Keelson/mast step;
- Ceiling boards/ceiling;
- Crossbeam constructions and/or transverse connections protruding through the shell, decks or platforms, deck layout and castles;
- Drainage for the rainwater and bilge water;
- Sails, rope and blocks;
- Most prominent items found from the ship's equipment, inventory and personal effects.

Nijkerk-II cog, Chapter 6.

- Large cog, built 2nd quarter/sank 3rd quarter of the 14th century, IJssel river basin;

- Did not founder at sea, but was scuttled/discarded in a peat trench ('harbour basin') and deposited. Probably the most important reason that high parts of the ship were also preserved;
- Presumably partly demolished;
- Situated near the 12th to 15th-century farming village of Ark;
- Dimensions and proportions of the reconstructed ship:
  A. 12.30 m
  B. 11.00 m
  C. 3.40 m
  D. 20.00 m
  E. 8.70 m
  F. 4.30 m
  G. 3.2:1
  H. 2.3:1
- Carvel-built/clinker-built bottom, clinker-built sides;
- Seams in the clinker-built planking sealed with sintel clamped moss caulking on the inside and outside. Carvel seams of the bottom sealed on the outside;
- Long keelson with integrated wider mast step with three riders on either side;
- Open ceiling;
- Crossbeams protruding through the shell. One beamhead guide found;
- Probably modified 10-20 years after construction (one strake higher, a continuous deck and evidence of castles);
- Drain channel for pump water/toilet. Small (closable) hole in shell into which the drain channel was inserted;
- Openings in the upper gunwale of the forebody (for gangway?);
- Fragment of a woven and tarred woollen sail;
- An exceptional red earthenware jug and an intact wooden shovel.

Nijkerk-I cog, Chapter 8.
- Large cog, built around the middle of/sunk in the 3rd/4th quarter of the 13th century, Baltic side of north-western Germany/Sleswig-Holstein and/or southern Jutland;
- Did not founder at sea, but sank against a peat coast. Probably the most important reason that high parts of the ship's hull were also preserved;
- Most likely partially demolished;
- Dimensions and proportions of the reconstructed ship:
  A. 13.20 m
  B. 12.90 m
  C. 2.70 m
  D. 18.30 m
  E. 6.90 m
  F. 3.40 m
  G. 3.2:1
  H. 2.7:1
- Fully carvel-built underwater body, rest of the sides clinker-built;
- Longue-shaped plank scarfs at several places on the outside of the sides, and several (closed-off) oval openings in the sides for pump water drainage. On the outside, it was determined that one oval hole very probably had a short leather spray hose nailed around the hole;
- Seams in the clinker-built planking sealed with sintel clamped moss caulking on the inside and outside. The carvel seams of the bottom were caulked on the outside, while sintel clamps were pounded into place on the inside at intervals of approx. 15 cm to seal the moss caulking in place. The triangular caulking seams covered the entire thickness of the planks;
- Long keelson with integrated wider mast step with three riders on either side and aft of that the hole for the base of the mast, with a knee-shaped mast support;
- Open ceiling;
- Crossbeams protruding through the shell and striking transverse connections in the afterbody and forebody;
- The pump water was probably drained to the side via a gutter and carried off through a hole in the shell and the short leather spray hose;
- Fragment of a woven and tarred woollen sail;
- Wooden chest filled with loam in use as a firebox;
- Striking iron (battle) axe.

Ens cog, Chapter 9.
- Small cog (for inland lakes and waterways), built 2nd/3rd quarter/sunk at findspot in the late 13th/early 14th century, IJssel river basin?;
- It is likely that the old wreck, which was already incomplete and had sunk elsewhere, ended up at the findspot when the river was high and the current was running strong;
- Dimensions and proportions of the reconstructed ship:
  A. 8.70 m
  B. 8.40 m
  C. 1.50 m
  D. 11.20 m
  E. 3.50 m
  F. 1.50 m
  G. 5.6:1
  H. 3.2:1
- The vessel was repaired during the period in which it was in use. On that occasion, the keel plank was probably replaced by a beechwood (!) keel beam. In addition, skegs were fastened into place under the stern

and stem hooks with thick beechwood (!) treenails, presumably with a plank or filler piece in between;
- The planks of the underwater body were entirely carvel-built from stern to stem and are only 2 cm thick. Only a few pieces of planking were found from the sides;
- The triangular carvel seams were either slightly open, or the planks only touched at the top, and were completely sealed with sintel clamped moss caulking;
- The frames were to the shell with wooden treenails 2 cm thick;
- Very sparing use of nails;
- Probably a short mast step;
- It was not clear whether the clinker-built planks of the sides were fastened together with double-clenched nails (folded twice and driven back into the wood) or with wooden treenails. In addition, it is not entirely certain that the vessel had clinker-built sides;
- Small frames 10-12 cm thick and wide;
- Almost certainly a bilge ceiling, probably also an open ceiling along the sides;
- No finds were discovered.

Rutten cog, Chapter 10.
- Large cog (largest from the Low Countries, possibly a military vessel [or a *Heerkogge*/war-cog?]), 4th quarter of the 13th century/shortly before or around 1300, German/Polish Baltic coast;
- Presumably stranded against or close to the coast (which was a peat coast);
- At low tide, the vessel jutted several metres above the water's surface, was easily accessible, and was probably largely demolished shortly after the disaster by e.g. the inhabitants of the nearby village of Lemmer;
- Dimensions and proportions of the reconstructed ship:
  A. 15.60 m
  B. 15.00 m
  C. 7.75 m
  D. 22.40 m
  E. 9.50 m
  F. 4.50 m
  G. 3.1:1
  H. 2.4:1
- The ship was extraordinarily robust in construction. The keel plank of 40-50 cm wide and approx. 15 cm thick. Bottom strakes of 50-60 cm wide and 5-8 cm thick. The plank ends at the >1 m long sloping scarfs were 4 cm thick and protruded beyond the shell. The frames were 22-30 cm wide and thick;
- Long keelson with integrated wider mast step section with three riders on either side;
- The vessel had an open ceiling and presumably a floor at approx. 1 m above the keel line;
- Under this floor, probably on a permanent basis (and from the ship's construction onwards), there were at least 5000-6000 kg of ballast stones. Approximately half of them consisted of Baltic flint;
- Between the ballast there were large bricks (and fragments of such bricks), possibly from a fireplace, which correspond in terms of their fabric to the bricks used to build the 13th-century churches in Schleswig-Holstein and along the south-western Baltic coast;
- It seems very probable that the ship had crossbeams protruding through the shell on two levels, as well several transverse connections at the stern and stem in the underwater body;
- The seams of the clinker-built planking were sealed with sintel clamped moss caulking on the inside and the outside. The carvel seams of the bottom and the cross-cut plank ends at the scarfs were sealed with sintel clamped moss caulking on the outside;
- There were a large number of nail hole plugs, indicating that temporary frames were used on a large scale during the construction;
- A complete skeleton of a sheep or goat and a piece of a thick hemp rope are the most striking finds.

Dronten-I cog, Chapter 11 (Reinders *et al.* 1980, 6-15 and 47-50).
- Small seagoing cog, shortly before or around 1300/ transition to 1st/2nd quarter of the 14th century, south-western Netherlands/Flanders?;
- Foundered on the east side of the (underwater) peat barrier between Schokland and Elburg, near one of the former estuaries (the Zuiderdiep) of the IJssel River;
- Dimensions and proportions of the reconstructed ship:
  A. 12.40 m
  B. 11.85 m
  C. 2.05 m
  D. 15.80 m
  E. 5.20 m
  F. 5.8:1
  G. 3.0:1
- Remarkably robust stem hook;
- Nails in the clinker-built fastenings of the planks were driven into the wood alternately from the outside to the inside and vice versa;
- Not a mast step but a recess in a wide, thick V-shaped floor timber for the base of the mast;
- Several small temporary frames left in place between the frames;
- Transverse connections in the underwater body at the stem and stern. Sail beam/transverse connection at deck height structurally reinforced with ham-

mer-shaped iron rosebolts;
- Presumably complete (or near-complete) inventory of ceramics and tools;
- Remarkable finds include the two lance tips, the hunting set of knives, a dagger knife with inlaid initials, and drinking and cooking utensils consisting of near-stoneware, *kugeltopf* and red earthenware pottery.

Marknesse cog, Chapter 12 (Modderman 1945).
- Small cog for navigating inland lakes and waterways and the Zuiderzee, 2nd quarter of the 14th century/ca 1375, northern Netherlands;
- First ship find identified as a cog (1945). This find was the reason for regularly conducted nautical archaeological excavations in the Netherlands, particularly in the IJsselmeer polders;
- The best, and most completely preserved, and oldest example of a small cog;
- Dimensions and proportions of the reconstructed ship:
  A. 11.80 m
  B. 11.30 m
  C. 2.85 m
  D. 14.80 m
  E. 4.65 m
  F. 1.60 m
  G. 3.1:1
  H. 3.2:1
- The vessel was most likely used for a relatively long period of time, and suffered significant damage in the vicinity of the mast during its period of operation. This damage was repaired. On that occasion, the mast was probably moved forward and placed in a new, short mast step that was placed on the frames. The mast was initially in a recess in a V-shaped floor timber, slightly farther back. This presumably was also related to the simultaneous transition from a square sail to a spritsail and the addition of the removable mast;
- Inventory with comparatively few tools;
- Cargo of 20 tonnes of large bricks.

Kuinre cog, Chapter 13.
- Small cog for navigating inland lakes and waterways and the Zuiderzee, 3rd/4th quarter of the 14th century/ca 1400, north-eastern Netherlands?;
- One of the first shipwrecks found in the Noordoostpolder and oldest known find of a cog in the Netherlands (1943);
- Only intact elements are stern hook plus sternpost. Two photos presumably show the excavated wreckage;
- The small vessel probably sank in the Kuinder or in an adjacent 'harbour basin' and lay between the locations of the two castles of Kuinre.
- Dimensions and proportions of the reconstructed ship:
  A. ?
  B. ?
  C. ?
  D. ca 13.00 m
  E. ?
  F. ?
  G. ?
  H. ?

Spakenburg-I cog, Chapter 14.
- Small cog for navigating inland lakes and waterways and the Zuiderzee, 4th quarter of the 14th century/1st-2nd quarter of the 15th century, south-eastern coast of the Zuiderzee?;
- Sank in the approach route from Spakenburg/Bunschoten. Discarded or deliberately scuttled after a dike breach?;
- Dimensions and proportions of the reconstructed ship:
  A. 13.00 m
  B. 11.80 m
  C. 3.10 m
  D. 15.20 m
  E. 4.80 m
  F. 1.70 m
  G. 3.8:1
  H. 3.2:1
- Closed ceiling/loadin floor;
- Short mast step and a housing for a removable mast;
- Load with (new) bricks in the middle of the hold;
- Several (striking) finds: part of a compass glass, a broken anchor blade that had penetrated through the shell on from the outside at the turn of the bilge, and the upper part of a discarded, half-height shoe.

Spakenburg-II cog, Chapter 15 (Van de Moortel 1991).
- Small seagoing cog/working schuyt, 1st-2nd quarter/2nd quarter or c. mid-15th century, south-eastern coast of the Zuiderzee?;
- Met with an accident/sank and ran aground on a high Pleistocene donk;
- Dimensions and proportions of the reconstructed ship:
  A. 9.30 m
  B. 9.05 m
  C. 2.80 m
  D. 12.40 m
  E. 4.30 m
  F. 1.25 m
  G. 3.2:1
  H. 2.9:1
- Remarkably 'round', low-lying vessel;

- Instead of a mast step, there were heavy chocks in the sides with a recess for vertically placed beams or stanchions. In any case, there was a chock on the starboard side and the port side. There may also have been two of these chocks at the back of the ceiling/work floor;
- The vessel may have been one of two on which a hoisting structure could have been built. Between the schuyts, an object could be hoisted from the seabed or, for example, piles could be driven into the seabed.

Almere-I cog, Chapter 16 (Hocker & Vlierman 1996).
- Small cog for navigating inland waterways and the Zuiderzee, shortly before or around 1400/early 2nd quarter of the 15th century, south-eastern coast of the Zuiderzee?;
- One of the small cogs that were most completely recovered;
- Short mast step enclosed by a closed ceiling;
- Striking iron rosebolts with large round, curved heads to reinforce the construction of two transverse beams connected to the sides;
- Contained the oldest disc block in the Netherlands to date;
- Beautifully crafted tools, a leather bag and a unique birchwood (lepers') clapper with the depiction of Saint Catherine of Alexandria, several coins and human bones.

Dronten-II cog, Chapter 17.
- Boat with cog-like features, 2nd/3rd quarter of the 15th century/2nd half of the 15th- early-16th century, Zuiderzee coastline of the provinces of Overijssel/ Gelderland;
- Old vessel that had presumably reached the end of its useful life and was discarded;
- Dimensions and proportions of the reconstructed ship:
  A. 7.50 m
  B. 7.10 m
  C. 1.35 m
  D. 9.45 m
  E. 2.35 m
  F. 0.65 m
  G. 5.25:1
  H. 4.00:1
- Only three strakes on each side of the keel plank, one for the bottom and two for the sides;
- Only sintel clamped on the outside of the seams between the keel plank and the first bottom strakes. The sintel clamps overlapped. No caulking and sintel clamps in seams between the clinker-built planking of the sides;
- Much of the rudder still present;
- Probably had a thwart through which the mast was inserted. There was probably originally a plank bridge above the bottom of the ship and on the frames (but none was found).

Swifterbant cog, Chapter 18.
- Small seagoing cog, probably similar to Dronten-I, early 14th century/c. 1325, south-western Netherlands/ Flanders? (field evaluation performed);
- Foundered in the middle of the sea;
- Dimensions and proportions of the reconstructed ship:
  A. ca 11.50 m
  B. ca 11.00 m
  C. ca 2.00 m
  D. ca 15.00 m
  E. ca 5.00 m
  F. ca 5.5:1
  G. ca 3.0:1
- Clinker-built seams were sintel clamped on the inside and outside, and bottom seams on the outside;
- Small temporary frame left in place at the turn of the bilge between the frames;
- Foreign(?) Burgundian (roof) tiles of the fireplace, bronze tripod pipkin and a lance tip.

Almere-II cog, Chapter 19.
- Small seagoing cog, 2nd quarter/2nd half of the 14th century?, north-eastern Netherlands (field evaluation performed);
- Foundered in the middle of the sea, turned turtle;
- Dimensions and proportions of the reconstructed ship:
  A. ca 12.20-13.20 m
  B. ca 11.50-12.50 m
  C. ca 2.00 m
  D. ca 16.50-17.50 m
  E. ca 6.30 m
  F. ca 3.20 m
  G. ca 4.0-4.2:1
  H. ca 2.6-2.8:1
- Stem hook plus what is presumably the complete and intact stempost still present;
- Sides start amidships at a fairly acute angle compared to the bottom of the ship;
- Clinker-built seams were sintel clamped on the inside and outside, and bottom seams on the outside.

Doel-I cog, Chapter 20.

- Large cog, shortly after 1326/3rd quarter of the 14th century, Baltic coast near Lower Saxony (salvaged);
- Probably capsized during tsunami-like tidal wave and sunk in the 'Deurganck' or 'Grote Gat' near Doel on the Scheldt. Was situated upside down. Hull structures deformed as a result of unnatural positioning and as a result of the iron fastenings that had been completely dissolved (by exposure to the salt water);
- Dimensions and proportions of the reconstructed ship:
  A. ca 14.00 m
  B. ca 13.30 m
  C. ca 3.10 m
  D. ca 20.00-21.00 m
  E. ca 7.00 m
  F. ca 4.40 - 4.50 m
  G. ca.3:1
  H. ca 2.85-3.00:1
- Ship's hull found in very complete state. Port side of forebody present as far as the top of the gunwale;
- Stern hook and stem hook plus stempost present and intact, sternpost, false stern and -stem missing;
- Crossbeams protruding through the shell were partly present, plus a number of beam-head guides;
- Shell strakes on the outside strikingly decorated on the outside with gouged seams in which the nails of the clinker-built connections had been placed;
- Opening in the gunwale of the forebody (see also Nijkerk-II cog);
- Shell seams made watertight using sintel-clamped moss caulking on the inside as well as the outside. Not sintel clamped on the outside at the level of the waterline from unloaded state to loaded state;
- Striking transverse connections in the underwater body at the stern and stem;
- Some wooden treenails had a protruding and slightly curved head on the outside of the ship;
- The ship had presumably originally been tarred fairly thickly on the inside and outside.

Doel-I cog, Chapter 21.

- Large cog, 2nd/3rd quarter of the 14th century, northern German/Polish Baltic coast (salvaged);
- Probably capsized simultaneously with the Doel-I cog during a tsunami-like tidal wave and sunk in the 'Deurganck' near the Scheldt;
- Dimensions and proportions of the reconstructed ship:
  A. ca 13.00 m
  B. ca 12.00-12.30 m
  C. ca 3.00 m
  D. ca 18.90-19.60 m
  E. ca 6.50 - 6.70 m
  F. ca 3.80 - 4.10 m
  G. ca 3.9-4.1:1
  H. ca 2.9:1
- Only a large part of the upside-down bottom of the ship plus frames, ceiling strakes and keelson;
- No stern- and stem hooks with a keel plank between them, but a stern- and stem hook with long keel plank sections that were fastened to each other amidships with a sloping scarf;
- Completely carvel-built bottom from stern to stem;
- Striking combination of moss and animal hair as caulking material for the sintel-clamped moss caulking;
- Some wooden treenails had a curved heads standing proud from the shell on the outside of the ship;
- The ship suffered considerable damage to the bottom during the period in which it was in use and was later repaired.

Kraggenburg cog, Chapter 22.

- Large, mid-sized or small cog, 1st/2nd quarter 14th century, northern or north-western Netherlands (exploratory study);
- Probably stranded and sunk with a south-western storm(?) in the shallows north of the original De Voorst peninsula.
- Keel beam comprised of two parts. Sintel clamps on the inside above the carvel seams. Reversed V-shaped carvel seams along the entire plank thicknesses.

Kampen (IJssel) cog, Chapter 23.

- Large cog, 1st half/3rd quarter of the 15th century (field evaluation performed, salvage 10 February 2016);
- The old ship, which had presumably reached the end of its useful life, was most likely deliberately scuttled in the IJssel River off Kampen in order to dam a stream channel of the IJssel and thus improve the water discharge in relation to the silting up of the IJssel estuaries;
- Dimensions and proportions of the reconstructed ship:
  A. ?
  B. ?
  C. ?
  D. ca 20.00 m
  E. ca 7.50 m
  F. ?
  G. ?
  H. ?
- The ship's hull is fairly intact up to the crossbeams protruding through the shell. Higher parts and decks were also found;

- The wreck is likely to provide additional/unknown information about the changing construction of the cogs during the period in question;
- A brick oven on the deck and remnants of a loading bed above the open ceiling in the bottom of the ship.
- No further data available yet.

Enkerhuizerzand cog, Chapter 24.
- Large cog, 3rd/4th and 4th quarter of the 15th century (field evaluation performed);
- Foundered in the middle of the sea;
- Ship's hull probably present at least up to deck/crossbeam height.
- No further data available.

Medemblik 'Zeebad', section 25.1 (Maarleveld 1984, 45-61).
- Large cog, 15th century (field evaluation performed);
- Probably stranded off the coast;
- Dimensions and proportions of the reconstructed ship:
  A. ?
  B. ca 15.00 m
  C. ca 3.60 m
  D. ?
  E. ?
  F. ?
  G. ca 4.15:1
  H. ?
- Almost entire bottom of the ship with floor timbers and keelson;
- Keel (or keel plank?) with the sternpost placed on top of it. Probably not a stern hook;
- Poles, possibly from a loading bed, and a large number of ballast stones.

Cargo vessel G 37, Noordoostpolder, section 25.2
- Mid-sized cog for inland waterways, Zuiderzee and Waddenzee, 15th century (excavated, data not developed further);
- Foundered in the middle of the sea;
- Entire underwater body and almost the entire starboard side, stern and stem;
- Keel plank in two parts, sternpost and stempost placed on the keel plank;
- Plank ends nailed in rabbets in the sternpost and stempost;
- Keelson/mast step consisting of three parts;
- Inside and outside seams of the clinker-built connections made watertight using sintel-clamped moss caulking;
- Approx. 275 large, heavy ballast stones.

Boat K 73/74 Eastern Flevoland, section 26.1. (Reinders *et al.* 1980, 17-29 and 47-50).
- Boat with cog-like features, 1st/2nd/3rd quarter of the 15th century, Kampen/IJssel delta?;
- Foundered in the middle of the sea/met with an accident and drifted away underwater to the site, where it settled to the bottom. Judging from the length, width and height of the gunwale, the vessel is not actually suitable for sailing on the Zuiderzee;
- Dimensions and proportions of the reconstructed ship:
  A. 13.55 m
  B. 13.35 m
  C. 1.90 m
  D. 15.80 m
  E. 2.40 m
  F. 0.80 m
  G. 7:1
  H. 6.6:1
- Hook-shaped stern and stem with rabbets for the bottom strakes and shell strakes in the sides. They were fastened onto the ends of the middle plank of the bottom;
- Mostly flat, carvel-built bottom and slightly everted sides consisting of two clinker-built strakes. Built with L-shaped frames in the flat part of the bottom. Aft and forward, the planks of the bottom transitioned into a clinker-built construction; the seams on the inside were sintel clamped;
- Small recesses for placing a mast/vertical posts in two floor timbers/L-shaped frames. In the latter case, possibly provided with a horizontal beam, over which a sail could be fitted as a canopy;
- The vessel was built as a small cog with a low gunwale with a piece of about 1/3 of the total length placed in between;
- Inventory including two stoneware jugs, tools, a beautiful dagger and a number of coins, amongst others several special coins from the city of Kampen;
- No fireplace;
- Could be rowed and poled. Predecessor of the track/tow boat intended for regional and interregional transport of passengers and goods.

Boat by the Kampen (IJssel) cog, section 26.2.
- Boat with cog-like features. 1st half/3rd quarter of the 15th century, Kampen/IJssel delta?
- In terms of construction, lines and execution, it is directly comparable to the boat of lot K 73/74 Eastern Flevoland (see section 26.1).
- Old vessel at the end of its useful life (many repairs). Together with the Kampen (IJssel) cog and a barge (Chapter 23 and Section 26.3), it was used to dam an old stream channel of the IJssel;

- Dimensions and proportions of the reconstructed ship:
  A. 13.60 m
  B. 13.40 m
  C. 1.95 m
  D. 15.80 m
  E. 2.35 m
  F. 0.75 m
  G. 6.9:1
  H. 6.7:1
- Rudder still present. So far oldest known example of a rudder on a shipwreck in the Netherlands;
- Rowlock/tholepin cut from a branch;

Barge near the Kampen (IJssel) cog, section 26.3.
- Barge built according to the cog-building tradition. 1st quarter/3rd quarter 15th century, Kampen/IJssel delta?;
- Part of an old vessel at the end of its useful life. Together with the Kampen (IJssel) cog and a boat (Chapter 23 and Section 26.2), it was used to dam an old stream channel of the IJssel;
- Dimensions and proportions of the reconstructed ship:
  A. ?
  B. ?
  C. ?
  D. ?
  E. ?
  F. ?
  G. ?
  H. ?
- Large eel buck (trap)

Large boat or small cog for inland waterways and Zuiderzee, L 89 Eastern Flevoland (excavation data not yet worked out). Section 26.4.
- Larger, but in terms of construction, lines and execution, it is directly comparable to the boats from lot K73/74 and near the Kampen (IJssel) cog, sections 26.1 and 26.2, ca middle/3rd quarter of 16th century, river basin of the IJssel?;
- 'Old-fashioned' (15th-century) model. Not moss laths, but split withies were used to apply the sintel clamped moss caulking in the seams. Until this shipwreck was found, split withies were only known in vessels predating the 12th century;
- Probably stranded on the west side of the (underwater) peat barrier between Schokland and Elburg and not far from the west-southwestern estuaries of the IJssel delta. The vessel broke in half on the end of this underwater 'bank';
- A keelson/mast step and a fireplace;
- Extensive equipment and inventory and several coins.
- Dimensions and proportions of the reconstructed ships:
  A. ?
  B. ?
  C. ?
  D. ca 20.0-21.0 m
  E. ca 4 m
  F. ca 0.8-0.9 m
  G. ?
  H. ca 5:1

Cargo vessel on lot B 55 Eastern Flevoland, section 26.5. (Reinders *et al.* 1980, 31-41 and 47-50).
- Large (inland) cargo vessel with completely flat, carvel-built bottom and clinker-built sides. Stern placed on middle plank of the bottom, late 15th century/ca 1500 - early 16th century. Not a cog;
- Dimensions and proportions of the reconstructed ship:
  A. ca 19.00 m
  B. ca 17.60 m
  C. ca 3.45 m
  D. ca 23.20-23.50 m
  E. ca 4.90 m
  F. 1.30 m
  G. ca 5.1:1
  H. ca 4.7:1
- Sintel clamped moss caulking in the seams in the bottom on the outside, and the seams in the clinker-built sides on the inside;
- Foundered in the middle of the sea;
- Incomplete inventory, exceptional finds: a macehead (from a weapon known as a *goedendag* or godendac), or in any case the upper part with iron knobs on it, and a pewter jug.

Antwerp, Lefèbvre Dock. Five different vessels and a crossbeam 6.5 m long and 0.3 m wide (possibly protruding through the hull) from a large vessel (cog?). Found at the end of the 19th century, section 26.6 (1-6) and fig. 26.4 (Hasse 1907; Vlierman 2002, 132).

Vessel 1. Barge-like planking boat with several cog-like features, presumably 12th/13th century, Antwerp region;
- Dimensions and proportions of the reconstructed ship:
  A. -
  B. ca 20.00 m
  (greatest length of bottom plus *heves*)
  C. ca 2.20 m
  D. ca 20.00 m
  (greatest length of bottom plus *heves*)

E. ca 4.00 m
F. ca 1.50 m
G. -
H. 5:1

- Carvel-built bottom and clinker-built sides;
- A *heve* (square bow/stern) about 80 cm wide aft and forward;
- Probably kept on course with a steering oar;
- Carvel seams completely open and filled with moss caulking. Fastened on the underside with triangular moss laths/sintel rods and sintel nails. On the inside only with sintel nails;
- Clinker-built fastenings with nails (possible blind?) that were driven into the wood alternately from the inside to the outside and vice versa;
- At least three transverse connections, probably originally five plus an after deck and forward deck. Decks connected by gangways.

*Vessel 2. **Barge-like planking boat with several cog-like features, presumably 12th/13th century, Antwerp region;***

- Dimensions and proportions of the reconstructed ship:
  A. -
  B. ca 15.00 m
  (greatest length of bottom plus *heves*)
  C. ca 1.00-1.20 m
  D. ca 15.00 m
  (greatest length of bottom plus *heves*)
  E. ca 3.00 m
  F. ca 1.20 m
  G. -
  H. 5:1
- Carvel-built bottom and clinker-built sides;
- A *heve* about 80 cm wide aft and forward;
- Probably kept on course with a steering oar;
- Carvel seams completely open and filled with moss caulking. Fastened on the underside with triangular moss laths/sintel rods and sintel nails. On the inside only with sintel nails;
- Clinker-built fastenings with nails (possible blind?) that were driven into the wood alternately from the inside to the outside and vice versa;
- Originally at least two, but more likely four transverse connections plus after and forward decks, possibly a ceiling strake or gangway against the sides.

*Vessel 3. **Boat with several cog-like features, presumably 12th/13th century, Antwerp region;***

- Dimensions and proportions of the reconstructed ship:
  A. ca 10.50 m
  B. ca 10.00 m
  C. ca 1.70 m
  D. ca 13.00 m
  E. ca 2.50 m
  F. ca 1.20 m?
  G. ca 5.9:1
  H. ca 5.2:1
- Carvel-built bottom and clinker-built sides;
- Stern and stem. Whether the vessel had stern- and stem hooks and a stern rudder was not determined;
- Carvel seams completely open and filled with moss caulking. Fastened on the underside with triangular moss laths/sintel rods and sintel nails. On the inside only with sintel nails;
- Clinker-built fastenings with nails (possible blind?) that were driven into the wood alternately from the inside to the outside and vice versa. A layer of moss caulking between the overlapping plank sections. It has not been determined whether this moss caulking was fastened with moss laths and sintel nails on the inside and/or on the outside;
- At least five (probably six) transverse connecting beams with rising knees on which the gunwale was fastened into place. Presumably also at the aft and forward ends of the after- and forward decks.

*Vessel 4. **Boat with cog-like features, presumably 12th/13th century, Antwerp region;***

- Dimensions and proportions of the reconstructed ship:
  A. ca 8.50 m
  B. ca 8.00 m
  C. ca 1.50 m
  D. ca 10.00 m
  E. ca 2.20 m
  F. ca 1.00 m?
  G. ca 5.3:1
  H. ca 4.5:1
- Carvel-built bottom and clinker-built sides;
- Stern and stem. Whether the vessel had stern- and stem hooks and a stern rudder was not determined;
- Carvel seams completely open and filled with moss caulking. Fastened on the underside with triangular moss laths/sintel rods and sintel nails. On the inside only with sintel nails;
- Clinker-built fastenings with nails (possible blind?) that were driven into the wood alternately from the inside to the outside and vice versa;

- At least two, possibly originally more transverse connecting beams with rising knees on which the gunwale was fastened into place.

Vessel 5. Small (seagoing) cog, 15th century, very probably older (13th?/14th century), Antwerp region;

- Dimensions and proportions of the reconstructed ship:
  - A. ca 11.50 m
  - B. ca 11.00 m
  - C. ca 1.70 m
  - D. ca 13.00 m
  - E. ca 5.00 m
  - F. ca 1.50 m
  - G. ca 6.5:1
  - H. ca 2.6:1;
- Carvel-built bottom and clinker-built sides;
- Whether the vessel had stern- and stem hooks and a stern rudder was not determined.

Ship wood, 's-Hertogenbosch, section 26.7 (Vlierman 1988, 97-101; idem 1996b, 107-110 and Appendix 13).

- Ship fragment used as revetment, mid-13th century, 's-Hertogenbosch region?;
- Possible fragment of a river barge or an inland cog;
- Sintel clamped moss caulking and clenched nails driven into the wood from the inside out.

Barge at Kessel (North Brabant), section 26.8 (Vlierman 1996b, 19-32).

- River barge, 15th century
- Dimensions and proportions of the reconstructed ship:
  - A. -
  - B. ca 15.50 m (greatest length of bottom plus *heves*)
  - C. ca 2.00 m
  - D. ca 15.50 m (greatest length of bottom plus *heves*)
  - E. ca 2.75 m
  - F. ca 0.85 m
  - G. -
  - H. ca5.6:1
- Sintel clamped moss caulking used to make the seams watertight: in the bilge seam on the inside, and the seams of the bottom on the outside; only moss caulking in clinker-built seam. Nails in the clinker-built fastenings driven into the wood alternately from the outside to the inside and vice versa;
- Presumably used as a ferry.

Several parts of (river) vessels from a different building tradition?, section 26.9 (respectively Reinders 1983, 15-20 and 40-41; Vlierman 1995, 91-104; Idem. 1996b, 115-119 and 137 and Van Holk 2001, 103-117).

- Meinerswijk, Amsterdam, Hattem and Rotterdam (ship 4), 13th century;
- Respectively probably foundered, used as a foundation platform under the walls of the reinforcement on the Nieuwezijds Kolk, discarded(?) and foundered, and parts used for a lock in the Rotte River;
- Instead of nails, wooden treenails were used to fasten the clinker-built planks;
- Vessels from the Rhine/Moselle/Meuse region;
- Sintel clamped moss caulking in the seams;
- Not examples of the cog-building tradition, but may possibly show influences of that tradition.

Small cargo vessel/boat from another shipbuilding tradition, Nieuwlande/Oosterschelde (Eastern Scheldt), section 26.10 (Vlierman 1996b, 33-52 and 132-133).

- Went down near the village of Nieuwlande near Krabbendijke in the Eastern Scheldt, which drowned ca 1500;
- Similarities to boat K 73/74 O. FL. (Section 26.1), but more to the (later) Zeeland shipbuilding industry, e.g. a strongly raking straight stem;
- Lanceolate, flat bottom with stern and stem placed on the ends of the middle plank of the bottom;
- Clinker-built sides consisting of two strakes;
- Mast step with a sail beam above it, and a broad thwart in the forebody;
- Inwale on which a splashboard could be placed;
- Transverse connection at gunwale height in the afterbody? and forebody;
- Not sintel clamped moss caulking in the seams, but moss covered with moss laths and fastened into place with wooden spikes (small wedges);
- Not (or no longer) built according to the cog-building tradition.

For more detailed information on the various vessels, please refer to the relevant chapters and sections.

29.4.4 Some early-15th-century miniatures with realistical depicted vessels

The boats in these two miniatures, also from the *Apocalips in Dietsche*,[32] can probably be reckoned among the earliest medieval examples with a very realistic rendering; they may almost serve as photographs of the boats in the cog-building tradition (as described in Chapter 26: K73/74 from Flevoland and the very similar boat found in the river IJssel beside the Kampen cog, as well as a boat of a different building tradition from the Oosterschelde marine inlet near Nieuwlande).

The first miniature (fig. 29.5a) shows two detailed, entirely realistically rendered, probably flat-bottomed vessels. The boat in full view has a relatively slender shape with straight, raked stern and stem and a stern rudder with a tiller. This boat has a reinforcing timber along the inside of the gunwale (an inwale, in small vessels like this one known as a covering inwale). This covering inwale covers the heads of the frames. This boat has a forward-deck and probably also an after -deck. Aft and fore, double tholepins are set into the covering inwale.

Of the other vessel, most probably of the same type, only the afterbody with the stern rudder and tiller, the covering inwale with tholepins and part of the flat floor are represented. The latter is not an underfoot ceiling, but a removable plank bridge. The planking of a ceiling is always laid longitudinally, across the frames. To show the planked hull to best effect, no nails are shown. Therefore it is impossible to tell whether this boat had a fully carvel-built, or a partially clinker-built bottom; and carvel- or clinker-built sides. The latter is the most likely, however.

The other miniature (fig. 29.5b) shows the afterbody of a similar boat. This too shows the covering inwale with pairs of tholepins, a stern rudder with a tiller and distinct rudder fittings. The after-deck is supported by a small deck beam; also the plank bridge on the bottom is visible. Further there is a thwart which in the visible part of the boat is supported by two stanchions - presumably there were three or four in total. The figure in the image shows that such vessels were not only rowed, but also punted.

Fig. 29.5a. Scenes from the life of St. John the Evangelist, 15th-century miniature in *Apocalips Dietsche*, folio 1 r. Bibliothèque nationale de France, Paris. From De Hommel-Steenbakkers 2001.

Fig. 29.5b. St. John's revelations, 15th-century miniature in *Apocalips in Dietsche* 1, folio 2 r. Bibliothèque nationale de France, Paris. From De Hommel-Steenbakkers 2001.

32. De Hommel-Steenbakkers 2001.

29.5 Sintel clamped moss caulking

29.5.1 A late-medieval caulking method as a means of dating

In ...*Van Zintelen, van Zintelroeden ende Mossen*...[33] a method of caulking wooden ships is described under the designation of '*gesinteld mosbreeuwsel*', i.e. 'sintel clamped moss caulking' (fig. 29.6). Its origin goes back at any rate to Carolingian times, more likely even further back in time.[34] It is a method that was part of the rivercraft building tradition within and north and immediately south of the Alpine region, an area that roughly corresponds with Lorraine, and the East-Francian realm of Louis the German, King of Bavaria, after the Empire was split up in 843. Sintel nails and sintel clamps from a 10th- to 14th-century context are known also from Novgorod.[35] The fifteen-year research preceding of that publication showed that the earliest examples of the sintel clamped-moss caulking method are to be found on rivercraft and especially barges. The title of the 1996 publication sums up the materials used in this method (sintels, laths of oak or split withies, and moss), as they were listed in a late-medieval source.[36]

Into the second half of the 12th century we see virtually no change in the shape of the sintel nails (narrow, thin, rectangular or slightly trapezoid clamps) by which the moss laths or withies and the moss caulking were secured into the shell seams of ships. The sintel nails were driven into the adjacent shell planks as an inverted U across the laths or withies and the moss-filled seams. With the emergence of the cog building tradition in the second half of the 12th century and in the region of the rising Hanseatic League the shape of the sintel nails slowly evolved from an oblong to a laurel-leaf outline (ca 1200), and then to a butterfly-shaped sintel, which had an oval centrepiece and two wings that were driven into the shell planks on either side of the seam. In the same period, the spacing (from centre to centre) of the sintel(nail)s became standardized to ca 5-10 cm. This process of change took place roughly between 1175 and 1250.

Very probably, it was in this period that sintel clamped moss caulking was first applied to seagoing vessels, which were built according to the then emerging cog-building tradition. After the mid 13th century the evolution of sintel shapes continued into the 16th century. The iron plate

Fig. 29.6. Part of the underside of the bottom of a small river barge of the late 14th- or early 15th century, the seams of which were closed with sintel clamped moss caulking. Oosterhout, province of Noord-Brabant.

33. Vlierman 1996a.
34. Arnold 1977, 293-297; Arnold 1992 part II, 87-94; 'The complex moss caulking of clearly Celtic origin survived in many regions....'.
35. Vlierman 1996a, 26.
36. Vlierman 1996a, 19 and 21; Verwijs & Verdam 1885, 609 (AD 1401).

that remained visible over the seam became an ever-longer oval. This change took place especially during the rise and heyday of the Hanseatic League in the European area where the League operated (fig. 29.7). This globally is the area of the European continent between the southern border of Denmark, the German North-Sea and Baltic coasts in the northwest, and the eastern Zuiderzee coast with the river IJssel, and from Flanders to the Lower-Rhine region in the south. From ca 1250 probably increasingly along the southeastern Baltic coast and the west coast of the Zuiderzee. With the demise of the Hanseatic League also the application of this caulking method disappeared. By ca 1550 sintel clamped moss caulking was no longer used for seagoing and larger inland vessels; at most for repairs following an old, familiar method. The continuous change in shape of the sintel is characteristic and exclusive to the cog-building tradition; and from the early 13th century onwards, sintels were probably employed for all large and small vessels built according to this tradition (distinguishing mark). The importance of these changing shapes lies in their datability. The six different types enable us to date the building of a vessel to within a period of 50 to 70 years, sometimes even more closely. The central parts of the sintels, progressively elongated to lengths of 35 to 50 mm, covered and thus protected increasing parts of the moss laths or withies, which were just a few mm thick. In 16th-century vessels, the sintels might even overlap to provide full cover.

In 'isolated' shipbuilding regions (the Alps), along the major rivers (Upper Rhine, Danube and their tributaries) and probably also along the rivers of (eastern) France and those flowing northward in Germany and Poland, the caulking method persisted almost unchanged in traditionally built vessels between ca 1200 and the mid-20th century. Along the Volga the same may apply in the 19th and 20th centuries. To this it may be added that in the Danube basin, vessels were built with gaping seams, in which the moss caulking was covered over with thick moss laths of trapezoid section. The laths were secured with broad or narrow sintels or sintel nails (*Klampfen*) with pointed wings. The closely spaced *Klampfen* also joined the planks together.[37] The thick moss lath, triangular in section, in the moss-filled crack in a 14th-century ship's timber from Novgorod also deviates from the thin split withies or moss laths used in the vessels from the Rhine and the (later) Hanseatic territory. We may be fairly certain that also along the rivers of eastern Europe the method goes back to the Middle Ages. Hence it seems plausible that rendering watertight the large and small rivercraft by means of sintel clamped moss caulking extended eastward to the

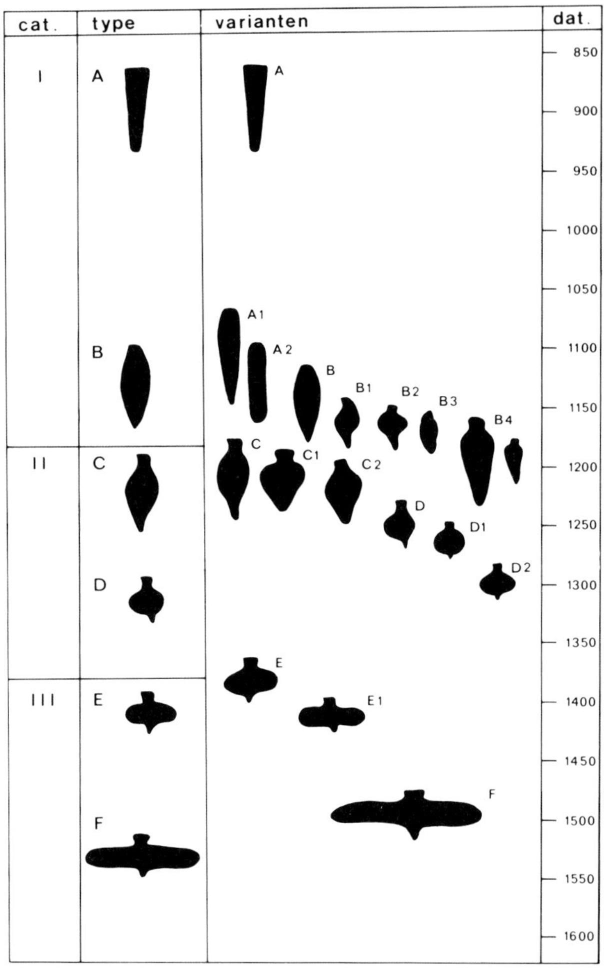

Fig. 29.7. Overview of the sintel nail and sintel types and their dating. After Vlierman 1996a.

rivers emptying into the Black Sea and Caspian Sea. The manufacture and application of sintels besides nails must, especially during the 14th to the mid-16th century, have constituted a substantial part, in terms of labour and time, of the activities of shipyard blacksmiths and shipbuilders. Building a 14th-century cog of 20 m length would easily require 5,000 nails and 15,000 sintels. That forging sintels was no job for an apprentice is evident from the Gdansk guild statutes of 4 December 1446 which ordain that a ship's blacksmith as his masterpiece should 'forge a rudder's fittings, a boathook, and a sintel ' (... *eyn Schute Rudil smeyde und eyn Bothoken und eyn Syntil'...*). Moll wonders whether *Syntil* should be interpreted as a sounding lead.[38] Yet this clearly is a reference to a sintel, and probably the model forged from two thin strips of iron (*...mit langen sterten...*,[39] *i.e.* with long tails; type E1). The evolution from straight, narrow strips (the sintel nails) to the butterfly-like shapes of sintels could be

37. Wöginger 2007.
38. Moll 1930, 23; Vlierman 1996a, 31.
39. Luns 1985, 31.

traced mainly thanks to the large number of Dutch shipwrecks investigated by NISA; by their inventories they could usually be dated to within a period of 25 years or even less.[40] The construction date of a ship can in many cases be established quite accurately by tree-ring analysis, which also may provide evidence of the provenance of the timber, and in vessels predating ca 1300 also of the ship. This is because 13th-century and earlier ships are most likely to have been built in the region where the timber had grown. After ca 1300 timber was increasingly imported from elsewhere. Also the moss used for caulking was of a different composition. Pre-1300 caulking was usually composed of different moss species some of which do not or no longer occur in The Netherlands. Ships of a later date are generally caulked with a single species (*Sphagnum*).

Even before 1980[41] it was established that there were six types of sintel nail or sintel (A to F; fig. 29.8). Finds in medieval cities, especially of reused ships' timbers (for example to revet quaysides) and foreign finds contributed to this research. In the summary of the earlier research it is concluded that most sintels and sintel nails can be dated by their shapes to within a period of 50 to 70 years, and in some cases even more closely. Also it is believed that new finds may allow further refinement in the near future, and that we may speak of a new archaeological method for dating shipbuilding in medieval northwestern Europe.[42]

Even shortly after the publication of Vlierman in 1996, it aroused a keen interest among maritime and indeed urban archaeologists. The iron sintel clamps had surfaced in numerous excavations, but in finds lists they had at best been described as "(probably) used in caulking ships". Regular requests arrived for coming over and examining, dating, and maybe publishing them. Also such finds were newly reported in ever-increasing numbers. This produced the predicted and increasingly distinct picture of the

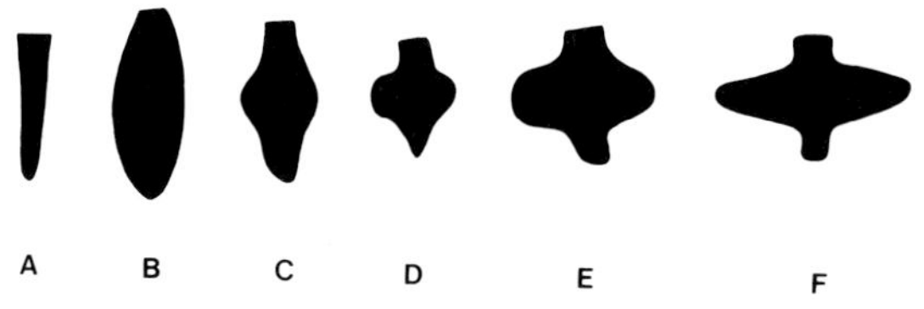

Fig. 29.8. Sintel nails and sintels: the six types.

distribution of types, associated datings and the regions where they had been manufactured and/or used. Besides, it became ever clearer that certain forms did not evolve simultaneously in every region, but that there were episodes of inertia in the process. This must have been the case especially in regions where vessels were well-suited for their function and range of operation, and gave no cause for change. The local blacksmith continued to forge his sintels in the way he had always done and this may have been continued by his successor(s). This possibility, also mentioned above, explains the survival (up to the mid-20th century) of authentic traditional shipbuilding, as for example in the lake district of northwestern Overijssel in The Netherlands, regions in northern Germany and along the above-mentioned major rivers: the *punters, Kahnen, Zillen, Plätten, Fähren*, etc. built there barely differ from their medieval predecessors found in archaeological contexts. Nowadays several transitional forms between sintel types have been identified, which must derive from a region where typological evolution did proceed.

Naturally, as the the study was concluded and published in 1996 there were some periods that could not yet be fully covered, owing to a dearth of dated finds. On the basis of the sintels then available, a typology of types and presumed transitional forms was drawn up. Finds at home and abroad have since almost fully confirmed both the typology and the chronology. The few deviating finds are those of sintel nails in ship's timbers reused in a lock in the river Rotte at Rotterdam[43] and in a broken-up ship used as a foundation under the wall of a fortification on the Nieuwendijk-Dirk van Hasseltsteeg site in Amsterdam.

J.M. Baart,[44] then Amsterdam's municipal archaeologist, confidently declared this structure to be the Castle of Aemstel. During the excavations in 1994 sintel nails were found in the foundation timbers under the uncovered walls. Invited by Baart, I visited the excavation and was able to establish that these were not just loose foundation beams and planks, but that under the north wall the bottom of a ship had been used as a foundation plate. On the basis of the data then available it was concluded that these sintel clamp models belonged in the periods ca 1175-1225 and ca 1200-1250.

40. Vlierman 1983, 1; 1997a, 15.
41. Vlierman & Van Dijk, 1980b.
42. Vlierman 1996a, 54.
43. Van Holk 2001, 93-104.
44. Baart 1995, 11-27.

A report of the subsequent research appeared in the book *Het "Kasteel van Aemstel", Burcht of Bruggehoofd?*[45] (the 'Castle of Aemstel', fort or bridgehead?). The preliminary conclusions in the report include a remark on these datings, to which other researchers in their criticism failed to pay attention. The remark reads: "Up to the writing of this contribution, no account was taken of the possibility that the vessel was built in the Upper Rhine region or another area in the margin of Hanseatic territory. If such was the case, a younger date cannot be ruled out, as sintel types from ca 1200 onwards in those parts barely changed in shape".[46] However, the suggested dates of ca 1175-1225 and ca 1200-1250 best suited Baart, and he stuck to them.

In the course of subsequent research into the fortifications during the 1998/'99 winter, the most promising wood samples were selected and sawn out of the uncovered ship's bottom under the north wall. Dendrochronological analysis of these samples showed that the youngest dated samples came from trees felled between May/June 1276 - February/March 1277 and September/October 1276 - February/March 1277. The datings were made possible by matching reference curves in the chronologies of central and southern Germany and southern Germany and the Meuse valley.[47] So the vessel was most probably built before 1280 in the Rhine/Meuse/Moselle region.[48] This confirmed the suggestion that the sintel nails might be younger than had been assumed in 1995, and that the ship had been built outside Hanseatic territory.

Tree-ring analysis of ship's timbers at Rotterdam produced a similar picture. In 2002 a fresh look was taken at the foundations of the fortification wall in Amsterdam. Beneath the now uncovered south wall more ship's timbers were found. The dendrochronological analysis performed shortly thereafter yielded an unsurprising date: AD 1276; the sequence matched the reference chronologies of the central and southern Netherlands, the Meuse valley, northern France or Germany. For good measure a beam was radiocarbon-dated: its outcome, AD 1250 quite closely corresponds with the more precise dendrochronological date.[49]

Recently also the log boat from Meinerswijk (sintel-nail types B and B/C) could be dendrochronologically matched to the reference chronology NLHist.1-zuid: felled in the first quarter of the 13th century.[50] A ship fragment from 's Hertogenbosch, made from silver fir (like the lock floor at Rotterdam) could soon after its recovery be dated by its sintel nails to around 1200, and provenanced to southern Germany or the southern Meuse watershed. Subsequent tree-ring analysis confirmed this: felling date after 1211 (and before 1220). Silver fir (*Abies alba*) does not naturally occur in the Netherlands, but in the lower European mountain ranges like the Baden-Württemberg region, Bavaria, the Black Forest or the adjacent part of France. The vessel therefore most probably originated in the southern watershed of the Meuse or in southwestern Germany.[51]

In the context of the construction of the Noord-Zuid metro line in Amsterdam two caissons were sunk into the Damrak at the former mouth of the river Amstel, near the Nieuwe Brug. The municipal Bureau Monumenten en Archeologie (BMA) simultaneously performed an archaeological investigation. This yielded over 450,000 finds, almost 1,000 of which were sintels. The BMA invited me to examine these finds; at the same time also the sintels from earlier excavations in Amsterdam were looked into. The report of this study[52] discusses not just the familiar sintel types, but also a large, as yet unknown one. These sintels can be reckoned to type D2/E with a probable date between ca 1375 and 1425. Instead of the usual oval to pointed-oval central parts, this part here first widens from the wings outward, and at the ends is almost straight and of roughly the same or even greater width than the centre (fig. 29.9). The largest recovered specimen (find no. 387-002) has a central part as much as 90 mm long and 40 mm wide.

Such sintels are unsuited for closing shell seams of ships. Yet they are useful for, for instance, fixing the caulking of seams of wooden drains or making watertight a cooling cask, like the one mentioned in a mid-16th-century invoice for repairing the cooling cask at Vredenburg castle in Utrecht. (.....*noch in't coelvat in de brouwerije. Daertoe gedaen tweehondert sintelen*. (...also in the cooling cask in the brewery; put in 200 sintels).[53] Judging by the model and the various findspots, these seem to have been specially made in Amsterdam for closing planked revetments of quaysides, piers and lock walls. Maybe the model was developed specifically for the wharves along the river IJ.[54] They were probably applied abutting or even

45. Vlierman 1995, 91-104.
46. Vlierman 1995, 99-101.
47. By Ms E. Hanraets (RING) and K. Vlierman; report on the dating research by RING: Amsterdam-Nieuwezijds Kolk; fortification February/March 1999.
48. Sintels in het Damrak, Vlierman, in prep. a.
49. Toebosch 2011, 50.
50. Reinders 1983, 21-35; Vlierman 1996a, 71-72; file on the logboat Meinerswijk-III, RCE Lelystad, rapportage RING, 11 april 2006.
51. Vlierman 2000, 118-125.
52. Vlierman in prep. a.
53. Vlierman 1996a, 40 (Hoekstra, 1984; RAU, RA 388, written comm.).
54. Vlierman, in prep. a.

Fig. 29.9. The unusual shape of sintel type D2/E. Amsterdam, Damrak, find number 00097-1.

overlapping. This would have prevented the constantly moving water from washing out the soil from behind the revetment.

The study of the sintel finds from the Damrak site further revealed that there were used as well as 'new', unused specimens, that they most probably dated from ca 1310-1500, and that in that same period they ended up in the then harbour at the spot where originally the river Amstel emptied into the river IJ. On the one hand, this suggests the dumping of urban debris, in this case coming from a probably nearby shipyard and maybe also a smithy. On the other hand, the sintels might derive from ship repairs in the harbour. Considering the demolition around 1300 of the fortification at the Nieuwendijk/Dirk van Hasselt-steeg site, and the presumably immediate levelling of the site (the spit of land situated between the present Dirk van Hasselsteeg (the then Boerenwetering canal), the Nieuwezijds Voorburgwal, the river IJ and the Damrak, then all open water), it seems quite likely that the site was straightaway given a new purpose. The establishment of a shipyard on the Damrak side with a nearby smithy seems an obvious use. In the case of the smithy, we might consider a relocation of the 13th-century smithy on the Nieuwendijk, whose activities ceased around this time. For comparison: in Bruges, excavations in the northernmost part of the old town uncovered a shipyard and a smithy in the Wulpenstraat, close to the Lange Rei stream (connecting the site with the Zwin marine inlet) and within the ramparts of 1297. Reexamination of the sintels and sintel nails found there has shown that these were of types B3, C2 and D1. The smithy appears to have been active throughout the 13th century, hence still outside the ramparts.[55]

Over half of the sintels recovered at the Damrak site can be classified as types D2 and E; in many cases the distinction is hard to make. This suggests that the model was barely evolving over an extended period, remaining essentially unchanged throughout the 14th century. Also the subsequent models, E1 and F, have poorly defined features. They all seem to come from the same workshop. The specimens that do match the more pronounced typological forms may derive from ships built and caulked elsewhere, and undergoing repairs in Amsterdam.[56]

Some of the sintels in the Marknesse cog (sections 12.2.2 and 12.8, type D) still bear some resemblance to the preceding 13th-century model. All other data, however, indicated that the vessel most probably was built in the second quarter of the14th century. The outcome of tree-ring analysis,[57] which became known only after the closing of the manuscript, confirmed this. In a small cargo vessel excavated north of Dronten[58] sintels of type E (1375-1450) had been used to secure split withies covering the moss caulking. Up till this discovery, withies were known only from 12th-century and older vessels. The ship is directly comparable to a mid-15th-century vessel excavated west of Dronten.[59] Tree-ring analysis indicated a felling date of 1545 ± 6 years, while one of the five recovered coins had been issued in 1556. So the vessel had probably foundered in the third quarter of the 16th century. Both wrecks seem to provide evidence of slowly evolving sintel forms from, for instance, an isolated or strongly regionally focused shipbuilding area. The converse might also occur. The sintels (type D2) of the Doel-I cog, which probably dates from the second quarter of the 14th century (Chapter 20), already display traits anticipating type E. In this wreck the seams at the fluctuating waterline on the outside (unloaded and loaded) were no longer closed with sintels. In the Marknesse cog even the entire clinker-built sides no longer have sintels externally. Up till the present study it had been assumed that this change only occurred after ca 1350.

The sintel typology[60] inspired a master's thesis at the Christian Albrechts University in Kiel. Siegloff[61] studied the sintel (nail) finds of the Schleswig isthmus and particularly those from the present village of Hollingstedt, a medieval settlement on the east bank of the river Treene.

55. Vlierman 1996a, 38-39 and fig. 20; Vlierman, in prep. a.
56. Vlierman, in prep. a.
57. Van Holk 2010, 138 (dated 1339 with a presumable provenance of the timber/the ship from the northern Netherlands).
58. In lot L 89 in Oostelijk Flevoland, file RCE-Lelystad; section 21.2, figs 21.2a en b.
59. K 73/74 O.Fl., Reinders *et al.* 1980, 17-29; section 21.1 and fig. 21.1.
60. Vlierman 1996a.
61. Siegloff 2004.

This is part of the Eider-Treene river system, which empties into the North Sea (fig. 29.10). The village lies at the western end of the Danevirke.This was a defensive bank across the Schleswig isthmus that since Viking days strategically closed off continental Denmark (Jutland) from the Frankish and German realms. Into Early Modern times, also the extensive marshy lowlands of the Eider-Treene floodplain formed a hard to overcome obstacle to commercial traffic as well as armies. To the east, the Schlei inlet, running 38 km inland, was difficult to cross, while the 15-km land bridge between Haithabu/Schleswig and Hollingstedt was protected by the earthworks and walls of the Danevirke.[62] The first mention of Hollingstedt appears in the oldest part of the statute book *Schleswiger Stadtrecht*, written probably shortly after 1150. In later sources too, Hollingstedt is regularly named as the western endpoint of the Viking-age transit road between the North Sea and the Baltic. Three 13th-century sources describe the crossing of the isthmus by Sven Grathe's army in 1151, which also carried along its warships.[63]

My study of 1996 discusses just 19 sintel nails of type A.[64] These were all recovered specimens known at the time. The three different models (A, A1 and A2) in the publication were just an indication of potential variants. Now 191 specimens are known from Hollingstedt, a highly valuable addition for achieving a better insight into any variation in especially the early types A and B. Nowadays, however, it increasingly seems to me that a classification of sintels and sintel nails boils down to six basic types (A to F) and that the initially distinguished possible variants are less relevant. Their design was strongly dependent on the blacksmith involved, and was (in part) determined by the presence or absence of local typological developments. In many cases the shape seems to have been more or less fortuitous and not forged to conform to any predetermined model. This at any rate goes for types A and B. The finds from Hollingstedt provide a more reliable insight into the shapes of sintel nails between ca 1100 and 1175. It is debatable whether this justifies a Hollingstedt typology.[65] To a researcher it is tempting to devise one. Yet it is most likely that any future finds of substantial assemblages will give rise to new typochronologies.

Besides the finds from Hollingstedt, Siegloff also describes and illustrates sintel(-nail) finds from Schleswig, Kolding, Haderslev, Lübeck, Stade, Greifswald, Alt-List/

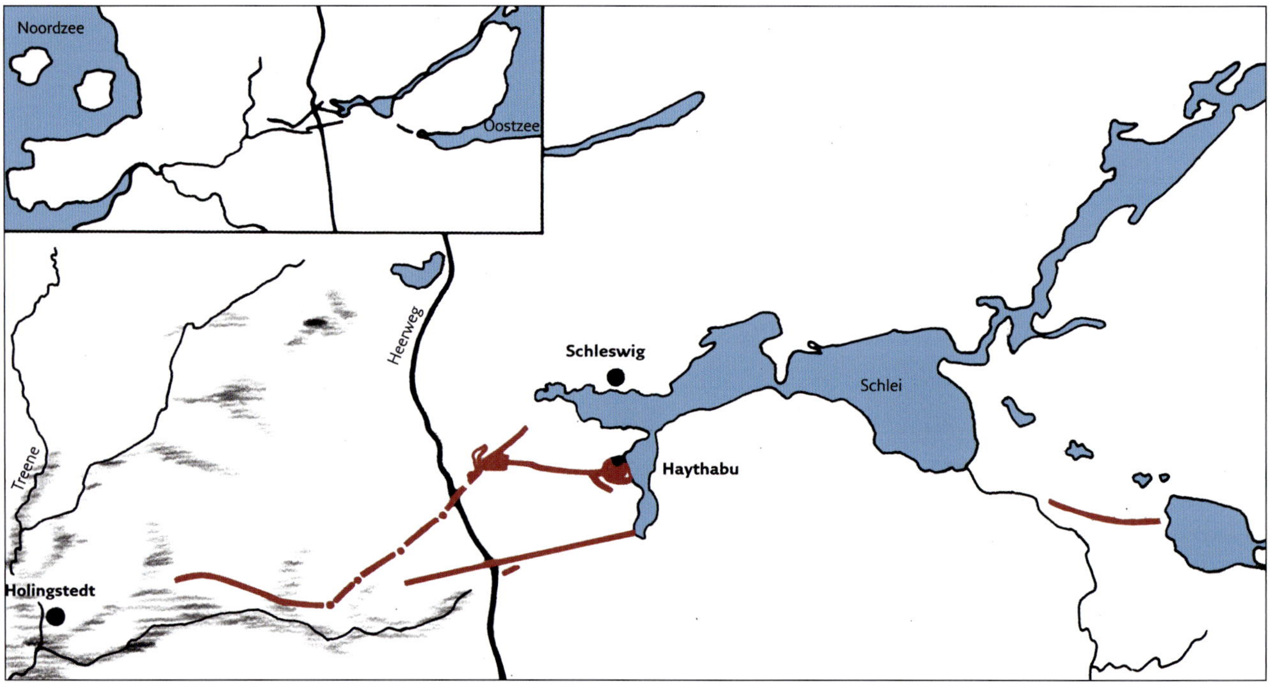

Fig. 29.10. Topography of the Schleswig isthmus with the Eider-Treene marshlands, the Danevirke with its successive extensions, the Military Road (Heerbaan), Haithabu (Hedeby) and Hollingstedt.

62. Siegloff 2004, 17.
63. Siegloff 2004, 22.
64. Vlierman 1996a (Siegloff 2004, 29).
65. Siegloff 2004, 41 Abb. 25.

Sylt, Grosser Schichtenberg (all Germany), Skänor (Sweden), Ribe and Dragør (both Denmark) and, as mentioned before, Novgorod ((Russia). In all cases the sintel nails and sintels are of types whose youngest specimens can be dated to the second half or end of the 13th century.[66] Remarkable and interesting is the repair to a cargo vessel built according to the Nordic tradition (riveted clinker-style) between ca 1130 and 1200, and found at Karschau on the Schlei. The timber for the ship came from western Denmark, the isle of Fünen and/or Schleswig-Holstein.[67] The ship was probably built in the same region. The repair was carried out probably in the second half of the 12th century, with the use of sintel clamped moss caulking.

A second ship find of interest in this regard is the clinker-riveted vessel (Ship 1) from the cutting of the Willemsspoortunnel (railway tunnel) in Rotterdam.[68] The ship, carrying ca 1000 kg of red bricks (28.5-30x14-15.5x7.5-8 cm), had been patched with thin planks in various places on the outside of the shell. The planks were fastened with double-clenched nails driven in from the outside. Sintel clamped moss caulking was used on a crack and around a replacement patch set into the shell planking (sintel-nail type B2/3). Tree-ring analysis showed that the composite curve of the sampled timbers produced a felling date between AD 1252 and 1264. The wreck must have sunk into the bottom between the building of the dam (around 1270) and the first expansion of the dam (in the early 14th century or a little later). Van Holk in his summary[69] observes that the ship displays features from two different shipbuilding traditions: the construction possibly representing an Anglo-Saxon tradition whereas the repairs reveal traits of the cog-building tradition. In combination with provenancing its timber, the origin of the ship might be sought in the southern or eastern coastal regions of the North Sea.

The finds from Hollingstedt also included 57 rivets. They belong in the Nordic or Viking shipbuilding tradition of fully clinker-built vessels. Siegloff suggests that his sintel (nail) type 1A (=A and A1) and sintel (nail) 1B are evidence of ship repairs in the first and second half of the 12th century, respectively.[70] The straight, and presumably unused specimens indeed prompt this idea. The others however may derive from parts of broken-up ships, which were reused as quayside revetment. The same may be the case with the rivets.[71]

In the harbour of Haithabu (Hedeby) a few 'clamps' (probably sintel nails), sintels and 'cog nails' were found. Further study may show whether there was, for instance, a shipyard.[72] In the final quarter of the 10th century the decline of Haithabu set in. This in part resulted from the increasing contacts between Germany and the Slavs in Holstein, which opened up a far shorter route for German trade between Hamburg and the Baltic, across the *Limes Saxoniae* and to Alt-Lübeck.[73] In the first half of the 11th century both Haithabu and Schleswig, on the German and Danish shores of the Schlei respectively, still were active merchant settlements; after which "Haithabu died a slow death".[74]

Summarizing, it may be concluded that the sintel(-nail) typology published in 1996 remains useful unaltered for globally dating ships' timbers to which sintel clamped moss caulking was applied. In the case of ships suspected to have been built in the Upper Rhine region this is somewhat different: there, type-C sintel nails saw barely any change in shape from ca 1200-1250 onward, allowing younger dates. In such cases dendrochronology may offer a way out. Research on the finds from Hollingstedt in Schleswig-Holstein, Germany on the one hand has shed more light on the sintel-nail forms dating from ca 1100-1175. On the other, it provided further confirmation that the method was not applied north of the German-Danish border. The repair with sintel clamped moss caulking to the Nordic-built cargo ship from Karschau (second half of the 12th century) is the earliest known example of a combination of characteristic procedures from two shipbuilding traditions.

The ship found in Rotterdam indicates that riveted vessels built clinker-style in the Nordic tradition, but with sintel clamped repairs, were (still) sailing along the North Sea coast even in the mid-13th century. Any sintel finds in Denmark, Norway or Sweden most probably directly relate to ships from the Hanseatic territories. The finds known so far of sintel clamped moss caulking predating ca 1200 in the future Hanseatic territory show that the carvel seams gaped and were fully plugged with moss. This was covered with split withies (*sintelroeden*) or moss laths (thin strips of wood). The sintel nails merely served to hold the withies/laths and moss in place. The sintel nails evolved from type A via B and C to sintels of type

66. With the exception of Pl. 4, Lübeck, number 14. This sintel dates from the 15th/16th century (type E1 or F). The same goes for Pl. 8, Skänor (Sweden), sintels 3 and 4; and for the sintel from Dragør (Denmark) on Pl. 9. The sintels from Grosser Schlichtenberg all date from the 13th century (Pl. 7).
67. Englert *et al.* 2001, 141-154; Siegloff 2004, 120.
68. Van Holk 2001, 71-84.
69. Van Holk 2001, 84.
70. Siegloff 2004, 53.
71. Vlierman 1996c; 1997d; 1998, 56-58; 2002, 119-148.
72. Crumlin-Pedersen 1999, 198.
73. Crumlin-Pedersen 1999, 47-48.
74. Crumlin-Pedersen 1999, 204.

D, which probably was used throughout the near-entire Hanseatic territory by ca 1250, and types E and F. The caulking method may, from the very origin of the cog, be regarded as a characteristic constituent of the associated shipbuilding method.

The sintels known so far from the Danube basin and presumably also the other rivers in eastern Europe (*e.g.* the Volga) have a different shape, which appears to have changed little through time. The seams are much wider, the moss laths thicker, and the sintels also play a role in joining the planks together. Not much is known about pre-1800 vessels in these parts because they, and especially the larger ones, remained in use for just 4 or 5 years. Along the Danube, written-off ships were bought up by the *Plättenschinder*, men who cut up the timbers for firewood.[75] Archaeological ship finds therefore will always be rare.

The finds from Hollingstedt, Schleswig and Stade show that the early sintel nail types were in use there for several centuries before the start of their evolution into sintels (type D). It is clear that early sintel clamped mos caulking (pre-1150) was practised along the rivers in a large part of Europe and there too may be considered a characteristic caulking method. Along the North Sea coasts (Flanders, western Netherlands, Friesland, Groningen, Ostfriesland, *i.e.* the former Frisia) and in England most probably at any rate well into the 11th century, mariners sailed in riveted, clinker-built seagoing vessels in which no clamped moss caulking was applied. The ship from Rotterdam shows that this was the case even well into the 13th century, while the sintel clamped-moss repair shows that the shipyard was familiar with that method. I consider it beyond any doubt that the sintel clamped moss caulking method did not have a Frisian origin, and that its application to seagoing vessels was derived from its centuries-old use on continental rivercraft. The rapid evolution from sintel nail to sintel between ca 1175 and 1250, and the further evolution of sintel shapes into the 16th century was directly linked to the cog-building tradition and the area in which the Hanseatic League operated.

29.5.2 Sintel clamped moss caulking featuring in a miniature in a medieval French edition of the *Roman de Troie*.

The following description of a medieval illumination is of significance because the image shows how the shell seams of two different vessels are being closed with sintel clamped moss caulking, together with the tools and materials involved. It expressly draws attention to this caulking method, which is an unusual method for such seagoing ships, as it is known only from cogs and medieval rivercraft.

The 15th-century miniature may well be a copy of a more ancient original image (as the *Roman* dates from the 12th century), so we may in fact be looking at an image of an actual vessel type that sat at the origin of the development of the cog. One of the vessels bears a strong resemblance to a cog. The miniature, and in particular the small scene top right, has been reproduced by other authors too: generally just as an example of a medieval ship. On rare occasions the caption includes the word 'cog'. To my knowledge, the miniature has never yet been examined through the lens of nautical archaeology. The historians who focused on the *Roman de Troie* merely remarked what a fine miniature this was.

Figure 29.11a and b shows the miniature, which precedes the *Roman de Troie*, a famous medieval romance, as an introductory illumination. The manuscript, written in Old French, is a 15th-century copy. It is part of a manuscript entitled *Histoire ancienne jusqu' à César*.[76] The full-page miniature, comprising three images, quite remarkably illustrates the closing of seams in a ship by means of sintel clamped moss caulking,[77] as well as some technical details. The vessels in the miniature almost certainly represent Crusaders' ships, which together with other details suggests that the original image was (much) older than this miniature. Medieval art generally has a narrative aspect, relating remarkable, usually religious stories. Might there be an unusual event taking place here too? The answer is: Yes, most probably there is!

75. Wöginger 2007, 16.
76. folio 31 r, MS Douce 353, Bodleian Library, University of Oxford. There are indications that it was compiled during the reign of the French king Charles V the Wise (1364-1380) (Chesney 1942, 46). This implies that the original was made in the same or an earlier period, Williams Jr. (1984, 68) writes that also references are possible to other copies of the second redaction. The royal Histoire de Troie he believes reflects much of the original story, more than the Roman de Troie en prose. He believes the fair copy of the story may be seen as a kind of ultimate History of Troy in the medieval French tradition, a work written in the vernacular and intended to surpass the more pretentious Latin story by Guido delle Colonne (written in 1272-1282). Guido delle Colonne was an Italian poet, belonging to the so-called Sicilian School, which flourished at the court of Emperor Frederick II (1220-1250) and his son Manfred. Chesney (1942, 66) closes his description of MS Douce 353 with the remark that his concise survey cannot be concluded without referring to the particularly charming miniatures. All, with the exception of two (folios 1 and 26) are illustrations of the history of Troy. There is one full page (the miniature described here) that is placed as an introduction to the Romance. At the end of the History, is a second, showing the rebuilt city with the wonders of the Chambres des Beautés (folio 43). Benoît de Sainte Maure wrote the Roman in the mid 12th century; in the 13th century Guido delle Colonne created an abridged, Latin version. This Latin version was much translated, including into French and thence into English.
77. The sintel nails that secure the withies into and onto the seams in the hull of the ship in the second scene of the medieval illumination suggest that they were applied widely spaced. In reality the interspaces will definiely have been smaller, with the sintel nails closer together. By representing a smaller number the miniaturist presumably wanted to draw attention to the process of sintel application. With reference to section 29.5.1 it seems that, taking into consideration the depicted tools and those described below, we are looking at a scene of the (late) 12th, or at the latest the early 13th century. Probably this illustrates a then novel manner of caulking seagoing ships.

Fig. 29.11a. Picture 2 of a miniature in a 15th-century French manuscript of a 13th-century Latin abridgement of the 12th-century *Roman de Troie*, in Histoire ancienne jusqu' à César. Bodleian Library, University of Oxford. folio 31 r MS Douce 353.

Scene 1: the commission
The first scene shows a ruler on his throne, surrounded by several (ecclesiastical) advisers, and one person in orange hose and a green jerkin, who receives a commission. This individual reappears in the second small image.

Scene 2: a ship being built/fitted out, made watertight and prepared for sailing
The second scene shows the man who received the commission on the aftercastle of a ship. He appears to be playing a game of chess with someone wearing a burgundy jerkin and an orange hat. There are soldiers on the deck and on the forecastle, while sailors climb into the rigging and sit on the yard: they are preparing to set sail. A green flag with a yellow cross is flying from the small mast on the forecastle. Three black men carry victuals(?) aboard, a barrel of wine(?) and a rod(?), while a fourth is drawing in the gangway. There are some very interesting details to be seen. Prominently large and central is a figure hammering sintel-nails across moss laths/withies on a seam in the shell. In the foreground is a man holding a two-handed saw, while beside him lies an axe with which the plank will later be finished to be fitted tightly into the hull. The standing figure on the left holds a caulking-mallet and a caulking iron, while his apron (fell or pelt) may well relate to the tarring of the caulking material before it was driven into a shell seam. To the left and right of the carpenter lie what probably are clumps of moss. The person on the lower right appears to be stirring the molten tar in a *Kugeltopf* over an open fire. The pictured T-shaped axe is a typical ship's carpenter's tool, certainly in the 11th century and before. Between the 11th century and the late -13th/early-14th century, a change in shape occurred.[78] This change mainly manifested itself in the altered length-width ratio of the blade (from 7:1 to 2-1.5:1).[79] If the dimensions in the picture are reliable, this might be a 12th/13th-century model. *Kugeltöpfe* were in general use up till the late 13th-or early-14th century.

Scene 3: the forest felled for the ship(s), the slaying of Netherworld monsters, and the ships' departure
The large third scene depicts a forest of tall trees, partially felled for building or refurbishing the ship(s). Also it shows a knight in armour slaying Netherworld monsters and hanging up the Golden Fleece on a tree before being the last to step aboard. In a tender, another civilian passenger or crew member is brought aboard. On the forecastle of the ship setting sail we see the man in the burgundy jerkin and orange hat issuing commands for departure, while on the aftercastle two men in burgundy jerkins with cream-coloured crosses blow horns to announce departure. On the sail is a cross in red. A spare yard or mast hangs alongside the port gunwale. The different hull shapes and other details of the two ships suggest that two separate episodes at different European locations are depicted. The large miniature shows not only the departing ship, but also a smaller one in the distance.

Crusaders' ships, Barbarossa and the Third Crusade
Apart from being very fine and colourful, and presenting remarkable and fascinating information about the caulking method, it shows several other compelling details. First, the red cross on the sail and on the outfits of the knights and soldiers. Together with the crosses on the hornblowers' jerkins and the flag, they leave no doubt that we are looking at one or more Crusaders' ships. Presumably these represent a complete fleet and a large army.
A large area of forest was cut down for it. Another fascinating detail is the ruler on his throne: he wears a prominent red beard. Almost certainly this is Emperor Frederick I, known as Barbarossa, who commissions the building or adaptation of ships to carry his troops to the Holy Land. This can only refer to the Third Crusade (1189-1192), the last significant European crusade, which was undertaken by the armies of Emperor Frederick I Barbarossa of the Holy Roman Empire, King Richard I the Lionheart of England, and King Philip II Augustus of France - and during which Barbarossa drowned, on 10 June 1190.

Emblems of the parties taking part in the Third Crusade
Given the globally datable details described above, the original image must have been made at the time of, or fairly shortly after this crusade. As a later copy, it has been copied very accurately with regard to several details, which tallies with the assessment of the miniature by Williams jr.[80] Otherwise it seems most unlikely that the miniaturist would have depicted a type of axe that almost certainly fell out of use in the late 13th century.[81] The red cross in the sail, the green and red jerkins of the commanders, the hornblowers' red jerkins with cream-coloured crosses and the green flag with a yellow cross probably refer to the parties taking part in the campaign. To visually distinguish the troops embarking on the Third Crusade, the three commanders organising the campaign (King Philip II Augustus of France, King Richard I the Lionheart of England (who in 1189 succeeded Henry II) and Philip of the

78. Flatman (2009) includes reproductions of several miniatures with ships, in which axes too are depicted. Those on pages 133 and 42 date from 1121-1161 and 1225-1275, respectively. They show the typical T-shaped ship's carpenter's axe. The miniatures on pages 11, 134, 135 and 136 show different axe models. These images date from 1325-1430. They match the axes from the cogs that feature in this study.
79. Vlierman 1985b, 33-39 and fig.8.
80. Williams Jr. 1984, 68.
81. See also the remarks about details in scene 2, described above.

Fig. 29.11b. The miniature in the *Roman de Troie*. Bodleian Library. University of Oxford, f.31r MS Douce 353.

Alsace, Count of Flanders), met at Gisors (France) in January 1189 and agreed they would bear different colours in their banners and emblems: red for the French, white for the English and green for the Flemish.[82] This may mean that the ship in the third (large) image represents a French one, and the other a Flemish vessel. The miniature as a whole in that case suggests that the French and the Flemish sailed for the Holy Land together.

Prior to the Third Crusade the plan was that the English and French troops under Richard the Lionheart and Philip II Augustus would embark from southern France. Richard assembled a substantial fleet in England, Normandy, Brittanny and Poitou, which in 1190 sailed to join the French king at Marseilles. Richard himself marched part of his troops over land to Marseilles. Their meet-up with the French there failed. The count of Flanders travelled over land to southern Italy, where he met up with the French and English kings and found himself having to mediate between the kings, who were at each other's throats. After wintering there, the ships in April 1191 finally left the port of Messina (Sicily) and the king of France and the count of Flanders sailed together. They arrived off Acre on 20 April 1191. King Richard's fleet joined them there on 6 June of that year.[83]

The composition of the original romance and the French version

The composition of the original romance and the painting of the miniature may well have been commissioned by Emperor/King Frederick II during his reign (1212-1250) as a tribute to his legendary grandfather Frederick I; but possibly even earlier. Maybe by Barbarossa himself. It is not surprising that such a fine French version should have been made, particularly with this splendid miniature to open with, as the French king and his army made the journey almost entirely by ship. King Philip II Augustus concluded the earliest contract that has come down to us about transporting crusader troops by ship. In February 1190, at a cost of 5850 Marks, he hired Genoese ships to carry 650 knights, 1300 armour-bearers and 1300 horses with provisions for eight months from the day of departure, plus wine for four months. The decision taken by kings Philip II Augustus and Richard I the Lionheart was quite forward-looking in travelling to the Holy Land by the Mediterranean Sea. The adventures of Emperor Frederick I Barbarossa during the Third Crusade in Asia Minor without doubt contributed to the development of waterborne transport.

The depicted vessel types

The original image or the information for it are almost certain to have been made or recorded by someone who was well-informed, indeed someone who intimately knew about the situation/the preparation of the voyage. Maybe he has portrayed himself as the figure climbing aboard from the tender. The four black men most probably indicate a Mediterranean location, as such dark-skinned people presumably had never been seen north of the Alps. Of the two depicted ships we can say that they both seem to be carvel-built, that they have a straight stern and a stern rudder, a curved stem, and an after- and a forecastle, one large mast with a square-rigged sail, and on the castles small square-rigged sails. Further the side elevations of the two ships are different. In the second (small) scene we see a high hull with a distinct sheer from stern to stem. The ship in the large image has a gunwale rising to the front of the aftercastle and the rear of the forecastle. So two deliberately different vessel types or variants have been depicted. The ship in the large scene shows a hull that one might expect in the Mediterranean or the English/French Atlantic region (fig. 4.5).[84] The hull shape of this possibly Atlantic ship and the castles on both vessels suggest a 15th-century date, a late-14th-century one at the earliest. So in this apparent copy of a much older original image, some details of the ships have been updated. One of these is an innovative (continental western European) method for rendering shell seams watertight, with sintel clamped moss caulking. Despite the curved stems, these ships - and especially the (Flemish?) vessel in the second scene - bears a close resemblance to the cog as we know it from civic seals and archaeological finds. If this indeed is a copy of a late-12th- or early-13th-century image, the historical evidence about four large pilgrims' ships quoted by Heinsius[85] also deserves mention. These departed from Cologne in 1188. Heinsius writes: "*Es ist naheliegend dass es sich (auch) hier schon um Koggen gehandelt haben wird*" (It is likely that these too would already have been cogs).[86]

29.5.3 Sculpture. A relief on the Sainte-Chapelle, Paris

Many forms of sculptures are known from the Roman period. Among them are many examples of ships of various large and small functional types for both maritime and inland navigation. The ships in the Roman sculptures generally are characterised by a very detailed rendering of various structural features. But few sculptures are known from the Early Middle Ages. European sculpture only

82. Lloyd 2004, 68; De Maesschalck 2012, 167-169.
83. De Maesschalck 2012, 169.
84. There is a remarkable similarity in hull shape to the ship on the 13th-century civic seal of Ipswich.
85. Heinsius 1956, 76.
86. Heinsius 1956, 247-248; Vlierman 1997b, 89.

truly revived when around the year 1000 suddenly an international wave of building activity occurred. A few exceptions apart, Romanesque sculpture is inextricably bound up with architecture, in a style mainly focused on relief art. In terms of content, Romanesque and the beginning of Gothic art presents innumerable (religious) images and figures that to us are hard to identify.[87] Ships are rarely portrayed, and like many other motifs should be regarded as symbolic rather than realistic.

The 13th-century relief of a ship on the Sainte-Chapelle therefore may be considered exceptional. Indeed quite remarkably so for its representation, as in the *Roman de Troie*, of the application of sintel clamped moss caulking. Although the vessel in question is just generically rendered, for geographical reasons a link to the clinker-built 'Viking ship'/the *nef* may be made. In this case too, this caulking method is remarkable given the type of vessel represented, probably indicative of a changing style in building crusader ships, changing towards cogs.

In the 1996 publication on sintel clamped moss caulking[88] attention was briefly given to this remarkably detailed sandstone relief on a portal of the Sainte-Chapelle in Paris. According to local information and Mme Courant-Vidol, the scene represents the building of Noah's Ark.[89] The Sainte-Chapelle was built between 1245 and 1247. To my knowledge this is the only sculpture representing the application of sintel clamped caulking (figs 29.12a and b). A potential explanation for the sintel clamping of the represented hull was as a possible 'foreign' style of building or caulking of an outlandish ship, maybe modelled on Flemish cogs. Meanwhile, over two decades have passed since that publication on sintel clamped moss caulking. The then suggested datings of the various types of sintel (nail) shapes have been corroborated and additional evidence on the caulking method has been gathered.[90] Also referring to the miniature discussed above from the *Roman de Troie*, we may understand somewhat more about the presumable meaning of the relief.

In the first place, about the vessel. Most probably this is not a particular type of ship, but, as said before, a symbolic, generic ship. Secondly, it probably does not refer to the construction of a particular ship (the Ark), despite the apparent scene. This may be the result of an erroneous interpretation. The original, eroded sculptures on the Sainte-Chapelle were largely replaced by newly made ones during restoration in the late 19th century. We cannot ascertain whether this also goes for this particular relief, but it is quite likely.[91] Still it seems plausible that here too, as in the miniature, we are seeing the building and tarring of a Crusaders' ship and attention being drawn to an innovative caulking method. The building almost certainly represents the Temple of Jerusalem. The closely spaced sintels tally with the nautical archaeological evidence from the mid 13th century.

Fig. 29.12a. Relief showing clamped moss caulking on the Sainte-Chapelle in Paris, 1245-1247.

87. Geese 1996, 256-257.
88. Vlierman 1996a, 38 and fig. 19.
89. Written comm., Mme Courant-Vidol, custodian of the Sainte-Chapelle, Paris, 1995.
90. See section 29.5.1.
91. Written comm., Mme Courant-Vidol, custodian of the Sainte-Chapelle, Paris, 1995.

Fig. 29.12b. Relief with sintel clamped moss caulking on the Sainte-Chapelle in Parijs 1245-1247.

29.5.4 Historical images explained by archaeology (and vice versa)

There are very many medieval iconographic sources featuring ships. Yet they do not all yield relevant information. Nor were they ever intended to be realistic. They had a symbolic meaning, and on top of this they might be (but usually were not) realistic. Realism tends on the whole to be limited to certain details and characteristic attributes or tools. Often the vessel type, the shape of the ship and its proportions are decidedly doubtful. For instance, people tend to be represented to a much larger scale than the ship that carries them. Also one wonders whether the lines of the hull might not have been adjusted to fit the round shape of civic seals. Such doubts apply to medieval art in general. Hence some restraint has traditionally been exercised in interpreting the details depicted on ships' hulls. Some features remained a total mystery. Thanks to archaeological finds, a number of such details are now explainable and it can be stated that through time, they have provided evidence at the time considered relevant or characteristic about the construction or for instance the dimensions of the ship (large or small), or an indication of their function. Such recognisable details may at times be very truthful representations.

In a number of seals and miniatures this is the case with the short vertical lines across the strakes, down the sides of hulls (sometimes two parallel lines, or with individual nailheads) and the rows of nails along the lower edges of the strakes.[92] They are not hard to interpret: the scarfs and nails of clinker fastenings. They match what has come to light through archaeology. On other seals and miniatures the scarfs are omitted, and just a few - if any - nails are shown to indicate the fastening method. Shipwrecks have also made clear what the narrow triangles at deck level on the outside of hulls represented. They are the heads of protruding crossbeams, with beam-head guides attached onto their forward faces. With these, the protruding beam heads were protected and the vessel when mooring would not have a beam head catching behind a post, but would rub and slip along it. Some were recovered on the Doel-I and the Nijkerk-II cogs.[93]

The late-13th-century seal of Harderwijk (Netherlands) shows three such beam heads.[94] The near-round 'beam heads', without the triangular beam-head guides, may also represent iron rosebolts with large, convex heads, as found in the Almere-I cog. Rosebolts are functionally similar to wall anchors securing tie-beams in buildings, and served to strengthen the connection between deck beams and the sides of a vessel.

In the course of the present research, an explanation was found also for the curious ca 8 cm long oval holes in the shell at deck level in the Nijkerk-I cog. They were found to be scuppers for discharging pumped-up water. Most had been closed with a piece of wood, presumably scuppers used in a previous phase. A still open scupper may have had a short leather 'hose' nailed to the outside, which allowed the water to be discharged, while waves could not come splashing in. Such 'scupper hoses' appear in some 16th century images. These are reproduced in the report on the Nijkerk-I cog.[95]

A parallel for the curious trapezoid gap in the bulwark of the forebody of the Nijkerk-II and Doel-I cogs appears (amidships) in a 14th-century wall painting in the cellar of the town hall of Wismar (Germany).[96] Only very rarely do artists show sintel clamped moss caulking in the shell seams.[97]

92. Images of seals and miniatures in Ewe 1972 and Flatman 2009.
93. Chapters 20 and 6, respectively.
94. See figs 16.18a and b and 16.35 b of the Almere-I cog, chapter 16.
95. Chapter 8, figs 8.23a and b, and 8.39a and b.
96. See chapter 6, plates 6.I and 6.V A and B; chapter 20, figs 20.15, 20.17 and plate 20.I; and chapter 6, fig. 6.21, respectively.
97. See sections 29.5.2 and 29.5.3.

29.6 Layout, function, inventory and the Hanseatic network

29.6.1 The layout, function and inventory of cogs

Layout and function

The large, 17-21 m long, 13th-15th-century cogs all have a similar hull shape and length-width ratio across the gunwales. The underwater body is sharp aft and fore, while the bottom of the hold is almost flat, both longitudinally and transversely. The parts between the bottom and the decks and also above have everted sides, and upwards have a progressively more walnut-shaped horizontal outline. This made for a spacious hold amidships. Stowage of cargo in the sharp after- and forebody is barely possible. The study found no or very little evidence to indicate that there were bulkheads separating the hold from the after- and/or forebody. About any other internal joinery there is little that we can say. Only in the Nijkerk-II cog could it be established that - presumably during a refurbishment - a second row of protruding crossbeams had been inserted, at any rate in the forebody. In the space between the two beam rows, ca 0.8 m high, the lower row of beams may have supported a between-deck. This may have served as a sleeping area (for soldiers or crew). Also it seems plausible that aft and fore in this vessel, floors had been inserted at about 1.5 to 2 m above the keel plank. It also was the only large cog that produced indications of both an after-castle and a forecastle.

The large cogs discussed in this study were without doubt principally intended for making maritime voyages and transporting voluminous cargoes. In the case of the Rutten cog, we might tentatively consider it being a warship or *Heerkogge*. The tonnage of cogs ranges from ca 70 to 150 gross tons. The large cogs had a single mast, a square sail and simple rigging. Rare textile finds in wrecks have demonstrated that they carried woven woollen sails. Rope fragments in a few cogs are the only remnants of their rigging.

Characteristic in this shipbuilding tradition is the consistent use of stern rudders, for the first time in northwestern Europe. This steering system had probably been introduced previously on the *nef* in western coastal regions of Flanders, northern France and England.

Whether the large cogs initially had just an after- and foredeck or were fully decked-over aft to fore even at an early stage, is hard to tell. Currently the former seems the more likely, with the closed deck possibly being gradually introduced in the course of the 14th century. Part of the deck planks and beams probably were removable, with few fastenings. They would be taken out to facilitate loading and unloading.

Several of the smaller cogs may also have been seagoing vessels, given their hull lines that virtually match those of the large cogs, their length-width ratio at the gunwales, and their horizontal outline of the gunwales. Their length ranges from ca 12 to 16 m. Yet the most marked difference is the height of the sides: is just a little over half that of the large cogs.

In the large ships it was necessary to tie together the tall, everted sides at deck level with massive protruding cross-beams, but in the smaller ships there was no need for these. Here transverse beams between the sides will suffice, sometimes with their fastening to the sides reinforced with rose-bolts. The gunwale was reinforced internally with a heavy ceiling strake, or a (covering) inwale. These vessels probably (all) had gangways connecting the after- and foredeck. They had a single mast and probably carried a square sail.

A second group of small cogs appears to have been primarily designed for inland waterways and the Zuiderzee. The somewhat more elongated shape in top view is reminiscent of the later cargo barges. Some of these vessels (the Marknesse and Spakenburg-I cogs) may have had a lowerable mast and carried a spritsail. The latter possibly also applies to the Almere-I cog.

In principle it can be said that the large and small cogs rarely if ever had any part designated as for instance 'crew's quarters' separate from the hold. In the (early) boats and small cogs there are the cuddies or decks aft and fore, at either end of the hold, that create a tripartition of sorts. Yet the sharply built stern and bow below the decks were hardly if at all accessible, and at best served for stowing cordage, tools, a spare sail or personal belongings and victuals for the voyage. When under sail, crew, passengers and cargo travelled in the midships section or the hold; the helmsman on the afterdeck. It appears that there almost never was a partition in the form of a transverse bulkhead between the three parts of the ship. Such seems to have been the usual layout of many inland craft in the Zuiderzee region right into the 18th century, and in some cases, well into the 19th.

Little can be said about the layout of the decks. Parts of the deck were found on the Bremen cog. Just a few protruding cross-beams were recovered of the afterdeck of the Nijkerk-II cog. Yet the early 16th century *waterschip* wrecks in lots MZ 22 in Zuidelijk Flevoland (fig. 27.2) and K 84.II in Oostelijk Flevoland (fig. 27.3) and the mid-16th-century *waterschip* from lot W10 in Oostelijk Flevoland did yield important evidence about the construction of decks in medium-sized (fishing) vessels of the Late Middle Ages. Only in late 2015 did parts of decks and bulkheads first come to light, in the Kampen(IJssel) cog.

The small cogs were suitable for transporting small quantities of cargo (capacity ca 8-25 gross tons), and without doubt served to distribute the goods arriving in the large sea-going ships across the inland regions. Some will have been specifically built and employed to carry, for instance, building materials.

The term 'boats' covers a range of craft with cog-like features, of various types and dimensions. Apart from transporting goods and passengers across the smaller waterways, these boats presumably also functioned as fishing vessels. Most wrecks, however, produced no concrete evidence regarding any particular function and/or use. Rowing and punting would have been the obvious manner of propulsion, possibly supported by a small mast and steadying sail. As for the mast steps in these small cogs and boats, these mostly seem to have been short ones fastened onto the frames. Only in the Marknesse cog was this found to have been a secondary solution: initially the mast in this ship had stood in a recess in one of the V-shaped floor timbers. The drawing of the vessels from Antwerp shows a mast step only in the cog (figs 3.2 and 26.14).

Equipment and inventory

The large and small cogs generally had onboard cooking facilities. Both in the seagoing and the inland craft, until late in the 16th century, the fireplace usually took the form of a low wooden box measuring ca 85x85 cm. This would be placed in the most stable part of the ship, immediately aft of the mast, on the ceiling or on the ballast. The earliest known example was filled with loam, later on filled with sand and paved with bricks; some even had fine tiles with inscriptions (figs 29.13a-c). The fuel was kindling and peat. Galley accoutrements such as fire-tending implements, cooking vessels and crockery, and maybe some personal belongings, were usually found close to this fireplace. In late-medieval ships there was no place for furniture such as a seat or a table. There was no question of (separate) crew quarters. Much of the day-to-day operations took place at deck level. So far no evidence has come to light of dedicated sleeping areas aboard late-medieval ships. The presumed between-deck in the Nijkerk-II cog and the double cross-beam rows apparent in some seal images may represent spaces for sleeping groups of passengers, such as soldiers.

Barely any items of the original inventory have been found in the wrecks of the large, seagoing ships. On the whole there are just sherds of earthenware and stoneware, or the odd tool. Hence little can be said with certainty about the amount and composition of the original equipment and inventory of a working cog. In the smaller cogs and the boats too, very few artefacts were found. Maybe they never carried much equipment at any time.

Among the small cogs there are a few (Dronten-I, Marknesse, Almere-I), and among the boats with cog features too (K73/74, L89, and the barge B55 in Oostelijk Flevoland), that contained a presumably (near-)complete inventory, or at any rate a cluster of items that offer insights into what may have been the 'standard' range of artefacts aboard a late-medieval ship. On the basis of these finds and the size of the ship in which they were found, we may tentatively form an idea of the number of crew. This is somewhat complicated by the probably multi-purpose nature of many items. Cogs like the Dronten-I and the Spakenburg-I could definitely be sailed by a crew of two or three; for most boats, this may have been one to three. A large cog like the Nijkerk-II is likely to have required at least 5 to 8 crew members.

Of all shipwrecks so far uncovered in the former Zuiderzee region, just a very small proportion can be classified as 'proper' (large) seagoing ships, including the large (Hanseatic) cogs and the smaller seagoing types. Besides fishing craft and some working boats, these on the whole were typical cargo vessels for inland navigation that sailed the Zuiderzee and now have been recovered. Large ships were less likely to founder in this relatively shallow sea. The same probably goes for the Scheldt estuary.

29.6.2 'Standard' ship's equipment and inventory in late-medieval ships

Establishing to what degree excavated artefacts are representative of the original ship's inventory and whether they were standard components of it, requires a great deal of research. Van Holk based his study of living and working aboard inland-vessels in the 1600-1900 period on both documentary and archaeological sources.[98] Some of his conclusions were:

- Most of the ships from the former Zuiderzee region foundered while actively working. The inventories of these wrecks therefore may be complete, but this must not be blindly assumed. The completeness of these ship's inventories with respect to other assemblages

98. Van Holk 1996, 241-259.

Fig. 29.13a. The 16th-century *waterschip* from lot P40 in the Noordoostpolder. In the foreground, the afterbody with ballast stones, and upon it, the firebox paved with tiles. Amidships is the fish-well cover with the trough.

Fig. 29.13b. The reconstructed tile plateau from the fire box in the *waterschip* in lot W10 in Oostelijk Flevoland. Most of the tiles bear the inscription "ALLE DINC HEEFT SINEN TIIT" (Everything has its time). Two tiles are inscribed with the date "1561" in reverse.

Fig. 29.13c. The wooden firebox from the *waterschip* from lot W10 in situ, with part of the tiles still in place; in the bottom of the box lay a flat piece of limestone to insulate the base. The tiled paving lay on a bed of sand.

obliges us to formulate precise questions. So caution is needed with respect to apparent completeness. By comparing several items per material category, Van Holk found that some of the investigated inventories were incomplete. On the one hand this is due to the structural integrity - or otherwise - of the hull (items may have been washed away); on the other, it may (in part) be due to a dearth of suitable treatments for waterlogged artefacts in the 1970s. Items of wood, leather, bone and textile, but also iron, which is poorly preserved in the former sea floor, may have decayed after excavation.[99]

- The overall number of artefacts, when considered in their contemporary context, will offer a good impression of the completeness of the assemblage.
- The research into written sources showed that these too are incomplete. For instance, inventories drawn up at a sale indicated only what was sold with the ship rather than what actually was aboard.
- Van Holk noted that in archaeological and written inventories there is a trend for the number of items appearing to grow from 1700 onwards, with an increasing emphasis on domestic rather than operational items. The number of items remained constant between 1650 and 1800. From 1850 on, a marked increase occurred.

In various other publications before and since, attention has been given to groups of items or tools from the late- and post-medieval ships' inventories from the former Zuiderzee region.[100]

99. Vlierman & Van Dijk 1980a, 7. From the mid-1970s on, NISA successfully worked on developing conservation treatments for many types of material and material combinations.
100. Vlierman 1983; 1985b; 1991; 1992a, b and c; 1997a; 2005; 2006; 2010; Oosting & Vlierman 1993, 87-106.

The equipment and inventory of a small 15th-century cog is the first comparative study of the inventory of late-medieval ships' inventories.[101] This study focused on five wreck finds from Flevoland: from lots N5 (Dronten-I cog, Chapter 11); M107 (Marknesse cog, Chapter 12); Almere W13 (Almere-I cog, Chapter 16); K73/74 (boat with cog features, section 26.1) and B 55 [1] (cargo vessel with sintel clamped moss caulking, section 26.5).[102] Important and reliable sources of evidence to test the completeness and the standard presence of particular items in these inventories are the inventories from a relatively large number of 16th-century *waterschip* vessels, excavated in the former Zuiderzee region (but as yet largely unpublished).[103] Documentary sources on the composition of late-medieval ships' inventories are few or nonexistent; for the Late Middle Ages we are fully dependent on nautical archaeological assemblages, particularly for small ships and rivercraft.

The implements from the Dronten-I, the Marknesse, the Almere-I and the K73/74 wrecks are newly discussed and illustrated in the respective chapters of the present study. These vessels foundered between ca 1325 and 1500. The items are classified according to function. This classification was designed especially for small 19th-century (cargo) vessels of the Zuiderzee.[104] The classification and comparison of the 14th-/15th-century inventories revealed differences from the post-medieval ones which probably are characteristic of the era. There is a difference in the nature of the items, in their numbers, and in certain cases there is multi-purposing in the medieval period. The conclusions from this study are:

- The smaller number of items in late-medieval ships in part results from multifunctional use. The absence of items with specific functions means fewer items (or even zero scores) per category.

- Despite this multifunctional use, items in three groups (categories) seem universal in small late-medieval vessels: cooking/eating/drinking; maintenance of the ship; and personal equipment/weaponry.

In that trial study it was a deliberate choice to focus on just these three groups, in order to highlight the multifunctional nature of the items within them. Other artefacts, that could be placed in the categories of Nautical, Operational and Military Equipment and Victuals, were then omitted. In Table 1 in the publication it was indicated which items in the Almere-W13 were associated with multifunctional use.[105] In the excavation report of the wreck the recovered items were still classified under the eleven headings proposed by Reinders.[106] In the comparison and debate on the report a provisional division into three 'standard' classes of implements aboard (small) late-medieval ships was presented.[107] In the same framework, attention was given to cooking and kitchen ware on ships (for inland navigation).[108] Van Holk adds the items that I deliberately provisionally omitted in 1992 to the objects from the three 'standard' groups and arrives at a division into seven categories:[109]

- Ship's equipment
- Operational equipment
- Galley equipment
- Eating and drinking
- Victuals
- Tools
- Personal equipment/weaponry

In the present study, the objects that I purposely omitted in 1992, have been included. In contrast to Van Holk's seven categories, listed above, I have consistently classified them under the eight headings below and in the same order. All recovered objects will be discussed in detail and illustrated in the respective research reports on the individual wrecks.

1. Ship's equipment
2. Operational equipment
3. Administration and navigation
4. Tools
5. Domestic utensils
6. Cooking, eating and drinking
7. Victuals
8. Personal equipment and weaponry

Groups 1, 2 and 4 are self-explanatory. It should be noted, though, that the items of ship's equipment that belong at deck level and above, usually are missing. Group 3 is a practical merger of Administration and Navigation, as generally very few if any objects in these categories are recovered. For this reason group 3 might be scrapped, but still such items

101. Vlierman 1992b, 2-22.
102. The vessels from lots K73/74 and B55 in Oostelijk Flevoland were published in Reinders *et al.* 1980,17-29 and 31-41, respectively, and are briefly discussed in sections 26.1 and 26.5.
103. Vlierman 2006, 151-169.
104. Reinders1985b. The classification system was first applied to the inventory of the cargo ship "Lutina" (Zwiers & Vlierman 1988).
105. Vlierman 1992b, 12.
106. Hocker & Vlierman 1996, 51-97.
107. Vlierman 1992b, 76-79.
108. Vlierman 1992a, 50-59.
109. Van Holk 1996, 122.

definitely belong in late-medieval ships' inventories. Group 5 is of debatable validity: aboard small late-medieval cogs, boats and other inland craft, it is as good as certain that there never were any such items. But they cannot be ruled out in the large, seagoing cogs. Apart from the candle-holder in the Nijkerk-II, none have been found so far. There may be more evidence yet to come from the Kampen (IJssel) cog and the Enkhuizerzand cog. Group 6 is again self-evident: it combines Van Holk's categories of Galley equipment and Eating and drinking. After all, the late-medieval ships lacked a separate space for food preparation; there are no indications of proper galleys.

Section 29.6.1 already mentioned the fireplace aboard late-medieval ships. It is the central point around which in most cases the objects relating to cooking, eating and drinking are found to cluster.[110] For this reason, the fireplace will be the first item to be listed in this category. Group 7 (Victuals) represents foodstuffs and their storage. Group 8, finally, again combines two of Van Holk's categories: Personal equipment and Military equipment. The reason is that in the (Late) Middle Ages not only one's clothes etc. were personal belongings, but so was any weaponry. This was the regular situation on land at the time, and so it must have been aboard the small(er) vessels. Only in the case of a 'true' warship or one fitted out for warfare can we speak of military armaments. So far just a single vessel in the former Zuiderzee area could be definitely typified as such.[111] This armed *tjalk* of the latter half of the 17th century had a proper galley in the forebody.

In the transition of the Middle Ages to the Early Modern era, the process of change in the equipment and inventory of ships progressed even more gradually than in actual shipbuilding. In the tens of excavated *waterschip* wrecks especially, it was evident that very little change occurred in the range of artefacts up till ca 1600. This may be due to the perennially mostly quite simple and purely functional nature of the items, and their low numbers in the usually very scanty inventories of these fishing vessels. Aboard the few excavated 16th-century cargo ships there is somewhat more evidence of change. From ca 1500 on, the amount of crockery at any rate seems to increase. This is due in part to the appearance on board of large and small plates, low skillets with or without feet, and the so-called *kromsteerten* ('curly tails', bowls with a curved handle). The earliest plate- or bowl-like piece of crockery aboard a ship in the Zuiderzee region was found in the early-15th-century Almere-I cog (section 16.5.2).
The movable fireplaces (figs 29.13a-c) remained in use up to the end of the 16th century. In some fishing vessels such wooden 'fireboxes' seem to have been used even into the early 20th century. They usually were placed aft of the mast above the keel; and in the *waterschip* vessels, on the ballast aft of the fish-well. Shortly before 1600, domestic settings in the Netherlands saw the appearance of hearth-backs of decorated tiling, despite the cast iron hearth-back already having been introduced. This was no different aboard ships. The earliest known example so far of a fixed hearth with a tiled back was found in the forebody of a *beurtschip* (ferry); it had foundered ca 1625 and must have been built around 1600.[112] The fixed hearth and the crew's quarters were not moved to the afterbody until much later.

Without here enlarging on the rapid development of the nature and number of onboard items after 1600, there is no doubt that the (late-)medieval nature of the ships' inventories remained clearly evident into the early 17th century. Even from Heemskerk and Barentsz' ship, which in 1596 sailed from Amsterdam with a crew of fifteen, the galley outfit, which in 1597 was left behind in the *Behouden Huys* on the island of Novaya Zemlya, can be typified as a typically medieval inventory.[113]

Considering the title of the present section, we can first establish that as yet we have relatively few archaeological sources from the 1250-1500 period. The three cogs mentioned, the boat and the cargo ship with cog-like features, are the only ones of which we can, be it cautiously, speak of a fairly complete inventory. By including the more numerous 16th-century ships, given the almost unchanged nature over this period of the composition of inventories, it does yet seem possible at least for small and medium-sized vessels to compile a hypothetical 'standard' list of ship's equipment and inventory. For the large ships this is more difficult. So far, too few artefacts have been found in the large cogs and other vessels. The Enkhuizerzand cog (Chapter 24) in this respect is likely to offer a better perspective. Below, the types and numbers of implements that may be expected aboard small to medium-sized late-medieval ships (with a crew of three to five) are described by category.

1 Ship's equipment

In general, little tends to be recovered of items in this category: ground tackle, sails, cordage, blocks, windlasses, pump, implements and boats. The main reason is that the parts of the ship where they belonged are only rarely found in shipwreck excavations. Almost always, such missing objects have become detached with the above-

110. Vlierman 1992a, 51-53.
111. Vlierman 1997, 157-166.
112. Vlierman 1992, 52.
113. Vlierman 2005.

deck timbers and been washed away. But on the basis of what has been recovered, and judging by documentary evidence, we know that 12th- to 16th-century vessels were equipped with simple rigging. Anything that in those days might be considered nonessential would be dispensed with; everything had to be functional.

2 ***Operational equipment***

The classes of Loading and unloading gear and Additional equipment are represented depending on the type of ship. Here too the rule is that only what was indispensable was carried aboard. Occasionally certain tools may reveal what kind of cargo the ship (regularly) transported.

3 ***Administration and navigation***

This combined category is usually represented by a few fragments of chalk (with which the cargo might be marked), sometimes parts of a compass, and very rarely cash that might be regarded as business capital. Late-medieval wrecks in the Netherlands have so far yielded no large amounts of coinage that may be regarded as such. Mostly the coins are just pocket money, i.e. personal belonging.

4 ***Tools***

The category of tools is the first one to be almost always represented. Usually one or more from the subgroups of carpentry, caulking, maintenance and splicing, and sailmaking tools is encountered. Despite the occasional absence of a particular tool, we can be fairly certain that the (medium-)sized late-medieval vessels carried a 'standard' toolkit comprising the following items.

Carpentry tools

An axe, one or more hammer-adzes, a spoon auger, and sometimes a handsaw or pincers are the most frequently found carpentry tools.

Caulking and maintenance tools

A caulking iron; with the increasing prevalence of carvel-built ships, the shape of the blades of these indispensable tools too changed. Very occasionally a tar pot or a deck swab is recovered. Curiously, the typical wooden caulking mallet, which sometimes turns up in later ships, is absent in the late-medieval wrecks. Presumably it was primarily a shipyard tool. For interim repairs, the hammer end of the hammer-adze might be used for jamming moss into the shell seams and for driving in treenails and iron nails: a multi-purpose tool. On rare occasions, spare sintels, nails, treenails, remnants of leather to repair or blocks of tar are encountered.

Tools for splicing and sailmaking

Another standard necessity is a marlinespike made of red-deer antler, sometimes two - indispensable tools for splicing rope. The more familiar iron ones seem to first appear around the middle of the 16th century. Sail needles too can be reckoned to the standard toolkit. However, none have yet been found in late-medieval shipwrecks - just a lump of grease with impressions of such a needle.

Miscellaneous tools

A whetstone and/or a grindstone may definitely be reckoned to the standard on-board toolkit. Sometimes there are more than one. Remarkable are the grind- and whetstones of quartz-phyllite, frequently encountered in medieval ships (and settlements), many of which must come from the quarries at Eidsborg in Telemark, Norway.

5 ***Domestic implements***

No domestic implements are to be expected aboard the smaller medieval vessels. A candle-holder, like the one from the Nijkerk-II cog (Chapter 6), would hardly be surprising in a large ship, to provide some light below decks. Chamberpots are not found in late-medieval ships. In the above-mentioned cog, the gutter for discharging pump water may have doubled as a latrine.

6 ***Cooking, eating and drinking:*** *Fireplace*

The smaller inland craft generally lack any kind of fireplace. Small seagoing cogs may contain 'fireboxes' or traces of their former presence, which is unsurprising, as these ships probably also made voyages taking several days. In some instances (Almere-I cog, Chapter 16) parts of cooking pots and chunks of peat suggest that at some time cooking had taken place on board, but by the time the ship went down this was no longer the case.
Another possibility may be in play with the Marknesse cog (Chapter 12). In this wreck cooking equipment was found, while it is well possible thet a fire was lit on top of the cargo of large bricks. At any rate it seems to be a fact that the large ships had a fireplace. Apart from a wooden box filled with loam or paved with bricks or floor tiles, generally some flints or a firesteel are found for lighting a fire. Firetongs and a trivet may also be regarded as standard equipment. Instead of a trivet, a few (half) bricks set around the fire might serve the same purpose of supporting the cauldron or cooking pot. The fuel almost always consists of peat, while some kindling (wood) is occasionally found as well. Ash pots, found widely in late-medieval homes, do not generally appear in ships until after ca 1600. The same goes for built-in fireplaces. Highly unusual is the discovery in 2015 of a brick-built oven aboard the 15th-century Kampen (IJssel) cog.

Galley metalware
In principle, a copper kettle may be found for preparing communal meals, sometimes two kettles. They are cylindrical in shape, two vertical copper loops projecting upwards from the rim, joined by an iron handle. The size tends to be quite uniform: ca 130-170 mm high and 220-260 mm across. This means a content of 5 to 7 litres.[114] Among the late-16th-century galleyware from the ship of Barentsz and Heemskerk, two different kettles were found. They are 260 and 370 mm high with diameters of 375 and 365 cm, respectively.[115] The bases of all of these kettles are somewhat convex, which makes them sit on the trivet more securely. Because of the ship's motion, it probably was too hazardous to suspend these small kettles over the fire. The type of kettle and its manufacture seems to have remained unchanged between ca 1300 and 1650. They are even found in 18th-century wrecks. Other 'standard' metalware to be encountered is the bronze *Grapen* or cooking pot with three feet. So far only small ones have been found in 14th-century ships, with a content of 0.5 to 1 litre. These too, albeit generally larger ones, still regularly turn up in 17th- and 18th-century shipwrecks. No metal spoons or skimmers have been found in pre-1600 wrecks. Presumably wooden spoons were used, which are rarely recovered, most having floated away.

Galley crockery
In the cog wrecks dating from the first half of the 14th century, not only copper and bronze containers were used, but also *Kugeltopf* cooking pots of small to average sizes. Besides, we see the first onboard appearance of small redware pots (Dronten-I cog, section 11.5.2, foundered ca 1320s). In the Almere-I cog (section 16.5.2, foundered ca 1420s) there are sherds of at least five to seven small pots and four skillets. They probably were not in use simultaneously. In the vessels dating from the first half of the 16th century (mainly of the *waterschip* type), we find a more-or-less stable number of cooking pots (5-7), including some that are a little larger. Also we then tend to see one or two skillets and a *kromsteert* (curved-handled) bowl on board.

Eating and drinking: *Eating utensils*
Of specific tableware, such as plates, dishes and bowls, there is virtually no evidence aboard 14th-century ships. Just a fragment of a small wooden dish in the Nijkerk-II cog. The earliest known dish- or bowl-like item was mentioned above and dates from the early 15th century. So far only very few ships of the 15th century have turned up in the Zuiderzee region. Yet it seems likely that tableware gradually became a more regular element in ships' inventories. In early-16th-century ships they already are a standard feature. Meals will have been eaten mainly with wooden spoons, but none of these are recovered, for the reason mentioned above. In the known wrecks of ships for inland navigation, pewter spoons appear only in the late 16th century. In the 14th century and before, people probably eat directly from the cooking pots. Roast meat was eaten with the fingers; it was cut off with a knife. Every man would generally carry his own knife.

Drinking ware
In wrecks of the 14th to the late 16th century, usually one or two, sometimes three stoneware drinking jugs are found. They have a content of 1 to 2.5 litres. Especially the 1-litre jugs can be regarded as typical drinking jugs; the larger, also as storage jugs or jars, especially in small river ships undertaking single-day voyages. In some ships even smaller stoneware jugs have come to light. They may have been used for drinking, but also for *e.g.* medicinal draughts.

7 ***Victuals:*** *Storage of provisions*
For storing drinking water on board, large (5- to 6-litre) stoneware jugs were used. Occasionally some are found with a content of 7 and 10 litres or even more.[116] It seems likely that aboard the large cogs during their voyages of weeks, a larger store of drinking water was carried in wooden barrels. In the wrecks of these ships staves have been found, though these may equally come from casks containing provisions of *e.g.* salted meat, or indeed from cargo. The oldest known example of specific water barrels in a ship in the former Zuiderzee region was from a *waterschip* that must have foundered around 1625.[117] Bones of various animals with butchering cutmarks show that the stowed provisions included meat.

*8 **Personal equipment and weaponry***
The most frequently encountered items of a personal nature are shoes and boots or parts of them. In the latter case, these generally are worn-out and maybe already cut-up parts which were used for various repairs on board. Occasionally fragments of textile garments are found. Single or a few coins which can be regarded as pocket money will also emerge now and then. On rare occasions there are unique items, such as the clapper from the Almere-I cog. Presumably the clapper with the image of St Catherine of

114. Vlierman 1992a, 55.
115. Braat *et al.* 1998, 235; Vlierman 2005, XV.
116. Vlierman 2010, 189-204.
117. The shipwreck in lot U86 in Oostelijk Flevoland. The two barrels lay on a support on top of the fish-well cover near the living quarters in the afterbody. One of the barrels had a copper-alloy tap.

Alexandria and a leather bag or pouch had been used by a misdiagnosed leper. The clapper's owner may have been registered as suffering from a particular form of incurable leprosy, which had been miraculously healed by supernatural intervention. Such recoveries, subsequently labelled as miracles, would occur quite often.[118] The clapper is discussed in detail in the research report (Chapter 16). Knives and knife sheaths too are recovered. The same goes for (bollock or kidney) daggers. Both knives and daggers may be considered multi-purpose tools. The knife in fact is the predecessor of the pocket-knife; the dagger, a personal weapon. In 17th- to 20th-century sailing ships often an intermediate form is found: the so-called seaman's or sailor's knife, also known as a *kortjan*. On rare occasions a set of knives comes to light, made up of a large knife or a dagger and a sheath which holds various smaller knives as well. Five such small knives belong to the inventory of the Dronten-I cog (section 11.5.2). A remarkable knife from the same ship has initials on the blade, inlaid with silver. A dagger or knife was usually carried on the belt, and therefore is likely to disappear when the owner leaves the ship as it goes down. A rapier and a halberd aboard a *waterschip* that was lost in the second half of the 16th century,[119] a macehead from a cargo ship lost around 1500,[120] two spearheads from the Dronten-I cog (section 11.5.2), one spearhead from the Swifterbant cog (section 18.5.2), and possibly also the axe from the Nijkerk-I cog (section 8.5.2), are true weapons of warfare or defence. Yet here they are classified as personal weaponry rather than military equipment, because there are just one or two of them.

As regards the ceramics recovered from the pre-1600 shipwrecks, it is notable that they generally were seconds or even third-rate products. Only after 1600 will there just occasionally be a standard- or top-quality stoneware jug or small majolica dish.

29.6.3 External connections, the Hanseatic network and Hanseatic-related finds

Having extensively discussed the cog wrecks of the Hanseatic era, we cannot ignore the external links, trade and the Hanseatic network. In the following, we shall give some attention to some aspects of these, which are mostly based on the work of Jahnke.[121] Besides, we shall examine to what extent these aspects have been tangibly established. In other words, whether any merchandise was encountered in the wrecks, that demonstrate a direct link to the Hanseatic league.

The Hanseatic League

In the Late Middle Ages, the Hanseatic League in fact was the great circulation pump of European trade. A pump that ceaselessly propelled huge flows of merchandise in every direction. So above all, the Hanseatic League served the trading among large production regions in far-flung parts of Europe. Apart from the wholesale, also lively internal trade went on among the Hanseatic cities themselves and retail trade between the towns and the countryside. The Hanseatic system was almost entirely dependent on the commercial activities of the city-based merchants. The territory where the League operated, the Hanseatic member cities and the foreign towns with Hanseatic trading posts lay scattered across a large area of northwestern Europe (fig. 29.14).

Therefore the Hanseatic League is best described by its cities, which can be divided into three categories. In the first place there were the seaports, such as Kampen, Zwolle and Deventer in the west, Bremen and Hamburg on the German North Sea coast, and the Baltic cities of Lübeck, Rostock, Stralsund, Gdansk and Königsberg (Kaliningrad), with Riga and Tallinn even further east. These cities formed the outer boundary of Hanseatic territory. After the seaports, Cologne and Thorn (Poland) were the principal river ports. Cities like Dortmund, Hannover, Magdeburg and Dorpat (Estonia) formed the connection between the seaports and the inland towns; here waterborne transportation met road transport. The principal land route ran between Hamburg/Stade on the North Sea and Lübeck on the Baltic. Also there were Hanseatic trading stations in many cities, including Bergen in Norway, London, and Bruges in Flanders.

The inland Hanseatic cities were the majority by far. The greatest revenue however came from the external trade that was run from the ports. Despite the diversity of size, importance and mercantile interests, all Hanseatic cities had one thing in common: the Low German dialect. Thus the Hanseatic territory can be equated with the Low German language area (fig. 29.15). Outside it, area people spoke another kind of German (Upper German) or a different language.
In the early 12th century the Hanseatic League did not yet exist and traders travelled along centuries-old, familiar routes. In transshipment ports such as Haithabu (Hedeby) /Schleswig and Soest, goods from *e.g.* Russia might be bought. Already Scandianavian merchants were active in southern France. Around the turn of the 13th century, the situation changed profoundly. More towns were granted

118. Koldeweij & Vlierman in: Hocker & Vlierman 1996, Annex A; chapter 16, the Almere-I cog.
119. Reinders *et al.* 1978, 27 and plate 13, shipwreck in lot W 10, Oostelijk Flevoland.
120. Reinders *et al.* 1980, 39 and plate 11, shipwreck in lot B 55, Oostelijk Flevoland.
121. Jahnke 2010, 45-49.

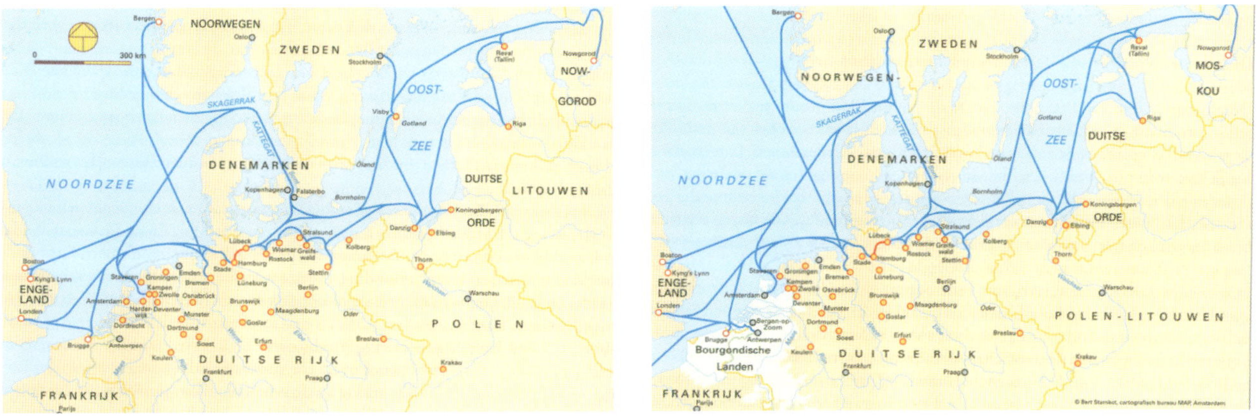

Fig. 29.14 The member cities of the Hanseatic League and the principal trading routes in 1370 (left) and 1474 (right). After Jahnke 2010, 48.

German city charters. As a result, the Baltic coast and its hinterland were progressively opened up. Some traders settled in these new areas and towns to promote their interests more effectively, or established collaborative Hanseatic trading stations there.

Hanseatic trade in the east opened up the region between the Baltic and the Urals. Here furs and beeswax were acquired. In the opposite direction went large amounts of woolen cloth, salt, fish and especially silver. To the southeast, the League linked the northwest-European to the Arabian - Asian trading area, where silk and spices were exchanged for *e.g.* fish. From the forests of Byelorussia and the mines of Slovakia, timber and ores, and later also grain were exported to western Europe via the Baltic. Also in the cities of Nuremberg, Frankfurt and Venice it was the Hanseatic League that linked northwestern Europe to the Mediterranean and Arabian worlds. The principal commodities traded between the Iberian peninsula and the north-European market were salt, fish and amber. Also the great European wine-growing regions were incorporated into the Hanseatic trading system. In the west, Flanders and England gained special importance. Here, beeswax, furs and in the 15th century also grain were sold, while these parts supplied the desirable woollen cloth in medium and top grades. Apart from iron and copper, especially fish (herring and stockfish) from Scandinavia was transported and sold by the Hanseatic traders. Fish was in high demand throughout Europe. Thus the Hanseatic League supplied consumers' requirements in every part of Europe.

Tangible evidence of the links between the cog wrecks, recovered artefacts and the Hanseatic League

It always is a tricky matter to match historical evidence directly to archaeological finds. Of the ca 450 recorded shipwrecks in the former Zuiderzee region, there are just a few, 19th-century examples for which this has succeeded. In a handful of other cases there is a fair degree of certainty that documentary evidence can be linked to a particular wreck. As the vessels become older, such connections are ever harder to establish. In the case of the cog finds there is no question of doing so. For the Kampen (IJssel) cog a link to the Hanseatic city of Kampen seems obvious, but proving this is a very different matter.

Starting with the ships, and the question of whether these were Hanseatic cogs, the first indication is the dating of the wrecks in the Low Countries. Then there is the provenance of the timber. In these respects, both would tally with a Hanseatic context in all cases discussed in the present study. Combining this with the fact that in the 1200-1500 period this area was marked by a single cog-building tradition, we can safely say that these vessels in one way or another, directly or indirectly, are likely to have been involved with Hanseatic trade and transport.

In some ships, certain constructions, procedures and materials seem to point to a provenance in the Baltic region. In the Nijkerk-I cog these are the remarkable transverse constructions in the underwater body at the stern and stem.[122] some longue-shaped scarfs on the outside of the shell amidships and the convex heads of the treenails that slightly protrude from the shell.[123] The Rutten cog probably also had such transverse connections. Analysis of the ship's timbers supports the idea that this ship, built before 1300, must come from the Baltic region. In the Doel-I vessel too such transverse connections have been observed in the underwater body, while the lower edges of the overlapping shell planks are provided with a gouged groove into which the nails of the clinker fastenings had

122. Crumlin-Pedersen 1979, fig. 2.9. The Vejby cog, ca 1375 (Crumlin-Pedersen 2000, 237).
123. Crumlin-Pedersen 1997, 120-125 and 224-251.

been driven - a form of decoration otherwise known only from the Nordic shipbuilding tradition. In the Doel-II cog not only moss but also animal hair was used in the caulking. Animal hair typically is a caulking of ships built in the Nordic tradition. In the case of this ship, it may in fact have undergone repairs in northern parts. Probably the 16th-century vessels briefly discussed in Chapter 27 also originated in the Baltic region. The fully clinker-built construction with rivets, comparable to the Nijkerk-I and Doel-I cogs transverse constructions and protruding treenail heads are indications of this. Considering the fully clinker-built ships with double-clenched nails in the strakes, curved stem and sintel clamped moss caulking in the shell seams, I believe quite firmly that the *waterschepen* were built on the western (Holland) shore of the Zuiderzee.

The above-mentioned features in some cog finds from the Low Countries not only deviate from the other cogs discussed in this study, but also from almost all of the hundreds of other shipwrecks excavated in the Zuiderzee region. Those can mostly be identified as characteristic regional vessel types of the northern and eastern Netherlands, which do not incorporate such features. Despite this not being definitive proof, it is hardly far-fetched to assume that most of the discussed cogs (after ca 1300) were built in the northeastern Netherlands and/or the northwest-German, Saxon region adjacent to the North Sea. Unfortunately nothing can be said with certainty about the early river cogs or boats/barges from Antwerp, but a Flemish provenance seems most likely. Of the early, other (river)craft mentioned in Chapter 26 it is quite certain that they were built in the Rhine, Moselle and Meuse areas.

Occasionally, ballast or a particular cargo will point to a specific provenance. Some examples of this follow below. The identification and assemblage of erratic boulders in the ballast found in the Nijkerk-I, Rutten and Nijkerk-II cogs make it clear that these definitely were not erratics gathered in the Netherlands. In the case of the Nijkerk-I, we may assume that the assemblage derives from southern Denmark at the latitude of the isle of Bornholm or more northerly. Over half of the originally probably 5000-6000 kg of ballast stones in the Rutten cog consists of Baltic flint. These flints must have been collected from a limestone coast on the Baltic, *e.g.* the isle of Møn. Unfortunately the crystalline rocks in the assemblage were not analysed. The brick fragments among the ballast in terms of fabric and dimensions match the contemporary building bricks used in Schleswig-Holstein and along the southwest Baltic coast up to the isle of Rügen. The bricks may indeed derive from a brick-paved fireplace. The ballast stones had presumably been brought aboard shortly after the ship was built and probably lay permanently in the underwater body. At about 1 m above the keel it seems the ballast was covered by a wooden floor. The erratic stones in the Nijkerk-II most probably were gathered on the southwestern coast of Sweden (around Malmø, Trelleborg). These ballast origins merely indicate that the ships probably visited those areas or regions, but do not necessarily reveal the ships' provenance.

The Marknesse cog was found to contain a 20-ton cargo of large bricks, as well as a few traces of earlier cargoes of bricks, seashells and maybe rooftiles. The Spakenburg-I cog contained a load of bricks; it is unclear whether these were part of a cargo. Maybe this batch had been put into the ship to help scuttle it. Here too there were bricks and brick fragments presumably belonging to a former cargo. In the Nijkerk-II cog, fragments of a curious, thin and round millstone(?) were recovered. The fresh grooves and scratches in its surface show that it had never been used. The schistose rock may well have a Scandinavian provenance.

Two items from the inventories seem to definitely indicate an origin outside the Low Countries. The axe from the Nijkerk-I, for which parallels are known only from Denmark, and the small drinking jug from the Nijkerk-II. This jug in terms of type and fabric is unknown in the Hanseatic territories, but strongly resembles products from the pottery centre of Saintonge, southeast of La Rochelle (France). The German stoneware found in the inventory of almost any cog was in use throughout the Hanseatic territories and therefore tells us nothing about ships' provenance or their voyages. The same goes for the recovered (sherds of) redware and globular *Kugeltöpfe* pots. In most cases these will be products from the Low Countries. Both types of crockery occur throughout the Hanseatic territories and were produced locally. Only some specific types, such as the highly decorated Flemish jugs of earthenware, are easy to identify. But no such items have been found in the cogs. Many of the grind-/whetstones of finely layered quartz phyllite must come from the quarries at Eidsborg in Telemark, Norway. These clearly are merchandise that was widely distributed over medieval western Europe, turning up in settlements as well. The finds from the cogs, however, are part of their toolkits, not their cargoes.

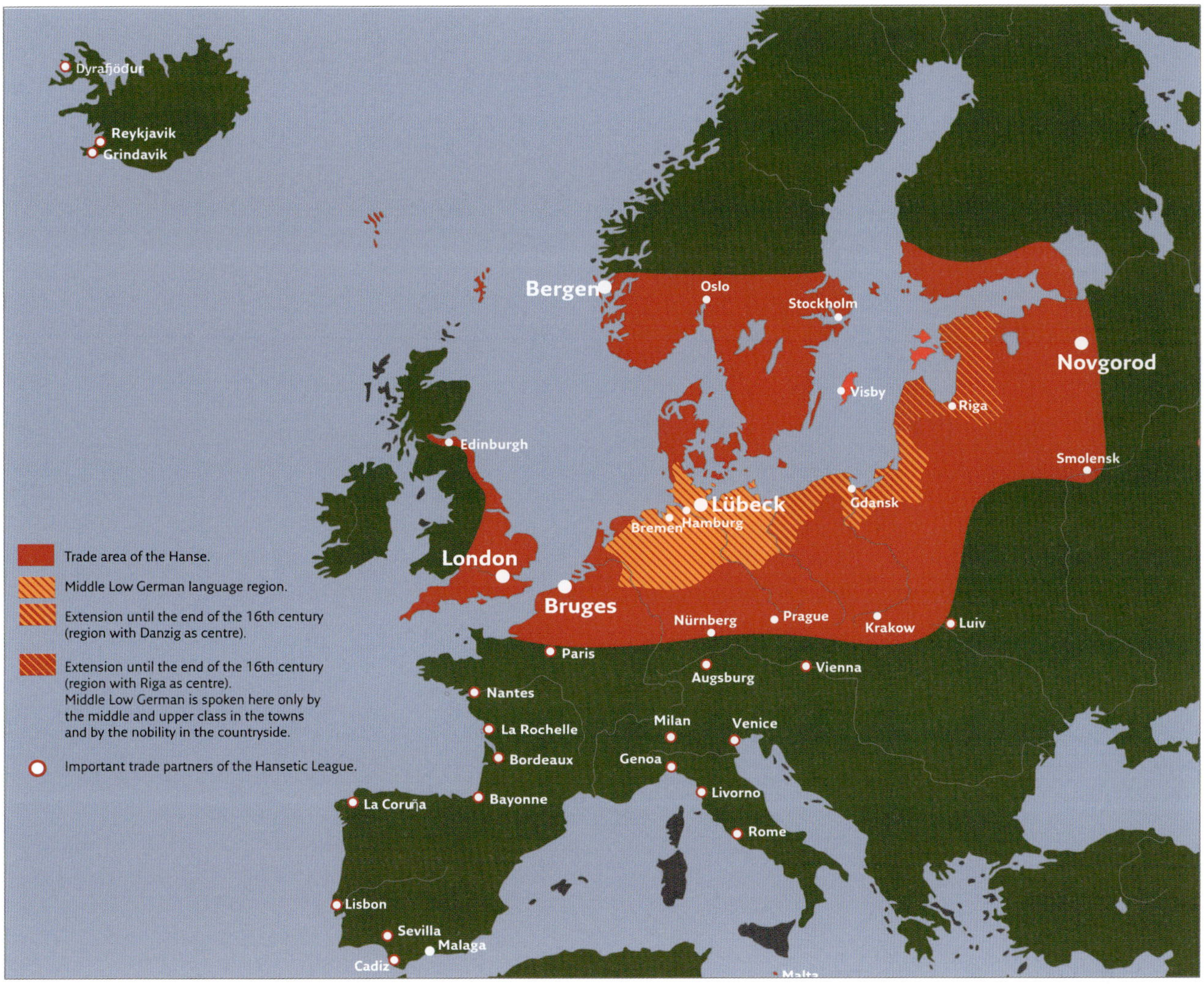

Fig. 29.15. After map in: Katalog Europäisches Hansemuseum Lübeck.

A good example of an early-14th-century cog (Dutch, 15th century, Bl. Add MS 11575, T65r) after Flatman 2009.

Origin and decline of the cog[1]

30.1 Origin

Until the discovery of the shipwreck in the Weser near Bremen in 1962, the cog, known from historical sources and from seals, had an almost mythical significance. Initially, historians were mainly interested in the large, seagoing late-medieval Hanseatic ship. It could be deduced from written sources that in the rapidly developing Western Europe there was quite suddenly also a new type of ship, with which larger quantities of goods could be transported and with which long sea voyages were made. In early (12th-/13th-century) sources the cog is also mentioned as a means of transporting crusade armies to and over the Mediterranean Sea and of transporting goods by inland waterways. All in all and certainly at that time, an intriguing and appealing type of ship.

In section 1.1, I have already indicated that, in north-western Europe, there were probably two rather different methods of building seagoing ships in the approximate period of 1000-1450. One was the Viking-/Nordic or Scandinavian building tradition, which had already existed for centuries. The other building tradition, regularly mentioned in historical sources was linked to the cog after the discovery of the large late-14th-century cargo ship near Bremen. A large vessel that must have had a relationship with the Hanseatic League in particular. The type probably originated in the second half of the 12th century and disappeared from the north-west European waters around 1500. In the Low Countries and in Scandinavia/the Baltic region about thirty shipwrecks have been found that match the characteristics of the underwater hull of the Bremen cog. More than two-thirds of these wrecks were found in the Netherlands.

Chapters 2 and 4 describe historical and archaeological data and ideas that provide a new or at least an additional (different) picture of the possible area and the period in which the cog was developed. Instead of talking about the origins of the cog, however, it is probably better to think about the origins of the cog-building tradition, in which the large seagoing ship was of course a leading figure.

I am not aware of any historical and/or archaeological sources that suggest that the (so far few) carvel-built vessels from Great Britain and from the North Sea- and Canal coasts in Flanders and France, dating from the Late Roman and Early Middle Ages, were built up to the Late Middle Ages as well.

(Late) 13th-century river craft from the Netherlands, with carvel-built bottoms and planks which transition to a clinker-built construction at the stern and stem, probably originate from the Rhine-, Meuse- and Moselle region. The sides are also clinker-built. The sintel clamped moss caulking in the seams is likewise similar to that of the cogs. However, the connections between the clinker-built planks are not made with nails folded over twice and driven back into the wood, but with thin (Ø approx. 2 cm) treenails. In my opinion, these vessels represent a different building tradition (from river vessels). The characteristics corresponding to the construction of cogs can be explained by assuming that the cogs were influenced by this.[2] Cautiously I would suggest that the corresponding construction details were already applied to these vessels for a long time. It is therefore also possible that we see here the start of the cog building features, which were also used elsewhere at the same time, but where they had already been further developed. I will come back to this later.

In section 29.2, in my opinion, the construction regions of the ships built according to the Nordic tradition are also to be found in Scandinavia, the Baltic region, the British Isles, along the Flemish/French Atlantic coasts and possibly also in the former Frisia. A number of finds from the British Isles are known, but to my knowledge, not yet from the Atlantic coasts. From the Netherlands we know of archaeological finds that belong in this building tradition. These British and Dutch finds suggest that at least until the middle of the 12th century, these were the typical types of ships in north-western Europe. The traders from England, Flanders, western Scandinavia and probably also from the coastal region from Flanders to Denmark sailed the North Sea, the Channel and the French Atlantic coast with these ships. The western Scandinavian

1. Translated by Sasja van der Vaart-Verschoof.
2. Förster (2008, 315) mentions the mid-13th-century Puck wreck, north-west of Gdansk. The connections of the vessel, which was entirely clinker-built, were also made with wooden treenails. He speaks of a Nordic/Slavic building tradition. It is not clear whether the wreck, which is 12 > m long and about 3.5 m wide, is a river vessel.

vessels undoubtedly already sailed to the Baltic Sea area for a long time. They were (initially) ships with a relatively low freeboard and not very suitable for carrying large cargoes. Because of their shallow draught they were suitable for sailing up the rivers as well. Based on the Dutch finds of river barges and barge-like plank boats from the period 1000-1200 it also seems very plausible that, in places not far from the sea coast, goods were already being transhipped on such vessels for transport further inland. As the seagoing vessels became larger and therefore more difficult to sail up rivers, this transhipping became more and more a matter of course. This has not changed to this day.

The finds of seagoing vessels in western Europe did not provide an answer to the question of whether, for example, in the 12th century there were already adapted, larger and higher vessels from this building tradition and whether they transported larger cargoes in the southern part of the North Sea, the Channel and along the Flemish/French Atlantic coasts. Asaert[3] believes that the *nef* (the Norseman's ship) from the 11th and later centuries was the large Atlantic cargo ship that sailed mainly between France and England, an area where the cog had not yet spread. He also writes that the difference between the two types of ships is not clear and that they were both round, low in shape and clinker-built. So far we have no concrete information about the *nef*, only that this type of ship was most probably built according to the Nordic tradition. A number of 13th-century English city seals, presumably depicting the *nef*, show protruding cross beams with fenders or fairings. This means that they probably represent relatively high vessels, comparable in height to the large cogs. We now know a lot more about the cog and it has become clear that the construction etc. is quite different from what Asaert assumed.

Crumlin-Pedersen[4] gives an overview of nine cargo ships from the Nordic tradition that have been found in Denmark, Sweden, Norway and Germany. The construction regions of six of these ships are the west Baltic and Kattegat areas and western Norway, of the remaining three this was/is not known. They date approximately from the period 1000-1200. The reconstructed lengths of these vessels vary between ca 16-30 m, the widths between ca 4-9.5 m. Six of the vessels are >20 m long. I do not know of any reconstructed midship board heights. The construction parts found of one of the vessels, from Bergen (N), suggest that the vessel was at least 30 m long. Archaeological evidence that around 1200 there were already large, completely clinker-built cargo ships in the north. Crumlin-Pedersen[5] points to the misconception that it was only after the creation of the Hanseatic League and the introduction of the cog that there was the first instance of the transport of bulk goods. The Gedesby wreck, of which a replica[6] has been sailing around since 1995, and the wreck of Vidby Hage[7], probably built on Falster around 1300 and on Sjaelland or Lolland-Falster around 1435, respectively, are examples of Nordic construction with typical construction elements that have clearly been copied from cog construction.[8] With the exception of repairs with sintel clamped moss caulking, the cargo ship from Karschau in the Schlei[9], built between ca 1130-1200 on Fünen or in Schleswig-Holstein according to the Nordic method, does not yet show an intensive exchange of structural elements between the two building traditions. It is reasonable to assume that the repair was most likely carried out in the final phase of the ship's operation. In addition to the aforementioned ship finds, there are other known wrecks in the Baltic region of vessels from the Nordic construction tradition. This also includes the Baltic cogs, entirely clinker-built from coniferous wood. Förster mentions them all in his *Katalog der Schiffsfunde*.[10] The vast majority of all (late-medieval) wreck finds in the Baltic Sea area and around Denmark were built according to the Nordic tradition.

In his article from 2000, Crumlin-Pedersen dates the construction of the oldest wreck with cog characteristics (Kollerup [DK]) around ca 1150, the wreck from Skagen (DK) around ca 1200 and the cog from Kolding (DK) ca 1250. Dendrochronological research has shown that they were built with wood from South Jutland and that at that time they were among the ships that sailed around Skagen to the Baltic Sea.[11] Crumlin-Pedersen furthermore believes that the area between the river Eider and Ribe has the greatest potential for the development of a hypothetical older type of proto-cog. I also think it is very likely that the construction of these ships took place in that region. Ribe was an important, if not the most important trading port on the Frisian/Danish North Sea coast during the period in question. Whether this means that the cog originated there is, in my opinion, a completely different matter. The first two (Kollerup and Skagen) are so far the

3. Asaert 1974, 111 and 112.
4. Crumlin-Pedersen 2000.
5. Crumlin-Pedersen 2000.
6. Bill 1991; Bill & Vinner 1995.
7. Gøthche & Myrhøj 1996.
8. Crumlin-Pedersen 2000, 241.
9. Englert *et al.* 2001, 141-154.
10. Förster 2008, 299-357.
11. Crumlin-Pedersen 2000, 234, 238 and 239; Daly, Eriksen & Englert 2000, 61.

oldest examples with a number of (somewhat) similar cog characteristics from later times. The planks of these ships are tangentially split from trunks instead of sawn, a type of material processing that is typical of the Nordic building tradition. The modern line drawings of the hull of the Kollerup cog furthermore show a symmetrically built ship of more than 20 m long, about 4.5 m wide and about 2.3 m high amidships. The construction of the other Danish cogs (Lille Kregme and Vejby) are dated to ca 1358 and ca 1372 respectively. Wood from Poland was used for both vessels. Presumably they were also built in the south-eastern Baltic area.[12]

The construction of the oldest cog-find in Sweden (Bossholmen) probably took place around 1275, the cog Helgeandsholm 2 around ca 1330 and that of Skanör around ca 1400. The first and last ones probably come from shipyards in the western Baltic region.

From Germany in addition to the Bremen-cog, built after 1380, we know the Darsser-cog (Vor-Pommern). A large vessel with a reconstructed length of approx. 21 m. The ship had five crossbeams, three of which have been preserved, and a long keelson. According to the dendrochronological research, the ship must have been built between 1298-1303/after 1335.[13] Moss and animal hair were used as caulking. The wood of the Darsser-cog comes from the area around Danzig and Elbing.[14]

For the origin of the cog, only the finds from Kollerup and Skagen are actually interesting. The others all date from around or after about 1250, the period in which the cogs sailed from the North Sea via the Kattegat to the Baltic Sea and vice versa. According to dendrochronological research, the mid-13th-century cogs from Kolding and Bossholmen probably originate from South Jutland and the western Baltic region, respectively. I suspect that the Dutch find Nijkerk-I (Chapter 8) was built in the same period and also in the western Baltic area. Unfortunately, no dendrochronological analysis has yet been carried out on this vessel. The large Rutten-cog (Chapter 10), probably built in the third quarter of the 13th century, is made of oak from Poland and Lower Saxony. The enormous amount of ballast, especially Baltic flint, was probably brought on board (shortly after construction) in the south-western part of the sea (Møn, Rügen?).

The cog wrecks found in the Baltic Sea area and around Denmark with a probable construction region in the west-Baltic area/Schleswig-Holstein, plus the Nijkerk-I cog from the Netherlands, date from the period ca 1150-1250. The other, younger ones were built elsewhere (in the Baltic area). From this it could be concluded that (the attempt to) build cogs in South Jutland/Holstein was of relatively short duration and subsequently developed more along the south-east coast of the Baltic Sea, whereby the whole clinker-built construction of the Nordic tradition in principle continued to be followed and maintained around Denmark and further north. The seals of Lübeck, Barth and Neustadt (in Holstein) from 1226, 1304 and 1351 respectively show vessels from this building tradition. The oldest city seals almost certainly depicting cogs date from after about 1300. Cogs of oak from the southern coastal region of the Baltic Sea, with the characteristics of the Bremen cog and the ship finds in the Netherlands and dating from around 1300 and later, were not necessarily built in the region of origin of the wood.

Hagedorn[15] is the first(?) to assume that the cog may have originated on the French coasts. Asaert[16] writes, among other things, that in the 13th century a distinction was still made between the *nef* and the *cog*, but that they evolved simultaneously and analogously. They are both mentioned in the same texts and sail in the same fleets. In my opinion, it seems very possible and perhaps even very obvious that during the explosive growth in all fields in Flanders and Artesia in the 11th-12th century, it was precisely there that the cog-building tradition arose. The Atlantic *nef* could then have been the starting point for the slowly developing larger cargo ship. The completely riveted (Nordic) clinker-built construction of these vessels, but possibly also the carvel-built vessels already known from the Merovingian/Carolingian period from this area, would have been the model for the combination of carvel/clinker-built construction of the cog hulls and the double-clenched nailing of the clinker-built connections. The combination of carvel-built bottom and clinker-built sides and double-clenched bottom planks was already known in Roman times from the large river vessels and the Celtic-Roman seagoing vessels (Sections 3.1 and 3.3), while the method of sealing the seams was almost certainly copied from the centuries-old tradition of river-ship building in the Rhine area on the western European continent. This caulking method (the sintel clamp shape) was adapted as required during the developing construction of the cogs. The rapidly growing population and cities in this (coastal) region and the there located organisation and probably also construction of a large part of the ships for

12. Crumlin-Perdersen 2000, 237 Table 1.
13. Förster 2008, 304.
14. Förster 2000, 53
15. Hagedorn 1914.
16. Asaert 1974, 111 and 112.

trade with England and for the Second and Third Crusade, kept pace with the initially dispersed (10th-11th century) and systematic reclamations (12th century) in the County of Flanders. As a result, large and small forest areas quickly disappeared.[17] The oldest known written designation of the cog type refers to two Flemish ships, which were used during the Second Crusade in 1147.[18]

Corresponding to the observation by Ellmers and Reinders that there is (as yet) no archaeological evidence for the creation of the cog from a Frisian *waddenschip*, it must be concluded that this also applies to the creation of the type of ship in the Flemish coastal area. Based on the currently available data, it seems almost certain that this occurred in the period 1125/1150-1225.

Within the explosive growth in all areas in Flanders and Artesia in the 12th and 13th centuries and with the preparations for the extensive troop transports for the Second and Third Crusade in mind, it therefore seems almost obvious that a maritime tradition developed in the Flemish coastal regions. At the same time, this was probably also the case in the Lower Rhine region. The aforementioned river vessels with cog characteristics from this area could be an indication of this. The most important reason for this was undoubtedly the (further) development and construction of large(er) seagoing vessels in order to be able to transport larger quantities of goods as efficiently as possible. The already mentioned Crusades undoubtedly put even more pressure on and created opportunities for developing and building large English, German, French and Flemish fleets as soon as possible in order to be able to transport the armies to the Holy Land and the goods they acquired there to western Europe at a certain point in time. This was undoubtedly stimulated in all sorts of ways by the monarchs involved, the clergy and, last but not least, the merchants.

The need to assemble these fleets out of large(er) vessels was partly because of the limited possibilities with regard to the intended long-distance voyages. People were very dependent on favourable winds and delays were common, which meant that voyages sometimes took a long time. In the winter months there was little or no sailing, especially to the north and the Baltic Sea area. In other words, the possibilities were seasonal and annually very limited. As a result, the existing capacity was unable to meet ever in creasing needs. The most obvious solution was to increase the capacity of the vessels or to increase their carrying capacity.

The knowledge, craftsmanship and experience built up since the Carolingian era in building ships for the trade between England and the continent and of the ships for the patrolling the French/Flemish Atlantic coasts and the rivers that flow into them, was fuelled by all kinds of (new) ideas with regard to reliably increasing the size and efficient operation of the vessels. This was particularly the case during the preparations for the Crusades around the middle and in the second half of the 12th century. The influences from the Mediterranean, probably already brought in first instance by the Normans, undoubtedly played an important role in this. Indeed, as early as around 1100, some of the major ports in northern Italy were already involved in striking activities relating to trade with the eastern Mediterranean and the large vessels used for this purpose, which had been made technically more suitable. The Norsemen of southern Italy, with their shipbuilding experience, will undoubtedly have made an important contribution to the latter. The initial adaptation of an English/Norman(/Flemish?) type of ship, the *nef* built according to the Nordic tradition and then very probably the slow development of a new, larger and higher type of ship (the cog) in the western European coastal area and along the Lower Rhine in the second half of the 12th century, fits within these developments.

This probably quickly developed into a shipbuilding tradition in the Flemish heartland. The centre of these activities and the further development of the cog-building tradition into the vessels as covered in this study, very probably moved more and more from the Flemish coast in a north-eastern direction towards the emerging Hanseatic region at the end of the 12th century. The already not or hardly available construction timber in Flanders and its surroundings at that time must have been a strongly contributing factor to this. The construction of sea-going vessels in the Lower Rhine area probably (also) moved almost simultaneously to the IJssel region, the eastern Zuiderzee region and north-west Germany. In these regions (eastern Netherlands, Westphalia and Lower Saxony), construction timber was abundantly available during the period in question.

Another, probably no less important reason for relocating shipbuilding and commercial activities within the boundaries of the German Empire to the north-east was probably due to the activities of Duke Hendrik de Leeuw in the Baltic Sea area, in particular around Lübeck. Emperor Frederick I did not like the fast-growing power of Hendrik. He successfully besieged Lübeck. After the surrender of the city and the deposition of Hendrik de Leeuw

17. Verhulst 1995, 129-133.
18. Bill *et al.* 1997, 158.

in 1180, Barbarossa confirmed the privileges of the now important trading city, where the origin of the Hanseatic League is also presumed.

Much of the trade with Westphalia and Lower Saxony had already been going through Deventer for centuries and for the most part by land. Western sea trade to and from the Baltic Sea also ran over land in Schleswig and via Bremen and Hamburg (Stade). The German emperor, the church and the merchants from the Rhineland must have been very interested in improving the connection with the Baltic Sea and getting a better handle on the situation there. The competition from the Flemish and English merchants was also felt by the Rhinelanders. Since the Second Crusade, Cologne had a large fleet with seafarers who had gained the requisite experience of sailing across the North Sea, the Channel, the Atlantic Ocean and the Mediterranean Sea. These sailors also had to be able to round up Jutland and sail to the Baltic Sea.

In order to shorten the long distance by water between the Rhineland and the Baltic Sea, the Lower Rhine merchants needed an 'intermediate station', preferably not too far from the North Sea. This was possible because as a result of storms, the Almere had been transformed into the Zuiderzee in the second half of the 12th century and within the German Empire a good navigable route from the Rhine via the IJssel over the Zuiderzee to the North Sea had been created. Approximately halfway the total distance, a suitable location was found in the IJssel delta, not far from the then coast of the Almere/the Zuiderzee. On the south-east side, the area of Kampen was naturally separated from the county of Gelre by an inlet or creek, De Enk, which ran from the Zuiderzee to the hardly known hamlet of Het Oenen or Onden. 12th-century sherds were found on some old river dunes in 1959 (and around 2000, also by me). From about 1300, the first dikes were constructed around Kampen and a piece of no man's land was created at De Enk. During the construction of the dike De Slaper in the middle of the 14th century, the hamlet was divided between Kampen (polders Broeken and Maten) and Kamperveen. According to research, De Enk is not an old IJssel branch, but a channel (*inbraakgeul* in Dutch) between two existing polders, as a result of which at high tide the Zuiderzee in particular - but also the IJssel on occasion - made its way. It is possible that the tidal channel was connected to the Oude IJssel during its largest expansion. In the Middle Ages, a ferry was present for the passage of the trench. In 1558 Philips II allowed the city of Kampen to build a bridge over 'a strand of water coming from the Zuiderzee, called den Enck, where they have a ferry that is unusable for part of the year due to storms and high tides'. During a heavy storm and ice drift in 1573, the bridge over the Enk river was destroyed, washed away and not rebuilt due to the costs involved. For decades the ferry was the most important connection between Kampen and Gelderland.[19]

In the context of 'Space for the river. IJssel Delta' ('*ruimte voor de rivier. IJsseldelta*' in Dutch), this old connection between the former Zuiderzee and the IJssel will now (2016-2019) be more or less restored to its former glory, in order to have an extra outlet for the IJssel in the event of high-water levels. Prior to the excavation and dike construction of this project, Isala Delta and ADC-ArcheoProjecten dug several test trenches. In two of them, wooden structures and posts were found on the north and south sides of De Enk in the vicinity of the medieval hamlet 'Het Oenen'. Presumably from jetties.[20] Perhaps the location of the aforementioned ferry. However, the results of dendrochronological research revealed that they are not late medieval, but rather date to the 16th/17th century.[21]

Kampen was most probably founded in the last quarter of the 12th century. A transfer station/trading settlement where the seagoing and river ships exchanged goods and where the ships could spend the winter. Since 2014 we can safely assume that Kampen had a shipyard in the 13th and in the first half of the 14th century where they at least repaired ships, and almost certainly also built small boats and barges and possibly also cogs.[22] The area along the IJssel up to and including the village of Zalk had long been part of the Oversticht (Overijssel) and not of Gelre, despite the fact that the whole is located on the south side of the IJssel. It stretched out to the Gelderse Kolk, a few kilometres south of Zalk. Not far from this whirlpool (*kolk* is Dutch for whirlpool) and on the IJssel was Buckhorst Castle, the only castle in the municipality of Kampen. The castle is historically known from 1224 on. The discovery of tuff and the inheritance-free past indicate a founding date in the 11th or 12th century. During the conflict in 1224 there was mention of a tower, which indicates that the Buckhorst was a motte castle, probably built for military purposes.[23] On the other side of the river, about 2 km from there, lay the infamous castle of Voorst (near Zwolle). Together they guarded the river and the border with Gelre and thus served the interests of the Bishop of Utrecht, but also those of the lords of the castle themselves.

19. Jager 2015, 90-92.
20. Pamphlet Rijkswaterstaat. Ministerie van Infrastructuur en Milieu. Ruimte voor de rivier 2016.
21. Van Daalen 2017.
22. See section 28.2.
23. Jager 2015, 387-406; Klomp 2005, 16.

It is therefore almost self-evident that the building tradition of the ships considered in this study was finalised in the first half of the 13th century in the Oversticht part of the river basin of the Overijsselse IJssel/the eastern coastal area of the Zuiderzee, possibly at the mouth of the Ems and along the lower reaches of the Weser and the Elbe to the areas around Bremen and Hamburg. In the same period there is probably also early cog construction in Flanders, Dordrecht and Vlaardingen. According to my personal opinion it seems more and more that in the Flemish, Zeeland and Dutch coastal area, the completely clinker-built construction of ships (the *nef*) does not disappear and possibly even regains the upper hand. The virtual absence of cogs or components thereof from the 13th and 14th centuries in the western part of the Netherlands is a not insignificant reason for this suspicion. The fishing and fish transport vessel, the *Hollandse waterschip*, which has been known from written sources since 1339, is a notable exception. The relatively large number of wrecks (more than thirty) in the former Zuiderzee area of this type of vessel, which appears to have many similarities with the small cogs (and the *nef*), supports this suspicion. The late medieval type of 16 m long is entirely (!) clinker-built, has a curved (!) stem, but the seams of the clinker-built planking are provided with sintel clamped moss caulking on the inside, while the clinker-built fastening is carried out with nails that are bent twice and driven back into the wood. Clearly under the influence of the cog-building tradition. Unfortunately, only late medieval *waterschip* built in the first half of the 16th century have been found so far. From around 1550 the shape and construction of the *waterschip* remained the same until the 19th century, except that from the middle of the 16th century the vessel was 20 m long and fully carvel-built. Heinsius and Asaert also indicate that the *nef* most likely managed to maintain its position alongside the cog for hundreds of years.[24] At the Batavia wharf in Lelystad, at the end of 2015, the reconstruction of the most complete late-medieval *waterschip* from Flevoland (from lot Mz 22), excavated by the NISA, was started.

After studying the data on cogs available in 1999, Weski concluded that there should (also) be an IJsselmeer type.[25] There are indeed differences with the Danish, Swedish and Baltic cogs built in the north. For example, it is not yet clear whether the Kollerup cog was guided using a stern rudder or a steering oar. The split rather than sawn planks of the Kollerup and Skagen cog indicate that these vessels were built in the Nordic tradition. It is remarkable that in the early 13th-century cogs from Denmark and the Netherlands the bottom strake and one or more of the side strakes were carvel-built from stern to stem. In the later cogs, the carvel-built planks have been put together over the last metres towards the stern and stem and the strakes from the sides are completely clinker-built. It should be noted, however, that only a few (3) shipwrecks dating from before approx. 1200 are known. In the case of the cogs built around 1250 and later, it can be said that there is an almost fixed construction and building sequence. The construction of other, smaller (inland) vessels derived from this. This standard formula was probably (partially) adopted from the end of the 13th century for the construction of cargo ships in the Baltic region.

If the construction regions shown above (Overstichtse en Nedersaksische river- and coastal areas) are correct, it is very likely that the name cog will suffice for the large and small vessels with the characteristics formulated in this study. The relatively large number of shipwrecks with similar construction characteristics (there are only a few finds from the Nordic tradition known, which moreover almost certainly come from elsewhere), the frequent occurrence of the name in written sources, the large number of ship hulls on city seals that correspond to those of the archaeological finds, and the fact that there is no actual alternative, are my most important reasons for this hypothesis. It is also possible to use the name Hanseatic-cog for the large, seagoing ships. It is very likely that almost all of the vessels were built in this core area of the Hanseatic League. The possibility, at least from the middle of the 13th century onwards, of using the larger and seaworthy vessels to sail via the Skagerrak and Kattegat to the Baltic Sea will have contributed to the flourishing of the Hanseatic towns along the river IJssel.

Importance of the cog for the Hanseatic League and other types of ships

In the previous paragraphs I hypothetically indicated that the cog-building tradition probably developed from about 1200 onwards, especially in the Overstichtse area along the IJssel and the lower reaches of the Weser and the Elbe. Within this building tradition the large sea-going cogs were undoubtedly the most important. From the middle of the 13th century, these vessels were used to transport bulk goods from the north to western European seaports. It goes without saying that the ships also took goods with them on the outward voyage to trade in the north. They also made trips to England and Flanders and to the French Atlantic coasts to obtain salt and wine. Based on the archaeological finds from the Netherlands and especially from the Zuiderzee area, I have been able to establish that all kinds of functional medium-sized and small types of vessels from the period 1200-1500/the Hanseatic period

24. Heinsius 1956, 251.
25. Weski 1999a.

were built in that area according to the cog-building tradition. From the northern countries only a few large(er) seagoing types are known. The small(er) vessels were built there according to the usual Nordic construction method.

In the historical sources, in addition to the name cog, other type names are also mentioned. As already noted, it is not possible to deduce from these references what these vessels looked like. Apart from the vessels from the cog-building tradition and a few vessels from the Nordic tradition, the archaeological finds from the Low Countries do not provide any indication that there were other vessels in western Europe/a different building tradition in the period from about 1200 to 1500. This is in fact a peculiar and hardly explainable observation. It therefore seems most likely to me, if there were other types that their origin should be sought in the developing Nordic building tradition along the Atlantic coasts (such as the *waterschip*) and/or in the (east) Baltic area. Förster[26] mentions several 15th- and 16th-century wrecks of commercial vessels from the (east) Baltic area. From Brittany the all clinker-built 15th-century vessel from the Aber Wrac'h river is known.[27] From a ship-archaeological point of view, nothing can be said about different types of seagoing vessels in service of the Hanseatic League. Only that in the Hanseatic period vessels built according to the Nordic tradition also carried out trade transports. The Dutch finds show that they also sailed to western Europe.

The name cog

In chapter 4, the same thing was established about the name cog. This name cannot be linked to a certain type of ship with 100% certainty. The difference with other ship type names is that a large number of vessels with similar hull shape and other characteristics are known. They all date from the Hanseatic period and were found in the Hanseatic area, some were found in areas where the Hanseatic merchants also came. The relatively large number of wrecks with these characteristics corresponds to the proportionally greater occurrence of the name cog in historical sources. In addition, the hull shape of the ship corresponds to the vessels depicted on the seals of the Hanseatic cities. It therefore seems perfectly acceptable to assume that an important or perhaps even most important part of the bulk transport by sea in north-western Europe in the late Middle Ages was carried out by the Hanseatic merchants with cogs. In addition, it seems obvious that the smaller inland transport vessels, especially in the regions around the seaports in the west of the core area of the Hanseatic League (the IJssel region/Zuiderzee area), were built according to the same tradition.

I therefore assume that the historical name *cog* refers to a vessel and that the group of large and smaller cargo ships in this study were referred to by this name in the Late Middle Ages. In this research it has been assumed that ships with these characteristics are in any case ships from the cog-building tradition. Until proven otherwise, I therefore agree with the aforementioned conclusion of Luns and the proposition of Crumlin-Pedersen, who is of the opinion that for vessels with the described characteristics of the Bremen cog, the name cog should/can be used as a general archaeological term.[28]

The three oldest (Danish) cog finds (Kollerup, Skagen and Kolding) were probably a (temporary?) 'northern response' to the growing trade with the north from England and Flanders since the 11th century and the large merchant ship on the Flemish coast that had been developing since the middle of the 12th century. The transport of goods from the seagoing vessels further inland was initially carried out in western Europe with barge-like plank boats and later with boats, barges and small cogs. In the Hanseatic region, the cog-building tradition was maintained as a barely evolving construction method until well into the 15th century. In some areas and for some small ships until at least the middle of the 16th century.

The research on the cogs from the Netherlands has also yielded a number of observations of another and different nature. Some can be reasonably explained. For others this is more difficult or I do not have the answer.

Shape of the amidships cross-sections

The cross-sections of the vessels studied are quite different amidships. The carvel built bottom is almost horizontal and has transitions of two to four strakes in the bilge to the sloping sides at an ever-changing angle. A second variant shows a slightly ascending surface from the keel plank towards both bilges. Another form is the plane that runs directly from the keel plank slowly 'round' upwards and merges smoothly into the sides. There are also ships that have a flat bottom and of which the bilge strakes are nailed at an oblique angle to the outer bottom planks. In the case of large vessels, the first two variants are usually used. A plausible explanation for the different designs seems to me to be that this is mainly due to the customs in the shipyard/region where the ship was built. On the other hand, in certain cases it may (also) have to do with a particular intended function that was already assigned to it during the construction contract. For example, for the transport of building materials (bricks), a flat bottom is by far preferable to a round or sharply constructed surface

26. Förster 2008, 299-257.
27. L' Hour & Veyrat 1994, 174-176.
28. Crumlin-Pedersen 2000, 240.

at the location of the hold. Remarkable is the construction of the river vessel/boat K 73/74 and the similar one found next to the Kampen (IJssel) cog.[29] These almost 16.00 m long, almost 2.50 m wide and amidships 0.75 m high vessels can be considered as the forerunners of the track boat (*trekschuit* in Dutch). They were probably used and built in (the region of) Kampen. They were rowed and were probably meant to carry passengers with hand luggage. The imaginary division of the vessels into three parts and the joining of the rear and front parts show that there is in fact a small, low-built cog, approximately 10-11 m long, between which a hold had been built with a flat bottom and almost vertical sides that were connected with each other by means of L-shaped timbers (*knikspanten* in Dutch).

Deck constructions

Little is known about the construction of decks in cogs. In the case of excavated wrecks, these parts are almost always missing. Of the Nijkerk-II from the first half of the 14th century, we can only be almost certain that there was probably a deck on the knee structures on the crossbeams 1 and 2 and between beam 1 and the stern beam. The planks of the rear, slightly lowered part lay in the longitudinal direction of the ship. Probably across the width between beams 1 and 2. The planks rested on slats nailed to the knee structures, but were not fixed to them. A curious and confusing observation.[30] Parts of the deck structures were found with the Bremen cog which presumably were mostly loose/removable.

The *waterschip* K 84 (fig. 27.6), built in the first quarter of the 16th century, has an aft deck with planks nailed to deck beams and lying in the longitudinal direction of the ship. In addition, the deck planks are held in place at the front of the aft deck by a deck beam. At the stem there was a small forward deck on which there was an anchor windlass. This deck consisted of transverse planks attached to the stringers and to the sides. At the back of the forward deck, the stringers rested on a transverse connection/deck beam. Between this forward and the beginning of the fish well (behind the mast) the stringers were connected to the sides by three short knee-shaped beams, which were attached to the stringers with a notch. The opening between the stringers, the forward deck and the sailing beam was open. Possibly meant to be able to put the sprit with the sail in the foreship and/or to lower the mast forwards. The mast was held in a square mast partner, which was let into the stringers. No indications were found that deck planks were also fitted transversely between the short knee-shaped beams. Above the fish well there were thick and relatively narrow gangways on the sides and between the aft and foredeck. Probably mainly meant to stand on when hauling in the fishing nets.

Somewhat comparable to this vessel are the also completely clinker-built *waterschepen* Mz 22 and Nz 42-I (figs 27.4 and 27.5 respectively).[31] They were both built in the first half of the 16th century. For the first vessel, it was established with certainty that there were loose deck planks between the stringers and the transverse knee-shaped beams. In the case of the *waterschip* Nz 42 I, there was a rear and foredeck connected to each other along the sides by thick, approximately 60 cm wide gangways which, as a stringer and as part of the decks, continued to the stern and stem. The deck planks lay in the length of the ship and were fixed to the deck beams with wooden pegs. There were two hatch openings in the foredeck. It can therefore be inferred from these three fishing vessels that in the first half of the 16th century there were clear constructional changes and that the decks from ca 1525 onwards formed a pegged-down element of these vessels. It should be noted, however, that these are fishing vessels. Trading vessels require relatively large hold/hatch openings. This may have been accommodated for a long time by making the deck planks and certain deck beams removable. This was assumed for the reconstruction drawings of the Nijkerk-II cog. The probably 15th-century Kampen (IJssel) cog could possibly shed more light on this peculiar issue.

Pumps

Another interesting subject is how the medieval shipmen got the rain and leakage water from their presumably partially open ships, back out of the ships. Wear marks on the keel plank/keel part of the stern hooks and some loosely found parts suggest that water flowed through limber holes in the underside of the floor timbers to the deepest point. It was then pumped up by a simple wooden pump barrel, into which a pump rod with handle was inserted and to which several round pieces of leather were attached. These pumps are directly comparable to our well-known bicycle pump and are the simplest device to pump water. They have regularly been found in wrecks of vessels in the IJsselmeer polders, dating back to well into the 19th century. In the Nijkerk-II cog a gutter was found that at approx. deck height ran from the centre of the ship diagonally to the ship's hull and through a small hole of 10x10 cm. Via the pump and this gutter the water was removed from the ship. In the 13th-century Nijkerk-I cog, the pump water was probably brought to the side of the ship via a gutter. In this case through an oval hole in the hull. On the outside there was probably a leather hose

29. Sections 26.1 and 26.2.
30. Chapter 6 and plates 6.IV A and B and 6.V A.
31. Respectively dossier Mz 22, RCE-Lelystad and Pedersen 1996, Enclosure 6.

nailed around that hole. This made it easy for the water to be drained, while the hose prevented any water from flowing in from the outside. Some 15th- and 16th-century images show such hoses. On lower, open ships, the water was thrown out of the boat with bailing spades. The completely carvel-built *waterschip* of lot W 10[32], which was built around 1550, is the oldest vessel to date with three wooden scuppers on both sides in the skin. The bilge and rain water and the water from the pump(s) were discharged through these. These scuppers were located on the lowest parts of the aft deck, the gangways and the foredeck. The gangways had a slanting coaming on the inside, while over the midship section above the fish well a completely closed canopy was built. This vessel was therefore completely closed off at deck height and above it.

Sails
A special observation made during the investigation was that pieces of coarsely woven woollen fabric were found in various cogs. Further investigation made it clear that it must have been pieces of sails. The pieces of fabric were treated with tar. Unique is the large piece of woollen tarpaulin, which was found in 2014, probably on the site of a 13th/14th-century shipyard in Kampen. Like the Viking ships, the cogs apparently sailed with tarred woollen sails. The sails were almost certainly composed of relatively narrow (up to approx. 90 cm wide) woven lengths of fabric (so-called *kleden* in Dutch). The piece of sail from Kampen may have been a tent sail and is made of 53 cm wide strips.

Cleaning the limber holes
In a number of cases it was established that in the longitudinal direction of the ship ropes had been attached through the limber holes in the underside of the floor timbers at the hold, some of which may have been knotted or tied in with pieces of wood. These ropes were probably pulled back and forth with some regularity to keep the limber holes free of any debris that could clog them up.

Divisions of hulls
In the case of the small cogs and boats, it could be established with a relatively high degree of certainty that, apart from the division of the ship's hull into an afterbody, a hold and a forward section, there was no further subdivision. This division is determined by the aft and foredeck and the hold in between. There has almost certainly never been any division between these parts of a vessel by, for example, fixed transverse bulkheads. In any case, no indications of this have ever been found. It is possible, for example, that certain parts of the hold could have been shielded with loose bulkheads or tarpaulins. In the case of the large cogs, however, a number of fastening pins/nails have been identified which may be related to transverse constructions. Perhaps the Kampen (IJssel) cog will provide more clarity about this.

Locks
A completely different observation is that the widths of the small 14th/15th-century (inland) cogs do not seem to correspond to the historically known widths of the Holland/Rhineland locks, such as those near Gouda. The vessels considered in this study would not have been able to pass through the locks at that time.

30.2 Decline of the cog

Until this year (2016), it is not possible from an archaeological point of view to establish a gradual transition from the construction of the large cogs, which had remained unchanged for about two centuries, to the larger, more sail-bearing merchant ships. It was noted above that there was a changing/evolving construction from the 12th to the beginning/middle of the 13th century. In the 14th century we can speak of different types of large and small cogs, of which the constructions, adapted to the size and probable function of a ship, were basically the same.
The Bremen cog from after about 1380 was not only the first, but until 2016 also the youngest large vessel of the cog-building tradition. Comparison with other, less completely preserved ships could only show small differences to a limited extent. This was due to the lack of sufficient comparative data from the same period. The next most completely preserved cogs Nijkerk-I, -II and Doel-I date from the middle of the 13th and the first half of the 14th century respectively. This is the first time that they have been fully described and depicted. From the 15th and the beginning of the 16th century only a few wrecks of large trading ships are known, but they have not yet been published. There is thus more than 100 years difference between these vessels and the Bremen cog. It is precisely the period (the 15th century) of which it is historically known that there were major changes in the construction of ships in western Europe at that time. The discovery of the Kampen (IJssel) cog, of which at this moment (end 2016) it is provisionally assumed that the ship was built in the first half of the 15th century, can provide additional information about possible changes in construction. After analysing the collected data on the construction of this ship we can probably say more about the further evolution of the construction of the youngest large cogs. After the en-

32. Reinders *et al.* 1978, plates 7, 9, 10 and 11.

tire investigation of all the cogs was completed, including the Analysis and the Summary and Conclusions, in the autumn of 2016 all the necessary data for reconstruction on paper of the Kampen (IJssel)-cog in Lelystad were documented. In the autumn of 2016, I received permission to (also) present the research results in this book about all cogs from the Netherlands, including the reconstruction drawings of the ship I made. The end result of the investigation into this exceptionally large and remarkably completely preserved vessel and the conclusions drawn from it with regard to the further development and/or decline of the cog are discussed again in the summary of the investigation report (Chapter 23). The wreck of the Enkhuizerzand cog probably offers even more perspective/clarity about these aspects and about the inventory and possible cargo of large cogs in time. This vessel was built in the third or fourth quarter of the 15th century.

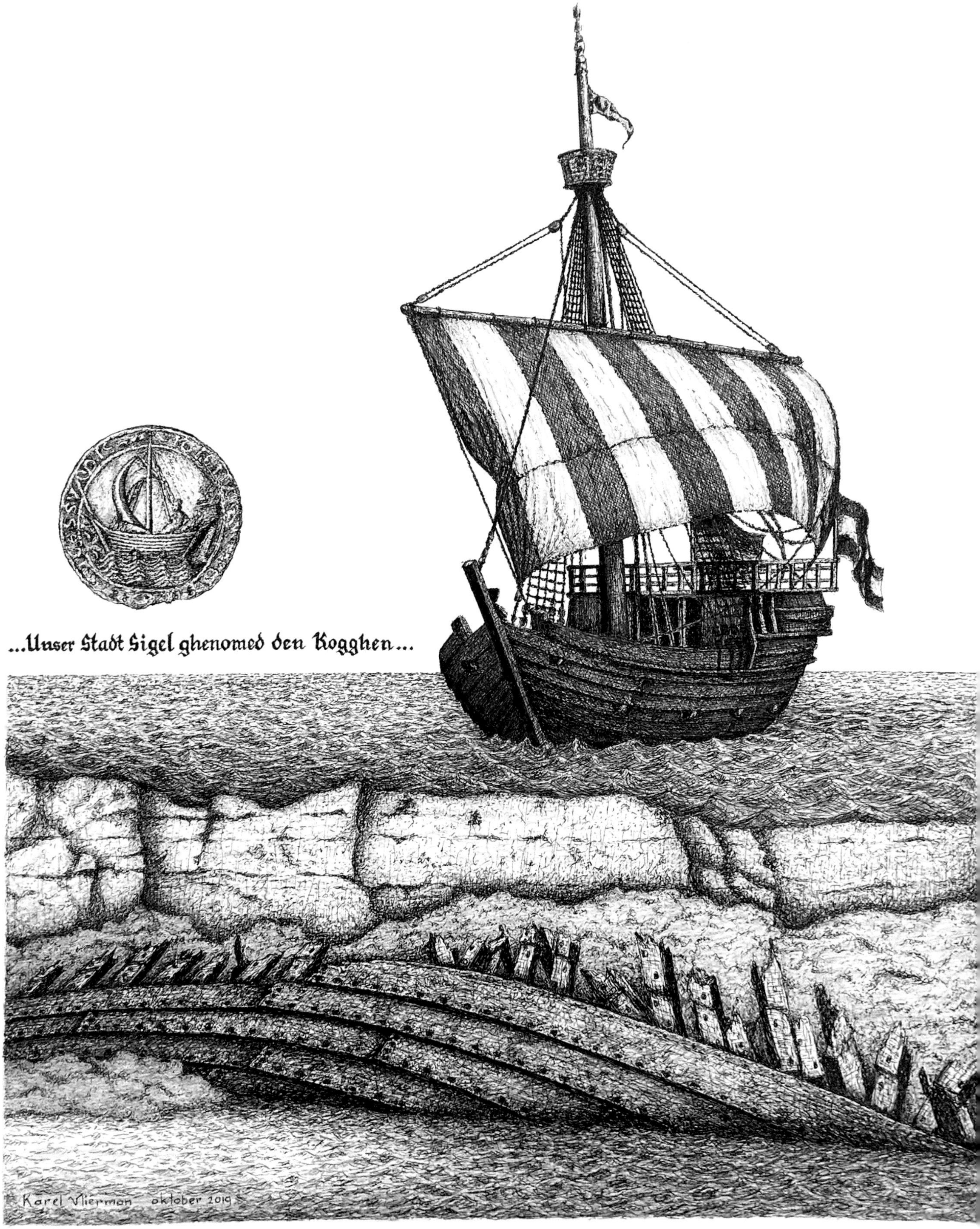
...Unser Stadt Sigel ghenomed den Kogghen...
Karel Vlierman oktober 2019

Summary and conclusions[1]

Until the recent discoveries in the Hanseatic city of Kampen (2014-2016), the conclusions and the summary are based on the results of the research until the end of 2014. The additional and new data and insights provided by the Kampen finds are expressed in the final sections of this concluding chapter.

Since the discovery in 1962 of the shipwreck in the Weser near Bremen, which was probably built shortly after 1380 and identified as a Hanseatic cog, it has been possible for the first time to identify other wreck finds on the basis of similar structural elements and the structure of the underwater hull as such. Previously, knowledge about this most evocative medieval trading vessel consisted of written sources and images, particularly on city seals. It was not known what the cog actually looked like or how the ship was built.

In the 1940s, a number of wrecks had already been found in Denmark and the Netherlands reminiscent of the ship type cog. One find in the Netherlands already gave rise to the suspicion that it concerned a small cargo cog. This 'Modderman cog' (the Marknesse cog) is not only the first wreck that was named as such, but also the best surviving example of this type. The vessel, which dates from the first half of the 14th century, was also the reason for routine archaeological research of ships in the Netherlands.

With the construction of the polders in the former Zuiderzee area - an important medieval and post-medieval part of a complex north-western European transport network - a worldwide unparalleled and more than 150,000 ha shipping cemetery has been drained. The Netherlands Institute for Ship and Underwater Archaeology (NISA) and its predecessors have so far recorded a total of approximately 450 shipwrecks in this area, more than 350 of which have been excavated over the years. They date from around 1250-1900. More than seventy of these wrecks date from before ca 1600. Among these vessels are sixteen or seventeen large and small vessels, which can be designated as a COG on the basis of the characteristics in the underwater hull of the wreck in Bremen. This concerns approximately two-thirds of all European finds. In 1985, on the basis of the then known wrecks, it was already established that there must have been a cog-building tradition.

Only four of the Dutch cog finds have been published since 1945. For two of them, the information presented is outdated and/or very brief. Over the years short articles have been published about the other finds and/or remarkable discoveries have received media attention. More than sufficient reason to examine the excavation data of all wreck finds with cog characteristics and to publish the research results in a complete overview. The vessels studied appear to represent the entire spectrum from small to large cogs from the period 1250-1500. Another conclusion is that almost the entire arsenal of late medieval wrecks from the former Zuiderzee area was built according to these characteristics, which further confirms that there was a building tradition.

Together with the two cog wrecks near Doel in Flanders (B) and the boats found in Belgium and in the Netherlands, where at least some characteristics of the cog have been found, there have been at least thirty archaeological finds in the Low Countries so far that can be attributed to the cog-building tradition. A unique and extensive source for research into the types, shape, dimensions, structure and construction (order) of the national late-medieval maritime heritage, in this case the cog. During the investigation it was not established whether the wrecks of Meinerswijk 2, Amsterdam, Hattem and Rotterdam 4 could also be considered as belonging to this group of vessels. The clinker-built connections were made using thin wooden pins instead of twice-folded nails, while a number of typical characteristics of cog construction were also found. The vessels from Amsterdam and Rotterdam were most probably built in the second half of the 13th century in the Rhine/Meuse/Moselle area.

One of the conclusions of the study is that almost certainly in the period 1200-1500 in the core of the Hanseatic region (along the Overijsselse IJssel/the eastern coastal region of the Zuiderzee, the coastal region of the Waddenzee and North Sea up to and along the lower reaches of the Weser and the Elbe) ships were only built using this method. There therefore must have been a cog-building tradition in this area, in which the large cog was dominant and possibly also used to determine the name. Initially, the first cogs were probably built on the Flemish coast and in the Lower Rhine area. Archaeological evidence is still lacking,

1. Translated by Sasja van der Vaart-Verschoof.

however. The vessels with cog characteristics (the Baltic cogs) built in the Baltic Sea area and mostly from coniferous wood and entirely clinker-built are not included in the study because there is too little (detailed) data available and because from a practical perspective it was almost impossible to include them in the present study.

Between 1981-1985, the wrecks of three large, seagoing cogs were excavated in Flevoland. The Nijkerk-II, built in the first half of the 14th century, was the most complete of these. The data from this ship were used between 1994 and 1997 to build a model scale 1:10 and to reconstruct the ship in full size (the Kamper Cog). The absence of especially large pieces of the sides, the deck and information about the rigging of the original ship were the reasons why the construction had to be carried out experimentally for a considerable part. This also provided an opportunity to gain more insight into the way in which cogs were built and the order in which this took place.

Direct involvement in the excavations of various cogs, the lack of fully detailed excavation data and publications of these and other European cog finds, and the completely unexpected occurrence of the possibility of writing a dissertation on a ship's archaeological subject led to the decision in 1999 and the proposal to elaborate and analyse all cog finds from the Netherlands. The data of the Nijkerk-II cog and the construction of the Kamper Cog were chosen as reference/research model for the shipwrecks to be investigated. This most complete large vessel plus the data on the reconstruction would in principle offer the best opportunities for answering the many outstanding questions concerning the shape, construction, layout, materials used, construction sequence and methods. Missing information on the original ship could probably be supplemented with data from the other archaeological finds. This was possible after the discovery of a large cog near Doel (B) in the autumn of 2000. A wreck from the Noordoostpolder and one in the IJsselmeer near Medemblik have not been included in the research plan on the basis of deviating constructions. In the end, it turned out that they should in fact have been included in the study. This was not carried out for practical reasons. The wrecks are briefly discussed.

The starting point for the investigation of all late medieval shipwrecks from the Low Countries that were built according to the characteristics of the cog, was to produce an extensive documentation as possible of relevant constructions and details with the ultimate aim of producing the most accurate possible reconstruction drawings of the ships. To this end, following on from the series of excavation reports published by the NISA in 1996 (Hocker & Vlierman 1996), a standard for the (re)construction drawings and a standard layout for the descriptions in research reports were designed.

The most completely recovered large cog in the Netherlands (the Nijkerk-II) and the Kamper Cog, reconstructed with the construction data of this ship, served as a research/reference model for the analysis of the data of all other wrecks. They also served as a control tool for these ships. In 2005 it became clear to me that no reconstruction drawings had yet been made of the Nijkerk-II and that De Boer & Sars had experimentally completed the Kamper Cog, building on the almost complete underwater hull found in its original form and with the 1:10 scale drawn parts from the sides. It is a beautiful and with the available data very well-built ship, that has been sailing for 20 years now. The 14th century shipbuilders of the original would recognize themselves very well in the ship. When I made the reconstruction drawings in 2005/2006, it became clear to me that the ship must have had a different shape from just below the protruding crossbeams upwards. In addition to the (re)construction drawings, the field drawings of all the cog finds have also been made suitable for publication and presented to serve as justification for the reconstructions. Anyone who wishes can use them to test the reconstructions. The texts describe the underlying assumptions.

The drawings presented on the plates are made on a scale of 1:10 in order to be able to reconstruct the shape as well as possible and to make the details stand out clearly. The aim of the scaling of the sintel clamps on the caulking seams is to approximate the original appearance of the ships as realistically as possible. The reconstructed course of the plank lines, the shape and the dimensions of the ships can be considered as certain for at least 95%. Presenting the vessels on the same scale allows for a good comparison and also reflects the differences between the large and small vessels. The detailed analysis of the wrecks made it possible, in most cases, to draw up a sound hull reconstruction and to determine the order of construction. The order of construction of the ships appeared to have been carried out according to a fixed pattern and in different phases. In this study this could be reconstructed as follows.

The number of building phases was the highest in the construction of the large, seagoing and high cogs. The planks of the largely carvel-built bottom were held in place with auxiliary frames and only fitted with floor timbers after the entire bottom had been assembled. The ends of the keel planks in the stern and foredeck were almost completely twisted (burnt) and clinker-built to each other and attached to the stern hooks. The subsequent strakes of the sides were completely clinker-built according to the shell first method. The fastening was carried out with nails folded twice ('double-clenched'), the tip of which was driven back into the wood. The frames were attached to the hull planks with treenails. Before the frame parts were placed, the clinker-built seams on the inside of the ship were made watertight with sintel clamped moss caulking. Initially, these seams were also caulked on the outside in the same way. The carvel-built surface seams and the drying cracks in the hull that occurred during construction were only sintel clamped on the outside. The sintel clamped moss caulking on the outside of the hull took place shortly before the launch.

A publication on the sintel clamped moss caulking was published in 1996 in which the shape development of the used sintel clampnails and sintel clamps and the period of use of the different models are discussed. This typo-chronological classification of the sintel clamps has been updated, but remains in force unchanged for the global dating of the construction period of late medieval vessels from the Hanseatic period and the area in which the Hanseatic League operated. It turned out that probably for the first time the application of the method for large seagoing vessels, which almost certainly did not happen before, was depicted on a miniature in the *Roman de Troie*. The original of the miniature was probably made relatively shortly after the Third Crusade. The moss caulking with sintel clamps (and sintel clampnails) was as a caulking method at least in use between the Carolingian times and the 20th century. It was a hardly changing part of the river ship construction in the area that roughly corresponds with Lorraine and the Empire of Lodewijk the German after the division of the Carolingian empire in 843. A slightly different version was in use in the catchment area of the Danube and along the north-eastern major rivers of Europe. Probably up to the Black and Caspian Sea.

In the entirely clinker-built and riveted Nordic or Viking building tradition of seagoing vessels along the Atlantic and North Sea coasts of France, England, Flanders, the west and east Frisian coastal region, in Denmark and the other Scandinavian countries, the method was not used for seagoing vessels. The caulking method characteristic of the cog-building tradition is undoubtedly derived/adopted from the continental European river ship building. The rapid development of form from sintel clamp nails to sintel clamps took place at the same time as the development of cog building in the period 1125/1150-1225 and only in the emerging Hanseatic region. With the disappearance of the Hanseatic/cog-building tradition around 1550 the caulking of seams of large seagoing and inland vessels ceases, of which the partial clinker-built construction is almost entirely replaced by the now more common carvel construction. The shape of the sintel clamps changes with the vessels of the Upper Rhine up to ca 1200 and stays almost the same after that (until the 20th century). In the case of vessels of the Moselle and the Lower Rhine, sintel clamps of the 15th/16th-century type F are still known, which remained in use until the 19th century, and in the case of some (small) vessels even until the 20th century. The clamped moss caulking was only applied at the underside of the bottom.

From historical written sources it was already clear that there were *Cogghen*, *Cleene cogghen ende Schuten*. The excavations in the former Zuiderzee area have shown over the years that there was indeed variation. The archaeological finds probably even represent the full spectrum of functional types (large and small seagoing cogs, small cogs for the Zuiderzee and [large] inland waters and all kinds of barges and boats [river cogs]) for regional transport and fishing. The analysis of the excavation data also made it clear that there is in fact no question of THE COG, but of vessels of one and the same ship-/cog building tradition in the period around 1200-1500. The final shape, length-width ratios of 3.2:1 or less, construction methods and construction sequence of the large and small seagoing cogs were probably already realised in the first half of the 13th century. With the small cogs for the Zuiderzee and (large) rivers, the length to width ratio changed slightly from the end of the 14th century onwards. Almost all of the wrecks in the former Zuiderzee area and the period in question that have been investigated so far can be considered to be part of this hardly evolving construction method. Exceptions are some entirely clinker-built and riveted 15th- and 16th-century vessels excavated in the IJsselmeer polders, the construction of which can be assumed in the Baltic Sea area and possibly on the (southern) North Sea coast.

An intriguing contemporary of the cogs is the late medieval *waterschip*, the most characteristic (fishing) ship of the Zuiderzee. The type of ship was already mentioned in historical sources in 1339 as a vessel for the transport of fresh fish in the large fish well, which usually consists of two compartments. In the 15th and 16th centuries it was also used as a fishing vessel. They sailed in convoy across the Zuiderzee with trawls in between. From about 1600 they were also used as a towing vessel to bring the large and heavily loaded VOC vessels (in ship's camels) to Amsterdam. This took place until the opening of the North Holland Canal (*Noord-Hollandskanaal* in Dutch) in the 19th century. The relatively large number of *waterschepen* found to date is the only type of ship with a little or no change in shape and layout during at least the 16th to 19th centuries. The wrecks studied so far date mainly from the 16th century, some from the 17th century. There are various paintings and prints, such as that of the battle on the Haarlemmermeer in 1572, but also from later times on which the type is depicted in an easily recognizable manner. It is also very remarkable that there are two types which were built in different ways. A late medieval, entirely clinker-built type of 16 m long and a completely carvel-built ship of 20 m long, which was first built in the middle of the 16th century. Both archeologically identified types illustrate the changes in shipbuilding in the Late Middle Ages. With regard to the decks of cogs, Chapter 30 also considered the changing deck structures of *waterschepen* in the first half of the 16th century. The hull shape, the length to width ratio, the construction (elements), the connections with nails twice folded and driven back into the wood and sintel clamped moss caulking in the seams is almost identical to the small, seagoing cogs. Different from the cogs, however, are the curved stem and the entirely clinker-built construction from keel plank to gunwale, characteristics whose origin is most likely to be found in the Atlantic *nef*. The similarities and differences between the cogs and the *waterschepen* and the origin of the *waterschip* form an interesting subject that requires further investigation. The present cog research was not the right place for this. The *waterschepen* deserve a study similar to that of the cogs.

The large (Hanseatic) cogs in the 13th and 14th century had a keel plank at both ends connected with stern- and stem hooks, straight stern- and stemposts and high sides, while the underwater body was built to a sharp angle at both ends and almost flat amidships. After the middle of the 15th century this was (gradually) replaced by a keel plank with stern and stem placed on it. They had protruding crossbeams and more than 1 m upper parts of the high sides. The sloping sides were 'held together' by heavy upright knees on the crossbeams. The top view has a half walnut-like shape at gunwale height, tapering towards the stern and stem and with a length to width ratio of 3.2:1 or less. The small seagoing cogs correspond to the large cogs in shape, construction and length to width ratio, but are about half as high and have transverse or cross-connections, which are attached to the inside of the sides. The small cogs for the Zuiderzee and rivers have a slightly more elongated shape with the result that the length to width ratio can be slightly greater than 3.2:1. However, this may also have to do with the evolution of the construction method. The different boats from this building tradition show a greater length/width difference and in general have very low sides.

Some large cogs on seals show a double row of protruding crossbeams. This was first archaeologically identified on the front half of the ship on the Nijkerk-II cog. In the space in between, which is less than 1 m high, there may have been a connecting deck, where for example soldiers had a place to sleep. None of the ships had a separate living area or partition between the hold and the aft or foreward. The living area was situated around the fire place, if it was present. The cooking, eating and drinking utensils, tools and personal belongings are usually also found around the fire place. In the cogs, however, no concrete indication of permanent berths has ever been found in any of the late- and post-medieval vessels from the former Zuiderzee area. The oldest examples of berth's found date from the second half of the 19th century. The Spakenburg-II was probably a workboat. It is possible that two of these vessels were linked together with large beams and used to lift large objects from the seabed or, for example, to pile posts.

Pieces or fragments of sails were found in six cogs. Remarkable and new is the fact that the cogs apparently had sails of coarsely woven brown wool and that, just like the Viking ships, they were tarred. The investigated cogs all had one mast and in principle they sailed with a square sail. Some small cogs for the Zuiderzee and inland waters were probably already equipped with a spritsail in the second half of the 14th century. In a few ships, indications have been found for a mast that could be lowered. Only in the Nijkerk-II are indications found for a rear and a fore-castle.

The items of equipment and inventory found in the wrecks have also been described according to a standard format. Hardly any objects were found in the large and a handful of small cogs. The reason for this is that certain ships were probably easily accessible at the wreckage site (near the coast) for a longer period of time and that everything of value had already been taken with the crew or had been taken away later. It is highly likely that some of the vessels were even partly scrapped. Some of them probably had hardly any objects on board originally. Other cogs sunk 'in the middle of the sea', where the objects from the equipment and inventory were found in the wrecks. Depending on the completeness of the equipment and inventories, these objects, together with the size of the ship, can be used to form an idea of the number of crew members. Some remarkable finds are the pieces of sail mentioned above. Furthermore, a clapper, a leather bag, three silver coins and some human bones in the Almere-I, the lance points, knives from a hunting set and an inlaid dagger with initials in the Dronten-I, the unique red earthenware jug, a fake coin of Duke Reinald II or III coined in Harderwijk, a wooden spade with iron fittings and a wooden candlestick in the Nijkerk-II cog, in the Spakenburg-I a piece of a compass glass and in the 13th-century Nijkerk-I a fireplace consisting of a square wooden box filled with loam and a strikingly shaped (battle) axe.

The archaeological finds of cog wrecks have been able to resolve a number of technical questions and explain some details on historical images.

- The initial assumption that the type of vessel would be completely clinker-built and built on frames proved to be wrong. There is a carvel-built bottom, which changes to a clinker-built construction and clinker-built sides towards the stern and stem. They are built according to the bottom first/'shell first'. For 13th-century vessels, the strakes from the bottom to the stern and stem are entirely carvel-built.

- The 'knobs' with long triangular, tapered and horizontally applied pieces on the front of them on the sides of some ships on seals and other pictures turned out to be the heads of protruding cross-beams, against which so called beam-head guides or fenders (*balkwillen* in Dutch) had been nailed. These guides were intended to prevent the vessel, for example, from getting stuck behind the protruding cross beams during mooring. These guides can also be seen on English 13th-century seals, on which the *nef* is probably depicted. Other, smaller 'bumps' on the ship's hulls probably stand for rose bolts.

- The openings in the high gunwale of the forebody of the Nijkerk-II and Doel-I cog can be seen on the seal of Vlaardingen from ca 1300 and on a fresco in the Town Hall of Wismar from the 14th/15th century. They were probably intended to be used as a gangway.

- A few oval or rectangular holes in the ship's hull at deck level on the Nijkerk-I, the Nijkerk-II and the Doel-I cog are most likely a kind of scuppers/holes for the discharge of rainwater and leaking water (via a gutter). The water was pumped up using a simple wooden pump tube and transported via a gutter to the hole in the hull of the ship. Remnants of nails around one of the holes on the outside of the hull (Nijkerk-I) indicate that a leather tube was most likely attached to the hull, allowing the water to drain away and preventing water from flowing in through the hole. The vessel on the title page of the *Waterrecht van Wisby,* from 1535 and on a painting with a late medieval *Hulk* show respectively one and two of these tubes. In the Nijkerk-II cog a complete drainage gutter has been found.

The historical data (naturally) did not provide any information about the layout, furnishing and construction on the inside of the cogs. It is only thanks to the archaeological finds that this can be clarified. Up to now, almost nothing can be said with certainty about the layout of a living or working area or about partitions below deck. Nor does it seem that this has been the case. The Bremen cog is the only wreck that provided information about the (rear part of) the deck, the castle deck and the layout below it. Of the Nijkerk-II cog only the rear two crossbeam constructions have been found, so that something can be said with certainty about the after decks of this vessel.

Research has shown that the oldest known vessels in the cog-building tradition in South(-west) Jutland were built between the middle and the end of the 12th century. However, this does not prove that the type was developed there. The three wrecks (from Kollerup, Kolding and Skagen) show typical Scandinavian construction methods and solutions, all clinker-built and riveted construction. It is possible that the construction of these ships was a (temporary) northern 'response' by the traders from Ribe to the Flemish, English and Dutch-German merchants who were more and more present in the north in their 'new', larger ships. The Flemish and Lower German merchants themselves travelled with their cogs to the Baltic Sea as early as the middle of the 13th century. The oldest mention of cogs dates from 1147 and refers to two Flemish vessels that were used during the Second Crusade. The oldest known cog finds from the Low Countries date from the second and third quarters of the 13th century.

The reason for building larger seagoing vessels with more loading capacity was undoubtedly the explosive economic growth on all terrains in the core area of the Flemish/Artesian coast in the first half and around the middle of the 12th century. The already lively and growing trade between England and the continent and between Flanders and northern Italy by land, forced the traders to increase the transport capacity. Over land this was not possible and it became more and more expensive as the number of tolls grew. Better shipping contacts between England/Flanders, the Lower Rhine region and Scandinavia and with the south-west coast of France and between the Italians and western Morocco made the Mediterranean Sea and the

Atlantic Ocean more attractive for transport. The initial fear of the ocean gradually disappeared. In particular, the preparations for and implementation of the Second and Third Crusade undoubtedly made a very important contribution to the creation of a new, larger type of vessel with more loading capacity suitable for distant sea voyages.

That this type of ship developed in the core area on the Flemish coast from an already existing building tradition there, complemented by technical, innovative improvements from the Mediterranean area and the adoption and adaptation of a continental caulking method of river vessels has been demonstrated in this study. The entirely clinker-built *Atlantic nef* with curved stem was almost certainly the starting point for this.

Another hypothetical conclusion of this study is that cogs such as the vessels described in this study were probably never built in the west of the Netherlands, and almost certainly not in the 13th/14th century or before. There is no archaeological evidence for this. However, it does seem quite conceivable that with the *Hollandse waterschip* in this part of the Netherlands and within the developing cog-building tradition, the complete clinker-built construction and curved stem of the *Atlantic nef* in the late 12th and 13th centuries was adopted and maintained via Flanders and Zeeland in Holland, while the construction and further development of the cog on the north-east side of the Zuiderzee, along the IJssel and in north-western Germany (with the partly carvel-built bottom and straight stern and stem) took place. The entirely clinker-built, but with straight stern and stem, wreck K 84-II from eastern Flevoland could be a *waterschip* of the eastern shore (east coast of the Zuiderzee, the island of Schokland) or the lower course of the IJssel, which was partly built in the beginning of the 16th century in the last phase of the cog-building tradition. Possibly it was the forerunner of the later Schokker. In this line of thought, the cog and the *waterschip* probably both originated from the *nef*. The functionally and in terms of size very different vessels from the cog-building tradition, and in particular the large seagoing cogs, in the Low Countries and other European regions in the Late Middle Ages, for more than 300 years, almost unchanged and for a very large part determined the look of transport by water.

By the end of the 15th and the beginning of the 16th century, the large cogs were probably no longer part of the sea fleets, but were gradually replaced by other types of ships. Some of them remained (initially) entirely clinker-built, others were entirely carvel-built. There are no archaeological indications yet that the cog ship 'merged' into one of these types. The large cog seems to disappear. The shift of economic power to other regions is probably one of the causes of the disappearance of the cog-building tradition. Trade grew in the 15th century and moved to Holland. At the same time, partly due to the silting up of the IJssel, the trade activities on the east side of the Zuiderzee declined strongly. In such a situation, people usually do not invest in new and/or innovative construction initiatives, as in this case with ships. Rather, people continue to sail for longer with an old ship and they continue with the construction method that has been used for centuries. In north-west Overijssel and east Frisian/Lower Saxony (the former Hanseatic region), the construction of small vessels has changed little or not at all since the Late Middle Ages and well into the 20th century. The cargo ship (or small [river]cog) of lot L 89, built around the middle of the 16th century according to the 'old-fashioned' tradition of the cogs, of which I suspect it was built in the region of the Lower IJssel, is probably an example of this. Unfortunately, this wreck came into view only after the finds at Kampen and for practical reasons it was impossible to investigate it and to add it to the research.

In the same period (the outgoing 15th and the first half of the 16th century) a fast-growing development took place in Holland with regard to trade and transport by sea. An ever-increasing shipbuilding industry also developed there, especially large seagoing ships, but also inland vessels. I believe that this happened within the then still mainly followed (adjusted) Nordic construction method. The contacts with Portugal and the Mediterranean area, where carvel-built ships were used at the time, will also have played a significant role in the introduction of this construction method in the west of the Netherlands. The mid-16th century and later carvel-built *Hollandse waterschepen* are a striking example of this. These developments and, at the same time, the displacement of the Hanseatic League from the international stage as a result of the idea, increasingly used in the 16th century, that the sovereign state had to control political action and thinking instead of the so-called intermediary powers (almost all of the Hanseatic Cities that remained at that time)[2] are probably the most important reasons for the disappearance of the great Hanseatic cogs around 1500. In the case of the small vessels of this building tradition, this process seems to be proceeding in the same way, but at a slower pace. The small cogs for the Zuiderzee and inland waterways evolved into inland cargo vessels and around 1600 probably had hardly any distinctive similarities with their late medieval predecessors.

The study showed how the various vessels were built and in what order. It also showed that the vessels discussed

2 . Hammel-Kiesow 2010.

were most probably all built in the core area of the Hanseatic League. Research questions that have not yet been answered sufficiently relate to the creation of the cog. Hopefully, 12th/13th-century archaeological finds will be made in the Flemish coastal area, for example, which will shed more light on this. The same applies to the disappearance of the construction method. In the case of the small 15th- and 16th-century vessels, the evolving structures etc. have become quite clear. The wrecks near Medemblik, on the Enkhuizerzand and in the IJssel by Kampen deserve special attention for the questions concerning the large cogs.

The cog plus the two small vessels for Kampen in the IJssel also form a unique collection of 15th-century wrecks, which have a direct relationship with water management (the silting up of the IJssel in the late Middle Ages) and the city. The site, together with the late 15th-century cog wreck on the Enkhuizerzand, has a heritage value of supra-national importance and should be treated as such (responsible recovery/research, archaeological/technical and public-oriented publications, conservation and the exhibition of the wrecks).

The recent discoveries in the Hanseatic city Kampen (2014-2016)

In the introduction it was stated that this research project on all cogs and cog-like vessels from the Low Countries was completed by the end of 2014. The shipwrecks dealt with in each chapter can be read as separate reports. In chapter 23 and sections 26.2 and 26.3 the text about the finds at Kampen was based on the report *Inventariserend Veldonderzoek Onderwater-Waarderend* (Underwater Assessment Fieldwork) by ADC-Maritiem from 2012. The remark about the supra-national importance and heritage value of these finds and the treatment that necessarily follows from them, as set out in the previous section, was subsequently completed by the authorities involved. After the salvage of the three wrecks between September 2015 and March 2016, the research and documentation on dry land could be continued.

Initially it was planned that, in addition to describing the vessels and their components, the documentation would be recorded in 3D-photogrammetrical images. In addition, it was assumed that 3D-reconstructions of the hull could be made from these images in order to perform all kinds of calculations about stability, payload, etc. The photogrammetric images provided beautiful, dimensionally stable images of the current situation of the cog wreck (Plates 23.I A and B, 23.II and 23.III). However, the deformed hull and the supports of the 600-year-old vessel made it impossible to extract the necessary data from these recordings to make a 3D-reconstruction. In order to achieve this, the additional data had to be collected manually and 2D-reconstruction drawings had to be made. At the same time as the instruction to carry out this work, the ADC promised that the reconstruction drawings and all other research data and images could be used for the (re)writing of chapter 23 and sections 26.2 and 26.3 of this manuscript. Thanks to this promise, it became possible to make a complete and conclusive overview of all cog finds from the Netherlands for the Hanseatic period (ca 1250-1550).

It was immediately decided to maintain unchanged the chapters already created, including chapters 29 and 31 (RESULTS AND ANALYSIS and SUMMARY AND CONCLUSIONS [up to the finds at Kampen]). An overview of the state of the cogs research by the end of 2014. The additional and new data and conclusions about the 15th-century finds at Kampen and their significance for our knowledge about the cog (-building tradition) are described and summarised in the chapter in question and in sections 26.2 to 26.4. The most important results, insights and conclusions that these wrecks from the last phase of the cog-building tradition have produced are given below.

The Kampen (IJssel)-cog

In the first instance, the Kampen (IJssel)-cog, built around 1420, still has the typical characteristics of the (large) cog from the 14th century and the hull dimensions are almost the same as the Bremen cog, built after 1380. The Nijkerk-II cog is one step smaller (fig. 23.40). The Kampen (IJssel) cog, however, appears after reconstruction not to have had a board height of approx. 4.25 m, but of 5.85 m. As a result, a length of approx. 25 m is reached at board height and a maximum width over the board of 8.5 m, due to the raked stern and stem. Attempts to determine by dendrochronological examination whether there had been a remodelling or enlarging of the ship during the period of operation could not be established. In the hull, however, changes in the construction were observed at various places which indicate that there had been an alteration/renovation. It has not become clear whether this took place immediately during construction, in other words whether the hull was enlarged immediately during construction and following on from the 14th century cog, or whether this was decided 10-20 years later, for example. In my opinion, the latter should certainly not be ruled out.

The highest parts of the heavy cross-beam constructions are virtually the same as those of the 14th century predecessors of the IJssel cog in terms of location in relation to the keel line. The strength and shape of the higher part of the sides was achieved by vertically affixed thick beams (vertical riders) on the inside against the ceiling board and on the outside against the clinker-built hull. In the longitudinal direction of the ship, the strength of the shape for this highest part of the ship's hull was obtained by applying wales to the outside of strakes N and P. This is the earliest archaeological example of this construction method to date. They were already known (in a slightly different version) from the large clinker-built cargo or warship of lot U 34 in Eastern Flevoland, the construction date of which has been set at 1528 + 9/- 6 by means of an examination of the year-rings, and from the Mary Rose, the flagship of Henry VIII, built in 1511 and remodelled in the period prior to its collapse in 1545 (file RCE-Lelystad; Marsden 2009).

Already at the Bremen and Nijkerk-II cog indications were found that certain parts of the deck may have been loose. During the investigation of the Kampen (IJssel) cog it was conclusively established that the large cogs had a 'main deck' of which the components were loose! Only the two long stringers on either side of the ship's centreline were secured on the knees of transverse beam construction 3 (halfway along the length of the vessel) with a single 30 mm thick wrought iron bolt. The deck planks were laid transversely in the rabbets of the long and short stringers. The probably two-part aft deck at a slightly higher and different level had longitudinal planks. These decks were primarily intended for the helmsman and for operating the large windlass.

In fact, the protruding cross-beam constructions, together with the sides built up from shell strakes, frames and ceiling strakes and the keelson, determine the shape strength of the ship's hull. The sideboards/stringers in the after- and forebody and the side stems in the bows also contribute to this. As a result, the deck sections between two cross-beam constructions and the sides can be made completely free for loading and unloading the ship without any problems. Long, heavy beam clambs/ceiling-boards below the protruding crossbeams and against the ceiling of the sides, with recesses in the top halfway between the crossbeam constructions, were most likely intended to create an extra and also completely loose deck at approximately 1 m below the 'main deck'. Probably intended to store vulnerable and/or valuable cargo. One could also think of sleeping places for soldiers if they were on board. A few traces on the inside of some vertical riders against the side ceiling strakes can be seen, which suggest that crossbeams could also be placed there. This would mean that an extra deck could also be installed at 1 m above the 'main deck'. The loading space between the crossbeam constructions 1 and 5 and the sides and the height of the loading floor below the 'main deck' is approx. 190 m^3. Together with a possible extra deck above this between crossbeam 2 and the bow, this would roughly mean a total gross cargo space of about 250 m^3 or a payload of about 250 gross tons.

The brick oven and fixed starboard fireplace on the 'main deck' between the crossbeam constructions 1 and 2, probably composed of loam mortar, are the earliest known examples on board of late medieval ships in the Netherlands. All other cogs with heating facilities were equipped with wooden boxes of approx. 80x80x20 cm filled with loam or sand with a tile or brick floor fixed in lime mortar. On top of this a fire was lit with peat, on top of which the cooking pot or kettle was placed on a trivet. These separate fireplaces were always located at the bottom of the ship on the ceiling and probably behind the mast. The deck area under the oven and fireplace consists of several layers of planks and small beams. They were all loose. On the bottom layer you can see traces of fire. This indicates that the fireplaces were initially built on the actual deck and later on the top layer of planks as they were found during the investigation. The deck has therefore been adjusted to take account of the weight of the brick structures. It also implies that in this section of the ship/deck there was not yet a fully enclosed galley section. Rather, it seems that the first signs of this can be seen with this ship. Another conclusion is that there must have been cooking and baking food for a larger group of people than just the crew. Here the possible presence of soldiers is again indicated. In Waldus & Vlierman 2018, food supply for the crew of convoy cogs is considered.

The oven and fireplace also indicate that they must have been protected at the top by a deck. The short stringer in front of the oven and fireplace has vertical openings for posts. A number of parts found loose were part of a structure in the afterbody. The shape of the rising part of a heavy knee found separate and next to the afterbody differs so much from the other knees that it must have been part of an eighth crossbeam construction. It was probably placed above the crossbeam construction 1 at the level of strake R and formed an important attachment point for the posts between which the large windlass was hung. All parts of the structure that were found suggest that an U-shaped castle/poop deck was probably built above the afterbody at some point, similar to the Bremen cog. The observed alterations in the ship's hull suggest that this did not take place immediately during construction. Some other structural parts found suggest that a roof construction was also found above a part of the rear castle. An S-shaped futtock from the forebody indicates that a roof structure was also installed above this part of the ship. It is possible that in the future, when the cog is reconstructed, more certainty can be obtained about this. For the idea formation of the possible construction above the back and forebody, the copper engraving of a *kraek* by Master WA from 1475 and the image of a cog in the Kamper *Digestum Vetus* in particular were examined.

Two heavy, knee-shaped pieces of wood originally formed the top of a frame of a loading and unloading opening on the port side between the crossbeam constructions 2 and 3. The underside was probably made in a similar way. The opening could be closed with thick planks that were clamped in place with two vertical posts and two horizontal beams. This special construction/opening has never been found in medieval ships before, but it has been explicitly depicted in two drawings in the aforementioned *Digestum Vetus* from Kampen. The images dating from 1455 may depict the Kampen (IJssel)-cog. In addition, it seems quite possible that the cog of Johan Witte is depicted.

In the Nijkerk-II cog and in several other late medieval wrecks, clues have been found for a loading-bed (*slietenvloer* in Dutch) on the ceiling. The parts of such a loading-bed found in the Kampen (IJssel) cog have made it clear that this was most likely standard practice in late medieval cargo ships. The loading beds were made up of pieces of split logs and pieces of thick branches in one or more layers. They were placed between and on the ceiling strakes. A layer of branches was laid across them, followed by one or more layers of twigs laid crosswise. The whole was covered with several layers of reed or straw, which were also laid crosswise over each other.

It is very likely that the cargo, mainly packed in barrels, was stowed on this springy floor of about 20-40 cm thick. A special and necessary feature of these loading-beds was that the rainwater that leaked through the deck seams could sink further down to between the frames. From there it ran through the limber holes at the underside of the frames to the pump in the afterbody. Through a wooden pump barrel the water was pumped up and poured into a cistern at deck level that was connected to a gutter which drained it off board. Scuppers in the boards of watertight decks have been archaeologically identified for the first time on a *waterschip* excavated in Flevoland, built in the middle of the 16th century. The permeable loading-beds were also intended to air the underwater body and to prevent decay. In some shipwrecks it has been established that these parts of the hull were also treated with extra tar for this reason.

The boat and the barge at the IJssel-cog, the boat K 73/74 and a larger variant L 89 from Flevoland.

The two 'mast steps' in the 15th-century barges of lot K 73/74 and next to the Kampen (IJssel) cog may have included posts on which a tarpaulin canopy was placed for passengers, for example. This may also have been the case with the barge found at the bow of the Kampen (IJssel) cog. They are probably precursors of the track boat (*trekschuit* in Dutch). They were almost certainly intended to transport passengers with hand luggage across the IJssel and beyond in the region, based on the findings in Kampen. In the approximately 16 m long, more than 2 m wide and only 80 cm high boats, clear indications have been found that they were rowed by at least two, but probably sometimes by three men. Probably one in the rear and two in the fore body. The large variant of the boats is the mid-16th-century wreck that was excavated on lot L 89 in Oostelijk Flevoland. This is probably a small, late (river)-/inland waterway cog, which was sailed. It seems very likely that all four of them belong to the area of the Lower IJssel/Kampen and were built there as well. Together with the IJssel-cog, the vessels provide a beautifully reconstructable overview of the ship types and movements on the IJssel in the 15th/16th century. Until now this image was mainly known from (later) maps and paintings of the Hanseatic city on the IJssel. A lesser-known drawing in the Kamper *Digestum Vetus* shows that some smaller cargo ships/cogs already had roof constructions or convex hatches above the hold in 1456.

Despite the relatively short period after the salvage in which the documentation and investigation of the deformed and in many places difficult to access wreck of the Kampen (IJssel) cog had to be carried out (summer and late summer of 2016) and the 2D-reconstruction drawings were realised (2017), it can be said that an image of a large Hanseatic cog from the last phase of the more than three centuries old building tradition could be obtained that was almost beyond expectation and accurate. It can justifiably be called a key find, which probably shows the maximum dimensions and payload that could be achieved within the cog-building tradition. Traces on the inside and outside, the wales and the use of a minimum of forty long, 30 mm thick wrought iron bolts to make crucial connections are 'new'. This also applies to the additional deck/cargo space, which may have been intended, for example, for special goods or deploying soldiers. On the other hand, the IJssel cog shows that it is still completely built in an old-fashioned way.

Despite the fact that no hard evidence has been found to date, it seems almost certain that the ship must have had a direct relationship with the Hanseatic city of Kampen. It is to be hoped that after a successful impregnation the wreck can be disassembled and reconstructed/constructed as much as possible in its original form and configuration. Kampen is the most suitable place for a Hanze museum/information centre with international allure to exhibit the IJssel cog and all Dutch historical and archaeological information about the cogs treated in this manuscript and about the Hanseatic League, and to further disseminate the collected knowledge. The Hanze Museum could have an extra crowd puller by keeping the cog-building tradition alive. Consideration can be given to building the 'Modderman-cog' (Chapter 12) and the boat that was found next to the IJssel cog, on the Koggewerf (Dutch for 'cog wharf') in Kampen. The schematics are ready. In addition to the already well-known Kamper Cog, Kampen has a unique small fleet of reconstructed 14th- and 15th-century vessels that can be used to make trips on the IJssel and further afield. This fleet could even be expanded by also building the IJssel cog.

A medieval shipyard on an old course of the IJssel in Kampen

Archaeological research on the site of the former St. Geertruidengasthuis (Guesthouse of St. Gertrude) in the oldest part of the city centre of Kampen has made it clear that there was a shipyard at least between the end of the 12th and the third quarter of the 14th century. Also, that there must have been a blacksmith on the same site or in the immediate vicinity to forge the nails and sintel clamps needed for shipbuilding and maintenance. After Bruges (Vlierman 1996a) it is the first time that a shipyard from the Hanseatic period has been located with certainty. The ships were hoisted up the slowly rising bank for maintenance with a windlass. It is not yet clear how wide and deep the Burgel was at that time, in other words whether large cogs could have sailed there as well. The pieces of ship's wood (dendro-date 1337 +/- 5) found in the foundation of a building on the site are in any case from a relatively large and a small to medium-sized vessel from the cog-building tradition, which may have been demolished at the shipyard. The shipyard site was bordered on the south side of the current Burgwalstraat by a ditch or moat that was approximately 15 m wide. This ran at least from the Boven Nieuwstraat to the Burgel. In the beginning of the 14th century, the Burgel was also the moat outside the first wall of the city. It is possible that the shipyard was on the outside of it. The elevation map of Kampen combined with multibeam data of the IJssel bed (Chapter 23, fig. 23.6) has made it clear that the channels of the IJssel in the immediate vicinity of the IJssel cog in the 15th century and before had a different course.

Despite the fact that there are no hard data at the moment, the current insights do not seem too daring to take into account the possibility that in the 11th and a large part of the 12th century the IJssel followed the current course of the Burgel from about the beginning on the south-east side at the IJssel/low landscape County, veered to the north at the Karpersteeg, ran diagonally across the Botervatsteeg and the Van Heutszplein, and then crossed over the current course of the IJssel into the Ganzendiep. The Burgel could then actually have been the current channel of the IJssel at that time. In this hypothetical situation, in my opinion, the foundation of the first settlement of Kampen would have taken place **not on the left, but on the right side of the IJssel**, i.e. in the Oversticht (Overijssel).

The swampy area on the south side of the settlement was bordered on the south-west side by a creek that almost ran up to the IJssel and flowed into the Zuiderzee, the Enk. It continued to the hamlet of Het Oenen/the current route of the N 50. The Enk formed the natural border with the county of Gelre and before 1300 could only be reached by land from Kampen with a ferry over the Enk near Het Oenen. The swamp area stretched out to the south-west to a few kilometres before present-day Dronten in Flevoland (see Chapter 11). The area around Zalk in the IJssel Valley was important to protect the (Over)stichtse/Kampen interests. At Zalk and 2 km from there on the other side of the IJssel near Westenholte there were probably already in the 11th century the military motte castles Buckhorst and Voorst. The development of the Zuiderzee in the second half of the 12th century and the associated delta formation of the IJssel ensured a continuous change in the area. Until the start of dike construction around 1300.

It is possible that during the creation of the Zuiderzee in the last quarter of the 12th century, the IJssel shifted via the north-eastern part of the moat around the settlement to the current course and that Kampen, which had already grown in the meantime, came to be situated, as it were,

Fig. 31.1. The navigable Reeve (in red), the watercourse connecting the Burgel/Geerstraat and the Zuiderzee, as shown on the 1724 map by Johan Muller.

on an island. The mid-15th-century shoring at the IJssel cog indicates that the left bank at that time was almost in the middle of the current IJssel. I think it is almost certain that the Burgel (the IJssel) at the level of the Geerstraat split and that one arm bowed off to the south/south-west, the long time well navigable Oude Trekvaart on the map by Johan Muller from 1724, which was formerly called the Reeve. By means of excavation or auger investigations (to be carried out when occasions arise) on the routes mentioned, it is probably possible to obtain more clarity as to whether this was indeed the case. If this is confirmed, the IJssel delta on the south-west side had the navigable Reeve, the watercourse on the map with IJssel estuaries from 1632, which runs from the city canal to the Camperzanden. The rightmost waterway is the Enk (Fig. 31.1).

Epilogue

The research on cogs presented in this two-volume book accompanied by technical drawings is, first and foremost, a study of primary sources, providing access to data which will be essential for interpretations based on the analyses recorded and depicted here. Without the ability to refer to the archaeological data interpreted in these pages, this research and further analysis of the cog is impossible, and can only be hypothetical or theoretical in nature. All eighteen cogs discovered in the Netherlands and Belgian Flanders to date, as well as several closely related shipwrecks, were subjected to systematic investigation, many of them right down to the very last nail and sintel clamp, sometimes even including the blind nailhole plugs. Within the framework of this extensive study, the data and analysis on one cog were used to supplement another, making it possible to arrive at a reconstruction even when faced with missing data for ship elements found separate from the main body of the shipwrecked vessel. The method used to present all the data here is universally verifiable, as is only appropriate for a proper publication of primary sources. It goes without saying that there may be differences in interpretation, in minor details, or in more sweeping themes, but that strengthens the scientific debate.
The vast majority of the research published here had already been concluded early in 2014.[1] It was already clear at that point that hardly any research had been done that examined all the aspects of cog-building and the structure of the cog in such depth as the analyses compiled here. This was all the more evident when a request was made in 2015/16 to create 2D reconstruction drawings of the Kampen (IJssel) cog and the boat found with it. During the advisory period in 2015, it had become clear that the documentation would be carried out by means of 3D photography, but the modern 3D images were inadequate for the purpose. Although they showed a field survey of the find at a specific moment in time, they offered almost nothing to support reconstruction, or in any case not enough. Reconstruction required first-hand knowledge on interpretation, manual collection of necessary data, and the drawing skills to produce 'old-fashioned' 1:10 drawings.[2] The reconstruction of the Kampen (IJssel) cog took into account the deformations that occurred during the vessel's extended stay under water, the disintegration and collapse of the hull and the wooden structural elements that had disappeared. All this was due to its location against the original sloping riverbank below the water's surface, the fluctuations in river currents and water levels over time (sometimes changing dramatically), and the excavation of the wreckage. This research on the Kampen (IJssel) cog was added to the data and analyses collected here, and is the culmination of the data and analyses collected here. Moreover, a 2014 excavation in the oldest part of the Hanseatic city of Kampen revealed finds clearly related to shipbuilding, and the analysis of that material has also been included in this study.[3]

Most studies on cogs to date are archaeological, historical, art-historical or theoretical in nature, sometimes also influenced by ideology. Ideological influence applies both to considerations —two examples of which are provided here, one German and the other Dutch under German occupation— and to earlier and later studies. In 1924, a group of writers, which included some members with far-right sympathies, founded a literary society known as *Die Kogge* in Bremen, which was disbanded ten years later amidst the tensions of the 1930s. *Die Kogge* was re-established in 1953 in Minden as an initiative to promote German literature, including the award of the *Kogge-Preis*. In 1974 the literary society was reformed and became the *Europäische Autorenvereinigung DIE KOGGE e.V.*: a European literary association. Very concisely summarised here, this history clearly shows the ideological associations that the term 'Kogge' or cog evokes, ranging from artistic and free to 'national socialist', and later to European cultural heritage.[4] From a very different perspective, the cog was promoted in the Netherlands as part of Dutch cultural heritage before and during the Second World War. In the supplement to the weekly *Schuttevaêr* magazine, an article on Dutch seafaring history was published in 1940 to mark the publication of J.C. Mollema's *Geschiedenis*

1. See Chapter Credits per Cog: supporting material for the chapters: 6 Nijkerk-II cog, October 2007; 14 Spakenburg-I cog, January 2008; 13 Kuinre cog, February 2008; 9 Ens cog, August 2008; 12 Marknesse cog, March 2009; 8 Nijkerk-I cog, August 2011; 11 Dronten-I cog, June 2012; 17 Dronten-II cog, November 2012; 10 Rutten cog, March 2013; 18 Swifterbant cog, March 2013; 19 Almere-II cog, April 2013; 20 Doel-I cog, April 2013; 21 Doel-II cog, May 2013; 24 Enkhuizerzand cog, August 2013; 15 Spakenburg-II cog, November 2014; 16 Almere-I cog, January 2015; 25 Cog shipwrecks, March 2018; 26 Schuyts and barges, March 2018; 27 Clinker-built vessels, March 2018; 28 Shipyards, March 2018; 22 Kraggenburg cog, October 2018; 23 Kampen (IJssel) cog, April 2018.
2. See for the combination of analogous documentation and photogrammetry 31-32, 35, 37, 44, 73-74 and the Plates.
3. See Chapter 23 Kampen (IJssel) cog; Chapter 28 Shipyards.
4. Behrens, Mark & Uli Rothfuss, n.d.

van Nederland ter zee. in vier deelen, Amsterdam: Uitgeversmaatschappij Joost van den Vondel, 1939.[5] The article opens heroically with the quote "Our ship in 900 AD was the Cog." Four years later, in 1944, the same magazine proudly wrote: "Nevertheless, the broad, flattened-oval ship of the Frisians, called *kogge* (cog), continued to predominate throughout the Middle Ages. ... the essential cog type emerged in Friesland. ... The cog remains ... the predominant type of ship in Western Europe. ... Following the form of its Frisian archetype, the cog is a heavy, short, stocky, flattened-oval ship ...".[6] Others also claimed that "the 'cog' was of Frisian or Saxon origin".[7] From a very different perspective, and equally imaginatively, Reinhard Paulsen's dissertation recently attributed the origins of the cog to the Celts, intended as a correction to the older German literature, which naturally attributed the origins to the Germanic peoples.[8] At the same time, Paulsen himself states that "*Im Kern sind nationalistische Forschung und Wissenschaftsethos nicht miteinander vereinbar*", and adds that his weighty tome has, among other things, become a "*Abrechnung mit deutschen Nationalismus*".[9] Similarly, Daniel Zwick (University of Kiel) recently stated in the introduction to his article on Southern European shipbuilding that his study primarily aims to bring together historical and archaeological sources in a meaningful way.[10] He notes that in academic treatises, shipwrecks labelled as 'cogs' have so far been discovered exclusively in northern waters, although there is rich iconographic and written evidence that they were also built on the Iberian Peninsula and in the north-western Mediterranean. Zwick looks at quite a few (known) historical sources. However, there are few relevant archaeological finds from the Mediterranean region, if any, and the shipwrecks he mentions do not provide any technical evidence of similarities with the cogs from northern regions, but rather with the Atlantic *nef*. In Figure 10, for instance, he clearly shows his lack of (detailed) data. The study is mainly theoretical. According to Zwick, however, his research provided historical evidence to support the fact that clinker-built and carvel-built vessels are also referred to as cogs in Southern European sources. In his opinion, seagoing vessels that were built 'bottom first' should therefore no longer be referred to as cogs in an archaeological sense, since he feels this would be inaccurate given the arguments gathered from a European perspective.

In this book, I have stayed far away from these theoretical, historiographical and ideological issues. The 'Nachleben' of the cogs is not the research topic addressed here, nor is this an analysis of the Hanseatic League in historical context – not because that approach and that type of research are not important, but because this specific study concentrates entirely on the hard data of all the recovered wrecks of ships that can be characterised as cogs. After all, the remains of the archaeologically investigated cogs and cog-like vessels that have now been reconstructed in coherent context and drawn in meticulous detail are the essential primary material source for further historiography of shipping in Northwestern Europe in the Late Middle Ages. This overview of the comprehensive array of underlying hard data, the compiled data on the analysis of excavated cogs, had been lacking until now.
After the conclusion of *Cogghen, kleene cogghen ende schuten* and the accompanying drawings, Chris Nicholson, director of the Centre for Digital Antiquity, wrote a blog post emphasising the long-term importance of recorded data above or at least equivalent to published interpretations: "I personally interpret this to mean that the data you produce are a more valuable contribution to your discipline [Archaeology] than are many of your interpretations." Nicholson here succinctly elaborated on an article published in 2019 by the team comprised of Sarah W. Kansa, Levent Atici, Eric C. Kansa and Richard H. Meadow presenting guidelines for archaeological data management, in which they state:[11] "... we strongly emphasize the importance of viewing data as a first-class research outcome that is as important as, if not more important than, the interpretive publications that result from their analysis". In essence, both these arguments retroactively justify the choice to publish *Cogghen, kleene cogghen ende schuten* and to defend it as an academic thesis, since this research project and its extensive sub-studies are primarily concerned with making the data available in as much detail as possible. Of course every effort has been made to present these data within the broader perspective of the Hanseatic era, looking at the origin of this very specific, time-limited, unique shipbuilding tradition, the variation primarily in size and thus a determining factor for inland or seagoing shipping, the -slight- development in design, the dating on the basis of sintel clamps, the ship inventory, etc. Given the number of cogs examined and the overarching similarities in hard data, it is possible to draw conclusions on that basis. Those conclusions are presented in this study; these interpretations can be part of the scientific debate, but the data cannot!

5. Anonymous, 1940. 17-18.
6. Anonymous, 1944. 11 March.
7. Visser, B., 1943, 194.
8. Ellmers, Detlev, 2010. 113-140; Paulsen, 2010. 19-112; idem, 2015 (verschenen 2016) 99-114; idem, 2016; Belasus, 2017, 222-241. Paulsen's dissertation was briefly and critically reviewed by Mike Belasus in VSWG. Vierteljahrschrift für Sozial- und Wirtschaftsgeschichte 105, 2018/1. 107-108.
9. Paulsen 2016, 7, 721.
10. Zwick, Daniel, 2016, 647-680: https://archaeologia-navalis.org/online-manuscripts.
11. Kansa *et al* 2019. 40-52: https://doi.org/10.1017/aap.2019.36.

A number of questions were formulated at the start of this study.[12] They have been answered as best as possible along the way. These answers are summarised concisely below:

1. *Can the shipwrecks investigated from the Zuiderzee region be categorised in the same structural type, and were the vessels built for a specific range of operation?*
Cogs can be categorised into a single structural type of vessel, which originated in the second half of the twelfth century and hardly evolved at all until 1400. The last large cogs were probably no longer seen by 1500. The shipbuilding tradition was established in the area that (later) belonged to the Hanseatic Empire, and the ships were used primarily in that area. They had a straight stern and stem, high sides, a single mast with a mainsail made of (roughly) woven woollen sailcloth that had been tarred, and a stern rudder.

2. *Can any particular (functional) types be deduced from e.g. the dimensions, lines and layout of the vessels?*
Larger cogs were shaped like a half-walnut in top view, with a sharply sloping underwater body and a length-width ratio of 3.2:1 or slightly less, and were approx. 20 m in length and more than 4 m high amidships. The mast stood in a recess in a long keelson, and the shape of the sides/hull was supported by crossbeams protruding through the shell with heavy knee constructions resting on the crossbeams. They had decks made up of almost completely detached elements. Before ca 1300, cogs were probably usually (somewhat) shorter and higher.
Small (seagoing) cogs had similar lines and a similar length-width ratio, were up to approx. 16 m long and had a gunwale height amidships of up to approx. 3 m. Crossbeams in these smaller vessels did not protrude through the sides. The mast stood in a short mast step or in a recess in a floor timber.
Presumably intended more for inland navigation, the smaller cogs were somewhat more elongated and would in principle have been no higher than approx. 2 m amidships. The elongated lines may possibly (also) be the result of evolution from ca 1400 onwards. A few vessels found so far may have been used for transporting certain kinds of cargo (e.g. bricks) if they had more robust structures or flatter bottoms. Probably from about the mid-14th century, a spritsail was sometimes used and the mast could be lowered.
Apart from the sharply tapered afterbody and forebody and the wide, flat midship section, there are no or hardly any indications of the interior layout of the large and small cogs.
Boats with cog-like features had a low gunwale and a variety of dimensions.

3. *Were these cogs all built (in a certain area) according to a generally accepted tradition, and what do the different types look like?*
The difference in types has already been indicated in response to the previous question. According to accepted tradition, the area in which shipbuilding took place between ca 1200 and 1500 can, based on the shipwrecks covered here, primarily be defined as the river basin of the Overijsselse IJssel River / the east side of the Zuiderzee and the Lower Saxon coastal area along the North Sea.

4. *How were these ships built?*
Cogs were built in phases according to the *bottom first/shell first* principle. This applies to all these vessels, from small cogs used for inland navigation to large seagoing cogs. The *shell first* principle can particularly be confirmed from the use of sintel-clamped moss caulking, which is characteristic of cogs. This is clear because the sintel clamps on the inside were applied to the seams before the frames were placed.

5. *Was a particular sequence of construction used?*
There is a very clear sequence of construction that applies to all cogs. There is hardly any deviation from this. Changes or additions may have been made later, but the original design followed a fixed pattern.

6. *In the past, the ships found in the Zuiderzee region and investigated here have been called cogs or described as cog-like on the basis of incomplete documentation. Was that designation justified?*
After analysing the data on most of the shipwrecks covered here, it could be confirmed that the building tradition suspected in 1985 must indeed have existed. All finds described at the time as cogs or cog-like vessels fit within the defined tradition, and there are several shipwrecked vessels that can most probably be added to that list.

7. *It is known from the literature that various shipwrecks found in the Baltic region have also been defined as cogs. Do the structural features correspond to those of the finds in the Low Countries?*
Hardly any cog finds from the Baltic region have been published, and insufficient data are available to say anything with certainty on that matter. There are a few shipwrecks that presumably do have some similarities to finds from the Netherlands, but most are vessels that almost certainly originated from the Baltic region.

12. See 31, Research questions.

8. Were these vessels demonstrably built in the Baltic region and, if so, are there e.g. any regional variations?
Following from the previous question and answer, it is not yet possible to say anything about regional differences. A few of the wrecks discussed here (Nijkerk-I, Rutten and Doel-I) may possibly have been built in the Baltic or Western Baltic region.

9. Why was the (large, seagoing) cog seemingly so dominant a type in historical sources and archaeological finds? Is that related to the type or volume of cargo, the seas to be navigated and/or the ports?
The most likely reason is related to the possibility of transporting large cargoes in the relevant period and the reliability of the cog. Otherwise, the type would never have remained in use and unchanged for some 300 years.

10. What are the dates of the ships (years/periods in which they were built and when they foundered) and is there any indication of evolution in the structure of the vessels? If the latter is the case, was this happening in different areas, for example simultaneously in Northwestern Europe and the Hanseatic region?
The ships covered here were built and perished in the period from ca 1250-1550. In the few cogs that date from before 1300, it has been established that the bottom planks/underwater body were completely carvel-built from stern to stem. The seams between the boards were V-shaped and initially covered the entire thickness of the boards. Then the aft and fore 2-3 m of the bottom sections and all the strakes of the sides were joined in a clinker-built construction. The rest of the bottom planks were applied in a carvel-built construction and only sintel-clamped from below (sometimes supplemented in 13th-century vessels by sintel nails on the inside). The youngest cog, the 15th-century Kampen (IJssel) cog, had elevated sides, but its hull was otherwise completely built the same as its 14th-century predecessors. A remarkable detail is the complete sintel-clamping of the clinker-built seams on both the inside and outside of the IJssel cog, while it has been established in other vessels and in vessels originating from other regions that sintel-clamping of the seams on the outside was partially (Doel-I) or completely omitted from the first half of the 14th century. As far as can be deduced from these ship finds, changes probably did not take place everywhere at the same time in the Hanseatic region.

11. Where did the ships found in the Zuiderzee region come from (where were they built)?
In Table 1 of Section 28.1, the cogs are arranged according to the period in which they were built, the year in which they were excavated and the provenance of the wood. Viewed in that format, it is striking that there is a correlation between the year in which the excavation took place and the availability of data on the area of origin. In the period when dendrochronological research was not yet possible, there was a corresponding lack of data on an area of origin. Where such data were available, it is not possible to pinpoint in all cases whether the provenance of the wood can be traced e.g. to the Netherlands or to Westphalia; in such cases, it is best to see the larger area as a single region. The IJssel River and the surrounding region, as a border river and 'transport channel' connected to the German hinterland[13], undoubtedly played an important role here. Some elements of the cogs that have not yet been subjected to dendrochronological dating have been preserved. Even for the wrecks that have been dated, recalibration and verification of the sintel clamp typology would also be a valuable addition. This perspective emerged gradually during the course of the cog research.[14] In ships built before ca 1300, the provenance of the wood probably also indicates the region where the vessel was built. This is supported by the moss species used for caulking. After that, wood was often brought in from elsewhere and only one moss species was used. It can be stated that all the shipwrecks covered here were almost certainly built in the Hanseatic region, possibly in the core of that region. The Marknesse cog was most likely launched in the Netherlands, quite possibly in the northern Netherlands.

12. Has more certainty been obtained about the previously expressed surmise that the dateable development in the shape of the sintel clamps and the application of sintel-clamped moss caulking as the caulking method used had a direct correlation to the origin, area and period in which the cogs sailed?
The centuries-old method of waterproofing the seams of river vessels with moss caulking, moss laths and sintel nails, particularly in the Rhine river basin, was probably first used when adapting an existing type of seagoing vessel and/or during the creation of the cog around the mid-12th century. Between ca 1175 and 1225, the sintel nail evolved into the sintel clamp. This and further changes in the shape of the sintel clamp only took place in the Hanseatic region (or what would become the Hanseatic region) and in ships built according to the cog-building tradition. It was a specific feature of the cogs. In any case most likely until the 14th century, the method was not used in the Nordic/Scandinavian shipbuilding industry, in which the vessels were completely clinker-built and riveted (as they were also in Great Britain and the area that was once Frisia), at most seen as minor repairs. The

13. Looper 2017, 9-32.
14. Daly 2007; Daly 2017, 334-338.

archaeological finds known to me so far from those areas and that period concern river vessels. There is certainly no indication of a possible Frisian caulking method. With the disappearance of the cogs and the end of the Hanseatic era, the sintel-clamped moss caulking also disappeared ca 1550. Sintel-clamping remained in use in the bottom of some (small) river vessels until the 20th century.

13. Earlier researchers have already wondered whether a new type of vessel emerged within a specific short time period or whether a long-established type of vessel was slowly developed in what would become the Hanseatic region. Has the present investigation provided answers or additional data and/or new insights?
Despite the lack (as yet) of archaeological finds in the Netherlands and especially in Flanders dating back to the 12th and early 13th century, the research at hand yielded new insights and established that there was almost certainly a slow, ongoing development of an existing type of vessel (see answer to next question).

14. The question of the region in which the cog itself and the cog-building tradition may have originated, when that took place, and the possible reasons for its emergence are also intriguing aspects and secondary aims of the research. Correlations to the Crusades and the Hanseatic League cannot be viewed separately from these questions.
Hypothetically, it could be surmised that the development of the cog around the mid-12th century may have been prompted primarily by the Crusades; that its development initially took place in the Lower Rhine region, but especially near the Artesian and Flemish Channel coast, and that the existing Atlantic *nef* was adapted for this purpose. Mainly due to the lack of suitable and sufficient building timber that soon became apparent, but also for economic and political reasons, the construction and actual further development of this type of ship and the associated shipbuilding method shifted to the north-east of the nascent Hanseatic region around 1200. Based on the data collected and analysed and the lack of archaeological finds in Friesland and Groningen, there is no reason whatsoever to think that the cog could be a Frisian type of vessel.

15. Does the standardised description of the objects found in the cogs offer better insight into the ship's equipment/inventory and does this provide more clarity about the number of crew members aboard the various vessels?
It is difficult to answer this question, since it was often still possible to access the vessel after it foundered, and people therefore removed everything that was still usable and could be reached. A complete or near-complete inventory was only found in ships that perished 'out to sea', were hastily abandoned, and were no longer accessible after they sank. This was the case for a few small cogs and boats; with all due caution, it is therefore possible to say something about the number of crew members. In any case, a regularly recurring feature is the fireplace and a limited number of implements or utensils, which were often multi-functional in nature.
The large group of *Hollandse waterschepen*, which can be categorised into two types in terms of structure, deserves the same sort of research as has been devoted to the cogs. First steps in this direction have already been taken.[15]

15. Maarleveld 1995, 3-7; Verwey e.a. 2012 65-93.

Glossary of ship terms

(Dutch terms in italics)

afterbody *achterschip*
The part of a ship's hull aft of the (main) mast.

aftercastle see castle

amidships see midship

anchor bitt see bitt

athwartships *dwarsscheeps*
Transverse(ly).

ballast *ballast*
Worthless, heavy material carried aboard a ship to enhance stability, draught and trim. In medieval ships, usually erratic stones.

barge, Rhine-barge, river-barge *(Rijn-)aak*
In Roman and medieval times, long, open (sailing) vessels used as ferries and for cargo transport on the major rivers. Flat-bottomed, square-ended, fore and aft ending in a trapezoid panel (*heve*). In Roman times with vertical sides and L-shaped floor timbers; in the Middle Ages with straight, everted sides. Such a craft built of straight planks is known as a plank boat (*plankboot*).

bargepole see punting-pole

beam clamb see shelf clamb

beam-head guide see guide

belaying pin see belaying rack

belaying rack *(nagel)bank, pennenbank*
A timber on the inwale, with holes into which wooden pins with a head (belaying pins *korvijnagels*) may be inserted. Ropes or lines could be secured around these. See also fife rail.

berth *kooi*
A bed or sleeping space aboard a ship.
Also a mooring space *(ligplaats, aanlegplaats)*.

beurtschip
Vessel providing a regular public-transport service for goods and passengers on inland waterways and the Zuiderzee.

(turn of the) bilge *kim*
The (construction at the) transition from the bottom to the side of a ship's hull. In transverse section it may be flowing, round or angular.

bilge ceiling board, bilge deal *kimweger*
An (extra-thick) ceiling board to cover the scarfs in the frames at the turn of the bilge.

bilge lath *kimlat*
A lath or plank, lozenge-shaped in section, placed inside an angular turn of the bilge, fastened with nails to both the bottom strake and the lowermost side strake. In occasional cogs, a plank on the transition from the carvel-laid bottom to the clinker-built side covers the seam, to counteract the working of the wood in these different constructions.

bilge strake *kimgang*
Any of the shell strakes forming the turn of the bilge.

bilge water *ruimwater, lenswater*
Water that has collected in the bottom of the ship, and will have to be pumped out.

bitt *(anker)beting*
A short, heavy post located on deck against the bulwark and used for securing lines and ropes. An anchor bitt secures the anchor cable. Cf. bollard.

blind bolt/nail/treenail *blinde bout/spijker/(houten) pen*
A bolt, nail or treenail that does not pass right through a plank or beam and is visible only on the side from which it was driven in.

block *blok*
A usually wooden pulley containing one or more rotating sheaves (*schijf/schijven*), or a wooden pulley with one or more holes in it (see deadeye).

boeitang
A washpeg-like, oakwood tool by which a plank in a clinker-built shell is held onto the shell plank beneath it while being fitted and fastened, comparable to a modern glueing cramp (see Chapter 7, Kampen Cog), figs 7.5a and b).

bollard *bolder*
A short, heavy post or pair of posts, the onshore equivalent of a bitt or bitts, used for mooring.

bottom *vlak*
The underside of a ship, in cogs constructed from a series of bottom strakes including the garboards, and the keel plank.

bow *boeg*
The foremost part of a ship above the waterline, especially the curved sides of the hull on either side of the stem.

brace *bras*
Either of the lines running from the ends of the yard to the deck, by which the sail can be rotated around the mast so that the ship can sail at a different angle to the wind.

brother-and-sister pattern *broeder-en-zusterpatroon*
An arrangement of planks in the hull of a ship, by which planks sawn adjacently from a tree are placed in equivalent positions on either side of the keel. Intended to evenly distribute the working of the wood.

bulkhead *schot*
A vertical partition, either fore-and-aft or athwartships.

bulwark *(op)boeisel, reling, verschansing*
The part of a ship's side above (the upper) deck level.

bumboat *parlevinker*
Any small craft from which goods are sold to ships in the harbour or on main rivers.

to burn, to char *branden*
Bending a plank into the desired curve over an elongated, open charcoal fire. The plank is secured and weighted during the procedure. Also it is kept wet, to stop the wood from catching fire or smouldering. The bent plank should be fastened into position immediately, before cooling.

'butterdish' compass *botervlootkompas*
A compass in a (lathe-turned) wooden bowl, covered with a glass pane and loose wooden lid.

buttress timber see rider

cable-laid, water-laid (rope) *kabelslag, (scheeps)tros*
Heavy cable for *e.g.* anchoring and mooring. Usually made out of three or four three-stranded ropes twisted together.

camel *scheepskameel*
A pair of lighters, connected by chains, which, when pumped dry, would lift a large ship across shallows (*e.g.* to reach Amsterdam from the Zuiderzee). [See also *waterschip*.]

capstan *kaapstander*
A winch with a vertical drum. See also windlass.

carvel-built, flush-laid *karveel gebouwd*
Said of a shell constructed from strakes meeting flush, producing smooth inboard and outboard surfaces. The strakes are held together by treenails or nails (and later, iron bolts) fastening them to the framing.

castle *kasteel*
In medieval ships, the castle is a raised platform built above the after- and/or foredeck (aftercastle *achterkasteel*, forecastle *voorkasteel*). In the 15th century the castle gradually becomes an integral part of the hull: the poop (*kampanje*).

caulking *breeuwsel*
Material used for sealing seams between planks, such as moss or other vegetable fibres, hemp fibres or picked rope (oakum *werk*), tarred paper, cotton, horse- or cowhair and pitch. The hulls of cogs were caulked with strands of *sphagnum* moss; the moss is held in place with laths or split withies, secured with sintel clamps.To caulk (*breeuwen*) is to make a ship's hull watertight.

caulking iron *breeuwijzer*
A chisel-like tool used with a mallet for driving caulking into the seams between shell planks.

ceiling, ceiling board, ceiling strake *wegering, weger, wegergang*
All fore-and-aft planking that lines the hull of a ship on the inboard side of the floor timbers and futtocks. Serving to strengthen the hull longitudinally and to provide flooring in the hold, protecting cargo from bilge water. A ceiling can be open (spar ceiling), with spaced strakes; or closed, to provide a floor or loadbed in the hold (*buikdenning, laadvloer*) underfoot. A beam may be part of the ceiling (*wegerbalk, balkweger*): see shelf clamb. Cogs and other late-medieval ships usually had an open ceiling, with narrow gaps of up to a plank's width between the strakes.

channel *rust*
A horizontal timber at deck level on the outboard side of a vessel, to which shrouds and rigging are fastened.

to clinch, to clench *klinken*
To join ship's timbers, especially overlapping (lapstrake *overnaads*) shell planks, by means of clinched/clenched nails (*klinknagels*), nails whose pointed ends were flattened or turned over, with or without a rove (*klinkplaatje*). A method typically employed in the Scandinavian (Nordic, Viking) shipbuilding tradition.

clinker-built, clincher-built, lapstrake *overnaads, klinkwerk, zoomwerk*
Said of a ship whose sides (and sometimes bottom) are built of planks overlapping at the seams.
See also clinch, land, and double-clenched nail.

coaming *den, denneboom, luikhoofd*
A raised frame around a deck opening such as a hatch (*luikhoofd*). In late-medieval vessels (small cogs) they are planks nailed obliquely along the inboard sides of the gangways bordering the hold, to prevent easy entry of water.

cog, cog-building tradition, *kogge, koggebouwtraditie*
The cog is one of the earliest named vessel types of the North Sea and Baltic regions known to us. The (large) cog was primarily a late-medieval cargo ship, also used in warfare. A cog-building tradition is evident throughout the Hanseatic territories between ca 1200 and 1550.

cordage *touw*
Any kind of rope or cord, especially that in a ship's rigging. See also hawser-laid and cable-laid.

covering inwale see inwale

crook *kromhout*
A timber naturally grown to form an angle, often part of a trunk and limb, and used for making *e.g.* knees or V-shaped floor timbers.

cross-sawn, cross-cut (wood) *kops (hout)*
(The end of) a plank or timber sawn or hewn through transversely, across the grain.

crotch see V-shaped floor timber

crutch see rider

cuddy *(voor)plecht*
A small foredeck or the space below it.

cutwater see gripe

deadeye *jufferblok*
A circular or semicircular wooden block through which a rope may be secured.

dead flat *stilstaand*
Said of an area of the ship's bottom that has no deadrise fore-and-aft or athwartships.

deadwood *slemphout*
A block of wood which on the inboard side reinforces the connection of the stern or stem with the keel. Not in cogs.

deck rider *dekligger*
A transverse beam on a deck, firmly fastened to the side or inwale, and to the underlying deck planking and deck beams. Applied in cogs and *waterschepen*, which often have deck riders with standing knees fastening them to the sides.

deck, afterdeck, foredeck *dek, achterdek, voordek*
A roof (and floor) that covers the hold of a ship, or just the foremost and/or after part of the hold. Some cogs (may have) had multiple deck levels.

deck beam *dekbalk*
A transverse beam that supports the deck (see also transverse beam and transverse connection, crossbeam).

deck plan *dekkenplan*
The layout of the deck or decks of a ship.

dinghy see tender

double-clenched nail, dead nail *naainagel, koggespijker, koggenagel*
A nail by which the planks of a cog's clinker-built shell were joined together, laterally and end-to-end. The chamfered surfaces (lands *landen*) were fitted together and the nails were driven in from the outboard side, passing through both. The tips of the nails were twice beaten sideways and thus driven back into the wood.

duckboard see plank bridge

to fair *stroken, uitlijnen*
To smooth and streamline a timber or a template.

false stem (or false stern), outer stem/stern *buitensteven*
In cogs, a plank or beam fastened in front of the stempost to cover and protect the hood ends of the shell strakes nailed to the mitred/double-bevelled stempost. Occasionally a similar construction on the stern.

fashionboard see washboard

fender see guide

fife rail, belaying rack, pin *rail mastbeting*
A rack around the mast, with belaying pins on which the lines running up the mast can be belayed.

filler/intermediate piece/beam *(opvulstuk, opvulbalk)*
A timber that fills up excess space between other timbers, *e.g.* a rider beam that fills the space between a crossbeam and a deck.

fireplace *stookplaats*
In cogs and *waterschip* vessels, a square wooden box, ca 80x80x20 cm, filled with loam (later with sand), and paved with mortar and bricks or tiles. On this floor an open peat fire was tended.

fish well *bun, kaar*
A bin in the hold of a fishing vessel for keeping and transporting live fish. Under the well, the ship's shell is pierced with hundreds of small holes (ca 1 cm across), allowing the water to circulate freely inside. Usually divided into two compartments. It is covered by the fish-well deck (*bundeken*).

flat-bottomed vessel *platbodem, platboomd schip*
A vessel built not with a keel beam but with a keel plank which may be somewhat thicker than the other bottom planking. Transversely, the bottom is quite flat or has very little rise towards the turn of the bilge. The bilge strakes are placed against the bottom at an angle. Flat-bottomed ships are generally used in shallow and in tidal waters where they can await the tide on the flats.

flat scarf see scarf

floor timber *legger, wrang*
A (straight or V-shaped) transverse timber connecting the strakes of the bottom and the keel, part of a frame. See also V-shaped floor timber.

fo'c'sle, forecabin *vooronder*
The space beneath the foredeck.

fore-and-aft *langsscheeps*
longitudinal(ly), from stern to stem

forebody *voorschip*
The part of a ship's hull forward of the (main) mast.

forecastle see castle

frame, rib *spant*
Any of the transverse sets of timbers that form part of a ship's skeleton. The frames sit on the shell at right angles to the keel or keel plank and extend up to the gunwale. Frames generally are composed of several parts: straight or V-shaped floor timbers and futtocks.

framing *inhouten*
Collective name for all the internal timbers making up the frames or skeleton of a ship and ensuring the transverse rigidity of the hull and the shell planks.

freeboard *vrijboord*
The vertical distance between the waterline of a ship and the main deck or gunwale.

futtock *oplanger*
Upright part of a frame. Futtocks are extensions of the (straight and V-shaped) floor timbers. A frame may have one or several futtocks fastened to the shell at either end of the floor timber.

galley *kombuis*
The kitchen in a ship.

gangway *gangboord*
A heavy plank connecting the after- and foredeck along the bulwark. Often in part integrated with the decks.

garboard, sand strake *zandstrook, kielgang*
Bottom strake, extending fore-and-aft immediately adjacent to the keel plank.

gripe, cutwater *loefbijter*
In some cogs, a small triangular piece projecting forward from the stem hook onto which the false stem was placed. A similar feature (heel) on the stern hook would serve to stop debris from getting caught and fouling the rudder.

gudgeon *vingerling*
The female part of a forged-iron rudder-suspension system, consisting of a ring with two brackets that are fastened to the sides of the sternpost. There will be several gudgeons on a stern.

guide, beam-head guide, fender *balkwil, geleider*
An elongated-triangular, tapering piece of wood, with a flat rear and underside, and rounded outer surface. It was nailed to the shell against the forward face of the protruding head of a crossbeam. Its function was to stop the crossbeam head from getting caught, e.g. during mooring manoevres. Made of beech or elm wood. Having suffered wear and tear it is easy to replace.

gunwale, gunnel *boord*
The upper edge of the side of a vessel.

hanging knee see knee

hawser-laid (rope) *wantslag*
A three-strand, right-hand laid rope. Also called plain-laid rope.

heel see mast heel, see gripe

heve
The upward-sloping stern- and/or bow-part and rectangular or trapezoid ended bottom of a barge or boat.

hogging *katterug*
A tendency of the ends of a hull to droop. See also sagging.

hold, cargo-hold *(laad)ruim, (vracht)ruim*
The space in a ship where the goods to be transported were stowed.

hood(ing) end *eind van een plank/gang*
The forward or after end of a shell strake.

hook scarf see scarf

hook see stern/stem hook

hook scarf *haaklas*
A scarf in which the table is stepped, so that the two halves cannot pull apart longitudinally; several varieties exist, all with their own names and peculiarities.

hulc *hulk*
A type of seagoing cargo vessel that in the Middle Ages was used by Scandinavian, German, Dutch, Flemish and English mariners.

inwale *binnenboord*
A piece of wood, often triangular in section, fastened inside the frame along the top strake of planking. Reinforces the gunwale. In smaller craft, it is usually square or oblong in section and as a covering inwale (*randgaarde*) sits atop the heads of the frames.

keel plank, plank keel *kielplank*
A usually fairly broad, fore-and-aft plank that amidships forms the underside of the ship and is the base for building up the stern and stem and the bottom and the hull. In the cog-building tradition the keel plank (up to three times the thickness of the shell planks) together with the horizontal parts of the stern and stem hooks forms the base for the construction of the hull. Later ships usually have a (much) thicker keel beam, onto which the stern- and stemposts are placed.

keelson *zaathout*
A heavy beam extending fore-and-aft inside the frames and above the keel. Serving to stiffen the underwater body. In cogs, usually engaging with the floor timbers, with an integral broader part to contain the mast step.

keep-net *leefnet, korf*
A round or square enclosure woven from withies, positioned in shallow water with stakes, for temporarily holding live caught fish.

killick *ankersteen*
A boulder provided with an iron eye, that serves as an anchor. Used for instance by fishermen to keep their boat in position while they put out or empty their creels.

knee *knie, kromhout*
A timber in the shape of a knee, naturally grown in an L shape. Used for connecting *e.g.* the deck and the bulwark or a transverse beam and the sides. Standing knees have one arm pointing upward; hanging knees have one arm pointing down.

L-shaped frame, hard chine *knikspant*
Naturally grown, L-shaped crook, shaped into a partial frame. Used mainly in boats, to join a flat bottom with (near-)vertical sides.

ladder *ladder, trap*
Stairway (*trap*) or steps, made of wood, metal or rope.

land *land, landing, overlap*
The part of a plank in a clinker-built shell which is covered by the next higher plank.

lapstrake see clinker-built

launching blocks *stapeling*
Piles of wooden blocks supporting the hull of a ship under construction. These are removed to launch the ship.

limber hole *loggat, kattegat*
One of several rectangular or triangular tunnels cut across the bottom surface of floor timbers to allow bilge water to flow towards the lowest part of the ship.

lines (drawing(s)) *lijnenplan*
A drawing or set of drawings delineating the shape of the hull.

load-bed, loading bed *buikdenning, laadvloer*
The underfoot part of the ceiling, of closely laid boards. In (large) late-medieval cargo ships such as cogs, this may be covered with a 20- to 40-cm layer of branches, twigs and straw. Occasionally there is a floor of closely-set, transverse thin stems and stakes (*slietenvloer*) laid on 30-cm posts set onto the ceiling planks. This is covered with wickerwork, sand and a cobbled surface (cargo or warhip U34 Oostelijk Flevoland).

marlinespike, fid *marlpriem, splitspen*
A pointed tool for lifting individual strands, used in splicing cordage. In the late Middle Ages (and after) made from a tine of antler. More recently, mostly of iron/steel.

mast beam see sail beam

mast heel *mastvoet, hiel/hieling van de mast*
Lowermost part of the mast, usually squared to fit into the similarly shaped hole in the mast step.

mast partner *mastvissing*
A reinforcement in the deck around the mast hole. The mast may be secured within it with wedges, and provided with a collar to stop water from entering. See also carling.

mast step *mastspoor*
A thicker and/or wider part of the keelson, into which a recess has been cut to hold the heel of the mast. The mast step distributes the mast's weight and pressure across several frames and riders and other structural parts. This construction is known from large Roman-period river-barges and large late-medieval cargo ships such as cogs. The recess may have a small channel carved out to drain off any rainwater or spray running down the mast.

mast support *maststeun*
An L-shaped timber set tightly against the rear of the mast and with the other arm treenailed to the keelson.

midship *middenschip*
The middle part of a vessel, midway between the bow and stern, and/or the broadest part of the ship. At this place: amidships (*midscheeps*).

mitred, double-bevelled *dakvormig afgeschuind*
Bevelled on both sides to that the oblique surfaces meet midway, like a rooftop. Often the shape of the forward surface of a cog's stempost, and less frequently of the rear surface of the sternpost. The port and starboard hood ends of the side strakes meet here and are nailed to the bevelled surfaces.

moss caulking *mosbreeuwsel*
In cogs, caulking was always done with bog moss, *Sphagnum*. This is the prime material found in all (probably Dutch-built) pre-1900 wooden ships, among the hundreds of wrecks in the Zuiderzee area. Oakum caulking, made of waste hemp fibres or picked old rope, has never been encountered there.

moss lath *moslat*
A lath or split withy used for covering moss caulking and securing it in the seam or crack. Also known as ribband or batten.

mould, template *plankuitslag*
A full-scale drawing of the true outline of a plank or timber.

nail header *nagelijzer*
A tool with which the heads of nails are forged. A hand-held 'anvil' with a hole into which the pointed shank is placed, after which the head is shaped on the surface.

nailhole plug *spijkerpen*
Pyramidal or conical wooden peg (ca 0.5 cm in diameter) to plug the hole left by a nail of *e.g.* temporary framing.

nef
A medieval type of seagoing vessel. This French name applies to ships of the Atlantic coasts, clinker-built and with a curved stem. Some (larger) *nefs* had protruding crossbeams and beam-head guides, like cogs.

oakum see caulking and moss caulking

outer stem (or stern) see false stem (or stern)

pairing marks *paringstekens*
Marks scratched or chiselled into adjacent planks or timbers so as to facilitate their correct repositioning after temporary removal. For instance, on parts during construction, or on removable deck planking.

pintle *roerpen*
On a rudder, a pin or set of pins fitting into the gudgeons on the stern, making a hinge on which the rudder can rotate.

plain-sawn (timber), through-and-through sawn (timber) *dosse gezaagd (hout)*
A log sawn into planks with the cuts all running parallel. Plain-sawn planks are more prone to warping (cupping) than quarter-sawn wood. *Cf.* quarter-sawn.

plain weave *effen binding, platbinding*
A simple weave in which the weft yarns pass alternatingly over and under the warp yarns.

plank boat see barge

plank bridge, duckboard *vlonder*
A loose, removable wooden floor in small craft, resting on the floor timbers.

poop see castle

promenade deck *wandelpier*
Either of the two (port and starboard) decks extending from the castle deck.

protruding crossbeam *door de huid stekende dwarsbalk*
One of several massive beams that on either side protrude through the ship's shell at deck level or somewhat lower. The beam heads have rabbets around them to engage with the enclosing shell strakes. Such crossbeams are a feature typical of (large) cogs. Also referred to as a through-beam.

punt *punter*
A narrow, flat-bottomed, open boat used in various parts of the Netherlands, *e.g.* at Giethoorn or Aalsmeer, which is propelled with a pole or sailed. To punt is to propel a small vessel by means of a punting-pole.

punting-pole, bargepole *vaarboom*
A pole with a wooden knob at one end and a hooked or forked, usually iron fitting at the other. Used for punting or for manoeuvring a (small) inland vessel.

quarter-sawn, rift-sawn, radially sawn (timber) *kwartiers gezaagd (hout)*
(Timbers/planks) sawn from a log in such a way that the cuts are as much as possible perpendicular to the annual rings. *Cf.* plain-sawn.

rabbet, rebate *sponning*
A groove or recess cut into a timber so that another piece fits in tightly to form a joint. Rabbets were made *e.g.* in the sides of stern and stem hooks to receive the hood ends of shell planks.

rake (of the stern/stem) *valling, stevenhoek*
The angle made by a ship's stern or stem with the perpendicular.

rebate see rabbet

rider (beam) *dekligger*
A beam seated atop another or on the decks. Also thick beams on the ceiling of large ships, usually to strengthen it.

riders, buttress timbers, crutches *kattesporen*
One or more pairs of short floor timbers abutting on either side of the mast-step part of the keelson, and treenailed to one or more of the underlying floor (or ceiling) timbers. A rider usually has a knee-shaped end where it meets the mast step, of the same height as the mast step. Riders take on the lateral pressure exerted by the mast and the sail, and distribute it more evenly over the hull.

Rine-barge see barge

rib see frame

rift-sawn see quarter-sawn

rigging *tuig(age), want*
Everything non-structural above decks, including sail(s), spars, mast and the cordage holding up the mast (shrouds, stays *want, verstaging*) and operating the sail(s).

rosebolt *rozenbout*
An iron bolt with a broad head that protrudes from a ship's side and that is fastened onto a deck beam or bulkhead to stiffen the hull laterally. Its inboard end is flat with holes in it and is nailed to the beam, or has lugs and is secured with clamps. It serves a similar function as a wall anchor.

rowlock, oarlock *dol, roeipen*
A device on the gunwale for holding an oar while it is pulled: a swivelling crutch, fork or ring, or a grommet on a tholepin. See also tholepin.

sagging *doorzakking*
An unintended tendency of the ship's aft and forebody.

sail beam, mast beam, supporting beam *zeilbalk, mastbalk*
A heavy transverse beam that supports the mast at deck level immediately before or aft of the mast.

sand strake see garboard

sapwood *spint(hout)*
The outermost, still unripe annual rings of a tree trunk: the light coloured, soft wood between the heartwood and the bark.

scarf *las*
An oblique joint by which two chamfered (cut obliquely, sometimes with steps or notches: hook scarf) timbers or planks are joined end-to-end without a change of width or thickness. In late-medieval ships the thin plank ends are usually left a few centimetres thick, resulting in locally increased thickness, the nib ends standing proud of the shell inside and outside. The scarf will be secured by means of some nails. A sloping flat scarf (*schuine las*) has plain surfaces.

schokker
A traditional type of sailing fishing vessel of the eastern shore of the Zuiderzee.

scupper *spuigat*
An opening in the bulwark at deck level for discharging water. In cogs it may be a square or oval hole in the shell, opening with a short leather hose on the outside.

to scuttle *wegbrengen, doorboren*
To sink a ship deliberately (by piercing the hull).

seam, caulking seam *naad, breeuwnaad*
A joint or slit between two adjacent planks. A caulking seam (*breeuwnaad*) has been reamed wider to allow caulking. In medieval vessels there are open seams, followed by seams of triangular section that cover the entire plank thickness; and finally seams of triangular section that cover just part of the plank thickness.

sheave see block

sheer *zeeg, sprong*
The upward curvature of a ship's main deck or gunwale from amidships to each end.

sheer strake *berghoutsgang*
The uppermost continuous shell strake.

shelf clamb, beam clamb *balkweger*
A heavy ceiling board with rabbets in which deck beams rest.

shell *huid*
The outer planking of the bottom and sides of a ship's hull.

shell-first see temporary framing

sideboard see waterway

side-stems *zijstevens*
A pair of long, broad planks or beams which in some cogs are fastened inside the frame on either side of the stem for the purpose of stiffening the bow and distributing pressure.

side stiffener *spoor*
A vertical timber on the inside of the ceiling or on the outside of the shell to enhance the stiffness of the upper part of the side.

sintel (clamp) *sintel*
A butterfly-shaped clamp, serving to secure a lath or split withy over the moss caulking in a seam or crack in a wooden hull. Evolved from the sintel nail.

sintel nail *sintelnagel*
A staple-like clamp without wings, serving to secure a lath or split withy over the moss caulking in a seam or crack in a wooden hull. Evolved into sintel (clamp).

sintel clamped moss caulking *gesinteld mosbreeuwsel, gesinteld werk*
A method for caulking wooden ships. *Sintel* is the Middle Dutch name of small forged-iron plates in the shape of staples with a broad, oval central part and two pointed extremities which were driven into the wood on either side of the caulked seam. The method is characteristic of European river craft in an area roughly coinciding with Lorraine and the realm of Louis the German (East Francia) after the division of the Empire in 843. At any rate from those days (and probably earlier) it continued to be practised, into the 20th century. The sintels used into the 12th century were narrow, staple-like strips of iron which first around 1175 gradually took on the shape of a laurel leaf. From the early 13th century onwards they barely changed in shape in some isolated parts of shipbuilding Europe. In the Danube basin and along other major rivers in eastern Europe, the method was applied in a slightly different form. Between ca 1150 and 1550, it was the typical caulking method for all ships, large and small, constructed according

to the cog-building tradition in the Hanseatic territories. Over the period 1150-1225 their shape in this region swiftly changed from the narrow sintel staple into the sintel clamp, whose central part successively was forged into the shape of a laurel leaf, then acquired a round, an oval and eventually an elongated-oval central part. The various shapes can be dated to periods of 50-75 years, some to shorter episodes. By around 1550, sintel-clamping was no longer applied to larger seagoing and inland vessels.

skeg *scheg*
A triangular piece fastened to the keel to improve sailing qualities of a flat-bottomed ship.

spar *rondhout*
A wooden pole to support the sail: *e.g.* mast, yard, sprit.

spar ceiling see ceiling

splashboard see washboard

spoon auger *lepelboor*
A sharp tool with a spoon-shaped bit, for cutting holes into timbers to receive treenails.

spritsail *sprietzeil*
A quadrilateral sail held up by a short mast and extended by a sprit (spar) between its lower forward corner at the lower part of the mast and its peak at the after top corner (sprit rigging *sprietuig*). Used on small craft, maybe including small cogs.

square bow/stern *heve*
The heve is the upward-sloping stern or bow of a square-ended barge or boat, from the waterline up to gunwale level. This part may be rectangular or trapezoid; in the latter case, with the sides converging or diverging upwards.

stanchion *stut*
A supporting post.

standing knee see knee

steering-oar *stuurriem*
A long oar, as used from Antiquity onwards, to steer river barges and rafts. Viking ships and ships built in the Nordic tradition into the 12th century also had steering oars.

stern/stem hook *stevenhaak*
A naturally grown (and squared) knee-shaped timber that in cogs forms an after or forward (horizontal) extension of the keel plank while its other (rising) arm forms the lower end of the stern- or stempost. Joined to this post and to the keel with flat scarfs. In the sides of the horizontal part and the rising arm are rabbets to accommodate the hood ends of the bottom strakes. This timber is a characteristic structural element of cogs.

stem *(voor)steven*
In cogs, the post that is placed on the stem hook with a sloping flat scarf. The hood ends of the side strakes are nailed to the double-bevelled/mitred front surface.

stern(post) *(achter)steven*
In cogs, the post that is placed on the stern hook with a sloping flat scarf. The hood ends of the side strakes are nailed to the double-bevelled/mitred rear surface.

stern rudder *stevenroer*
A rudder suspended with forged-iron fittings and pintles in gudgeons on the rear of the stern. The cog is the first medieval vessel type in Europe to have been fitted with a stern rudder from its beginning.

stocks *stapeling*
A structure to support the hull under construction.

strake *gang*
In wooden ships, a series of planks joined end-to-end, running fore-and-aft, forming the ship's shell (bottom and sides) or ceiling. The planks are joined by flat scarfs, to make up garboard, bilge and wale strakes, etc. Some strakes may not run the full length of the ship (drop strake, stealer *insteker*).

stringer *schaarbalk*
A longitudinal timber serving to reinforce the structure of the hull (*e.g.* the bilge, the side). In cogs, usually in parallel pairs 0.5 to 1 m apart; resting on transverse beams above the keel, or short stringers resting in rabbets in such beams. Some stringers have rabbets into which the loose, transverse deck planks were laid. These planks and short stringers were removable for loading and unloading the ship.

sweeping stern, overhanging stern *geveegde kont*
A stern with a sharply converging underwater body and widely everted sides with an S-shaped profile, overhanging the rudder.

template see mould

temporary floor timber *bouwlegger, hulplegger*
A (light) floor timber to temporarily hold together the planks in a ship's bottom during construction. Eventually to be replaced by a permanent floor timber.

temporary frame *bouwspant*
A full frame erected on the keel, to which the shell planks are fixed.

temporary framing *noodspanten(raam), bouwspanten*
A structure comprising the keel (plank), stern and stem, and light frames or templates of the outer surface of the frames, to which the shell planks are fitted and temporarily fastened during construction. In due course, the temporary frames are replaced by the actual frames.

tender, dinghy *werkschuit, bijboot*
A small open boat, usually rowed, and serving a larger vessel, used for the ship's maintenance and for carrying goods and people to and from the shore.

tholepin(s) *(dubbele) dol*
An upright pin or pair of pins on the gunwale, by which an oar is held in place in a rowing boat. See also rowlock.

thwart *doft*
A plank or bench at right angles to the keel in a boat, to seat rowers or passengers. Sometimes with a notch or hole to support a mast.

tiller, helm *helmstok*
A horizontal lever by which the rudder can be moved.

tjalk
A flat-bottomed sailing barge of variable size, with a round bow and usually one mast, sailing inland and coastal waters.

to tow, to haul *jagen*
Pulling a vessel along an inland waterway by a horse or people walking on a towpath. The towline would be secured on a bitt on the ship's afterbody, then fastened higher up on the mast, before running forward to the harness worn by the horse or person(s).

track boat, tow barge *trekschuit*
A horse- or human-drawn vessel formerly used on inland waterways. It often had a short mast from the top of which the towline ran.

treenail, trenail *pen*
A (hardwood) peg, used as a fastening in wooden ships. Driven into holes prepared with a spoon auger, and often secured with a wedge. See also wedge.

truss band *rakband*
A line wound around the mast, strung with wooden beads (*kloten*), to facilitate the hoisting of the yard with the sail by reducing friction.

turn of the bilge see bilge

underwater body *onderwaterschip*
The underwater part of the ship's hull.

ventjager
A fast sailing vessel with a fish-well for transporting and landing of fresh fish and for provisioning of skippers.

vertical rider *spoor*
A separate vertical timber against the side, that extends the support of a frame to the uppermost side strakes on the in- and outside.

victuals *victualie, leeftocht*
Provisions carried onboard to feed the crew during a voyage.

V-shaped floor timber, crotch (timber) *wrang*
In the after- and forebody, a slightly to strongly angled (part of a) frame, holding together the keel (plank), bottom strakes and side strakes above the keel plank or stern and stem. Those furthest into the bow and stern (*stevenwrang*) may be fastened only to shell strakes.

wale *berghout*
A heavy strake fastened to the outside of the shell to protect the shell of 15th/16th century large vessels when in contact with e.g. a quaside

washboard, splashboard, fashionboard *zetboord*
Removable plank or (partial) strake that may be fitted onto the gunwale or low bulwark of a vessel to temporarily increase the freeboard and thus to keep out waves and spray.

waterschip
A type of fishing vessel, also used for transporting live fish. First mentioned in 1339. Later also used as tugs for moving East-Indiamen across the Pampus shallows into the port of Amsterdam. See also camel.

waterway *watergang, lijfhout*
The outermost plank of a deck serving to guide any water towards the scuppers.

windlass *spil, windas*
In medieval ships such as cogs, a windlass was a solid wooden drum (*spilrol*) held horizontally between a pair of cheeks with ratcheting, and rotated by means of hand spokes. Often placed on, or against the forward edge of the afterdeck, it served for hoisting the yard or sprit with the sail. Also, on the foredeck, as an anchor winch. See also capstan.

wedge, treenail wedge *wig, ark, deutel, keg*
A tapering piece of hardwood, driven into a slit in the outer end of a treenail (usually on the outside of a ship's shell) to secure it in place. They may be pyramidal (*deutels*). Also, a small wooden wedge used in some late-medieval ships and some more recent fishing vessels for securing moss laths and caulking instead of sintels (*prikken*).

yard *ra*
A long spar suspended transversely from a mast, from which a sail hangs.

Abbrevations

ADC
Archeologisch Diensten Centrum, (Archaeological Service Centre), Amersfoort

ADW
Archeologische Dienst Waasland, (Waasland Archaeological Service), 1987-2016, Sint-Niklaas, Belgium

BAI/GIA
Biologisch-Archeologisch Instituut van de Rijksuniversiteit Groningen, (Biological-Archaeological Institute of the University of Groningen)/Groninger Instituut voor Archeologie (Groningen Institute of Archaeology)

NISA
Nederlands Instituut voor Scheeps- en onderwaterArcheologie, (Netherlands Institute for Ship and underwater Archaeology), Ketelhaven/Lelystad

RIJP
Rijksdienst voor de IJsselmeerpolders, (IJsselmeerpolders Development Authority), 1968-1989, Lelystad

ROB/RACM/RCE
Rijksdienst voor het Oudheidkundig Bodemonderzoek, (State Service for Archaeological Investigation), Amersfoort; as of 2006 Rijksdienst voor Archeologie, Cultuur en Monumenten (State Service for Archaeology, Culture and Monuments); as of 2009 Rijksdienst voor het Cultureel Erfgoed (Cultural Heritage Agency), Amersfoort

RWS
Rijkswaterstaat, (Ministry of Public Works), The Hague

The 'open depot' of ship inventories with the National Reference Collection in the building of the Netherlands Institute for Ship- and underwaterArchaeology (NISA) in Lelystad, 2001.

List of finds

L lenght
W width
H height
D diameter
(all measurements in cm)

* not anymore in the collection
** does not belong to the ship

6 The Nijkerk-II cog

numbers	find	size
1 t/m 4	jug, stoneware (Siegburg, sherds)	
5	jug, redware	H 18.2
6 t/m 8	cooking pot, redware (sherds)	
9	tar-pot, blue greyware, tar (sherds)	
10	*Kugeltopf*, greyware, (sherds)	neck D 12.5
11	idem	neck D12
12	pot, blue greyware (sherd)	
13	pot, earthenware (Pingsdorf, sherd)	
14	*Kugeltopf* earthenware (Paffrath-like, sherd)	
15	pot, earthenware (sherd)	
16	jug, proto-stoneware (sherd)	
17	pot, earthenware (Paffrath-like, sherd)	
18	jug, proto-stoneware (sherd)	
19	idem	
20	pot, earthenware (sherd)	
21	*Kugeltopf* earthenware (sherd)	
22	idem	
23	*Kugeltopf*, blue greyware (sherd)	
24 t/m 27	pot, earthenware (prehistoric,sherd)	
28 t/m 30	pot, earthenware (Karolingian? sherd)	
31	pot, earthenware (Andenne? sherd)	
32	pot, earthenware (Karolingian, sherds)	
33	pot, earthenware (Andenne? sherd)	
34	idem	
35	*Kugeltopf* earthenware (sherd)	
36	pot/jug earthenware (Andenne? sherd)	
37 t/m 40	jug, earthenware (Pingsdorf, sherd)	
41 t/m 46	jug, proto-stoneware (sherds)	
47	pot, (pipkin) blue-grey ware (Paffrath? sherd)	
48 t/m 50	pot, blue-grey ware (Paffrath? sherds)	
51	pot, earthenware (Karolingian? sherd)	
52	*Kugeltopf* pot-ware (sherd)	
53 t/m 70	idem	
71	pot, blue-grey ware (sherd)	
72	pot, redware (sherd)	
73	pot, earthenware (prehistoric, sherd)	
74 t/m 81	idem	
82	candlestick, wood and iron	H 19/ 22
83	idem (copy)	
84	small plate, wood (fragment)	
85	knife-handle, wood/iron (fragment)	
86	knife-handle, wood	L 11
87	handle? wood (fragment)	
88	ring?, wood (fragment)	D 5.5
89	stick, wood (fragment)	
90	small dish, stave built, wood (fragments)	
91	small sticks, wood	D 0.5-1.2
92 t/m 97	idem	D 0.2-0.9
98	barrel, wood (fragments)	
99	idem	
100	plank (repair?), wood	L 68
101	plug, wood	D 5
102 t/m 106	idem	D 0.3-1.7
107	spade, wood and iron	L 89
108	idem (copy)	
109	grain shovel, wood (fragments)	
110	idem	
111	sleg (big hammer), wood	L 21
112	idem	L 29
113	sheave, wood	D 7.5
114	lid, wood	13.2x13
115	small plank, wood	L12.8
116	lid of a pump-cylinder?, wood	D 16
117	drain spout, wood	L 170
118	idem (copy)	
119	plug or pen, wood (fragment)	D 4
120	idem	D 2
121	pen, wood	L 36
122 t/m 126	idem	D 3.2-4.2
127	small plank, wood	L 20
128	small beam, wood (fragment)	
129	floor of branches, twigs and straw	
130	knee, wood	72x44
131	knee or anchor stone?, wood	H 60
132	handspike?, wood	L 55.5
133	piece of wood from a load bed of branches and twigs	L 60
134	idem	L 21.5
135	idem	
136	branches of a load bed of thin stems	
137	squaring axe, iron/wood	L 21
138	spoon-auger, iron (fragment)	D 1.5
139	knife-blade, iron (fragment)	
140	load hook, iron	L 14.5
141	grapnel, iron (fragment)	
142	strip, iron	L 6.8
143	gudgeon, iron (fragment)	
144	bolt with eyelet (suspension for rudder), iron (fragment)	
145	bolt, iron	L 42,5

146	nails, iron	L 17.5
147 t/m 154	idem	L -16
155	small bar, iron	L 12
156 t/m 158	sintel clamps, iron	
159	knife of ditchcutter (recent), iron	
160	compass ? needle, copper (fragment)	L 7.3
161	small plate, copper	D 3.3
162	idem	4x3
163	button, silver?	D 1.3
164	tripod pipkin, bronze (fragment)	
165	small plate, tin	3x3
166	small plate (rolled up), lead	8,5x7
167	marlinspike, antler tine	L 19
168	idem	L 22
169	small bar, bone	D 0.4-0.7
170	needle, bone (fragment)	L 10.5
171 t/m 186	shoe, leather (fragments)	
187	boot or jacket, leather (fragment)	
188	boot, leather (fragment)	
189	shoe, leather (fragment)	
190	boot?, leather (fragment)	
191 t/m 211	shoe, leather (fragment)	
212	knife-sheath, leather (fragments)	
213	knife-sheath, leather (fragments)	
214	idem(?)	
215	bag?, leather (fragment)	
216 t/m 222	shoe, leather (fragments)	
223	lace, leather	L 11
224	belt?, leather (fragment)	
225	shoe?, leather (fragments)	
226	sail, wool and tar (two pieces)	ca 70x50
227	idem (fragment)	ca 70x60
228 t/m 230	rope (fragments)	
231 t/m 268	bone (fragments)	
269	horn pit, bone	
270	molar	
271	bone (fragment)	
272	canine	
273	bone (fragment)	
274	idem	
275	molar	
276 t/m 283	bone (fragments)	
284-1 t/m 284-18	idem	
285-1 t/m 285-116	idem	
286-1 t/m 286-12	idem	
287-1 t/m 287-56	idem	
288-1 t/m 288-19	idem	
289 t/m 292	slate	
293	ballast?, stone	
294	flint	
295	idem	
296	brick, red (fragment)	
297	idem	
298	killick or anchor stone, granite	40x22x17
299 t/m 301	coal (pieces)	
302	charcoal (piece)	
303	net for keeping fish?, sample wicker-worck	
304	hazelnut, (shell and shell fragments)	
305	moss caulking, sample from the drain spout	
306	sail, wool and tar (piece)	
307	sand, sample from the covering of the load bed of branches, twigs and straw	
308	coin (coin catalogue no. Zo36-M1), Reinald II of III (1327-1343 of 1343-1361), zilver	

8 The Nijkerk-I cog

numbers	find	size
1 a en b	jug, proto-stoneware, base/sherd	D 5,8, H 4
2 a - c	jug, proto-stoneware, rim sherds	3, 2, 5
3 a - c	jug, proto-stoneware, three sherds	L 2, 2, 1.5
4	jug, stoneware (Siegburg), neck sherd (does not belong to ship)	L 6
5 a - l	*Kugeltopf*, grey brown earthenware, native (Nord-Netherlands), rim-/wall sherd, bottom sherds, soot/burning traces	H 19, D neck 16
6	*Kugeltopf*, sherd	L 3.2
7	grapnel with ring, wrought iron	L 51, D ring 12.7
8 a en b	axe with handle (fragments)	L 20, steel L > 30
9	flat piece of iron/ironoxyde	L 19, B 4.5
10	nails (21)	L 8-0, D kop 2-3
11 a	gudgeon, wrought iron	H ca. 43
11 b en c	nails (2), wrought iron	L 7.5, .2
12	sintel nail clamps (13) (several fragments), wrought iron	L 2.1 - 2.8
13	clamp of firebox?, wood	L 24, B 8, H 5
14	treenail, wood	L 33.5
15	branche, wood (wrought)	L 20
16	treenail, wood	L 20
17	treenail, wood	L 20.6
18	treenail, wood	L 18.6
19	treenail, wood	L 16.5
20	treenail, wood	L .2
21	treenail, wood	L 13.5
22	treenail, wood	L 10.3
23	treenail, wood (4)	L 8.7, 8.5, 6.5, 5
24	treenail with wedge from ceiling board/floor timber, wood	L 20
25	treenail with wedge, wood	L 10.4
26	stick with notch (fragment), wood	L 9.1
27	stick, wood	L 14.5
28	plug?, wood	L 8
29	small plank, wood (from compass box)	L 14.3
30	small plank, wood (fragment)	L 17
31 a en b	plank fragment with remains of textile, wood/wool	L 14.2
32	'cog sticks', wood (12 + 19 fragments)	L 16-20
33	stick, wood	L 36

34	piece of leather	L 24, B 11-13
35	treadsole, leather	L 15.5, B 11.5
36	upper, part of a shoe, leather (fragments)	
37*	shoe, leather	
38	whetstone, quartz phyllite (4 fragments of)	L 8.8,
39	whetstone, granite (boulder)	L 11.5, B 9.5, h 6
40*	flint (3), 2 lack	D 2
41	skullcap cattle, bone	L 26
42	*atlas* cattle, bone	L 7.5
43	*pelvis* cattle, bone	L 9
44	*femur* sheep/goat, bone	L 6
45 a en b	*humerus* and *radius* (r) sheep/goat, bone	L 11
46	*pelvis* sheep/goat, bone	L 6.8
47	rib cattle, bone	L 2.5
48	rib cattle, bone	L 6.5
49	rib sheep/goat, bone	L 7.5
50	rib cattle, bone	L 11.5
51	rib sheep/goat, bone	L10.5
52	rib sheep/goat, bone	L 12.5
53	rib sheep/goat, bone	L 9.2
54	rib cattle, bone	L 12
55	rib sheep/goat, bone	L 7.7
56	rib sheep/goat, bone	L 7.7
57	*thoracic vertebra* sheep/goat, bone	L 6.3
58	rib sheep/goat, bone	L 14
59	rib sheep/goat, bone	L 7.7
60	medium mammal, bone	L 7
61	*thoracic vertebra* sheep/goat	L 5.6
62	molar sheep/goat	L 3.7
63 a en b	*radius* cattle, bone	L 14.5
64	sail (fragment), wool (woven)	ca. 60 x 80
65	hazelnut shells	
66	wood chips/hazelnut shells (sample)	
67	peat (9 pieces of peat bog)	L 7 - 15
68	loam/ river sludge from fire box (sample)	
69	frame head, wood (cut off)	L 40, B 11
70	rope (fragment)	D 1.3
71	rope (fragments)	D 1.1
72	rope (fragments)	D 1.8
73*	rope (fragment)	
74	rope (fragment)	D 1.6
75	rope (fragments)	D 1.7
76	rope (fragments)	
77	rope (fragment)	D 1.8
78	rope (fragment)	D 1.8
79	rope (leftovers)	
80	rope (fragment)	D 1.8
81	rope (fragments)	
82	rope (fragments)	D 1.8
83	moss caulking (sample) from seam stern hook/ garboard	
84	idem from seam keel plank/garboard port side	
85	idem from seam garboard/B strake port side	
86	idem from seam B strake/C strake starboard side	
87	idem (and fragment moss lathe) from seam J strake port side outside	
88	idem from seam D stake/E strake starboard side outside	
89	idem from scarf between planks F1 and F2 starboard side	
90	idem scattered from the wreck	

9 The Ens cog

numbers	find	size
1	five nails, folded over twice and four incomplete heads	L 5-5.5 d 1.6-2
2	eight sintel nail clamp fragments	
3	ten pieces of moss lathe, oak	
4	ca 20 triangular strands of moss caulking (*Sphagnum*) from the bottom seams	
5	four clods of moss caulking, probably from plank scarfs	
6	wedge or keg (oak) from pen made of beech wood	
7	small boulder, unworked flint	4x3,5

10 The Rutten cog

numbers	find	findspot
1	*Kugeltopf* sherd with '*Besenstrich*' decoration	ground level
2	glacier millstone	near the ditch
3	sherd, red earthenware, post- medieval	in top soil
4	idem	idem
5	idem	idem
6	idem	idem
7	pen, iron, from a harrow?, recent	idem
8	screw bolt, iron, recent	idem
9	pen, iron, from a harrow?, recent	on top of stem
10	majolica dish, 3 sherds and fragment of leather, post-medieval	in top soil
11	3 sherds, red earthenware, post-medieval	idem
12	stone, recent	idem
13	gudgeon?, iron fragment	drainage slit, afterbody
14	sherd, red earthenware, post-medieval	top soil, afterbody
15	rim sherd, red earthenware, post-medieval	ground level
16	gudgeon?, iron fragment	idem at the stern
17	?, iron fragment	idem
18	bone, cattle	in topsoil
19	file, iron, recent	idem
20	flat piece of iron, fittings?, recent?	Idem
21	large brick (*kloostermop*), fragment	idem
22	sherd, red earthenware, post-medieval	idem
23	brick, fragment, post-medieval	idem
24	?, iron, post-medieval	idem
25	nail/pen, iron, post-medieval?	idem
26	bone, fragment, 19th/20th century	in topsoil
27	bone and piece of leather, post-medieval?	idem
28	sherds, red earthenware, post-medieval	idem
29	sherd, proto-stoneware	above midship
30	idem	idem
31	small stone	idem
32	sherds, red earthenware, post-medieval	idem
33	sticks (22)	idem

34	small ribs, nail, piece of leather	midship
35	sherds, proto-stoneware en blue-grey (Elmpter ware)	idem
36*	pieces of pine wood, trimmed	idem
37	bone of bird?	idem
38	a lot of sticks (22)	idem
39	large brick (*kloostermop*), fragment and pieces	idem
40	idem, fragment	forbody
41	sherd, *Kugeltopf*	idem
42	rib	idem
43	small stone?, iron?	Idem
44	sherd, proto-stoneware and sintel clamp	midship
45	sherds, yellow grey 'soft' earthenware and near-stoneware	afterbody
46	rope (8) and sherd, red earthenware, post-medieval	midship
47	staves of a small dish (21 A), wood	idem
48	skeleton, sheep or goat	idem
49	sherd, proto-stoneware	idem
50	sticks (22)	midship
51	sherds, proto-stoneware and *Kugeltopf*	idem
52	rib, nail and some sherds, post-medieval	idem
53	sherds, near-stoneware (3)	forbody
54	small bones	on stem hook
55	sticks (22)	idem
56	large brick (*kloostermop*), fragment	idem
57	sherds, proto-stoneware	forbody
58	small bone	idem
59	peat, fragment	idem
60	pieces of wood, wrought	idem
61	nail and sintel clamp	idem
62	sticks (22)	idem
63	sintel clamp	idem
64	sherd, *Kugeltopf* (4)	midship
65	sticks (22)	forbody
66	rope, fragments (10)	idem
67	sintel clamp	idem
68	bone	idem
69	sherd, *Kugeltopf*	idem
70	sintel clamps (16) and small bone	idem
71	nails	idem
72	small bones	idem
73	moss caulking?	idem
74	sherd, near-stoneware	idem
75	sticks (22)	idem
76	lumps of peat (28)	idem
77	wooden 'button' (14)	idem
78	small repair(?) planks	midship
79	sherds, proto-stoneware and *Kugeltopf*	idem
80	sticks (22) and small bone	idem
81	gravel and pieces of peat (28)	idem
82	sherd, *Kugeltopf* pot	idem
83	bone, cattle	idem
84	peat, fragments (28)	idem
85	moss caulking, strands	idem
86	sherd, *Kugeltopf*	midship
87	idem (4) and near-stoneware (3)	idem
88	large brick (*kloostermop*)	idem
89	piece of iron, stone and peat (28)	forbody
90	sticks (22), pieces of wood, sherd	idem
91	peat, fragment	idem
92	button?, small bone and pieces of wood	idem
93	gravel	idem
94	pieces of wood	idem
95	moss caulking	midship
96	peat (28)	forbody
97	sherds, proto-stoneware, nails (11)	idem
98	sticks (22), wrought pieces of wood	idem
99	rope, fragments (10)	idem
100	sintel clamps (16)	idem
101	rivets with plate (15)	midship
102	small bone	idem
103	piece of leather (26), sticks (22), sherd, *Kugeltopf* (4), small stave (21 B)	idem
104	piece of peat (28)	idem
105	sherds and small bones	idem
106	gravel	idem
107	splitwood box?, fragments (20)	idem
108	sample wood	idem
109	nail?	idem
110	pieces of wood	midship
111	sherd, *Kugeltopf*	idem
112	boulder	idem
113	small square plank	idem
114	sticks (22)	idem
115	sample gravel/plant remains	idem
116*	leather, fragment	midship
117	peat (28)	forbody
118	nails (11), small bone, sherd near-stoneware	idem
119	rib, cattle	midship
120	jaw, sheep/goat	idem
121	sherd, stoneware	idem
122	sample wood	idem
123	sintel clamps, nails (11)	midship
124	sticks (22), small bone	idem
125	fishbones	idem
126	seeds	idem
127*	leather, fragment	idem
128A	sherds	idem
128B	nails (11)	idem
129	sherds, stoneware and *Kugeltopf*	idem
130	bone fragments	idem
131*	pieces of peat	idem
132	seeds and sticks	idem
133	fishbones	idem
134*	charcoal	idem
135	bone, fragment	forbody
136	sherds near-stoneware (3)	idem
137	sticks (22)	idem
138	bone	midship
139	nail and sintel clamps (16)	idem
140	pieces of chalk	idem
141	nails (12)	idem
142	sticks (22)	idem

143	sherds, proto-stoneware	idem
144	small bone	idem
145	nail with eye?	midship
146	small stave, small keg and stick	forbody
147	flint with iron slag?	idem
148	sintel clamps (16), nails	idem
149	bone from human (29)	idem
150	pieces of thick rope (7)	under forbody
151	sherds, stoneware	midship
152	nail?	idem
153	piece of iron	idem
154	sticks (22)	idem
155	sample of seeds?	idem
156	peat (28)	idem
157	nail	idem
158	split wood box, bottom?	idem
159	sample of seeds?	idem
160*	wooden button	idem
161	bone fragments	idem
162	wood fragments	idem
163	idem	idem
164	sherd	forbody
165	pieces of peat	idem
166	sticks (22)	idem
167	small staves?	idem
168	peat (28)	next to forbody
169	sherd, near-stoneware (2)	forbody
170	small bones (burned)	idem
171	sherds, *Kugeltopf* and large brick, fragment	idem
172	sintel clamp, nail and iron pin?	idem
173	sticks (22)	idem
174	piece of iron	midship
175	bones and piece of charcoal*	idem
176	sticks (22), nails (11) and sintel clamps	afterbody
177	neck sherd, proto-/near-stoneware (1)	idem
178	rope, fragment (9)	idem
179	sticks (22), beechnut, hazelnuts (29)	idem
180	bottom sherd, proto-/near-stoneware	idem
181	bone, fragment	idem
182	wooden button? wound up with rope	idem
183	bone	forbody
184	knob of flagpole?	afterbody
185	three sherds	idem
186	decorative button?, tin?	idem
187	rope fragment with knot	idem
188	sintel clamp	idem
189	pieces of chalk from ballast	idem
190	fish hook* and hazelnuts (29)	idem
191	stick (22) and keg	idem
192	rope, fragments	idem
193	piece of iron	midship
194	bone, fragment of skull	
195	sticks (22)	afterbody
196	sherd, proto-stoneware and peat (28)	idem
197	sintel clamps (16), nails and sherd of *Kugeltopf*	idem

198	two nails	idem
199*	two pieces of leather and button	midship
200	sherds	idem
201	piece of iron	idem
202	keg, piece of flat iron*, piece of moss lath	idem
203	bone fragments	idem
204	fish bones and hazelnuts (29)	idem
205	sticks (22)	idem
206	peat (28)	idem
207	piece of wood with fire traces	idem
208	small stave	idem
209	bone fragments	idem
210	sherds	idem
211	sintel clamp, nails	idem
212	sticks (22)	idem
213	four sherds, proto-stoneware, sticks (22) and nails	idem
214	piece of lead	midship
215	sherds, *Kugeltopf* and sintel clamp	idem
216*	piece of rope with a wooden ring at the end	idem
217	small bones, nails	idem
218	sherds, pieces of amber	idem
219	large brick, bones, sherds, *Kugeltopf* (5) and nails	idem
220	sherds	idem
221	sintel clamps	idem
222	sherds	idem
223	sherds	idem
224	pen or handle, pieces of wrought wood	idem
225	small bones	idem
226	fishbones	idem
227	nails	idem
228	sherds	idem
229	charcoal	idem
230	pieces of amber	forbody
231	coin	idem
232	sherds, stoneware and *Kugeltopf*	midship
233	bone fragments	idem
234	hazelnut shells (29)	idem
235	sample tar	forbody
236	wooden spoon, fragments (18)	idem
237	pieces of wood	idem
238	rim sherd, stoneware	near stern
239	foot of a trivet?, iron (19) and sintel clamp	under keel afterbody
240	iron fittings, fragment, of stem post	near stem
241	sand stone and nail	midship
242	sintel clamps	under wreck
243	nails	idem
244	sherds, earthenware	idem
245	piece of wood	idem
246	pieces of iron corrosion	idem
247	fragments of iron fittings	near stem
248	leather fragments	under wreck
249	sample ballast stones, flint and granite	midship
250	sample of caulking and moss laths	A1/keel plank star board
251	idem	but of scarf A1/ A2 star board

252	bundle moss laths and caulking	seam A1/B1 star board
253	sample of caulking	scarf keel plank/ stern hook
254	idem	scarf B1/B2 star board
255	idem	stern hook
256	moss caulking and moss lath	scarf B1/C1 star board
257	caulking	stern hook port side
258	idem	seam C1/D1 star board
259	idem	seam D1/E1 star board

11 The Dronten-I cog

numbers	find	size
1	storage jar, near-stoneware (Langerwehe or Brühl)	H 31.5
2	storage jar, near-stoneware (Siegburg)	H 33.5
3	drinking jug, near-stoneware	H 20.5
4	*Kugeltopf*, '*Besenstrich*' decoration	H 13
5	*Kugeltopf*, '*Besenstrich*' decoration	H 14.5
6	*Kugeltopf*, '*Besenstrich*' decoration, sherd	L 8.5
7	storage pot or bowl, blue grey ware (rim sherd)	L 3
8	*Kugeltopf*, '*Besenstrich*' decoration (two sherds)	L 6 en 2
9	cooking pot, red earthenware (sherd)	L 3.5
10	spearhead, iron	L 28
11	spearhead, iron	L 18.2
12	knife, iron with wooden handle	L 20.5
13	knife/dagger with part of a handle (burr walnut?), iron blade with inlaid letters and 'stars', silver	L 17.5
14	knife with sickle-shaped blade, handle is missing, iron	L 14
15	knife, handle is missing, iron, copper, silver	L 14
16	small knife, handle is missing, iron, copper	L 9
17	axe, iron	L 26.8
18	adze, iron	L 20.3
19	caulking iron, iron	L 18.5
20	fire tongs, iron (two pieces)	L 32 and 16
21	gudgeon, iron (on stern hook)	
22	rosebolt, iron (incomplete)	L 44.8
23	rosebolt, iron (incomplete)	L 21.5
24	strip with nail, iron (fragment)	L 8.5
25	sintel clamp, iron	L 3.5
26	nail, on the bottom of the nail head four notches, iron	L 14.7
27	idem	L 8.5
28	nail, iron	L 9
29	wetstone, quartz phyllite	L 32.8
30	idem	L 17.3
31	floor tile, red earthenware	13.8x13.8x 2.8
32	large brick	29x13.8x6.5
33	idem (fragment)	18.5x14x6.5
34	idem (fragment)	18x13.5x6.5
35	idem (fragment)	13.5x13.5x6.5
36	idem (fragment)	9x13.5x7

37	idem (fragment)	6x12x6
38	lump of cracked flint (six pieces)	
39	marline spike or split pin, antler tine from red deer	L 14.2
40	cattle, *lumbar vertebra* ve	
41	cattle, *calcaneum* (r) ve	
42A	cattle, *femur* (r) d ve	
42B	cattle, *astragalus* (r) belongs to 41 (chop mark)	
42C	cattle, *lumbar vertebra* onv (chop mark)	
42D	cattle, rib (chop marks)	
42E	cattle, rib (chop marks)	
42F	cattle, rib (chop marks)	
42G	pig, *humerus* (r) p+d onv	
43+A	shoe fragments of at least one high and probably one low model, leather (fragments)	
44	sample moss caulking	
45	idem with moss lath fragment and sintel clamp	
46	sample moss caulking	
47	sample moss caulking with lath fragment and sintel clamp	
48	sample of tar of the inside stem hook	
49	idem of inside stern hook	
50	sample of loam remains	
51	stem hook	L 290
52	stern hook	L 200
53	V-shaped floor timber of frame 18 with hole for the mast heel	L 305
54*	clamp, wood (fragment)	L 21.5
55*	handle of pump, wood (three fragments)	L 14
56	point of a stem hook	
57	nail, iron	L 9
58	idem	L 9
59	idem	L 9
60	idem	L 9
61	idem	L 9
62	idem	L 9
63	idem	L 9
64*	peat (fragments)	

12 The Marknesse cog

numbers	find	size
1	drinking jug, stoneware (Siegburg)	H 23
2	small drinking jug (Meuse valley)	H 9.5
3*	rim sherd, stoneware, smooth with brown patch of glaze	
4*	sherd	
5	cooking pot (Grapen), bronze (iron handle missing)	H 13
6*	cooking pot (Grapen), bronze (casting)	H 15
7	kettle, copper (iron handle and ears and one is missing)	
8	gudgeon, wrought iron wings	L 23
9*	bolt, wrought iron	
10*	nail, wrought iron	
10-1	nail, wrought iron	L 8.5 mm
11*	sintel clamps, wrought iron	
11-1	sintel clamps, wrought iron	

11-2	sintel clamps, wrought iron	
11-3	sintel clamps, wrought iron	
11-4	sintel clamps, wrought iron	
11-5	sintel clamps, wrought iron	
12*	anchor fluke	L 23.5
13*	fire tongs	L 34.5
14	axe, wrought iron	L 22.3
15*	forked tool, wrought iron	
16	boat hook, wrought iron	L 17
17*	knife, wrought iron and wooden(?) hilt	L 20.5
18 A en B	small reparation board, oak and piece of caulking, wool	42.5x13
19 A en B	small reparation board, oak and piece of caulking, wool	32x9.5
20	small piece of caulking, wool	23x14
21	rope, fragments	
22*	rope out of channel	
23*	upper leather of shoe	
24	whetstone	L 32.5
25	brick (cargo)	ca L 28, W 13.5, H 6.5
26	brick (cargo)	ca L 28, W 13.5, H 6.5
27	brick (cargo)	ca L 28, W 13.5, H 6.5
28	brick (misfire with soot, cargo?)	ca L 28, W 12, H 6.5
29	6 brick fragments (cargo)	
30*	5 roof tile fragments with splotches of glaze cargo?)	
31*	sample of shells of cargo	
32	peat, fragments (drag-peat)	ca 13.5x10.5x 6-8

14 The Spakenburg-I cog

numbers	find	size
1	jug, stoneware (Siegburg, sherd)	
2	jug, stoneware (sherd)	
3	fire pan, red earthenware (sherd; does not belong to ship)	
4	compass glass (fragment)	D 8
5	roof tile red earthenware with sooth (fragment)	
6	idem	
7	ring with octagonal diameter, wrought iron	D 9
8	nail, wrought iron	L 12.5
9	5 nails, wrought iron	L 10.5-12.5
10	5 nails, wrought iron	L 10.5-11.5
11	2 nails, wrought iron	L 8 en 10
12	2 nails, wrought iron	L 2.5 en 5
13	sintel clamps, wrought iron	L 4-5
14	anchor blade, wrought iron (broken off; dos not belong to the ship	ca 50x40
15	sail, wool (woven; 2 fragments)	ca 50x40 en 15x15
16	sail, wool (woven and sewn; selvage and threads)	L 15
17	bottom of barrel, oak (fragment)	L 39
18	stopper? wood	L 5, D 3.5
19	broom, heather	L 30-40
20	shoe, leather (upper part; restored and reconstructed, sole	L 26
21	shoe, leather (fragment)	
22	shoe, leather (fragment)	
23	shoe?, leather (fragment)	
24	shoe, leather (fragment)	
25	shoe?, leather (fragment)	
26	rib, bone	
27	rib, bone	
28	small bone	
29	rib, bone	
30	rib, bone	
31	*scapula/pelvis*, bone	
32	rib, bone	
33	rib, bone	
34	*vertebra*, bone (fragment)	
35	small bone	
36	skull of rat, bone	
37	walnut-/hazelnut shells	
38	rope (fragment)	L 13
39	9 long peat bricks (partly fragments)	L 18-24
40	seed grains (do not belong tot he ship)	
41	writing chalk (pieces)	
42	sample moss caulking (moss, moss lath and sintel clamp)	
43	idem	
44	idem	
45	sample moss caulking	
46	sample moss caulking	
47	idem	
48	idem	
49	idem	
50	brick	21.5x11x5
51	idem	
52	piece of stone, white quartz (used as flint for fire striking?)	10x9x5
53	flint, flint stone	6.5x6x4.5
54	2 pieces of brick (worn)	

15 The Spakenburg-II cog

numbers	find	size
1	sherds, red earthenware	
2**	bottom sherd, Pingsdorf ware(?)	
3**	handle-/rim sherd, red earthenware (17th century)	
4**	rim-/body sherd, red earthenware (17th/18th century)	
5**	sherd, red earthenware	
6	plate, iron	
7	bolt with square head (3 fragments), iron	
8	bolt (fragments), iron	
9	?, iron	
10	treenails (7 fragments), wood	
11	piece of leather	
12**	heel of a shoe, leather (post-medieval)	
13	piece of woven sail, wool	

14**	bone
15	*vertebra*, bone
16	rib(?), bone
17	brick (fragment)
18**	idem
19	idem
20	idem
21	idem
22	idem, rests of mortar
23**	(fish) net weight, lime stone
24**	idem
25	sample caulking, moss
26	idem
27	Idem
28	? (fragments), iron
29	pen (with wedge), wood
30	handle of ?, wood
31	sawed off end of a crossbeam, wood
32	sample tar
33	sample caulking, moss
34	idem
35	sample, tar
36	sample caulking, moss
37	treenails with wedges, wood

16 The Almere-I cog

numbers	find	size
1	jug, stoneware	H 31.5
2	jug, like 1 (fragments)	
3	jug, stoneware (fragments)	
4	small (funnel) jug, stoneware (fragment)	
5	cooking pot, red earthenware (fragments)	
6	idem	
7	idem	
8	idem	
9	idem	
10	tripod bowl, red earthenware (fragments)	
11	idem	
12	idem	
13	frying-pan, red earthenware (handle of)	
14	idem	
15	idem	
16	frying-pan, red earthenware (fragments)	
17	tripod pipkin, red earthenware (fragments)	
18	cooking pot, red earthenware (fragments)	
19	*Kugeltopf*?, earthenware (fragment)	
20	shovel/spade with fragment of handle, iron and wood	B 22.5
21	adze with fragment of handle, iron and wood	L 21, handle, L 44
22	drawbar with fragment handle, iron and wood	L 20.5
23	caulking iron, iron	L 22
24	kidney dagger (blade), iron	L 22.7
25	knife, iron and wood	L 20
26	handle (of handsaw), wood and iron	L 17.5
27	spur, iron	L 15.5
28	buckle(?), iron	L 3.6
29	rosebolt (broken), iron	L 27.3
30	idem	L 23.5
31	idem	L 15.3
32	flat of the stem construction, iron	L 14.8-41
33	ring, iron	L 4.6
34	nails, iron (3)	L 5.9-7.5
35	idem	L 7.2
36*	idem	
37	nails (4), iron	L 9
38	nails (10 different examples), iron	L 5.7-8.9
39	nail with fragment leather, iron and leather	L 9.7
40	sintel clamps (6), iron	L 4.5-6
41*	idem	
42*	idem	
43	clapper, wood and leather	L 25.8
44	brace-block, wood	L 32.5
45	handle of tool(?)/part of anchor(?), wood	L 29
46	stirring- or churn stick(?), wood	L 37.5
47	stopper, wood	L 9.5
48	marlinspike, point of antler from red deer	L 18
49	human bones (8, A-H)	
50	rib, bone	L 9
51	bone fragment	L 10.5
52	small bone (fragment)	L 3.2
53	bone fragment	L 7
54	boulder (wind-abraded rock), quartzite	10x10
55	writing chalk, limestone	6x5
56	roof tile, red earthenware (fragment)	L 9.8
57	flint	L 2.8
58.	shell from mussel	
59	iron(?) slag	L 7
60	clod of fat	D 7.5
61	peat (8)	
62	rope (fragment)	D 1.8
63	shoe (fragment), leather	
64	sheath (fragment), leather	L 10
65	belt (small fragment), leather	
66	idem (6), leather	l 0.7-0.9
67	idem (small fragment), leather	L 17.5
68	idem (3 small fragments; one with silver? button), leather and silver?	L 4-7.5
69	bag/pouch (fragments), leather	H 30
70	coin (Gelre, Willem I), silver	
71	idem (Gelre, Reinald IV), silver	
72	idem (Utrecht, Frederik van Blankenheim), silver	
73-77	caulking material (5 samples), moss	
78-82	caulking material with moss lathe (5 samples), moss and laths	
83*	sintel clamp, iron	

17 The Dronten-II cog

numbers	find	size
1	gudgeon, wrought iron	L 16.2
2	rudder fitting with pin, wrought iron	L 14.6, d pen 2
3 A, B en C	three sintel clamps (two still sticking together), wrought iron	L 3.5, 3,5 and 4
4 A en B	two sintel clamps (sticking together), wrought iron	L 3.8 en 4
5 A en B	sintel clamp en moss filling from seam, wrought iron/moss	L 5, moss 1.5x1.5
6 A en B	two sintel clamps, wrought iron	L 5 en 5.2
7	sintel clamp, wrought iron	L 3.8
8 A en B	two sintel clamps, wrought iron	L 4.4, en 5.4
9	sintel clamp, wrought iron	L ca 5
10 A en B	two sintel clamps, wrought iron	L 5.8
11	nail, wrought iron	L 9, kop D 2.5
12	nail, wrought iron	L 7, kop D 2.2
13	nail, wrought iron	L ca 8, kop D 2.5 14
14	nail, wrought iron	L ca 8, kop D 2.5
15	nail, wrought iron	l ca 7, kop D 2.5 16
16	nail, wrought iron	l 5.5
17	nail, wrought iron	l ca. 3
18	130 boulders and four half red bricks, some fragments and one misfire. One big boulder, the other ones have the size of a fist.	
19	sample moss caulking	
20	sample moss caulking	

18 The Swifterbant cog

numbers	find	size
1	storage jar, near-stoneware, not complete (find no. OG 77/7)	H 29.3
2	small beaker, near-stoneware, not complete (find no. OG 77/18 + 3x 28)	H 8, D voet 6
3	drinking jug, near-stoneware, 7 sherds (find no. OG 77/23 en 28)	
4	floor tile, fragment, red earthenware, mainly streaky white engobe and lead glaze (find no. OG 77/10 en 12)	17x14x 1.7
5	idem, fragment (find no. OG 77/ 12)	10x12.5 x 1.7
6	floor tile, fragment, red earthenware, mainly dark brown glaze (find no. OG77/5 en 14)	11x8x1.7
7	spindle-whorl, grey yellow earthenware (find no. OG77/21)	D 3.5
8*	kettle, copper, small model with pouring lip and iron handle	
9	small cooking pot (grape), bronze with iron handle (find no.OG77/9).	H 15.2
10	spoon augur, fragment, wrought iron (find no. OG77/8)	L 36.2
11	spearhead, shaft is mainly missing, wrought iron/ wood (find no. OG77/27)	L 42
12	nail, wrought iron (find no. OG77/19)	L 9.5, head D 3
13	three nails, wrought iron (find no. OG77/25)	L > 4, head D 2,2
14	sintel clamp, wrought iron (find no. OG77/1)	L 3.8
15	sintel clamp, wrought iron (find no. OG77/30)	L ca. 4.5
16	whetstone quartz phillyte (find no. OG77/4)	L 64.5
17	whetstone?, coarse-grained sand stone (find no. OG77/16)	14x9x5
18 A-V	writing chalk, 22 pieces and pieces of chalk (find nos. OG77/6, 17 en 20	L 1-14
19	brick, fragment, (find no. OG77/13)	L 10
20	cattle, skull fragment with hornpit, bone (find no. OG77/26)	
21 A en B	duck, 2 bones, (find no. OG77/11)	
22	sheep/goat, rib, bone (find no. OG77/31)	
23	clamp, L-shaped, oak (find no. OG77/15)	11x5
24*	caulking, moss (sample M-1) (find no. OG77/2)	
25*	shells, *Mya arenaria* (sample M-2) (find no. OG77/3)	
26*	unknown material (sample M-3) (find no. OG77/22)	
27*	iron concretion, different pieces (find no. OG77/24)	
28	tar, piece (find no. OG77/29)	14x7.5x5
29 A t/m U	ballast stones, 21 pieces (find no. OG77/32)	
30*	flint	

22 The Kraggenburg cog

numbers	find	size
1*	flint, stone	12x10x6
2*	flint, stone	10x1.5x4
3*	rope, ?	D 2
4	2 sintel clamps, 6 nails, flint, iron/stone	4.5, 4.4
5*	caulking, lath and sintel clamps, moss/wood/iron	
6	split pin, point of antler of red deer	l 19,2
7	sintel clamp, iron (fragment)	3.5x2.5
8	2 lath, wood (fragments)	
9	rope (fragment)?	D 3.5
10	3 nails and 2 sintel clamps, iron (fragments)	
11*	flint, stone	7x1.5-4x3
12	jug, near-stoneware sherd (+ 16, 23,24,39, 45, 50, 51, 2x 52 and 55)	
13	jug, blue grey earthenware, sherd (+ 34, 43 en 59)	
14	2 red brick (fragments)	
15	granite, stone (fragments)	
16	sherd, near-stoneware (see 12)	
17	nail, iron	L 5
18	brick (fragment)	
19	bone (fragment)	L 2
20	sintel clamp, iron (fragment)	3x2
21	3 red brick (fragments)	
22	coal, piece	2,5x2
23	sherd, near-stoneware (see 12)	
24	sherd, near-stoneware (see 12 en 54)	
25	jug, near-stoneware, sherd (+58)	
26	large brick ('kloostermop' with soot traces)	28x14x6,5
27	nail, iron	L 4
28*	pen, wood	L 14, D 3
29	nail, iron	L 3.5
30	*Kugeltopf* sherds, earthenware and small bone	
31	nail, iron	L 3.5
32	axe with handle, iron/wood	23.5, handle L 33
33	large brick ('kloostermop' with soot traces)	27.5x13.5x6,5

34	*Kugeltopf* sherd, earthenware	6x4
35	L-shaped piece of wood	5x4
36	nail, iron	L 3
37	nail, iron	L 4
38	nail, iron	L 3.5
39	jug, near-stoneware, sherd (+44)	
40	nail, iron	L 2.5
41	whetstone, possible soot traces, red sand stone	15x12x9
42	peat, peat bog	31x10x7
43	sherd, blue grey earthenware (see 13)	
44	sherd, near-stoneware (see 39)	
45	sherd, near-stoneware (see 12)	
46	adze, iron	L 20
47	large brick, ('kloostermop', fragment)	15x14x6
48	2 nails, iron	
49	bone fragment	L 4.5
50	sherd, near-stoneware (see 12)	
51	sherd, near-stoneware (see 12)	
52	sherd, near-stoneware (see 12)	
53	peat log	23x12x3-7
54	sherd, near-stoneware (see 24)	
55	sherds, near-stoneware (see 12)	
56*	sherds, near-stoneware and *Kugeltopf*?	
57	nails, iron	
58	sherd, near-stoneware (see 25)	
59	2 sherds, blue gray and *Kugeltopf* ware (see 13)	
60	fire tongs, iron (2 fragments)	

24 The Enkhuizerzand cog

numbers	find	size
1*	three pieces of handmade red bricks	22.5x10.5x5.5
2*	brick fragment	
3*	idem	
4*	idem	
5*	idem	
6*	idem	
7*	idem	
8*	idem	
9*	idem	
10*	idem	
11*	eight fragments of brick	
12*	boulder, red granite	
13*	flint	
14*	(fish) net weight, limestone	
15*	two sherds of red earthenware from two different small pots	
16*	fragment rib, bone	
17*	fish net weight, lead	
18*	eight nails, wrought iron	
19*	two sintel clamps, wrought iron	

Appendices

6 The Nijkerk-II cog

Pollen spectrum of the moss caulking

E. Mook-Kamps

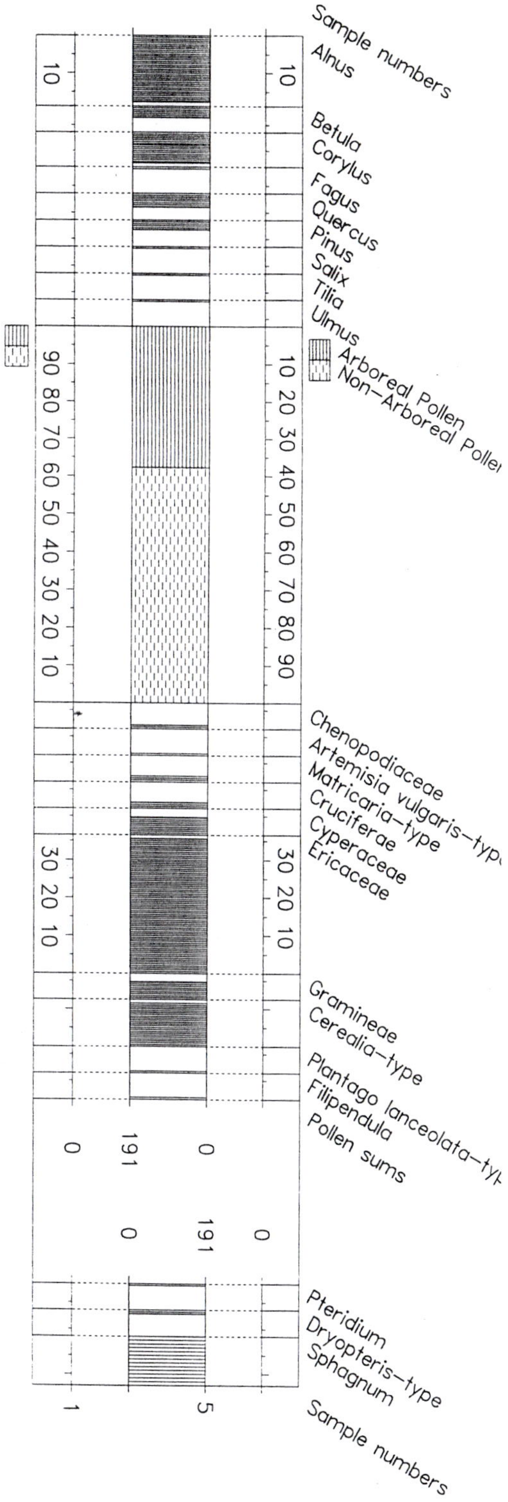

Interpretation

For the moss caulking of the Nijkerk-II cog (sample 17) a pure mix of moss has been used. The condition of the investigated sample was bad. The sample was very corroded and contained few pollen.
Trees 38%, herbaceous plants 62%
Alder 18%, hazel 8%, birch 3%, oak 4%, beech, willow, lime and elm 0,5%.
Heather 36%, cereals 11%, grasses en cyperus grasses 5%, arable weeds 4%, e.g. mugwort, *Cruciferae* and narrowleaf plantain.
Local: peat moss 120%, bracken and dryopteris ferns 1%.
Indetermined: 0%.
Oligotropic conditions. A large peatland, adjacent to reclaimed land, heathland and trees of many types growing in an open landscape.

Examination of the bones

F. Laarman

Introduction

This study covered the bone-material that has been found during the excavation of a number of cogs in the IJsselmeerpolders (Zuidelijk Flevoland: lot Zo 36 [chapter 6], lot Zo 43 [chapter 8], lot Zn 42 II [chapter 14], lot Zn 43 [chapter 15], lot Zw 13 [chapter 16]. Oostelijk Flevoland: lot On 5 [chapter 11], lot Og 77 [chapter 18] and Noordoostpolder: lot Na 57 [chapter 10]). A complete sheep has been found and also 406 bones from mammals, 17 bones from birds and 116 bones from fishes with a total weight of roughly 10.5 kg. The preservation of the bones varies from good to excellent.
If we compare the bones with the four classes for a morphologic assessment for the quality of bone[1], then all the bones belong to class 1. The bones are determined with the help of the reference collection of the RCE. In this study attention has been given to traces of chopping, cutting, bite marks and burning as well as traces of rounding. The description/determination of the bones and the results of this study are presented per cog in the appendices.

Expectation pattern

The bones expected aboard cogs mainly sailing between Hanseatic towns, would be from pieces of meat ready for consumption during the voyage and in addition from complete animals (pets and/or cargo/living stock). Luxurious food is not to be expected, instead less expensive pieces of ox, sheep/goat and pig and possibly fowl or fish.
In this study we assume that the remains of the bones represent

1. Huisman *et al.*, 2006.

the meat provisions, unless we have reason to assume another conclusion, as the refuse from meals was thrown overboard. It is difficult to distinguish the bones from sheep and goat (except for the skull), so they have been put into a single category called sheep/goat.

Conclusion

In this study the bones from eight cog wrecks only partly confirmed the expected pattern of meat ready for consumption expected on board ship. Especially in the bigger complexes (the Nijkerk II cog, chapter 6 and the Rutten cog, chapter 10), other bones also had drifted into the ship. We can conclude that a wreck is not always a closed find complex, a conclusion already evident in studies of ship wrecks from the Dutch Waddenzee[2], as ship wrecks often protrude above the seabed for a long time or become partly visible due to the current. Bones, sherds and other debris from elsewhere may be washed onto the irregular seabed above the wreck by the current and may be found later in the upper layers of the disturbed soil.

Results Nijkerk-II cog

During the study of the remains of the Nijkerk-II cog, 260 bones were collected. The researcher classified the bones into seven categories, based on the find circumstances. The results of the study of animal species are presented per find category as seen in the tables 1 and 2.

Both table 1 and 2 show that category 4 is the largest and contains a lot of different species, many more than one would expect as food on a ship and also quite unexpected ones as a ration on a cog, such as horse, red deer, black-throated diver and crane. The bones in the categories 2, 3, 4 and 6 also contain bones of pets (dog/cat). When a cog perishes with a pet on board, one would expect a more or less complete skeleton. However, this is not the case here. The diversity of category 4 and the presence of dogs and cats indicate that the bones from the cog must have become mixed with material from a settlement. In table 3 we see some parameters of the bone-study per category.

The high percentage of rounded material in category 2 of table 3 is remarkable and also that rounded material is present in each category. Chop marks are relatively dominant in category 1 and are the lowest in category 7. This is the reason why the average weight in category 7 is the largest by far. This is not an indication for meat ready for consumption, as would be expected on board ship. Tabel 4 gives an overview of the position of cattle bones in the skeleton.

The lower legs, and to a lesser extent, the head skeleton are considered offal. One does not expect offal on a ship, unless the cargo consisted of offal. In this case, the cattle bones of category 1 may belong to the ship. However, that is definitely not the case for the cattle bones of category 2, because this is exclusively offal. In the other categories one can see a mix of consumption and offal remains.

To summarise, the bones of category 1 may have been on board of the ship and the bones from the categories 2 and 7 cannot have been. The bones from the other categories are probably a mix of bones from the ship and bones from a settlement that had been washed into the wreck.

Table 1. Number of bones per animal species per find category

	Category 1	Category 2	Category 3	Category 4	Category 5	Category 6	Category 7
cattle	5	6	10	47	7	22	27
pig	-	-	3	14	3	5	6
sheep/goat	2	-	1	23	-	9	-
horse	-	1	-	4	1	5	9
red deer	-	-	-	-	1	-	-
dog	-	1	-	2	-	5	-
cat	-	-	1	8	-	4	-
large mammal	-	-	2	4	-	3	-
medium mammal	-	-	-	5	-	1	-
small mammal	-	-	-	1	-	-	-
mammal	-	-	-	2	-	-	-
chicken	-	-	1	1	-	-	-
duck	-	-	-	2	-	1	-
goose	-	-	-	2	-	-	-
black-throated diver	-	-	-	1	-	-	-
crane	-	-	-	1	-	-	-
bird indet.	-	-	-	1	-	-	-

2. Laarman & Lauwerier, 2006.

Table 2. Weight in grams of the bones per animal species per find category

	Category 1	Category 2	Category 3	Category 4	Category 5	Category 6	Category 7
cattle	123.7	122.0	89.9	737.9	266.7	422.1	1967.8
pig	-	-	20.5	139.1	84.6	44.2	220.9
sheep/goat	6.6	-	1.9	136.7	-	35.8	-
horse	-	20.7	-	373.4	49.9	355.4	1444.5
red deer	-	-	-	-	37.4	-	-
dog	-	39.2	-	4.6	-	24.4	-
cat	-	-	3.7	16.2	-	14.8	-
large mammal	-	-	7.4	23.1	-	14.3	-
medium mammal	-	-	-	16.5	-	3.1	-
small mammal	-	-	-	1.7	-	-	-
mammal	-	-	-	3.6	-	-	-
chicken	-	-	1.3	3.6	-	-	-
duck	-	-	-	1.4	-	0.6	-
goose	-	-	-	5.2	-	-	-
black-throated diver	-	-	-	3.7	-	-	-
crane	-	-	-	7.8	-	-	-
bird indet.	-	-	-	2.1	-	-	-

Table 3. Some bone-data per find category

	N bones	rounded (%)	chop marks (%)	average weight
Category 1	7	43	71	18.6
Category 2	8	75	13	22.7
Category 3	18	33	39	6.9
Category 4	118	36	16	12.5
Category 5	12	50	17	36.6
Category 6	55	38	13	16.6
Category 7	42	52	5	86.5

Table 4. The number of cattle bones per element category per find category

	head skeleton	axial skeleton	front leg	hind leg	lower legs
Category 1	-	4	1	-	-
Category 2	3	-	-	-	3
Category 3	-	4	4	-	2
Category 4	13	14	5	4	11
Category 5	2	3	-	1	1
Category 6	1	4	7	4	6
Category 7	9	6	5	3	4

r	- right	PI	- phalanx 1
l	- left	PII	- phalanx 2
p	- proximal	PIII	- phalanx 3
d	- distal	Gl	- greatest length
ve	- fused	Bp	- width proximal
onv	- non fused	Bd	- width distal
juv	- juvenile	mnd.	- months

List of determinations of the bones from the Nijkerk-II cog

invent. no Zo36-	description	weight in grams
231	horse 1x *tibia* (l) d ve Bd = 67 mm, lightly rounded	332.6
232	horse 1x *tibia* (r) d onv, rounded	2646
233	cattle 1x *radius* (l) p ve Bp = 60 mm, rounded	88.4
234	horse 1x *tibia* (l) d onv, rounded (belongs to 232)	134.6
235	horse 1x *metacarpus* (l) ve Gl = 222 mm, rounded	170.6
236	cattle 1x *metacarpus* (l) ve Gl = 179 mm	136.2
237	cattle 1x *metatarsus* (r)	84.2
238	cattle 1x *metatarsus* (l) juv, rounded	36.3
239A	cattle 1x *femur* (l), rounded	58.6
239B	cattle 1x tibia (r) d onv (traces of burning)	25.2
240	horse 1x *astragalus* (r), rounded (fossil appearance)	69.2
241	cattle 1x *humerus* (l) p+d ve, rounded	298.7
242A	cattle 1x *humerus* (l) d ve, rounded	240.8
242B	cattle 1x *humerus* p ve	32.1
243	cattle 1x *humerus* (l) d ve, rounded	150.2
244	cattle 1x rib weathered, rounded	35.0
245	cattle 1x rib, rounded	24.9
246A	cattle 1x rib, (chop mark), lightly rounded	18.1
246B	sheep/goat 1x rib, (chop mark), lightly rounded	6.0
247A	cattle 1x rib, (chop mark)	12.9
247B	sheep/goat 1x rib	0.6
248	cattle 1x rib, (chop mark)	14.3
249	cattle 1x rib (chop mark)	39.0
250	cattle 1x *metacarpus*, rounded	12.1
251	pig 1x *scapula* (r) ve	51.4
252	horse 1x *scapula* (l), rounded	201.6
253	pig 1x *scapula* (l) onv	16.1
254	pig 1x *scapula* (l) (chop mark) wild boar?	25.0
255	cattle 1x *pelvis* (r)	63.7
256	horse 1x *calcaneum* (l) (traces of bite marks), rounded	50.1
257	horse 1x *ulna* (l), rounded	41.1
258	cattle 1x *ulna* (r) p onv, rounded	47.6
259	cattle 1x *thoracic vertebra* (chop mark traces on or traces of bite marks dog?)	30.8
260	cattle 1x halswervel onv, rounded	24.6
261	cattle 1x *lumbar vertebra* onv, rounded (hole in protuberance)	54.5
262	horse 1x turner ve, rounded	180.1
263	cattle 1x *sacrum* weathered	54.1
264	horse 1x *pelvis* (r) (fossil appearance, strangely warped)	252.4
265	cattle 1x *cranium*	15.0
266	pig 1x *cranium* wild boar?	107.0
267	cattle 1x *cranium* (not smashed)	178.9
268A+B	cattle 1x *cranium* (not smashed) horn pits cut off	147.7
269	cattle 1x horn pit (r) outline base 102 mm, outside length 140 mm	66.4
270A	cattle 1x lower jaw tooth (looks older than the ship)	6.3
270B	cattle 1x fragment tooth (looks older than the ship, fossile?), rounded	8.0
271	cattle 1x *mandibula* (l) calf, rounded	39.0
272	pig 1x *canine* ♂	11.7
273	cattle 1x *mandibula* (r) incl. P4 M1	52.6
274	cattle 1x *mandibula* (r) incl. M3 (chop mark)	48.8
275	cattle 1x upper jaw tooth	12.9
276	pig 1x *cranium* (l)	9.7
277	cattle 1x *centrotarsale* (r) ve small model, rounded	23.3
278	horse 1x PIII small model, rounded	20.7
279	cattle 1x PII ve	9.3
280	cattle 1x PII ve small model, rounded	5.6
281	cattle 1x *cranium* (jaw joint) (traces of cutting), rounded	10.8
282	dog 1x *tibia* (r) G1 = 191 mm, fibula stuck -> old, rounded	39.2
283	cattle 1x *maxilla* (l), rounded	6.6
284-1	cattle 1x *ulna* (r) (chop mark), rounded	29.7
284-2	cat 1x *humerus* (l) (chop mark), lightly rounded	3.7
284-3	pig 1x *humerus* (r) traces of bite marks, rounded	10.9
284-4	chicken 1x *tibiotarsus* (l)	1.3
284-5	cattle 1x *thoracic vertebra* (protrusion) (chop mark)	14.4
284-6	cattle 1x *thoracic vertebra* (protrusion) weathered	10.6
284-7	pig 1x rib	4.7
284-8	sheep/goat 1x rib (chop mark)	1.9
284-9	cattle 1x rib (chop mark)	2.5
284-10	cattle 1x rib (chop mark)	9.5
284-11	cattle 1x *humerus*, lightly rounded	4.9
284-12	large mammal 1x pipe bone	2.6
284-13	cattle 1x *tarsale*	3.1
284-14	cattle 1x *scapula*, lightly rounded	5.5
284-15	cattle 1x *scapula*	3.3
284-16	large mammal 1x pipe bone	4.8
284-17	pig 1x *pelvis* (r) (chop mark)	4.9
284-18	cattle 1x PIII weathered, lightly rounded	6.4
285-1	horse 1x *femur* (l) juv, lightly rounded	81.2
285-2	cattle 1x *femur* (l) weathered	65.6
285-3	paard 1x *humerus* (r) weathered, lightly rounded	68.3
285-4	cattle 1x *tibia* (l) juv, lightly rounded	23.6
285-5	horse 1x PI ve Gl = 75 mm, rounded	44.2
285-6	cattle 1x *humerus* (r) d ve (chop mark), rounded	77.4
285-7	cattle 1x *astragalus* (l) (chop mark)	26.2
285-8	cattle 1x *atlas*, rounded	24.5
285-9	cattle 1x turner weathered, rounded	23.3
285-10	cattle 1x *calcaneum* juv	27.2
285-11	sheep/goat 1x *scapula* (l) (traces of cutting)	33.4
285-12	horse 1x *pelvis* (l), rounded	179.7
285-13	pig 1x *pelvis* (r) (chop mark), lightly rounded	20.3
285-14	pig 1x rib (chop mark)	8.8
285-15	cattle 1x rib (chop mark)	11.8
285-16	cattle 1x rib weathered	6.9
285-17	cattle 1x rib (chop mark)	15.0
285-18	cattle 1x *lumbar vertebra* (chop mark) weathered	15.7
285-19	pig 1x *lumbar vertebra* onv	55.1
285-20	sheep/goat 1x upper jaw tooth adult	27.6
285-21	cattle 1x *cranium*, lightly rounded	29.3

285-22	cattle 1x *maxilla* (r) (fits on 24A)	32.5
285-23	d 1x *cranium*	10.9
285-24A	cattle 1x *maxilla* (r) (fits on 22)	69.9
285-24B	cattle 1x upper jaw tooth adult	27.6
285-25	cattle 1x *cranium*	7.8
285-26	cattle 1x *scapula* (r), lightly rounded	17.2
285-27	cattle 1x *cranium*	8.4
285-28	sheep/goat 1x rib	7.1
285-29	dog 1x rib	3.4
285-30	dog 1x rib	1.2
285-31	small mammal 1x rib	1.7
285-32	goose 1x *furcula*	1.1
285-33	sheep/goat 1x rib	1.5
285-34	sheep/goat 1x rib	1.9
285-35	cattle 1x rib (chop mark)	11.8
285-36	pig 1x rib, rounded	4.8
285-37	sheep/goat 1x rib, rounded	5.7
285-38	medium mammal 1x *thoracic vertebra* (protrusion)	3.4
285-39	cattle 1x rib (chop mark)	6.7
285-40	sheep/goat 1x rib	2.6
285-41	sheep/goat 1x rib, rounded	2.3
285-42	cattle 1x rib, rounded	6.9
285-43	cattle 1x rib, rounded	2.2
285-44	cattle 1x *hyoïd*	1.5
285-45	sheep/goat 1x *scapula* (chop mark)	2.8
285-46	sheep/goat 1x rib (chop mark)	1.0
285-47	goose 1x *tarsometatarsus* (r), rounded	4.1
285-48A	pig 1x *metapode* onv. rounded	5.6
285-48B	pig 1x *metapode* onv.	6.3
285-48C	sheep/goat 1x PI onv.	1.0
285-49	duck 1x *carpometacarpus* (r)	0.6
285-50+51	sheep/goat 1x *metatarsus ve*	9.1
285-52	sheep/goat 1x *metapode* ve	1.1
285-53	cat 1x *fibula*	0.9
285-54	black-throated diver 1x *tibiotarsus* (r), rounded	3.7
285-55	cat 1x *femur* (l) Gl = 92 mm, rounded	3.9
285-56	cat 1x *humerus* (l), rounded	4.5
285-57	bird indet. 1x pipe bone	2.1
285-58	chicken 1x *femur* (l) Gl = 73 mm, rounded	3.6
285-59	cat 1x *tibia* (l) juv	1.3
285-60	duck 1x *ulna* (l) Gl = 64 mm	0.8
285-61	cat 1x *tibia* juv, rounded	0.8
285-62	cat 1x *femur* (l) juv, rounded	1.2
285-63	crane 1x *humerus* (r), rounded	7.8
285-64	medium mammal 1x pipe bone weathered, rounded	4,8
285-65	sheep/goat 1x *tibia* (l), rounded	6,5
285-66	medium mammal 1x pipe bone	3.9
285-67	large mammal 1x pipe bone	3.8
285-68	large mammal 1x pipe bone	2.5
285-69	medium mammal 1x pipe bone	1.8
285-70	cattle 1x PII ve	9.6
285-71	cattle 1x PII ve weathered, rounded	6.3
285-72	sheep/goat 1x *radius* (l) p ve (chop mark)	1.3
285-73	sheep/goat 1x PII ve	0.7
285-74	sheep/goat 1x PI ve, rounded	1.8
285-75	pig 1x PI ve, rounded	4.1
285-76	sheep/goat 1x PI ve, rounded	0.8
285-77	pig 1x PI onv, rounded	2.0
285-78	cattle 1x PIII, rounded	6.2
285-79	cattle 1x PIII	10.6
285-80	cattle 1x PIII weathered	8.5
285-81	cattle 1x *centrotarsale* (l) (past bij 108)	2.9
285-82	pig 1x *scapula* (r), rounded	13.0
285-83	cattle 1x *ulna* (r) (chop mark)	5.8
285-84	cattle 1x *centrotarsale* (r) muis! (maatje van 81)	11.9
285-85	cattle 1x *cranium* weathered	16.6
285-86	sheep/goat 1x halswervel juv	7.4
285-87	sheep/goat 1x *lumbar vertebra*	6.3
285-88	cat 1x *thoracic vertebra* juv	0.5
285-89	cattle 1x *thoracic vertebra*, rounded	10.3
285-90	cattle 1x *thoracic vertebra* (chop mark)	14.2
285-91	cattle 1x *cranium* (l) (oogboog)	9.2
285-92	sheep/goat 1x *cranium* (l)	5.3
285-93	cat 1x *pelvis* (l), rounded	3.1
285-94	cattle 1x *pelvis* (r)	9.8
285-95	sheep/goat 1x *calcaneum* (r) juv	1.9
285-96	cattle 1x *tibia* (r), rounded	9.3
285-97	cattle 1x rib (chop mark)	8.9
285-98	pig 1x *mandibula* (r), rounded	9.2
285-99	medium mammal 1x pipe bone juv	2.6
285-100	cattle 1x *ulna*, lightly rounded	14.0
285-101	large mammal 1x pipe bone, rounded	6.4
285-102	cattle 1x *scapula*	4,6
285-103	mammal 1x *scapula*	2.1
285-104	mammal 1x indet., rounded	1.5
285-105	cattle 1x *cranium*	3.7
285-106	cattle 1x horn pit	5.0
285-107	cattle 1x halswervel (chop mark)	6.4
285-108	cattle 1x *tarsale* (past bij 81)	3.5
285-109	pig 1x *pelvis* (r) (chop mark)	2.7
285-110	large mammal 1x pipe bone, rounded	10.0
285-111	cattle 1x *mandibula*	6.0
285-112	pig 1x *radius* (r) d onv, rounded	3.4
285-113	sheep/goat 1x *femur* p onv	1.8
285-114	pig 1x *femur* p onv (chop mark)	2.2
285-115	cattle 1x *carpale* weathered (chop mark)	4.7
285-116	pig 1x *vertebra* (disc) onv	1.6
286-1	pig 1x *humerus* (r) d ve, rounded	39.1
286-2	pig 1x *humerus* (r) d ve, rounded	30.8
286-3	cattle 1x *astragalus* (l) (chop mark), rounded	40.0
286-4	cattle 1x *tibia* (l), rounded	75.7
286-5	horse 1x *tibia* (l) weathered	49.9
286-6	pig 1x *pelvis* (r) (chop mark)	14.7
286-7	cattle 1x *cranium*	9.8
286-8	cattle 1x *cranium*	3.8
286-9	red deer 1x rib	37.4
286-10	cattle 1x rib, rounded	48.5
286-11	cattle 1x turner, rounded	28.4
286-12	cattle 1x *lumbar vertebra* onv traces of bite marks	60.5
287-1	horse 1x *ulna* (r) weathered, rounded	73.0
287-2	cattle 1x *sacrum* weathered, rounded	27.4

287-3	horse 1x rib, rounded	30.7
287-4	cattle 1x rib (chop mark)	10.5
287-5	dog 1x rib	2.1
287-6	sheep/goat 1x rib	3.0
287-7	sheep/goat 1x rib (chop mark)	3.8
287-8	sheep/goat 1x rib (chop mark)	5.2
287-9	dog 1x rib	1.7
287-10	sheep/goat 1x rib (chop mark)	1.6
287-11	sheep/goat 1x rib (chop mark)	1.2
287-12	cattle 1x *scapula* (r) (chop mark)	42.7
287-13	cat 1x *scapula* (r)	1.7
287-14	cat 1x *humerus* (l) d ve, rounded	5.2
287-15	dog 1x *radius* (r) p ve, lightly rounded	13.4
287-16	dog 1x *metapode* ve	3.3
287-17	horse 1x *humerus* (l) weathered, rounded	151.2
287-18	horse 1x *femur* (l) p onv weathered, rounded	23.4
287-19	pig 1x *humerus* (l)	14.8
287-20	pig 1x *metapode* d onv, rounded	6.2
287-21	cat 1x *tibia* (r) p ve	6.3
287-22	cattle 1x *tibia* (r) onv, rounded	28.4
287-23	cattle 1x *radius* (r) d ve	21.0
287-24	cattle 1x *radius* (l) d ve, rounded	27.9
287-25	cattle 1x *radius* (r) d onv	26.4
287-26+28	cattle 1x *femur* juv	16.0
287-27	cattle 1x *tibia*	31.7
287-29	medium mammal 1x pipe bone	3.1

287-30	sheep/goat 1x *metatarsus*	6.4
287-31	cattle 1x *astragalus* (r), rounded	22.6
287-32	cattle 1x *carpale* (r)	4.4
287-33	pig 1x *astragalus* (r) (chop mark)	5.2
287-34	sheep/goat 1x *astragalus* (r) weathered and rounded-> game?	3.6
287-35	dog 1x *metapode* ve	3.9
287-36	horse 1x PI + PII ve pathologically fused, rounded	77.1
287-37	pig 1x PI onv	3.2
287-38	cattle 1x PIII, rounded	5.2
287-39	cattle 1x PIII	9.6
287-40	cattle 1x PIII	7.0
287-41	pig 1x *atlas*, rounded	14.8
287-42	cattle 1x *thoracic vertebra* ve, rounded	41.4
287-43	cattle 1x *thoracic vertebra* onv (chop mark), rounded	34.5
287-44	cattle 1x *scapula*	8.0
287-45	cattle 1x *pelvis* (l) juv bite marks hond, rounded	17.4
287-46	sheep/goat 1x *pelvis* (r) juv	4.4
287-47	cat 1x *pelvis* (r), rounded (does not fit on 285-93!)	1.6
287-48	cattle 1x *radius* (l) p ve damaged (hole) weathered, rounded	18.9
287-49	cattle 1x horn pit	9.7
287-50	sheep/goat 1x horn pit	6.6
287-51	large mammal 1x pipe bone	7.3
287-52	large mammal 1x pipe bone	2.5
287-53	cattle 1x *carpale* (r), afgerond	8.1
287-54	cattle 1x *scapula*	3.3
287-55	large mammal 1x pipe bone	4.5
287-56	duck 1x *coracoïd* (r)	0.6

Dendrochronological report RING
E. Jansma & A.E.M. Hanraets

ring

Stichting Nederlands Centrum voor Dendrochronologie
Kerkstraat 1, 3811 CV Amersfoort
Tel (033) 63 42 33
Fax (033) 65 32 35
E-mail: ring @ archis.nl

RAPPORTAGE

Object/vindplaats: Ketelhaven, kavel: OZ-36 (kogge).

Datum monstername :
Datum aanvraag : december 1993

Aanvrager:
Naam : R. Oosting
Instantie : Museum voor Scheepsarcheologie
Stad : Lelystad

Onderzoekers:
Naam : E. Jansma, E. Hanraets
Instantie : RING
Datum onderzoek : maart 1994

Vondstno./object	**Dendronaam**	**kern**	**spint**	**wankant**	**n**
WD/SB, m.7	kko070	+(5-10)	-	-	68
OO, m. 10	kko100	+2	25	-	77
CL oplanger, m.11	kko110	+ ±10	-	-	93
J oplanger, m.12	kko120	+2	-	-	92
WC2/SB, m.13	kko130	+2	-	-	61
WB/SB, m.14	kko140	-	24	-	135
M-15, m.15	kko151	-	19	-	79
N oplanger, m.16	kko160	-	-	-	110
M wrang, m.17	kko171	+ ±4	26	-	86
QQ spant, m.18	kko180	+5	24	-	150
CT oplanger, m.19	kko190	+3	19	-	90

kern = (afstand tot) de eerstgevormde (oudste) jaarring in de stam
spint = aantal ringen spinthout
wankant = laatstgevormde jaarring (direkt onder bast), nodig voor een absolute datering van de veldatum
n = totaal aantal jaarringen in houtmonster

DATERING
Geen van de monsters kon afzonderlijk gedateerd worden met een van de ons ter beschikking staande referentiechronologieen. Alle monsters behalve m. 12 en 13 konden echter onderling gedateerd worden en worden gemiddeld tot een objectcurve, <u>kkostd</u>. Deze werd vervolgens absoluut gedateerd:

Dendronaam	Referentie-Chronologie	Datering	%PV	t	P
kkostd	Apeldoorn	1173-1334AD	68.7	5.53	0.001
kkostd	O. Nederland/ Westfalen	1173-1334AD	61.8	4.91	0.01

Toelichting op de statistische variabelen:

%PV = "Gleichlaufigkeit" (Duitse term) of "Percentage Agreement" (Engelse term); het percentage van de ringen in het onderzochte jaarringpatroon die aan de referentiechronologie identieke toe- en afnames van de breedte vertonen op de door de datering van het patroon aangegeven positie t.o.v. de referentiechronologie. De significantie van dit percentage is een functie van de lengte in jaren van het onderzochte jaarringpatroon (zie bijgevoegd documentatiemateriaal).

t = De waarde die resulteert uit een Student_ t-test op de kruiscorrelatie die behoort bij de beste "match" tussen het onderzochte jaarringpatroon en de referentiechronologie.

P = Het significantieniveau behorend bij de gevonden waarde van gl. Wanneer dit niveau bijvoorbeeld 0.001 is, betekent dit dat de datering juist is met een waarschijnlijkheid van 99.9%

De datering van de afzonderlijke monsters t.o.v. de referentiechronologieen is:

Dendronaam	Referentie-chronologie	Datering
kko070		1238-1305AD
kko100		1258-1334AD
kko110		1220-1312AD
kko120		niet dateerbaar
kko130		niet dateerbaar
kko140		1199-1333AD
kko151		1247-1325AD
kko160		1202-1311AD
kko171		1233-1318AD
kko180		1173-1322AD
kko190		1236-1325AD

INTERPRETATIE VAN DE DATERINGEN

Object	Einddatering	veldatum
OZ-36 (kogge)	1334 + 1 (de laatste ring wordt nooit gemeten)	1335/1336 AD

Gezien de positie van de spintgrens (zie bijgevoegde figuur) is het onwaarschijnlijk, dat de houtmonsters afkomstig zijn uit gelijktijdig gevelde bomen. In volgorde van ouderdom schatten we de veldatum van de houtmonsters die spinthout bevatten, en uitgaande van een gemiddeld aantal spintringen van 27 ± 7, als volgt:

Dendronaam	**Einddatering**	**veldatum**
kko170	1319	1319-1326 AD
kko180	1323	1323-1331 AD
kko190	1326	1326-1339 AD
kko150	1326	1326-1339 AD
kko140	1334	1334-1342 AD
kko100	1335	1335-1342 AD

Hoewel deze schattingen ten dele overlappen, is duidelijk dat de veldatum van kko170 niet kan samenvallen met die van kko100 en kk0140, en waarschijnlijk evenmin samenvalt met die van kko150 en kko190, hetgeen betekent dat we hier met ten minste twee 'bouwfasen', of tenminste 1 reparatie, te maken hebben, en misschien zelfs met meer.

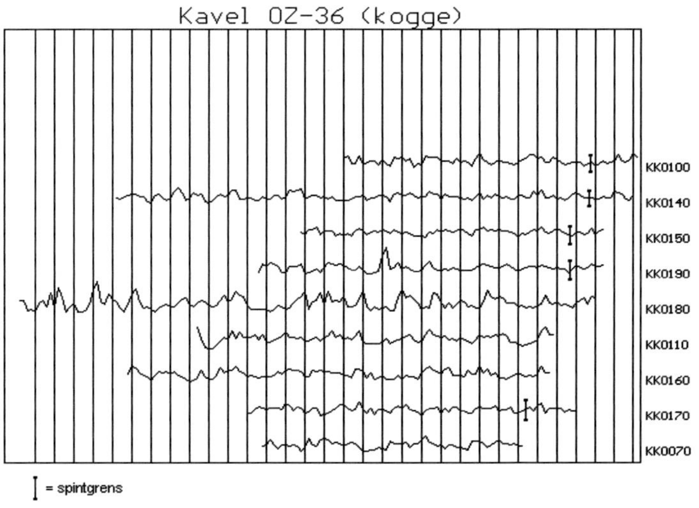

---

8 The Nijkerk-I cog

Examination of the bones

F. Laarman

For the introduction, the pattern of expectations and the conclusion of the study of bones from eight cogs see appendix. of the Nijkerk-II cog.

Results Nijkerk-I cog

During the investigation of the Nijkerk-I cog, 24 bones are counted. Table I shows the number of bones and the weight in grams.

With the exception of the *cranium* of cattle (41) and a *molar* from the upper jaw of a sheep/goat (62) the bone collection meets the expectations of bones found aboard a ship. They belong to meat ready for consumption with a lot of chop marks. Five out of seven cattle bones show chop marks, as do ten out of fourteen sheep/goat bones. There were no bones at all that were rounded. The weight of the bones, without the numbers 41 and 62 is 273 grams, which represents 3,5 kg meat by a conversion factor. If we assume that a person eats 250 grams meat per day, this amount is enough for four persons for 3,5 days.

Table 1. Number of bones per animal species and weight

	N	Grams
cattle	8	443.3
sheep/goat	15	82.6
medium mammal	1	2.2

r	- right	PI	- phalanx 1
l	- left	PII	- phalanx 2
p	- proximal	PIII	- phalanx 3
d	- distal	Gl	- greatest length
ve	- fused	Bp	- width proximal
onv	- non fused	Bd	- width distal
juv	- juvenile	mnd.	- months

List of determinations of the bones from the Nijkerk-I cog

Invent. no. Zo 43-	description	weight in grams
41	cattle 1x *cranium* + horn pit (r) outline base = 130 mm	255.5
42	cattle 1x *atlas* (chop marks)	27.6
43	cattle 1x *pelvis* (l) (chop marks)	62.2
44A+B	sheep/goat 1x *femur* (l) d onv	19.3
45A	sheep/goat 1x *humerus* (l) d ve	7.7
45B	sheep/goat 1x *radius* (r) p ve	11.5
46	sheep/goat 1x *pelvis* (l) (chop marks)	7.7
47	cattle 1x rib	5.4
48	cattle 1x rib (chop marks)	11.4
49	sheep/goat 1x rib (chop marks)	4.1
50	cattle 1x rib (chop marks)	19.8
51	sheep/goat 1x rib (chop marks)	4.0
52	sheep/goat 1x rib (chop marks)	5.7
53	sheep/goat 1x rib (chop marks)	4.2
54	cattle 1x rib (chop marks)	15.2
55	sheep/goat 1x rib (chop marks)	3.5
56	sheep/goat 1x rib (chop marks)	2.0
57	sheep/goat 1x point of *thoracic vertebra* (chop marks)	1.8
58	sheep/goat 1x rib (chop marks)	1.6
59	sheep/goat 1x rib	1.4
60	medium mammal 1x *scapula*	2.2
61	sheep/goat 1x point of *thoracic vertebra*	2.1
62	sheep/goat 1x *molar*	6.5
63A+B	cattle 1x *radius* (r)	46.2

Acta Bot. Neerl. **38**(4), December 1989, p. 477–491

Alder carr, growth and drowning in the IJsselmeer region, an aspect of the Dutch coastal development

D. G. VAN SMEERDIJK

Hugo de Vries-Laboratory, University of Amsterdam, Kruislaan 318, 1098 SM Amsterdam, The Netherlands

SUMMARY

A peat layer and the underlying mineral soil in a section of the Zuidelijke Flevopolder in the central Netherlands has been studied for pollen analysis. The pollen spectra from the mineral soil reflect a Late-Atlantic forest development under relatively dry soil conditions. Podzolizing has started, but as a consequence of a rise of the groundwater table the soil became wetter and an alder carr could develop. A further rise of the water table is thought to be responsible for the decline of the alder carr after 3780 ± 60 BP. Eutrophic peat growth at site OZ-43 is synchronous with the deposits belonging to the 'Cardium transgression' phase. The development in the IJsselmeer region was strongly correlated with the coastal development of Noord-Holland. The oldest 'Cardium deposits' are thought to be transported via the Oer-IJ and the youngest via tidal channels in the north near Hoorn. The clay sediments covering the peat belong to the Zuiderzee deposits and indicate a hiatus between the peat and the clay deposits of at least 3500 calendar years.

Key-words: alder carr, Cardium transgression, coastal development, palaeoecology, peat, radiocarbon dating.

INTRODUCTION

In the autumn of 1980 the wreck of a ship was recorded on parcel OZ-43 in the polder of Zuidelijk Flevoland. The ship had fallen into pieces, and lay on top of the Pleistocene sandy subsoil; its deepest parts had sunk into the sand. At some distance from the wreck there was a natural peat layer of *c.* 30 cm thick covering the sand, so that the wreck must have sunk through the peat into the Pleistocene sand. The wreck was covered with disturbed peat and clay material and overlain by humic clay. At the start of the excavation of the wreck the age of the ship was unknown, and neither was the age of the peat layer. Presumably there is a relationship between the age of the ship and the age of the peat, and in this case the peat deposit would be rather young. An alternative assumption is that the peat dates from much older times. In that case it can be of great value for dating and correlating other sediments in the IJsselmeer region, because most of the former peat deposits in the area have been eroded away.

This paper is dedicated to Professor Dr T. van der Hammen on the occasion of his 65th birthday.

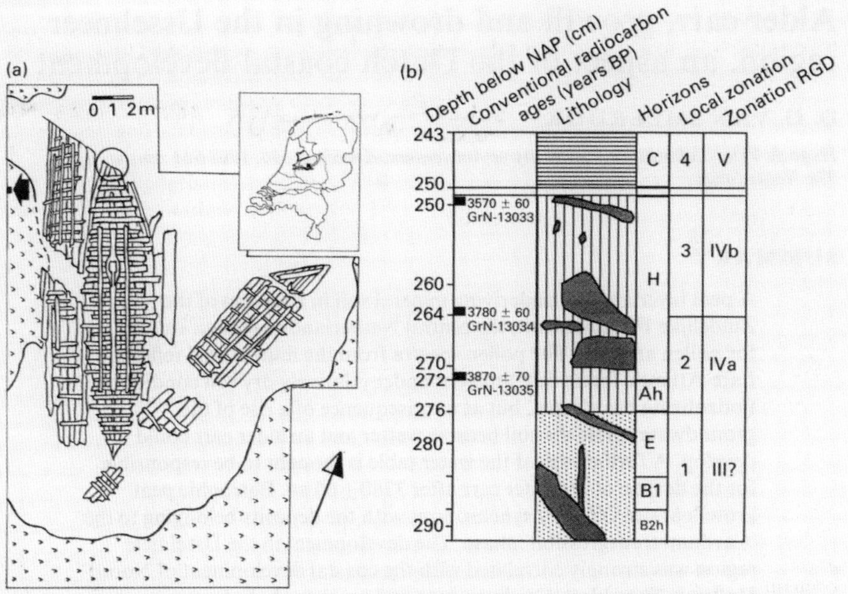

Fig. 1. (a) Excavation pit of the medieval shipwreck at parcel OZ-43 with the limits of the peat *in situ* and location of the section OZ-43. (○) sand samples, (▨) peat *in situ*, (←) location of section OZ-43. (b) Lithology, zonation, pedogenic horizons and conventional ^{14}C ages. (▤) Clay, (▦) sand, (■) peat and (▩) wood.

MATERIAL AND METHODS

Sampling had been done in 1981 during an excavation of a wreck, dating from the 13th or 14th century (Vlierman 1985), on parcel OZ-43 in the polder of Zuidelijk Flevoland, the Netherlands (Fig. 1a). The excavation was carried out under the supervision of the IJsselmeerpolders Development Authority (Department of Nautical Archaeology) in Ketelhaven. The samples were collected by means of a metal box, 50 × 10 × 15 cm, which was pushed into the exposed W profile of the excavation pit (on 27th May 1981) and subsequently cut out. From each centimetre, except from the top and bottom one, subsamples were taken for micro- and macrofossil analyses. From a depth of 250 cm, the clay and peat were collected separately. The exact depth of the section with regard to the Dutch ordnance Data (NAP) could not be determined, and is taken as accurate to within 15 cm (Koopstra, 1988 personal communication). The top of the Pleistocene sand was podzolized, and a description of the pedogenic horizons follows (Fig. 1b):

I C: humic grey clay with many shell fragments; 244–250 cm below NAP;

II H (histic horizon): brown peat with wood remains; 250–272 cm below NAP;

III Ah (histic A horizon): strongly decomposed black peat, with sand; 273–276 cm below NAP;

III E (= A2): humic blackish sand; 277–280 cm below NAP; dark grey sand; 281–284 cm below NAP;

III B1: grey sand; 285–288 cm below NAP; and

III B2h: brownish sand: 289–293 cm below NAP.

Pollen and other microfossils

Subsamples of known volumes for pollen analyses were treated according to the standard method of Erdmann (Faegri & Iversen 1975). The clay- and sand-containing subsamples were also treated with bromoform (sp.gr. 2·0). Subsequently the residues were treated according to the method of Middeldorp & Mijzen (1986). This is a new method for establishing pollen concentrations by embedding the total amount of microfossils of the subsample in a film of polyvinyl alcohol. The relative contributions of the microfossils are expressed as percentages of the tree pollens ($\Sigma AP \geqslant 300$). The results are shown in Figs 2 and 3 and Table 1.

Macrofossils

Subsamples of *c.* 10 cm^3 were treated with a 5% KOH solution and rinsed with tap-water over a sieve with 177 μm pores. The results are given as percentages of the residue on the sieve or as numbers (Fig. 4; Table 1). At the site several litres of sand from the inside of the wreck (see Fig. 1) were sieved and yielded a few *Pinus* cones and a handful of *Corylus* nut fragments.

Diatom analysis

Diatom analyses of the upper 6 cm of the clay deposits and of the underlying 8 cm of the peat deposits have been carried out by J.A.E. Kooyman (method according to Van der Werff 1960). At least 100 diatoms per sample were counted. The results (Table 2) have been interpreted according to the methods used by the Rijks Geologische Dienst (De Wolf 1982). The peat samples did not contain any diatoms.

Organic carbon, nitrogen, hydrogen and C/N ratio

A selection of subsamples were used for the determination of total carbon and nitrogen and the C/N ratio. The untreated material was dried at 105°C for 24 h and ground. Aliquots of 2 mg were incinerated using a Perkin-Elmer 240 element analyser. The results are shown in Fig. 5. The amounts of total carbon and nitrogen are more or less constant in the main part of the peat (44·5–46·7% C; 2·4–2·6% N) and subsequently decrease to 38·3% and 1·7%, respectively, at a depth of 272 cm.

In the Ah horizon, where sand is incorporated in the peat, both values fall off sharply but then stay at a more or less constant level (10·6–15·4% C; 0·7–0·8% N). In the deeper horizons they decline further to a very low level (1·3% C; 0·2% N). The C/N-ratio does not fluctuate very much in either the peat or the mineral soil (18–22 in the peat and 13–21 in the mineral soil). The C/N values in the peat point to a high rate of decomposition. The values in the mineral soil point to moder, a characteristic of rather dry soils.

In all diagrams, 49 spectra have been shown, namely, 48 spectra corresponding with 48 cm of sediment, and an additional one extra corresponding with the clay from depth 250 cm. The space between the two samples from depth 250 cm represents a hiatus. The scale distances in the pollen and macrodiagrams represent 5 plot units (short bars) and 10 plot units (long bars).

Radiocarbon dating

Initially the age of the peat of the section OZ-43 was unclear. Palynological dating only yields a rough estimate of the age: Late-Atlantic/Subboreal. Radiocarbon dating, carried out by the Centre for Isotope Research of the University of Groningen, placed the peat growth in the middle of the Subboreal (Table 3).

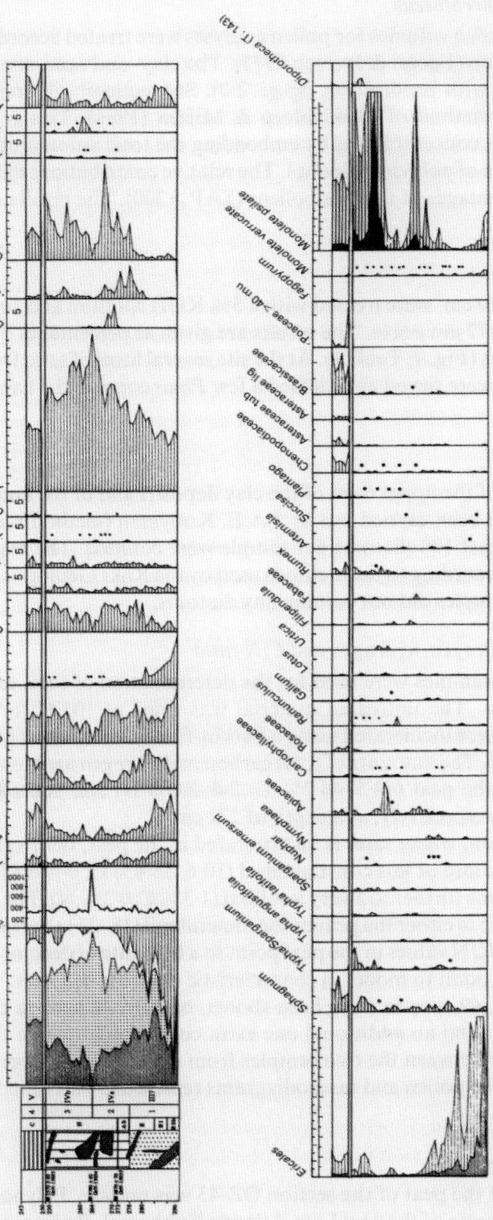

Fig. 2. Percentages diagram of pollen and spores, section OZ-43. (▦) Trees, excluding *Alnus* and *Salix*, (⊞) *Alnus* and *Salix*, (□) Poaceae and Cyperaceae, (▤) *Ericales*.

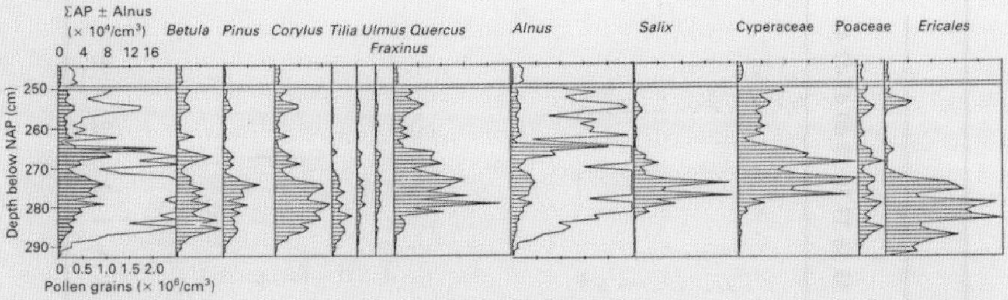

Fig. 3. Pollen concentration diagram, section OZ-43. The upper scale for ΣAP excluding *Alnus* (□), *Alnus* (□) and all other pollen types. The lower scale for ΣAP includes *Alnus* (▤) and *Alnus* (▤).

RESULTS AND DISCUSSION

Peat accumulation and pollen concentration

In this section the selection of samples for radiocarbon dating was primarily lithostratigraphical: the topmost peat sample and the first that does not contain sand from the subsoil. The third sample was taken at the point where the cumulative arboreal pollen curve plotted against depth shows a remarkable turn (Fig. 6a). The same trend, although less sharp, can be observed in the cumulative arboreal-*Alnus* pollen curve. In Fig. 6b the calibrated ^{14}C-dates are compared with the depth. On the basis of this relationship the peat accumulation rate is calculated. Figure 6c shows a slightly decreasing peat accumulation rate with decreasing depth, a mean of 0·068 cm year^{-1} between depths 272 and 264 cm, and of 0·047 cm year^{-1} between depths 264 and 250 cm.

The highest pollen concentrations (Fig. 3) were found in the top layer of the Pleistocene sand, and for peat in the samples from depths 276 to 272 cm (Ah horizon) and in the samples from depths 269 to 264 cm. In the last series these high pollen concentrations are caused by *in-situ Alnus* production (Brock *et al.* 1989). The former high concentrations point to favourable preserving conditions for the locally produced pollen, attributable to a low biological activity. This may also point to a rather high humidity of the soil, in particular of the Ah horizon where peat formation took place. In the Ah horizon, representing the initial phase of peat growth, the high pollen concentrations reflect a slow rate of peat accumulation. The decrease of the pollen concentrations in the peat from a depth of 272 cm (4302 cal. BP = 3870 BP) onward is primarily caused by the decline of the pollen production of all the forest elements. In the period from 4302 cal. BP to 4184 cal. BP (3780 BP), i.e. the depths 272 to 264 cm, the arboreal pollen production recovers somewhat, but after 4184 cal. BP a total decline of all forest elements takes place. This points to a gradual disappearance of the forests.

Reconstruction of the local vegetational succession

Zone 1: 292–277 cm below NAP (mineral soil). Pollen analyses of soil profiles have been subject to many discussions in the last 10 years (e.g. Havinga 1962; 1974; Dimbleby 1985; Gaudreau *et al.* 1989). Several processes, dependent on the local soil-forming conditions, are thought to have been responsible for the pollen content of mineral soils.

In the present study the mineral soil had a well-developed podzol profile. Pollen occurs to at least 16 cm below the top of the mineral soil and shows a characteristic distribution pattern similar to that of a peat deposit. However, it is generally accepted that once pollen

Table 1. Less frequently occurring micro- and macrofossils, restricted to water

Microfossil (m) or macrofossil (M)* Species	Sample number† 1	2	3	4	5	6	8	9	10	11	12	13	14	16	17	18	19	22	23	24	26	29	30	31	32	33	34	35	36	37	38	39	40	41	42	43
M *Mya arenarea*	++	++		+	+	+	++																													
M *Cardium glaucum*	1	1				1																														
M *Hydrobia ulvae*		2	1																																	
M Ostracoda	3		3	3	1	5																														
M *Balanus* sp.		1																																		
m *Pediastrum*	7	3	7	5	3	8																														
m *Botryococcus*	4	5	2	1	3	5																														
m Type 128	+	+		+		+		+		+																	+	+								
m *Spirogyra* (T.132)		+	+			2	+		+	2	2	+	+									+			+		+									
M *Chara*						25	48	7	3		4																									
M *Batrachium*								2																												
m *Spirogyra* (T.131)				+							+																									
m *Mougeotia*							+																	+												
M *Plumatella fungosa*						1						1																								
m *Spirogyra* (T.315)										+	+																									
m *Spirogyra* (T.130)											+																									
M *Daphnia pulex*-type (T.189C)											1				1			1		1																
M *Moina* (189B)												1						2																		
m Zygnemataceae														+	+	+	+						+				+	+				1				
M *Lophopus crystallinus* (T.192)																		2	1		1	1		1												
M *Cristatella mucedo* (T.390)																							2													
m *Debarya*																														+						
m Type 314		+			+	+																+	+	+	1	+	1	1	+	1	2	1	2	1	+	+

*Microfossils (m) given in percentages or presence (+), macrofossils (M) given in numbers or presence (+).
†Sample number 1 correlates with a depth of 245 cm, sample 49 correlates with 293 cm.

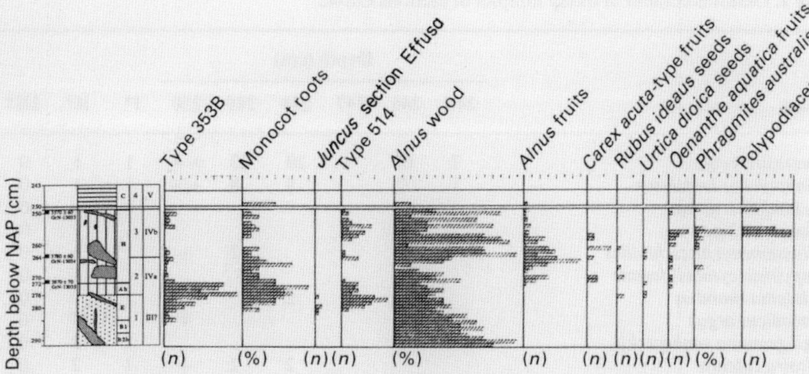

Fig. 4. Macrofossil diagram, section OZ-43. In number (*n*) or percentage (%).

has become incorporated in the deposit it moves neither upwards nor downwards. In mineral soils, on the other hand, the course of events responsible for incorporation of pollen in the soil are quite different from that in peat. Pollen diagrams from soils must therefore be interpreted in a different way.

Pollen in soils is not freely dispersed but is incorporated in fossil excrements of the soil fauna (modex). The infiltration of pollen into deeper layers of the soil is strongly related to the movement of modexial material (Van Mourik 1986). In podzols the absolute pollen frequencies are high in the topmost layer of the soil and decrease with depth. The quantities of pollen of the various taxa in the soil are not only related to the depth but also to the age. Each depth level may contain pollen of different ages.

In the mineral part of the soil several distribution patterns in the absolute pollen frequencies can be observed. (a) The pollen frequencies of *Pinus*, *Betula*, *Corylus*, *Tilia* and Poaceae increase from a depth of 292 to 284 cm and reach maximum values between 282 and 276 cm; *Corylus* is the most abundant. (b) The pollen frequencies of Ericales increase from 292 to 282 cm and reach maximum values between 282 and 278 cm and subsequently decrease rather rapidly. (c) The pollen frequencies of *Fraxinus* and *Ulmus* are low and increase very slightly to reach maximum values between 280 and 274 cm. (d) The pollen frequencies of the total arboreal pollen and of *Alnus* increase from 292 to 276 cm and attain very high values between 280 and 276 cm. (e) A sudden rise of the pollen frequencies of *Salix* and Cyperaceae from 282 cm onward. (f) The pollen frequencies of *Quercus* show low values until 282 cm and subsequently increase strongly.

It is evident that the spectra from the mineral soil are mixed. Conceivably *Quercus* is under-represented in the lowest 12 cm because of the rather high susceptibility of its pollen to corrosion (Havinga 1962; 1974; 1984). Havinga (1962) suggested that high *Tilia* pollen values may point to *in-situ* stands of *Tilia*. Keatinge (1982) has shown that a high concentration of *Tilia* pollen is found in a narrow zone around trunks of *Tilia*. This has been interpreted as follows.

The first vegetation phase was an open *Betula–Quercus* forest with local stands of *Pinus* and *Tilia* and heathers, grasses and *Pteridium* in the understorey. *Corylus* co-dominated in the shrub layer. *Alnus* was not yet an element of the local stand of vegetation so its pollen must have infiltrated from a younger horizon. The podzolizing of the relatively dry soil

Table 2. Diatoms content of 6 clay samples of sections OZ-43

Species	Depth (cm)								
	245	246	247	248	249	250	I*	II*	III*
Cymatosira belgica	7	14	1	20	20	++	1	1	0
Actinoptychus undulatus	1	2		8	8	++	1	1	0
Coscinodiscus nitidus	2						1	1	0
Podosira stelliger	2						1	1	0
Thalassionema nitzschioides		2			2	+	1	1	0
Campylosia cymbelliformis		1					1	1	0
Biddulphia rhombus				2			1	1	0
Aulacodicus argus						+	1	1	0
Plagiogramma vanheurckii		1					1	2	0
Melosira culcata				2	2	+	1	2	0
Dimerogramma minor				2			1	2	0
Diploneis bombus						+	1	2	0
Rhaphoneis surirella	1	1	1	4	2	+	1	3	0
Rhaphoneis amphiceros		5			2		1	3	0
Cyclotella striata	5						2	1	3
Coscinodiscus lacustris	1						2	1	2
Achnanthes hauckiana	2						2	3	0
Fragilaria inflata	8	3			2		3	1	0
Cyclotella menighiniana		2					3	1	2
Nitzschia hungarica	1						3	2	2
Rhopalodia gibberula				2	6		3	3	0
Navicula mutica		1					3	4	3
Fragilaria construens	2	2	1		2		4	1	2
Fragilaria virescens		1		2			4	1	2
Stephanodiscus astraea					2		4	1	1
Synedra ulna		1					4	1	3
Fragilaria pinnata	50	54	79	48	20	+	4	3	2
Fragilaria pinnata var. *lancettula*	15	6	13	4	10		(4	3	2)
Fragilaria brevistriata				2	2		4	3	2
Hantzschia amphioxys					4		4	4	3
Nitzschia debilis	1						4	4	2
Achnanthes lanceolata		1					4	4	2
Cocconeis diminuta			2				5	3	1
Navicula insociabilis			1				5	4	3
Amphora sabyii	1	3		6	8	+			
Cocconeis thumensis	1					+			
Cocconeis sp.			2	2	4				
Navicula mutica var. *nivalis*				2					
Nitzschia frustulum					2				
Licmophora abbreviata						+			

*Ecofactors.

I: Salinity: (1) polyhalobions, (2) mesohalobions, (3) oligohalobions (halophyl), (4) oligohalobions (indifferent), (5) holophobes.

II: Life form: (1) plankton, (2) benthos, (3) epiphyts, (4) aerophil, (5), eu-terrestrial.

III: pH spectrum: (1) alkalibiontic, (2) alkaliphilous, (3) indifferent, (4) acidophilous, (5) acidobiontic.

probably took place in this period. In the second vegetation phase a brook forest developed with *Alnus*, *Salix* and Cyperaceae.

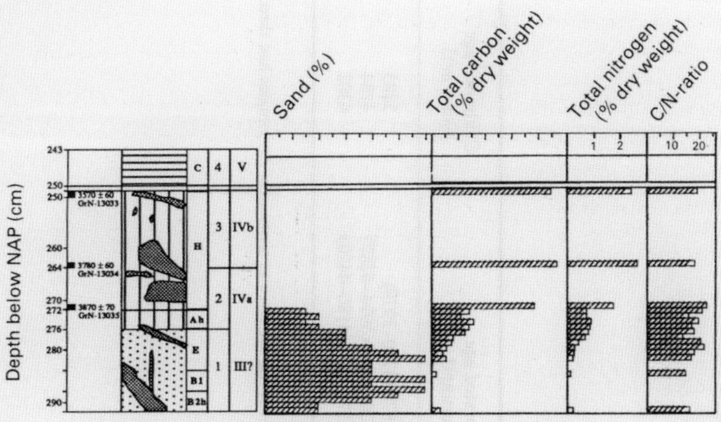

Fig. 5. Contents of sand, total carbon, total nitrogen and C/N-ratio.

The dating of pollen spectra in mineral soils does not yield satisfactory results. The pollen spectra in the mineral soil of this section are thought to have been formed in the Late-Atlantic or Early-Subboreal period. In summary, the pollen spectra from the mineral soil reflect an Atlantic/Subboreal forest development starting on a relatively dry soil. As a consequence of a rising groundwater table the soil became wetter and an alder carr could develop, which finally resulted in a peat layer on top of the mineral soil.

Zone 2: 276–265 cm below NAP (Early-Subboreal, pollen zone IVa). Subzone 2a: (276–273 cm). The sediment consists of a mixture of the mineral soil material and peat. Roots of monocots and woody material of alder are the most important macroscopic remains. This subzone is rather poorer in microfossils than in other tree pollen. The presence of pollen of *Nuphar, Nymphaea* and Typhaceae, and of algae (*Mougeotia* and Type 132, 314), Bryozoa (*Cristatella mucedo, Lophopus crystallinus*) and cocoons of flat-worms (T.353B, described by Van Geel *et al.* 1980/81) and Type 514 (described by Ran 1989) is noteworthy (see Table 1, Fig. 2) because it points to local freshwater conditions. Some of these fossils have penetrated into the underlying sand deposits. Under the influence of a rising groundwater table local pools were formed and subsequently the present forest became submerged. Simultaneously a *Carex* mire developed which later became overgrown by *Salix* and *Alnus*.

The relatively high values of *Corylus, Quercus* and Ericales point to the former vegetation cover and must have arisen from the sandy subsoil, incorporated into this peat layer.

Subzone 2b (272–265 cm; 4301–4200 cal. BP). The peat is strongly degraded and contains much woody material of *Alnus* and roots of monocots. The vegetation type is an alder carr with a predominance of *Alnus*, which reaches its maximum extension towards the end of this subzone, and with some local stands of *Salix*. Cyperaceae and ferns are the most important herbs in the understorey. The ecological conditions point to a rather eutrophic environment with the ground-water table a little above the surface, this is supported by the presence of several aquatic organisms (see Table 1).

Zone 3: 264–250 cm below NAP (Late-Subboreal, pollen zone IVb). The sediment is similar to that of subzone 2b. The pollen percentages curve of *Alnus* decreases strongly,

Table 3. Radiocarbon dating of the peat

GrN-No	Depth below NAP* (cm)	Conventional ^{14}C age (years BP)	Calibrated age (years BP)†	Calibrated age (years BP)‡
13033	250	3570 ± 60	3845;3846;3870 (3780–3979)	3883
13034	264	3780 ± 60	4154;4173;4176;4206;4219 (3994–4287)	4184
13035	272	3870 ± 70	4300,4330;4346;4374;4377 (4180–4420)	4302

*Depth is approximate.
†Ages (calendar years) calibrated according to Stuiver & Becker (1986); between parentheses the upper and lower age are given at sigma one. All samples had three calibrated ages.
‡Ages (calendar years) calibrated according to Van der Plicht *et al.* (1986) at 50% probability.
The calibrated ages were used for all calculations and for final dating of the zone boundaries.

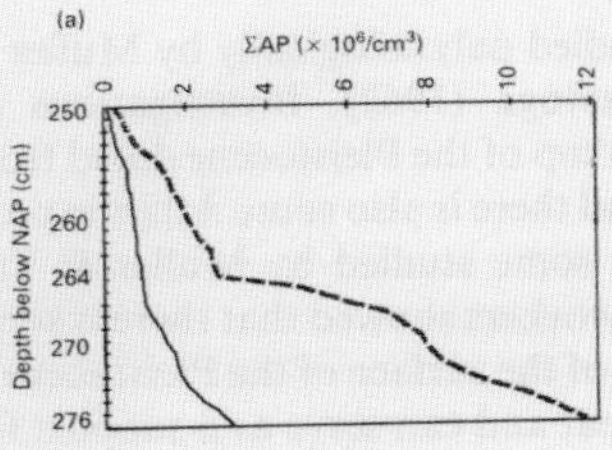

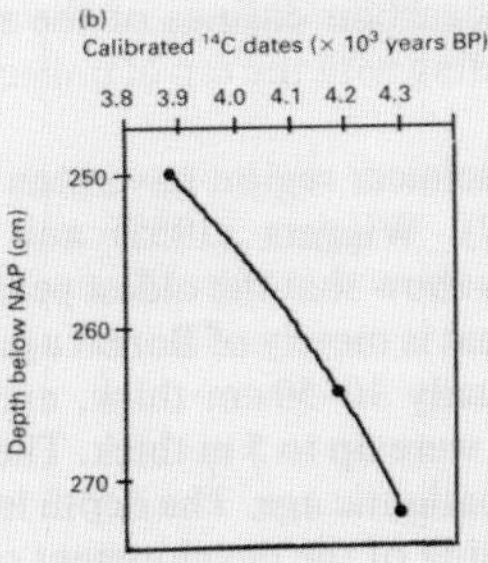

(c)

Depth (cm)	Age (cal.year BP)	Age/cm peat	Peat accumulation (cm/cal. year)
250	3881	25	0.04
251	3907	26	0.038
252	3932	25	0.04
253	3956	24	0.042
254	3980	24	0.042
255	4003	23	0.043
256	4025	22	0.045
257	4048	23	0.043
258	4069	21	0.048
259	4089	20	0.05
260	4109	20	0.05
261	4129	20	0.05
262	4147	18	0.056
263	4165	18	0.056
264	4183	18	0.056
265	4200	17	0.059
266	4217	17	0.059
267	4232	15	0.067
268	4247	15	0.067
269	4261	14	0.071
270	4275	14	0.071
271	4288	13	0.077
272	4201	13	0.077

Fig. 6. (a) Cumulative $\Sigma AP/cm^3$ (– – –) and cumulative $\Sigma AP - Alnus/cm^3$ (———) plotted versus depth for peat samples of section OZ-43. (b) Calibrated radiocarbon dates plotted versus depth. Polynomal (2°): $x = -21\,742 + 179{\cdot}20455y - 0{\cdot}30681818y^2$ ($R = 1{\cdot}0000$). (c) Calculated ages per depth in calendar years BP, age per cm peat and peat accumulation in cm/calendar year.

while the absolute frequences are more or less constant at a rather low level. In this subzone Cyperaceae and Monolete spores reach their second maximum, the fern spores in particular being very abundant. The *Carex acuta*-type, Apiaceae (such as *Oenanthe aquatica*), *Phragmites australis*, and ferns dominated the understorey. Towards the end of this zone there is some evidence for *in-situ* wet conditions, such as an increase of pollen of Typhaceae and Sparganiaceae, spores of *Spirogyra*, and the regular occurrence of oosporangia of *Chara* and other aquatic organisms (Table 1). The top of the peat is eroded and the aquatic organisms pointing to open fresh water probably date from the erosion phase and have been washed into the peat. It is also possible that they were brought into the alder carr by innundations, heralding an approaching wet phase. The arboreal pollen except that of *Alnus* and *Salix* now represents the regional forest. They occur in low quantities, and this points to a restricted forest development in the IJsselmeer region in the period between 4183 cal. BP and 3881 cal. BP.

Zone 4: 250–245 cm below NAP (Subatlantic, pollen zone Vb2). The sediment consists of a dark-grey clay with many shell fragments. Next to *Mya arenarea*, the most important among the shell fragments, *Cerastoderma glaucum*, *Hydrobia ulvae* and Ostracoda are found. *Pediastrum* and *Botryococcus* are the most frequently occurring microfossils. The clay sample immediately above the peat also contains some micro- and macrofossils (such as *Spirogyra*, *Chara* and *Plumatella fungosa*) pointing to fresh water.

Diatom analyses of the clay samples (Table 2) point to a saline or saline–brackish environment with freshwater influences. The presence of *Mya arenarea* in these sediments no doubt indicates that these sediments belong to the Zuiderzee-deposits (Ente *et al.* 1986) which were formed between 1600 AD and 1932 AD which implies that there is a hiatus of *c.* 3500 calendar years between the top of the peat and the clay deposits.

Peat growth in the IJsselmeer region

The palynological dating of this section is hampered by the predominance of alder pollen and by the scarcity of such characteristic markers as pollen of *Fagus* and *Carpinus*. Only a rough dating could be made: 'Subboreal'. This is an interesting period in the development of the Dutch coast. Radiocarbon datings of the section place it more precisely in the Subboreal time corresponding with the top of pollen zone IVa and the lower part of pollen zone IVb.

Peat deposits in the IJsselmeer region have been studied palynologically by Muller & Van Raadshooven (1947), Wiggers (1955) and Havinga (1962). Investigations of Noordoostpolder deposits show that the oldest peat on top of the Pleistocene dated from Preboreal times, but the peat is mostly of Boreal age, and there is also some Atlantic peat. These peat layers are usually 10–50 cm thick, except some studied by Muller & Van Raadshooven (1947), that were up to 5 m thick. These workers showed that there is some peat of Subboreal and Subatlantic age. The depth level of the surface of the Pleistocene is very important for the dating of the development of peat and can serve as a support for other datings. Most of the latter are palynologically estimated, and of course may have to be re-interpreted. Much of the peat is eroded away, and therefore it is hardly possible to reconstruct the Subboreal peat growth. In the south-western and north-eastern parts of the Noordoostpolder the Subboreal peat is found at a depth of 3·5–2 m below NAP. Wiggers (1955) proposed that during the Subboreal the Noordoostpolder was covered with mesotrophic and oligotrophic peat, with some eutrophic alder peat in the south along the gullies. From the other IJsselmeerpolders there is little information on peat formation available. Havinga (1962) interpreted the thin layers of peat on top of the Pleistocene (*c.* 3·2–3·0 m below NAP) to be of Atlantic age.

If we accept that eutrophic peat growth takes place at the ground-water level, and is thus directly correlated with the mean sea level (MSL) at 3780 ± 60 BP the palaeo-MSL was 264 ± 15 cm below NAP at site OZ-43. Roep and Van Regteren Altena (1988) reconstructed palaeo-tidal levels at about 3700 BP in north Holland and showed that their palaeo-MSL is *c.* 2·30 m below NAP; 10 cm higher than the one based on the MSL trend curve of the barriers.

Cardium transgression (see Fig. 7)

The peat of site OZ-43 is more or less synchronous with deposits belonging to the 'Cardium transgression', but ever since the first description of these sediments (Muller & Van Raadshooven 1947) their age has been disputed (Zagwijn 1973; Koopstra 1981; Ente *et al.* 1986). The sediments belonging to the 'Cardium deposits' partly consist of clay or clay–sand and partly of organic (detritus-gyttja) material, with a characteristic marine and brackish fauna, in particular *Cerastoderma glaucum* Poiret (*Cardium glaucum*). The shells were initially identified as *Cardium edule*, but taxanomic studies have proved that these shells have to be conceived as *Cerastoderma glaucum* Poiret (Janssen & Van der Slik 1978). Koopstra (1981) gives an overview of the radiocarbon ages of *Cerastoderma* shells from the Noordoostpolder (Tollebeek II: 3920 ± 60 BP) and Oostelijk Flevoland (G42c: 3995 ± 40 BP). De Jong (1988) studied two sections rather close to that of site OZ-43, in one, KZ-58, *c.* 10 km NW from section OZ-43, *Cerastoderma* shells were dated at 4220 ± 90 BP. Palynological dating (boundary of pollen zone IVa and IVb) of the *Cerastoderma* level of this section does not fit with the radiocarbon age of the shells.

The 'Cardium deposits' in the Noordoostpolder were dated palynologically in the Subboreal, particularly at the boundary of the pollen zones IVa and IVb (Florschütz in

Wiggers 1955). This boundary is not synchronous in the Netherlands (De Jong 1982) but is assumed to be at 3800 BP. Without local ^{14}C-dating this boundary cannot be used for exact dating purposes. The radiocarbon date of section OZ-43 (3780 ± 60 BP) of the beginning of the decline of the alder carr in this area could be used for dating the transition between pollen zones IVa and IVb, although the usual marker for this transition, the first continuous occurrence of *Fagus* pollen, is not discernible in the pollen diagrams of site OZ-43. The start of the 'Cardium transgression' is usually placed at this boundary, but on the other hand the time span in which the 'Cardium deposits' are found is a global one (Ente *et al.* 1986: 3900–3200 BP; Zagwijn, 1986: 4000–3400 BP). There is little doubt about the reliability of the radiocarbon dating of the shells (Koopstra 1981; De Jong, 1988); they are thought to be too old (3920–3995–4220 BP).

Radiocarbon dating of shells in general implies more uncertainties than those of peat (Mook & Streurman 1983) and for that reason radiocarbon datings of shells are used without any reserve. In the Noordoostpolder a clay–peat layer correlating with the 'Cardium clay' (Urk: 3505 ± 120 BP) and an eutrophic peat layer, shaded off into a 'Cardium clay' layer (Schokland: 3315 ± 90 BP) were used for dating this 'Cardium clay' (Wiggers, 1955). Although these dates do not really reflect the age of the 'Cardium deposits' they are the best approximation, if no direct datings are available.

There are arguments pleading for accuracy of the older datings. Comparison of the data of site J51 (Zagwijn 1973) and G42c (Koopstra 1981) *c.* 5 km apart shows that the datings of the shell-containing layers lay very close to each other. The boundary of pollen zone IVa dn IVb in section J51 can be set at *c.* 3800 BP, the *Cerastoderma* shells found just below that boundary must then be older than 3800 BP. The *Cerastoderma* shells of site G42c were dated at 3995 ± 40 BP.

If the datings of the *Cerastoderma* shells are accepted the conclusion must be that the oldest deposits belonging to the 'Cardium transgression' phase lay in Zuidelijk Flevoland, the somewhat younger deposits lay in Oostelijk Flevoland and in the Noordoostpolder and the youngest lay in the Noordoostpolder.

The development of the IJsselmeer region is strongly correlated with the coastal development of north Holland (Ente *et al.* 1986; Zagwijn 1986; Westerhoff *et al.* 1987, Westerhoff & Cleveringa 1989). The peat growth of the IJsselmeer region in the coastal plain moved further to the east under the influence of a continuous rise of the sea level during the period 7000–3000 BP. In the beginning of this period a series of W–E orientated tidal channels developed. In the west most of these channels were filled with sediments, at *c.* 5300 BP two were left which ran eastwards into the Noordoostpolder and Flevopolders. The position of the two tidal channels were primary influenced by the relief of the Pleistocene surface (Zagwijn 1986; paleogeographical map 1).

At the start of the Subboreal (5000 BP) beach barriers developed for the first time, this means an enormous change in the coastal development. The open coast is getting closed. Since then there has been a lack of sediments to fill the more eastward plains, and in the central part of the IJsselmeer region a lagoonal system could establish. The northern tidal channel (via Heer hugowaard-Hoorn) silted up, but was active until *c.* 3200 BP, the inlet near Bergen is closed indefinitely.

In the lagoon the influence of the sea decreased, the ebb/flood movements were reduced and probably stopped, and in the lagoon a new water management system established. In view of the radiocarbon datings the Oer-IJ, as the southern-most connection to the sea, is thought to be the main transport system for sediments with *Cerastoderma* shells in the Flevopolders, and the northern tidal channel mentioned above (originated from the inlet near Bergen) was the transport system for these sediments in the Noordoostpolder. The sediments in which the *Cerastoderma* shells occur give rise to the idea that the *Cerastoderma* shells were not deposited continuously but transported as spatefall (Westerhoff *et al.* 1984). This spatefall transport could have taken place several times during the period that the lagoonal system was connected to the sea.

Eutrophic peat growth had already started before 3870 BP in the section OZ-43. The eastern limits of the eutrophic peat growth in Zuidelijk Flevoland (shown on the paleogeographical map of the Mid-Subboreal by Zagwijn 1986) should be diverted to the southeast for at least 5 km. This implies that peat growth had already taken place at a higher level on the Pleistocene surface, and that the ground-water table was also somewhat higher. The eutrophic peat growth at site OZ-43 is synchronous with the 'Cardium deposits' in the IJsselmeer region, and should have taken place at ground-water level, which is nearly the same as the water level in the lagoon. However, the deposition of the shells themselves, which is thought to have taken place during the frequent storm surges, did not influence the peat growth at site OZ-43. Flooding of this part of the IJsselmeer region was responsible for hampering peat growth and ultimately also for the decline of the alder carr. The flooding is thought to be caused by a continuous rise of the water level in the lagoon, which in turn was depending on the continuous rise of the sea level.

ACKNOWLEDGEMENTS

The IJsselmeerpolders Development Authority, in particular Mr K. Vlierman (Department of Nautical Archeology) is acknowledged for permission to join the excavation of the wreck of the ship and to work out the ecological data. I would also like to thank Mrs E. Beglinger and Mr R. Bregman for their technical assistance, Mrs J.A.E. Kooyman for the diatom analyses, Mr P. Cleveringa (RGD, Haarlem) for discussion on the geology, Mr J.M. van Mourik (FRW, Amsterdam) for his help in interpreting the soil profile, Mr J.P. Pals (IPP, Amsterdam) for collecting the samples and Mr T. Meijer (RGD, Haarlem) for the identification of the molluscs.

REFERENCES

Brock, T.C.M., Jongerhuis, R., Van der Molen, P.C. & Ran, E.T.H. (1989): A comparison of the history and present state of an *Alnus glutinosa* and a *Betula pubescens* dominated patch of wetland forest in the nature reserve 'Het Molenven', The Netherlands. *Acta Bot. Neerl.* **38:** 425–437.

De Jong, J. (1982). Chronostratigraphic subdivision of the Holocene in the Netherlands. In: Mangerud, J., Birks, H.J. and Jäder, D.K., (eds): *Chronostratigraphic subdivision of the Holocene. Striae,* **16:** 71–74.

— (1988) Pollenanalytisch onderzoek en een ^{14}C-ouderdomsbepaling aan afzettingen van holocene ouderdom uit Zuidelijk Flevoland. *Intern Rapport Rijks Geol. Dienst,* **1051:** 1–6.

De Wolf, H. (1982): Method of coding of ecological data from diatoms for computer utilization. *Med. Rijks Geol. Dienst,* **36:** 95–98.

Dimbleby, G.W. (1985): *The Palynology of Archaeological Sites.* Academic Press, London.

Ente, P.J., Koning, J. & Koopstra, R. (1986): De bodem van oostelijk Flevoland. *Flevobericht* **258:** 3–181.

Faegri, K. & Iversen, J. (1975): *Textbook of Pollen Analysis.* Munksgaard, Copenhagen.

Gaudreau, D.C., Jackson, S.T. & Webb, T. (1989): Spatial scale and sampling strategy in paleoecological studies of vegetation patterns in mountainous terrain. *Acta Bot. Neerl.* **38:** 369–390.

Havinga, A.J. (1962): *Een Palynologisch Onderzoek van in Dekzand Ontwikkelde Bodemprofilen.* Thesis, University of Wageningen.

— (1974): Problems in the interpretation of the pollen diagrams of mineral soils. *Geol. Mijnbouw,* **53:** 449–453.

— (1984): A 20-year experimental investigation into the differential corrosion susceptibility of pollen and spores in various soil types. *Pollen Spores,* **26:** 541–558.

Janssen, A.W. & Van der Slik, L. (1978): De fossiele schelpen van de Nederlandse stranden en zeegaten, tweede serie 7. *Basteria*, **42:** 49–72.

Keatinge, T.H. (1982): Influence of stemflow on the representation of pollen of *Tilia* in soils. *Grana*, **21:** 171–174.

Koopstra, R. (1981): De problematiek van de 'Cardiumtransgressie' in het IJsselmeergebied. *Flevobericht*, **206:** 55–59.

Middeldorp, A.A. & Mijzen, P. (1986): Embedding pollen in a film of polyvinyl alcohol, a method for establishing pollen concentrations. *Pollen Spores*, **28:** 451–456.

Mook, W.G. & Streurman, H.J. (1983): Physical and chemical aspects of radiocarbon dating. *Pact*, **8:** 31–55.

Muller, J. & Van Raadshooven, B. (1947): Het Holoceen in de Noordoostpolder. *KNAG 2° Series 1*, **44**(2): 153–185.

Ran, E.T.H. (1989): *Dynamics of Vegetation and Environment During the Middle Pleniglacial in the Dinkelvalley (The Netherlands)*. Thesis, University of Amsterdam.

Roep, Th. B. & Van Regteren Altena, J.F. (1988): Paleotidal levels in tidal sediments (3800–3635 BP): compaction, sea level rise and human occupation (3275–2620 BP) at Bovenkarspel, NW Netherlands. In: De Boer, P.L. *et al.* (eds): *Tide-Influenced Sedimentary Environments and Facies*. 215–231. Reidel Publishing Comp., Dordrecht.

Stuiver, M. & Becker, B. (1986): High-precision decadal calibration of the radiocarbon time scale, AD 1950–2500 BC. *Radiocarbon*, **28**(2B): 863–910.

Van der Plicht, J., Mook, W.G. & Hasper, H. (1987): An automatic calibration program for radiocarbon dating. In: *Abstracts of the Second International Symposium on Archaeology and* ^{14}C. 38. Groningen.

Van der Werff, A. (1960): Die Diatomeen des Dollart-Ems Gebietes. Symposium Ems-estuarium. *Verh. Kon. Ned. Geol. Mijnb. Genoot.* **29:** 153–201.

Van Geel, B., Bohncke, S.J.B. & Dee, H. (1980/81): A palaeoecological study of an upper Late-Glacial and Holocene sequence from 'De Borchert', The Netherlands. *Rev. Palaeobot. Palynol.* **31:** 367–448.

Van Mourik, J.M. (1986): Pollen profiles of slope deposits in the Galician Area (N.W. Spain). *Ned. Geogr. Stud.*, **12:** 1–171.

Vlierman, K. (1985): Neolithische en middeleeuwse vondsten op de kavels OZ 35 en OZ 36 in Zuidelijk Flevoland. *RIJP Rapport*, **51:** 1–39.

Westerhoff, W.E. & Cleveringa, P. (1989): Sea-level rise and coastal sedimentation in central Noord-Holland (The Netherlands) around 5000 BP: a case of change in sedimentation dynamics and sediment distribution patterns. In: Beukema, J.J. (ed.): *Proceedings Workshop on Climatic Effects on Coastal Ecosystems*. Balkema, Rotterdam.

—, Cleveringa, P. & Mücher, H.J. (1984): Development of Dunkirk III deposits near Alkmaar, The Netherlands. *Geol. Mijnb.* **63:** 277–286.

—, De Mulder, E.F.J. & De Gans, W. (1987): *Toelichting bij de Geologische kaart van Nederland 1:50.000. Blad Alkmaar (19 O+W)*. Rijks Geol. Dienst, Haarlem.

Wiggers, A.J. (1955): *De Wording van het Noordoostpoldergebied*. Tjeenk Willink, Zwolle.

Zagwijn, W. (1973): Pollenanalytisch onderzoek van lacustriene afzettingen uit het gebiedvan de IJsselmeerpolders. *Intern Rapport Rijks Geol. Dienst*, **697:** 1–5.

Zagwijn, W. (1986): *Nederland in het Holoceen. Geologie van Nederland, Deel I.* Rijks Geol. Dienst, Haarlem.

9 The Ens cog

Pollen spectrum of the moss caulking

E. Mook-Kamps

Interpretation

The conservation of the moss caulking (sample 19) is bad and contains a lot of charcoal
Trees 48%, herbaceous plants 52%.
Birch 20%, alder and oak 11%, juniper 3%, hazel, hornbeam, beech, willow and bog mirtle 1%.
Heather 22%, grasses 14%, cereals 7%, arable weeds 5% (*Cruciferae,* mugwort, dock, narrowleaf plantain and black-bindweed).
Goosefoot and sea plantain may indicate saline conditions.
Local: horsetail 22%, burr-reed and Dryopteris fern 1%, *Spagnum* moss 5%.
Indeterminates: 1,3%.
Oligotrophic peatland. The pollen spectrum looks like sample 17 (the Nijkerk-II cog) but the relative percentages of the species are very different. This sample must have come from a less open landscape, and a smaller area of peat.

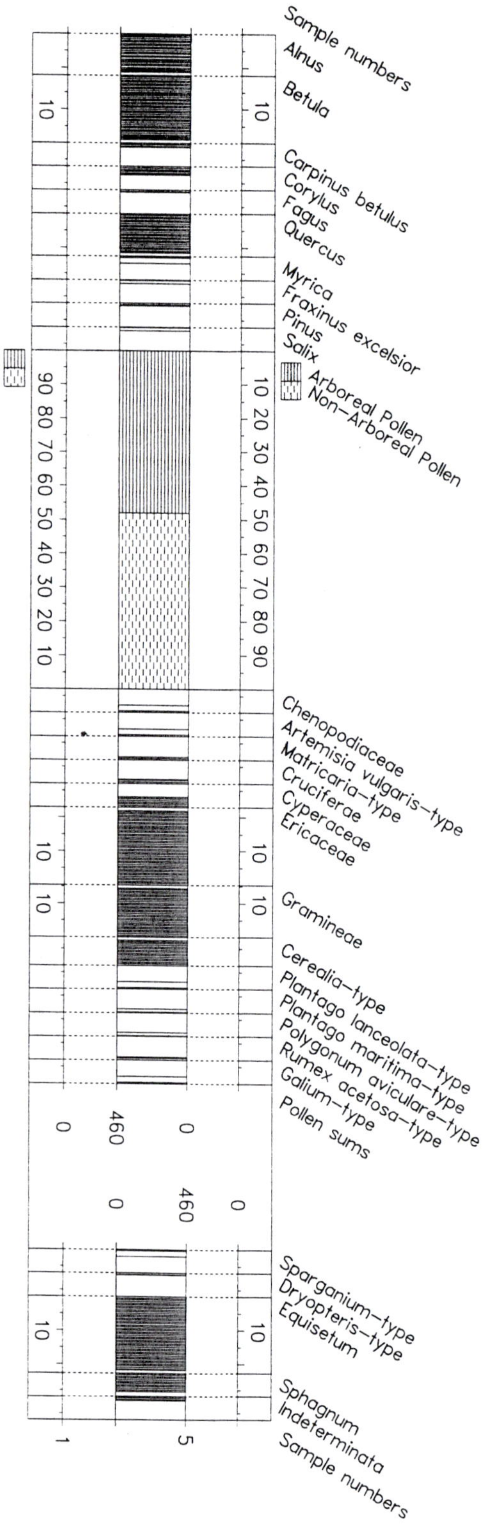

10 The Rutten cog

Pollen spectrum of the moss caulking

E. Mook-Kamps

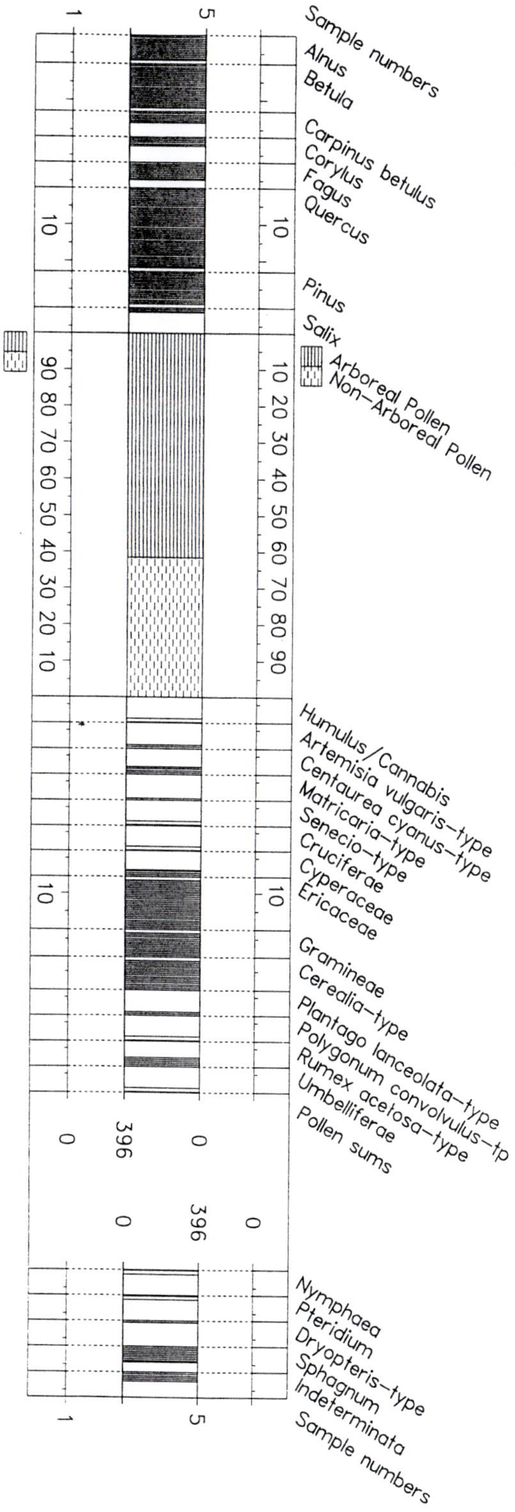

Interpretation

Pure peat moss was used to caulk the Rutten cog (sample 6); the conservation of the moss was reasonable.
Trees 60%, herbs 40%.
Oak 22%, birch 12%, elder 7%, beech 5%, hornbeam 3%, hazel 2,5%, willow 1,3% and den 9%.
Heather 14%, grains 8%, grasses 7%, cyperus grasses 2%, field weeds 6%, among which cornflower, small plantain, sorrel and knotgrass.
Locally small percentages for water lilly, *Pteridium aquilinum*, dryopteris ferns and peat moss (4%).
Indeterminates: 2,5%.
Reasonable nutrient-rich circumstances. It looks like a disturbed area with a small moor, fields and a forest with oak, beech and hornbeam. The cornflower may be an indication for the cultivation of rye.

Examination of the bones

F. Laarman

Results Rutten cog

For the introduction, the pattern of expectations and the conclusion of the study of bones from the Rutten cog see appendix of the Nijkerk-II cog.

It seems as if this ship was excavated more carefully, because even small fish remains and bone splinters were collected. In this wreck, 106 bones and a nearly complete sheep skeleton were found. With the exception of the complete sheep, all bones and their weight (in grams) are presented in table 1.

Table 2 shows that bones from all parts of the skeleton are present, but the emphasis lies with the axial part, especially ribs. Six out of twenty seven cattle bones and 24 out of 43 sheep/goat bones have chop marks. In addition traces of rounding have been noted on four cattle bones. In addition to waste from consuming meat, offal is also present in this cog. In addition to the bones mentioned in the tables, the remains of a complete sheep have been found with 87 bones and a weight of 1126,4 grams.
Find number 48 contained nearly all the bones of this sheep, but small bones have also been found at the numbers 181, 219, 225 en 226. According to the dental information, combined with the fusion phases of the pipe bones, this sheep was about 12 months old. It may have been brought on board as livestock and drowned during the sinking, but since the tail and hoof bones are absent, it is possible that it was a skinned carcass.

Apart from the bones of mammals and birds, a number of fish bones and scales were found. The species and numbers are presented in table 3.

It is of course the question if these fish remains were food supply or entered the wreck after it sunk in one way or another. According to Dorleijn[3], it was not possible to fish coldfish and whiting in the former Zuiderzee. This is probably related to the period that the Zuiderzee was salty (ca 1600-1932). How the fishing conditions were at the time of the wrecking of the Rutten cog (second part of the 13th century) in the then 'sweet water' Almere sedimentation phase is hard to tell. It seems likely that these species were aboard as food supply. The other species may have come into the wreck later, after it had sunk, as was determined with the Biddinghuizer Colfschip[4]. How, finally, a human calf bone ended up in this wreck is a puzzle. Is it the last remnant of a drowned seaman or passenger?

Table 1. Number of bones per animal species and their weight

	N	Grams
cattle	27	446.3
sheep/goat	43*	161.3
pig	3	5.2
large mammal	3	17.8
medium mammal	1	2.0
indet. mammal	23	23.2
human	1	2.8
chicken	4	3.9
duck	1	0.5

* find number 48, a complete sheep, is not included in the table.

Table 2. Number of bones per element category for cattle en sheep/goat

	head skeleton	axial skeleton	front legs	hind legs	lower legs
cattle	3	16	4	1	3
sheep/goat	2	30	4	4	3

Table 3. Number of fish bones per species

	bones	scales
codfish	7	-
whiting	1	-
pike	9	-
perch	17	29
bream	1	-
indet. fish	12	40

3. Dorleijn 1987-
4 Lauwerier & Laarman, 2006

r	- right	PI	- phalanx 1
l	- left	PII	- phalanx 2
p	- proximal	PIII	- phalanx 3
d	- distal	Gl	- greatest length
ve	- fused	Bp	- width proximal
onv	- non fused	Bd	- width distal
juv	- juvenile	mnd.	- months

List of determinations of the bones from the Rutten cog

Findno. A57-	Description	weight in grams
18	sheep/goat 1x *cervical vertebra* ve	25.8
27	cattle 1x *pelvis* (r) (chop- and gnaw marks)	43.7
34	sheep/goat 1x rib (chop mark)	4.9
34	sheep/goat 1x rib (chop mark)	1.7
37	chicken 1x *scapula* (r)	0.6
42	cattle 1x rib (complete), rounded	35.4
48	sheep (complete, see end of list)	
52	cattle 1x rib (chop marks) lightly burned	18.1
54	cattle 1x tail vertebra	1.2
54	cattle 1x *carpale* (l)	8.3
58	sheep/goat 1x rib	3.8
68	cattle 1x *thoracic vertebra* onv, rounded	28.1
70	cattle 1x *thoracic vertebra* cap	1.9
72	cattle 1x *carpale* (l)	14.2
72	cattle 1x *thoracic vertebra* onv	15.0
72	cattle 1x *vertebra* ve	3.6
80	sheep/goat 1x rib (chop marks)	3.0
83	cattle 1x *mandibula* (l)	68.6
92	sheep/goat 1x *thoracic vertebra*	1.4
92	sheep/goat 1x PII	0.9
92	sheep/goat 1x *vertebra* (chop mark)	1.6
92	medium mammal 1x fragment *cranium*	2.0
102	sheep/goat 1x rib (chop mark)	2.7
105	2x indet.	4.9
109	pig 1x upper jaw *molar*	0.9
118	1x indet. piece of *cranium*	0.9
119	cattle 1x rib (chop marks)	32.9
120	sheep/goat 1x *mandibula* (l) ± 8 mnd	9.5
124	sheep/goat 1x rib	1.2
125	cattle 1x *cervical vertebra* (chop mark)	5.0
125	sheep/goat 1x rib (chop mark)	0.9
125	schaap/goat 1x *calcaneum* (l)	2.7
125	sheep/goat 1x *vertebra* (burned)	0.5
125	whiting 1x *vertebra*	
125	sheep/goat 1x rib	0.6
130	sheep/goat 1x rib (burned)	0.8
130	sheep/goat 1x *humerus*	3.1
130	1x indet.	0.9
132	perch 29 scales	
133	indet. fish 40 scales	
133	perch 1x *epihyale*, 2x *interoperculum*	
133	pike 1x *suboperculum*	
133	indet. 1x small bone	0.6
133	perch 1x *supracleithrale*, 1x *frontale*, 2x *operculum*, 1x *dentale*,	
	2x *preoperculare*, 5x *vertebra*	
135	sheep/goat 1x *metatarsus* (l)	12.1
138	cattle 1x *thoracic vertebra* onv, rounded	40.3
144	sheep/goat 1x rib	5.7
149	human 1x calf bone (l) d	2.8
161	sheep/goat 1x rib (chop mark)	1.4
161	sheep/goat 1x rib (chop mark)	3.1
161	sheep/goat 1x piece of *mandibula* (fits on nr. 120)	
161	sheep/goat 1x *pelvis* (r)	6.5
161	large mammal 1x pipe bone	4.5
170	1x indet. (burned)	0.5
170	sheep/goat 1x rib (burned)	1.5
175	cattle 1x rib	3.4
175	cattle 1x rib (lightly burned)	4.1
176	pig 1x *fibula*	2.1
181	sheep/goat 1x rib (chop mark)	4.1
181	sheep/goat 1x rib (chop mark)	5.2
181	sheep/goat 1x rib (chop mark)	3.2
181	cattle 1x *scapula* (r)	13.3
181	sheep/goat 1x *lumbar vertebra* (belongs to 48!)	0.5
183	cattle 1x *mandibula* (l), rounded	38.6
190	fish jaw	
194	cattle 1x *cranium*	13.2
203	cattle 1x rib	5.7
203	sheep/goat 1x *atlas* juv (cutting trace)	1.8
203	chicken 1x *pelvis*	0.7
203	1x indet.	0.9
204	pike 1x *palatinum*, 2x *vertebra*	
204	indet. 12 fish bones	
209	duck 1x *carpometacarpus* (r)	0.6
209	sheep/goat 1x rib (hakspoor), 1x PIII, 1x *hyoid*	2.8
209	codfish 1x *postcleithrum*	
209	indet. 13x bone splinters	8.7
217	sheep/goat 1x *carpale* (belongs to nr. 48!)	0.4
217	1x indet.	0.9
217	large mammal 1x pipe bone	4.8
219	sheep/goat 1x *scapula* (r)	4.5
219	sheep/goat 1x *humerus* p onv (belongs to nr. 48!)	3.3
225	pig 1x *fibula*	2.2
225	cattle 1x *carpale* (l)	6.9
225	sheep/goat 4x rib (chop marks)	9.2
225	sheep/goat 1x *vertebra* cap onv (belongs to nr. 48!)	0.4
225	codfish 1x *vertebra*	
225	indet. 1x small bone	3.6
226	pike 1x *dentale* (l), 2x *vertebra*	
226	codfish 3x *vertebra*	
226	sheep/goat 1x *lumbar vertebra* ve	4.7
226	sheep/goat 1x PII, 1x *carpale* (r) (belongs to 48!)	2.5

226	indet. 1x small bone	1.3
233	bream 1x *operculum*	
233	perch 1x *interoperculum*	
233	codfish 1x *vertebra*	
233	sheep/goat 2x rib, 1x *lumbar vertebra* (chop marks)	6.8
233	sheep/goat 1x *tibia*	3.2
233	cattle 1x rib (chop marks)	5.9
233	large mammal 1x indet.	8.5
233	sheep/goat 1x *humerus* (r) juv	4.2
233	duck 1x *sternum*	0.5
233	chicken 1x *coracoïd* (r), 1x *tarsometatarsus* (r)	2.6
248	pike 1x *basioccipitale* (very big), 1x *vertebra*	
248	codfish 1x *vertebra*	
248	perch 1x *maxillare*	
248	cattle 1x *thoracic vertebra* (chop mark)	20.6
248	cattle 1x tail *vertebra* onv	1.9
248	cattle 3x *scapula* (looks like turned to stone)	16.4
248	sheep/goat 1x *scapula* (r), 1x *femur* (l) (chop marks)	16.2

Complete sheep, find number 48

Present: 1x skull, 2x *madibula*, 7x *cervical vertebra*, 12x *thoracic vertebra*, 6x *lumbar vertebra*, 1x *sacrum*, 2x *scapula*, 2x *humerus*, 2x *radius*, 2x *ulna*, 2x *metacarpus*, 3x *carpale*, 24x rib, 2x *pelvis*, 2x *femur*, 2x *tibia*, 2x *astragalus*, 2x *calcaneum*, 2x *centrotarsale*, 2x *metatarsus*, 3x *sternum*, 7x PI, 2x PII.

Total weight: 1136.4 gram.

Fused: 2x *humerus* d ve, 2x *radius* p ve, 7x PI p in the proces of fusion.

Age based on fusions: 13-16 mnd.

Age based on teeth: ± 10 mnd.

Striking absent: tail, 1x PI, 6x PII, 8x PIII.

Dendrochronological report RING
A.E.M. Hanraets & E. Jansma

ring

Stichting Nederlands Centrum voor Dendrochronologie
Kerkstraat 1, 3811 CV Amersfoort
Tel (033) 63 42 33
Fax (033) 65 32 35
E-mail: ring @ archis.nl

RAPPORTAGE

Vindplaats: Ketelhaven, kavel: A-57

Datum monstername :
Datum aanvraag : dec. 1993

Aanvrager:
Naam : R. Oosting
Instantie : Museum voor Scheepsarcheologie
Stad : Lelystad

Onderzoekers:
Naam : E. Hanraets, E. Jansma
Instantie : RING
Datum onderzoek : april 1994

Vondstno.	Dendronaam	kern	spint	wankant	n
m. 6	ka5 06	-	-	-	125
WB1/SB, m.7	ka5 07	-	-	-	139
huidgang R, m.8	ka5 08	-	grens	-	159
GC1/BB, m.24	ka5 24	-	-	-	197

DATERING

Dendronaam	Referentie-chronologie	Datering	%PV	t	P
ka5 06	Polen	1113-1237AD	63.7	4.59	0.01
ka5 07	Polen	1109-1247AD	65.2	4.57	0.001
ka5 08	Nedersaxen	1091-1249AD	57.9	4.27	0.05
ka5 24	Polen	1046-1242AD	63.8	6.67	0.001

Deze monsters konden onderling gedateerd worden en gemiddeld tot een objectchronologie, ka5all:

Dendronaam	Referentie-chronologie	Datering	%PV	t	P
ka5all	Polen	1046-1249AD	64.3	6.86	0.001
	Nedersaxen	1046-1249AD	60.6	4.88	0.1

INTERPRETATIE VAN DE DATERINGEN

Dendronaam	**datering**	**veldatum**
ka5 08	1249 + 20 ± 6jaar	1263-1275 AD

Alleen bij monster ko5 08 is een duidelijke spintgrens te herkennen. Aangenomen dat de veldatum voor dit monster geldig is voor de hele partij, ligt deze veldatum dus tussen 1263 en 1275 AD.

---

11 The Dronten-I cog

Pollen spectrum of the moss caulking

E. Mook-Kamps

Interpretation

Pure peat moss was used in the Dronten-I cog (sample 9). The conservation was reasonable. The sample contained charcoal.
Trees 34%, herbs 66%.
Alder 16%, birch and hazel 7%, oak 2%, bog myrtle 2%, beech and elm 0,5%.
Heather 46%, grassen 7%, cyperus grasses 4%, grains 4%, field weeds 3% (small plantain and sorrel), common sundew 0,4%.
Local: small percentages for wood fern, *Pteridium aquilinum* and horse tail. Peat moss 95%.
Interdeterminates: 2,9%.
Nutrient poor area. A large moor. Little agriculture in the environment.

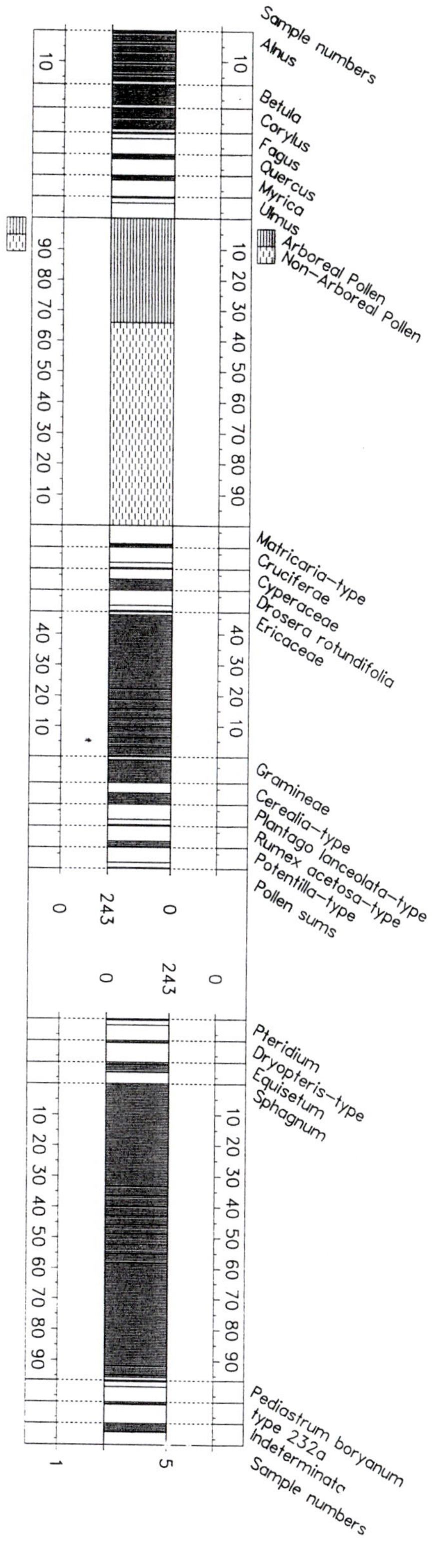

Examination of the bones

F. Laarman

For the introduction, the pattern of expectations and the conclusion of the study of bones from the eight cogs see appendix of the Nijkerk-II cog.

Results Dronten-I cog

The bones of the Dronten-I cog match the pattern of expectations. Except for a marline spike, made from an antler tine from a red deer, all the bones are from ready made lumps of meat. One bone from a pig and eight bones from cattle represent about 6 kilo of meat. Chop marks are found on five out of eight cattle bones; traces of rounding were absent.

Table 1. Number of bones per animal species and their weight

	N	grams
Red deer	1	53.0
Pig	1	24.6
Cattle	8	467.5

r	- right	PI	- phalanx 1
l	- left	PII	- phalanx 2
p	- proximal	PIII	- phalanx 3
d	- distal	Gl	- greatest length
ve	- fused	Bp	- width proximal
onv	- non fused	Bd	- width distal
juv	- juvenile	mnd.	- months

List of determinations of bones from the Dronten-I cog

Invent nr. On 5-	description	weight in grams
39	red deer 1x antler point made into a marline spike	53.0
40	cattle 1x *lumbar vertebra* ve	43.4
41	cattle 1x *calcaneum* (r) ve	120.8
42A	cattle 1x *femur* (r) d ve	122.5
42B	cattle 1x *astragalus* (r) chop mark) belongs to 41	60.9
42C	cattle 1x *lumbar vertebra* onv (chop mark)	38.2
42D	cattle 1x rib (chop mark)	7.7
42E	cattle 1x rib (chop marks)	16.6
42F	cattle 1x rib (chop marks)	57.4
42G	pig 1x *humerus* (r) p+d onv	24.6

12 The Marknesse cog

Dendrochronological report RING

M. Dominguez

Stichting RING, p/a RACM Postbus 1600, 3800 BP Amersfoort; 0334217545; m.dominguez@racm.nl; www.archis.nl/ringnieuw

ring

Nederlands Centrum voor Dendrochronologie
Stichting RING
p/a RACM Amersfoort
Postbus 1600 3800 BP Amersfoort
Tel. 033-4227545
E-mail m.dominguez@racm.nl
www.archis.nl/ringnieuw

Aan: Dhr. A. Van Holk en mevrouw A. Overmeer
Nieuw Land Erfgoedcentrum
Oostvaardersdijl 01-13
8200 AB Lelystad

Uitslag dateringsonderzoek
***RING* Intern Rapport nummer: 2008080**
Datum: 3 november 2008

Geachte heer Van Holk en mevrouw Overmeer,

Wij onderzochten voor u 5 eikenmonsters (*Quercus sp.*) afkomstig uit de kavel NM 107 bij Emmeloord, Noordoostpolder in Flevoland (offerte O2008058). De volgende dateringen zijn het resultaat (zie met name de vijfde kolom van tabel 1):

Tabel 1. Uitslag dendrochronologisch onderzoek

Uw omschrijving	RINGs Dendrocode	Datering van de laatste gemeten ring	Zekerheid van de datering (*probability*)	Periode waarin/ waarna de boom is omgehakt	Gebruikte Referentie-chronologieën
NM 107 / monster 4, los hout; werkput onbekend, put 2A	EML00011	1331 n.Chr.	>99,0%	1336 n.Chr. ±4	NLNOOR1E
NM 107 / monster 5, los hout; spant 1B / BB, put 1	EML00020	1339 n.Chr.	>99,5%	zomer/winter 1339 n.Chr.	NLTWEN03
NM 107 / monster 6, los hout; spant 2B / BB, put 1	EML00030	1339 n.Chr.	>97,5%	zomer/winter 1339 n.Chr.	NLNOOR1B
NM 107 / monster 7, los hout; los hout / gang, put 2-2A	EML00040	1305 n.Chr.	>98,0%	ná 1320 n.Chr.	NLNOOR02
NM 107 / monster 8, los hout; los hout / binnenboord, put 2-2A	EML00050	1301 n.Chr.	>99,5%	1318 n.Chr. ±5	NLNOOR1B

De monsters van meetreeksen EML00011 en EML00040 (monsters 4 en 7 respectievelijk) synchroniseren met elkaar (zie bijlage 1, tabel 2 en bijlage 2, afbeelding 2). Dit wijst er op dat de bomen van deze monsters in hetzelfde gebied gegroeid zijn.

Terwijl de individuele meetreeksen lage synchronisaties met de referentiekalenders tonen, laat de gemiddelde curve van alle de meetreeksen (EML5MMMM) een goede synchronisatie met de referentiekalender NLNOOR02 zien (zie bijlage 1, tabel 3 en bijlage 2, afbeelding 4).

3/9

Aanvullende informatie over de laboratoriumresultaten, de gebruikte statistiek en/of de gebruikte referentiekalenders, vindt u in de bijlagen.

RING Intern Rapport nummer:	Laboratoriumnummer, verwijzing naar de analyse.
Zekerheid van de datering:	De kans dat de gevonden match met de referentiechronologie niet op toeval berust. Deze waarde is gebaseerd op de 'Gleichlaufigkeit' tussen de twee vergeleken reeksen, ook wel %PV genoemd (*percentage of parallel variation*; Jansma 1995).
Verantwoording van de dateringen:	Dendrochronologische dateringen door RING zijn gebaseerd op een combinatie van waarnemingen: (a) vergelijking en relatieve datering (ten opzichte van elkaar) van de jaarringpatronen binnen een vindplaats/bouwfase; (b) vergelijking van deze jaarringpatronen met *meerdere* absoluut gedateerde referentiekalenders. Deze vergelijkingen zijn statistisch onderbouwd en worden visueel gecontroleerd. Wanneer observaties elkaar ondersteunen en bevestigen, wordt de datering geaccepteerd als zijnde correct.

RING-rapport 2008080, Emmeloord, Noordoostpolder, Flevoland, Kavel NM 107

Bijlage 1. Statistische resultaten van het dendrochronologisch onderzoek

Tabel 1. Statistische resultaten van de monsters.

Uw omschrijving	RINGs Dendrocode	n	Kern	Spint	Wankant	1e jaar	ne jaar	Kapdatum*	t	%PV	p	Kalender
NM 107 / monster 4, los hout; werkput onbekend, put 2A	EML00011	90	+?	14	5±4	1242	1331	1336 n.Chr. ±4	2,11	63,9	0,01	NLNOOR1E
NM 107 / monster 5, los hout; spant 1B / BB, put 1	EML00020	56	+1	25	WK	1284	1339	zomer/winter 1339 n.Chr.	2,92	70,5	0,005	NLTWEN03
NM 107 / monster 6, los hout; spant 2B / BB, put 1	EML00032	85	+1	22	WK	1254	1339	zomer/winter 1339 n.Chr.	3,7	62,2	0,025	NLNOOR1B
NM 107 / monster 7, los hout; los hout / gang, put 2-2A	EML00040	111	+?	-	>15	1195	1305	ná 1320 n.Chr.	4,8	62,2	0,02	NLNOOR02
NM 107 / monster 8, los hout; los hout / binnenboord, put 2-2A	EML00050	102	+1	1	17±5	1200	1301	1318 n.Chr. ±5	2,83	66,0	0,005	NLNOOR1B

* Kapdatum geschat volgens Jansma, E. 2007 voor Nederland.

Tabel 2. Relevante statistische resultaten tussen meetreeksen

Meetreeksen	Houtsoort	Overlap	t	%PV	p
EML00011 vs. EML00040	eik	64	8,52	64,8	0,02

Tabel 3. Gemiddelde curven

Meetreeksen	Houtsoort	RINGs Dendrocode	n	1e jaar	ne jaar	t	%PV	p	Kalender
EML00011 EML00020 EML00030 EML00040 EML00050	eik	EML5MMMM	140	1195	1339	5,28	63,2	0,005	NLNOOR02

RING-rapport 2008080, Emmeloord, Noordoostpolder, Flevoland, Kavel NM 107

Bijlage 2. Afbeeldingen van de gedateerde meetreeksen

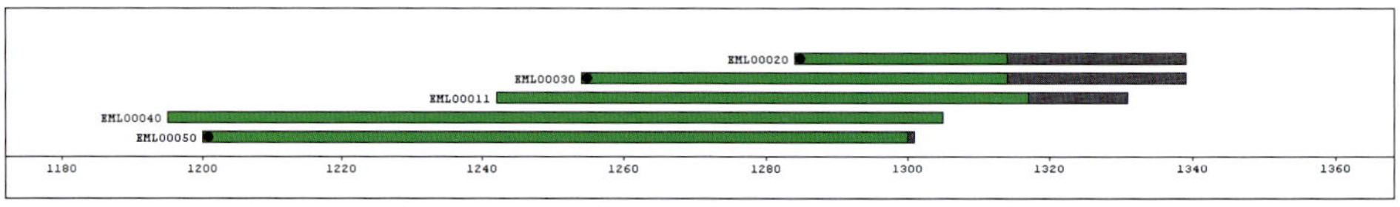

Afbeelding 1. Tijdspanne van de gedateerde meetreeksen. Deze zijn gesorteerd op de laatst gemeten jaarring. De schatting van de veldatum is niet weergegeven op deze afbeelding. Het grijze gebied geeft de spintringen weer. De zwarte punt wijst erop dat de kern in het monster aanwezig is. x-as: kalenderjaar.

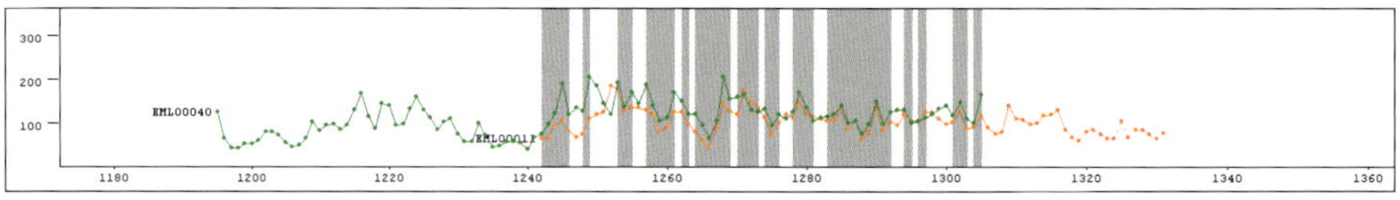

Afbeelding 2. Visuele synchronisatie tussen de meetreeksen EML00011 (oranje) en EML00040 (groen); y-as: jaarringbreedte in mm*10^{-2}; x-as: kalenderjaar. Het grijze gebied toont de *Gleichläufigkeit* (percentage van parallelle ringbreedte variaties (%PV) van beide patronen). De bomen van deze monsters zijn in hetzelfde gebied gegroeid.

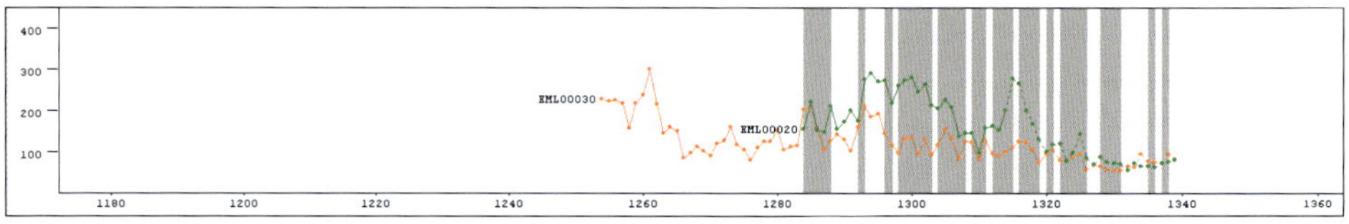

Afbeelding 3. Visuele synchronisatie tussen de meetreeksen EML00020 (groen) en EML00030 (oranje); y-as: jaarringbreedte in mm*10^{-2}; x-as: kalenderjaar. Het grijze gebied toont de *Gleichläufigkeit* (percentage van parallelle ringbreedte variaties (%PV) van beide patronen).

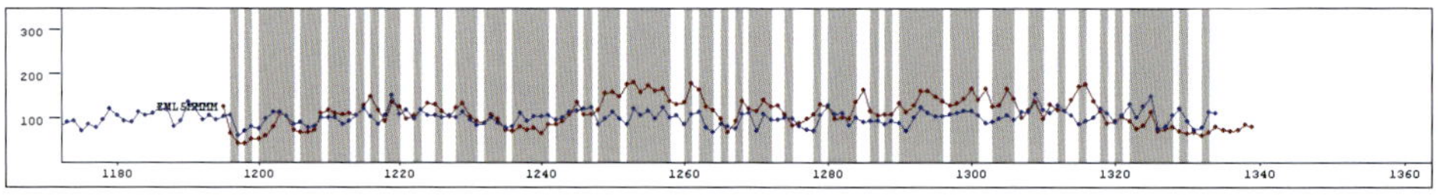

Afbeelding 13. Bruin: gemiddelde curve EML5MMMM; blauw: referentiekalender NLNOOR02; y-as: jaarringbreedte in mm*10^{-2}; x-as: kalenderjaar. Het grijze gebied toont de *Gleichläufigkeit* (percentage van parallelle ringbreedte variaties (%PV) van beide patronen).

RING-rapport 2008080, Emmeloord, Noordoostpolder, Flevoland, Kavel NM 107

Bijlage 3. Toelichting op de resultaten van de dendrochronologische analyse

Houtsoort = Het hout wordt door ons enkel gedetermineerd ten behoeve van de datering. Alleen de *genus*, bijv. Den (*Pinus*), wordt bepaald. Verdere soortbepaling, zoals bijv. grove den (*Pinus sylvestris*), blijft in principe achterwege, tenzij deze eenvoudig vastgesteld kan worden. Een uitzondering hierop is *Abies alba* (Zilverspar), de enige soort *Abies* die in het verleden in Nederland is toegepast.

Kern = (Geschatte afstand tot) de eerstgevormde (oudste) jaarring in de stam.

Spint = Aantal gemeten ringen spinthout. Volgens Hollstein (1980) heeft eik een gemiddeld aantal spintringen van 16 ± 5 bij een boom tot 100 jaar oud, 20 ± 6 bij een boom van 100 tot 200 jaar oud, en 26 ± 8 bij een boom ouder dan 200 jaar. Wij gebruiken een nieuwe, bijgestelde spinthoutberekening voor archeologisch/ historisch constructiehout dat dateert met Nederlandse en Duitse chronologieën (Jansma 2007). Bij eikenhout uit het Baltische gebied is het gemiddelde aantal spintringen iets lager dan in West Europa, 15 (+9/-6) (Wazny, 1990). Grove den, (*Pinus sylvestris*) heeft weliswaar ook duidelijk zichtbaar spinthout, maar doordat het aantal spinthoutringen onregelmatig is, is een schatting van de velddatum niet mogelijk. Fijnspar, (*Picea abies*) heeft geen spinthout. Uiteraard geeft een aanwezige wankant wel de precieze kapdatum van de boom.

Wankant = Het geschatte aantal jaarringen tot de wankant, d.w.z. tot de laatstgevormde jaarring (direct onder de bast), nodig voor een absolute datering van de veldatum.

Veldatum = De datum waarop de boom geveld is. Als er wankant aanwezig is, is er een absolute datering mogelijk. Als er spintringen aanwezig zijn, of zelfs alleen spintgrens, wordt de veldatum berekend door het aantal ontbrekende spintringen te berekenen. Als er bij een eik van 100 tot 200 jaar oud b.v. 4 spintringen gemeten zijn, is het geschatte aantal ontbrekende spintringen dus 16 ± 6. Dit getal wordt bij de datering opgeteld. Als er geen spintringen meer op het monster aanwezig zijn, is het onbekend hoeveel *kernhoutringen* er nog ontbreken . De veldatum ligt dan een onbekend aantal jaren ná de datering van de laatste (jongste) ring + de schatting van het ontbrekende aantal spinthoutringen. Bij een boom, die 100 tot 200 jaar oud is, is de veldatum dus xxxx AD + 20 (± 6) + X.

n = Totaal aantal jaarringen in het houtmonster.

x = Geschat aantal missende ringen (kernhout en/of spinthout) tot de wankant.

%PV = "Gleichlaufigkeit" (Duitse term) of "Percentage of Parallel Variation" (Engelse term); het percentage van de ringen in het onderzochte jaarringpatroon die aan de referentiechronologie identieke toe- en afnames van de breedte vertonen op de door de datering van het patroon aangegeven positie t.a.v. de referentiechronologie. De significantie van dit percentage is een functie van de lengte in jaren van het onderzochte jaarringpatroon en de referentie chronologie.

t = De waarde die resulteert uit een Students t-test op de kruiscorrelatie die behoort bij de beste "match" tussen het onderzochte jaarringpatroon en de referentiechronologie.

P = De kans (uitgedrukt als een fractie van 1) dat de gevonden waarde voor %PV per toeval optreedt, dus niet op een datering duidt.

RING-rapport 2008080, Emmeloord, Noordoostpolder, Flevoland, Kavel NM 107

Bijlage 4. Gebruikte referentiechronologieën

NLNOOR1B/E	Hout toegepast in Noordwest-Nederland (*Jansma, 1995*).
NLNOOR02	Cluster X (*Van Daalen, n.p.*).
NLTWEN03	Twente 2 (*De Vries, n.p.*).

RING-rapport 2008080, Emmeloord, Noordoostpolder, Flevoland, Kavel NM 107

Bijlage 5. Literatuur

Hollstein, E., 1980. *Mitteleuropäische Eichenchronologie.* Verlag Phillipp von Zabern, Mainz am Rhein.

Jansma, E., 1995. *Rememberings, The development and application of local and regional tree-ring chronologies of oak for the purposes of archaeological and historical research in the Netherlands.* Diss. UvA (Nederlandse Archeologische Rapporten 19), 150 pp.

Jansma, E., 2007: *Datering, herkomst en bouwvolgorde van De Meern 4. In: T. de Groot & J.-M.A.W. Morel (red.), 2007: Het schip uit de Romeinse tijd De Meern 4 nabij boerderij de Balije, Leidsche Rijn, gemeente Utrecht. Waardestellend onderzoek naar de kwaliteit van het schip en het conserverend vermogen van het bodemmilieu.* RACM.

Wazny, T., 1990. *Aufbau und Anwendung der Dendrochronologie fuer Eichenholz in Polen.* Dissertatie Universiteit van Hamburg.

14 The Spakenburg-I cog

Pollen spectrum of the moss caulking

E. Mook-Kamps

Interpretation

The sample of moss caulking from the Spakenburg-I cog produced few pollen and a lot of charcoal.
Trees: 42%, herbs 58%.
Alder 16%, oak 9%, hazel 6%, birch 5%, hornbeam 2%, beech 1%, willow 1%, elm 0,5% and pine tree 2%
Heather 18%, grasses 11%, cyperus grasses 3%, grains 17% (buckwheat 0,3%), field weeds 2% (ribwort plantain and sorrel).
Local: wood fern and *Pteridium aquilinum* in small percentages and 15% peat moss.
Interdeterminates: 1,7%.
Circumstances: nutrient poor till nutrient rich, the beech and hornbeam will have grown on nutrient rich grounds. A peat bog bordering on cultivated land. Probably cultivation of buckwheat

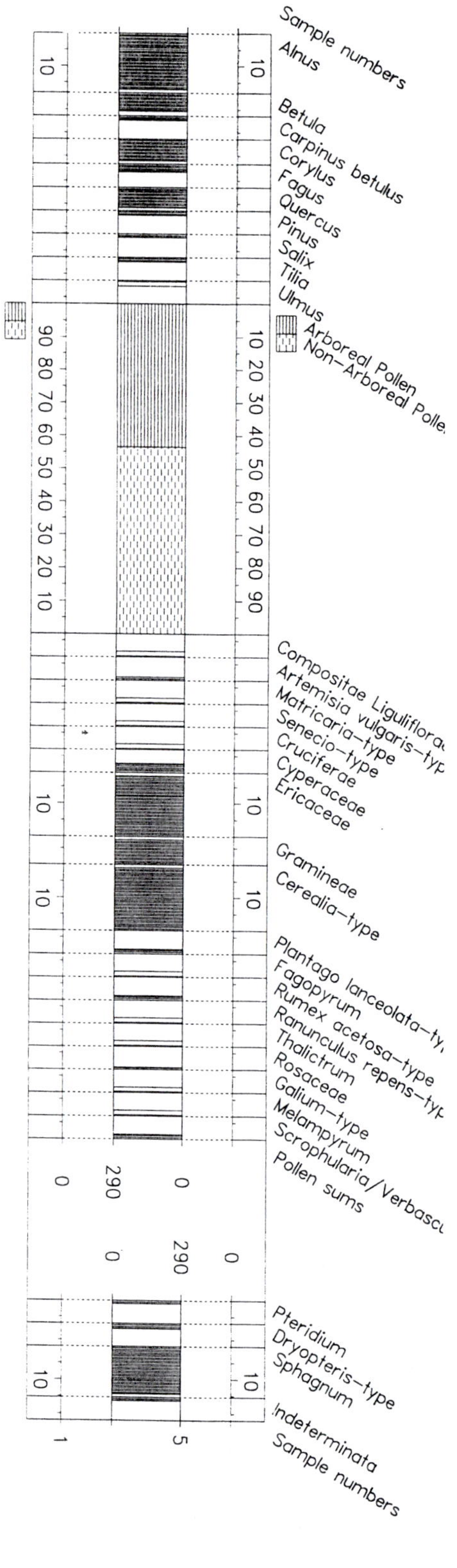

Examination of the bones

F. Laarman

For the introduction, the pattern of expectations and the conclusion of the study of bones from the eight cogs see appendix of the Nijkerk-II cog..

Results Spakenburg-I cog
Eleven bones have been found in this wreck. The numbers and their weight are presented in table 1.

This is a collection of bones that match the expectations of meat on board ship. Six out of eleven bones show chop marks and - with the exception of a chicken head - the bones are from the meat parts of the animals, and ready for consumption. It is possible that the chicken was taken on board alive and slaughtered during the voyage. The four cattle bones and the chicken head show traces of rounding. It is possible that this wreck has lain for some time without a sediment cover on the bottom of the Zuiderzee.

Table 1. Number of bones per species and their weight

	N	grams
cattle	4	115.8
pig	1	14.9
sheep/goat	3	13.8
chicken	1	3.2
goose	2	5.7

r - right
l - left
p - proximal
d - distal
ve - fused
onv - non fused
juv - juvenile
PI - phalanx 1
PII - phalanx 2
PIII - phalanx 3
Gl - greatest length
Bp - width proximal
Bd - width distal
mnd. - months

List of determinations of bones from the Spakenburg I-cog

Invent- no. ZN42 II-	description	weight in grams
26	cattle 1x rib (complete), lightly rounded	71.6
27	cattle 1x rib (chop mark), rounded	11.7
28	goose 1x *tibiotarsus* (l) d ve	4.7
29	sheep/goat 1x rib (chop mark)	5.1
30	sheep/goat 1x rib (chop mark)	3.2
31	goose 1x *pelvis*	1.0
32A+B	cattle 1x rib (chop marks), rounded	14.9
33A+B	pig 1x rib (chop marks)	10.2
34	sheep/goat 1x *lumbar vertebra* onv (chop mark)	5.5
35	cattle 1x *patella* (r), rounded	17.6
36	chicken 1x *cranium*, rounded	3.2

15 The Spakenburg-II cog

Pollen spectrum of the moss caulking

E. Mook-Kamps

Interpretation

The sample of caulking moss (11) from the Spakenburg-II cog is of pure peat moss and the conservation is good.
Trees: 21%, herbs 79%.
Birch 7%, alder 5%, oak 3%, hazel 3% and small percentages for hornbeam, beech, walnut, ash, pine, willow, linden and elm .
Heather 68%, grasses 3%, cyperus grasses 2%, grains 2%, buckwheat 0,1%, field weeds 2% (sandspurrys and sorrel).
Local: wood fern and *Pteridium aquilinum* both 1%, peat moss 14%.
Interdeterminates: 1%.
Fields of heather, peat and a rich diversity of trees. For instance tree growth on the richer grounds of a river valley. Some agriculture of rye and buckwheat in the near environment.

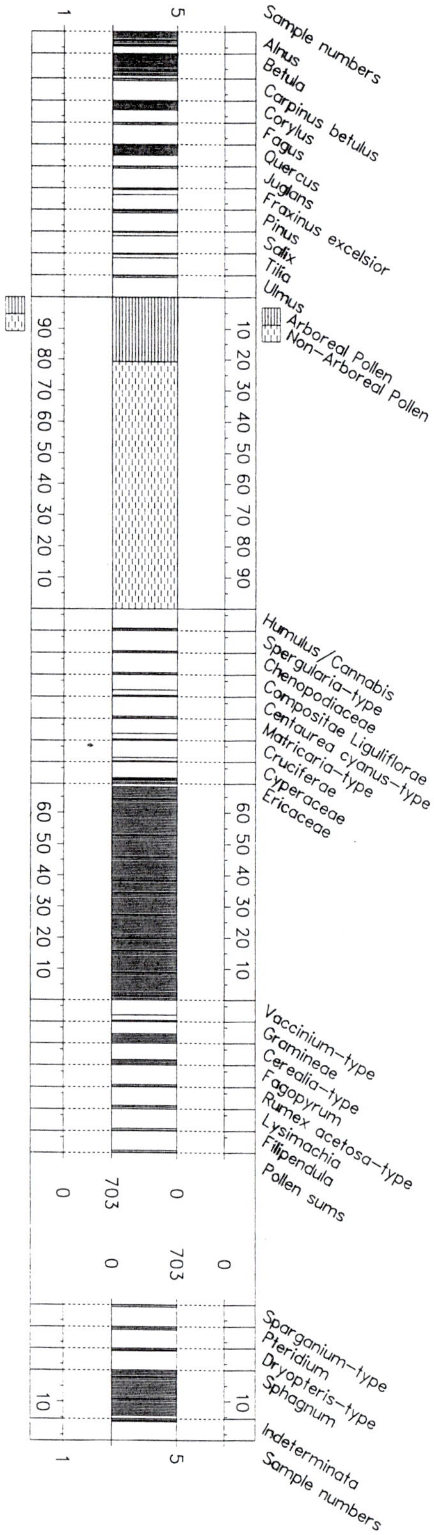

Examination of the bones

F. Laarman

For the introduction, the pattern of expectations and the conclusion of the study of bones from the eight cogs see appendix of the Nijkerk-II cog.

Results Spakenburg-II cog

Only three cattle bones were collected during the excavation of the Spakenburg-II cog. Two of these may be considered consumption waste and the third bone is a *metatarsus* from a meatless lower leg. This bone is part of offal, but is often used as raw material for bone working. This unworked bone has a length of 209 mm, so the height of the animal at the withers was 114 cm.[5] This size is comparable with 14th-century cattle from Deventer.[6] The bones show no traces of rounding.

r	- right	PI	- phalanx 1
l	- left	PII	- phalanx 2
p	- proximal	PIII	- phalanx 3
d	- distal	Gl	- greatest length
ve	- fused	Bp	- width proximal
onv	- non fused	Bd	- width distal
juv	- juvenile	mnd.	- months

List of determinations of bones from the Spakenburg-II cog

Invent-no. ZN43-	description	weight in grams
14	cattle 1x *metatarsus* (r) Gl = 209 mm Bp = 41 mm Bd = 49 mm	141.4
15A+B	cattle 1x *cervical vertebra* onv	85.6
16	cattle 1x boned cartilage, rib end	25.5

5. Van den Driesch en Boesneck 1974.
6. IJzereef and Laarman 1986.

Dendrochronological results RING

E. Hanraets & E. Jansma

ring

Stichting Nederlands Centrum voor Dendrochronologie
Kerkstraat 1, 3811 CV Amersfoort
Tel (033) 63 42 33
Fax (033) 65 32 35
E-mail: ring @ archis.nl

RAPPORTAGE

Vindplaats: Ketelhaven kavel:NZ-43

Datum monstername : -
Datum aanvraag : december 1993

Aanvrager:
Naam : Ing. R. Oosting
Instantie : Museum voor Scheepsarcheologie
Stad : Lelystad

Onderzoekers:
Naam : E. Hanraets, E. Jansma
Instantie : RING
Datum onderzoek : juni 1994

Vondstno.	Dendronaam	kern	spint	wankant	n
G2C/BB, m.15	kn4 015	-	11	-	112
G5A, m.31	kn4 031	-	-	-	127
SPT 15, m.38	kn4 038	+3	-	-	89
G1C/BB, m.42	kn4 042	-	-	-	152
G10B/SB, m.45	kn4 045	-	-	-	109

DATERING

Dendronaam	Referentie-chronologie	Datering	%PV	t	P
kn4 015	kn4 045	1288-1399AD	75.8	7.27	0.001
kn4 031	niet dateerbaar				
kn4 038	niet dateerbaar				
kn4 042	niet dateerbaar				
kn4 045	NL/Westfalen	1246-1354AD	69.0	7.66	0.01

de middelcurve van beide monsters dateert als volgt:

Dendronaam	**Referentie-chronologie**	**Datering**	%PV	t	P
kn4 2x0	NL/Westfalen	1246-1399 AD	62.1	7.15	0.01

INTERPRETATIE VAN DE DATERINGEN

Dendronaam	**datering**	**veldatum**
kn4 2x0	1399 + 9 jaar ± 6	1402-1414 AD

De schatting van de ontbrekende ringen spinthout is gebaseerd op een totaal aantal spintringen van 20 ± 6 (bomen de 100 en 200 jaar oud; Hollstein 1980). De 3 overige monsters, kn4 031, kn4 038 en kn4 042, konden noch relatief (ten opzichte van elkaar), noch absoluut (t.o.v. een referentiechronologie) gedateerd worden.

----

16 The Almere-I cog

Pollen spectrum of the moss caulking

E. Mook-Kamps

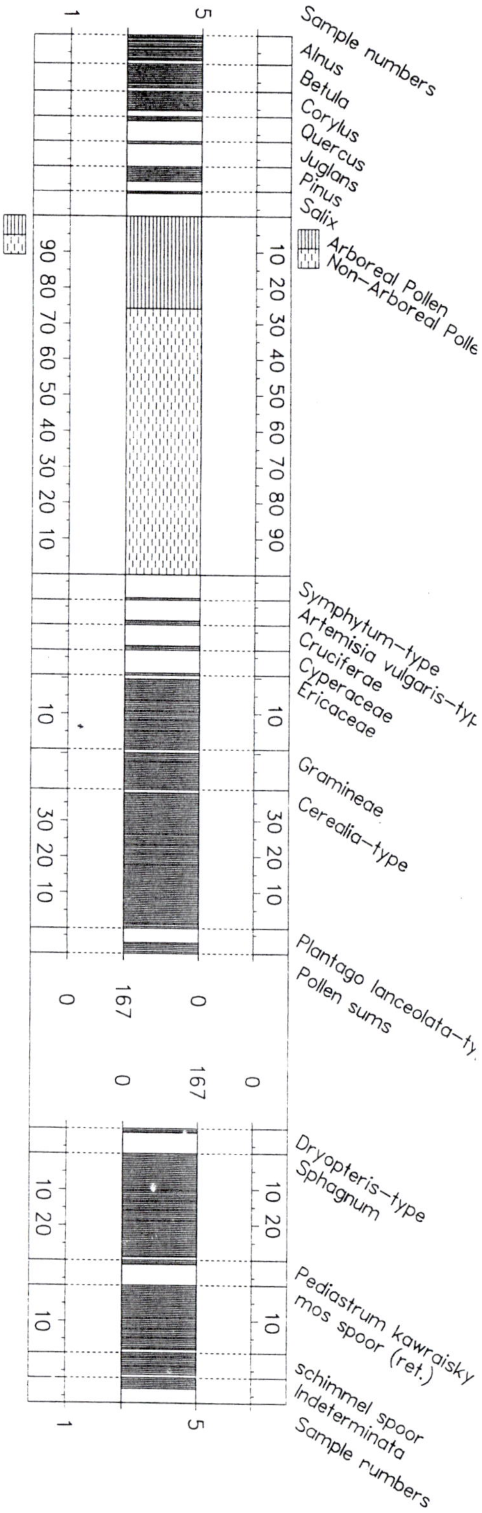

Interpretation

The caulking from the Almere-I cog is peat moss, the conservation of the sample (1) is between mediocre and bad.
Trees 25%, herbs 75%.
As hand birch 7%, hazel 5%, oak 1% and willow 0,5%.
Grains: 38%, heather 20%, grasses 10%, cyperus grasses 1% (probably underrepresented because of the bad conservation) field weeds 5% (Brassicaceae, common mugwort and ribwort plantain).
Local: peat moss 30%, reticulate moss traces 18%, mould spores 6%.
Indeterminates: 3%.
Nutricient poor area. A peat area bordering uncultivated land.
Heather on the higher parts of the peat area.

Examination of the bones

F. Laarman

For the introduction, the pattern of expectations and the conclusion of the study of bones from the eight cogs see appendix of the Nijkerk-II cog.

During the excavation of the Almere-I cog five bones were collected. In table 1, the amount of bones and their weight are presented.

Only a few bones were found in the wreck. Except for a marlinspike made from the end of an antler of red deer, the bones are the consumption waste of cattle and sheep/goat. The bone from sheep/goat has a chop mark. Traces of rounding were not seen.

Note: A number of human bones were found in the wreck. These are not subject of this study and were studied earlier by G.J.R. Maat.[7]

Table1. The number of bones per animal and their weight

	N	Grams
red deer	1	65.9
sheep/goat	1	2.5
cattle	3	111.7

r	- right	PI	- phalanx 1
l	- left	PII	- phalanx 2
p	- proximal	PIII	- phalanx 3
d	- distal	Gl	- greatest length
ve	- fused	Bp	- width proximal
onv	- non fused	Bd	- width distal
juv	- juvenile	mnd.	- months

Determination list of the bones from the Almere I cog

Invent no. ZW13-	Omschrijving	weight in grams
48	red deer 1x end of antler (worked) marlinspike	65.9
50	sheep/goat 1x rib (chop mark)	2.5
51	cattle 1x *tibia* (l)	44.1
52	cattle 1x rib (burned)	2.0
53	cattle 1x *radius* p ve	65.6

7 Maat in: Hocker & Vlierman 1996a, 99-103.

ring

Stichting Nederlands Centrum voor Dendrochronologie
Kerkstraat 1, 3811 CV Amersfoort
Tel (033) 63 42 33
Fax (033) 65 32 35
E-mail: ring @ archis.nl

RAPPORTAGE

Object/vindplaats: Almere Wijk 13; scheepswrak

Datum monstername : -
Datum aanvraag : nov. 1996

Aanvrager:
Naam : Dhr. R. Oosting
Instantie : NISA
Straat : Vossemeerdijk 21
Stad : 8251 PM Dronten

Onderzoekers:
Naam : Dr. E.Jansma, Drs. E.Hanraets
Instantie : RING
Datum onderzoek : nov. 1996

Meetgegevens RING

Vondstno./object:	Vondstomschrijving (vondstnummer, balklaag enz.)
Dendronaam:	bestandsnaam RING
kern:	(afstand tot) de eerstgevormde (oudste) jaarring in de stam
spint:	aantal ringen spinthout
wankant:	laatstgevormde jaarring (direkt onder bast), nodig voor een absolute datering van de veldatum
n:	totaal aantal jaarringen in houtmonster

Vondstno./object	**Dendronaam**	**kern**	**spint**	**wankant**	**n**
GB3/SB, m.3	kaw011	+40±20	-	-	68
GG3/BB, m.6	kaw021	+40±20	-	-	53
buikdenning, m.11	kaw031	+20±10	-	-	36
buikdenning, m.12	kaw041	+20±10	-	-	38
buikdenning, m.13	kaw051	+20±10	-	-	31
GE/SB, m.1	kaw061	+40±20	-	-	40

1

DATERING

Toelichting:

Referentiecurve:	De standaardkalender(s) die de beste 'match' opleverde(n). De literatuurverwijzingen staan in Jansma 1995.
Datering:	De kalenderjaren waarin respectievelijk de eerst- en laatstgemeten ring in het houtmonster zijn gevormd.
%PV:	"Gleichlaufigkeit" (Duitse term) of "Percentage Agreement" (Engelse term); het percentage van de ringen in het onderzochte jaarringpatroon die aan de referentiechronologie identieke toe- en afnames van de breedte vertonen op de door de datering van het patroon aangegeven positie t.o.v. de referentiechronologie. De significantie van dit percentage is een functie van de lengte in jaren van het onderzochte jaarringpatroon (zie bijgevoegd documentatiemateriaal).
t:	De waarde die resulteert uit een Student_ t-test op de kruiscorrelatie die behoort bij de beste "match" tussen het onderzochte jaarringpatroon en de referentiechronologie.
α:	De overschrijdingskans behorend bij de gevonden waarde van %PV. Wanneer deze kans bijvoorbeeld 0.001 is, betekent dit dat er een kans is kleiner dan of gelijk aan 1 op de 1000 dat de gevonden waarde van %PV toeval is, dus niet samenhangt met een werkelijk verband tussen de vergeleken series (de kans dus dat de datering onjuist is).

Dendronaam	**Referentie-curve**	**Datering**	**%PV**	**t.**	**α**
kaw1tmt6	geen datering				

INTERPRETATIE VAN DE DATERINGEN

De monsters zijn met al onze chronologieën vergeleken, dus ook met de Franse, Engelse en Poolse, maar een datering was helaas niet mogelijk.
De monsters hebben erg weinig ringen, en zijn vrij onregelmatig. Daardoor was zelfs een onderlinge correlatie niet duidelijk.

Dendrochronological report RING

E. Jansma & A.E.M. Hanraets

ring

Stichting Nederlands Centrum voor Dendrochronologie
Kerkstraat 1, 3811 CV Amersfoort
Tel (033) 63 42 33
Fax (033) 65 32 35
E-mail: ring @ archis.nl

RAPPORTAGE

Object/vindplaats: kavel M.61/OFL (scheepswrak)

Datum monstername : 10-1-1996
Datum aanvraag : 22-8-1996

Aanvrager:
Naam : Dhr. R. Oosting
Instantie : NISA
Straat : Vossemeerdijk 21
Stad : 8251 PM Dronten
Telefoon : 0321-313287

Onderzoekers:
Naam : Dr E.Jansma, Drs. E.Hanraets
Instantie : RING
Datum onderzoek : aug. 1996

Meetgegevens RING

Vondstno./object:	Vondstomschrijving (vondstnummer, balklaag enz.)
Dendronaam:	bestandsnaam RING
kern:	(afstand tot) de eerstgevormde (oudste) jaarring in de stam
spint:	aantal ringen spinthout
wankant:	laatstgevormde jaarring (direkt onder bast), nodig voor een absolute datering van de veldatum
n:	totaal aantal jaarringen in houtmonster

Vondstno./object	**Dendronaam**	**kern**	**spint**	**wankant**	**n**
GA1/SB, m.4 km6041	+40±20	-	-	121	
G1B/SB, m.6 km6061	+40±20	-	-	120	

1

DATERING

Toelichting:

Referentiecurve:	De standaardkalender(s) die de beste 'match' opleverde(n). De literatuurverwijzingen staan in Jansma 1995.
Datering:	De kalenderjaren waarin respectievelijk de eerst- en laatstgemeten ring in het houtmonster zijn gevormd.
%PV:	"Gleichlaufigkeit" (Duitse term) of "Percentage Agreement" (Engelse term); het percentage van de ringen in het onderzochte jaarringpatroon die aan de referentiechronologie identieke toe- en afnames van de breedte vertonen op de door de datering van het patroon aangegeven positie t.o.v. de referentiechronologie. De significantie van dit percentage is een functie van de lengte in jaren van het onderzochte jaarringpatroon (zie bijgevoegd documentatiemateriaal).
t:	De waarde die resulteert uit een Student_ t-test op de kruiscorrelatie die behoort bij de beste "match" tussen het onderzochte jaarringpatroon en de referentiechronologie.
α:	De overschrijdingskans behorend bij de gevonden waarde van %PV. Wanneer deze kans bijvoorbeeld 0.001 is, betekent dit dat er een kans is kleiner dan of gelijk aan 1 op de 1000 dat de gevonden waarde van %PV toeval is, dus niet samenhangt met een werkelijk verband tussen de vergeleken series (de kans dus dat de datering onjuist is).

Dendronaam	**Referentie-curve**	**Datering**	**gl.**	**t.**	**α**
km6041	DLS9	1155-1275 AD	71.0	5.38	0.0001
km6061	DLS9	1157-1276 AD	63.8	5.53	0.005

De beide monsters zijn van dezelfde boom afkomstig. Ze zijn gemiddeld tot:

km646	DLS9	1155-1276 AD	68.2	5.67	0.0001

INTERPRETATIE VAN DE DATERINGEN

Object	**Dendronaam**	**Laatste ring**	**+ X =**	**Veldatum**
GA1/SB + G1B/SB				
	km646	1276 AD	+ tenminste 20±6	na 1296 AD±6

X = het ontbrekende aantal spintringen. Dit is geschat op 20 ± 6 bij een boom van 100-200 jaar oud (Hollstein 1978). Er was echter geen spintgrens te zien, zodat er ook nog een onbekend aantal kernringen kan ontbreken. De datering valt dus een onbekend aantal jaren nà 1296 AD ± 6.

---

18 The Swifterbant cog

Examination of the bones

F. Laarman

For the introduction, the pattern of expectations and the conclusion of the study of bones from the eight cogs see appendix of the Nijkerk-II cog.

During the first survey of the Swifterbant cog, only four bones were found. Table 1 shows the bones and their weight.

The four bones, of which two are from ducks, may have been washed into the wreck and the cattle skull fragment probably also is not part of the ship's provisions. In contrast, the fragment of a sheep/goat rib could well have been from a piece of meat for consumption. The two bones from mammals show chop marks, but all four bones are not rounded.

Table 1. Number of bones and their weight

	N	Grams
duck	2	1.2
cattle	1	158.2
sheep/goat	1	0.9

r	- right	PI	- phalanx 1
l	- left	PII	- phalanx 2
p	- proximal	PIII	- phalanx 3
d	- distal	Gl	- greatest length
ve	- fused	Bp	- width proximal
onv	- non fused	Bd	- width distal
juv	- juvenile	mnd.	- months

List of determinations of the bones from the Swifterbant cog

Invent.- no. Og 77-	description	weight in grams
20	cattle 1x *cranium* + hornpit (r) circumference base 173 mm	158.2
21A	duck 1x *tibiotarsus* (l) (small species)	0.5
21B	duckl 1x *radius* (l) G1 = 74 mm (chop mark)	0.7
22	sheep/goat 1x rib (chop mark)	0.9

Dendrochronological report RING
A.E.M. Hanraets

ring

Nederlands Centrum voor Dendrochronologie/Stichting RING
Rijksdienst voor het Oudheidkundig Bodemonderzoek (ROB)
Kerkstraat 1, 3811 CV (Postbus 1600, 3800 BP) Amersfoort
Tel (033) 422 75 13; Fax (033) 422 77 99
E-mail RING@ARCHIS.NL

RAPPORTAGE DATEREND ONDERZOEK

Object: kavel: OG-77; scheepswrak   coörd.:

Aanvrager
Naam : Mevrouw L. Koehler
Instantie : NISA
Stad : Lelystad

Onderzoekers
Naam : drs A. E. M. Hanraets
Instantie : RING/ROB
Datum onderzoek : juli 2001

Resultaten van het onderzoek

Vondstno./ Object	Dendro-naam	kern	spint	wankant	n	referentiecurve	datering	t	%PV	p
m. 1	kog 01.1	+2	26	+ca.10	109	NLWestfalen *(Tisje)*	1187-1295	4.9	64.4	0.005
m. 3	kog 02.0	-	2	-	153	NLHist_3 *(Jansma)*	1125-1277	8.5	71.1	0.0001
m. 2	kog 02.0	+ca.15	-	-	69	NSS 7*	1212-1280	3.3	66.9	0.01

*NSS 7 is de referentiechronologie van Niedersachsen, subregio Niedersächsischer Küstenraum *(Leuschner)*.

Toelichting op de resultaten
kern = (afstand tot) de eerstgevormde (oudste) jaarring in de stam
spint = aantal ringen spinthout.
wankant = laatstgevormde jaarring (direct onder de bast), nodig voor een absolute datering van de veldatum
n = totaal aantal jaarringen in het houtmonster
%PV = "Gleichlaufigkeit" (Duitse term) of "Percentage of Parallel Variation" (Engelse term); het percentage van de ringen in het onderzochte jaarringpatroon die aan de referentiechronologie identieke toe- en afnames van de breedte vertonen op de door de datering van het patroon aangegeven positie t.a.v. de referentiechronologie. De significantie van dit percentage is een functie van de lengte in jaren van het onderzochte jaarringpatroon en de referentie chronologie.
t = De waarde die resulteert uit een Students t-test op de kruiscorrelatie die behoort bij de beste "match" tussen het onderzochte jaarringpatroon en de referentiechronologie.
P = De kans (uitgedrukt als een fractie van 1) dat de gevonden waarde voor %PV per toeval optreedt, dus niet op een datering duidt.

Interpretatie van de datering

Objectcode	Einddatering	Schatting ontbrekend spinthout	Veldatum
kog 01.1	1295 AD	+ ca. 10	ca. 1305 AD
kog 02.0	1277 AD	+ 24 +/- 8	1301 AD +/- 8
kog 03.0	1280 AD	+ $\geq$ 20 $\pm$ 6	na 1300 AD $\pm$ 6

Voor de berekening van het geschatte aantal ontbrekende spintringen gaan we uit van een gemiddeld aantal spintringen van 16 ± 5 bij een boom tot 100 jaar oud, 20 ± 6 bij een boom van 100 tot 200 jaar oud, en 26 ± 8 bij een boom ouder dan 200 jaar (Hollstein, 1980). Als er helemaal geen spintringen meer op het monster aanwezig zijn, is het onbekend hoeveel *kernhoutringen* er ook nog ontbreken, en ligt de veldatum dus een onbekend aantal jaren nà de einddatering + de schatting van het ontbrekende aantal spinthoutringen.

Literatuur

Hollstein, E., 1980. *Mitteleuropäische Eichenchronologie.* Verlag Phillipp von Zabern, Mainz am Rhein.
Jansma, E., 1995. *RemembeRINGs: The development and application of local and regional tree-ring chronologies of oak for the purposes of archaeological and historical research in the Netherlands.* Diss. UvA (Nederlandse Archaeologischerapporten 19), 150 pp.

19 The Almere-II cog

Dendrochronological report RING

A.E.M. Hanraets

ring

Nederlands Centrum voor Dendrochronologie/Stichting RING
Rijksdienst voor het Oudheidkundig Bodemonderzoek (ROB)
Kerkstraat 1, 3811 CV (Postbus 1600, 3800 BP) Amersfoort
Tel (033) 422 75 13; Fax (033) 422 77 99
E-mail RING@ARCHIS.NL

RAPPORTAGE DATEREND ONDERZOEK

Object: kavel: CZ-46/Oostvaardersplassen, scheepswrak

Aanvrager

Naam : Laura Köhler
Instantie : NISA
Stad : Lelystad

Onderzoekers

Naam : Drs. A. E. M. Hanraets
Instantie : RING/ROB
Datum onderzoek : september 1999

Resultaten van het onderzoek

Vondstno./ Object	Dendro-naam	kern	spint	wankant	n	referentiecurve	datering	t	%PV	p
m.1/kiel	kcz 01.0	+±10	-	-	46	-	GEEN	-	-	-
m.2/gang	kcz 02.0	-	-	-	100	-	GEEN	-	-	-
GC2/BB	kcz 03.0	+±30	-	-	86	NLHist_2 (*Jansma*)	1222-1307	5.1	73.5	0.0001
G2B/BB	kcz 04.1	+±20	-	-	65	NLWestfalen (*Tisje*)	1230-1294	7.0	77.3	0.0001
G2A/BB	kcz 05.0	+±20	-	-	49	NLWestfalen	1242-1290	5.7	74.0	0.001

Toelichting op de resultaten

kern = (afstand tot) de eerstgevormde (oudste) jaarring in de stam
spint = aantal ringen spinthout.
wankant = laatstgevormde jaarring (direct onder de bast), nodig voor een absolute datering van de veldatum
n = totaal aantal jaarringen in het houtmonster
%PV = "Gleichlaufigkeit" (Duitse term) of "Percentage of Parallel Variation" (Engelse term); het percentage van de ringen in het onderzochte jaarringpatroon die aan de referentiechronologie identieke toe- en afnames van de breedte vertonen op de door de datering van het patroon aangegeven positie t.a.v. de referentiechronologie. De significantie van dit percentage is een functie van de lengte in jaren van het onderzochte jaarringpatroon en de referentie chronologie.
t = De waarde die resulteert uit een Students t-test op de kruiscorrelatie die behoort bij de beste "match" tussen het onderzochte jaarringpatroon en de referentiechronologie.
P = De kans (uitgedrukt als een fractie van 1) dat de gevonden waarde voor %PV per toeval optreedt, dus niet op een datering duidt.

Interpretatie van de datering

Object/code	Einddatering	Schatting ontbrekend spint	Veldatum
kcz 03.0	1307 AD	+ ≥ 20 ± 6	NA 1327 AD ± 6
kcz 04.1	1294 AD	+ ≥ 20 ± 6	NA 1314 AD ± 6
kcz 05.0	1290 AD	+ ≥ 20 ± 6	NA 1310 AD ± 6

Voor de berekening van het geschatte aantal ontbrekende spinthoutringen gaan we uit van een gemiddeld aantal spintringen van 16 ± 5 bij een boom tot 100 jaar oud, 20 ± 6 bij een boom van 100 – 200 jaar oud, 26 ± 8 bij een boom ouder dan 200 jaar (Hollstein, 1980). Als er helemaal geen spinthout meer op het monster aanwezig is, is het onbekend hoeveel ringen *kernhout* ook nog ontbreken, en ligt de veldatum een onbekend aantal jaren na de einddatering + de schatting van het aantal ontbrekende spintringen.

Monster kcz 02.0 heeft een zwakke statistische en visuele overeenkomst met de drie daterende monsters, maar deze is te zwak voor een betrouwbare datering. Met monster kcz 01. 1 is er helemaal geen overeenkomst.

Literatuur

Hollstein, E., 1980. *Mitteleuropäische Eichenchronologie.* Verlag Phillipp von Zabern, Mainz am Rhein.
Jansma, E., 1995. *RemembeRINGs: The development and application of local and regional tree-ring chronologies of oak for the purposes of archaeological and historical research in the Netherlands.* Diss. UvA (Nederlandse Archaeologischerapporten 19), 150 pp.

20 The Doel-I cog

Dendrochronological report RING

A.E.M. Hanraets

ring

Nederlands Centrum voor Dendrochronologie/Stichting RING
Rijksdienst voor het Oudheidkundig Bodemonderzoek (ROB)
Kerkstraat 1, 3811 CV (Postbus 1600, 3800 BP) Amersfoort
Tel (033) 422 75 13; Fax (033) 422 77 99

RAPPORTAGE DATEREND ONDERZOEK

Object: Antwerpen, Doel; kogge

Aanvrager

Naam	: Prof. Dr. Guy Deboe
Instantie	: Instituut voor het Archaeologisch Patrimonium
Straat	: Doornveld Industrie Asse 3 nr. 11, bus 30
Postcode, Stad	: B-1731 Asse-Zellik, België

Onderzoekers

Naam	: Drs. A. E. M. Hanraets
Instantie	: ROB/RING
Datum onderzoek	: november/december 2000

Resultaten van het onderzoek

Vondstno./ Object	Dendro- naam	kern	spint	wankant	n	referentiecurve	datering	t	%PV	p
GN2 - 1SB	kka 01.0	+ca.5	-	-	95	NLWestfalen *(Tisje)*	1145-1239	5.1	70.2	0.0002
						NSS 4 *		5.1	69.7	0.0002
GL2 - 1SB	kka 02.1	+ca.10	-	-	106	NSS 4	1151-1256	6.2	69.0	0.0002
						NLWestfalen		5.0	67.1	0.0005
GK2 - 1SB	kka 03.0	+ca.3	-	-	67	kka 07.1	1197-1263	9.2	81.0	0.0001
S1 - 3SB	kka 04.0	+1	20	-	159	NSS 4	1164-1322	5.4	65.8	0.0002
						NSS 5*		6.0	65.5	0.0002
GR1 - 3BB	kka 05.1	-	3	-	91	NSS 5	1220-1310	4.9	68.9	0.0005
GQ1 - 2BB	kka 06.1	+ca.5	1(3)**	-	159	NSS 5	1150-1308	7.2	68.4	0.0001
						NSS 4		6.5	68.7	0.0001
GJ1 - 2SB	kka 07.1	+ca.15	-	-	61	NSS 9*	1213-1273	5.1	71.1	0.001
						NSS 1*		5.1	71.1	0.001
GM4 - 1BB	kka 08.2	-	-	-	79	NSS 1	1223-1301	5.4	67.9	0.005
GE3 - 2SB	kka 09.1	+ca.15	-	-	46	-	GEEN	-	-	-
GJ1 - 1BB	kka 10.1	+2	-	-	101	NSS 4	1201-1301	4.8	65.5	0.005
						NSS 5		4.6	675	0.0005
GG1 - 1BB	kka 11.1	+ca.10	13	-	83	kka 3710***	1240-1322	4.6	76.2	0.0001
GA2 - 2SB	kka 12.0	+ca.10	-	-	49	-	GEEN	-	-	-
GF4 - 2BB	kka 13.0	-	1	-	46	-	GEEN	-	-	-
GB2 - 1SB	kka 14.1	+ca.10	-	-	84	NSS 9	1198-1281	6.9	74.1	0.0001
						NSS 1+3		6.8	74.1	0.0001
S13A - BB	kka 15.0	+2	19	**+1=wk,z/w**	93	NSS 4	1232-1324	4.4	66.3	0.005
S6 - SB	kka 16.0	+1	14	**+1=wk,z/w**	104	NSS 7	1221-1324	4.6	62.6	0.02
						O.Friesl. *(Leuschner)*		4.6	63.6	0.01
S8 - SB	kka 17.1	+1	-	-	57	NSS 5	1220-1276	4.6	64.3	0.04

GC2 - 2BB Dit monster is ongeschikt. Het heeft minder dan 35 ringen.

*

NSS 1 is de refentiechronologie van de regio Niedersachsen, subregio Südniedersächsisches Bergland (Leuschner)
NSS 4 is de refentiechronologie van de regio Niedersachsen, subregio Nordwestdeutsche Bergland Schwelle (Leuschner)
NSS 5 is de refentiechronologie van de regio Niedersachsen, subregio Ostniedersächsisches Tiefland (Leuschner)
NSS 9 is de refentiechronologie van de regio Niedersachsen, subregio Westniedersächsisches Tiefland (Leuschner)

**

Drie spintringen zijn niet mee gemeten omdat ze beschadigd zijn.

***

kka 3710 is de middelcurve van kka 03.0, 07.1,10.1. De overlap tussen kka 3710 en kka 11.1 is 62 ringen.

De monsters *kka 01.0 (GN2-1SB) en kka 02.1 (GL2-1SB)* hebben een zeer grote onderlinge overeenkomst: t= 16.3, %PV=85.2. Deze beide monsters moeten van dezelfde boom afkomstig zijn. Monster *kka 14.1 (GB2-1S*B) heeft met beide een zeer goede overeenkomst: t=7.1, %PV=68.5.
Ook *kka 05.1 (GR1-3BB) en kka 06.1 (GQ1-2BB)*: t=11.3, %PV- 78.9 zijn zeker afkomstig van één boom.
Waarschijnlijk zijn ook *kka 03.0 (GK2-1SB) en kka 07.1 (GJ1-2SB)* t= 9.2, %PV= 81.0 afkomstig van één boom. Monster kka 10.1 (GJ1-1BB) heeft met deze twee monsters een zeer goede overeenkomst: t= 6.7. %PV=75.0.

Toelichting op de resultaten

kern	=	(afstand tot) de eerstgevormde (oudste) jaarring in de stam
spint	=	aantal ringen spinthout.
wankant	=	laatstgevormde jaarring (direct onder de bast), nodig voor een absolute datering van de veldatum
n	=	totaal aantal jaarringen in het houtmonster
%PV	=	"Gleichlaufigkeit" (Duitse term) of "Percentage of Parallel Variation" (Engelse term); het percentage van de ringen in het onderzochte jaarringpatroon die aan de referentiechronologie identieke toe- en afnames van de breedte vertonen op de door de datering van het patroon aangegeven positie t.a.v. de referentiechronologie. De significantie van dit percentage is een functie van de lengte in jaren van het onderzochte jaarringpatroon en de referentie chronologie.
t	=	De waarde die resulteert uit een Students t-test op de kruiscorrelatie die behoort bij de beste "match" tussen het onderzochte jaarringpatroon en de referentiechronologie.
P	=	De kans (uitgedrukt als een fractie van 1) dat de gevonden waarde voor %PV per toeval optreedt, dus niet op een datering duidt.

Interpretatie van de datering

Object/code	Einddatering	Schatting ontbrekend spint	Veldatum
kka 01.0+02.1	1256 AD	+ ≥ 20 ± 6	NA 1276 AD ± 6
kka 03.0	1263 AD	+ ≥ 20 ± 6	NA 1283 AD ± 6
kka 04.0	1322 AD	+ max.6	tussen 1323 en 1328 AD
kka 051+06.1	1310 AD	+ 17 ± 6	1327 AD ± 6
kka 07.1	1273 AD	+ ≥ 20 ± 6	NA 1293 AD ± 6
kka 08.2	1301 AD	+ ≥ 20 ± 6	NA 1321 AD ± 6
kka 10.1	1301 AD	+ ≥ 20 ± 6	NA 1321 AD ± 6
kka 11.1	1322 AD	+ 7 ± 6	1329 AD ± 6
kka 14.1	1281 AD	+ ≥ 20 ± 6	NA 1301 AD ± 6
kka 15.0	1324 AD	+ 1	1325 AD
kka 16.0	1324 AD	+ 1	1325 AD
kka 17.1	1276 AD	+ ≥ 20 ± 6	NA 1296 AD ± 6

Voor de berekening van het geschatte aantal ontbrekende spinthoutringen gaan we uit van een gemiddeld aantal spintringen van 16 ± 5 bij een boom tot 100 jaar oud, 20 ± 6 bij een boom van 100 – 200 jaar oud, 26 ± 8 bij een boom ouder dan 200 jaar (Hollstein, 1980). Als er helemaal geen spinthout meer op het monster aanwezig is, is het onbekend hoeveel ringen *kernhout* ook nog ontbreken, en ligt de veldatum een onbekend aantal jaren na de einddatering + de schatting van het aantal ontbrekende spintringen.

De monsters kka 15.0 en kka 16.0 hebben beide wankant. De bomen waarvan de monsters afkomstig zijn, zijn gekapt in de loop van de zomer of in de winter, maar vóór het opnieuw uitlopen van de bomen in het daarop volgende voorjaar, d.w.z. **tussen zomer 1325 AD en voorjaar 1326 AD**. Deze kapdatum is voor alle daterende monsters gemeenschappelijk mogelijk.

De monsters dateren het beste met verschillende referentiechronologieen uit Niedersachsen (NSS 1,4,5,9). De herkomst van het hout uit dit gebied is dan ook waarschijnlijk.

Literatuur

Hollstein, E., 1980. *Mitteleuropäische Eichenchronologie.* Verlag Phillipp von Zabern, Mainz am Rhein.
Jansma, E., 1995. *RemembeRINGs: The development and application of local and regional tree-ring chronologies of oak for the purposes of archaeological and historical research in the Netherlands.* Diss. UvA (Nederlandse Archaeologischerapporten 19), 150 pp.

21 The Doel-II cog

Dendrochronological report
K. Haneca

Gent, donderdag 20 maart 2003.

Laboratorium voor Houttechnologie

UNIVERSITEIT GENT

Verslag dendrochronologisch onderzoek: Kogge schepen - Deurganckdok

Verwerking & verslag: Kristof Haneca
Universiteit Gent – FLTBW, Laboratorium voor Houttechnologie
Coupure Links 653, 9000 Gent
Tel 09-264.61.24 Fax 09-264.62.33
e-mail: kristof.haneca@rug.ac.be

BESCHRIJVING DATASET:

Tijdens de aanleg van het nieuwe Deurganckdok te Doel zijn tot twee maal toe de resten gevonden van een kogge schip. Van beide schepen werden monsters ter beschikking gesteld voor dendrochronologisch en houtbiologisch onderzoek. Naast het zoeken naar een sluitende datering zal ook een poging ondernomen worden om het oorspronggebied van het hout nader te bepalen.

Tijdens de maand augustus 2002 werden 7 monsters genomen van de resten van de romp van een kogge schip. Al deze planken bleken eik (*Quercus sp.*) te zijn. De labels op de monsters waren soms vrij lang waardoor een kortere dendro-code werd toegekend. In onderstaande tabel wordt een overzicht gegeven van de onderzochte monsters en hun dendro-codes. In de uiterst rechtse kolom wordt tevens omschreven hoe het zaag/klief-patroon van elke plank er uit ziet. Bij kwartiers (= radiaal) gezaagd/gekloven hout staat het zaag/kliefvlak loodrecht op de groeiringen. Indien loodrecht op deze richting werd gezaag/gekloven spreekt men van dosse (= tangentiaal).

Dendro-code	Label	Zaag/klief-schema
GAI-ISB	GAI-ISB Dendro 3	Dosse + merg
GBI-2SB	GBI-2SB Dendro 2	Kwartiers + merg
GC2-ISB	GC2-ISB Dendro 1	Dosse
KP-2-1	Kielplank 2-1 Dendro 5	Dosse
SLOS-1B	SLOS-1B Dendro 6	Dosse + merg
SLOS-2B	SLOS-2B Dendro 7	Kwartiers + merg
WAI-2SB	WAI-2SB Dendro 4	Dosse

DENDROCHRONOLOGISCH ONDERZOEK:

Datering:

Van al deze monsters werd het groeiringpatroon opgemeten. Op één monster (GC2-ISB) blijken 13 spinthoutringen aanwezig te zijn. Nadien werden de individuele groeiringpatronen met elkaar

1

vergeleken. De correlaties tussen de verschillende monsters is niet al te hoog, maar na visuele controle blijkt het berekenen van een middelcurve, *Kogge1m*, te verantwoorden.
Zowel de middelcurve *Kogge1m* als de afzonderlijke curven werden op hun beurt vergeleken met bestaande referentiechronologieën. Deze referentiechronologieën zijn afkomstig uit de eigen databank en de databank van het Nederlands Centrum voor Dendrochronologie (Stichting RING; Rijksdienst voor het Oudheidkundig Bodemonderzoek (ROB), Amersfoort, Nederland). Er werd een goede overeenkomst gevonden met referentiechronologieën die zijn opgesteld met archeologisch hout, afkomstig uit het Baltisch gebied. In onderstaande tabel wordt een overzicht gegeven van de correlaties met de referentiecurven en de daaruit volgende dateringen.

Dendro-code	**Merg**	**Spint**	**Aantal ringen**	**Referentiecurve**	**Datering (BC/AD)**	**t_{BP}**	**GLK**
GAI-ISB	+	-	83	66STD6	1230 – 1312	3,7	69,8
				EUBig7st	1230 – 1312	4,9	71,6
				GBI-2SB	1230 – 1312	5,0	73,1
GBI-2SB	+	-	77	66STD6	1209 – 1285	3,6	69,8
				EUBig7st	1209 – 1285	3,7	71,3
				SLOS-1B	1209 – 1285	4,1	56,0
GC2-ISB	-	13	109	66STD6	1218 – 1326	5,1	59,3
KP-2-1	-	-	136	66STD6	1169 – 1304	8,7	72,3
SLOS-1B	+	-	145	66STD6	1164 – 1308	5,9	69,2
SLOS-2B	+	-	107	66STD6	1193 – 1299	5,0	66,2
WAI-2SB	-	-	136	66STD6	1161 –1296	4,9	67,5
Alle deze monsters werden gemiddeld tot *Kogge1m*							
Kogge1m			166	66STD6	1161 – 1326	8,7	72,3
				Baltic 1	1161 – 1326	5,1	67,1
				EUBig7st	1161 – 1326	10,1	74,7

Toelichting bij de tabel:

Dendro-code: Code toegekend aan de opgemeten groeiringcurven.
Merg: - : merg (= kern van de boom) niet aanwezig / + : merg aanwezig
Spint: Aantal ringen spinthout / - : geen spinthoutringen aanwezig
Aantal ringen: Totaal aantal ringen die op het monster aanwezig zijn
Referentiecurve: EUBig7st = Referentiecurve van eikenhout afkomstig uit het huidige Polen.
66STD6 = Referentiecurve van eikenhouten monsters uit Oost-Pommeren, Noord-Polen (Wazny, 1990)
Baltic 1 = Engelse referentiecurve opgesteld met geïmporteerd Baltisch eikenhout dat gebruikt werd voor schilderijpanelen (Hillam & Tyers, 1995)
t_{BP}: Statistische maat die resulteert uit een Students t-test op de Pearson correlatie tussen monster en referentiecurve, na transformatie volgens het algoritme van Baillie & Pilcher (1973). Deze t-waarden zijn significant voor waarden boven 3,5.
GLK: "Gleichlaufigkeit" of "percentage of parallel variation", d.i. het percentage van de ringen uit het monster die een gelijkaardige toename/afname vertonen t.o.v. het voorgaande jaar zoals de referentiecurve, voor één welbepaalde positie op de tijdsas.

In onderstaande figuur worden de gedateerde groeiringreeksen en hun middelcurve t.o.v. elkaar op een tijdsas gepositioneerd.

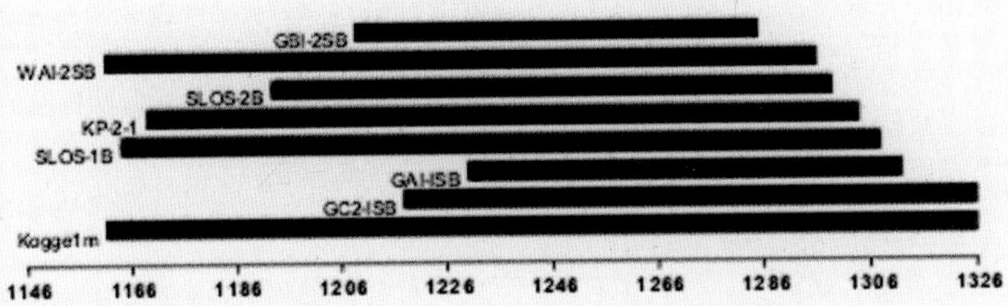

Fig. 1: Situering van gedateerde groeiringreeksen en hun middelcurve *Kogge1m* op een tijdsas.

Interpretatie van de datering:

Om tot een correcte datering te komen van het jaar waarin de boom werd geveld, moeten we het aantal ontbrekende groeiringen tot aan de bast bij de opgemeten groeiringen tellen. Indien nog een deel van het spinthout aanwezig is, kan een schatting gemaakt worden van het aantal onbrekende ringen. In West-Europa komen er gemiddeld bij een boom tot 100 jaar oud, 16 ± 5 spinthoutringen voor. Bij een boom van 100 tot 200 jaar oud zijn dit er gemiddeld 20 ± 6 en bij een boom ouder dan 200 jaar, 26 ± 8 spinthoutringen (Hollstein, 1980). Voor het Baltisch gebied ligt het gemiddelde aantal spinthoutringen iets lager, nl. 15 (+9/-6), waar 15 de mediaan is en +9 en –6 het interval beschrijven waarin 90% van de gevallen zich situeren (Wazny, 1990). Indien geen spinthoutringen meer aanwezig zijn op het monster, moet er nog rekening gehouden worden met een onbekend aantal ontbrekende kernhoutringen. De veldatum van de boom ligt dan een ongekend aantal jaar na de einddatering, vermeerderd met de schatting van het aantal spinthoutringen.

In dit geval zijn er 13 spinthoutringen aanwezig op het staal GC2-ISB. Aangezien zowel de middelcurve *Kogge1m* als de individuele curven de hoogste correlaties vertonen met Baltische referentiechronologieën, wordt het gemiddelde aantal spinthoutringen op 15 (+9/-6) geschat. Het aantal ontbrekende spinthoutringen voor *Kogge1m* bedraagt dan 15 – 13 = 2 (+9/-6). De uiteindelijke veldatum (= kapdatum van de boom waaruit de planken/balken zijn gezaagd) ligt dan na 1328 AD +9/-6.

Dendro-code	**Einddatering**	**Schatting ontbrekend aantal spinthoutringen**	**veldatum**
Kogge1m	1326	+ 2 (+9/-6)	Na 1328 AD +9/-6

Oorspronggebied van het constructiehout:

Na het daterend onderzoek werden zowel de individuele groeiringpatronen als de middelcurve *Kogge1m* vergeleken met Europese regionale eikenchronologieën. Deze regionale chronologieën worden verondersteld het gemiddelde groeiringpatroon te weerspiegelen van een welbepaalde regio. Alhoewel er uitgebreid kan gediscussieerd worden over de vraag hoe 'regionaal' deze chronologieën wel zijn, kan vergelijking van het groeiringpatroon van een stuk eikenhout met ongekend oorsprong, met deze regionale chronologieën een indicatie geven in welke regio de boom waaruit dat stuk hout gekliefd werd, groeide.
Het gebied waarop de regionale chronologie betrekking heeft, wordt voorgesteld a.d.h.v. een punt op de kaart van Europa (rode punten Fig. 2). Deze punten, één voor elke regionale chronologie, worden

verondersteld het centrum aan te duiden van de regio waarin de bomen groeiden die hebben bijgedragen tot het berekenen van de regionale chronologie. Nadien werd een correlatiematrix gegenereerd met de groeiringcurven afkomstig van het kogge schip aan de ene zijde en de regionale chronologieën aan de andere zijde. De berekende correlatiecoëfficiënten (t-waarden) werden daarna gevisualiseerd door vanuit elk punt op de kaart een zwarte cirkel te teken met een straal die evenredig is met de correlatie tussen de regionale chronologie en de betreffende groeiringcurve van het kogge schip. In figuur 2a worden de correlaties voorgesteld tussen de middelcurve *Kogge1m* en de regionale chronologieën. Daarin is te zien dat de correlaties consequent toenemen langs de ZW-NO as. De t-waarden bereiken hun hoogste waarden (t_{BP} = 8,7) wanneer de middelcurve wordt vergeleken met een chronologie voor Noord-Polen. Een alternatieve manier om vanuit dezelfde basisgegevens een idee te krijgen over het oorspronggebied van het gebruikte hout is het genereren van iso-lijnen. Dit zijn lijnen die op een gelijkaardige manier worden berekend als hoogtelijnen. In dit geval bakenen deze (hypothetische) iso-lijnen echter een gebied af waarbinnen de correlaties hoger liggen dan de waarde toegekend aan de lijn zelf (Fig. 2b).
In beide gevallen suggereren de figuren dat het oorspronggebied van het hout te situeren is in het Baltisch gebied. De hoogste correlaties worden gevonden met regionale chronologieën uit het huidige Polen.

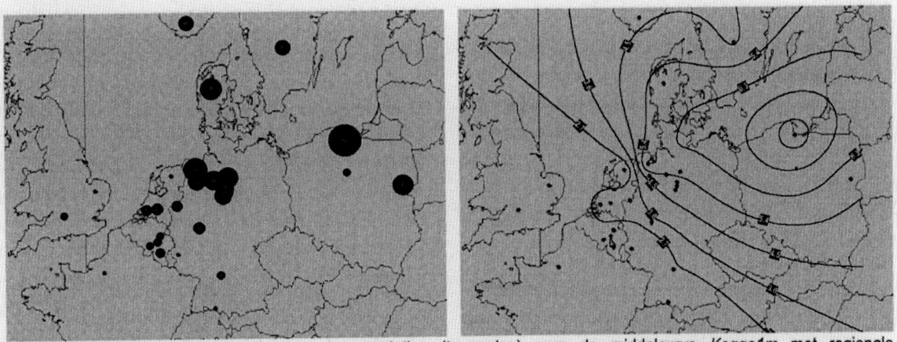

Fig. 2: Geografische interpretatie van de correlaties (t-waarden) voor de middelcurve *Kogge1m* met regionale eikenchronologieën (rode punten). In Figuur 2a worden vanuit deze punten zwarte cirkels getrokken waarvan de straal evenredig is met de graad van correlatie. Vanuit dezelfde dataset worden in Figuur 2b iso-lijnen gegenereerd.

DATEREND ONDERZOEK OP RESTANTEN KOGGE SCHIP (Stichting RING; november 2000):

Situering onderzoek:

In november 2000 werd door Stichting RING daterend onderzoek uitgevoerd op 18 monsters van een eerder opgegraven kogge schip uit het Deurganck dok. Bij 17 van die monsters werd een sluitende datering bekomen. Op twee stalen (S13A-BB en S6-SB) was de laatste spinthoutring aanwezig voor de bast. Dit laat toe om een zeer accurate datering van de veldatum te verkrijgen. Deze veldatum werd vastgelegd op 1325 AD. Vermoedelijk werd de boom dus gekapt tussen de zomer van 1325 AD en het voorjaar van 1326 AD (Hanraets, 2000).
In figuur 3 worden de gedateerde groeiringreeksen t.o.v. elkaar op een tijdsas gepositioneerd. De einddatum aangegeven op de tijdsas is 1324. Dit komt omdat de laatste groeiring niet is opgemeten, maar wel in de analyses is opgenomen.

Gent, donderdag 20 maart 2003.

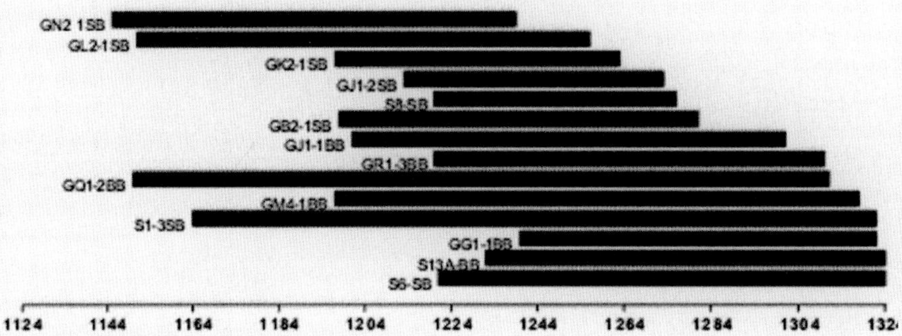

Fig. 3: Situering van gedateerde groeiringreeksen op een tijdsas.

Oorspronggebied van het constructiehout:

Ook hier werd een gelijkaardige analyse uitgevoerd als bij het dendrochronologisch onderzoek op de restanten van het kogge schip dat werd opgegraven in augustus 2002. De individuele, gedateerde curven werden opnieuw vergeleken met de dataset van regionale chronologieën.
De onderlinge correlaties liggen, op enkele uitzonderingen na, vrij laag. Op statistische basis is het berekenen van een middelcurve dan niet te verantwoorden. Aangezien alle fragmenten wel uit een zelfde context komen, kan er wel een zogenaamde object-chronologie berekend worden, i.e. alle groeiringcurven die nu uit een zelfde context komen middelden tot één chronologie. Voor deze kogge krijgt de object-chronologie de naam *Kogge2m*. Met deze chronologie werden de hoogste correlaties (t_{BP} = 7,0 - 7,8) gevonden bij de vergelijking met regionale chronologieën uit Niedersachsen (H. Leuschner). Dit suggereert dat dit ook het gebied van oorsprong kan zijn waar de bomen groeiden waarvan het hout gebruikt werd voor de constructie van het schip. Gelijkaardige resultaten worden gevonden als de individuele curven vergeleken worden met de regionale chronologieën.

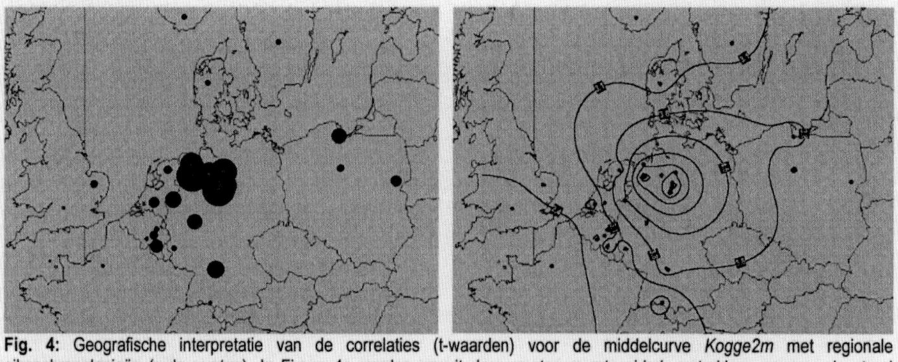

Fig. 4: Geografische interpretatie van de correlaties (t-waarden) voor de middelcurve *Kogge2m* met regionale eikenchronologieën (rode punten). In Figuur 4a worden vanuit deze punten zwarte cirkels getrokken waarvan de straal evenredig is met de graad van correlatie. Vanuit dezelfde dataset worden in Figuur 4b iso-lijnen gegenereerd

Literatuur

Baillie, M.G.L. & Pilcher, J.R. (1973). A simple cross-dating program for tree-ring research. *Tree Ring Bulletin*, 33: 7-14.

Hanraets, A.E.M. (2000). Rapportage Daterend Onderzoek – ROB/RING. Amersfoort, pp. 3

Hillam, J., & Tyers, I. (1995). Reliability and Repeatability in Dendrochronological Analysis: Tests Using the Fletcher Archive of Panel-Painting Data. *Archaeometry*, 37(2): 395-405.

Hollstein, E. (1980). Mitteleuropäische Eichenchronologie. Verlag Phillipp von Zabern, Mainz am Rhein.

Wazny, T. (1990). Aufbau und Anwendung der Dendrochronologie für Eichenholz in Polen. Diss. Univ. Hamburg.

22 The Kraggenburg cog

Dendrochronological report RING

M. Dominguez

Stichting RING, p/a Rijksdienst voor het Cultureel Erfgoed; Postbus 1600, 3800 BP Amersfoort; 0334217545; m.dominguez@cultureelerfgoed.nl

ring

Nederlands Centrum voor Dendrochronologie
Stichting Ring
p/a Rijksdienst voor het Cultureel Erfgoed
Smallepad 5, 3811 MG Amersfoort
Postbus 1600, 3800 BP Amersfoort
Tel: 033-4217545
E-mail: m.dominguez@cultureelerfgoed.nl

Aan: Dhr. Van Holk
Nieuw Land Erfgoedcentrum
Postbus 73
8200 Lelystad

Betreft: uitslag dateringsonderzoek kogge NT25 uit Kraggenburg, Noordoostpolder
***RING* Intern Rapport nummer: 2010077**

Datum: 8 december 2010

Geachte heer Van Holk,

Wij onderzochten voor u 7 eikenmonsters (*Quercus sp.*) afkomstig uit kogge NT25 uit Kraggenburg, Noordoostpolder (offerte O2010069).

Het dendrochronologische onderzoek op het monster leverde het volgende resultaat op (zie met name de vijfde kolom in tabel 1):

Tabel 1. Uitslag dendrochronologisch onderzoek

Monsternr / onderdeel / spoor	RINGs Dendrocode	Datering van de laatste gemeten ring	Zekerheid van de datering (*probability*)	Periode waarin/ waarna de boom is omgehakt	Gebruikte Referentie-chronologie
M1 / vlakplank / -	NKR00011	1294 n.Chr.	>99,99%	na 1307 n.Chr.	NLNOOR10
M6 / vlakplank / -	NKR00020	1268 n.Chr.	>99,95%	na 1281 n.Chr.	NLHOOR01
M2 / vlakplank / -	NKR00030	1259 n.Chr.	>97%	na 1281 n.Chr.	NLNOOR1E
M5 / spant / S16	NKR00040	1293 n.Chr.	>90%	1310 n.Chr ±5	NLNOOR10
M4 / spant / S11	NKR00050	-	-	-	-
M3 / spant / S10	NKR00060	-	-	-	-
M7 / spant / -	NKR00070	-	-	-	-

Het hout van monster 2 (NKR00030) en 6 (NKR00020) is afkomstig uit dezelfde boom (zie bijlage 1, tabel 2 en bijlage 2, afbeelding 5). De meetreeksen zijn gemiddeld tot de curve NKR2_3T. Deze curve dateert met de referentiekalender NLNOOR1E (zie bijlage 1, tabel 3 en bijlage 2, afbeelding 6).

Het hout van de gedateerde monsters is waarschijnlijk afkomstig uit het noordwesten van Nederland.

3/10

Helaas was het onmogelijk om de monsters 3, 4 en 7 te dateren. Deze monsters beschikken over weinig ringen en ze tonen geen duidelijke synchronisatie met elkaar of met de andere monsters.

Aanvullende informatie over de laboratoriumresultaten, de gebruikte statistiek en/of de gebruikte referentiekalenders, vindt u in de bijlagen.

RING Intern Rapport nummer:	Laboratoriumnummer, verwijzing naar de analyse.
Zekerheid van de datering:	De kans dat de gevonden match met de referentiechronologie niet op toeval berust. Deze waarde is gebaseerd op de 'Gleichlaufigkeit' tussen de twee vergeleken reeksen, ook wel %PV genoemd (*percentage of parallel variation*; Jansma 1995).
Verantwoording van de dateringen:	Dendrochronologische dateringen door RING zijn gebaseerd op een combinatie van waarnemingen: (a) vergelijking en relatieve datering (ten opzichte van elkaar) van de jaarringpatronen binnen een vindplaats/bouwfase; (b) vergelijking van deze jaarringpatronen met *meerdere* absoluut gedateerde referentiekalenders. Deze vergelijkingen zijn statistisch onderbouwd en worden visueel gecontroleerd. Wanneer observaties elkaar ondersteunen en bevestigen, wordt de datering geaccepteerd als zijnde correct.

RING-rapport 2010077, houtmonsters kogge NT25, Kraggenburg, Noordoostpolder

Bijlage 1. Statistische resultaten van het dendrochronologisch onderzoek

Tabel 1. Statistische resultaten van de monsters

Monsternr / onderdeel / spoor	RINGs Dendrocode	n	Kern	Spint	Wankant	1e jaar	ne jaar	Kapdatum*	t	%PV	p	Kalender
M1 / vlakplank / -	NKR00011	102	+?	-	>13	1193	1294	na 1307 n.Chr.	7,22	70,8	0,0001	NLNOOR10
M6 / vlakplank / -	NKR00020	78	+?	-	>13	1191	1268	na 1281 n.Chr.	5,85	70,5	0,0005	NLHOOR01
M2 / vlakplank / -	NKR00030	100	+5	-	>13	1160	1259	na 1281 n.Chr.	5,63	61	0,03	NLNOOR1E
M5 / spant / S16	NKR00040	108	+1	1	17±5	1186	1293	1310 n.Chr ±5	4,95	53,7	0,1	NLNOOR10
M4 / spant / S11	NKR00050	67	+1	19	WK (zomer/winter)	-	-	-	-	-	-	-
M3 / spant / S10	NKR00060	83+1	+1	14+1	-	-	-	-	-	-	-	-
M7 / spant / -	NKR00070	63	+1	-	-	-	-	-	-	-	-	-

*Kapdatum geschat volgens Jansma, 2007.

Tabel 2. Relevant statistische resultaten tussen de meetreeksen

Meetreeksen	Ol	Gl	p	THO
NKR00020 vs. NKR00030	69	80,4	0,0001	15,4

Tabel 3. Gemiddelde curve

Meetreeksen	RINGcode gemiddelde curve	n	1e jaar	ne jaar	t	%PV	p	Kalender
NKR00020 NKR00030	NKR2_3T	109	1160	1268	6,32	63,8	0,005	NLNOOR1E

RING-rapport 2010077, houtmonsters kogge NT25, Kraggenburg, Noordoostpolder

Bijlage 2. Afbeeldingen van de meetreeksen

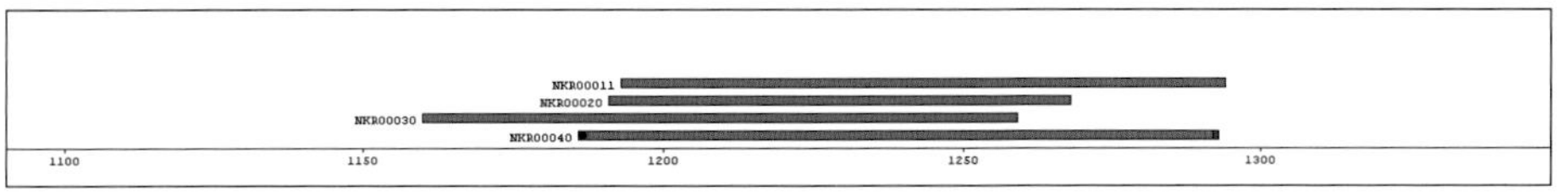

Afbeelding 1. Tijdspanne van de gedateerde monsters. Deze zijn gesorteerd op de laatst gemeten jaarring. De schatting van de veldatum is niet weergegeven op deze afbeelding. Het grijze gebied geeft de spintringen weer. x-as: kalenderjaar.

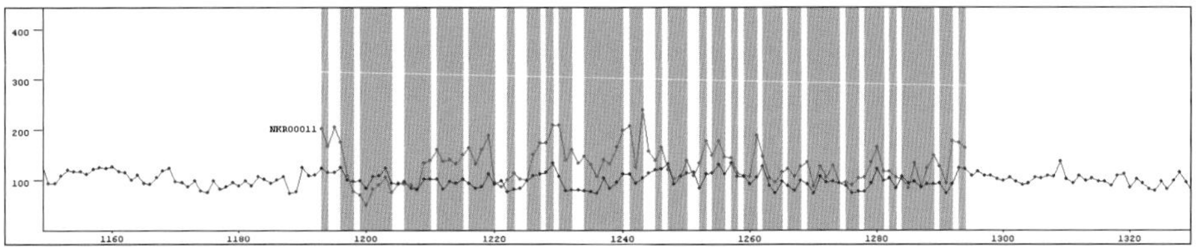

Afbeelding 2. Groen: gedateerde meetreeks NKR00011; blauw: referentiechronologie NLNOOR10; y-as: jaarringbreedte in mm*10^{-2}; x-as: kalenderjaar. Het grijze gebied toont de *Gleichläufigkeit* (percentage van parallelle ringbreedte variaties (%PV) van beide patronen).

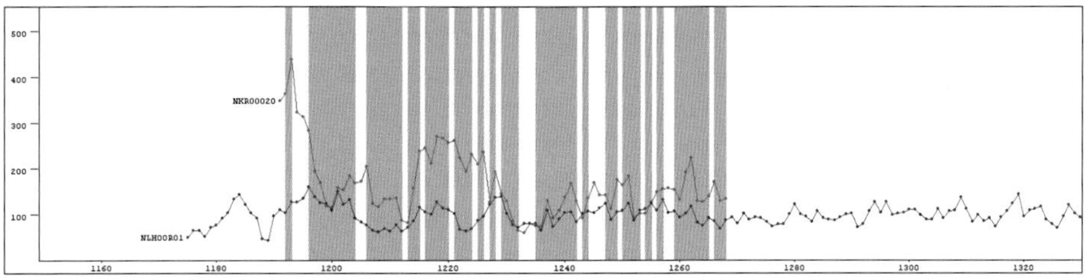

Afbeelding 3. Groen: gedateerde meetreeks NKR00020; blauw: referentiechronologie NLHOOR01; y-as: jaarringbreedte in mm*10^{-2}; x-as: kalenderjaar. Het grijze gebied toont de *Gleichläufigkeit* (percentage van parallelle ringbreedte variaties (%PV) van beide patronen).

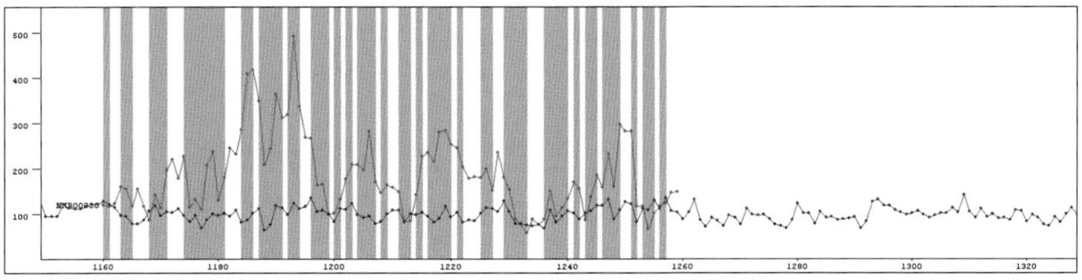

Afbeelding 4. Groen: gedateerde meetreeks NKR00030; blauw: referentiechronologie NLNOOR1E; y-as: jaarringbreedte in mm*10^{-2}; x-as: kalenderjaar. Het grijze gebied toont de *Gleichläufigkeit* (percentage van parallelle ringbreedte variaties (%PV) van beide patronen).

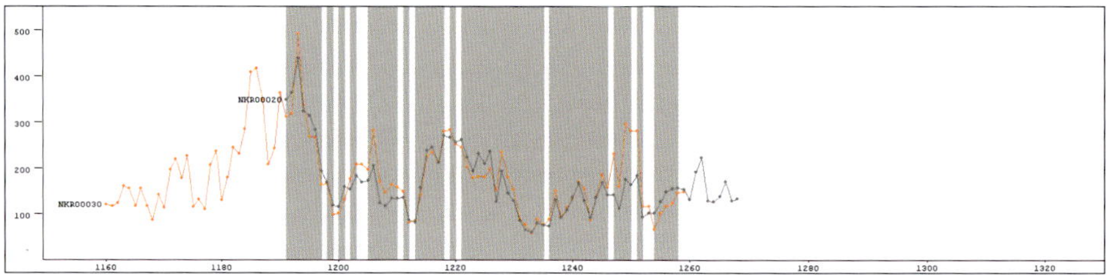

Afbeelding 5. Visuele synchronisatie tussen de gedateerde meetreeksen NKR00020 (grijs) en NKR00030 (oranje). Deze monsters zijn afkomstig uit dezelfde boom. y-as: jaarringbreedte in mm*10^{-2}; x-as: nummer van jaarringen. Het grijze gebied toont de *Gleichläufigkeit* (percentage van parallelle ringbreedte variaties (%PV) van beide patronen).

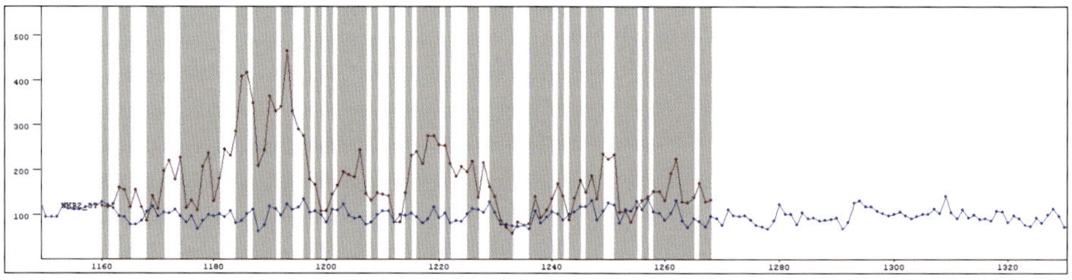

Afbeelding 6. Bruin: gemiddelde curve NKR2_3T; blauw: referentiechronologie NLNOOR1E; y-as: jaarringbreedte in mm*10^{-2}; x-as: kalenderjaar. Het grijze gebied toont de *Gleichläufigkeit* (percentage van parallelle ringbreedte variaties (%PV) van beide patronen).

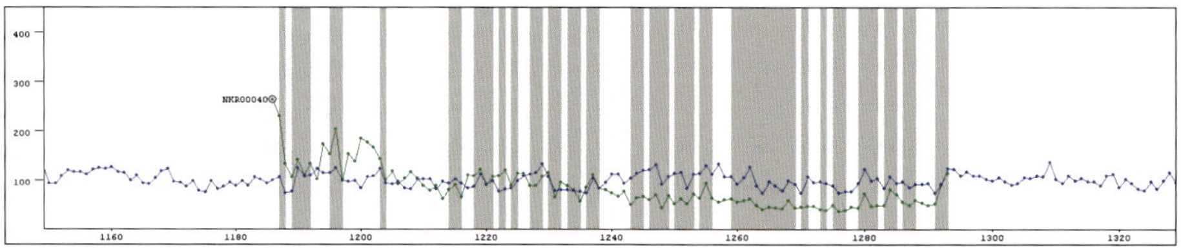

Afbeelding 7. Groen: gedateerde meetreeks NPS00040 (de gestippelde lijn geeft de spintring weer, de punt wijst erop dat de kern in het monster aanwezig is); blauw: referentiechronologie NLNOOR10; y-as: jaarringbreedte in mm*10^{-2}; x-as: kalenderjaar. Het grijze gebied toont de *Gleichläufigkeit* (percentage van parallelle ringbreedte variaties (%PV) van beide patronen).

RING-rapport 2010077, houtmonsters kogge NT25, Kraggenburg, Noordoostpolder

Bijlage 3. Toelichting op de resultaten van de dendrochronologische analyse

Houtsoort = Het hout wordt door ons enkel gedetermineerd ten behoeve van de datering. Alleen de *genus*, bijv. Den (*Pinus*), wordt bepaald. Verdere soortbepaling, zoals bijv. grove den (*Pinus sylvestris*), blijft in principe achterwege, tenzij deze eenvoudig vastgesteld kan worden. Een uitzondering hierop is *Abies alba* (Zilverspar), de enige soort *Abies* die in het verleden in Nederland is toegepast.

Kern = (Geschatte afstand tot) de eerstgevormde (oudste) jaarring in de stam.

Spint = Aantal gemeten ringen spinthout. Volgens Hollstein (1980) heeft eik een gemiddeld aantal spintringen van 16 ± 5 bij een boom tot 100 jaar oud, 20 ± 6 bij een boom van 100 tot 200 jaar oud, en 26 ± 8 bij een boom ouder dan 200 jaar. Wij gebruiken een nieuwe, bijgestelde spinthoutberekening voor archeologisch/ historisch constructiehout dat dateert met Nederlandse en Duitse chronologieën (Jansma 2007). Bij eikenhout uit het Baltische gebied is het gemiddelde aantal spintringen iets lager dan in West Europa, 15 (+9/-6) (Wazny, 1990). Grove den, (*Pinus sylvestris*) heeft weliswaar ook duidelijk zichtbaar spinthout, maar doordat het aantal spinthoutringen onregelmatig is, is een schatting van de velddatum niet mogelijk. Fijnspar, (*Picea abies*) heeft geen spinthout. Uiteraard geeft een aanwezige wankant wel de precieze kapdatum van de boom.

Wankant = Het geschatte aantal jaarringen tot de wankant, d.w.z. tot de laatstgevormde jaarring (direct onder de bast), nodig voor een absolute datering van de veldatum.

Veldatum = De datum waarop de boom geveld is. Als er wankant aanwezig is, is er een absolute datering mogelijk. Als er spintringen aanwezig zijn, of zelfs alleen spintgrens, wordt de veldatum berekend door het aantal ontbrekende spintringen te berekenen. Als er bij een eik van 100 tot 200 jaar oud b.v. 4 spintringen gemeten zijn, is het geschatte aantal ontbrekende spintringen dus 16 ± 6. Dit getal wordt bij de datering opgeteld. Als er geen spintringen meer op het monster aanwezig zijn, is het onbekend hoeveel *kernhoutringen* er nog ontbreken . De veldatum ligt dan een onbekend aantal jaren ná de datering van de laatste (jongste) ring + de schatting van het ontbrekende aantal spinthoutringen. Bij een boom, die 100 tot 200 jaar oud is, is de veldatum dus xxxx AD + 20 (± 6) + X.

n = Totaal aantal jaarringen in het houtmonster.

x = Geschat aantal missende ringen (kernhout en/of spinthout) tot de wankant.

%PV = "Gleichlaufigkeit" (Duitse term) of "Percentage of Parallel Variation" (Engelse term); het percentage van de ringen in het onderzochte jaarringpatroon die aan de referentiechronologie identieke toe- en afnames van de breedte vertonen op de door de datering van het patroon aangegeven positie t.a.v. de referentiechronologie. De significantie van dit percentage is een functie van de lengte in jaren van het onderzochte jaarringpatroon en de referentie chronologie.

t = De waarde die resulteert uit een Students t-test op de kruiscorrelatie die behoort bij de beste "match" tussen het onderzochte jaarringpatroon en de referentiechronologie.

P = De kans (uitgedrukt als een fractie van 1) dat de gevonden waarde voor %PV per toeval optreedt, dus niet op een datering duidt.

RING-rapport 2010077, houtmonsters kogge NT25, Kraggenburg, Noordoostpolder

Bijlage 4. Gebruikte referentiechronologieën

NLNOOR10	Hout toegepast in Noordwest-Nederland (*Jansma 1995*).
NLHOOR01	Hoorn, Nederland (*Hanraets, E. n.p.*).
NLNOOR1E	Hout toegepast in Noordwest-Nederland (*Jansma 1995*).

RING-rapport 2010077, houtmonsters kogge NT25, Kraggenburg, Noordoostpolder

Bijlage 5. Literatuur

Hollstein, E., 1980. *Mitteleuropäische Eichenchronologie.* Verlag Phillipp von Zabern, Mainz am Rhein.

Jansma, E., 1995. *Rememberings, The development and application of local and regional tree-ring chronologies of oak for the purposes of archaeological and historical research in the Netherlands.* Diss. UvA (Nederlandse Archeologische Rapporten 19), 150 pp.

Jansma, E., 2007: *Datering, herkomst en bouwvolgorde van De Meern 4. In: T. de Groot & J.-M.A.W. Morel (red.), 2007: Het schip uit de Romeinse tijd De Meern 4 nabij boerderij de Balije, Leidsche Rijn, gemeente Utrecht. Waardestellend onderzoek naar de kwaliteit van het schip en het conserverend vermogen van het bodemmilieu.* RACM.

Wazny, T., 1990. *Aufbau und Anwendung der Dendrochronologie fuer Eichenholz in Polen.* Dissertatie Universiteit van Hamburg.

23 The Kampen (IJssel) cog

Pollen analysis of the moss caulking

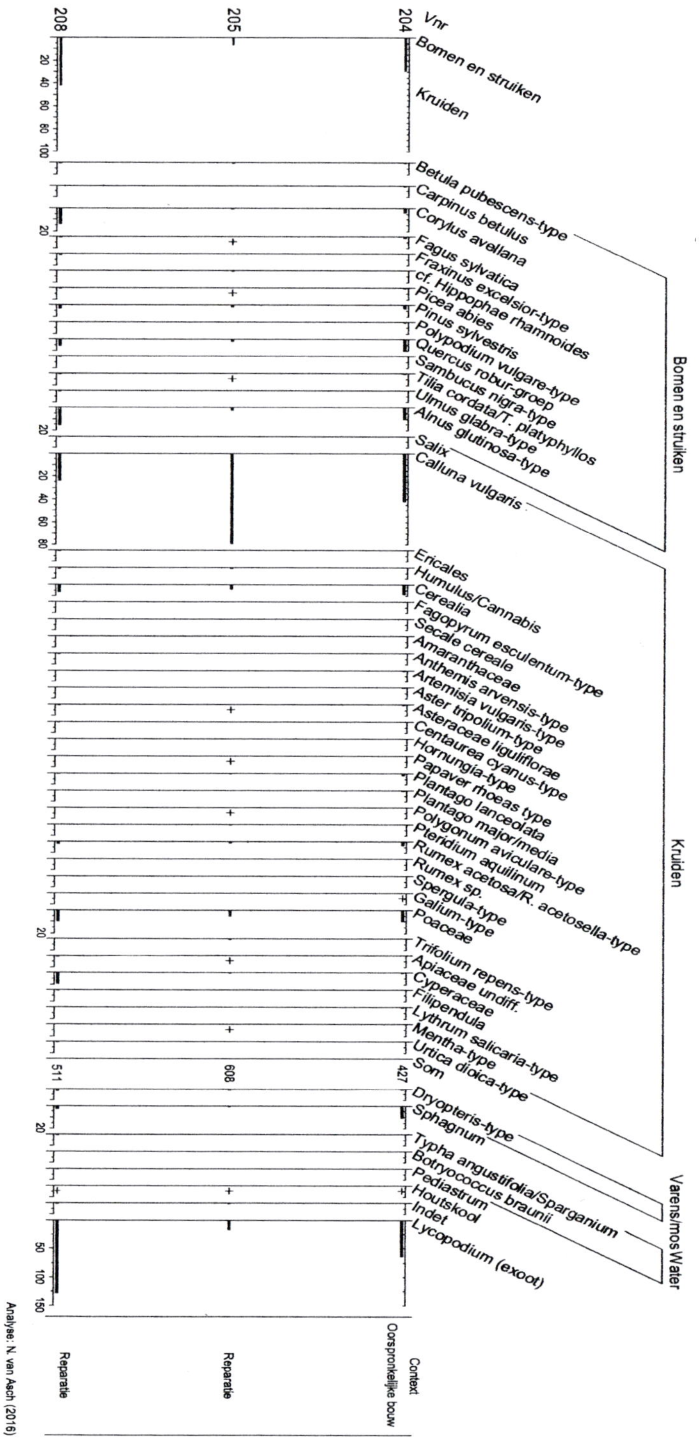

Stichting RING, p/a Rijksdienst voor het Cultureel Erfgoed; Postbus 1600, 3800 BP Amersfoort; 0334217545; m.dominguez@cultureelerfgoed.nl

ring

Nederlands Centrum voor Dendrochronologie
Stichting Ring
p/a Rijksdienst voor het Cultureel Erfgoed
Smallepad 5, 3811 MG Amersfoort
Postbus 1600, 3800 BP Amersfoort
Tel: 033-4217545
E-mail: m.dominguez@cultureelerfgoed.nl

Aan: Dhr. W. Waldus
ADC ArcheoProjecten
Nijverheidsweg Noord 114
3800 BM Amersfoort

Betreft: uitslag dateringsonderzoek houtmonsters uit scheepswrak van vermoedelijk een kogge in Kampen

***RING* Intern Rapport nummer: 2012063**

Amersfoort, 29 oktober 2012

Geachte heer Waldus,

Wij onderzochten voor u 12 eiken monsters (*Quercus* sp.) (tabel 1) afkomstig uit een scheepswrak van vermoedelijk een kogge, gevonden in de haven van Kampen (KAMN-12) (offertenummer O2012054).

Tabel 1. Gegevens van onderzochte monsters

Monsternr.	Werkput	Objecttype	Opmerkingen
17	Put 1	Duik 26	Gezaagd van LSH1, een plank van het dek of van de scheepshuid
24	Put 1	Duik 31	Gezaagd van LSH3, knie die tegen de wegering aan BB aanzat
26	Put 1	Duik 31	Gezaagd van LSH2, knie aan BB zijde
28	Put 3	Duik 37	Gezaagd als monsters van inhout uit voorschip aan BB thv balk 4
29	Put 3	Duik 37	Gezaagd als monsters van inhout uit voorschip aan BB thv balk 4
31	Put 1	Duik 35	Gezaagd van LSH5, een plank van de wegering
32	Put 1	Duik 40	Gezaagd van LSH7, een losse balk aan BB bij balk 3, mogelijk een schot deel
35	Object 5	Duik 46	Gezaagd van LSH8, een overnaadse huidplank van object 5
36	Put 1	Duik 46	Gezaagd van een huidplank van de kogge tussenbalk 2 en 3, losliggend
37	Put 2	Duik 46	Gezaagd van een weger bij de achtersteven van de kogge
38	Put 2	Duik 46	Gezaagd van een weger bij de achtersteven van de kogge
39	Put 2	Duik 46	Gezaagd van een weger bij de achtersteven van de kogge

Uit het vooronderzoek blijkt dat vier van de monsters niet geschikt zijn voor dendrochronologisch onderzoek zijn omdat het hout over onvoldoende jaarringen beschikt (tabel 2). Deze monsters zijn afkomstig uit onderdelen die tangentiaal uit snel gegroeide bomen zijn gezaagd:

Tabel 2. Uitslag vooronderzoek

Monsternr.	N	Kern	Spinthout	Opmerkingen
32	48	ca. 15	9	Snel gegroeid; tangentiaal (gezaagd?)
37	27	ca. 15	-	Snel gegroeid; tangentiaal (gezaagd?)
38	20	ca.15	-	Snel gegroeid; tangentiaal (gezaagd?)
39	10	ca. 15	-	Snel gegroeid; tangentiaal (gezaagd?)

Het dendrochronologische onderzoek op de andere monters leverde het volgende resultaat op (zie met name de vijfde kolom in tabel 3):

Tabel 3. Uitslag dendrochronologisch onderzoek

Monsternr. / opmerkingen	RINGs Dendrocode	Datering van de laatste gemeten ring	Zekerheid van de datering (*probability*)	Periode waarin/ waarna de boom is omgehakt	Gebruikte Referentie-chronologie
17 / radiaal (gespleten?)	KAK00010	1555 n.Chr.	>99,99%	Na 1568 n.Chr.	FRQUSP_Meuse
24 / vierkante kwart boom	KAK00020	-	-	-	-
26 / vierkante kwart boom	KAK00030	-	-	-	-
28 / rechthoekige kwart boom	KAK00040	1379 n.Chr.	>99,99%	Na 1393 n.Chr.	NLNOOR10
29 / rechthoekige kwart boom	KAK00050	1386 n.Chr.	>99,95%	Na 1399 n.Chr.	NLNOOR10
31 / tangentiaal (gezaagd?)	KAK00060	1276 n.Chr.	>99,99%	Na 1288 n.Chr.	DESUDE02
35 / radiaal (gespleten?)	KAK00070	1378 n.Chr.	>99,99%	Na 1386 n.Chr.	EUPLNW01
36 / radiaal (gezaagd/gespleten?)	KAK00080	1265 n.Chr.	>99,99%	Na 1279 n.Chr.	NLNOMIDD01

Van de acht dendrochronologisch onderzocht monsters konden zes worden gedateerd. Deze groep monsters is zeer heterogeen, zoals de verschillende dateringen, groeipatronen en herkomsten duidelijk laten zien.

Het hout van monster 17 hoort waarschijnlijk niet bij de andere monsters, aangezien deze boom meer dan anderhalf eeuw later dan de rest is gekapt. Deze boom groeide ergens in het gebied van de Meuse in Frankrijk.

De meetreeksen van de monsters 28 en 29 synchroniseren zeer goed met elkaar (zie bijlage 1, tabel 2 en bijlage 2, afbeelding 3), wat er op wijst dat het hout van beide elementen uit hetzelfde gebied afkomstig is. De metingen zijn gemiddeld tot de curve KAK4_5M, die een goede overeenkomst toont met de referentiekalender NLNOOR10 (zie bijlage 1, tabel 3 en bijlage 2, afbeelding 4). Het hout is waarschijnlijk afkomstig uit Nederland, hoewel het niet mogelijk is om de herkomst nauwkeuriger te bepalen. Het hout van het monster 36 heeft ook waarschijnlijk een lokale herkomst, terwijl het hout van monster 31 afkomstig is uit het zuiden van Duitsland. Het hout van monster 35 is waarschijnlijk afkomstig uit het noordwesten van Polen.

Aangezien in geen van de gedateerde monsters spinthout aanwezig heeft, kunnen de kapdatums alleen als *post quem* dateringen worden gegeven. De vroegerer *post quem* dateringen van monsters 31 en 36 hoeft niet te betekenen dat deze elementen tot een andere fase behoren. Het is mogelijk dat deze monsters uit het binnenste deel van de bomen afkomstig zijn, terwijl de monsters 28, 29 en 35 uit de buitenste deel van de bomen kunnen zijn genomen. De *post quem* dateringen van deze drie monsters lijken de bouwperiode van het schip te plaatsen in de eerste helft van de vijftiende eeuw.
Het heterogene karakter van het materiaal kan erop wijzen dat het bouwhout van het schip is gekocht op een grote houtmarkt. Echter, de kleine hoeveelheid onderzochte monsters staat dergelijke sterke conclusies niet toe. Wij bevelen daarom aan om meer houten elementen van het schip te laten onderzoeken, in combinatie met een

grondige verkenning van de constructie en historisch onderzoek. Hopelijk kan op deze manier meer informatie worden verkregen over de herkomst en datering van het schip.

Aanvullende informatie over de laboratoriumresultaten, de gebruikte statistiek en/of de gebruikte referentiekalenders, vindt u in de bijlagen.

RING Intern Rapport nummer:	Laboratoriumnummer, verwijzing naar de analyse.
Zekerheid van de datering:	De kans dat de gevonden match met de referentiechronologie niet op toeval berust. Deze waarde is gebaseerd op de 'Gleichlaufigkeit' tussen de twee vergeleken reeksen, ook wel %PV genoemd (*percentage of parallel variation*; Jansma 1995).
Verantwoording van de dateringen:	Dendrochronologische dateringen door RING zijn gebaseerd op een combinatie van waarnemingen: (a) vergelijking en relatieve datering (ten opzichte van elkaar) van de jaarringpatronen binnen een vindplaats/bouwfase; (b) vergelijking van deze jaarringpatronen met *meerdere* absoluut gedateerde referentiekalenders. Deze vergelijkingen zijn statistisch onderbouwd en worden visueel gecontroleerd. Wanneer observaties elkaar ondersteunen en bevestigen, wordt de datering geaccepteerd als zijnde correct.
N:	Totaal aantal jaarringen in het houtmonster.
Kern	(Geschatte afstand tot) de eerstgevormde (oudste) jaarring in de stam.
Spinthout	Aantal spinthout ringen aanwezig in het monster.
Wankant	Het geschatte aantal jaarringen tot de wankant, d.w.z. tot de laatstgevormde jaarring (direct onder de bast), nodig voor een absolute datering van de veldatum.

RING-rapport 2012063, scheepswrak, vermoedelijk kogge, Kampen (KAMN-12)

Bijlage 1. Statistische resultaten van het dendrochronologisch onderzoek

Tabel 1. Statistische resultaten van de monsters

Monsternr.	RINGs Dendrocode	n	Kern	Spint	Wankant	1e jaar	ne jaar	Kapdatum	t	%PV	p	Kalender
17 / radiaal (gespleten?)	KAK00010	106	+?	-	>13	1450	1555	Na 1568 n.Chr.*	7,28	73,1	0,0001	FRQUSP_Meuse
24 / vierkante kwart boom	KAK00020	57+1	+1	-	-	-	-	-	-	-	-	-
26 / vierkante kwart boom	KAK00030	108	+?	-	-	-	-	-	-	-	-	-
28 / rechthoekige kwart boom	KAK00040	130	Ca. 15	-	>14	1250	1379	Na 1393 n.Chr.*	6,77	69,6	0,0001	NLNOOR10
29 / rechthoekige kwart boom	KAK00050	94+3	+?	-	>13	1293	1386	Na 1399 n.Chr.*	3,78	68,6	0,0005	NLNOOR10
31 / tangentiaal (gezaagd?)	KAK00060	144	Ca. 15	-	>12	1133	1276	Na 1288 n.Chr.*	11,8	76,8	0,0001	DESUDE02
35 / radiaal (gespleten?)	KAK00070	247	+?	-	>8	1132	1378	Na 1386 n.Chr.**	8,32	64,0	0,0001	EUPLNW01
36 / radiaal (gezaagd/gespleten?)	KAK00080	124	Ca. 15	-	>14	1141	1265	Na 1279 n.Chr.*	7,43	74,0	0,0001	NLNOMIDD01

* Kapdatum geschat volgens Jansma, E. 2007.
** Kapdatum geschat volgens Wazny, 1990.

Tabel 2. Relevant statistische resultaten tussen meetreeksen

Meetreeksen		OI	%PV	p	t
KAK00040	KAK00050	87	78,7	0,0001	87

Tabel 3. Gemiddelde curve

Meetreeksen	RINGcode gemiddelde curve	n	1e jaar	ne jaar	t	%PV	p	Kalender
KAK00040 KAK00050	KAK4_5M	137	1250	1386	7,54	75,2	0,0001	NLNOOR10

RING-rapport 2012063, scheepswrak, vermoedelijk kogge, Kampen (KAMN-12)

Bijlage 2. Afbeeldingen van de gedateerde meetreeksen

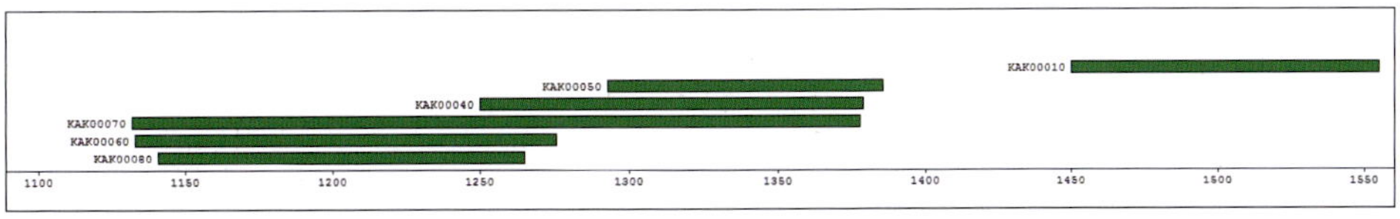

Afbeelding 1. Tijdspanne van de gedateerde monsters. Deze zijn gesorteerd op de laatst gemeten jaarring. x-as: kalenderjaar.

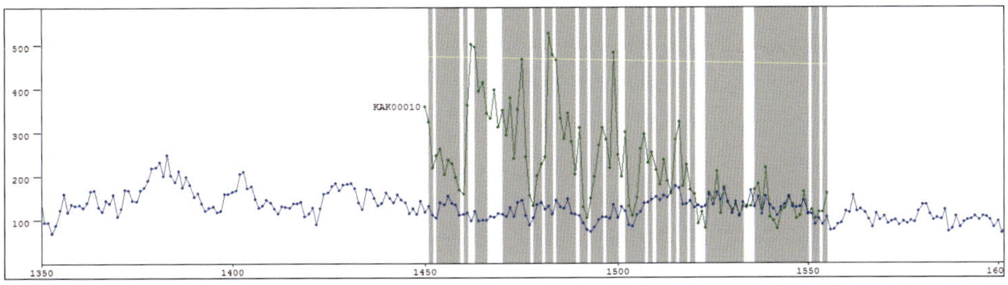

Afbeelding 2. Groen: gedateerde meetreeks KAK00010; Blauw: referentiechronologie FRQUSP_Meuse; y-as: jaarringbreedte in mm*10^{-2}; x-as: kalenderjaar. Het grijze gebied toont de *Gleichläufigkeit* (percentage van parallelle ringbreedte variaties (%PV) van beide patronen).

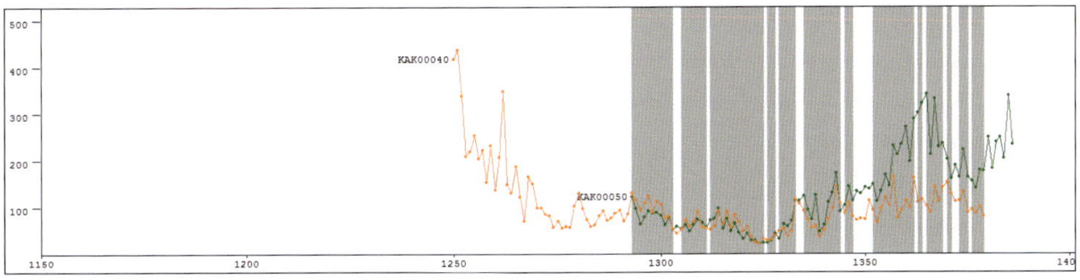

Afbeelding 3. Visuele synchronisatie tussen de monsters KAK00040 (oranje) en KAK00050 (groen). Het hout van deze monsters is afkomstig uit dezelfde gebied. Y-as: jaarringbreedte in mm*10^{-2}; x-as: kalenderjaar. Het grijze gebied toont de *Gleichläufigkeit* (percentage van parallelle ringbreedte variaties (%PV) van beide patronen).

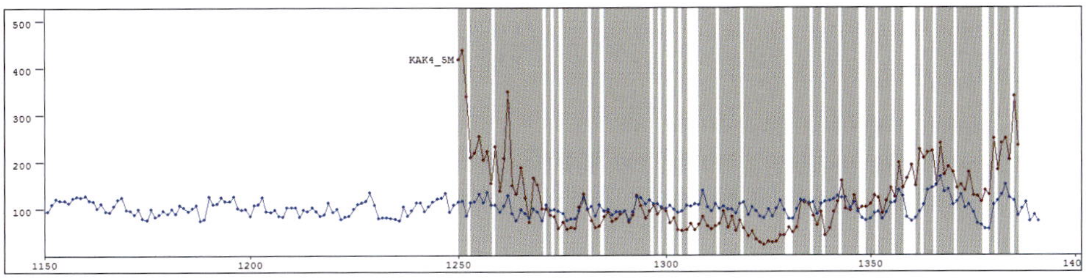

Afbeelding 4. Bruin: gemiddelde curve KAK4_5M; Blauw: referentiechronologie NLNOOR10; y-as: jaarringbreedte in mm*10^{-2}; x-as: kalenderjaar. Het grijze gebied toont de *Gleichläufigkeit* (percentage van parallelle ringbreedte variaties (%PV) van beide patronen).

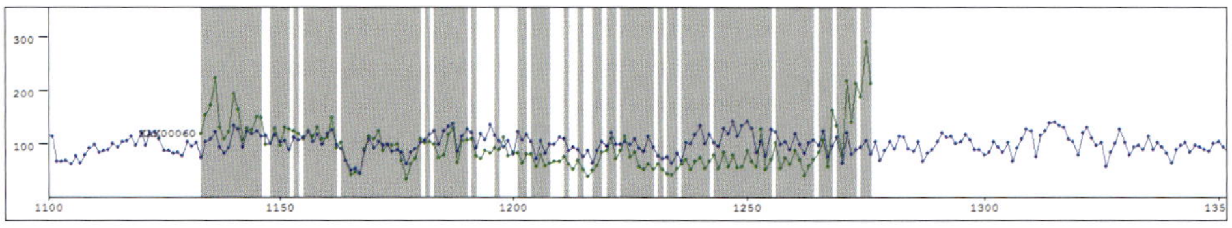

Afbeelding 5. Groen: gedateerde meetreeks KAK00060; Blauw: referentiechronologie DESUDE02; y-as: jaarringbreedte in mm*10^{-2}; x-as: kalenderjaar. Het grijze gebied toont de *Gleichläufigkeit* (percentage van parallelle ringbreedte variaties (%PV) van beide patronen).

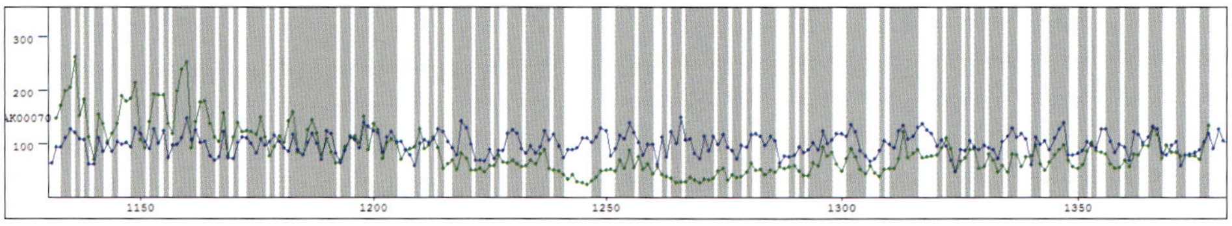

Afbeelding 6. Groen: gedateerde meetreeks KAK00070; Blauw: referentiechronologie EUPLNW01; y-as: jaarringbreedte in mm*10^{-2}; x-as: kalenderjaar. Het grijze gebied toont de *Gleichläufigkeit* (percentage van parallelle ringbreedte variaties (%PV) van beide patronen).

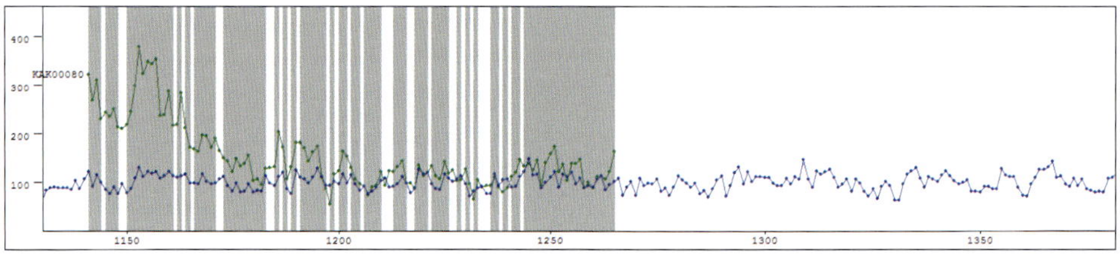

Afbeelding 7. Groen: gedateerde meetreeks KAK00080; Blauw: referentiechronologie NLNOMIDD01; y-as: jaarringbreedte in mm*10^{-2}; x-as: kalenderjaar. Het grijze gebied toont de *Gleichläufigkeit* (percentage van parallelle ringbreedte variaties (%PV) van beide patronen).

RING-rapport 2012063, scheepswrak, vermoedelijk kogge, Kampen (KAMN-12)

Bijlage 3. Toelichting op de resultaten van de dendrochronologische analyse

Houtsoort = Het hout wordt door ons enkel gedetermineerd ten behoeve van de datering. Alleen de *genus*, bijv. Den (*Pinus*), wordt bepaald. Verdere soortbepaling, zoals bijv. grove den (*Pinus sylvestris*), blijft in principe achterwege, tenzij deze eenvoudig vastgesteld kan worden. Een uitzondering hierop is *Abies alba* (Zilverspar), de enige soort *Abies* die in het verleden in Nederland is toegepast.

Kern = (Geschatte afstand tot) de eerstgevormde (oudste) jaarring in de stam.

Spint = Aantal gemeten ringen spinthout. Volgens Hollstein (1980) heeft eik een gemiddeld aantal spintringen van 16 ± 5 bij een boom tot 100 jaar oud, 20 ± 6 bij een boom van 100 tot 200 jaar oud, en 26 ± 8 bij een boom ouder dan 200 jaar. Wij gebruiken een nieuwe, bijgestelde spinthoutberekening voor archeologisch/ historisch constructiehout dat dateert met Nederlandse en Duitse chronologieën (Jansma 2007). Bij eikenhout uit het Baltische gebied is het gemiddelde aantal spintringen iets lager dan in West Europa, 15 (+9/-6) (Wazny, 1990). Grove den, (*Pinus sylvestris*) heeft weliswaar ook duidelijk zichtbaar spinthout, maar doordat het aantal spinthoutringen onregelmatig is, is een schatting van de velddatum niet mogelijk. Fijnspar, (*Picea abies*) heeft geen spinthout. Uiteraard geeft een aanwezige wankant wel de precieze kapdatum van de boom.

Wankant = Het geschatte aantal jaarringen tot de wankant, d.w.z. tot de laatstgevormde jaarring (direct onder de bast), nodig voor een absolute datering van de veldatum.

Veldatum = De datum waarop de boom geveld is. Als er wankant aanwezig is, is er een absolute datering mogelijk. Als er spintringen aanwezig zijn, of zelfs alleen spintgrens, wordt de veldatum berekend door het aantal ontbrekende spintringen te berekenen. Als er bij een eik van 100 tot 200 jaar oud b.v. 4 spintringen gemeten zijn, is het geschatte aantal ontbrekende spintringen dus 16 ± 6. Dit getal wordt bij de datering opgeteld. Als er geen spintringen meer op het monster aanwezig zijn, is het onbekend hoeveel *kernhoutringen* er nog ontbreken . De veldatum ligt dan een onbekend aantal jaren ná de datering van de laatste (jongste) ring + de schatting van het ontbrekende aantal spinthoutringen. Bij een boom, die 100 tot 200 jaar oud is, is de veldatum dus xxxx AD + 20 (± 6) + X.

N = Totaal aantal jaarringen in het houtmonster.

x = Geschat aantal missende ringen (kernhout en/of spinthout) tot de wankant.

%PV = “Gleichlaufigkeit” (Duitse term) of “Percentage of Parallel Variation” (Engelse term); het percentage van de ringen in het onderzochte jaarringpatroon die aan de referentiechronologie identieke toe- en afnames van de breedte vertonen op de door de datering van het patroon aangegeven positie t.a.v. de referentiechronologie. De significantie van dit percentage is een functie van de lengte in jaren van het onderzochte jaarringpatroon en de referentie chronologie.

t = De waarde die resulteert uit een Students t-test op de kruiscorrelatie die behoort bij de beste “match” tussen het onderzochte jaarringpatroon en de referentiechronologie.

P = De kans (uitgedrukt als een fractie van 1) dat de gevonden waarde voor %PV per toeval optreedt, dus niet op een datering duidt.

RING-rapport 2012063, scheepswrak, vermoedelijk kogge, Kampen (KAMN-12)

Bijlage 4. Gebruikte referentiechronologieën

DESUDE02	Zuid-Duitsland (Huber & Becker n.p.).
EUPLNW01	Noodwest-Polen (EU Contractnr. ENV4-CT95-0128).
FRQUSP_Meuse	Frankrijk, Meuse gebied (Tegel, W., Dendronet, n.p.).
NLNOMIDD01	Hout toegepast in Noord- en Midden-Nederland (Hanraets, n.p.).
NLNOORD10	Hout toegepast in Noordwest-Nederland (Jansma, 1995).

RING-rapport 2012063, scheepswrak, vermoedelijk kogge, Kampen (KAMN-12)

Bijlage 5. Literatuur

Hollstein, E., 1980. Mitteleuropäische Eichenchronologie. Verlag Phillipp von Zabern, Mainz am Rhein.

Jansma, E., 1995. Rememberings, The development and application of local and regional tree-ring chronologies of oak for the purposes of archaeological and historical research in the Netherlands. Diss. UvA (Nederlandse Archeologische Rapporten 19), 150 pp.

Jansma, E., 2007: Datering, herkomst en bouwvolgorde van De Meern 4. In: T. de Groot & J.-M.A.W. Morel (red.), 2007: Het schip uit de Romeinse tijd De Meern 4 nabij boerderij de Balije, Leidsche Rijn, gemeente Utrecht. Waardestellend onderzoek naar de kwaliteit van het schip en het conserverend vermogen van het bodemmilieu. RACM.

Wazny, T., 1990. Aufbau und Anwendung der Dendrochronologie fuer Eichenholz in Polen. Dissertatie Universiteit van Hamburg.

Kampen, IJsselkogge

Dendrochronologisch onderzoek

Van Daalen Dendrochronologie
Projectnummer: 16.075
Uitgevoerd: november 2016
Auteur: ir. S. van Daalen
Opdrachtgever: ADC ArcheoProjecten

Contact:
H.G. Gooszenstraat 1, kamer 15, 7415 CL Deventer
vandaalen@dendro.nl
www.dendro.nl
tel: +31 (0)630114237

INLEIDING

Bij archeologisch onderzoek op de locatie van de IJsselkogge bij Kampen zijn door ADC ArcheoProjecten de resten van een punterachtig wrak, een aakachtig wrak en een (vermoedelijke) beschoeiing aangetroffen. Voor de verschillende contexten zijn monsters aangeleverd voor dendrochronologisch onderzoek om de ouderdom vast te stellen.
Het onderzoek is uitgevoerd in opdracht van ADC ArcheoPorjecten en vond plaats in oktober 2016 op het laboratorium van Van Daalen Dendrochronologie te Deventer.

METHODE

Selectie en vooronderzoek
Voor ieder monster is nagegaan of het een dateerbare houtsoort betrof, of het voldoende jaarringen leek te hebben (minimaal 70) en of het jaarringpatroon vrij was van verstoringen. Waar mogelijk wordt voorkeur gegeven aan monsters met spinthout of wankant (zie hieronder). Voor monsters waarvan de houtsoort niet met het blote oog bepaald kon worden is aan de hand van microscopische coupes en een determinatiesleutel[1] de houtsoort bepaald.

Meting(en)
Geschikt bevonden monsters hebben elk een unieke metingcode toegekend gekregen en zijn volgens standaard methodes langs één of meerdere radiale trajecten geprepareerd.[2] Langs ieder radiaal traject zijn de jaarringbreedtes ingemeten met een daartoe ingerichte meetopstelling.[3] Waar meerdere metingen aan hetzelfde monster verricht zijn, zijn deze gemiddeld tot één meting zodat ieder individueel element altijd door één meting vertegenwoordigd wordt (zie tabel 2).

Bij het inmeten is gelet op aanwezigheid van spinthout of wankant.[4] Deze informatie wordt gebruikt voor het schatten van een kapjaar of kapinterval. Hierbij worden de volgende situaties onderscheiden (zie tabel 1). De codering is gebaseerd op Baillie (1982, p.61) en wordt toegelicht in bijlage 1.

---

[1] Schweingruber, 1990.
[2] Pilcher 1990.
[3] Een Velmex meetopstelling met Acu-Rite QV10-V lineaire codeerder met een nauwkeurigheid van 10 μm gekoppeld aan een Euromex binoculair microscoop met een vergroting van 10 en 30 maal.
[4] De termen spinthout en wankant worden toegelicht in bijlage 1.

Tabel 1. Verschillende schattingsmethoden voor kapintervallen voor een datering in het jaar x.

code	omschrijving	notatie
A	wankant aanwezig, kapinterval vastgesteld buiten groeiseizoen van laatste jaar.	herfst/winter x/x+1
A1	wankant aanwezig, kapinterval vastgesteld tijdens groeiseizoen van laatste jaar.	zomer x
A2	wankant aanwezig; kapinterval vastgesteld in aanvang van volgend groeiseizoen.	lente x+1
B	geen wankant, spinthout deels aanwezig; Bayesiaanse schatting van een kapinterval (alleen voor eik)	mediaan, (2•δ interval)
C	alleen spinthoutgrens aanwezig; schatting van een kapinterval (alleen voor eik)	mediaan, (2•δ interval)
D	geen spinthout aanwezig (alleen voor eik)	na x+min. aantal spinthout
E	geen spinthout aanwezig	na x

Dateringsonderzoek

De metingen zijn met behulp van dendrochronologische software[5] met elkaar vergeleken. Voor iedere positie tussen de metingen zijn twee parameters berekend:

1. Student t-waarde. De t-waarde beschrijft de overeenkomst tussen twee getallenreeksen voor een gegeven positie. Hoe hoger deze waarde, hoe sterker de gelijkenis is; een t-waarde hoger dan 5 komt grofweg neer op een kans van 1 op 10.000 dat de gevonden uitslag op toeval berust en kan als een indicatie voor een datering beschouwd worden. Voorafgaand aan het berekenen van de t-waarde worden de jaarringbreedtes logaritmisch getransformeerd[6] zodat deze een normale verdeling benaderen.
2. *Gleichläufigkeit* (GLK); het percentage van de intervallen tussen twee jaren waarin de meting en referentiecurve gelijktijdig een stijging of daling in het jaarringpatroon laten zien. In de praktijk wordt een GLK van minder dan 62 als zwak beschouwd.

Synchronisaties die aan de statistische vereisten voldoen zijn door de dendrochronoloog visueel beoordeeld. De synchronisatie is vervolgens geaccepteerd of verworpen. Onderlinge dateringen zijn uitgevoerd om metingen uit dezelfde boom te identificeren en/of één of meerdere middelcurven samen te stellen die het dateren faciliteren.

[5] PAST4. Uitgegeven door SCIEM, Wenen (Oostenrijk). www.sciem.com

[6] De zogeheten transformatie van Hollstein (Hollstein 1980).

RESULTATEN

Selectie en vooronderzoek

In totaal zijn 17 monsters aangeleverd. Hiervan zijn 2 monsters komen te vervallen omdat deze niet voldoende jaarringen bevatten (zie tabel 2).

In de wrakken is eik (*Quercus sp.*) en grove den (*Pinus sylvestris* L.) aangetroffen. Voor de beschoeiing is beuk (*Fagus sylvatica* L.) gebruikt.

Metingen

Tabel 2. Overzicht van de meetgegevens. n*:aantal jaarringen,* $n_{(s)}$*: aantal spintringen, type: schattingswijze voor het kapinterval conform tabel 1.*

structuur	vondstnr.	omschrijving	houtsoort	meting	*n*	$n_{(s)}$	type
kogge	M173	balk	eik	16.075.001	90	5	B
kogge	M186	plank binnenbetimmering	grove den	16.075.002	110	-	D
kogge	M187	plank binnenbetimmering	eik	16.075.003	232	-	D
kogge	M188	plank binnenbetimmering	grove den	16.075.004	205	-	D
aak	M83	plank	eik	16.075.005	138	-	D
aak	M84	plank	eik	16.075.006	148	-	D
aak	M87	spant	eik	16.075.007	127	15	B
kade	M06	plank	beuk	16.075.008	180	n.v.t	A
kade	M08	balk	beuk	16.075.009	207	n.v.t	A
kade	M11	plank	beuk	16.075.010	142	n.v.t	A
punter	M13	balk	eik	16.075.011	84	16	B
punter	M22	plank	eik	16.075.012	65	-	D
punter	M29	balk	eik	16.075.013	85	7	B
punter	M37	plank	eik	16.075.014	69	-	D
punter	M46	plank	eik	-			
punter	M71	plank	eik	16.075.015	62	-	D
punter	M79	plank	eik	-			

Dateringsonderzoek

Onderlinge synchronisatie van de metingen leverde resultaten op voor de planken van het punterachtige wrak. Met enige moeite kan hieruit een middelcurve gemaakt worden waarvoor goede resultaten gevonden kon worden die voor de individuele metingen in mindere mate behaald werden.

Daarnaast konden de metingen van 2 beuken palen onderling gedateerd worden. Hiervoor werden (zwakke) resultaten met een referentiecurve voor eik gevonden. Deze dateringen werden bij aanvullend onderzoek door S. Cremer[7] bevestigt.

Enkele van de resterende metingen konden op individuele basis gedateerd worden (zie tabel 3). De overige eiken metingen zijn voor aanvullend onderzoek naar dr. A. Daly[8] gestuurd. Dit leverde voor M13 van het punterachtige wrak resultaat op.

De vermelde referentiecurven staan in tabel 4 toegelicht.

Tabel 3. Overzicht van de dateringen met statistische onderbouwing. De grafische weergave van de metingen met de onderstreepte referentiecurve staat in bijlage 2. eind$_{(m)}$/eind$_{(r)}$*: positie van de laatste jaarring van de meting/referentie.*

meting	eind$_{(m)}$	referentie	eind$_{(r)}$	overlap	GLK	t-waarde	middelcurve
16.075.003	1385	BALTIC1	1597	230	66,3	7,47	
16.075.004	1372	K500090M	1990	199	60,8	6,17	
16.075.005	1321	BELG003	1989	138	61,2	5,07	
16.075.006	1357	NLTWWF01	1972	148	64,5	5,64	
16.075.007	1376	NLTWWF01	1972	127	65,0	5,06	
16.075.008	1451	DEWEBE01	1970	180	62,8	4,06	
16.075.010	1451	16.075.008	1451	142	66,2	7,34	
16.075.012	1406	NL.VME-NT	1835	65	67,7	4,70	
		16.075.014	1400	59	69,5	4,26	16.075.M2
16.075.014	1400	NLTWWF01	1972	69	67,4	5,43	
		16.075.012	1406	59	69,5	4,26	16.075.M2
16.075.015	1426	DEWEFA01	1801	62	73,4	4,13	
16.075.M2	1426	NL.VME-NT	1835	95	68,9	6,05	

[7] Het Koninklijk Instituut voor het Kunstpatrimonium (KIKIRPA), Brussel (B).

[8] dendro.dk, Brønshøj (DK)

Tabel 4. Overzicht van vermelde referentiecurven.

referentie	omschrijving
BALTIC1	Paneelschilderijen (import uit Balticum of Oost-Europa). Referentiecurve voor eik (1156 - 1597). Hillam, Tyers, 1995.
BELG003	België, Meuse Valley Archaeological. Referentiecurve voor eik (672 - 1986). Hoffsummer, via ITRDB.
DEWEBE01	Duitsland, Weser Bergland. Referentiecurve voor eik (1004 - 1970). Delorme, 1972.
DEWEFA01	Duitsland, Westfalen. Referentiecurve voor eik (1060 - 1801). Hollstein, 1980.
K500090M	Midden-Polen. Referentiecurve voor grove den (1990 - 817). Wazny, niet gepubliceerde data.
NL.VME-NT	Nederland, algemeen. Referentiecurve voor eik (282 - 1835). Van Daalen, niet gepubliceerde data.
NLTWWF01	Nederland, Duitsland (Twente, Westfalen). Referentiecurve voor eik (1040 - 1972). Tisje, niet gepubliceerd data.

INTERPRETATIE

Het onderzoek is deels geslaagd in het dateren van de monsters. Hoewel voor iedere context één of meerdere dateringen beschikbaar zijn (zie tabel 5) bleek het niet mogelijk voor iedere context een kapinterval vast te stellen. In deze gevallen kon alleen een ondergrens van het kapinterval bepaald worden.

Tabel 5. Schatting van de kapintervallen. Het type is de schatting volgens tabel 1.

structuur	monster	meting	eind	kapinterval	type
kogge	M173	16.075.001	-		B
kogge	M186	16.075.002	-		D
kogge	M187	16.075.003	1385	na 1393	D
kogge	M188	16.075.004	1372	na 1372	D
aak	M83	16.075.005	1321	na 1327	D
aak	M84	16.075.006	1357	na 1363	D
aak	M87	16.075.007	1376	rond 1380 (1376 – 1393)	B
kade	M06	16.075.008	1451	herfst/winter 1451/52	A
kade	M08	16.075.009	-		A
kade	M11	16.075.010	1451	herfst/winter 1451/52	A
punter	M13	16.075.011	1431	rond 1435 (1431 – 1448)	B
punter	M22	16.075.012	1406	na 1412	D
punter	M29	16.075.013	-		B
punter	M37	16.075.014	1400	na 1406	D
punter	M71	16.075.015	1426	na 1432	D

Het hout is afkomstig van verschillen bronnen. In de kogge is Pools eiken en grenen verwerkt. De overige metingen zijn ofwel van een lokale herkomst, ofwel van nabijgelegen gebieden in Duitsland afkomstig. De resultaten zijn is veel gevallen niet sterk of talrijk genoeg voor een degelijke herkomstbepaling.

LITERATUUR

Baillie, M.G.L., 1982: *Tree-ring dating and Archaeology.* ISBN 0-7099-0613-7. Croom Helm Ltd. London.

Bronk Ramsey, C., 2009: Bayesian analysis of radiocarbon dates. In: Radiocarbon, 51(1), pp. 337-360.

Delorme, A., 1972: Dendrochronologisch Untersuchingen an Eichen des Südlichen Weser- und Leineberglandes. Doctoraal dissertatie, Universiteit van Göttingen.

Hillam, J., Tyers, I., 1995: Reliability and repeatability in dendrochronological analysis: tests using the Fletcher archive of panel-painting data. Archaeometry, 37(2), pp. 395-405

Hoffsummer, P.: BELG003. Data set id: noaa-tree-3501. Ontleend aan http://hurricane.ncdc.noaa.gov/pls/paleox/f?p=519:1::::P1_STUDY_ID:3501 op 4-2-2014.

Hollstein, E., 1980: Trierer Grabungen und Forschungen. Band XI, Rheinisches Landesmuseum Trier. ISBN 3-8053-0096-4. Verlag Philipp von Zabern, Mainz am Rhein.

Pilcher, J.R., Sample preparation, Cross-dating, and Measurement. In: Cook, E.R., Kairiukstis, L.A., (eds) 1990: *Methods of Dendrochronology, Applications in the Environmental Sciences.* Kluwer Academic Publishers. ISBN 0-7923-0586-8.

Schweingruber, F.H., 1990: Mikroskopische Holzanatomie. Formenspektren mitteleuropäischer Stamm- Und Zweigölzer zur Bestimmung von recentem und subfossilem Material. 226 pp. Zürcher AG. ZugOxf.: 811.1 __ 016 : 810 : 814.7 (4). 3e druk.

BIJLAGE 1

A. Wankant aanwezig: De jaarringgrens van de buitenste jaarring direct onder de bast maakt het mogelijk het seizoen te bepalen waarin de boom gekapt is. Aanwezigheid van de wankant betekent per definitie dat het spinthout volledig aanwezig is. Het seizoen waarin de boom gekapt is volgt uit de mate waarin de buitenste ring gevormd is:

1. A: De buitenste jaarring is volledig gevormd. Het kapinterval valt buiten het groeiseizoen van de laatste (gedateerde) jaarring.
2. A1: De buitenste jaarring is niet volledig gevormd. Het kapinterval valt in het groeiseizoen van de laatste (gedateerde) jaarring.
3. A2: Alleen de aanzet tot de buitenste jaarring is aanwezig. Deze jaarring wordt niet ingemeten. Het kapinterval valt aan het begin van het groeiseizoen volgend op de laatste (ingemeten) jaarring.

B. Spinthout aanwezig: Het spinthout is de buitenste zone van de stam waar het hout nog niet is omgezet in kernhout. Niet alle houtsoorten vormen kernhout en alleen bij eik is het aantal jaarringen in het spinthout statistisch te omschrijven zodat een schatting gemaakt kan worden van het aantal ontbrekende jaarringen tot de wankant. Voor het bereken van het kapinterval wordt OxCal[9] gebruikt met door de auteur samengestelde spinthoutstatistieken. Hieruit volgt een jaartal dat het meest waarschijnlijk is (de mediaan), met daarom heen een 2·δ (95,4%) betrouwbaarheidsinterval. Spinthoutstatistieken verschillen zijn niet voor alle herkomstgebieden hetzelfde, waardoor naar gelang de herkomst van het hout andere spinthoutstatistieken toegepast kunnen worden.

C. Spinthoutgrens aanwezig: Als (een deel van) de contouren van een monster één en dezelfde jaarring volgen dan kan dit geïnterpreteerd worden als de overgang tussen het kernhout en het (niet meer aanwezige) spinthout. Hierbij wordt op dezelfde wijze als hierboven een kapinterval berekend. Hierbij moet de kanttekening geplaatst worden dat dit alleen met redelijke zekerheid vastgesteld kan worden als dit langs een voldoende groot deel van de contouren van het monster zichtbaar is.

D. Geen spinthout aanwezig: Hierbij is het niet mogelijk een kapinterval te schatten en kan alleen gesteld worden dat in ieder geval een klein aantal spinthoutringen (6 stuks) volgt op het kernhout. De vroegst mogelijke datering wordt dan met een corresponderend aantal jaarringen gecorrigeerd. Dit geldt alleen voor eik.

E. Geen spinhoutstatistieken beschikbaar of geen kernhoutvorming: Hierbij is het niet mogelijk een kapinterval te schatten en kan alleen gesteld worden dat het kapjaar ná de datering van de buitenste ring valt. Dit wordt zowel toegepast voor houtsoorten die geen kernhout vormen, of waarvoor het aantal spinthoutringen niet rekenkundig te omschrijven is.

---

[9] Bronk Ramsey 2009.

BIJLAGE 2

Hier onder staan de metingen afgebeeld met de in tabel 3 aangegeven referentie. Op de x-as staan de jaartallen, op de y-as de ringbreedtes op een logaritmische schaal, uitgedrukt in 1/100 mm. Het spinthout is gestippeld aangegeven. De grijze banen geven intervallen met een positieve GLK aan.

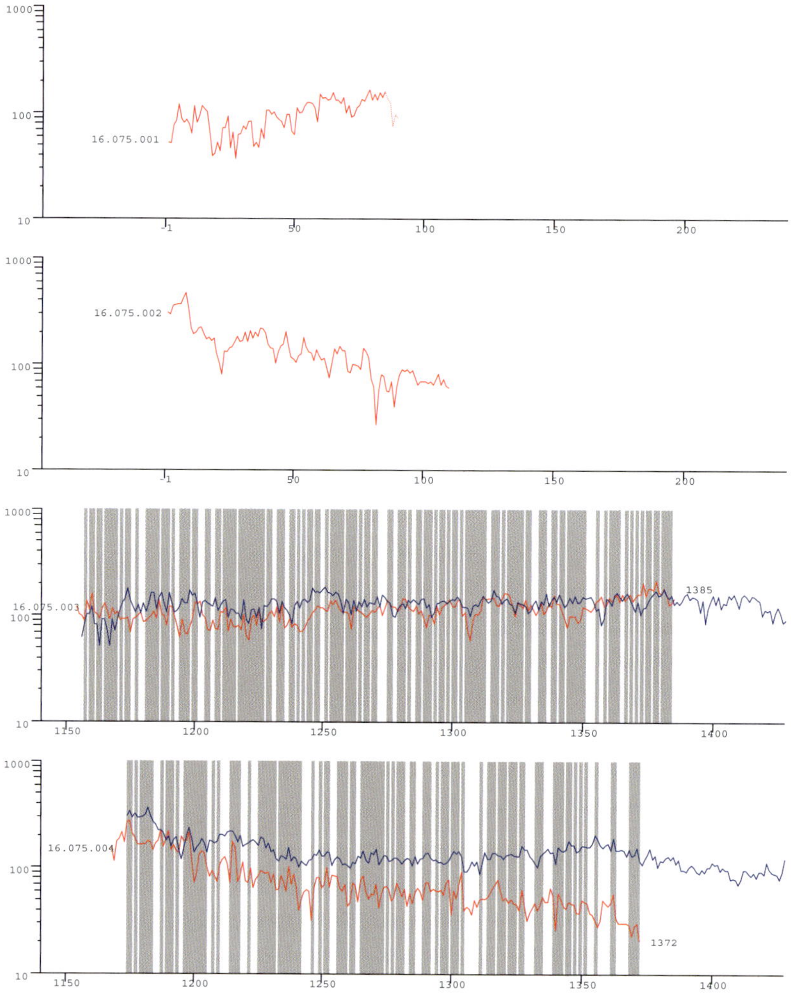

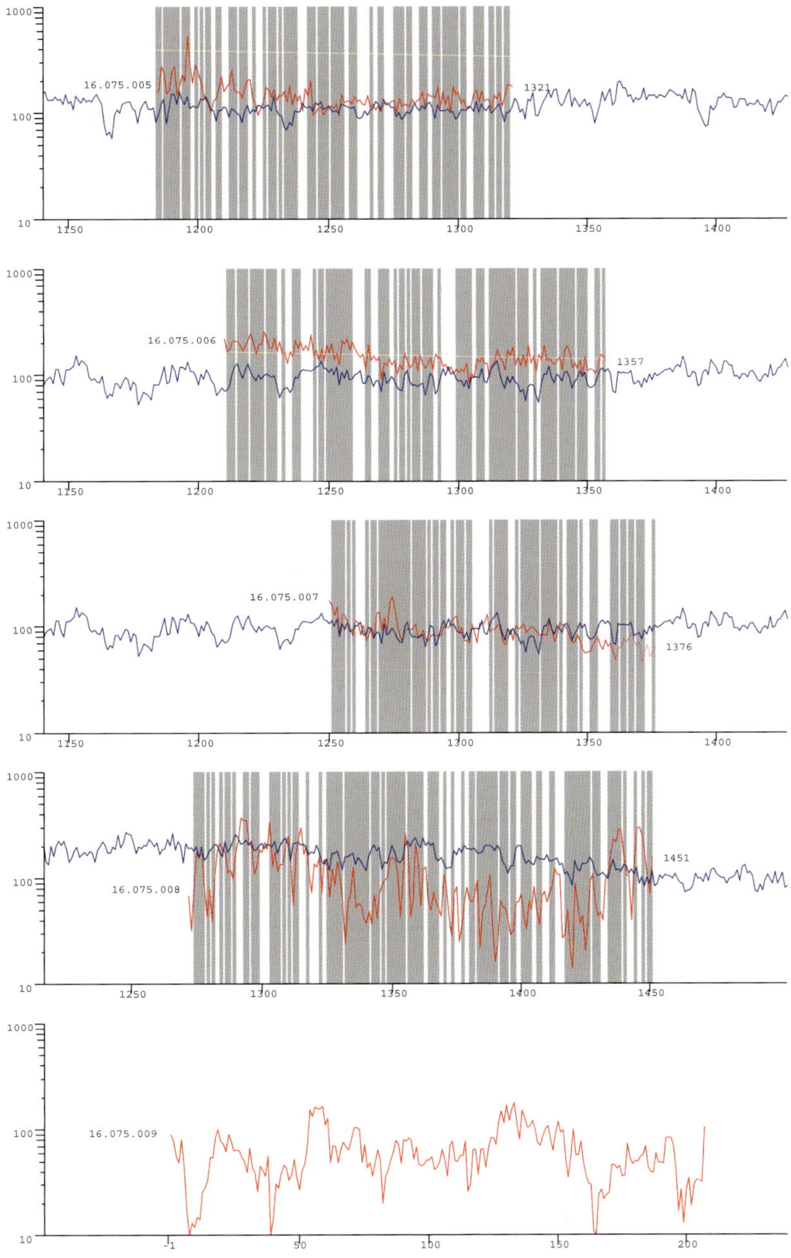

16.075.005
1321
16.075.006
1357
16.075.007
1376
16.075.008
1451
16.075.009

10

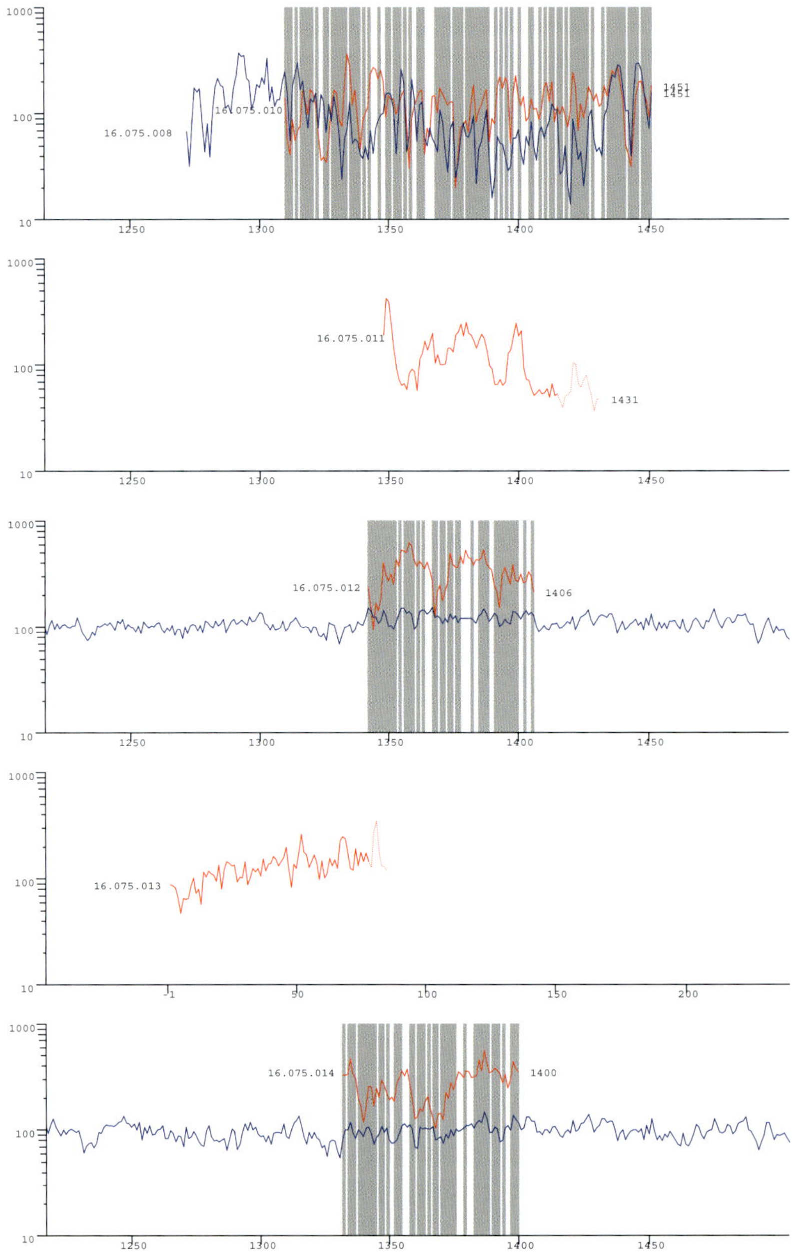
16.075.008
16.075.010
1451
1451
16.075.011
1431
16.075.012
1406
16.075.013
16.075.014
1400

11

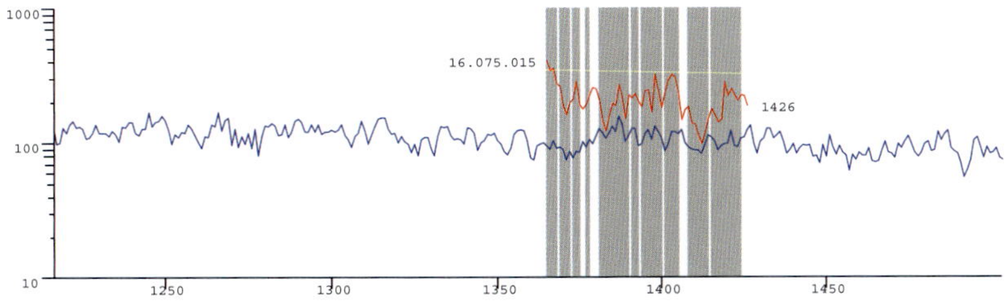
1000
100
10
16.075.015
1426
1250
1300
1350
1400
1450

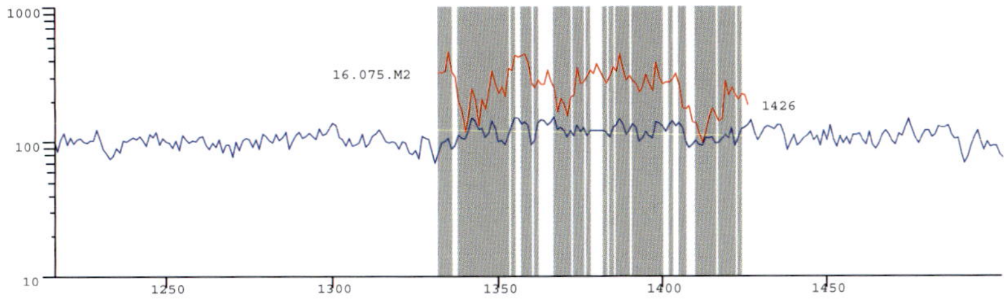
1000
100
10
16.075.M2
1426
1250
1300
1350
1400
1450

24 The Enkhuizerzand cog?

Dendrochronological report RING

E. Jansma & E. Hanraets

ring

Nederlands Centrum voor Dendrochronologie/Stichting RING
Rijksdienst voor het Oudheidkundig Bodemonderzoek (ROB)
Kerkstraat 1, 3811 CV (Postbus 1600, 3800 BP) Amersfoort
Tel (033) 422 75 13; Fax (033) 422 77 99
E-mail RING@ARCHIS.NL

RAPPORTAGE DATEREND ONDERZOEK

Object: kavel: YM-1; scheepswrak

Aanvrager

Naam	: de heer K. Vlierman
Instantie	: NISA
Straat	: Oostvaardersdijk 01-04
Postcode, stad	: 8242 PA Lelystad
Telefoon	: 0320-269 700

Onderzoekers

Naam	: dr. E. Jansma, drs. E. Hanraets
Instantie	: RING/ROB
Datum onderzoek	: maart 1999

Resultaten van het onderzoek

vondstno.	dendro-naam	kern	spint	wankant	n	referentiecurve	datering	t	%PV	p
huidpl./m.1	kym 01.0	-	13	-	64	NlWestfalen (Tisje)	1395-1458	5.86	70.6	0.001
						EU1(EU project 96)*		3.68	74.6	0.0002
huidpl./m.2	kym 02.1	-	-	-	41	-	geen			
huidpl./m.3	kym 03.0	+ca.15	-	-	58	-	geen			
huidpl./m.4	kym 04.1	-	-	-	52	-	geen			
legger?/m.5	kym 05.0	+ca.5	grens	-	64	-	geen			
dekb.?/m.6	kym 06.0	+3	24	-	172	-	geen			
spant?/m.7	kym 07.0	K	-	-	146	NlWestfalen	1291-1436	5.11	62.1	0.005
						C.Duitsland(Hollstein)		5.28	63.5	0.005

* EU1 is de regio Nederland en Noord Duitsland

Toelichting op de resultaten

kern	=	(afstand tot) de eerstgevormde (oudste) jaarring in de stam
spint	=	aantal ringen spinthout
wankant	=	laatstgevormde jaarring (direkt onder de bast), nodig voor een absolute datering van de veldatum
n	=	totaal aantal jaarringen in het houtmonster
%PV	=	"Gleichlaufigkeit" (Duitse term) of "Percentage of Parallel Variation" (Engelse term); het percentage van de ringen in het onderzochte jaarringpatroon die aan de referentiechronologie identieke toe- en afnames van de breedte vertonen op de door de datering van het patroon aangegeven positie t.o.v. de referentiechronologie. De significantie van dit percentage is een functie van de lengte in jaren van het onderzochte jaarringpatroon.
t	=	De waarde die resulteert uit een Students t-test op de kruiscorrelatie die behoort bij de beste "match" tussen het onderzochte jaarringpatroon en de referentiechronologie.
P	=	De kans (uitgedrukt als een fractie van 1) dat de gevonden waarde voor %PV per toeval optreedt, dus niet op een datering duidt.

Interpretatie van de datering

Code	Einddatering	Schatting ontbrekend spint	Veldatum
kym 01.0	1458 AD	+ 7 ± 6	1465 AD ± 6
kym 07.0	1436 AD	+ ≥ 20 ± 6	NA 1456 AD ± 6

****

TSAP JOB OUTPUT (DATE: 1999.4.1 TIME: 19.42): CORRELATION TABLE
Time interval refers to absolute dates
CORRELATION VALUE:
TVH = T-Value with Hollstein-Standardization, 'Wuchswert' (0<=TVH<=100)

No/Len:TVH	P.	KYM00010/ 64	KYM00070/ 146
EUBIG1ST/	2966	0.9/ 63	0.1/ 145
EUBIG2ST/	2675	1.4/ 63	3.0/ 145
EUBIG3ST/	2983	0.7/ 63	1.0/ 145
EUBIG4ST/	2181	0.0/ 63	1.0/ 145
EUBIG5ST/	1785	2.1/ 63	1.0/ 145
EUBIG6ST/	1332	1.6/ 63	0.4/ 145
EUBIG7ST/	1183	1.2/ 63	1.7/ 145
EUBIG8ST/	1226	0.7/ 63	1.2/ 145
FRFB1351/	449		
FRNO1274/	706	1.6/ 63	3.2/ 145
FROO1360/	625	1.5/ 63	1.7/ 76
FRPARDST/	750	1.7/ 63	2.6/ 145
FRPARDST/	750	1.7/ 63	2.6/ 145
FRPAREST/	367		
FRWESBST/	820	0.6/ 63	2.8/ 145
FRWESBST/	820	0.6/ 63	2.8/ 145
GBBALTI1/	442	0.8/ 63	0.7/ 145
GBBALTI1/	442	0.8/ 63	0.7/ 145
GBLONDON/	1316	1.6/ 63	1.7/ 145
GBNORTHC/	472	0.5/ 63	1.0/ 145
GBNORTHM/	1303	0.4/ 63	1.2/ 145
GBSOUTHM/	1189	0.8/ 63	0.2/ 145
HAM1080M/	888	3.9/ 63	0.2/ 145
MAAS672M/	1315	1.6/ 63	4.9/ 145
NBH1036M/	937	2.6/ 63	3.9/ 145
NKU1082M/	891	2.2/ 63	1.0/ 145
NLMIDDEN/	**644**	**5.0/ 63**	3.6/ 145
NLNOORDM/	306		1.6/ 55
NLWF1040/	**933**	* **5.8/ 63**	* **5.1/ 145**
NLZEELAN/	428		
NLZUIDMM/	1326	1.9/ 63	4.6/ 145
NSSUB1HL/	1077	3.1/ 63	3.6/ 145
NSSUB3HL/	1077	3.2/ 63	3.6/ 145
NSSUB4HL/	1118	2.6/ 63	2.9/ 145
NSSUB5HL/	1113	3.5/ 63	2.8/ 145
NSSUB6HL/	1112	2.9/ 63	1.9/ 145
NSSUB7HL/	1128	3.3/ 63	2.2/ 145
NSSUB9HL/	1117	3.2/ 63	3.7/ 145
OFRLAATT/	993	3.4/ 63	2.6/ 145
POL457MM/	1267	1.5/ 63	1.6/ 145
POL996MM/	990	0.6/ 63	0.6/ 145
SCH1115M/	529	0.1/ 63	0.5/ 145
SH436MMM/	1533	2.4/ 63	0.5/ 145
WBL1004M/	967	3.0/ 63	3.2/ 145
WD400STD/	1576	2.4/ 63	* **5.2/ 145**
ZD631MMM/	1320	3.0/ 63	4.0/ 145
ZDLLAATT/	951	3.1/ 63	3.9/ 145

****

```
**************************************************************************
TSAP JOB OUTPUT (DATE: 1999.4.1  TIME: 19.42): CORRELATION TABLE
Time interval refers to absolute dates
CORRELATION VALUE:
Glk = Gleichlaeufigkeit, % of equivalent slope intervals (0<=Glk<=100)
```

No/Len:Glk	P.	KYM00010/ 64	KYM00070/ 146
EUBIG1ST/ 2966		42.7/ 64	44.4/ 146
EUBIG2ST/ 2675		35.5/ 64	43.4/ 146
EUBIG3ST/ 2983		45.2/ 64	49.7/ 146
EUBIG4ST/ 2181		51.6/ 64	50.7/ 146
EUBIG5ST/ 1785		69.4/ 64	51.4/ 146
EUBIG6ST/ 1332		66.1/ 64	52.1/ 146
EUBIG7ST/ 1183		65.3/ 64	49.7/ 146
EUBIG8ST/ 1226		53.2/ 64	50.0/ 146
FRFB1351/ 449			
FRNO1274/ 706		53.2/ 64	56.9/ 146
FROO1360/ 625		49.2/ 64	52.0/ 77
FRPARDST/ 750		68.5/ 64	60.8/ 146
FRPARDST/ 750		68.5/ 64	60.8/ 146
FRPAREST/ 367			
FRWESBST/ 820		46.8/ 64	60.1/ 146
FRWESBST/ 820		46.8/ 64	60.1/ 146
GBBALTI1/ 442		50.0/ 64	52.1/ 146
GBBALTI1/ 442		50.0/ 64	52.1/ 146
GBLONDON/ 1316		55.6/ 64	50.0/ 146
GBNORTHC/ 472		55.6/ 64	53.1/ 146
GBNORTHM/ 1303		56.5/ 64	58.3/ 146
GBSOUTHM/ 1189		55.6/ 64	51.4/ 146
HAM1080M/ 888		63.7/ 64	47.6/ 146
MAAS672M/ 1315		59.7/ 64	56.9/ 146
NBH1036M/ 937		53.2/ 64	53.8/ 146
NKU1082M/ 891		67.7/ 64	53.8/ 146
NLMIDDEN/ 644	*	**70.2/ 64**	62.2/ 146
NLNOORDM/ 306			61.1/ 56
NLWF1040/ 933	*	**72.6/ 64**	62.2/ 146
NLZEELAN/ 428			
NLZUIDMM/ 1326		59.7/ 64	57.6/ 146
NSSUB1HL/ 1077	*	**76.6/ 64**	63.5/ 146
NSSUB3HL/ 1077	*	**74.2/ 64**	62.5/ 146
NSSUB4HL/ 1118		66.1/ 64	54.5/ 146
NSSUB5HL/ 1113	*	**75.8/ 64**	54.9/ 146
NSSUB6HL/ 1112		61.3/ 64	52.8/ 146
NSSUB7HL/ 1128	*	**71.8/ 64**	50.7/ 146
NSSUB9HL/ 1117	*	**73.4/ 64**	62.8/ 146
OFRLAATT/ 993	*	**74.2/ 64**	51.0/ 146
POL457MM/ 1267		58.9/ 64	50.0/ 146
POL996MM/ 990		61.3/ 64	49.3/ 146
SCH1115M/ 529		49.2/ 64	51.4/ 146
SH436MMM/ 1533		66.1/ 64	47.2/ 146
WBL1004M/ 967	*	**73.4/ 64**	61.5/ 146
WD400STD/ 1576		59.7/ 64	63.5/ 146
ZD631MMM/ 1320		58.9/ 64	60.1/ 146
ZDLLAATT/ 951		58.9/ 64	60.1/ 146

```
**************************************************************************
```

Credits per cog

6 The Nijkerk-II cog

Discovery and Notification:
23 September 1981; K. Vlierman

Reconnaissance:
28 September - 10 October 1981; L. van Dijk and J. van der Land.

Excavation:
27 April - 28 October 1983; G. Bek, L. Van Dijk, W. Folkertsma (trainee), R. Hageraats, A.F.L. van Holk (stagiair), J. van der Land, A. Luns (stagiair), R. Oosting, H.R. Reinders, K. Vlierman (day-to-day archaeological management), members of the Archeologische Werkgemeenschap voor Nederland (AWN) members of the Oudheidkundige Vereniging Arent toe Boecop of Elburg.

Post-field research, augmentation of the field data, report and (reconstruction) drawings:
K. Vlierman. For the Appendices 2.III A-C the drawings of J.-M.A.W. Morel are used as a basis.

Description of the finds and restoration of objects:
K. Vlierman and T.J.C. van Zuidam.

Pollen spectrum of the moss caulking:
E. Mook-Kamps (RUG).

Examination of the bones:
F. Laarman (RCE)

Dendrochronological report:
E. Jansma and E. Hanraets (RING).

Determination of wood samples:
L. Koehler and J. Nientker (RCE).

Find numbers:
Zo 36-1 to Zo 36-308.

photos:
L. van Dijk, T. Penders, H. Schraal (RCE) and K. Vlierman.

Photo numbers:
84938-84947
90658-90718
91107-91121
92461
Zo 36-1 to Zo 36-307
90721-90768
91123-91142
model Oz 36 ZFL
85131-85144
90784-90791
91155
100292
Zo 36 [-13]
91281-91307
100799
Zo 36-M1 and -M1a
91326-91357
100809
91413-91415
K. Vlierman private ownership

Slide numbers:
69306-69316
70108-70127
71225
87793-87888
88069-88070
91007-91026
69962-69966
70236-70277
87942-87948
88079-88087
69970-69981
70597-70604
69987-69993
70609-70613

Manuscript completed in October 2007.

8 The Nijkerk-I cog

Notification:
4 September 1980; J. Abrahamse

Reconnaissance:
3 November 1980; J. van der Land and H. van Veen

Excavation:
1 May - 10 November 1981; L. van Dijk, R. van Fucht (trainee), R. Hulst (trainee), J. van der Land, R. Oosting, H.R. Reinders, K. Vlierman (day-to-day archaeological management), P.B. Zwiers, members of the Archeologische Werkgemeenschap voor Nederland (AWN) and members of the Oudheidkundige Vereniging Arent toe Boecop of Elburg.

Post-field research, augmentation of the field data, report and (reconstruction)drawings:
K. Vlierman

Description of the finds:
K. Vlierman

Pollen spectrum of the moss caulking:
E. Mook-Kamps (RuG)

Examination of the bones:
F. Laarman (RCE)

Study of peat:
D.G. van Smeerdijk

Find numbers:
Zo 43-1 to Zo 43-90

Photos:
L. van Dijk and T. Penders (RCE), L. Potuyt (RIJP),

Photo numbers:
84284-84337
84759-84815
model COG
Zo43-1 to Zo43-90
84621-84625
84925-84937
model Nijkerk-I
84720-84750
85109-85130
model Nijkerk-I [-001]

Slide numbers:
68356-68359
70062-70066
105049-105050
68468-68482
70082-70084
CSA00422-CSA00425
68233-69245
88058-88061
69323-69335

Manuscript completed in August 2011

9 The Ens cog

Notification:
1946. Superintendent Brandsma?

Reconnaissance: ?

Excavation:
28 May-3 June and 11-24 June 1949.
An unknown number of camp inmates under leadership of Th.J. Remmerswaal under direction of G.D. van der Heide.

Field drawings:
G.D. van der Heide / Th.J. Remmerswaal.

Preliminary excavation report:
G.D. van der Heide.

Post-field research, augmentation of the field data, report and (reconstruction)drawings:
K. Vlierman.

Description of the finds:
K. Vlierman

Pollen spectrum of the moss caulking:
E. Mook-Kamps RuG.

Photos:
Hiemstra and Klaar (Directie Wieringermeer), L. van Dijk, T. Penders and H. Schraal (RCE) and K. Vlierman.

Find numbers:
Nq 75-1 to Nq 75-7.

Photo numbers:
7188, 7424, 7428 to7430, 97.234
NQ75-1 to NQ75-7, Z1950-I 58, Z1950-I 59 and Z1950-I 60.

Manuscript completed in August 2008.

10 The Rutten cog

Notification:
23 October 1952; R. Terpstra, 616 Kamp Rutten.

Reconnaissance:
20 September 1982; J. van der Land and K. Vlierman.

Excavation:
3 June-21 August 1985; L. van Dijk, J. van der Land, R. Loos, R. Oosting, H. Post, K. Vlierman (day-to-day archaeological management), J. van der Zee (RIJP), J. Cupido and A. Vos (trainees), pupils and two teachers LTS-class of the Copernicus Scholengemeenschap in Dronten and members of the Oudheidkundige vereniging Arent Thoe Boecop of Elburg (volunteers).

Post-field research, augmentation of the field data, report and (reconstruction) drawings:
K. Vlierman.

Description of the finds:
K. Vlierman.

Pollen spectrum of the moss caulking:
E. Mook-Kamps (RuG).

Examination of the bones:
F. Laarman (RCE).

Dendrochronological report:
RING; A.E.M. Hanraets and E. Jansma.

Find numbers in the field:
A57/1 to A57/259; fictive inventory numbers 1 to 29

Photos:
L. van Dijk and T. Penders (RCE).

Photo numbers:

85.997
71346-71440
93.043-93.071
71487-71503
93.133-93.228
93.232-93.258
93.300-93.323
95.179-95.187
A57-1 to A57-261
A57-[-001]
A57-[-002]

Slide numbers:
71346-71440
71487-71503

Manuscript completed in March 2013

11 The Dronten-I cog

Notification:
Oktober 1973, B. Haayer.

Reconnaissance:
29 October 1973, L. van der Land, H. van Veen and K. Vlierman.

Excavation/research:
2 August -3 September 1976, L. van Dijk, J. van der Land, H.R. Reinders, F.J. Tjallingii, H. van Veen, K. Vlierman and P.B. Zwiers.

Excavation report 1980:
H.R. Reinders, H. van Veen, K. Vlierman and P.B. Zwiers.

Post-field research 2012:
Augmentation of the field data, report and (reconstruction) drawings: K. Vlierman

Description of the finds:
K. Vlierman.

Restoration ceramics:
K. Vlierman.

Conservation iron objects:
Van der Woude.

Pollen spectrum of the moss caulking:
E. Mook-Kamps (RuG).

Inventory numbers:
On 5-1 to On 5-64.

Photos:
L. van Dijk and T. Penders (RCE), J.H. Kamps (Delft), J.U. Potuyt (RIJP) and K. Vlierman.

Photo numbers:
54865-54892
54943-54946
55201-55213
59213-59214
97219-97232
DSCF 3186-3195
ON5-1 to ON5-63

Slide numbers:
62248-62264
88059-88060
88062-88065
88085-88086

Manuscript completed in June 2012.

12 The Marknesse cog

Notification:
?

Excavation and re-burial of the wreck in 1944:
P.J.R. Modderman, D. Huismans and J.P.G. Kloppenburg (Nederlands Openluchtmuseum), J. Lanting and R. Woudstra (Biologisch Archeologisch Instituut), camp inmates (Directie Wieringermeer) and employees of the Amsterdamsche Scheepwerf (De Vries Lentsch).

Post-field research in the field, August 2008:
A.F.L. van Holk and H.R. Reinders with students of the International Fieldschool for Maritime Archaeology Flevoland (IFMAF); K. Vlierman.

Post-field research, report and (reconstruction) drawings:
K. Vlierman.

Description of the finds:
P.J.R. Modderman and K. Vlierman

Dendrochronological report:
RING; M. Dominguez

Find numbers:
Nm 107-1 to -32.

Photos:
Managing board Wieringermeer; T. Penders (RCE); K. Vlierman

Photo numbers:
5685, 5686, 9172, 9184, 9192, 16643 and 16644
Photos (Nom):
13114 to 13132
13144 to 13150
Model M 107 NOP
M107-1 to M107-23
Z1944-II 2
Z1944-II 3
Z1944-II 5
Z1944-II 6
Z1944-II 7
Z1944-II 9
Z1944-II 9 [-001]
Z1944-II 11
Z1944-II 14
Z1944-II 15
Z1944-II 15a
Z1944-II 20
Z1944-II 21

Slide numbers:
9172, 9184, 19557, 75007-750016, 88064, 88073, 88078, CSA 00418-CSA 00421, 16298-106326.

Manuscript completed in March 2009.

13 The Kuinre cog

Notification:
?

Notes:
P.J.R. Modderman and G.D. van der Heide

Research, report and drawings:
K. Vlierman

Find number:
Z1950/ I-96

Photos:
Managing board Wieringermeer; T. Penders (RCE); K. Vlierman

Photo numbers:
10534 and 10535.
(NM133) Z1950-I 96

Manuscript completed in February 2008

14 The Spakenburg-I cog

Notification:
B. Proem.

Reconnaissance:
29 october 1975. J. van der Land and H. van Veen.

Excavation:
30 July - 11 August 1979. L. and M. Boot, R. Bouwmans, M. Brinkman, M. Eijkman, N. de Goeje, A. Krekel, A. Pronk, H. van Vastenhoven, M. van Veen, J. Verschoor and M. van der Vlugt (N.J.B.G.), M. Casals, R. Le Conte, Ch. De Malglaive and P. Perrin (French guests).
14 - 24 August 1979. L. van Dijk, J. van der Land, H. R. Reinders and K. Vlierman.

Post-field research, field drawings, report and (reconstruction) drawings:
K. Vlierman.

Description of the finds:
K. Vlierman.

Pollen spectrum of the moss caulking:
E. Mook-Kamps (RUG).

Examination of the bones:
F. Laarman (RACM).

Find numbers:
Zn 42 II-1 to-54.

Photos:
J.U. Potuyt, L. Van Dijk, T. Penders (RCE) and K. Vlierman.

Photo numbers:
80606-80647
97235-97243
M.D. 3 and 45
K. Vlierman private
ZN42 II-1 to ZN42 II-56
Z42 II-[-001]

Slide numbers:
67078-67092
67095-67105
67124-67125
74979-74981
74989-74991
88080.

Manuscript completed in January 2008.

15 The Spakenburg-II cog

Notification:
Agricultural supervisor H.D. Engberts, July 1971.

Reconnaissance:
J. Kuipers and K. Vlierman, 19 July 1971.

Excavation:
L. van Dijk, J. van der Land, H.R. Reinders, H. van Veen, K. Vlierman, J. van der Zee and P.B. Zwiers, 27 April - 31 May 1979.

Excavation report 1991:
A. van de Moortel, 1991.

Post-field research, report and drawings:
K. Vlierman.

Pollen spectrum of the moss caulking:
R.T.J. Cappers and E. Mook-Kamps (RuG).

Dendrochronological report:
E. Hanraets and E. Jansma, RING.

Photos:
L. van Dijk and T. Penders (RCE).

Photo numbers:
80247-80281
91422-91426
80683-80698
96490-96494
84156-84158
Nz 43 Z. Fl. (model)

Slide numbers:
66524-66553
70614-70615
69950-69961
70617-70618
70085-70089

Manuscript completed in November 2014.

16 The Almere-I cog

Notification:
2 April 1986. Brandhorst (supervisor from the Rijksdienst voor de IJsselmeerpolders).

Reconnaissance:
2 April, 1986. F. Folkertsma and R.Loos.

Excavation and salvage:
2-23 April, 1986. L. van Dijk, F. Folkertsman, J. van Helden, J. van der Land, R. Loos, W. Oosterhof, R. Oosting, H,J. Post, H.R. Reinders, P. Stassen and K. Vlierman.

Drawing of the ship parts scale 1:10:
15 June-17 August 1988. M. Fitzgerald, F.M. Hocker, S. Mark and R. Neyland (Texas A & M University, USA).

Excavation report 1996:
F.M. Hocker and K. Vlierman with contributions of D. van der Broek, A.M. Koldeweij, G.J.R. Maat and J. van der Zee.

Post-field research, report and drawings:
K. Vlierman.

Determination of wood samples:
A. Wevers.

Pollen spectrum of the moss caulking:
R.T.J. Cappers and E. Mook-Kamps (RuG).

Examination of the bones:
F. Laarman

Dendrochronological report
E. Jansma and E. Hanraets (RING, november 1996).

Photos:
L. van Dijk (Rijksdienst voor de IJsselmeerpolders).

Photo numbers:
93.759-93.785
93.886-93-911
93.918-93.923
93.931-93-964
95.189-95.194
95.300-95-305
96.219-96.241
96.452-96.453
96.482-96.489
99.048-99.112

Slide numbers:
71874-71877
72249-72301
72381
73587-74889
75535-75574
75599-75614
89130
105669-105693
105697
950094

Manuscript completed in January 2015.

17 The Dronten-II cog

Notification:
Late summer 1959, G.J. Rook, map-maker of the department Bodemkunde of the Directory board of the Wieringermeer.

Reconnaissance:
26 Oktober 1959, G.D. van der Heide, B. Van Dalen and H. Van Veen, department Oudheidkundig Bodemonderzoek of the Directory board of the Wieringermeer..

Reconnaissance:
4 and 5 Oktober 1982, K. Vlierman, department Oudheidkundig Bodemonderzoek van de Rijksdienst voor de IJsselmeerpolders and R. Kuiper (volunteer).

Excavation:
19 June - 11July 1995, G.G.W. Schreurs, H.J.Post (NISA), R. Hilgers and D. Zeldenrust (volunteers).

Excavation report:
G.G.W. Schreurs, in: NISA-Jaarboek 2 1995.

Post-field research 2012:
Augmentation field data, report and (reconstruction) drawings, K. Vlierman.

Dendrochronological report
RING, E. Jansma and E. Hanraets.

Photos:
G.G.W. Schreurs (RCE) and K. Vlierman.

Photo numbers:
Black/white 85.977-85.983
Color 100.139-100.170

Video recordings:
G.G.W. Schreurs.

Manuscript completed in November 2012.

18 The Swifterbant cog

Notification:
June 1962, by agricultural supervisor Bakker.

Reconnaissance:
7 June 1962, G.D. van der Heide
13 - 25 Juni 2001, A.F.L. van Holk, G. Schreurs, H. Post (NISA), Ch. Peen, G. Wilgerson and E. Weijters (students of Leiden University, Archaeological faculty).

Research 2013:
Augmentation of the field drawings, report and drawings K. Vlierman.

Description of the finds:
K. Vlierman.

Dendrochronological report
A.E.M. Hanraets RING.

Inventory numbers:
OG 77-1 to OG 77-30.

Photos:
G. Schreurs and T. Penders (RCE)

Photo numbers:
Rol 1-2, -7,-10, -13, -14, -15, -18, -19, -24, -25, -27, -28, -29, -30, -31, -32, -33 and -34
Rol 2-OA, -1A, -3A, -4A, -6A, -7A, -9A, -10, -14, -15, -16, -17 and -18

Og77-4, Og77-5-10-12-14, Og77-7-18-28, Og77-8, Og77-9, Og77-21, Og77-27.

Manuscript completed in March 2013.

19 The Almere-II cog

Notification:
R.S.R. Thijssen, supervisor Beheerseenheid Oostvaardersplassen.

Reconnaissance:
24 and 25 August and 1 - 3 September 1999, G. Schreurs and H. Post (NISA).

Research 2013:
Augmentation of the field drawings, report and drawings: K. Vlierman.

Dendrochronological report
A.E.M. Hanraets (RING).

Photos:
G. Schreurs (RCE).

Photo numbers:
Zc 46/ 32-1 to 32-22.
Zc46 [-001].

Manuscript completed in April 2013.

20 The Doel-I cog

Discovery:
In the beginning of September 2000; employees of the Archeologische Dienst Waasland (ADW), Sint-Niklaas under supervision of J.-P. Van Roeyen.

Request for advice to the Nederlands Instituut voor Scheeps- and onderwater Archeologie (NISA/ ROB):
R. Van Hove, director / archaeologist of the ADW.

Advisory visit at the site:
28 September 2000; A.F.L. van Holk and R. Oosting (NISA).
16 October 2000; K. Vlierman (NISA).

Excavating and cleaning of the ship's hull:
29 September - 5 October 2000; D. Boel, J. Van Cauter, S. Delamelle, J.-P. Van Roeyen (project leader) and G. De Wilde (all employees of the ADW).

Field documentation:
K. Vlierman; Photos, slides, sketches, written information and concept research report.
J.-P. Van Roeyen; Photos and slides, sketches of the position of cross beams/soil profile and ceiling.
Photographer of the Instituut voor het Acheologisch Patrimonium (IAP); Photos.
3D Solutions & Technics, G. De Meurichy, Temse/ Bornem ; 3 D-scans of the ship's hull.

Disassembly, salvage and packing of the ship's parts:
D. Boel, J. Van Cauter, S. Delamelle, B. Klinck, L. Meesschaert, N. Van der Putten, J.-P. Van Roeyen (projectleider), Van Roeyen Sr., G. De Wilde (ADW/ Universiteit van Gent), K. Vlierman (advisor).

Augmentation field data, report and drawings:
K. Vlierman.

Dendrochronological report
A.E.M. Hanraets (RING).

Photos and slides:
H. Schraal and K. Vlierman (NISA); slides fig. 16.1 and 16.24 J.-P. Van Roeyen (ADW).

Photo numbers:
28-1 to 28-333.

Slide numbers:
29-1 to 29-303.

Manuscript completed in April 2013.

21 The Doel-II cog

Discovery:
July 2002; employees of the Archeologische Dienst Waasland (ADW) under supervision of J.-P. Van Roeyen.

Request for advice to NISA/ROB:
R. Van Hove, director / archaeologist of the ADW.

Excavating and drawing of the remains of the ship:
Employees ADW under supervision of J.-P. Van Roeyen.

Advisory visit, describing and sketching the bottom of the cog:
K. Vlierman.

Disassembly, salvage and storing:
Employees of the ADW under supervision of J.-P. Van Roeyen.

Augmentation field data, drawings in appendix I (after the field drawings of J.-P. Van Roeyen) and report:
K. Vlierman.

Dendrochronological report
K. Haneca. Laboratorium voor Houttechnologie, Universiteit-Gent.

Photos and slides:
K. Vlierman; figures 17.6A , J.-P. Van Roeyen.

Photo numbers:
Doel II-1 to -24.

Slide numbers:
Doel II-100 to -155.

Manuscript completed in May 2013

22 The Kraggenburg cog

Notification:
16 May 2007. R. Crébas (Brennels Buiten) to D. Velthuizen, Nieuw Land Erfgoedcentrum, Lelystad. D. Velthuizen to R. Oosting, Maritiem Erfgoed RACM, Amersfoort.
14 June 2007. Second notification: PR-employee Brennels Buiten to RACM, location Lelystad.

Reconnaissance:
23 May 2007: J.-M.A.W. Morel, A. Vos, L. Koehler and G. Schreurs (RACM, Lelystad).
24 May 2007: L. Koehler and G. Schreurs.
15 June 2007: J.-M.A.W. Morel and A. Vos.

Exploratory research:
July 2010. International Fieldschool for Maritime Archaeology Flevoland (IFMAF).

Description and preliminary conclusions and drawing:
K. Vlierman.

Description of the finds:
K. Vlierman.

Dendrochronological report
M. Dominguez (RING).

Field find numbers :
NT 25-1 to -60

Photos:
T. Penders (RCE) and K. Vlierman.

Photo numbers:
NT 25.

Manuscript completed in August 2013.
Mid 2018 supplemented with several measurements and other data from the report: '*Het vlak van een COG-achtig schip bij Kraggenburg (Flevoland). Verkennend onderzoek aan een laatmiddeleeuws scheepswrak op kavel NT 25*' by K. Blok & M. van de Hoek, 2017.

23 The Kampen (IJssel)-cog[1]

Discovery:
April 2011. During the preliminary investigation in the context of the planned deepening of the summer-bed of the Beneden IJssel commissioned by Rijkswaterstaat Programmadirectie 'Ruimte voor de Rivier' carried out by ADC Maritiem under the direction of W.B. Waldus.

Reconnaissance / inventory field survey under water:
15 to 31 October 2012. ADC Maritiem: W.B. Waldus (senior KNA underwater archaeologist), K. Van Campenhout (KNA underwater archaeologist) and J. Verweij (maritime archaeologist), K. Vlierman (counsel).

Documentation on land:
10 May-29 October 2016. ADC Maritiem: W.B. Waldus (project leader and KNA underwater archaeologist), T. Van Damme (specialist 3D-Photography), G. Dijkstra (student maritime archaeology University of Groningen), M. van der Linden (maritime archaeologist), J. Vermeersch (KNA underwater archaeologist), J. Verweij (maritime archaeologist) and K. Vlierman (senior counsel ship archaeology and specialist cogs).

Publications and media:
'*De IJsselkogge*'. ADC Maritiem Rapport 3300 by Waldus *et al.* 2012;
'*De vijftiende-eeuwse IJsselkogge: onderdeel van een lange scheepsbouwtraditie uit de Hanzetijd*'. By Vlierman & Waldus 2017;
'*De opgraving en lichting van de 15de-eeuwse IJsselkogge*' by Waldus (red.) *et al.* 2018;
- Wide coverage in foreign, national and regional press;
- Various radio and TV programmes.

Post-field research, augmentation of the documentation, reconstruction drawings and report:
K. Vlierman. Note: The report draws on the content of ADC Maritime Report 3300, 2012 and ADC Monograph 24, 2018.

All 3D drawings, reconstruction drawings, photos and maps were made available by the ADC, except for:

- Figure 23.2, 23.12, 23.17: Photos RCE;
- Figure 23.4, 23.20, 23.29: Photos K. Vlierman;
- Figure 23.10 B, 23.11, 23.32 B, 23.35 A and B, 23.37: after Dijkstra 2018, fig.14.13,
- Figure 23.39, 23.44: drawings K. Vlierman;
- Figure 23.32 B and C and 23.42: Stadsarchief Kampen.
- Figure 23.43: Photo after Asaert 1974, 110.

The post -field research was carried out after completion of the reconstruction drawings at the end of November 2017. The manuscript was completed on 15 April 2018.

1. In this list of the persons that are responsible for the research, only the archaeological co-workers and the 3D-photographer are mentioned. They have been involved with the documentation of the wreck and the ship's timber on dry land and the reconstruction drawings. You can find the names of all the other co-workers who were involved with the underwater research and the salvage (Isalacogghe-project) and the specialist in Waldus 2018 (see references).

24 The Enkhuizerzand cog

Notification:
22 March 1999. Rijkswaterstaat Directie IJsselmeergebied. Dienstkring IJsselmeer-Markermeer/M. House of the family. Van Oord.

Evaluation visit cutter suction dredger:
23 March 1999. K. Vlierman and J. van den Akker (NISA). Report: K. Vlierman.

Dendrochronological report
March 1999. RING, E. Jansma & E. Hanraets.

Sonar mapping and diving survey:
26 March 1999. ADT/SNAP-combination.

Advice concerning the ship finds Noordelijke Aanloop Houtribsluizen ROB/NISA, 30 May 1999:
Th.J. Maarleveld, letter number NISA U991111, 4 June 1999.

Manuscript completed in August 2013.

25 Shipwrecks classified as cogs

Reports:
K. Vlierman

Manuscript completed in March 2018.

26 Dutch barges and boats with characteristics of cogs

Reports:
K. Vlierman

Manuscript completed in March 2018.

27 15th- and 16th- century clinker built ships

Reports:
K. Vlierman

Manuscript completed in March 2018.

28 Medieval shipyards

Research/description of the finds/report:
K. Vlierman

Illustrations:
Amersfoort, Archeologie gemeente Zwolle, Stadsarchief gemeente Kampen, K. Vlierman.

Manuscript completed in March 2018

References

Anonymous, 1940. 'Uit de geschiedenis van Nederland ter zee', *Voor het gezin*. Supplement to *"Schuttevaêr", weekblad voor binnen- en Rijnvaart* 52 no. 3, 13 April.

Anonymous, 1944. 'Scheepvaartaaangelegenheden. Van holle boomstam tot motorschip II', *"Schuttevaêr", hoofdorgaan der Schippersvereniging "Schuttevaêr"* 55 no. 50, 11 March.

Aa, A.J. van der, 1839-1851. *Aardrijkskundig Woordenboek der Nederlanden.* Deel A-Z. Gorinchem.

Aalders, M., 2018. Dierlijk botmateriaal. In: Waldus W.B. (red.). *De opgraving en lichting van de 15e-eeuwse IJsselkogge.* ADC Monografie 24. 309-326.

Adrian, K. & R. Mielke, 1903. Klappergeräte. In: *Zeitschrift des Vereins für Volkskunde 13.* 436-438.

Åkerlund, H., 1951. *Fartygsfynden i den forna hamnen i Kolmar.* Uppsala.

Allema, J. & A. Hubregtse, 2008. Een kogge langs de hydromechanische lat. In: SWZ/ Maritime. Maritiem Technisch Vakblad. Jaargang 18, september 2008 9.

Amkreutz, C. & J. Brattinga, 2018. Twee geweibijlen. In: Waldus, W.B. (red.) *De opgraving en lichting van de 15e-eeuwse IJsselkogge.* ADC Monografie 24. 241-246.

Andersen, P.K., 1983. *Kollerupkoggen*. Museet forthy og vester hanherred. Denmarken.

Andree, R., 1910. Ratschen, Klappern und das verstummen der Karfreitagsglocken. In: *Zeitschrift für Volkskunde 20.* 250-264.

Andree, R., 1985. Die Hillebille. In:*Zeitschrift des Vereins für Volkskunde 5.* 103-107.

Arnold, B., 1977. Some remarks on caulking in Celtic boat constructions and its evolution in areas lying northwest of the Alpine area. In: *The International Journal of Nautical Archaeology and Underwater Exploration 6.* 293-297.

Arnold, B., 1992. *Battelerie Gallo-Romaine sur le Lac de Neuchâtel, tome 1 et 2, Archéologie Neuchâteloise 12.* St. Blaise.

Asaert, G., 1974. *Westeuropese scheepvaart in de middeleeuwen.* Bussum.

Baart, J.M., 1995. Het Kasteel van Aemstel. In: Roever, M.B. de (red.) *Het "Kasteel van Amstel". Burcht of bruggehoofd?* Amsterdam. 11-27.

Baart, J.M., W. Krook, A. Lagerweij et.al. 1977. *Opgravingen in Amsterdam. 20 jaar stadskernonderzoek.* Haarlem.

Bakker, J.A., 2004. *Kanttekeningen bij mijn publicaties en enige andere zaken.* Baarn.

Bartels, M.H., 2001. Verkennend archeologisch onderzoek Noordenbergstraat 10, 'De Gymzalen', Deventer (Rapportage Archeologie Deventer 4). Gemeente Deventer.

Barwasser, M., 1997. De oudste leervondsten uit Kampen. In: Barwasser, M. & M. Smit (red.) *Acht eeuwen tussen twee stegen. Archeologisch, historisch en bouwhistorisch onderzoek in Kampen.* Archeologie IJssel/ Vechtstreek. 85-89.

Baykowski, U, 1991. *Die Kieler Hanzekogge. Der Nachbau eines historischen Siegelschiffes von 1380.* Kiel.

Behrens, Mark & Uli Rothfuss, n.d. 'Geschichte - Die KOGGE' on the website Die Kogge. *Europäische Autorenvereinigung e.V.*: https://www.diekogge.com/geschichte/.

Belasus, Mike, 2017. Historical Ship Archaeology in the Shadow of Historism and Nationalism. A German Perspective, in: Brooks, Alasdaire & Natascha Mehler, 2017. *The Country Where My Heart Is. Historical Archaeologies of Nationalism and National Identity,* Gainesville e.a.: University Press of Florida.

Berg, G. van den, J. Boerhof, R. van 't Hul, J. van der Steeg, J. Spruit, I. Streekstra & M. van Ommen, 2007. *Kamper Kogge. Van Oz 36 naar OLV04.* Stichting Kamper Kogge. Kampen.

Berk, G.L., 1984. *De punter.* Weesp.

Besançon, J., ca. 1950. *Flämische Zeichnungen des XV. Und XVI. Jahrhunderts.* Hyperto Verlag.

Beuningen, H.J.E. van& A.M. Koldeweij, 1993. *Heilig en Profaan. 1000 laat-middeleeuwse insignes uit de collectie H.J.E. van Beuningen.* Rotterdam Papers VIII. Cothen.

Beylen, J. van, 1985. *Zeilvaart Lexicon.* Weesp.

Beylen, J. van, P.A. de Groote, A. van Kampen, J.A.M. Kramer, L.L. von Münching, W.P.L. Spruit & A. de Vos, 1970 en 1972. *Maritieme Encyclopedie I en V.* Bussum.

Bill, J., 1991. Gedesbyskibet-Middelalterlig skude-og faergeart fra Falster. *Natinalmuseets Arbeidsmark 1991.* 188-198. København.

Bill, J. & M. Vinner, 1995. The Gedesby ship under sail. *Maritime Archaeology Newsletter from Roskilde, Denmark 5.* 3-8.

Bill, J. B. Poulsen, F. Rieck & O. Ventegodt, 1997. *Fra stammebåd til skib.* Dansk Søfarts Histori I. Indtil 1588. København.

Bitter, P., 1985. Archeologisch Bodemonderzoek in het bouwterrein van C & A Nederland aan de Haarlemmerstraat en Spijkerboorsteeg te Leiden. In: *Archeologisch Jaarverslag 1985.* Leiden.

Blockmans, W., 2010. *Metropolen aan de Noordzee. De geschiedenis van Nederland, 1100-1560.* Amsterdam.

Blok, K. & M. van den Broek, 2017. Het vlak van een kogge-achtig schip bij Kraggenburg (Flevoland). Verkennend onderzoek van een laatmiddeleeuws scheepswrak op kavel NT 25. *Grondsporen 27. Opgravings- onderzoeksrapporten van het Groninger Instituut voor Archeologie.* Groningen.

Blom, M.C., 2018. De leervondsten. In: Waldus, W.B. (red.) *De opgraving en lichting van de 15e-eeuwse IJsselkogge.* ADC Monografie 24. 283-288.

Blommesteijn, C.M, H.A. Heidinga, H.H & C.L. Verkerk, 1977. *De Veluwe. Archeologisch-historische verkenning van de bewoningsgeschiedenis tot 1200. Project Middeleeuwse Geschiedenis* Van Regteren Altena *1975-1976.* Universiteit van Amsterdam. A.E. van Giffen Instituut voor Pre- en Protohistorie (I.P.P.). Working paper 4, Historisch Seminarium, werkschrift 14. Amsterdam.

Bolin, G., 1933. *Stockholms upkomst.* Uppsala.

Boockman, H., 1980. *Die Stadt im späten Mittelalter.* München.

Borsboom,H., 2002. Scheepswrak IJHI-VAL2: *IJsselmeer_vaargeul Amsterdam-Lemmer; Archeologische begeleiding inventariserend veldonderzoek.* RAAP-rapport 755.

Bouma, N., 2012. *Kampen Burgwal. Een inventariserend veldonderzoek in de vorm van profielsleuven.* ADC-Rapport 3017. Amersfoort.

Braat, J., J.H.G. Gawronski, J.B. Kist, A.E.D.M. van de Put & J.P. Sigmond, 1998. *Behouden uit het Behouden Huys.* Catalogus van de voorwerpen van de Barentsexpeditie (1596) gevonden op Nova Zembla. De Rijksmuseumcollectie aangevuld met Russische en Noorse vondsten. Amsterdam.

Brand, H. & E. Knol (red.), 2010. *Koggen, Kooplieden en Kantoren.* Hilversum/Groningen.

Brouwers, W., E. Jansma & M. Manders, 2015. Middeleeuwse scheepsresten in Nederland. De Vroege-Middeleeuwen 500-1050. In: *Archeobrief jaargang 19 nummer 3. September 2015. Onderzoek.* Utrecht. 6-24.

Bruijn, A., 1962-'63. Die mittelalterliche Keramische Industrie in Südlimburg. *Berichten ROB 12-13.* 357-459.

Bruijn, A., 1966. *De middeleeuwse pottenbakkerijen in Zuid-Limburg.* Vrienden van de Nederlandse Ceramiek 44. Amsterdam.

Bruijn, A., 1979. *Pottersvuren langs de Vecht. Aardewerk rond 1400 in Utrecht. Rotterdam Papers III.* Rotterdam.

Buisman, J., 1995 en 1996. *Duizend jaar weer, wind en water in de Lage Landen.* Deel I (tot 1300) deel II (1300-1450). Franeker.

Cappers, R.T.J., E. Mook-Kamps, S. Bottema, B.O. van Zanten & K. Vlierman, 2000. The Analysis of caulking material in the study of shipbuilding technology. In: *Paleohistoria. 30/40. 1997/1998.* Rotterdam/Brookfield. 577-590.

Cederlund, C.O., 1977. Recording the remains of a sewn ship found at Skeppargatan 4 in Stockholm. In: McGrail, S. (ed.) *Sources and Techniques in Boat Archaeology. BAR Supplementary Series 29.* Greenwich. 191-201.

Cederlund, C.O., 1978. Ett Fartyg byggt med syteknik. *Statens Sjohistorika Museum Rapport 7.* Stockholm.

Chesney, K., 1942. A neglected prose version of the Roman de Troie. In: *Medium Ñvum/Society for the study of Medieval Languages and Literature. Volume 00011.* Oxford. 46-67.

Comis, S.Y., 1998. Kleding. In: Braat, J., J.H.G. Gawronski, J.B. Kist, A.E.D. van de Put & J.P. Sigmond (red.) *Behouden uit het Behouden Huys.* Amsterdam. 243-257.

Comis, S.Y., 2013. Onderzoek naar textielvondsten. In: Ploegaert, P.H.J.I. *Rotterdam markthal, archeologische onderzoek 2. Rotterdam rapporten 469-deel 2.* Rotterdam. 415-428.

Cremers, W., 2017. Houtskeletbouw. in : Klomp, M. (red.) *Myosotis,in de schaduw van het Sint Geertruidengasthuis. Archeologisch onderzoek tussen Boven Nieuwstraat en Burgwal.* Zwolle. 53-92.

Crone, G.C.E., 1926. *Nederlandsche jachten, binnenschepen, visschersvaartuigen en daarmee verwante kleine zeeschepen.* Amsterdam.

Crumlin-Pedersen, O.,1965. *Cog-Kogge-Kaag, Handels-Søfarts-museet par Kronborg, Arborg.* 1-144.

Crumlin-Pedersen, O., 1967. The Skuldelev Ships: The Ships. *Acta Archaeologica 38.*

Crumlin-Pedersen, O., 1969. *Das Haithabuschiff. Berichte über die Ausgrabungen in Haithabu 3.* Neumünster.

Crumlin-Pedersen, O., 1977. Some principles for the recording and presentation of ancient boat structures. In: McGrail, S. (ed.) *Sources and Techniques in Boat Archaeology. BAR Supplementary Series 29.* Greenwich. 163-179.

Crumlin-Pedersen, O., 1979. Danish cog-finds. In: McGrail, S. (ed.) *Medieval Ships and Harbours in Northern Europe. BAR International Series 66.* Greenwich. 17-34.

Crumlin-Pedersen, O., 1983. *From Viking Ships to Hanseatic cogs. Third Paul Johnstone Lecture.* National Maritime Museum Greenwich.

Crumlin-Pedersen, O., 1997. Large and small warships of the North. In: *Military aspects of Scandinavian society.*

Crumlin-Pedersen, O., 1997. *Viking-Age Ships and Shipbuilding in Hedeby/Haythabu and Schleswig. Ships and Boats of the North. Volume 2.* Schleswig/Roskilde.

Crumlin-Pedersen, O., *2000.* To be or not be a cog: The Bremen Cog in perspective. In: *The International Journal of Nautical Archaeology (2000).* 230-246.

Daly, A., O.H. Eriksen & A. Englert, 2000. New dendro dates for Danish medieval ships from Eltang and Kollerup. In: *Roskilde Newsletters 14, Denmark.* 61.

Daly, Aoife, 2007. *Timber, Trade and Tree-rings. A dendrochronological analysis of structural oak timber in Northern Europe, ca AD 1000 to ca AD 1650.* PhD thesis University of Southern Denmark, Esbjerg: Microsoft Word - Timber, Trade and Tree-rings june 07.doc (sdu.dk)

Daly, Aoife. 2017, Timber-regionality and temporality in Northern Europe's shipbuilding resource, in: Gawronski, Jerzy & André van Holk, Joost Schokkenbroek, *Ships and maritime landscapes: Proceedings of the Thirteenth International symposium on boat and ship archaeology*, Amsterdam 2012, ISBSA 13, Eelde: Barkhuis Publishing.

Daalen, S. van, 2017. *Kampen Reevediep vindplaats 4.* Dendrochronologisch onderzoek in opdracht van ADC-Archeoprojecten. Deventer.

De Maesschalck, E., 2012. *De Graven van Vlaanderen; 861-1384.* Leuven.

Dinter, M. van, 2018. Fysische geografie. In: Waldus, W.B. (red.) *De opgraving en lichting van de 15e-eeuwse IJsselkogge.* ADC Monografie 24. 119-134.

Doesburg, J. van & P. C de Boer, 2001. Burchten op de bodem van de zee: Aanvullend Archeologisch Onderzoek (AAO) naar de burchten van Kuinre. *Rapportage Archeologische Monumentenzorg 91.* ROB. Amersfoort.

Dooren, F. van, 1986. *Landschappen van Nijkerk-Arkemheen.* Nijkerk.

Dorleijn, P., 1987. *Zuiderzeevisserij in beeld.* De Walburg Pers.

Driesch, A. von den & J. Boessneck, 1974. Kritische Anmerkungen zur Widerristhöhenberechnung aus Längenmassen vor- und frühgeschichtlicher Tierknochen, *Säugetierkundliche Mitteilungen 22.* 325-348.

Duroselle, D., 2002. *Europa. Geschiedenis van zijn bewoners.* Utrecht/Antwerpen.

Ehbrecht, W., 1989. Fulda II, Stadt. In: *Lexicon des Mittelalters.* München-Zürich. Kol. 1022-1023.

Ellmers, D., 1972. *Frühmittelalterliche Handelsschiffahrt in Mittel- und Nordeuropa.* Neumünster.

Ellmers, D., 1975. Neue Ergebnisse der Kogge-Forschung am Deutschen Schiffahrtsmuseum. Förderverein Information 3.

Ellmers, D., 1979. The Cog of Bremen and Related Boats. In: *Medieval Ships and Harbours in Northern Europe. BAR International Series 66.* 1-15.

Ellmers, D., 2002. Mittelalterliche Koggensegel - Ein Discussionsbeitrag. In: *Schriften des Deutschen Schiffahrtsmuseum. Herausgegeben von U. Schnall. Band 58.* Hamburg. 160-164.

Ellmers, D., 2010. Koggen Kontrovers, *Hansische Geschichtsblätter,* 128

Engelen, C. & M. Marx, 1993. *Middeleeuws erfgoed. Beelden voor namen - namen voor beelden.* (cataloque). Hoogstraten.

Englert, A., J. Fischer, S. Hartz, H.J. Kühn & O. Nakoinz, 2001. Ein Nordischer Frachtschiff der 12. Jahrhunderts in Schleswig-Holstein. In: *Archäologisches Korrespondenzblad 31-2001.* 141-154.

Ente, P.J., 1973. De IJsseldelta. Overdruk uit *de Kamper Almanak, uitgave 1973.* Rijksdienst voor de IJsselmeerpolders. Lelystad. 137-164.

Ente, P.J., J. Koning & R. Koopstra, 1986. *De bodem van Oostelijk Flevoland.* Flevobericht 258. Lelystad.

Erbstösser, M., 1976. *Die Kreuzzüge. Eine Kulturgeschichte.* Leipzig.

Es, W.A. van & W.J.H. Verwers, 1980. *Excavations at Dorestad I. The Harbour: Hoogstraat I.* Nederlandse Oudheden 9. Amersfoort.

Es, W.A. van & W.J.H. Verwers, 2009. *Excavations at Dorestad 3. Hoogstraat, II-IV.* Nederlandse Oudheden 16. Amersfoort.

Evans, A.C.,1977. Recording and presentation of finds from nautical contexts. In: McGrail, S. (ed.) *Sources and Techniques in Boat Archaeology. BAR Supplementary Series 29.* Greenwich. 179-191.

Ewe, H., 1972. *Schiffe auf Siegeln.* Bielefeld/Berlin.

Fenwick, V. 1978. The Graveney Boat: a tenth-century Find from Kent. *BAR British Series 53.*
Flatman, J. 2009. *Ships & Shipping in Medieval Manuscripts.* London.

Fliedner, S., 1964. Die Bremer Kogge. *Hefte des Focke-Museum. Bremen Nr. 2.* Bremen.

Formsma, W.J. 1934. De polders van Kuinre en Blankenham. *Verslagen en Mededeelingen Overijsselsch Regt en Geschiedenis 50/ 2e reeks 26.* Zwolle. 22-34.

Förster, Th., 2000. Stand der schiffsarchäologischen Untersuchungen in der Wismarbucht und Neue Wrackfunde zwischen Rügen und Darsser ort. *NAU band 7-2000.* Freiburg i. Br. 46-54.

Förster, Th., 2008. Grosse Handelsschiffe der Spätmittelalters. Untersuchungen an zwei Wrackfunden des 14. Jahrhunderts vor der Insel Hiddensee und der Insel Poel. *Des Deutschen Schiffahrtsmuseums Bremerhaven. Band 67.* Bremerhaven/Kuden.

Fremery, G. de, 1878. *Schip en tuig, artillerie, reglementen, kompas, vlaggen enz.* Rotterdam.

Frutaz, A.P., 1961. Katharina. In: *Lexikon für Theologie und Kirche* (herdruk 1986). Deel 6. Freiburg i. Br. Kol. 60-62.

Gardiner, R., 1996. *The earliest ships. The evolution of boats into ships.* Conway's History of the ship. London.

Geese, U., 1996. Romaanse beeldhouwkunst. In: *Romaanse Kunst. Architectuur,Beeldhouwkunst, Schilderkunst.* Keulen. 256-323.

Gelder, H.E. van, 1972. Muntvondsten Oostelijk Flevoland. *Jaarboek voor Munt- en Penningkunde 60/61.* 186-191.

Giffen, A.E. van, 1946. Oudheidkundige aanteekeningen over Drentsche vondsten (XIII). Een vernield Hunebed, D LIIa, het zogenaamde Pottiesbargien, in het (vroegere) Wapserveld bij Diever, Gem. Diever. In: *Nieuwe Drentsche Volksalmanak 64, 1946.* 61-71, afb. 1 en 2.

Goodman, W.L., 1964. *The History of Woodworking tools.* London.

Gøthche, M. & H.M. Myrhoy, 1996. The wreck from Vedby Hage. A late medieval wreck from Starstrømmen. In: *Roskilde Newsletter 7, Denmark.* 12-15.

Gottschalk, M.K.E., 1971. *Stormvloeden en rivieroverstromingen in Nederland. I De periode vóór 1400*. Assen.

Goubitz, O., 1983. De ledervondsten. In: H.L. Janssen (red.) *V Bos tot Stad. Opgravingen in 's Hertogenbosch*. 's Hertogenbosch. 274-283.

Goubitz, O., C. van Driel-Murray & W. Groenman-Van Wateringe, 2001. *Stepping through time. Archaeological footwear from Prehistoric times until 1800*. Stichting Promotie Archeologie. Zwolle.

Groten, M., 1990. Forms of economic life in the High Middle Ages. In: Toman, R. (ed.) *The High Middle Ages in Germany*. Köln. 66-85.

Hagendorn, B., 1914. *Die Entwicklung der wichtigsten Schiffstypen bis ins 19. Jahrhundert*. Berlin.

Hammel-Kiesow, R, 2010. Hoe de Hanze verdween en op de drempel van de twintigste naar eenentwintigste eeuw weer opleeft. In: Brand, H.& E. Knoll (red.) *K, Kooplieden & Kantoren. De Hanze een praktisch netwerk*. Hilversum/Groningen. 191-203.

Hanke, F., 1995. Traditionelle Boote in Deutschland 3: Holzbootbau am Mittelmain. In: U. Schnall & U. Feldkamp (eds.) *Deutsches Schiffahrtsarchiv 18. 1995*. Altenburg. 261-294.

Hasse, G., 1907. Les barques de pêche trouvées à Anvers en 1884 et 1904-1905. In: *Annales de la Fédération archéoloque et historique de Belgique. Comtes endres et memoires des Congrès, 20e congrès, Gand*. 1-18.

Heide, G.D. van der, 1951. De burcht van Kuinre, Overijssel. *Jaarboek voor Cultuur en Historie 5*. 18-27.

Heide, G.D. van der, 1951. *Scheepsopgravingen in het Zuiderzeegebied*. Intern rapport Afd. Oudheidkundig Bodemonderzoek van de Directie voor de Wieringermeer. Zwolle.

Heide, G,D. van der, 1951/52. Opgraving in Kampen. In: *Kamper Almanak 1951/52*. Kampen.

Heide, G.D. van der, 1965. *Van Landijs tot Polderland. 2000 eeuwen Zuiderzeegebied*. Amsterdam.

Heide, G.D. van der, 1974. *Scheepsarcheologie. Scheepsopgravingen in Nederland en elders in de wereld*. Naarden

Heide, G.D. van der, ca. 1974. *Archäologie auf den Meeresboden. Archäologischen Untersuchungen im Gebiet der Zuidersee. Schiffsarchäologisches Museum Ketelhaven und Ärchäologisches Museum Schokland im Rijksdienst voor de IJsselmeerpolders*. Kampen.

Heidinga, H.A., 1984. *De Veluwe in de Vroege Middeleeuwen, aspecten van de nederzettingsarcheologie van Kootwijk en zijn buren*. Proefschrift. Amsterdam.

Heinsius, P., 1956. *Das Schiff der Hansischen Frühzeit*. Weimar.

Heinsius, P., 1986. *Das Schiff der Hansischen Frühzeit 2. Verbesserte Auflage*. Köln/Wien.

Hellinga, W. Tj., 1984. Onderzoek ballastmateriaal scheepswrak Oz 36. Schriftelijke rapportage.

Henderikx, P.A., 1995. De ringwalburgen in het mondingsgebied van de Schelde in Historisch perspectief. In: Heeringen, R.M. van, P.A. Henderikx & A. Mars (red.) *Vroeg-Middeleeuwse ringwalburgen in Zeeland*. Amersfoort. 71-113.

Henn, V., 2010. Het ontstaan van de Hanze. In: Brand, H. & E. Knoll (red.) *Koggen, Kooplieden en Kantoren*. Hilversum/ Groningen. 11-25.

Hillewaert, B., 2011. Wat de archeologie vertelt over de Brugse binnenstad. In: Hillewaert, B., Y. Hollevoet & M. Ryckaert (red.) *Op het raakvlak van twee landschappen. De vroegste geschiedenis van Brugge*. Brugge.

Hocker, F.M., 2000. Relocating the Kolding cog. *Maritime Archaeology Newsletter from Roskilde, Denmark. 14*. 50-55.

Hocker, F.M., 2001. Cog-finds from Danish waters. In: *Newsletter No. 17 December 2001. Maritime Archaeology from Roskilde, Denmark.*

Hocker, F.M. & K. Vlierman, 1996. *A small cog wrecked on the Zuiderzee in the early fifteenth century*. Excavation report 19. Flevobericht 408. Almere.

Hockman, O., 1994. Das Schiff. In: Hellenkemper Salis, G., H.H. von Prittwitz und Gafreh & G. Bauchhenss (Herausgeber) *Das Wrack. Der Antike Schiffsfund von Madhia*. Köln. 53-81.

Hoek, 1899. *De vischtuigen volgens de bestaande reglementen in ons vaderland geoorloofd, gerangschikt naar provinciën.*

Hoekstra, T.J., 1976. Schepen met geld. Utrecht.

Hoffmann-Krayer, E., 1932. Hille. In: H. Bächtold-Stäubli (ed.) *Handwörterbuch des Deutschen Aberglaubens 4*. Berlin. K. 1.

Höhlbaum, K. & H. Künze, 1879. *Hansisches Urkundenbuch (1876-1904) I-IV HUB V nr. 509*. Halle.

Holck, P., 1986. Scurvy as a paleopathological problem. In: Capecci, V. & Rabino Massa (eds.) *Proceedings Paleopathology Association. 5th European Meeting. Sienna 1984. Paleopathology Association*. Sienna.

Holk, A.F.L. van, 1986. Jaarringonderzoek van scheepsresten. Doctoraalscriptie. Groningen.

Holk, A.F.L. van, 1987. Jaarringonderzoek van scheepsresten. In: Reinders, R. (red.) *Raakvlakken tussen scheepsarcheologie, maritieme geschiedenis en scheepsbouwkunde*. Flevobericht 280. Lelystad. 75-81

Holk, A.F.L. van, 1996. *Archeologie van de binnenvaart. Wonen en werken aan boord van binnenvaartschepen (1600-1900). Scheepsarcheologie IV. Flevobericht 410*. Almere.

Holk, A.F.L. van, 2001. Vier 13e eeuwse schepen in de dam van Rotterdam. In: Carmiggelt, A., A.J. Guiran & M.C. van Trierum (red.) *BOORbalans 4. Archeologisch onderzoek in het trace van de Willemsspoortunnel te Rotterdam. Sluizen en schepen in de dam van de Rotte*. Rotterdam. 71-123.

Holk, A.F.L. van, 2010. Maritieme archeologie van de kogge. In: Brand, H. & E. Knol (red.) *Koggen, Kooplieden en Kantoren. De Hanze een praktisch netwerk*. Hilversum/ Groningen. 122-144.

Holk, A.F.L. van, D. Velthuizen, D. Biewinga & B. van Rosmalen, 2008. Een schip dat tot tweemaal toe zoekraakte, in: *Westerheem 6*. Jaargang 57, december 2008. 353-367.

Hommel-Steenbakkers, N. de, 2001. *Een openbaring: Paris Bibliothèque Nationale ms. Néerlandais 3*. Nijmegen.

Hove, R. Van, 2005. De Doelse kogge(n). Maritiem erfgoed van Europees formaat. In: *Monumenten, Landschappen & Archeologie 24/4 juli/augustus 2005*. Brussel. 50-69.

Huisman, D.J., R.C.G. M. Lauwerier, M.M.E., Jans, A.G.F.M. Cuijpers & F.J. Laarman, 2006. *Degradatie en bescherming van archeologisch bot, Praktijkboek Instandhouding Monumenten Deel II - 11 Overige onderwerpen 14*.

Hulst, R.S., 1985. Schriftelijke mededeling.

Hulst, R., 1987. Iconografische aspecten van de 'Kogge.' In: Reinders, R. (red.) *Raakvlakken tussen scheepsarcheologie, maritieme geschiedenis en scheepsbouwkunde. Inleidingen gehouden tijdens het Glavimans Symposion 1985*. Flevobericht 280. Lelystad. 49-56.

Hupperetz, W. & E. Nijhof, 1995. Ceramiek uit twee 14e-eeuwse beerputten aan het O.L.V. plein te Maastricht. *Corpus van Middeleeuwse Aardewerk uit gesloten vondstcomplexen in Nederland en Vlaanderen (CMA). Afleveringen 11 & 12*. 's-Hertogenbosch.

Hurst, J.G., 1997. Langerwehe Stoneware of the Fourteenth and Fifteenth centuries. In: Apted, M.R., R. Gyard-Beer & A.D. Saunders (eds.) *Ancient Monuments and their interpretation. Essays presented to A.J. Taylor*. London. 219-238.

Jacobi, H.W., 1983. Determinatie munt Oz 36. Schriftelijke rapportage.

Jacoby, A., 1932. Hille. In: Bächtold-Stäubli (ed.) *Handwörterbuch der deutschen Aberglaubens 4*. Berlin. K. 2.

Jager, A., 2012. De IJsselkogge, een werkpaard van de Hanze in de rivierbodem bij Kampen. In: Clevis, H. & S. Wentink (red.) *Overijssels Erfgoed. Archeologische en Bouwhistorische Kroniek 2012*. Assen. 131-200.

Jager, A., 2015. *Middeleeuws Kampen. De ruimtelijke en economische structuur van de stad aan de hand van archeologische, bouwhistorische en historische bronnen*. Zwolle.

Jahnke, C., 2010. De Hanze en de Europese economie in de Middeleeuwen. In: Brand, H. & E. Knol (red.) *Koggen, kooplieden en Kantoren. De Hanze een praktisch netwerk*. Hilversum / Groningen. 44-62.

Jansma, E. & E. Hanraets, 1994. RING. Dendrochronologisch onderzoek Oz 36. Schriftelijke rapportage.

Janssen, H.L., 1981. Later Medieval pottery production in The Netherlands. In: *Ceramics and Trade*. Department of Prehistory and Archaeology. University of Sheffield. Sheffield. 121-187.

Janssen, H.L., 1983. Later Medieval Pottery Production in the Netherlands. I: Hodges, R. & P. Davey (eds.) *Ceramics and Trade. The production and distribution of Later Medieval Pottery in Northwest Europe*. Sheffield. 121-185.

Janssen, H.L., 1988. The dating and typology of the earliest Siegburg stoneware in the Netherlands, in: Gaimster, D.M.R., M. Redknap & H.H. Wegner (eds.) *Zur Keramiek des Mittelalters und der beginnenden Neuzeit im Rheinland. Medieval and later pottery from the Rhineland and its markets. BAR International Series 440*. Oxford.

Jaspers, N.L., 2018. Het (post-)middeleeuws aardewerk. In: Waldus, W.B. (red.). 253-262.

Johnstone, P., 1974. *The Archaeology of Ships*. London, Sydney, Toronto.102-114.

Kansa, Sarah W. & Levent Atici, Eric C. Kansa, Richard H. Meadow, 2019. Archaeological Analysis in the Information Age: Guidelines for Maximizing the Reach, Comprehensiveness, and Longevity of Data, *Advances in Archaeological Practice* 8. 40 - 52: https://doi.org/10.1017/aap.2019.36.

Kars, H., 1995. De voorwerpen van natuursteen, in: Heeringen. R.M. van, P.A. Henderikx & A. Mars (red.) *Vroeg-Middeleeuwse ringwalburgen in Zeeland*. Goes/Amersfoort. 185-191.

Ketting, G.N.A., 1922. *Bijdrage tot de geschiedenis van lepra in Nederland*. Amsterdam.

Keweloh, H.W., 1983. Kalfathölzer - Zur interpretation eines Beifundes des Hansekogge. In: Kludas, A. & U. Schnall (eds.) *Deutsches Schiffahrtsarchiv 6. 1983*. Oldenburg. 7-10.

Kipp, F., 1990. Water en Vuur. Brandpreventie en het Middeleeuwse dak. In: Groot, H.L. de (red.) *Het vuur beschouwd*. Archeologisch Bouwhistorisch Centrum Utrecht. Utrecht.

Klomp, M., 2014. Stationskwartier/Niersallee. *Archeologische Rapporten Kampen I*. Zwolle.

Klomp, M., (red.), 2018. *Myosotis, in de schaduw van het Sint Geertruiden gasthuis. Archeologisch onderzoek tussen Boven Nieuwstraat en Burgwal*. Archeologische Rapporten Kampen 4. Zwolle, 137.

Koenen, M.J. & J. Endepols, 1977. Verklarend handwoordenboek der Nederlandse taal. Zevenentwintigste druk 1977.

Kok, H.J., 1958. *Proeve van een onderzoek van de patrocinia in het middeleeuwse bisdom Utrecht*. Assen.

Koldeweij, A.M., 1987. Teekenen van Mijnheer St. Lenaert: pelgrimsinsignes uit Dudzele. In: *Biekorf, West-Vlaams Archief voor Geschiedenis, Archeologie, Taal en Volkskunde 87*. 305-317.

Koning, J. & P.J. Ente, 1986. *De bodemkundige code- en profielenkaart van Zuidelijk Flevoland*.
Flevobericht 271. Lelystad.

Koopstra, R., 1982. Schriftelijke rapportage bodemprofiel scheepswrak M 61 Oostelijk Flevoland.

Körber-Grohne, U., 1989. Botanische Untersuchungen an Mittelalterlichem tauwwerk aus Schleswig. Ausgrabungen in Schleswig. In: *Berichte und Studien 7*. Schleswig. 101-135.

Kossmann, F.J., 1990. Het verbond van de Hanzesteden. In: M. van Rooyen (red.) *Steden des tijds. Historische stadstypen in de Nederlanden*. Utrecht.

Krause, G. 2018. Zwischen Überlieferung und Propaganda. Zur Geschichte der Duisburg-Ruhrorter Häfen und ihrer Rezeption. In: Rhein-Maas. *Geschichte, Sprache, Kultur*, Bd. 8 (2018), S. 153-175.

Kruijer, J.D. & E.H.P. Leusink, 2018. Botanisch onderzoek. In: Waldus, W.B. (red.) *De opgraving en lichting van de 15e-eeuwse IJsselkogge*. ADC Monografie 24. 297-308.

Laarman, F.J. & R.C. Lauwerier, 2006. Bot uit scheepswrakken in de Waddenzee (16e - 18e eeuw). *Rapportage G. Archeologische Monumentenzorg 132.*

La Cour, V., 1961. *Naesholm, Nationalmuseet.* Kopenhavn.

Lahn, W., 1992. *Die Kogge von Bremen.* Bremen.

Langelaar, J. & W.B. 2018. Voorwerpen van metaal. In: Waldus, W.B. (red.) *De opgraving en lichting van de 15e-eeuwse IJsselkogge.* ADC Monografie 24. 289-296.

Lanting, J.N., 1994. C14-onderzoek. Schriftelijke rapportage.

Lauwerier, R.C.G.M. & F.J. Laarman, 2006. Hollandse Nieuwe en de mythe van Willem Beukelszoon. *Holland.* 38ste jaargang nr. 3. 19-161.

Lenselink, G. & R. Koopstra, 1994. *Paleographical Map of the Zuiderzee region and North-Holland in the Early Middle Ages (Mid-Subatlanticum, 850 AD and recent coast-line).* Lelystad. fig. 5.21d.

Lenselink, G. & U. Menke, 1995. *Geologische en Bodemkundige Atlas van het Markermeer. Blad 6, dikte holocene afzettingen.* Directoraat Rijkswaterstaat Directie IJsselmeergebied. Lelystad.

L'Hour, M. & E. Veyrath, 1994. The French Medieval Clincker Wreck from Aber Wrac'h. In: Westerdahl, Ch. (ed.) *Crossroads in ancient Shipbuilding. Proceedings of the sixth International Symposium on Boat and Ship Archaeology. Roskilde 1991.* Oxford. 165-180.

Litwin, J., 1979. Studium Konstrukcji i proba klasyfikatcji 'miedziowska'. *Kwartalnik Historii Kultury Materialnej 3.* 321-343.

Litwin. J., 1980. 'The Copper Wreck'. The wreck of a medieval ship raised by the Central Maritime Museum in Gdansk, Poland. *International Journal of Nautical Archaeology 9.* London. 217-225.

Lloyd, S., 1999. Die Kreuzzugsbewegung 1096 bis 1274. In: *Illustrierte Geschichte der Kreuzzüge.* Herausgegeben von J. Riley-Smith. Frankfurt/New York.

Looper, B., 2010. De Nederlandse Hanzesteden: Scharnieren in de Europese economie 1250-1550. In: H. Brandt & E. Knol (red.) *Koggen, Kooplieden en Kantoren.* Hilversum/Groningen.

Looper, B., 2017. Aan de grenzen van de Hanze. De IJsselregio als economisch systeem in de Middeleeuwen. In: Dijk, E.A. van (red. et al) *Overijsselse Historische Bijdragen. 132e stuk 2017. De Hanze. Verslagen en mededelingen van de Vereeniging tot beoefening van Overijsselsch Regt en Geschiedenis.* Hilversum. 9-32.

Lübben, A. & Chr. Walther, 1888. *Mittelniederdeutsches Handwörterbuch.* Nachdruck 1993. Darmstadt.

Luns, A., 1985. Item van den Cogghen .... . Doctoraalscriptie Middeleeuwse Geschiedenis, Rijksuniversiteit Leiden. Leiden. Niet gepubliceerd.

Maarleveld, Th. J., 1984. Archeologie in troebel water. Een nieuwe werkwijze voor scheepsonderzoek. Twee schepen onderzocht. *Archeologie onder water, 1e onderzoeksrapport. Ministerie van Welzijn, Volksgezondheid en Cultuur.* Rijswijk.

Maarleveld, Th.J., 1995. *'Type or technique.* Some thoughts on boat and ship finds as indicative of cultural traditions', *International Journal of Nautical Archaeology 24.1 (1995) 3-7*

Maarleveld, Th.J., 20**, *Het land, het water en de binnenvaart. Een schets van de middeleeuwse binnenvaart in de Lage Landen.*

Maat, G.J.R., 1984a. Dating and rating of Harris's lines. In: *American Journal of Physical Anthropology 63.* 291-299.

Maat, G.J.R., 1984b. A search for secular growth changes in The Netherlands proceeding 1850. In: J. Borms, R. Hauspi, A. Sand, C. Susanne & M. Hemmelinck (eds.) *Human growth and Development.* New York. 185-191.

Maat, G.J.R., 1984c. Scurvy in Dutch whalers burried at Spitsbergen. In: G.T. Haneveld, W.R.K. Perisonius & P.J. Janssens (eds.) *Proceedings Paleopathology Association, 4th European meeting. Middelburg/Antwerpen 1982. Paleopathology Association.* Utrecht. 82-92.

Maat, G.J.R., 1996. Healed survey, a Dutch example from the fifteenth century. In: Hocker, F.M. & K. Vlierman *A small cog wrecked on the Zuiderzee in the early fifteenth century. Excavation report 19.* Nds Instituut voor Scheeps- en onderwaterArcheologie /ROB (NISA). Annex B.

Marsden, P., 2009. Reconstruction of the Mary Rose: her Design and Use. In: Marsden, P. (ed.) *Mary Rose: your Noblest Shippe. Anatomy of a Tudor Warship. The Archaeology of the Mary Rose Volume 2.* 369-390.

McGrail, S., 1981. *The Ship. Rafts, boats and ships from Prehistoric times to the Medieval Era.* London.

McKee, J.E.G., 1977. Hypothetical Reconstructions in Boat Archaeology. In: McGrail, S. (ed.) *Sources and Techniques in Boat Archaeology. BAR Supplementary Series 29.* Greenwich. 205-215.

McLaughin-Neyland, K. & B. Neyland, 1993. *Two prams wrecked on the Zuiderzee in the late eighteenth century. Excavation reports 15 and 16. Flevobericht 383.* Lelystad.

Melkert, M.J.A., 2018. Keramisch bouwmateriaal en Natuursteen. In: Waldus, W.B. (red.) *De opgraving en lichting van de 15e-eeuwse IJsselkogge.* ADC Monografie 24. 263-274.

Meulemeester, J.L., 1987-1988. De melaatsenklepper in het Museum O.-L.-Vrouw-ter-Potterie te Brugge. In: *Antiek 22ste Jaargang no. 7.* Lochem. 393-399.

Modderman, P.J.R., 1945. *Over de wording en de betekenis van het Zuiderzeegebied.* Proefschrift.

Groningen - Batavia.
Mook-Kamps, E., 1997. Stuifmeelonderzoek aan breeuwsels. RUG. Groningen. (niet gepubliceerd).

Moolhuizen, C. & N. van Asch, 2018. Botanisch onderzoek. In: Waldus, W.B. (red.) *De opgraving en lichting van de 15e-eeuwse IJsselkogge.* ADC Monografie 24. 297-308.

Moortel, A. van de, 1991. *A cog-like vessel from The Netherlands.* Flevobericht 331. Lelystad.

Moortel, A. van de, 2011. Medieval boats and ships of Germany, The Low Countries, and Northeast France - archaeological evidence for shipbuilding traditions, shipbuilding resources, trade and communications. In: *Siedlungs - und Küstenforschung im Südlichen Nordseegebiet. Settlement and Coastal research in the Southern North Sea Region.* Niedersachsisches Institut für historisches Küstenforschung./ Lower Saxony Institute for historical Coastal Research. Wilhelmshaven. Rhaden/Westphalen. 67-104.

Morel, J.-M.A.W., *1993.* Het Biddinghuizer Colfschip. In: Tiesinga, G.H.L. (eindred.) *Drooggelegd land- blootgelegd verleden. Cultuurhistorisch Jaarboek voor Flevoland.* Lelystad. 25-37.

Neugebauer, W.,1975. Arbeiten der Böttcher und Drechsler aus mittelalterliche Bodenfunden Hansestadt Lübeck. In: Renaud, J.G.N. (red.) *Rotterdam Papers II. A contribution to medieval archaeology.* Rotterdam. 117-137.

Neyland, R.S. & K. *McLaughin-Neyland,* 1996. *A late-sixteenth century freither from the Workumer Nieuwlandpolder in Workum, Friesland. Excavation report 18. Flevobericht 407.* Almere.

Neyland, R.S. & B. Schröder, 1996. *A late seventeenth century Dutch freither wrecked on the Zuiderzee. Excavation report 20. Flevobericht 409.* Almere.

Nie, M. van & M. Smit, 1997. Houten huizen aan de Oudestraat. In: M. Barwasser & M. Smit (red.) *Acht eeuwen tussen twee stegen. Archeologisch, historisch en bouwhistorisch onderzoek in Kampen.* Stichting Archeologie IJssel/ Vechtstreek.

Oosting, R., 1985. Opgraving van het vlak van een kogge bij Rutten (Noordoostpolder, kavel A 57). In: *Jaarverslag Onderafdeling Scheepsarcheologie 1985.* Lelystad. 7-14.

Oosting.R., 1987a. De opgraving van een laat middeleeuws vrachtschip op kavel U 34 in Oostelijk Flevoland. In: *Jaarverslag 1987 Onderafdeling Scheepsarcheologie.* Rijksdienst voor de IJsselmeerpolders 1987, Lelystad. 9-18.

Oosting, R., 1987b. De opgraving van het vlak van een kogge bij Rutten. In: *Raakvlakken tussen scheepsarcheologie, maritieme geschiedenis en scheepsbouwkunde.* Lelystad. 57-63.

Oosting, R. & K. Vlierman, 1993. Een 17de-eeuws vistransportschip gevonden in Swifterbant. In: *Drooggelegd land-blootgelegd verleden.* Cultuur Historisch Jaarboek voor Flevoland. Lelystad. 87-106.

Oppers, V.M., 1966. The secular trend in growth and maturation in The Netherlands. *Nederlands Tijdschrift voor Sociale Geneeskunde 44.* 539-548.

Overvoorde, J.C., 1924. Catalogus van voorwerpen in het Stedelijk Museum "De Lakenhal" te Leiden.
(2de editie). Wassenaar.

Paulsen, Reinhard, 2010. Die Koggendiskussion in der Forschung. Methodische Probleme und ideologische Verzerrungen, *Hansische Geschichtsblätter* 128.

Paulsen, Reinhard, 2015 (verschenen 2016). The History of the "Hanse Cog". A ship type between Science and Politics, *Hansische Geschichtsblätter,* 133.

Paulsen, Reinhard, 2016. *Schiffahrt, Hanse und Europa im Mittelalter. Schiffe am Beispiel Hamburgs, europäischen Entwicklungslinien und die Forschung in Deutschland,* Köln/ Weimar/Wien: Böhlau (Quellen und Darstellungen zur hansischen Geschichte 73);.

Pedersen, R.K., 1996. *Waterschip ZN 42 I: a glenched-lap fishing vessel from Flevoland, The Netherlands. Excavation report 17. Flevobericht* 406. Almere.

Perkmann, A., 1932. Klapper/Hillebille. In: H. Băchtold-Stăubli (ed.) *Handwörterbuch des deutschen Aberglaubens 4.* Berlin. K. 2-3 ud K. 1443.

Pol, H.J. van der, 1951. De periodisering van de geschiedenis. In: Oosthoeks Encyclopedie 1968. Deel 10,167. Utrecht.

Pryor, J.H., 1992. *Geography, technology and war. Studies of Maritime History of the Mediterranean 641-1571.* Cambridge.

RAAP-Notitie 43, 1992. Weerstandsonderzoek naar kasteel 'Hulkestein'.

Raczynski-Henk, Y., 2018. Vuursteen. In: Waldus, W.B. (red.) *De opgraving en lichting van de 15e-eeuwse IJsselkogge.* ADC Monografie 24. 247-252.

Reinders, R., 1979. Medieval ships: Recent finds in the Netherlands. In: McGrail, S. (ed.) *Medieval ships and Harbours in Northern Europe.* National Maritime Museum Greenwich. Archeological Series no. 5. BAR International Series 66.

Reinders, H.R., 1983. *Drie middeleeuwse rivierschepen. Opgravingverslagen 5, 6 en 7.* Flevobericht 221. Lelystad.

Reinders, R., 1985a. *Cog finds from the IJsselmeerpolders.* Flevobericht 248. Lelystad.

Reinders, H.R., 1985b. The inventory of a cargo vessel, wrecked in 1888. In: Cederlund, C.O. (ed.) *Postmedieval boat and shiparchaeology. British Archaeological Reports. BAR International Series 256.* Oxford. 81-99.

Reinders, R., 1986. *Maritime Archaeological Heritage: responsibility and management. Paper presented at a meeting concerning the VOC-ship Amsterdam,* organised by the Stichting Centrum Historische Instellingen. Haarlem, november 1985. Werkdocument 1986-24Abw. Lelystad.

Reinders, R., 1987. Inleiding/Cornelis Jan Glavimans. In: Reinders, R. (red.) *Raakvlakken tussen scheepsarcheologie, maritieme geschiedenis en scheepsbouwkunde. Inleidingen gehouden tijdens het Glavimans Symposion 1985.* lelystad.

Reinders, H.R., H. van Veen, K. Vlierman & P. B. Zwiers, 1978. *Het wrak van een 16e eeuws vissersschip in Flevoland. Het onderzoek van een vissersschip gevonden op kavel W 10 in Flevoland.* Opgravingverslag I. Lelystad.

Reinders, H.R., H. van Veen, K. Vlierman en P.B. Zwiers, 1980. *Drie schepen uit de late middeleeuwen. Opgravingverslagen 2, 3 en 4.* Flevobericht 166. Lelystad.

Reinders, H.R., H. van Veen, K. Vlierman & P.B. Zwiers, 1984. *Vier werkschuiten uit de zeventiende eeuw.* Opgravingsverslagen 8, 9, 10 en 11. Flevobericht 235. Lelystad.

Reinders. R. & R. Oosting, 1986. De opgraving van een overnaads gebouwd vrachtschip op kavel U 34 in Oostelijk Flevoland. In: *Jaarverslag Onderafdeling Scheepsarcheologie 1986.* Rijksdienst voor de IJsselmeerpolders. Lelystad. 10-13.

Reinders, H.R., & R. Oosting, 1989. Mittelalterliche Schiffsfunde in de IJsselmeerpoldern. In: *Wilhelmshavener Tage nr. 2. Ländliche und Städtische Küstensiedlungen im 1. und 2. Jahrtausend.* Wilhelmshaven. 106-122.

Reinders, H. R., 2009. 'The Sailing Route from Uffelte to Utrecht' in: R. Bockius (red.), *Between the Seas* (Mainz 2009) 337-348.

Renaud, J.G.N., 1973. *Rhodesteyn, schatkamer der middeleeuwse ceramiek*. Lochem.

Renaud, J.G.N., 1976. Middeleeuwse Ceramiek. *AWN-Monografie nr. 3*. Den Haag.

Reineking-Von Bock, G., 1976. *Steinzeug*. Köln.

Rieth, E., 1979. Pirogue monoxyle et port fluvial de Port-Berteau (Charente-Maritime, France). In : McGrail, S. (ed.) *Medieval Ships and Harbours in Northern Europe. BAR Archaeological Series No. 5. BAR International Series 66*. Greenwich. 117-129.

Rieth, E., 2010. L'Epave du XVe siècle de Beutin, Canche (Pas de Calais). Archéologie en Nord-Pas-de Calais. L'Ètat et la Patrimoine Archéologique. Le Ministèr de la Culture Republique Française.

Roede, M.J. & J.C. Van Wieringen, 1985. Growth diagrams 1980. Netherlands third nation-wide survey. *Tijdschrift voor Sociale Geneeskunde 63, supplement*. 1-34.

Rule, M., 1982. *The Mary Rose. The excavation and raising of Henry VIII's flagship*. London.

Schilder, K., 1983. *Digestum Vetus 1448-1478*. Kamper Genealogische en Historische Bronnen. Deel 3. Oud archief Kampen nr. 8. Kampen.

Schreurs, G.G.W. & H.J. Post, 1995. Oostelijk Flevoland, kavel M 61. In: *NISA-Jaarboek 2 1995. Nederlands Instituut voor Scheeps- en onderwaterArcheologie/ ROB/NISA*. 36-38.

Schreurs, G.G.W. & H.J. Post, 1995. Oostelijk Flevoland, kavel H 61. *In NISA-Jaarboek 2 1995. Nederlands Instutuut voor Scheeps- en onderwaterArcheologie/ROB/NISA*. 38-42..

Scholkmann, B. 1992. *Dass Fallbeispiel Esslingen, Stadtluft, Hirsebrei und Bettelmönch. Die Stadt um 1300*. (catalogus). Zürich-Stuttgart. 451-463.

Schoute, N., 2018. Duigen. In: Waldus, W.B. (red.) *De opgraving en lichting van de 15e-eeuwse IJsselkogge*. ADC Monografie 24. 327-338.

Schröder, B., 1994. *Aardewerk uit de inventaris van een laat-17e-eeuwse praam in Oostelijk Flevoland*. Corpus van Middeleeuws Aardewerk uit gesloten vondstcomplexen in Nederland en Vlaanderen (CMA). Complex 10. Amersfoort.

Seiler, R., 1992. *Medizin und Gesundheitsfürsorge, Stadtluft, Hirsebrei und Bettelmönch. Die Stadt um 1300*. (catalogus). Zürich-Stuttgart. 469-471.

Siegloff, E. 2004. *Studien zu den Sintelfunden an der Schleswiger Landenge*. Schriftliche Hausarbeit zur erlangung des Grades einen Diplom- Prähistorikers. Kiel.

Smeerdijk, D.G. van, 1989. Alder carr, growth and drowning in the IJsselmeer region, an aspect of the Dutch coastal development. In: *Acta Botanica Neerlandica. 38 (4), december 1989*. 477-491.

Smeyers, M. (ed.), 1993. Vlaamse Miniaturen van Van Eyck (ca. 1380-ca. 1420). *Corpus van Verluchte Handschriften. Vol. 6. Low Countries Series 4*. Catalogus. Leuven. 191-196 cat. 61 en 62.

Smit, M. m.m.v. J.E. Dilz, 1987. *Ceramiek uit een laat-13e-eeuwse vulling van een immuniteitssloot van de Sint Paulusabdij te Utrecht*. Corpus van Middeleeuws Aardewerk uit gesloten vondsten in Nederland en Vlaanderen (CMA). Aflevering 4. Amersfoort/ Laarne.

Smits, J.G., 1995. *Vorst en onderdaan. Studies over Holland en Zeeland in de Late Middeleeuwen*. Leuven.

Sopers, P.J.V.M., 1974. *Schepen die verdwijnen*. Amsterdam.

Speet, B.J.M. & Th. Rothfusz, 1986. *Historische Stedenatlas van Nederland*. Herwijnen, G. van, G. van der Kieft, J.C. Visser & J.G. Wegner (red.). *Aflevering 4 KAMPEN*. Delft.

Stallaert, K. 1890. Glossarium van verouderde rechtstermen, kunstwoorden en andere uitdrukkingen 1. *Woordenboek der Nederlandsche Taal 6*. Leiden.

Stassen, P., 1992. Touwwerk en Schepen. Touwtechnologie; een onderzoek naar touwfragmenten afkomstig uit een 16e-eeuwse koopvaarder. Bijlage 5. Scheepstouwwerk uit de Late Middeleeuwen. Afstudeerproject (niet gepubliceerd).

Stephan, H.G., 1980. The development and production of medieval stoneware. In: Davey, P. & R. Hodges (eds.) *Ceramics and Trade*. Sheffield. 95-120.

Schwartz, W., 1896. Noch einmal die Hillebille. In: *Zeitschrift des Vereins für Volkskunde 6*. 444-445.

Tanner, P., 2017. The Bremen Kogge. Still a cog but not as we knew it. 3D scanning Id Limited & Traditional Boats of Ireland project. On behalf of German Maritime Museum. Leibnitz Institute for German Maritime History. Bremerhaven.

Toebosch, Th., 2011. *De Nieuwezijds Kolk en de Nieuwendijk in dertiende-eeuws Amsterdam. Een archeologische speurtocht*. Amsterdam.

Torsy, J., 1976. *Der grosze Namenstagskalender*. Breisgau.

Trotter, M. & G.C. Gleser, 1952. Estimation of stature from the log bones of American whites and negroes. *American Journal of Physical Anthropology. N.S. 10*. 463-514.

Trotter, M. & G.C. Gleser, 1985. A re-evaluation of estimation of stature based on measurements of stature during live of long bones after death. *American Journal of Physical Anthropology. N.S. 16*. 79-123.

Van Damme, T., 2018. 3D-fotogrammetrie. In: Waldus, W.B. (red.). *De opgraving en lichting van de 15e eeuwse IJsselkogge*. ADC monografieën 24. 87-92.

Varenius, B., 1985. Medieval ships from the center of Stockholm. In: Papers based on those presented to the International Symposium on Boat and Ship Archaeology in Stockholm in 1982. *BAR International Series 256*. 437-440.

Verhoeven, A.A.A., 1998. *Middeleeuws gebruiksaardewerk in Nederland (8ste-13de eeuw). Archaeological Studies 3*. Amsterdam.

Verhulst, A., 1995. *Landschap en landbouw in Middeleeuws Vlaanderen*. Gent.

Vermeersch, J.L., J.P.F. Verweij, K. Vlierman & W.B. Waldus, 2018. De schuit met kenmerken van de koggetraditie (object 14, de punter). In: Waldus, W.B. (red.) *De opgraving en lichting van de 15e-eeuwse IJsselkogge.* ADC Monografie 24. 159-174.

Vermeersch, J.L., G. Dijkstra, K. Vlierman & W.B. Waldus, 2018. De IJsselkogge. In: Waldus, W.B. *De opgraving en lichting van de 15e-eeuwse IJsselkogge.* ADC Monografie 24. 191-224

Verweij, J., W. Waldus en A. van Holk, 2012. 'Continuity and change in Dutch shipbuilding in the Early Modern period. The case of VAL7 and the watership in general', *Journal of Archaeology in the Low Countries 4 (2012)*, 65-93

Verweij, J.P.F., J.L. Vermeersch & W.B. Waldus, 2018. Het aak-achtige wrak. In: Waldus, W.B. (red.) *De opgraving en lichting van de 15e-eeuwse IJsselkogge.* ADC Monografie 24. 175-190.

Verwijs, E. & J. Verdam, 1885. *Middelnederlandsch Woordenboek.* 's Gravenhage.

Visser, B., 1943. De Europeesche zee- en kustvisscherij in den loop der eeuwen, *De visscherijwereld; officieel orgaan van de Vakgroep Vischverwerkende industrie en van de Federatie van Organisaties in het Vischbedrijf* 3 nr. 17, 30 april. 194.

Vlek, R., 1987. *The Medieval Utrecht Boat. BAR International Series 382.* National Maritime museum Greenwich. Oxford.

Vlierman, K., 1981. Archeologie in de kruidentuin. In: Verenigingsblad *Arent thoe Boecop, april 1981. Nr. 18.* Elburg. 1-10.

Vlierman, K., 1983. *Aardewerk uit de inventaris van twee laatmiddeleeuwse vrachtschepen in Oostelijk Flevoland.* Corpus van Middeleeuwse Aardewerk uit gesloten vondstcomplexen in Nederland en Vlaanderen (CMA). Complex 1 en 2. Lelystad.

Vlierman, K., 1985a. *Neolithische en Middeleeuwse vondsten op de kavels Oz 35 en Oz 36 in Zuidelijk Flevoland.* RIJP-Rapport 1985-51Abw. Lelystad.

Vlierman, K., 1985b. *Bijlen uit laatmiddeleeuwse schepen.* RIJP-Rapport 1985-55Abw. Lelystad.

Vlierman, K., 1985c. *Vuurstenen werktuigen op kavel M 132 in de Noordoostpolder.* RIJP-Rapport 1985-Abw. Lelystad.

Vlierman, K., 1985d. De vondst van een kogelpot in Flevoland, sectie OZ. In: *Jaarverslag Onderafdeling Scheepsarcheologie 1985.* Rijksdienst voor de IJsselmeerpolders. 29, 31 en 32.

Vlierman, K., 1988. Scheepshout als oeverbeschoeiing uit een 13e-eeuwse sloot aan de Postelstraat in 's Hertogenbosch. In: Boekwijt, H.W. & H.L. Janssen (red.) *Kroniek Bouwhistorisch en Archeologisch onderzoek 's Hertogenbosch nr. 1.* 's Hertogenbosch. 97-101.

Vlierman, K., 1991. Een zeventiende eeuwse ventjager. In: Reinders, R. & R. Oosting (red.) *Scheepsarcheologie: prioriteiten en lopend onderzoek. Inleidingen gehouden tijdens de Glavimans symposia in 1986 en 1988.* Lelystad. 95-104.

Vlierman, K., 1992a. Koken en kookgerei op (binnenvaart-) schepen 1300-1900. In: Ruempol, A & A. Van Dongen (eindred.) *Quintessens. Wederwaardigheden over acht eeuwen kookgerei.* Museum Boymans-Van Beuningen. Rotterdam. 50-59.

Vlierman, K., 1992b. De uitrusting en inventaris van een 15e-eeuwse kogge. In: Reinders, R. (red.) *Scheepsuitrusting en -inventaris. Inleidingen gehouden tijdens het vijfde Glavimans symposion Amsterdam, 5 april 1991.* Groningen. 10-22.

Vlierman, K., 1992c. Ceramiek uit een vroeg- 15e-eeuwse kogge in Almere. *Corpus van Middeleeuws Aardewerk uit gesloten vondstcomplexen in Nederland en Vlaanderen (CMA). Aflevering 7.* Amersfoort.

Vlierman, K., 1993. Late medieval pottery on Dutch shipwrecks and a well-dated inventory of the early 15th century. In: Blackmore, L. & M. Redknap (eds.) *Medieval Ceramics Volume 17.* Warwickshire. 69-76.

Vlierman, K., 1995. Scheepshout: mos en sintelnagels. Datering van middeleeuws scheepshout aan de hand van een breeuwmethode. In: Roever, M.B. de (red.) *Het "Kasteel van Amstel". Burcht of bruggehoofd.* Amsterdam. 91-104.

Vlierman, K., 1996a. *'...Van Zintelen, van Zintelroeden ende Mossen...". Een breeuwmethode als hulpmiddel bij het dateren van scheepswrakken uit de Hanzetijd. Scheepsarcheologie I.* Flevobericht 386. Almere.
Vlierman, K., 1996b. *Kleine bootjes en middeleeuws scheepshout met constructiedetails. Scheepsarcheologie II.* Flevobericht 404. Almere.

Vlierman, K., 1996c. De schepen. In: Sarfatij, H. & A. Mars (red.) De Middeleeuwse haven van Tiel. Archeologie in Tiel 3. Tiel.

Vlierman, K., 1997a. The IJsselmeerpolders: a 'source book' for late medieval and early post-medieval wreck inventories en The galley, galley utensils and cooking, eating and drinking vessels from an armed *'Tjalck'* wrecked on the Zuiderzee in 1673: a preliminary report. In: Redknap, M. (ed.) *Artefacts from wrecks. Dated assemblages from the Late Middle Ages to the Industrial Revolution.* Oxbow Monograph 84. Exeter. 15-37 en 157-166.

Vlierman, K., 1997b. De middeleeuwse kogge; scheepsarcheologische vondsten in het Zuiderzeegebied. In: Boer, D.E.H. de, E.H.P. Cordfunke & H. Sarfatij (red.) *Holland en het water in de middeleeuwen. Strijd* tegen *het water en gebruik* van *het water.* Hilversum. 71-95.

Vlierman, K., 1997c. A caulking method used as an aid to dating shipwrecks from the Hanseatic period. In: Boe, G. De & F. Verhaeghe (eds.). *Travel Technology & Organisation in Medieval Europe. Papers of the 'Medieval Europe Brugge 1997' Conference. Volume 8.* I.A.P. Rapporten 8. Zellik. 41-52.

Vlierman, K., 1997d. Scheepsarcheologisch onderzoek. In: Dijkstra, J. & K. Vlierman. *De Middeleeuwse haven van Tiel (2). Archeologie in Tiel 5.* Tiel.

Vlierman, K., 1997e. Sintels: ijzeren voorwerpjes met een eigen verhaal. In: Barwasser, M. & M. Smit (red.) *Acht eeuwen tussen twee stegen. Archeologisch, historisch en bouwhistorisch onderzoek in Kampen.* Stichting Archeologie IJssel/ Vechtsreek. 113-121.

Vlierman, K., 1997f. A 'reference book' for late medieval and post-medieval shipwrecks and artefactual ship's inventories. In: De Boe, G. & F. Verhaeghe (eds) *Travel Technology & Organisation in Medieval Europe, Papers of the 'Medieval Europe Brugge 1997' Conference Volume 8.* Zellik 1997, 57-68.

Vlierman, K., 1998. Scheepsfragmenten uit de Viking-scheepsbouwtraditie (voorlopig verslag). In: Dijkstra, J. *Archeologisch onderzoek in de binnenstad van Tiel. Juni t/m september 1996.* Amersfoort. 56-58.

Vlierman, K., 2000. Unieke vondst: dertiende-eeuwse kadebeschoeiing van scheepshout. In: Vos, A, J. Buiks, H. Janssen & T. Hoogbergen (red.) *Bossche Bladen 2000-4.* 's Hertogenbosch. 118-125.

Vlierman, K., 2002. Scheeps- en stadsarcheologie. De betekenis van scheeps(hout)vondsten in Nederlandse middeleeuwse steden. In: Woltering, P.J., W.J.H. Verwers & G.H. Scheepstra (red.). *Middeleeuwse toestanden. Archeologie, geschiedenis en monumentenzorg.* Heerhugowaard, Amersfoort, Hilversum. 119-148.

Vlierman, K., 2003. *Naar een 'Vlaams Centrum voor Scheepsarcheologie en houtconservering'.* Dronten (niet gepubliceerd).

Vlierman, K., 2004. Scheepsfragmenten uit de Viking-scheepsbouwtraditie in Vlaardingen. In: Vredenbregt, A.H.L. & T. de Ridder (red.) *VLAK-verslag 15.2 Gat in de Markt 1.101. Houtgebruik in 11e-eeuwse graven.* Vlaardingen. 22-29.

Vlierman, K., 2005. *Ceramiek uit het Behouden Huys op Nova Zembla (1596). Ceramics from the Behouden Huys on Novaya Semlya (1596).* Corpus van Middeleeuws Aardewerk uit gesloten vondstcomplexen in Nederland en Vlaanderen (CMA). Aflevering 17. 's-Hertogenbosch.

Vlierman, K., 2006. Laat- en post-middeleeuwse vissersschepen van de Zuiderzee. In: Pieters, M., F. Verhaeghe & G. Gevaert (eds.) *Fishery, Trade and Piracy. Fishermen and fishermen's settlements in and around the North Sea area in the Middle Ages and later. Papers from the colloquium at Oostende-Raversijde. Provinciaal Museum Walraversijde, Belgium, 21-23 november 2003.* Brussel. 151-169.

Vlierman, K., 2007. Houten schepen. In: Hurkmans, W, N. Köster-Constandse, P. Voermans & G. van der Velde (red.) *De Houtverzamelaar. De Nederlandse Vereniging van Houtsoorten Verzamelaars (NEHOSOL). Jaargang 60, september 2007 Nr. 77. 16-29.*

Vlierman, K., 2010. An overview of the dated stoneware(groups) from ships in the Netherlands (Zuiderzeeregion) and Novaya Zemlya (1200-1600). In: De Groote, K., D. Tys & M. Pieters (eds.). *Exchanging Medieval Material Culture. Studies on Archaeology and history presented to Frans Verhaeghe. Relicta Monografiën 4.* Brussel. 189-204.

Vlierman, K., 2018. Sintels en andere aan schepen gerelateerde vondsten. In: M. Klomp (red) *Myosothis, in de schaduw van het Sint Geertuidengasthuis. Archeologisch onderzoek tussen Boven Nieuwstraat en Burgwal.* Zwolle. 257-268.

Vlierman, K., in voorbereiding a. Sintels uit het Damrak. (Vondsten uit de Noord-Zuidlijn in Amsterdam; werktitel).

Vlierman, K., in voorbereiding b. The boat of Bruges reviewed. (International Journal of Nautica archaeology 2021).

Vlierman, K. & L. van Dijk, 1980a. Conservering en restauratie van kleine voorwerpen bij het Museum voor Scheepsarcheologie te Ketelhaven. Werkdocument WD 1980-149 Abw. Lelystad (niet gepubliceerd).

Vlierman, K. & L. van Dijk, 1980b. Sintels. Werkdocument 1980-304Abw. Lelystad (niet gepubliceerd).

Vlierman, K. m.m.v. P. Kleij, 1990. Ceramiek uit de inventaris van een klein 17de-eeuws vrachtschip in Zuidelijk Flevoland. *Corpus van Middeleeuws Aardewerk uit gesloten vondstcomplexen in Nederland en Vlaanderen (CMA). Aflevering 5.* Amersfoort.

Vlierman, K. & A.M. Koldeweij, 1992-1993. Een bijzondere klepper in de inventaris van een 15e-eeuwse kogge in Flevoland. In: *Antiek 27.* Lochem. 323-333.

Vlierman, K. & A.M. Koldeweij, 1996-1997. Een melaatsenklepper met een handgreep in de vorm van 'Hilgi nin'. In: *Antiek 31.4.* Zwolle. 155-162.

Vlierman, K. & T. Tahey, 1998. 15.000 sintels en een fles bier. Herbouw van een hanzeschip. In: *Natuur & Techniek 3, 66e jaargang maart 1998.*

Vlierman, K. & W. Waldus, 2017. De vijftiende-eeuwse IJsselkogge: onderdeel van een lange scheepsbouwtraditie uit de Hanzetijd. In: Dijk, E.A. van (red. *et al.*) *Overijsselsche Historische Bijdragen. Verslagen en mededelingen van de Vereeniging tot beoefening van Overijssels Regt en Geschiedenis. 132e stuk 2017. De Hanze.* Hilversum. 77-91.

Vlierman, K. & W.B. Waldus, 2018. De reconstructie van de punter en de IJsselkogge. In: Waldus, W.B. (red.) *De opgraving en lichting van de 15e-eeuwse IJsselkogge.* ADC Monografie 24. 225-240.

Vogel, W., 1915. *Geschichte der deutschen Seeschiffahrt.* Berlijn.

Vos, A., 1987. Een bijzonder grote kogge in de Noordoostpolder, het onderzoek ven een scheepswrak op kavel A 57. Werkdocument 1987-26Cbw. Rijksdienst voor de IJsselmeerpolders. Lelystad (niet gepubliceerd).

Vos, A., 2012. Onderwaterarcheologie op de rede van Texel. Waardestellende onderzoeken in de westelijke Waddenzee (Burgzand). *Nederlandse Archeologische Rapporten 041.* Rijksdienst voor het Cultureel Erfgoed. Ministerie van Onderwijs, Cultuur en Wetenschap. Almere.

Waldus, W.B. (red.), 2018. *De opgraving en lichting van de 15e-eeuwse IJsselkogge.* ADC Monografie 24. Amersfoort.

Waldus, W.B., 2018. Voorwerpen van hout en touw en een intacte turf. In: Waldus, W.B. (red.) *De opgraving en lichting van de 15e-eeuwse IJsselkogge.* ADC Monografie 24, 275-282.

Waldus, W.B., J.F. Verweij, H.M. van der Velde, A.F.L van Holk & S.E. Vos, 2019. The IJsselcog project: from excavation to 3D reconstruction. In: *The International Journal of Nautical Archaeology (2019).* 48.2: 466-494.

Waldus, W.B., K. Camphout & J. Verweij, 2012. De IJsselkogge. *ADC-rapport 3300.* Amersfoort.

Waldus, W.B., K. Vlierman 2018. Synthese. In: Waldus, W.B. (red.) *De opgraving en lichting van de 15e-eeuwse IJsselkogge.* ADC Monografie 24, 339-354.

Wahl, R., 1959. *Barbarossa. Eine Historie.* München.

Weinhold, K. 1898. Zur Hillebille. In: *Zeitschrift des Vereins für Volkskunde 8.* 347.

Weinhold, K. & J., 1895. Zur Hillebille. In: *Zeitschrift des Vereins für Volkskunde 5.* 327-329.

Weski, T., 1999a. The IJsselmeer type: some thoughts on Hanseatic cogs. In: *The International Journal of Nautical Archaeology 28.4.* 360-379.

Weski, T., 1999b. Fiktion oder Realität. Anmerkungen zum archäoligischen Nachweis Spätmittelalterliche Schiffsbezeichnungen. In: *Skyllis. Zeitschrift für Unterwasserarchäologie. 2.Jahrgang 1999. Heft 2.*

Westerhoff, W.E., T.E. Wong & M.C. Geluk, 2003. De opbouw van de ondergrond. In: Mulder, E.F.J. de, M.C. Geluk, I. E. Wong Ritsema, W.E. Westerhoff & T.E. Wong (eds.) *De ondergrond van Nederland. Nederlands Instituut voor Toegepaste Geowetenschappen TNO, Geologie van Nederland 7.* Utrecht. 247-352.

Westphal, V., 1999. Die Kollerup-kogge. Ein Unikum oder ein Schlüsselfund zur Schiffsgeschichte? *In: Das Logbuch. Zeitschrift für Schiffbaugeschichte und Schiffsmodelbau. Heft 1999-3.* 103-115

Westphal, V., 2004. Schiffsfund Port Berteau II. Ältestes Zeugnis einer mittelalterlichen Kraweelbauweise nördlich der Alpen. *DAS LOCHBUCH 40. Zeitschrift für Schiffbaugeschichte und Schiffsmodelfbau.Heft 2004-1.* 9-21.

Weststrate, J., 2008. *In het kielzog van de moderne markten. Handel en Scheepvaart op de Rijn, Waal en IJssel, ca. 1360-1560.* Hilversum.

Wevers, A., 1990. Treatment of waterlogged rope. ICOM. Bremerhaven.

Wiechell, H., 1978. Das Schiff auf der Lübecker Siegeln der 13. Jahrhunderts-Ein frühe Kogge? *Zeitschrift des Vereins für Lübeckischer Geschichte und Altertumskunde 58.*

Wiggers, A.J., 1955. De wording van het Noordoostpoldergebied. In: *Van Zee tot Land nr. 14.* Zwolle.

Wijsenbeek, A., 1996. *Vuur- en bakenloodjes voor een veilige vaarweg op de Zuiderzee. Scheepsarcheologie III. Flevobericht 405.* Almere.

Williams Jr., C.C., 1984. A case of mistaken identity: still an other Trojan narrative in old French
prose. In: *Medium Ñvum/ Society for the Study of Medieval Languages and Literature. Vol. 00053.* Oxford. 59-72.

Winsen, M.R. van & H. J. van Velzen, 2012. *Binnen de Poorten. Ontstaan en ruimtelijke ontwikkeling van de binnenstad van Kampen.* IJsselacademie nr. 227. Kampen.

Witsen, N., 1671. *Aeloude en hedendaegsche scheepsbouw en bestier.* Amsterdam. (facsimile 1979).

Wöginger, Chr., 2007. *Stromaufwärts, Stromabwärts. Eine Reise durch die spannende Geschichte der historischen Donauschiffahrt in Barocksloss Erlahof. Schiffahrtsmuseum Spitz.* Spitz.

IJzereef, G.F. & F.J. Laarman, 1986. The Animal Remains from Deventer (8th-19th) Centuries AD). *B.R.O.B.* 36. 405-444.

Zaretsky, O., 1905. Zur Hillebille. In: *Zeitschrift des Vereins für Volkskunde 15.* 93-94.

Zeeuw, J.Gzn., P. de, 1974. *Nijkerk door de eeuwen heen.* Nijkerk.

Zeller, G., 1902. Die Klebern. In: *Zeitschrift des Vereins für Volkskunde 12.* 214-215.

Zimmermann, W., 1982. *Nef der cinque ports. Das Normannenschiff des 13. Jahrhunderts.* München.

Zwick, D., 2016. Bayonese cogs, Genoese carracks, English chomons and Iberian carvels: tracing technology transfer in medieval Atlantic shipbuilding. In: *Itsas Memoria. Revista de Estudios Maritimos del Pais Vasco* 8.

Zwiers, B.P. & K. Vlierman, 1988. *De 'Lutina'. Een Overijssels vrachtschip, vergaan in 1888.* Flevobericht 292. Lelystad.

Curruculum vitae[1]

Karel Vlierman (Steenwijk {Overijssel}, 28 May 1942)

When, in the summer holidays of 1955, a school friend suggested going to find decorated potsherds in a sandy knoll on the Berkenheuvel estate at Wapse (prov. Drenthe), this meant the beginning of a lifelong contamination with the 'archaeology bug'. Not much later, an exhibition was being set up in a church at Steenwijk, to mark the 750th anniversary of the town's city charter, I identified the recovered sherds as Funnel-Beaker (TRB) ware.
For the then 13-year-old, the immediate results were a home visit from a real professor (H.T. Waterbolk), and subsequently the tutelage and introduction to (amateur) archaeology and geology by Mr M. Gerrits, headmaster at Eesveen and school doctor Dr F.D. Carels. A special highlight was their present for my eighteenth birthday: a five-week tour of Italy to acquaint myself also with the legacy of the Etruscans, Greeks, Romans and medieval Italians. During an excursion of the Drents Prehistorische Vereniging (Prehistoric Society of Drenthe) in 1955, 'youth membership' of this society - then composed mainly of (elderly) notables - was introduced especially for my benefit. Family circumstances obliged me to leave secondary school early and find employment. My great dream of one day professionally working in archaeological research especially in the province of Drenthe, was never to be realized

During a home visit by emeritus professor A.E. van Giffen and his wife, I showed him the potsherds that as a boy I had collected from the '*Pottiesbarchien*' at Wapse (the site of the destroyed megalithic passage grave D52a, excavated by Van Giffen in 1927), and he asked me to lend them to him for examination and publication. After Van Giffen died (in 1973), it was Dr J.A. Bakker (University of Amsterdam) who took over the project. In 1982, in consultation with Dr Bakker and Dr J.N. Lanting (University of Groningen) all finds from this passage grave (from Van Giffen's excavation and the potsherds subsequently gathered by private individuals, kept at the Drents Museum in Assen), were transferred to Dronten, so that I could investigate them myself. In his publication[2] Van Giffen had presented sherds from just a fairly small number of pots. By early 1987 this investigation, which unfortunately remained uncompleted, had resulted in drawings of parts of at least 149 different vessels. I also predicted and indicated where I was sure that thousands of sherds were still to be found in the spoil left by two stone-robbers from Steenwijk who in 1739 had dismantled the then probably still largely intact passage grave. Since Van Giffen had uncovered the site, a lot of illicit sherd-collecting had gone on at the site. Therefore Dr Lanting of the University of Groningen, in consultation with the Forestry Commission, decided to perform a second investigation in July 1988. The predicted amount of recovered material was very much exceeded (15,800 potsherds, 88 arrowheads, 9 amber beads, a stone axe and some scrapers). It was the only time that I ever took part in an archaeological excavation in Drenthe.
In 1958 I had submitted an open application to G.D. van der Heide, Head of the Archaeological Department of the *Rijksdienst voor de IJsselmeerpolders* (Development Authority) at Schokland, and after eleven years, on 1 September 1969, this resulted in my appointment as the supervisor of a newly formed excavation team. This second team was needed to excavate the fast-growing number of ship finds in the new polders Oostelijk Flevoland and Zuidelijk Flevoland, and to staff the Museum for Maritime Archaeology to be opened at Ketelhaven in 1971.
In over thirty years of practicing maritime archaeology, I was involved in assessing, excavating, documenting and salvaging over a hundred shipwrecks in and outside the Flevoland polders, ranging in age from Roman times to the 20th century. The discovery and salvage of the six Roman-period ships at Zwammerdam and two medieval ones at Utrecht in 1973/'74 were occasion for the Development Authority to create facilities for conserving large, waterlogged archaeological timber from across the Netherlands. This necessitated making an inventory and selection of the huge amount of previously excavated and preserved ships' timbers, which were being kept wet on rinsing floors and stored in large sheds. The colleagues appointed to carry out this formidable task backed out. But I accepted the challenge, on condition that I should be permitted subsequently to also make accessible for research and to provide better storage for the collection of objects gathered since the Second World War, excavated from maritime and terrestrial contexts, from prehistoric and medieval sites.

1. Translated by Xandra Bardet
2. Van Giffen 1946, 61-71 and figs 1 and 2.

I took care of the day-to-day supervision of excavations, restored pottery, designed the fitting-out of a restoration workshop and initiated responsible restoration and conservation methods for finds of every kind of material[2] designed and arranged permanent displays and temporary exhibitions on prehistoric, medieval, geological, palaeozoological and numismatic themes, introduced improvements to the storage of the various collections in the buildings and the depots of the Schokland Museum (in connection with its transfer to the Municipality of Noordoostpolder), (co-)wrote research reports and gave lectures at home and abroad, while also working on a new inventory of the Maritime Museum/NISA and Schokland Museum collection etc. from 1974 until 2002.
Linked to the new inventory, the archaeological assessment of the terrestrial, mostly prehistoric finds, for which up till then it was in fact assumed that the Flevo polders were unlikely to yield any new sites, resulted in a completely new picture; this led to a degree project of J.W.H. Hogestijn and ultimately, on 1 September 1987, to appointment as the first Provincial Archaeologist of Flevoland. The Department of Maritime Archaeology had been subsumed under the State Archaeological Service (ROB) as of April 1992, but in 2002 reorganisation at the ROB resulted in my leaving the Service early; this meant that I was unable to complete the depot I had designed to be fully publicly accessible and which by then had been partly realised. In this depot, the National Reference Collection of ships' inventories of the period ca 1300-1900, comprising ca 30,000 items, was to be presented ship-by-ship.[3] My plan of completing the depot before reaching retirement age would certainly have succeeded.
The new depot was designed to be housed in the allocated part of the new building of the Department of Maritime Archaeology, which on 12 April 1995[4] had been renamed 'Netherlands Institute for Ship and underwater Archaeology' (NISA). The new building, at Lelystad, was opened in 1998.

At the transfer of the Department of Maritime Archaeology from the Directorate-General of Rijkswaterstaat (IJsselmeer-Region Directorate) to the ROB, the future NISA was afforded a 'dowry'. The maritime archaeologists were given the opportunity and funding to prepare for publication by the end of 1995 ten research projects of excavated ships. Besides, *Rijkswaterstaat* (the Ministry of Waterways and Public Works) made available the funds for appointing five full-time academics to expand and enhance NISA's activities. The research reports were completed towards the end of 1996.[5] In 2016 the NISA was closed down.

2. Vlierman & Van Dijk 1980a.
3. Vlierman 2002, 121.
4. NISA-Jaarboek (1) 1994, 5-6 en NISA-Jaarboek (2) 1995, 5-10.
5. McLaughin-Neyland & Neyland 1993; Vlierman 1996a; Vlierman 1996b; Wijzenbeek 1996; Pedersen 1996; Neyland & McLaughin-Neyland 1996; Hocker & Vlierman 1996; Neyland & Schröder 1996; Van Holk 1996.